CALCIUM *as a* CELLULAR REGULATOR

Edited by
Ernesto Carafoli
Claude Klee

New York   Oxford
Oxford University Press
1999

Oxford University Press

Oxford   New York
Athens   Auckland   Bangkok   Bogotá   Buenos Aires   Calcutta
Cape Town   Chennai   Dar es Salaam   Delhi   Florence   Hong Kong   Istanbul
Karachi   Kuala Lumpur   Madrid   Melbourne   Mexico City   Mumbai
Nairobi   Paris   São Paulo   Singapore   Taipei   Tokyo   Toronto   Warsaw

and associated companies in
Berlin   Ibadan

Published by Oxford University Press, Inc.
198 Madison Avenue, New York, New York 10016

Oxford is a registered trademark of Oxford University Press

Library of Congress Cataloging-in-Publication Data
Calcium as a cellular regulator / edited by Ernesto Carafoli and Claude Klee.
p. cm
Includes bibliographical references and index.
ISBN 0-19-509421-2
1. Calcium—Physiological effect.   2. Calcium channels.
3. Cellular signal transduction.   I. Carafoli, Ernesto.   II. Klee, Claude B.
QP535.C2C2624   1998
572'.516—dc21   97–48843

1   3   5   7   9   8   6   4   2

Printed in the United States of America
on acid-free paper

CALCIUM *as a* CELLULAR REGULATOR

Preface

Since S. Ringer our understanding of the broad role of calcium in the regulation of cellular activity has advanced in quantum steps. It took forty years for L. V. Heilbrunn to revisit this concept, and another twenty years to obtain convincing evidence that a calcium signal preceded muscle contraction and for S. Ebashi to show how Ca^{2+} works in skeletal muscle. Thinking of the great pioneers one must also mention O. Loewy. His famous statement in 1959 ("calcium is everything") is a tribute to his clairvoyance; considering what we know today, it wasn't off the mark by much. But it was not until the discovery of calmodulin by G. Cheung, and most importantly of its dependence on calcium so elegantly shown by S. Kakiuchi, that it became firmly established that Ca^{2+} was a universal second messenger in eukaryotic cells. Calmodulin is now the recognized prime actor in the decoding of the Ca^{2+} signal, and the understanding of the structural aspects of its function was greatly aided by the landmark contribution of R. H. Kretsinger on the structure of parvalbumin. His identification of the EF-hand motif and his generalization of it to an ever increasing number of Ca^{2+}-modulated proteins, including calmodulin, is another striking example of clairvoyance.

The following ten years saw the continuous expansion of the list of calmodulin-regulated enzymes in eukaryotic cells from yeast to man. The list gradually grew to cover the phosphodiesterases, adenylate cyclase, calcium pumps, many kinases, the unique phosphatase and more recently the NO synthase and an array of cytoskeletal proteins.

At this point it was of course obvious that such a variety of Ca^{2+}-regulated enzymes should attract the attention of molecular and structural biologists and their comprehensive efforts to unravel the complex interplay of these enzymes in the regulation of cellular processes. The past few years have indeed seen a flurry of reports dealing with the role of Ca^{2+} as a second messenger and with its major role in the regulation of gene expression, cell differentiation, secretion, and a number of neuronal functions. Again, this takes us back to Ca^{2+} binding proteins. The EF-hand proteins are now numbered in dozens, some of them even extracellular. Why do we need so many? And, to compound the problem, why do we need so many diverse Ca^{2+}-binding proteins, from annexins to gamma-carboxyglutamate containing proteins?

An obvious corollary to the discoveries on Ca^{2+}-regulated enzymes/cellular processes is the necessity to finely regulate Ca^{2+} in the cellular compartments: that is, the existence of the understanding of the mechanisms to transport Ca^{2+} across membrane phases. The area of Ca^{2+} transporters, including the Ca^{2+} channels, has now expanded enormously to include dozens of isoforms of pumps and exchangers, and numerous subtypes of plasma membrane and intracellular channels. Although we now understand a lot about their function, structural studies are sorely needed: the recent solution of the 3D structure of a K^+ channel raises hopes that the 3D structure of Ca^{2+} channels/transporters may not be too far away. The recent results by D. Stokes on the pump of sarcoplasmic reticulum are certainly promising in this respect. And in thinking of ways and means to control cellular Ca^{2+} it is appropriate to mention mitochondria; prominent actors in the early times of cellular Ca^{2+} transport, they went into oblivion during the era of sarco(endo)plasmic reticulum dominance, but are now enjoying a robust revival as important controllers of cytosolic Ca^{2+}.

Naturally, the study of Ca^{2+} homeostasis depends critically on dependable means to measure intracellular free Ca^{2+}. Here too, phenomenal progress has been made from the early days of metallochromic indicators, thanks essentially to the introduction and development of fluorescent indicators by R. Y. Tsien.

These compounds have become everyday tools in calcium research, and are now being joined by ingenious new molecules that have greatly expanded our ways to measure intracellular Ca^{2+}, from targeted aequorin to chameleons.

In conclusion, then, Ca^{2+} is alive and well, and promises to continue to be a prime actor in cell biology for a long time to come. This book, even if it does not cover all possible aspects of the subject (it would have been impossible to do so within the borders of a normal-size book), offers a reasonably comprehensive appraisal of its role. We hope it will become a reference point for Ca^{2+} fans everywhere.

Contents

Contributors

Yasuhiro Anraku
Department of Biological Sciences
Graduate School of Science
University of Tokyo
Tokyo
Japan

George J. Augustine
Department of Neurobiology
Duke University Medical Center
Durham, NC
USA

Ad Bax
Laboratory of Chemical Physics
National Institute of Diabetes and Digestive and Kidney
  Diseases
National Institutes of Health
Bethesda, MD
USA

Joseph A. Beavo
Department of Pharmacology
University of Washington
Seattle, WA
USA

Jörg Benz
Max-Planck-Institut für Biochemie
Abteilung für Strukturforchung
Martinsried
Germany

Stephen Bolsover
Department of Physiology
University College London
London
UK

Andrew Braun
Department of Neurobiology
Stanford University School of Medicine
Stanford, CA
USA

Edward M. Brown
Endocrine-Hypertension Division
Department of Medicine
Brigham and Women's Hospital
Harvard Medical School
Boston, MA
USA

Ernesto Carafoli
Institute of Biochemistry
Swiss Federal Institute of Technology (ETH)
Zürich
Switzerland
Present address: Department of Biochemistry
University of Padova
Italy

Giorgio Carmignoto
Department of Biomedical Sciences and CNR Center for
    Biomembranes
University of Padova
Padova
Italy

Patrice Catty
Unité de Biochimie Physiologique
Université Catholique de Louvain
Louvain-La-Neuve
Belgium
Present address: Département de Biologie Moléculaire et
    Structurale
Centre d-Etudes Nucléaires de Grenoble
Grenoble
France

Guy C. K. Chan
Department of Pharmacology
University of Washington
Seattle, WA
USA

Donna G. Crenshaw
Department of Pharmacology and Cancer Biology
Duke University
Durham, NC
USA

Anna Crivici
Ligand Pharmaceuticals
San Diego, CA
USA

Jennifer S. Dayton
Department of Pharmacology and Cancer Biology
Duke University
Durham, NC
USA

Richard M. Denton
Department of Biochemistry
School of Medical Sciences, University of Bristol
Bristol
UK

Torbjörn Drakenberg
Department of Physical Chemistry
Lund University
Lund
Sweden

Setsuro Ebashi
National Institute for Physiological Sciences
Okazaki
Japan

Makoto Endo
Saitama Medical School
Saitama
Japan

David E. Evans
School of Biological and Molecular Sciences
Oxford Brookes University
Oxford
UK

Sture Forsén
Department of Physical Chemistry
Lund University
Lund
Sweden

Teiichi Furuichi
Department of Molecular Neurobiology
Institute of Medical Science
University of Tokyo
Tokyo
Japan

André Goffeau
Unité de Biochimie Physiologique
Université Catholique de Louvain
Louvain-La-Neuve
Belgium

Danilo Guerini
Institute of Biochemistry
Swiss Federal Institute of Technology (ETH)
Zürich
Switzerland

Robert Huber
Max-Planck-Institut für Biochemie
Abteilung für Strukturforschung
Martinsried
Germany

Mitsuhiko Ikura
Division of Molecular and Structural Biology, Ontario
    Cancer Institute, and Department of Medical Biophysics
University of Toronto
Toronto, Ontario
Canada

Christina R. Kahl
Department of Pharmacology and Cancer Biology
Duke University
Durham, NC
USA

Claude Klee
Laboratory of Biochemistry
National Cancer Institute
National Institutes of Health
Bethesda, MD
USA

Susanne Liemann
Max-Planck-Institut für Biochemie
Abteilung für Strukturforschung
Martinsried
Germany

Julian Loke
Banting and Best Department of Medical Research
University of Toronto
Charles H. Best Institute
Toronto, Ontario
Canada

David H. MacLennan
Banting and Best Department of Medical Research
University of Toronto
Charles H. Best Institute
Toronto, Ontario
Canada

David J. McConkey
Department of Cell Biology
The University of Texas M.D. Anderson Cancer Center
Houston, TX
USA

James G. McCormack
Target Cell Biology
Novo Nordisk,
Bagsvaerd
Denmark

Anthony R. Means
Department of Pharmacology and Cancer Biology
Duke University
Durham, NC
USA

Edon Melloni
Institute of Biological Chemistry
University of Genoa
Genoa
Italy

Takayuki Michikawa
Department of Molecular Neurobiology
Institute of Medical Science
University of Tokyo
Tokyo
Japan

Katsuhiko Mikoshiba
Department of Molecular Neurobiology
Institute of Medical Science
University of Tokyo
Tokyo
Japan

Baruch Minke
Department of Biological Chemistry,
    Physiology and the Kuhne Minerva
    Center for Studies of Visual Transduction
Hebrew University of Jerusalem
Jerusalem
Israel

Alex Odermatt
Banting and Best Department of Medical Research
University of Toronto
Charles H. Best Institute
Toronto, Ontario
Canada

Iwao Ohtsuki
Faculty of Medicine
Kyushu University
Fukuoka
Japan

Yoshikazu Ohya
Department of Biological Sciences
Graduate School of Science
University of Tokyo
Tokyo
Japan

Sten Orrenius
Institute of Environmental Medicine
Division of Toxicology
Karolinska Institutet
Stockholm
Sweden

Lucia Pasti
Department of Biomedical Sciences and CNR Center for
    Biomembranes
University of Padova
Padova
Italy

Kenneth D. Philipson
Departments of Physiology and Medicine and the
    Cardiovascular Research Laboratory
UCLA School of Medicine
Los Angeles, CA
USA

Sandro Pontremoli
Institute of Biological Chemistry
University of Genoa
Genoa
Italy

Tullio Pozzan
Department of Biomedical Sciences and CNR
    Center for Biomembranes
University of Padova
Padova
Italy

Stephen M. Quinn
Endocrine-Hypertension Division
Department of Medicine
Brigham and Women's Hospital
Harvard Medical School
Boston, MA
USA

Hao Ren
Laboratory of Biochemistry
National Cancer Institute
National Institutes of Health
Bethesda, MD
USA

Franca Salamino
Institute of General Physiology and Biological Chemistry
University of Sassari
Sassari
Italy

Luigia Santella
Laboratory of Cell Biology
Stazione Zoologica "A. Dohrn"
Napoli
Italy

Howard Schulman
Department of Neurobiology
Stanford University School of Medicine
Stanford, CA
USA

Zvi Selinger
Departments of Biological Chemistry, Physiology and
    the Kuhne Minerva Center for Studies of Visual
    Transduction
Hebrew University of Jerusalem
Jerusalem
Israel

Carolyn M. Slupsky
Department of Biochemistry
University of Alberta
Edmonton, Alberta
Canada

Johan Stenflo
Department of Clinical Chemistry
Lund University
University Hospital
Malmö
Sweden

Daniel R. Storm
Department of Pharmacology
University of Washington
Seattle, WA
USA

Brian D. Sykes
Department of Biochemistry
University of Alberta
Edmonton, Alberta
Canada

Isei Tanida
Department of Biological Sciences
Graduate School of Science
University of Tokyo
Tokyo
Japan

Nico Tjandra
Laboratory of Chemical Physics
National Institute of Diabetes and Digestive and
    Kidney Diseases
National Institutes of Health
Bethesda, MD
USA

Richard W. Tsien
Department of Molecular and Cellular Physiology
Stanford University School of Medicine
Stanford, CA
USA

Roger Y. Tsien
Howard Hughes Medical Institute and Department of
 Pharmacology
University of California, San Diego
La Jolla, California
USA

Peter M. Vassilev
Endocrine-Hypertension Division
Department of Medicine
Brigham and Women's Hospital
Harvard Medical School
Boston, MA
USA

Samuel S.-H. Wang
Department of Neurobiology
Duke University Medical Center
Durham, NC
USA
Present address: Biological Computation Research
 Department
Bell Labs
Murray Hill, NJ
USA

Xutong Wang
Laboratory of Biochemistry
National Cancer Institute
National Institutes of Health
Bethesda, MD
USA

David B. Wheeler
Department of Molecular and Cellular Physiology
Stanford University School of Medicine
Stanford, CA
USA

Robert J. P. Williams
University of Oxford
Inorganic Chemistry Laboratory
South Parks Road
Oxford
UK

Chen Yan
Department of Pharmacology
University of Washington
Seattle, WA
USA

PART I

General Aspects

1

Calcium: The Developing Role of its Chemistry in Biological Evolution

Robert J. P. Williams

The theme of this chapter will be the evolving role of the calcium ion as the major communicator, and later almost the controller, of the conditions outside cells relative to those inside cells as eukaryotic cells developed from prokaryotes to become multicellular organisms. Stages in this evolution range from calcium being a rejected element from very primitive cells to being a major part of the integrated homeostasis and the triggered activity of a multicellular eukaryotic organism. In order to appreciate the reason for the selection of calcium in these different roles during evolution, we need a background review of the character of calcium as an inorganic ion. I have given much detail of this chemistry elsewhere (Williams, 1976), but it is probably good to have present an outlined account here to remind the reader of salient points. I shall provide this summary in two parts while describing the development of the changing use of calcium from prokaryote to eukaryote single cells. The first concerns the evolution of calcium chemistry in the environment, the sea, which is given in the next section, while the second concerns the selective interaction of calcium with organic molecules, especially proteins, which is given in a later section, "Selection for Calcium by Proteins". The final use of calcium in multicellular organisms is a development from that in single-cell eukaryotes.

The Inorganic Chemistry of Calcium in the Sea

Calcium is a plentiful element in nature and the sea contains some 3 mM today while it also contains 10 mM magnesium. We do not know the condition of the original ocean 4 billion years ago, when life began, but it is likely that the sea was more acidic and that the atmosphere above it was 1000 times higher in carbon dioxide (Henderson, 1986). Acidity slowly dissolved silicates and the pH rose

$$M_4SiO_4 + 4H^+ \leftrightarrows 4M^+ + Si(OH)_4 \qquad (1)$$
$$M_2SiO_4 + 4H^+ \leftrightarrows 2M^{2+} + Si(OH)_4 \qquad (2)$$

which gave a sea rich in Na^+, K^+, and Mg^{2+}. The reduced acidity and the high carbon dioxide worked in opposite directions as far as free calcium is concerned. At any pH, there are the calcium-dependent equilibria

$$Ca_2SiO_4 \downarrow +4H^+ \leftrightarrows 2Ca^{2+} + Si(OH)_4 \qquad (3)$$
$$2Ca^{2+} + 2H_2O + 2CO_2 \leftrightarrows 2CaCO_3 \downarrow +4H^+ \qquad (4)$$

Given the fact that CO_2 fell, the consequences were that, eventually, the ocean became saturated in $CaCO_3$ and $Si(OH)_4$ at a pH around 8.0. The Ca^{2+} concentration has not been too far from constant during all the later periods of life's evolution, but life may well have started under somewhat different calcium conditions of acidity and mineral content of the sea.

The Primitive Cell

The early sea also contained chloride, sulphide, phosphate, organic anions, and trace elements, as well as sodium, potassium, magnesium, and calcium (Williams and Frausto da Silva, 1996a). Cellular life, separated from the sea by a membrane, is based on organic chemicals, and required that most of these organic molecules were anions in order to maintain solubility in water. Any metabolic chart shows these anions, often as phosphates or carboxylates, to dominate cells to this day. To keep a reasonably low internal osmotic pressure and a reasonable electrical neutrality, there was then demanded a rejection from the cell cytoplasm of both of the components of the dominant salt of the sea, NaCl (Fig. 1.1). To maintain stable conditions and the neutralisation of charge (internally, and a low osmotic pressure), the only sufficiently available cation which could be allowed in the cell and did not bind to organic molecules was potassium. Thus, Na^+ and Cl^- were pumped out of cells while K^+ was admitted. Note that this requires a potassium channel. Now even quite low calcium was (and is) also disadvantageous in such a cell since it precipitates organic anions rather readily. To prevent this precipitation, yet to allow acid catalysis for synthesis, Ca^{2+} had to be rejected and Mg^{2+}, which forms more soluble phosphates, accepted by cells. There is then a further requirement for a calcium pump in addition to those for sodium and chloride (Fig. 1.1). So it is in all cells to this day—cells are rich in K^+, Mg^{2+}, phosphates, and carboxylates and poor in Na^+, Ca^{2+}, and Cl^-—and it has to be so. This division of elements was more critical to the very beginnings of a protected metabolism inside a membrane enclosure than was the need for any code, e.g. RNA or DNA.

Of course, such separation of elements to stabilise a cell required energy and while we shall not consider the origin of energy capture here, we notice that all metabolism, synthesis, and pumping is related to the energy source, through ATP (Fig. 1.1). We shall assume in this chapter that ATP could be made very early in life's development. Given that the cell is full of such phosphate anions and HCO_3^-, there is one more major step needed in the description of the homeostatic cytoplasm before we return to calcium functioning. It was clearly necessary to control pH of the cell. Roughly, this control was helped by the buffered external pH of the sea but, internally, cellular metabolism always created an internal/external out of balance, a pH gradient, due to metabolism. If the reactions of carbon assimilation started from CO_2, then reduction was required and the pH rose in the cell:

$$CO_2 + 4H^+ + 4e \rightarrow (HCHO) + H_2O \qquad (5)$$

If CH_4 was the starting material, then oxidation must have been active:

$$H_2O + CH_4 \rightarrow (HCHO) + 4H^+ + 4e \qquad (6)$$

and the pH would have fallen. (Note that electron sources and sinks and energy for these redox reactions are assumed to be present possibly through iron/sulphur chemistry) (Williams, 1965; Wachterhauser, 1988).

Since the pH affects many phosphate anions, e.g. $R\text{-}OPO_3^{2-} + H^+ \rightleftarrows ROPO_3H^-$, including ATP used in maintaining the above ion gradients, the require-

ment is clearly for a coupling of H^+/OH^- movement with the electrostatic balances of Na^+, Cl^-, HCO_3^-, K^+, Mg^{2+}, Ca^{2+}, and phosphates. The link was partly by exchange and partly that of ATP to proton gradients (Fig. 1.1), which may have been the primary connection to the basic redox source of energy in cells (Williams, 1961; Wachterhauser, 1988; Williams and Frausto da Silva, 1996a). As far as we know, these two forms of energy—proton gradients and ATP—have always been intimately connected in all cells. There are ATP syntheses from metabolically generated H^+ gradients and excess H^+ pumping out using ATP (Fig. 1.1). Thus pH homeostasis is linked to virtually all activities of primitive cells.

In effect, two energy sources, the proton gradient from redox reactions [e.g. equations (5) and (6), or light-driven, or due to iron sulphide reactions] and energy directly from the disproportionation of carbon compounds, glycolysis [e.g. of $(HCHO)_n$, sugars, etc.], were in controlled connection with synthesis through a relationship with ATP (Fig. 1.2). There is, in fact, a requirement for such dynamic homeostasis if a cell is to develop and be reproduced in a precise way.

We must see, too, that even an early cell had also to operate as a metabolising system for *balanced* synthesis of several organic molecules by different pathways, sugars, fats, proteins, and nucleotides. Such balance requires for its homeostasis some common *internal* cell messengers between and within reaction paths so as to control and regulate each path proportionally. Path (a) is then in feed-back or feed-forward connection with path (b) and so on. The best messengers between different paths are substances which do not metabolise but do exchange quickly, e.g. Mg^{2+}, Na^+, K^+, or Cl^-, but they must also bind quite well, which leaves Mg^{2+} as an extremely valuable internal messenger. An alternative general messenger is a substance which only metabolises in one or two simple chemical ways, with relatively fast kinetics, but that is required in many reaction sequences. Obviously carbon/nitrogen compounds are not the most suitable since they can undergo many different transformations, have slow kinetics, and do not link directly either to any simple substance which is required in several pathways, or in a simple way to energy which must also be carried to every pathway in a balanced way. [Note that of course co-enzymes are required for H, C, N, O, transfer and can act as limited messengers]. The obvious further element which meets the criteria for a good messenger is phosphorus, as phosphate; phosphate reactions are required in many pathways and they are readily made rapid, phosphate is neither oxidised nor reduced in cells, and phosphate as pyrophosphate (ATP) carries energy and is essential for all

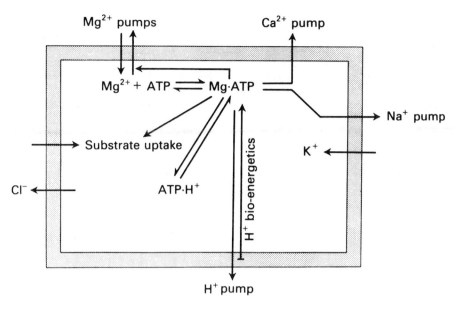

Figure 1.1 The major movements of the simple cations and anions in all cells at rest.

pathways of synthesis. Many (primitive) prokaryote cell messengers, pathway controllers, are, in fact, phosphates. Again, many phosphates are linked with both H^+ and Mg^{2+} concentrations through binding so that these three homeostatic networks are linked. Figure 1.2 shows many phosphate esters and anhydrides in cells acting as messengers and energy distributors between paths, e.g. cNMP (mes-senger only) and NTP (messengers and energy distri-butors), where N is any nucleotide. Since ATP is connected to the pumps for Na^+, K^+, Cl^- and Mg^{2+}, these ions can assist in the control of metabo-lism as their levels, quite high in cells, are related to energy (ATP) available and they can act as allosteric messengers in liaison with phosphates. As far as cal-cium is concerned in this, the earliest, stage of cellular

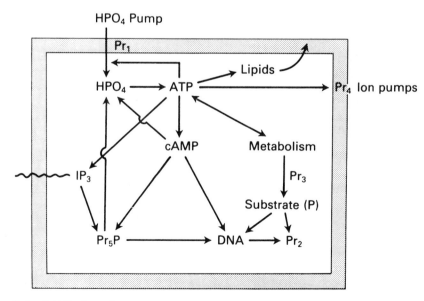

Figure 1.2 The link of ATP, through many proteins (Pr), to metabolism, to controls via transcription factors binding to DNA and using cAMP or phosphorylation, to messengers such as IP_3, and to ion pumps.

development (prokaryotes), the calcium-selective system was the ATPase pump alone, which had to select Ca^{2+} against Na^+, K^+, and Mg^{2+}, while pumping out Ca^{2+} to a very low level, and we presume that calcium had no messenger role inside primitive cells. As far as the author knows, anaerobic prokaryotes to this day depend largely on K^+, H^+, Mg^{2+}, and phosphate internal controls while showing little dependence on Na^+, Cl^- or Ca^{2+} all of which were rejected. (Undoubtedly today some *aerobic* prokaryotes use calcium in limited triggers very different from those of eukaryotes).

With this quick dash through the description of the cytoplasm of the primitive cell and its homeostasis, we see that calcium had virtually no role to play except outside the cell where it could act so as to stabilise the cell membrane and wall, and perhaps to assist enzyme breakdown of large molecules for food. This is not a bad description of the role of calcium in the simplest anaerobic bacteria to this day. Lowering (or increasing) extracellular calcium does them little harm except that it can destabilise (or stabilise) the outermost membrane, and, internally, early cellular systems just maintain free calcium at a low but not very fixed level.

The Increase in Cell Complexity

It is believed that a relatively early step of great consequence in evolution was the development of the nucleated eukaryote cell (Margulis and Schwartz, 1988). Our first interest is not so much in the nucleus itself but in the variety of vesicles and filamentous proteins which also appeared and, in particular, the reticula and then the organelles (Fig. 1.3). In effect, there were generated new aqueous solution *compartments*, in addition to the cytoplasm and the sea, and not in equilibrium with either of them. These compartments, vesicle solutions, served as chemical depositories, as buffers for internal solutions, as zones for isolated activities such as degradative digestion locked away from the synthesis in the cytoplasm. They became used for endocytosis and exocytosis of wanted and unwanted small or large molecules. Since they could be at a different pH or a different redox potential from the cytoplasm, different reactions were carried out in each one of them, including synthesis steps such as protein glycosylation. The vesicles were also used to prepare materials for export—biominerals, wall or periplasmic proteins, and polysaccharides. Increase in spatial separation, isolation of activities, has many advantages for the success of chemical processes, as is obvious in a laboratory or factory. These vesicles had to be held in place, of course, and this was achieved by the filaments. There is a prerequisite, however, which is that

the regions which are spatially separated must remain in communication with one another and with the cytoplasm to ensure homeostasis, and this is especially so since a cell must reproduce exactly. In a laboratory, this objective of devising communicating compartments is achieved by the chemist's apparatus. To understand the communication in a single biological eukaryote cell, we need to note new chemical features of vesicles, filaments, reticula, and organelles, all of which must now be in homeostatic connection.

In large part, vesicles are more like the electrolyte solution outside the cell, the sea. They are usually quite high in Na^+ and Cl^-, though low in pH. Almost invariably, they are also high in calcium, since the pumps which face the external waters on the outer membrane now face the inside of the vesicle. The vesicles do not usually contain high levels of energised phosphate since NTP (where N is a nucleotide) does not pass through membranes except when pumped. (There are some exceptions, as in the synthesis of glycoproteins in the Golgi.)

Observations on single eukaryote cells, such as yeasts, show that, unlike prokaryotes, they now have many internal *cytoplasmic* calcium-receptor proteins, such as calmodulins (Table 1.1). An immediate suggestion is, therefore, that a major additional messenger for coordinating the eukaryote cell's activity was calcium, this calcium coming either from outside the cell or from vesicles. Thus, we shall consider the possibility that calcium became a major messenger to the cytoplasm from all aqueous compartments external to the cytoplasm, interacting with vesicles and filaments, as well as cytoplasmic systems at low concentrations. Remember that calcium has all the prerequisites to act as a messenger—no metabolism, fast kinetics, easy recognition, it can bind to proteins of many pathways, and an energised gradient of it had already been made in prokaryotes by its rejection to protect the cytoplasm. Simultaneously, with the need for development of *internal* communication between cytoplasm, filaments and vesicles, (homeostasis) the larger eukaryote cell needed sensors and a messenger on its *external* surface, which had become flexible, to allow it to react to the environment to its advantage, e.g. to change shape or to locate chemical gradients. Given the presence of calmodulin (troponin) as a receptor, on acto-myosin it seems very probable that calcium became this messenger too. The calcium message in this and other cases clearly demanded, now, a controlled calcium membrane channel, allowing calcium to enter the cytoplasm, and, of course, the channel had to be switched on or off by conditions external to the cell. Of necessity, the calcium pump already present in prokaryotes removed the excessive calcium after the signal had been received and acted upon. Many

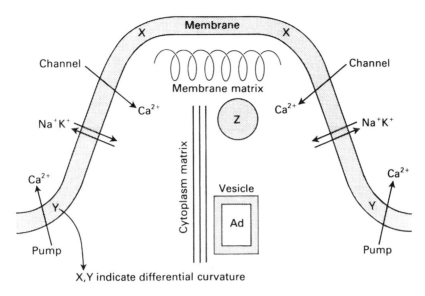

X,Y indicate differential curvature

Figure 1.3 The link between membrane, membrane-associated protein structures, calcium, and vesicle compartments. The diagram indicates that the whole is a unit of activity in that tension, curvature, positioning of pumps and channels, and movement of vesicles *all* depend on the activity of a cell in an interactive manner (see text). Cell shape is a steady-state property. Ad, adrenaline; Z, a vesicle which could contain calcium.

eukaryotes also build extensive $CaCO_3$ units internally in vesicles, to make external shells later, so that control over calcium became extremely important.

Now, as already stated, the maintenance of and control over cell shape and over the local disposition of vesicles within a cell required the simultaneous development of filaments in managed tension. Filaments are fundamental to the process of division, even in prokaryotes where they depend on the phosphate metabolism for contractile energy. The activity of the filamentous system, connecting different parts of the outer membrane and the vesicles to this membrane, obviously had to be linked to (phosphate) energy, ATP, allowing the cell to maintain steady state, to respond to the outside environment, to change cell shape, or to reject material from vesicles for example (Fig. 1.3). All these activities came then to be dependent on the calcium and the phosphate message systems.

In summary, in order that calcium should work as a message there had to be control over input to the cytoplasm using receptors on membranes The new message system developed as a control of (triggered) Ca^{2+} input channels, from outside or from internal vesicles, and utilised the existing calcium pumps or exchangers to maintain or restore rest-state homeos-

Table 1.1 Some Internal Calcium Proteins of Eukaryotes

Protein	Function	Structure
Calmodulin	Trigger of metabolism	α-Helical
Troponin C	Trigger of contraction	α-Helical
Calbindins	Transport and/or buffer	α-Helical
Annexins	Trigger of exocytosis	α-Helical
S-100	?	α-Helical
Parvalbumin	Calcium buffer	α-Helical
Calcineurin B	Phosphatase	α-Helical
Neuroproteins (neurocalcin)	?	α-Helical
C-2 domains	?	β-sheet
Phospholipases	Hydrolysis	β-sheet

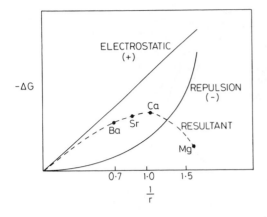

Figure 1.4 The contributions to binding energy, $-\Delta G$, from electrostatic forces plotted against $1/r$ and the effect of increasing ligand–ligand repulsion as the ligands crowd around the central ion. The dashed line shows the resultant binding free energy with a maximum at calcium.

tasis. We need to appreciate three parts of this calcium chemistry in the isolated eukaryote cell: (1) the nature of the protein chemistry which allows the input channel, export pump, and receptor to be selective, especially over Mg^{2+}; and (2) the relationship of the new calcium message system to the pre-existing phosphate prokaryote messenger system in the cell, since clearly the calcium and phosphate control systems had to be dove-tailed; (3) The synthesis of the new proteins required for the new calcium activities. Only when we have such a picture can we turn to multicellular organisms.

We have assumed this all-embracing role of calcium in the evolution of eukaryotes without absolute proof, and it would be difficult to give any very convincing discussion, but we note that the extensive distribution of calcium and its functioning as a control ion in cells is today common to a huge range of eukaryote cell types from animal, plant, or fungal sources, which diverged billions of years ago. It even extends into the organelles (captured bacteria?) where calcium controls dehydrogenases of mitochondria and the dioxygen-evolving enzyme of chloroplasts. We must see how the functional value of calcium could have so changed from its simple rejection and use outside prokaryote cells. This involves a second look at calcium chemistry (Linse and Forsén, 1995).

The Possible Evolution of Calcium Pumps

Before examining this calcium biochemistry in eukaryotes, I present an intriguing possibility of the step which allowed the importance of calcium to evolve as eukaryotes appeared. We must ask first, what is the concentration level of calcium in cells which a prokaryote can manage without damage? An analysis of calcium precipitation of phosphate and carboxylate compounds suggests that a value of 10^{-5} M would be

protective. The activity of organelles such as mitochondria, as well as that of present-day bacteria, does not seem to be affected by such an internal concentration. This level of internal calcium would be managed by the then existing (3×10^{-9} million years ago) calcium pumps. The use of calcium as a fast trigger of signalling is then difficult, since a very considerable influx of calcium, raising its concentration to 10^{-4} M, would be necessary. One great advance in the eukaryotes that we may propose is that the calcium ATPase pump became more effective by mutation, so as to lower internal calcium to 10^{-7} M. At this level of internal calcium, the calcium cation could be used in a fast signalling system. Of course, a channel was required too. Note that this change requires a change in homeostasis. To this day, it may well be that anaerobic prokaryotes are not linked to calcium pulses between 10^{-7} M and 10^{-5} M. We now turn to the evolution of calcium binding proteins in eukaryotes.

Selection for Calcium by Proteins

The affinity of an organic matrix, protein or some other, for simple ions such as calcium is due to electrostatic interaction with negative charge or dipolar centres (Fig. 1.4). This would be purely a matter of the charge and size dependence of affinity, ΔG, which for divalent ions falls in the order

$$Mg^{2+} > Ca^{2+} > Sr^{2+} > Ba^{2+}$$

but for the effect of steric constraints amongst ligands around the central ion (Figs. 1.4 and 1.5). It is observed that Mg^{2+} fits almost perfectly into the hole created by six octahedral water molecules, i.e. six oxygen donor atoms, which collapse around the cation to the oxygen–oxygen close-packed distance. Any organic group tends to fit less well around this

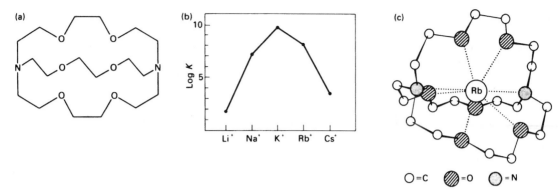

Figure 1.5 The formation of cavity ligands here to distinguish between group 1 metal ions. (a) The cavity ligand; (b) binding constants; (c) fitting by collapse around an ion.

small ion so that $[Mg(H_2O)_6]^{2+}$ prefers to exchange water with very few, small, simple (organic) anions either in complexes or precipitates. The salts of its larger anions are then soluble, e.g. those of SO_4^{2-} and of most organic anions. Chelation, as in oxalate or pyrophosphate (as in ATP), increases magnesium binding in complexes, but they remain somewhat soluble. Organic groups of lower polarity, e.g. carbonyl and ether oxygen, even in chelating ligands, do not readily displace water from $[Mg(H_2O)_6]^{2+}$, though alcohol —OH groups can do so (Birch, 1993). Calcium, which is considerably larger in size (diameter 2 Å, compare magnesium 1.2 Å), has a hydrate of ill-defined coordination number, probably fluctuating dynamically between seven-, eight- and nine-coordinate with H_2O molecules held at various distances. Not surprisingly, then, Ca^{2+} has a very adaptable coordination sphere in its complexes and can accept larger (organic) anionic groups more easily than Mg^{2+}. Hence, quite simple calcium salts of anions such as SO_4^{2-} and organic phosphates tend to precipitate. Again, this ion readily accepts the carbonyl, ether, and alcohol O-donor atoms in place of water in multidentate chelates where close distance of approach is not demanded. Then, Ca^{2+} has a very flexible geometry both with relationship to bond angle and bond distance. In solution or in crystals, we observe that both the first and second coordination sphere are usually irregular, and both contribute to overall stability. Very little can be predicted quantitatively. However, it is easily seen, generally, how selection in favour of calcium rather than magnesium association is achieved in complexes, e.g. by EGTA, or in cross-linking, or in precipitates, e.g. of carbonate. It is simply a matter of easier fitting at larger distances and the lower energy of hydration of the calcium as opposed to the magnesium ion. When we move on to Sr^{2+}, and Ba^{2+} ions, the electrostatic interactions with anions are weaker (Fig. 1.4) and

the cations are larger, so that while geometry remains flexible, a usual binding series to a complex organic chelating ligand, such as a protein is

$$Ca^{2+} > Sr^{2+} > Ba^{2+}$$
$$> Mg^{2+}$$

However, while strontium and barium are rare elements, magnesium is often present in much higher concentrations than calcium. There are now many conditions in which the competition between Mg^{2+} and Ca^{2+} for protein binding sites must be managed.

Notice that the flexibility of the calcium coordination sphere allows very fast exchange of water for other binding groups but this is not true for magnesium (Eigen, 1960). We shall see that in fact eukaryotes have evolved a very large number of calcium-binding proteins not found in any prokaryotes both of low and high binding strength and different exchange rates.

Calcium-Binding Selection Outside Cells

Selection for Ca^{2+} is now seen to be relatively easily managed, but there are two different concentration ranges in which this had to happen, even in unicellular eukaryotic systems. The first is outside the cell, applicable equally to prokaryotes, where $[Ca^{2+}]$ is, say, 3×10^{-3} M in the sea while $[Mg^{2+}]$ is some 10^{-2} M. Here, organic constructs from polysaccharides or proteins must have binding constants or solubility products such that the value for Ca^{2+} is at least 10^3 and exceeds that of Mg^{2+} by 10^2. Undoubtedly the best way of achieving this objective is to make a matrix from lowly charged groups that generate a large cavity and fits, or can collapse *locally* to fit, the Ca^{2+} ion but will not collapse further to fit the small Mg^{2+} ion (Williams, 1976). The best donor groups are oxygen atoms of carbonyls, ethers, or alcohols with one or two carboxylates. Too low a charge centre could favour Na^+, which has the same size as Ca^{2+}. Table 1.2 indicates how this is managed in lattices and in proteins. Notice that, apart from bacterial wall stabilisation, there are very important cases of protein–protein and mixed saccharide–protein complexes of their kind, e.g. the lectins (of multicellular organisms) which use Ca^{2+} as a binding cross-link (see below). These considerations apply outside cells (in the sea), inside vesicles, and later in evolution in the controlled extracellular fluids of higher multicellular organisms (see below). In all cases, free calcium exceeds 10^{-3} M.

A particularly stable construct outside cells uses a β-sheet protein with a fixed cavity for calcium that cross-links the system in an almost rigid fashion since dynamics are not required for these structures. Examples quite close to this description are various

extracellular enzymes, such as phosphatases (Fig. 1.6) and saccharases. There are many more examples in multicellular organisms, as we shall see. There are also some very flexible proteins which bind calcium in external fluids and serve a very different purpose (see section on "The nature of multicellular organisms").

Calcium Recognition Inside Cells

The general problem of binding Ca^{2+} selectively inside a eukaryote cell is quite different since the competition from Mg^{2+} is much more severe (Frausto da Silva and Williams, 1991). (Note that later we shall consider specific classes of cells which differ in Ca^{2+} ion dependence). The cell contains 1×10^{-3} M Mg^{2+}. A calcium message has to operate from an observed baseline of intracellular Ca^{2+} concentration between 10^{-7} M and 10^{-8} M due to out-

Table 1.2 Calcium Sites in some proteins

Protein	Sequence	Continuous[a]
Phospholipase A_2	Tyr-28, Glu-30, Gly-32, Asp-49, $2H_2O$	No
Trypsin	Glu-70, Asn-72, Glu-80, $2H_2O$	Yes
Staphylococcal nuclease	Asp-19, Asp-21, Asp-40, Thr-41, Glu-43, $1H_2O$	No
Con A	Asp-10, Tyr-12, Asn-14, Asp-19, $2H_2O$	Yes
Thermolysin	Asp-138, Glu-177, Asp-185, Glu-187, Glu-190, $1H_2O$	No
	Glu-177, Asn-183, Asp-185, Glu-190, $2H_2O$	No
	Asp-57, Asp-59, Glu-61, $3H_2O$	Yes
	Tyr-193, Thr-194, Ile-197, Asp-200, $2H_2O$	Yes
Parvalbumin		
CD	Asp-51, Asp-53, Ser-55, Phe-57, Glu-59, Glu-62	Yes
EF	Asp-90, Asp-92, Asp-94, Lys-96, Glu-101, $1H_2O$	Yes
Troponin C, rabbit skeletal		
I	Asp-27, Asp-29, Asp-33, Ser-35, Glu-38, $1H_2O$	Yes
I	Asp-63, Asp-65, Ser-67, Thr-69, Asp-71, Glu-74, $1H_2O$	Yes
III	Asp-103, Asp-105, Asp-107, Tyr-109, Asp-111, Glu-114, $1H_2O$	Yes
IV	Asn-139, Asn-141, Asp-143, Arg-145, Asp-147, Glu-150, $1H_2O$	Yes
Calmodulin		
I	Asp-20, Asp-22, Asn-24, Thr-26, Thr-28, Glu-31, $1H_2O$	Yes
II	Asp-56, Asp-58, Asn-60, Thr-62, Asp-64, Glu-67, $1H_2O$	Yes
III	Asp-93, Asp-95, Asn-97, Tyr-99, Ser-101, Glu-104, $1H_2O$	Yes
IV	Asn-129, Asp-131, Asp-133, Glu-135, Asn-137, Glu-140, $1H_2O$	Yes
S-100	Asp-61, Asp-63, Asp-65, Glu-67, Asp-69, Glu-72, $1H_2O$	Yes
Calbindin		
I	Ala-15, Glu-17, Asp-19, Glu-21, Leu-23, Glu-26, $1H_2O$	Yes
II	Asp-54, Asn-56, Asp-58, Glu-60, Ser-62, Glu-65, $1H_2O$	Yes

[a]"Continuous" implies that the binding groups are close together in a loop of sequence.

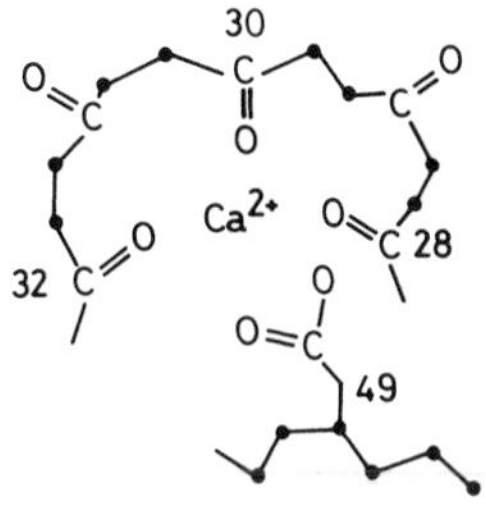

Figure 1.6 The cavity formed mainly by carbonyls for the calcium ion of phospholipase A_2.

ward pumping and cannot be allowed to increase above 10^{-5} M since the cell must recover quickly (pump out calcium) to avoid destruction by cross-linking and then to be ready for the next message. It is observed that simple single-pulse activity in the eukaryote cytoplasm due to Ca^{2+} develops at around 10^{-6} M. Thus, the binding constant for calcium must be $\log K = 6$ while that for Mg^{2+} must be $\log K < 10^3$ to avoid competition. There are several known designs of single protein binding sites based on a moderately loose cavity, formed by α-helices, which, by adjustment of the protein *fold*, as well as by *local* changes, can contract to fit Ca^{2+} but cannot contract further to fit Mg^{2+}. The cavity is usually made from five to two or three protein carbonyl groups plus from one to four carboxylates, respectively, and an odd water molecule (Table 1.2). The Ca^{2+} is, as expected, seven- or eight-coordinate. Notice that I have stated, in this case, that the cavity *contracts* through *fold* changes. It is this contraction which forces a conformational change in the protein matrix and generates a localised (mechanical) message in the cell system. This demands a protein design for conformational change and not just a relatively rigid body recognition (see below). It is possible to design a cavity to fit calcium precisely without contraction but while selectivity would be improved, the kinetics involving binding, i.e. on/off rates, would be slow. While such a slow system makes a good device for calcium binding alone, say outside a cell, or for calcium transport, it is useless for *fast* transmission of information since the protein does not adjust. We observe that the *fast* mechanical devices inside cells are all α-helical proteins while such proteins are sel-

dom seen outside cells. Now we must stress that the way in which this triggering works usually involves the removal of inhibition by calcium-binding to a single protein site. It is not to be confused with calcium activities where calcium acts as a cross-linking agent between proteins or between a protein and a negative surface which we describe next.

Alternatively it is possible to use combinations of ligands inside cells, such as a protein cavity plus a phospholipid, say, of a membrane to achieve the selective binding of $\log K = 6$ for Ca^{2+}. Here the Ca^{2+} binding to the protein itself has a low affinity, $\log K = 4$. The protein may be an α-helix or a β-sheet and sometimes several Ca^{2+} ions are bound e.g. in the C_2-domains. Triggering may or may not be of the protein conformation but of protein/phospholipid association. In this case calcium-binding kinetics are slower and of considerable complexity. The parallel is with lectin cross-links outside cells, (see Fig. 10). The types of cavity are illustrated in Table 1.3.

This illustrates the general control over calcium selective binding *at equilibrium* but it has been observed that the rates of reaction of sites for calcium in different proteins are very different.

Calcium Kinetics

The kinetics of calcium reaction control the response/relaxation and residence times (Frausto da Silva and Williams, 1991). Different message times are required for a fast on/off response and for a sustained reactivity.

Consider first the simple reaction

Table 1.3 Types of Calcium Cavity

1. Relatively rigid, generated by tight protein β-fold, e.g. phospholipase A_2 and C_2-domains
2. More loosely folded protein, e.g. calmodulin, α-fold
3. "Random" protein folded by calcium, e.g. osteocalcin
4. More loosely folded protein plus phospholipid, e.g. annexin, α-fold
5. Relatively rigid protein plus polysaccharide, e.g. lectin, β-fold
6. Relatively rigid pair of proteins, e.g. gelsolin, β-fold

$$Ca^{2+} + L \underset{k_{off}}{\overset{k_{on}}{\rightleftarrows}} CaL^{2+}$$

The ratio of rates k_{on}/k_{off} is the binding constant K. Now the fastest on-rates are limited by the diffusion rate of calcium and/or the rate of water loss from calcium on collision. Water exchange rate from Ca^{2+} is 10^9 per second, one thousand times faster than for Mg^{2+} which makes Ca^{2+} an excellent messenger for a fast response when we remember that its divalent charge and its large size allow it to bind much better than magnesium to many centres. The off-rate inside a cell for this on-rate has the maximal value of 10^3 per sec if $\log K = 6$. Thus, here calcium action leads to millisecond single-pulse turnover of cell machinery as in fast muscles. This fast machinery has to operate concommitantly with slow changes in the cell's activities. The slow changes are protected from calcium by its fast rejection. Now the kinetics can be made slower if the properties of the ligand, L, are such that the cavity for calcium binding changes relatively slowly to accept calcium. Here, the rate-limiting on-step is not the rate at which water leaves calcium but the relaxation of L, the protein, which itself is initially not in the best state for calcium binding. A simple way of achieving this objective is to make the protein stiffer for access of calcium. We can write

$$L' \underset{k'_{off}}{\overset{k'_{on}}{\rightleftarrows}} L + Ca^2 \underset{}{\overset{}{\leftrightarrows}} CaL^{2+}$$

The equilibrium binding constant of L' may remain 10^6 but the overall on-rate can have any value, say 10^6 seconds^{-1} if k'_{on} is slow when the calcium off-rate is 1.0 per second. The residence time of calcium is now 1 second not 10^{-3} seconds. Another way of looking at the equation is to say that L is less stable (k'_{on}/k'_{off}) than L' by 10^3. Since calcium binds to L' with a constant of 10^6 it now has to bind to L with a constant of 10^9. A variety of changes of L' to L allow such slow kinetics for calcium. For example (1) L' is an Mg^{2+}-bound protein, (2) L' is in the wrong conformation to bind Ca^{2+}, and (3) L' is a composite of say a lipid and a protein, where *the combination* must associate and change shape to bind Ca^{2+}. (It must be made expressly clear again that the most selective process is the fitting of two rigid objects, dye/mould, but this cannot be the fastest and cannot give a signal through mechanical changes of structure since the mould is rigid). In general calcium *signalling* demands the mechanical rearrangement of proteins but the *transport* of calcium or its long-term *buffering* do not necessarily require such mechanical changes, see below.

Given such variable on-rates it is easy to see that the single fastest pulse (on and off) can be completed in 10^{-3} seconds with no other consequence than say contraction for a cell, but this will require removal of excess calcium to prevent slower processes. Typically calmodulins respond in this way. An alternative method of stimulation is to alter the duration of the calcium fall in the cell by pulsing, see next section. A slower response is that of protein/phospholipid association due to calcium. Typically annexins and C_2 domains operate in this second mode where they themselves bind weakly.

Fast Single Pulses and Multiple Sustained Pulsing

We now must distinguish in some detail fast contractile (muscle) responses of filaments of 10^{-3} seconds from general sustained action (> 1 second), although both are known to be governed by calcium in-put. Frequently they are observed in different types of cell. The simplest case of fast action is that induced by a single calcium pulse, for example that caused by the total cell-depolarisation of the Na^+/K^+ membrane potential, a switch which is imposed on a fast muscle cell or at nerve synapse. All action is over in 10^{-3} seconds or thereabouts. The depolarisation causes calcium channel openings. Now if this contractile process of calcium activation, by pulsing via a nerve depolarisation of the whole cell, is repeated rapidly (even artifically) over a long period, then intracellular calcium never returns fully to resting levels. For example if the resting in-cell calcium level is approximately 10^{-8} M and the muscle triggering is at 10^{-6} M, then repeated fast pulsing may well raise the new "resting" concentration to, say, 10^{-7} M, (Fig. 1.7). Let us assume there are present, proteins which accept calcium slowly with binding constants of close to 10^{-7} M, while the rapid contractile system responds to 10^{-6} M calcium (Chin et al., 1998). The rapid pulsing will elevate the new "resting" state so as to activate the slower processes while the repeated twitch activities of the fast single pulse remain. This is observed in the differentiation of slow muscles to fast muscles on rapid artificially imposed pulsing and, although it is known as an artificial example, it may occur during differentiation of muscle cells in the foetal state. A fast muscle system could not develop until fast nerve signalling was in place.

Consider next, the impact of a messenger that, unlike the Na^+/K^+ message of depolarisation, which is very fast (10^{-6} seconds), is sustained. An example would be the impact of a hormone molecule which bound to a receptor on the outside of the cell, with a binding constant of 10^9 and an on-rate of 10^8 - the off-rate is 10 seconds. Let this binding open a

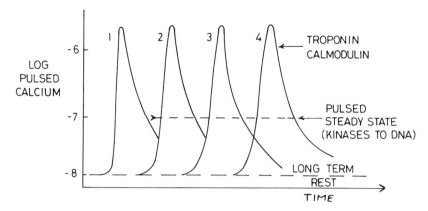

Figure 1.7 The levels of free calcium in a cell following a train of fast pulses (1) to (4). The pulses must be fast enough so that there is never complete recovery from a fast pulse before the next pulse is generated. Note the "resting" level is raised from 10^{-8} M to 10^{-7} M.

calcium channel. Now the channel must not remain open for a long period since this would flood the cell with calcium. A solution to the problem is to allow the calcium which first enters the cell to switch off the calcium input channel when the calcium reaches 10^{-6} M. As the calcium falls due to pumping or exchange the calcium input channel linked to the hormone re-opens, say at 10^{-7} M Ca^{2+} and calcium entry is switched on again. The calcium enters the cell in a series of pulses or waves just like an artificially pulsed depolarisation. This again raises the "resting" calcium level somewhat and may allow many (slow) processes which require say 10^{-7} M calcium. Thus there can be different types of calcium-dependent activities in single cells (Williams, 1995).

1. Those which respond to *general* cell depolarisation—fast contracting muscle cells.
2. Those which respond *locally* to hormones at receptors of which there can be many types. (the responding cells may have no contractile systems). Here the calcium in *local* vesicles may be extremely valuable, (Fig. 1.3).
3. Combinations of (2) then (1) to alter the prepared state of a cell at "rest" for very rapid action e.g. adrenaline tuning.
4. Slower changes due to protein/phospholipid association when relaxation especially is slow, e.g. vesicle discharge.

Cooperative Binding of Calcium Itself and With Other Messengers

The cooperativity of calcium proteins is well-illustrated by calmodulin which binds four calcium ions, (Linse and Forsén, 1995). Cooperativity with a Hill constant >2 gives a tighter range of activation by calcium than a simple sigmoidal binding curve ($n = 1$). The advantage is that a pulse acts over a more limited range of concentration in order to go from rest to 100% action.

A second cooperative mode is for the original event at the surface of the cell to release both calcium and another (organic) messenger. The two then stimulate differently at different enzymes and bring about a combined effect in the cell. Such complex situations are well known in the releases of IP_3 and of lipids, from a lipid/IP_3 combination, both of which may act on calcium sensitive systems. Thus IP_3 activates internal calcium stores while the released lipid, e.g. arachadonic acid, may well work on an outer membrane channel for calcium admittance. The complexity of calcium signalling includes also calcium monitoring of calcium release.

Clearly cells need different kinds of proteins in order to respond differently to calcium pulses. We now turn to a more detailed description of calcium-binding proteins which are found both inside and outside *single* and multicellular eukaryote cells.

The Proteins Which Bind Calcium

Proteins have two major secondary structure motifs α-helices and β-sheets. The difference between the two lies in their mobility. The α-helical segments are to be thought of as rigid rods, which can move relative to one another to a variable degree depending on the protein under consideration or on a rigid surface of say a β-sheet, or within a lipid bilayer perpendicular to the plane of the bilayer. A β-sheet is a fixed structure which can bend around to form a barrel. Barrels can move around one another but individual staves of the barrel are fixed. Thus, a β-sheet has the great advantage of presenting an almost fixed surface, while a set of α-helices present a variable surface (Williams, 1995). Helices then allow the construction of mechanical devices such as a fulcrum, a hinge, a rising hinge, a variable channel, a pump and so on, but they are not of the highest selectivity in their interaction with substrates. Moreover, the fact that upon binding a substrate, a set of helices may be adjusted, allosteric change, removes the possibility of utilising the substrate-binding energy optimally to strain the bound object. There follows three observations (see Frausto da Silva and Williams, 1991):

1. A helix bundle cannot act as the optimal catalyst, enzyme, since very good specific recognition and generated strain in the substrate are of the essence of enzyme action.
2. A helix bundle cannot induce optimal attacking power in a bound local group or cofactor e.g. a metal ion, the so-called entatic condition (Vallee and Williams, 1968; Williams, 1996).
3. A helix bundle readily alters conformation, tertiary structure, on binding ions, molecules or other proteins.
4. A β-sheet can engender optimal recognition.
5. A β-sheet can induce optimal catalytic ability, the most powerful *entatic state*, but cannot effect mechanical transmission of energy. Note that calcium-dependent enzymes are usually made from β-sheets sometimes cross-linked by —S—S— bridges to helices.
6. A β-sheet may move as a whole, as can an α-helix, to associate with a surface of a phospholipid as calcium neutralises negative charge on both.

Overwhelmingly calcium ions are messengers for the induction of mechanical change in helical proteins when admitted *into cells*, and many known *cytoplasmic* calcium proteins are α-helical bundles (Table 1.1). (Calcium is not known to be a catalyst inside any cell.) This does not imply that they all undergo equivalent conformation changes or that their rates of change are all similar (see below). However, recently a number of C_2-β-sheet domains which bind Ca^{2+} in association with phospholipids has been discovered. As stated before, they act so as to trigger association and not by protein conformational change. *Outside cells*, calcium binding is often used in structural cross-linking or is used to induce new structures (Table 1.4), where some proteins are rejected from cells, and so activate enzymes. These latter uses are common to prokaryotes. Now and then, calcium outside cells aids catalysis directly, but it is a very poor Lewis acid and, more generally, it assists proenzyme-to-enzyme conversion. Some of these proteins contain random conformations which are organised by calcium, while others are α/β structures (Table 1.3). Few, if any, are simple α-helices.

In *membranes*, calcium channels are adjustable multihelical units. An agonist which generates calcium entry adjusts the helical bundle to allow channel opening: an antagonist blocks this mechanical activity. The outward calcium membrane pump (Carafoli, 1991) is similar to a channel in its membrane unit, but has attached an ATPase which adjusts the channel so that the energy of ATP hydrolysis pumps Ca^{2+} ions out (Fig. 1.8). The ATPase is based on a rising hinge of α-helices which is energised by the relative motion of two β-sheet domains, which hydrolyse ATP, linked to it. The energisation is the binding energy plus hydrolysis of the substrates. The rising hinge acts to transfer energy from the ATPase β-sheets to the helical membrane domain in pumping (Fig. 1.9). Note that a rising hinge stores energy so that once the perturbing activity is removed, it returns to rest automatically. The activity of these single domain pumps cannot be related to that of ATP-synthetases.

The messenger calcium α-helical proteins, calmodulins, which function in the cytoplasm, act on enzymes (including the ATPase pump), which are like the ATPase part of a pump. In the calcium-bound state, the calmodulins drive a conformation

Table 1.4 Calcium-Binding Proteins Outside Cells[a]

Protein	Function	Structure without Calcium
Osteocalcin	Bone-binding protein	Random[b]
Osteonectin	Bone-binding protein	Random
Calsequestrin	Vesicle protein for calcium storage	Random
Chromogranin A	Storage in adrenal vesicles	Random

[a] "Outside cells" includes in vesicles.
[b] "Random" implies without long-range globular or domain structure.

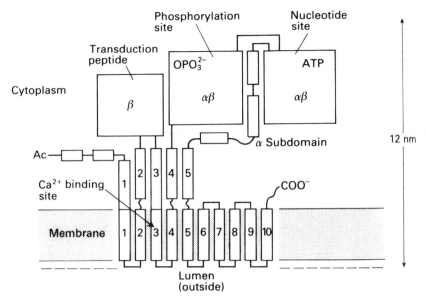

Figure 1.8 The calcium pump in outline. The mechanism is very like that of the activation of a kinase (see Fig. 1.9).

change of an enzyme, so removing inhibition say of a kinase, through a rising hinge, i.e. in the reverse sense to that of a pump. The kinase, like the ATPase, has two β-sheet domains attached to the rising hinge (Fig. 1.9). A similar principle applies to calcium/troponin activation of muscle, which is obviously a direct mechanical activity using ATP hydrolysis. The precise function of other α-helical proteins such as annexins and S-100 proteins is not known. Again the function of C_2 domains is not fully explained.

Calcium Buffers and Transport

The buffer and transport calcium proteins (Heizmann, 1984) are very like the messenger proteins, calmodulins and troponins, in that they are four-helix bundles but now, although the binding is similar in strength, the on and off rates are slower. The high concentration of such proteins as calbindin and parvalbumin indicates those cells in which a maintained high *free calcium* has to be

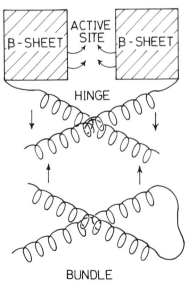

Figure 1.9 In the kinases case, a helical bundle (calmodulin lower protein) is changed in conformation by ion binding, X, which forces a more distant conformation change upon the disposition of two β-sheet domains (upper protein) so as to create an enzyme reaction, phosphorylation, often by removing inhibition. The calcium pump operates in the reverse sense when the reaction, hydrolysis of ATP, forces two β-sheet domains to alter relative disposition in a membrane, which changes a helical bundle, and so forces a cation through the membrane. The enzyme in each case connects to the helical bundle, middle domain, via a rising hinge also made from helices. Energy capture, the formation of ATP, from ion gradients works in the opposite sense from a pump.

prevented, or in which a slightly raised free calcium could be useful, while a relatively high *bound calcium* has been established. The choice rests in the strength of the buffer. If the buffer protein, slow calcium uptake and release, has a conditional binding constant for calcium, i.e. a constant in certain conditions of Mg^{2+}, pH, etc., of $> 10^7$, then the buffer will effectively remove free calcium, but relatively slowly compared with calmodulin or troponin, while allowing later leisurely removal of calcium to a low level by the (still slower) calcium pumps. This is so for parvalbumin. It is the parvalbumin-bound Mg^{2+} which slows the binding step. If, on the other hand, a buffer made from a *slow-release* protein has a binding constant of 10^6, then once it has been heavily saturated it will maintain a higher total bound calcium for a longer period of time than would be allowed by fast pumping out. At the same time, a slow diffusing trickle of low-level calcium will arise. An example is calbindin. During the delayed release time, the calcium bound by either buffer can be transported along distances in the cell. More details are given below.

Finally, where Ca^{2+} activates only in the presence of an additional chemical, e.g. a phospholipid, the activation process is now a relatively slow relay of a more complex kind since it may (annexins) or may not (C_2 domains) involve change of conformation of (1) the protein by mechanical motion of helices, as in calmodulin (Weng et al., 1993); and of (2) a phospholipid; and (3) cooperative dynamics of the phospholipid and the annexin or C_2 domain. This mechanism could engender vesicle release—endocytosis. (See also gelsolin below.)

In summary, the internal calcium proteins allow mechanical message transmission in a single eukaryote cell, thus connecting the filament and vesicular internal structures, absent in prokaryotes, and the environment to the cytoplasm acting through a variety of devices with different time constants:

1. Directly on filaments to adjust cell shape and contact, pseudo-pod formation, in one initially fast switch, as in muscle cells.
2. On the disposition of vesicles—release of digestive proteins, building of extracellular or intracellular structures, ejecting poisons. This is a slower activity.
3. On the metabolism of the cell so as to provide enzymes and energy to meet environmental conditions. This is also a slower activity.
4. On transcription systems to adjust the protein complement of the cell to new conditions outside it. This is the slowest switch (see pp. 18–19).

Calcium "Buffers" in Different Cells

Conventional[3] thermodynamic calcium buffering is achieved in the flowing liquids outside cells, the sea, in unicellular organisms, or in higher multicellular organisms by equilibration with bone (p. 20). Inside vesicles, the calcium concentration is similarly buffered by the storage proteins, calsequestrin and its relatives. These proteins are described in later sections. In the vesicles, calcium is maintained constant somewhat above 10^{-3} M. The background buffering of calcium inside most cells is due to the *steady state* flow, including the exchange binding by proteins, giving balance of inward leak and outward pumps with $[Ca^{2+}] \ll 10^{-7}$ M. This is basic homeostasis (see the later section on this topic). Temporary buffering of calcium at a somewhat higher level in the cell cytoplasm, as described above, is the different process of maintenance, *for some time, of bound calcium* slowly and continuously released after a pulse of fixed concentration. [The ability to hold a concentration of an ion constant is traditionally called (thermodynamic) buffering, but the sense of calcium "buffering" in a cell is the process of transient retention of calcium by proteins before rejection. Remember that in the conventional treatment of buffers, time is not considered and the buffer helps to keep an "instantaneous" thermodynamic equilibrium concentration as, for example, in the use of a pH buffer.] Now, differently differentiated cells can use calcium differently by utilising the above "buffering" times of retention and binding strengths.

One example is the action of parvalbumin as a relaxation factor, able to mop up free calcium faster than the calcium pump, and is very valuable, particularly in very *fast muscles* as follows:

$$\text{input to cell} \rightarrow \text{activity} \rightarrow \text{parvalbumin} \rightarrow \text{pump out}$$
$$\text{troponin C} \qquad \text{"buffer"}$$

Thus, the action of parvalbumin is to reduce the degree to which free calcium remains high enough in the cytoplasm to maintain some tension while total calcium in the cell remains quite high. It prevents also any triggering of slower acting strongly binding proteins, since it itself has a high calcium-binding constant. The mechanism of slow parvalbumin uptake of calcium is the "slow" release of Mg^{2+}, as described above (Heizmann, 1984).

Calbindin works in a different way (Linse et al., 1991). While it closely resembles the EF-hand calcium-binding proteins, it does not bind Mg^{2+}, and it undergoes only a slight conformation change on binding Ca^{2+}. It has slower uptake and release kinetics, presumably due to the stiffness of the protein. Calbindin will not function like parvalbumin

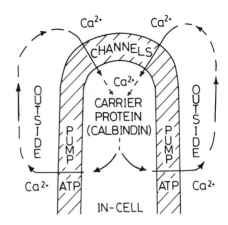

Figure 1.10 Circulating calcium currents around a cell tip due to the disposition of calcium pumps and channels.

since while both will allow the concentration of free Ca^{2+} to rise rapidly to 10^{-6} M, it binds Ca^{2+} with only equal strength to that of calmodulin. Thus, both calbindin and parvalbumin hold on to bound calcium for some time for different reasons: parvalbumin through a high binding, calbindin through slow protein relaxation. The function of any hold-up of ions in such a calbindin "buffer" is to establish a slow trickle of Ca^{2+} in the cell, which can activate other calcium-binding proteins, e.g. in memory cells. Additionally, protein can diffuse distributing calcium at a low level away from any local region of admittance and, hence, can be used to distribute calcium across epithelial cells, or in circulating Ca^{2+} currents (Fig. 1.9). In a memory cell, the effect of calbindin-like proteins could therefore be to give a *sustained* very low pulse level, i.e. slow return to the full resting state, which can affect phosphate transcription systems and then generate long-term memory through protein production. In this sense, such proteins, including S-100, assist the activities due to repeated pulses (Fig. 1.7). Additionally, any calcium protein, such as calmodulin, can be made to act slowly if it is bound to a protein, say a kinase, which holds the calmodulin in the bound state for a long period of time. Different cells can also be different, of course, in the number of calcium pumps (and exchangers) and their location around cells (Fig. 1.3 and 1.10) (Harold, 1990; Frausto da Silva and Williams, 1991). Control over filament networks can also have different time constants for similar reasons.

Intracellular Filament Network

While a variety of cellular activities have been discussed, it has been assumed that the cell network internally is intact to a large degree. The network is made from proteins, such as actin, which form a permeable (to molecules) gel, especially under the membrane. In some circumstances, this gel is broken down to a solution or sol condition. One protein active in assisting this gel–sol transition is gelsolin, together with its relatives villin and severin. These proteins are calcium-activated in cells. Here, the role of the calcium when its concentration rises for a longish period in a cell is to cross-link gelsolin to actin chain-ends, so breaking up a continuous actin network. This inhibitory action is very different from the description given above of a required conformational muscle triggering. In fact, the trigger proteins, calmodulins, are quite different from gelsolins in their structures. Here, binding of calcium does not give a long-range structural change of the protein but just local binding, as in extracellular proteins. In fact, like these extracellular proteins, gelsolin is a β-sheet protein with a surface binding site for calcium (McLaughlin et al., 1993). This intracellular activity is to be compared with certain extracellular activities of calcium in multicellular organisms (see adherins and lectins) where calcium makes cross-links so as to join cells together (Iobst et al., 1994), and with the functions of C_2 domains.

The Coordination of Intracellular Functions in Eukaryotes: Summary

The development of the eukaryote cell demanded first and foremost a permanent, though adjustable, filamentous structure to keep the nucleus in place. This structure connected the nuclear compartment membrane with the outer membrane. The rearrangement of the membrane, the nucleus, and the filaments had to be coincidental on cell division. Thus, new signalling was required above that in the prokaryote cell and in addition to fast responses. Again, the new filamentous structure had to maintain the cell shape and all vesicle positions internally and yet make it adaptable on contact with other cells to form colonies (*not* yet multicellular organisms) or allow the cell to attach to a substratum. Again, a new signalling device was required. It is very probable that calcium entry into the cytoplasm was this signal since all these activities are today associated with calcium in many single eukaryote cell types of widely different species. It is then necessary to postulate the development in eukaryotes of four new kinds of protein in addition to the calcium pumps of prokaryotes and apart from any buffering or transport activity described above. They are calcium channels; calcium receptors on the filaments, troponins, for fast action; fast receptors on enzyme-coupled systems, calmodulins; and a series of calcium receptors for slower action. Subsequently, in the unicellular eukaryote this signalling had to be increased in complexity to allow calcium modulation of metabolism so as to generate extra energy opposite the calcium-induced action. To be able to use the calcium signal discriminately in cell space, the calcium input channels could also be associated with special receptor zones of outer membrane curvature (Figs. 1.3 and 1.10), and this function could be refined further by utilising calcium stores (with additional new proteins) in new vesicles close to the channel entry points (Fig. 1.11). Calcium-induced calcium release and calcium-controlled calcium entry from outside became useful so that triggering was then localised and not general to the whole cell. All these developments link the outside world to the cell activity, so that we can imagine that the various calcium entry channels were the means by which the single eukaryote cell coordinated inside activity with vesicles and different outside stimuli, both chemical and physical. However, all of this activity met with one extreme problem. There was inside the prokaryote cell from which the eukaryote evolved a pre-existing signalling network based on phosphate (Fig. 1.12), which had to be kept intact but now of moderated activity. The essential character of this phosphate network has already been described as controlling the homeostasis of ions such as H^+, Na^+, K^+, Mg^{2+}, and including Ca^{2+} by pumping and that this network linked through ATP to most (all synthetic) metabolic pathways, as well as through phosphorylation, dephosphorylation, and through cyclic nucleotides to many transcription factor activities. There was an obvious demand that the new signalling pathway due to calcium, which informed the eukaryote cell about the world outside itself, should intercommunicate with the internal network based on phosphate. Before we enter into the description of this intermeshing of networks in all eukaryote single cells, we need to note that the two networks, calcium and phosphate, are profoundly different. The calcium network is based on calcium binding and

exchange using gradients (described above) not on chemical enzymic transformation, which is the basis of phosphate messages. Calcium triggering is essentially faster than phosphate triggering. The essence of linking the two networks lies in their time constants.

The Phosphate Network

Phosphate signalling is due to covalent bond formation, esterification, or destruction, hydrolysis. Now the control of such chemical bond changes rests upon mechanisms using catalysts (enzymes), which are called phosphorylations (kinases), dephosphorylations (phosphatases), and reactions such as cyclic ester, i.e. cyclic nucleotide, formation (cyclases), and their decomposition (cyclic esterases). Moreover, the energetics of reaction are not based on a gradient of phosphate (as is the case for calcium gradient triggering) but on storage of the phosphate in an unstable energised condition as pyrophosphate, ATP (Fig. 1.2). Central to all cells is this storage of energy as pyrophosphate (not only as a signalling device):

$$ATP + X \rightarrow X\text{-}P + ADP$$
$$ATP \rightarrow cAMP + P_2$$
$$cAMP + Y \rightarrow Y(cAMP)$$

where X and Y may be internal receptors and can be transcription factors.

It is all these activities *inside* the cell, linked to polymer synthesis and regulation through transcription factors, which became linked to the environment of the cell through calcium signals. The further achievement in eukaryotes was therefore to make the phosphate network dependent, in part, on calcium by controlling the enzymes of much of the phosphate system using calcium binding. A major calcium protein here is, once again, calmodulin, which affects many kinases, pumps, and phosphatases. Some of the interlinking of the phosphate (internal) and the calcium (external/internal) signalling of a single eukaryote cell is shown in Fig. 1.11. However, there developed also the variety of S-100, annexins and C_2 domain proteins, the complexity of which is only slowly being recognised.

Calcium Links to DNA

The functions of sustained calcium rise (Fig. 1.7) are obviously linked to the synthesis of proteins since differentiation due to such calcium changes is observed. There is then the major consideration of the control of protein synthesis, apart from that of the control over expressed proteins, which is the con-

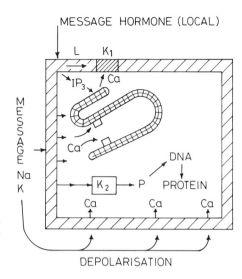

Figure 1.11 The effect of localised hormone receptors is to deliver a regional or local calcium pulse. The pulse can be used to activate calcium from internal vesicles locally. The effect is very different from a general cell depolarisation leading to general calcium admittance.

sideration of regulatory genes and the transcription factors for the production of these new (to eukaryotes) calcium-binding proteins. To date, no calcium-binding protein has been discovered that binds directly to DNA so as to act in transcription. Very frequently, the regulation of protein transcription acts through

1. Phosphorylation/dephosphorylation or cyclic nucleotide systems, known in both prokaryotes and eukaryotes.
2. Zinc fingers, often activated by organic molecules (hormones), such as sterols, and which mechanism is absent in prokaryotes.

The suggestion is, then, that the more primitive system of regulation is (1), which in eukaryotes became associated with calcium levels, and that (2) only arose with the advent of eukaryotes and was developed in multicellular organisms (see below). It is then of interest to see what part of the synthesis of calcium proteins is under the control of zinc-finger transcription rather than that of phosphate-dominated transcription. At the present time, there are insufficient data to allow any generalisation but some indication of possibilities will be given below.

Assuming that the above description of the involvement of calcium in the evolution of the single cell from the simple prokaryote to the complex eukaryote condition is correct, I will now attempt to show how the calcium signalling developed further in multicellular organisms (Table 1.5). This will give us a better insight into the problems of transcription factors linked to calcium.

Table 1.5 Calcium-Controlled Events in Multi-cellular Organisms

Activity	Control
Photosynthesis	Dioxygen release (Chloroplasts)
Oxidative phosphorylation	Dehydrogenases (Mitochondria)
Receptor responses	(a) Nerve synapse
	(b) IP$_3$-linked reactions
Contractile devices	(a) Muscle triggering (actomyosin)
	(b) Cell filament controls (tubulins)
Digestion	Activation of hydrolases
Adehesion and cell association	Surface glycoproteins
Immune reaction and clotting system	Complement reactions and Gla-proteins
Membrane/filament organisation	Calpactin-like proteins provide tension
Cell division	S-100 proteins(?)

Multicellular Organisms

If the above description of the calcium message system in eukaryotes is accepted, then the development of the multicellular system could be based on external cell-to-cell signalling using the *self-environment* which can be created in the multicellular organism, coupling it now to both calcium and phosphate signals (Williams and Frausto da Silva, 1996b). However, this signalling can be truly effective only if extracellular structures and solutions are fixed, both chemically and in space. Evolution of multicellular life depended on making the cell environment more and more constant while diversifying cells within it. In such an environment, a new set of cell-to-cell messengers is required and these came to be organic chemicals, deliberately released by cell A, positioned purposefully some distance from cell B. The organic messenger molecules flowed inside the fixed structures of the organism as transmitters or hormones so as to reach targets on the outside surface of another cell B membrane or so as to reach DNA transcription factors directly. Taking the membrane receptor first, once bound by an agonist the organic messenger molecule had to open a calcium channel or communicate directly with phosphate-dependent systems. For obvious reasons, the free calcium outside the cell, and thence the pulse size, should now have as fixed a value as possible since this would allow refinement of the message pulse system. The requirement is for a bulk channel pump system of organs to control circulating calcium. However, before any of this refinement could be managed the extracellular structure had to be put into place, which requires us to give a description of the nature of the extracellular filaments and spaces that separate organs. In one sense, the switch in complexity from prokaryotes to unicellular eukaryotes, already described, is the same as the switch from unicellular to multicellular eukaryotes. Both arise from increases in numbers of compartments, both demand new filamentous structure, and both demand new coordinating message systems which must dove-tail with pre-existing systems, i.e. phosphate and calcium messages, the first seen in the prokaryote and the combination observed in the eukaryote single cell. It is organic messengers which were added to both of these messenger systems in the multicellular organisms. The new message systems and the new filamentous systems of multicellular systems are largely based on oxidative organic chemistry which dioxygen introduced, after $1–2 \times 10^9$ years of evolution. It was dioxygen chemistry which also allowed the synthesis of the necessary chemicals for the construction of the required, more rigid, extracellular filaments. Here, some new calcium chemistry became possible, assisting the constancy of the extracellular solutions and structures.

The Nature of Multicellular Organisms

External to the cells of multicellular organisms, a number of differently *cross-linked matrices* are formed, e.g. collagen, sulphated saccharides, chitins, and lignins. These polymers were bound together, in part, by disulphide bridges, but dominantly by newly synthesised side-chains (Table 1.6) which form cross-links using calcium. They all required oxidative chemistry. Through this matrix, the external fluids flow carrying calcium ions. The levels of calcium could be buffered by external inorganic precipitates, fixed in place within the organism by the external filaments (Fig. 1.12). These precipitates also further strengthened the extracellular matrices. The composite materials are the shells and bones of modern organisms. Apart from the fact that they are marvellous construction (structural) materials, they act as buffers of Ca^{2+}, H^+, and carbonate ($CaCO_3$), or of Ca^{2+}, H^+, and phosphate [$Ca_2(OH)PO_4$, bone] in the extracellular fluids of the organisms (Fig. 1.12).

To establish the constancy of the animal cell environment during growth requires that calcium is constantly taken into the organism to balance any loss to the environment. Now, only part of a complex organism is in contact with the environment through digestive epithelial cells (input) and filtering organs such as the kidney (output). It is essential that these units or compartments communicate with one another to keep free calcium constant while transporting it to the areas of construction, shell or bone. Excess calcium, as well as a deficiency of it, is hazardous. A set of calcium proteins for transport in epithelial cells was required and it was necessary for the organism to regulate this uptake much as single cells control calcium channels and pumps. A new set of calcium-binding proteins was also needed to give the mineral–protein composite stability. Finally, where bone controls calcium levels the phosphate concentration in the extracellular fluid has to be fixed too. We find control proteins for both calcium and phosphate levels and we find novel calcium-binding proteins in shells, bones, and teeth (Fig. 1.12).

The use of these new calcium proteins is a new management system for calcium and the production

Table 1.6 Modified Amino-Acids for Calcium Binding

Amino acid	Protieins
γ-Carboxyglutamate	Osteocalcin, prothrombin
Hydroxyproline	Collagen
Hydroxyaspartate	EGF (growth factor) hands

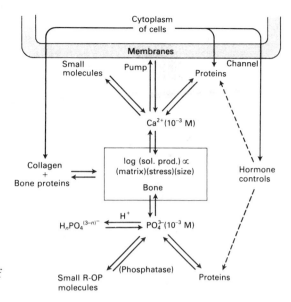

Figure 1.12 The relationship between calcium (phosphate and pH) buffering by bone and the functions of bone proteins and hormones such as vitamin D.

of the proteins required new transcription systems within the cell cytoplasm. It would have been difficult to build the transcription on phosphate or the connected calcium itself, given the involvement of these messages in the processes in single cells described already, so that a new nonredox element was incorporated as a messenger. One element generated in large amounts by oxidation of sulphides, following the rise of O_2 in the atmosphere, was zinc. We find zinc-finger transcription factors in multicellular systems (Fig. 1.13), closely connected to mineral metabolism outside cells, calcium transport, calbindins,

and phosphate release, alkaline phosphatase, and thence to bone formation via steroid messages which bypass phosphate and calcium message systems and go directly to the DNA. Thus, there is a new signalling system, including vitamin D, concerned with extracellular homeostasis based on a pathway independent from phosphates or calcium but necessary to control *extracellular* phosphate and calcium. This new set of transcription factors also controls the other minerals in *extracellular* fluids, e.g. Na^+, K^+, and Cl^-, through similar sterol hormones, the glucocorticoids.

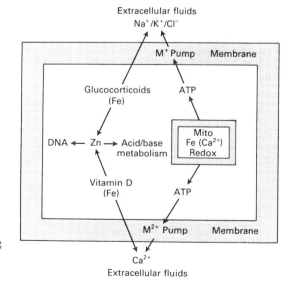

Figure 1.13 A scheme for the tight interrelationships between mineral ion concentration gradients, including that of calcium, the functioning of hormones (using zinc and iron) and the use of energy ATP.

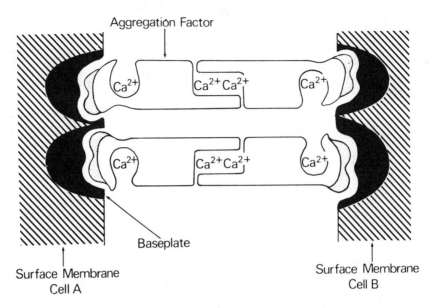

Figure 1.14 Lectin cross-linking of cells via calcium bridges.

Of course, much of the new extracellular signalling had to connect to the phosphate transcription systems too. Here, we find that new organic compounds are released (by calcium), often aided by zinc enzymes, not zinc fingers (e.g. formation of peptides) or copper and iron oxidative action (e.g. formation of some amidated peptides and hormones such as adrenaline). These compounds, even down to transmitters such as acetylcholine, are frequently not only released by calcium from cell A, using annexins(?), but also afterward, through receptors on cell B surfaces, they connect (via a second calcium entry) messages to phosphate systems such as kinases. Thus, calcium, through the extracellular environment of multicelluar organisms, links the activities of different organs, connecting their intracellular phosphate homeostasis to the battery of new organic messengers. We need to observe very carefully where these message systems are dependent on chosen time constants of the internal calcium pulses as described above. Thus, a single Na^+/K^+ depolarisation at nerve junctions or at nerve muscle junctions must lead to only a fast calcium action (10^{-3} seconds), while a hormone such as adrenaline must lead to more sustained activity, and a slow calcium-maintained rise (Fig. 1.7) may even lead to differentiation.

Finally, cell–cell recognition was essential in multicellular organisms and here calcium is utilised in lectins which bind glycosylated protein surfaces of one cell to proteins (lectins) of another (Fig. 1.14). A gigantic matrix based on specific lectin recognition

and other calcium proteins was achieved (Table 1.7). Some of these proteins are poorly structured in the absence of calcium e.g. osteocalcin.

The Brain and Calcium

The final step in evolution was that of a quite new organ, the brain. A connection between the environment and the interior of an organism can be of immediate value but a memory of environmental features assists in quite other ways. To develop a memory, the organism must change in some structured feature so that a part of it reflects observation of the environment at a given time. This is not a useful feature of the structure of a whole organism. The simplest way to imagine this is to create literally a localised growth pattern so that the morphology of part of the organism, but only a part of it, represents the environmental experience. The basic requirement is for a message from the environment to stimulate local cell growth. In one sense, this requires homeostasis, the maintainence of a given structure present in all other parts of the organism, to be replaced by a limited plasticity in a specific region, which needs to be isolated. The autonomous nervous system grows into a fixed pattern but the brain cells form an evolving, ever-changing growth connection. Here, the role of calcium is as the second mediator of the external event which is first generated as a Na^+/K^+ current before it becomes a calcium in-flow current Bolsover and others (1995). It is the value of calcium

Table 1.7 Structure Roles of Calcium

Protein	Role	Reference(s)
Gelsolin (villin, severin)	Binds to ends of actin Breaks actin gels	1
EGF-like domains	Domain-linking Virus coat cross-linking	2 3
Cadherin	Transmembrane protein giving cell–cell adhesion	4, 5
Integrin	Links extracellular matrix to actin cytoskeleton	6
Lectins	Glycoproteins linkage to cell surface receptors	7
Blood-clotting factors	Linking domains so as to activate enzymes	8

1. McLaughlin, P., Gooch, J. T., Mannhoerz, M.-G., and Weeds, A. G. (1993) *Nature* 364: 685–692.
2. Rao Z., Handford, P., Mayhew, M., Knott, V., Brownlee, G. G. and Stuart, D. 1995 *Cell* 82: 131–141.
3. Bloomer, A. C., Champress, J. N., Bricogne, G., Staden, R., and Klug, A. (1978) *Nature* 276: 362–368.
4. Overduin M., Harvey, T. S., Bagby, S., Tong, K. I., Yau, P., Takaichi, M., and Ikuri, M. (1995) *Science* 267: 386–389.
5. Shapiro, L., Fannon, A. M., Kwong, P. D., Thompson, A., Lehmann, M. S., Grubel, G., Legrand, J.-F., Al-Nielsen, J., Colman, D. R., and Hendrickson, W. A. (1995) *Nature* 374: 327–337.
6. Graves, B. J. (1995) *Struct. Biol.* 2: 181–183.
7. Iobst, S. T., Wormald, M. R., Weiss, W. I., Dwek, R. A., and Drickamer, K. (1994) *J. Biol. Chem.*. 269: 15505-15511.
8. Jackson, C. M. and Nemerson, Y. (1980) *Annu. Rev. Biochem.* 49: 765–811.

that it can modulate plasticity (growth) indirectly through phosphates (Williams, 1996). At present, we can only pose the proposition that it is the impact of the environment on calcium input to the cell, coupled to phosphate reactions in the cell, which allows growth at a synapse to produce a memory in the brain.

Poorly Structured Calcium-Binding Proteins

I have stressed in this chapter the great advantage of two kinds of calcium-binding protein—the β-sheet, or highly structured and relatively rigid proteins largely linked to enzymes with highly selective recognition, or to condensation of surfaces upon one another and the α-helical proteins, which are dynamic mechanical machine components. There is a third group of calcium-binding protein which has a much poorer structural definition and it is often even termed "random" (Levine and Williams, 1982). Such proteins or protein domains are common outside cells (Table 1.8) and are associated with such processes as blood clotting, calcium-storage (in vesicles), and mineralisation. The reason for the looseness of the non-calcium-bound proteins is not obvious. Clearly, they are readily hydrolysed so that, in

some cases, they are destroyed shortly after release. Again, they are often very highly charged through phosphate or γ-carboxyglutamate groups, as well as carboxylates, which causes their "random" structure but also generates binding of many calcium atoms. In such a situation, they store calcium compactly or they may provide surfaces which correspond almost epitaxially to those of minerals. Invariably, the proteins exchange calcium very rapidly.

Homeostasis

One of the great difficulties in understanding the role of any element in living organisms is the haphazard manner in which evidence accumulates. A considerable number of years ago the calmodulin series of proteins were uncovered. Due to gene technology we now know there are some five calmodulin-like proteins in yeast. Somewhat more recently, there have been discovered the series of S-100 proteins—there are probably twenty in mice—and the annexins, where again there may be some ten to twenty in higher organisms. There are no annexins in yeast. The number of C_2 domain proteins, which are the most recently discovered, is in excess of five. Some of the proteins, such as calmodulins, have at least ten

Table 1.8 Loosely Folded Calcium-Binding Proteins

Protein	Function
Chromagranin A	Peptide storage in chromaffin granule
Osteocalcin	Bone-binding protein
Peptide A	Component of prothrombin
Calciquestrins	Calcium storage in vesicles
Phosphoproteins	Components of bone matrix
Shell proteins	Highly acidic proteins of shells

functions. To these cytoplasmic proteins we must add a variety of channels, pumps and exchangers in plasma and vesicle membranes and the proteins of the endoplasmic reticulum and the golgi—many are now known. As this knowledge has grown, the perception has strengthened that the function of calcium is triggering at a variety of rates. Finally, there are the calcium-binding proteins outside cells for use in structures and activities; the list of them is now very long. Here again, triggering by rejection of proteins to fixed calcium is seen as the major calcium action. However, it is obvious that the external concentration of calcium is highly controlled in a dynamic fashion, the most refined system being associated with bone. Here external homeostasis is a very dominant feature. The sum of all this knowledge raises the question: is there a more general continuous role of calcium in homeostasis inside all eukaryotes cells?

If instead of seeing the resting cell internally as an inactive or passive state of fixed energy and material, here calcium concentration, we may have to see it as a system in constant flow (of both energy and material) when we need a better concept of intracellular dynamic homeostasis. We know in fact that all of the organic content of cells, substrates and proteins, turns over constantly in the resting state. Is this also true of calcium, so that the resting state is one of fixed flux to which (fast) extensive triggering is an added feature to adjust to a changed environment? It is largely this selective triggering which I have addressed in this chapter, where a calcium flow into a cell initiates certain activities or, in multicellular organisms, organic chemicals are ejected from cells to a fixed calcium concentration which activates them. We know, however, that in the resting states of cells the same activities at a lower level are maintained; consider muscle tone, hormone circulation and brain rhythms for example, and clearly they must be so maintained, by background flows quite similar in effect to extensive triggering. Given that cells have a constant energy and material demand, these flows need control. Hence I am led to ask—do all eukaryotic cells experience constant calcium currents which, with their connections to phosphate-based chemical

turnover, monitor resting state homeostasis? If so, calcium is flowing in and out of all cells at fixed background levels (Fig. 1.15). Calcium could then be viewed as a current carrier in a cell connecting together its various activities at "rest" (in fact an active condition), much as background currents run in computers. External messengers trigger more selectively. If this is a correct analysis of dynamic homeostasis inside cells, then we need experimental approaches to calcium currents and circuits different from those used at present (Williams, 1999). The importance of such calcium homeostasis could lead us to reconsider certain diseased conditions as largely a consequence of its loss or persistence at the wrong level. Typical effects could be looked for in cancer. Again, morphological patterns in lower organisms are known to be related to calcium flow patterns and apoptosis appears to be connected to altered calcium levels at "rest". Perhaps we are still some way from a full appreciation of calcium activity in organisms. Finally a parallel constant current flow of calcium connects the cells in multicellular organisms.

Conclusion

In reviewing the "fitness of calcium", we have seen that this "fitness" has evolved from a fundamental requirement for organic synthesis in water—the need to concentrate free anions in space within a robust container free from calcium. Calcium cross-links anions readily and its "fitness" was at first, and is still ideal for stabilising organic structures outside the cell (i.e. in prokaryotes), but this makes it extremely "unfit" for the interior of cells. Calcium was initially rejected. This rejection made the generated $[Ca^{2+}]$ gradient into a large external energy store which could be used in eukaryotes as a signal of all kinds of impacts, chemical or physical, upon the outside of a cell. Calcium ions are both large and doubly charged and are selectively bound—a part of their fitness—since it is possible to devise binding sites which greatly prefer Ca^{2+} over Mg^{2+}. Moreover, Ca^{2+} diffuses rapidly. Thus, once an increased gradi-

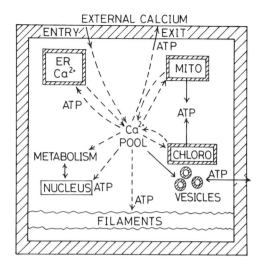

Figure 1.15 The flow of calcium is shown in the "resting" state. It continuously monitors all the cell's activities so as to assist coordination. The external homeostasis is also due to continuous flow linked to bone, for example.

ent of Ca^{2+} had been established across cell membranes, Ca^{2+} could be utilised as the current carrier, by incorporating entry channels in the membrane, reflecting environmental circumstances, and then as the activator of transformation inside the cell. This activity had to be dove-tailed with the pre-existing network of internal cell messages, based on phosphate, in prokaryotes. However, eukaryotes have the extra necessity of managing at "rest", the complications of a flexible cell, filaments and internal compartments. I propose that the resting state of the cytoplasm of each and every eukaryote cell is in dynamic homeostasis, including its relationship to its more or less permanent environment, its internal vesicular compartments, and its filaments, and that this condition is continuously inspected by calcium currents (Fig. 1.15). Thus, calcium flow acts as a monitor of the resting state, adjusting by triggering all cellular parts so that they act cooperatively. This condition may have developed from one of a fixed static pattern and shape to one of slow development, as is known in acetobularia, where until growth is complete there is no true constant resting state. Calcium currents around many cells are well documented.

Given that Ca^{2+} has the ability, when in high concentration, to cross-link organic and inorganic acidic groups, it later became an invaluable aid, outside cells, to cell–cell filamentous organisation, multicellular organisms, and to the formation of biological hard structures, minerals. This fitness of calcium could not be demonstrated fully until oxidation reactions became possible outside cells with the advent of a dioxygen atmosphere. This atmosphere is required to give a stiffening of connective tissue of all kinds by organic cross-links and to provide a more acidic surface for calcium binding to these tissue fibres.

As stated, the fitness of calcium for cross-linking depends on the maintenance of a high calcium concentration outside cells, just as its fitness as a messenger depended on its very low concentration inside cells. Hence, when multicellular plants and animals left the sea, it was essential for them to develop their own external fluids high in calcium. Thus, calcium in high organisms is rejected from cells, only to be accumulated in the circulating fluids and internal vesicles. Eventually, ways were found for precipitating locally the calcium as carbonate (shell) or phosphate (bone) and, as a consequence, the Ca^{2+} concentration in body fluids could be held at a very fixed "buffered" value. This homeostasis gave the external fluid/internal fluid of cells a fixed gradient and thereby greatly enhanced the refinement of the external organic message systems devised using calcium and oxidative chemistry. Messages passed now from cell to cell or organ to organ in one organism, as well as from the external environment to the internal cytoplasm. Several of the qualities of the calcium ion, its "fitness", were now combined. In particular, new organic message systems from cell to cell had to interlock with the calcium gradient and the internal phosphate messenger systems. Much of this interlocking was achieved by using water-soluble organic messengers to open membrane calcium channels. Much of the additionally needed control over external calcium and that of other minerals was linked to a new transcription system independent of phosphate and based on zinc fingers. The zinc fingers first arose in the eukaryotes. The fingers are activated by a different class of lipid-soluble messenger, e.g. sterols, which

Table 1.9 Summary of the Evolution of Calcium Functions

Prokaryotes: Calcium Rejected	Eukaryotes: Calcium Rejected	Multicellular Organisms: Calcium Maintained High in Extracellular Fluids but Rejected from Cells
Calcium rigidifies outer matrix	Calcium pulses link external and vesicular systems	Calcium linked to phosphate transcription systems as in eukaryotes
		Calcium pulses and organic molecular pulses link extracellular to cellular systems and cells to cells
Phosphate signals control transcription	Calcium linked to phosphate transcription systems	Calcium supports extracellular structures Organic messenger molecules linked to new calcium protein production via new transcription factors, zinc fingers

act directly at the DNA level, passing through membranes, and which do not concern the calcium message systems directly. However these very sterols, in the form of vitamin D, act so as to control expression of the calcium proteins essential for homeostasis of external flowing calcium. It is seen that calcium is a dominant, if not *the* dominant, element in the connection of the cell cytoplasm to perhaps all external, including environmental, events in higher organisms. An outline scheme of calcium in evolution is given in Table 1.9.

References

Birch, N. J. (ed), (1993) *Magnesium and the Cell.* Academic Press, London.

Bolsover, S. and others (1995) In Symposium on Calcium Signalling Mechanisms: Implications for Neuronal Function. *Trans. Biochem. Soc.* 23: 627–655.

Carafoli, E. (1991) *The Membrane Calcium ATP-ase in Cellular Metabolism* (McCormack, J. G. and Cobbold P. H., eds.). IRL Press, Oxford.

Chin, E. R., Olson, E. N., Richardson, J. A., Yang, Q., Humphries, C., Shelton, J. M., Wu, H., Bassel-Duby, R., and Sanders Williams, R. (1998) A calcineurin-dependent transcriptional pathway controls skeletal muscle fiber type. *Genes Dev.* 12: 2499–2509.

Eigen, M. (1960) Exchange of water from around cations. *Z. Electrochem.* 64: 115–130.

Frausto da Silva, J. J. R. and Williams R. J. P. (1991) *The Biological Chemistry of the Elements.* Oxford University Press, Oxford, pp. 268–298.

Harold, F. M. (1990) Morphogenesis in micro-organisms. *Microbiol. Rev.* 54: 381–431.

Heizmann, C. W. (1984) Parvalbumin: an intracellular calcium-binding protein. *Experimentia* 40: 910–921.

Henderson, P. (1986) *Inorganic Geochemistry.* Pergamon Press, Oxford.

Iobst, S. T., Wormald, M. R., Weis, W. I., Dwek, R. A, and Drickamer, K. (1994) Binding of sugar ligand to calcium dependent lectins. *J. Biol. Chem.* 269: 15505–15511.

Levine, B. A. and Williams, R. J. P. (1982) Calcium-binding to proteins. In *Calcium and Cell Function* (Cheung, W. V., ed.). Academic Press, New York, pp. 1–38.

Linse, S. and Forsén, S. (1995) Determinants that govern high-affinity calcium binding. In *Advances in Second Messenger and Phosphoprotein Research* Means, A. R., ed. Raven Press, New York, pp. 89–151.

Linse, S., Johansson, C., Brodin, P., Grundstrom, T., Drakenburg, T., Forsén S. (1991) Electrostatic contributions to the binding of calcium in calbindin D_{9k}. *Biochemistry* 30: 154–162.

Margulis, L. and Schwartz K. V. (1988) *Five Kingdoms.* Freeman, New York.

McLaughlin, P., Gooch, J. T., Mannhoerz, M.-G., and Weeds A. G. (1993) The Structure of gelolsin 1-actin complex. *Nature* 364: 685–692.

Vallee, B. L. and Williams R. J. P. (1968) The entatic state. *Proc. Natl. Acad. Sci.* USA 59: 498–505.

Wachterhauser, G. (1988) Evolution of life. *Microbiol. Rev.* 52: 452–484.

Weng, X., Luecke, H., Song, I. S., Kang, D. S., Kim, S. H., and Huber R. (1993) Crystal structure of human annexin I. *Protein Sci.* 2: 448–458.

Williams, R. J. P. (1961) The functions of chains of catalysts. *J. Theoret. Biol.* 1: 1–13.

Williams, R. J. P. (1965) Electron migration in iron compounds. In *Non-haem Iron Proteins* (San Pietro, A. ed.). Antioch Press, Yellow Springs, pp. 7–22.

Williams, R. J. P. (1976) *Calcium Chemistry and its relation to biological function.* In *Calcium in Biological Systems, Symposia of the Society for Experimental Biology,* No. XXX. Cambridge University Press, Cambridge, pp. 1–18.

Williams, R. J. P. (1995) Energised (entatic) states of groups and of secondary structure in proteins and metalloproteins. *Eur. J. Biochem.* 234: 363–381.

Williams, R. J. P. (1996) Calcium binding proteins in normal and transformed cells. *Cell Calcium* 20: 87–93.

Williams, R. J. P. and Frausto da Silva J. J. R. (1996a) *The Natural Selection of the Chemical Elements.* Oxford University Press, Oxford, pp. 289–322.

Williams, R. J. P. and Frausto da Silva J. J. R. (1996b) *The Natural Selection of the Chemical Elements.* Oxford University Press, Oxford, pp. 415–504.

Williams, R. J. P. (1999) Calcium Inside/Outside Homeostasis and Signalling. *Biochem. Biophys Acta* (Accepted).

2

Monitoring Cell Calcium

Roger Y. Tsien

Significance

To understand cellular Ca^{2+}, one must be able to measure it. Cellular fluxes of Ca^{2+} bear some analogy to the flow of water in a plumbing system, the movement of charge in an electrical circuit, or even the distribution of goods in an economy. In each case, one needs to measure two distinct but related quantities, which for these examples are the water flow rate and the pressure, the electrical current and the voltage, or the quantity of goods and their unit price, respectively. For Ca^{2+}, the corresponding quantities are the concentration of total Ca^{2+} and the chemical potential of Ca^{2+}, which we would like to know in each relevant compartment of the cell. The chemical potential measures how much the thermodynamic free energy would change upon addition on an infinitesimal unit of Ca^{2+}. It therefore describes the "pressure" of Ca^{2+} or its propensity to drive Ca^{2+}-binding reactions. Most biologists find the chemical potential too abstract a parameter and prefer to think about and measure the free concentration of Ca^{2+}, perhaps because that has the familiar units of moles/liter. The chemical potential μ equals $RT \ln c$, where R is the gas constant, T is the absolute temperature, and c is the free concentration, so the chemical potential and the free concentration carry the same information on logarithmic vs. linear scales. Newcomers to the field often wonder why we care about the free concentration, since, by definition, it measures the Ca^{2+} that is not bound, whereas the bound Ca^{2+} is what actually triggers biochemical responses. The answer to this query is that the free Ca^{2+} is just another way of expressing the local driving force available for Ca^{2+}-binding reactions in general (Campbell, 1983; Tsien, 1983). Naturally, the best way to measure free Ca^{2+} is to provide a sensitive reference reaction in which Ca^{2+} can participate and then measure how far the equilibrium proceeds.

Methods for Measuring Total Ca^{2+} or Ca^{2+} Fluxes

Methods for the absolute measurement of total cellular Ca^{2+} (Campbell, 1983) generally involve destruction of the tissue and liberation of bound Ca^{2+}. For example, the tissue can be ashed or extracted with acid. The Ca^{2+} content of the ash or extract is then determined by atomic absorption spectrophotometry, which measures the characteristic absorptions of vaporized calcium ions at extremely high temperatures in a flame or graphite furnace (Sanui and Rubin, 1982). Alternatively, if the tissue can be incubated with radioactive Ca^{2+} for long enough to reach tracer equilibrium, then scintillation counting of total radioactivity and specific activity yield the total cellular Ca^{2+}. Neither of these methods is used frequently nowadays, perhaps because cells usually contain large quantities of statically bound Ca^{2+}. Therefore, the changes in total Ca^{2+} that accompany signal transduction are usually buried in the experimental error, which includes variations in the amount of tissue in each successive sample.

Methods that measure *changes* of total Ca^{2+} as influx or efflux of Ca^{2+} across the plasma membrane are much more practical and common. If $^{45}Ca^{2+}$ is included in the bathing medium, unidirectional influxes can be measured as the initial rates of uptake of the isotropic tracer. Sequential samples of tissue must be taken and subjected to scintillation counting. The main technical problem is how to wash away, very rapidly, the large amount of radioactivity bound to the exterior of the cells without letting a significant amount of intracellular Ca^{2+} escape. Unidirectional effluxes may be measured by prelabeling the tissue to isotopic equilibrium, then counting the radioactivity released back into successive samples of supernatant medium. For many years, these techniques (see Borle, 1981, for a review) were among the most popular types of experiments on cellular Ca^{2+}, perhaps because they require no specialized reagents or equipment. They are much less fashionable nowadays because they demand much skilled but repetitive manual manipulation of samples containing a short-lived hazardous isotope, and their spatial and temporal resolution is poor. Nevertheless, if unidirectional fluxes are to be quantified, isotopic methods are essential.

Several more methods are available for measuring *net* fluxes, especially across the plasma membrane. If the Ca^{2+} flows through channels whose ionic selectivity is known, measurement of the electrical current through the channel yields the flux of Ca^{2+}. Likewise, if a carrier or pump is electrogenic, if the stoichiometry of the transport cycle is known, and if the associated current can be resolved from all the other currents across the plasma membrane, then the flux is easily deduced. A good example is the resolution of a current component representing Na^+-Ca^{2+} exchange (Yau and Nakatani, 1984). However, in small nonexcitable cells, the currents associated with important Ca^{2+} influxes and effluxes are often minuscule and difficult to measure (Penner et al., 1988).

Another class of methods detects changes in extracellular free Ca^{2+} just outside the cell(s) under study; increases and decreases in extracellular free Ca^{2+} reflect net cellular extrusion and uptake, respectively. The extracellular concentration changes can be detected by low-affinity Ca^{2+} indicators of Ca^{2+}-selective electrodes. The fundamental difficulty in this approach is that the fractional change in extracellular free Ca^{2+} concentration is small. It can be increased by lowering the background level of Ca^{2+} so that cellular fluxes cause bigger percentage changes, though such low-Ca^{2+} media are likely to depress Ca^{2+} influxes. Decreasing the volume of extracellular medium being sampled, for example by pressing the cells against a Ca^{2+}-selective electrode or dispersing them in aqueous microdroplets under oil, is also necessary (Miller and Korenbrot, 1987; Tepikin et al., 1994; Belan et al., 1996). A Ca^{2+} indicator with an octadecyl tail, adsorbed to the outside of osteoblasts, is reported to detect hormone-stimulated Ca^{2+} efflux into a low-Ca^{2+} buffer (Lloyd et al., 1995). Mechanical vibration of Ca^{2+}-selective electrodes helps them to detect small local differences in extracellular Ca^{2+} next to sites of net Ca^{2+} uptake or extrusion (Smith et al., 1994). The tiny changes in electrode potential reflecting the local diffusion gradient of Ca^{2+} are synchronized with the mechanical motion of the electrode and can be selectively amplified to increase discrimination against random drift and noise.

A complementary strategy is to load the cell with a fluorescent indicator at a concentration high enough to become the dominant Ca^{2+} buffer and to clamp the intracellular free Ca^{2+} ($[Ca^{2+}]_i$) nearly constant. Most of the Ca^{2+} entering or leaving the cell binds to, or comes from, the dominant buffer, which optically reports how much Ca^{2+} it has bound (Tsien et al., 1982; Tsien and Rink, 1983). A variant of this method is to replace the fluorescent indicator by EGTA, which release two protons per Ca^{2+} bound, plus a pH indicator (Jong et al., 1995). Obviously, methods in which $[Ca^{2+}]_i$ is nearly clamped are more suited to measuring Ca^{2+} fluxes into rather than out of the cytoplasm.

Yet another common need is to measure spatial variations in Ca^{2+} content, particularly to localize compartments of high Ca^{2+} within cells. Several methods are available for microscopic quantitation of total Ca^{2+} within fixed or quick-frozen tissue sections, including laser-activated mass spectrometry and various electron microscopic techniques (Somlyo et al., 1988; Baumann et al., 1991; Wendt-Gallitelli and Isenberg, 1991; Chandra et al., 1994; Grohovaz et al., 1996). Perhaps the most challenging aspect of these methods is the problem of preventing redistribution of Ca^{2+} during sample preparation for electron microscopy.

Free Ca^{2+} Monitoring by Aequorin, Metallochromic Indicators, Ion-Selective Electrodes, and Bioassay

Most Ca^{2+} measurements nowadays focus on the concentration of intracellular free Ca^{2+}, abbreviated as $[Ca^{2+}]_i$. Such measurements require placing a known Ca^{2+} binding sensor inside the cell(s) and measuring the extent to which the sensor becomes Ca^{2+}-occupied. The first practical sensor molecule was a chemiluminescent protein, aequorin, extracted from bioluminescent jellyfish (*Aequorea forskalea* or *victoria*) that live in Puget Sound. Once aequorin binds its full complement of Ca^{2+} ions, three per protein molecule, it emits blue light. The actual emitting chromophore is a small organic cofactor, coelenterazine, which is destroyed in the process. The uses, advantages, and disadvantages of aequorin have been extensively reviewed previously (Campbell et al., 1979; Blinks et al., 1982; Thomas, 1982; Cobbold and Rink, 1987; Cobbold and Lee, 1991; Miller et al., 1994; Rizzuto et al., 1994b). The main disadvantages of aequorin are as follows: (1) it is irreversibly destroyed by high Ca^{2+}; (2) the light output is very low, because on average twelve Ca^{2+} ions have to bind to and consume six aequorin molecules to cause the emission of just one photon (Shimomura, 1995), so that imaging of individual small cells is difficult (Rutter et al., 1996) or impossible; (3) the light output is a steep 2.5th power law with respect to $[Ca^{2+}]_i$, so that when aequorin is exposed to spatially nonuniform $[Ca^{2+}]_i$, its signal is heavily biased towards the highest local $[Ca^{2+}]_i$ values; (4) it is technically quite difficult to introduce the intact preformed protein into cells without exposing either to high Ca^{2+}, even momentarily. The main new development in the last few years (Rizzuto et al., 1994b) has been the use of the aequorin gene and molecular biological techniques to cause the protein to be synthesized in situ and targeted to specific organelles within cells such as mitochondria (Rizzuto et al., 1994a), nuclei (Brini et al., 1994), and endoplasmic reticulum (Montero et al., 1995; Button and Eidsath, 1996; Kendall et al., 1996). Fortunately, coelenterazine is commercially available, membrane-permeant, and spontaneously able to bind to the apoprotein to form functional aequorin as a Ca^{2+} sensor in the desired location. Aequorin is thus particularly advantageous for assessing free Ca^{2+} levels in specific organelles or sites to which it can be targeted by molecular biological signals. It is also useful in organisms such as bacteria, yeast (Nakajima-Shimada et al., 1991), slime molds (Cubitt et al., 1995a), and plants (Campbell et al., 1996), whose cells are easy to transfect but difficult to load with small-molecule fluorescent indicators, and in subpopulations of cells that are cotransfected with other genes of interest (Brini et al., 1995). Aequorins of reduced Ca^{2+} sensitivity have been engineered to extend the measurement range to higher concentrations (Shimomura et al., 1993). However, Ca^{2+} levels reported by aequorin in the lumen of the endoplasmic reticulum are contentious (Montero et al., 1995; Button and Eidsath, 1996; Kendall et al., 1996), probably because that organelle is heterogeneous and because the high Ca^{2+} not only stimulates light output but also kills the protein.

Nonfluorescent metallochromic dyes such as murexide, arsenazo III, antipyrylazo III used to be very popular indicators of intracellular Ca^{2+} (Tsien and Rink, 1983; Thomas, 1991). The Ca^{2+} changes their absorbance spectra mainly by displacing H^+, which makes their Ca^{2+}-sensitivity intrinsically pH-sensitive. Other problems include low affinity and complicated stoichiometry for Ca^{2+}, low selectivity against Mg^{2+}, the insensitivity and difficulty of calibrating absorbance signals in thin cells, the lack of membrane-permeant ester derivatives, and their tendency to react with NAD(P)H and sulfhydryls to generate free radicals and superoxide (Beeler, 1990; Docampo et al., 1993). Currently, the main area where metallochromic dyes have significant advantages is in the measurement of very fast $[Ca^{2+}]_i$ transients in amphibian skeletal muscle. Modified murexides have Ca^{2+} dissociation constants just below 1 mM, so that they can respond quickly and linearly to $[Ca^{2+}]_i$ elevations of tens of micromolar. Their low molecular weight

minimizes binding to cytosolic components-(Hirota et al., 1989).

The two major nonoptical methods for quantifying $[Ca^{2+}]_i$ are Ca^{2+} selective microelectrodes and nuclear magnetic resonance (NMR) of fluorinated indicators. Ca^{2+}-selective microelectrodes (Thomas, 1982; Orchard et al., 1991; Baudet et al., 1994) consist of a glass micropipet plugged with a viscous organic matrix doped with a Ca^{2+}-selective ionophore. Because of the ionophore, the membrane is selectively permeable to Ca^{2+}, so that the potential across the membrane approaches the Nernst potential for Ca^{2+} and changes by about 29 mV per decade change in $[Ca^{2+}]_i$. The unique advantage of Ca^{2+}-selective microelectrodes is their wide dynamic range. from 10^{-8} to $> 10^{-2} M[Ca^{2+}]$, thanks to the inherently logarithmic response. However, they have many difficult aspects. Each microelectrode needs to be individually fabricated and calibrated, which takes considerable skill and effort. The selectivity of the ionophore is not absolute, so the response levels off in the region of 10^{-7} to 10^{-8} M. Response times are small fractions of seconds at high $[Ca^{2+}]$, but slow down to seconds or tens of seconds at typical resting $[Ca^{2+}]_i$. It is hard to make tips below $0.5 \mu m$ in diameter, which limits impalements to relatively large and sturdy cells. Even when an impalement is achieved, the absolute potential of the electrode then includes the membrane potential across the plasma membrane. The plasma membrane potential needs to be measured with a separate electrode and subtracted. Thus, two simultaneous high-quality membrane impalements are required. Any leak at either electrode can let Ca^{2+} into the cell or depolarize the membrane and thereby distort the results. The $[Ca^{2+}]_i$ measurement comes from a single point inside the cell, which could give valuable spatial resolution if one knew exactly where the tip is relative to other cellular landmarks, but this is often not the case. For all these reasons, intracellular Ca^{2+}-selective microelectrodes are rarely used nowadays.

The NMR indicators are chemical relatives of the fluorescence indicators and will be discussed after the latter.

Because cells have so many Ca^{2+}-sensitive functions, it is sometimes possible to use a relatively well-characterized endogenous response as a nonperturbing internal bioassay for $[Ca^{2+}]_i$. This approach is particularly attractive for monitoring $[Ca^{2+}]_i$ in specific locations that are difficult for exogenous indicators, such as the immediate vicinity of plasma membrane Ca^{2+} channels or secretory vesicles (Chow et al., 1994). The problems lie in independently calibrating the endogenous Ca^{2+} sensor, for example using Ca^{2+} buffers and caged Ca^{2+}, and in ensuring that its Ca^{2+}-sensitivity remains the same during the physiological phenomena under study.

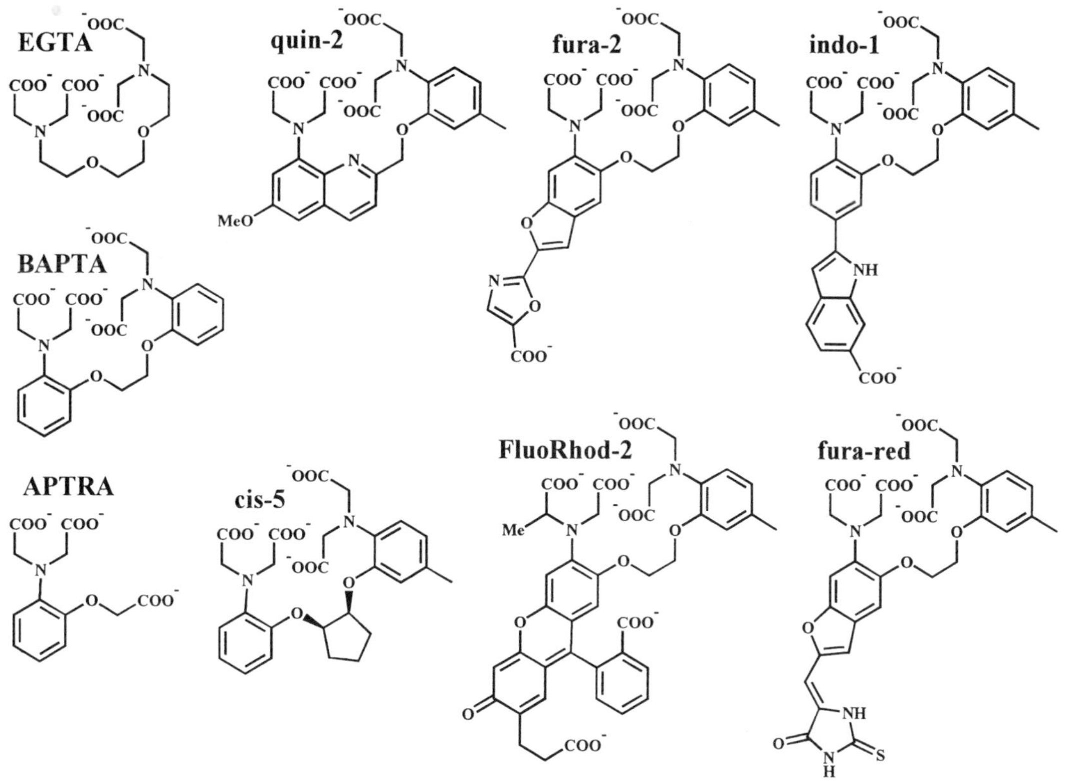

Figure 2.1 Structures of Ca^{2+} chelators EGTA, BAPTA, APTRA, and cis-5, plus fluorescent indicators whose excitation or emission wavelengths are altered by Ca^{2+} binding: quin-2, fura-2, indo-1, FluoRhod-2, and fura-red.

Polycarboxylate Chelators and Indicators for Ca^{2+}

The use of fluorescent polycarboxylate indicators for Ca^{2+} has been reviewed many times (Cobbold and Rink, 1987; Tsien, 1988, 1989a, 1989b, 1992; Tsien and Pozzan, 1989; Williams and Fay, 1990; Callaham and Hepler, 1991; Moreton, 1991; Thomas and Delaville, 1991; Duchen, 1992; Poenie, 1992; Kao, 1994). The current discussion will focus on the general principles and molecular mechanisms rather than catalog the dye properties and biological applications. More practical experimental advice can be found in the above reviews, as well as in the primary literature on the particular tissue or cell type of interest.

BAPTA

Modern Ca^{2+} indicators for biological use are descended from BAPTA (Tsien, 1980), which, in turn,

was designed from EGTA (structures in Fig. 2.1), the first chelator shown to have high selectivity for Ca^{2+} over Mg^{2+}. BAPTA has about the same high selectivity ($\sim 10^5$) because it has eight ligand groups in much the same steric arrangement as EGTA. The benzene rings in BAPTA have many beneficial effects. They reduce the proton ionizations (pK_a values) to well below 7, compared with 9.58 and 8.96 for EGTA. Both proton and Ca^{2+} affinities may be further modulated by ring substituents, which do not alter the geometry of the ligand groups. Finally, the aromatic rings constitute primitive chromophores whose absorbance and fluorescence spectra are strongly affected by Ca^{2+}. The Ca^{2+}-sensitive spectra of BAPTA itself are too deep in the ultraviolet to be useful inside cells, though they can be quite useful in checking the concentration and degree of Ca^{2+} saturation of buffer solutions in vitro. The main Ca^{2+} indicators are obtained by elaborating the benzene rings into chromophores and fluorophores with longer wavelengths. Therefore, the design of indicators and chelators is highly modular; the ligand groups responsible for cation affinity and selectivity interact in a limited but well-defined manner with th chromophore or fluorophore.

APTRA

London, Levy, and colleagues sought to create fluorescent and NMR indicators for Mg^{2+} (Levy et al., 1988; Raju et al., 1989). Because Mg^{2+} is smaller than Ca^{2+} and can accomodate fewer ligands, they truncated BAPTA to reduce its number of chelating groups from eight to five. The result is a chelator called APTRA (Fig. 2.1) for o-aminophenol-N, N, O-triacetic acid, whose K_d values for Mg^{2+} and for Ca^{2+} are about 1.45 mM and 18 μM, respectively (S. R. Adams and R. Y. Tsien, unpublished observations). The truncation of the BAPTA binding site has therefore increased the Mg^{2+} affinity by about 1 order of magnitude while weakening the Ca^{2+} affinity by about 2 orders of magnitude. Fluorophores can be attached to APTRA and show spectral responses closely analogous to their BAPTA relatives (Raju et al., 1989). If cytosolic Ca^{2+} remains at values of $< 1 \mu$M as in resting cells, APTRA-based dyes can indeed be used to measure cytosolic Mg^{2+}, hence their commercial nicknames (Haughland, 1996) "mag-fura," "mag-indo," and "Magnesium Green," etc. Biologists desperate for fluorescent Ca^{2+} indicators of low affinity have seized upon the APTRA dyes to measure micromolar or higher levels of Ca^{2+}, despite their poor selectivity. In this reviewer's opinion, such use of the APTRA dyes should be approached with care; in particular, it is important to get independent confirmation of Mg^{2+} levels or at least determine that they remain constant.

Effect of Substituents on Ca^{2+} Affinities

The simplest way to modify the cation affinities of BAPTA or APTRA in a rational and controllable way is to add electron-donating or -withdrawing substituents, which, strengthen and weaken cation binding, respectively. Chemists have long had parameters that measure how strongly different substituents donate or attract electron density (Perrin et al., 1981). These parameters predict reasonably accurately the effect of the substituents on Ca^{2+} affinities (Tsien, 1980; Pethig et al., 1989). Most of the fluorescent indicators based on BAPTA have a methyl substituent on the benzene ring that lacks the fluorophore. The main purpose for this methyl is to block any further substitution and to ensure that only one fluorophore can be attached to BAPTA, but it also slightly increases the Ca^{2+} affinity. Replacement of this weakly electron-donating methyl by a strongly electron-withdrawing nitro group, as in "Calcium

Green-5N" and "Calcium Orange-5N," weakens the affinity by about 70–400 fold (Pethig et al., 1989; Haugland, 1996). A roughly similar weakening results from addition of two fluorine substituents (London et al., 1994).

Another way to modify the cation affinities is to alter the bridge connecting the two halves of BAPTA. Replacement of one of the ether oxygens by an sp^2-hybridized nitrogen, as in the quin-1 and quin-2, has relatively little effect on Ca^{2+} affinity but strengthens Mg^{2+} binding by about 5-fold (Tsien, 1980), one of the major disadvantages of quin-2. Inclusion of the central $-CH_2-CH_2-$ group into a five- or six-membered ring greatly constraints its geometry and can either raise or lower the K_d for Ca^{2+} (Adams et al., 1988). Ca^{2+} binding is strengthened by a five-membered ring in which both oxygens sprout from the same side of the ring, i.e., a cis-1,2-cyclopentane-diyl linkage as in "cis-5" (Fig. 2.1). When the oxygens are attached to opposite sides of the ring, i.e., a $trans$-1,2-cyclopentanediyl linkage, then the K_d is greatly increased. In six-membered (cyclohexanediyl) rings, both stereochemistries tend to weaken Ca^{2+} binding, but by smaller factors. These modification are relatively pure steric effects on the connection geometry between the otherwise unchanged halves of BAPTA, and their respective effects were surprises. They would be good challenges for any theoretical chemist who feels able to predict affinities of cations for ligands in aqueous solution. Nevertheless, the cis- and $trans$-1,2-cyclopentanediyl analogs of BAPTA are quite useful, in that the cis forms give the highest affinities and Ca^{2+}:Mg^{2+} selectivities of any BAPTA derivatives, and the $trans$ forms are useful isomers that provide low-affinity chelators that retain strong rejection of Mg^{2+}.

The remaining locations where substituents have been added are on the chelating arms themselves. Extra methyl groups hanging from the carboxymethyl side arms, as in FluoRhod-2 (fig. 2.1), considerably increase both Ca^{2+} and H^+ affinities. The increase in pK_a brings those values over 7 and therefore makes the chelators pH-sensitive unless electron-withdrawing substituents are also present on the benzene rings, as is the case in FluoRhod-2 (Clarke et al., 1993; Smith et al., 1993). Also, the considerable increase in hydrophobicity will make the dyes difficult to load by means of hydrolyzable esters. Therefore, such dyes have not yet been useful in practical biological measurements.

Effect of Environmental Factors: Ionic Strength, pH and Macromolecules

Aside from molecular substituents, several environmental factors affect Ca^{2+} affinities. Increasing ionic strength weakens Ca^{2+} binding, i.e., raises the K_d as would be expected for the reaction of a tetra-anion with a divalent cation (Grynkiewicz et al., 1985; Pethig et al., 1989). Temperature per se has relatively modest effects (Lattanzio and Bartschat, 1991), because complexation has only a small enthalpy of reaction. The K_d values for Ca^{2+} are typically 1.6–2.2-fold higher at 37°C in $\sim$ 150 mM ionic strength at 20°C in 100 mM KCl (Grynkiewicz et al., 1985; Merritt et al., 1990), but most of this difference is due to the ionic strength rather than the temperature. Increasing acidity or Mg^{2+} concentrations generally have modest effects on BAPTA-based indicators until pH < 6.5 or $[Mg^{2+}]$ > 10 mM are attained. Furthermore, even when these competing cations bind, they alter the spectra of most indicators relatively little, because they primarily bind to just the half of the BAPTA that does not carry the fluorophore (Minta et al., 1989; Tsien, 1980). By contrast, Ca^{2+} engages hasboth halves of BAPTA simultaneously and therefore has both a higher affinity and larger spectral effect. Of course, the APTRA-based indicators are much more vulnerable to Mg^{2+} perturbation. Finally, binding to certain macromolecules is reported to weaken the indicator's Ca^{2+} affinities considerably (Konishi et al., 1988; Blatter and Wier, 1990; Bancel et al., 1992; Zhao et al., 1996), especially in skeletal muscle. The seriousness of this problem, in practical biological applications is controversial (Westerblad and Allen, 1994).

Reaction Kinetics

Because the BAPTA nucleus is already fully ionized, Ca^{2+} can bind to an empty BAPTA moiety at essentially diffusion-controlled rates, i.e., rate constants of $10^8 - 10^9 \, M^{-1} \, s^{-1}$. The rate constants for dissociation vary in order of magnitude from 10^2 to $10^4 \, s^{-1}$ for k_d values of 0.1 to 100 μM (Quast et al., 1984; Kao and Tsien, 1988; Eberhard and Erne, 1991; Lattanzio and Bartschat, 1991). Higher K_d values result mainly from increased dissociation rate constants, with only modest decreases in association rate constants.

Fluorescence Characteristics ad How They are Affected by Ca^{2+} and Other Cations

In order for Ca^{2+} to bind to all the ligand groups simultaneously, the bond between the nitrogen and the aromatic ring has to twist by about 90° so that the lone pair of electrons from the nitrogen can point toward the Ca^{2+} placed symmetrically between the two ether oxygens. This twist, together with the electrostatic attraction of the Ca^{2+} for the electrons, prevents them from conjugating with the rest of the benzene ring or any attached chromophore. Therefore, the effect of Ca^{2+} binding can be generally

predicted by conceptually disconnecting the nitrogens from the rest of the molecule. This steric twist was initially predicted by model building (Tsien, 1980) but later confirmed by x-ray crystallography (Gerig et al., 1987). An additional effect of the Ca^{2+} is an overall rigidification of the ligand configuration, which tends to inhibit vibrational deactivation and thereby makes the fluorescence quantum efficiency of the Ca^{2+} complex generally higher than that of the metal-free indicators.

Quin-2

Quin-2 was the first fluorescent tetracarboxylate indicator that was of practical biological use as an intracellular Ca^{2+} indicator. Like most tetracarboxylate indicators from the Tsien lab, its name reflects the first four letters of the name of the fluorophore, followed by a number giving its chronological order within its chemical series. Quin-2 was thus the second completed quinoline derivative of BAPTA (Tsien, 1980). Its structure is shown in Fig. 2.1. The chromophore consists of a 6-methoxyquinoline, whose extinction coefficient and fluorescence wavelengths are rather similar to dansyl groups. The quinoline fluorophore includes a ring nitrogen, which takes the place of one of the ether oxygens of BAPTA. The double function of the quinoline keeps the molecule very compact, but lowers the $Ca^{2+}:Mg^{2+}$ selectivity by almost an order of magnitude. Fortunately, the excitation spectra of the free dye and of the Mg^{2+} complex cross at 337 nm, so, at this wavelength, Mg^{2+} binding happens not to alter the fluorescence, although it is still a silent inhibitor of Ca^{2+} binding. The Ca^{2+} binding causes a larger spectral shift, but the most obvious effect is a 6-fold increase in excitation amplitude a this wavelength, which is the traditional setting for working with quin-2. The usual estimate of 115 nM for the effective K_d for Ca^{2+} at 37°C assumes a free Mg^{2+} of 1 mM (Tsien et al., 1982). Quin-2 is little used nowadays because it has a relatively small extinction coefficient, low quantum yield even when bound to Ca^{2+}, and is relatively photobleachable, all of which make it almost impossible to image microscopically. Other drawbacks include the poor selectivity for Ca^{2+} over Mg^{2+} and other divalent cations and the difficulty of deducing Ca^{2+} from the ratio of the fluorescences obtained at two wavelengths (Grynkiewicz et al. 1985). The lack of ratiometric readout is because the quantum yield of the Ca^{2+} complex is so much greater than the free dye that their spectra do not have a useful crossover point. The main advantage of quin-2 is that because it is the smallest tetracarboxylate indicator that is sufficiently fluorescent to be detected intracellularly, its permeant acetoxymethyl (AM) ester is the most efficient at depositing high concentrations of quin-2 inside cells. Therefore, quin-2 is particularly useful (Tsien et al., 1982; Tsien and Pozzan, 1989) when (1) millimolar or higher concentrations of intracellular chelator need to be attained by the permeant ester technique and (2) those dye concentrations and the efficacy of Ca^{2+} buffering need to be quantified by fluorescence measurements, which are impractical with BAPTA.

Fura-2

Fura-2 (structure in Fig. 2.1) is probably the most popular Ca^{2+} indicator (Grynkiewicz et al., 1985; Tsien, 1989a) because it combines convenient excitation ratioing, fairly good photostability, and relatively easy loading via its acetoxymethyl ester fura-2/AM, which made it the first that could be readily imaged at the single-cell level (Tsien and Poenie, 1986). The name "fura-2" reflects its origin as the second member of a family of indicators containing benzofuran groups. Free fura-2 has an excitation peak at 362 nm, which shifts to 335 nm and increases in amplitude upon binding of Ca^{2+} (Fig. 2.2). The K_d for Ca^{2+} is 135 nM in 100 mM KCl at 20°C, vs. 224 nM in buffer-simulating mammalian cytoplasm at 37°C. The emission peak, at 518 nm for the free dye, shifts only to 510 nm when Ca^{2+} binds (Grynkiewicz et al., 1985), probably because in the excited state the amino group disengages from the Ca^{2+}. The evidence for such excited-state relaxation is that when fura-2 solutions are made highly viscous, for example by addition of 70% sucrose and chilling to −10°C, Ca^{2+} binding then causes about a 55 nm shift to shorter wavelengths (R. Tsien, unpublished observations).

Indo-1

Indo-1 (Fig. 2.1) is unique among the commercially available Ca^{2+} indicators in shifting not only its excitation but also its emission wavelengths (from 482 to 398 nm) upon Ca^{2+} binding (fig. 2.2). The K_d for Ca^{2+} in buffers mimicking mammalian cytoplasm at 37°C is 250 nm (Grynkiewicz et al., 1985). A probable reason for the emission shift is that the indole fluorophore is unusually electron-rich, so that it is less prone than the oxazole of fura-2 to steal electron density from the BAPTA amino nitrogen in the excited state. Therefore, the Ca^{2+} is retained in the binding site and keeps the emission wavelengths short. Emission ratioing is particularly valuable for instruments such as flow cytometers and confocal microscopes that use laser excitation, because it is relatively inconvenient to alternate between two UV excitation lasers at 340 and 380 nm. It is much easier to excite with a single UV laser at 351 or 364 nm and to ratio the two emission bands that are simulta-

neously produced, especially when a pair of inexpensive photomultipliers or other nonimaging detector can be used instead of imaging cameras. Emission ratioing requires no wavelength-selecting moving parts and is not limited in speed by the rate of alternating excitation wavelengths.

One of the most exciting trends in laser-based microscopy is multiphoton excitation, the use of very brief but high-intensity pulses of focused infrared light to excite dyes that normally require UV photons (Xu et al., 1996). If two or even three IR photons hit the dye molecule essentially simultaneously, which means within a femtosecond or so, their energies may add together to produce the same result as a single photon of twice or treble the energy, i.e. $\frac{1}{2}$ or $\frac{1}{3}$ the wavelength. Once the excited state is reached, the subsequent fluorescence properties are the same regardless of whether the excitation energy had been delivered by single UV or multiple IR photons. The main advantages of multiphoton excitation are that (1) the excitation beam is infrared, which penetrates biological tissue much further and with less attenuation than ultraviolet, and which is less subject to chromatic aberration from microscope optics; (2) the very high instantaneous photon fluxes necessary for the low-energy photons to cooperate are achieved only at the focus point of the laser, not above or below that focus. Therefore, dye molecules in out-of-focus planes are not excited and not subjected to photodegradation, and a confocal aperture is not necessary for depth resolution. By contrast, out-of-focus molecules are being wastefully excited and degraded in ordinary confocal microscopy, even though their emission is filtered out by the confocal pinhole that creates depth resolution. The main disadvantage of multiphoton excitation is the major expense and complexity of infrared lasers with femtosecond pulsewidths. Indo-1 is currently the only dye that permits ratiometric measurements of $[Ca^{2+}]_i$ with multiphoton excitation (Szmacinski et al., 1993). The two-photon excitation spectrum of indo-1 is unfortunately peaked at wavelengths somewhat shorter than twice the ordinary (i.e., one-photon) excitation spectrum; the necessary wavelengths, especially for the Ca^{2+} complex, are at or below 700 nm (Xu and Webb, 1996), which is at the very limits of the tuning range of currently available pulsed lasers. Three-photon excitation may actually be somewhat more convenient (Xu et al., 1996).

Miscellaneous Longer Wavelength, Wavelength-Shifting Indicators

Yet longer wavelengths of excitation and emission are possible while retaining wavelength shifts, though tradeoffs must be accepted. For example, DeMarinis et al. (1990) developed an analog of fura-2 in which

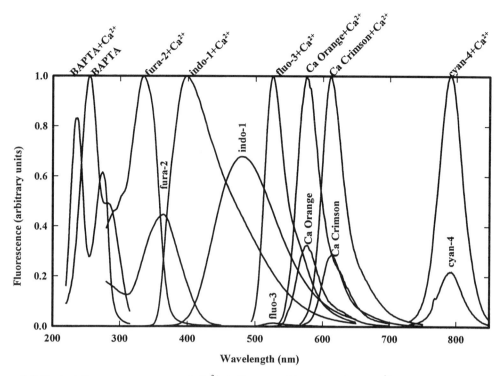

Figure 2.2 Typical fluorescence spectra of Ca^{2+} indicators with and without Ca^{2+} bound, normalized by the maximal amplitude within each pair. Therefore, the relative amplitudes of the free vs. bound forms are shown realistically, but not the relative brightnesses of different indicators. Excitation spectra are shown for BAPTA and fura-2, emission spectra for all the other indicators. Because indo-1 shifts both its excitation and emission wavelengths upon binding Ca^{2+}, the relative amplitudes of the free vs. bound emission spectra depend on excitation wavelength, which was 339 nm for these curves. Data for Ca-Orange and Ca-Crimson courtesy of M. Kuhn and P. Hewitt, Molecular Probes.

the oxazole is replaced by yet longer and more electron-withdrawing groups. The main commercially available example is "fura-red" (Fig. 2.1). Another styryl-based Ca^{2+} indicator without the furan bridge was reported by Akkaya and Lakowicz (1993). Unfortunately, such dyes, which are really styryl merocyanines, tend to have very low quantum yields in aqueous solution. This dimness is probably because the major redistribution of electron density in the excited state from the amino group to the electron-withdrawing end of the chromophore causes a large change in dipole moment, which causes massive redistribution of the surrounding water dipoles, which promotes vibrational deactivation before the excited state can emit a photon. Likewise, several BAPTA-based oxazones and carbazones were developed, but their Ca^{2+} affinities were greatly depressed by the chromophores's strong electron withdrawal, and the quantum yields were disappointing (Tsien, 1983). A later development was FluoRhod-2 (Fig. 2.1), in which the BAPTA is made an integral part

of a rhodamine chromophore (Smith et al., 1993). This dye changes from rhodamine-like (excitation and emission maxima at 537 and 566 nm, respectively) to fluorescein-like spectra (480 and 537 nm) upon binding Ca^{2+} with a K_d of $1.07\,\mu M$. Unfortunately, no biological results in live cells have been reported yet. Along the same lines, incorporation of BAPTA into a coumarin chromophore generates "BTC," which binds Ca^{2+} with a K_d near 10^{-5} M, while shifting its excitation maximum from 462 to 401 nm (Iatridou et al., 1994). In skeletal muscle, BTC appears to be particularly strongly bound to myoplasmic constituents and to sense some signal other than $[Ca^{2+}]_i$ (Zhao et al., 1996).

Non-Wavelength-Shifting Indicators: General Principles

For many purposes, it is very helpful to be able to excite the indicator fluorescence with visible light. The most common reason is that laser-based

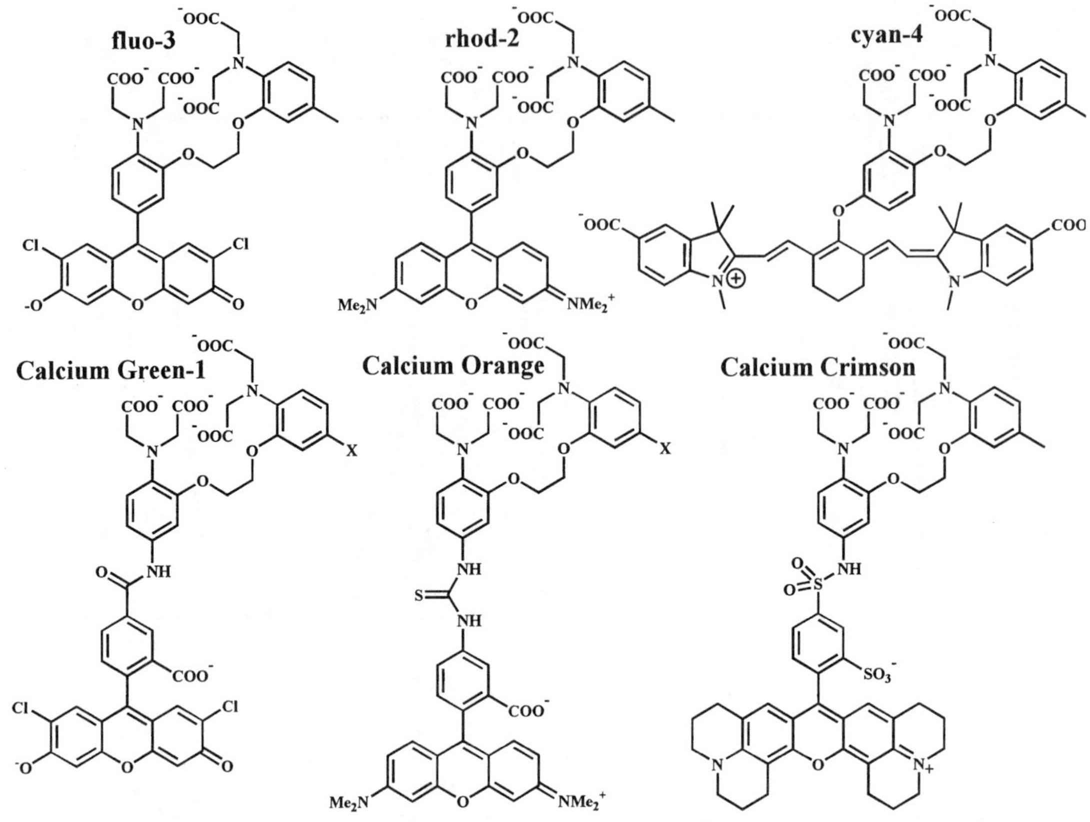

Figure 2.3 Structures of some longer wavelength Ca^{2+} indicators that work by modulation of photoinduced electron transfer: fluo-3, rhod-2, cyan-4, Calcium Green-1, Calcium Orange, and Calcium Crimson.

instruments, such as flow cytometers and confocal microscopes, are often supplied only with visible excitation, most commonly 488 nm from an argon-ion laser. Ultraviolet lasers are certainly available but consume more power, cooling capacity, and initial expense. Sometimes the instrument optics or sample chambers use glasses or plastics that do not pass ultraviolet or that emit fluorescence at those wavelengths. Occasionally, the biological preparation genuinely demands relatively long wavelengths, perhaps because of excessive autofluorescence or light sensitivity at short wavelengths. The most commonly used visible-excitation Ca^{2+} indicators, such as fluo-3, rhod-2, Ca-Green, Ca-Orange, and their analogs (Fig. 2.3), do not shift excitation or emission wavelengths significantly between Ca^{2+}-free and Ca^{2+}-bound forms. Instead, Ca^{2+} binding merely increases the quantum yield and hence the intensity of emission.

All of these dyes contain a fluorophore that is linked to, but clearly distinct from, the BAPTA moiety. The mechanism by which Ca^{2+} occupancy affects the quantum yield of the separate fluorophore is a phenomenon called photoinduced electron transfer (Rettig, 1986; Huston et al., 1988; Bisell et al., 1993; Tsien, 1993). Excitation of the fluorophore involves promotion of an electron from the highest occupied molecular orbital (HOMO) to the lowest unoccupied molecular orbital (LUMO) of the fluorophore. Naturally, this leaves a vacancy in he HOMO. In the absence of Ca^{2+}, the BAPTA portion of the indicator is quite electron-rich and is able to donate an electron to the fluorophore, which fills the vacancy in the HOMO (Fig. 2.4A). Now the originally promoted electron is marooned in the LUMO with no way to emit a photon while returning directly to the orbital from which it came. Therefore, Ca^{2+}-free BAPTA quenches the fluorescence of the attached fluorophore. Eventually, the electron should return to the BAPTA in a relatively slow, radiationless process that restores the ground state, but the net effect is that the excitation energy is dissipated by two

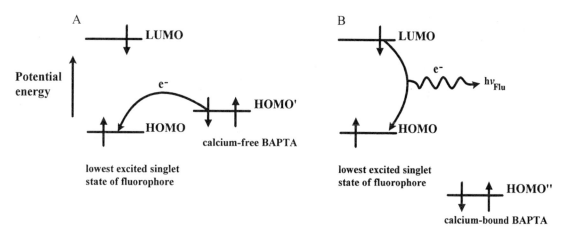

Figure 2.4 Energy level diagrams to explain why nonratiometric visible Ca^{2+} indicators increase their fluorescence upon Ca^{2+} binding. (A) In the absence of Ca^{2+}, the BAPTA unit quenches the attached fluorophore by photoinduced electron transfer. (B) When Ca^{2+} is bound, photoinduced electron transfer becomes energetically unfavorable.

successive electron transfers to and from the fluorophore. By contrast, Ca^{2+}-bound BAPTA cannot donate an electron to the HOMO (Fig. 2.4B) because the Ca^{2+} both electrostatically attracts electron density and decouples the nitrogen lone pair electrons from the rest of the molecule by the steric twist mechanism. Therefore Ca^{2+} inhibits the quenching, i.e., enhances the fluorescence. Photoinduced electron transfer and the consequent quenching are naturally more efficient the more direct and close the linkage between BAPTA and the fluorophore. Therefore, dyes such as fluo-3 and rhod-2, in which the linkage between the fluorophore and the BAPTA is only one bond, show the lowest quantum yields in their Ca^{2+}-free forms and the highest degrees of enhancement (> 100-fold) by Ca^{2+} binding (Haugland, 1996). Early estimates (Minta et al., 1989) giving lower enhancement factors were probably artifacts of contamination by Ca^{2+}-insensitive fluorescent impurities contributing to the fluorescence in the absence of Ca^{2+}.

Fluo-3 and Rhod-2

In fluo-3 (Minta et al., 1989), th fluorophore attached to BAPTA is a fluorescein-like xanthene substituted with two chlorine atoms, whose purpose is to shift the pK_a of the fluorophore down to about 5, safely remote from physiological pH. The chloro substituents also place wavelengths of excitation and emission to 506 and 526 nm, slightly longer than those of fluorescein itself. The main spectroscopic effect of Ca^{2+} binding, as stated above, is a ~ 100-fold enhancement in fluorescence amplitude (Fig. 2.2).

Fluo-3's K_d for Ca^{2+} is about 400 nm at 20°C in 100 mM KCl (Minta et al., 1989) and 864 nM at 37°C in the ionic strength of mammalian cytosol. These numbers are distinctly higher than those of BAPTA, fura-2, or indo-1, indicating that the xanthene chromophore is somewhat electron-withdrawing. Rhod-2 has a tetramethylrhodamine-like xanthene chromophore, which is positively charged so that rhod-2 has a yet higher K_d for Ca^{2+}, 570 nM (Haugland, 1996). The previous estimate of 1 μM (Minta et al., 1989) was probably perturbed by impurities.

Calcium Green, Calcium Orange, and Calcium Crimson

Michael Kuhn (1993) at Molecular Probes developed a general approach to produce fluorophore-conjugated BAPTAs, in which 5-amino-5′-methyl BAPTA is conjugated to any of a wide variety of amine-reactive fluorescent probes such as 2,7-dichlorofluorescein isothiocyanate, 5-carboxytetramethylrhodamine N-hydroxysuccinimidyl ester, and Texas red sulfonyl chloride. These three produce "Calcium Green-1," "Calcium Orange," and "Calcium Crimson," respectively (Haugland, 1996), in which the chromophores are separated from the BAPTA units by an extra benzene ring and an amide, thiourea, or sulfonamide linkage compared with the direct single-bond connections in fluo-3 and rhod-2. The increased spacer length weakens the interaction between the chromophores and the BAPTA, so that Calcium Green-1 and Calcium Orange show higher quantum efficiencies and Ca^{2+} affinities than fluo-3 and rhod-2 (Haugland, 1996). However, the spacers also greatly reduce the degree of fluorescence enhancement due to Ca^{2+} binding, and their extra molecular weight may hinder loading via permeant esters. Although Calcium Green-1 is not ratiometric, it works well with two-photon excitation to detect local $[Ca^{2+}]_i$ elevations (Denk et al., 1995).

Infrared Indicators Based on Cyanine Fluorophores

The approach of tying BAPTA to an independent fluorophore to produce Ca^{2+}-modulated PET appears to be fairly general and has been exploited to produce both shorter wavelength indicators, using fluorophores like anthracene and pyrazoline (Bissell et al., 1993), and very long-wavelength dyes, using heptamethine cyanines (S. R. Adams and R. Y. Tsien, in preparation). The cyanines yield infrared excitation and emission wavelengths (770 and 790 nm, respectively) that should be very valuable for particular biological problems, such as measurement of $[Ca^{2+}]_i$ in photoreceptors or other cells sensitive to UV and visible wavelengths, in highly pigmented or autofluorescent tissues like blood or green plants, or deep inside scattering tissues. Their K_d values are around 280 nM. However, the very large size and hydrophobicity of the heptamethine cyanine chromophore has hindered purification and successful intracellular delivery of the molecules by permeant esters or patch clamp techniques, though pressure injection through sharp microelectrodes works.

Effect of Competing Cations: Mg^{2+}, Sr^{2+}, and Ba^{2+}

The basis for discrimination between Ca^{2+} and Mg^{2+} is built into BAPTA and has already been discussed. Discrimination of Sr^{2+} and Ba^{2+} from Ca^{2+} is more difficult, not only for synthetic chelators but also for biological processes, which explains why Sr^{2+} and Ba^{2+} are often partial to excellent surrogates for Ca^{2+}. Fura-2 binds Sr^{2+} and Ba^{2+} with K_d values of 780 nM and 2.62 μM, respectively at 37°C (Schilling et al., 1989), i.e., Ca^{2+} is preferred by factors of 3.5 and 11.7, respectively. The spectral effect of Sr^{2+} and Ba^{2+} binding is basically similar to that of Ca^{2+} binding, so in order to use fura-2 to measure free $[Sr^{2+}]$ or $[Ba^{2+}]$, $[Ca^{2+}]_i$ needs to be known and preferably kept constant. Explicit tests of the effect of Sr^{2+} and Ba^{2+} on other BAPTA derivatives have not been reported, but one would expect the $Ca^{2+}:Sr^{2+}$ and $Ca^{2+}:Ba^{2+}$ selectivity ratios to be roughly preserved.

Transition metal ions also bind to BAPTA derivatives, but here the binding is usually much stronger than that of Ca^{2+} and the spectral effect more unpredictable. The most extensive data are for fura-2 and fluo-3. Fura-2 is quenched to practically zero quantum yield by Mn^{2+}, Co^{2+}, and Ni^{2+} (Grynkiewicz et al., 1985), whose paramagnetism presumably promotes intersystem crossing from the excited singlet state to the triplet, and, in turn, from the triplet to the singlet ground state. All three ions are commonly used as blockers for Ca^{2+} channels but Mn^{2+} and Co^{2+} may also permeate such channels to some extent (Shibuya and Douglas, 1992). The ability of Mn^{2+} to sneak through Ca^{2+} channels and quench quin-2 and fura-2 is the basis for a popular method for semiquantitative assessment of plasma membrane Ca^{2+} permeability (Hallam and Rink, 1985). The Mn^{2+} is added extracellularly, typically at 0.1–1 mM. The rate of quenching of the fluorescence is an index of plasma membrane permeability to divalent cations. Such quenching can, at least initially, be measured separately from the free $[Ca^{2+}]_i$ measured by ratioing, because the latter should not be affected by losses of fluorescence at both wavelengths. However, the validity of the method assumes that (1) Mn^{2+} is neither stored inside internal organelles nor pumped back out of the cell and (2) incoming Mn^{2+} does not remain free or bind to cellular buffers but binds essentially exclusively to the dye. Assumption (1) is generally accepted; assumption (2) is more problematic. Very little is known about cellular buffering of Mn^{2+} and its competition with Ca^{2+}. Although fura-2 binds Mn^{2+} about 42-fold more tightly than Ca^{2+} (Grynkiewicz et al., 1985), one must be concerned that when $[Ca^{2+}]_i$ is high, some of the fura-2 may remain bound to Ca^{2+} and unquenched.

Binding of Mn^{2+} to fluo-3 does not quench the fluorescence completely (Minta et al., 1989). Presumably, the metal both prevents photoinduced electron transfer (as does Ca^{2+}) and contributes paramagnetic quenching instead, but in fluo-3 the paramagnetic ion is much further away from the fluorophore than it is in fura-2. The net effect is that the fluo-3–Mn^{2+} complex is about one fifth as fluorescent as the Ca^{2+} complex and simulates about 100 nM free $[Ca^{2+}]_i$. This partial quenching is the basis for one of the few methods for calibrating intracellular fluo-3 signals in terms of absolute $[Ca^{2+}]_i$ without lysing the cell. A Ca^{2+} ionophore, such as ionomycin, is used first to elevate $[Ca^{2+}]_i$ to levels that saturate fluo-3, then the external Ca^{2+} is replaced by Mn^{2+}. The fluorescence levels observed when the fluo-3 is Ca^{2+}- and Mn^{2+}-saturated, together with the K_d of fluo-3 for Ca^{2+}, are enough to determine $[Ca^{2+}]_i$ (Kao et al., 1989; Merritt et al., 1990).

Heavy divalent cations without unpaired electrons, such as Zn^{2+}, Cd^{2+} and Pb^{2+}, also bind fura-2 and fluo-3 quite avidly. The K_d values are in the low nanomolar range for Zn^{2+} and picomolar for Cd^{2+} and Pb^{2+} (Tsien, 1980; Minta et al., 1989; Jefferson et al., 1990; Tomsig and Suszkiw, 1990; Hinkle et al., 1992). The effects on the spectra are roughly similar to those of Ca^{2+}, usually with somewhat less enhancement of quantum yield. Lanthanides bind with even higher affinity, which results in dissociation rates on the time scale of seconds to minutes.

Fluorescence Lifetimes

Lifetime Ratioing vs. Wavelength Imaging

The excited-state lifetimes of fluorescent indicators are in the low nanosecond range. Generally, the Ca^{2+}-complex has a longer lifetime than the free dye, as would be expected from the higher quantum efficiency of the former. The difference in lifetimes offers an alternative to dual-wavelength ratioing for measuring the percent Ca^{2+}-occupancy of the dye separately from the total amount of dye, pathlength, lamp intensity, or detector efficiency. The dye with the easiest-to-resolve lifetimes is quin-2, which changes from 1.35 to 11.6 ns lifetime upon binding Ca^{2+} (Lakowicz et al., 1992b, 1994). Unfortunately, quin-2, as mentioned previously, is not very bright, photostable, or Ca^{2+}-selective. Fura-2 and indo-1 have rather modest changes in lifetime (Szmacinski et al., 1993; Szmacinski and Lakowicz, 1995) and would be more conveniently quantified using their excitation or emission shifts. Fluo-3 ought to have a huge increase in lifetime to match its > 100-fold increase in quantum efficiency upon Ca^{2+} binding, but its excited-state photophysics are rather more complicated. The free dye gives rise to at least two excited-state populations: a species that decays too fast to be measured and another species with a lifetime not so different from the Ca^{2+} complex. The detailed structural or conformational changes responsible for the heterogeneity are not understood. The Calcium-(Green, Orange, Crimson) series would seem to be the most promising for lifetime imaging, given that they contain efficient visible-wavelength chromophores, show significant lifetime changes in response to Ca^{2+}, and cannot be ratioed by dual-wavelength techniques (Lakowicz et al., 1992a; Sanders, 1995). However, the lifetimes observed in intact cells seem rather different from those in vitro, so that interpretation and calibration of intracellular signals has remained difficult.

Two-Photon Excitation

Analysis of excited-state lifetimes requires excitation that is modulated at tens to hundreds of megahertz (MHz) or pulsed with nanosecond or shorter durations. The detector must have comparable frequency bandwidth or time resolution, which is not too difficult for a nonimaging detector but quite challenging for a camera-type system. The expense and commercial nonavailability of such imaging systems have greatly limited the exploration of lifetime detection. In my view, the most promising way to develop lifetime imaging is in conjunction with laser-scanning two-photon excitation, which inherently creates pulses of ~ 0.1 ps duration and 80 MHz repetition rate. The detector can be a nonimaging type because the spatial resolution is supplied by the excitation scanning. The outputs from two phase-sensitive detectors, respectively tuned to report photons emitted immediately (say less than 2 ns) vs. later (2–12 ns) after the pulse, could be ratioed (Sanders, 1995), just as signals at two different wavelengths are ratioed at present.

NMR Indicators

Binding of Ca^{2+} to BAPTA- or APTRA-derived chelators affects not only their optical spectra but also their NMR spectra, as expected from the ability of the cation to attract electron density. To be biologically useful, the chelator must be enriched in an isotope detectable by NMR against a low endogenous background. One approach (Robitaille and Jiang, 1992) has been ^{13}C-enrichment, which affects the ^1H spectrum, but a much more popular isotope is ^{19}F, the only natural isotope of fluorine. A variety of fluorinated chelators have been prepared (Metcalfe and Smith, 1991, London et al., 1994). Their major advantage is complete indifference to tissue opacity or autofluorescence, so that $[Ca^{2+}]_i$ can be monitored in brain slices (Badar-Goffer et al., 1990) and intact beating heart (Harding et al., 1993). Disadvantages include the expensive specialized NMR spectrometers required, the need to keep the preparation alive deep inside a superconducting magnet cavity, and the insensitivity of NMR detection. The latter necessitates relatively high concentrations of intracellular chelator ($\sim 100\,\mu M$) and long acquisition times (> 10 min), while preventing much spatial resolution. If the tissue undergoes repetitive Ca^{2+} transients that can be synchronized to the NMR pulses, millisecond time resolution can be achieved by stroboscopically varying the time delay (Harding et al., 1993).

Fluorescently Labeled Ca^{2+}-Sensitive Proteins

The two best characterized Ca^{2+}-sensitive proteins, troponin C and calmodulin, have each been labeled by an environmentally sensitive fluorophore, dansylaziridine and a merocyanine, respectively. Their fluorescence thereby becomes sensitive to the Ca^{2+} occupancy of the protein. Although preparation and microinjection of the labeled proteins are too cumbersome for routine monitoring of cellular Ca^{2+}, these probes can give useful insights into the physiological interactions of these two important Ca^{2+} sensors (Ashley et al., 1991; Hahn et al., 1992).

Targeting of Chelators/Indicators Inside Cells

Microinjection, Iontophoresis, Internal Perfusion, and Electroporation

The most direct techniques for putting the indicators inside cells all involve disruption of the plasma membrane: pressure microinjection, ionthophoresis, internal perfusion with a patch pipet, or electroporation. The first three techniques, which all use micropipets, deliver the dye to one cell at a time, which can be quite helpful in singling out one cell from a complex tissue, but can also be severely restrictive when one would prefer to monitor a larger population at once. Other advantages are that the dye is initially localized solely in the cytoplasm, that its concentration and degree of initial Ca^{2+} loading can often be controlled relatively accurately, and that no chemical modifications or enzymatic activities are required. Major disadvantages are the invasive nature of the procedures and their potential for lasting membrane damage, which would particularly perturb subsequent $[Ca^{2+}]_i$ signals because of the large electrochemical gradient favorig Ca^{2+} entry. Internal perfusion and electroporation cause major replacement of the native intracellular constituents with those supplied by the experimenter, which can cause severe rundown of physiological processes. All the micropipet techniques require considerable skill and specialized equipment.

AM Esters as Membrane-Permeant, Intracellularly Hydrolyzable Derivatives

To avoid disruption of the plasma membrane and the requirement for single cells robust enough for micropipet penetrations, membrane-permeant lipophilic derivatives of the dyes were devised. The idea to mask all the highly polar features, especially the negative charges of the carboxylates, with temporary protecting groups that permit diffusion through the plasma membrane and that fall off inside cell (Tsien, 1981). Similar concepts had been used for delivery of drugs across membrane barriers, but they had, at most, one carboxylate per molecule and delivery into a whole organism was the desired end result (Jansen and Russell, 1965; Ferres, 1980). It was unknown whether molecules with four or more masked carboxylates could cross membranes and whether each individual cell (as opposed to a specialized organ, like the liver) had the capacity to split off the protecting groups intracellularly. Fortunately, AM ester groups proved to confer the desired properties of adequate extracellular stability, membrane permeability, and susceptibility to cytoplasmic esterases. The AM esters are mixed esters of methylene glycol, $CH_2(OH)_2$, in which one hydroxyl is esterified to acetic acid and the other to the carboxylate of the chelator. Sometimes, AM esters have been termed "acetomethoxy," but this name is chemically incorrect. Most of the specificity of intracellular esterases seems to center on the carboxylic acid moeity rather than the alcohol. Simple esters, e.g., methyl or ethyl, or the chelators seem to be poor esterase substrates because the latter is a foreign carboxylic acid. However, acetyl esters are fairly common in biochemistry, so they are readily cleaved by esterases. The free hydroxyl thus released chemically destabilizes the remaining ester, so AM esters are a nice way to transfer the hydrolyzability of acetyl esters onto other carboxylic acid esters. Simple incubation of most mammalian cells with an AM ester results in intracellular accumulation of the parent carboxylic acid over tens of minutes to an hour (Tsien, 1981; Tsien et al., 1982). Because the hydrolysis of the ester groups seems to be due to intracellular esterases, the polyanion reaction products are trapped inside the cells and can accumulate to much higher concentrations than that of the extracellular ester. It must be remembered that all the carboxylates must be hydrolyzed to regenerate the indicator properly. Although the commercially available compounds are the fully anionic and fully esterified forms, intracellular hydrolysis is not an all-or-nothing process but proceeds through an entire set of partially de-esterified intermediates. Even after the initial ester is gone, sufficient time must be allowed for adequate hydrolysis of those intermediates.

The AM esters of BAPTA and quin-2 fairly readily load mammalian cells to final intracellular concentrations of several millimolar. As the indicators were optimized for stronger and longer wavelength fluorescence, the ease of loading generally decreased. Our casual impression, not backed by any quantitative comparisons, is that the loading efficiency with AM esters decreases in the order quin-2 > indo-1 > fura-2 > fluo-3 > Ca Green > cyan-4. This order also happens to fit with increasing molecular weight (except for indo-1 and fura-2, which are almost the same size), so the > 1 kDa moledular weight of the larger AM esters may obstruct permeation through the membrane. Another possibility is that the aqueous solubility of the esters is also decreasing as a steep function of molecular weight. For example, the true aqueous solubilities of BAPTA/AM and fura-2/AM are 15 and $0.11\,\mu M$, respectively (Kao et al., 1990). Therefore, in typical loading protocols using concentrations of fura-2/AM well above the above-quoted solubility, most of the ester is precipitated or bound, and not immediately available to diffuse across the membrane.

The AM esters that have formed precipitates visible under the microscope are probably ineffective at loading cells, but ester that is still colloidal or bound to certain amphiphilic macromolecules still seems able to be transferable to the plasma membrane and from there to reach the cytoplasm. Because its true solubility is so low, it is common practice to disperse the ester as much as possible, for example by gradually squirting the DMSO solution of the ester into a stirred suspension of cells. If the cells are adherent, the DMSO stock solution can be stirred into medium, sonicated briefly, then added to the culture dish.

Ester loading is often improved (Poenie et al., 1986) by including macromolecular amphiphiles such as serum albumin, Pluronic F-127, Cremophor EL, and cyclodextrins, which probably act by reversibly binding the ester, buffering its free concentration, and obstructing precipitation. Albumin is well known to carry hydrophobic molecules in the bloodstream and would be the most physiological of the above adjuvants. Pluoronic F-127 is a block copolymer of propylene oxide and ethylene oxide in which a single macromolecule probably forms a sort of micelle, with the relatively hydrophobic propylene units as the core and the ethylene glycol ethers as the skin. It is a sort of detergent that is relatively nontoxic, perhaps because it cannot split into momoners that make holes in membranes. Two important facts about Pluronic F-127 should be noted. It is not an acid, so the phrase "Pluronic acid" sometimes seen in the biological literature is an oxymoron that has somehow become commonplace through ignorant repetition. "Pluronic" is a tradename from BASF Wyandotte for an entire family of related copolymers. Pluronic F-127 is just one particular member that helps deliver hydrophobic substances like tetrahydrocannabinol and voltage-sensitive dyes to cells (see Lojewska and Loew, 1987, and references therein). It was chosen for trials with AM ester loading merely because it happened to be available in the author's laboratory (Poenie et al., 1986). Other Pluronics might be as good or better. Cremophor EL is another macromolecular detergent, which has been claimed to enable fura-2/AM to load adult brain slices (Kudo et al., 1991). This valuable action has not yet been reproducible in our hands (V. Lev-Ram and R. Tsien, unpublished observations).

Loading of the higher molecular weight indicators via AM esters is often more difficult, or impossible, in

microorganisms or plants with cell walls. The AM esters of indicators with big bright fluorophores have molecular masses of 1 kDa or greater, which are probably too high for ready permeation through the cell walls. Also, many of these cell types seem not to have much acetylesterase activity, yet have relatively strong autofluorescence. Even when the dyes are successfully introduced into the cytoplasm, they are often rapidly sequestered into vacuoles (Callaham and Hepler, 1991). For these reasons, successful applications of AM esters in such species are far less common than in vertebrate, especially mammalian, cells.

Alternative Esters

In principle, many derivatives other than AM esters might serve to mask the carboxylates until they get through the membrane. For example, anhydrides or substituted phenyl esters are relatively nonpolar yet hydrolyzable back to the carboxylic acids. Unfortunately, such derivatives are also commonly used to label proteins. Aside from potentially damaging cellular proteins, those dye molecules would be fluorescent but relatively insensitive to Ca^{2+} due to replacement of one or more carboxylates by amides. Alkoxymethyl, silyl, and dialkulaminoethyl esters (Ferres, 1980) spontaneously hydrolyze by different mechanisms that are unlikely to cause labeling of cellular constituents, but have not yet proven effective for getting Ca^{2+} indicators into cells (L. R. Makings, S. R. Adams, and R. Tsien, unpublished observations). t-Butyldimethylsilyl esters hydrolyze in 0.5–1 h, a convenient time course, but the partially hydrolyzed intermediates cause lysis of the erythrocytes used as a test system (R. Tsien and V. L. Lew, unpublished observations). A general concern with any substrate that spontaneously and nonenzymatically hydrolyzes quickly enough to load cells is that it will also be sensitive to moisture during storage and will probably have to be prepared freshly before use.

An acetoxymethyl ester is just one of a large family of acyloxyalkyl esters that should all load cells by the same mechanism. Formyloxymethyl would be a very attractive alternative, because it would reduce both the molecular weight and the hydrophobicity, and formyl groups are generally considered more labile than acetyl groups. Unfortunately, formyloxymethyl esters of tetracarboxylate chelators have so far proven difficult or impossible to synthesize (A. Minta and R. Tsien, unpublished observations). Various other esters such as 1-(acetoxy)ethyl, (methoxycarbonyloxy)methyl, and (methoxyacetoxy)methyl, have been prepared (M. Poenie, S. R. Adams, R. Tsien, unpublished observtions) but none has shown any advantage over AM esters.

A final possibility would be to use photolabile ester groups such as *o*-nitrobenzyls or related caging groups. The ester molecules would be loaded into the tissue in the absence of UV, then photolyzed to trap some of the carboxylates intracellularly. The problems with this approach would be the high molecular weight and hydrophobicity of esters carrying four or five caging groups; the need to photolyze every ester group in order to regenerate full Ca^{2+}-binding affinity; the lack of enzymatically driven intracellular accumulation and the tendency of BAPTA chromophores to quench the photolysis reaction, probably by photoinduced electron transfer (Adams et al., 1989).

Leakage and Compartmentation

The first trials with quin-2/AM loading indicated that the quin-2 was deposited mainly in the cytoplasm and nucleus, as judged by selective permeabilization of the plasma membrane with digitonin or electroporation (Tsien et al., 1982). When fura-2/AM was introduced, microscopic imaging of the dye in single cells became possible. It soon became evident that a significant amount of the dye could be taken up into organelles (Almers and Neher, 1985; Malgaroli et al., 1987; Connor, 1994), especially if the loaded cells were kept at 37°C for long periods. In principle, such compartmentation could occur either by the ester crossing two membranes and becoming hydrolyzed inside the organelles, or by transport of partially or fully anionic dye from the cytoplasm into the organelle. The former possibility might seem more likely, and is probably the dominant mechanism. However, in some cell types, sequestration of fura-2 or fluo-3 into organelles and extrusion across the plasma membrane can be inhibited by drugs such as probenecid and sulfinpyrazone (Di Virgilio et al., 1988; Merritt et al., 1990), which are used clinically to inhibit uric acid transport in the kidney. Unfortunately, probenecid and sulfinpyrazone are sometimes ineffective or have unwanted side effects on cell function. Curiously, one well-defined transporter, the multidrug resistance protein MDR1, seems able to extrude the AM esters but not the free acid forms of the dyes (Homolaya et al., 1993).

Ideally, one would alter the dye structure to reduce its ability to be transported. Vorndran et al. (1995) have made significant progress in this direction by linking a positively charged piperazine to fura-2 on the ring that is not attached to the fluorophore. The resulting "fura-PE3" seems to show much less leakage and compartmentation than fura-2, perhaps because the dye is now a zwitterion and should be less susceptible to nonspecific anion transporters.

The most thoroughgoing way to prevent compartmentation and leakage of the dye is to attach it covalently to a macromolecule such as dextran. Conjugates of fura-2, indo-1, and Calcium Green to dextran (Haugland, 1996) are now commercially available and have proven quite valuable in cases where compartmentation or leakage are particularly severe or in which observations over many hours to days are desired. Of coarse, such indicators can no longer be loaded as membrane-permeant AM esters and probably are unsuitable for ionrophoresis due to the reduction in charge per unit mass. Therefore, pressure microinjection is typically employed.

Targeting to Particular Intracellular Sites or Organelles

There is widespred interest in the likelihood that open Ca^{2+} channels in the plasma membrane or endoplasmic reticulum generate steep Ca^{2+} gradients within the first few nanometers to micrometers distance from its mouth. For a review, see Augustine and Neher (1992). The evidence for such gradients at submicroscopic dimensions arises mainly from theoretical modeling and from discrepancies between the amplitude and time course of bulk cytoplasmic $[Ca^{2+}]_i$ signals and the behavior of Ca^{2+}-sensing channels and secretory events (Chow et al., 1994). Several factors contribute to the difficulty in directly observing such local signals. Obviously, their submicroscopic spatial extent and high speed pose tremendous problems in achieving adequate sensitivity and signal-to-noise ratio. The high cooperativity of aequorin's response helps it to detect local transients (Llinás et al., 1992) but also makes quantification trickier. The very high Ca^{2+} amplitudes expected, up to hundreds of micromolar, would locally saturate conventional indicators. High concentrations of diffusible buffers would themselves tend to blur gradients by raising the effective diffusibility of intracelular Ca^{2+} (Neher, 1986; Nowycky and Pinter, 1993). Dextran conjugation should reduce this problem by lowering the diffusibility of the exogeneous buffering sites. However, the better long-term solution would be to target the Ca^{2+} indicators by biochemical affinity or molecular biological sorting to the appropriate locations, rather than relying entirely on the limited spatial resolution of visible light imaging.

A first attempt at such localization has been the attachment of a 12-carbon tail to a derivative of fura-2 (Vorndran et al., 1995). This tail serves to anchor the indicator to membranes such as liposomes, organelles, or plasma membranes. A hydrophilic spacer in the form of a piperazine amide seems necessary between the dodecyl tail and the rest of the dye so that the latter is not pulled too deep into the membrane. Unfortunately, AM ester loading appears poorly effective, so that the free acid has to be microinjected. Preliminary reports indicate that some selectivity for near-plasma-membrane Ca^{2+} may be obtainable.

Delocalized Positive Charge for Mitochondrial Uptake

Mitochondria can be loaded with Ca^{2+} indicators by two strategies. The first relies on the organelles' large negative membrane potential, which strongly accumulates delocalized cations such as the permeant AM ester of rhod-2 (Minta et al., 1989; Tsien and Bacskai, 1995). As the four esters hydrolyze and the Ca^{2+}-binding site is unmasked, the net charge shifts from +1 to −3, but by then the dye should no nonger be membrane-permeant. Not all the dye winds up in the mitochondria, presumably because some hydrolysis may occur in the cytoplasm before the esters reach the mitochondria, but the usual appearance is of bright dots or worms, readily distinguished from the dimmer diffuse background of cytoplasmic fluorescence. The identity of the organelles is demonstrated by the costaining with mitochondrial markers and the ability of mitochondrial uncouplers to release the Ca^{2+}. An alternative strategy is to reduce the rhod-2 ester to a colorless nonfluorescent dihydro derivative (Hajnóczky et al., 1995). Mitochondria and peroxisomes, perhaps because they specialize in mediating redox processes, accelerate the reoxidation back to the fluorescent dye. Unfortunately, the dihydro derivative also reoxidizes spontaneously and has to be made just before use (Haugland, 1996); also, it is uncharged and cannot itself be accumulated in a potential-dependent manner. For those seriously wishing to image mitochondrial Ca^{2+}, both methods should probably be tried.

Targeting to Particular Neurons

The use of fluorescent Ca^{2+} indicators in intact neural circuits poses special opportunities and problems. The ubiquity of Ca^{2+} channels as the main transducers that mediate biochemical responses to depolarization means that $[Ca^{2+}]_i$ transients are common markers of neuron activity. Under favorable circumstances, the optical signals that result from such $[Ca^{2+}]_i$ transients are much larger than those obtainable from current voltage-sensitive indicators. Therefore, $[Ca^{2+}]_i$ imaging is valuable not just to analyze the biophysics and biochemistry of an important messenger but also as a technique to eavesdrop on ensemble coding in neural networks. However, a major problem has been that adult central nervous system (CNS) tissue tends to load poorly or not at all

with the conventional AM esters (Yuste and Katz, 1991). The probable cause is that the esters cannot penetrate the highly tortuous and compacted extracellular spaces in the mature vertebrate CNS. Those cells that do load are usually on the surface of the brain slice, where physiological recording is suspect because of the high percentage of damaged cells. Probably, there is no fundamental defect in membrane permeability or intracellular esterase activity, because healthy neurons in immature slices or dissociated cultures load satisfactorily. Unfortunately, the neuronal connectivity and signal processing in such preparations are different from intact adult brain. Furthermore, even if all the neurons in a mature slice could be loaded, it would often be quite difficult to distinguish individual cells due to their close packing.

Remote loading of AM esters (Regehr and Tank, 1991; Wu and Saggau, 1994; Backsai et al., 1995) or dextran conjugates (McClellan et al., 1994) provides at least a partial solution to these challenges when the neurons of interest extend processes to some remote location. A high concentration of AM ester or dye dextran conjugage is adminisered at that remote site. If necessary, membranes are locally disrupted to promote loading of the free acid or conjugate. Over the next few hours to days, the indicator gradually diffuses or is transported to the imaging zone, where the only neurons that show up are those whose processes pass through the loading zone. The loading site contains much extraneous fluorescence and may be damaged, so the major requirement is that the indicator should spread further and faster than the injury that enabled it to gain entry to the cells.

Chemical and Biological Problems

Like all complex techniques, the use of Ca^{2+} indicators is subject to a number of pitfalls. Some of the more common ones are discussed below, starting with the most chemical problems and ending with the biological concerns.

Chemical Purity and Stability of Nonesterified Chelators

Almost all biological users buy the chelator free acids, esters, or conjugates from a commercial source and are therefore insulated from the nontrivial problems of multistep organic syntheses and quality control. But when such reagents fail to work or seem to have undesirable Ca^{2+}-binding, pharmacological, or fluorescence properties, it is always possible that impurities are at fault (Zucker, 1992) rather than the molecule whose name is on the label. Sometimes, these problems are obvious because material from different batches or different vendors behaves differently. But often, such comparisons are not available or the impurity is a reproducible artifact, so biological customers must be aware that purity not only should but actually can be checked when problems arise.

BAPTA-family Ca^{2+} chelators are highly polar molecules that generally do not crystallize well, especially as salts. The salts typically contain varying amounts of water and inorganic salts left over from the reagents used to hydrolyze the esters. Such non-fluorescent impurities may not affect the final use of the chelators in solution but mean that the total weight may not yield an accurate estimate of moles of chelator content. The free acids are less likely to contain much inorganic salts, but unfortunately have some tendency to decarboxylate. This reaction converts $—CH_2COOH$ groups to CH_3 groups plus CO_2, which is lost. Such decarboxylation does not greatly affect the spectra but does reduce the affinity for Ca^{2+} significantly. The form of the dye that is most stable and easily purified is usually the ester (e.g., methyl, ethyl, or t-butyl, not AM ester) from which the free acid is first prepared. Quantitative hydrolysis of a known weight of such a pure ester is usually the best way to prepare a chelator or indicator solution of known molarity, from which the extinction coefficient of the dye can be measured at a suitable absorbance maximum. Once that value is published, subsequent solutions can be assayed by absorption spectrophotometry. Absolute chelator concentrations are most important when trying to make solutions of defined percent Ca^{2+}-saturation, either to buffer Ca^{2+} or to measure the K_d of the equilibrium. A useful alternative to knowing the absolute concentrations of chelator and Ca^{2+} is to divide the chelator stock solution into two aliquots, carefully titrate one aliquot to Ca^{2+} saturation as detected by no further change in the spectral properties, then mix the nontitrated aliquot with the Ca^{2+}-saturated stock solution in the desired proportions.

Deterioration of fura-2 salts upon standing in solution at room temperature is particularly easy to notice because addition of excess Ca^{2+} no longer produces a high ratio of 340 nm to 380 nm excitation amplitudes. Really fresh fura-2, saturated with Ca^{2+} and measured on a spectrofluorometer with narrow excitation bandwidths, can give a ratio as high as 35 (Grynkiewicz et al., 1985), but this easily declines to < 20 upon careless storage. The loss of ratio is mainly because Ca^{2+} no longer suppresses the 380 nm excitation as effectively as on a fresh sample. The chemical basis for this deterioration has not been determined.

A somewhat distinct problem is the detection of impurities that are spectroscopically similar to the indicator itself. These include incompletely hydro-lyzed esters, decarboxylation and certain photodecomposition products, leftover impurities from the synthesis, etc. Unlike water, excess salts, and buffers, such contaminants are detectable by high-performance liquid chromatography (HPLC) or capillary electrophoresis. The main caution is that HPLC in stainless steel equipment is problematical because the chelators tend to leach variable amounts of heavy metals from the steel, which leads to broad peaks and irreproducible retention times. In our lab, we use nonmetallic, "biocompatible" HPLC equipment or fused silica electrophoresis capillaries. Typical HPLC conditions are on octadecyl reverse-phase silica, pH 5–7, a few millimolar EGTA, with methanol/water or acetonitrile/water gradients. Capillary electrophoresis works in capillaries of 50 cm length, 50–75 μm bore, filled with 50 mM borate, pH 8.5, or 10 mM phosphate buffer and electrophoresed at 20 kV (S. R. Adams, unpublished observations).

AM Ester Problems

The membrane-permeant acetoxymethyl esters can also have significant impurities, which, in the worst case, can actually prevent the detection of real biological signals in ester-loaded cells. Because only the contaminants with chromophores matter, AM ester purity can be checked by thin-layer chromatography on silica gel or by reverse-phase HPLC. Reverse-phase HPLC has the advantage that it is more tolerant of DMSO, which is the most common solvent for stock solutions of AM esters. Heavy metal leaching from stainless steel HPLC apparatus is not important because the AM esters have negligible chelating power until hydrolyzed. Even partial hydrolysis makes the compounds very much more polar, i.e., slower-migrating on normal-phase silica or faster-migrating on reverse-phase columns. It is also worth mentioning that partial hydrolysis is unlikely if the AM ester has been reasonably freshly prepared in dry DMSO, which should be exposed to atmospheric moisture as little as possible. Moisture in the DMSO is most readily noticed by checking the melting point as the stock solution comes out of the freezer. Dry DMSO melts at about 18°C, i.e., only just below normal room temperature. A solution that remains liquid at 0–4°C probably contains substantial amounts of water and may begin to hydrolyze the AM ester groups over a few days.

Yet another problem is how to determine whether the AM esters have fully hydrolyzed within the tissue. This should be checked if the apparent $[Ca^{2+}]_i$ signals are much smaller than expected. With some of the dyes, the excitation spectra change somewhat during the hydrolysis process, but seeing some spectral alteration is not a reliable proof that the hydrolysis has proceeded to completion. The most rigorous test is to do HPLC or capillary electrophoresis on a tissue lysate (Kawanishi et al., 1989; Tran et al., 1995), but often the quantity of material is insufficient or the apparatus may not be available in a physiological laboratory. The next best measure is to use EGTA buffers to titrate a tissue lysate over a range of free Ca^{2+} concentrations near the reported K_d for the fully hydrolyzed dye. If a Scatchard or other binding plot shows a single binding component with the right K_d, then the indicator must be substantially satisfactory (Kao et al., 1989). But if there is a significant component with a much weaker Ca^{2+} affinity, then incomplete hydrolysis should be suspected.

One of the most notorious problems with AM ester loading of cells is that not all the indicator molecules end up in the cytosol or nucleus. A significant fraction can become loaded into intracellular organelles, such as endoplasmic reticulum or mitochondria, where they obviously encounter a different ambient free Ca^{2+}. For most applications, such compartmentation is undesirable. It is minimized by avoiding ester loading altogether, for example by microinjecting or internally perfusing the indicator as a salt or a dextran conjugate. If ester loading is necessary, for example because large populations of cells are to be examined or the cells are too small to tolerate membrane puncture, then compartmentation is generally reduced by exposing cells to the ester for relatively short times and using them fairly quickly without prolonged periods of incubation at higher temperatures, such as 37°C. In some but not all cell types, compartmentation, as well as active extrusion from the cytoplasm, can be reduced by treatment with clinically useful inhibitors of anion transport, such as probenecid or sulfinpyrazone. Also, the zwitterionic indicators, such as Fura-PE3, are supposed to suffer less from compartmentation and extrusion. Compartmentation can often be recognized simply by the blotchy or punctate distribution of fluorescence. A more quantitative criterion is to permeabilize the plasma membrane with saponin, digitonin, or a patch pipet in whole-cell configuration. The truly cytosolic dye should diffuse away rapidly; the remainder, assuming the membrane-delimited organelles remain intact, represents the compartmentalized fraction. For an interesting analysis of compartmentalization, see Connor (1994).

There is considerable interest in measuring Ca^{2+} levels in the Ca^{2+}-accumulating organelles. In this case, the fraction of dye that enters the organelles is to be maximized. A lower-affinity indicator should be used, and prolonged incubation at 37°C may encourage active transport into the organelles (Hofer and Machen, 1993; Short et al. 1993; Hofer and Schulz, 1996). Cytoplasmic dye can be removed by permeabilization or patch pipet dialysis, or it can be simply quenched by dialysis with Mn^{2+}-containing intracel-

lular solutions (Tse and Hille, 1994) as long as one is sure than Mn^{2+} does not alter the Ca^{2+} transport processes of the organelles.

Bleaching and Formation of Fluorescent Ca^{2+}-Insensitive Degradation Products

All fluorescent dyes bleach at some nonzero rate under intense illumination. Obviously, illumination should be curtailed to the minimum necessary to do the experiment. Certainly, the excitation source should be shuttered off whenever one is not collecting data from the specimen. A ratiometric dye provides an additional level of protection from simple bleaching upon illumination, whose effect is simply to lower the amount of fluorophore remaining, without affecting the ratio of the two wavelengths or decay time components. Of course, the signal-to-noise will gradually degrade, but at least the apparent $[Ca^{2+}]_i$ should not be affected by the bleaching. Unfortunately, it has gradually become clear that the main ratiometric dyes to some extent under photodegradation give products that are still fluorescent but have reduced Ca^{2+} responsivity and affinity. This problem was most fully analyzed in the case of indo-1 (Scheenen et al., 1996). When indo-1-loaded cells are subjected to bright excitation, not only does the dye bleach gradually, but also the apparent resting $[Ca^{2+}]_i$ value and amplitude of $[Ca^{2+}]_i$ transients calculated from the emission ratio decrease. Thus, conversion to a fluorescent but relatively Ca^{2+}-insensitive photoproduct competes with overall bleaching. Overall bleaching requires the presence of O_2 but is relatively independent of Ca^{2+} levels. Conversion to the fluorescent, Ca^{2+}-insensitive product does not require O_2 but is inhibited by high Ca^{2+}. Mass spectroscopic analysis shows that the fluorescent, Ca^{2+}-insensitive product results from loss of a $-CH_2COO^-$ chelating arm from indo-1. Fortunately, this photodealkylation can be substantially inhibited by a few micromolar Trolox, which is a commercially available, water-soluble analog of vitamin E. Although this mechanism has been studied in greatest detail with indo-1, a similar phenomenon also occurs, to some extent, with fura-2 (Becker and Fay, 1987) and with Ca-Green (Scheenen et al., 1996). Because Ca-Green is nonratiometric, the loss of Ca^{2+}-sensitivity superficially resembles simple bleaching, which also occurs. The difference is that simple bleaching does not alter the ability of the remaining fluorescence to respond to a Ca^{2+} increase, whereas formation of a Ca^{2+}-insensitive product means that a bigger $[Ca^{2+}]_i$ increase is required to increase the remaining fluorescence by a given percentage. It has not yet been demonstrated whether Trolox inhibits formation of fluorescent Ca^{2+}-insensitive products from fura-2 and Ca-Green.

Calibration

Calibration of intracellular Ca^{2+} indicators requires observation of their readouts (fluorescence intensity, ratio of intensities at two excitation or two emission wavelengths, or excited-state lifetime) at several defined free Ca^{2+} concentrations, under conditions mimicking the normal intracellular milieu as closely as possible. The ideal would be to clamp the $[Ca^{2+}]_i$ in situ at the end of every experiment, but currently available Ca^{2+} ionophores, such as A23187 or ionomycin, do not mediate enough Ca^{2+} flux to guarantee equality of submicromolar Ca^{2+} concentrations across the plasma membrane except when cell ATP has been depleted (Lew et al., 1982; Chused et al., 1987). Permeabilization of the plasma membrane does equalize $[Ca^{2+}]$ but usually releases the indicator as well. A really powerful Ca^{2+}-specific ionophore or insertable channel would greatly aid calibration.

Single-wavelength fluorescence signals F are calibrated in terms of $[Ca^{2+}]_i$ by the equation (Tsien and Pozzan, 1989)

$$[Ca^{2+}]_i = K_d \left(\frac{F - F_{min}}{F_{max} - F} \right)$$

where F_{min} and F_{max} are the respective fluorescences observed at saturating and zero Ca^{2+} and the same concentration of indicator, optical pathlength, source brightness, and overall detector sensitivity. These four parameters are usually so variable that each experiment must be individually calibrated at its conclusion. Autofluorescence cancels out as long as it is a constant. However, any leaked dye contaminates F and should be removed by washing the cells or quenched by extracellular Mn^{2+} or other paramagnetic ion. In a cell suspension in a cuvette, lysis of the cells does not change the macroscopic average dye concentration or pathlength, so permeabilization is usable as a way of clamping $[Ca^{2+}]$ very high and very low. In cells under a microscope or in a flow cytometer, diffusion of already leaked dye into the bathing solution usually keeps its fluorescence contribution down, but the calibration procedure must not let any additional dye out of the microscopic zone of observation.

The great advantage of ratios R of fluorescences at two wavelengths λ_1 and λ_2, $R = F(\lambda_1)/F(\lambda_2)$, is that they cancel out changes in the concentration, pathlength, and instrumental sensitivity, assuming that background autofluorescence has been properly subtracted and that the molecules are not optically shielding one another. Obviously, the dye's spectral properties and Ca^{2+} affinity and the instrument's spectral bias for λ_1 vs. λ_2 will still affect the ratios, but these factors are much easier to keep constant.

The calibration equation then becomes (Grynkiewicz et al., 1985)

$$[Ca^{2+}]_i = K_d \left(\frac{R - R_{min}}{R_{max} - R} \right) \left(\frac{S_{f2}}{S_{b2}} \right)$$

where R_{min} and R_{max} are the respective ratios at zero and saturating Ca^{2+}. The new factor of (S_{f2}/S_{b2}), sometimes abbreviated to β, is the intrinsic ratio of intensities of free vs. Ca^{2+}-bound indicator observed at λ_2. The main residual problem is that dyes inside cells behave somewhat differently from dyes in buffers, whose $[Ca^{2+}]$ is easily controlled. Much of the effect can be traced to viscosity-induced alterations in R_{min}, R_{max}, and β, which, in turn, can be minimized by judicious choice of wavelengths (Busa, 1992). However, the possibility of more fundamental protein-induced perturbations, including alterations in K_d, should not be forgotten (see "Effect of Environmental Factors: Ionic Strength, pH, and Macromolecules" above).

Reports of Toxicity

Toxicity refers to biological effects that are not explained by the obvious ability of the chelators to bind and buffer Ca^{2+} and other metal ions such as Zn^{2+}. Because Zn^{2+} is important for many proteins, such as protein kinase C and zinc fingers, the strong (approximately nanomolar K_d) Zn^{2+} affinity of BAPTA-series chelators should not be forgotten. A relatively simple control (Arslan et al., 1985) is to test TPEN, which stands for N,N'-tetrakis(2-pyridylmethyl)ethylenediamine, a commercially available chelator with extremely high affinity for Zn^{2+} ($K_d = 10^{-15.6}$ M) and other transition metals (Anderegg et al., 1977), but negligible affinity for Ca^{2+} ($K_d \sim 40\,\mu M$) and Mg^{2+} ($K_d \sim 20\,mM$) (Arslan et al., 1985). Because TPEN is membrane-permeant, 20–100 μM extracellular concentration is generally sufficient. If TPEN mimics the effect of the Ca^{2+} chelator, heavy metal chelation should be suspected. If TPEN has no or a different effect, then heavy metal involvement is unlikely. Likewise, TPEN easily strips any heavy metal away from a Ca^{2+} indicator and reveals whether the latter was being perturbed by heavy metals. One recently discovered complication is that TPEN can buffer Ca^{2+} inside internal stores and thereby activate capacitative Ca^{2+} entry (Hofer et al., 1998).

There was an early report that quin-2 loading could alter lymphocyte mitogenic activity (Hesketh et al., 1983), but no comparable complaints about indo-1 and fura-2 have appeared, despite extensive use in many laboratories. The inositol-1,4,5-trisphosphate receptor may be inhibited by high concentrations of BAPTA, though estimates of the severity of the effect vary (Combettes et al., 1994; Bootman et al., 1995; Patel and Taylor, 1995; Scheenen et al., 1996). This receptor is also subject to strong feedback (both positive and negative) from the Ca^{2+} that it releases, so it is hard to be sure which effects are toxicity vs. very efficient buffering of the submicroscopic local Ca^{2+} (Horne and Meyer, 1995). Nevertheless, direct pharmacological effects of BAPTA are a general concern with the common practice of using very high concentrations of BAPTA to check for local gradients of Ca^{2+}. Obviously, such experiments require stringently high purity of the BAPTA.

The hydrolysis of each AM ester group releases one molecule each of formaldehyde and acetic acid, and one additional proton from the ionization of the chelator free acid. Up to five AM esters need to hydrolyze per indicator molecule loaded. Both the formaldehyde and the acidity (Spray et al., 1984) are potential sources of toxicity. Fortunately, the formaldehyde is gradually generated at low concentrations far below those that fix tissue and not so different from those produced endogenously from demethylation reactions. Also, one can give effective antidotes, such as ascorbate and pyruvate, typically included in the extracellular medium at millimolar concentrations. Ascorbate may work by reversing the reaction of formaldehyde with lysines (Trezl et al., 1983); pyruvate may fuel oxidative phosphorylation and counteract formaldehyde's tendency to inhibit glycolysis (Garcia-Sancho, 1984). These antidotes have even been able to protect the retina (Ratto et al., 1988), which is probably the tissue most sensitive to formaldehyde. As for the acidity, the pH drop (if any) can be measured by various methods, such as the pH indicator BCECF. Once measured, the acidification can be mimicked by extracellular application of appropriate concentrations of permeant acids, such as acetic or propionic acids. The pH can also be clamped by mixtures of weak bases and acids. Finally, useful control compounds for both the formaldehyde and acidification are methylene diacetate and anis-1/AM (Tsien and Rink, 1983), which are AM esters that release the same byproducts but do not generate Ca^{2+} chelators. Methylene diacetate is available from Aldrich Chemical Co.; anis-1/AM, which represents an isolated half of BAPTA, was formerly available from Molecular Probes as N-(o-methoxyphenyl)iminodiacetic acid, AM.

Despite these concerns about BAPTA and AM ester loading, there is some surprising evidence that BAPTA/AM, far from being toxic to whole animals, is actually protective against neuronal ischemia (Tymianski et al., 1993). The mechanism appears not to be brute force buffering of Ca^{2+} elevations but rather diffusional dissipation of $[Ca^{2+}]_i$ microdomains and damping of glutamate release (Tymianski et al., 1994).

Measuring Free $[Ca^{2+}]_i$ vs. Buffering

The Ca^{2+} indicators must inevitably buffer and perturb Ca^{2+} to some extent in the process of measuring it. To detect changes in $[Ca^{2+}]_i$ with as little perturbation as possible, the most important consideration is to use the lowest concentration of indicator that gives an adequate optical signal-to-noise ratio. This requires that the binding vs. unbinding of Ca^{2+} should make the largest possible change in the fluorescence per molecule of indicator. If absolute levels of $[Ca^{2+}]_i$ are to be measured, they should be comparable to the K_d, because it is at the K_d that the fractions of Ca^{2+}-bound and Ca^{2+}-free dye are jointly most sensitive to Ca^{2+}. If a range of $[Ca^{2+}]_i$ levels, from basal to the peak of a transient, are to be accurately quantified, the mathematically ideal K_d is the geometric mean of the extreme concentrations.

The inevitable buffering has three types of possible consequences: depression of basal Ca^{2+}, blunting of transient increases, and blurring of spatial gradients. Depression of basal $[Ca^{2+}]_i$ by AM-ester loaded chelators or indicators is hardly ever significant in cells exposed to normal extracellular Ca^{2+} concentrations. Low $[Ca^{2+}]_i$ is self-correcting because it increases Ca^{2+} influx via capacitative Ca^{2+} entry and starves Ca^{2+} efflux mechanisms. Only if a cell is heavily loaded with Ca^{2+} buffers in the absence of extracellular Ca^{2+}, or if it is continuously perfused with low-Ca^{2+} buffers, is it possible to lower basal $[Ca^{2+}]_i$ to a significant and sustained manner. For example, when several millimolar quin-2 was loaded into cells in zero external Ca^{2+}, the $[Ca^{2+}]_i$ could be lowered to 10–20 nM. However, on restoration of normal external Ca^{2+}, the $[Ca^{2+}]_i$ rapidly rose back to a normal resting value (Tsien et al., 1982). Although the mechanism for the fast $[Ca^{2+}]_i$ restoration was unknown at the time, in retrospect this must have been one of the first manifestations of capacitative Ca^{2+} influx.

The $[Ca^{2+}]_i$ transients are obviously more severely buffered the higher the total concentration of chelator, the closer the $[Ca^{2+}]_i$ to the K_d, and the briefer the Ca^{2+} fluxes. Because the free and Ca^{2+}-bound chelators are more mobile than Ca^{2+} bound to endogenous buffers, the chelators also considerably raise the effective diffusibility of the Ca^{2+} and smooth out spatial gradients of $[Ca^{2+}]_i$. Such effects have been the subject of considerable theoretical and some experimental investigation.

Buffering of $[Ca^{2+}]_i$ is highly desirable for many types of experiments, for example to determine the biological function of $[Ca^{2+}]_i$ transients or gradients. If these spatiotemporal fluctuations are necessary for a cell function, introduction of enough chelator should prevent the relevant physiology. Even if the $[Ca^{2+}]_i$ transient is too brief or microscopically localized to be directly monitored, an indirect idea of its magnitude can often be obtained by trying chelators of varying K_d values and seeing which are the most effective, e.g. (Speksnijder et al., 1989; Ranganathan et al., 1994). If some or all of the chelator is a fluorescent indicator, its signal helps to determine the concentration of chelator and the extent to which $[Ca^{2+}]_i$ has been stabilized. Often, $[Ca^{2+}]_i$ is crucially involved in complex feedback loops; inclusion of sufficient buffering power serves as a sort of $[Ca^{2+}]_i$ clamp that can help disentangle the individual reactions, just as voltage clamping separates ionic permeabilities. Finally, once the exogenous chelator becomes the dominant intracellular buffer, changes in its Ca^{2+}-loading become measures of Ca^{2+} fluxes. Often, it is desirable to vary the concentration or K_d and to extrapolate to infinite chelator concentration (Tsien and Rink, 1983). A particularly useful application has been to measure the fraction of an ionic current that is carried by Ca^{2+}; the total charge (integral of the current) is compared with the Ca^{2+} flux measured by the fluorescent indicator under conditions where it buffers the $[Ca^{2+}]_i$ well enough to shield endogenous buffers (Schneggenburger et al., 1993).

Prospects for the Future

What are the most likely areas for major future progress in the monitoring of intracellular Ca^{2+}? It is always risky to make such predictions, but my guess would point toward improved targeting of fluorescent indicators, especially by molecular genetic tricks roughly analogous to those already demonstrated for aequorin. Because fluorescent indicators give 10^3 to 10^5 more photons per molecule than aequorin can, and are not destroyed by Ca^{2+}, they will always have major advantages once they too are constructable by molecular biology. This dream is well within possibility. For example, Ca^{2+} indicators that shift emission wavelengths have been constructed (Miyawaki et al., 1997) from color-shifted mutants of the Green Fluorescent Protein (GFP) of *Aequorea victoria* (Cubitt et al., 1995b) fused to calmodulin and a calmodulin-binding peptide. Simple expression of the genetic construct synthesizes the fluorescent indicator in situ without addition of any small molecules such as coelenterazine. Such protein-based indicators are easily targeted to specific organelles (Miyawaki et al., 1997), and should be fused to other key proteins in Ca^{2+} signaling and incorporated into transgenic organisms. Such harnessing of

molecular biology should answer many of today's remaining intractable questions about $[Ca^{2+}]_i$.

Acknowledgments   The research in my laboratory is supported by the Howard Hughes Medical Institute and by the National Institutes of Health (NS27177).

References

Adams, S. R., Kao, J. P. Y., Grynkiewicz, G., Minta, A., and Tsien, R. Y. (1988) Biologically useful chelators that release Ca^{2+} upon illumination. *J. Am. Chem. Soc.* 110: 3212–3220.

Adams, S. R., Kao, J. P. Y., and Tsien, R.Y. (1989) Biologically useful chelators that take up Ca^{2+} upon illumination. *J. Am. Chem. Soc.* 111: 7957–7968.

Akkaya, E. U. and Lakowicz, J. R. (1993) Styryl-based wavelength-ratiometric probes: a new class of fluorescent calcium probes with long wavelength emission and a large Stokes' shift. *Anal. Biochem.* 213: 285–289.

Almers, W. and Neher, E. (1985) The Ca signal from fura-2 loaded mast cells depends strongly on the method of dye-loading. *FEBS Lett.* 192: 13–18.

Anderegg, G., Hubmann, E., Podder, N.G., and Wenk, F. (1977) Pyridinderivate als Komplexbildner. XI. Die Thermodynamik der Metallkomplexbildung mit Bis-, Tris, und Tetrakis[(2-pyridyl)methyl]-aminen. *Helv. Chim. Acta* 60; 123–140.

Arslan, P., Di Virgilio, F., Beltrame, M., Tsien, R. Y., and Pozzan, T. (1985) Cytosolic Ca^{2+} homeostasis in Ehrlich and Yoshida carcinomas. *J. Biol. Chem.* 260: 2719–2727.

Ashley, C. C., Griffiths, P. J., Lea, T. J., Mulligan, P., Palmer, R. E., and Simnett, S. J. (1991) Use of fluorescent TnC derivatives and 'caged' compounds to study cellular Ca^{2+} phenomena. In *Cellular Calcium* (McCormack, J. G. and Cobbold, P. H. eds.). Oxford, Oxford University Press, pp. 177–203.

Augustine, G. J. and Neher, E. (1992) Neuronal Ca^{2+} signalling takes the local route. *Curr. Opin. Neurobiol.* 2: 302–307.

Bacskai, B. J., Wallén, P., Lev-Ram, V., Grillner, S., and Tsien, R. Y. (1995) Activity-related calcium dynamics in lamprey motoneurons as revealed by video-rate confocal microscopy. *Neuron* 14: 19–28.

Badar-Goffer, R. S., Ben-Yoseph, O., Dolin, S. J., Morris, P. G., Smith, G. A., and Bachelard, H.S. (1990) Use of 1,2-bis(2-amino-5-fluorophenoxy)ethanane-N, N, N', N'-tetraacetic acid (5FBAPTA) in the measurement of free intracellular calcium in the brain by ^{19}F-nuclear magnetic resonance spectroscopy. *J. Neurochem.* 55: 878–884.

Bancel, F., Salmon, J. M., Vigo, J., Vo-Dinh, T., and Viallet, P. (1992) Investigation of noncalcium interactions of fura-2 by classical and synchronous fluorescence spectroscopy. *Anal. Biochem.* 204: 231–238.

Baudet, S., Hove-Madsen, L., and Bers, D.M. (1994) How to make and use calcium-specific mini- and microelectrodes. *Methods Cell Biol.* 40: 93–113.

Baumann, O., Walz, B., Somlyo, A. V., and Somlyo, A.P. (1991) Electron probe microanalysis of calcium release and magnesium uptake by endoplasmic reticulum in bee photoreceptors. *Proc. Natl. Acad. Sci. USA* 88: 741–744.

Becker, P. L. and Fay, F. S. (1987) Photobleaching of fura-2 and its effect on determination of calcium concentrations. *Am. J. Physiol.* 253: C613–C618.

Beeler, T. (1990) Oxidation of sulfhydryl groups and inhibition of the $(Ca^{2+} + Mg^{2+})$-ATPase by arsenazo III. *Biochim. Biophys. Acta* 1027: 264–267.

Belan, P. V., Gerasimenko, O. V., Tepikin, A. V., and Petersen, O.H. (1996) Localization of Ca^{2+} extrusion sites in pancreatic acinar cells. *J. Biol. Chem.* 271: 7615–7619.

Bissell, R. A., de Silva, A. P., Gunaratne, H. Q. N., Lynch, P. L. M., McCoy, C. P., Maguire, G. E. M., and Sandanayake, K. R. A. S. (1993) Fluorescent photoinduced electron-transfer sensors: the simple logic and its extensions. In *Fluorescent Chemosensors for Ion and Molecule Recognition.* (Czarnik, A. W. ed.) Washington, DC, American Chemical Society, pp. 45–58.

Blatter, L. A. and Wier, W. G. (1990) Intracellular diffusion, binding, and compartmentalization of the fluorescent calcium indictors indo-1 and fura-2. *Biophys. J.* 58: 1491–1499.

Blinks, J. R., Wier, W. G., Hess, P., and Predergast, F. G. (1982) Measurement of Ca^{2+} concentrations in living cells. *Prog. Biophys. Mol. Biol.* 40: 1–114.

Bootman, M. D., Missiaen, L., Parys, J. B., De Smedt, H., and Casteels, R. (1995) Control of inositol 1,4,5-trisphosphate-induced Ca^{2+} release by cytosolic Ca^{2+}. *Biochem. J.* 306: 445–451.

Borle, A. B. (1981) Control, modulation, and regulation of cell calcium. *Rev. Physiol. Biochem. Pharmacol.* 90: 13–153.

Brini, M., Marsault, R., Bastianutto, C., Pozzan, T., and Rizzuto, R. (1994) Nuclear targeting of aequorin. A new approach for measuring nuclear Ca^{2+} concentration in intact cells. *Cell Calcium*, 16: 259–268.

Brini, M., Marsault, R., Bastianutto, C., Alvarez, J., Pozzan, T., and Rizzuto, R. (1995) Transfected aequorin in the measurement of cytosolic Ca^{2+} concentration ($[Ca^{2+}]_c$). A critical evaluation. *J. Biol. Chem.* 270: 9896–9903.

Busa, W. B. (1992) Spectral characterization of the effect of vicosity on fura-2 fluorescence: excitation wavelength optimization abolishes the viscosity artifact. *Cell Calcium* 13: 313–319.

Button, D. and Eidsath, A. (1996) Aequorin targeted to the endoplasmic reticulum reveals heterogeneity in luminal Ca^{++} concentration and reports agonist- or InsP$_3$-induced release of Ca^{++}. *Mol. Biol. Cell* 7: 419–434.

Callaham, D. A. and Hepler, P. K. (1991) Measurement of free calcium in plant cells. In *Cellular Calcium* (McCormack, J. G. and Cobbold, P. H. eds.). Oxford, Oxford University Press, pp. 383–410.

Campbell, A. K. (1983) *Intracellular Calcium.* John Wiley and Sons, Chichester.

Campbell, A. K., Lea, T. J., and Ashley, C. C. (1979) Coelentereate photoproteins. In *Detection and Measurement of Free Ca^{2+} in Cells* (Ashley, C. C. and Campbell, A. K. eds.). Amsterdam, Elsevier, pp. 13–72.

Campbell, A. K., Trewavas, A. J., and Knight, M. R. (1996) Calcium imaging shows differential sensitivity to cooling and communication in luminous transgenic plants. *Cell Calcium* 19: 211–218.

Chandra, S., Fewtrell, C., Millard, P. J., Sandison, D. R., Webb, W. W., and Morrison, G. H. (1994) Imaging of total intracellular calcium and calcium influx and efflux in individual resting and stimulated tumor mast cells using ion microscopy. *J. Biol. Chem.* 269: 15168–15194.

Chow, R. H., Klingauf, J., and Neher, E. (1994) Time course of Ca^{2+} concentration triggering exocytosis in neuroendocrine cells. *Proc. Natl. Acad. Sci. USA* 91: 12765–12769.

Chused, T. M., Wilson, H. A., Greenblatt, D., Ishida, Y., Edison, L. J., Tsien, R. Y., and Finkelman, F.D. (1987) Flow cytometric analysis of cytosolic free calcium in murine splenic B lymphocytes: responses to anti-IgM and anti-IgD differ. *Cytometry* 8: 396–404.

Clarke, S. D., Metcalfe, J. C,. and Smith, G. A. (1993) Design and properties of new ^{19}F NMR Ca^{2+} indicators: modulation of the affinities of BAPTA derivatives *via* alkylation. *J. Chem. Soc., Perkin Trans.* 2: 1187–1209.

Cobbold, P. H. and Lee, J. A. C. (1991) Aequorin measurements of cytoplasmic free calcium. In *Cellular Calcium* (McCormack, J. G. and Cobbold, P. H. eds.) Oxford, Oxford University Press, pp. 55–81.

Cobbold, P. H. and Rink, T. J. (1987) Fluorescence and bioluminescence measurement of cytoplasmic free calcium. *Biochem. J.* 248: 313–328.

Combettes, L., Champeil, P., Finch, E. A., and Goldin, S. M. (1994) Calcium and inositol 1,4,5-trisphosphate-induced Ca^{2+} release. *Science* 265: 813–815.

Connor, J. A. (1994) Dissecting signals from fura-2 in cytosol, nucleus, and other organelles in living cells. *Neuroprotocols* 5: 25–34.

Cubitt, A. B., Firtel, R. A., Fischer, G., Jaffe, L. F., and Miller, A. L. (1995a) Patterns of free calcium in multicellular stages of Dictyostelium expressing jellyfish aequorin. *Development* 121: 2291–2301.

Cubitt, A. B., Heim, R., Adams, S. R., Boyd, A. E., Gross, L. A., and Tsien, R. Y. (1995b) Understanding, using and improving green fluorescent protein. *Trends Biochem. Sci.* 20: 448–455.

DeMarinis, R. M., Katerinopoulos, H. E., and Muirhead, K. A. (1990) New tetracarboxylate compounds as fluorescent intracellular calcium indicators. *Biochem. Methods* 112: 381

Denk, W., Holt, J. R., Shepherd, G. M., and Corey, D. P. (1995) Calcium imaging of single stereocilia in hair cells: localization of transduction channels at both ends of tip links. *Neuron* 15: 1311–1321.

Di Virgilio, F., Steinberg, T. H., Swanson, J. A., and Silverstein, S.C. (1988) Fura-2 secretion and sequestration in macrophages. *J. Immunol.* 140: 915–920.

Docampo, R., Moreno, S. N. J., and Mason, R. P. (1993) Generation of free radical metabolites and superoxide anion by the calcium indicators arsenazo III, antipyrylazo III, and murexide in rat liver microsomes. *J. Biol. Chem.* 258: 14920–14925.

Duchen, M. R. (1992) Fluorescence—monitoring cell chemistry *in vivo*. In *Monitoring Neuronal Activity* (Stamford, J. A. ed.). Oxford, Oxford University Press, pp. 231–260.

Eberhard, M. and Erne, P. (1991) Calcium binding to fluorescent calcium indicators: calcium green, calcium orange and calcium crimson. *Biochem. Biophys. Res. Commun.* 180: 209–215.

Ferres, H. (1980) Pro-drugs of β-lactam antibiotics. *Chem. Ind.* 435–440.

Garcia-Sancho, J. (1984) Inhibition of glycolysis in the human erythrocyte by formaldehyde and Ca-chelator esters. *J. Physiol.* 357: 60P.

Gerig, J. T., Singh, P., Levy, L. A., and London, R. E. (1987) Calcium complexation with a highly calcium selective chelator: crystal structure of Ca(CaFBAPTA)·5H$_2$O. *J. Inorg. Biochem.* 31: 113–121.

Grohovaz, F., Bossi, M., Pezzati, R., Meldolesi, J., and Tarelli, F.T. (1996) High resolution ultrastructural mapping of total calcium: electron spectroscopic imaging/electron energy loss spectroscopy analysis of a physically/chemically processed nerve-muscle preparation. *Proc. Natl. Acad. Sci. USA* 93: 4799–4803.

Grynkiewicz, G., Poenie, M., and Tsien, R.Y. (1985) A new generation of Ca^{2+} indicators with greatly improved fluorescence properties. *J. Biol. Chem.* 260: 3440–3450.

Hahn, K., DeBiasio, R., and Taylor, D. L. (1992) Patterns of elevated free calcium and calmodulin activation in living cells. *Nature* 359: 736–738.

Hajnóczky, G., Robb-Gaspers, L. D., Seitz, M. B., and Thomas, A. P. (1995) Decoding of cytosolic calcium oscillations in the mitochondria. *Cell* 82: 415–424.

Hallam, T. J. and Rink, T. J. (1985) Agonists stimulate divalent cation channels in the plasma membrane of human platelets. *FEBS Lett.* 186: 175–179.

Harding, D. P., Smith, G. A., Metcalfe, J. C., Morris, P. G., and Kirschenlohr, H. L. (1993) Resting and end-diastolic [Ca^{2+}]$_i$ measurements in the Langendorff-perfused ferret heart loaded with ^{19}F NMR indicator. *Magn. Reson. Med.* 29: 605–615.

Haugland, R. P. (1996) *Handbook of Fluorescent Probes and Research Chemicals,* 6th edn. Molecular Probes, Eugene, OR.

Hesketh, T. R., Smith, G. A., Moore, J. P., Taylor, M. W., and Metcalfe, J.C. (1983) Free cytoplasmic calcium concentration and the mitogenic stimulation of lymphocytes. *J. Biol. Chem.* 258: 4876–4882.

Hinkle, P. M., Shanshala, E. D. II, and Nelson, E. J. (1992) Measurement of intracellular cadmium with fluorescent dyes. Further evidence for the role of calcium channels in cadmium uptake. *J. Biol. Chem.* 267: 25553–25559.

Hirota, A., Chandler, W. K., Southwick, P. L., and Waggoner, A.S. (1989) Calcium signals recorded from two new purpurate indicators inside frog cut twitch fibers. *J. Gen. Physiol.* 94: 597–631.

Hofer, A. M. and Machen, T. E. (1993) Technique for *in situ* measurement of calcium in intracellular inositol 1,4,5-trisphosphate-sensitive stores using the fluorescent indicator mag-fura-2. *Proc. Natl. Acad. Sci. USA* 90: 2598–2602.

Hofer, A. M. and Schulz, I. (1996) Quantification of intraluminal free [Ca] in the agonist-sensitive internal calcium store using compartmentalized fluorescent indicators: some considerations. *Cell Calcium* 20: 235–242.

Hofer, A. M., Fasolato, C., and Pozzan, T. (1998) Capacitative Ca^{2+} entry is closely linked to the filling state of internal Ca^{2+} stores: a study using simultaneous measurements of I_{CRAC} and intraluminal [Ca^{2+}]. *J. Cell Biol.* 140: 325–334.

Homolya, L., Hollo, Z., Germann, U. A., Pastan, I., Gottesman, M. M., and Sarkadi, B. (1993) Fluorescent cellular indicators are extruded by the multidrug resistance protein. *J. Biol. Chem.* 268: 21493–21496.

Horne, J. H. and Meyer, T. (1995) Luminal calcium regulates the inositol trisphosphate receptor of rat basophilic leukemia cells at a cystolic site. *Biochemistry* 34: 12738–12746.

Huston, M. E., Haider, K. W., and Czarnik, A.W. (1988) Chelation-enhanced fluorescence in 9,10-bis(TMEDA)anthracene. *J. Am. Chem. Soc.* 110: 4460–4462.

Iatridou, H., Foukaraki, E., Kuhn, M. A., Marcus, E. M., Haugland, R. P., and Katerinopoulos, H.E. (1994) The development of a new family of intracellular calcium probes. *Cell Calcium* 15: 190–198.

Jansen, A. B. A. and Russell, T. J. (1965) Some novel penicillin derivatives. *J. Chem. Soc.* 2127–2132.

Jefferson, J. R., Hunt, J. B., and Ginsburg, A. (1990) Characterization of indo-1 and quin-2 as spectroscopic probes for Zn^{2+}-protein interactions. *Anal. Biochem.* 187: 328–336.

Jong, D. S., Pape, P. C., Baylor, S. M., and Chandler, W. K. (1995) Calcium inactivation of calcium release in frog cut muscle fibers that contain millimolar EGTA or fura-2. *J. Gen. Physiol.* 106: 337–388.

Kao, J. P. Y. (1994) Practical aspects of measuring [Ca^{2+}] with fluorescent indicators. *Methods Cell Biol.* 40: 155–181.

Kao, J. P. Y. and Tsien, R. Y. (1988) Ca^{2+} binding kinetics of fura-2 and azo-1 from temperature-jump relaxation measurements. *Biophys. J.* 53: 635–639.

Kao, J. P. Y., Harootunian, A. T., and Tsien, R. Y. (1989) Photochemically generated cytosolic calcium pulses and their detection by fluo-3. *J. Biol. Chem.* 264: 8179–8184.

Kao, J. P. Y., Alderton, J. M., Tsien, R. Y., and Steinhardt, R. A. (1990) Active involvement of Ca^{2+} in mitotic progression of Swiss 3T3 fibroblasts. *J. Cell Biol.* 111: 183–196.

Kawanishi, T., Blank, L. M., Harootunian, A. T., Smith, M. T., and Tsien, R. Y. (1989) Ca^{2+} oscillations induced by hormonal stimulation of individual fura-2-loaded hepatocytes. *J. Biol. Chem.* 264: 12859–12866.

Kendall, J. M., Badminton, M. N., Sala-Newby, G. B., Campbell, A. K., and Rembold, C. M. (1996) Recombinant apoaequorin acting as a pseudo-luciferase reports micromolar changes in the endoplasmic reticulum free Ca^{2+} of intact cells. *Biochem. J.* 318: 383–387.

Konishi, M., Olson, A., Hollingworth, S., and Baylor, S. M. (1988) Myoplasmic binding of fura-2 investigated by steady-state fluorescence and absorbance measurements. *Biophys. J.* 54: 1089–1104.

Kudo, Y., Nakamura, T., and Ito, E. (1991) A 'macro' image analysis of fura-2 fluorescence to visualize the distribution of functional glutamate receptor sybtypes in hippocampal slices. *Neurosci. Res.* 12: 412–420.

Kuhn, M. A. (1993) 1,2-Bis(2-aminophenoxy)ethane-N,N,N',N'-tetraacetic acid conjugates used to measure intracellular Ca^{2+} concentration. In *Fluorescent Chemosensors for Ion Molecule Recognition.* (Czarnik, A. W. ed.). Washington, DC, American Chemical Society, pp. 147–161.

Lakowicz, J. R., Szmacinski, H., and Johnson, M. L. (1992a) Calcium imaging using fluorescence lifetimes and long-wavelength probes. *J. Fluoresc.* 2: 47–62.

Lakowicz, J. R., Szmacinski, H., Nowaczyk, K., and Johnson, M. L. (1992b) Fluorescence lifetime imaging of calcium using quin-2. *Cell Calcium* 13: 131–147.

Lakowicz, J. R., Szmacinski, H., Nowaczyk, K., Lederer, W. J., Kirby, M. S., and Johnson, M. L. (1994) Fluorescence lifetime imaging of intracellular calcium in COS cells using quin-2. *Cell Calcium* 15: 7–27.

Lattanzio, F. A. Jr. and Bartschat, D. K. (1991) The effect of pH on rate constants, ion selectivity and thermodynamic properties of fluorescent calcium and magnesium indicators. *Biochem. Biophys. Res. Commun.* 177: 184–191.

Levy, L. A., Murphy, E., Raju, B., and London, R. E. (1988) Measurement of cytosolic free magnesium ion concentration by ^{19}F NMR. *Biochemistry* 27: 4041–4048.

Lew, V. L., Tsien, R. Y., Miner, C., and Bookchin, R. M. (1982) Physiological $[Ca^{2+}]_i$ level and pump-leak turnover in intact red cells measured using an incorporated Ca chelator. *Nature* 298: 478–481.

Llinás, R., Sugimori, M., and Silver, R. B. (1992) Microdomains of high calcium concentration in a presynaptic terminal. *Science* 256: 677–679.

Lloyd, Q. P., Kuhn, M. A., and Gay, C. V. (1995) Characterization of calcium translocation across the plasma membrane of primary osteoblasts using a lipophilic calcium-sensitive fluorescent dye, calcium green C_{18}. *J. Biol. Chem.* 270: 22445–22451.

Lojewska, Z. and Loew, L. M. (1987) Insertion of amphiphilic molecules into membranes is catalyzed by a high molecular weight non-ionic surfactant. *Biochim. Biophys. Acta* 899: 104–112.

London, R. E., Rhee, C. K., Murphy, E., Gabel, S., and Levy, L.A. (1994) NMR-sensitive fluorinated and fluorescent intracellular calcium ion indicators with high dissociation constants. *Am. J. Physiol.* 266: C1313 -C1322.

Malgaroli, A., Milani, D., Meldolesi, J., and Pozzan, T. (1987) Fura-2 measurement of cytosolic free Ca^{2+} in monolayers and suspensions of various types of animal cells. *J. Cell Biol.* 105: 2145–2155.

McClellan, A. D., McPherson, D., and O'Donovan, M. J. (1994) Combined retrograde labeling and calcium imaging in spinal cord and brainstem neurons of the lamprey. *Brain Res.* 663: 61–68.

Merritt, J. E., McCarthy, S. A., Davies, M. P. A., and Moores, K. E. (1990) Use of fluo-3 to measure cytosolic Ca^{2+} in platelets and neutrophils: loading cells with the dye, calibration of traces, measurements in the presence of plasma, and buffering of cytosolic Ca^{2+}. *Biochem. J.* 269: 513–519.

Metcalfe, J. C. and Smith, G. A. (1991) NMR measurement of cytoplasmic free calcium concentration by fluorine labelled indicators. In *Cellular Calcium* (McCormack, J. G. and Cobbold, P. H. eds.), Oxford, Oxford University Press, pp. 123–132.

Miller, A. L., Karplus, E., and Jaffe, L. F. (1994) Imaging $[Ca^{2+}]_i$ with aequorin using a photon imaging detector. *Methods Cell Biol.* 40: 305–338.

Miller, D. L. and Korenbrot, J. I. (1987) Kinetics of light-activated Ca fluxes across the plasma membrane of rod outer segments: a dynamic model of the regulation of cytoplasmic Ca concentration. *J. Gen. Physiol.* 90: 397–425.

Minta, A., Kao, J. P. Y., and Tsien, R. Y. (1989) Fluorescent indicators for cytosolic calcium based on rhodamine and fluorescein chromophores. *J. Biol. Chem.* 264: 8171–8178.

Miyawaki, A., Llopis, J., Heim, R., McCaffery, J. M., Adams, J. A., Ikura, M., and Tsien, R. Y. (1997) Fluorescent indicators for Ca^{2+} based on green fluorescent proteins and calmodulin. *Nature* 388: 882–887.

Montero, M., Brini, M., Marsault, R., Alvarez, J., Sitia, R., Pozzan, T., and Rizzuto, R. (1995) Monitoring dynamic changes in free Ca^{2+} concentration in the endoplasmic reticulum of intact cells. *EMBO J.* 14: 5467–5475.

Moreton, R. B. (1991) Optical techniques and Ca^{2+} imaging. In *Cellular Neurobiology* (Chad, J. and Wheal, H. eds.). Oxford, Oxford University Press, pp. 205–222.

Nakajima-Shimada, J., Iida, H., Tsuji, F. I., and Anraku, Y. (1991) Monitoring of intracellular calcium in *Saccharomyces cerevisiae* with an apoaequorin cDNA expression system. *Proc. Natl. Acad. Sci. USA* 88: 6878–6882.

Neher, E. (1986) Concentration profiles of intracellular calcium in the presence of a diffusible chelator. *Exp. Brain Res.* 14: 80–96.

Nowycky, M. C. and Pinter, M. J. (1993) Time courses of calcium and calcium-bound buffers following calcium influx in a model cell. *Biophys. J.* 64: 77–91.

Orchard, C. H., Boyett, M. R., Fry, C. H., and Hunter, M. (1991) The use of electrodes to study cellular Ca^{2+} metabolism. In *Cellular Calcium: A Practical Approach* (McCormack, J. G. and Cobbold, P. H. eds.). Oxford, Oxford University Press, pp. 83–113.

Patel, S. and Taylor, C. W. (1995) Quantal responses to inositol 1,4,5-trisphosphate are not a consequence of Ca^{2+} regulation of inositol 1,4,5-trisphosphate receptors. *Biophys. J.* 312: 789–794.

Penner, R., Matthews, G., and Neher, E. (1988) Regulation of calcium influx by second messengers in rat mast cells. *Nature* 334: 499–504.

Perrin, D. D., Dempsey, B., and Serjeant, E. P. (1981) *pKa Prediction for Organic Acids and Bases.* Chapman and Hall, New York.

Pethig, R., Kuhn, M., Payne, R., Adler, E., Chen, T.-H., and Jaffe, L. F. (1989) On the dissociation constants of BAPTA-type calcium buffers. *Cell Calcium* 10: 491–498.

Poenie, M. (1992) Measurement of intracellular calcium with fluorescent calcium indicators. In *Neuromethods vol. 20: Intracellular Messengers*, (Boulton, A., Baker, G., and Taylor, C. eds.). Humana Press, Totowa, NJ, pp. 129–170.

Poenie, M., Alderton, J., Steinhardt, R., and Tsien, R. (1986) Calcium rises abruptly and briefly throughout the cell at the onset of anaphase. *Science* 233: 886–889.

Quast, U., Labhardt, A. M., and Doyle, V. M. (1984) Stopped-flow kinetics of the interaction of the fluorescent calcium indicator quin-2 with calcium ions. *Biochem. Biophys. Res. Commun.* 123: 604–611.

Raju, B., Murphy, E., Levy, L. A., Hall, R. D., and London, R. E. (1989) A fluorescent indicator for measuring cytosolic free magnesium. *Am. J. Physiol.* 256: C540- C548.

Ranganathan, R., Bacskai, B. J., Tsien, R. Y., and Zuker, C. S. (1994) Cytosolic calcium transients: spatial localization and role in *Drosophila* photoreceptor cell function. *Neuron* 13: 837–848.

Ratto, G. M., Payne, R., Owen, W. G., and Tsien, R. Y. (1988) The concentration of cytosolic free calcium in vertebrate rod outer segments measured with fura-2. *J. Neurosci.* 8: 3240–3246.

Regehr, W. G. and Tank, D. W. (1991) Selective fura-2 loading of presynaptic terminals and nerve cell processes by local perfusion in mammalian brain slice. *J. Neurosci. Methods* 37: 111–119.

Rettig, W. (1986) Charge separation in excited states of decoupled systems—TICT compounds and implications regarding the development of new laser dyes and the primary processes of vision and photosynthesis. *Angew. Chem., Int. Ed. Engl.* 25: 971–988.

Rizzuto, R., Bastianutto, C., Brini, M., Murgia, M., and Pozzan, T. (1994a) Mitochondrial Ca^{2+} homeostasis in intact cells. *J. Cell Biol.* 126: 1183–1194.

Rizzuto, R., Brini, M., and Pozzan, T. (1994b) Targeting recombinant aequorin to specific intracellular organelles. *Methods Cell Biol.* 40: 339–358.

Robitaille, P. M. and Jiang, Z. (1992) New calcium-sensitive ligand for nuclear magnetic resonance spectroscopy. *Biochemistry* 31: 12585–12591.

Rutter, G. A., Burnett, P., Rizzuto, R., Brini, M., Murgia, M., Pozzan, T., Tavare, J. M., and Denton, R. M. (1996) Subcellular imaging of intramitochondrial Ca^{2+} with recombinant targeted aequorin: significance for the regulation of pyruvate dehydrogenase activity. *Proc. Natl. Acad. Sci. USA* 93: 5489–5494.

Sanders, R. (1995) Fluorescence lifetime as a contrast mechanism in confocal imaging. Ph.D. thesis, Universiteit Utrecht.

Sanui, H. and Rubin, H. (1982) Atomic absorption measurement of cations in cultured cells. In *Ions, Cell Proliferation, and Cancer.* (Boynton, A. L., McKeehan, W. L., and Whitfield, J. F. eds.). Academic Press, New York, pp. 41–52.

Scheenen, W. J. J. M., Makings, L. R., Gross, L.R ., Pozzan, T., and Tsien, R. Y. (1996) Photodegradation of indo-1 and its effect on apparent Ca^{2+} concentrations. *Chem. Biol.* 3: 765–774.

Schilling, W. P., Rajan, L., and Strobl-Jager, E. (1989) Characterization of the bradykinin-stimulated calcium influx pathway of cultured vascular endothelial cells. Saturability, selectivity, and kinetics. *J. Biol. Chem.* 264: 12838–12848.

Schneggenburger, R., Zhou, Z., Konnerth, A., and Neher, E. (1993) Fractional contribution of calcium to the cation current through glutamate receptor channels. *Neuron* 11: 133–143.

Shibuya, I. and Douglas, W. W. (1992) Calcium channels in rat melanotrophs are permeable to manganese, cobalt, cadmium, and lanthanum, but not to nickel: evidence provided by fluorescence changes in fura-2 loaded cells. *Endocrinology* 131: 1936–1941.

Shimomura, O. (1995) Luminescence of aequorin is triggered by the binding of two calcium ions. *Biochem. Biophys. Res. Commun.* 211: 359–363.

Shimomura, O., Musicki, B., Kishi, Y., and Inouye, S. (1993) Light-emitting properties of recombinant semi-synthetic aequorins and recombinant fluorescein-conjugated aequorin for measuring cellular calcium. *Cell Calcium* 14: 373–378.

Short, A. D., Klein, M. G., Schneider, M. F., and Gill, D. L. (1993) Inositol 1,4,5-trisphosphate-mediated quantal Ca^{2+} release measured by high resolution imaging of Ca^{2+} within organelles. *J. Biol. Chem.* 268: 25887–25893.

Smith, G. A., Metcalfe, J. C., and Clarke, S. D. (1993) The design and properties of a series of calcium indicators which shift from rhodamine-like to fluorescein-like fluorescence on binding calcium. *J. Chem. Soc., Perkin Trans.* 2: 1195–1204.

Smith, P. J. S., Sanger, R. H., and Jaffe, L. F. (1994) The vibrating Ca^{2+} electrode: a new technique for detecting plasma membrane regions of Ca^{2+} influx and efflux. *Methods Cell Biol.* 40, 115–134.

Somlyo, A. V., Bond, M., Broderick, R., and Somlo, A .P. (1988) Calcium and magnesium movements through sarcoplasmic reticulum, endoplasmic reticulum, and mitochondria. *Adv. Exp. Med. Biol.* 232: 221–229.

Speksnijder, J. E., Miller, A. L., Weisenseel, M. H., Chen, T.-H., and Jaffe, L. F. (1989) Calcium buffer injections block fucoid egg development by facilitating calcium diffusion. *Proc. Natl. Acad. Sci. USA* 86: 6607–6611.

Spray, D. C., Nerbonne, J., Campos de Carvalho, A., Harris, A. L., and Bennett, M. V. L. (1984) Substituted benzyl acetates: a new class of compounds that reduce gap junctional conductance by cytoplasmic acidification. *J. Cell Biol.* 99: 174–179.

Szmacinski, H. and Lakowicz, J. R. (1995) Possibility of simultaneously measuring low and high calcium concentrations using fura-2 and lifetime-based sensing. *Cell Calcium* 18: 64–75.

Szmacinski, H., Gryczynski, I., and Lakowicz, J.R. (1993) Calcium-dependent fluorescence lifetimes of indo-1 for one- and two-photon excitation of fluorescence. *Photochem. Photobiol.* 58: 341–345.

Tepikin, A. V., Llopis, J., Snitsarev, V. A., Gallacher, D. V., and Petersen, O. H. (1994) The droplet technique: measurement of calcium extrusion from single isolated mammalian cells. *Pflügers Arch.* 428: 664–670.

Thomas, A. P. and Delaville, F. (1991) The use of fluorescent indicators for measurements of cytosolic-free calcium concentration in cell populations and single cells. In *Cellular Calcium* (McCormack, J. G. and Cobbold, P. H. eds.). Oxford University Press, Oxford, pp. 1–54.

Thomas, M. V. (1982) *Techniques in Calcium Research.* Academic Press, New York.

Thomas, M. V. (1991) Metallochromic indicators. In *Cellular Calcium*, (McCormack, J. G. and Cobbold, P. H. eds.). Oxford University Press, Oxford, pp. 115–122.

Tomsig, J. L. and Suszkiw, J. B. (1990) Pb^{2+}-induced secretion from bovine chromaffin cells: fura-2 as a probe for Pb^{2+}. *Am. J. Physiol.* 259: C762–C768.

Tran, N. N. P., Leroy, P., Bellucci, L., Robert, A., Nicolas, A., Atkinson, J., and Capdeville-Atkinson, C. (1995) Intracellular concentrations of fura-2 and fura-2/AM in vascular smooth muscle cells following perfusion loading of fura-2/AM in arterial segments. *Cell Calcium* 18: 420–428.

Trezl, L., Rusznak, I., Tyihak, E., Szarvas, T., and Szende, B. (1983) Spontaneous N^ϵ-methylation and N^ϵ-formylation reactions between L-lysine and formaldehyde inhibited by L-ascorbic acid. *Biophys. J.* 214: 289–292.

Tse, F. W. and Hille, B. (1994) Cyclic Ca^{2+} changes in intracellular stores of gonadotropes during gonadotropin-releasing hormone-stimulated Ca^{2+} oscillations. *Proc. Natl. Acad. Sci. USA* 91: 9750–9754.

Tsien, R. Y. (1980) New calcium indicators and buffers with high selectivity against magnesium and protons: design, synthesis, and properties of prototype structures. *Biochemistry* 19: 2396–2404.

Tsien, R. Y. (1981) A non-disruptive technique for loading calcium buffers and indicators into cells. *Nature* 290: 527–528.

Tsien, R. Y. (1983) Intracellular measurements of ion activities. *Annu. Rev. Biophys. Bioeng.* 12: 94–116.

Tsien, R. Y. (1988) Fluorescent measurement and photochemical manipulation of cytosolic free calcium. *Trends Neurosci.* 11: 419–424.

Tsien, R. Y. (1989a) Fluorescent indicators of ion concentrations. *Methods Cell Biol.* 30: 127–156.

Tsien, R. Y. (1989b) Fluorescent probes of cell signaling. *Annu. Rev. Neurosci.* 12: 227–253.

Tsien, R.Y. (1992) Intracellular signal transduction in four dimensions: from molecular design to physiology (1992 Bowditch Lecture). *Am. J. Physiol.* 263: C723 - C728.

Tsien, R. Y. (1993) Fluorescent and photochemical probes of dynamic biochemical signals inside living cells. In *Fluorescent Chemosensors for Ions and Molecule Recognition.* (Czarnik, A. W. ed.). pp. 130–146. American Chemical Society, Columbus, OH, pp. 130–146.

Tsien, R. Y. and Bacskai, B. J. (1995) Video-rate confocal microscopy. In *Handbook of Biological Confocal Microscopy.* 2nd edn. (Pawley, J. B. ed.). Plenum Press, New York, pp. 459–478.

Tsien, R. Y. and Poenie, M. (1986) Fluorescence ratio imaging: a new window into intracellular ionic signaling. *Trends Biochem. Sci.* 11: 450–455.

Tsien, R. Y. and Pozzan, T. (1989) Measurement of cytosolic free Ca^{2+} with quin2: practical aspects. *Methods Enzymol.* 172: 230–262.

Tsien, R. Y. and Rink, T.J. (1983) Measurement of cytoplasmic free Ca^{2+}. In *Current Methods of Cellular Neurobiology*, 3rd edn. (Barker, J. and McKelvey, J. F. eds.). Wiley, New York.

Tsien, R. Y., Pozzan, T., and Rink, T.J. (1982) Calcium homeostasis in intact lymphocytes: cytoplasmic free Ca^{2+} monitored with a new, intracellularly trapped fluorescent indicator. *J. Cell Biol.* 94; 325–334.

Tymianski, M., Wallace, M. C., Spigelman, I., Uno, M., Carlen, P. L., Tator, C. H., and Charlton, M. P. (1993) Cell-permeant Ca^{2+} chelators reduce early excitotoxic and ischemic neuronal injury *in vitro* and *in vivo*. *Neuron* 11: 221–235.

Tymianski, M., Spigelman, I., Zhang, L., Carlen, P. L., Tator, C. H., Charlton, M. P., and Wallace, M. C. (1994) Mechanism of action and persistence of neuroprotection by cell-permeant Ca^{2+} chelators. *J. Cereb. Blood Flow Metab.* 14: 911–923.

Vorndran, C., Minta, A., and Poenie, M. (1995) New fluorescent calcium indicators designed for cytosolic retention or measuring calcium near membranes. *Biophys. J.* 69: 2112–2124.

Wendt-Gallitelli, M. F. and Isenberg, G. (1991) X-ray microanalysis. In *Cellular Calcium: A Practical Approach.* (McCormack, J. G. and Cobbold, P. H. eds.). Oxford University Press, Oxford, pp. 133–157.

Westerblad, H. and Allen, D. G. (1994) Methods for calibration of fluorescent calcium indicators in skeletal muscle fibers. *Biophys. J.* 66: 926–928.

Williams, D. A. and Fay, F. S. (1990) Imaging of cell calcium: collected papers and reviews. *Cell Calcium* 11: 55–249.

Wu, L. G. and Saggau, P. (1994) Presynaptic calcium is increased during normal synaptic transmission and paired-pulse facilitation, but not in long-term potentiation in area CA1 of hippocampus. *J. Neurosci.* 14: 645–654.

Xu, C. and Webb, W. W. (1996) Measurement of two-photon excitation cross sections of molecular fluorophores with data from 690 to 1050 nm. *J. Opt. Soc. Am. B* 13: 481–491.

Xu, C., Ziipfel, W., Shear, J. B., Williams, R. M., and Webb, W. W. (1996) Multiphoton fluorescence excitation: new spectral windows for biological nonlinear microscopy. *Proc. Natl. Acad. Sci. USA* 93: 10763–10768.

Yau, K. W. and Nakatani, K. (1984) Electrogenic Na–Ca exchange in retinal rod outer segment. *Nature* 311: 661–663.

Yuste, R. and Katz, L. C. (1991) Control of postsynaptic Ca^{2+} influx in developing neocortex by excitatory and inhibitory neurotransmitters. *Neuron* 6: 333–344.

Zhao, M., Hollingworth, S., and Baylor, S. M. (1996) Properties of tri- and tetracarboxylate Ca^{2+} indicators in frog skeletal muscle fibers. *Biophys. J.* 70: 896–916.

Zucker, R. S. (1992) Effects of photolabile calcium chelators on fluorescent calcium indicators. *Cell Calcium* 13: 29–40.

3

Intracellular Calcium Pools and Calcium Oscillations in Cells from the Central Nervous System

Giorgio Carmignoto
Lucia Pasti
Tullio Pozzan

The development of fluorescent Ca^{2+} indicators that can be trapped in the cytosol of practically all cell types via application of intracellularly hydrolizable esters represents a milestone in the study of cellular Ca^{2+} homeostasis (Tsien, 1980; Tsien et al., 1989). The first of these indicators, quin-2, because of its relatively low fluorescence, allowed mainly studies on cell populations, which, however, unraveled some key characteristics of the intracellular Ca^{2+} response which were later confirmed by more sophisticated single-cell analysis. This latter technique was made possible by the new, more strongly fluorescent Ca^{2+} probes, such as fura-2 and indo-1, allowing the dissection of inter- and intra-cellular heterogeneities in Ca^{2+} handling (Grynkiewicz et al., 1985; Tsien, 1989). Ironically, however, probably the most intriguing new phenomenon concerning Ca^{2+} homeostasis, i.e., the oscillations of cytosolic Ca^{2+} concentration ($[Ca^{2+}]_i$), were not first described using the new sophisticated technique of single-cell image analysis, but with the most traditional, old-fashioned Ca^{2+} probe, the photoprotein aequorin (Woods et al., 1986). Soon after this seminal observation in rat hepatocytes, $[Ca^{2+}]_i$ oscillations and waves were detected in practically all cell types using different fluorescent probes and stimuli (Berridge and Galione, 1988). It is now well established that periodic $[Ca^{2+}]_i$ changes represent not only an ubiquitous response to cells to a variety of agonists, but probably the typical response of cells to physiological agonist concentrations. Numerous excellent reviews have been published over the last decade about the mechanism and role of $[Ca^{2+}]_i$ oscillations in eukaryotes (Berridge and Galione, 1988; Jacob, 1990; Tsien and Tsien, 1990; Jaffe, 1991; Berridge and Dupont, 1994). Rather than simply adding a new

contribution to this too crowded field, we have concentrated our attention on two aspects of this phenomenon which have not been extensively considered (1) the $[Ca^{2+}]_i$ oscillations in cells of the central nervous system; (2) the correlation between $[Ca^{2+}]_i$ oscillations and cellular functions. Because of space limitations, numerous important observations will not be mentioned and we apologize in advance to these colleagues.

$[Ca^{2+}]_i$ Oscillations and Waves: General Considerations

The idea that the $[Ca^{2+}]_i$ changes during cellular activity are not represented by simple, dose-dependent increases, but rather by more complex periodic events, goes back to over 20 years ago. Since the early work of Jaffe and collaborators on *Medaka* eggs injected with aequorin (Jaffe, 1991), it was clear that, at least in this cell type, the $[Ca^{2+}]_i$ increases induced by fertilization are organized in the form of spatially oriented waves. Similarly, in *Xenopus* oocytes (Berridge, 1988), as well as in other cell types (Woods et al., 1986; Ambler et al., 1988; Gray, 1988; Jacob et al., 1988; Foskett et al., 1991), the application of hormones or neurotransmitters such as acetylcholine induces repetitive oscillations of Ca^{2+}-activated Cl^- currents. Last, but not least, it was well known since the 1950s that neuronal activity is often composed of a burst of action potentials. Given that depolarization leads to opening of voltage-operated Ca^{2+} channels (VOCs), it was almost obvious to expect an oscillatory behavior of $[Ca^{2+}]_i$ in relation to oscillations of membrane potential. Nonetheless, it was not until 1986 that $[Ca^{2+}]_i$

oscillations were measured directly in an intact mammalian cell. Surprisingly, these oscillations were observed in a nonexcitable cell, the rat hepatocyte, (Woods et al., 1986) and were not dependent on periodic opening and closing of plasma membrane Ca^{2+} channels, but rather on cycles of Ca^{2+} release and reuptake from an intracellular compartment sensitive to the second messenger inositol(1,4,5)-trisphosphate (InsP3). Since then, hundreds of papers have appeared describing similar oscillatory phenomena in a variety of cell types and the reader is referred to the original contributions or to the numerous reviews on this topic published in the last years (Berridge and Galione, 1988; Jacob, 1990; Tsien and Tsien, 1990; Jaffe, 1991; Berridge and Dupont, 1994). Here, it is sufficient to say that the phenomenon of $[Ca^{2+}]_i$ oscillations is characterized by periodic increases of the cytosolic Ca^{2+} concentration, whose amplitude is rather constant while their frequency varies according to the cell type, the stimulus, and other experimental conditions. A crude first distinction of $[Ca^{2+}]_i$ oscillations can be based on the origin of the Ca^{2+} spike. In some cell types, essentially excitable cells, the source of Ca^{2+} is extracellular, via plasma membrane Ca^{2+} channels; in other, mainly nonexcitable cells, it is intracellular and depends on the release of Ca^{2+} from endoplasmic reticulum or one of its subcompartments. This distinction is, however, largely artificial, since most often a contribution of intracellular stores is evident also in cells which oscillate mainly through opening and closing of plasma membrane channels; vice versa, in cells which rely primarily on intracellular stores, plasma membrane Ca^{2+} channels can play a key permissive role (see below). The classical approach to distinguish between the intra- or extracellular origin of $[Ca^{2+}]_i$ oscillations is to apply the stimulus in Ca^{2+}-free medium. However, while in most cases this approach gives unambiguous answers, in some it can lead to erroneous interpretations. An obvious example is that of cardiac myocytes (Fabiato, 1983). No doubt exists about the fact that most of the Ca^{2+} which sustains the periodic contraction of the cells is released from the sarcoplasmic reticulum (SR). However, it is also well established that the release of Ca^{2+} from the SR is triggered by the localized influx of Ca^{2+} through VOCs that, in turn, triggers the opening of the SR Ca^{2+} channels, a process known as Ca^{2+}-induced Ca^{2+} release (CICR). Thus, in this tissue, the inhibition of $[Ca^{2+}]_i$ oscialltions upon removal of Ca^{2+} from the medium could be erroneously interpreted as evidence for an extracellular origin of the $[Ca^{2+}]_i$ oscillations. In many instances, more sophisticated pharmacological approaches are needed to draw firm conclusions about the origin of the $[Ca^{2+}]_i$ oscillations, such as the use of sarco-endoplasmic reticulum Ca^{2+} ATPase (SERCA) inhibitors, channel blockers, or activators.

In addition to the periodicity of the $[Ca^{2+}]_i$ increases, another key characteristic of this phenomenon is its intracellular spatial complexity. Depending on the cell type and stimulus, $[Ca^{2+}]_i$ oscillations may appear in the form of a single wave originating from one specific cellular region, often referred to as the "oscillator" (Berridge and Gallione, 1988); or in the form of multiple radially moving waves, as in the case of oocytes (Lechleiter and Clapman, 1992; Bootman and Berridge, 1995). Finally, abortive oscillations may occur in specific cellular hot spots, as in the case of pancreatic acinar cells (Thorn et al., 1993), the cardiac myocytes (Cannell et al., 1995), PC12 cells (Grohovaz et al., 1991), or the glamorous phenomenon of Ca^{2+} "puffs" recently described in the *Xenopus* oocytes (Yao et al., 1995). As discussed below in more detail, neither the mechanism nor the physiological function of $[Ca^{2+}]_i$ oscillations has been completely elucidated. As to the mechanism, several unifying hypotheses have been proposed (Berridge and Galione, 1988, Meyer and Streyer, 1988; Petersen and Wakui, 1990; Friel, 1995), but no consensus has been reached yet, probably because no such unique mechanism exists. As to the physiological role, the most traditional and, in our opinion, still the most convincing hypothesis is that which sees in the $[Ca^{2+}]_i$ oscillations a digital coding of cell activation (see below).

Mechanism of $[Ca^{2+}]_i$ Oscillations: One or Multiple Mechanisms?

From what is briefly discussed above, it is clear that at least two major mechanisms exist: one due to periodic opening and closing of plasma membrane Ca^{2+} channels, the other due to cycles of Ca^{2+} release and reuptake from intracellular stores.

As to the first, in a cell firing repetitive action potentials, and endowed with Ca^{2+} VOCs, and behavior of $[Ca^{2+}]_i$ is predicted to be, and has been shown to be, oscillatory. This applies to both neuronal and neuroendocrine cell types. However, even in this relatively simple model, the situation is often not straightforward inasmuch as a contribution by intracellular Ca^{2+} stores in modulating the amplitude or kinetics of the oscillations can be demonstrated, at least in some cell types (see, for example, Kuba and Takeshita, 1981; Friel and Tsien, 1992; Nohmi et al., 1992; and below).

Is there any evidence that $[Ca^{2+}]_i$ oscillations can be sustained by Ca^{2+} influx through plasma membrane channels other than Ca^{2+} VOCs? At least two examples have been presented in which it is clear that $[Ca^{2+}]_i$ oscillations can be triggered by an oscillatory influx of Ca^{2+} through a novel type of Ca^{2+} channels, often referred to as "store-operated channels" (SOCs). These channels are activated when the Ca^{2+} content of intracellular stores decreases and accordingly, the current has been named I_{CRAC} (Penner et al., 1988), for Ca^{2+} release activated Ca^{2+} current (for review, see Fasolato et al, 1994). The first example, quite artificial, is that of voltage-clamped mast cells in which, after depletion of intracellular stores, a periodic change of membrane potential, imposed via the patch pipette, generates oscillations of $[Ca^{2+}]_i$ through SOCs (Penner et al, 1988; Fasolato et al., 1993). Though this example of $[Ca^{2+}]_i$ oscillations is clearly imposed by the experimental protocol, it is at least theoretically reasonable to predict that in an intact cell a prolonged stimulation of these channels, paralleled by oscillations of membrane potential, could result in similar $[Ca^{2+}]_i$ oscillations. Indeed, in many cell types it had been shown that $[Ca^{2+}]_i$ oscillations induced by InsP3 generating agonists either subside imediatley upon removal of Ca^{2+} ions from the extracellular medium (see, for example, Cornell-Bell et al., 1990) or significantly decrease their frequency (Igusa and Miyazaki, 1983). The effect of extracellular Ca^{2+} removal on $[Ca^{2+}]_i$ oscillations, however, has been interpreted as evidence for a secondary role of Ca^{2+} infux in the phenomenon, i.e., for preventing either the unloading of the stores or the decrease of the $[Ca^{2+}]_i$ below a permissive level. Though this may indeed be the case, the possibility has not been considerd, at least to our knowledge, that oscillations are, instead, entirely due to Ca^{2+} influx through SOCs. In other words, it is possible to imagine the following scenario: (1) the agonist, via InsP3 production, causes a complete and prolonged emptying of the stores; (2) I_{CRAC} is activated; (3) membrane potential oscillates and/or I_{CRAC} itself undergoes cycles of activation/inactivation, possibly modulated by $[Ca^{2+}]_i$. A model of this type could explain not only the inhibition of $[Ca^{2+}]_i$ oscillations by removal of extracellular Ca^{2+}, but also the oscillatory behavior of cells treated with suboptimal doses of thapsigargin (Dolmetsch and Lewis, 1994).

By far the most extensively studied model of $[Ca^{2+}]_i$ oscillations is, however, that involving the periodic Ca^{2+} release from intracellular stores. The main actors in this type of oscillations are the intracellular Ca^{2+} channels localized on the endo-sarcoplasmic reticulum membrane. Two classes of Ca^{2+}-release channels have been identified, though others may exist. The first channel is gated by the binding of InsP3 and is thus named InsP3 receptor (InsP3R) (Ma et al., 1988; Furuichi et al., 1989; Mignery et al., 1990; Niggli and Lederer, 1990; Berridge, 1993). The second, the ryanodine receptor (RyRs), was initially described in striated muscle (Otsu et al., 1990; for

a recent review, see Pozzan et al., 1994) and derives its name from the plant alkaloid ryanodine, which, because of its high affinity binding, was instrumental in the purification of the protein (Inui et al., 1987). These two receptors show a 5- to 10-fold preference for Ca^{2+} over Na^+ and K^+, share structural and functional similarities, and can both be expressed by the same cell. It is noteworthy that they can both be modulated by Ca^{2+}, ATP, phosphorylation, and by interaction with other proteins (for reviews, see Henzi and MacDermott, 1992; Sorrentino and Volpe, 1993).

To summarize and simplify a vast literature, the models proposed fall into three main categories: models in which the oscillatory release of Ca^{2+} depends on oscillations of the signal generated at the receptor level (model 1), models which take into account the existence of multiple independent stores and channels influencing each other (model 2), and models which rely on feedback mechanisms between Ca^{2+} and InsP3 at the intracellular channels responsible for the release (model 3). It would be too lengthy and beyond the purpose of this chapter to discuss in detail the evidence for or against each model. We will, therefore, summarize the main features of each model, also mentioning the most relevant arguments in favor of or against each of them.

In model 1 (see, for example, Meyer and Streyer, 1988), the key point is the periodic variation of the concentration of the second messenger, primarily InsP3, leading to release of Ca^{2+}. The latter could vary either because of a negative feedback at the receptor level carried out, for example, by protein kinase C-dependent phosphorylation (Bird et al., 1993; Nishizuka, 1988), or by a positive feedback caused by Ca^{2+} itself on phospholiphase C (Cockroft and Thomas, 1992). The advantage of this model is that it is simple; the main disadvantage is that it is, at present, impossible to prove (or disprove) it directly, given that kinetic measurements of the InsP3 level in single cells are not possible yet. The strongest, though not conclusive, argument against this proposal is the finding that $[Ca^{2+}]_i$ oscillations can be induced also by intracellular perfusion with nonmetabolizable analogs of InsP3, i.e., under conditions in which the concentration of the second messenger is presumed to be constant (Wakui et al., 1989).

In model 2 (see, for example, Berridge, 1993), oscillations are due to a complex interplay betwen InsP3-sensitive and Ca^{2+}-sensitive stores. In other words, upon generation of InsP3, a small catalytic release of Ca^{2+} is induced through InsP3-gated channels; in turn, the increase in $[Ca^{2+}]$ close to RyRs results in the opening of the latter channels, thereby causing a much more robust Ca^{2+} release. The Ca^{2+} then diffuses away, RyRs close, and the cycle can start again. In favor of this model are the findings that, in several cell types, ryanodine, which keeps the channels in an open conformation (Fleischer and Inui, 1989), often abolishes $[Ca^{2+}]_i$ oscillations, thus offering a plausible explanation for the coexistence in the same cell type of two differently regulated Ca^{2+}-release channels (Goldbeter et al., 1990). The main argument against this model is the finding that ryanodine is often inefficient at inhibiting $[Ca^{2+}]_i$ oscillations (see, for example, Miyazaki et al., 1993; Uneyama et al., 1993).

In model 3, oscillations are attributed to the molecular characteristics of the InsP3Rs themselves, which could periodically open and close in the presence of a constant level of InsP3. The essential features of this model are in the existence of (1) a synergistic action between Ca^{2+} and InsP3 in opening the channel (Meyer et al., 1988), (2) an inhibitory action of high $[Ca^{2+}]$ on channel opening (Bezprozvanny et al., 1991), and (3) a refractory period of the InsP3-gated channels after their opening, induced by InsP3 itself, or by cytosolic and/or lumenal $[Ca^{2+}]$ (Missiaen et al., 1992; Hajnoczky and Thomas, 1994). The main attractions of this model are: (1) it can explain also $[Ca^{2+}]_i$ oscillations in cells expressing only InsP3 receptors; (2) it accounts for a few established features of InsP3-receptor gating and binding characteristics, for example the biphasic activation by Ca^{2+} (Bezprozvanny and Erlich, 1994; Hajnoczky and Thomas, 1994), the inhibitory action on oscillations of elevated $[Ca^{2+}]_i$ (Marshall and Taylor, 1993), and the effects of agents affecting the affinity for InsP3 (Bootman et al., 1992). Against this model is the fact that it does not account for the inhibitory action of ryanodine, when it occurs. A conservative attitude would be that none of the models is entirely correct or incorrect and that, depending on the cell type, one or another mechanism predominates. It is the biased opinion of the authors of this chapter that the third model is currently the most popular.

$[Ca^{2+}]_i$ Oscillations in Cells of the Central Nervous System (CNS)

Neurons

Almost all cell types tested are capable, under selected experimental conditions, of undergoing cyclic changes of $[Ca^{2+}]_i$. From hepatocytes to lymphocytes, from neutrophils to epithelial cells, $[Ca^{2+}]_i$ oscillations have been repetitively reported, all responding to a common theme, that is cycles of $[Ca^{2+}]_i$ transients of similar amplitudes whose frequency is modulated by several factors, such as the extracellular Ca^{2+} concentration, the intensity of the stimulation, the nature of the agonist, etc. As discussed above, both InsP3Rs and RyRs appear to be involved in these oscillations, while the role of Ca^{2+} influx is quite variable. In nonexcitable cells, in most cases, Ca^{2+} influx from the extracellular space usually plays a secondary role with respect to Ca^{2+} mobilization from the stores, either preventing the depletion of the intracellular pools or participating in the maintenance of a permissive $[Ca^{2+}]_i$ level. On the contrary, in excitable neuroendocrine cells, i.e., primary cell cultures of the hypophysis (Cheek et al., 1993) and endocrine pancreas (Stojikovic and Catt, 1992), as well as cell lines derived from them, such as GH3 (Schlegel et al., 1987), RINm5F (Dunne et al., 1990), and INS-1 (Rutter et al., 1993), spontaneous action potentials can generate oscillatory $[Ca^{2+}]_i$ changes which depend entirely on Ca^{2+} influx through plasma membrane VOCs. Most of these cell types, however, can undergo $[Ca^{2+}]_i$ oscillations also driven by intracellular Ca^{2+} release. Quite surprisingly, however, neurons, either in culture or in semi-intact tissue preparations (such as brain slices), appear, with a few exceptions, to be incapable of oscillating via intracellular stores and undergo $[Ca^{2+}]_i$ oscillations only in synchrony with changes of the membrane potential (Cherubini et al., 1991; Van de Pol et al., 1992; Lawrie et al., 1993; Gu et al., 1994). These oscillations in neuronal $[Ca^{2+}]_i$ can be sustained by plasma membrane channels gated by either membrane voltage or ligand binding. In both cases, the increase in $[Ca^{2+}]_i$ is a consequence of channel openings and the repetitive oscillations are the result of the oscillatory nature of neuronal activity. For example, cortical and cerebellar neurons in primary cultures exhibit spontaneous oscillations that, to a large extent, occur in synchrony among adjacent neurons (Fields et al., 1991; Gu et al., 1994). These Ca^{2+} transients are dependent on synaptic activity being blocked by both glutamate receptor antagonists and tetrodotoxin (TTX). Spontaneous $[Ca^{2+}]_i$ oscillations, highly correlated among subgroups of ganglion and amacrine cells, have been observed also in the developing mammalian retina (Wong et al., 1995). In these cells, $[Ca^{2+}]_i$ oscillations are not blocked by glutamate receptor antagonists and are supposed to depend on spontaneous openings of sodium channels as no $[Ca^{2+}]_i$ increase was observed in the presence of TTX (Wong et al., 1995). Although $[Ca^{2+}]_i$ transients in neurons rely, in general, on Ca^{2+} influx through plasma membrane channels, it has been demonstrated that conditions that prevent Ca^{2+} release from intracellular stores can modify the pattern of the $[Ca^{2+}]_i$ transients and the effects of the Ca^{2+} changes (Regehr and Tank, 1990; Holliday et al., 1991; Gu et al., 1994). For example, it has been found that the pattern of differentiation of spinal cord neurons is altered by agents affecting Ca^{2+} mobilization, suggesting that not only Ca^{2+} influx but also Ca^{2+} release is necessary to promote a correct maturation of these cells (Regehr and Tank, 1990).

A major difference exists between $[Ca^{2+}]_i$ oscillations driven by cycles of Ca^{2+} release-uptake from intracellular stores and $[Ca^{2+}]_i$ oscillations dependent on synaptic activity or spontaneous openings of Ca^{2+} channels. In the first case, the stimulus is constant; in the second, the trigger itself is oscillatory. Thus, although apparently the end result, i.e., the behavior of $[Ca^{2+}]_i$ is similar, the underlying mechanism is intrinsically different in the two situations. It goes beyond the purpose of this chapter to deal in detail with the dynamics of the $[Ca^{2+}]_i$ changes that follow synaptic transmission. One aspect of this phenomenon, however, will be briefly dealt with here: the expression and possible relevance of $[Ca^{2+}]_i$ changes highly localized in small neuronal compartments, such as the dendritic spine.

Localized $[Ca^{2+}]_i$ Rises in Neurons

The development of confocal fluorescence microscopy and its use for the study of $[Ca^{2+}]_i$ homeostasis has not only led to significant improvements of image quality, but also has expanded the possibility of investigating cells in more physiological conditions, such as tissue slices. It is now possible to investigate, with unprecedented spatial and temporal resolution, the features of the $[Ca^{2+}]_i$ changes at the level of very small compartments, less than 1 μm in diameter, and in semiphysiological situations. Numerous recent studies have described $[Ca^{2+}]_i$ rises, following synaptic activation, that are restricted to individual dendritic spines (Regehr et al., 1990; Gurthrie et al., 1991; Müller and Connor, 1991; Denk et al., 1995; Eilers et al., 1995). For example, by modulating the intensity of the electrical stimulus applied to the presynaptic input, the Ca^{2+} increase in CA3 hippocampal neurons has been observed to be confined to the postsynaptic spine (Guthrie et al., 1991; Müller and Connor, 1991; Konnerth et al., 1992; Denk et al., 1995; Eilers et al., 1995), demonstrating that spines could regulate $[Ca^{2+}]_i$ independent of the parent dendrite. Indeed, the spine and its parent dendrite are believed to represent the key region in the cell for the functional and morphological changes that control synaptic plasticity (Konnerth et al., 1992; Hosokawa et al., 1995). Along this line, long-term changes in synaptic efficacy mediated by Ca^{2+} transients restricted to the dendrites have been recently described in Purkinje cells (Hartell, 1996).

The changes of $[Ca^{2+}]_i$ occurring in the spine are large and long lasting, exceeding by orders of magnitude the duration of the presynaptic activity (Murphy

et al., 1995). Increases in the $[Ca^{2+}]_i$ that persist after the decay of the Ca^{2+} current that has triggered them have also been described in other models, for example in bullfrog sympathetic neurons (Hernández-Cruz et al., 1990) and rat cerebellar Purkinje neurons (Llano et al., 1994). These kinetic discrepancies depend, on the one hand, on the different rates of Ca^{2+} fluxes across channels (10^6-10^7 ions/sec/channel) and extrusion mechanisms (10–100 ions/sec/pump or transporter), and, on the other hand, on Ca^{2+} release from intracellular stores (Furuichi et al., 1989). As to the first, a short opening of a Ca^{2+} channel, e.g., 20 msec, will let in an amount of Ca^{2+} ions which requires several hundreds of milliseconds to be pumped out of the cytosol. As to the second, the $[Ca^{2+}]_i$ transient in the dendrite and dendritic spines of CA1 hippocampal neurons triggered by N-methyl-D-aspartate receptor (NMDAR) activation has been observed to be substantially depressed by either thapsigargin or ryanodine (Alford et al., 1993). In particular, approximately 65% of the $[Ca^{2+}]_i$ increase induced in these neurons by tetanic stimulation has been shown to be due to a CICR process (Alford et al., 1993). From a functional point of view, the release from intracellular stores is of major functional relevance since, at least in some neurons, for example hippocampal CA1 neurons, depletion of internal Ca^{2+} pools by dantrolene and thapsigargin was observed to interfere with the induction of LTP (Obenaus et al., 1989; Bortolotto and Collingridge, 1993).

The mechanism underlying $[Ca^{2+}]_i$ mobilization may vary according to the neuronal type and/or cellular region. For example, in rat Purkinje neurons, InsP3Rs are expressed at high concentrations in the soma, dendrites and spines (Ross et al., 1989; Satoh et al., 1990; Nakanishi et al., 1991; Walton et al., 1991), while RyRs in the same cell type have been identified at the level of the soma and dendritic tree, but not the spines (Walton et al., 1991). In contrast, in hippocampal neurons, RyRs appear to be concentrated in dendritic spines (Sharp et al., 1993a). The InsP3Rs and RyRs subtypes are, therefore, differentially expressed in neurons from different brain regions and, within the same neuron, in different locations, thereby providing a sophisticated mechanism for regulating the $[Ca^{2+}]_i$ level in restricted and functionally relevant regions.

In spite of the technical improvement provided by the confocal microscope, the resolution of the available methods might be insufficient to reveal the small signal of localized oscillations. The study of Swandulla et al. (1991) provided a convincing demonstration of this point by showing that injection of EGTA into the squid presynaptic terminals eliminates all diffuse $[Ca^{2+}]_i$ rises. In spite of this, synaptic transmission was unaffected, demonstrating that the $[Ca^{2+}]_i$ rise that triggers the release of the neurotransmitter is smaller in size and/or faster than the spatial and temporal resolution of the presently available imaging systems. Different experimental approaches, therefore, must be developed to investigate further the presence and features of very rapid $[Ca^{2+}]_i$ oscillations in extremely restricted neuronal compartments. Fast millisecond confocal imaging and targeting of dyes or Ca^{2+}-sensitive photoproteins may be techniques for the future in this field.

$[Ca^{2+}]_i$ Oscillation from Stores in Neurons

As mentioned above, some neurons, such as bullfrog sympathetic neurons (Kuba and Takeshita, 1981; Friel and Tsien, 1992; Nohmi et al., 1992) and chick ciliary ganglion neurons (Rathouz et al., 1995), do show $[Ca^{2+}]_i$ oscillations driven by intracellular Ca^{2+} stores. These neurons have been shown to undergo repetitive cycles of Ca^{2+} release and uptake from intracellular pools endowed with RyRs and/or InsP3Rs.

Why are neurons so reluctant to oscillate using intracellular stores as the source of Ca^{2+}? Certainly, lack of InsP3Rs or RyRs cannot be advocated as the cause of this, given that neurons are known to express, at relatively high concentrations, most isoforms of both intracellular channels (McPherson and Campbell, 1990; Satoh et al., 1990; McPherson et al., 1991; Furuichi et al., 1994). Similarly, receptors coupled to InsP3 generation are quite abundant in neurons (Worley et al., 1989; Sharp et al., 1993b) and increases in InsP3 levels have been documented in both intact tissues and primary neuronal cultures upon stimulation with a variety of agonists (Berridge and Irvine, 1984). As discussed above, Ca^{2+} release from intracellular stores, at the level of either the cell body or the dendritic tree, has also been widely documented in neurons (for a review, see Simpson et al., 1995). The main difference between neurons and cells of different embryological origin that display $[Ca^{2+}]_i$ oscillations from stores is the lack, in the former cell type, of Ca^{2+} currents (or Ca^{2+} influxes) activated by store depletion (I_{CRAC}-like currents). There is no example, to our knowledge, that either receptor agonists or SERCA inhibitors (such as thapsigargin), induce sustained increases in $[Ca^{2+}]_i$ in neuronal cells of mammals, arguing that this pathway of Ca^{2+} influx is not expressed in neurons. This striking difference cannot, however, be the only explanation for the lack of $[Ca^{2+}]_i$ oscillations (from stores) in neurons, given that $[Ca^{2+}]_i$ oscillations in many cell types can be elicited also in the absence of extracellular Ca^{2+}. Another intriguing possibility is that $[Ca^{2+}]_i$ oscillations driven by Ca^{2+} release from intracellular Ca^{2+} stores appear in neurons rather late during development. Indeed, most experiments of

this type have been carried out in cells isolated from embryos or from neonatal animals in which the oscillatory mechanism may not yet be operative, while clear Ca^{2+} oscillations have been obtained in sympathetic ganglion neurons from adult bullfrogs (Kuba and Takeshita, 1981; Hernández-Cruz et al., 1990; Friel and Tsien, 1992; Nohmi et al., 1992). Further indirect support for this idea comes from the observations that spontaneous $[Ca^{2+}]_i$ oscillations, presumably originating from intracellular Ca^{2+} stores, have been observed in neurons from the visual cortex of rats at postnatal day 0–7 (Yuste et al., 1992), and $[Ca^{2+}]_i$ oscillations, in hippocampal and visual cortical neurons, following stimulation of metabotropic glutamate receptors, can be more easily induced in slices derived from animals older than 10 days (L. Pasti et al., unpublished).

Astrocytes

A second type of central nervous system cell that, unlike neurons, displays consistent and regular $[Ca^{2+}]_i$ oscillations that are dependent on intracellular stores is the astrocyte. Astrocytes represent the major class of glial cells in the mammalian brain and serve a series of important functions, participating in the regulation of the ionic composition of the extracellular space, the formation of the blood–brain barrier and guiding the migration of neurons in the developing embryo (Cserr and Bundgaard, 1986; Hatten, 1990). Astrocytes also possess an efficient uptake system for several neurotransmitters (Barbour et al., 1988; Atwell et al., 1993), thereby contributing to their removal from the synaptic cleft, voltage-dependent ion channels, and a variety of neurotransmitter receptors, including ionotropic and metabotropic glutamate receptor subtypes (Bormann and Kettenmann, 1988; Sontheimer et al., 1988; Bevan, 1990; Wyllie et al., 1991). The function of these receptors is still unclear but their activation, both in vitro (Cornell-Bell et al., 1990; Glaum et al., 1990; Charles et al., 1991) and in brain slices (Dani et al., 1992; Duffy and MacVicar, 1995; Porter and McCarthy, 1995), causes a rapid elevation in the $[Ca^{2+}]_i$. In particular, L-glutamate, i.e., the major excitatory neurotransmitter in the brain, as well as its analogs quisqualate and kainate, can trigger in astrocytes repetitive Ca^{2+} spikes (Cornell-Bell et al., 1990; Glaum et al., 1990; Charles et al., 1991; Dani et al., 1992; Porter and McCarthy, 1995) These responses are due to the production of InsP3 (Pearce et al., 1986; Milani et al., 1989) following activation of the metabotropic receptor subtype mGluR5 (Romano et al., 1995). Numerous studies demonstrate that many other stimuli can produce $[Ca^{2+}]_i$ oscillations in astrocytes, for example ATP, acetylcholine, γ-aminobutyric acid, noradrenaline, or mechanical stimulation (Charles et al., 1991; Wilkin et al., 1991; Nilsson et al., 1992). The response of these cells to InsP3-producing stimuli appears to be particularly complex in terms of spatiotemporal behavior: ranging from sustained $[Ca^{2+}]_i$ increases (at high stimulus concentration) to prolonged episodes of $[Ca^{2+}]_i$ oscillations. Quite interestingly, $[Ca^{2+}]_i$ oscillations induced by histamine can be observed in discrete regions of type-2 astrocyte processes (Inagaki et al., 1991) The presence of multiple sites of $[Ca^{2+}]_i$ elevations that are not synchronized, even within the same cell, suggests that each hot spot may represent an independent compartment of $[Ca^{2+}]_i$ signaling (Inagaki et al., 1991).

The $[Ca^{2+}]_i$ oscillations often trigger waves that can propagate from one cell to the other for hundreds of micrometers (Igusa and Miyazaki, 1983; Dani et al., 1992; Finkbeiner, 1992). One of the most intriguing possibilities is that glutamate-mediated $[Ca^{2+}]_i$ oscillations and propagating waves may represent a Ca^{2+}-based form of excitability that functions as a nonsynaptic, long-range signaling system (Igusa and Miyazaki, 1983; Smith, 1994). Indeed, it has been demonstrated that $[Ca^{2+}]_i$ oscillations and waves in astrocytes from either neuron–astrocyte mixed cultures (Murphy et al., 1993), organotypic hippocampal slices (Dani et al., 1992) or acute brain slices (Pasti et al., 1997) can be induced by neuronal stimulation, indicating that these cells are sensitive to glutamate released by synaptic activity (see below). The long and elaborate astrocyte processes that are in intimate association with nerve terminals in the synaptic cleft (Peters et al., 1991) may account for their sensitivity to detect sudden local changes of glutamate concentration.

The mechanisms underlying $[Ca^{2+}]_i$ oscillations and waves are still poorly defined in this cell type also. Both types of changes can be induced even in the absence of extracellular Ca^{2+} (Igusa and Miyazaki, 1983; Jensen and Chiu, 1990), demonstrating that astrocytes possess an organized and efficient system of Ca^{2+} release from intracellular stores. Though I_{CRAC}-type currents have not been directly measured in these cells, sustained increases in the $[Ca^{2+}]_i$ induced by high concentration of the agonist-mediating $[Ca^{2+}]_i$ oscillations or SERCA inhibitors, represent a typical behavior of astrocytes (L. Pasti et al., unpublished observation). No evidence for functional expression of RyRs has yet been obtained in astrocytes, and ryanodine does not affect either the expression or the pattern of glutamate-induced $[Ca^{2+}]_i$ oscillations (Pasti et al., 1995). This finding suggests that $[Ca^{2+}]_i$ oscillations in astrocytes depend on InsP3-gated channels only. Quite unique, at least in terms of its spreading for long distances, is the phenomenon of intercellular $[Ca^{2+}]_i$ waves. These waves are represented by repetitive $[Ca^{2+}]_i$ increases

originating in one cell and then spreading to other cells at distance as long as hundreds of micrometers. A number of experimental findings suggests that intercellular Ca^{2+} waves propagate via gap junctions: (1) inhibitors of gap junctional coupling, such as halothane and octanol, almost completely block wave propagation between astrocytes, but not Ca^{2+} waves within individual astrocytes (Finkbeiner, 1992); (2) C6 glioma cells show clear intercellular Ca^{2+} waves only when they are transfected with connexin 43 (Charles et al., 1992); (3) in an elegant study, Wade et al. (1986) showed that when injected with a fluorescent probe confluent, but not isolated, astrocytes showed not only a rapid diffusion of the dye from the injected cell, but also a rapid recovery of fluorescence after photobleaching of the dye in an individual cell (Wade et al., 1986; Anders, 1988). Gap junctions may play a crucial role in this long-range, nonsynaptic communication (see below for a discussion on its possible physiological meaning). Last, but not least, intercellular coupling can be modulated by activation of neurotransmitter and neuropeptide receptors, probably via kinases A and C (Stagg and Fletcher, 1990). In particular, activation of α- and β-adrenergic receptor subtypes in astrocytes has been shown to decrease and increase gap junctional coupling, respectively (Giaume et al., 1991).

Role of $[Ca^{2+}]_i$ Oscillations: General Considerations

A plausible hypothesis concerning the physiological role of $[Ca^{2+}]_i$ oscillations is that they represent the code of the Ca^{2+} signal transduction mechanism. In other words, the action of Ca^{2+} as a second messenger may be based on a frequency-dependent rather than an amplitude-dependent code (Woods et al., 1986; Berridge and Galione, 1988; Jacob et al., 1988). This mechanism ensures that the information carried by the intensity of the stimulus is preserved and converted into a defined frequency of oscillations. In favor of this hypothesis are the observations that the frequency of oscillations in hepatocytes (Woods et al., 1986), endothelial cells (Jacob et al., 1988), *Xenopus* oocytes (Berridge, 1988), insulinoma cells (Prentki et al., 1988), and astroglia (Cornell-Bell et al., 1990) is dependent on the concentration of the agonist and that the frequency of $[Ca^{2+}]_i$ oscillations in several cell types, for example pituitary (Rapp and Berridge, 1981) and salivary gland (Holl et al., 1988) cells, is positively correlated with the secretion process.

Why should cells prefer a digital type of coding for Ca^{2+} signals, rather than the apparently simpler analogic type of stimulation, such as in a Michaelis

Menten kinetic model? The reader is referred to the numerous reviews for discussion of the advantages of a digital coding. Below are a few speculations that we would like to propose to our colleagues for discussion.

Two types of established cellular phenomena may take advantage from an oscillatory behavior of $[Ca^{2+}]_i$. In particular, only through oscillations it is possible to generate the local large $[Ca^{2+}]_i$ increases which are necessary to couple $[Ca^{2+}]_i$ to secretion and mitochondrial metabolism. As to secretion, it is now clear that release of neurotransmitters relies on local increases of $[Ca^{2+}]_i$, up to concentrations of at least several tens of micromoles (Adler et al., 1991; Swandulla et al., 1991; Llinas et al., 1992; Heidelberger et al., 1994). Such increases of $[Ca^{2+}]_i$ would be incompatible with cell survival if prolonged even for a few seconds in the whole cytosol (Choi, 1995). On the contrary, local and large oscillations of $[Ca^{2+}]_i$ close to the secretory sites ensure a repetitive efficient stimulus–secretion coupling without flooding the entire cell with toxic doses of Ca^{2+} ions. In more general terms, in order to trigger a process regulated by Ca^{2+} with an affinity in the 10–100 μM range, the most economical and safe way is that of generating $[Ca^{2+}]_i$ oscillations. A similar reasoning applies also to the coupling of cell activation to mitochondrial metabolism. In particular, for a long time it has been mysterious how the low-affinity mitochondrial uptake system could sense the modest increases of $[Ca^{2+}]_i$ induced by cell activation and thus couple them to the activation of the Ca^{2+}-dependent dehydrogenases of their matrix. This paradox has been recently solved by the demonstration that mitochondria are transiently exposed, during a Ca^{2+} spike, to local domains of high $[Ca^{2+}]_i$ in the vicinity of the InsP3-gated channels (Rizzuto et al., 1993, 1994). Furthermore, it has been demonstrated that, depending on the frequency of the $[Ca^{2+}]_i$ oscillations, the mitochondria can respond with either an oscillatory activation or with a prolonged activation of their metabolism, thus offering a way to turn a digital into an analogic signal (Hajnoczky, et al., 1995). Interestingly, such local microdomains of very high $[Ca^{2+}]_i$ close to the mitochondria can be generated also in the proximity of plasma membrane Ca^{2+} channels in cells, such as neurons, relying primarily on Ca^{2+} influx (T. Pozzan et al., unpublished).

Another advantage of an oscillatory vs. a steady-state increase in $[Ca^{2+}]_i$ may derive from the characteristics of some of Ca^{2+}-activated enzymes. For example, the Ca^{2+}-calmodulin-dependent kinase II (CaM kinase II), highly expressed in the CNS, is known to require an increase in $[Ca^{2+}]_i$ for activation, but after its autophosphorylation it becomes Ca^{2+} independent (Miller and Kennedy, 1986). The activation of the enzyme is, therefore, triggered by

the initial $[Ca^{2+}]_i$ spike, a sustained increase largely representing, in this case, a waste of metabolic energy. Interestingly, it has recently been shown that the autonomous activation of CaM kinase II in vitro increased steeply as a function of $[Ca^{2+}]_i$ oscillation frequency, thus suggesting that CaM kinase II can act as a frequency decoder of $[Ca^{2+}]_i$ oscillations (De Koninck and Schulman, 1998). The observations that $[Ca^{2+}]_i$ oscillations can increase the efficiency and specificity of gene expression give further support to the idea that the frequency of $[Ca^{2+}]_i$ oscillations represents the code of the Ca^{2+} signal transduction system (Dolmetsch et al., 1998; Li et al., 1998).

Functional Significance of $[Ca^{2+}]_i$ Oscillations in the CNS

Neurons

The $[Ca^{2+}]_i$ oscillations in neurons are mainly due to sequential openings of plasma membrane Ca^{2+} VOCs and/or NMDARs (Cherubini et al., 1991; Murphy et al., 1992; Lawrie et al., 1993; Gu et al., 1994). These changes in membrane conductance reflect the rhythmic oscillatory firing, i.e., one of the principal characteristic of neuronal cells. The functional relevance of this mode is, therefore, linked to the peculiar role of the basic synaptic communication within and among neuronal circuitries and will be not discussed here.

As mentioned above, $[Ca^{2+}]_i$ oscillations that involve Ca^{2+} release from intracellular stores are not commonly observed in neuronal cells. The discussion on their possible role in the activation of specific Ca^{2+}-sensitive biochemical pathways is necessarily limited to the few reports present in the literature. Nevertheless, the general considerations discussed above about the functional significance of Ca^{2+} oscillations hold true for the CNS cells as well. Can (other) specific functions of neurons be directly correlated to the slow oscillations of Ca^{2+}? Of particular interest along this line are the spontaneous and synchronous $[Ca^{2+}]_i$ oscillations that occur in neurons of developing rat visual cortex at an early stage of development (Berridge and Irvine, 1984). In contrast to those observed at the level of the retina (Wong et al., 1995) and in neuronal cultures (Murphy et al., 1992), the oscillations in rat visual cortex neurons are not dependent on synaptic activity and are not blocked by TTX application, suggesting that they most likely result from Ca^{2+} release from intracellular stores. These $[Ca^{2+}]_i$ oscillations probably involve also gap junctions (Yuste et al., 1992). Katz and collaborators (Yuste et al., 1992) suggested that this coordinated pattern of activity may drive cells from the same domain to a common developmental stage and thus define modular functioning units during development, such as the ocular dominance columns of the primary visual cortex (Levay et al., 1978). These $[Ca^{2+}]_i$ oscillations may also be involved in the phenomenon of synaptic strengthening. According to Hebb (1949), the stabilization and strengthening of a synapse depends on the time coincidence of pre- and postsynaptic activities. Since the time of its formulation, this hypothesis has been validated by numerous experimental observations (Stent, 1973; Stryker et al., 1990; Singer, 1993). The pattern of activity underlined by the synchronous $[Ca^{2+}]_i$ oscillations may represent a powerful source of correlation that serves to coordinate innervation of incoming axonal terminals and the strengthening of synapses among neurons from the same domain. This remains a working hypothesis that needs experimental evidence to prove or disprove it.

As to the localized repetitive $[Ca^{2+}]_i$ changes occurring at the level of the spines and dendrites, their most obvious role is in the modulation of the specific events occurring at these cellular locations. For example, a rise in the dendritic $[Ca^{2+}]$ mediated by activation of the NMDA receptor has been found to be crucial for the generation of LTP (Alford et al., 1993; Collingridge and Bliss, 1995), though it may not be sufficient to generate stable LTP (Kullmann et al., 1992) and LTD (Hosokawa et al., 1995). However, an intriguing observation published recently by Tsien and collaborators (Deisseroth et al., 1996) demonstrates how these extremely localized $[Ca^{2+}]_i$ changes may have effects at very long distances. These authors showed that phosphorylation of the transcription factor CREB requires postsynaptic $[Ca^{2+}]_i$ changes so strictly localized that their functional consequences can be abolished by the fast Ca^{2+} chelator BAPTA but not by EGTA, despite the fact that the latter completely abolishes the $[Ca^{2+}]_i$ changes measurable with indicators.

Astrocytes

As discussed above in the CNS, $[Ca^{2+}]_i$ oscillations driven by cycles of Ca^{2+} release and uptake from intracellular stores are typical of astrocytes. Results from two studies performed in primary mixed cultures from the cerebral cortex demonstrated that $[Ca^{2+}]_i$ elevations in astrocytes can trigger $[Ca^{2+}]_i$ increases in neighboring neurons (Nedergaard, 1994; Parpura et al., 1994). Nedergaard (1994) showed that the $[Ca^{2+}]_i$ increase in neurons triggered by astrocyte oscillations is not prevented by kynurenic acid, which, in control experiments, inhibited glutamate-mediated responses of the neurons. The conclusion that glutamate released by oscillating

astrocytes cannot account for the $[Ca^{2+}]_i$ change in neurons was further supported by the observation that the application of gap junction inhibitors prevented the occurrence of the phenomenon (Nedergaard, 1994). In contrast, Parpura et al. (1994) showed that, in their experimental conditions, the increase in neuronal $[Ca^{2+}]_i$ following stimulation of astrocyte oscillations with bradykinin is prevented by NMDA receptor antagonists and enhanced by the NMDA receptor cotransmitter glycine, thereby indicating that glutamate, probably released by stimulated astrocytes, is responsible for the $[Ca^{2+}]_i$ changes in neurons. Beside the conflicting conclusions, both studies reported that the stimulus-induced $[Ca^{2+}]_i$ increase in cultured astrocytes can trigger significant $[Ca^{2+}]_i$ increases in neighboring neurons, thereby suggesting that active interactions may exist between neurons and astrocytes, at least in culture. An active role for astrocytes in the modulation of $[Ca^{2+}]_i$ in neurons was then demonstrated in the CA1 hippocampal region and the visual cortex from acute brain slices (Pasti et al., 1997) . In this study, we showed that astrocytes from both regions can respond to stimulation of the mGluR, performed either by exogenous application of the mGluR agonist t-ACPD or glutamate released by synaptic activity, with $[Ca^{2+}]_i$ oscillations. These oscillations resulted in the release of glutamate which triggers repetitive $[Ca^{2+}]_i$ increases in neurons (Pasti et al., 1997). We also found that the neurotransmitter release in astrocytes in situ, as already reported for astrocytes in culture (Parpura et al., 1994), is a Ca^{2+}-dependent event and is mediated by prostaglandin formation (Bezzi et al., 1998). Together, these results demonstrated that glutamate-mediated $[Ca^{2+}]_i$ oscillations in astrocytes represent a signaling system that underlies the bidirectional form of communication between neurons and astrocytes. Importantly, two recent studies by Haydon and collaborators convincingly demonstrated that astrocytes in co-cultures with neurons from the hippocampus can modulate spontaneous excitatory and inhibitory synaptic transmission by increasing the probability of transmitter release through activation of NMDA receptors (Araque et al., 1998a, 1998b).

The sensitivity of astrocytes to glutamate is of particular interest since glutamate represents the main excitatory neurotransmitter in the CNS and it is responsible for phenomena as diverse as neurodegeneration (Choi, 1988) and long-term potentiation-depression of synaptic connections (Bliss and Lomo, 1973; Linden, 1994). These latter phenomena are thought to represent a form of memory function at the cellular level (Singelbaum and Kandel, 1991; Bliss and Collingridge, 1993). In this scenario, the sensitivity of astrocytes to glutamate is generally believed to be of secondary importance, at least for the proces-

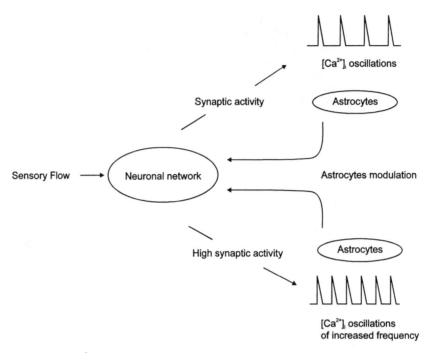

Figure 1 Possible role of $[Ca^{2+}]_i$ oscillations in the modulation of information processing in the neuron astrocyte network. The key points of the schema can be summarized as follows: (1) glutamate released in the synaptic cleft as a result of synaptic activity induces $[Ca^{2+}]_i$ oscillations in astrocytes that are located in the proximity of the active synapses; (2) oscillating astrocytes, in turn, modulate neuronal responses, for example by inducing significant $[Ca^{2+}]_i$ increases in neighboring neurons; (3) astrocytes increase the frequency of their oscillations in the presence of repetitive synaptic activity at the same synaptic sites, thereby inducing a quantitatively and/or qualitatively different neuronal modulation.

sing of information in the neural network. However, the data mentioned above, and a recent finding obtained by us, suggest a revision of the role played by the glial cells in this phenomenon. We found that astrocytes from the visual cortex in culture display a surprising form of cellular memory: in response to glutamate these cells show periodic oscillations in their $[Ca^{2+}]_i$; when exposed to repetitive stimulations with glutamate at the same concentration, the frequency of these oscillations increased dramatically (Pasti et al., 1995, 1997). The potentiation of the response is long-lasting, does not depend on Ca^{2+} entry from the extracellular space, and is mediated by activation of a metabotropic glutamate receptor. The inhibition of the nitric oxide (NO) synthase was also found to abolish the increase in oscillation frequency otherwise observed following repetitive stimulations with glutamate, without significantly affecting the amplitude of the $[Ca^{2+}]_i$ spikes. A change in the frequency of oscillations might modulate a number of functional features of astrocytes, such as glutamate and other neurotransmitters (Pin

and Bockaert, 1989; Szatkowski et al., 1990; Gallo et al., 1991; Pasti et al., 1997), NO (Garthwaite, 1993), cytokines (Martin, 1992), and neurotrophins (Lindsay, 1979; Zafra et al., 1992), which can affect neuronal migration, proliferation, development, and plasticity of synapses. Interestingly, NO represents also one of the intercellular messengers likely involved in the induction of long-term potentiation in hippocampal neurons (Schuman and Madison, 1993; Dawson and Snyder, 1994). The possibility for a new role of astrocytes in the neuronal network is schematically illustrated in Fig. 3.1. Astrocytes might participate in the processing of information by detecting high neuronal activity and responding to the release of glutamate with a progressive increase in the frequency of oscillations. These $[Ca^{2+}]_i$ oscillations and waves may, in turn, trigger the release of glutamate and other neuroactive compounds, including NO, neurotrophins, and perhaps other neurotransmitters. Noteworthy, NO itself appears involved in the induction of the potentiation of the astrocyte response to glutamate (Pasti et al., 1995).

The recent findings described above add new information on the interactions between neurons and astrocytes and, more interestingly, point to important and unexpected roles of astrocytes in the function of the synapse.

References

Adler, E. M., Augustine, G. J., Duffy, S. N. and Charlton, N. T. (1991) Alien intracellular calcium chelators attenuate neurotransmitter release at the squid giant synapse. *J. Neurosci.* 11: 1496–1507.

Alford, S., Frenguelli, B. G., Schfield, J. G., and Collingridge, G. L. (1993) Characterization of Ca^{2+} signals induced in hippocampal CA1 neurones by the synaptic activation of NMDA receptors. *J. Physiol.* 469: 693–716.

Ambler, S. K., Poenie, M., Tsien, R. Y., and Taylor, P. (1988) Agonis-stimulated oscillations and cycling of intracellular free calcium in individual cultured muscle cells. *J. Biol. Chem.* 263: 1952–1959.

Anders, J. J. (1988) Lactic acid inhibition of gap junctional intercellular communication in in vitro astrocytes as measured by fluorescence recovery after laser photobleaching. *Glia* 1: 371–379.

Araque, A., Parpura, V., Sanzgiri, R. P., and Haydon, P. G. (1998a) Glutamate-dependent astrocyte modulation of synaptic transmission between cultured hippocampal neurons. *Eur. J. Neurosci.* 10: 2129–2142.

Araque, A., Sanzgiri, R. P., Parpura, V., and Haydon, P. G. (1998b) Calcium elevation in astrocytes causes an NMDA receptor-dependent increase in the frequency of miniature synaptic currents in cultured hippocampal neurons. *J. Neurosci.* 18: 6822–6829.

Attwell, D., Barbour, B., and Szatkowski, M. (1993) Nonvesicular release of neurotransmitter. *Neuron* 11: 401–407.

Barbour, B., Brew, H., and Attwell, D. (1988) Electrogenic glutamate uptake in glial cells is activated by intracellular potassium. *Nature* 335, 433–435.

Berridge, M.J. (1988) Inositol trisphosphate induced membrane potential oscillations in *Xenopus* oocytes. *J. Physiol.* 403: 589–599.

Berridge, M. J. (1993) Inositol trisphosphate and calcium signalling. *Nature* 361: 315–325.

Berridge, M. J. and Dupont, G. (1994) Spatial and temporal signalling by calcium. *Curr. Biol.* 6: 267–274.

Berridge, M. J. and Galione, A. (1988) Cytosolic calcium oscillators. *FASEB J.* 2: 3074–3082.

Berridge, M. J. and Irvine, R. K. (1984) Inositol trisphosphate, a novel second messenger in cellular signal transduction. *Nature* 312: 315–320.

Bevan, S. (1990) Ion channels and neurotrasmitter receptors in glia. *Sem. Neurosci.* 2: 467–481.

Bezprozvanny, I. and Ehrlich, B. E. (1994) Inositol (1,4,5)-trisphosphate (InsP3)-gated Ca channels from cerebellum: conduction properties for divalent cations and regulation by intraluminal calcium. *J. Gen. Physiol.* 104: 821-856.

Bezprozvanny, I., Watras, J., and Ehrlich, B. E. (1991) Bell-shaped calcium-response curves of Ins(1,4,5)P3- and calcium-gated channels from endoplasmic reticulum of cerebellum. *Nature* 351: 751–754.

Bezzi P., Carmignoto G., Pasti L., Vesce S., Rossi D., Lodi Rizzini B., Pozzan T., and Volterra A. (1998) Prostaglandins stimulate calcium-dependent glutamate release in astrocytes. Nature 391: 281-285.

Bird, G. S. T. W., Rossier, M. F. Obie, G. F., and Pitney, J. Y. Jr. (1993) Sinusoidal oscillations in intracellular calcium requiring negative feedback by protein kinase C. *J. Biol. Chem.* 268: 8425–8428.

Bliss, T. V. P. and Collingridge, G. L. (1993) A synaptic model of memory: long-term potentiation in the hippocampus. *Nature* 361: 31–39.

Bliss, T. V. P. and Lomo, T. (1973) Long-term potentiation of synaptic transmission in the dentate area of anaesthetized rabbit following stimulation of the perforant path. *J. Physiol.* 232: 331–356.

Bootman, M. D. and Berridge, M. J. (1995) The elemental principles of calcium signaling. *Cell* 83: 675–678.

Bootman, M. D., Taylor, C. W., and Berridge, M. J. (1992) The thiol reagent, thimerosal, evokes Ca^{2+} spikes in HeLa cells by sensitizing the inositol 1,4,5-trisphosphate receptor. *J. Biol. Chem.* 267: 25113–25119.

Bormann, J. and Kettenmann, H. (1988) Patch-clamp study of gamma-aminobutyric acid receptor Cl^- channels in cultured astrocytes. *Proc. Natl. Acad. Sci. USA* 85: 9336–9340.

Bortolotto, Z. A., and Collingridge, G. L. (1993) Characterisation of LTP induced by the activation of glutamate metabotropic receptors in area CA1 of the hippocampus. *Neuropharmacology* 32: 1–9.

Cannell, M. B., Cheng, H., and Lederer, W. J. (1995) The control of calcium release in heart muscle. *Science* 268: 1045–1059.

Charles, A. C., Merrill, J. E., Dirksen, E. R., and Sanderson, M. J. (1991) Intercellular signaling in glial cells: calcium waves and oscillations in response to mechanical stimulation and glutamate. *Neuron* 6: 983–992.

Charles, A. C., Naus, C. C., Zhu, D., Kidder, G. M., Dirksen, E. R., and Sanderson, M. J. (1992) Intercellular calcium signaling via gap junctions in glioma cells. *J. Cell Biol.* 118: 195–201.

Cheek, T. R., Morgan, A., O'Sullivan, A., Moreton, R. B., Berridge, M. J., and Burgoyne, R. D. (1993) Spatial localization of agonist-induced Ca^{2+} entry in bovine adrenal chromaffin cells. Different patterns induced by histamine and angiotensin II, and relationship to catecholamine release. *J. Cell Sci.* 105: 913–921.

Cherubini, E., Ben-Ari, Y., Ito, S. and Krnjevic, K. (1991) Persistent pulsatile release of glutamate induced by N-methyl-D-aspartate in neonatal rat hippocampal neurones. *J. Physiol.* 436: 531–547.

Choi, D. W. (1988) Glutamate neurotoxicity and diseases of the nervous sytem. *Neuron* 1: 623–634.

Choi, D. W. (1995) Calcium and excitotoxic neuronal injury. *Ann. N.Y. Acad. Sci.* 747: 162–171.

Cockroft, S. and Thomas, G. M. H. (1992) Inositol-lipid-specific phospholipase C isoenzymes and their differential regulation by receptors. *Biochem. J.* 288: 1–14.

Collingridge, G. L. and Bliss, T. V. P. (1995) Memories of NMDA receptors and LTP. *Trends Neurosci.* 18: 54–56.

Cornell-Bell, A. H., Finkbeiner, S. M., Cooper, M. S., and Smith, S. J. (1990) Glutamate induces calcium waves in cultured astrocytes: long-range glial signaling. *Science* 247: 470–473.

Cserr, H. F. and Bundgaard, M. (1986) The neuronal microenvironment: a comparative view. *Ann. N.Y. Acad. Sci.* 481: 1–6.

Dani, J. W., Chernjavsky, A., and Smith, S. J. (1992) Neuronal activity triggers calcium waves in hippocampal astrocyte network. *Neuron* 8: 429–440.

Dawson, T. M. and Snyder, S. H. (1994) Gases as biological messengers: nitric oxide and carbon monoxide in the brain. *J. Neurosci.* 14: 5147–5159.

Deisseroth, K., Bito, H. and Tsien, R. W. (1996) Signaling from synapse to nucleus: postsynaptic CREB phosphorylation during multiple forms of hippocampal synaptic plasticity. *Neuron* 16: 89–101.

De Koninck, P. and Schulman, H. (1998) Sensitivity of CaM kinase II to the frequency of Ca^{2+} oscillations. *Science* 279: 227–230.

Denk, W., Sugimori, M., and Llinas, R. (1995) Two types of calcium response limited to single spines in cerebellar Purkinje cells. *Proc. Natl. Acad. Sci. USA* 92: 8279–8282.

Dolmetsch, R. and Lewis, R.S. (1994) Signaling between intracellular Ca^{2+} stores and depletion-activated Ca^{2+} channels generates $[Ca^{2+}]_i$ oscillations in T lymphocytes. *J. Gen. Physiol.* 103: 365–388.

Dolmetsch, R. E., Xu, K., and Lewis, R. S. (1998) Calcium oscillations increase the efficiency and specificity of gene expression. *Nature* 392: 933–936.

Duffy, S. and MacVicar, B. A. (1995) In vitro ischemia promotes calcium influx and intracellular calcium release in hippocampal astrocytes. *J. Neurosci.* 15: 5535–5550.

Dunne, M. J., Yule, D. I.,Gallacher, D. V., and Petersen, O. H. (1990) Comparative study of the effects of cromakalim (BRL 34915) and diazoxide on membrane potential, $[Ca^{2+}]_i$ and ATP-sensitive potassium currents in insulin-secreting cells. *J. Membr. Biol.* 114: 53–60.

Eilers, J., Augustine, G. J. and Konnerth, A. (1995) Subthreshold synaptic Ca^{2+} signalling in fine dendrites and spines of cerebellar Purkinje neurons. *Nature* 373: 155–158.

Fabiato, A. (1983) Calcium-induced release of calcium from the cardiac sarcoplasmic reticulum. *Am. J. Physiol.* 245: C1–C14.

Fasolato, C., Hoth, M., Matthews, G., and Penner, R. (1993) Ca^{2+} and Mn^{2+} influx through receptor-mediated activation of nonspecific cation channels in mast cells. *Proc. Natl. Acad. Sci. USA* 90: 3068–3072.

Fasolato, C., Innocenti, B., and Pozzan, T. (1994) Receptor-activated Ca^{2+} influx: how many receptors for how many channels? *Trends Pharmacol. Sci.* 15: 77–83.

Fields, R. D., Yu, C., and Nelson, P. G. (1991) Persistent pulsatile release of glutamate induced by N-methyl-D-aspartate in neonatal rat hippocampal neurones. *Neuroscience* 11: 134–146.

Finkbeiner, S. (1992) Calcium waves in astrocytes–filling in the gaps. *Neuron* 8: 1101–1108.

Fleischer, S. and Inui, M. (1989) Biochemistry and biophysics of excitation–contraction coupling. *Annu. Rev. Biophys. Biophys. Chem.* 18: 333–364.

Foskett, J. K., Roifman, C. N., and Wong, D. (1991) Spatial and temporal signalling by calcium. *J. Biol. Chem.* 266: 2778–2782.

Friel, D. D. (1995) $[Ca^{2+}]_i$ oscillations in sympathetic neurons: an experimental test of a theoretical model. *Biophys. J.* 68: 1752–1766.

Friel, D. D. and Tsien, R. W. (1992) Phase-dependent contributions from Ca^{2+} entry and Ca^{2+} release to caffeine-induced $[Ca^{2+}]_i$ oscillations in bullfrog sympathetic neurons. *Neuron* 8: 1109–1125.

Furuichi, T., Kohda, K., Miyawaki, A., and Mikoshiba, K. (1994) Intracellular channels. *Curr. Opin. Neurobiol.* 4: 294–303.

Furuichi, T. S., Yoshikawa, A., Miyawaki, A., Mikoshiba, K., and Okawa, K. (1989) Primary structure and functional expression of the inositol 1,4,5-trisphosphate-binding protein P400. *Nature* 342: 32–38.

Gallo, V., Patrizio, M., and Levi, G. (1991) GABA release triggered by the activation of neuron-like non-NMDA receptors in cultured type 2 astrocytes is carrier-mediated. *Glia* 4: 245–255.

Garthwaite, J. (1993) Nitric oxide signalling in the nervous system. *Sem. Neurosci.* 5: 171–180.

Giaume, C., Marin, P., Cordier, J., Glowinski, J., and Premont, J. (1991) Adrenergic regulation of intercellular communications between cultured striatal astrocytes from the mouse. *Proc. Natl. Acad. Sci. USA* 88: 5577–5581.

Glaum, S. R., Holzwarth, J. A. and Miller, R. J. (1990) Glutamate receptors activate Ca^{2+} mobilization and Ca^{2+} influx into astrocytes. *Proc. Natl. Acad. Sci. USA* 87: 3454–3458.

Goldbeter, A., Dupont, G., and Berridge, M. J. (1990) Minimal model for signal-induced Ca^{2+} oscillations and for their frequency encoding through protein phosphorylation. *Proc. Natl. Acad. Sci. USA* 87: 1461–1465.

Gray, P. A. (1988) Oscillations of free cytosolic calcium evoked by cholinergic and catecholaminergic agonist in rat parotid acinar cells. *J. Physiol.* 406: 35–53.

Grohovaz, F., Zacchetti, D., Clementi, E., Lorenzon, P., Meldolesi, J., and Fumagalli, G. (1991) $[Ca^{2+}]_i$ imaging in PC12 cells: multiple response patterns to receptor activation reveal new aspects of transmembrane signaling. *J. Cell Biol.* 113: 1341–1350.

Grynkiewicz, G., Poenie, M., and Tsien, R. Y. (1985) A new generation of Ca^{2+} indicators with greatly improved fluorescent properties. *J. Biol. Chem.* 260: 3440–3450.

Gu, J., Olson, E. C., and Spitzer, N. C. (1994) Spontaneous neuronal calcium spikes and waves during early differentiation. *J. Neurosci.* 14: 6325–6335.

Guthrie, P. B., Segal, M., and Kater, S. B. (1991) Independent regulation of calcium revealed by imaging dendritic spines. *Nature* 354: 76-79.

Hajnoczky, G. and Thomas, A. P. (1994) The inositol trisphosphate calcium channel is inactivated by inositol trisphosphate. *Nature* 370: 474–477.

Hajnoczky, G., Robb-Gasper, L. D., Seitz, M. B., and Thomas, A. P. (1995) Decoding of cytosolic calcium oscillations in the mitochondria. *Cell* 82: 415–424.

Hartell, N. A. (1996) Strong activation of parallel fibers produces localized calcium transients and a form of LTD that spreads to distal synapses. *Neuron* 16: 601–610.

Hatten, M. E. (1990) Riding the glial monorail: a common mechanism for glial-guided neuronal migration in different regions of the developing mammalian brain. *Trends Neurosci.* 13: 179–184.

Hebb, D. (1949) *The Organization of Behavior. A Neuropsycological Theory.* Wiley, New York.

Heidelberger, R., Heinemann, C., Neher, E., and Matthews G. (1994) Calcium dependence of the rate of exocytosis in a synaptic terminal. *Nature* 371: 513–515.

Henzi, V. and MacDermott, A. B. (1992) Characteristics and function of Ca^{2+}- and inositol 1,4,5-trisphosphate-releasable stores of Ca^{2+} in neurons. *Neuroscience* 46: 251–274.

Hernández-Cruz, A., Sala, F. and Adams, P. R. (1990) Subcellular calcium transients visualized by confocal microscopy in a voltage-clamped vertebrate neuron. *Science* 247: 858–862.

Holl, R. W., Thorner, M. O., Mandell, G. L., Sullivan, J. A., Sinha, Y. N. and Leong, D. A. (1988) Spontaneous oscillations of intracellular calcium and growth hormone secretion. *J. Biol. Chem.* 263: 9682–9685.

Holliday, J., Adams, R. J., Sejnowski, T. J. and Spitzer, N. C. (1991) Calcium-induced release of calcium regulates differentiation of cultured spinal neurons. *Neuron* 7: 787–796.

Hosokawa, T., Rusakov, D. A., Bliss, T. V. P. and Fin, A. (1995) Repeated confocal imaging of individual dendritic spines in the living hippocampal slice: evidence for changes in length and orientation associated with chemically induced LTP. *J. Neurosci.* 15: 5560–5573.

Igusa, Y. and Miyazaki, S. I. (1983) Effects of altered extracellular and intracellular calcium concentration on hyperpolarizing responses of the hamster egg. *J. Physiol.* 340: 611–632.

Inagaki, N., Fukui, H., Ito, S, Yamatodani, A., and Wada, H. (1991) Single type-2 astrocytes show multiple independent sites of Ca^{2+} signaling in response to histamine. *Proc. Natl. Acad. Sci. USA* 88: 4215–4219.

Inui, M., Saito, A., and Fleischer, S. (1987) Purification of the ryanodine receptor and identity with feet structures of junctional terminal cisternae of sarcoplasmic reticulum from fast skeletal muscle. *J. Biol. Chem.* 262: 1740–1747.

Jacob, R. (1990) Calcium oscillations in electrically non-excitable cells. *Biochim. Biophys. Acta* 1052: 427–438.

Jacob, R., Merritt, J. E., Hallam, T. J., and Rink, T. J. (1988) Repetitive spikes in cytoplasmic calcium evoked by histamine in human endothelial cells. *Nature* 335: 40–45.

Jaffe, L. (1991) The path of calcium in the cytosolic calcium oscillations: a unifying hypothesis. *Proc. Natl. Acad. Sci. USA* 88: 9883–9887.

Jensen, A. M. and Chiu, S. Y. (1990) Fluorescence measurement of changes in intracellular calcium induced by excitatory amino acids in cultured cortical astrocytes. *J. Neurosci.* 10: 1165–1175.

Konnerth, A., Dreessen, J. and Augustine, G. J. (1992) Brief dendritic calcium signals initiate long-lasting synaptic depression in cerebellar Purkinje cells. *Proc. Natl. Acad. Sci. USA* 89: 7051–7055.

Kuba, K. and Takeshita, K. (1981) Stimulation of intracellular Ca^{2+} oscillations in a sympathetic neurone. *J. Theor. Biol.* 93: 1009–1031.

Kullmann, D. M., Perkel, D .J., Manabe, T., and Nicoll, R. A. (1992) Ca^{2+} entry via postsynaptic voltage-sensitive Ca^{2+} channels can transiently potentiate excitatory synaptic transmission in the hippocampus. *Neuron* 9: 1175–1183.

Lawrie, A. M., Graham, M. E ., Thorn, P., Gallacher, D. V. and Burgoyne, R. D. (1993) Synchronous calcium oscillations in cerebellar granule cells in culture mediated by NMDA receptors. *Neuroreport* 4: 539–542.

Lechleiter, J. D. and Clapman, D. E. (1992) Molecular mechanism of intracellular calcium excitabiltty in X. Láevis oocytes. *Cell* 69: 283–294.

Levay, S., Stryker, M. P. and Shatz, C. J (1978) Ocular dominance columns and their development in layer IV of the cat's visual cortex. *J. Comp. Neurol.* 179: 559–576.

Li, W., Llopis, J., Whitney, M., Zlokarnik, G., and Tsien, R.Y. (1998) Cell-permeant caged InsP3 ester shows that Ca^{2+} spikes frequency can optimize gene expression. *Nature* 392: 936–941.

Linden, D. J. (1994) Long-term synaptic depression in the mammalian brain. *Neuron* 12: 457–472.

Lindsay, R. M. (1979) Adult rat brain astrocytes support survival of both NGF-dependent and NGF-insensitive neurones. *Nature* 282: 80–82.

Llano, I., Di Polo, R., and Marty, A. (1994) Calcium-induced calcium release in cerebellar Purkinje cells. *Neuron* 12: 663–673.

Llinas, R., Sugimori, M. and Silver, R. B. (1992) Microdomains of high calcium concentration in a presynaptic terminal. *Science* 256: 677–679.

Ma, J., Fill, M., Knudson, C. M., and Campbell, K. P. (1988) Ryanodine receptor of skeletal muscle is a gap junction-type channel. *Science* 242: 99–102.

Marshall, I. C. B. and Taylor, C. W. (1993) Biphasic effects of cytosolic Ca^{2+} on Ins(1,4,5)P3-stimulated Ca^{2+} mobilization in hepatocytes. *J. Biol. Chem.* 268: 13214–13220.

Martin, D. L. (1992) Synthesis and release of neuroactive substances by glia cells. *Glia* 5: 81–94.

McPherson, P. S. and Campbell, K. P. (1990) Solubilization and biochemical characterization of the high affinity [3H]ryanodine receptor from rabbit brain membranes. *J. Biol. Chem.* 265: 18454–18460.

McPherson, P. S., Kim, Y.-K., Valdivia, H., Knudson, M., Takekura, H., Franzini-Amstrong, C., Coronado, R., and Campbell, K. P. (1991) The brain ryanodine receptor: a caffeine-sensitive calcium release channel. *Neuron* 7: 17–25.

Meyer, T. and Stryer, L. (1988) Molecular model for receptor-stimulated calcium spiking. *Proc. Natl. Acad. Sci. USA* 85: 5051–5055.

Meyer, T., Holowka, D. and Stryer, L. (1988) Highly cooperative opening of calcium channels by inositol 1,4,5-trisphosphate. *Science* 240: 653–656.

Mignery, G. A., Newton, C. L., Archer, B. T. III, and Sudhof, T. C. (1990) Structure and expression of the rat inositol 1,4,5-trisphosphate receptor. *J. Biol. Chem.* 265: 12679–12685.

Milani, D., Facci, L., Guidolin, D., Leon, A., and Skaper, S. D. (1989) Activation of polyphosphoinositide metabolism as a signal-transducing system coupled to excitatory amino acid receptors in astroglial cells. *Glia* 2: 161–169.

Miller, S. G. and Kennedy, M. B. (1986) Regulation of brain type II Ca^{2+}/calmodulin-dependent protein kinase by autophosphorylation: a Ca^{2+}-triggered molecular switch. *Cell* 44: 861–870.

Missiaen, L., De Smedt, H., Droogmans, G. and Castells, R. (1992) Ca^{2+} release induced by inositol 1,4,5-trisphosphate is a steady-state phenomenon controlled by luminal Ca^{2+} in permeabilized cells. *Nature* 357: 599–602.

Miyazaki, S., Shirakawa, H., Nakada, K., and Honda, Y. (1993) Essential role of the inositol 1,4,5-trisphosphate receptor/Ca^{2+} release channel in Ca^{2+} waves and Ca^{2+} oscillations at fertilization of mammalian eggs. *Dev. Biol.* 158: 62–78.

Müller, W. and Connor, J.A. (1991) Dendritic spines as individual neuronal compartments for synaptic Ca^{2+} responses. *Nature* 354: 73–76.

Murphy, T. H., Baraban, J. M. and Wier, W. G. (1995) Mapping miniature synaptic currents to single synapses using calcium imaging reveals heterogeneity in postsynaptic output. *Neuron* 15: 159–168.

Murphy, T. H., Blatter, L. A., Wier, W. G., and Baraban, J. M. (1992) Rapid communication between neurons and astrocytes in primary cortical cultures. *J. Neurosci.* 12: 4834–4845.

Murphy, T. H., Blatter, L. A., Wier, W. G. and Baraban, J. M. (1993) Rapid communication between neurons and astrocytes in primary cortical cultures. *J. Neurosci.* 13: 2672–2679.

Nakanishi, S., Maeda, N., and Mikoshiba, K. (1991) Immunohistochemical localization of an inositol 1,4,5-trisphosphate receptor, P400, in neural tissue: studies in developing and adult mouse brain. *J. Neurosci.* 11: 2075–2086.

Nedergaard, M. (1994) Direct signaling from astrocytes to neurons in cultures of mammalian brain cells. *Science* 263: 1768–1771.

Niggli, E. and Lederer, W. J. (1990) Voltage-independent calcium release in heart muscle. *Science* 250: 565–568.

Nilsson, M., Hansson, E., and Rönnbäck, L. (1992) Agonist-evoked Ca^{2+} transients in primary astroglial cultures–modulatory effects of valproic acid. *Glia* 5: 201–209.

Nishizuka, Y. (1988) The molecular heterogeneity of protein kinase C and its implications for cellular regulation. *Nature* 334: 661–665.

Nohmi, M., Hua, S. Y. and Kuba, K. (1992) Basal Ca^{2+} and the oscillation of Ca^{2+} in caffeine-treated bullfrog sympathetic neurones. *J. Physiol.* 450: 513–528.

Obenaus, A., Mody, I., and Baimbridge, K. G. (1989) Dantrolene-Na (Dantrium) blocks induction of long-term potentiation in hippocampal slices. *Neurosci. Lett.* 98: 172–178.

Otsu, K., Willard, H. F., Khanna, V. K., Zorzato, F., Green, N. M., and MacLennan, D. H. (1990) Molecular cloning of c-DNA encoding the Ca^{2+} release channel (ryanodine receptor) of rabbit cardiac muscle sarcoplasmic reticulum. *J. Biol. Chem.* 265: 13472–13483.

Parpura, V., Basarky, T. A., Liu, F., Jeftinija, K., Jeftinija, S., and Haydon, P. G. (1994) Glutamate-mediated astrocyte-neuron signalling. *Nature* 369: 744–747.

Pasti, L., Pozzan, T., and Carmignoto, G. (1995) Long-lasting changes of calcium oscillations in astrocytes. A new form of glutamate-mediated plasticity. *J. Biol. Chem.* 270: 15203–15210.

Pasti, L., Pozzan, T., and Carmignoto, G. (1997) Intracellular calcium oscillations in astrocytes: a highly plastic, bidirectional form of communication between neurons and astrocytes in situ. *J. Neurosci.* 17: 7817–7830.

Pearce, B., Albrecht, J, Morrow, S. and Murphy, S. (1986) Astrocyte glutamate receptor activation promotes inositol phospholipid turnover and calcium flux. *Neurosci. Lett.* 72: 335–340.

Penner, R., Matthews, G. and Neher, E. (1988) Regulation of calcium influx by second messengers in rat mast cells. *Nature* 344: 499–504.

Peters, A., Palay, S. L. and Webster, H. F. (1991) *The Fine Structure of the Nervous System: Neurones and Their Supporting Cells.* Oxford University Press, New York, pp. 276–295.

Petersen, O. H. and Wakui, M. (1990) Oscillating intracellular Ca^{2+} signals evoked by activation of receptors linked to inositol lipid hydrolysis. Mechanism of generation. *J. Membr. Biol.* 118: 93–105.

Pin, J. P. and Bockaert, J. (1989) Two distinct mechanisms, differentially affected by excitatory amino acids, trigger GABA release from fetal mouse striatal neurons in primary culture. *J. Neurosci.* 9: 648–656.

Porter, J. T. and McCarthy, K. D. (1995) GFAP-positive hippocampal astrocytes in situ respond to glutamatergic neuroligands with increases in $[Ca^{2+}]_i$. *Glia* 13: 101–112.

Pozzan, T., Rizzuto, R., Volpe, P., and Meldolesi, J. (1994) Molecular and cellular physiology of intracellular calcium stores. *Physiol. Rev.* 74: 595–636.

Prentki, M., Glennon, M. C., Thomas, A. P., Morris, R. L., Matschinsky, F. M., and Corkey, B. E. (1988) Cell-specific pattern of oscillating free Ca^{2+} in carbomycholine-stimulated insulinoma cells. *J. Biol. Chem.* 263: 11044–11047.

Rapp, P. E. and Berridge, M. J. (1981) The control of transepithelial potential oscillations in the salivary gland of Calliphora erythrocephala. *J. Exp. Biol.* 93: 119–132.

Rathouz, M. M., Vijayaraghavan, S., and Berg, D. K. (1995) Acetylcholine differentially affects intracellular calcium via nicotinic and muscarinic receptors on the same population of neurons. *J. Biol. Chem.* 270: 14366–14375.

Regehr, W. G. and Tank, D. W. (1990) Postsynaptic NMDA receptor-mediated calcium accumulation in hippocampal CA1 pyramidal cell dendrites. *Nature* 345: 807–810.

Rizzuto, R., Brini, M., and Pozzan, T. (1994) Targeting recombinant aequorin to specific intracellular organelles. *Methods Cell. Biol.* 40: 339–358.

Rizzuto, R., Brini, M., Murgia, M., and Pozzan, T. (1993) Microdomains with high Ca^{2+} close to IP3-sensitive channels that are sensed by neighboring mitochondria. *Science* 262: 744–746.

Romano, C., Sesma, M. A., MacDonald, C., O'Malley, K., Van de Pol, A. N., and Olney, J. W. (1995) Distribution of metabotropic glutamate receptor mGluR5 immunoreactivity in rat brain. *J. Comp. Neurol.* 355: 455–469.

Ross, C. A., Meldolesi, J., Milner, T. A., Satoh, T., Supattupone, S., and Snyder, S.H. (1989) Inositol 1,4,5-trisphosphate receptor localized to endoplasmic reticulum in cerebellar Purkinje neurons. *Nature* 339: 468–470.

Rutter, G. A., Theler, J.-M., Murgia, M., Wolheim, C. B., Pozzan, T., and Rizzuto, R. (1993) Stimulated Ca^{2+} influx raises mitochondrial free Ca^{2+} to supramicromolar levels in a pancreatic beta-cell line. Possible role in glucose and agonist-induced insulin secretion. *J. Biol. Chem.* 268: 22385–22390.

Satoh, T., Ross, C. A., Villa, A., Supattapone, S., Pozzan, T., Snyder, S. H. and Meldolesi, J. (1990) The inositol 1,4,5,-trisphosphate receptor in cerebellar Purkinje cells: quantitative immunogold labeling reveals concentration in an ER subcompartment. *J. Cell Biol.* 111: 615–624.

Schlegel, W., Winiger, B. P., Mollard, P., Vacher, P., Warin, F., Zahnd, G. R., Wollheim, C. B., and Dufy, B. (1987) Oscillations of cytosolic Ca^{2+} in pituitary cells due to action potentials. *Nature* 329: 719–721.

Schuman, E. M. and Madison, D. V. (1993) Nitric oxide as an intercellular signal in long term potentiation. *Sem. Neurosci.* 5: 207–215.

Sharp, A. H., Dawson, T. M., Ross, C. A., Fotuhi, M., Mourey, R. J. and Snyder, S. H. (1993a) Inositol 1,4,5-trisphosphate receptors: immunohistochemical localization to discrete areas of rat central nervous system. *Neuroscience* 53: 927–942.

Sharp, A. H., McPeterson, P. S., Dawson, T. M., Aoki, C., Campbell, K. P., and Snyder, S. H. (1993b) Differential immunohistochemical localization of inositol 1,4,5-trisphosphate- and ryanodine-sensitive Ca^{2+} release channels in rat brain. *J. Neurosci.* 13: 3051–3063.

Simpson, P. B., Challiss, R. A. J., and Nahorski, S. R. (1995) Neuronal Ca^{2+} stores: activation and function. *Trends Neurosci.* 18: 299–306.

Singelbaum, S. A. and Kandel, E. R. (1991) Learning-related synaptic plasticity: LTP and LTD. *Curr. Opin. Neurobiol.* 1: 113–120.

Singer, W. (1993) Neuronal representations, assemblies and temporal coherence. *Prog. Brain Res.* 95: 461–474.

Smith, S. J. (1994) Neuromodulatory astrocytes. *Curr. Biol.* 4: 807–810.

Sontheimer, H., Kettenmann, H., Backus, K. H. and Schachner, M. (1988) Glutamate opens Na^+/K^+ channels in cultured astrocytes. *Glia* 1: 328–336.

Sorrentino, V. and Volpe, P. (1993) Ryanodine receptors. How many, where and why? *Trends Pharmacol. Sci.* 14: 98–103.

Stagg, R. B. and Fletcher, W. H. (1990) The hormone-induced regulation of contact-dependent cell–cell communication by phosphorylation. *Endocrine Rev.* 11: 302–325.

Stent, G. S. (1973) A physiological mechanism for Hebb's postulate of learning. *Proc. Natl. Acad. Sci. USA* 70: 997–1001.

Stojilkovic, S. S. and Catt, K. J. (1992) Calcium oscillations in anterior pituitary cells. *Endocrine Rev.* 13: 256–280.

Stryker, M. P., Chapman, B., Miller, K. D., and Zahs, K. R. (1990) Experimental and theoretical studies of the organization of afferents to single orientation columns in visual cortex. *Cold Spring Harbor Symp. Quant. Biol.* 55: 515–527.

Swandulla, D., Hans, M., Zipser, K., and Augustine, G. (1991) Role of residual calcium in synaptic depression and posttetanic potentiation: fast and slow calcium signaling in nerve terminals. *Neuron* 7: 1–20.

Szatkowski, M., Barbour, B., and Atwell, D. (1990) Nonvesicular release of glutamate from glial cells by reversed electrogenic glutamate uptake. *Nature* 348: 443–446.

Thorn, P., Lawrie, A. M., Smith, P. M., Gallacher, D. V., and Petersen, O. H. (1993) Local and global cytosolic Ca^{2+} oscillations in exocrine cells evoked by agonist and inositol trisphosphate. *Cell* 74: 661–668.

Tsien, R. W. and Tsien, R. Y. (1990) Calcium channels, stores and oscillations. *Annu. Rev. Cell Biol.* 6: 715–760.

Tsien, R. Y. (1980) New calcium indicators and buffers with high selectivity against magnesium and protons: design, synthesis and properties of prototype structures. *Biochemistry* 19: 2396–2404.

Tsien, R. Y. (1989) Fluorescent indicators of ion concentration. *Methods Cell. Biol.* 30: 127–156.

Tsien, R. Y., Pozzan, T., and Rink, T. J. (1982) Calcium homeostasis in intact lymphocytes: cytoplasmic free calcium monitored with a new, intracellulary trapped fluorescent indicator. *J. Cell Biol.* 94: 325–334.

Uneyama, H., Uneyama, C., and Akaike, N. (1993) Intracellular mechanisms of cytoplasmic Ca^{2+} oscillation in rat megakaryocyte. *J. Biol. Chem.* 268: 168–174.

Van den Pol, A. N., Finkbeiner, S. M., and Cornell-Bell, A. H. (1992) Calcium excitability and oscillations in suprachiasmatic nucleus neurons and glia in vitro. *J. Neurosci.* 12: 2648–2664.

Wade, M. H., Trosko, J. E., and Shindler, M. (1986) A fluorescence photobleaching assay of gap junction-mediated communication between human cells. *Science* 232: 525–528.

Wakui, M., Potter, B. V. L., and Petersen O. H. (1989) Pulsatile intracellular calcium release does not depend on fluctuations in inositol trisphosphate concentration. *Nature* 339: 317–320.

Walton, P. D., Airey, J. A., Sutko, J. L., Beck, C. F., Mignery, G. A., Sudhof, T. C., Deerinck, T. J., and Ellisman, M. H. (1991) Ryanodine and inositol trisphosphate receptors coexist in avian cerebellar Purkinje neurons. *J. Cell Biol.* 113: 1145–1157.

Wilkin, G. P., Marriott, D. R., Cholewinski, A. J., Wood, J. N., Taylor, G. W., Stephens, G. J., and Djamgoz, M. B. A. (1991) Receptor activation and its biochemical consequences in astrocytes. *Ann. N.Y. Acad. Sci.* 633: 475–488.

Wong, R. O. L., Chernjavsky, A., Smith, S. J. and Shatz, C. (1995) Early functional neural networks in the developing retina. *Nature* 374: 716–718.

Woods, N. M., Cuthbertson, K. S. R., and Cobbold, P. H. (1986) Repetitive transient rises in cytoplasmic free calcium in hormone-stimulated hepatocytes. *Nature* 319: 600–602.

Worley, P. F., Baraban, J. M., and Snyder, S. H. (1989) Inositol 1,4,5-trisphosphate receptor binding: autoradiographic localization in rat brain. *J. Neurosci.* 9: 338–346.

Wyllie, D. J. A., Mathie, A., Symonds, C. J., and Cull-Candy, S. G. (1991) Activation of glutamate receptors and glutamate uptake in identified macroglial cells in rat cerebellar cultures. *J. Physiol.* 432: 235–258.

Yao, Y., Choi, J., and Parker, I. (1995) Quantal puffs of intracellular Ca^{2+} evoked by inositol trisphosphate in *Xenopus* oocytes. *J. Physiol.* 482: 533–553.

Yuste, R., Peinado, A., and Katz, L. C. (1992) Neuronal domains in developing neocortex. *Science* 257: 665–669.

Zafra, F., Lindholm, D., Castren, E., Hartikka, J., and Thoenen, H. (1992) Regulation of brain-derived neurotrophic factor and nerve growth factor mRNA in primary cultures of hippocampal neurons and astrocytes. *J. Neurosci.* 12: 793–4799.

PART II

Calcium-Modulated Proteins

4

The Structural Basis of Regulation by Calcium-Binding EF-Hand Proteins

Carolyn M. Slupsky
Brian D. Sykes

Perspective and Overview

Calcium is essential for many processes, such as nerve transmission, muscle contraction and cell motility, metabolic regulation, as well as cell division and growth. Calcium is the fifth most abundant element on earth, and is ubiquitous in biological organisms, processes, and structures (McPhalen et al., 1991). Most calcium-dependent intracellular functions are regulated in some manner by calcium-binding proteins. Calcium-binding proteins comprise a large class of regulatory proteins which may be subdivided into several groups, each with distinctive structural features. The two largest groups are the calcium-modulated proteins which utilize a helix–loop–helix calcium-binding site referred to as the EF-hand (Moews and Kretsinger, 1975), which occur most often in pairs, and the annexin protein family which contain sequences resembling mutated EF-hands (Heizmann and Hunziker, 1991). In this review chapter, the focus will be on the proteins of the EF-hand family which are involved in regulation of cellular processes.

A calcium signal begins with the activation of a cell surface receptor or channel protein by an extracellular stimulus, whereupon the cell surface receptor communicates with intracellular calcium stores located in the endoplasmic reticulum or sarcoplasmic reticulum (Falke et al., 1994). The signal from the cell surface results in the opening of the calcium channels, thereby releasing calcium into the cytoplasm. The rise in free calcium concentrations modulates the calcium-regulatory proteins at key control points in the esstential physiological pathways until calcium is pumped out of the cytoplasm by a calcium ATpase. The EF-hands of calcium-binding proteins function in a variety of ways. In some cases, such as calmodulin, troponin C, or recoverin, calcium acts to modulate these proteins such that they can interact with other proteins or membranes. In other cases, such as calbindin or parvalbumin, the free calcium concentrations in the cell are buffered.

There have been many reviews of EF-hand-type calcium-binding proteins (Kretsinger, 1980; Means and Dedman, 1980; Levine and Dalgarno, 1983; Persechini et al., 1989; Strynadka and James, 1989; Heizmann and Hunziker, 1990, 1991; Marsden et al., 1990; McPhalen et al., 1991; Falke et al., 1994; Kawasaki and Kretsinger, 1994). The review by Kawasaki and Kretsinger (1994), for example, is particularly complete and contains nearly 7000 references on calcium-binding proteins. Since the helix–loop–helix calcium-binding or EF-hand site is comprised of contiguous residues from the amino acid sequence, calcium-binding proteins are easily recognized from sequences alone. Again, the Kawasaki and Kretsinger reference lists more than 1000 sequences, divided into 32 subfamilies. Thus, most of the reviews to date have centered primarily on the structure of the calcium-binding site, the structural relationships between calcium-binding sites, and the determinants of calcium affinity, specificity, and kinetics for the EF-hand calcium-binding site. The reviews of Falke et al. (1994) and Linse and Forsén (1995) are particularly complete and up to date in this regard on all aspects of the chemistry, thermodynamics, and kinetics of calcium-binding.

Until recently, only a limited number of structures of calcium-binding proteins have been available. In particular, very few pairs of structures (apo and calcium-saturated) were complete. Recently, there has been a tremendous increase in the number of structures available, including several where the pair of structures is available and the calcium-induced conformational change is exposed. We have decided, for the purposes of this review, to take the position that the determinants of calcium-binding have been well studied and reviewed, and, in any case, merely serve

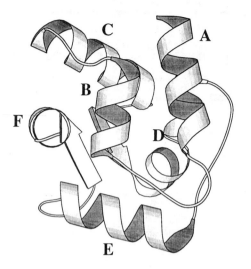

Figure 4.1 Ribbon diagram of calcium-saturated β-parvalbumin (pike pI 4.10) (Declercq et al., 1991). The helices are lettered in boldface type, starting from the N-terminus. Calcium binds in the loops between helices C and D, as well as between E and F. The ribbon diagram was prepared using MOLSCRIPT (Kraulis, 1991).

to poise the calcium affinity in the range of interest relative to the biological system wherein they serve as calcium sensors. Further, in some cases, such as the myosin light chains, the calcium-binding properties per se have been lost. The main focus of this review will be in the structures of these calcium-binding domains which contain at least pairs of EF-hand calcium-binding sites, and, in particular, to look at the conformational change that is induced in the two-site domain upon calcium-binding and the mechanism of coupling of calcium-binding to the structural change. This structural change is fundamental to the role of calcium-binding proteins in regulatory biological processes which they carry out via interaction with other proteins. We hope to set the stage for viewing calcium-binding proteins as generalized allosteric proteins.

General Features

What Constitutes an EF-Hand Calcium-Binding Site?

The term "EF-hand protein" was originally derived from the first EF-hand calcium-binding protein to have its structure solved—parvalbumin (Moews and Kretsinger, 1975). The structure of parvalbumin revealed six α-helical regions, designated with the letters A through F, with a calcium ion bound between helices C and D and another between helices E and F (Fig. 4.1). The two helices E and F, joined by a calcium-binding loop, resemble a hand where the extended forefinger, clenched middle finger, and extended thumb of the right hand represent helix E,

the EF-loop, and helix F, respectively (Fig. 4.2). The EF-hand calcium-binding site virtually always occurs tightly coupled with another EF-hand calcium-binding site to form the basic structural/functional unit where the pair of EF-hands are related by an approximate 2-fold axis of symmetry.

The major feature of the calcium-binding sites is a contiguous sequence of approximately 30 amino acids that form a helix–(calcium-binding)loop-helix (HLH) structural motif (Kretsinger and Nockolds, 1973). The helices are amphiphilic, and the calcium-binding loops all start with a γ-turn followed by a three-residue β-strand and end with the three amino-terminal residues of the second α-helix (Strynadka and James, 1989). Several studies employing 34-residue helix–loop–helix peptides that represent calcium-binding sites in the muscle protein have demonstrated that the hydrophobic residues of the amphiphilic helices are important for formation and stabilization of the EF-hand domain (Reid et al., 1981; Gariépy et al., 1982; Shaw et al., 1990, 1991a, 1991b, 1992c, 1994; Monera et al., 1992). A helix–loop–helix peptide that represents calcium-binding site III of TnC (residues 93–126 of chicken skeletal TnC) in the presence of calcium spontaneously forms a symmetric dimer which associates in a head-to-tail arrangment (Shaw et al., 1990, 1991a, 1992c). The same occurs for a peptide that represents calcium-binding site IV (121–159 of rabbit skeletal TnC) (Kay et al., 1991). When two 34-residue peptides (one that represents calcium-binding site III and the other that represents calcium-binding site IV) are mixed, a heterodimer which is more stable than either of the homodimers is preferentially formed (Shaw et al., 1991c, 1992a, 1992b, 1994). These results may be explained by the

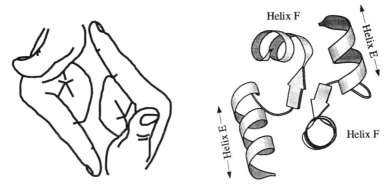

Figure 4.2 Schematic illustrating the EF-hand, helix–loop–helix calcium-binding motif. Calcium binds in the loop formed between helices E and F. The two helices E and F, joined by the calcium-binding loop, resemble a hand where the extended forefinger, clenched middle finger, and extended thumb of the right hand represent helix E, the EF-loop, and helix F, respectively. These EF-hands virtually always occur in pairs to form an "EF-handshake". The ribbon diagram was prepared using MOLSCRIPT (Kraulis, 1991). (Adapted from Kretsinger and Nockolds, 1973.)

optimization of hydrophobic interactions, as mutation of nonliganding residues in the 34-residue peptides results in increased or decreased calcium affinity with no alteration of secondary structure (Shaw et al., 1991b; Monera et al., 1992).

Calcium or magnesium may be coordinated to the 12-residue loop region via the carboxyl, hydroxyl, or amide side chains of five amino acid residues at positions $1\,(x)$, $3\,(y)$, $5\,(z)$, $9\,(-x)$, and $12\,(-z)$ and to the main chain carbonyl of the residue at position $7\,(-y)$. Seven oxygen ligands (the residue at position 12 is involved in a bidentate coordination with calcium) result in a pentagonal bipyramid coordination geometry where each vertex is approximately 2.4 Å from the central calcium ion (Strynadka and James, 1989). Not all of the seven ligands are necessarily from the protein. There are cases where the calcium-binding sites have one or more water molecules in the coordination sphere of the calcium ion (Strynadka and James, 1989). Studies of the calcium-binding EF-hands reveal invariant residues (Kanellis et al., 1983; Boguta and Bierzynski, 1988; Marsden et al., 1990). These residues are located at positions 1 (Asp), 3 (Asp/Asn), 6 (Gly), 8 (Ile), and 12 (Glu) (Marsden et al., 1990). Positions 6 and 8 are highly conserved due to structural constraints. A glycine at position 6 allows $[(\phi, \psi) = (90°, 0°)]$, thus enabling the side chain at position 5 and the carbonyl main chain at position 7 to coordinate calcium. The residue at position 8 is involved in a β-sheet H-bonding interaction between adjacent calcium-binding loops, which is needed for the construction of the hydrophobic core of the domain (Herzberg and James, 1985b). The stability of these calcium-binding sites is determined primarily by the hydrophobic interactions between the two sites (Monera et al., 1992), as well

as by the helices that surround the calcium-binding loops (Marsden et al., 1990).

There are some calcium-binding sites which bind only calcium, while others appear to bind both calcium and magnesium. There have been studies on the importance of metal charge and radius on metal-binding affinity (Reid and Hodges, 1980; Gariépy et al., 1985; Snyder et al., 1990); however, those sites which can bind both calcium and magnesium at physiological calcium and magnesium concentrations do so by their increased affinity for divalent ions rather than specifically by an increase in affinity for magnesium.

Sequences of EF-Hand Calcium-Binding Proteins

There have been many sequences found which have "typical" EF-hand metal-ion-binding sites. In a review by Kawasaki and Kretsinger (1994), over 1000 sequences are divided into 32 subfamilies. The proteins which will be covered here will include parvalbumin, oncomodulin, calcyclin (an S100 protein), calmodulin, troponin C, calbindin, recoverin, sarcoplasmic calcium-binding protein, calcineurin B, and the myson light chains.

Structures of EF-Hand Calcium-Binding Proteins

Parvalbumin

Parvalbumin is a protein found in the cytosol of muscle cells, and is thought to act as a calcium buffer to ensure fast response for calcium sensitivity.

Parvalbumin is of low molecular weight (approximtely 12 kDa) and is characterized by a high calcium affinity ($K°_a \sim 10^8 M^{-1}$). Although the affinity of parvalbumin for calcium is greater than that for TnC or calmodulin, upon release of calcium from the sarcoplasmic reticulum, calcium binds first to TnC and calmodulin, followed by parvalbumin. This is because at physiological concentrations of magnesium and during the resting state of muscle, parvalbumin's metal-ion-binding sites are occupied by magnesium. Magnesium has a slow off rate from parvalbumin, therefore allowing calcium to bind to TnC and calmodulin first. Parvalbumin then acts like a sponge, pulling calcium away from TnC and calmodulin—thus, quickly reducing calcium concentrations and ensuring that contraction does not reoccur. Parvalbumins are essentially skeletal muscle proteins and, since they are not found in cardiac or smooth muscle, they are not indispensable components of the contractile mechanism (Strynadka and James, 1989). Parvalbumins have also been found in nervous tissue where synaptic transmission is triggered by calcium influx (Levine and Dalgarno, 1983).

Comparison of parvalbumin amino acid sequences from different species indicates the presence of two distinct phylogenetic lineages, termed α (pI 5.0) and β (pI 4.1). The crystallographic structures of calcium-saturated β-parvalbumin (Moews and Kretsinger, 1975; Swain et al., 1989; Kumar et al., 1990; Declercq et al., 1988, 1991) and α-parvalbumin (Roquet et al., 1992, 1995; McPhalen et al., 1994; Padilla et al., 1988) reveal globular molecules which are ellipsoid in shape (Fig. 4.3). The two structures are virtually identical, as can be seen in Fig. 4.3. There are six α-helices designated by the letters A through F. Calcium binds between the loops of helices C/D and E/F. The N-terminal extension, which comprises helices A and B separated by an eight-residue loop, does not bind calcium and it lies over the tightly packed CD and EF-hand domains to cover the hydrophobic core. In vivo, the two calcium-binding sites are occupied by either calcium or magnesium ions. To date, no metal-free structure of parvalbumin has been solved and therefore no comparison between apo- and calcium-saturated parvalbumin may be made. It should be noted, however, that under physiological conditions the metal-ion sites are occupied by metal ions (either calcium or magnesium) and therefore the structure of the apo form of parvalbumin may not be relevant.

Oncomodulin

Oncomodulin is a member of the EF-hand family of calcium-binding proteins that shares 50% identity of amino acid sequence with rat parvalbumin (Golden et al., 1989). Oncomodulin is a protein which is expressed in prenatal development in the placenta and in the majority of tumors (Golden et al., 1989; Ahmed et al., 1990; Palmer et al., 1990). Oncomodulin is capable of activating enzymes such as phosphodiesterase and calcineurin in a calcium-dependent manner and can inhibit enzymes such as glutathione reductase (MacManus, 1981; Palmer et al., 1990).

Rat oncomodulin, isolated from rat tumors, consists of 108 residues and contains six α-helices (A through F) and two calcium-binding loops (between helices C and D, as well as between helices E and F) as does parvalbumin (Ahmed et al., 1990). The structure of oncomodulin was solved to 1.85-Å resolution and reveals a close similarity to parvalbumin (backbone root-mean-square deviation (RMSD) is 0.62 Å for residues 5–108 between the two structures) Ahmed et al., 1990). Although oncomodulin and parvalbumin are very similar structurally, they have very different functions. To date, no structure of apo-oncomodulin has been accomplished, and it remains to be determined if this form of oncomodulin is physiologically relevant.

Calbindin (Intestinal Calcium Binding Protein)

There are two types of calbindin: calbindin D_{9K}, and calbindin D_{28K}. These proteins bind calcium with high-affinity ($K_a \sim 2 \times 10^6$) and are widely distributed in various species and organs. These proteins are hormonally controlled by vitamin D (hence the "D" in the name) and the number refers to the molecular weight. Although they are vitamin D dependent in tissues such as the intestine and kidney, there is evidence that they are not vitamin D dependent in tissues of the central nervous system (Gross and Kumar, 1990). The highest concentrations of calbindin D_{9K} and calbindin D_{28K} are found in calcium-transporting tissues, such as the intestine, kidney, and placenta, and smaller concentrations are found in nonepithelial tissues that do not transport calcium, such as bone, parathyroid, and brain (Gross and Kumar, 1990). Thus, calbindin has a role mainly in calcium transport but may also be multifunctional.

Calbindin D_{28K} is present in all vertebrate species and some invertebrates (for a review, see Heizmann and Hunziker, 1990). It is thought that the protein may function as a mediator of vitamin D-dependent calcium transport. This protein contains six putative calcium-binding sites but binds only three to four calcium ions with high-affinity (Heizmann and Hunziker, 1990). Calbindin D_{9K} has been studied more extensively. It is present mainly in mammalian intestine (and was originally called intestinal calcium-binding protein) and appears to be involved in the translocation of calcium (Heizmann and Hunziker,

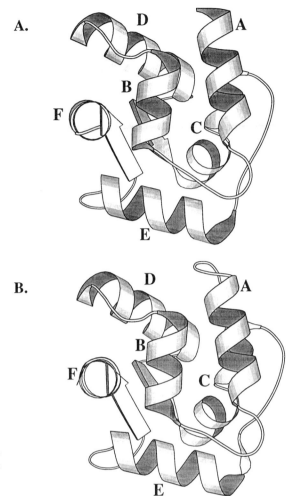

Figure 4.3 Ribbon diagrams of (A) Calcium-saturated β-parvalbumin (pike pI 4.10) (Declercq et al., 1991) and (B) calcium-saturated α-parvalbumin (pike pI 5.0) (J. P. Declercq et al., unpublished data). The helices are lettered in boldface type, starting from the N-terminus. The ribbon diagram was prepared using MOLSCRIPT (Kraulis, 1991).

1990). Calbindin D_{9K} has two EF-hand calcium-binding sites and binds two calcium ions strongly. The small size of calbindin D_{9K} (75 residues) means it is ideal for study using NMR methods.

Calbindin was the second calcium-binding protein to have its structure solved by x-ray crystallographic methods (Szebenyi et al., 1981). This calcium-saturated structure of calbindin was later refined to 2-Å resolution (Szebenyi and Moffat, 1986), and is shown in Fig. 4.4. The three-dimensional calcium-saturated (Kördel et al., 1993) and apo- (Skelton et al., 1994) calbindin D_{9K} structures have been determined using NMR methods, and are shown to be very similar (Fig. 4.5). Calbindin D_{9K} is comprised of four helices, two ion-binding loops, and a linker loop, regardless of the calcium state. It has been determined that binding of calcium results in little change in the secondary structure of the protein and that

reorientation of side chains occurs at the helical interface formed by a pair of helix–loop–helix regions (Dalgarno et al., 1983b). Table 4.1 compares the interhelical angles and helix midpoint distances of the calcium and apo-calbindin structures. The interhelical angles are virtually the same between helix A and helix B regardless of the ion-binding state. The interhelical angle is 124° in the apo structure and 135° in the $2Ca^{2+}$ structure. Between helices C and D, the interhelical angle is 117° for the apo structure and 115° for the $2Ca^{2+}$ structure. The distance between the midpoints of the helices is 12 Å between helix A and helix B, and 13 Å between helix C and helix D, and for the $2Ca^{2+}$ form of calbindin, 11 Å between helix A and helix B, and 12 Å between helix C and helix D. Comparison of the apo and $2Ca^{2+}$ structures in detail reveals a root-mean-square deviation of 1.7 Å between backbone atoms for residues 5–73.

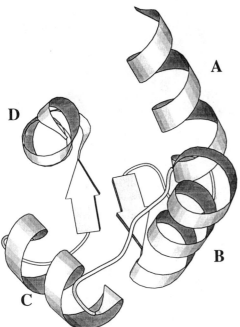

Figure 4.4 Calcium-saturated structure of calbindin D_{9K} (intestinal calcium-binding protein) refined to 2.0-Å resolution (Szebenyi and Moffat, 1986). The helices are lettered in bold-face type, starting from the N-terminus. Calcium binds in the loop formed between the helices A and B, as well as between C and D. The ribbon diagram was prepared using MOLSCRIPT (Kraulis, 1991).

This difference may be related to three primary effects associated with calcium-binding: a change in the backbone conformation of helix IV from regular α-helix in the apo state to a mixture of 3_{10} and a-helix in the $2Ca^{2+}$ state, a rigid body movement of helix III of 1.5 Å with respect to the rest of the protein, and a reorganization of the packing of helices III and IV into each other and onto helices I and II (Skelton et al., 1994). The reason for the slight structural change in calbindin when calcium binds may be due to side

chains in the C-terminal EF-hand (helices III and IV) repositioning for chelation of the calcium ion (Skelton et al., 1994).

Sarcoplasmic calcium-binding protein

Sarcoplasmic calcium-binding proteins are important in invertebrate muscle, and appear to have a similar function to parvalbumin in vertebrates (Cook et al., 1993). Two recent structures have

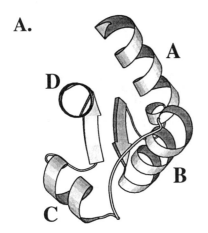
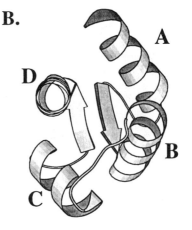

Figure 4.5 Ribbon diagrams of the NMR structures of (A) calcium-saturated calbindin D_{9K} (Kördel et al., 1993) and (B) apo-calbindin D_{9K} (Skelton et al., 1994). The helices are labeled in boldface type. The ribbon diagrams were generated using MOLSCRIPT (Kraulis, 1991).

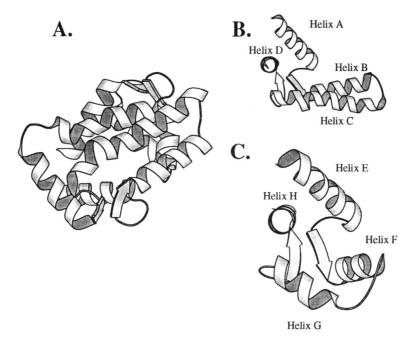

A.

B.

Helix A

Helix D

Helix B

Helix C

C.

Helix E

Helix H

Helix F

Helix G

Figure 4.6 Ribbon diagram of the structure of the sarcoplasmic calcium-binding protein from *Nereis diversicolor* (Vijay-Kumar and Cook, 1992). (A) Intact sarcoplasmic calcium-binding protein, (B) the N-terminal domain of the sarcoplasmic calcium-binding protein and (C) the C-terminal domain of the sarcoplasmic calcium-binding protein. Calcium binds between helices A and B, helices E and F, and helices G and H. The ribbon diagrams were generated using MOLSCRIPT (Kraulis, 1991).

appeared for sarcoplasmic calcium-binding protein: one from *Nereis diversicolor* (Vijay-Kumar and Cook, 1992) and the other from *Amphioxus* (Cook et al., 1993). These structures are shown in Figs. 4.6, and 4.7, respectively. Each structure is compact, and consists of two domains which contain two EF-hand calcium-binding sites. The linker region between the two domains contains a tight turn which is responsible for the compact structure of these proteins, as compared with TnC or calmodulin (Vijay-Kumar and Cook, 1992; Cook et al., 1993). Of the four EF-hand calcium-binding sites, only three bind calcium in each structure. The structure of *Nereis diversicolor* sarcoplasmic calcium-binding protein indicates that calcium binds to calcium-binding sites I, III, and IV (Vijay-Kumar and Cook, 1992), whereas in the structure of *Amphioxus* sarcoplasmic calcium-binding protein, calcium is bound to calcium-binding sites I, II, and III (Cook et al., 1993). The interhelical angles for the saturated calcium-binding domains are around 110° and 102°, which indicates a structure that is similar to the C-terminal domain of TnC in terms of "openess" of the hydrophobic pocket (see Table 4.1). This may

indicate that this domain of the sarcoplasmic calcium-binding protein has a structural role (i.e. no regulatory function associated). The domain of the sarcoplasmic calcium-binding protein that binds only one calcium ion has interhelical angles of approximately 58° and 82°. The interhelical angles are similar to those of the regulatory N-terminal domain of TnC and both domains of calmodulin (Table 4.1) indicating that this domain appears to have more of a regulatory role.

Myosin regulatory and essential light chain structures

There are two light chains associated with myosin in muscle, and these are referred to as the essential light chain and the regulatory light chain. In the crystal structure of myosin S1, the essential light chain was shown to interact with the C-terminal long helical portion of S1 that connects to the α-helical coiled coil backbone (Rayment et al., 1993). This light chain has conformational flexibility in the crystal structure. The regulatory light chain interacts with the region further to the C-terminal end of S1. The

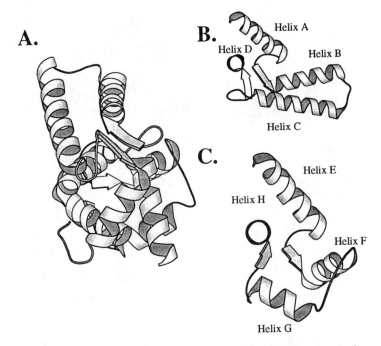

Figure 4.7 Ribbon diagram of the structure of the sarcoplasmic calcium-binding protein from *Amphioxus* sarcoplasmic calcium-binding protein (Cook et al., 1993). (A) Intact sarcoplasmic calcium-binding protein, (B) the N-terminal domain of the sarcoplasmic calcium-binding protein and (C) the C-terminal domain of the sarcoplasmic calcium-binding protein. Calcium binds between helices A and B, helices C and D, and helices E and F. The ribbon diagrams were generated using MOLSCRIPT (Kraulis, 1991).

two light chains make contact with one another over a very limited region. The contact area involves the first loop of the essential light chain with the linker region in the C-terminal domain of the regulatory light chain (G117) (Xie et al., 1994). Both classes of light chain subunits appear to have evolved from four domain EF-hand proteins, as they share sequence homology to calmodulin and TnC, but it has been shown that the essential light chains do not bind calcium, whereas the regulatory light chains can bind one calcium ion (Kretsinger 1980; Tollemar et al., 1986; Persechini et al., 1989).

The structures of these proteins have been solved in complex with myosin (Rayment et al., 1993; Xie et al., 1994). The proteins consist of eight α-helices and two antiparallel β-sheets, with each domain exhibiting folds similar to other EF-hand calcium-binding proteins. In the crystal structure of a fragment of scallop myosin that contains the light chains with part of the heavy chain associated, it was determined that calcium binds to the first calcium-binding loop of the regulatory light chain, as well as to the first calcium-binding loop of the essential light chain (Xie et al., 1994). In the structures, the regulatory light

chain is extended with a kink in the interconnecting helix region so as to reorient the two globular lobes appropriately, whereas the structure of the essential light chain is similar to calmodulin bound to the myosin light chan kinase (MLCK) peptides (Figs. 4.8 and 4.9). Except for the N-terminal lobe of the essential light chain, the contacts made by the light chains with the heavy chain are similar to those found in calmodulin–peptide complexes (Xie et al., 1994).

The structure of the C-terminal domain (which does not bind calcium) has interhelical angles of 108° and 90° respectively, for the E/F and G/H helices of the essential light chain, and 125° and 115°, respectively, for the E/F and G/H helices of the regulatory light chain. These values are somewhat smaller than those for apo-TnC or apo-calmodulin (see Table 4.1), indicating a slightly more open structure in the C-terminal domain of the light chains than was observed for TnC or calmodulin. This slightly more open structure makes sense since a hydrophobic patch is needed to complex the light chains to the heavy chain. The N-terminal domain structures of the essential and regulatory light chains (each with one calcium ion

Table 4.1 Interhelical Angles and Distances in Various Calcium-Binding Proteins

Protein	Interhelical Angles[a] (degrees)	Interhelical Distance[a] (Å)
Apo EF-hands		
$2Ca^{2+}$-TnC[b] I/II	138°	12.3 Å
$2Ca^{2+}$-TnC[b] III/IV	144°	12.3 Å
Apo-NTnC[c] I/II	124°	13.8 Å
Apo-NTnC[c] III/IV	124°	14.0 Å
Apo-calmodulin[d] I/II	130°	10.6 Å
Apo-calmodulin[d] III/IV	133°	12.3 Å
Apo-calmodulin[d] V/VI	144°	10.4 Å
Apo-calmodulin[d] VII/VIII	134°	12.6 Å
1-Ca^{2+}-recoverin[e] V/VI	107°	18.3 Å
1-Ca^{2+}-recoverin[e] VII/VIII	127°	12.0 Å
Apo-myristoylated recoverin[f] I/II	167°	14.6 Å
Apo-myristoylated recoverin[f] III/IV	141°	10.8 Å
Apo-myristoylated recoverin[f] V/VI	126°	13.7 Å
Apo-myristoylated recoverin[f] VII/VIII	97°	14.6 Å
Apo-calbindin[g] I/II	124°	11.7 Å
Apo-calbindin[g] III/IV	117°	13.0 Å
Ca^{2+}-*bound EF-hands*		
$2Ca^{2+}$-TnC[b] V/VI	104°	14.8 Å
$2Ca^{2+}$-TnC[h] VII/VIII	111°	13.8 Å
$4Ca^{2+}$-TNC[h] I/II	79°	18.9 Å
$4Ca^{2+}$-TNC[h] III/IV	67°	16.3 Å
$4Ca^{2+}$-TnC[h] V/VI	83°	16.6 Å
$4Ca^{2+}$-TnC[h] VII/VIII	106°	15.3 Å
$2Ca^{2+}$-NTnC[i] I/II	87°	18.8 Å
$2Ca^{2+}$-NTnC[i] III/IV	69°	17.4 Å
$4Ca^{2+}$-calmodulin[j] I/II	88°	17.5 Å
$4Ca^{2+}$-calmodulin[j] III/IV	86°	15.6 Å
$4Ca^{2+}$-calmodulin[j] V/VI	100°	14.5 Å
$4Ca^{2+}$-calmodulin[j] VII/VIII	94°	14.0 Å
$2Ca^{2+}$-calbindin[k] I/II	135°	11.3 Å
$2Ca^{2+}$-calbindin[k] III/IV	115°	12.2 Å
$1Ca^{2+}$-recoverin[e] I/II	99°	13.2 Å
$1Ca^{2+}$-recoverin[e] III/IV	99°	14.8 Å

[a]Calculated using the in-house program written by Kyoko Yap and Mitsuhiko Ikura (University of Toronto). Helices I/II, III/IV, V/VI, and VII/VIII correspond to helices that form calcium-binding sites I, II, III, and IV, respectively. A decrease in the value of I/II, III/IV, V/VI, or VII/VIII, indicates an opening of the structure. [b]PDB accession code: ITOP; [c]PDB accession code: ITNP; [d]coordinates were a kind gift from Dr. M. Ikura (University of Toronto); [e]PDB accession code: IREC; [f]coordinates were a kind gift from Dr. M. Ikura (University of Toronto); [g]PDB accession code: ICBI; [h]PDB accession code: ITNW, ITNX; [i]PDB accession code: ITNQ; [j]PDB accession code: ICII; [k]PDB accession code: 2BcB.

bound) are quite different from one another. The essential light chain has interhelical angles of 139° and 129°, respectively, for the A/B and C/D helix pairs, whereas the regulatory light chain has angles of 91° and 92°, respectively, for the A/B and C/D helix pairs. The essential light chain has interhelical angles that indicate a closed structure, as may be observed for apo-calmodulin and apo-TnC, whereas the regulatory light chain has interhelical angles that indicate an open structure—such as the calcium-saturated forms of calmodulin and TnC (see Table 4.1). Without the heavy chain and the regulatory light

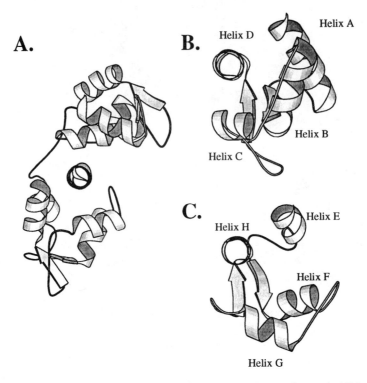

Figure 4.8 Ribbon diagram of the structure of the essential light chain (Xie et al., 1994). (A) Intact structure when bound to the heavy chain of myosin, (B) the N-terminal domain of the essential light chain and (C) the C-terminal domain of the essential light chain. In this structure, calcium binds to calcium-binding site I (between helices A and B). The ribbon diagrams were generated using MOLSCRIPT (Kraulis, 1991).

chain, the essential light chain does bind calcium. The ternary complex stabilizes binding of calcium to the first loop of the essential light chain which is proposed to be the specific calcium-binding or triggering site that is responsible for myosin activation (Xie et al., 1994).

Calcyclin

Calcyclin is a member of the S100 family of proteins. The S100 family of proteins represents another major subfamily of calcium-binding proteins. The S100 proteins have been primarily associated with human disease, and, in particular, several S100 genes or gene products have been shown to exhibit deregulated expression in human cancer, and have been suggested to participate in tumor progression (Potts et al., 1995). Calcyclin is among the genes in the cluster of S100 genes on human chromosome 1q21 that has shown deregulated expression in association with tumor transformation (Potts et al., 1995). The structure exhibits a similarity in global fold to apo-calbindin D_{9K}, with four helices and two calcium-binding

sites. The structure of the calcyclin homodimer is shown in Fig. 4.10A. The dimer interface is mediated by hydrophobic interactions between side chains, and involves primarily helices D and D', as well as residues at the C-terminal of helices A and A'. The structure of a monomer of calcyclin is shown in Fig. 4.10B. This structure is similar in global fold to calbindin. It will be interesting to see if the calcium-saturated structure of calcyclin is similar to the calcium-saturated structure of calbindin, which would suggest that calcyclin acts as a buffer protein. Other members of the S100 family of calcium-binding proteins undergo large changes in their chemical shifts upon calcium-binding, which suggests large tertiary structure rearrangements (G. S. Shaw, personal communication).

Troponin C

Troponin C (TnC) is the calcium-binding regulatory protein of the troponin complex in muscle tissue. Recombinant chicken skeletal TnC has a molecular weight of 18,257 and utilizes four EF-hand helix–

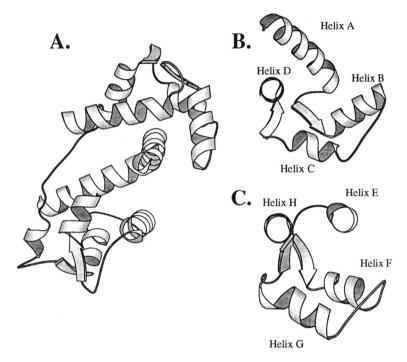

Figure 4.9 Ribbon diagram of the structure of the regulatory light chain (Xie et al., 1994). (A) Intact structure when bound to the heavy chain of myosin, (B) the N-terminal domain of the regulatory light chain and (C) the C-terminal domain of the regulatory light chain. In this structure, calcium binds to calcium-binding site I (between helices A and B). The ribbon diagrams were generated using MOLSCRIPT (Kraulis, 1991).

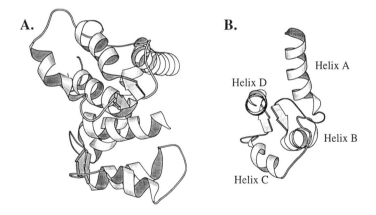

Figure 4.10 Structure of (A) the apo-calcyclin homodimer and (B) one of the calcyclin monomers. The helices are lettered starting with helix A. The preliminary coordinates for calcyclin were kindly provided by Dr. Walter Chazin. This figure was prepared using MOLSCRIPT (Kraulis, 1991).

loop–helix motifs for binding calcium. Troponin C is a very acidic protein with a pI of 4.1–4.4 (Hartshorne and Driezen, 1972) . It was realized in the mid-1970s that TnC contained four homologous regions in the sequence (Collins et al., 1973; Kretsinger and Barry, 1975), each of which presumably contained one of the four calcium-binding sites in a helix–loop–helix arrangement as described by Potter and Gergely (1975). Since then, there have been many studies on TnC that have tried to delineate its calcium-binding properties, its structure, and the conformational change that occurs upon binding calcium.

It was first suggested by Ebashi et al. (1968) that troponin has four calcium-binding sites, two of which are of high-affinity and two of which are of low-affinity. Potter and Gergely (1975) and Potter et al. (1976) determined that the calcium-binding component of troponin, TnC, has two high-affinity sites which bind calcium and magnesium competitively ($K_{Ca} = 2.1 \times 10^7 M^{-1}$ and $K_{Mg} = 5 \times 10^3 M^{-1}$) and two low-affinity sites which selectively bind calcium over magnesium ($K_{Ca} \sim 3 \times 10^5 M^{-1}$). It was shown that in complexes of TnI/TnC and TnI/TnT/TnC, all four calcium-binding sites have approximately the same affinity for calcium in the presence of 2 mM Mg^{2+}. Since the change in free magnesium concentration by approximately 2 mM did not affect calcium sensitivity of the myofibrillar ATPase activity, it was reasoned that the low-affinity sites were the ones related to the regulation of muscle contraction (Potter and Gergely, 1975). Analysis of the kinetics of calcium removal from TnC using stopped flow fluorimetric studies revealed that the release of calcium from the low-affinity sites was much greater than that from the high-affinity sites (Johnson et al., 1979). Therefore, since the events which regulate the skeletal muscle contraction–relaxation cycle must be complete within approximately 50 ms after excitation, only the calcium-specific sites exchange rapidly enough to be involved in regulation.

To understand the complex series of events upon calcium release in the muscle, several NMR studies of calcium-binding to TnC were undertaken in the 1970s. It was determined that calcium-binding to the low-affinity N-terminal domain sites did not alter the backbone of these sites significantly, but changed the hydrophobic interactions (Levine et al., 1977). A comparison of magnesium and calcium-binding to the C-domain demonstrated differences in the degree of backbone folding of this domain and altered interactions between hydrophobic resides when magnesium was bound (Seamon et al., 1977; Levine et al., 1978); however, magnesium was able to induce almost the same conformation as calcium (Seamon et al., 1977; Drakenberg et al., 1987; Tsuda et al., 1990).

The crystal structure of skeletal TnC initially was solved with the high-affinity calcium-binding sites filled and the low-affinity calcium-binding sites unfilled to 2.0-Å resolution (Herzberg and James, 1988; Satyshur et al., 1988) and, more recently, to 1.78-Å resolution (Satyshur et al., 1994). The structure reveals that TnC is a 70-Å-long dumbbell-shaped molecule with approximately 66% of the structure being helical. The domains have a mean radius of approximately 17 Å, with their centers separated by a 31-residue α-helix that corresponds to 44 Å (Herzberg and James, 1988). Each domain of TnC has two helix–loop–helix motifs tightly coupled by a short segment of β-sheet which helps to form the calcium-binding loops. The N-terminal domain consists of helix A/calcium-binding loop I/helix B and helix C/calcium-binding loop II/helix D. The C-terminal domain consists of helix E/calcium-binding loop III/helix F, and helix G/calcium-binding loop IV/helix H. The N-terminal domain has an additional helix termed the N-helix, which is apparently unique to the TnCs (Strynadka and James, 1989). In the crystal structure, the central connecting region is a continuous helix from helix D to helix E. Figure 4.11 illustrates the major tertiary structural differences between the apo N-terminal domain and the $2Ca^{2+}$ C-terminal domain of the crystal structure. On a superficial level, it appears that the N-terminal domain adopts a more closed structure whereas the C-terminal domain adopts a more open structure. An analysis of (ϕ, ψ) angles in the crystal structure reveals that one residue in helix B, E41, is irregular. This residue possesses the angles [(ϕ, ψ) = ($-90°$, $-7°$)] which produce a kink in helix B at about residue 41. This irregularity appears for the modified structure of TR_1C (Findlay and Sykes, 1993; Findlay et al., 1994; Gagné et al., 1994) and the apo N-domain (residues 1–90) hereafter referred to as NTnC (Gagné et al., 1994).

Since 1985, there was a relative standstill in the understanding of the structural changes which accompany calcium-binding to the N-terminal domain. Over the next few years, calcium-binding properties were studied. In particular, it was discovered what residues are required for liganding a metal ion (see previous section), and the hydrophobic interactions in peptide homodimer domains were studied. It turned out that the structure of these peptide homodimer domains was similar to the C-terminal domain of TnC (Shaw et al., 1990, 1992c). A synthesized 34-residue peptide corresponding to residues 93–126 of chicken skeletal troponin C was shown to form a homodimer in the presence of calcium that exhibited an approximate 2-fold rotational symmetry with hydrophobic residues and a three-residue β-sheet that stabilized the protein domain at the interface between the two sites (Shaw et al., 1990,

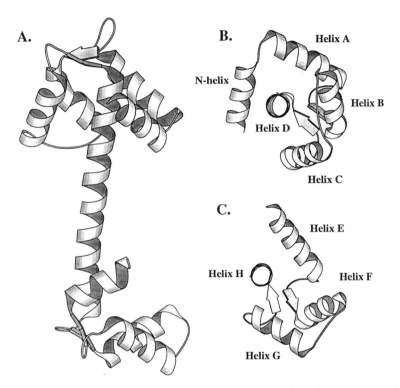

Figure 4.11 Ribbon diagram of the structure of half-saturated chicken skeletal muscle troponin C (at 1.78-Å resolution) (Satyshur et al., 1994). (A) Intact troponin C, (B) the N-terminal domain of troponin C and (C) the C-terminal domain of troponin C. Calcium is bound at calcium-binding sites III and IV (between helices E and F, as well as between helices G and H). The ribbon diagrams were generated using MOLSCRIPT (Kraulis, 1991).

1992c). Figure 4.12 illustrates the homodimer domain. Interhelical angles between the two helices (E and F) in the symmetric homodimer are 124°. These results are somewhat similar to the apo domains of TnC and calmodulin, thus indicating a relatively closed structure. These homodimers bind only one calcium with high-affinity (Shaw et al., 1991b, 1992b, 1992c; C. M. Slupsky, unpublished data). The exact reason for this is unknown, but hydrophobic interactions could play a role since two peptides, one representing calcium-binding site III and the other representing calcium-binding site IV, bind two calcium ions in the micromolar range (Shaw et al., 1992d) and have a ^{1}H NMR spectrum which is very similar to the spectrum of TR$_2$C (a proteolytic fragment of TnC that represents residues 92–162 of chicken skeletal TnC) (Shaw et al., 1992d).

During this period, attempts were also made to study the conformational change of TnC. A model of calcium-saturated TnC was formulated (Herzberg et al., 1986) where the N-terminal sites were modeled after the C-terminal domain sites. The key feature of the model was a movement of the B/C helix pair of

helices away from the A/D pair to expose a patch of hydrophobic residues. This model gained support from several lines of evidence including cysteine reactivity (Ingraham and Hodges, 1988; Fuchs et al., 1989; Putkey et al., 1993), NMR (Gagné et al., 1994; Lin et al., 1994), and site-directed mutagenesis (Fujimori et al., 1990; Grabarek et al., 1990; Gusev et al., 1991; Pearlstone et al., 1992); for a minireview see daSilva and Reinach, 1991).

Structures of the apo N-terminal domain (Fig. 4.13) were shown to be very similar to the crystal structure of the N-terminal domain. Apo-TR$_1$C (a proteolytic fragment that represents the N-terminal domain) (Findlay et al., 1994; Gagné et al., 1994) and apo-NTnC (Gagné et al., 1994, 1995) both have interhelical A/B and C/D angles which are similar to the apo N-terminal crystal structure of TnC (TR$_1$C A/B interhelical angle is 133° and C/D interhelical angle is 137°; NTnC A/B interhelical angle is 124° and C/D interhelical angle is 124°; and the N-terminal crystal structure interhelical angles are 138° and 144° for A/B and C/D, respectively). The structures of the apo N-terminal domain (x-ray crystal

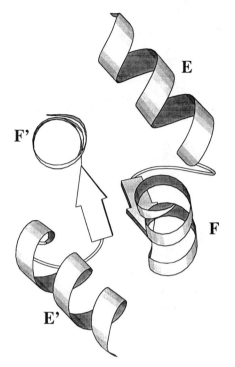

Figure 4.12 Ribbon diagram of the structure of the SCIII homodimer (a homodimer which is formed from a synthetic peptide
comprising residues 93–126 of chicken skeletal troponin C) (Shaw
et al., 1992c). The helices are indicated with boldface type. One
calcium ion is bound in this structure. This diagram was prepared
using MOLSCRIPT (Kraulis, 1991).

structure, TR_1C, and apo-NTnC) are all very similar.
They all have similar helix lengths (except for TR_1C,
which has no N-helix) and a kink at residue E41 that
results in a closed structure.

Recently, the structures of calcium-saturated TnC
(Slupsky and Sykes, 1995; Slupsky et al., 1995b) and
NTnC (Gagné et al., 1994, 1995) were solved. Figure
4.14 shows the structures of NTnC and TnC. As can

be observed from the figure, there are five helices, N,
A, B, C, and D, and they have approximately the
same interhelical angles between helices A/B and C/
D. For calcium-saturated TnC, the A/B interhelical
angles are 79° and the C/D interhelical angles are 67°
and for calcium-saturated N-terminal domain, the
interhelical angles are 87° and 69° for A/B and C/
D, respectively. These results suggest that NTnC

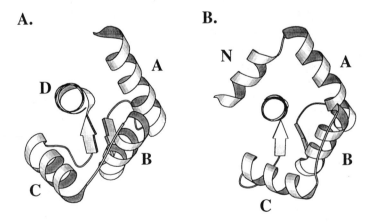

Figure 4.13 Ribbon diagram of the structures of (A) apo-TR_1C (residues 12–87 of turkey skeletal troponin C)
(Findlay et al., 1994) and (B) apo-NTnC (residues 1–90 of cloned chicken skeletal troponin C) (Gagné et al., 1995).
The ribbon diagrams were prepared using MOLSCRIPT (Kraulis, 1991).

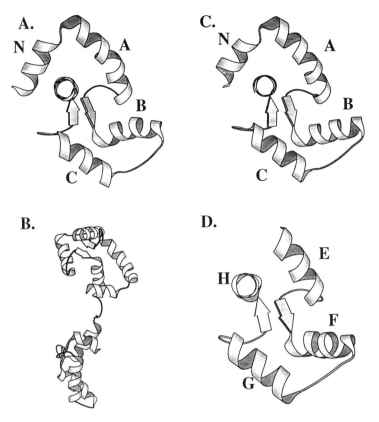

Figure 4.14 Ribbon diagrams of the structures of (A) calcium-saturated NTnC (residues 1–90 of cloned chicken skeletal troponin C) (Gagné et al., 1995), (B) calcium-saturated cloned chicken skeletal troponin C—intact (Slupsky and Sykes, 1995), (C) the N-terminal domain of B and (D) the C-terminal domain of B. Calcium is bound between helices A and B, and helices C and D of NTnC, and between helices A and B, helices C and D, helices E and F, and helices G and H. The ribbon diagrams were prepared using MOLSCRIPT (Kraulis, 1991).

may represent the whole TnC molecule in terms of conformational changes, and that communication between the N and C domains is not required for full movement of the helices to expose the hydrophobic pocket which may then interact with TnI. The structures of the calcium-saturated N-terminal domain (NTnC and TnC) are also very similar, with the same helix lengths and a straight B-helix [E41 has helical (ϕ, ψ) angles]. These structures are very different from the apo N-terminal domain (RMSD = 5.45 Å) and are different from the calcium-saturated C-terminal domain in that the structures are more open. The N-terminal domain hydrophobic pocket is so exposed that TnC dimerizes N-domain to N-domain at saturating calcium and pH 7.0 conditions (Slupsky et al., 1995a). Table 4.1 illustrates the inter-helical angles and the distance between the midpoints of the helices for calcium-saturated and apo structures. In the table, the larger the angle, the more closed the structure; whereas the smaller the angle, the more open the structure and the further apart the midpoints of the helices. In general, the calcium-saturated structures are more open than the apo structures.

Calmodulin

Calmodulin regulates a number of fundamental cellular activities, such as cyclic nucleotide and glycogen metabolism, intracellular motility (microtubules and microfilaments), calcium transport, and calcium-dependent protein kinases (Means and Dedman, 1980). It also mediates the activation of a number of different intracellular enzymes, including phosphodiesterase, MLCK, calcineurin, erythrocyte calcium ATPase, brain adenylate cyclase, phosphorylase kinase, and nicotinamide dinucleotide kinase (Strynadka and James, 1989). Calmodulin exists as a monomer of approximate molecular weight 17,000 and is very similar in tertiary structure

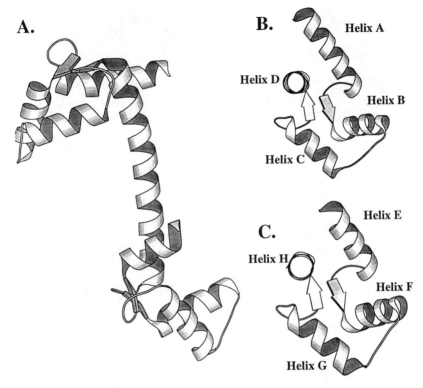

A.

B.

Helix A

Helix D

Helix B

Helix C

C.

Helix E

Helix H

Helix F

Helix G

Figure 4.15 Ribbon representation of the crystal structure of calmodulin refined to 1.7-Å resolution (Chattopadhyaya et al., 1992). (A) Intact calmodulin, (B) the N-terminal domain and (C) the C-terminal domain. Calcium binds between helices A and B, helices C and D, helices E and F, and helices G and H. The ribbon diagrams were prepared using MOLSCRIPT (Kraulis, 1991).

to skeletal TnC. The major differences between calmodulin and TnC are the presence of an N-terminal helix in TnC of between 7 and 8 residues, three extra amino acid residues in the linker between the two domains (between 78 and 79 in calmodulin), and an extra residue at the C-terminus in TnC. There is approximately 45% direct homology between the two proteins and approximately 78% homology for conservative replacement upon alignment of the calcium-binding loops.

The crystal structure was initially solved by Babu et al. (1985), and was subsequently refined to 2.2-Å resolution (Babu et al., 1988), and 1.7-Å resolution (Chattopadhyaya et al., 1992) in the presence of saturating calcium (both domains in the calcium-bound form), and reveals an elongated protein that looks like a dumbbell (Fig. 4.15). The two domains are connected by a long α-helical linker, with each globular domain consisting of two EF-hands in a helix–loop–helix conformation and a short stretch of antiparallel β-sheet between each pair of calcium-binding loops. The helices are designated with

the letters A through H; where four calcium ions bind in each of the loops formed between the helices A/B, C/D, E/F, and G/H. Both domains of calmodulin are comparable in structure (RMSD = 0.751 Å) (Strynadka and James, 1989).

Recently, there have been several structural studies on the apo form of calmodulin (Finn et al., 1993, 1995; Kuboniwa et al., 1995; Zhang et al., 1995). Apo-calmodulin consists of two small globular domains separated by a flexible linker (Kuboniwa et al., 1995; Zhang et al., 1995) (Fig. 4.16) and is analogous to calcium-saturated calmodulin in terms of secondary structure (Finn et al., 1993, 1995; Kuboniwa et al., 1995; Zhang et al., 1995). The structure reveals eight helices lettered A through F and two antiparallel β-sheets. At residue 31, helix B has a kink [$(\phi, \psi) = (-75°, -10°)$] which is similar to residue 41 found in the structurally homologous protein troponin C (Kuboniwa et al., 1995; Zhang et al., 1995). The NMR amide hydrogen exchange experiments have indicated that the helices in calcium-free calmodulin are less stable than those in calcium-satu-

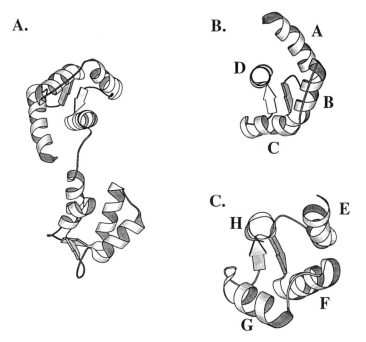

Figure 4.16 Ribbon representation of the solution NMR structure of apo-calmodulin (M. Ikura, personal communication). (A) Intact apo-calmodulin, (B) the N-terminal domain and (C) the C-terminal domain. The ribbon diagrams were prepared using MOLSCRIPT (Kraulis, 1991).

rated calmodulin (Kuboniwa et al., 1995; Zhang et al., 1995), with the C-terminal domain being severely affected by conformational averaging (Kuboniwa et al., 1995). The net effect of removing calcium is to cause the interhelical angles of four EF-hand motifs to increase by 36° to 44°, which creates major changes in surface properties, including a closure of the deep hydrophobic cavity that is essential for target protein recognition. The interhelical angles of apo-calmodulin are very similar to those obtained for the apo N-terminal domain and are shown in Table 4.1. It may therefore be deduced that TnC and calmodulin share similar structures and that, since calcium-binding causes these proteins to interact with other proteins, they may be classified as regulatory proteins. Minor differences between TnC and calmodulin such as the N-helix in TnC (daSilva et al., 1993; Gulati et al., 1993; Chandra et al., 1994; Ding et al., 1994; Smith et al., 1994), differences in the helical linker (Gulati et al., 1990, 1993; Dobrowolski et al., 1991; Babu et al., 1993), and the high-affinity of calcium-binding sites in the C-terminal domain of TnC, ensure that neither can totally substitute for the other.

Before the apo structure was solved, the structures of calmodulin in complex with peptides and drugs were solved—shedding light as to how these regula-

tory proteins bind to their targets (for a review, see Crivici and Ikura, 1995). The first complex structure to be solved was that between calmodulin and the MLCK peptide (Ikura et al., 1992; Meador et al., 1992). The structure of this protein is shown in Fig. 4.17. It was determined that the two domains (residues 6–73 and 83–146) remained essentially unchanged in structure upon complexation, with the central helix connecting the two domains (85–93) being disrupted into two helices connected by a flexible loop (74–82) (Ikura et al., 1992). Thus, the hydrophobic pocket forms upon calcium-binding and then seeks an amphiphilic helix to bind. The interaction of calmodulin and the MLCK peptide encompasses residues 3–21 of the MLCK peptide with the stabilization of the complex by hydrophobic interactions, including a large number of methionines (Ikura et al., 1992). The structure of calmodulin was also determined in complex with a peptide that represents the calmodulin-binding domain of calmodulin-dependent protein kinase, and was shown to be similar to the calmodulin MLCK peptide complex structure (Meador et al., 1993) (Fig 4.17). These studies have shown that the central helix is highly flexible and bends to allow the hydrophobic regions at each end of the molecule to interact with enzymes. Calmodulin interaction may be inhibited by a num-

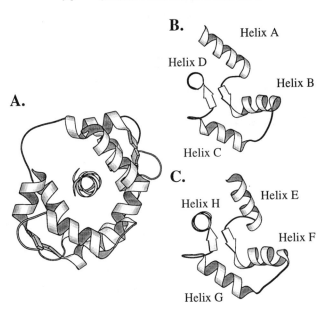

Figure 4.17 Ribbon representation of the structure of calmodulin in complex with the myosin light chain kinase peptide (Ikura et al., 1992; Meador et al., 1992). (A) Calmodulin in complex with the myosin light chain kinase peptide. (B) the N-terminal domain of calmodulin and (C) the C-terminal domain of calmodulin. Calcium is bound to all four calcium-binding sites in this structure. The structure is similar to the one obtained with calmodulin in complex with the calmodulin-binding domain of calmodulin-dependent protein kinase (Meador et al., 1993). This diagram was prepared using MOLSCRIPT (Kraulis, 1991).

ber of pharmacological agents. One of the strongest inhibitors is trifluoperazine (TFP), and the structure of calmodulin in a 1:1 complex with TFP has been solved and is shown in Fig 4.18 (Cook et al., 1994). This structure is very similar to the structures where calmodulin is in complex with peptides (Ikura et al., 1992; Meador et al., 1992, 1993). Residues 75–80 were not observed in the crystal structure and therefore were assumed to be flexible (Cook et al., 1994).

Recoverin

Recoverin is a member of the EF-hand superfamily, and differs from calmodulin and troponin C in having additional residues at the amino and carboxy termini, plus an insertion between the third and fourth EF-hand domains. The recoverin family of proteins all appear to exhibit a calcium–myristoyl switch for membrane targeting that is not found in calmodulin or troponin C (Ames et al., 1994). Recoverin is a 23-kDa calcium-binding protein in retinal rod and cone cells, the calcium-bound form of which prolongs the photoresponse—most likely by blocking the phosphorylation of photoexcited rhodopsin (Ames et al., 1994). In effect, it modulates the calcium-sensitive deactivation of rhodop-

sin. The larger size of this protein (in comparison with TnC and calmodulin) may be accounted for by additional residues at the amino and carboxy termini, plus an insertion between EF3 and EF4. This retinal protein has a covalently attached myristoyl or related N-acyl group at its amino terminus (Dizhoor et al., 1992) and calcium-binding causes recoverin to partition into lipid bilayers (Zozulya and Stryer, 1992; Dizhoor et al., 1993), suggesting that the protein has a calcium–myristoyl switching mechanism. The crystal structure of unmyristoylated recoverin with one calcium ion bound to EF3 has been solved to 1.9-Å resolution (Flaherty et al., 1993) (Fig 4.19), and reveals four EF-hand domains in a linear array with three disordered regions: residues 2–8, 199–202, and 74–78. Calcium is bound to only one EF-hand (calcium-binding site III) in the crystal form. Recoverin is shown to be a compact protein made of two domains separated by a narrow cleft, with each domain containing a pair of EF-hands. The protein does not contain an extended α-helix that separates the two domains, but rather a U-shaped linker. The protein is also different from sarcoplasmic calcium-binding protein (Cook et al., 1993) where the two pairs of EF-hands are on opposite faces of a

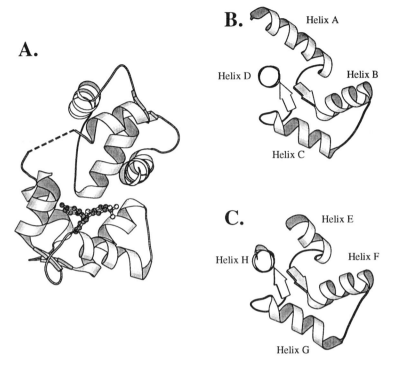

Figure 4.18 Ribbon representation of the structure of calmodulin in complex with one molecule of trifluoperazine (Cook et al., 1994). (A) Calmodulin in complex with trifluoperazine, (B) the N-terminal domain and (C) the C-terminal domain. Calcium is bound to all four calcium-binding sites. The diagram was prepared using MOLSCRIPT (Kraulis, 1991).

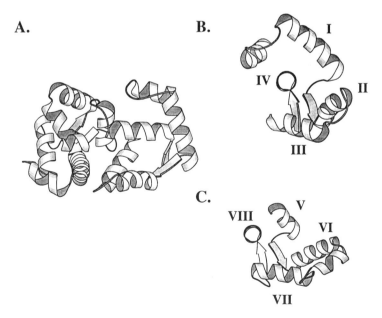

Figure 4.19 Crystal structure of recoverin with one calcium ion bound (Flaherty et al., 1993). (A) Intact recoverin, (B) the N-terminal domain of recoverin and (C) the C-terminal domain of recoverin. One calcium ion is bound to calcium-binding site III (between helices V and VI). The diagram was prepared using MOLSCRIPT (Kraulis, 1991).

compact molecule. Recently, the study of myristoylated recoverin in the apo form, using NMR, has been undertaken (Ames et al., 1994). The secondary structure reveals 11 helical segments and two pairs of antiparallel β-sheets. It was determined that the N-terminal helix is longer and more flexible in this form of recoverin in contrast to that obtained with the crystal structure of the unmyristoylated calcium-bound form. The three-dimensional structure of this form of recoverin was also solved (Tanaka et al., 1995), and is shown in Fig. 4.20. The structure contains eleven α-helices with two pairs of short antiparallel β-sheets. Residues 92–97 form a sharply bent linker between the two domains to allow the four EF-hands to form a compact linear array that contrasts with the dumbbell arrangement seen in calmodulin and TnC. The myristoyl group adopts an extended conformation that projects into a deep hydrophobic pocket in the N-terminal domain. Helices B, C, E, and F surround the myristoyl group and provide a frame for the binding pocket, and helix A serves as a lid on top of the other four helices. Residues G2 and N3 form a tight hairpin turn to position the myristoyl group inward and antiparallel to the N-terminal helix (Tanaka et al., 1995).

Comparison of myristoylated and nonmyristolated recoverin reveals a fairly similar C-terminal domain (backbone RMSD = 2.2 ± 0.1 Å), whereas

the N-terminal domains are very different (6.9 ± 0.1 Å). Comparison of the C-terminal domain interhelical angles reveals very similar values for both proteins (for the E/F and G/H helices, respectively, the interhelical angles are 107° and 127° for the nonmyristoylated form of recoverin and 126° and 96° for the myristoylated form of recoverin), whereas comparison of the N-terminal domain reveals very different interhelical angles (for the A/B and C/D pairs of helices, respectively, the interhelical angles are 167° and 141° for the myristoylated form of recoverin and 99° and 99° for the nonmyristoylated form of recoverin). The eight-residue loop (17–24) between EF1 and the N-terminal helix A allows helix A to adopt different positions with respect to helix B. Helix A is stabilized by the myristoyl group in the apo structure, whereas the first seven residues of helix A are unstructured in the crystal. Furthermore, the relative orientation of the N-terminus and C-terminus is different between the two forms of recoverin (EF2 is rotated by 45° with respect to EF3) (Tanaka et al., 1995). Further evidence has shown the myristoyl group of recoverin to be sequestered into the hydrophobic core when the protein is in the apo form and extruded upon calcium-binding (Ames et al., 1995), thus confirming recoverin as having a calcium–myristoyl switch.

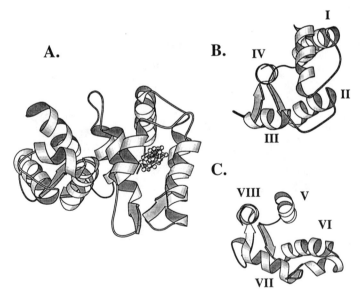

Figure 4.20 Solution NMR structure of apo, myristoylated recoverin (M. Ikura, personal communication). (A) Intact recoverin, (B) the N-terminal domain of recoverin and (C) the C-terminal domain of recoverin. The myristoyl group is attached to the N-terminal helix, and projects into a deep hydrophobic pocket in the N-terminal domain formed by helices B, C, E, F, and A (helices II, III, V, VI, and I, respectively). The diagram was prepared using MOLSCRIPT (Kraulis, 1991).

Calcineurin B

Calcineurin is a type 2B protein phosphatase whose major substrates in brain appear to be the RII regulatory subunit of cAMP-dependent protein kinase and protein phosphatase 1 inhibitors, DARPP-32 and G protein (for a review see Persechini et al., 1989). Calcineurin has a broad specificity, causing phosphate groups to be removed from both protein and nonprotein substrates (see Heizmann and Hunziker, 1991; chapter 14, for reviews). It has been identified as the intracellular target of the immunosuppressant–immunophilin complexes FKBP–FK506 and cyclophilin–cyclosporin A (Liu et al., 1991). Calcineurin is a heterodimer consisting of subunit A (60,000 molecular weight) and subunit B (19,000 molecular weight). Subunit A has the catalytic and calmodulin-binding sites (Heizmann and Hunziker, 1991). Subunit B has approximately 35% sequence homology with calmodulin and contains four high-affinity calcium-binding sites (Heizmann and Hunziker, 1991). The calcium-binding loops share 54% sequence identity with the calcium-binding loops of calmodulin; however, outside these four loops, the degree of sequence homology is only 20% (Anglister et al., 1994). Recently, the secondary structure and NOE (Nuclear Overhauser Enhancement) exchange rates were determined for calcineurin B (subunit B of calcineurin) using NMR spectroscopy (Anglister et al., 1993, 1994; Grzesiek and Bax, 1993). It was determined, in the presence of CHAPS to prevent protein aggregation, that calcineurin B has eight helices distributed in four EF-hand calcium-binding domains. The secondary structure appeared to be highly homologous to calmodulin (Anglister et al., 1994), except that helices B and C are shorter and helix G is longer in calcineurin than in calmodulin. It was also determined that the central helix is flexible (residues 84–88, which are analogous to residues 77–81 of calmodulin) (Grzesiek and Bax, 1993; Anglister et al., 1994).

The x-ray structure of calcineurin B complexed with a fragment of calcineurin A and the immunosuppressive FKBP-12–FK506 complex has recently been determined (Griffith et al., 1995). The structure shows the expected four EF-hand, two-domain structure in close agreement with the NMR structure of uncomplexed calcineurin B. The manner in which calcineurin B interacts with the binding helix portion (BBH) of calcineurin A is, however, very strikingly different from the manner in which calmodulin binds to the MLCK α-helical peptide. The BBH is a five-turn amphipathic α-helical peptide with one surface being completely nonpolar. This surface binds to a complementary hydrophobic glove in calcineurin B, which thereby covers only one side of the peptide and leaves the other side available to contact the FKBP-12–FK506 complex. In this structure, the N-terminal glycine of calcineurin B is covalently linked to a myristate group which is shown to be located at the end of the calcineurin complex and near the end of the BBH.

This structure reveals the complexities of larger complexes that involve regulatory calcium-binding proteins, and is undoubtedly an indicator of exciting new features to be seen in future structures.

Classification

The Calcium-Induced Conformational Change

The first structures to be solved were those of calcium-saturated ($2Ca^{2+}$) parvalbumin (Moews and Kretsinger, 1975), calcium-saturated ($2Ca^{2+}$) calbindin D_{9K} (Szebenyi et al., 1981), half-saturated ($2Ca^{2+}$) troponin C (Herzberg and James, 1985a,b; Sundaralingam et al., 1985), and calcium-saturated ($4Ca^{2+}$) calmodulin (Babu et al., 1985). At this time, there was a significant delay in the field of calcium-binding proteins (in terms of solving structures). The major problem with x-ray crystallography was the inability to obtain suitable crystals in the absence and presence of calcium. During this period, a lot of attention was paid to the calcium-binding sites themselves and models of calcium-saturated troponin C (Herzberg et al., 1986) and apo-calmodulin (Strynadka and James, 1988) were formulated. Recently, with the advent of multidimensional heteronuclear NMR experiments, the structures of calcium-saturated TnC (Gagné et al., 1995; Slupsky and Sykes, 1995), apo-calmodulin (Finn et al., 1995; Kuboniwa et al., 1995; Zhang et al., 1995), and apo-calbindin (Skelton et al., 1994) were solved. Comparison of these structures with either their calcium-saturated or apo counterparts suggested that there are different calcium-induced changes depending on the function of the proteins. Calmodulin and troponin C appear to be regulatory proteins and induce large changes in interhelical angles as calcium is bound, thereby causing some residues to move as much as 19 Å.

Table 4.1 is a compilation of interhelical angles with the distances between midpoints of helices for apo and calcium-saturated structures. For the apo structures, the interhelical angles indicate very closed structures of TnC, calmodulin, and recoverin, and a slightly more open structure for calbindin. The distance between the midpoints of the helices ranges from 10 Å (for calmodulin) to 14 Å for NTnC and 18 Å for recoverin. For the calcium-saturated structures, the interhelical angles are somewhat smaller-ranging from 59° for the C-terminal domain of calmodulin to 135° for calbindin. The distance between the midpoints of the helices is somewhat larger than for the apo structures and ranges from 11 Å for calbindin to 19 Å for TnC. The larger calcium-induced opening causes TnC to dimerize via its N-terminal domain (Slupsky et al., 1995a). The smallest change (upon calcium-binding) occurs for calbindin, and the largest change occurs for the N-terminal domain of TnC. In general, it appears that all regulatory domains have open structures (interhelical angles < 100°) in the calcium-saturated state, whereas the more structural proteins (proteins which require calcium for their structure) or buffer proteins have interhelical angles > 100° in their calcium-saturated state. In the apo form, the distinction is not as clear-cut.

Summary

In this review chapter, the structures of several calcium-binding proteins have been studied. Several of the proteins have been solved only in the metal-bound state, and some which do not bind metal ions have been solved in the apo state.

It is clear from the data presented that regulatory proteins, such as calmodulin and the N-terminal domain of TnC, incur large changes in tertiary structure as calcium is bound. Buffer-type proteins, such as calbindin, do not incur large changes in structure as calcium is bound. As all of these proteins contain the EF-hand pair of calcium-binding sites, it is clear that they have all most likely evolved from a common ancestor. Slight alterations in the hydrophobic interactions between the sites—and, indeed, changes to calcium-liganding residues—have led to differences in calcium affinity and possibly to differences in structural rearrangement. Future work in this area will prove interesting in terms of whether it is possible to synthesize genetically engineered calcium-binding proteins to elicit certain functions.

Note Added in Proof

Since this article was written, a large number of new structures of calcium-binding proteins have been solved. For example, the structure of calcineurin subunit-B (Griffith et al., 1995) reveals an open structure similar to calmodulin or troponin C. The structures that have been determined for the apo-S100B dimer (Drohat et al., 1996; Kilby et al., 1996) reveal a structure similar to calcyclin (Potts et al., 1995).

At this stage, it is appropriate to ask what new insights have been obtained or concepts developed based upon these new structures. Most of these have focused on the calcium-induced conformational change, and subsequent binding to target peptides

and/or proteins. First, it is apparent that a variety of structures are possible in terms of the "closed/open" states of the paired EF-hand domains. Structures have been observed in the closed, semiopen, and open states. Perhaps one of the more interesting features that has arisen is that not all calcium-bound structures exhibit the open conformation (as defined by an almost perpendicular orientation of the two helices involved in the EF-hand calcium-binding loop). Two recent structures, one of cardiac troponin C (Sia et al., 1997) and the other of calpain (Lin et al., 1997), illustrate this. In cardiac troponin C, the regulatory N-terminal domain has lost the ability to bind a calcium ion in site I. As a result, the regulatory domain of cardiac troponin-C remains closed upon calcium-binding, with only a slight opening of site II (Sia et al., 1997). In calpain, a cytosolic cysteine protease, the third EF-hand binds calcium in the "closed" conformation. The fact that these proteins bind calcium and remain "closed" does not preclude these proteins from binding to their targets. As indicated earlier, the essential and regulatory chains of myosin bind to the heavy chain in the "closed" and "semi-open" conformations. Ikura and coworkers (Swindells and Ikura, 1996) have recently addressed the question of the preformation of the "semiopen" conformation of apo-calmodulin and its implication for the binding of IQ motifs.

The second major feature of calcium-binding proteins which is now just starting to be understood is the mechanism that links calcium-binding to the induced conformational change in some regulatory domains. By mutating the glutamic acid in position 12 of the first calcium-binding loop of the regulatory domain of skeletal troponin C to an alanine, it was shown that calcium-binding does not lead to an "open" structure but results in only slight rearrangements of helices, similar to what is found for cardiac troponin C (Gagné et al., 1997). The authors show that calcium binds to troponin C in a stepwise manner, first to site II of the N-terminal domain—this does not significantly alter the structure but sets the stage for a larger structural change to take place upon calcium-binding to site I. Similarly, the mutant E140Q (also position 12, this time in loop IV) in the C-terminal domain of calmodulin leaves the protein in a 65/35 equilibrium between the "closed" and "open" states (Evenäs et al., 1997). It is interesting to note that the domains of the sarcoplasmic calcium-binding protein discussed above, which are more open than is typical, represent the only proteins that have an aspartic acid residue in place of the glutamic acid in position 12. In terms of the developing mechanism, this would imply that the domain must open more to allow the carboxyl group of the aspartic acid residue to chelate to the bound calcium ion. Thus, it appears as if the residue in position 12 places a crucial role in the balance of forces which govern the opening structural change of these domains.

Enormous strides have been taken in the past few years toward the understanding of the nature of calcium-binding proteins. Hopefully, the next few years will bring us closer to understanding even more about calcium-binding proteins, and how drugs can act to sensitize or inhibit these proteins in binding calcium.

Acknowledgment The authors are indebted to Dr. Mitsu Ikura for providing the apo-calmodulin and myristoylated recoverin structures and related papers in advance of publication, and for providing the program to calculate interhelical angles, and to Dr. Water Chazin for providing the calcyclin coordinates in advance of publication.

References

Ahmed, F. R., Przybylska, M., Rose, D. R., Birnbaum, G. I., Pippy, M. E., and MacManus, J. P. (1990) Structure of oncomodulin refined at 1.85 Å resolution. *J. Mol. Biol.* 216: 127–140.

Ames, J. B., Tanaka, T., Ikura, M., and Stryer, L. (1995) NMR evidence for Ca^{2+}-induced extrusion of the myristoyl group of recoverin. *J. Biol. Chem.*, 270: 30909–30913.

Ames, J. B., Tanaka, T., Stryer, L., and Ikura, M. (1994) Secondary structure of myristoylated recoverin determined by three-dimensional heteronuclear NMR: implications for the calcium–myristoyl switch" *Biochemistry* 33: 10743–10753.

Anglister, J., Grzesiek, S., Ren, J., Klee, C. B., and Bax, A. (1993) *J. Biomol. NMR* 3: 121–126, 1993.

Anglister, J., Grzesiek, S., Wang, A. C., Ren, H., Klee, C. B., and Bax, A. (1994) ^1H, ^{13}C, ^{15}N nuclear magnetic resonance backbone assignments and secondary structure of human calcineurin B. *Biochemistry* 33: 3540–3547.

Babu, A., Rao, V. G., Su, H., and Gulati, J. (1993) Critical minimum length of the central helix in troponin C for the Ca^{2+} switch in muscular contraction. *J. Biol. Chem.* 268: 19232–19258.

Babu, Y. S., Bugg, C. E., and Cook W. J. (1988) Structure of calmodulin refined at 2.2 Å resolution. *J. Mol. Biol.* 204: 191–204.

Babu, Y. S., Sack, J. S., Greenhough, T. J., Bugg, C. E., Means, A. R., and Cook, W. J. (1985) Three-dimensional structure of Calmodulin *Nature* 315: 37–40.

Boguta, G. and Bierzynski, A. (1988) Conformational properties of Ca^{2+}-binding segments of proteins from the troponin C superfamily. *Biophys. Chem.* 31: 133–137.

Chandra, M., daSilva, E. F., Sorenson, M. M., Ferro, J. A., Pearlstone, J. R., Nash, B. E., Borgford, T., Kay, C. M., and Smillie, L. B. (1994) The effects of N helix deletion and mutant F29W on the Ca^{2+} binding and functional properties of chicken skeletal muscle troponin C. *J. Biol. Chem.* 269: 14988–14994.

Chattopadhyaya, R., Meador, W. E., Means, A. R., and Quiocho, F. A. (1992) Calmodulin structure refined at 1.7 Å resolution. *J. Mol. Biol.* 228: 1177–1192.

Collins, J. H., Potter, J. D., Horn, M. J., Wilshire, G., and Jackman, N. (1973) The amino acid sequence of rabbit skeletal muscle troponin C. Gene replication and homology with calcium-binding proteins from carp and hake muscle. *FEBS Lett.* 36: 268–272.

Cook, W. J., Jeffrey, L. C., Cox, J. A., and Vijay-Kumar, S. (1993) Structure of a sarcoplasmic calcium-binding protein from Amphioxus refined at 2.4 Å resolution. *J. Mol. Biol.* 229: 461–471.

Cook, W. J., Walter, L. J., and Walter, M. R. (1994) Drug binding by calmodulin: crystal structure of a calmodulin-trifluoperazine complex. *Biochemistry* 33: 15259–15265.

Crivici, A. and Ikura, M. (1995) Molecular and structural basis of target recognition by Calmodulin. *Annu Rev. Biophys. Biomol. Struct.* 24: 85–116.

Dalgarno, D. C., Levine, B. A., Williams, R. J. P., Fullmer, C. S., and Wasserman, R. H. (1983) Proton-NMR studies of the solution conformations of vitamin-D induced bovine intestinal calcium-binding protein. *Eur. J. Biochem.* 137: 523–529.

daSilva, A. C. R. and Reinach, F. C. (1991) Calcium binding induces conformational changes in muscle regulatory proteins. *Trends Biochem. Sci.* 16: 53–57.

daSilva, E., Sorenson, M. M., Smillie, L. B., Barrabin, H., and Scofano, H. M. (1993) Comparison of calmodulin and troponin C with and without its amino-terminal helix (residues 1 to 11) in the activation of erythrocyte Ca^{2+}-ATPase. *J. Biol. Chem.* 268: 26220–26225.

Declercq, J. P., Tinant, B., Parello, J., Etienne, G., and Huber, R. (1988) Crystal structure determination and refinement of pike 4.10 parvalbumin (minor component from *Esox lucis*). *J. Mol. Biol.* 202: 349–353.

Declercq, J. P, Tinant, B., Parello, J., and Rambaud, J. (1991) Ionic interactions with parvalbumins. Crystal structure determination of pike 4.10 parvalbumin in four different ionic environments. *J. Mol. Biol.* 220: 1017–1039.

Ding, X., Akella, A. B., Su, H., and Gulati, J. (1994) The role of glycine (residue 89) in the central helix of EF-hand protein troponin-C exposed following amino-terminal α-helix deletion. *Protein Sci.* 3: 2089–2096.

Dizhoor, A. M., Chen, C. K., Olshevskaya, E., Sinelnikova, V. V., Phillipov, P., and Hurley, J. B. (1993) Role of acylated amino terminus of recoverin in Ca^{2+}-dependent membrane interaction. *Science* 259: 829–832.

Dizhoor, A. M., Ericsson, L. H., Johnson, R. S., Kumar, S., Olshevskaya, E., Zozulya, S., Neubert, T. A., Stryer, L., Hurley, J. B., and Walsh, K. A. (1992)

The NH_2 terminus of retinal recoverin is acylated by a small family of fatty acids. *J. Biol. Chem.* 267: 16033–16036.

Dobrowolski, Z., Xu, G., Chen, W., and Hitchcock-Degregori, S. E. (1991) Analysis of the regulatory and structural defects of troponin C central helix mutants. *Biochemistry* 30: 7089–7096.

Drakenberg, T., Forsén, S., Thulin, E., and Vogel, H. J. (1987) The binding of Ca^{2+}, Mg^{2+} and Cd^{2+} to tryptic fragments of skeletal muscle troponin C. Cadmium-113 and proton NMR studies. *J. Biol. Chem.* 262: 672–678.

Drohat, A. C., Amburgey, J. C., Abildgaard, F., Starich, M. R. Baldisseri, D., and Weber, D. J. (1996) Solution structure of rat Apo-S100B ($\beta\beta$) as determined by NMR spectroscopy. *Biochemistry* 35: 11577–11588.

Ebashi, S., Kodama, A., and Ebashi, F. (1968) Troponin. I. Preparation and physiological function. *J. Biochem.* 64: 465–477.

Evenäs, J., Thulin, E., Malmendal, A., Forsen, S., and Carlstrom, G. (1997) NMR studies of the E140Q mutant of the carboxyl-terminal domain of calmodulin reveal global conformational exchange in the Ca^{2+}-saturated state. *Biochemistry* 36: 3448–3457.

Falke, J. J., Drake, S. K., Hazard, A. L., and Peersen, O. B. (1994) Molecular tuning of ion binding to calcium signalling proteins. *Quart. Rev. Biophys.* 27: 219–290.

Findlay, W. A. and Sykes, B. D. (1993) ^{1}H-NMR resonance assignments, secondary structure, and global fold of the TR_1C fragment of turkey skeletal troponin C in the calcium-free state. *Biochemistry* 32: 3461–3467.

Findlay, W. A., Sönnichsen, F. D., and Sykes, B. D. (1994) Solution structure of the TR_1C fragment of skeletal muscle troponin C. *J. Biol. Chem.* 269: 6773–6778.

Finn, B. E., Drakenberg, T., and Forsén, S. (1993) The structure of apo-calmodulin. A ^{1}H NMR examination of the carboxy-terminal domain. *FEBS Lett.* 336: 368–374.

Finn, B. E., Evenäs, J., Drakenberg, T., Waltho, J. P., Thulin, E., and Forsén, S. (1995) Calcium-induced structural changes and domain autonomy in calmodulin. *Nat. Struct. Biol.* 2: 777–783.

Flaherty, K. M., Zozulya, S., Stryer, L., and McKay, D. B. (1993) Three-dimensional structure of recoverin, a calcium sensor in vision. *Cell* 75: 709–716.

Fuchs, F., Liou, Y. M., and Grabarek, Z. (1989) The reactivity of sulfhydryl groups of bovine cardiac troponin C. *J. Biol. Chem.* 264: 20344–20349.

Fujimori, K., Sorenson, M., Herzberg, O., Moult, J., and Reinach, F. C. (1990) Probing the calcium-induced conformational transition of troponin C with site-directed mutants. *Nature* 345: 182–184.

Gagné, S. M., Li, M. X., and Sykes, B. D. (1997) Mechanism of direct coupling between binding and induced structural change in regulatory calcium binding proteins. *Biochemistry* 36: 4386–4392.

Gagné, S. M., Tsuda, S., Li, M. X., Chandra, M., Smillie, L. B., and Sykes, B. D. (1994) Quantification of the calcium-induced secondary structural changes in the regulatory domain of troponin C. *Protein Sci.* 3: 1961–1974.

Gagné, S. M., Tsuda, S., Li, M. X., Smillie, L. B., and Sykes, B. D. (1995) Structures of the troponin C regulatory domains in the apo and calcium-saturated states. *Nat. Struct. Biol.* 2: 784–789.

Gariépy, J., Kay, C. E., Kuntz, I. D., Sykes, B. D., and Hodges, R. S. (1985) Nuclear magnetic resonance determination of the metal–proton distances in a synthetic calcium-binding site of rabbit skeletal troponin C. *Biochemistry* 24: 544–550.

Gariépy, J., Sykes, B. D., Reid, R. E., and Hodges, R. S. (1982) Proton nuclear magnetic resonance investigation of synthetic calcium-binding peptides. *Biochemistry* 21: 1506–1512.

Golden, L. F., Corson, D. C., Sykes, B. D., Banville, D., and MacManus, J. P. (1989) Site-specific mutants of oncomodulin. *J. Biol. Chem.* 264: 20314–20319.

Grabarek, Z., Tan, R. Y., Wang, J., Tao, T., and Gergely, J. (1990) Inhibition of mutant troponin C activity by an intra-domain disulphide bond. *Nature* 345: 132–135.

Griffith, J. P., Kim, J. L., Kim, E. E., Sintchak, M. D., Thomson, J. A., Fitzgibbon, M. J., Fleming, M. A., Caron, P. R., Hsiao, K., and Navia, M. A. (1995) X-ray structure of calcineurin inhibited by the immunophilin-immunosuppressant FKBP–12–FK506 complex. *Cell* 82: 507–522.

Gross, M. and Kumar, R. (1990) Physiology and biochemistry of vitamin D-dependent calcium-binding proteins. *Am. J. Physiol.* 259: F195–F209.

Grzesiek, S. and Bax, A. (1993) Measurement of amide proton exchange rates and NOEs with water in $^{13}C/^{15}N$-enriched calcineurin B. *J. Biomol. NMR* 3: 627–638.

Gulati, J., Babu, A., Su, H., and Zhang, Y. (1993) Identification of the regions conferring calmodulin-like properties to troponin C. *J. Biol. Chem.* 268: 11685–11690.

Gulati, J., Persechini, A., and Babu, A. (1990) Central helix role in the contraction–relaxation switching mechanisms of permeabilized skeletal and smooth muscles with genetic manipulation of calmodulin." *FEBS Lett.* 263: 340–344.

Gusev, N. B., Grabarek, Z., and Gergely, J. (1991) Stabilization by a disulfide bond of the N-terminal domain of a mutant troponin C (TnC48/82). *J. Biol. Chem.* 266: 16622–16626.

Hartshorne, D. J., and Driezen, D. (1972) Studies on the subunit composition of troponin. *Cold Spring Harbor Symp. Quant. Biol.* 37: 225–234.

Heizmann, C. W. and Hunziker, W. (1990) Intracellular calcium-binding molecules. In *Intracellular Calcium Regulation* (Bronne, F., ed.). Alan R. Liss, New York, pp. 211–248.

Heizmann, C. W. and Hunziker, W. (1991) Intracellular calcium-binding proteins: more sites than insights. *Trends Biochem. Sci.* 16: 98–103.

Herzberg, O. and James, M. N. G. (1985a) Structure of the calcium regulatory muscle protein troponin C at 2.8 Å resolution. *Nature* 313: 653–659.

Herzberg, O. and James, M. N. G. (1985b) Common structural framework of the two Ca^{2+}/Mg^{2+} binding loops of troponin C and other Ca^{2+} binding proteins. *Biochemistry* 24: 5298–5302.

Herzberg, O. and James, M. N. G. (1988) Refined crystal structure of troponin C from turkey skeletal muscle at 2.0 Å resolution. *J. Mol. Biol.* 203: 761–779.

Herzberg, O., Moult, J., and James, M. N. G. (1986) A model for the Ca^{2+}-induced conformational transition of troponin C. *J. Biol. Chem.* 261: 2638–2644.

Ikura, M., Clore, G. M., Gronenborn, A. M., Zhu, G., Klee, C. B., and Bax, A. (1992) Solution structure of a calmodulin-target peptide complex by multidimensional NMR. *Science* 256: 632–638.

Ingraham, R. H. and Hodges, R. S. (1988) Effects of Ca^{2+} and subunit interactions on surface accessibility of cysteine residues in cardiac troponin." *Biochemistry* 27: 5891–5898.

Johnson, J. D., Charlton, S. C., and Potter, J. D. (1979) A fluorescence stopped flow analysis of Ca^{2+} exchange with troponin C. *J. Biol. Chem.* 254: 3497–3502.

Kanellis, P., Yang, J., Cheung, H. C., and Lenkiski, R. E. (1983) Synthetic peptide analogs of skeletal troponin C: fluorescence studies of analogs of the low-affinity calcium-binding site II. *Arch. Biochem. Biophys.* 220: 530–540.

Kawasaki, H., and Kretsinger, R. H. (1994) Calcium-binding proteins. I: EF-hands. *Protein Profile* 1: 343–391.

Kay, L. E., Forman-Kay, J. D., McCubbin, W. D., and Kay, C. M. (1991) Solution structure of a polypeptide dimer comprising the fourth Ca^{2+}-binding site of troponin C by nuclear magnetic resonance spectroscopy. *Biochemistry* 30: 4323–4333.

Kilby, P. M., Van Edkik, L. J., Roberts, G. C. K. (1996) The solution structure of the bovine S100B protein dimer in the calcium-free state. *Structure* 4: 1041–1052.

Kördel, J., Skelton, N. J., Akke, M., and Chazin, W. J. (1993) High-resolution solution structure of calcium-loaded calbindin D_{9K}. *J. Mol. Biol.* 231: 711–734.

Kraulis, P. J. (1991) MOLSCRIPT: a program to produce both detailed and schematic plots of protein structures. *J. Appl. Crystallogr.* 24: 946–960.

Kretsinger, R. H. (1980) Structure and evolution of calcium-modulated proteins. *CRC Crit. Rev. Biochem.* 8: 119–174.

Kretsinger, R. H. and Barry, C. D. (1975) The predicted structure of the calcium-binding component of troponin. *Biochim. Biophys. Acta.* 405: 40–52.

Kretsinger, R. H. and Nockolds, C. E. (1973) Carp muscle calcium-binding protein. II. Structure determination and general description. *J. Biol. Chem.* 248: 3313–3326.

Kuboniwa, H., Tjandra, N., Grzesiek, S., Ren, H., Klee, C. B., and Bax, A. (1995) Solution structure of calcium-free calmodulin. *Nat. Struct. Biol.* 2: 768–776.

Kumar, V. D., Lee, L., and Edwards, B. F. P. (1990) Refined crystal structure of calcium-liganded carp parvalbumin 4.25 at 1.5 Å resolution. *Biochemistry* 29: 1404–1412.

Levine, B. A., and Dalgarno, D. C. (1983) The dynamics and function of calcium-binding proteins. *Biochim. Biophys. Acta* 726: 187–204.

Levine, B. A., Mercola, D., Coffman, D., and Thornton, J. M. (1977) Calcium binding by troponin C. A proton magnetic resonance study. *J. Mol. Biol.* 115: 743–760.

Levine, B. A., Thornton, J. M., Fernandes, R., Kelly, C. M., and Mercola, D. (1978) Comparison of the calcium- and magnesium-induced structural changes of troponin C. A proton magnetic resonance study. *Biochim. Biophys. Acta* 535: 11–24

Lin, G., Chattopadhyay, D., Masatoshi, M., Wang, K. K. W., Carson, M., Jin, L., Yuen, P., Takano, E., Hatanaka, M., DeLucas, L. J., and Narayana, S. V. L. (1997) Crystal structure of calcium bound domain VI of calpain at 1.9 Å resolution and its role in enzyme assembly, regulation and inhibitor binding. *Nat. Struct. Biol.* 4: 539–547.

Lin, X., Krudy, G. A., Howarth, J., Brito, R. M. M., Rosevear, P. R., and Putkey, J. A. (1994) Assignment and calcium dependence of methionyl ϵC and ϵH resonances in cardiac troponin C. *Biochemistry* 33: 14434–14442.

Linse, S. and Forsén, S. (1995) *Advances in Secondary Messenger Phosphoprotein Research*, Vol. 30 (Means, A. R., ed.). Raven Press, New York, pp. 89–151.

Liu, J., Farmer, J. D. Jr., Lane, W. S., Friedman, J., Weissman, I., and Schreiber, S. (1991) Calcineurin is a common target of cyclophilin–cyclosporin A and FKBP–FK506 complexes. *Cell* 66: 807–815.

MacManus, J. P. (1981) The stimulation of cyclic nucleotide phosphodiesterase by a M_r 11500 calcium-binding protein from hepatoma. *FEBS Lett.* 126: 245–249.

Marsden, B. J., Shaw, G. S., and Sykes, B. D. (1990) Calcium binding proteins. Elucidating the contributions to calcium affinity from an analysis of species variants and peptide fragments. *Biochem. Cell Biol.* 68: 587–601.

McPhalen, C. A., Sielecki, A. R., Santarsiero, B. D., and James, M. N. G. (1994) Refined crystal structure of rat parvalbumin, a mammalian α-lineage parvalbumin, at 2.0 Å resolution. *J. Mol. Biol.* 235: 718–732.

McPhalen, C. A., Strynadka, N. C. J., and James, M. N. G. (1991) Calcium-binding sites in proteins: a structural perspective. *Adv. Protein Chem.* 42: 77–146.

Meador, W. E., Means, A. R., and Quiocho, F. A. (1992) Target enzyme recognition by calmodulin: 2.4 Å structure of a calmodulin-peptide complex. *Science* 257: 1251–1255.

Meador, W. E., Means, A. R., and Quiocho, F. A. (1993) Modulation of calmodulin plasticity in molecular recognition on the basis of x-ray structures. *Science* 262: 1718–1721.

Means, A. R. and Dedman, J. R. (1980) Calmodulin—an intracellular calcium receptor. *Nature* 285: 73–77.

Moews, P. C. and Kretsinger, R. H. (1975) Refinement of the structure of carp muscle calcium-binding parvalbumin by model building and difference Fourier analysis. *J. Mol. Biol.* 91: 201–228.

Monera, O. D., Shaw, G. S., Zhu, B. Y., Sykes, B. D., Kay, C. M., and Hodges, R. S. (1992) Role of interchain α-helical hydrophobic interactions of Ca^{2+} affinity, formation, and stability of a two-site domain in troponin C. *Protein Sci.* 1: 945–955.

Padilla, A., Cave, A., and Parello, J. (1988) Two-dimensional ^1H nuclear magnetic resonance study of pike pI 5.0 parvalbumin. Sequential resonance assignments and folding of polypeptide chain. *J. Mol. Biol.* 204: 995–1017.

Palmer, E. J., MacManus, J. P., and Mutus, B. (1990) Inhibition of glutathione reductase by oncomodulin. *Arch. Biochem. Biophys.* 277: 149–154.

Pearlstone, J. R., Borgford, T., Chandra, M., Oikawa, K., Kay, C. M., Herzberg, O., Moult, J., Herklotz, A., Reinach, F. C., and Smillie, L. B. (1992) Construction and characterization of a spectral probe mutant of troponin C: application to analyses of mutants with increased Ca^{2+} affinity. *Biochemistry* 31: 6545–6553.

Persechini, A., Moncrief, N. D., and Kretsinger, R. H. (1989) The EF-hand family of calcium-modulated proteins. *Trends Neurosci.* 12: 462–467.

Potter, J. D. and Gergely, J. (1975) The calcium and magnesium binding sites on troponin and their role in the regulation of myofibrillar adenosine triphosphate. *J. Biol. Chem.* 250: 4628–4633.

Potter, J. D., Seidel, J. C., Leavis, P., Lehrer, S. S., and Gergely, J. (1976) Effect of Ca^{2+} binding on troponin C. *J. Biol. Chem.* 251: 7551–7556.

Potts, B. C. M., Smith, J., Akke, M., Macke, T. J., Okazaki, K., Hidaka, H., Case, D. A., and Chazin, W. J. (1995) The structure of calcyclin reveals a novel homodimeric fold for S100 Ca^{2+}-binding proteins. *Nat. Struct. Biol.* 2: 790–796.

Putkey, J. A., Dotson, D. G., and Mouawad, P. (1993) Formation of inter- and intramolecular disulfide bonds can activate cardiac troponin C. *J. Biol. Chem.* 268: 6827–6830.

Rayment, I., Rypniewski, W. R., Schmidt-Bäse, K., Smith, R., Tomchick, D. R., Benning, M. M., Windelmann, D. A., Wesenberg, G., and Holden, H. M. (1993) Three-dimensional structure of myosin subfragment-1: a molecular motor. Science 261: 50–58.

Reid, R. E., and Hodges, R. S. (1980) Co-operativity and calcium/magnesium binding to troponin C and muscle calcium-binding parvalbumin: an hypothesis. *J. Theor. Biol.* 84: 401–444.

Reid, R. E., Gariépy, J., Saund, A. K., and Hodges, R. S. (1981) Calcium-induced protein folding. *J. Biol. Chem.* 256: 2742–2751.

Roquet, F., Declercq, J. P., Tinant, B., Rambaud, J., and Parello, J. (1992) Crystal structure of the unique parvalbumin component from muscle of the leopard shark (*Triakis semifasciata*). The first x-ray study of an α-parvalbumin. *J. Mol. Biol.* 223: 705–720.

Satyshur, K. A., Pyzalska, D., Greaser, M., Rao, S. T., and Sundaralingam, M. (1994) Structure of chicken skeletal muscle troponin C at 1.78 Å resolution. *Acta Crystallogr.* D50: 40–49.

Satyshur, K. A., Rao, S. T., Pyzalska, D., Drendel, W., Greaser, M., and Sundaralingam, M. (1988) Refined structure of chicken skeletal muscle troponin C in the two-calcium state at 2 Å resolution. *J. Biol. Chem.* 263: 1628–1647.

Seamon, K. B., Hartshorne, D. J., and Bothner-By, A. A. (1977) Ca^{2+} and Mg^{2+} dependent conformations of troponin C as determined by ^1H and ^{19}F nuclear magnetic resonance. *Biochemistry* 16: 4039–4046.

Shaw, G. S., Findlay, W. A., Semchuk, P. D., Hodges, R. S., and Sykes, B. D. (1992a) Specific formation of a heterodimeric two-site calcium-binding domain from synthetic peptides. *J. Am. Chem. Soc.* 114: 6258–6259.

Shaw, G. S., Golden, L. F., Hodges, R. S., and Sykes, B. D. (1991a) Interactions between paired calcium-binding sites in proteins: NMR determination of the stoichiometry of calcium-binding to a synthetic troponin C peptide. *J. Am. Chem. Soc.* 113: 5557–5563.

Shaw, G. S., Hodges, R. S., Kay, C. M., and Sykes, B. D. (1994) Relative stabilities of synthetic peptide homo- and heterodimeric troponin-C domains. *Protein Sci.* 3: 1010–1019.

Shaw, G. S., Hodges, R. S., and Sykes, B. D. (1990) Calcium-induced peptide association to form an intact protein domain: ^1H NMR structural evidence. *Science* 249: 280–283.

Shaw, G. S., Hodges, R. S., and Sykes, B. D. (1991b) Probing the relationship between α-helix formation and calcium affinity in troponin C: ^1H NMR studies of calcium-binding to synthetic and variant site III helix–loop–helix peptides. *Biochemistry* 30: 8339–8347.

Shaw, G. S., Hodges, R. S., and Sykes, B. D. (1991c) Synthetic calcium-binding peptides which form heterodimeric two-site domains. In *Peptides: Chemistry and Biology* (Smith, J. A. and Rivier, J. E., eds). ESCOM Science Publishers, Leiden, The Netherlands, pp. 209–212.

Shaw, G. S., Hodges, R. S., and Sykes, B. D. (1992b) Calcium-induced folding of troponin C: formation of homodimeric and heterodimeric two-site domains from synthetic peptides. *Techniques in Protein Chemistry III* (Hogue Angeletti, R. ed.). Academic Press, New York, pp 347–353.

Shaw, G. S., Hodges, R. S., and Sykes, B. D. (1992c) Determination of the solution structure of a synthetic two-site calcium-binding homodimeric protein domain by NMR spectroscopy. *Biochemistry* 31: 9572–9580.

Shaw, G. S., Hodges, R. S., and Sykes, B. D. (1992d) Stoichiometry of calcium-binding to a synthetic heterodimeric troponin-C domain. *Biopolymers* 32: 391–397.

Sia, S. K., Li, M. X., Spyracopoulos, L., Gagné, S. M., Liu, W., Putkey, J. A., and Sykes, B. D. (1997) Structure of cardiac muscle troponin C unexpectedly reveals a closed regulatory domain. *Biochemistry* 272: 18216–18221.

Skelton, M. J., Kördel, J., Akke, M., Forsén, S., and Chazin, W. J. (1994) Signal transduction versus buffering activity in Ca^{2+}-binding proteins. *Nat. Struct. Biol.* 1: 239–245.

Slupsky, C. M. and Sykes, B. D. (1995) NMR solution structure of calcium-saturated skeletal muscle troponin C. *Biochemistry*, 34: 15953–15964.

Slupsky, C. M., Kay, C. M., Reinach, F. C., Smillie, L. B., and Sykes, B. D. (1995a) Calcium-induced dimerization of troponin C: mode of interaction and use of trifluoroethanol as a denaturant of quaternary structure. *Biochemistry* 34: 7365–7375.

Slupsky, C. M., Reinach, F. C., Smillie, L. B., and Sykes, B. D. (1995b) Solution secondary structure of calcium saturated troponin C monomer determined by multidimensional heteronuclear NMR spectroscopy. *Protein Sci.* 4: 1279–1290.

Smith, L., Greenfield, N. J., and Hitchcock-Degregori, S. E. (1994) The effects of deletion of the amino-terminal helix on troponin C function and stability. *J. Biol. Chem.* 269: 9857–9863.

Snyder, E. E., Buoscio, B. W., and Falke, J. J. (1990) Calcium (II) site specificity: effect of size and charge on metal ion binding to an EF-hand-like site. *Biochemistry* 29: 3937–3943.

Strynadka, N. C. J. and James, M. N. G. (1988) Two trifluoperazine binding sites on calmodulin predicted from comparative molecular modeling with troponin C. *Proteins: Struct. Funct. Genet.* 3: 1–17.

Strynadka, N. C. J. and James, M. N. G. (1989) Crystal structures of the helix–loop–helix calcium-binding proteins. *Annu. Rev. Biochem.* 58: 951–958.

Sundaralingam, M., Bergstrom, R., Strasburg, G., Rao, S. T., Roychowdhury, P., Greaser, M., and Wang, B. C. (1985) Molecular structure of troponin C from chicken skeletal muscle at 3 Å resolution. *Science* 227: 945–948.

Swain, A. L., Kretsinger, R. H., and Amma, E. L. (1989) Restrained least squares refinement of native (calcium) and cadmium-substituted carp parvalbumin using x-ray crystallographic data at 1.6 Å resolution. *J. Biol. Chem.* 264: 16620–16628.

Swindells, M. B., and Ikura, M. (1996) Pre-formation of the semi-open conformation by the apo-calmodulin C-terminal domain and implications for binding IQ-motifs. *Nat. Struct. Biol.* 3: 501–504.

Szebenyi, D. M. E. and Moffat, K. (1986) The refined structure of vitamin D-dependent calcium-binding protein from bovine intestine. Molecular detail, ion binding, and implications for the structure of other calcium-binding proteins. *J. Biol. Chem.* 261: 8761–8777.

Szebenyi, D. M. E., Obendorf, S. K., and Moffat, K. (1981) Structure of vitamin D-dependent calcium-binding protein from bovine intestine. *Nature* 294: 327–332.

Tanaka, T., Ames, J. B., Harvey, T. S., Stryer, L., and Ikura, M. (1995) Sequestration of the membrane-targeting myristoyl group of recoverin in the calcium-free state. *Nature*, 376: 444–447.

Tollemar, U., Cunningham, K., and Shriver, J. W. (1986) Lack of communication between LC2 light chain and the SH_1 region of myosin S-1 studied by ^{19}F NMR. *Biochim. Biophys. Acta* 873: 243–251.

Tsuda, S., Ogura, K., Hasegawa, Y., Yagi, K., and Hikichi, K. (1990) ^1H NMR study of rabbit skeletal muscle troponin C: Mg^{2+}-induced conformational change. *Biochemistry* 29: 4951–4958.

Vijay-Kumar, S. and Cook, W. J. (1992) Structure of a sarcoplasmic calcium-binding protein from *Nereis diversicolor* refined at 2.0 Å resolution. *J. Mol. Biol.* 224: 413–426.

Xie, X., Harrison, D. H., Schlichting, I., Sweet, R. M., Kalabokis, V. N., Szent-Györgi, A. G., and Cohen, C. (1994) Structure of the regulatory domain of scallop myosin at 2.8 Å resolution. *Nature* 368: 306–312.

Zhang, M., Tanaka, T., and Ikura, M. (1995) Solution structure of apo calmodulin: conformational transition induced by calcium-binding. *Nat. Struct. Biol.* 2: 758–767.

Zozulya, S. and Stryer, L. (1992) Calcium–myristoyl protein switch. *Proc. Natl. Acad. Sci. USA* 89: 11569–11573.

5

Calcium-Binding Proteins: Annexins

Susanne Liemann
Robert Huber
Jörg Benz

Although it has been known for a long time that calcium plays a central role in many physiological processes, such as nervous stimulation and muscle contraction, the discovery of calcium as a second messenger in stimulus response coupling led to an intense search for calcium-binding proteins in order to elucidate its complex signal transduction pathways. Distinct from the well-known "EF hand" proteins, a novel family of structurally related proteins has been characterized, termed the annexins, which share the property of calcium-dependent binding to negatively charged phospholipids. About 13 members have been identified so far on the basis of sequence homology, related biochemical properties, and phospholipid binding assays. With the exception of yeast and prokaryotes, annexins are found ubiquitously in a wide range of tissues and cell types in lower and higher eukaryotes, including mammals (except for erythrocytes) (for a review, see Moss 1992; Raynal and Pollard, 1994), birds (Erikson and Erikson, 1980; Horseman, 1989; Radke et al., 1980), fish (Ivanenkov et al., 1994; Walker 1982), *Xenopus* (Gerke et al., 1991), *Drosophila* (Johnston et al., 1990), *Dictyostelium* (Doring et al., 1991; Gerke, 1991), and plants (Andrawis et al., 1993; Boustead et al., 1989; Smallwood et al., 1990). Annexins clearly differ from the "EF hand" family of calcium-binding proteins in their sequence and geometry of the calcium-binding sites, as well as in their affinity for calcium. Although the biochemical properties of the annexins have been extensively investigated and their molecular structures, in crystalline and membrane-bound forms, have been elucidated in detail, their in vivo functions still remain unknown.

Annexins: A Novel Family of Calcium-Binding Proteins

Nomenclature and Characteristics of the Annexins

During the last two decades various annexins have been identified independently, stemming from various research fields, which led to a rather confusing nomenclature within the family. The different members were called lipocortins, endonexins, chromobindins, calelectrins, placental anticoagulant proteins (PAP), vascular anticoagulants (VAC), calcimedins, calphobindins, etc., before a new nomenclature was adopted by Crumpton and Dedman (1990) based on the suggestion by Geisow (1986) and the lipocortin numbering system of Pepinsky et al. (1988). The new generic term "annexins" summarizes their main feature to bind and therefore to "annex" to cellular membranes in a calcium-dependent manner.

Annexins are characterized structurally by their typical four-repeat structure and biochemically by their ability to bind to preferably negatively charged phospholipids in a calcium-dependent manner (for a review, see Burgoyne and Geisow, 1989; Crompton et al., 1988; Geisow and Walker, 1986; Klee, 1988; Liemann and Lewit-Bentley, 1995; Moss et al., 1991; Raynal and Pollard, 1994; Smith et al., 1990; Swairjo and Seaton, 1994). Except for annexin VI, which is composed of eight repeats (probably due to gene duplication), all annexins are composed of four homologous repeats of about 70 amino acids (Fig. 5.1) building up the conserved "core" of the protein (Barton et al., 1991). The repeats share a sequence identity of 25–35% within an individual protein (Haigler et al., 1989) and of 45–50% among

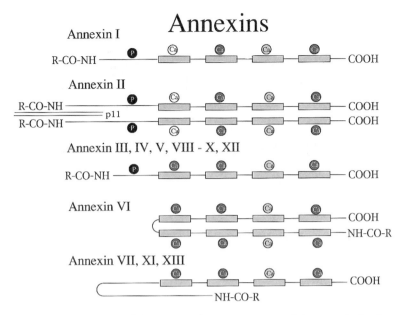

Figure 5.1 Schematic representation of the conserved repeat-structure of the annexin family. Each grey rectangle represents one homologous repeat of approximately 70 amino acid residues. The calcium ions bound to the consensus sequence G-X-G-T-{38}-D/E are shown in grey. Calcium bound to the slightly different sequence in domain III is depicted in light grey and that bound to a sequence only found in the first repeat of annexin I and II in white. The black circles represent N-terminal phosphorylation sites on tyrosine, serine, or threonine residues.

the different family members (Smith and Moss, 1994). They all harbour a characteristic calcium- and phospholipid-binding site within a 17 amino acid consensus sequence called the "endonexin fold" (Geisow et al., 1986; Kretsinger and Creutz, 1986). In contrast, their N-termini are diverse in sequence and length, ranging from 12 to 196 residues, and may confer specific functions upon each annexin type. This regulatory region is subjected to various post-translational modifications: phosphorylation by tyrosine and serine/threonine kinases in the case of annexins I and II (Fava and Cohen, 1984; Gerke and Weber, 1984; Glenney and Tack, 1985; Greenberg and Edelmann, 1983; Johnsson et al., 1988; Schlaepfer and Haigler, 1988), binding to proteins of the S100-type such as annexin II to p11 (Gerke and Weber, 1985) and annexin XI to calcylin (Tokumitsu et al., 1992), glycosylation of annexins I and II (Goulet et al., 1992), N-myristoylation of annexins II and XIII (Fiedler et al., 1995; Soric and Gordon, 1985; Wice and Gordon, 1992), intracellular cross-linking of annexin I by tissue transglutaminase (Ando et al., 1991), proteolysis (Chuah and Pallen, 1989), and acylation (Pepinsky et al., 1988).

In comparison to the "EF hand" proteins, annexins display only weak calcium affinity with K_d values in the micro- to millimolar range, except for annexin VI with $K_d \sim 1\,\mu$M, whereas the binding to phospholipids is of high affinity with a K_d value around 10^{-10} M at calcium concentrations of $\sim$10–100 μM. The ability to mediate membrane binding is highly specific for calcium and shows a decreased order for Cd > Zn > Co > Ba > Mg in the case of annexin V, whereas Zn ions have a synergistic effect on calcium-dependent binding (Andree et al., 1990). In general, annexins can be released reversibly from the membrane in a soluble form by the removal of calcium, indicating a peripheral binding of the protein to the lipids. In addition, nevertheless, membrane-bound forms resistant to EDTA extraction have been described in lung, liver, heart, and brain that are soluble only after detergent treatment (Bianchi et al., 1992; Boustead et al., 1993; Pula et al., 1990; Trotter et al., 1995b; Valentine-Braun et al., 1987), but the structural difference between these two forms remains unclear.

Overview of the Proposed In Vivo Functions

On the basis of many various, intensively investigated biochemical in vitro functions of the annexins, a wide variety of in vivo functions have been proposed, but the diversity of these functions led to confusion rather than to an understanding of the annexins' biological role. Annexins are nearly ubiquitous in nature. They are usually cytoplasmic proteins found either in the cytosol or associated with membranes or the cytoskeleton. They have been identified in a wide range of different cell types and tissues, such as chromaffin cells, fibroblasts, muscle cells, glial cells, hepatocytes, keratinocytes, epithelial cells, B-lymphocytes, placenta, lung, aorta, liver, intestine, kidney, brain, synaptic vesicles, and others (Moss, 1992; Raynal and Pollard, 1994). Some annexins have also been found in the extracellular compartment (Christmas et al., 1991; Croxtall and Flower, 1992; Flower and Rothwell, 1994; Jacquot et al., 1990; Pepinsky et al., 1986; Russo-Marie, 1992; Solito et al., 1991; Thorin et al., 1995; Violette et al., 1990), including blood plasma (Flaherty et al., 1990; Römisch et al., 1992), but since a hydrophobic signal peptide for secretion is lacking (Muesch et al., 1990), their mechanism of movement to and through the plasma membrane is not yet clear (Haigler and Schlaepfer, 1992). Annexins I (Jindal et al., 1991; Raynal et al., 1992), V (Koster et al., 1993; Sun et al., 1992), and XI (Mizutani et al., 1992) have been immunolocalized in the nucleus, although, again, missing a signal sequence. Annexin II is reported to stimulate activity of DNA polymerase α, suggesting a role in cell differentiation (Jindal et al., 1991). Annexin II also acts as a DNA-binding protein, interacting with ubiquitous dispersed repeats of the Alu family (Boyko et al., 1994). For annexin XI, the long N-terminal tail is described as being responsible for nuclear localization (Mizutani et al., 1995).

In general, annexins bind reversibly to phospholipid membranes mediated by calcium ions, but the discovery of EDTA-resistant forms suggested a link to the lipids via an annexin-binding protein. Recently, a 85 kDa complex containing annexin V has been observed using bifunctional cross-linking agents on platelet membranes (Trotter et al., 1995a), suggesting that complex formation or receptor binding is important for annexin V relocation to platelet membranes following physiological stimulation or exposure of intracellular annexin V to calcium. Other direct protein–protein interactions have been described for annexin II which binds to p11, a small 11 kDa protein from the "EF-hand" family, in a calcium-independent manner (Gerke and Weber, 1984; Weber, 1992). The so-formed heterotetramer (annexin II–p11)$_2$ exists both in vitro and in vivo (Gerke and Weber, 1985) and is probably the predominant species in most cells (Gerke, 1992). In the presence of calcium, annexin II, VI (Zeng et al., 1993), and XI (Tokumitsu et al., 1992) can associate with calcyclin, a growth-regulated S100 protein assumed to play a role in cell proliferation. For annexin, I a calcium-dependent interaction with

S100C, purified from porcine heart, has been observed (Naka et al., 1994).

The first annexin to be identified was annexin VII (synexin), from the bovine adrenal medulla, by Creutz et al. (1978). As most other annexins in vitro, it binds to phospholipid membranes and promotes aggregation and fusion of vesicles in a calcium-dependent manner, first shown with chromaffin secretory granules (Creutz et al., 1982; Zaks and Creutz, 1991). Also, for annexin I and II, an involvement in the secretion of adrenal chromaffin cells has been reported (Ali et al., 1989). These results strongly suggest an involvement of the annexins in membrane traffic processes such as endo- and exocytosis (Creutz, 1992; Nakata et al., 1990). Phosphorylation of annexins I and II alter their membrane-binding and aggregation properties significantly. Phosphorylated annexin I requires less calcium for binding (Ando et al., 1989; Schlaepfer and Haigler, 1987) and reveals an enhanced degradation by proteolysis (Chuah and Pallen, 1989), but needs an increased calcium concentration for vesicle aggregation (Johnstone et al., 1993; Wang and Creutz, 1992). The opposite has been observed for annexin II, revealing a lower affinity for phospholipids after phosphorylation (Powell and Glenney, 1987). Its membrane interactions are further determined by p11-binding and the heterotetramer (annexin II–p11)$_2$ displays the lowest calcium concentration needed for vesicle aggregation (1 μM) (Drust and Creutz, 1988; Weber, 1992). Annexin II directly participates in calcium-evoked exocytosis of permeabilized chromaffin cells, but requires phosphorylation by protein kinase C (Sarafian et al., 1991). For phosphorylated annexin XI, a complete loss of membrane binding has been observed (Mizutani et al., 1993).

Annexins II and VI appear to be involved in endocytosis (Burgoyne and Clague, 1994). Annexin II is transferred between early endosomes upon fusion in vitro, and immunogold labelling studies confirmed that annexin II is present on early, fusogenic endosomes in vivo (Emans et al., 1993). Annexin VI plays a role in budding of clathrin-coated pits (Lin et al., 1992). Budding activation of annexin VI requires ATP, calcium, and cytosol, indicating that full activity probably depends on additional cytosolic components. Annexin VI appears to be not only an active compound in the detachment of coated pits from the membrane, but also a site for regulating the formation of coated vesicles (Lin et al., 1992). In contrast, it has been recently demonstrated that endocytosis occurs independently of annexin VI in human A3431 cells (Smythe et al., 1994). Transfection with annexin VI had no influence on the endocytotic pathway and annexin VI failed to exert any influence on internalization and recycling of the transferrin receptor.

Annexins display anti-inflammatory properties. Early studies showed a decrease of the production of eicosanoids, a class of inflammatory mediators and derivatives of arachidonic acid, after treatment with glucocorticoids, hence suggesting a glucocorticoid induced expression of an anti-inflammatory drug. This observation led to the term "lipocortin" (Blackwell et al., 1980; Flower and Blackwell, 1979; Hirata et al., 1980; Russo-Marie, 1992; Wallner et al., 1986). In further studies, the direct correlation between glucocorticoids and annexin expression was a matter of controversy (Nakano et al., 1990; Piltch et al., 1989). The inhibition of phospholipase A_2, and therefore blocking the release of the inflammatory mediator arachidonic acid, could be shown in vitro for all annexins tested so far (Russo-Marie, 1992). Since enzyme inhibition can be abolished by increasing the relative phospholipid concentration (Davidson et al., 1987, 1990), annexins probably do not interact with phospholipase A_2 directly (as suggested by Kim et al., 1994), but compete with it for membrane binding and thereby reduce the substrate availability by a process called "phospholipid substrate depletion" (Ahn et al., 1988; Russo-Marie, 1992). This mechanism has been confirmed by the additional in vitro inhibition of phospho-inositide-specific phospholipase C (Machoczek et al., 1989) and phospholipase A_1/lysophospholipase from hepatic lipase (Bohn et al., 1992).

The anticoagulant properties as a general feature of all annexins can be explained by membrane sequestering in an analogous manner (Reutlingsperger et al., 1985). Annexins bind to membranes with high affinity and therefore might displace the phosphatidyl serine- and calcium-dependent coagulation factors needed for thrombin activation and final fibrin clot formation (Andree et al., 1992; Funakoshi et al., 1987; Romisch et al., 1990).

Annexins are expressed in a growth-dependent manner and are targets of various cellular kinases and, therefore, might play a role in cell proliferation, cell differentiation, and signal transduction (Haigler et al., 1987). Annexin I and II are substrates of the epidermal growth factor receptor (EGFR) kinase (Fava and Cohen, 1984), and the rous sarcoma virus-encoded pp60[v src] tyrosine kinase (Erikson and Erikson, 1980; Glenney and Tack, 1985), the insulin receptor tyrosine kinase (Karasik et al., 1988) and protein kinase C (PKC) (Gould et al., 1986; Schlaepfer and Haigler, 1988). Annexin V, which is not a substrate of PKC, has been shown to inhibit phosphorylation of annexin I by PKC (Schlaepfer et al., 1992), probably due to membrane sequestering (Raynal et al., 1993), although a direct interaction between PKC and annexin V has been postulated (Schlaepfer et al., 1992).

Some annexins bind to cytoskeletal proteins (Gerke and Weber, 1985; Glenney et al., 1987; Khanna et al., 1990; Mangeat, 1988). (Annexin II–p11)$_2$ can bundle actin filaments at near physiological calcium concentrations (Ikebuchi and Waisman, 1990) but is inhibited by a nonapeptide of helix IVB of annexin II (Jones et al., 1992). Annexins V and VI are major components of matrix vesicles which initiate mineral deposition in cartilage (Genge et al., 1989, 1991; Mollenhauer and von der Mark, 1983; Wu et al., 1991, 1993). They bind to types II and X collagen, the major collagens of hypertrophic cartilage, in a calcium-independent manner (Kirsch and Pfäffle, 1992; Wu et al., 1991) thereby mediating cell–matrix interaction and calcium loading of matrix vesicles (Rojas et al., 1992).

Molecular Structure of the Annexins

Crystal Structures

The first annexin to be structurally characterized by its crystal structure was annexin V (Huber et al., 1990a, 1990b) . So far, the three-dimensional structure of human (Huber et al., 1992; Lewit-Bentley et al., 1992; Sopkova et al., 1993), chicken (Bewley et al., 1993), and rat (Concha et al., 1993) annexin V have been solved to high resolution, as well as the structures of an N-terminally truncated form of human annexins I (Weng et al., 1993) and II (Burger et al., 1996). In summary, these structures document that all annexins have the same topology in accord with their high primary sequence homology. Supporting the results from circular dichroism (Geisow, 1986; Sopkova et al., 1994), the annexins are shown to be almost entirely α-helical. The polypeptide chain is folded into four compact domains of similar structure, consistent with the four highly conserved amino acid repeats. Each domain consists of five α-helices (labelled A–E) of 7–16 amino acids length that are wound into less than two turns of a right-handed superhelix. Four of the helices are oriented approximately (anti-)parallel, whereas the connecting helix C lies almost perpendicular to them (Fig. 5.2). The domains I and II, and III and IV, respectively, are connected by short interhelical turns, whereas the connector between the domains I and II is extended.

The four domains are arranged in an almost planar, cyclic array. The molecule has an overall slightly flat, curved shape with a convex and a concave face. All calcium ions defined in the various crystal forms of the annexins are located on the convex site of the molecule, whereas the N- and C-termini are placed on the concave site. The domains II and III, and I and IV, respectively, have tight contacts mediated by

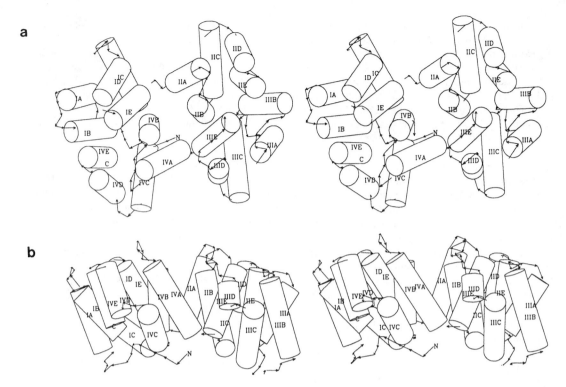

Figure 5.2 Stereoviews of the annexin structure with the helical segments (A–E) by cylinders. The four repeats are indicated by Roman numbers. (a) View from the flat face from the membrane. The central pore appears widely open. (b) Side view. The convex face on top is the calcium- and membrane-binding side.

hydrophobic residues, thus generating two tight modules with approximately 2-fold symmetry. These two modules interact less tightly, mostly by polar or charged contact residues and are related by another 2-fold local dyad. This axis marks the centre of the molecule and a very prominent hydrophilic pore which has been identified as the calcium-selective ion pathway (Berendes et al., 1993b; Burger et al., 1994; Huber et al., 1992). The framework of the funnel-shaped channel is formed by a four-helix bundle of the helices IIA, IIB, IVA, and IVB and coated with highly conserved charged residues. A passing ion would require at least side-chain rearrangement and, in this respect, particular impediments are the two invariant saltbridges within the pore. Buried water molecules in the structure are located predominantly in the channel, evident by their well-defined electron density.

The structure of the short N-terminus of annexin V containing 15 amino acid residues is extended (Huber et al., 1992) and, so far, no crystal structure including a naturally long tail has been reported. For annexin VII (synexin) that possesses an N-terminal sequence of 167 amino acids with a characteristic GYP motif repeated eight times, a β-turn and β-sheet structure has been predicted (Burns et al., 1989).

Calcium-Binding Sites

The annexins have calcium-binding sites within a 17-amino acid consensus sequence in each of the homologous repeats called the "endonexin fold" (Geisow et al., 1986; Kretsinger and Creutz, 1986) that are located on the convex site of the molecule shown to be the membrane-binding site. After crystallization under high calcium concentrations (2–20 mM), the calcium is bound to a loop between the helices A and B with the consensus sequence G-X-G-T-{38}-(D/E), in which the glycines enable a tight loop geometry. The calcium is coordinated by three main-chain carbonyl oxygen atoms of the loop, a bidentate carboxylate group from an acidic side chain 38 residues down the sequence in the helix D–E loop, and additional water molecules or a sulfate from the precipitating agent, respectively (Berendes et al., 1993b; Concha et al., 1993; Huber et al., 1990b; Sopkova et al., 1993). The coordination sphere is a pentagonal bipyramid with a main-chain carbonyl and a water molecule at its vertexes. The calcium–oxygen distances are between 2.4 and 2.7 Å. These characteristic annexin calcium-binding sites, also termed type II sites (Huber et al., 1992), are closely related to the binding site in the catalytic centre of phospholipase A_2, which shows a comparable calcium affinity but has localized the coordinating bidentate acidic side chain only 16 residues away (Dijkstra et al., 1981). Different phospholipases A_2 have been crystallized with substrate-analogous inhibitors in order to emulate the tetrahedral transition state of the phospholipid hydrolysis (Scott et al., 1990; Thunnissen et al., 1990; White et al., 1990). These structures revealed that two water molecules in the coordination sphere of the calcium can be replaced by the inhibitors phosphate or phosphonate group and an acyl carbonyl oxygen, suggesting that a sulfate or an equivalent water molecule in the annexin binding site might mimic a polar phosphoryl headgroup of a phospholipid upon membrane binding. In contrast, the "EF hand" calcium-binding sites are distinguished by their higher calcium affinity and their different structure, presenting a loop of 12–14 amino acids which contributes 6–8 coordinating oxygen atoms and is flanked by two α-helices (Tufty and Kretsinger, 1975).

In all annexin structures, except for the hexagonal crystal form of human annexin V in which the metal-binding sites are not well defined (Huber et al., 1990a), a second type of calcium-binding site, type III (Weng et al., 1993), has been found. The calcium is bound to the D–E loop by one or two main chain carbonyl oxygen atoms, a bidentate carboxylate from the E helix, and several water molecules. This type of binding site displays a lower affinity to calcium ions but a higher affinity to lanthanides and is therefore called a lanthanum-binding site (Huber et al., 1990b).

Two different conformations have been observed for annexin V for the third repeat whose calcium-binding sequence is G-E-L-K-W-G-T-{38}-E and therefore slightly deviates from the other repeats (Fig. 5.3). Crystallization under low calcium concentrations showed a structure with three canonically bound calcium ions in domains I, II, and IV (Bewley et al., 1993; Huber et al., 1990b; Lewit-Bentley et al., 1992) in which the affinity of the A–B loops decreases in the order of domain I > IV > II with respect to the exchange for lanthanum (Lewit-Bentley et al., 1992). Under high calcium concentrations, a new crystal form exists in which the normally buried tryptophan in the A–B loop of domain III becomes exposed to the molecular surface and is thereby moved a distance of about 18 Å (Berendes et al., 1993b; Concha et al., 1993; Sopkova et al., 1993). Due to this drastic conformational change, a new calcium-binding site within domain III is formed, structurally very similar to the canonical

sites in the other three repeats. These results indicate that the buried or exposed conformation within this domain depends on crystal packing in a highly sensitive calcium-dependent manner. The same conformational change has been observed in solution by fluorescence studies in the presence of calcium and phospholipids, suggesting a direct contact between the exposed tryptophan and the ester-bond region of the membrane (Meers, 1990; Meers and Mealy, 1993a).

In comparison, annexins I and II display a different calcium-binding sequence in domain I, lacking the acidic amino acid side chain for bidentate coordination (Burger et al., 1996; Weng et al., 1993). Nevertheless, it has been shown by crystallography that both annexins bind calcium there, although in a rather different way with a neighbouring glutamate as a bidentate ligand. Both annexins also bind calcium in domain III with a sequence similar to annexin V, but since the tryptophan is replaced by a lysine here, the A–B loop is exposed in both molecules to render the charged side chain solvent-accessible (Fig. 5.4).

Annexin Membrane Interactions

Phospholipid Specificity and Binding Mode

Annexins bind to phospholipid membranes with high affinity (Tait et al., 1989) and are preferrably localized adjacent to the plasma membrane of the cells (Geisow et al., 1984; Gerke and Weber, 1984; Giambanco et al., 1991), indicating that the membrane-bound state seems to be of eminent importance with regard to the annexin role in vivo. The composition and physical state of the membrane is a crucial factor for the calcium-dependent formation of the ternary annexin–calcium–phospholipid complex. The typical concentrations for half-maximal binding to phosphatidylserine are in the range of 1–200 μM calcium (Bazzi et al., 1992). These values depend on the annexin and membrane type and are higher for small unilamellar vesicles than for large vesicles or planar bilayers, indicating an influence of the membrane curvature on binding (Andree et al., 1992; Swairjo et al., 1994). Annexins preferably bind to negatively charged phospholipids, bind zwitterionic lipids at higher calcium concentrations (Andree et al., 1990; Bazzi et al., 1992; Junker and Creutz, 1994), but do not bind to phosphatidyl choline except in the presence of negatively charged amphiphiles (Meers and Mealy, 1993b) or in the lipid-bound state (Fuente and Parra, 1995). For annexin V, the calcium concentration needed for half-maximal binding decreases in the order of cardiolipin > dioleoyl-phosphatidyl glycerol > dioleoyl-phosphatidyl serine

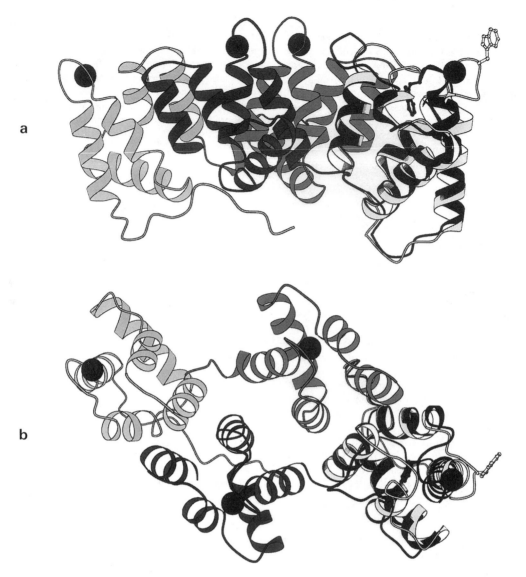

Figure 5.3 Superposition of the high-calcium (Berendes et al., 1993b) and low-calcium (Huber et al., 1992) forms of human annexin V, showing the conformational change in domain III. The structure is seen from the side in (a) and from the top in (b). Calcium ions are indicated as spheres in all four domains and each domain is shown in a different grey scale [domain I: light grey; domain II: grey; domain III: white (high-calcium form) and black (low-calcium form); domain IV: dark grey]. The side chain of Trp187 (ball-and-stick representation) is buried in the low-calcium form and exposed to the surface by the calcium-induced conformational change in the high calcium-form [Figure drawn with MOLSCRIPT (Kraulis, 1991)].

> phosphatidyl inositol > phosphatidic acid > dio-leoyl-phosphatidyl ethanolamine > sphingomyelin, whereby adsorption was independent of the overall surface charge of the membrane (Andree et al., 1992). Annexin V has a high affinity for human platelets ($k_d \sim 7$ nM) (Thiagarajan and Tait, 1990), and some annexins also bind to erythrocytes (Bandorowicz et al., 1992). Interestingly the affinity to neutral or anionic phospholipid monomers is very low ($K_d > 2 \times 10^{-5}$ M) (Tait et al., 1989).

On the basis of the three-dimensional annexin structure with the conserved calcium-binding sites located on the convex surface of the molecule, this site has been suggested to be the membrane-binding

Figure 5.4 Stereo view of the C^{α}-traces of the different annexin types. The calcium ions are depicted as balls; the N- and C-termini are numbered. The following annexin structures are superimposed: annexin V: high-calcium form (Berendes et al., 1993b): light grey (residues 4–316); low-calcium form (Huber et al., 1990b): grey (residues 3–318); annexin I (Weng et al., 1993): dark grey (residues 6–318); annexin II (Burger et al., 1996): black (residues 33–339). The cystine 297-316 of annexin I and the cystine 133–262 of annexin II are indicated as small spheres [Figure drawn with MOLSCRIPT (Kraulis, 1991)].

site where binding is mediated by calcium ions. This is supported by the structural similarity of the calcium-binding sites to the phospholipase A_2-binding site in its catalytic centre (Dijkstra et al., 1981), by the results of the fluorescence measurement of the unique tryptophan-187 in human annexin V in the presence of calcium and phospholipids (Meers, 1990; Meers and Mealy, 1993a) which are in accord with the crystallographic observation of an exposed calcium-binding loop in domain III after crystallizing of annexin V under very high calcium concentrations (Berendes et al., 1993b; Concha et al., 1993; Sopkova et al., 1993), and has recently been confirmed by electron microscopy and image processing of two-dimensional crystals of annexin V on phospholipid monolayers (Voges et al., 1994).

In general, the annexin crystal structures reveal between 2 and 6 bound calcium ions (with the exception of the hexagonal crystal form of human annexin V (Huber et al., 1990a), whereas 10–12 ions have been determined for membrane binding by affinity chromatography on immobilized phospholipids (Evans and Nelsestuen, 1994). For the eight-repeat annexin VI, 15 or 16 calcium ions have been detected by this method, whereas by equilibrium dialysis a value of 8 was obtained (Bazzi and Nelsestuen, 1991).

Lipids in the Annexin-Bound State

The number of phospholipid molecules involved in annexin V binding has been estimated by different methods. According to simple geometrical considerations, each annexin molecule should cover an area of about 26 phospholipids (Huber et al., 1992). By ellipsometry, the number of 42 lipid molecules has been determined (Andree et al., 1990), which agrees well

with the value of 59 molecules found by the annexin-binding effect on the excimer-to-monomer fluorescence ratio of pyrene-phosphatidyl choline (Meers et al., 1991) and with a value of 50 molecules found by titration of FITC–labelled annexin V with vesicles at different calcium concentrations (Tait and Gibson, 1992). The minimum average number of phospholipid molecules for complete covalent binding has been determined to be 3–5 (Meers and Mealy, 1993b) which suggests that one phospholipid-binding site is associated with each of the four domains. Binding to calcium and membranes has been shown to be highly cooperative (Meers and Mealy, 1993a).

In a detailed study, Meers and Mealy showed that the relatively shallow nature of the annexin V phospholipid-binding sites is reflected by the nearly equivalent binding to D and L lipid molecules and the low influence of the headgroup specificity (except for phosphatidyl inositol) (Meers and Mealy, 1994). In contrast, binding depends strongly on the presence of an sn-3 phosphate group, and an electronegative group at the sn-2 position followed by an sn-2 acyl chain of sufficient length. This indicates important polar interactions with the phosphate group and possibly the ester carbonyl oxygen, in parallel with hydrophobic interactions at the sn-2 acyl chain. The latter has been confirmed by quenching the intrinsic tryptophan-187 fluorescence with a nitroxide-labelled phosphatidyl choline derivative at position 5, in human annexin V revealing an interfacial location of the single tryptophan residue (Meers and Mealy, 1994). The weaker quenching effect at spin label positions 12 and 16 suggests only peripheral binding of the annexin molecule (Meers, 1990).

Annexin IV membrane binding has been shown to reduce the lateral diffusion coefficient about 35-fold by fluorescence recovery after photobleaching, thus leading to an extremely strong immobilization of the membrane. In addition, two populations of the fluorescence probes were found, as if annexin IV were able to induce two phases of different composition in these lipid bilayers (Gilmanshin et al., 1994). Different studies demonstrated that the bound phospholipid molecule can rapidly exchange between the sites on membrane-bound annexin V (Meers and Mealy, 1993b). The effect of annexin binding on fluidity of phosphatidyl serine/phosphatidyl choline membranes was investigated by electron spin resonance. The decrease of the order parameter induced by calcium was completely abolished by annexins IV and VI, with no significant effect on the fatty acid chains (Sobota et al., 1993b). The occurrence of phospholipid phase separation and clustering upon annexin binding is still a matter of controversy (Bazzi and Nelsestuen, 1991; Bazzi et al., 1992; Junker et al., 1993; Meers et al., 1991). Recently, the effects of annexin binding to phospholipids have been investigated by NMR experiments (Swairjo and Seaton, 1994). ^{1}H-NMR T_1 relaxation measurements on small unilamellar vesicles showed no influence on the hydrocarbon-chain segmental motions, while ^{31}P-NMR spectra revealed a shift of the outer- and inner-leaflet phosphoryl headgroups. These data propose that protein binding occurs only peripherally but affects the environment of the inner vesicular surface, probably due to a protein-induced change in vesicle morphology (Swairjo and Seaton, 1994).

Theoretical studies of the annexin V membrane interaction on the basis of simple dielectric models suggested that binding to phospholipids is regulated by the electrostatic potential of the protein (Karshikov et al., 1992). The calculations show that a strong local gradient of the electrostatic potential exists at the protein–membrane interface, which may result in an increased disorder and enhanced permeability of the lipid bilayer comparable to the phenomenon of electroporation (Neumann, 1988).

Structure of Membrane-Bound Annexins

Annexins are water-soluble proteins with highly charged hydrophilic surfaces. No evidence exists for a refolding in the presence of lipids, going along with the observation that the calcium-dependent binding is rapidly reversible by addition of calcium chelators. Although the occurrence of EDTA-resistant annexins has been described, their distribution and structure is not understood yet. In the case of annexin VII a "TIM barrel" model postulated for voltage-gated sodium, calcium, and potassium channels (Durell and Guy, 1992; Guy and Conti, 1990) has been suggested in which the tertiary structure of the molecule

is reorganized for possible hydrophobic membrane contacts (Guy et al., 1991). The observation of capacity currents (Rojas and Pollard, 1987) proposed a penetration of the molecule into the lipid bilayer, but, so far, no experiments could confirm this major reorganization (Plager and Nelsestuen, 1994). Instead, a disulfide bridge in annexins I and II has been detected by sequence analysis and crystallography. It should consolidate the tertiary fold within one or two domains, respectively, such that a refolding of the protein is very unlikely (Burger et al., 1996; Weng et al., 1993). Annexin V forms calcium-induced dimers in aqueous solution (Ahn et al., 1988; Neumann et al., 1994), while in the presence of phospholipids trimers, hexamers and higher aggregates have been observed (Concha et al., 1992). Self-association on membrane surfaces in the presence of calcium has been reported for different annexins. Various techniques were applied to obtain structural information about the membrane-bound protein. Low-angle neutron scattering experiments were conducted on small unilamellar vesicles in the absence and presence of annexin V (Ravanat et al., 1992). After addition of protein, the radius of gyration increased consistent with the formation of a protein monolayer shell of 3.5 nm thickness around the vesicles, with little or no insertion.

The membrane-bound structures of annexin V (Brisson et al., 1991; Mosser et al., 1991; Newman et al., 1991; Olofsson et al., 1994; Voges et al., 1994) and annexin VI (Driessen et al., 1992; Newman et al., 1989) have been investigated by electron microscopy. More recently, negatively stained two-dimensional crystals of annexin V bound to phospholipid monolayers have been analyzed (Fig. 5.5) and image processing of recorded tilt series allowed the three-dimensional structure of the annexin-membrane complex to be reconstructed (Voges et al., 1994). The trimers found in the crystal structure (Huber et al., 1990a, 1990b) were also present on the lipid and might act as nucleation centres for the two-dimensional crystal. The mass coverage per unit area corresponded well with the value obtained previously by ellipsometry (Andree et al., 1990). This electron microscopic study could unambiguously show that annexin V binds to the model membrane with its convex side harbouring the calcium sites as proposed by Huber et al. (1990b). By slightly altering the relative orientation of the two modules (composed of domains II/III and I/IV, respectively), the structure determined by x-ray crystallography fits well into the three-dimensional reconstruction of the membrane-bound form (Fig. 5.6), thus revealing that the Ca-binding sites in all four domains become coplanar with the membrane. The domain structure of the annexin molecule is completely conserved on the membrane and the stain excluding height is about 3

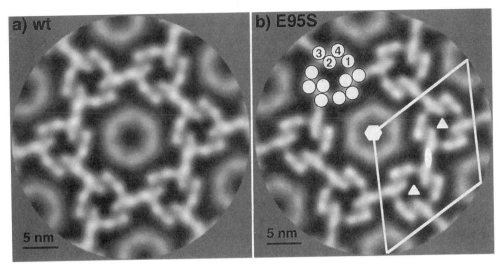

Figure 5.5 Electron microscopy images of membrane-bound (a) wild-type and (b) a Glu95→Ser mutant at 20 Å resolution. The images are views from the membrane side after image averaging. The negatively stained crystals are isomorphous, with a p6 lattice and a periodicity of 18.3 nm. The unit cell, outlined by the white diamond in (b), contains two trimers of annexin V and a central ring located on the head, which is a translationally and rotationally disordered annexin V trimer (Voges et al., 1994).

nm. This shows that the protein binds interfacially without penetrating the membrane, which is consistent with proposed models (Huber et al., 1992; Karshikov et al., 1992; Meers, 1990; Ravanat et al., 1992; Swairjo et al., 1994).

Annexins: *Janus-Faced* Proteins with Ion Channel Activity

Annexins share structural and functional properties with integral and soluble proteins and are therefore called *Janus-faced* proteins. The crystal structure, in coincidence with electron microscopy images, showed that the central axis of pseudosymmetry is normal to the membrane plane (Brisson et al., 1991). For the annexins, this central axis defines a polar pore. The same feature has also been observed for several integral membrane proteins with channel activity, as, e.g. the gap junctions (Milks et al., 1988), the acetylcholine receptor (Changeux, 1990; Unwin, 1995), bacteriorhodopsin (Henderson et al., 1990), and halorhodopsin (Havelka et al., 1995), whereby most of them are characterized by a high α-helical content.

Although the annexins seem to be different from other channel proteins since they are not supposed to penetrate the membrane upon binding, ion channel activity has so far been reported for annexins I, II, V, VI, and VII (Berendes et al., 1993b; Burger et al., 1994, 1996; Pollard and Rojas, 1988; Pollard et al., 1992; Rojas et al., 1990). When annexin VII was

bound to a bilayer of a patch pipette, a significant increase in the size and the time constant of the displacement current, a 5-fold decrease of the resistance, and a 10-fold increase of the capacitance, were detected (Rojas and Pollard, 1987). Further studies showed that annexin VII acts as a voltage-gated ion channel in phosphatidyl serine bilayer membranes highly selective for calcium (Pollard and Rojas, 1988). Annexin V displays voltage-dependent ion channel activity on acidic phospholipid bilayer membranes but has been shown to interact also with other di- and monovalent cations (Rojas et al., 1990).

Although various annexins display ion channel activity in vitro, their in vivo role is not understood in detail. Annexin V is supposed to act as an calcium channel in matrix vesicles (MV), thus indicating an involvement in bone formation. Activity might be modulated by binding of type II and X collagens to MV so that attachment of MV to the collagen network in hypertrophic cartilage could facilitate rapid calcium influx (Kirsch and Wuthier, 1994; Kirsch et al., 1994). Other studies demonstrated that annexin V is overexpressed independently from other annexins in tracheal epithelial cell lines with cystic fibrosis (CF), at the mRNA and protein levels showing a mutation in the gene for the cystic fibrosis transmembrane conductance regulator (CFTR), (Della Gaspera et al., 1995). These findings suggest that annexin V, which shows structural similarity to CFTR, might be used to compensate for the defective

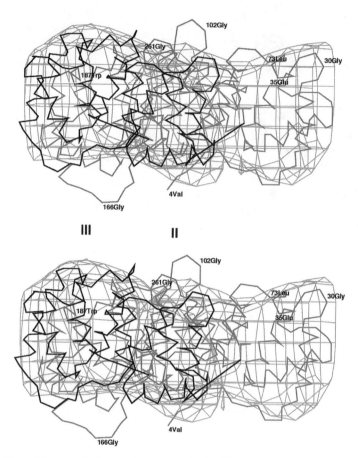

Figure 5.6 Superposition of the annexin V crystal structure obtained by x-ray crystallography (Huber et al., 1990b), with the EM structure as seen from the side. The EM density (light grey) is shown around one monomer together with the original annexin V x-ray structure (grey) and the proposed model [rotated module (II/III) shown in black, unchanged module (I/IV) in grey]. The proposed model is created by rotating the module (II/III) separately so that the molecule is oriented exactly parallel to the membrane. The lipid layer lies on top and some residues involved in calcium binding are labelled (Voges et al., 1994).

cAMP-dependent regulation of Cl⁻ channels by calcium channelling, thus activating calcium signalling by calcium-dependent Cl⁻ channel regulation in CF cells. For annexin VI, a concentration-dependent release of ^{45}Ca from chromaffin granules has been reported (Jones et al., 1994). Since annexin VI is predominantly localized to the apical plasmalemma, its participation in the secretory event by calcium channelling is suggested.

For annexin VI also, modulation of the gating properties of the ryanodine-sensitive calcium-release channel from the sarcoplasmic reticulum has been reported (Diaz-Munoz et al., 1990; Hazarika et al., 1991a, 1991b). The protein causes an increase of the open probability by 2.7-fold and of the mean open time by 82-fold, whereas annexins I to V and the four-domain N- and C-terminal annexin VI halves did not show any channel regulation (Diaz-Munoz et al., 1990; Hazarika et al., 1991a, 1991b). Annexin IV has recently been shown to inhibit the calmodulin-dependent protein kinase II (CaM KII)-activated chloride channel (Kaetzel et al., 1994). Introduction of exogenous annexin IV into colonic T84 cells specifically prevented calcium-dependent chloride current activation, which was again enhanced by further introduction of anti-annexin IV antibodies. Since annexin IV does not inhibit or interact with CaM KII in vitro, it appears that annexin IV inhibits phosphorylation-dependent anion conductance activation by preventing CaM KII–ion channel interaction, thus proposing a novel mechanism by which calcium-binding proteins, cytoplasmic kinases, and ion channels interact synergistically to regulate membrane conductance (Chan et al., 1994).

Functional and Structural Investigation of the Annexins by Mutagenesis

Picture of the Ion Channel Function Revealed by Mutagenesis

As described in the preceding subsection, annexins interact, calcium mediated, with membranes and upon this intimate contact, the annexins exhibit calcium activity in phosphatidyl serine bilayers formed at the tip of a patch pipette. Based on the crystal structure of annexin V (Huber et al., 1990a, 1992) the electrophysiological properties were elucidated by extensive mutational analysis (Berendes et al.,

1993a, 1993b; Burger et al., 1994; Demange et al., 1994), whereby important residues for the ion selectivity and for the voltage dependence of the channel could be identified.

On the basis of the three-dimensional structure, the central four-helix bundle of human annexin V was suggested as the ion pathway (Huber et al., 1992) (Fig. 5.7). In this region, 12 highly conserved charged or polar residues and two saltbridges are found which are preferred targets for site-directed mutagenesis with regard to the elucidation of the ion channel activity. Subsequently, we substituted glutamic acid-95 for a serine residue in the centre of the pore (Fig. 5.8). Compared with wild-type annexin V, the mutated protein exhibited a reduced single-channel conductance for calcium and a highly increased conductance for sodium and potassium (Berendes et al., 1993b; Burger et al., 1994). Furthermore, the mutant lost selectivity of calcium versus sodium (Fig. 5.9). This indicates that Glu95 is essential for the selectivity and at least one of the crucial components of the ion-selectivity filter within

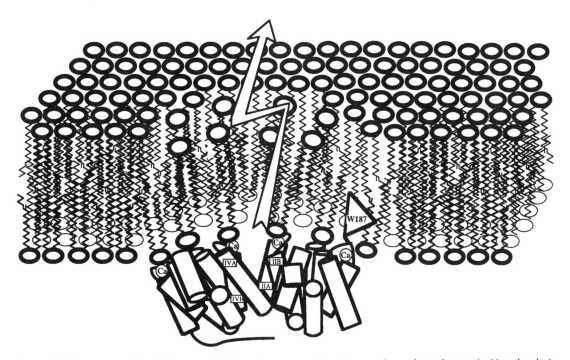

Figure 5.7 Current model of the annexin V membrane complex. The membrane-bound annexin V molecule is shown with cylinders representing α-helices, the pore-forming four-helix bundle (helix assignment as in Huber et al., 1992) in grey, and the calcium ions labelled. The membrane-penetrating Trp187 (depicted as a triangle), the electroporation phenomenon (represented by a flash of lightning), and the interactions between the calcium-binding loops and the phospholipid headgroups are proposed to lead to a local disorder of the phospholipids adjacent to the annexin V molecule (see Demange et al., 1994).

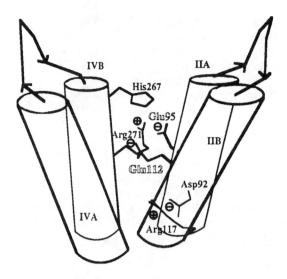

Figure 5.8 Cylinder representation of the central four-helix bundle forming the hydrophilic pore through annexin V. Some of the mutated residues are shown as derived from the high-resolution crystal structure.

the pore. The crystal structure of the mutant revealed only minor differences around the mutation site compared with the wild-type annexin V.

The ion channel activity of annexin V is clearly voltage dependent. Hodgkin and Huxley (1952) developed the concept that an electric field can act by moving charges in the channel protein, the so-called gating charges or voltage sensor. The models proposed for the known Na, K, and Ca channels suggest a highly charged helix called S4 as the voltage sensor (Greef, 1992; Papazian et al., 1991; Stühmer et al., 1989). In the case of annexin V, Karshikov et al. (1992) postulated that the conserved amino acid residues Glu112–Arg271 and Asp92–Arg117, which form saltbridges across the pore, may be involved in the voltage-sensing gates of the channel based on stereochemical and energetic considerations. This is now supported by mutational analysis. We changed

the glutamate into glycine in the saltbridge Glu112–Arg271 (Liemann et al., 1996). The electrophysiological analysis showed the loss of the voltage dependence in the channel opening and closing for the mutant Glu112→Gly compared with the wild-type annexin V (Fig. 5.10). This clearly indicates a role of this saltbridge in the control of the channel gating and the participation of the residue 112 in the voltage sensor of annexin V. The role of Arg271 and the second saltbridge remains to be established and is now under investigation in our laboratory. The result of the Glu112→Gly mutant led to the first picture of a voltage sensor at atomic resolution (Liemann et al., 1996).

In order to examine the macro-dipole of human annexin V, mutations were introduced on the surface of the protein (Burger et al., 1994). These mutations led to a new, lower conductance level. As no changes

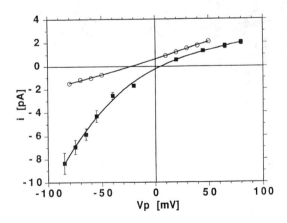

Figure 5.9 The current-voltage relationship for wild-type annexin V (open circles) and the Glu95Ser mutant (filled squares) resulting from single current recordings (V_p, pipette potential; pipette solution, 50 mM Ca^{2+}; bath solution, 100 mM Na^+). The continuous lines represent nonlinear least-squares fits (Berendes et al., 1993b).

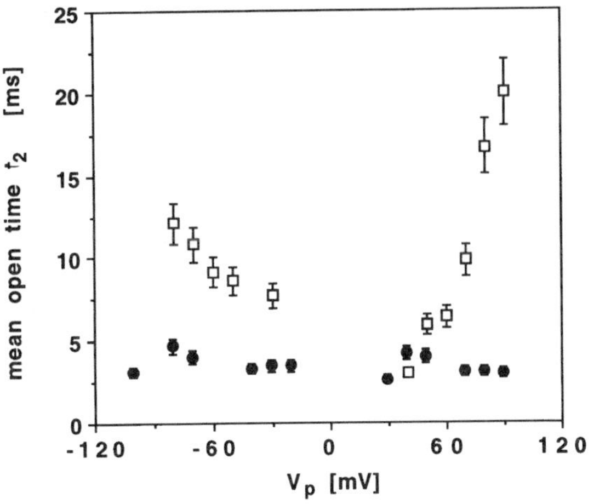

Figure 5.10 Voltage-dependence of the large mean open time (t_2) of wild-type annexin V (open squares) and the large conductance level of the Glu112Gly mutant (filled squares). The mutant has clearly lost the voltage dependence of its channel gating (V_p, pipette potential; recordings were made using symmetrical pipette and bath solutions, both containing Ca^{2+}).

in the crystal structures of the mutants compared with the wild type were observed, the additional conductance level may result from a change in the dynamic behaviour of the two modules of annexin V. Relative mobility of the two modules is apparent from the slightly, but significantly different angle in various crystal forms of human annexin V (Huber et al., 1992; Lewit-Bentley et al., 1992). By using different intermodule angles, the pore width can thus be controlled. The surface mutations are not located in the same part of the molecule and are far away from the central pore, although the electrophysiological properties of the mutated proteins are very similar (Burger et al., 1994). Therefore, we propose an influence of the charge distribution on the intermodule angle that determines the pore width. This may result in different conformers of annexin V with discrete conductance levels.

The Influence of the N-terminus on the Function of the Annexins

The high amino acid sequence homology in the typically four annexin domains is responsible for the common biochemical properties of all annexin members as well as for their calcium and membrane binding. In contrast, their N-terminal regions differ strongly in length, amino acid diversity, and in the sites of secondary modifications like phosphorylation, glycosylation, or acetylation. Taking this into account, the N-terminus was thought to give the annexins their distinct biological functions. Large efforts have been made in the last few years to investigate this hypothesis by site-directed mutagenesis. Berendes et al. (1993a) investigated a deletion mutant of the whole N-terminus of annexin V from an electrophysiological point of view by a calcium influx

assay into fura-2-loaded liposomes and by single-channel experiments. The truncated mutant has completely lost ion channel activity in contrast to experiments done on N-terminally truncated potassium channels (Armstrong and Bezanilla, 1977; Stühmer et al., 1989; Vassilev et al., 1988). In potassium channels the N-terminus is responsible for a fast inactivation by blocking the pore as proposed in the "ball-and-chain" model (Armstrong and Bezanilla, 1977). According to this model, the N-terminal region hinders sterically the ion flux by a direct ineraction with the pore region (Hoshi et al., 1990; Jan and Jan, 1992; Zagotta et al., 1990). Therefore, deletion of the N-terminus results in a permanently open channel. In the case of annexin V, a possible explanation for the channel inactivation upon truncation of the N-terminus could be an influence of the N-terminus on the equilibrium of different conformers for the open and closed state.

The role of the N-terminus in the process of membrane fusion and aggregation was elucidated by several groups with different approaches. Reutelingsperger and coworkers produced a chimera consisting of the N-terminus of annexin I and the C-terminal core of annexin V. In contrast to annexin V, annexin I is able to mediate membrane fusion and aggregation (Meers et al., 1992). Since chimera is also able to produce vesicle aggregation, it was concluded that the N-terminus is crucial for this function (Andree et al., 1993) and that a direct interaction of the N-terminus with phospholipid vesicles occurs (Hoekstra et al., 1993). Another approach to investigate the influence of the N-terminus on the aggregation behaviour on chromaffin granules was taken by the group of Creutz by a stepwise truncation of the N-terminus of annexin I and characterizing these mutants by chromaffin binding and aggregation assays (Wang and Creutz, 1994). The truncations led to significant changes in the calcium sensitivity of the proteins in aggregation but not in membrane-binding capabilities.

Mutagenesis in the Calcium-Binding Sites

Most of the proposed functions of the annexins are linked to their binding to membranes and thus to their binding to calcium. The biochemical character- ization of the calcium-binding sites was carried out by mutagenesis in the last few years. Jost et al. (1992) identified three acidic residues in human annexin II as being involved in the formation of the type II cal- cium-binding sites type II. Replacement of these resi- dues by alanine resulted in a mutant protein which showed a reduced affinity for calcium in liposome binding assay but binding at higher calcium concen- trations was not influenced. The additional distortion of the type III calcium-binding sites together with the mutations in the type II sites abolished the mem- brane-binding properties of the protein (Jost et al., 1994). Furthermore, alteration of the type II sites led to changes in the intracellular distribution of the pro- tein, but no influence on the distribution of the type III mutants could be observed (Jost et al., 1994). Therefore, the type II calcium sites seem to be most important for the physiological, intracellular activ- ities of the annexins.

Similar studies were done on annexin IV by Nelson and Creutz (1995). In all four domains they changed the acidic residue which participates in the calcium binding by alanine and also made combina- torial mutagenesis with all mutants. The investigation of the mutated proteins in chromaffin granule aggre- gation and binding assays revealed that when increas- ing the number of mutated domains, a higher amount of calcium is needed, for membrane binding repre- senting the same result as shown previously by Jost et al., (1994). Furthermore, the authors showed that the single mutations had different effects on the bind- ing or aggregation behaviour, in that the alteration of the calcium-binding site in the second domain had a greater effect on membrane aggregation than on membrane binding.

Trave et al., (1994a, 1994b) mutated a conserved lysine and the two glycines into glutamates, in the endonexin fold of the second domain. The mutants resulted also in different binding and membrane aggregation behaviour. In summary, systematic in- vestigation of all calcium-binding sites, the character- ization of specificities of the single domains against defined phospholipids and the influence on the dis- tribution in the cell upon modification in the calcium- binding sites presents hints on the in vivo function of the annexins.

Conclusion and Perspectives

The annexins are a novel family of calcium-binding proteins whose structure in solution and in the mem- brane-bound form has been elucidated to high reso- lution. Nevertheless, their biological role still remains unclear although a wide range of in vitro properties has been intensively studied. The best documented include aggregation and fusion of membranes and an involvement in endo- and exocytosis, inhibition of phospholipase A_2, and thus anti inflammatory effects, anticoagulation, ion channel activity, interac- tion with cytoskeletal proteins as well as a role in cell differentiation, cell proliferation, and mitogenic sig- nal transduction. Due to their uncertainties, for the following decade a clear understanding of the annex- ins in vivo function will be an interesting and, of course, challenging aim. Experiments of gene dele- tions in higher vertebrates and mammals will prob- ably play a decisive role.

In vitro, the voltage-dependent ion channel activ- ity has been investigated in detail. The combination of x-ray crystallography, electron microscopy, and electrophysiology on the basis of site-directed muta- genesis on human annexin V led to a detailed picture of the location of amino acid residues which are important for the ion channel properties. The central hydrophilic pore was identified as the ion conduction pathway since crucial constituents of the ion channel selectivity filter and the voltage sensor has been loca- lized within the pore. In addition, the mutational exchange of charged for uncharged residues on the annexin surface led to different conductance levels of the channel, indicating that the charge distribution in the molecule might be important for ion permeabil- ity. Nevertheless, further studies are necessary in order to understand the selectivity and voltage gating of the channel precisely. A major goal also should be the detailed characterization of the membrane in the annexin-bound and ion channelling state. Annexins as peripherally bound membrane proteins are rather unusual ion channels, but might be a paradigm of a variety of proteins existing in water-soluble and membrane-bound states, in which they generate pores, as for example the toxins. Since the annexin ion pore is made up of an α-helical bundle, it bears resemblance to some models of integrated membrane ion channels and thus some of the features elucidated in annexins might apply to those proteins, too.

References

Ahlmers, W. and McClesky, E. W. (1984) Non-selective conductance in calcium channels of frog muscle: cal- cium selectivity in a single file pore. *J. Physiol.* 353: 565–583.

Ahn, N. G., Teller, D. C., Bienkowski, M. J., McMullen, B. A., Lipkin, E. W., and C. de Häen (1988) Sedimentation equilibrium analysis of five lipocortin-related phospholipase A$_2$ inhibitors from human placenta. Evidence against a mechanistically relevant association between enzyme and inhibitor. *J. Biol. Chem.* 263: 18657–18663.

Ali, S. M., Geisow, M. J., and Burgoyne, R. D. (1989) A role for calpactin in calcium-dependent exocytosis in adrenal chromaffin cells. *Nature* 340: 313–315.

Ando, Y., Imamura, S., Owada, M. K., Kakunaga, T., and Kannagi, R. (1989) Cross-linking of lipocortin I and enhancement of its Ca^{2+} sensitivity by tissue transglutaminase. *Biochem. Biophy. Res. Commun.* 163: 944–951.

Ando, Y., Imamura, S., Owada, M. K., and Kannagi, R. (1991) Calcium-induced intracellular cross-linking of lipocortin I by tissue transglutaminase in A431 cells. Augmentation by membrane phospholipids. *J. Biol. Chem.* 266: 1101–1108.

Andrawis, A., Solomon, M., and Delmer, D. P. (1993) Cotton fiber annexins: a potential role in the regulation of callose synthase. *Plant J.* 3: 763–772.

Andree, H. A., Reutelingsperger, C. P., Hauptmann, R., Hemker, H. C., Hermens, W. T., and Willems, G. M. (1990) Binding of vascular anticoagulant alpha (VAC alpha) to planar phospholipid bilayers. *J. Biol. Chem.* 265: 4923–4928.

Andree, H. A., Stuart, M. C., Hermens, W. T., Reutelingsperger, C. P., Hemker, H. C., Frederik, P. M., and Willems, G. M. (1992) Clustering of lipid-bound annexin V may explain its anticoagulant effect. *J. Biol. Chem.* 267: 17907–17912.

Andree, H. A., Willems, G. M., Hauptmann, R., Maurer-Fogy, I., Stuart, M. C., Hermens, W. T., Frederik, P. M., and Reutelingsperger, C. P. (1993) Aggregation of phospholipid vesicles by a chimeric protein with the N-terminus of annexin I and the core of annexin V. *Biochemistry* 32: 4634–4640.

Armstrong, C. M. and Bezanilla, F. (1977) Inactivation of the sodium channel. II. Gating current experiments. *J. Gen. Physiol.* 70: 567–590.

Bandorowicz, J., Pikula, S., and Sobota, A. (1992) Annexins IV (p32) and VI (p68) interact with erythrocyte membrane in a calcium-dependent manner. *Biochim. Biophys. Acta* 1105: 201–206.

Barton, G. J., Newman, R. H., Freemont, P. S., and Crumpton, M. J. (1991) Amino acid sequence analysis of the annexin super-gene family of proteins. *Eur. J.Biochem.* 198: 749–760.

Bazzi, M. D. and Nelsestuen, G. L. (1991) Extensive segregation of acidic phospholipids in membranes induced by protein kinase C and related proteins. *Biochemistry* 30: 7961–7968.

Bazzi, M. D., Youakim, A., and Nelsestuen, G. L. (1992) Importance of phosphatidylethanolamine for association of protein kinase C and other cytoplasmic proteins with membranes. *Biochemistry* 31: 1125–1135.

Berendes, R., Burger, A., Voges, D., Demange, P., and Huber, R. (1993a). Calcium influx through annexin V ion channels into large unilamellar vesicles measured with fura-2. *FEBS Lett.* 317: 131–134.

Berendes, R., Voges, D., Demange, P., Huber, R., and Burger, A. (1993b). Structure–function analysis of the ion channel selectivity filter in human annexin V. *Science* 262: 427–430.

Bewley, M. C., Boustead, C. M., Walker, J. H., Waller, D. A., and Huber, R. (1993) Structure of chicken annexin V at 2.25-Å resolution. *Biochemistry* 32: 3923–3929.

Bianchi, R., Giambanco, I., Ceccarelli, P., Pula, G., and Donato, R. (1992) Membrane-bound annexin V isoforms (CaBP33 and CaBP37) and annexin VI in bovine tissues behave like integral membrane proteins. *FEBS Lett.* 296: 158–162.

Blackwell, G. J., Carnuccio, R., di Rosa, M., Flower, R. J., Parente, L., and Persico, P. (1980) Macrocortin: a polypeptide causing the anti-phospholipase effect of glucocorticoids. *Nature* 287: 147–149.

Bohn, E., Gerke, V., Kresse, H., Loffler, B. M., and Kunze, H. (1992) Annexin II inhibits calcium-dependent phospholipase A$_1$ and lysophospholipase but not triacyl glycerol lipase activities of rat liver hepatic lipase. *FEBS Lett.* 296: 237–240.

Boustead, C. M., Brown, R., and Walker, J. H. (1993) Isolation, characterization and localization of annexin V from chicken liver. *Biochem. J.* 291: 601–608.

Boustead, C. M., Smallwood, M., Small, H., Bowles, D.J., and Walker, J. H. (1989) Identification of calcium dependent phospholipid-binding proteins in higher plant cells. *FEBS Lett.* 244: 456–460.

Boyko, V., Mudrak, O., Svetlova, M., Negishi, Y., Ariga, H., and Tomilin, N. (1994) A major cellular substrate for protein kinases, annexin II, is a DNA-binding protein. *FEBS Lett.* 345: 139–142.

Brisson, A., Mosser, G., and Huber, R. (1991) Structure of soluble and membrane-bound human annexin V. *J. of Mol. Biol.* 220: 199–203.

Burger, A., Berendes, R., Liemann, S., Benz, J., Hofmann, A., Göttig, P., Huber, R., Gerke, V., Römisch, J., and Weber, K. (1996) The crystal structure and ion channel activity of human annexin II, a peripheral membrane protein. *J. Mol. Biol.* 257: 839–847.

Burger, A., Voges, D., Demange, P., Perez, C. R., Huber, R., and Berendes, R. (1994) Structural and electrophysiological analysis of annexin V mutants. Mutagenesis of human annexin V, an in vitro voltage-gated calcium channel, provides information about the structural features of the ion pathway, the voltage sensor and the ion selectivity filter. *J. Mol. Biol.* 237: 479–499.

Burgoyne, R. D. and Clague, M. J. (1994) Annexins in the endocytotic pathway. *Trends Biochem. Sci.* 19: 231–232.

Burgoyne, R. D. and Geisow, M. J. (1989) The annexin family of calcium-binding proteins. *Cell Calcium* 10: 1–10.

Burns, A. L., Magendzo, K., Shirvan, A., Srivastava, M., Rojas, E., Alijani, M. R., and Pollard, H. B. (1989)

Calcium channel activity of purified human synexin and structure of the human synexin gene. *Proc. Nat. Acad. Sci. USA* 86: 3798–3802.

Chan, H. C., Kaetzel, M. A., Gotter, A. L., Dedman, J. R., and Nelson, D. J. (1994) Annexin IV inhibits calmodulin-dependent protein kinase II-activated chloride conductance. A novel mechanism for ion channel regulation. *J. Biol. Chem.* 269: 32464–32468.

Changeux, J. P. (1990) The nicotinic acetylcholine receptor: an allosteric prototype of ligand-gated ion channels. *Trends Pharmaceut. Sci.* 11: 485–492.

Christmas, P., Callaway, J., Fallon, J., Jones, J., and Haigler, H. T. (1991) Selective secretion of annexin 1, a protein without a signal sequence, by the human prostate gland. *J. Biol. Chem.* 266: 2499–2507.

Chuah, S. Y. and Pallen, C. J. (1989) Calcium-dependent and phosphorylation-stimulated proteolysis of lipocortin I by an endogenous A431 cell membrane protease. *J. Biol. Chem.* 264: 21160–21166.

Concha, N. O., Head, J. F., Kaetzel, M. A., Dedman, J. R., and Seaton, B. A. (1992) Annexin V forms calcium-dependent trimeric units on phospholipid vesicles. *FEBS Lett.* 314:159–162.

Concha, N. O., Head, J. F., Kaetzel, M. A., Dedman, J. R., and Seaton, B. A. (1993) Rat annexin V crystal structure: Ca(2+)-induced conformational changes. *Science* 261: 1321–1324.

Creutz, C. E. (1992) The annexins and exocytosis. *Science* 258: 924–931.

Creutz, C. E., Pazoles, C. J., and Pollard, H. B. (1978) Identification and purification of an adrenal medullary protein (synexin) that causes calcium-dependent aggregation of isolated chromaffin granules. *J. Biol. Chem.* 253: 2858–2866.

Creutz, C. E., Scott, J. H., Pazoles, C. J., and Pollard, H. B. (1982) Further characterization of the aggregation and fusion of chromaffin granules by synexin as a model for compound exocytosis. *J. Cell. Biochem.* 18: 87–97.

Crompton, M. R., Moss, S. E., and Crumpton, M. J. (1988) Diversity in the lipocortin/calpactin family. *Cell* 55: 1–3.

Croxtall, J. D. and Flower, R. J. (1992) Lipocortin 1 mediates dexamethasone-induced growth arrest of the A549 lung adenocarcinoma cell line. [Erratum *Proc. Natl. Acad. Sci. USA* (1992) 89(17): 8408.] *Proc. Nat. Acad. Sci. USA* 89: 3571–3575.

Crumpton, M. J. and Dedman, J. R. (1990). Protein terminology tangle. *Nature* 345: 212.

Davidson, F. F., Dennis, E. A., Powell, M., and Glenney, J. R. Jr. (1987) Inhibition of phospholipase A₂ by "lipocortins" and calpactins. An effect of binding to substrate phospholipids. *J. Biol. Chem.* 262: 1698–1705.

Davidson, F. F., Lister, M. D., and Dennis, E. A. (1990) Binding and inhibition studies on lipocortins using phosphatidylcholine vesicles and phospholipase A₂ from snake venom, pancreas, and a macrophage-like cell line. *J. Biol. Chem.* 265: 5602–5609.

Della Gaspera, B., Weinman, S., Huber, C., Lemnaouar, M., Paul, A., Picard, J., and Gruenert, D. C. (1995)

Overexpression of annexin V in cystic fibrosis epithelial cells from fetal trachea. *Exp. Cell Res.* 219: 379–383.

Demange, P., Voges, D., Benz, J., Liemann, S., Gottig, P., Berendes, R., Burger, A., and Huber, R. (1994) Annexin V: the key to understanding ion selectivity and voltage regulation? *Trends Biochem. Sci.* 19: 272–276.

Diaz-Munoz, M., Hamilton, S. L., Kaetzel, M. A., Hazarika, P., and Dedman, J. R. (1990) Modulation of Ca^{2+} release channel activity from sarcoplasmic reticulum by annexin VI (67-kDa calcimedin). *J. Biol. Chem.* 265: 15894–15899.

Dijkstra, B. W., Kalk, K. H., Hol, W. G. J., and Drenth, J. (1981) Structure of bovine pancreatic phospholipase A₂ at 1.7 Å resolution. *J. Mol. Biol.* 147: 97–123.

Doring, V., Schleicher, M., and Noegel, A. A. (1991) Dictyostelium annexin VII (synexin). cDNA sequence and isolation of a gene disruption mutant. *J. Biol. Chem.* 266: 17509–17515.

Driessen, H. P., Newman, R. H., Freemont, P. S., and Crumpton, M. J. (1992) A model of the structure of human annexin VI bound to lipid monolayers. *FEBS Lett. 306:* 75–79.

Drust, D. S. and Creutz, C. E. (1988) Aggregation of chromaffin granules by calpactin at micromolar levels of calcium. *Nature* 331: 88–91.

Durell, S. R. and Guy, H. R. (1992). Atomic scale structure and functional models of voltage gated potassium channels. *Biophys. J.* 62: 238–250.

Emans, N., Gorvel, J. P., Walter, C., Gerke, V., Kellner, R., Griffiths, G., and Gruenberg, J. (1993) Annexin II is a major component of fusogenic endosomal vesicles. *J. Cell Biol.* 120: 1357–1369.

Erikson, E. and Erikson, R. L. (1980) Identification of a cellular substrate phosphorylated by the avian sarcoma virus-transforming-gene product. *Cell* 21: 829–836.

Evans, T. C. Jr. and Nelsestuen, G. L. (1994) Calcium and membrane-binding properties of monomeric and multimeric annexin II. *Biochemistry* 33: 13231–13238.

Fava, R. A. and Cohen, S. (1984) Isolation of a calcium dependent 35-kilodalton substrate for the epidermal growth factor receptor/kinase from A-431 cells. *J. Biol. Chem.* 259: 497–509.

Fiedler, K., Lafont, F., Parton, R. G., and Simons, K. (1995) Annexin XIIIb: a novel epithelial specific annexin is implicated in vesicular traffic to the apical plasma membrane. *J. Cell Biol.* 128: 1043–1053.

Flaherty, M. J., West, S., Heimark, R. L., Fujikawa, K., and Tait, J. F. (1990) Placental anticoagulant protein-I: measurement in extracellular fluids and cells of the hemostatic system. *J. Lab. Clin. Med.* 115: 174–181.

Flower, R. J. and Blackwell, G. J. (1979) Anti-inflammatory steroids induce a biosynthesis of a phospholipase A₂ inhibitor which prevents prostaglandin generation. *Nature* 278: 456–459.

Flower, R. J. and Rothwell, N. J. (1994) Lipocortin-1: cellular mechanisms and clinical relevance [see comments]. *Trends Pharmacol. Sci.* 15: 71–76.

Fuente, M. de la and Parra, A. V. (1995) Vesicle aggregation by annexin I: role of secondary membrane binding site. *Biochemistry* 34: 10393–10399.

Funakoshi, T., Hendrickson, L. E., McMullen, B. A., and Fujikawa, K. (1987) Primary structure of human placental anticoagulant protein. *Biochemistry* 26: 8087–8092.

Geisow, M. J. (1986) Common domain structure of Ca^{2+} and lipid-binding proteins. *FEBS Lett.* 203: 99–103.

Geisow, M. J. and Walker, J. H. (1986) New proteins involved in cell regulation by Ca^{2+} and phospholipids. *Trends Biochem. Sci.* 11: 420–423.

Geisow, M. J., Childs, J., Dash, B., Harris, A., Panayotou, G., Südhof, T., and Walker, J. (1984). Cellular distribution of three mammalian calcium-binding proteins related to Torpedo calelectrin. *EMBO J.* 3: 2969–2974.

Geisow, M. J., Fritsche, U., Hexham, J. M., Dash, B., and Johnson, T. (1986) A consensus amino-acid sequence repeat in Torpedo and mammalian Ca^{2+}-dependent membrane-binding proteins. *Nature* 320: 636–638.

Genge, B. R., Wu, L. N., Adkisson, H. D. T., and Wuthier, R. E. (1991) Matrix vesicle annexins exhibit proteolipid-like properties. Selective partitioning into lipophilic solvents under acidic conditions. *J. Biol. Chem.* 266: 10678–10685.

Genge, B. R., Wu, L. N., and Wuthier, R. E. (1989) Identification of phospholipid-dependent calcium-binding proteins as constituents of matrix vesicles. *J. Biol. Chem.* 264: 10917–10921.

Gerke, V. (1991). Identification of a homologue for annexin VII (synexin) in *Dictyostelium discoideum*. *J. Biol. Chem* 266: 1697–1700.

Gerke, V., Koch, W., and Thiel, C. (1991) Primary structure and expression of the *Xenopus laevis* gene encoding annexin II. *Gene*: 104: 259–264.

Gerke, V. (1992) Evolutionary conservation and three-dimensional folding of the tyrosine kinase substrate annexin II. In *The Annexins* Moss, S. E. (ed.). Portland Press, London.

Gerke, V. and Weber, K. (1984) Identity of p36 phosphorylated upon Rous sarcoma virus transformation with a protein purified from brush borders; calcium-dependent binding to non-erythroid spectrin and F-actin. *EMBO J.* 3: 227–233.

Gerke, V. and Weber, K. (1985) Calcium dependent conformational changes in the 36 kDa subunit of intestinal protein I to the cellular 36 kDa target of Rous sarcoma virus tyrosine kinase. *J. Biol. Chem.* 260: 1688–1695.

Giambanco, I., Pula, G., Ceccarelli, P., Bianchi, R., and Donato, R. (1991). Immunohistochemical localization of annexin V (CaBP33) in rat organs. *J. Histochem. Cytochemistry* 39: 1189–1198.

Gilmanshin, R., Creutz, C. E., and Tamm, L. K. (1994) Annexin IV reduces the rate of lateral lipid diffusion and changes the fluid phase structure of the lipid bilayer when it binds to negatively charged membranes in the presence of calcium. *Biochemistry* 33: 8225–8232.

Glenney, J. R. and Tack, B. F. (1985) Amino-terminal sequence of p36 and associated p11: identification of the site of tyrosine phosphorylation and homology with S100. *Proc. Nat. Acad. Sci. USA* 82: 7884–7888.

Glenney, J. R. Jr. Tack, B., and Powell, M. A. (1987) Calpactins: two distinct Ca^{++}-regulated phospholipid- and actin-binding proteins isolated from lung and placenta. *J. Cell Biol.* 104, 503–511.

Gould, K. L., Woodgett, J. R., Isacke, C. M., and Hunter, T. (1986) The protein tyrosine kinase substrate p36 is also a substrate for protein kinase C in vitro and in vivo. *Mol. Cell. Biol.* 6: 2738–2744.

Goulet, F., Moore, K. G., and Sartorelli, A. C. (1992) Glycosylation of annexin I and annexin II. *Biochem. Biophys. Res. Commun.* 188: 554–558.

Greef, N. G. (1992) Molecular structure-function relations in voltage-gated ion channels of excitable membranes. *Jerusal. Symp. Quant. Biochem.* 25: 279–296.

Greenberg, M. E. and Edelmann, G. M. (1983) Comparison of the 34,000-Da pp60scr substrate and a 38,000-Da phosphoprotein identified by monoclonal antibodies. *J. Biol. Chem.* 258: 8497–8502.

Guy, H. R. and Conti, R. (1990) Pursuing the structure and function of voltage-gated channels. *Trends Neurosci.* 13: 201–206.

Guy, H. R., Rojas, E. M., Burns, A. L., and Pollard, H. B. (1991) A TIM barrel model for synexin channels. *Biophys. J.* 59: 372a.

Haigler, H. T. and Schlaepfer, D. D. (1992) Annexin I phosphorylation and secretion. In *The Annexins* Moss, S. E., (ed.). Portland Press, London, pp. 11–22.

Haigler, H. T., Fitch, J. M., Jones, J. M., and Schlaepfer, D. D. (1989) Two lipocortin-like proteins, endonexin II and anchorin CII, may be alternate splices of the same gene. *Trends Biochem. Sci.* 14: 48–50.

Haigler, H. T., Schlaepfer, D. D., and Burgess, W. H. (1987) Characterization of lipocortin I and an immunologically unrelated 33-kDa protein as epidermal growth factor receptor/kinase substrates and phospholipase A_2 inhibitors. *J. Biol. Chem.* 262: 6921–6930.

Havelka, W. A., Henderson, R., and Oesterhelt, D. (1995) Three-dimensional structure of halorhodopsin at 7 Å Resolution. *J. Mol. Biol.* 247: 726–738.

Hazarika, P., Kaetzel, M. A., Sheldon, A., Karin, N. J., Fleischer, S., Nelson, T. E., and Dedman, J. R. (1991a) Annexin VI is associated with calcium-sequestering organelles. *J. Cell. Biochem.* 46: 78–85.

Hazarika, P., Sheldon, A., Kaetzel, M. A., Diaz-Munoz, M., Hamilton, S. L., and Dedman, J. R. (1991b) Regulation of the sarcoplasmic reticulum Ca(2+)-release channel requires intact annexin VI. *J. Cell. Biochem.* 46: 86–93.

Henderson, R., Baldwin, J. M., Ceska, T. A., Zmelin, F., Beckmann, E., and Downing, K. H. (1990) Model for the structure of bacteriorhodopsin based on high-resolution electron microscopy. *J. Mol. Biol.* 213: 899–929.

Hess, P. and Tsien, R. W. (1984) Mechanism of ion permeation through calcium channels. *Nature* 309: 453–456.

Hirata, F., Schiffmann, E., Venkatasubramanian, K., Salomon, D., and Axelrod, J. (1980) A phospholipase A$_2$ inhibitory protein in rabbit neutrophils induced by glucocorticoids. *Proc. Natl. Acad. Sci. USA* 77: 2533–2536.

Hodgkin, A. L. and Huxley, A. F. (1952) A quantitative description of membrane current and its application to conduction and excitation in nerve. *J. Physiol.* 117: 500–544.

Hoekstra, D., Buist-Arkema, R., Klappe, K., and Reutelingsperger, C. P. (1993) Interaction of annexins with membranes: the N-terminus as a governing parameter as revealed with a chimeric annexin. *Biochemistry* 32: 14194–14202.

Horseman, N. D. (1989) A prolactin-inducible gene product which is a member of the calpactin/lipocortin family. *Mol. Endocrinol.* 3: 773–779.

Hoshi, T., Zagotta, W. N., and Aldrich, R. W. (1990) Biophysical and molecular mechanism of Shaker potassium channel inactivation. *Science* 250: 533–538.

Huber, R., Berendes, R., Burger, A., Schneider, M., Karshikov, A., Luecke, H., Römisch, J., and Paques, E. (1992) Crystal and molecular structure of human annexin V after refinement. Implications for structure, membrane binding and ion channel formation of the annexin family of proteins. *J. Mol. Biol.* 223: 683–704.

Huber, R., Römisch, J., and Paques, E. P. (1990a) The crystal and molecular structure of human annexin V, an anticoagulant protein that binds to calcium and membranes. *EMBO J.* 9: 3867–3874.

Huber, R., Schneider, M., Mayr, I., Römisch, J., and Paques, E. P. (1990b) The calcium binding sites in human annexin V by crystal structure analysis at 2.0 Å resolution. Implications for membrane binding and calcium channel activity. *FEBS Lett.* 275: 15–21.

Ikebuchi, N. W. and Waisman, D. M. (1990) Calcium-dependent regulation of actin filament bundling by lipocortin-85. *J. Biol. Chem.* 265: 3392–3400.

Ivanenkov, V. V., Weber, K., and Gerke, V. (1994) The expression of different annexins in the fish embryo is developmentally regulated. *FEBS Lett.* 352: 227–230.

Jacquot, J., Dupuit, F., Elbtaouri, H., Hinnrasky, J., Antonicelli, F., Haye, B., and Puchelle, E. (1990) Production of lipocortin-like proteins by cultured human tracheal submucosal gland cells. *FEBS Lett.* 274: 131–135.

Jan, L. Y. and Jan, Y. N. (1992) Structural elements involved in specific K$^+$ channel functions. *Ann. Rev. Physiol.* 54: 537–555.

Jindal, H. K., Chaney, W. G., Anderson, C. W., Davis, R. G., and Vishwanatha, J. K. (1991) The protein-tyrosine kinase substrate, calpactin I heavy chain (p36), is part of the primer recognition protein complex that interacts with DNA polymerase alpha. *J. Biol. Chem.* 266: 5169–5176.

Johnsson, N., Marriott, G., and Weber, K. (1988) p36, The major cytoplasmic substrate of src tyrosine protein kinase, binds to its p11 regulatory subunit via a short amino-terminal amphiphatic helix. *EMBO J.* 7: 2435–2442.

Johnston, P. A., Perin, M. S., Reynolds, G. A., Wasserman, S. A., and Südhof, T. C. (1990) Two novel annexins from *Drosophila melanogaster*. Cloning, characterization, and differential expression in development. *J. Biol. Chem.* 265: 11382–11388.

Johnstone, S. A., Hubaishy, I., and Waisman, D. M. (1993) Regulation of annexin I-dependent aggregation of phospholipid vesicles by protein kinase C. *Biochem. J.* 294: 801–807.

Jones, P. G., Fitzpatrick, S., and Waisman, D. M. (1994) Chromaffin granules release calcium on contact with annexin VI: implications for exocytosis. *Biochemistry* 33: 8180–8187.

Jones, P. G., Moore, G. J., and Waisman, D. M. (1992) A nonapeptide to the putative F-actin binding site of annexin-II tetramer inhibits its calcium-dependent activation of actin filament bundling. *J. Biol. Chem.* 267: 13993–13997.

Jost, M., Thiel, C., Weber, K., and Gerke, V. (1992). Mapping of three unique Ca(2+)-binding sites in human annexin II. *Euro J. Biochem.* 207: 923–930.

Jost, M., Weber, K., and Gerke, V. (1994) Annexin II contains two types of Ca(2+)-binding sites. *Biochem. J.* 298,: 553–559.

Junker, M. and Creutz, C. E. (1993). Endonexin (annexin IV)-mediated lateral segregation of phosphatidylglycerol in phosphalidylglycerol/phosphatidylcholine membranes. *Biochemistry* 32: 9968–9974.

Kaetzel, M. A., Chan, H. C., Dubinsky, W. P., Dedman, J. R., and Nelson, D. J. (1994) A role for annexin IV in epithelial cell function. Inhibition of calcium-activated chloride conductance. *J. Biol. Chem.* 269: 5297–5302.

Karasik, A., Pepinsky, R. B. Shoelson, S. E., and Kahn, C. R. (1988) Lipocortin 1 and 2 as substrates for the insulin receptor kinase in rat liver. *J. Biol. Chem.* 263: 11862–11867.

Karshikov, A., Berendes, R., Burger, A., Cavalie, A., Lux, H.-D., and Huber, R. (1992) Annexin V membrane interaction: an electrostatic potential study. *Eur. Biophys. J.* 20: 337–344.

Khanna, N. C., Helwig, E. D., Ikebuchi, N. W., Fitzpatrick, S., Bajwa, R., and Waisman, D. M. (1990) Purification and characterization of annexin proteins from bovine lung. *Biochemistry* 29: 4852–4862.

Kim, K. M., Kim, D. K., Park, Y. M., Kim, C. K., and Na, D. S. (1994). Annexin-I inhibits phospholipase A$_2$ by specific interaction, not by substrate depletion. *FEBS Lett.* 343: 251–255.

Kirsch, T. and Pfäffle, M. (1992) Selective binding of anchorin CII (annexin V) to type II and X collagen and to chondrocalcin (C-propeptide of type II collagen). Implications for anchoring function between matrix vesicles and matrix proteins. *FEBS Lett.* 310: 143–147.

Kirsch, T. and Wuthier, R. E. (1994) Stimulation of calcification of growth plate cartilage matrix vesicles by binding to type II and X collagens. *J. Biol. Chem.* 269: 11462–11469.

Kirsch, T., Ishikawa, M. F., and Wuthier, R. E. (1994) Roles of the nucleational core complex and collagens (types II and X) in calcification of growth plate cartilage matrix vesicles. *J. Biol. Chem.* 269: 20103–20109.

Klee, C. B. (1988) Ca^{2+}-dependent phospholipid- (and membrane-) binding proteins. *Biochemistry* 27: 6645–6653.

Kraulis, P. J. (1991) MOLSCRIPT: a program to produce both detailed and schematic plots of protein structures. *J. Appl. Crystallogr.* 24: 946–950.

Kretsinger, R. H. and Creutz, C. E. (1986) Consensus in exocytosis. *Nature* 302: 573.

Lewit-Bentley, A., Morera, S., Huber, R., and Bodo, G. (1992) The effect of metal binding on the structure of annexin V and implications for membrane binding. *Eur. J. Biochem.* 210, 73–77.

Liemann, S. and Lewit-Bentley, A. (1995) Annexins: a novel family of calcium- and membrane-binding proteins in search of a function. *Structure* 3: 233–237.

Liemann, S., Benz, J., Berendes, R., Burger, A., Demange, P., Voges, D., Huber R., and Göttig, P. (1996) Structural and functional characterization of the voltage-sensor in the ion channel human annexin V. *J. Mol. Biol.* 258: 555–561.

Lin, H. C., Südhof, T. C., and Anderson, R. G. (1992) Annexin VI is required for budding of clathrin-coated pits. *Cell* 70: 283–291.

Machoczek, K., Fischer, M., and Soling, H. D. (1989) Lipocortin I and lipocortin II inhibit phosphoinositide- and polyphosphoinositide-specific phospholipase C. The effect results from interaction with the substrates. *FEBS Lett.* 251: 207–212.

Mangeat, P.-H. (1988) Interaction of biological membranes with the cytoskeletal framework of living cells. *Bio. Cell* 64: 261–281.

Meers, P. (1990) Location of tryptophans in membrane-bound annexins. *Biochemistry* 29: 3325–3330.

Meers, P. and Mealy, T. (1993a) Relationship between annexin V tryptophan exposure, calcium, and phospholipid binding. *Biochemistry* 32: 5411–5418.

Meers, P. and Mealy, T. (1993b) Calcium-dependent annexin V binding to phospholipids: stoichiometry, specificity, and the role of negative charge. *Biochemistry* 32: 11711–11721.

Meers, P. and Mealy, T. (1994) Phospholipid determinants for annexin V binding sites and the role of tryptophan 187. *Biochemistry* 33: 5829–5837.

Meers, P., Daleke, D., Hong, K., and Papahadjopoulos, D. (1991). Interactions of annexins with membrane phospholipids. *Biochemistry* 30: 2903–2908.

Meers, P., Mealy, T., Pavlotsky, N., and Tauber, A. I. (1992) Annexin I-mediated vesicular aggregation: mechanism and role in human neutrophils *Biochemistry* 31: 6372–6382.

Milks, L. C., Kumar, N. M., Houghton, R., Unwin, N., and Gilula, N. B. (1988) Topology of the 32-kD liver gap junction protein determined by site-directed antibody localizations. *EMBO J.* 7: 2967–2975.

Mironov, S. L. (1992) Conformational model for ion permeation in membrane channels: a comparison with multi-ion models and applications to calcium channel permeability. *Biophys. J.* 63: 485–496.

Mizutani, A., Tokumitsu, H., Kobayashi, R., and Hidaka, H. (1993) Phosphorylation of annexin XI (CAP-50) in SR-3Y1 cells. *J. Biol. Chem.* 268: 15517–15522.

Mizutani, A., Usuda, N., Tokumitsu, H., Minami, H., Yasui, K., Kobayashi, R., and Hidaka, H. (1992) CAP-50, a newly identified annexin, localizes in nuclei of cultured fibroblast 3Y1 cells. *J. Biol. Chem.* 267: 13498–13504.

Mizutani, A., Watanabe, N., Kitao, T., Tokumitsu, H., and Hidaka, H. (1995) The long amino-terminal tail domain of annexin XI is necessary for its nuclear localization. *Arch. Biochem. Biophys.* 318: 157–165.

Mollenhauer, J. and von der Mark, K. (1983) Isolation and characterization of a collagen-binding glycoprotein from chondrocyte membranes. *EMBO J.* 2: 45–50.

Moss, S. E. (ed.) (1992) *The Annexins.* Portland Press, London.

Moss, S. E., Edwards, H. C., and Crumpton, M. J. (1991) Diversity in the annexin family. In *Novel-Calcium Binding Proteins* Heizmann, C. W. (ed.) Springer Verlag, Berlin, pp. 535–566.

Mosser, G., Ravanat, C., Freyssinet, J. M., and Brisson, A. (1991) Sub-domain structure of lipid-bound annexin-V resolved by electron image analysis. *J. Mol. Biol.* 217: 241–245.

Muesch, A., Hartmann, E., Rohde, K., Rubartelli, A., Sitia, R., and Rapoport, T. A. (1990) A novel pathway for secretory proteins. *Trends Biochem. Sci.* 15: 86–88.

Naka, M., Qing, Z. X., Sasaki, T., Kise, H., Tawara, I., Hamaguchi, S., and Tanaka, T. (1994) Purification and characterization of a novel calcium-binding protein, S100C, from porcine heart. *Biochim. Biophys. Acta* 1223: 348–353.

Nakano, T., Ohara, O., Teraoka, H., and Arita, H. (1990) Glucocorticoids supress group II phospholipase A_2 production by blocking mRNA synthesis and post-transcriptional expression. *J. Biol. Chem.* 265: 12745–12748.

Nakata, T., Sobue, K., and Hirokawa, N. (1990) Conformational change and localization of calpactin I complex involved in exocytosis as revealed by quick-freeze, deep-etch electron microscopy and immunocytochemistry. *J. Cell Biol.* 110: 13–25.

Nelson, M. R. and Creutz, C. E. (1995) Combinatorial mutagenesis of the four domains of annexin IV: effects on chromaffin granule binding and aggregating activities. *Biochemistry* 34: 3121–3132.

Neumann, E. (1988). The electroporation hysteresis. *Ferroelectrics* 86: 325–333.

Neumann, J. M., Sanson, A., and Lewit-Bentley, A. (1994) Calcium-induced changes in annexin V behaviour in solution as seen by proton NMR spectroscopy. *Eur. J. Biochem.* 225: 819–825.

Newman, R., Tucker, A., Ferguson, C., Tsernoglou, D., Leonard, K., and Crumpton, M. J. (1989) Crystallization of p68 on lipid monolayers and as three-dimensional single crystals. *J. Mol. Biol.* 206: 213–219.

Newman, R. H., Leonard, K., and Crumpton, M. J. (1991) 2D crystal forms of annexin IV on lipid monolayers. *FEBS Lett.* 279: 21–24.

Olofsson, A., Mallouh, V., and Brisson, A. (1994) Two-dimensional structure of membrane-bound annexin V at 8 Å resolution. *J. Struct. Biol.* 113: 199–205.

Papazian, D. M., Timpe, L. C., Jan, Y. N., and Jan, L. Y. (1991) Alteration of voltage-dependence of Shaker potassium channel by mutation in the S4 segment. *Nature* 349: 305–310.

Pepinsky, R. B., Sinclair, L. K., Browning, J. L., Mattaliano, R. J., Smart, J. E., Chow, E. P., Falbel, T., Ribolini, A., Garwin, J. L., and Wallner, B. P. (1986). Purification and partial sequence analysis of a 37-kDa protein that inhibits phospholipase A2 activity from rat peritoneal exudates. *J. Biol. Chem.* 261: 4239–4246.

Pepinsky, R. B., Tizard, R., Mattaliano, R. J., Sinclair, L. K., Miller, G. T., Browning, J. L., Chow, E. P., Burne, C., Huang, K. S., and Pratt, D. (1988) Five distinct calcium and phospholipid binding proteins share homology with lipocortin I. *J. Biol. Chem.* 263: 10799–10811.

Piltch, A., Sun, L., Fava, R. A., and Hayashi, J. (1989) Lipocortin-independent effect of dexamethasone on phospholipase activity in a thymic epithelial cell line. *Biochem. J.* 261: 395–400.

Plager, D. A. and Nelsestuen, G. L. (1994) Direct enthalpy measurements of the calcium-dependent interaction of annexins V and VI with phospholipid vesicles. *Biochemistry* 33: 13239-13249.

Pollard, H. B. and Rojas, E. (1988). Ca^{2+}-activated synexin forms highly selective, voltage-gated Ca^{2+} channels in phosphatidylserine bilayer membranes. *Proc. Natl. Acad. Sci. USA* 85: 2974–2978.

Pollard, H. B., Guy, H. R., Arispe, N., de la Fuente, M., Lee, G., Rojas, E. M., Pollard, J. R., Srivastava, M., Zhang-Keck, Z.-Y., Merezhinskaya, N., Caohuy, H., Burns, A. L., and Rojas, E. (1992) Calcium channel and membrane fusion activity of synexin and other members of the annexins gene family. *Biophys. J.* 62: 15–18.

Powell, M. A. and Glenney, J. R. (1987) Regulation of calpactin I phospholipid binding by calpactin I light-chain binding and phosphorylation by p60v-src. *Biochem J.* 247: 321–328.

Pula, G., Bianchi, R., Ceccarelli, P., Giambanco, I., and Donato, R. (1990) Characterization of mammalian heart annexins with special reference to CaBP33 (annexin V). *FEBS Lett.* 277: 53–58.

Radke, K., Gilmore, T., and Martin, G. S. (1980) Transformation by Rous sarcoma virus: a cellular substrate for transformation specific phosphorylation contains phosphotyrosine. *Cell* 21: 821–828.

Ravanat, C., Torbet, J., and Freyssinet, J. M. (1992) A neutron solution scattering study of the structure of annexin-V and its binding to lipid vesicles. *J. Mol. Biol.* 226: 1271–1278.

Raynal, P. and Pollard, H. B. (1994) Annexins: the problem of assessing the biological role for a gene family of multifunctional calcium- and phospholipid-binding proteins. *Biochim. Biophys. Acta* 1197: 63–93.

Raynal, P., Hullin, F., Ragab-Thomas, J. M., Fauvel, J., and Chap, H. (1993) Annexin 5 as a potential regulator of annexin 1 phosphorylation by protein kinase C. In vitro inhibition compared with quantitative data on annexin distribution in human endothelial cells. *Biochem. J.* 292: 759–765.

Raynal, P., van Bergen en Henegouwen, P. M., Hullin, F., Ragab-Thomas, J. M., Fauvel, J., Verkleij, A., and Chap, H. (1992) Morphological and biochemical evidence for partial nuclear localization of annexin 1 in endothelial cells. *Biochem. Biophys. Res. Commun.* 186: 432–439.

Reutlingsperger, C. P. M., Hornstra, G., and Hemker, H. C. (1985) Isolation and partial purification of a novel anticoagulant from arteries of human umbilical cord. *Euro. J. Biochem.* 151: 625–629.

Rojas, E. and Pollard, H. B. (1987) Membrane capacity measurements suggest a calcium-dependent insertion of synexin into phosphatidylserine bilayers. *FEBS Lett.* 217: 25–31.

Rojas, E., Arispe, N., Haigler, H. T., Burns, A. L., and Pollard, H. B. (1992) Identification of annexins as calcium channels in biological membranes. *Bone and Miner.* 17: 214–218.

Rojas, E., Pollard, H. B., Haigler, H. T., Parra, C., and Burns, A. L. (1990) Calcium-activated endonexin II forms calcium channels across acidic phospholipid bilayer membranes. *J. Biol. Chem.* 265: 21207–21215.

Römisch, J., Schorlemmer, U., Fickenscher, K., Paques, E. P., and Heimburger, N. (1990) Anticoagulant properties of placenta protein 4 (annexin V). *Thrombosis Res.* 60: 355–366.

Römisch, J., Schuler, E., Bastian, B., Burger, T., Dunkel, F. G., Schwinn, A., Hartmann, A. A., and Paques, E. P. (1992) Annexins I to VI: quantitative determination in different human cell types and in plasma after myocardial infarction. *Blood Coag. Fibrinol.* 3: 11–17.

Russo-Marie, F. (1992) Annexins, phospholipase A_2 and glucocorticoids. In *The Annexins* (Moss, S. E. ed.). Portland Press, London, pp. 35–46.

Sarafian, T., Pradel, L. A., Henry, J. P., Aunis, D., and Bader, M. F. (1991) The participation of annexin II (calpactin I) in calcium-evoked exocytosis requires protein kinase C. *J. Cell Biol.* 114: 1135–47.

Schlaepfer, D. D. and Haigler, H. T. (1987) Characterization of Ca^{2+}-dependent phospholipid binding and phosphorylation of lipocortin I. *J. Biol. Chem.* 262: 6931–6937.

Schlaepfer, D. D. and Haigler, H. T. (1988) In vitro protein kinase C phosphorylation sites of placental lipocortin. *Biochemistry* 27: 4253–4258.

Schlaepfer, D. D., Jones, J., and Haigler, H. T. (1992) Inhibition of protein kinase C by annexin V. *Biochemistry* 31: 1886–1891.

Scott, D., Otwinowsky, Z., Gelb, M. H., and Sigler, P. B. (1990) Crystal structure of bee-venom phospholipase A$_2$ in a complex with a transition-state analogue. *Science* 250: 1563–1566.

Smallwood, M. F., Gurr, S. J., Choudhari, U., and Bowles, D. J. (1990) Characterization of plant annexin gene expression. *Biochem. Soc. Trans.* 18: 1116.

Smith, P. D. and Moss, S. E. (1994) Structural evolution of the annexin supergene family. *Trends Genet.* 10: 241–246.

Smith, V. L., Kaetzel, M. A., and Dedman, J. R. (1990) Stimulus-response coupling: the search for intracellular calcium mediator proteins. *Cell Reg.* 1: 165–172.

Smythe, E., Smith, P. D., Jacob, S. M., Theobald, J., and Moss, S. E. (1994). Endocytosis occurs independently of annexin VI in human A431 cells. *J. Cell Biol.* 124: 301–306.

Sobota, A., Bandorowicz, J., Jezierski, A., and Sikorski, A. F. (1993) The effect of annexin IV and VI on the fluidity of phosphatidylserine/phosphatidylcholine bilayers studied with the use of 5-deoxylstearate spin label. *FEBS Lett.* 315: 178–182.

Solito, E., Raugei, G., Melli, M., and Parente, L. (1991) Dexamethasone induces the expression of the mRNA of lipocortin 1 and 2 and the release of lipocortin 1 and 5 in differentiated, but not undifferentiated U-937 cells. *FEBS Lett.* 291: 238–244.

Sopkova, J., Gallay, J., Vincent, M., Pancoska, P., and Lewit-Bentley, A. (1994) The dynamic behavior of annexin V as a function of calcium ion binding: a circular dichroism, UV absorption, and steady-state and time-resolved fluorescence study. *Biochemistry* 33: 4490–4499.

Sopkova, J., Renouard, M., and Lewit-Bentley, A. (1993) The crystal structure of a new high-calcium form of annexin V. *J. Mol. Biol.* 234: 816–825.

Soric, J. and Gordon, J. A. (1985) The 36-kilodalton substrate of pp60v-src is myristylated in a transformation-sensitive manner. *Science* 230: 563–566.

Stühmer, W., Conti, F., Suzuki, H., Wang, X., Noda, M., Yahagi, N., Kubo, H., and Numa, S. (1989) Structural parts involved in activation and inactivation of the sodium channel. *Nature* 339: 597–603.

Sun, J., Salem, H. H., and Bird, P. (1992) Nucleolar and cytoplasmic localization of annexin V. *FEBS Lett.* 314: 425–429.

Swairjo, M. A. and Seaton, B. A. (1994) Annexin structure and membrane interactions: a molecular perspective. *Ann. Rev. Biophys. Biomol. Struct.* 23: 193–213.

Swairjo, M. A., Roberts, M. F., Campos, M. B., Dedman, J. R., and Seaton, B. A. (1994) Annexin V binding to the outer leaflet of small unilamellar vesicles leads to altered inner-leaflet properties: ^{31}P- and ^{1}H-NMR studies. *Biochemistry* 33: 10944–10950.

Tait, J. F. and Gibson, D. (1992) Phospholipid binding of annexin V: effects of calcium and membrane phospha-tidylserine content. *Arch. Biochem. Biophys.* 298: 187–191.

Tait, J. F., Gibson, D., and Fujikawa, K. (1989) Phospholipid binding properties of human placental anticoagulant protein-I, a member of the lipocortin family. *J. Biol. Chem.* 264: 7944–7949.

Thiagarajan, P. and Tait, J. F. (1990) Binding of annexin V/placental anticoagulant protein I to platelets. Evidence for phosphatidylserine exposure in the procoagulant response of activated platelets. *J. Biol. Chem.* 265: 17420-17423.

Thorin, B., Gache, G., Dubois, T., Grataroli, R., Domingo, N., Russo-Marie, F., and Lafont, H. (1995) Annexin VI is secreted in human bile. *Biochem. Biophys. Res. Commun.* 209: 1039–1045.

Thunnissen, M. M., Ab, E., Kalk, K. H., Drenth, J., Dijkstra, B. W., Kuipers, O. P., Dijkman, R., de Haas, G H., and Verheij, H. M. (1990) X-ray structure of phospholipase A$_2$ complexed with a substrate-derived inhibitor. *Nature* 347: 689–691.

Tokumitsu, H., Mizutani, A., Minami, H., Kobayashi, R., and Hidaka, H. (1992) A calcyclin-associated protein is a newly identified member of the Ca^{2+}/phospholipid-binding proteins, annexin family. *J. Biol. Chem.* 267: 8919–8024.

Trave, G., Cregut, D., Lionne, C., Quignard, J. F., Chiche, L., Sri Widada, J., and Liautard, J. P. (1994a). Site-directed mutagenesis of a calcium binding site modifies specifically the different biochemical properties of annexin I. *Protein Eng.* 7: 689–696.

Trave, G., Quignard, J. F., Lionne, C., Sri Widada, J., and Liautard, J. P. (1994b) Interdependence of phospholipid specificity and calcium binding in annexin I as shown by site-directed mutagenesis. *Biochim. Biophys. Acta* 1205: 215–222.

Trotter, P. J., Orchard, M. A., and Walker, J. H. (1995a). Ca^{2+} concentration during binding determines the manner in which annexin V binds to membranes. *Biochem. J.* 308: 591–598.

Trotter, P. J., Orchard, M. A., and Walker, J. H. (1995b) EGTA-resistant binding of annexin V to platelet membranes can be induced by physiological calcium concentrations. *Biochem. Soc. Trans.* 23: 37S.

Tufty, R. M. and Kretsinger, R. H. (1975) Troponin and parvalbumin calcium binding regions predicted in myosin light chain and T4 lysozyme. *Science* 187 167–169.

Unwin, N. (1995) Acetylcholine receptor channel image at the open state. *Nature* 373: 37–43.

Valentine-Braun, K. A., Hollenberg, M. D., Fraser, E., and Northup, J. K. (1987) Isolation of a major human placental substrate for the epidermal growth factor (urogastrone) receptor kinase: immunological cross-reactivity with transducin and sequence homology with lipocortin. *Arch. Biochem. Biophys.* 259: 262–282.

Vassilev, P. M., Scheuer, T., and Catterall, W. A. (1988) Identification of an intracellular peptide segment involved in sodium channel inactivation. *Science* 241: 1658–1661.

Violette, S. M., King, I., Browning, J. L., Pepinsky, R. B., Wallner, B. P., and Sartorelli, A. C. (1990) Role of lipocortin I in the glucocorticoid induction of the terminal differentiation of a human squamous carcinoma. *J. Cell. Physiol.* 142: 70–77.

Voges, D., Berendes, R., Burger, A., Demange, P., Baumeister, W., and Huber, R. (1994) Three-dimensional structure of membrane-bound annexin V. A correlative electron microscopy–X-ray crystallography study. *J. Mol. Biol.* 238: 199–213.

Walker, J. H. (1982) Isolation from cholinergic synapses of a protein that binds to membranes in a calcium dependent manner. *J. Neurosci.* 39: 815–823.

Wallner, B. P., Mattaliano, R. J., Hession, C., Cate, R. L., Tizard, R., Sinclair, L. K., Foeller, C., Chow, E. P., Browing, J. L., Ramachandran, K. L. et al. (1986) Cloning and expression of human lipocortin, a phospholipase A2 inhibitor with potential anti-inflammatory activity. *Nature* 320: 77–81.

Wang, W. and Creutz, C. E. (1992) Regulation of the chromaffin granule aggregating activity of annexin I by phosphorylation. *Biochemistry* 31: 9934–9939.

Wang, W. and Creutz, C. E. (1994) Role of the aminoterminal domain in regulating interactions of annexin I with membranes: effects of amino-terminal truncation and mutagenesis of the phosphorylation sites. *Biochemistry* 33: 275–282.

Weber, K. (1992). Annexin II: interaction with p11. In *The Annexins* (Moss, S. E. ed.). Portland Press, London, pp. 61–68.

Weng, X., Luecke, H., Song, I. S., Kang, D. S., Kim, S. H., and Huber, R. (1993) Crystal structure of human annexin I at 2.5 A resolution. *Protein Sci* 2: 448–458.

White, S. P., Scott, D. L., Otwinowsky, Z., Gelb, M. H., and Sigler, P. B. (1990). Crystal structure of cobra-venom phospholipase A$_2$ in a complex with a transition-state analogue. *Science* 250: 1560–1563.

Wice, B. M. and Gordon, J. I. (1992) A strategy for isolation of cDNAs encoding proteins affecting human intestinal epithelial cell growth and differentiation: characterization of a novel gut-specific N-myristoylated annexin. *J. Cell Biol.* 116: 405–422.

Wu, L. N., Genge, B. R., Lloyd, G. C., and Wuthier, R. E. (1991) Collagen-binding proteins in collagenase-released matrix vesicles from cartilage. Interaction between matrix vesicle proteins and different types of collagen. *J. Biol. Chem.* 266: 1195–1203.

Wu, L. N. Y., Yoshimori, T., Genge, B. R., Sauer, G. R., Kirsch, T., Ishikawa, Y., and Wuthier, R. E. (1993) Characterization of the nucleational core complex responsible for mineral induction by growth plate cartilage matrix vesicles. *J. Biol. Chem.* 268: 25084–25094.

Zagotta, W. N., Hoshi, T., and Aldrich, R. W. (1990) Restoration of inactivation mutants of Shaker potassium channels by a peptide derived from ShB. *Science* 250: 568–571.

Zaks, W. J. and Creutz, C. E. (1991) Ca(2+)-dependent annexin self-association on membrane surfaces. *Biochemistry* 30: 9607–9615.

Zeng, F. Y., Gerke, V., and Gabius, H. J. (1993) Identification of annexin II, annexin VI and glyceraldehyde-3-phosphate dehydrogenase as calcyclin-binding proteins in bovine heart. *Int. J. Biochem.* 25: 1019–1027.

6

Gamma-Glutamic Acid-Containing Proteins

Torbjörn Drakenberg
Sture Forsén
Johan Stenflo

Although the requirement of calcium for blood coagulation has been known for almost a century, it is only in the last two decades that we have begun to gain insight into the nature of the calcium-binding sites and the role of the individual calcium ions. Extracellular proteins exist in an environment where the concentration of free ionized calcium is approximately 1.2 mM. Yet, these proteins have binding constants for calcium ranging from 10^3 to $10^9 \, M^{-1}$. In most instances, the calcium-binding sites appear to stabilize the fold of proteins but calcium ions may also be directly involved in enzyme activation, enzyme–cofactor and domain–domain interactions, as well as in membrane binding. Coagulation factors containing the Ca^{2+}-binding γ-carboxyglutamic acid (Gla) will be discussed in this chapter, with an emphasis on the calcium binding of the proteins, on the nature of the calcium sites, and on properties induced by Ca^{2+} binding.

In all Gla-containing proteins, the N-terminal 9–12 glutamic acid residues have been carboxylated to Gla in postribosomal, vitamin K-dependent carboxylations (Furie and Furie, 1990; Mann et al., 1990; Stenflo and Suttie, 1977; Suttie, 1985). The Gla-containing coagulation factors are prothrombin (factor II), factor VII, factor IX, and factor X (collectively referred to as the "vitamin K-dependent coagulation factors"), and the anticoagulant regulatory proteins: protein C and protein S. Protein Z, a phylogenetically related protein of unknown function, also contains Gla. In addition, two proteins in mineralized tissues contain Gla: osteocalcin (or bone Gla protein) and matrix Gla protein, which contain three and five Gla residues, respectively (Hauschka et al., 1989). Gla has also been found in certain very potent neurotoxins from molluscs (Olivera et al., 1990). These proteins will not be discussed here. To facilitate an appreciation of the involvement of calcium in blood coagulation proteins, a brief recapitulation of the blood coagulation process is warranted.

Blood Coagulation

Vascular injury triggers a series of chemical reactions resulting in activation of the zymogen, prothrombin, to the corresponding active enzyme, thrombin, which is a trypsin-like serine protease (Davie et al., 1991; Jackson and Nemerson, 1980; Mann et al., 1990; Stenflo and Dahlbäck, 1994). Thrombin then cleaves four peptide bonds in fibrinogen, converting it to fibrin monomers which spontaneously assemble into a branching network of fibrils, fibrin, which is the primary material of blood clots. The reactions that resulted in thrombin formation are sequentially ordered in a so-called cascade, where an active serine protease activates the next zymogen in the cascade, an arrangement that ensures amplification of a small initial stimulus.

Blood coagulation is initiated by binding of factor VII or factor VIIa (factor VII is a zymogen whereas the "a" in factor VIIa denotes the active serine protease) to its cofactor, called tissue factor (TF) (Rapaport, 1991). Unlike free factor VII, factor VII bound to TF is rapidly activated by extremely low concentrations of factor VIIa, factor Xa, thrombin, and/or other as yet unidentified proteases that are released upon tissue injury (Fig. 6.1). Free factor VIIa is not inhibited by serine protease inhibitors, such as antithrombin III, and circulates in low concentration in plasma with a half-life identical with that of factor VII. The TF is an integral membrane protein expressed on fibroblasts in the adventitia of blood vessels, but not by the endothelial cells that line the vessel wall. Tissue injury exposes TF which

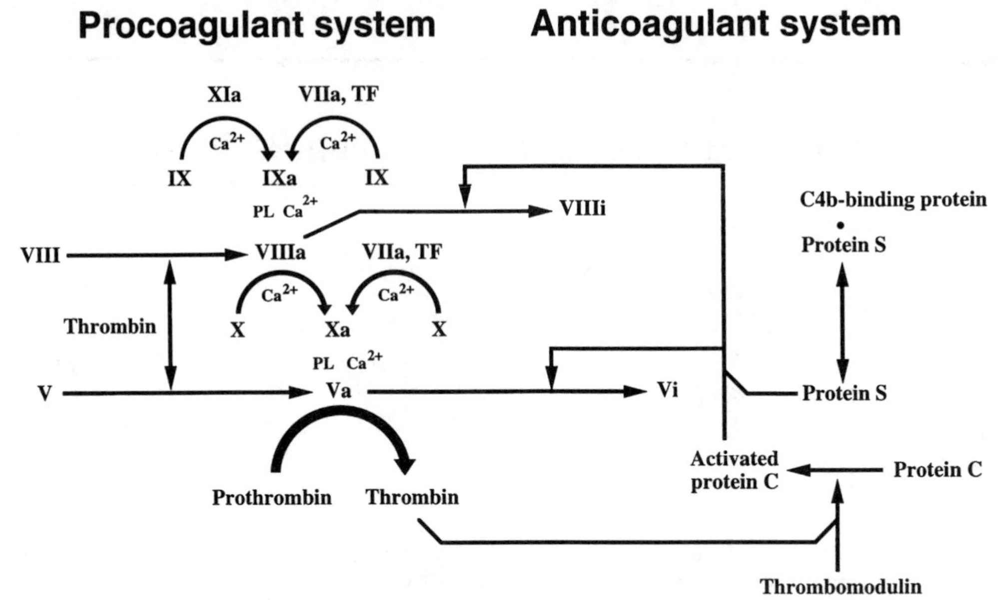

Procoagulant system   Anticoagulant system

Figure 6.1 Schematic representation of blood coagulation. The cascade of reactions is initiated by Factor VIIa in complex with an integral membrane protein, tissue factor (extrinsic pathway), or by factor XIa (intrinsic pathway). The factor VIIa–tissue factor complex activates either factor IX or factor X. In vivo, the coagulation cascade is initiated via the extrinsic pathway as a result of tissue damage and the exposure of blood to tissue factor. The active forms of the serine proteases and the two coactors V and VIII are indicated by lower case letter a, e.g. X denotes the zymogen factor X and Xa the active enzyme factor Xa respectively. The activation of factors V and VIII by thrombin is indicated. Thrombin can either convert fibrinogen to fibrin (procoagulant pathway, not shown) or form complex with the integral membrane protein, thrombomodulin, and activate protein C (anticoagulant pathway). Activated protein C degrades factors Va and VIIIa in the presence of phospholipid and turns off the blood coagulation process.

binds factors VII/VIIa, resulting in activation of factors IX and/or X.

A characteristic feature of the coagulation enzymes is that they are essentially inactive against their physiological substrates in the absence of suitable phospholipid and cofactor (Kane and Davie, 1988; Mann et al., 1990). To be biological active, factor IXa must form complex with factor VIIIa (the active form of the cofactor, factor VIII) on a cell membrane. The active membrane-bound macromolecular complex activates factor X to factor Xa, which forms a macromolecular complex with membrane-bound factor Va. Both the protein–phospholipid interactions, e.g. factor Xa and phospholipid, and the enzyme–cofactor interactions, e.g. factor Xa and factor Va, require calcium to be bound by both enzyme and cofactor. The macromolecular complexes are preferentially assembled on phosphatidyl serine-containing membranes. Phosphatidyl serine is enriched on the inner leaflet of biological membranes and is thus exposed upon cell injury; activation of platelets also results in exposure of phos-

phatidyl serine on the outer leaflet of the cell membrane (Swords and Mann, 1993).

The reversible membrane-binding of Gla-containing coagulation factors increases the probability of productive collisions between enzyme and cofactor on one hand, and between macromolecular enzyme complex and its substrate on the other. The two cofactors, factors Va and VIIIa, are specific high-affinity binding proteins ($K_d \approx 10^{-9}$ M) for the active enzymes, factors IXa and Xa, respectively, but have no measurable affinity for the corresponding zymogens (Dahlbäck and Stenflo, 1978; Miletich et al., 1977). It is noteworthy that factors V and VII, unlike the integral membrane protein, TF, circulate in plasma as inactive precursors of the cofactors; i.e. they only become activated upon limited thrombin- or factor Xa-mediated proteolysis and insertion in the appropriate biological membrane (Kane and Majerus, 1982). The rate of activation of prothrombin by factor Xa in the presence of calcium increases approximately 100-fold if phospholipid is present (reduces K_M for prothrombin). If factor Va is added instead of phospholipid, the rate increases approximately 1000-fold (increases V_{max}). Phospholipid and factor Va together increase the rate of prothrombin activation 10^5–10^6-fold (Mann et al., 1990).

Regulation of blood coagulation is obtained by several mechanisms; either by inactivation of the active enzymes by protease inhibitors (among which the serine protease inhibitors have been studied in most detail, particularly antithrombin III) or by degradation of the cofactor-binding proteins. Degradation of the cofactors—factor Va and factor VIIIa—by limited proteolysis, is mediated by activated protein C, resulting in down-regulation of the blood coagulation cascade (Dahlbäck and Stenflo, 1994; Esmon, 1989; Stenflo, 1988). Protein C is a Gla-containing serine protease. It is activated by thrombin in complex with thrombomodulin (TM), an integral membrane protein expressed by endothelial cells, particularly in the capillaries (Esmon, 1989, 1993). When in complex with TM, the activity of thrombin is modulated; it loses its procoagulant activities (conversion of fibrinogen to fibrin by limited proteolysis), but its ability to activate protein C is enhanced several thousand-fold.

The importance of Gla in calcium binding is illustrated by the well-known fact that vitamin K-antagonistic drugs, such as warfarin, inhibit carboxylation of the Glu residues in factors VII, IX, and X, prothrombin, protein C, and protein S (Stenflo and Suttie, 1977). The Glu-containing forms of the coagulation factors have lost ten to twelve calcium-binding sites and do not interact with biological membranes. For approximately 50 years, warfarin has been used in clinical medicine for the treatment of thrombotic disease. For more detailed information about blood coagulation, the reader is referred to several recent reviews (Dahlbäck and Stenflo, 1994; Davie et al., 1991; Furie and Furie, 1988; Jackson and Nemerson, 1980; Mann et al., 1990; Stenflo and Dahlbäck, 1994).

Modular Structure of γ-Carboxyglutamic Acid-Containing Coagulation Factors

The Gla-containing plasma proteins are glycoproteins that contain from 406 to 635 amino acids. Based on their modular structure, three types of Gla-containing proteins can be identified (Fig. 6.2). Factors VII, IX and X, and protein C, form one group, whereas both prothrombin and protein S have unique structures (Furie and Furie, 1988; Stenflo and Dahlbäck, 1994). The common denominator of the proteins is the NH_2-terminal module, the Gla module, which contains 9 to 12 Gla residues. In factors VII, IX, and X, and protein C, the Gla mod-

ule is followed by two modules that are homologous to the EGF precursor, whereas the serine protease module is C-terminal. In prothrombin, a hexadecapeptide with a disulfide loop has been inserted C-terminal of the Gla module. This region is followed by two so-called kringle modules, while the COOH-terminal half of prothrombin is occupied by the serine protease module. In protein S, the Gla module is followed by a small module which has an internal disulfide bond, and two arginyl bonds that are susceptible to cleavage by thrombin. The Gla module is followed by four EGF-like modules, whereas the COOH-terminal part consists of a domain that is homologous to plasma steroid hormone binding proteins.

The structure of vitamin K-dependent plasma proteins is consistent with the view that many genes in eukaryotes that encode extracellular and membrane proteins have been assembled via intron-mediated exon shuffling (Baron et al., 1991; Campbell and Bork, 1993; Patthy, 1985, 1991). According to this view, the exons are remnants of primordial genes which in the course of evolution have been shuffled between genes and duplicated, giving rise to proteins of complex modular design. In the Gla-containing proteins, the propeptide and the major NH_2-terminal part of the Gla module are encoded on the same exon. This is noteworthy, as the propeptide (which binds vitamin K-dependent carboxylase) and the Gla module together constitute a functional unit (Furie and Furie, 1988; Suttiet, 1993; Vermeer, 1990).

In the above scenario, the coagulation proteins have been derived from simple primordial serine proteases that contained a signal peptide required for secretion and a COOH-terminal serine protease part (Patthy, 1985, 1991, 1994). Exons were recruited to the genes encoding these simple proteins and inserted into the intron that separated the exons which encoded the signal peptide and the serine protease module, respectively. Recruitment of exons and exon duplications resulted in an increase in size of the NH_2-terminal, noncatalytic region, that accounts for approximately half of each of the Gla-containing serine protease zymogens. The noncatalytic modules have important regulatory functions relating to protein secretion, postribosomal modification, and regulation of the coagulation response at the site of injury.

Gla-Containing Modules

Calcium Binding to Gla-Containing Modules

Conversion of 9–12 Glu residues to Gla in the N-terminal domain is a prerequisite for calcium-binding

Factor X

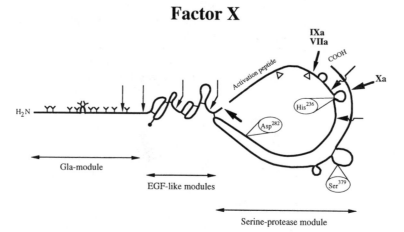

Prothrombin

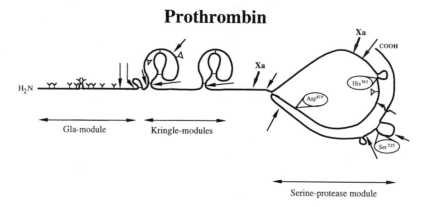

Protein S

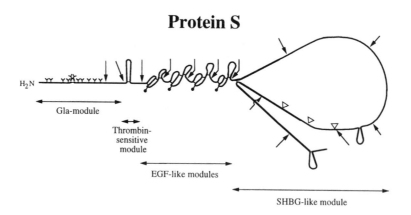

Figure 6.2 Modular structure of the vitamin K-dependent plasma proteins. The structure of factor X also represents that of the closely related factors VII and IX and protein C. The pre–pro leader sequences are not shown. Open triangles denote carbohydrate side chains. Cleavages in factor Xa and prothrombin that are mediated by factors VIIa and IXa and by factor Xa, respectively, are denoted by thick arrows. Small arrows denote the location of introns in the corresponding genes. The residues in the active sites of factor X and prothrombin are given in ovals. The symbol Y denotes γ-carboxyglutamic acid, a "lollipop" symbol on an EGF-like module denotes a β-hydroxyaspartic or β-hydroxyasparagine residue.

and biological activity of vitamin K-dependent coagulation factors. The dicarboxylic side chains of Gla residues inevitably have higher calcium affinity than the monocarboxylic Glu side chains. A monocarboxyblic acid, such as acetic acid, binds calcium ions with $K_d \approx 300$ mM, whereas malonic acid (a dicarboxylic acid similar to Gla) binds calcium ions with higher affinity, $K_d \approx 30$ mM. As the total calcium concentration in blood plasma is 2.2 ± 0.2 mM, and that of free ionized calcium is 1.1 ± 0.1 mM, it is evident that a native three-dimensional structure of the Gla domain is required to create sites with affinity sufficiently high to ensure at least partial site saturation at physiological calcium concentrations. However, it should also be noted that very strong calcium-binding sites of the EF-hand type consist of only unmodified amino acids. In fact, in one of the calcium sites in calbindin, all ligands except one (a glutamic acid residue) are backbone carbonyls (Szebenyi and Moffat, 1986). It is therefore clear that Gla residues are not necessarily a requirement for strong calcium binding, though they may be needed to obtain a conformation switch necessary for the function of these proteins.

The first steps toward a characterization of the calcium-binding properties of prothrombin containing Gla with those of uncarboxylated prothrombin were made in the early 1970s (Henriksen and Jackson, 1975; Nelsestuen and Suttie, 1972a; Stenflo, 1973; Stenflo and Ganrot, 1972). Since then, numerous studies of the calcium-binding properties of Gla-containing proteins have been performed, in most cases using equilibrium dialysis and ^{45}Ca, though spectroscopic methods, such as difference spectroscopy, fluorescence spectroscopy, circular dichroism and NMR spectroscopy, as well as calorimetry and ion-selective electrodes, have also been used (Jackson, 1988; Marsh et al., 1979; Nelsestuen, 1976; Öhlin et al., 1988b, 1990; Persson et al., 1991b; Prendergast and Mann, 1977).

Results of binding studies using equilibrium dialysis have indicated 6–12 calcium binding sites to exist in all vitamin K-dependent plasma proteins (Jackson, 1988). For several reasons, however, attempts to make accurate determinations of stoichiometric calcium-binding constants have been fraught with difficulty. The low affinity of the calcium-binding sites (K_d values between 0.1 and 2 mM) necessitates the use of high protein concentrations and requires high calcium concentrations in equilibrium dialysis experiments to attain a reasonable degree of site saturation. The cooperativity that characterizes binding of the first two to four calcium ions is also a complication, and has been difficult to evaluate as it may also reflect aggregation of the protein (Henriksen and Jackson, 1975; Stenflo, 1973). There have been few studies where attempts have been made to control this factor, for instance by performing binding measurements at varying protein concentrations. It should also be borne in mind that, as all Gla-containing coagulation factors in plasma, except protein S, are zymogens or serine proteases, control of proteolytic activity is crucial. For these reasons, information about binding constants, limiting calcium-binding stoichiometry, and degree of cooperativity is generally less accurate for Gla-containing coagulation factors than for intracellular calcium-binding proteins, such as calmodulin and related proteins, that bind few calcium ions with high affinity. Nevertheless, there seems to be a consensus that there are 6–12 calcium-binding sites in the Gla-containing coagulation factors, and that there is cooperative binding of the first two to four calcium ions (Bajaj et al., 1975; Borowski et al., 1985; Henrikens and Jackson, 1975; Jackson, 1988; Nelsestuen and Suttie, 1972b; Nelsestuen et al., 1981; Stenflo and Ganrot, 1973). The crystal structure of the calcium form of prothrombin fragment 1 has provided further support for inferences made about cooperative calcium binding (Soriano Garcia et al., 1992).

Studies of calcium-induced alterations of the intrinsic protein fluoresence have shed light on some of the metal ion-binding properties of Gla-containing proteins. In bovine prothrombin, calcium induces a slow ($t_{12} \approx 100$ min) fluorescence quenching, thought to be due to a *cis–trans* isomerization of Pro 22, a residue that is not conserved in human prothrombin or in the other vitamin K-dependent proteins (Fig. 6.3) (Nelsestuen, 1976). Calcium binding to factor X-GlaEGF$_N$ (fX-GlaEGF$_N$), consisting of the Gla domain and the first EGF-like domain, has also been followed with measurements of the fluoresence emission (Persson et al., 1991a). In this

									10										20										30										40				
Prothrombin	A	N	T	-	F	L	γ	γ	V	R	K	G	N	L	γ	R	γ	C	V	γ	γ	T	C	S	Y	γ	γ	A	F	γ	A	L	γ	S	-	S	T	A	T	D	V	F	W	A	K	Y	T
Factor VII	A	N	A	-	F	L	γ	γ	L	R	P	G	S	L	γ	R	γ	C	K	γ	γ	Q	C	S	F	γ	γ	A	R	γ	I	F	K	D	-	A	γ	R	T	K	L	F	W	I	S	Y	S
Factor IX	Y	N	S	G	K	L	γ	γ	F	V	Q	G	N	L	γ	R	γ	C	M	γ	γ	K	C	S	F	γ	γ	A	R	γ	V	F	γ	N	-	T	γ	R	T	T	γ	F	W	K	Q	Y	V
Factor X	A	N	S	-	F	L	γ	γ	M	K	K	G	H	L	γ	R	γ	C	M	γ	γ	T	C	S	Y	γ	γ	A	R	γ	V	F	γ	D	-	S	D	K	T	N	γ	F	W	N	K	Y	K
Protein C	A	N	S	-	F	L	γ	γ	L	R	H	S	S	L	γ	R	γ	C	I	γ	γ	I	C	D	F	γ	γ	A	K	γ	I	F	Q	N	-	V	D	D	T	L	A	F	W	S	K	H	V
Protein S	A	N	S	-	L	L	γ	γ	T	K	Q	G	N	L	γ	R	γ	C	I	γ	γ	L	C	N	K	γ	γ	A	R	γ	V	F	γ	N	D	P	γ	-	T	D	Y	F	Y	P	K	Y	L

Figure 6.3 Amino acid sequences in the Gla-modules of human vitamin K-dependent coagulation factors. Residues are shaded when at least three of six are identical.

fragment, fluorescence quenching can be linked to a single conserved Trp residue (position 41; Fig. 6.3) in the aromatic cluster that, in the three-dimensional structure, is adjacent to the disulfide bond linking Cys 19 and 22 (Soriano Garcia et al., 1991). In prothrombin, and presumably also in factor X, the orientation of the Trp indole is altered upon calcium binding, which presumably accounts for the metal ion-induced changes in intrinsic protein fluorescence (Soriano Garcia et al., 1992). The metal ion-induced fluoresence quenching in fX-GlaEGF$_N$ occurs in the same range of calcium concentrations as in intact factor X, indicating the fragment to contain sufficient structural information for normal folding of the Gla domain. Gla domains, cleaved in the aromatic cluster of the C-terminal α-helical peptide, can be isolated after limited chymotryptic digestion of factors IX and X and protein C (Johnson et al., 1983; Morita and Jackson, 1980; Morita and Kisiel, 1985; Morita et al., 1984; Sugo et al., 1984) and from factor VII after digestion with cathepsin G (Nicolaisen et al., 1992). The isolated Gla modules from factor X (residues 1–44), factor IX (residues 1–43), and prothrombin, bind calcium, but with lower affinity than does the Gla domain in the intact protein or that in the GlaEGF fragments (Astermark et al., 1991; Persson et al., 1991a). Other properties of the isolated Gla domains, e.g. their comparatively low solubility in the presence of calcium, also indicate that they do not have a native conformation. On the other hand, a synthetic Gla domain containing the entire C-terminal α-helix appears to have normal Ca^{2+}-binding properties (Colpitts and Castellino, 1994; Jacobs et al., 1994; Martin et al., 1993). It thus appears that an intact C-terminal α-helix is required for folding of the Gla domain to a native conformation. The adjacent domain, i.e. the tetradecapeptide disulfide loop (residues 47–60 in prothrombin fragment 1) and the N-terminal EGF-like domain in factors IX and X and protein C, is in close contact with the Gla domain but may not be necessary for its folding, particularly in the presence of Ca^{2+} (Sunnerhagen et al., 1995a, 1995b). The Gla domain with calcium bound protects susceptible peptide bonds in the C-terminal α-helix in the Gla domain of factors IX and X and protein C from proteolysis (Johnson et al., 1983; Morita and Jackson, 1980). Presumably, this is due to a reorientation of the helix relative to the adjacent EGF-like domain upon calcium binding (see below). Interaction between the Gla domain and the adjacent domains has also been inferred from studies of the calcium-binding properties of protein S in which the thrombin-sensitive bond had been cleaved. In cleaved protein S, the Gla domain is linked to the remainder of the molecule by a disulfide bond (Fig. 6.2) (Dahlbäck, 1983; Dahlbäck et al., 1986). The Gla domain in the cleaved protein has lower calcium affi-

nity than that in intact protein S, and the cleaved molecule is not a cofactor of activated protein C.

Calcium binding to Gla-containing proteins is clearly quite different from calcium binding to many intracellular proteins, such as calmodulin. In calmodulin, calcium binds to a more or less preformed pocket and induces a conformational alteration that exposes a hydrophobic pocket, thereby triggering a biological response (Strynadka and James, 1991). The cooperative calcium binding to Gla-containing proteins serves a primarily structural purpose (Henriksen and Jackson, 1975; Soriano Garcia et al., 1992; Stenflo, 1973). It induces folding of the N-terminal part of the domain, connects Gla residues that are remote in the linear sequence, and provides the energy required to force side chains of three hydrophobic amino acids into the aqueous solvent (Phe 5, Leu 6, and Val 9; Fig. 6.3), a prerequisite for the interaction with biological membranes (Sunnerhagen et al., 1995a).

The vitamin K-dependent plasma proteins also bind other divalent cations, such as Mg^{2+}, Mn^{2+}, Sr^{2+}, and lanthanide ions (Borowski et al., 1986; Deerfield et al., 1987; Jackson, 1988; Prendergast and Mann, 1977). However, these ions do not substitute for Ca^{2+} in blood coagulation. The structure of the Sr^{2+} form of prothrombin fragment 1 has been solved by x-ray crystallography (Seshadri et al., 1994). It was found to be very similar to that of the Ca^{2+} form although the coordination of certain Sr^{2+} ions was found to be different from that of the corresponding Ca^{2+} ion. Based on findings in experiments with conformation-specific antibodies, two classes of metal ion-binding sites have been postulated in factor IX (Borowski et al., 1986). One class of binding sites was found to be specific for Ca^{2+}, whereas the other class had lower specificity and could accommodate both Ca^{2+} and Mg^{2+}. Full biological activity was obtained only with Ca^{2+}, which could not be substituted by other cations. Recently, this hypothesis was modified as it was demonstrated that factor IX has at least one Mg^{2+}-binding site that does not interact with Ca^{2+} and that is saturated at physiological Mg^{2+} (Sekiya et al., 1995). Moreover, the rate of activation of factor IX to IXa was increased by Mg^{2+}, even in the presence of an excess of Ca^{2+}. This Mg^{2+}-binding site appears to be unique to factor IX.

Structure of Gla-Containing Modules

Gla modules contain approximately 45 amino acid residues and are encoded by two exons. As can be seen in Fig. 6.3, the sequence similarity indicates a phylogenetic relationship between these domains. All 9–12 Glu residues in the domain, but no other residues in the protein, have been carboxylated to Gla. The carboxylase thus appears to be unaffected by which amino acid is adjacent to a Glu residue that will be carboxylated. There are three pairs of Gla residues (6,7; 19,20; 25,25) that are conserved in all vitamin K-dependent proteins, as are the single Gla residues in positions corresponding to 14, 16, and 29 in human prothrombin. Position 32 is conserved in most Gla domains.

The pronounced sequence similarity among Gla modules indicates them to have similar properties and structures (Fig. 6.3). Calcium ion-induced spectral changes observed in Gla domain-containing proteins and in fragments of these proteins are quite similar (Astermark et al., 1991; Nelsestuen, 1976; Persson et al., 1991a; Öhlin et al., 1990; Prendergast and Mann, 1977). Moreover, all these proteins yield similar results in chemical modification experiments (Schwalbe et al., 1989; Welsch and Nelsestuen, 1988). Collectively, this evidence indicates the Gla modules in vitamin K-dependent coagulation factors to have a similar fold, which is now also becomming evident from recent structural studies. However, interactions of the Gla modules with adjacent modules in factors VII, IX, and X, and protein C are bound to be different from those in prothrombin and protein S, as the adjacent modules are different: in EGF-like domain in factors VII, IX, and X, and protein C, a short disulfide loop in prothrombin, and a thrombin-sensitive module in protein S (Fig. 6.2).

Although no intact Gla-containing protein has yet been crystallized, an N-terminal 156 amino acid residue long fragment from prothrombin, fragment 1, has been isolated from tryptic digests of bovine thrombin and crystallized in the presence of calcium (Soriano Garcia et al., 1992). Fragment 1 contains the Gla domain and the first kringle domain (Fig. 6.2). The structure of fragment 1 has been determined with x-ray diffraction methods (Soriano Garcia et al., 1989, 1992) to a resolution of 2.25 Å. The solution structure of an isolated Gla module from factor IX has been determined by NMR both in the apo form (Freedman et al., 1995b) and in the Ca^{2+} form (Freedman et al., 1995a). Moreover, the solution structure of the Gla–EGF domain pair from factor X has been determined in the apo form by NMR (Sunnerhagen et al., 1995a). Prothrombin fragment 1 has also been crystallized in the absence of calcium (Tulinsky et al., 1988). The C-terminal α-helix (residues 36–45) was well defined but, as there was no electron density N-terminal of this helix, it was assumed that this part of the Gla module did not have an ordered structure.

All forms, except the apo form of the isolated Gla module of factor IX, have essentially the same secondary structure consisting of three α-helices. In the N-terminal part of the module, there are two helices, with one and two turns, respectively (Fig. 6.4). The C-terminus is a three-turn helix (residues 36–45) organized in such a way that the side chains of Phe 40, Trp 41, and Tyr 44 form an aromatic cluster (Freedman et al., 1995a; Soriano Garcia et al., 1992).

The structure of the Gla module from fragment 1 as determined by x-ray diffraction methods shows the carboxylate groups from the Gla residues to be ligands to the calcium ions, apparently cross-linking parts of the module that are remote in the linear sequence. For example, Gla 6, Gla 16, and Gla 20 are ligands to the same Ca^{2+} ion as Gla 7, Gla 26, and Gla 30. The most striking structural feature is that carboxylate groups of Gla residues 6, 7, 16, 26, and 29 form a negatively charged "channel" with an array of five calcium ions with approximately 4 Å between neighbouring ions (Soriano Garcia et al., 1992). Gla residues 14, 19, and 20 form a negatively charged cluster adjacent to the hexapeptide disulfide

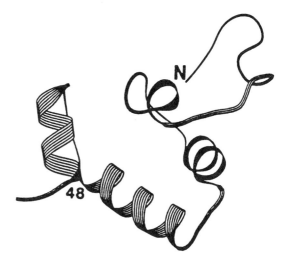

Figure 6.4 Ribbon drawing of the calcium form of prothrombin fragment 1. (Reproduced with permission from Soriano Garcia et al., 1992.)

loop peptide. In this region, two calcium ions have been identified. The geometry of the coordination around the calcium is not well defined and does not correspond to any of the idealized polyhedra. There are both unidentate interactions between carboxylate oxygens and calcium and chelate coordinations utilizing both carboxylates of Gla residues. Of the seven calcium ions identified, four to five seem to be inaccessible to solvent. An important structural feature is that the 9–12 N-terminal residues in the Gla module are folded in a horseshoe-like fashion, such that the amino group of Ala 1 forms a hydrogen-bonded ion pair with carboxylate groups of Gla 16 and 26. Another important feature of the structure in this region of fragment 1 is that the hydrophobic residues surrounding the first pair of Gla residues (i.e. Phe 4, Leu 5, and Val 8) are oriented such that their hydrophobic side chains appear to be exposed to the aqueous solvent (Figs. 6.3 and 6.4). The Gla residues in positions 15 and 20 also appear to form ion pairs with Arg 55 in the tetradecapeptide disulfide loop.

The solution structure of the Ca^{2+} form of the synthetic Gla module from factor IX has been determined by NMR (Freedman et al., 1995a). It has a global fold that is very similar to that of the Gla module of prothrombin but with significant differences in the N-terminal 11 residues. However, it should be borne in mind that due to solubility problems, the structure was determined with the protein dissolved in 3 M urea and 2.5 M guanidine HCl. It is somewhat surprising that the module is at all structured in this solvent. It is therefore debatable whether the differences observed reflect the influence of the solvent rather than a difference between the two proteins that would occur at physiological pH and ionic strength.

All Gla modules, except that in human factor IX, have an Ala as the N-terminus, and according to the x-ray structure a larger side chain cannot be accommodated. The N-terminal loop (residues 1–11) in the Gla module of factor IX therefore appears to adopt a fold that differs from that in prothrombin. In prothrombin, Ala 1 is completely buried in the interior of the protein (Soriano Garcia et al., 1992). This is in agreement with the observation that the amino-termini of vitamin K-dependent proteins are protected from chemical modification in the presence, but not absence, of Ca^{2+} ions (Schwalbe et al., 1989; Welsch and Nelsestuen, 1988). The only modification of Ala 1 that does not interfere with phospholipid binding is trinitrophenylation (Welsch and Nelsestuen, 1988a). It thus appears as if the N-terminal residues may adopt either of two alternative conformations, both of which allow membrane binding: one is characteristic of prothrombin and factors VII, X, and proteins C and S which have N-terminal Ala, whereas the other one is found in factor IX (which has an N-

terminal Tyr residue) and perhaps in trinitrophenylated factor X (Freedman et al., 1995a; Schwalbe et al., 1989; Soriano Garcia et al., 1992; Sunnerhagen et al., 1995a; Welsch et al., 1988b).

Attempts have been made to determine the solution structure of the Gla–EGF module pair from factor X in the presence of calcium (Sunnerhagen et al., 1995a). However, aggregation posed an insurmountable obstacle, and no suitable solvent mixture was found, though the urea/guanidine mixture was not tried.

As mentioned above, crystals from the apo form of prothrombin fragment 1 showed no electron density from amino acids 1–35, and this portion of the protein was therefore assumed to lack an ordered structure. Similarly, NMR studies of the synthetic Gla module with factor IX sequence (residues 1–47) resulted in a poorly defined structure (Freedman et al., 1995b). The C-terminal helix found in the crystal structure was also present in the solution structure of the Gla module of factor IX, as well as a loop containing residues 5–8 and the Cys-linked loop with residues 17–22. As these three structural elements were completely independent of each other, it seems fair to state that N-terminal to amino acid 35 there is no well-defined structure, in complete agreement with the x-ray studies. On the other hand, it has been found that the proteolytic fragment from factor X, which contains the intact Gla and the N-terminal EGF modules (fX-GlaEGF$_N$) (Sunnerhagen et al., 1995), has a fairly well-defined structure, even in the absence of Ca^{2+} ions. It was thus found that the EGF domain of fX-GlaEGF$_N$ has a structure that is virtually identical to that of the isolated EGF$_N$ domain, which has been determined previously (see below), while the Gla module was less well defined, indicating the presence of a more flexible polypeptide chain. Nevertheless, elements of secondary structure were identified (Fig. 6.5). The C-terminal α-helix is well defined, as in the calcium-free form of prothrombin fragment 1. In addition, the two N-terminal helices, comprising one and two turns, respectively, are present (residues 13–17 and 24–29). The major difference between the calcium-free and calcium-loaded forms of the fragment involves residues 1–9. In the calcium-loaded form, they form the horsehoe loop with the side chains of Phe 4, Leu 5, and Val 8 exposed to solvent and the side chains of Gla 6 and 7 coordinating Ca^{2+} ions in the interior of the protein. In the apo form, however, the three hydrophobic side chains face toward the interior of the Gla module, whereas the side chains of Gla 6 and 7 are exposed to solvent. there are no NOEs that identify the position of residues 1–3, probably due to flexibility of the N-terminal residues, an interpretation which is consistent with the previous observation that chemical modification of the

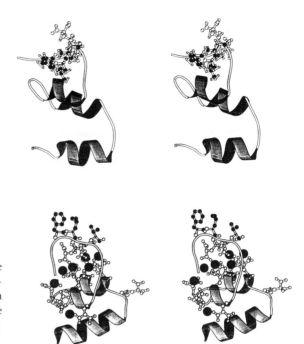

Figure 6.5 Stereo view of the structure of (upper) apo and (lower) calcium forms of a Gla module. The carbon atoms in the side chains of the three hydrophobic residues, Phe 4, Leu 5, and Val 8, are shown as filled circles, whereas those from Gla residues are shown as open circles. This figure was drawn using the MOLSCRIPT software (Kraulis, 1991).

N-terminus, e.g., by reductive methylation, occurs readily in the absence, but not in the presence, of calcium (Schwalbe et al., 1989).

The results described above demonstrate that even in the apo form the Gla module has a reasonably well-defined structure. The most striking difference between the apo and calcium-loaded forms is that the fold of the nine N-terminal residues is entirely different. This may explain why, in crystals of fragment 1 formed in the absence of calcium, the Gla module appears to have a disordered structure N-terminal of the long α-helix. The well-structured Gla module in fX-GlaEGF$_N$ thus contrasts to the poorly defined structure of the synthetic Gla domain of factor IX. Although the differences in structure may be a consequence of the difference in sequence, it appears reasonable at this stage to hypothesize that the isolated Gla module is unable to fold properly in the Ca^{2+}-free form but that the adjacent module is needed as a folding scaffold, whereas in the Ca^{2+}-saturated form the isolated Gla module will fold into its native conformation. This has recently been confirmed for factor X where even the reasonably well-characterized C-terminal helix, in the Gla module, shows a remarkable increase in flexibility upon removal of the EGF module (T. Drakenberg, unpublished results).

Recently, the crystal structure of prothrombin fragment 1 was solved with Ca^{2+} replaced by Sr^{2+} (Seshadri et al., 1994). Strontium is the only ion, other than calcium, that will restore the phospholipid-binding properties of Gla domains. The structure also turned out to be very similar to the Ca^{2+} structure. The root-mean-square deviation between the backbone atoms (Ca, C, N) of the two structures is 0.45 Å. The placement of the metal ions is very similar, even though the distances between metal ions are somewhat longer for Sr^{2+} than for Ca^{2+}.

Function of Gla Modules

For their interaction with membranes, it is an obligatory requirement that the vitamin K-dependent proteins contain Gla rather than Glu (Stenflo and Suttie, 1977). Phosphatidyl serine-containing phospholipid membranes (in the presence of calcium) increase the rate of prothrombin activation by factor Xa approximately 100-fold, due to binding of both enzyme and zymogen substrate (Kane and Davie, 1988; Mann et al., 1990). This increases the probability of productive collisions between macromolecular enzyme complexes and substrate.

Recently, each individual Gla residue in prothrombin and protein C has been mutated to Asp, and the mutant proteins expressed in eukaryotic cells (Ratcliff et al., 1993; Zhang and Castellino 1992; Zhang et al., 1992). The proteins were characterized with respect to biological activity and calcium- and phospholipid-binding properties. Results obtained in the two proteins were similar. Mutation of residues

equivalent to Gla 6, 7, 16, 26, and 29 in bovine prothrombin resulted in loss of biological activity and
calcium- and phospholipid-binding. In the structure
of bovine fragment 1, the carboxylate groups of these
Gla residues form a negatively charged "channel"
with an array of five calcium ions in its interior
(Fig. 6.4). Considering the differences in structure
of fX-GlaEGF$_N$ with vs. without calcium ions, the
site mutagenesis experiments clearly demonstrate the
importance of Gla residues 16, 26, and 29 on one
hand and Gla 8 on the other in ligating calcium
ions and imposing a native structure on the calcium-form of the protein (Ratcliff et al, 1993;
Zhang and Castellino, 1992; Zhang et al., 1992).
Properties of prothrombin in which Gla in position
14 or 19 had been mutated to Asp indicate that these
residues and calcium ions bound to them are not of
prime importance for the prothrombin/protein C–
phospholipid interaction. Mutation of other Gla residues gives results that are less clearcut.

Recently, rapid reaction techniques have been
used to study prothrombin activation by factor Xa–
factor Va on phospholipid vesicles (Krishnaswamy,
1990; Mann et al., 1990). Assembly of the prothrombinase complex on phospholipid membranes was
demonstrated to proceed through independent binding of factor Xa and factor Va to the membrane
(Giesen et al., 1991; Krishnaswamy, 1990).
Moreover, binding studies have indicated that factors
Va and Xa interact on the membrane surface with a
dissociation constant of approximately 1×10^{-9} M
(Krishnaswamy, 1990; Mann et al., 1990;
Nelsestuen et al., 1978; Nesheim et al., 1979; Tracy
et al., 1979; Ye and Esmon, 1995), whereas the dissociation constant for the interaction between factors
Va and Xa in solution has been estimated to be
$\approx 1 \times 10^{-6}$ M (Guinto and Esmon, 1984; Mann et
al., 1990; Persson et al, 1993). The increased affinity
on the membrane surface could be accounted for by
phospholipid-mediated conformational change(s) in
the proteins. Alternatively, the formation of binary
protein-phospholipid complexes and translational
diffusion of the proteins in the plane of the membrane may suffice to enhance the frequency of productive collisions between membrane-bound proteins
and account for the increase in binding affinity
(Mann et al., 1990).

Optimal procoagulant membrane surfaces, e.g.
activated platelets, contain 10–20% anionic phospholipids (Gerads et al., 1990; Mann et al., 1990).
In addition, the chemical structure of the polar
head groups of the phospholipid molecules is important, as vesicles which contain the anionic phospholipid phosphatidyl serine are most effective (Govers-
Riemslag et al., 1992). However, other anionic phospholipid membranes also promote the prothrombin-
converting activity, albeit less effectively than those

containing phosphatidyl serine Our limited insight
into the interaction between vitamin K-dependent
proteins and membranes is illustrated by the fact
that the prothrombin-converting activity in the presence of phosphatidyl serine-containing membranes
is relatively unaffected by high ionic strength
whereas, when phospholipid membranes containing
other head groups are used, high ionic strength
reduces the rate of prothrombin activation (Gerads
et al., 1990). Based on these observations, it has been
hypothesized that the interaction of Gla-containing
proteins with phosphatidyl serine-containing membranes is due to the formation of coordinated complexes with calcium as the central ion with ligands
provided by Gla residues in factors Xa or prothrombin on the one hand, and polar head groups of the
phospholipid on the other. Interaction between other
membranes with anionic phospholipid and Gla-containing proteins is thought to be electrostatic in nature. In the light of the structural information now at
hand, this is a less likely model for the membrane
binding.

The rate of prothrombin activation by the prothrombinase complex was recently shown to be
enhanced by neutral, phosphatidyl choline-contain-
ing membranes (Govers-Riemslag et al., 1994).
Also, in this case, calcium was required as well as
Gla domain on both enzyme and substrate. High
ionic strength inhibited the reaction and, most important, phospholipids that contained unsaturated
hydrocarbon side chains were far more effective
than those with saturated side chains. This is a crucial
observation as unsaturated hydrocarbon side chains
are known to facilitate hydrophobic interactions
between proteins and membranes. In this context, it
is also noteworthy that affinity chromatography
experiments have suggested the interaction between
the factors IX and X and biological membranes to be
both electrostatic and hydrophobic (Atkins and
Ganz, 1992; Lundblad, 1988). Evidence suggesting
that hydrophic interactions contribute significantly
to the interaction between vitamin K-dependent proteins and phospholipid membranes was recently
obtained by mutation of Leu 5 in protein C (Zhang
and Castellino, 1994). Leucine in this position is conserved in all vitamin K-dependent clotting factors.
On its mutation to Gln, while calcium binding was
normal, the interaction with anionic phospholipid
was virtually lost. Moreover, mutant activated protein C had lost its ability to inactivate factor Va in the
presence of phospholipid. These results should be
viewed in light of the structures of the Gla domain
in factor X (Sunnerhagen et al., 1995). In the apo
form, Phe 4, Leu 5, and Val 8 are located on the
hydrophobic interior of the domain. In the calcium
form of the domain, Gla 6 and 7 ligate the array of
calcium ions, thereby forcing the hydrophobic side

chains of the Phe, Leu, and Val residues into the aqueous solvent, where they presumably make hydrophic contact with suitable phospholipid membranes. Again, it should be borne in mind that vitamin K-dependent proteins synthesized after administration of vitamin K antagonistic drugs, such as warfarin, contain Glu rather than Gla (Senflo and Suttie, 1977). They do not have any calcium-binding sites in the "Gla" domain and they do not interact with biological membranes. A tentative model for the interaction of Gla modules and phospholipid membranes can now be presented, Fig. 6.6 (Arni et al., 1994; Christiansen et al., 1994; Li et al., 1995; Sunnerhagen et al., 1995; Zhang and Castellino, 1994).

Reduction of the dimensionality from three to two, such as occurs for blood coagulation when reactants are bound on a surface, is crucial for effective assembly of enzymatically active macromolecular complexes (Mann et al., 1990). However, this applies to the entire blood coagulation cascade (Fig. 6.1). It thus appears that the product of one reaction, e.g. factor Xa, rather than dissociating from and rebinding to the surface is "channelled" in the plane of the surface where in binds factor Va, forming a macromolecular complex which activates prothrombin to thrombin. More detailed discussions of the interaction between Gla-containing proteins and biological membranes have been published elsewhere (Abbott and Nelsestuen, 1988; Krishnaswamy, 1990; Lentz

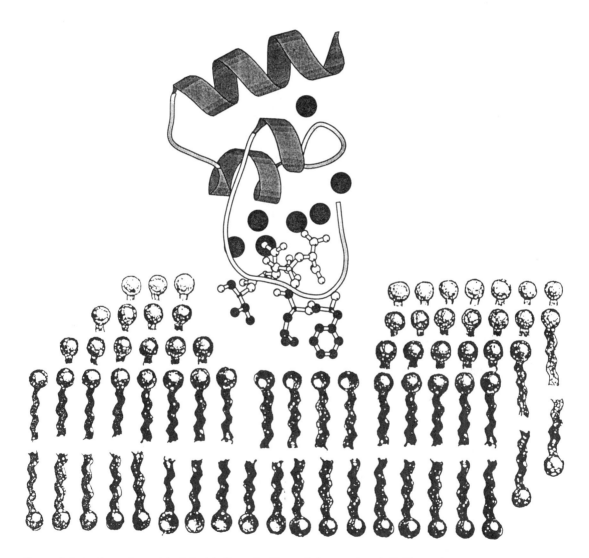

Figure 6.6 A schematic model of the binding of a Gla module to a phospholipid membrane.

et al., 1994; Mann et al., 1990; Plager and Nelsestuen, 1994)

Gla domains are important not only for the interaction of the vitamin K-dependent proteins with biological membranes but also for the interaction of factor VII/VIIa with tissue factor (Clarke et al., 1992; Krishnaswamy, 1992; Ruf et al., 1991; Toomey et al., 1991), and that of protein C with thrombomodulin (Galvin et al., 1987; Olsen et al., 1992; Suzuki et al., 1989; Zushi et al., 1989). In both cases, the interaction appears to be mediated by the Gla domain and the EGF-like domain(s). No detailed structural information is yet available.

EFG-Like Modules

Calcium Binding to EGF Like Modules

A Gla-containing domain binds approximately 10 calcium ions, and for some time it was thought to account for all calcium binding to vitamin K-dependent coagulation factors. With time it became apparent, however, that factors IX and X and protein C, from which the Gla domain has been removed by limited proteolysis, still have one or two calcium binding sites (Esmon et al., 1983; Morita et al., 1984; Sugo et al., 1984). Studies of the Ca^{2+}-binding properties of intact EGF-like modules, isolated after limited proteolysis of protein C, demonstrated that there is one site in a fragment containing the two EGF-like modules (Öhlin and Stenflo, 1987; Öhlin et al., 1988a, 1988b). Subsequently, the calcium-binding site was found to be in the N-terminal EGF-like module in protein C, as well as in factors IX and X (Handford et al, 1990; Persson et al, 1989; Öhlin et al., 1988b). The C-terminal EGF-like module does not bind calcium in these proteins. The second Gla-independent site is in the serine protease module and corresponds to that identified in trypsin (see below) (Bajaj et al., 1992; Bode and Schwager, 1975a, 1975b; Persson et al., 1993).

Calcium binding to the N-terminal EGF-like module has been studied using ^1H NMR by correlating the metal ion-induced chemical shift of the tyrosine multiplet (Tyr 68 in factor X; Fig. 6.7A) to the calcium concentration (Handford et al., 1990; Persson et al., 1989). This approach has allowed accurate determination of the calcium affinity of this site in $fIX-EGF_N$, $fX-EGF_N$, and $fX-GlaEGF_N$. The affinity of the site in $fX-GlaEGF_N$ has also been measured with fluorescence spectroscopy (Fig. 6.7B) (Valcarce et al, 1993). Calcium binding to the site in EGF_N enhances the intrinsic protein fluorescence from the single Trp residue (position 41) in $fX-GlaEGF_N$, whereas calcium binding to sites in the Gla domain is associated with a quenching of the protein fluorescence (Astermark and Stenflo, 1991; Öhlin et al., 1990; Persson et al, 1991a; Valcarce et al., 1993).

The calcium affinity of $fX-EGF_N$ is low ($K_d \approx 2.2 \pm 0.6\,mM$) irrespective of whether Asp 63 is hydroxylated or not (Selander-Sunnerhagen et al., 1993). The Ca^{2+} affinity of the site in the EGF-module site is approximately 20-fold higher in $fX-GlaEGF_N$ ($K_d \approx 124 \pm 20\,\mu M$) than in the isolated module (Valcarce et al., 1993). Decarboxylation of Gla to Glu in $fX-GlaEGF_N$ ($fX-desGlaEGF_N$) does not affect the calcium affinity of the site in the EGF module. Moreover, it is apparent that the C-terminal α-helical region in the Gla module is important for the calcium affinity, whereas the N-terminal part of the Gla module appears to be of less importance.

Calcium-binding sites in the EGF-like modules discussed so far all have dissociation constants for calcium of around $10^{-4}\,M$ in the intact proteins, and of around $10^{-3}\,M$ in isolated EGF-like modules. There is, however, another class of very high-affinity calcium-binding sites found in protein S (Dahlbäck et al., 1990). Protein S is unique among the vitamin K-dependent plasma proteins in that it contains four EGF-like modules, all with the consensus sequence required for Asp/Asn-β-hydroxylation (Dahlbäck, et al., 1986; Lundwall et al., 1986). The N-terminal EGF-like module has a Hya residue, whereas the other three domains each have one partially hydroxylated Asn/Hyn residue (Stenflo et al., 1988). Intact protein S has been found to contain four very high-affinity Gla-independent calcium-binding sites, $K_d\,10^{-7}$ to $10^{-9}\,M$, i.e. four to six orders of magnitude stronger than the EGF sites in factors IX and X (Dahlbäck et al., 1990). High-affinity calcium binding is retained in protein S fragments consisting of two or more EGF-like modules. It has thus been shown that the minimum requirement for the very high-affinity site is a protein fragment containing EGF modules three and four; other pairs of EGF modules or individual EGF modules show much lower Ca^{2+} affinity (Stenberg et al., 1997a, 1997b, and unpublished results). At present, with the lack of three-dimensional structure information there is no explanation for this unusual Ca^{2+} affinity.

Factors VII, IX, and X, and protein C all have the sequence Asp-Gly-Asp-Gln-Cys (corresponding to positions 46–50 in factor X; Fig. 6.8). Two of the EGF-like modules of protein S, however, have the sequence Asp-Ile/Val-Asp-Glu-Cys in the corresponding positions, and contain Hyn rather than Hya (Stenflo et al., 1989). Substitution of Glu for Gln and Hyn for Hya leaves the net charge in the calcium-binding region unaltered. Although the exchange of Gly for Ile/Val with hydrophobic side chains tends to increase the metal ion affinity, it cannot account for the large difference in calcium affi-

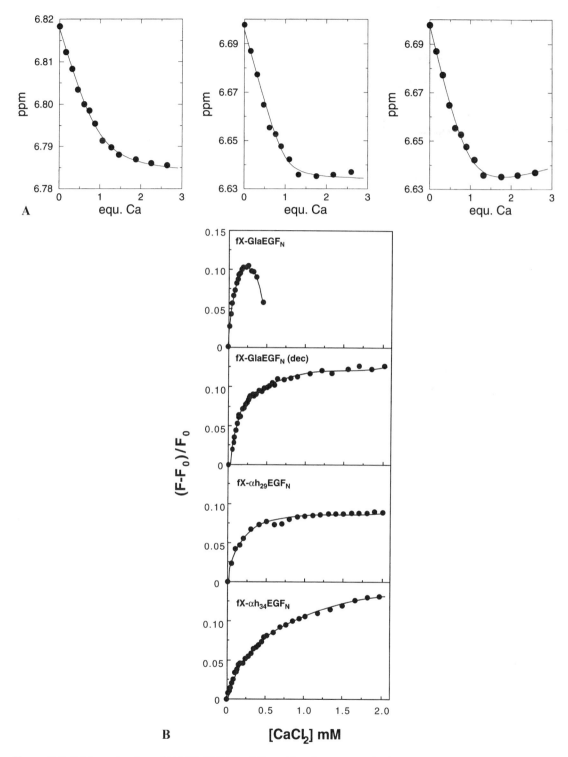

Figure 6.7 Calcium titration of (A) fX-GlaEGF$_N$ followed by observation of calcium-dependent chemical shifts by NMR and (B) fX-GlaEGD$_N$, fX-desGlaEGF$_N$, fX-ah29GlaEGF$_N$, and fX-ah34GlaEGF$_N$ followed by observation of changes in the intrinsic tryptophan fluorescence. F and F_0 are the emission intensities in the presence and absence of calcium, respectively. (Reproduced with permission from Valcarce et al., 1993).

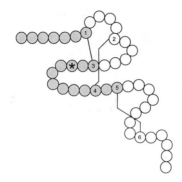

Factor			1.							
Factor	VII	Res. 44-50	Y	S	D	G	D	Q		C
Factor	IX	45-51	Y	V	D	G	D	Q		C
Factor	X	44-50	Y	K	D	G	D	Q		C
Protein	C	44-50	H	V	D	G	D	Q		C
Protein	S 1.	74-80	N	A	I	P	D	Q		C
	2.	114-120	E	F	D	I	N	E		C
	3.	158-164	C	K	D	V	D	E		C
	4.	200-206	C	E	D	I	D	E		C

	3.				*						4.	5.
Res. 61-72	C	K	D	Q	L	Q	S	Y	I	C	F	C
62-73	C	K	D	D	I	N	S	Y	E	C	W	C
61-72	C	K	D	G	L	G	E	Y	T	C	T	C
69-80	C	I	D	G	I	G	S	F	S	C	D	C
93-104	C	K	D	G	K	A	S	F	T	C	T	C
134-145	C	D	N	T	P	G	S	Y	H	C	S	C
176-187	C	K	N	I	P	G	D	F	E	C	E	C
215-228	C	V	N	Y	P	G	G	Y	T	C	Y	C

Figure 6.8 Alignment of the amino acid sequences of the NH$_2$-terminal EGF-like modules from factors VII, IX, and X, protein C, and the four EGF modules of protein S. The sequences amino-terminal to the first Cys residue are shown to the left and the sequences between the third, fourth, and fifth Cys residues are shown to the right. Asp/Hya/Glu residues are shaded. The hydroxylated Asp/Asn residues are denoted by * (the residue in factor VII is not hydroxylated).

nity between the two types of sites (Handford et al., 1991). The 10-fold difference in calcium affinity between free fX-EGF$_N$ and EGF$_N$ in fX-GlaEGF$_N$ suggests that adjacent modules have a profound effect on the metal ion affinity of the site. Presumably, this is also the case for protein S, although the structural basis of this effect is still unknown. The N-terminal EGF-like module in factor IX has been chemically synthesized, and the amino acids involved in calcium binding altered to make the site resemble sites that appear to have very high calcium ion affinity (Handford et al, 1991). However, none of the variant modules manifested calcium affinity much higher than that of the parent molecule. Similar experiments have been performed with the N-terminal EGF-like domain in factor X. In this instance, a 10-fold increase in calcium affinity was obtained when the module had Ile rather than Gly in position 47 and Glu rather than Gln in position 49 (Fig. 6.7) (K. Julenius, unpublished results).

It should be pointed out that the calcium-binding sites in the EGF-like modules of protein S resemble the sites at non-vitamin K-dependent proteins more closely than the sites in factors VII, IX, and X, and protein C (Selander-Sunnerhagen et al., 1992; Stenflo, 1991). Gly between the two Asp residues seems to be confined to the clotting proteases, whereas Ile/Val occurs frequently in many EGF-like modules in fibrillin (Lee et al, 1991; Maslen et al, 1991), NOTCH and DELTA (Kidd et al., 1986; Vässin et al., 1987; Wharton et al., 1985), TAN 1 (Ellisen et al., 1991), the low-density lipoprotein receptor, and thrombomodulin (Südhof et al., 1985). Moreover, the occurrence of Gln before the first Cys residue is uncommon, whereas Glu occurs frequently (Selander-Sunnerhagen et al., 1992).

Previously, the structure of pairs of protein modules was determined in fibronectin and in urokinase (Hansen et al., 1994; Williams et al., 1993). In fibronectin, close contact was observed between two adjacent type 1 modules. In contrast, the adjacent EGF-like and kringle modules in urokinase are structurally independent, with very little interaction between them. It has now become increasingly obvious that calcium binding to the N-terminal EGF-like module in factor X, and presumably to the corresponding domains in factors VII and IX, and protein C, determines the orientation of the adjacent Gla and EGF modules with respect to each other. Calcium titrations of fX-GlaEGF$_N$ have shown the calcium site in the EGF-like module to have higher affinity than any of the sites in the Gla module (Valcarce et al, 1993). Moreover, the fact that the calcium affinity of the site is low in fX-EGF$_N$ but approximately 20-fold higher in fX-GlaEGF$_N$, presumably due to interaction between the calcium ion and at least one residue in the C-terminal α-helix, attests to the interaction between the two modules Recent experiments, where the reactivity of disulfide bonds in fX-GlaEGF$_N$ has been probed with the low-molecular-weight thiol-containing protein thioredoxin, have also demonstrated an interaction to occur between the Gla module and the EGF-like module in fX-GlaEGF$_N$ in the presence of calcium, but it is not obvious in its absence (Valcarce et al., 1994).

It should also be mentioned that two of the EGF-like modules in thrombomodulin are of the calcium-binding type. Activation of protein C by the thrombin–thrombomodulin complex certainly requires calcium binding to protein C (Esmon, 1993; Hill and Castellino, 1987a). To what extent calcium binding to the EGF-like modules in thrombomodulin is a prerequisite for protein C activation is not clear.

Finally, it should be emphasized that the function of the hydroxyl group in Hya is still enigmatic. Although it does not affect the calcium affinity of fX-EGF$_N$ to any significant degree (Selander-Sunnerhagen et al., 1993), it may interact with neighbouring modules in the intact protein, or it may be important in the interaction with some other protein or receptor (Valcarce et al., 1993).

Structure of EGF-Like Domains

Nuclear-magnetic asonance has been used extensively to study the structure of EGF modules (Campbell and Bork 1993; Campbell et al, 1989; Cooke et al., 1987; Khoda et al., 1988, 1991; Kline et al., 1990; Montelione et al., 1987, 1992; Moy et al., 1994). All these structures are quite similar, manifesting a protein module consisting of two relatively independent submodules. The N-terminal submodule constitutes about 2/3 of the module, and its major structure is a triple-standard β-sheet in EGF and TGF-α where strands two and three are six amino acids long and connected by a somewhat distorted bend. The first strand, which is only two residues long (amino acids 6 and 7 in EGF), is not always present in the β-sheet and is not well defined. The C-terminal submodule of TGF-α has been described both as a small antiparallel double hairpin structure (Moy et al.,

1994) and as two loops connected by a short β-sheet and a flexible C-terminus (Kline et al., 1990). More recently, the structure of EGF-like modules has also been determined by x-ray diffraction methods (Graves et al., 1994; Padmanabhan et al., 1993; Picot et al., 1994; Rao et al., 1995). Only one of these studies deals with EGF modules of the Ca^{2+}-binding type (Rao et al., 1995). In the following discussion, we will restrict ourselves to those few structures.

The solution structure of the N-terminal EGF-like module of factor IX (fIX-EGF$_N$) and factor X (fX-EGF$_N$), both of which contain Hya, was recently determined by means of two-dimensional NMR (Baron et al, 1992; Huang et al., 1991; Selander et al, 1990; Ullner et al., 1992). Figure 6.9 depicts the structure of fX-EGF$_N$ in its apo form and provides a comparison with the structure with murine EGF. The structure, as well as that of fIX-EGF$_N$, is similar to the EGF and TGF-α-structure, except that the short N-terminal strand is absent from the main β-sheet.

The structure of fX-EGF$_N$ has now also been determined in the presence of calcium (Selander-Sunnerhagen et al., 1992). In this study, the calcium concentration was sufficiently high to ensure more than 90% saturation of the single calcium-binding site. As the calcium exchange is fast on the NMR time scale, only one species will be seen by NMR whatever the calcium concentration. The changes in the NMR spectra caused by calcium ion binding are localized to residues 46–51 and 62–68 (Fig. 6.10). A comparison of the structures of apo and calcium forms shows no major differences, as was to be expected since the structure is highly restrained by the three disulfide bridges. There are, however, some minor but important changes. In the calcium form, the N-terminus is more well defined and located on top of the main β-sheet (Selander-Sunnerhagen et al., 1992). Moreover, the turn connecting the two strands in the main β-sheet is bent slightly towards the N-terminus. Even though the metal ion cannot be seen directly from NMR restraints, the calcium site can readily be identified (Fig. 6.10). In the calcium form of fX-EGF$_N$, a cavity of appropriate size for a calcium ion is lined by four oxygens oriented towards the centre, where electrostatic considerations require a positive charge. There are several well-defined ligands; the backbone carbonyls of Gly 47 and Gly 64, the side-chain carboxyl of Asp 46 and Hya 63, and the side-chain carbonyl of Gln 49. The side chain of the Hya residue is rotated $120°$ around its $C^\alpha — C^\beta$ bond such that the hydroxyl group of Hya points away from the calcium ion. The hydroxyl group is not a ligand to the calcium ion, a finding in accord with the observation that hydroxylation only has a minor effect on the calcium affinity (Selander-Sunnerhagen et al., 1993). Moreover, the

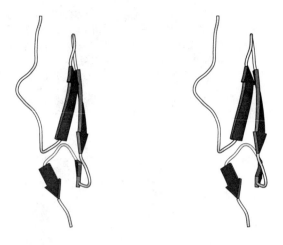

Figure 6.9 Stereo view of a ribbon drawing of the apo form of the N-terminal EGF module from factor X. This figure has been drawn using the MOLSCRIPT software (Kraulis, 1991).

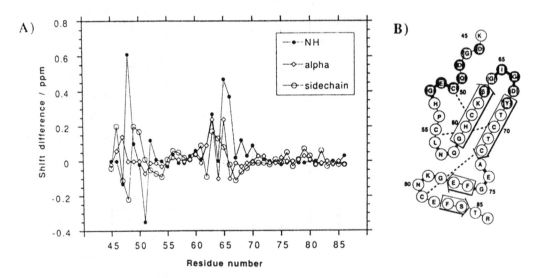

Figure 6.10 (A) ^{1}H NMR shift differences between calcium and apo forms of fX-EGF$_N$ displayed by residue. Shift changes larger than 0.04 ppm are shown. (B) Secondary structure of fX-EGF$_N$ with the largest shift differences for each residue indicated: $\Delta\delta > 0.3$ ppm (solid black), $\Delta\delta = 0.30$–0.20 ppm (checkered), and $\Delta\delta = 0.20$–0.10 ppm (wavy). Arrows indicate β-sheet structure and β-turns are shown with thick lines. Disulfide bonds are displayed as broken lines. (Reproduced with permission from Selander-Sunnerhagen et al., 1992.)

plane of the aromatic ring of Tyr 68 is displaced 2 Å in the calcium form, as compared with the metal-free form. The function of the hydroxyl group of Hya is unknown.

Prior to the determination of the structure of the calcium form of fX-EGF$_N$, it was proposed that the Hya residue and two Asp residues (corresponding to Asp 46 and 48 in factor X) are calcium ligands (Cooke et al., 1987). This hypothesis was based on the structure of EGF, determined by two-dimensional NMR, and assumes a similar fold for N-terminal EGF-like domains in coagulation factors. Site-directed mutagenesis studies performed in factor Ix and in protein C supported this notion (Handford et al., 1991; Öhlin et al., 1998b; Rees et al., 1988). In the structure described above, however, neither of the two Asp residues can unequivocally be designated as a calcium ligand (Selander-Sunnerhagen et al., 1992). Asp 46, however, points in the general direction of the metal ion even though the oxygens are within coordination distance in only a few of the calculated structures (Fig. 6.9). This carboxylate groups has nevertheless been assumed to be a ligand, as there are no NMR constraints in the structure calculations that will force the carboxylate group to point in the right direction. Moreover, no electrostatic potentials were used in the calculations. For Asp 48, on the other hand, the conclusion to be drawn from the structure calculations is that it is not a metal ion ligand. Nevertheless, the negative charge is necessary for a high-affinity calcium ion site, as demonstrated by site-directed mutagenesis studies (Handford et al., 1991; Rees et al., 1988). It should be noted that even negatively charged side chains in the neighbourhood of a Ca^{2+} site may have a significant effect on the Ca^{2+} affinity It has, for example, been demonstrated for calbindin D_{9k}, an EF-hand calcium-binding protein, that neutralization of a surface charge close to the calcium site reduces the affinity 5–10-fold (Linse et al., 1988). It is also noteworthy that, in most instances, Asn and Gln were found to be almost as good calcium ligands as Asp and Glu. However, if the Glu residue coordinates of the calcium ion with both its side-chain oxygens, it cannot be replaced by a Gln without a significant loss in calcium affinity (Maune et al., 1992).

The structure of the calcium form of fX-EGF$_N$ accounts for five calcium ligands, leaving room for one or two water molecules. The calcium affinity of the EGF module is approximately 20-fold higher in fX-GlaEGF$_N$ than in isolated EGF$_N$ (see below). Only tentative conclusions can be drawn regarding which residues are calcium ligands, based on the structure of isolated EGF$_N$, as any such conclusion may have to be modified once the structure of fX-GlaEGF$_N$ has been refined to high resolution.

Indeed, it is entirely feasible that the side chain of Asp 48 is a ligand to the calcium ion in fX-GlaEFG$_N$, but not in the isolated domain. A more accurate definition of the calcium ion ligands will probably be possible once the crystal structure has been determined.

The recently determined crystal structure of the N-terminal EGF domain of factor IX has largely confirmed the conclusions that could be drawn from the NMR study of factor X (Rao et al., 1995; Selander-Sunnerhagen et al, 1992). Not only were all the suggested Ca^{2+} ligands verified in the crystal structure, but additional ligands could also be identified. Asp 64 was found to contribute with both its side-chain oxygens, which is somewhat puzzling since in many Ca^{2+}-binding EGF modules there is an Asn in this position. In calcium-binding proteins of the EF type, it has been shown that when the Glu that uses both its side-chain oxygens as ligands is mutated to a Gln the calcium binding is dramatically decreased. The last calcium ligand is supplied by the other molecule in the asymmetric unit; a finding which suggested to the authors that the function of the calcium binding to EGF modules may be to orient the domains relative to each other, as was also suggested based on the NMR structure (Rao et al, 1995).

Role of Calcium Binding to EGF Modules

The Ca^{2+}-binding EGF-like modules often occur in tandem, and in some proteins there are up to 40 adjacent modules. A closer look at the organization of adjacent modules and the localization of the Ca^{2+}-binding site shows that a module preceding the Ca^{2+}-binding module may very well donate a missing ligand to fill the coordinating sphere of the Ca^{2+} ion. For steric reasons, a domain on the C-terminal side can hardly contribute to the Ca^{2+} binding. The crystal structure of the N-terminal module from factor IX consists of an asymmetric unit with two molecules that are packed in a manner similar to what might be expected for a module pair. It is also found that the "preceding" module donates one Ca^{2+} ligand to the site in the other domain (Rao et al., 1995) However, in factors IX and X the Ca^{2+}-binding EGF-like module is preceded by a Gla module and not an EGF-like module. Calcium binding to an isolated EGF-like module is at least one order of magnitude weaker than to an Gla-EGF module pair. In fact, only a minor part, residues 29–45, of the Gla module is needed to restore full Ca^{2+} binding; and a decarboxylated Gla module also restores the Ca^{2+} binding (Valcarce et al., 1993). It is therefore likely that the extra Ca^{2+} ligand is to be found in this peptide segment. Modelling studies by Rao et al., (1995) suggest that Gla 40 may donate a side-chain oxygen as a ligand. A model for the Ca^{2+} binding to a pair of EGF modules in fibrillin has recently been proposed after noting that many of these EGF-like modules have Asn/Asp in a position 4 or 6 residues before the last Cys (Hill and Castellino, 1987b). Modelling studies have shown that the two modules may well be oriented in such a way that a side-chain oxygen from this conserved Asn/Asp can participate in Ca^{2+} coordination. We have noted that also in protein S, the second and third EGF-like modules have this amino acid which may contribute to the very high-affinity Ca^{2+}-binding site in this protein (Lundwall et al., 1986). However, in the recently determined three-dimensional structure of a pair of EGF modules from fibrillin, no indication of an extra ligand could be detected even though the Ca^{2+} affinity of the second EGF module was more than 10-fold higher than for the first module (Downing et al., 1996).

The recent determination of the structure of the calcium-free form of fX-GlaEGF$_N$ by two-dimensional NMR has clearly established that the Gla and EGF-like modules are mobile relative to each other, with little contact between the two modules (Sunnerhagen et al., 1996). Indeed, the region between the Gla and EGF-like module seems to function as a hinge. Binding of the single calcium ion to the first EGF-like module bends the entire Gla module towards the EGF-like module and appears to lock the two modules relative to each other (Fig. 6.11). It also promotes the contact between the C-terminal α-helix in the Gla module and the top of the β-sheet in the EGF-like domain. The fact that mutant factor IX molecules, where amino acids shown to be calcium ligands in the EGF-like module have been mutated (corresponding to positions 46, 49, and 63 in factor X; Fig. 6.9), have very low biological activity suggests that calcium binding to the site in the EGF-like module is functionally important, presumably to orient the two modules relative to each other (Stenflo and Dahlbäck 1994).

Calcium Binding to Serine Protease Modules

The Gla-containing coagulation factors VII, IX, and X, prothrombin, and protein C all have a C-terminal serine protease module. The sequence identity between protein C on the one hand, and chymotrypsinogen, trypsinogen, and thrombin on the other, is 32%, 33%, and 38%, respectively (Fisher et al., 1994). Among Gla-containing coagulation factors, the three-dimensional structure of the serine protease module has been determined by x-ray crystallography for thrombin (Bode et al., 1989), factor IX (Hill et al., 1988), and factor X (Padmanabhan et al., 1993), the structure of thrombin having been determined at the highest resolution (1.9 Å). Each

of the three modules consists of two interacting six-stranded barrel-like domains that are linked by trans-domain straps. In thrombin, there are four of these transdomain straps and, in addition, five helical segments. Most of these structural elements are conserved in the other serine proteases. The active site residues Ser 195, His 57, and Asp 102 (using the chymotrypsin nomenclature of Bode et al., 1989) are located at the junction between the two barrels (Bode et al., 1989). The α-carbon structure of human α-thrombin is shown in Fig. 6.12, superimposed with that of the archetype serine protease, chymotrypsin, showing a topological equivalence within an r.m.s. of 078 Å between backbone residues in the two proteins (Bode et al., 1992). The molecule is shown in the so-called standard orientation with the A-chain in a groove at the back of the B-chain and with the active site in the centre and the substrate-binding cleft running "east–west".

The thrombin B-chain is 27 residues longer than chymotrypsin and most of the insertions are in the immediate vicinity of the active site and position the enzymatic reaction center in the bottom of a deep and narrow cleft (Bode et al., 1992). Moreover, the large active site region has an uneven distribution of surface charges in that a centrally positioned negatively charged surface patch is sandwiched between two positively charged areas towards the "northwest" and the "east" (the anionic-binding exosite). This large canyon-like active site that widens towards "east" and "west" is designed to accommodate a select number of macromolecular substrates, such as fibrinogen, factors V and VIII, as well as the so-called thrombin receptor, inhibitors such as antithrombin III and hirudin, and regulatory proteins such as thrombomodulin. The latter protein alters the topology of the active site such that thrombin can no longer accommodate and cleave fibronogen but is converted to an activator of protein C. Thrombomodulin thus transforms thrombin from procoagulant to an anticoagulant enzyme which is crucial for the regulation of blood coagulation (Esmon, 1993).

Calcium Binding Site in the Serine Protease Module of Trypsin

It has long been known that trypsinogen contains two calcium-binding sites, one of low affinity in the activation peptide and one with a higher affinity ($K_d = 2.6 \times 10^{-5}$ M) that is also present in the active enzyme (Bode and Schwater, 1975a, 1975b). Calcium binding to this site inhibits autolysis of trypsin and increases its thermal stability. This site is the archetype of similar calcium-binding sites in chymotrypsin, factors VII, IX, and X, and protein C (Chiancone et al, 1985; Persson et al., 1993; Rezaie and Esmon,

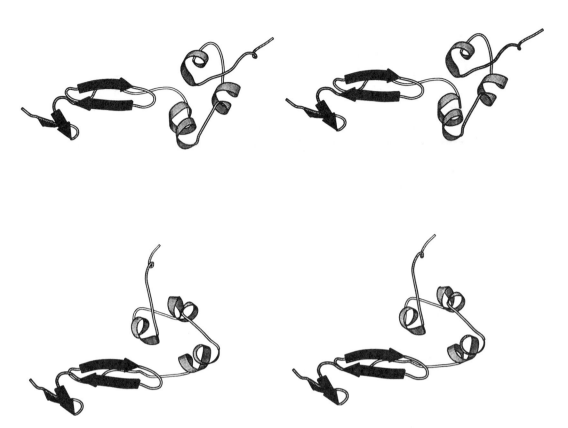

Figure 6.11 Stereo view of the (A) apo and (B) 1Ca forms of the fX-GlaEGF$_N$ fragment. This figure has been drawn using the MOLSCRIPT software (Kraulis 1991).

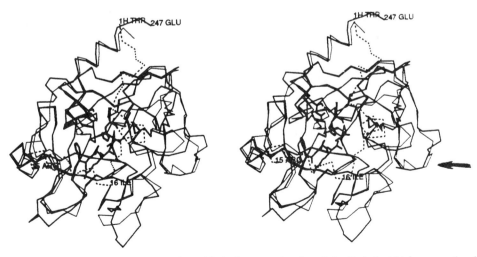

Figure 6.12 α-Carbon structure of the A-chain (dashed connections) and the B-chain (thick connections) super-imposed with bovine trypsin. The PPACK molecule is overlaid and the standard view is used. The arrow is pointing toward the Ca^{2+}-binding loop. (Reproduced with permission from Bode et al., 1992.)

		70								80		
B. Trypsin	L	G	E	D	N	I	N	V	V	E	G	N	E	Q	F
H. Thrombin	I	G	K	H	S	R	T	R	Y	E	R	N	I	E	K	I
H. Factor VII	L	G	E	H	D	L	S	E	H	D	G	D	E	Q	S
H. Factor IX	A	G	E	H	N	I	E	E	T	E	H	T	E	Q	K
H. Factor X	V	G	D	R	N	T	E	Q	E	E	G	G	E	A	V
H. Protein C	L	G	E	Y	D	L	R	R	W	E	K	W	E	L	D

Figure 6.13 Amino acid sequences in the calcium-binding region of bovine trypsin and human factors VII, IX, and X, and protein C and in the corresponding region of thrombin. In trypsin, the side chains of Glu 70 and Glu 80 have been identified as calcium ligands together with the backbone carbonyls of Asn 72 and Val 75.

1994; Rezaie et al., 1993; Wildgoose et al., 1993). In factors IX and X, however, a conspicuous absence of a calcium ion in the site was observed on determination of the crystal structure, presumably due the fact that crystals were obtained in the presence of high concentrations of SO_4^{2-} and at PH 5.5 or lower, conditions that preclude formation of stable calcium–protein complexes (Brandstetter et al., 1995; Padmanabhan et al., 1993). However, biophysical studies using fluorescence energy transfer and ^{43}Ca and ^{113}Cd NMR and lanthanide-induced NMR relaxation suggested it to be highly likely that a calcium ion occupies the same position in chymotrypsin as in trypsin (Chiancone et al., 1985). The calcium-binding site in factors VII and X, and protein C will be discussed in some detail below, as will the notable absence of a calcium-binding site in thrombin. The pronounced sequence identity of serine proteases in this region suggests that several of them, that are not involved in blood coagulation, also contain similar sites, although this has not yet been experimentally verified (Fig. 6.13).

The structure of the calcium-binding site in trypsin has been determined to a resolution of 1.8 Å and the ligands have been identified (Figs. 6.14 and 6.15). The Ca^{2+} ion is complexed by six ligands positioned at the edges of an octahedron. Each ligand is approximately 2.5 Å distant from the Ca^{2+} ion. Glu 70 and 80 provide side-chain ligands and Asn 72 and Val 75 provide backbone carbonyls as ligands. Two of the ligands are water molecules, one of which is sandwiched between the Ca^{2+} ion and the side chain of Glu 77. For the sake of clarity, the Ca^{2+}-binding site in factors VII, IX, X, and protein C are discussed separately below.

Calcium Binding to the Serine Protease Module of Factors IX and X

A Ca^{2+}-binding site has been identified in the serine protease module of factor IX. Its dissociation con-

stant was estimated to be $\approx 1.2 \times 10^{-4}$ M (Bajaj et al., 1992). Calcium binding to a synthetic peptide covering the sequence that corresponds to the Ca^{2+} site in trypsin was used to identify the site together with a monoclonal antibody and recognized a Ca^{2+}-dependent epitope in this region. The function of this Ca^{2+}-binding site has not yet been elucidated.

Immunochemical experiments have demonstrated a Ca^{2+}-induced conformational change in the serine protease module of factor X (Persson et al, 1993). Equilibrium dialysis showed the Ca^{2+} dissociation constant of the site to be 6.5×10^{-4} M (Rezaie et al., 1993). The structure of factor X, determined by x-ray crystallography, demonstrates that the Ca^{2+}-binding loop with the residues that ligate Ca^{2+} in trypsin is conserved (Padmanabhan et al., 1993). However, there was no Ca^{2+} in the site as the crystallization conditions did not allow complex formation.

Localization of the Ca^{2+}-binding site was attained by site-directed mutagenesis, i.e. mutation of Asp 70, a Ca^{2+} ligand in trypsin, to Lys (Rezaie and Esmon, 1994). The mutant protein, which also lacked the Gla domain and the first EGF-like domain, had no Ca^{2+} affinity. By mutating the Ca^{2+} site in a delection mutant of factor X, the effects of Ca^{2+} binding to the site in the serine protease module could be observed without interference from the Ca^{2+} sites in the Gla module and in the first EGF-like module The mutant protein retained functional activity but, unlike the wild-type protein, Ca^{2+} did not enhance the amidolytic activity. Retention of functional activity in the absence of a Ca^{2+} binding site was proposed to be due to an internal salt link between Lys 70 and Glu 80. Despite the absence of a Ca^{2+}-binding site in the mutant factor Xa, its rate of activation by factor VIIa in complex with tissue factor was Ca^{2+} dependent. Moreover, the rate of activation of prothrombin and prethrombin 1 (prothrombin lacking the Gla module) by the mutant factor Xa in complex with factor Va (without phospholipid) was Ca^{2+}-dependent. These effects presumably reflect a Ca^{2+}-

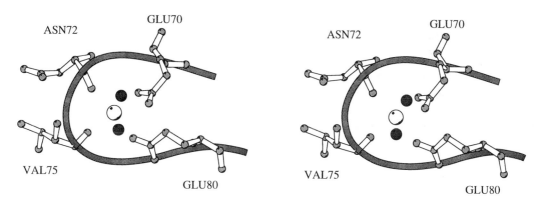

Figure 6.14 Stereo view of the calcium-binding site of the serine protease, trypsin. The side chains are shown only for the ligating residues. Note that V75 and N72 coordinate to the calcium ion via the backbone carbonyls. The two coordinating water molecules are shown as filled circles. This figure has been drawn using the MOLSCRIPT software (Kraulis, 1991).

dependent recognition of the macromolecular substrate (mutant factor X) by the factor VIIa–tissue factor complex and the effect of Ca^{2+} on the interaction between factor Va and factor Xa in the macromolecular complex that activates prothrombin. To resolve the function of calcium in these very complicated protein–protein interactions in atomic detail, the three-dimensional structure of the factor Va–factor Xa complex will have to be determined.

Calcium Binding to the Serine Protease Module of Factor VII

Recently, factor VII has been shown to have a Ca^{2+}-binding site in the serine protease module similar to that in trypsin and factor X (Sabharwal et al., 1995; Wildgoose et al, 1993). The Ca^{2+}-binding constant of the site is $\approx 1.5 \times 10^{-4}$ M. In addition to Ca^{2+}, this site in factor VII (and in protein C, see below) binds terbium, which was used as a fluorescent probe to characterize the site (the Ca^{2+}-binding sites in the Gla and EGF-like modules do not bind terbium). Moreover, a 37 amino acid residue-long synthetic peptide covering the putative Ca^{2+}-binding site was found to bind Ca^{2+}. Mutating Glu 220, corresponding to Glu 80—a Ca^{2+} ligand in trypsin, inactivates the site, i.e. Ca^{2+} does not enhance the amidolytic activity of factor VIIa in the mutant protein as it does in the wild-type protein (Wildgoose et al., 1993).

Figure 6.15 Schematic representation of the protease domain of factor VIIa. the polypeptide backbone is shown as ribbon with the N- and C-termini marked with N and C, respectively. The solid peptide ribbon represents the VIIa-TF inhibitory peptide. The Ca^{2+} is shown as an open circle and the two glutamic acid ligands, Glu 10 and Glu 220, are shown as stick models. (Modified after Sabharwal et al., 1995, with permission.)

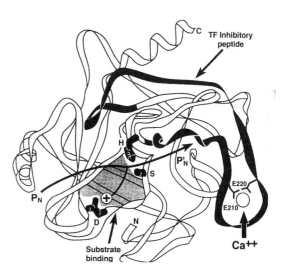

The function of the Ca^{2+}-binding site in the serine protease module in the interaction between factor VII/VIIa and tissue factor (TF) is attracting considerable interest. One approach to study the function of the various modules in factor VII/VIIa in the interaction with TF has been to make chimeric factor VII proteins using recombinant DNA techniques (Chang et al., 1995; Toomey et al., 1991). In these mosaic proteins, modules were exchanged between factor VII and factor IX. These, and other studies, have shown the N-terminal EGF-like module of factor VII to be crucial to the interaction. However, the serine protease module also appears to interact with TF, albeit more weakly. The Gla module and the second EGF-like module are also necessary for the TF–factor VII interaction, the Gla module by providing Ca^{2+} and phospholipid affinity to factor VII, and the second EGF-like module by imparting a native structure to the serine protease module. Calcium binding to the site in the serine protease module not only increases the amidolytic activity of factor VIIa but greatly increases the affinity of factor VII for TF (Sabharwal et al, 1995). Recently, the three-dimensional structure of the complex between factor VIIa and TF was determined by x-ray crystallography. The structure clearly shows the first EGF-like module of factor VII to be directly involved in the interaction, though a full report of this work is not yet available (Banner et al., 1995).

Calcium Binding to the Serine Protease Module of Protein C

The Ca^{2+}-binding site in the serine protease module of protein C has been localized to the same loop as in factors VII and X by site-directed mutagenesis (Rezaie and Esmon, 1992; Rezaie et al., 1992, 1994). To shed light on the complex effects of Ca^{2+} binding to this site in protein C, a battery of methods have been used. A monoclonal antibody recognizing an epitope in the activation peptide region and spanning the scissile bond has been useful (Le Bonniec et al, 1992). Moreover, in activated protein C, as well as in factors VII and IX, binding of Ca^{2+} to the site in the serine protease module increases the amidolytic activity of the protein (Rezaie et al., 1994). Calcium binding to the site in protein C (but not in activated protein C) can be monitored by measurements of the quenching of the intrinsic protein fluorescence intensity (Rezaie et al, 1992). Recently, the quenching was demonstrated to be due to perturbation of tryptophan residues 231 and 234 (Comp et al., 1984). Finally, in protein C, as in factor VII, terbium (Tb^{3+}) binds to the Ca^{2+} site with high affinity. The terbium binding results in increased energy transfer from tryptophan (Trp 231 and/or Trp 234) that results in creased intensity of fluorescence emission from the bound terbium,

making it a useful probe to monitor Ca^{2+} binding (Rezaie and Esmon, 1992; Wildgoose et al., 1993).

Activation of protein C by thrombin in solution is very slow and the rate of activation is reduced by Ca^{2+}. If, however, protein C is activated by thrombin in complex with thrombomodulin, the rate of activation is very rapid. Addition of Ca^{2+} increases the rate even further, giving a total increase in the k_{cat}/K_m that is approximately 1000-fold (Esmon, 1989, 1993). Careful studies have shown that both effects are due to Ca^{2+} binding to the site in the serine protease module of protein C. Occupancy of this site in protein C induces a conformational change in the activation peptide that is required for rapid activation by the thrombin–thrombomodulin complex. In the absence of thrombomodulin, this conformational change results in a reduced rate of activation of protein C, presumably due to a repulsive interaction between one or two Asp residues (positions 167 and 172) in the activation peptide region in protein C and Glu 192 in thrombin (Esmon, 1993; Rezaie and Esmon, 1992). Occupancy of EGF-like domains 5 and 6 of thrombomodulin in the anionic binding site in thrombin reduces this repulsion and results in an enhanced rate of activation of protein C.

The Ca^{2+}-binding sites in the serine protease modules of factors VII and X, and protein C have several properties in common. Saturation of the site increases the amidolytic activity of the protein. Moreover, binding of Ca^{2+} to the site in factors VII and IX, and protein C results in a quenching of the intrinsic protein fluorescence, but this does not occur in factor X. In factor VII/VIIa, Ca^{2+} binding to the site affects the interaction with tissue factor, and in protein C this binding has dramatic effects on the rate of activation of the protein by thrombin, whether in complex with thrombomodulin or not. Finally, the conspicuous absence of the site in thrombin appears to be due to exchange of a Glu residue to Lys which results in the formation of a salt bridge between the Lys residue and an opposing glu residue (Fig. 6.13).

Addendum

Since this chapter was written, important progress has been made relating to several aspects of the subject. New insight has been gained into the structure and calcium binding of pairs of EGF-like modules (Downing et al., 1996) and into the high-affinity calcium binding of EGF modules in protein S (Stenberg et al., 1997a, 1997b). These results are mentioned only briefly in the text. Moreover, the crystal structure of the complex of coagulation factor VIIa with tissue factor has been reported. This important structure also sheds light on the calcium-binding properties of this group of molecules (Banner et al., 1996).

References

Abbott, A. J. and G. L. Nelsestuen (1988) The collisional limit: an important consideration for membrane-associated enzymes and receptors. *FASEB J* 2: 2858–2866.

Arni, R. K., K. Padmanabhan, K. P. Padmanabhan, T.-P. Wu, and A. Tulinsky (1994) Structure of non-covalent complex of prothrombin kringle 2 with PPACK-thrombin. *Chem. Phys Lipids* 67/68: 59–66.

Astermark, J. and J. Stenflo (1991) The epidermal growth factor-like domains of factor IX. Effect on blood clotting and endothelial cell binding of a fragment containing the epidermal growth factor-like domains linked to the gamma-carboxyglutamic acid region. *J. Biol Chem.* 266: 2438–2443.

Astermark, J., I. Björk, A. K. Öhlin, and J. Stenflo (1991) Structural requirements for Ca^{2+}-binding to the gamma-carboxyglutamic acid and epidermal growth factor-like regions of factor IX. Studies using intact domains isolated from controlled proteolytic digests of bovine factor X. *J. Biol Chem.* 266: 2430–2437.

Atkins, J. S. and P. R. Ganz (1992) The association of human coagulation factors VIII, IXa and X with phospholipid vesicles involves both electrostatic and hydrophobic interactions. *Mol. Cell. Biochem.* 112: 61–71.

Bajaj, S. P., R. L. Butkowski, and K. G. Mann (1975) Prothrombin fragments. Ca^{2+} binding and activation kinetics. *J. Biol. Chem.* 250: 2150–2156.

Bajaj, S. P., A. K. Sabharwal, J. Gorka, and J. J. Birktoft (1992) Antibody-probed conformational transitions in the protease domain of human factor IX upon calcium binding and zymogen activation: putative high affinity Ca^{2+}-binding site in the protease domain. *Proc Natl. Acad. Sci. USA* 89: 152–156.

Banner, D. W., A. D'Arcy, C. Chène, F. Vilbois, W. H. Konigsberg, A. Guha, Y. Nemerson, and D. Kirchhofer (1995) The crystal structure of the complex of human factor VIIa with human soluble tissue factor. *Thromb. Haemostasis* 73: 1183.

Banner, D. W., A. D'Arcy, C. Chène, F. K. Winkler, A. Guha, W. H. Konigsberg, Y. Nemerson and D. Kirchhofer (1996) The crystal structure of the complex of blood coagulation factor VIIa with soluble tissue factor. *Nature* 380: 41–46.

Baron, M., D. G. Norman, and I. D. Campbell (1991) Protein modules. *Trends Biochem. Sci.* 16: 13–17.

Baron, M., D. G. Norman, T. S. Harvey, P. A. Handford, M. Mayhew, A. G. D. Tse, G. G. Brownlee, and I. D. Campbell (1992) The three-dimensional structure of the first EGF-like module of human factor IX: comparison with EGF and TGF-a. *Protein Sci.* 1: 81–90.

Bern, M. M., K. Suzuki, K. Mann, P. Tracy, L. Hoyer, W. Jensen, M. Gallivan, C. Arkin, and G. Davis (1984) Response of protein C and protein C inhibitor to warfarin therapy in patient with combined deficiency of factors V and VIII. *Thromb. Res.* 36: 485–495.

Bode, W. and P. Schwager (1975a) The refined crystal structure of bovine beta-trypsin at 1.8 Å resolution II.

Crystallographic refinement, calcium binding site, benzamidine binding site and active site at pH 7.0 *J. Mol. Biol.* 98: 693–717.

Bode, W. and P Schwager (1975b) The single calcium-binding site of crystalline bovine beta-trypsin. *FEBS Lett.* 56: 139–143.

Bode, W., I. Mayr, Y. Bauman, R. Huber, S. R. Stone, and J. Hofsteenge (1989) The refined 1.9 Å crystal structure of human α-thrombin: interaction with D-Phe-Pro-Arg chloromethylketone and significance of the Tyr-Pro-Pro-Trp insertion segment. *EMBO J.* 88: 3467–3475.

Bode, W., D. Turk, and A. Karshikov (1992) The refined 1.9-Å X-ray crystal structure of D-Phe-Pro-Arg chloromethylketone-inhibited human α-thrombin: structure analysis, overall structure, electrostatic properties, detailed active-site geometry, and structure–function relationships. *Protein Sci.* 1: 426–471.

Borowski, M., B. C. Furie, S. Bauminger, and B. Furie (1986) Prothrombin requires two sequential metal-dependent conformational transitions to bind phospholipid. Conformation-specific antibodies directed against the phospholipid-binding site on prothrombin. *J. Biol. Chem.* 261: 14969–14975.

Borowski, B., B. C. Furie, G. H. Goldsmith, and B. Furie (1985) Metal and phospholipid binding properties of partially carboxylated human prothrombin variants. *J. Biol. Chem.* 260: 9258–9264.

Brandstetter, H., M. Bauer, R. Huber, P. Lollar, and W. Bode (1995) X-ray structure of clotting factor IXa: active site and module structure related to Xase activity and hemophilia B. *Proc. Natl. Acad. Sci. USA* 92: 9796–9800.

Campbell, I. D. and P. Bork (1993) Epidermal growth factor-like modules. *Curr. Opin. Struct. Biol.* 3: 385–392.

Campbell, I. D., R. M. Cooke, M. Baron, T. S. Harvey, and M. J. Tappin (1989) The solution structure of epidermal growth factor and transforming growth factor alpha. *Progr. Growth Factor Res.* 1: 13–22.

Chang, J.-Y., D. W. Stafford, and D. L. Straight (1995) The roles of factor VII's structural domains in tissue factor binding. *Biochemistry* 34: 12227–12232

Chiancone, E., T. Drakenberg, O. Teleman, and S Forsén (1985) Dynamic and structural properties of the calcium binding site of bovine serine proteases and their zymogens. *J. Mol. Biol.* 185: 201–207.

Christiansen, W. T, A. Tulinsky, and F J. Castellino (1994) Functions of individual γ-carboxyglutamic acid residues of human protein C. Determination of functionally nonessential Gla residues and correlation with their mode of binding to calcium. *Biochemistry* 33: 14993–15000.

Clarke, B. J., F. A. Ofosu, S. Sridhara, R. D. Bona, F. R. Rickles, and M. A. Blajchman (1992). The first epidermal growth factor domain of human coagulation factor VII is essential for binding with tissue factor. *FEBS Lett.* 298: 206–210.

Colpitts, T. L. and F. J. Castellino (1994) Calcium and phospholipid binding properties of synthetic gamma-carboxyglutamic acid-containing peptides with sequence counterparts in human protein C. *Biochemistry* 33: 3501–3508.

Comp, P. C., R. R. Nixon, and C. T. Esmon (1984) Determination of functional levels of protein C, an antithrombotic protein, using thrombin–thrombomodulin complex. *Blood* 63: 15–21.

Cooke, R. M., A. J. Wilkinson, M. Baron, A. Pastorc, M. J. Tappin, I. D. Campbell, H. Gregory, and B. Sheard (1987) The solution structure of human epidermal growth factor. *Nature* 327: 339–341.

Dahlbäck, B. (1983) Purification of human vitamin K-dependent protein S and its limited proteolysis by thrombin. *Biochem. J.* 209: 837–846.

Dahlbäck, B. and J. Stenflo (1978) Binding of bovine coagulation factor Xa to platelets. *Biochemistry* 17: 4938–4945.

Dahlbäck, B. and J. Stenflo (1994) The protein C anticoagulant system. In *The Molecular Basis of Blood Diseases* (Stamatoyannopoulos, G., Nienhuis, A. W., Majerus, P. W., and Varmus, H., eds). W. B. Saunders, Philadelphia.

Dahlbäck, B., B Hildebrand, and S. Linse (1990) Novel type of very high affinity calcium binding sites in beta-hydroxyasparagine-containing epidermal growth factor-like domains in vitamin K-dependent protein S. *J. Biol. Chem.* 265: 18481–18489.

Dahlbäck, B., Å. Lundwall, and J. Stenflo (1986) Primary structure of bovine vitamin K-dependent protein S. *Proc. Natl. Acad. Sci. USA.* 83: 4199–4203.

Davie, E. W., K. Fujikawa, and W. Kisiel (1991) The coagulation cascade: initiation, maintenance, and regulation *Biochemistry* 30: 10363–10370.

Deerfield, D. W., D. L. Olson, P. Berkowitz, P. A. Byrd, K. A. Koehler, L. G. Pedersen, and R. G. Hiskey (1987) Mg(II) binding by bovine prothrombin fragment 1 via equilibrium dialysis and the relative roles of Mg(II) and Ca(II) in blood coagulation. *J. Biol. Chem.* 262: 4017–4023.

Downing, A. K., V. Knott, J. W. Wermer, C. M. Cardy, I. D. Campbell, and P. A. Handford (1996) Solution structure of a pair of calcium-binding epidermal growth factor-like domains: implications for the Marfan syndrome and other genetic disorders. *Cell* 85: 597–605.

Ellisen, L. W., J. Bird, D. C. West, A. L. Soreng, T. C. Reynolds, S. D. Smith, and J. Sklar (1991) TAN-1, the human homolog of the *drosophila* notch gene, is broken by chromosomal translocations in T lymphoblastic neoplasms. *Cell* 66: 649–661.

Esmon, C. T. (1989) The roles of protein C and thrombomodulin in the regulation of blood coagulation. *J. Biol. Chem.* 264: 4743–4746.

Esmon, C. T. (1993) Molecular events that control the protein C anticoagulant pathway. *Thromb Haemostasis* 70: 29–35.

Esmon, N. L., L. E. DeBault, and C. Y. Esmon (1983) Proteolytic formation and properties of γ-carboxyglutamic acid domainless protein C. *J. Biol. Chem.* 258: 5548–5553.

Fisher, C. L., J. S. Greengard, and J. H. Griffin (1994) Models of the serine protease domain of the human antithrombotic plasma factor activated protein C and its zymogen. *Protein Sci.* 3: 588–599.

Freedman, S. J., B. C Furie, B. Furie, and J. D. Baleja (1995a) Structure of the calcium ion-bound γ-carboxyglutamic acid-rich domain of factor IX. *Biochemistry* 34: 12126–12137.

Freedman, S. J., B. C. Furie, B. Furie, and J. D. Baleja (1995b) Structure of the metal-free γ-carboxyglutamic acid-rich membrane binding region of Factor IX by two-dimensional NMR spectroscopy. *J. Biol. Chem.* 270: 7980–7987.

Furie, B. and B. C Furie (1988) The molecular basis of blood coagulation. *Cell* 53: 505–518.

Furie, B. and B. C. Furie (1990) Molecular basis of vitamin K-dependent gamma-carboxylation. *Blood* 75: 1753–1762.

Galvin, J. B., S Kurosawa, K. Moore, C. T. Esmon, and N. L. Esmon (1987) Reconstitution of rabbit thrombomodulin into phospholipid vesicles. *J. Biol. Chem.* 262: 2199–2205.

Gerads, I., J. W. P Govers-Riemslag, G. Tans, R. F. A. Zwaal, and J Rosing (1990) Prothrombin activation on membranes with anionic lipids containing phosphate sulfate, and/or carboxyl groups *Biochemistry* 29: 7967–7974.

Giesen, P. L. A., G. M. Willems, and W. Th. Hermens (1991) Production of thrombin by the prothrombinase complex is regulated by membrane-mediated transport of prothrombin. *J. Biol. Chem.* 266: 1379–1382.

Govers-Riemslag, J. W. P., M. P. Janssen, R. F. A. Zwaal, and J. Rosing (1992) Effect of membrane fluidity and fatty acid composition on the prothrombin-converting activity of phospholipid vesicles. *Biochemistry* 31: 10000–10008.

Govers-Riemslag, J. W. P., M. P. Janssen, R. F. A. Zwaal, and J. Rosing (1994) Prothrombin activation on dioleoylphosphatidylcholine membranes. *Eur J. Biochem.* 220: 131–138.

Graves, B. J., R. L. Crowther, C. Chandran, J. M. Rumberger, S. Li, K.-S. Huang, D. H. Presky, P. C. Familletti, B. A. Wolitzky, and D. K. Burns (1994) Insight into E-selectin/ligand interaction from the crystal structure and mutagenesis of the lec/EGF domains. *Nature* 367: 532–538.

Griffin, J. H., D. F. Mosher, T. S. Zimmerman, and A. Kleiss (1982) Protein C, an antithrombotic protein, is reduced in hospitalized patients with intravascular coagulation. *Blood* 60: 261–264.

Guinto, E. R. and C. T. Esmon (1984) Loss of prothrombin and factor Xa–factor Va interactions upon inactivation of factor Va by activated protein C. *J. Biol. Chem.* 259: 13986–13992.

Handford, P. A., M. Baron, M. Mayhew, A. Wills, T. Beesley, G. G. Brownlee, and I. D. Campbell (1990) The first EGF-like domain from human factor IX contains a high-affinity calcium binding site. *EMBO J.* 9: 475–480.

Handford, P. A., M. Mayhew, M. Baron, P. R. Winship, I. D. Campbell, and G. G. Brownlee (1991) Key residues involved in calcium-binding motifs in EGF-like domains. *Nature* 351: 164–167.

Hansen, A. P., A. M. Petros, R. P. Meadows, D. G. Nettesheim, A. P. Mazar, E. T. Olejinczak, R. X. Xu, T. M. Pederson, J. Henkin, and S. W. Fesik (1994) Solution structure of the amino-terminal fragment of urokinase-type plasminogen activator. *Biochemistry* 33: 4847–4864.

Hauschka, P. A., J. B. Lian, D. E. C. Cole, and C. M. Gendberg (1989) Osteocalcin and matrix Gla protein: vitamin K-dependent proteins in bone. *Physiol. Rev.* 69: 990–1047.

Henriksen, R. A. and C. M. Jackson (1975) Cooperative calcium binding by the phospholipid binding region of bovine prothrombin: a requirement for intact disulfide bridges. *Arch. Biochem. Biophys.* 170: 149–159.

Hill, K. A. W. and F. J. Castellino (1987a) The binding of Mn^{2+} to bovine plasma protein C, des(1-41)-light chain protein C, and activated des(1-41)-light chain activated protein C. *Arch. Biochem. Biophys.* 254: 196–202.

Hill, K. A. W. and F. J. Castellino (1987b) Topographical relationships between the monovalent and divalent cation binding sites of des-1-41-light chain bovine plasma protein C and des-1-41-light chain-activated bovine plasma protein C. *J. Biol. Chem.* 262: 7105–7108.

Hill, K. A. W., S. A. Steiner and F. J. Castellino (1988) Estimation of the distance between the divalent cation binding site of des-1-41-light chain-activated bovine plasma protein C and a nitroxide spin label attached to the active-site serine residue. *Biochem. J.* 251: 229–236.

Huang, L. H., H. Cheng, A. Pardi, J. Tam, and W. V. Sweeny (1991) Sequence-specific ^{1}H NMR assignments, secondary structure, and location of the calcium binding site in the first epidermal growth factor like domain of blood coagulation factor IX. *Biochemistry* 30: 7402–7409.

Jackson, C. M. (1988) Calcium ion binding to gamma-carboxyglutamic acid-containing proteins from the blood clotting system: What we still don't understand. In *Current Advances in Vitamin K Research* (Suttie, J. W. ed.). Elsevier, New York.

Jackson, C. M. and Y. Nemerson (1980) Blood coagulation. *Annu. Rev. Biochem.* 49: 765–811.

Jacobs, M., S. J. Freedman, B. C. Furie, and B. Furie (1994) Membrane binding properties of the factor IX gamma-carboxyglutamic acid-rich domain prepared by chemical synthesis. *J. Biol. Chem.* 269: 25494–25501.

Johnson, A. E., N. L. Esmon, T. M. Laue, and C. T. Esmon (1983) Structural changes required for activation of protein C are induced by Ca^{2+} binding to a high affinity site that does not contain γ-carboxyglutamic acid. *J. Biol. Chem.* 258: 5554–5560.

Kane, W. H. and E. W. Davie (1988) Blood coagulation factors V and VIII: structural and functional similarities and their relationship to hemorrhagic and thrombotic disorders. *Blood* 71: 539–555.

Kane, W. H. and P. W. Majerus (1982) The interaction of human coagulation factor Va with platelets. *J. Biol. Chem.* 257: 3963–3969.

Khoda, D., N. Go, K. Hayashi, and F. Inagaki (1988) Tertiary structure of mouse epidermal growth factor determined by two-dimensional ^{1}H NMR. *J. Biochem.* 103: 741–743.

Khoda, D., T. Sawada, and F. Inagaki (1991) Characterization of pH titration shifts for all the non-labile proton resonances in a protein by two-dimensional NMR: the case of mouse epidermal growth factor. *Biochemistry* 30: 4896–4900.

Kidd, S., M. R Kelley, and M. W. Young (1986) Sequence of the Notch locus of *Drosophila melanogaster*: Relationship of the encoded protein to mammalian clotting and growth factors. *Mol. Cell. Biol.* 6: 3094–3108.

Kline, T. P., F. K. Brown, S. C. Brown, P. W. Jeffs, K. D. Kopple, and L. Mueller (1990) Solution structure of human transforming growth factor α derived from ^{1}H NMR data. *Biochemistry* 29: 7805–7813.

Kohda, D., N. Go, K. Hayashi, and F. Inagaki. 1988. Tertiary structure of mouse epidermal growth factor determined by two-dimensional ^{1}H NMR. *J. Biochem.* 103: 741.

Kraulis, P. J. (1991) MOLSCRIPT: a program to produce both detailed and schematic plots of protein structures. *J. Appl. Crystallogr.* 24: 946–950.

Krishnaswamy, S. (1990) Prothrombin complex assembly. Contributions of protein–protein and protein–membrane interactions toward complex formation. *J. Biol. Chem.* 265: 3708–3718.

Krishnaswamy, S. (1992) The interaction of human factor VIIa with tissue factor. *J. Biol. Chem.* 267: 23696–23706.

Le Bonniec, B. F., E. R. Guinto, and C. T. Esmon (1992) Interaction of thrombin des-ETW with antithrombin III, the kunitz inhibitors, thrombomodulin and protein C. *J. Biol. Chem.* 267: 19341–19348.

Lee, B., M. Godfrey, E. Vitale, H. Hori, M. G. Mattei, M. Sarfarzai, P. Tsipouras, F. Ramirez, and D. W. Hollister (1991) Linkage of Marfan syndrome and a phenotypically related disorder to two different fibrillin genes. *Nature* 352: 330–334.

Lentz, B. R., C.-M. Zhou, and J. R. Wu (1994) Phosphatidylserine-containing membranes alter the thermal stability of prothombin's catalytic domain: a differential scanning calorimetric study. *Biochemistry* 33: 5460–5468.

Li, L., T. Darden, C. Foley, R. Hiskey, and L. Pedersen (1995) Homology modelling and molecular dynamics simulation of human prothrombin fragment 1. *Protein Sci.* 4: 2341–2348.

Linse, S., C. Johansson, P. Brodin, T. Grundström, E. Thulin, and S. Forsén (1988) The role of protein surface charges in ion binding. *Nature* 335: 651–652.

Lundblad, R. L. (1988) A hydrophobic site in human pro-thrombin present in a calcium-stabilized conformer. *Biochem. Biophys. Res Commun.* 30: 295–300.

Lundwall, Å., W. R. Dackowski, E. H. Cohen, M. Shaffer, A. Mahr, B. Dahlbäck, J. Stenflo, and R. M. Wydro (1986) Isolation and sequence of the cDNA for human protein S, a regulator of blood coagulation. *Proc. Natl. Acad. Sci. USA.* 83: 6716–6720.

Mann, K. G., M. E. Nesheim, W. R. Church, P. Haley, and S. Krishnaswamy (1990) Surface-dependent reactions of the vitamin K-dependent enzyme complexes. *Blood* 76: 1–16.

Marsh, H. C., M. E. Scott, R. G. Hiskey, and K. A. Koehler (1979) The nature of the slow metal ion-dependent conformational transition in bovine prothrombin. *Biochem. J.* 183: 513–517.

Martin, D. M. A., D. P. O'Brien, E. G. D. Tuddenham, and P. G. H. Byfield (1993) Synthesis and characterization of wild-type and variant gamma-carboxyglutamic acid containing domains of factor VII. *Biochemistry* 32: 13949–13955.

Maslen, C. L., G. M. Corson, B. K. Maddox, R. W. Glanville, and L. Y. Sakai (1991) Partial sequence of a candidate gene for the Marfan syndrome. *Nature* 352: 334–337.

Maune, J. F., C. B. Klee, and K. Beckingham (1992) Calcium-binding and conformational change in two series of point mutations to the individual calcium-binding sites of calmodulin. *J. Biol. Chem.* 267: 5286–5295.

Miletich, J. P., C. M. Jackson, and P. W. Majerus (1977) Interaction of coagulation factor Xa with human platelets. *Proc. Natl. Acad. Sci. USA.* 74: 4033–4036.

Montelione, G. T., K. Wüthrich, A. W. Burgess, E. C. Nice, G. Wagner, K. D. Gibson, and H. A. Scheraga (1992) Solution structure of murine epiderman growth factor determined by NMR spectroscopy and refined by energy minimization with restraints. *Biochemistry* 31: 236–249.

Montelione, G. T., K. Wüthrich, E. C. Nice, A. W. Burgess, and H. A. Scheraga (1987) Solution structure of murine epidermal growth factor: determination of the polypeptide backbone chain-fold by nuclear magnetic resonance and distance geometry. *Proc. Natl. Acad. Sci. USA.* 84: 5226–5230.

Morita, T. and C. M. Jackson (1980) Structural and functional characteristics of a proteolytically modified. "Gla domain-less" bovine factor X and Xa (des light chain residues 1–144). In *Vitamin K metabolism and vitamin K-dependent proteins.* Suttie, J. W., (ed.). MD, University Park Press, Baltimore, pp. 124–128.

Morita, T. and W Kisiel (1985) Calcium binding to a human factor IXa derivative lacking gamma-carboxyglutamic acid: evidence for two high-affinity sites that do not involve beta-hydroxyaspartic acid. *Biochem. Biophys. Res. Commun.* 130: 841–847.

Morita, T., B. S. Isaacs, C. T. Esmon, and A. E. Johnson (1984) Derivative of blood coagulation factor IX containing a high affinity Ca^{2+}-binding site that lacks gamma-carboxyglutamic acid. *J. Biol. Chem.* 259: 5698–5704.

Moy, F. J., Y.-C. Li, P. Rauenbuehler, M. E. Winkler, H. A. Scheraga, and G. T. Montelione (1994) Solution structure of human transforming type-alpha transforming growth factor determined by heteronuclear NMR spectroscopy and refined by energy minimization with restraints. *Biochemistry* 32: 7334–7353.

Nelsestuen, G. L. (1976) Role of γ-carboxyglutamic acid. An unusual protein transition required for the calcium-dependent binding of prothrombin to phospholipid. *J. Biol. Chem.* 25: 5648–5656.

Nelsestuen, G. L. and J. W. Suttie (1972a) Mode of action of vitamin K. Calcium binding properties of bovine prothrombin. *Biochemistry* 11: 4961–4964.

Nelsestuen, G. L. and J. W. Suttie (1972b) The purification and properties of an abnormal prothrombin protein produced by Dicoumarol-treated cows. Comparison to normal prothrombin. *J. Biol. Chem.* 247: 8176–8182.

Nelsestuen, G. L., W. Kisiel, and R. G. Discipio (1978) Interaction of vitamin K dependent proteins with membranes. *Biochemistry* 17: 2134–2138.

Nelsestuen, G. L., R. M. Resnick, G. J. Wei, C. H. Pletcher, and V. A. Bloomfield (1981) Metal ion interaction with prothrombin and prothrombin fragment 1. Stochiometry of binding. Protein self-association and conformational change induced by a variety of metal ions. *Biochemistry* 20: 351–358.

Nesheim, M. E., J. B. Taswell, and K. G. Mann (1979) The contribution of bovine factor V and factor Va to the activity of prothrombinase. *J. Biol. Chem.* 254: 10952–10962.

Nicolaisen, E. M., P. C. Petersen, L. Thim, J. K. Jacobsen, M. Christensen, and U. Hedner (1992) Generation of Gla-domainless FVIIa by cathepsin G-mediated cleavage. *FEBS Lett.* 306: 157–160.

Öhlin, A. K. and J. Stenflo (1987) Calcium-dependent interaction between the epidermal growth factor precursor-like region of human protein C and a monoclonal antibody. *J. Biol. Chem.* 262: 12798.

Öhlin, A. K., I. Björk, and J. Stenflo (1990) Proteolytic formation and properties of a fragment of protein C containing the gamma-carboxyglutamic acid rich region and the EGF-like region. *Biochemistry* 29: 644–651.

Öhlin, A. K., G. Landes, P. Bourdon, C. Oppenheimer, R. M. Wydro, and J. Stenflo (1988a) Beta-hydroxyaspartic acid in the first epidermal growth factor-like domain of protein C. Its role in Ca^{2+} binding and biological activity. *J. Biol. Chem.* 263: 19240–19248.

Öhlin, A. K., S. Linse, and J. Stenflo (1988) Calcium binding to the epidermal growth factor homology region of bovine protein C. *J. Biol. Chem.* 263: 7411–7417.

Olivera, B. M., J. Rivier, C. Clark, C. A. Ramillo, G. P. Corpuz, F. C. Abogadie, E. E. Mena, S. R. Woodward, D. R. Hillyard, and L. J. Cruz (1990) Diversity of Conus neuropeptides. *Science* 249: 257–263.

Olsen, P. H., N. L. Esmon, C. T. Esmon, and T. M. Laue (1992) Ca^{2+} dependence of the interactions between protein C, thrombin, and the elastase fragment of thrombomodulin. Analysis of ultracentrifugation. *Biochemistry* 31: 746–754.

Padmanabhan, K., K. P. Padmanabhan, A. Tulinsky, C. H. Park, W. Bode, R. Huber, D. T. Blankenship, A. D. Cardin, and W. Kisiel (1993) Structure of human des (1–45) factor Xa at 2.2 Å resolution. *J. Mol. Biol.* 232: 947–966.

Patthy, L. (1985) Evolution of the proteases of blood coagulation and fibrinolysis by assembly from modules. *Cell* 41: 657–663.

Patthy, L. (1991) Modular exchange principles in proteins. *Curr. Opin. Struct. Biol.* 1: 351–361.

Patthy, L. (1994) Introns and exons. *Curr. Opin. Struct. Biol.* 4: 383–392.

Persson, E., I. Björk, and J. Stenflo (1991a) Protein structural requirements for Ca^{2+}-binding to the light chain of factor X. Studies using isolated intact fragments containing the gamma-carboxyglutamic acid region and/or the epidermal growth factor-like domains *J. Biol. Chem.* 266: 2444–2452.

Persson, E., P. J. Hogg, and J. Stenflo (1993) Effects of calcium binding on the protease module of factor Xa and its interaction with factor Va. Evidence of two Gla-independent calcium-binding sites in factor Xa. *J. Biol. Chem.* 268: 22531–22539.

Persson, E., M. Selander, T. Drakenberg, A. K. Öhlin, and J. Stenflo (1989) Calcium binding to the isolated β-hydroxyaspartic acid-containing epidermal growth factor containing domain of bovine factor X. *J. Biol. Chem.* 264: 16897–16904.

Persson, E., C. Valcarce, and J. Stenflo (1991) The gamma-carboxyglutamic acid and epidermal growth factor-like domains of factor X. Effects of isolated domains on prothrombin activation and endothelial cell binding of factor X. *J. Biol. Chem.* 266: 2453–2458.

Picot, D., P. J. Loll, and M. Garavito (1994) The X-ray crystal structure of the membrane protein prostaglandin H2 synthase-1. *Nature* 367: 243–249.

Plager, D. A. and G. L. Nelsestuen (1994) Direct enthalpy measurements of factor X and prothrombin association with small and large unilamellar vesicles. *Biochemistry* 33: 7005–7013.

Prendergast, F. G. and K. G. Mann (1977) Differentiation of metal ion-induced transitions of prothrombin fragment 1. *J. Biol. Chem.* 252: 840–850.

Rao, L. V. M., S. I. Rapaport, and A. D. Hoang (1993) Binding of factor VIIa to tissue factor permits rapid antithrombin III/heparin inhibition of factor VIIa. *Blood* 81: 2600–2607.

Rao, Z., P. Handford, M. Mayhew, V. Knott, G. G. Brownlee, and D. Stuart (1995) The structure of a Ca^{2+}-binding epidermal growth factor-like domain: its role in protein–protein interactions. *Cell* 82: 131–141.

Rapaport, S. I. (1991) The extrinsic pathway inhibitor: a regulator of tissue factor-dependent blood coagulation. *Thromb. Haemostosis* 66: 6–15.

Ratcliff, J. V., B. Furie, and B. C. Furie (1993) The importance of specific γ-carboxyglutamic acid residues in prothrombin. *J. Biol. Chem.* 268: 24339–24345.

Rees, D. J. G., I. M. Jones, P. A. Handford, S. J. Walter, M. P. Esnouf, K. J. Smith, and G. G. Brownlee (1988) The role of beta-hydroxyaspartate and adjacent carboxylate residues in the first EGF domain of human factor IX. *EMBO J.* 7: 2053–2061.

Rezaie, A. R. and C. T. Esmon (1992) The function of calcium in protein C activation by thrombin and the thrombin-thrombomodulin complex can be distinguished by mutational analysis of protein C derivatives. *J. Biol. Chem.* 267: 26104–26109.

Rezaie, A. R. and C. T. Esmon (1994) Asp-70 → lys mutant of factor X lacks high affinity Ca^{2+} binding site yet retains function. *J. Biol. Chem.* 269: 21495–21499.

Rezaie, A. R., N. L. Esmon, and C. T. Esmon (1992) The high affinity calcium-binding site involved in protein C activation is outside the first epidermal growth factor homology domain. *J. Biol. Chem.* 267: 11701–11704.

Rezaie, A. R., T. Mather, F. Sussman, and C. T. Esmon (1994) Mutation of Glu 80 to Lys results in a protein C mutant that no longer requires Ca^{2+} for rapid activation by the thrombin–thrombomodulin complex. *J. Biol. Chem.* 269: 3151–3154.

Rezaie, A. R., P. F. Neuenschwander, J. M. Morrissey, and C. T. Esmon (1993) Analysis of the functions of the first epidermal growth factor-like domain of factor X. *J. Biol. Chem.* 268: 8176–8180.

Ruf, W., M. W. Kalnik, T. Lund-Hansen, and T. S. Edgington (1991) Characterization of factor VII association with tissue factor in solution. High and low affinity binding sites in factor VII contribute to functionally distinct interactions. *J. Biol. Chem.* 266: 15719–15725.

Sabharwal, A. K., J. J. Birktoft, J. Gorka, P. Wildgoose, L. C. Petersen, and S. P. Bajaj (1995) High affinity Ca^{2+}-binding site in the serine protease domain of human factor VIIa and its role in tissue factor binding and development of catalytic activity. *J. Biol. Chem.* 270: 15523–15530.

Schwalbe, R. A., J. Ryan, D. M. Stern, W. Kisiel, B. Dahlbäck, and G. L. Nelsestuen (1989) Protein structural requirements and properties of membrane binding by γ-carboxyglutamic acid-containing plasma proteins and peptides. *J. Biol. Chem.* 264: 20288–20296.

Sekiya, F., T. Yamashita, H. Atoda, Y. Komiyama, and T. Morita (1995) Regulation of the tertiary structure and function of coagulation factor IX by magnesium (II) ions. *J. Biol. Chem.* 270: 14325–14331.

Selander, M., E. Persson, J. Stenflo, and T. Drakenberg (1990) 1H NMR assignment and secondary structure of the Ca^{2+}-free form of the amino-terminal epidermal growth factor like domain in coagulation factor X. *Biochemistry* 29: 8111–8118.

Selander-Sunnerhagen, M. E. Persson, I. Dahlqvist, T. Drakenberg, J. Stenflo, M. Mayhew, M. Robin, P. Handford, J. W. Tilley, I. D. Campbell, and G. G.

Brownlee (1993) The effect of aspartate hydroxylation on calcium binding to epidermal growth factor-like modules in coagulation factors IX and X. *J. Biol. Chem.* 268: 23339–23344.

Selander-Sunnerhagen, M., M. Ullner, E. Persson, O. Teleman, J. Stenflo, and T. Drakenberg (1992) How an epidermal growth factor (EGF)-like domain binds calcium. High resolution NMR structure of the calcium form of the NH_2-terminal EGF-like domain in coagulation factor X. *J. Biol. Chem.* 267: 19642–19649.

Seshadri, T. P., E. Skrzypczak-Jankun, M. Yin, and A. Tulinsky (1994) Differences in the metal ion structure between Sr- and Ca-prothrombin fragment 1. *Biochemistry* 33: 1087–1092.

Soriano Garcia, M., C. H. Park, A. Tulinsky, K. G. Ravichandran, and E. Skrzypczak-Jankun (1989) Structure of Ca^{2+} prothrombin fragment 1 including the conformation of the Gla domain. *Biochemistry* 28: 6605–6610.

Soriano Garcia, M. W., K. Padmanabhan, A. M. deVos, and A. Tulinsky (1992) The Ca^{2+} ion and membrane binding structure of the Gla-domain of Ca^{2+}-prothrombin fragment 1. *Biochemistry* 31: 2554–2566.

Stenberg, Y., K. Julenius, I. Dahlqvist, T. Drakenberg, and J. Stenflo (1997a) Calcium-binding properties of the third and fourth epidermal-growth-factor-like modules in vitamin K-dependent protein S. *Eur. J. Biochem.* 248. 163–170.

Stenberg, Y., S. Linse, T. Drakenberg, and J. Stenflo (1997b) The high affinity calcium-binding sites in the epidermal growth factor module region of vitamin K-dependent protein S. *J. Biol. Chem.* 272: 23255–23260.

Stenflo, J. (1973) Vitamin K and the biosynthesis of prothrombin. 3. Structural comparison of an NH_2-terminal fragment from normal and from dicoumarol-induced prothrombin. *J. Biol. Chem.* 248: 6325–6332.

Stenflo, J. (1988) The biochemistry of protein C. In *Protein C and related proteins* Bertina, R. M., (ed.). Churchill Livingstone, Edinburgh.

Stenflo, J. (1991) Structure–function relationships of epidermal growth factor modules in vitamin K-dependent clotting factors. *Blood* 78: 1637–1651.

Stenflo, J. and B. Dahlbäck (1994) Vitamin K-dependent proteins. In *The Molecular Basis of Blood Diseases* Stamatoyannopoulos, G., Nienhuis, A. W., Majerus, P. W., and Varmus, H., (eds.). Saunders, Philadelphia.

Stenflo, J. and P. Ganrot (1972) Vitamin K and the biosynthesis of prothrombin. 1. Identification and purification of a dicoumarol-induced abnormal prothrombin from bovine plasma. *J. Biol. Chem.* 247: 8160–8166.

Stenflo, J. and P. Ganrot (1973) Binding of Ca^{2+} to normal and dicoumarol-induced prothrombin. *Biochem. Biophys. Res. Commun.* 50: 98–104.

Stenflo, J. and J. W. Suttie (1997) Vitamin K-dependent formation of gamma-carboxyglutamic acid. *Annu. Rev. Biochem.* 46: 157–172.

Stenflo, J., E. Holme, S. Lindstedt, N. Chandramouli, L. H. Tsai Huang, J. Tam, and R. B. Merrifield (1989)

Hydroxylation of aspartic acid in domains homologous to the epidermal growth factor precursor is catalyzed by a 2-oxoglutarate-dependent dioxygenase. *Proc. Natl. Acad. Sci. USA* 86: 444–447.

Stenflo, J., A. K. Öhlin, W. G. Owen, and W. J. Schneider (1988) Beta-hydroxyaspartic acid or beta-hydroxyasparagine in bovine low density lipoprotein receptor and in bovine thrombomodulin. *J. Biol. Chem.* 263: 21–24.

Strynadka, N. C. J. and M. N. G. James (1991) Towards an understanding of the effects of calcium on protein structure and function. *Curr. Opin. Struct. Biol.* 1: 905–914.

Südhof, T. C., J. L. Goldstein, M. S. Brown, and D. W. Russel (1985) The LDL receptor gene: a mosaic of exons shared with different proteins. *Science* 228: 815–822.

Sugo, T., I. Björk, A. Holmgren, and J. Stenflo (1984) Calcium-binding properties of bovine factor X lacking the gamma-carboxyglutamic acid-containing region. *J. Biol. Chem.* 259: 5705–5710.

Sunnerhagen, M., S. Forsén, A.-M. Hoffrén, T. Drakenberg, O. Teleman, and J. Stenflo (1995) Structure of the Ca^{2+} free Gla domain sheds light on membrane binding of blood coagulation proteins. *Nat. Struct. Biol.* 2: 504–509.

Sunnerhagen, M., G. Olah, J. Stenflo, T. Drakenberg, and J. Trewhella (1996) The relative orientation of Gla and EGF domains in coagulation factor X is altered by Ca^{2+} binding to the first EGF domain. A combined NMR-small angle X-ray scattering study. *Biochemistry* 35: 11547–11559.

Suttie, J. (1985) Vitamin K-dependent carboxylase. *Annu. Rev. Biochem.* 54: 459–477.

Suttiet, J. W. (1993) Synthesis of vitamin K-dependent proteins. *FASEB J.* 7: 445–452.

Suzuki, K., Y. Deyashiki, J. Nishioka, and K. Toma (1989) Protein C inhibitor: structure and function. *Thromb. Haemostosis* 61: 337–342.

Swords, N. A. and K. G. Mann (1993) The assembly of the prothrombinase complex on adherent platelets. *Arterioscl. Thromb.* 13: 1602–1612.

Szebenyi, D. M. E. and K. Moffat (1986) The refined structure of vitamin D-dependent calcium-binding protein from bovine intestine. *J. Biol. Chem.* 261(19): 8761–8777.

Toomey, J. R., K. J. Smith, and D. W. Stafford (1991) Localization of the human tissue factor recognition determinant of human factor VIIa. *J. Biol. Chem.* 266: 19198–19202.

Tracy, P. B., J. M. Peterson, M. E. Nesheim, F. C. McDuffie, and K. G. Mann (1979) Interaction of coagulation factor V and factor Va with platelets. *J. Biol. Chem.* 254: 10354–10361.

Tulinsky, A., C. H. Park, and E. Skrzypczak-Jankun (1988) Structure of prothrombin fragment 1 refined at 2.8 Å resolution. *J. Mol. Biol.* 202: 885–901.

Ullner, M., M. Selander, E. Persson, J. Stenflo, O. Teleman, and T. Drakenberg (1992) Three-dimensional structure of the NH_2-terminal EGF-homologous module of blood coagulation factor X as determined by NMR spectroscopy and simulated folding. *Biochemistry* 31: 5794–5983.

Valcarce, C., A. Homgren, and J. Stenflo (1994) Calcium-dependent interaction between Gla-containing and N-terminal EGF-like modules in factor X. *J. Biol. Chem.* 269: 26011–26016.

Valcarce, C., M. Selander-Sunnerhagen, A.-M. Tämlitz, T. Drakenberg, I. Björk, and J. Stenflo (1993) Calcium affinity of the NH₂-terminal epidermal growth factor-like module of factor X. Effect of the gamma-carboxyglutamic acid-containing module. *J. Biol. Chem.* 268: 26673–26678.

Vermeer, C. (1990) Gamma-carboxyglutamate-containing proteins and the vitamin K-dependent carboxylase. *Biochem. J.* 266: 625.

Vässin, H., K. A. Bremer, E. Knust, and J. A. Campus Ortega (1987) The neurogenic gene delta of *Drosophila melanogaster* is expressed in neurogenic territories and encodes a putative transmembrane protein with EGF-like repeats. *EMBO J.* 6: 3431–3440.

Welsch, D. J. and G. L. Nelsestuen, (1988a) Carbohydrate-linked asparagine-101 of prothrombin contains a metal ion protected acetylation site. Acetylation of this site causes loss of metal ion induced protein fluoresence change. *Biochemistry* 27: 4946–4952.

Welsch, D. J., Pletcher, C. H. and Nelsestuen, G. L. (1988) Chemical modification of prothrombin fragment 1: documentation of sequential, two-stage loss of protein function. *Biochemistry* 27: 4933–4938.

Wharton, K. A., K. M. Johansen, T. Xu, and S. Artavanis Tsakonas (1985) Nucleotide sequence for the neurogenic locus notch implies a gene product that shares homology with proteins containing EGF-like repeats. *Cell* 43: 567–581.

Wildgoose, P., D. C. Foster, J. Schiodt, F. C. Wiberg, J. J. Birktoft, and L. C. Petersen (1993) Idenfitication of a calcium binding site in the protease domain of human blood coagulation factor VII: evidence for its role in the factor VII–tissue factor interaction. *Biochemistry* 32: 114–119.

Williams, M. J., I. Phan, M. Baron, P. C. Driscoll, and I. D. Campbel (1993) Secondary structure of a pair of fibronectin type 1 modules by two-dimensional nuclear magnetic resonance. *Biochemistry* 32: 7388–7395.

Ye, J. and C. T. Esmon (1995) Factor Xa–factor Va complex assembles in two dimensions with unexpectedly high affinity: an experimental and theoretical approach. *Biochemistry* 34: 6448–6453.

Zhang, L. and F. J. Castellino (1992) Influence of specific gamma-carboxyglutamic acid residues on the integrity of the calcium-dependent conformation of human protein C. *J. Biol. Chem.* 267: 26078–26084.

Zhang, L. and F. J. Castellino (1994) The binding energy of human coagulation protein C to acidic phospholipid vesicles contains a major contribution from leucine 5 in the gamma-carboxyglutamic acid domain. *J. Biol. Chem.* 269: 3590–3595.

Zhang, L., A. Jhingan, and F. J. Castellino (1992) Role of individual gamma-carboxyglutamic acid residues of activated human protein C in defining its in vitro anticoagulant activity. *Blood* 80: 942–952.

Zushi, M., K. Gomi, S. Yamamoto, I. Maruyama, T. Hayashi, and K. Suzuki (1989) The last three consecutive epidermal growth factor-like structures of human thrombomodulin comprise the minimum functional domain for protein C-activating cofactor activity and anticoagulant activity. *J. Biol. Chem.* 264: 10351–10353.

7

Calmodulin Structure and Target Interaction

Nico Tjandra
Ad Bax
Anna Crivici
Mitsuhiko Ikura

Of all EF-hand-type calcium-binding proteins, calmodulin (CaM) has been studied by the largest possible variety of biophysical and biochemical methods (Cohen and Klee, 1988). This 148-residue protein is found in virtually all eukaryotic cell types and its amino acid sequence is highly conserved. It plays a major role in transducing the influx of Ca^{2+} resulting from an extracellular signal, into a cellular response. In the resting state of an unstimulated cell, the Ca^{2+} concentration, is ca. 6×10^{-8} M, whereas following stimulation this rises to ca. 1.6×10^{-6} M. At this higher Ca^{2+} concentration, CaM binds four Ca^{2+} ions and forms tight complexes (K_d 10^{-7} to 10^{-11} M) with a large variety of target proteins (Klee, 1988). Binding of CaM to its targets remains an area of active research, and, thus far, detailed structural information is available only for complexes between CaM and peptide fragments of a few enzymes. A number of natural peptides also bind CaM with high affinity, including several hormones, neurotransmitters, and venoms (Malencik and Anderson, 1982, 1983a, 1983b). However, as these are either exogenous or secreted peptides, the physiological relevance of these interactions remains uncertain.

Many CaM-dependent proteins are involved either in cell signaling through phosphorylation or dephosphorylation of intracellular proteins or participate in signal transduction by modulating levels of intracellular second messengers such as Ca^{2+}, cAMP, cGMP, and nitric oxide (Greenlee et al., 1982; Guerini et al., 1987; Charbonneau, 1990; Lowenstein and Snyder, 1992; Schmidt et al., 1992). Another important function of CaM relates to the regulation of cytoskeletal elements (Pollard et al., 1991; Aderem, 1992; Scaramuzzino and Morrow, 1993), and new cellular targets of CaM are continuously being identified, although for many of these the function is not yet known.

Although no single consensus sequence for CaM recognition exists, the interactions with its target enzymes are highly specific and of high affinity, and activation of the majority of CaM-dependent proteins requires Ca^{2+}. Because most of the target proteins are large and multimeric, CaM–target complexes have been difficult to study at the molecular level. Consequently, much of the structural information available to date has been obtained from the study of CaM complexed with peptide venoms (Malencik and Anderson, 1982, 1983b; Kataoka et al., 1989), model peptides (Cox et al., 1985; O'Neil and DeGrado, 1989; Clore et al., 1994), and peptide fragments of the CaM-binding domains of several target proteins (Blumenthal and Krebs, 1987; Roth et al., 1991; Ikura et al., 1992; Meador et al., 1992, 1993; Trewhella, 1992. Functional studies (Putkey et al., 1986; Small and Anderson, 1988; Kataoka et al., 1989; Kowluru et al., 1989; Persechini et al., 1989; Izumi et al., 1992; Gao et al., 1993) using structurally modified CaM have also contributed to the elucidation of the different modes of recognition and complex formation found in different types of targets. Numerous comprehensive reviews describing structural (O'Neil and DeGrado, 1990; Strynadka and James, 1990; Clore et al., 1994; Weinstein and Mehler, 1994; Crivici and Ikura, 1995; Ikura, 1995, 1996) and functional (Persechini and Kretsinger, 1988b; Means et al., 1991) aspects of CaM are available, including collections edited by Means and Conn (1987), Cohen and Klee (1988), and Carafoli and Klee (1992). This chapter focuses on the structural features of CaM, as revealed in recent years by x-ray crystallography and nuclear magnetic resonance (NMR) techniques. Details of the CaM structure are now available for all three states: apo, Ca^{2+}-ligated, and complexes with different target peptides.

Historically, the crystal structure of Ca^{2+}-CaM was solved first (Babu et al., 1985, 1988; Kretsinger et al., 1986; Chattopadhyaya et al., 1992), followed by that of Ca^{2+}-CaM complexed with peptide fragments of three different target enzymes (Ikura et al., 1992; Meador et al., 1992, 1993), and only very recently the structure of the apo form has been derived from NMR data (Kuboniwa et al., 1995; Zhang et al., 1995). In our discussion, we will follow the more logical order, starting with the apo state of the protein, followed by the Ca^{2+}-ligated state, and interactions between Ca^{2+}-CaM and its targets. Structural information on the potentially interesting state in which only the C-terminal CaM domain, which has the highest Ca^{2+} affinity, is ligated remains sketchy and this is still an area of active research.

Apo-calmodulin

Determination of the three-dimensional structure of the apo form of CaM has been hampered by problems of growing crystals suitable for x-ray crystallography. The CaM amino acid sequence (Fig. 7.1) reveals that the protein contains four Ca^{2+} binding sites of the EF-hand type. This type of Ca^{2+}-binding motif was first described for Ca^{2+}-ligated parvalbumin and the EF-hand has been named after helices E and F in this protein (Kretsinger and Nockolds, 1973). It consists of a helix (E), followed by a 12-residue Ca^{2+}-binding "loop" which includes a short β-strand and the first few residues of helix F, and helix F itself. In the absence of Ca^{2+}, the protein is very sensitive to proteolysis, particularly at the Arg^{106}–His^{107} peptide bond. Remarkably, these residues are located near the middle of the the the F-helix of CaM's third EF-hand. Together with microcalorimetric results (Tsalkova and Privalov, 1985), which indicated the absence of the usual clear melting transition found for globular proteins, this had led to some speculation that CaM in the apo state does not adopt a well-defined structure but would be more similar to a "molten globule". Circular dichroism (CD) studies suggested that the apo form contained a somewhat smaller fraction of helix relative to the Ca^{2+}-ligated form (Hennessey et al., 1981; Martin and Bayley, 1986; Török et al., 1992), and small-angle x-ray scattering data indicated a maximum length vector of 59 Å, only slightly shorter than the value of 63 Å found for Ca^{2+}-CaM (Seaton et al., 1985; Heidorn and Trewhella, 1988). A model for the apo-CaM had been proposed (Herzberg et al., 1986; Strynadka and James, 1988) on the basis of the x-ray structure of the homologous protein troponin C (Herzberg and James, 1988), for which the two C-terminal EF-hands were Ca^{2+}

ligated, whereas the two EF-hands in the N-terminal domain were not.

Dynamic Features

A preliminary characterization by NMR indicates that the eight α-helices of CaM's four EF-hands are, indeed, also present in the apo form (Tjandra et al., 1995b). However, as can be seen from Fig. 7.2, the rates at which the backbone amide protons exchange with the solvent are much faster compared with what is typically found for stable globular proteins. The C-terminal half of the protein, in particular, shows rather rapid exchange for many of its backbone amides, including those located in the trypsin-sensitive helix F. Exchange of an amide hydrogen with solvent can occur only when it is not hydrogen-bonded, and these exchange data are therefore compatible with a situation where the helices are present most of the time, but also adopt a random coil or other nonhelical conformation for a small fraction of the time. Indeed, NMR relaxation data provide clear evidence for a conformational exchange process involving most of the residues in the C-terminal half of CaM which occurs on a time scale of a few hundred microseconds (Tjandra et al., 1995b). These relaxation data also indicate that the "alternate" conformation is present for only a very small fraction ($\ll 10\%$) of the time. For the N-terminal half of the protein, the amide hydrogen exchange data are, on average, also faster than what is typically found for small globular proteins but, with the exception of an equilibrium localized at residue Val^{55}, NMR relaxation data did not provide any evidence for measurable population of a second conformation.

The three-dimensional structure of apo-CaM in aqueous solution was derived independently and simultaneously in two laboratories, yielding very similar results (Kuboniwa et al., 1995; Zhang et al., 1995). In addition, the three-dimensional structure for the C-terminal fragment, TR_2C, containing residues 76–148, was also determined using multidimensional NMR techniques (Finn et al., 1993, 1995). Somewhat surprisingly, the conformation of this "half-calmodulin" appears to differ beyond the uncertainty in its coordinates from the conformation of the same domain in intact apo-CaM. At this stage, it remains unclear whether this difference can be attributed to differences in the ionic strength and temperatures at which these data were recorded or whether it is caused by some interaction between the N- and C-terminal halves of the protein. As will be discussed later, the NMR data for intact apo-CaM do not provide any evidence for such an interaction, but its presence nevertheless cannot be excluded entirely.

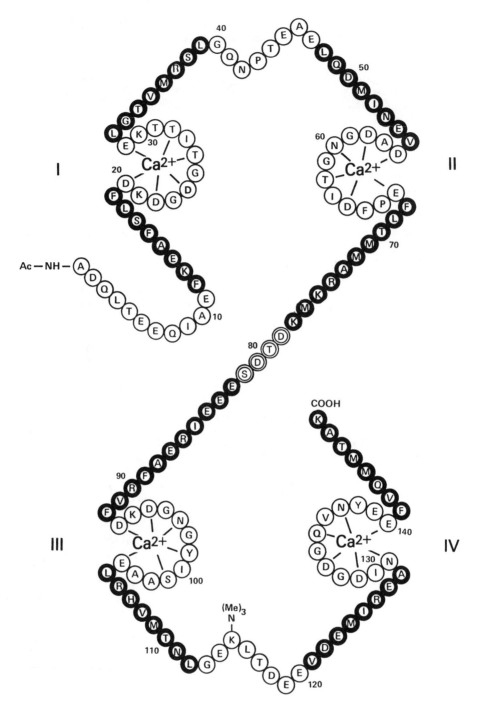

Figure 7.1 Amino acid sequence of mammalian calmodulin. The EF-hand type Ca^{2+}-binding sites are marked I–IV, and residues in the helices of each EF-hand are marked by bold circles. The last three residues of each Ca^{2+}-binding site are also part of the exiting "F-helix", with the conserved glutamate (E) in the 12th position of each site forming an N-cap hydrogen bond with the residue in position 9. The N-terminus of mammalian calmodulin is post-translationally N-acetylated, and residue Lys^{115} is trimethylated. (Courtesy of C. B. Klee.)

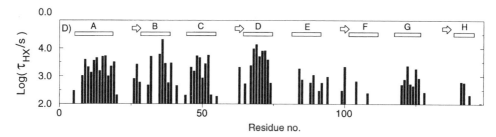

Figure 7.2 Graphical representation of hydrogen exchange times (τ_{HX}) and secondary structure in apo-CaM at pH 6.3, 100 mM KCl, 23°C. Open bars correspond to α-helices, open arrows correspond to the short β-strands. Only residues for which $\tau_{HX} > 100$ s were quantitatively evaluated and are shown in the figure. (From Tjandra et al., 1995b).

Fig. 7.3 shows a superposition of the set of structures which are all compatible with the NMR data. When superposition of the structures is optimized for the N-terminal half of CaM (Fig. 7.3A), the C-terminal half appears very "fuzzy", and vice versa (Fig. 7.3B). The reason for this fuzziness is that the orientation of the N-terminal relative to the C-terminal domain is ill-determined by the NMR data. In itself, this does not prove that the linker region is flexible; however, in principle at least, it could also result from incomplete analysis of the NMR data. Positive information regarding the flexibility of the linker region is provided by the NMR relaxation data of the ^{15}N nuclei of the polypeptide backbone. These provide detailed information on the orientation and magnitude of the rotational diffusion tensor (Barbato et al., 1992; Brüschweiler et al., 1995; Tjandra et al., 1995a). In the case of apo-CaM, results indicate that the rotational diffusion is, to a good approximation, axially symmetric, but the ratio of the diffusion rates parallel and perpendicular to the long axis is only 1.8, whereas a ratio of ~ 2.5 is predicted if the conformation were a rigid, extended dumbbell. If the two domains were packed against each other, and their relative orientations were fixed, this could possibly lead to an anisotropy as

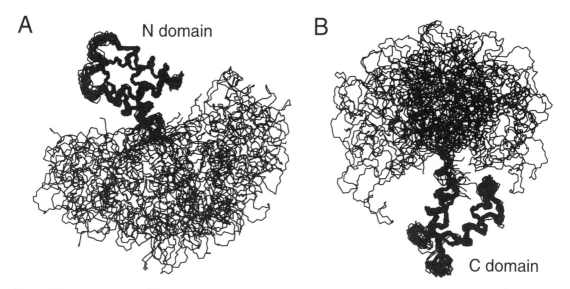

Figure 7.3 Superposition of the 30 solution structures of apo-CaM derived from NMR data by Zhang et al. (1995). (A) Only the N-terminal domain is used in best fitting the structures to one another and (B) only the C-terminal residues are used. The structures indicate that the conformations of the N- and C-terminal domains are well determined by the NMR data, but that their relative orientation is undefined.

low as 1.8, but, in this case, direct interactions between the N- and C-terminal domains should have been observed, which was not the case. In addition, the magnetic field dependence of the ^{15}N longitudinal relaxation times indicates large-amplitude domain motions on a time scale of several nanoseconds (Tjandra et al., 1995b), which also excludes the possibility of stable contacts between the two domains.

The N-terminal Domain

The structure of the N-terminal domain is determined with high precision by the NMR data, whereas the precision of the C-terminal domain is somewhat lower as a result of the extensive conformational exchange process mentioned earlier, which occurs in this domain on a time scale of several hundred microseconds (Tjandra et al., 1995b). Ribbon models of the N- and C-terminal domains are shown in Fig. 7.4. Although it is clear from Fig. 7.4 that the two domains are very similar, there are some interesting differences too, which will be discussed below.

The first helix in the N-terminal domain starts with an N-cap at residue Thr^5 and extends through Leu^{18}. This helix is followed by an extension of two residues, Phe^{19} and Asp^{20}, which form 3_{10}-type hydrogen bonds to Phe^{16} and Ser^{17}, respectively. Residues Thr^{26}–Thr^{28}, paired with Asp^{64}–Thr^{62}, are part of a very short antiparallel β-sheet. The second helix, B, corresponding to the F-helix of the EF-hand, initiates at Thr^{29}. This helix shows a remarkable kink at residue 31, which is responsible, in part, for the difference in A/B-interhelix angle relative to the Ca^{2+}-ligated state. A similar kink was observed for the homologous residue in both the solution and x-ray structures of the N-terminal troponin C domain in the apo state. The loop connecting helices B and C shows an increase in flexibility relative to the helices, but its average conformation remains reasonably well determined. The second EF-hand is virtually a repeat of the first one; initiating helix C with an N-cap at Thr^{44}, and ending helix D at Met^{76}. As shown in Fig. 7.4, the four helices form a tightly packed, highly twisted bundle, capped by the short antiparallel β-sheet. The bundle is held together

by numerous hydrophobic interactions, including a tight cluster of four phenylalanine residues at positions 16, 19, 65, and 68. The surface of the domain does not include any extensive hydrophobic patches and contains numerous negatively charged side chains.

The C-terminal Domain

As can be seen from Fig. 7.4, the C-terminal domain adopts a conformation very similar to that of the N-terminal domain, and, indeed, the backbone atoms of the two domains can be superimposed with a root-mean-square (rms) difference of 2.0 Å. The small differences observed are mainly caused by slightly different packing of the hydrophobic residues. For example, as can be seen in Fig. 7.4, Phe^{141} in helix H adopts an orientation very different from the homologous Phe^{68} in helix D. Other interesting differences relate to the absence of the N-cap in helix H and its presence in helix D, and the absence of the pronounced kink in helix F. The effects of the slow conformational exchange, mentioned above, are most pronounced in the Ca^{2+}-binding sites themselves, and make it difficult to compare quantitatively the conformation of these residues with the corresponding ones in the N-terminal domain.

Ca^{2+}-Ligated Calmodulin

Early NMR studies showed clear evidence for a two-step conformational change in CaM upon binding Ca^{2+}, with the C-terminal domain exhibiting an order of magnitude higher affinity for Ca^{2+} relative to the N-terminal domain (Andersson et al., 1983; Forsen et al., 1983; Ikura et al., 1984; Klevit and Vanaman, 1984). The Ca^{2+}-binding affinity and cooperativity are affected by target complex formation and K_d values for Ca^{2+} can increase significantly in the presence of target peptides (Klee, 1988), presumably because binding to its target stabilizes CaM's Ca^{2+}-ligated conformation.

In addition to measuring changes in Ca^{2+}-binding properties, solution experiments using small angle x-ray scattering (Seaton et al., 1985; Heidorn and

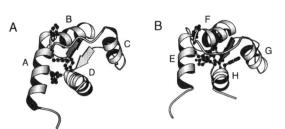

Figure 7.4 Ribbon diagram of (A) the N-terminal and (B) the C-terminal domain of apo-CaM. Phenylalanine side chains and tyrosine side chains are also shown. The two domains are connected by a flexible linker (Met^{76}–Ser^{81}) and their relative orientation is ill-determined (see Figure 7.3). All protein ribbon diagrams were created using the MOLSCRIPT program (Kraulis, 1991). (Adapted from Kuboniwa et al., 1995.)

Trewhella, 1988; Matsushima et al., 1989) and fluorescence anisotropy measurements of photo-cross-linked CaM (Small and Anderson, 1988) indicate that CaM becomes somewhat more extended in the Ca^{2+}-ligated state, with an increase in the maximum length vector from 59 to 63 Å. Circular dichroism measurements were interpreted as an indication that the helical content of CaM also increases upon ligation of Ca^{2+} (Hennessey et al., 1981; Martin and Bayley, 1986; Kowluru et al., 1989).

The crystal structure of Ca^{2+}-CaM was first solved by Babu et al. (1985), and subsequently by Kretsinger et al. (1986), and further refined by several groups (Babu et al., 1988; Chattopadhyaya et al., 1992; Rao et al., 1993). In the crystalline state, Ca^{2+}-CaM adopts a dumbbell shape, approximately 65 Å long, with the N- and C-terminal domains, of dimensions $25 \times 20 \times 20$ Å, separated by an interconnecting helix of approximately eight turns (Fig. 7.5). The two EF-hands in each domain adopt their usual conformation, with the helices preceding and following each Ca^{2+}-binding "loop" nearly orthogonal to one another. Next to the so-called "central helix", connecting the two domains, the most characteristic features are the cup-shapes of the individual domains, with a concave hydrophobic surface in the center and negatively charged residues at the rims (Fig. 7.6). In the crystalline state, the second helix of Ca^{2+}-binding site II and the first helix of Ca^{2+}-binding site III are connected by an α-helical spacer, consisting of residues 76–82. Together with this linker, these two helices constitute the so-called "central helix". Although residues 76–82 adopt an α-helical geometry in the crystalline state, they are fully solvent-exposed and show relatively high temperature

factors, indicative of increased flexibility of these residues.

Flexibility of the "Central Helix"

The question of whether the linker region is α-helical in solution or not has been the subject of considerable debate, and various experimental techniques have yielded different answers. For example, small-angle x-ray scattering data were interpreted by different groups as support for a rigid and a flexible dumbbell (Seaton et al., 1985; Heidorn and Trewhella, 1988; Matsushima et al., 1989), and fluorescence anisotropy decay measurements of photo-cross-linked CaM were interpreted as support for the rigid dumbbell model (Small and Anderson, 1988). Optical measurements of unmodified CaM indicated a decrease in effective correlation time when the pH is raised from 4.5 to 7, which has been explained by a decrease in rigidity of the "central helix" (Török et al., 1992). The high sensitivity of the linker to proteolysis suggested that, at least for part of the time, the linker adopts a nonhelical structure (Walsh et al., 1977; Newton et al., 1984). The question of central helix flexibility was finally settled by NMR. First, the interproton distances (measured from nuclear Overhauser effects or NOEs) and ^{13}C chemical shifts observed for the linker region are much closer to values expected for a random coil structure than for an α-helix (Ikura et al., 1991). Second, the backbone amide peptides undergo large amplitude motions on a subnanosecond time scale, considerably faster than the overall tumbling of the protein. Third, the rotational diffusion of the individual domains exhibits only a small degree of anisotropy, in contrast

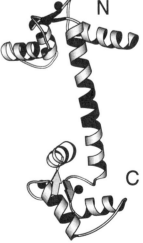

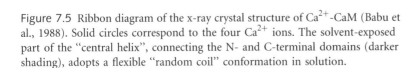

Figure 7.5 Ribbon diagram of the x-ray crystal structure of Ca^{2+}-CaM (Babu et al., 1988). Solid circles correspond to the four Ca^{2+} ions. The solvent-exposed part of the "central helix", connecting the N- and C-terminal domains (darker shading), adopts a flexible "random coil" conformation in solution.

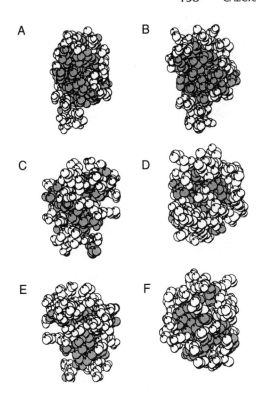

Figure 7.6 Comparison of surface structures of Ca^{2+}-free (Kuboniwa et al., 1995) and Ca^{2+}-saturated CaM (Babu et al., 1988). The "front" surfaces of (A) the N- and (B) the C-terminal domains of Ca^{2+}-saturated CaM, with hydrophobic residues shown shaded. The same views of the (C) N- and (D) C-terminal domain of apo-CaM. (E, F) The "back" views of the same structures shown in (C) and (D), respectively. (Adapted from Zhang et al., 1995.)

to the very large degree of anisotropy expected for a nearly rigid dumbbell. Fourth, the correlation times for rotational diffusion of the two domains, also derived from NMR data, differ by a statistically significant amount; the larger N-terminal half shows an effective rotational correlation time of 7.3 ns whereas the smaller C-terminal half yields 6.5 ns (Barbato et al., 1992). These data therefore indicate that rotational diffusion over angles of ca.1 radian occur on different time scales for the two domains, i.e. the linker region has a high degree of flexibility. Although this linker region was also found to be flexible in apo-CaM, quantitative analysis of the NMR data indicates that the linker in Ca^{2+}-CaM is even more flexible. As will be discussed later, this flexibility of the linker region is key to CaM's ability to bind to a large variety of target sites.

Ca^{2+}-Induced Conformational Change

Figure 7.7 compares the conformation of the apo- and Ca^{2+}-ligated states of the N-terminal CaM domain. As can be seen, the relative position of helices A and D remains nearly unchanged upon Ca^{2+} binding and the same applies for helices B and C. However, there is a large change in the relative orientations of helices A and B, and C and D. In the Ca^{2+}-ligated state, helices B and C are pulled

away from A and D, giving rise to the deep hydrophobic cleft observed in Fig. 7.6A. As will be discussed later, this hydrophobic cleft and a similar one in the C-terminal domain are critical for binding to many of CaM's target enzymes. In contrast to an increase in α-helicity upon Ca^{2+} binding, suggested by CD measurements (Hennessey et al., 1981; Martin and Bayley, 1986; Török et al., 1992), the secondary structure of CaM remains virtually unchanged upon Ca^{2+} ligation. The same is found to be true for troponin C, and it has been suggested that the increase in negative ellipticity observed by CD is not due to a change in secondary structure but to a change in tertiary structure (Gagne et al., 1994, 1995).

It is interesting to analyze in some more detail which changes in the Ca^{2+}-binding "loop" give rise to the switch in conformation upon Ca^{2+} binding. Figure 7.8 compares the conformation of the first Ca^{2+} binding loop in the apo- and in the Ca^{2+}-ligated states. Although a superposition of the entire 12-residue Ca^{2+}-binding site yields a poor fit (not shown), the first eight residues (Asp^{20}–Thr^{27}) adopt very similar conformations in both states, with a rms difference between the backbone atom coordinates of ca. 0.5 Å. Similarly, the last four residues also adopt the same structure in the two states, and the very short antiparallel β-sheet functions as a hinge. Thus, the polypeptide backbone of the first part of the Ca^{2+}-

Figure 7.7 Change in the relative orientation of the α-helices in the N-terminal domain of CaM upon ligation of Ca^{2+}. The figures superimpose the helices of the x-ray structure of the N-terminal domain of Ca^{2+}-CaM (Babu et al., 1988) (light shading) with those of the NMR structure of apo-CaM (dark shading), (a) with optimal alignment of helices A and D, and (b) with optimal alignment of helices B and C. A similar change upon ligation of Ca^{2+} occurs in the C-terminal domain. (Adapted from Zhang et al., 1995).

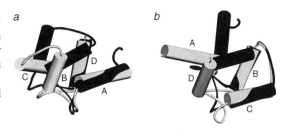

binding "loop" (Asp^{20}–Thr^{27}), which contains four of the Ca^{2+}-ligation sites (the side-chain carboxyls of Asp^{20}, Asp^{22}, and Asp^{24}, and the backbone carbonyl oxygen of Thr^{26}) is preformed in the absence of Ca^{2+}. In the apo state, the two carboxyl oxygens of Glu^{31}, which constitute a bidentate ligand for Ca^{2+}, are too far away from the position where Ca^{2+} would coordinate to its other four ligation sites. This side chain needs to be pulled by about 4 Å towards the center of the loop formed by the first eight residues in order to form the classical EF-hand type Ca^{2+}-coordination. In these sites, the coordination of Ca^{2+} is not octagonal, but together with a water molecule, the six protein ligands form a coordination geometry which resembles a pentagonal bipyramid, in which one of the Asp^{20} carboxyl oxygens corresponds to the apex, and the water oxygen is the apex at the opposite side of the pentagon. The movement of the side chain of Glu^{31} relative to residues 20–27 is primarily responsible for the large change of the relative orientation of helices A and B. For the second Ca^{2+}-binding site, consisting of Asp^{56}–Glu^{67}, which is linked to site I via the short antiparallel β-sheet, the hinge motion occurs at Ile^{63}, located opposite the hinge in site I. Residues 56–61 and 63–67 again

adopt conformations very similar to those found in the Ca^{2+}-ligated state. The precision at which the structure of the Ca^{2+}-binding loops of apo-CaM's C-terminal domain can be determined is adversely affected by the conformational exchange process mentioned earlier. However, the homology in the chemical shift patterns observed for residues in Ca^{2+}-binding sites of the N- and C-domains indicates that their structures must be quite similar, with the exception of the last few residues of site IV.

Cooperativity of Ca^{2+} Binding

In part, the change in the conformation of one of the two adjacent Ca^{2+}-binding sites is transmitted to the other site via the short antiparallel β-sheet connecting the two sites. A possibly even more important contribution relates to the concerted movement of helices B and C relative to A and D, and the changes in the interactions within each helix pair are relatively small. A similar pairing of helices has also been noted for troponin C (Strynadka and James, 1991). It thus appears that this helix pairing is a key structural feature for the cooperativity in Ca^{2+} binding to the adjacent Ca^{2+}-binding sites (Zhang et al., 1995).

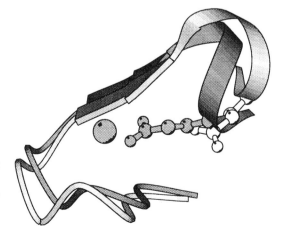

Figure 7.8 Comparison of the conformation of the first Ca^{2+}-binding site in the crystal structure of the Ca^{2+}-ligated state (darker shade) (Babu et al., 1988) and in the solution structure of the apo state (lighter shade). (Adapted from Kuboniwa et al., 1995.)

Calmodulin–Target Interaction

Over two dozen cellular target proteins have been identified for CaM (Kawasaki and Kretsinger, 1994; Crivici and Ikura, 1995). The majority of these are CaM-dependent protein kinases and one phosphatase, most of which appear to share a common mode of CaM recognition and regulation. The best studied members of this group include skeletal and smooth muscle myosin light-chain kinases (MLCKs) (Lukas et al., 1986; Blumenthal and Krebs, 1987; Persechini and Kretsinger, 1988b), multifunctional CaM-dependent protein kinase II (CaMKII) (Payne et al., 1988; Colbran et al., 1989; Colbran and Soderling, 1990; Hanson and Schulman, 1992), CaM-dependent protein phosphatase, also called calcineurin and protein phosphatase 2B (Stewart et al., 1982; Klee et al., 1988; Cohen, 1989), and phosphorylase kinase (Dasgupta et al., 1989). These enzymes all appear to share a similar mode of regulation by CaM (Fig. 7.9): in the absence of Ca^{2+}-CaM, an inhibitory domain of the enzyme blocks access to its active site. Constitutive Ca^{2+}-CaM-independent activity can be induced in vitro by proteolytic cleavage of the regulatory domain, which frequently overlaps with, or is immediately adjacent to, the CaM-binding site (Cohen and Klee, 1988). The binding of Ca^{2+}-CaM to its target site on the enzyme is believed to be responsible for removal of the neighboring autoinhibitory region of the enzyme, exposing its active site. This picture presumably oversimplifies reality, however, as evidenced by CaM mutagenesis studies which show no effect on target peptide affinity but which lower CaM's ability to activate the enzymes (George et al., 1990; VanBerkum and Means, 1991; Su et al., 1994, 1995). Also, it remains unclear how CaM can bind to its target peptide as high-affinity interaction requires the peptide to be clamped between the two domains of CaM (see below). Clearly, such interaction with CaM requires that the target peptide is first pulled away from the enzyme, and, at present, there is no detailed experimental information on how CaM could accomplish this.

Calmodulin also regulates a number of enzymes involved in signal transduction through generation or regulation of intracellular second messengers. These include cyclic nucleotide phosphodiesterase (PDE) (Charbonneau, 1990; Novack et al., 1991), type I adenylyl cyclase (Ladant, 1988; Vorherr et al., 1993), plasma membrane Ca^{2+}-ATPase (Vorherr et al., 1990; Falchetto et al., 1992), and nitric oxide synthase (Lowenstein and Snyder, 1992; Vorherr et al., 1993). There is little experimental data on the molecular mechanism by which CaM regulates these proteins but it is likely that for this class of proteins CaM regulation also occurs by removal of

an intramolecular inhibitory peptide fragment that blocks access to the active site. Indeed, PDE was the first CaM-regulated enzyme to be activated by proteolysis and whose activation was proposed to be mediated by the displacement of an autoinhibitory peptide (Cheung, 1969, 1970; Charbonneau et al., 1992).

A third category of proteins modulated by CaM are involved in regulation of cytoskeletal proteins. Again, the molecular basis of CaM regulation and the functional relevance of CaM interaction with these proteins and with other regulatory proteins has not yet been fully established. The CaM-binding proteins thought to participate in various cytoskeleton-mediated events, such as motility, cell growth and development, and morphogenesis, include spectrin (Stromqvist et al., 1988; Steiner et al., 1989), β-adducin (Scaramuzzino and Morrow, 1993), caldesmon (Ikebe, 1990; Marston et al., 1994; Zhang and Vogel, 1994), brush-border myosin I (BBMI) (Collins et al., 1990; Swanljung-Collins and Collins, 1991), and the myristoylated alanine-rich C kinase substrate (MARCKS) (Aderem, 1992; Blackshear, 1993) and a related protein, F52 (Blackshear et al., 1992; Blackshear, 1993).

Calmodulin Target Sites

The identification of CaM-target proteins among Ca^{2+}-dependent proteins has largely been based on binding affinity and a functional requirement for CaM (Cohen and Klee, 1988; O'Neil and DeGrado, 1990). Calmodulin dependence is typically characterized by activation constants in the nanomolar range (Klee, 1988), by a sensitivity to inhibition by CaM-binding drugs (Jarrett, 1984; Cohen and Klee, 1988), and inhibition by peptide fragments of the target protein which exhibit high affinity for CaM (Andreasen et al., 1981; DeGrado et al., 1985). These peptide fragments have generally been mapped using limited proteolysis, or by means of deletion, truncation, and site-directed mutagenesis. Extensive mapping is also carried out using synthetic peptides and proteolytic fragments that incorporate full or partial sequences of the protein segment of interest. Positive identification of a sequence as a CaM-binding domain requires that the synthetic peptide retains a high affinity for CaM, with a K_d value comparable to those of the intact target protein, and with an ability to inhibit CaM binding to the native target protein (Cohen and Klee, 1988).

Examination of the CaM-binding sequences of a number of target proteins reveals that the CaM-binding domain is limited to a short region of 14–26 residues that has a propensity to form a basic amphiphilic α-helix (reviewed by O'Neil and DeGrado, 1990). DeGrado et al. (1985) have devel-

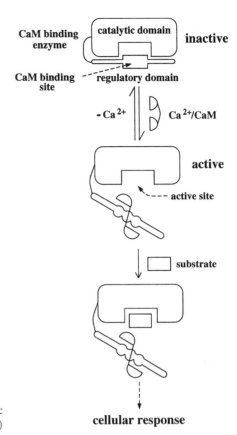

Figure 7.9 Model of regulation of a target enzyme by intrasteric inhibition and activation by CaM. (From Crivici and Ikura, 1995.)

oped a computer algorithm to screen cDNA libraries or protein sequences for potential CaM-binding domains. The algorithm evaluates a sequence on the basis of its electronic and hydrophobic properties, and its α-helical propensity. This approach has correctly predicted several CaM-binding domains (Buschmeier et al., 1987; Alexander et al., 1988; Chapman et al., 1992; Vorherr et al., 1993) but has also identified regions that do not have high CaM affinity when reproduced within a synthetic peptide (Dasgupta et al., 1989; Kataoka et al., 1991; Marston et al., 1994). However, residues in the primary sequence of known CaM-binding domains do not always exhibit a propensity to form an amphipathic α-helix (Table 7.1). In phosphorylase kinase, for example, one segment of the CaM-binding domain, reproduced in the synthetic peptide PhK13, is predicted to form an extended β-turn/β-sheet structure (Dasgupta et al., 1989).

Structure of Calmodulin Complex with MLCK Peptides

Although a substantial variety of models have been proposed for the interaction between CaM and its target sites (Persechini and Kretsinger, 1988a, 1988b; Strynadka and James, 1990; Vorherr et al., 1992), the definitive answer was provided only when the three-dimensional structures of CaM, complexed with target peptides of both skeletal and smooth muscle myosin light-chain kinase (skMLCK and smMLCK), were solved independently by NMR and x-ray crystallography (Ikura et al., 1992; Meador et al., 1992, 1993). A ribbon view of the NMR structure of CaM complexed with the 26-residue target peptide of skMLCK, often referred to as the M13 peptide (Blumenthal and Krebs, 1987), is shown in Figure 7.10. The complex is approximately ellipsoidal, with the peptide located in a hydrophobic channel that passes through the center of the ellipsoid at an angle of ca. 45° relative to its long axis. The complex appears to be stabilized primarily by hydrophobic interactions which, from the CaM side, involve an unusually large number of methionine residues. Key residues on the peptide side are a tryptophan in position 4, and a phenylalanine in position 17, which extend to the bottoms of the hydrophobic cusps of CaM's C- and N-terminal domains, respectively. Indeed, many of the target peptides display this common feature, and contain either aromatic residues or

Table 7.1 Primary Sequences of Some Known and Putative Calmodulin-Binding Domains of Protein and Peptide Calmodulin Targets

Target[a]	Sequence[b]	Reference
skMLCK (M13)[c]	K R R **W** K K N F I A V S A A N **R** F K K I S S S G A L	1
smMLCK (smMLCKp)	A R R K **W** Q K T G H A V R A I G **R** L S S	2
CaMKII	L K K F N A R R K **L** K G A I L T T M **L** A T	3
Caldesmon	G **V** R N I K S M W E K G N V **F** S S	4
Calspermin	A R R **L** K A A V K A V V A S S **R** L G S	5
PFK (M11)	F M N N W E V **Y** K L L A H I R P P A P **K** S G S Y T V	6
Calcineurin	A R K E V **I** R N K I R A I G K M A **R** **V** F S V L R	7
PhK (phK5)	L R R L I D A **Y** A F R I Y G H **W** V K K G Q Q Q N R G	8
(phK13)	R G K **F** K V I C L T V L A S V **R** **I** Y Y Y Q Y R R V K P G	9
Ca²⁺-ATPase (C28W)	L R R G Q I L **W** F R G L N R I Q T Q I **K** **V** V N A F S S S	10
59-kDa PDE	R R K H **L** Q R P I F R L R C L V **K** Q L E K	11
60-kDa PDE	T E K M **W** Q R L K G I L R C L V **K** Q L E K	12
NOS (NO-30)	K R R A I G **F** K K L A E A V K F S A **K** **L** M G Q	13
Type I AC (AC-28)	I K P A K R M K **F** K T V C Y L L V Q L M **H** C R K M F K A	14
Bordetella pertussis AC	**I** D L L **W** K I A R A G A **R** S A V G T E A	15
Neuromodulin	K A H **K** A A T K I Q A S F R G H I T R K **K** **L** K G E K K	16
Spectrin	K T A S P **W** K S A R L M V H T V A T **F** N S I K E	17
MARCKS	K K K K K **R** F S F K K S F K L S G F S **F** K K S K K	18
F52 or MacMARCKS	K K K K K **F** S F K K P F K L S G L S **F** K R N R K	19
β-Adducin	K Q Q K E K T R **W** L N T P N T Y L R V N V **A** D E V Q R N M G S	20
HSP90α	K D Q **V** A N S A F Q E R L R K **H** G L E V I	21
HIV-1 gp160	**Y** H R **L** R D L L L I V K **R** **I** V E L L G R R	22
BBMHCI	Q Q **L** A T L **I** Q K T Y R G W R C R T **H** **Y** Q L M	23
Dilute MHC	R A A C I R **I** Q K T I R G W L L R K **R** **Y** L C M Q	24
Mastoparan	**I** N L K A L A A L A K K I **L**	25
Melittin	G I G A V **L** K V L T T G L P A L I S **W** I K R K R Q Q	26
Glucagon	H S Q G T F T T S D **Y** S K Y L D S R R A Q D F **V** Q W L M N T	27
Secretin	S D G T F T S E **L** S R L R D S A R L Q R L **L** Q G L V	28
VIP	S D A V F T D N **Y** T R L R K Q **M** A V K K Y **L** N S I L N	29
GIP	A D G T F I S D **Y** S A I **M** N K I R Q Q D F **V** N W L L A Q Q Q K	30
	S	
Model peptide CBP2	K L **W** K K L L K L L K K L L **K** L G	31

[a]Abbreviations:AC, adenylyl cyclase; BBMHCI, brush-border myosin heavy chain I; CaMKII, calmodulin kinase II; CBP2, calmodulin-binding peptide 2; GIP, gastrin-inhibitory peptide; HIV-1 gp 160, human immunodeficiency virus envelope glycoprotein 160; HSP, heat-shock protein; MARCKS, myristoylated alanine-rich C kinase substrate; MHC, myosin heavy chain; NOS, nitric oxide synthase; PDE phosphodiesterase; PFK, phosphofructokinase; PhK, phosphorylase kinase; sk- and smMLCK, skeletal muscle- and smooth myosin light chain kinase; IP, vasoactive intestial peptide.

[b]Alignment of the CaM domains was made by visual inspection based on alignment of the putatively conserved major (bold and underlined) and minor (bold) hydrophobic anchors that interact with the hydrophobic patches of the C- and N-terminal domains of CaM (Ikura et al., 1992), and on the alignment of the conserved basic residue (bold and italicized) analogous to that residue of MLCK that is required for activation by CaM (Meador et al., 1992, 1993). Precise boundaries of the CaM-binding domain are not known for all targets.

[c]Names in parentheses are those used in the literature for the synthetic peptides containing the sequences listed.

References:1, Blumenthal & Krebs, 1987; 2, Lowenstein & Snyder, 1992; 3, Novack et al., 1991; 4, Zhang & Vogel, 1994; 5, Payne et al., 1988; 6, Buschmeier et al., 1987; 7, Kincaid et al., 1988; 8, Dasgupta et al., 1989; 9, Dasgupta et al., 1989; 10, Vorherr et al., 1990; 11, Vorherr et al., 1993; 12, Novack et al., 1991; 13, Vorherr et al., 1993; 14, Vorherr et al., 1993; 15, Ladant et al., 1989; Craescu et al., 1995; Oldenburg et al., 1992; 16, Chapman et al., 1991; 17, Leto et al., 1989; 18, Graff et al., 1991; 19, Blackshear et al., 1992; 20, Scaramuzzino & Morrow, 1993; 21, Minami et al., 1993; 22, Srinivas et al., 1993; 23, Mercer et al., 1991; 24, Mercer et al., 1991; 25, Malencik & Anderson, 1983b; 26, Malencik & Anderson, 1983b; 27, Malencik & Anderson, 1983a; 28, Malencik & Anderson, 1983a; 29, Malencik & Anderson, 1983a; 30, Malencik & Anderson, 1983a; 31, Degrado et al., 1985.

 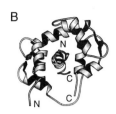

Figure 7.10 Ribbon diagram presentations of the complex between Ca^{2+}–CaM and the M13 peptide (Ikura et al., 1992), generated by the program RIBBONS. The N- and C-termini of both the peptide and protein are marked "N" and "C", respectively.

long-chain hydrophobic ones separated by a stretch of 12 residues (Table 7.1), suggesting similar modes of binding. Besides Trp^4 and Phe^{17}, numerous other hydrophobic peptide M13 residues, Phe^8, Ile^9, Ala^{10}, Val^{11}, Ala^{13}, and Ala^{14}, also contribute significantly to the hydrophobic contacts with CaM's globular domains (Fig. 7.11).

In the complex, the conformations of the globular domains remain basically identical to those observed in the crystal structure of Ca^{2+}-CaM. However, it is interesting to note that each of CaM's hydrophobic cusps contains four methionine residues. These residues are highly flexible and therefore provide a "soft" surface which can adapt to its target. Interestingly, most of these methionines retain considerable mobility upon complexation with target peptide (Bax et al., 1994). The importance of electrostatic contributions to peptide–target interactions is best defined by the x-ray structure of the complex between CaM and the smMLCK-peptide, revealing the presence of salt bridges primarily involving the acidic residues near the rims of CaM's hydrophobic cusps, and positively charged arginine and lysine side chains of the peptide (Fig. 7.11B). In particular, the cluster of basic residues near the peptide's N-terminus is involved in numerous salt bridges and hydrogen bonds, and it has been suggested that this cluster may be responsible for the orientation of the peptide relative to CaM (Meador et al., 1993).

Although, in the crystalline state, the linker region between CaM's N- and C-terminal domains in the absence of target peptide is helical, studies in aqueous solution have shown this linker region to be essentially a random coil. It is therefore not surprising that in the complex with target peptide, this region remains disordered and forms a loop connecting the two domains, both in the NMR structure of M13-CaM and in the x-ray structure of CaM complexed with the skeletal muscle MLCK peptide. Although the peptide fragments in these two studies have substantial homology, details regarding the length and precise location of the nonhelical region differ. In the X-ray structure, the region is extended, spans from residues 73–77, and is quite well defined. In the NMR-structure of M13-CaM, it extends from residues 74–82 and shows a high degree of disorder. The

location and high degree of flexibility of this loop in the M13-CaM structure has been subsequently confirmed by independent hydrogen exchange and ^{15}N NMR relaxation measurements (Grzesiek and Bax, 1993). The possibility that the longer and more flexible loop observed in solution may be an artifact of the structure calculation therefore can be safely excluded.

Calmodulin Complexed with CaMKII Peptide

The structures of CaM complexed with skMLCK and smMLCK peptides demonstrated the importance of four of the key hydrophobic residues, and reasonable predictions could be made regarding interactions with many other enzymes containing similar patterns. However, for a number of sequences it remained difficult to predict the precise mode of interaction. For example, for the target peptide of CaM-dependent kinase II (CaMKII), the pattern of hydrophobic residues does not fit the standard pattern (Table 7.1). The crystal structure of the complex between CaM and a 25-residue CaMKII peptide (Table 7.1) shows that the N-terminal half of the peptide binds in the same manner to the C-terminal CaM domain as do the MLCK peptides (Meador et al., 1993). Leu^{10} and, to a lesser extent, Ile^{14} occupy the space where Trp^4 of the skMLCK peptide (Trp^5 of the smMLCK) peptide was found. The difference in the size of these hydrophobic residues in CaMKII relative to their counterparts in MLCK is accommodated in part by small changes in the E/F and G/H interhelix angles of the C-terminal CaM domain. The N-terminal CaM domain shifts toward its C-terminal domain by approximately one turn of the α-helical CaMKII peptide, relative to its position in the complex with the MLCK peptides, and Leu^{19} of the CaMKII peptide takes the position of Phe^{17} (M13) and Leu^{18} (smMLCK peptide). For the N-terminal domain, there is also a small change in the interhelix angles of its two-EF-hands, which optimizes the interaction with the target peptide. These changes in interhelix angles are much smaller than the very large differences in angles between apo-CaM and Ca^{2+}-CaM, discussed earlier. Nevertheless, it reflects CaM's plasticity and allows the domains to "custo-

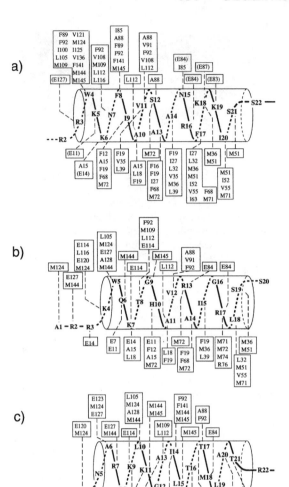

Figure 7.11 (a) Schematic summary of the intermolecular nuclear Overhauser effect (NOE) interactions (<5 Å distances) between residue pairs in M13 (residues indicated in the schematic helix) and Ca^{2+}-CaM (residues listed above and below the helix) observed in the NMR solution structure (Ikura et al., 1992). Potential electrostatic interactions between M13 and glutamic acid residues of CaM are inferred from the three-dimensional structure. (b) Schematic summary of the interactions (<4 Å distances) between residues of the smMLCK peptide and Ca^{2+}-CaM, as observed in the crystal structure of the complex (Meador et al., 1992). (c) Summary of the interactions (<4 Å) between residues of the CaMKII peptide fragment and Ca^{2+}-CaM, as observed in the crystal structure of the complex (Meador et al. 1993). Major hydrophobic anchors of the peptide targets: (a) Trp^4 and Phe^{17}, (b) Trp^5 and Leu^{18}, and (c) Leu^{10} and Leu^{19}.

mize" their fit to the target. For the two CaM-like light chains of myosin, three distinct conformational states are observed, namely "open", "semiopen", and "closed" (Houdusse and Cohen, 1995), which suggests that CaM's shape could possibly span a similar range of conformations. In addition to modulation of interhelix angles in the EF-hands, the high degree of flexibility of the side chains of the four methionines in each of CaM's two domains plays a major role in shaping the hydrophobic surfaces. Besides their high degree of flexibility, a second important reason why methionine residues are found is presumably related to the need for CaM to change from a closed (apo) state to an open state upon binding of Ca^{2+}. Hydration of methionines is energetically less costly relative to other hydrophobic residues (Makhatadze

and Privalov, 1993) and therefore facilitates the conformational change occurring upon Ca^{2+} ligation.

In the complex between CaM and the CaMKII peptide, the N-terminal CaM domain is shifted significantly relative to its C-terminal domain, and in the crystal structure this is accompanied by a nonhelical linker region between the two domains, which in this case extends from residues 73 to 83. According to Meador et al. (1993), this region of the "central helix" can be considered as a flexible extension joint that can "unravel" by a variable amount in order to accommodate the required positioning of CaM's two domains. On the other hand, residues 74–82 were found to be highly flexible under physiologically relevant conditions in the CaM–M13 complex (Ikura et al., 1992), and it has been confirmed that the definition of the loop region in solution does not result from incomplete NMR data (Grzesiek and Bax, 1993). Moreover, relative to the CaM–M13 complex, the highly flexible region in Ca^{2+}-CaM is only marginally shorter; for free Ca^{2+}-CaM, a gradual increase in backbone dynamics, starting at residue 74, is observed, with highest flexibility for residues 77–81 (Barbato et al., 1992). Therefore, it remains an open question as to what extent interactions observed for residues in this linker region in the complex with smMLCK peptide (Meador et al., 1992) are functionally significant.

Conclusions

Over the past 5 years, structural studies have contributed considerably to our increased understanding of how CaM functions at the molecular level. Experimental data obtained for the N-terminal domain of apo-CaM are generally in excellent agreement with a model proposed on the basis of the x-ray structure of the homologous N-terminal lobe of troponin C. For the C-terminal domain, significant discrepancies are found between the model and the solution structure, but the precision and thereby the accuracy of the solution structure is adversely affected by conformational averaging in the apo state. Optical spectra of apo-CaM show a clear dependence on the salt concentration (Török et al., 1992), and it has been suggested that the degree of conformational heterogeneity in the C-terminal domain is salt-dependent and that the short antiparallel β-sheet only forms at high salt concentration (Urbauer et al., 1995). Clear evidence for the presence of this sheet was observed by others at 100 mM salt (Kuboniwa et al., 1995; Zhang et al., 1995); but these latter studies also indicate conformational averaging in the β-sheet region and provide no quantitative information on what fraction of time this β-sheet is intact.

There have been contradictory interpretations of the cooperativity of CaM's Ca^{2+}-binding data. One group concluded that the data could be described adequately without requiring an interaction between the N- and C-terminal domains (Linse et al., 1991). Others argued that the optimal fitting of the data requires a degree of cooperativity (Klee, 1988), consistent with the negative interaction between the two halves of the protein revealed by thermodynamic studies of the CaM domain organization (Tsalkova and Privalov, 1985). Quantitative thrombin footprinting experiments indicate that proteolysis of the Arg^{37}–Ser^{38} peptide bond is sensitive to the Ca^{2+}-ligation state of the C-terminal domain, supporting the notion that the N-terminal half of the protein senses the Ca^{2+}-ligation state of the C-terminal domain (Shea et al., 1996). The NMR studies of apo-CaM

have shown no evidence for direct interdomain contacts, and therefore provide no support for the presence of stable, direct interactions between the two domains. However, neither the previous backbone dynamics study of apo-CaM nor the solution structures of apo-CaM exclude the possibility of transient interactions between the two domains. In this respect, it is intriguing to note that the C-terminal domain alone (Finn et al., 1995) adopts a structure somewhat different from the same domain in intact apo-CaM (Kuboniwa et al., 1995; Zhang et al., 1995). Clearly, more research is needed to fully clear up these apparent contradictions.

In the Ca^{2+}-ligated state, the CaM structure has been characterized in detail by a series of crystallographic studies (Babu et al., 1985, 1988; Kretsinger et al., 1986; Taylor et al., 1991). The N- and C-terminal domains adopt very similar conformations, and their backbone atoms can be superimposed to within a pairwise rms difference of 0.75 Å. As determined probably most convincingly by NMR (Barbato et al., 1992), the solvent-exposed part of the "central helix" observed in the crystalline state has random coil characteristics in aqueous solution at pH 6.3, and is therefore highly flexible, even more so than in the apostate of the protein. Again, there is no evidence from the NMR studies for any direct interaction between the N- and C-terminal domains, but transient interaction cannot be excluded. A fluorescence anisotropy study by Török et al. (1992) suggests that the rigidity of CaM's central helix increases when the solution pH is lowered from 7 to the crystallization pH of 4.5. In addition, it is conceivable that nonaqueous additives such as 2-methyl-2,4-pentanediol used in the solution from which CaM was crystallized may have contributed to the helical character of the linker region in the central helix.

The NMR and x-ray crystal structures of Ca^{2+}-CaM complexed with target peptides of skeletal and smooth MLCK and the x-ray structure for a complex with a CaMKII peptide confirm the importance of the hydrophobic pockets on Ca^{2+}-CaM's N- and C-terminal lobes for target binding. Extensive modeling studies had predicted the importance of these pockets and had emphasized the propensity of the target peptides to adopt amphipathic α-helical conformations (Persechini and Kretsinger, 1988a, 1988b; O'Neil and DeGrado, 1990; Strynadka and James, 1990; Vorherr et al., 1992). Details of the mode of binding in the experimental studies, however, differ substantially from all proposed models. The structures of the complexes highlight the importance of the flexibility of the "central helix", and, in particular, the complex with CaMKII peptide demonstrates the ability of each individual domain to readjust its conformation to optimally fit its target. A most striking example of this plasticity is demonstrated by CaM's ability to bind to variants of melittin and the smMLCK peptide synthesized with all D-amino acids (Fisher et al., 1994). The affinity for these D-form peptides is considerably lower relative to the natural L-form, however, and it appears that these peptides can only interact with either the C- or N-terminal CaM domain at a given time. A careful study of the hydrogen exchange rates of the backbone amides of the regular L-form smMLCK peptide complexed with CaM provides details regarding the energetics of the protein–peptide interaction and indicates a total free energy change of 5.5 kcal/mol upon complex formation (Ehrhardt et al., 1995).

Despite the information on the structure of several CaM–target peptide complexes, it would be premature to state that the mystery of activation of CaM-dependent enzymes has been solved. First, there are other proteins that do not appear to fit the pattern expected for the target peptide binding in an α-helical conformation to CaM, and there are proteins, such as phosphorylase kinase, where binding involves two target sites. Second, it remains unclear how CaM can recruit a pseudo-substrate site from the catalytic site of the enzyme. Finally, mutation data indicate that a number of CaM residues that are not involved in target peptide binding are important for enzyme activation (George et al., 1990; VanBerkum and Means, 1991; Su et al., 1995). Therefore, the model as depicted in Fig. 7.9 presumably represents an oversimplification, albeit a useful one, of the activation mechanism.

Besides the interaction between Ca^{2+}-CaM and its targets, there are also other proteins, such as neuromodulin and neurogranin, which bind to apo-CaM. A preliminary study of a neuromodulin peptide complexed with apo-CaM suggests that it interacts primarily with the C-terminal domain (Urbauer et al., 1995). Other information on how this type of interaction may occur can be gleaned from the crystal structure of the regulatory domain of myosin, where the two light chains, both members of the CaM superfamily, bind to the regulatory chain in a variety of modes, ranging from an open-lobe configuration, similar to Ca^{2+}-CaM, to a closed one, similar to apo-CaM (Xie et al., 1994; Houdusse and Cohen, 1995). It is anticipated that further structural studies of apo-CaM, semi-Ca^{2+}-ligated CaM, and its interaction with targets will continue to further unravel the details of the CaM puzzle.

Acknowledgments We thank Claude Klee for many useful comments during the preparation of this chapter. Work in the authors' laboratories is supported by grants from the Medical Research Council of Canada (MI) and the Intramural AIDS Targeted Anti-Viral Program of the Office of the Director of the National Institutes of Health (AB).

References

Aderem, A. (1992) Signal transduction and the actin cytoskeleton: the roles of MARCKS and profilin. *Trends Biochem. Sci.* 17: 438.

Alexander, K. A., Wakim, B. T., Doyle, G. S., Walsh, K. A., and Storm, D. R. (1988) Identification and characterization of the calmodulin-binding domain of neuromodulin, a neurospecific calmodulin-binding protein. *J. Biol. Chem.* 263: 7544.

Andersson, A., Drakenberg, T., Thulin, E., and Forsen, S. (1983) A ^{113}Cd and ^{1}H NMR study of the interaction of calmodulin with D600, trifluoperazine and some other hydrophobic drugs. *Eur. J. Biochem.* 134: 459.

Andreasen, T. J., Keller, C. H., LaPorte, D. C., Edelman, A. M., and Storm, D. R. (1981) Preparation of azidocalmodulin: a photoaffinity label for calmodulin-binding proteins. *Proc. Natl. Acad. Sci. USA* 78: 2782.

Babu, Y. S., Bugg, C. E., and Cook, W. J. (1988) Structure of calmodulin refined at 2.2 Å resolution. *J. Mol. Biol.* 204: 191.

Babu, Y. S., Sack, J. S., Greenhough, T. J., Bugg, C. E., Means, A. R., and Cook, W. J. (1985) Three dimensional structure of calmodulin. *Nature* 315: 37.

Barbato, G., Ikura, M., Kay, L. E., Pastor, R. W., and Bax, A. (1992) Backbone dynamics of calmodulin studied by ^{15}N relaxation using inverse detected two-dimensional NMR spectroscopy: the central helix is flexible. *Biochemistry* 31: 5269.

Bax, A., Delaglio, F., Grzesiek, S., and Vuister, G. W. (1994) Resonance assignment of methionine methyl groups and X_3 angular information from long range proton-carbon J correlation in a calmodulin–peptide complex. *J. Biomol. NMR* 4: 787.

Blackshear, P. J. (1993) The MARCKS family of cellular protein kinase C substrates. *J. Biol. Chem.* 268: 1501.

Blackshear, P. J., Verghese, G. M., Johnson, J. D., Haupt, D. M., and Stumpo, D. J. (1992) Characteristics of the F52 protein, a MARCKS homologue. *J. Biol. Chem.* 267: 13540.

Blumenthal, D. K. and Krebs, E. G. (1987) Preparation and properties of the calmodulin binding domain of skeletal muscle myosin light-chain kinase. *Methods Enzymol.* 139: 115.

Brüschweiler, R., Liao, X., and Wright, P. (1995) Long-range motional restrictions in a multidomain zinc-finger protein from anisotropic tumbling. *Science* 268: 886.

Buschmeier, B., Meyer, H. E., and Mayr, G. W. (1987) Characterization of the calmodulin-binding sites of mucle phosphofructokinase and comparison with known calmodulin binding domains. *J. Biol. Chem.* 262; 9454.

Carafoli, E. and Klee, C., eds. (1992) New developments in the calmodulin field. *Cell Calcium* 13: 353.

Chapman, E. R., Au, D., Alexander, K. A., Nicolson, T. A., and Storm, D. R. (1991) Characterization of the calmodulin binding domain of neuromodulin *J. Biol. Chem.* 266: 207.

Chapman, E. R., Alexander, K., Vorherr, T., Carafoli, E., and Storm, D. R. (1992) Fluorescence energy transfer analysis of calmodulin-peptide complexes. *Biochemistry* 31: 12819.

Charbonneau, H. (1990) Structure–function relationships among cyclic nucleotide phosphodiesterases. In *Cyclic Nucleotide Phosphodiesterases: Structure, Regulation and Drug Action* (Beavo, J., Houslay, M. D., eds.). Wiley and Sons, New York, pp. 267–296.

Charbonneau, H., Kumar, S., Novack, J. P., Blumenthal, D. K., and Griffin, P. R., Shabanowitz, J., Hunt, D. F., Beavo, J. A., and Walsh, K. A. (1992) Evidence for domain organization within the 61-kDa calmodulin dependent cyclic nucleotide phosphodiesterase from bovine brain. *Biochemistry* 30: 7931.

Chattopadhyaya, R., Meador, W. E., Means, A. R., and Quiocho, F. A. (1992) Camodulin structure refined at 1.7 Å resolution *J. Mol. Biol.* 228: 1177.

Cheung, W. Y. (1969) Cyclic 3′,5′-nucleotide phosphodiesterase: preparation of a partially inactive enzyme and its subsequent stimulation by snake venom. *Biochem. Biophys. Acta* 191:303.

Cheung, W. Y. (1970) Cyclic 3′,5′-nucleotide phosphodiesterase: demonstration of an activator. *Biochem. Biophys. Res. Commun.* 38: 533.

Clore, G. M., Bax, A., Ikura, M., and Gronenborn, A. M. (1994) Structure of calmodulin–target peptide complexes *Curr. Opin. Struct. Biol.* 3; 838.

Cohen, P. (1989) The structure and regulation of protein phosphatases *Annu. Rev. Biochem.* 58: 453.

Cohen, P. and Klee, C. B., eds. (1988) *Calmodulin.* Elsevier, Amsterdam.

Colbran, R. J. and Soderling, T. R. (1990). Calcium/calmo-dulin-dependent protein kinase II. *Curr. Topics Cell. Reg.* 31: 181.

Colbran, R. J., Schworer, C. M., Hashimoto, Y., Fong, Y. L., and Rich, D. P., Smith, M. K., and Soderling, T. R. (1989) Calcium/calmodulin-dependent protein kinase II. *Biochem. J.* 258: 313.

Collins, K., Sellers, J. R., and Matsudaira, P. (1990) Calmodulin dissociation regulates brush border myosin I (110-kD-calmodulin) mechanochemical activity in vitro. *J. Cell. Biol.* 110: 1137.

Cox, J. A., Comte, M., Fitton, J. E., and DeGrado, W. F. (1985). The interaction of calmodulin with amphiphilic peptides. *J. Biol. Chem.* 260: 2527.

Craescu, C. T., Bouhss, A., Mispelter, J., Diesis, E., Popescu, A., Chiriac, M., and Barzu, O. (1995) Calmodulin binding of a peptide derived from the regulatory domain of *Bordetella pertussis* adenylate cyclase. *J. Biol. Chem.* 270: 7088.

Crivici, A. and Ikura, M. (1995) Molecular and structural basis of target recognition by calmodulin. *Ann. Rev. Biophys. Biomol. Struct.* 24: 85.

Dasgupta, M., Honeycutt, T., and Blumenthal, D. K. (1989) The γ-subunit of skeletal muscle phosphorylase kinase contains two noncontiguous domains that act in concert to bind calmodulin. *J. Biol. Chem.* 264: 17156.

DeGrado, W. F., Prendergast, F., Wolfe, H. R., and Cox, J. A. (1985) The design, synthesis, and characterization of tight-binding inhibitors of calmodulin. *J. Cell. Biochem.* 29: 83.

Ehrhardt, M. R., Urbauer, J. L., and Wand, J. A. (1995) The energetics and dynamics of molecular recognition by calmodulin *Biochemistry* 34: 2731.

Falchetto, R., Vorherr, T., and Carafoli, E. (1992) The calmodulin-binding site of the plasma membrane Ca^{2+} pump interacts with the transduction domain of the enzyme. *Protein Sci.* 1: 1613.

Finn, B. E., Drakenberg, T., and Forsen, S. (1993) The structure of apo-calmodulin: A ^{1}H NMR examination of the carboxy-terminal domain. *FEBS Lett.* 336: 368.

Finn, B. E., Evenas, J., Drakenberg, T., Waltho, J. P., Thulin, E., and Forsen, S. (1995) Calcium-induced structural changes and domain autonomy in calmodulin. *Nat. Struct. Biol.* 2: 777.

Fisher, P. J., Prendergast, F. G., Ehrhardt, M. R., Urbauer, J. L., Wand, J. A., Sedarous, S. S., McCormick, D. J., and Buckley, P. J. (1994) Calmodulin interacts with amphiphilic peptides composed of all D-amino acids. *Nature* 368: 651.

Forsen, S., Andersson, A., Drakenberg, T., Teleman, O., Thulin, E., and Vogel, H. J. (1983) ^{25}Mg, ^{43}Ca and ^{113}Cd NMR studies of regulatory calcium binding proteins. In *Calcium-Binding Proteins* (de Bernard, B., Sottocasa, G. L., Sandri, G., Carafoli, E., and Taylor, A. N., eds.). Elsevier, Amsterdam, pp. 121–31.

Gagne, S. M., Tsuda, S., Li, M. X., Chandra, M., Smillie, L. B., and Sykes, B.D. (1994) Quantification of the calcium-induced secondary structural changes in the regulatory domain of troponin-C. *Protein Sci.* 3: 1961.

Gagne, S. M., Tsuda, S., Li, M. X., Smillie, L. B., and Sykes, B. D. (1995) Structures of the troponin C regulatory domains in the apo and calcium-saturated states. *Nat. Struct. Biol.* 2: 784.

Gao, Z. H., Krebs, J., VanBerkum, M. F. A., Tang, W. J., Maune, J. F., Means, A. R., Stull, J. T., and Beckingham, K. (1993) Activation of four enzymes by two series of calmodulin mutants with point mutations in individual Ca^{2+} binding sites. *J. Biol. Chem.* 268: 20096.

George, S. E., VanBerkum, M. F. A., Ono, T., Cook, R., Hanley, R. M., Putkey, J., and Means, A. R. (1990) Chimeric calmodulin-cardiac troponin C proteins differentially activate calmodulin target enzymes. *J. Biol. Chem.* 265: 9228.

Graff, J. M., Rajan, R. R., Randall, R. R., Nairn, A. C., and Blackshear, P. J. (1991) Protein kinase C substrate and inhibitor characteristics of peptides derived from the myristoylated alanine-rich C kinase substrate (MARCKS) protein phosphorylation site domain. *J. Biol. Chem.* 266: 14390.

Greenlee, D. V., Andreas, T. J., and Storm, D. R. (1982) Calcium-independent stimulation of *Bordetella pertusis* adenylate cyclase by calmodulin. *Biochemistry* 21: 2759.

Guerini, D., Krebs, J., and Carafoli, E. (1987) Stimulation of the erythrocyte Ca^{2+}-ATPase and of bovine brain cyclic nucleotide phosphodiesterase *Eur. J. Biochem.* 170: 35.

Grzesiek, S. and Bax, A. (1993) The importance of not saturating H_2O in protein NMR. Application to sensitivity enhancement and NOE measurements. *J. Am. Chem. Soc.* 115: 12593.

Hanson, P. I. and Schulman, H. (1992) Neuronal Ca^{2+}/calmodulin-dependent protein kinases. *Annu. Rev. Biochem.* 61: 559.

Heidorn, D. B. and Trewhella, J. (1988) Comparison of the crystal and solution structures of calmodulin and troponin C. *Biochemistry* 27: 909.

Hennessey, J. P., Parthasarathy, M., and Johnson, W. C. (1981) Conformational transitions of calmodulin as studied by vacuum-UV CD. *Biopolymers* 26: 561.

Herzberg, O. and James, M. N. G. (1988) Refined crystal structure of troponin C from turkey skeletal muscle at 2.0Å resolution. *J. Mol. Biol.* 203; 761.

Herzberg, O., Moult, J., and James, M. N. G. (1986) A model for the Ca^{2+}-induced conformational transition of troponin C. *J. Biol. Chem.* 261: 2638.

Houdusse, A. and Cohen, C. (1995) Target sequence recognition by the calmodulin superfamily: implications from light chain binding to the regulatory domain of scallop myosin. *Proc. Natl. Acad. Sci. USA* 92: 10644.

Ikebe, M. (1990) Phosphorylation of smooth muscle caldesmon by calmodulin dependent protein kinase II. *J. Biol. Chem.* 265: 17607.

Ikura, M. (1995) *Encyclopedia of Nuclear Magnetic Resonance* (Grand D. M. and Harris, R. K., eds.).Wiley, London, pp. 1100–1106.

Ikura, M. (1996) Calcium binding and conformational response in EF-hand proteins. *Trends Biochem. Sci.* 21: 14.

Ikura, M., Clore, G. M., Gronenborn, A. M., Zhu, G., Klee, C. B., and Bax A. (1992) Solution structure of a calmodulin–target peptide complex by multidimensional NMR *Science* 256: 632

Ikura, M., Hiraoki, T., Hikichi, K., Mikuni, T., Yazawa, M., and Yagi, K. (1983) Nuclear magnetic resonance studies on calmodulin: calcium-induced conformational change. *Biochemistry* 22: 2573.

Ikura, M., Hiraoki, T., Minowa, O., Yamaguchi, H., Yazawa, M., and Yagi, K. (1984) Nuclear magnetic resonance studies on calmodulin: calcium-dependent spectral change of proteolytic fragments *Biochemistry* 23: 3124.

Ikura, M., Kay, L. E., Krinks, M., and Bax, A. (1991) Triple-resonance multidimensional NMR study of calmodulin complexed with the binding domain of skeletal muscle myosin light-chain kinase: indication of a conformational change in the central helix *Biochemistry* 30: 5498.

Izumi, Y., Wakita, M., Yoshino, H., and Matsushima, N. (1992) Structure of the proteolytic fragment F34 of calmodulin in the absence and presence of mastoparan as revealed by solution X-ray scattering. *Biochemistry* 31; 12266.

Jarrett, H. W. (1984) The synthesis and reaction of a specific affinity label for the hydrophobic drug-binding domains of calmodulin. *J. Biol. Chem.* 259: 10136.

Kataoka, M., Head, J. F., Seaton, B. A., and Engelman, D. M. (1989) Melittin binding causes a large calcium dependent conformational change in calmodulin. *Proc. Natl. Acad. Sci. USA* 86: 6944.

Kataoka, M., Head, J. F., Vorherr, T., Krebs, J., and Carafoli, E. (1991) Small-angle X-ray scattering study of calmodulin bound to two peptides corresponding to parts of the calmodulin-binding domain of the plasma membranc Ca^{2+} pump. *Biochemistry* 30: 6247.

Kawasaki, H. and Kretsinger, R. (1994) Calcium-binding proteins 1: EF-hands. *Protein Profile* 1: 343.

Kincaid, R. L., Nightingale, M. S., and Martin, B. M. (1988) Characterization of a cDNA clone encoding the calmodulin binding domain of mouse calcineurin. *Proc. Natl. Acad. Sci. USA* 85: 8983.

Klee, C. B. (1988) Interaction of calmodulin with Ca^{2+} and target proteins. In *Calmodulin* (Cohen, P. and Klee, C. B., eds.). Elsevier, Amsterdam, pp. 35–46.

Klee, C. B., Draetta, G. F., and Hubbard, M. J. (1988) Calcineurin. *Adv. Enzymol.* 61: 149.

Klevit, R. E. and Vanaman, T. C. (1984) Azidotyrosylcalmodulin derivatives: specific probes for protein-binding domains. *J. Biol. Chem.* 259: 15414.

Kowluru, R. A., Heidorn, D. B., Edmonson, S. P., Bitensky, M. W., and Kowluru, A (1989) Glycation of calmodulin: chemistry and structural and functional consequences. *Biochemistry* 28: 2220.

Kraulis, P. J. (1991) MOLSCRIPT: a program to produce both detailed and schematic plots of protein structures. *J. Appl. Cryst.* 24: 945.

Kretsinger, R. H. and Nockolds, C. E. (1973) Carp muscle calcium-binding protein: structure determination and general decription. *J. Biol. Chem.* 248: 3313.

Kretsinger, R. H., Rudnick, S. E., and Weisman, L. J. (1986) Crystal structure of calmodulin. *J. Inorg. Biochem.* 28: 289.

Kuboniwa, H., Tjandra, N., Grzesiek, S., Ren, H., Klee, C. B., and Bax, A. (1995) Solution structure of calcium-free calmodulin. *Nat. Struct. Biol.* 2: 768.

Ladant, D. (1988) Interaction of *Bordetella persussis* adenylate cyclase with calmodulin. *J. Biol. Chem.* 263: 2612.

Ladant, D., Michelson, S., Sarfati, R., Gilles, A.-M., Predeleanu, R., and Barzu, O. (1989) Characterization of the calmodulin-binding and of the catalytic domains of *Bordetella pertussis* adenylate cyclase. *J. Biol. Chem.* 264: 4015.

Leto, T. L., Pleasic, S., Forget, B. G., Benz, E. J., and Harchesi, V. T. (1989) Characterization of the calmodulin-binding site of nonerythroid α-spectrin. *J. Biol. Chem.* 264: 5826

Linse, S., Helmersson, A., and Forsen, S. (1991) Calcium binding to calmodulin and its globular domains *J. Biol. Chem.* 266: 8050.

Lowenstein, C. J. and Snyder, S. H. (1992) Nitric oxide, a novel biologic messenger. *Cell* 70; 705.

Lukas, T. J., Burgess, W. H., Prendergrast, F. G., Lau, W., and Watterson, D. M. (1986) Calmodulin binding domains: characterization of a phosphorylation and calmodulin binding site from myosin light chain kinase. *Biochemistry* 25: 1458.

Makhatadze, G. I. and Privalov, P. L. (1993) Contribution of hydration to protein folding thermodynamics. *J. Mol. Biol.* 232: 639.

Malencik, D. A. and Anderson, S. R. (1982) Binding of simple peptides, hormones, and neurotransmitters by calmodulin. *Biochemistry* 21: 3481.

Malencik, D. A. and Anderson, S. R. (1983a) Binding of hormones and neuropeptides by calmodulin. *Biochemistry* 22: 1995.

Malencik, D. A. and Anderson, S. R. (1983b) High affinity binding of the mastoparans by calmodulin. *Biochem. Biophys. Res. Commun.* 114: 50.

Marston, S. B., Fraser, I. D. C., Huber, P. A. J., Pritchard, K., Gusev, N. B., and Török, K. (1994) Location of a two contact sites between human smooth muscle caldesmon and Ca^{2+}-calmodulin. *J. Biol. Chem.* 269: 8134.

Martin, S. R. and Bayley, P. M. (1986) The effects of Ca^{2+} and Cd^{2+} on the secondary and tertiary structure of bovine testis calmodulin. *Biochem. J.* 238: 485.

Matsushima, N., Izumi, Y., Matsuo, T., Yoshino, H., Ueki, T., and Miyake, Y. (1989) Binding of both Ca^{2+} and mastoparan to calmodulin induces a large change in the tertiary structure *J. Biochem. (Tokyo)* 105: 883.

Meador, W. E., Means, A. R., and Quiocho, F. A. (1992) Target enzyme recognition by calmodulin: 2.4 Å structure of a calmodulin–peptide complex *Science* 257: 1251.

Meador, W. E., Means, A. R., and Quiocho, F. A. (1993) Modulation of calmodulin plasticity in molecular recognition on the basis of X-ray structures *Science* 262: 1718.

Means, A. R. and Conn, P. M., eds. (1987) Cellular regulator. Part A. Calcium- and calmodulin-binding proteins. *Methods Enzymol.* Vol. 139. Academic Press, Orlando.

Means, A. R., VanBerkum, M. F. A., Bagchi, I., Lu, K. P., and Rasmussen, C. D. (1991) Regulatory functions of calmodulin. *Pharmacol. Ther.* 50: 255.

Mercer, J. A., Seperack, P. K., Strobel, M. C., Copeland, N. G., and Jenkins, N. A. (1991) Novel myosin heavy chain encoded by murine dilute coat colour locus. *Nature* 349: 709.

Minami, Y., Kawasaki, H., Suzuki, K., and Yahara, I. (1993) The calmodulin-binding domain of the mouse 90-kDa heat shock protein *J. Biol. Chem.* 268: 9604.

Newton, D. L., Oldewurtel, M. D., Krinks, M. H., Shiloach, J., and Klee, C. B. (1984) Agonist and antagonist properties of calmodulin fragments. *J. Biol. Chem.* 259: 4419.

Novack, J. P., Charbonneau, H., Bentley, J. K., Walsh, K. A., and Beavo, J. A. (1991) Sequence comparison of the 63-, 61-, and 59-kDa calmodulin-dependent cyclic nucleotide phosphodiesterases. *Biochemistry* 30: 7940.

O'Neil, K. T. and DeGrado, W. F. (1989) The interaction of calmodulin with fluorescent and photoreactive model peptides: evidence for a short interdomain separation. *Proteins* 6: 284.

O'Neil, K. T. and DeGrado, W. F. (1990) How calmodulin binds its targets: sequence independent recognition of amphiphilic α-helices. *Trends Biochem. Sci.* 15: 59.

Oldenburg, D. J., Gross, M. K., Wong, C. S., and Storm, D. R. (1992) High-affinity calmodulin binding is required for the rapid entry of *Bordetella pertussis* adenylyl cyclase into neuroblastoma cells. *Biochemistry* 31: 8884.

Payne, M. E., Fong, Y. L., Ono, T., Colbran, R. J., Kemp, B. E., Soderling, T. R., and Means, A. R. (1988) Calcium/calmodulin-dependent protein kinase II. *J. Biol. Chem.* 263: 7190.

Persechini, A. and Kretsinger, R. H. (1988a) The central helix of calmodulin functions as a flexible tether. *J. Biol. Chem.* 263: 12175.

Persechini, A. and Kretsinger, R. H. (1988b) Toward a model of the calmodulin–myosin light-chain kinase complex: implications for calmodulin function. *J. Cardiovasc. Pharmacol.* 12:, 501.

Persechini, A., Blumenthal, D. K., Jarrett, H. W., Klee, C. B., Hardy, D. O., and Kretsinger, R. H. (1989) The effects of deletions in the central helix of calmodulin on enzyme activation and peptide binding. *J. Biol. Chem.* 264: 8052.

Pollard, T. D., Doberstein, S. K., and Zot, H. G. (1991) Myosin-I. *Annu. Rev. Physiol.* 53: 653.

Putkey, J. A., Draetta, G. F., Slaughter, G. R., Klee, C. B., Cohen, P., Stull, J. T., and Means, A. R. (1986) Genetically engineered calmodulins differentially activate target enzymes. *J. Biol. Chem.* 261: 9896.

Rao, S. T., Wu, S., Satyshur, K. A., Ling, K. Y., Kung, C., and Sundaralingam, M. (1993) Structure of *Paramecium tetraurelia* calmodulin at 1.8 Å resolution. *Protein Sci* 2: 436.

Roth, S. M., Schneider, D. M., Strobel, L. A., VanBerkum, M. F. A., Means, A. R., and Wand, A. J. (1991) Structure of the smooth muscle myosin light-chain kinase calmodulin-binding domain peptide bound to calmodulin. *Biochemistry* 30: 10078.

Scaramuzzino, D. A. and Morrow, J. S. (1993) Calmodulin-binding domain of recombinant erythrocyte-adducin. *Proc. Natl. Acad. Sci. USA* 90: 3398.

Schmidt, H. H. H. W., Pollock, J. S., Nakane, M., Forstermann, U., and Murad, F. (1992) Ca^{2+}/calmodulin-regulated nitric oxide synthases. *Cell Calcium* 13; 427.

Seaton, B. A., Head, J. F., Engelman, D. M., and Richards, F. (1985) Calcium induced increase in the radius of gyration and maximum dimension of calmodulin measured by small angle scattering. *Biochemistry* 24: 6740.

Shea, M. A., Verhoeven, A. S., and Pedigo, S. (1996) Calcium-induced interactions of calmodulin domains revealed by quantitative thrombin footprinting of Arg 37 and Arg 106. *Biochemistry* 35: 2943.

Small, E. W. and Anderson, S. R. (1988) Fluorescence anisotropy decay demonstrates calcium-dependent shape changes in photo-cross-linked calmodulin. *Biochemistry* 27; 419.

Srinivas, S. K., Srinivas, R. V., Anantharamaiah, G. M., Compans, R. W., and Segrest, J. P. (1993) Cytosolic domain of the human immunodeficiency virus envelope glycoproteins binds to calmodulin and inhibits calmodulin-dependent proteins. *J. Biol. Chem.* 268; 22895.

Steiner, J. P., Walke, H. T. Jr., and Bennett, V. (1989) Calcium/calmodulin inhibits direct binding of spectrin to synaptosomal membranes. *J. Biol. Chem.* 264: 2783.

Stewart, A. A., Ingebritsen, T. S., Manalan, A., Klee, C. B., and Cohen, P. (1982) Discovery of a Ca^{2+} and calmodulin-dependent protein phosphatase: probable identity with calcineurin (CaM-BP$_{80}$). *FEBS Lett.* 137: 80.

Stromqvist, M., Berglund, A., Shanbhag, V. P., and Backman, L. J. (1988) Influence of calmodulin on the human red cell membrane skeleton. *Biochemistry* 27: 1104.

Strynadka, N. C. J. and James, M. N. G. (1988) Two trifluoroperazine-binding sites on calmodulin predicted from comparative modelling with troponin C. *Proteins: Struct. Funct. Genet.* 3: 1.

Strynadka, N. C. J. and James, M. N. G. (1990) Model for the interaction of amphiphilic helices with troponin C and calmodulin. *Proteins: Struct. Funct. Genet.* 7: 234.

Strynadka, N. C. J. and James, M. N. G. (1991) Towards an understanding of the effects of calcium on protein structure and function. *Curr. Opin. Struct. Biol.* 1: 905.

Su, Z., Blazing, M. A., Fan, D., and George, S. E. (1995) The calmodulin–nitric oxide synthase interaction. Critical role of the calmodulin latch domain in enzyme activation. *J. Biol. Chem.* 270: 29117.

Su, Z., Fan, D., and George, S. E. (1994) Role of domain 3 of calmodulin in activation of calmodulin-stimulated phosphodiesterase and smooth muscle myosin light chain kinase. *J. Biol. Chem.* 269: 16761.

Swanljung-Collins, H. and Collins, J. H. (1991) Ca^{2+} stimulates the Mg^{2+}-ATPase activity of brush border myosin I with three or four calmodulin light chains but inhibits with less than two bound. *J. Biol. Chem.* 266: 1312.

Taylor, D. A., Sack, J. S., Maune, J. F., Beckingham, K., and Quiocho, F. (1991) Structure of a recombinant calmodulin from *Drosophila melanogaster* refined at 2.2 Å resolution. *J. Biol. Chem.* 266: 21375.

Tjandra, N., Feller, S. E., Pastor, R. W., and Bax, A. (1995a) Rotational diffusion anisotropy of human ubiquitin from ^{15}N NMR relaxation. *J. Am. Chem. Soc.* 117: 12562.

Tjandra, N., Kuboniwa, H., Ren. H., and Bax, A. (1995b) Rotational dynamics of calcium-free calmodulin studied by ^{15}N NMR relaxation measurements. *Eur. J. Biochem.* 230: 1014.

Török, K., Lane, A. N., Martin, S. R., Janot, J.-M., and Bayley, P. M. (1992) Effects of calcium binding on the internal dynamic properties of bovine brain calmodulin, studied by NMR and optical spectroscopy *Biochemistry* 31: 3452.

Trewhella, J. (1992) The solution structures of calmodulin and its complexes with synthetic peptides based on target enzyme binding domains *Cell Calcium* 13; 377.

Tsalkova, T. S. and Privalov, P. L. (1985) Thermodynamic study of domain organization in troponin C and calmodulin. *J. Mol. Biol.* 181: 533.

Urbauer, J. L., Short, J. H., Dow, L. K., and Wand, J. A. (1995) Structural analysis of a novel interaction by calmodulin: high-affinity binding of a peptide in the absence of calcium. *Biochemistry* 34: 8099.

VanBerkum, M. F. A. and Means, A. R. (1991) Three amino acid substitutions in domain I of calmodulin prevent the activation of chicken smooth muscle myosin light chain kinase. *J. Biol. Chem.* 266: 21488.

Vorherr, T., James, P., Krebs, J., Enyedi, A., McCormick, D. J., Penniston, J. T., and Carafoli, E. (1990) Interaction of calmodulin with the calmodulin binding domain of the plasma membrane Ca^{2+} pump. *Biochemistry* 29: 355.

Vorherr, T., Kessler, O., Mark, A., and Carafoli, E. (1992) Construction and molecular dynamics simulation of calmodulin in the extended and in a bent conformation. *Eur. J. Biochem.* 204: 931.

Vorherr, T., Knopfel, L., Hofmann, F., Mollner, S.,. Pfeuffer, T., and Carafoli, E. (1993) The calmodulin binding domain of nitric oxide synthase and adenylyl cyclase. *Biochemistry* 32: 6081.

Walsh, M., Stevens, F. C., Kuznicki, J., and Drabikowski, W. (1977) Characterization of tryptic fragments obtained from bovine brain protein modulator of cyclic nucleotide phosphodiesterase. *J. Biol. Chem* 252: 7440.

Weinstein, H. and Mehler, E. L. (1994) Ca^{2+}/calmodulin-dependent protein kinase II is phosphorylated by protein kinase C in vitro. *Annu. Rev. Physiol.* 56: 213.

Xie, X., Harrison, D. H., Schlichting, I., Sweet, R. M., Kalabokis, V. N., Szent-Gyorgyi, A. G., and Cohen, C. (1994) Structure of the regulatory domain of scallop myosin at 2.8 Å resolution. *Nature* 368: 306.

Zhang, M. and Vogel, H. J. (1994) The calmodulin binding domain of caldesmon binds to calmodulin in an α-helical conformation. *Biochemistry* 33: 1163.

Zhang, M., Tanaka, T., and Ikura, M. (1995) Calcium-induced conformational transition revealed by the solution structure of apo calmodulin. *Nat. Struct. Biol.* 2: 758.

8

Voltage-Gated Calcium Channels

Richard W. Tsien
David B. Wheeler

Generic Properties and Structure of Voltage-Gated Ca^{2+} Channels

Voltage-gated Ca^{2+} channels regulate Ca^{2+} entry in a potential-dependent manner and thereby contribute to Ca^{2+}-signaling in a wide variety of cell types, including nerve, endocrine, and muscle cells. As signal transduction molecules par excellence, these channels act as links between the realms of electrical signaling and intracellular messengers (Hille, 1992). A single opening of a Ca^{2+} channel can allow many hundreds or thousands of Ca^{2+} ions to flow into the cytoplasm, thus generating a rise in $[Ca^{2+}]_i$ that may control vital functions such as excitability, rhythmicity, transmitter or hormone release, contraction, metabolism, and gene expression (Tsien and Tsien, 1990). To initiate such events effectively, Ca^{2+} channels have evolved as very efficient and highly regulated enzymes to catalyze the downhill flow of Ca^{2+} across membranes (Tsien et al., 1987b). Some key features of this enzymatic activity are as follows. *Activation* of Ca^{2+} channels is steeply voltage-dependent. The opening of individual channels occurs more quickly and is more complete with larger depolarizations, similar to voltage-gated Na^{2+} and K^+ channels. Typically, Ca^{2+} channels open within one or a few milliseconds after the membrane is depolarized from rest, and close (deactivate) within a fraction of a millisecond following repolarization. *Inactivation*, the closing of channels during maintained depolarization, strongly influences the cytosolic Ca^{2+} signal that arises from cellular electrical activity. While inactivation is a general property of Ca^{2+} channels, the speed of inactivation varies widely, ranging from very slow (requiring second-long depolarizations) to relatively rapid (tens of milliseconds). *Selectivity* of voltage-gated Ca^{2+} channels for Ca^{2+} ions is remarkably high, so that Ca^{2+} is the main charge carrier even when Ca^{2+} is greatly outnumbered by other

ions, as under normal physiological conditions. *Permeation* of Ca^{2+} through a single open Ca^{2+} channel can achieve rates of millions of ions per second when the electrochemical gradient is large. At driving forces reached physiologically, the flux rate is more modest, but sufficient to cause a large increase in $[Ca^{2+}]_i$ ($>1\,\mu M$) in a very localized domain ($\sim 1\,\mu m$) near the mouth of the open channel. The basic features of channel opening and closing (collectively referred to as "gating") and Ca^{2+} ion selectivity and permeation are intrinsic properties of all voltage-gated Ca^{2+} channels, evidently highly conserved in evolution. Our newly increased understanding of how these physiological characteristics come about is reviewed at the end of the chapter.

The powerful functional capabilities of Ca^{2+} channels are rooted in their molecular architecture. As far as we know, all voltage-gated Ca^{2+} channels are comprised of multiple components, that come together to form a large macromolecular complex ($\sim 500\,kDa$). The generic structure contains four subunits called α_1, α_2, δ, and β (Fig. 8.1). The first examples of each of these subunits were originally isolated from skeletal muscle transverse tubules by biochemical techniques more than a decade ago (Catterall and Curtis, 1987; Campbell et al., 1988; Catterall et al., 1988; Glossmann and Striessnig, 1990). Each subunit has been cloned in one or more forms within the last dozen years (Tanable et al., 1987; Ellis et al., 1988; Ruth et al., 1989). The α_1 subunit is a large (200–260 kDa) transmembrane protein that contains the channel pore, the voltage-sensor, and the gating machinery. Because the α_1 subunit appears to be able to form a functional Ca^{2+} channel on its own, the other subunits are sometimes referred to as auxiliary or ancillary subunits. The $\alpha_2\delta$-subunit (175 kDa) is a dimer, consisting of glycosylated α_2- and δ-proteins linked together by disulfide bonds, derived by post-translational processing of a single

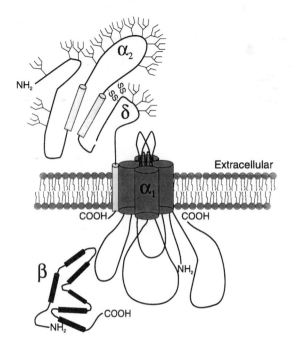

Figure 8.1 Structural organization of the subunits comprising a generic voltage-gated Ca^{2+} channel. (Reproduced with permission from Gurnett et al., 1996, copyright Cell Press.)

parent polypeptide (Ellis et al., 1988; De Jongh et al., 1990; Williams et al., 1992b). In current models (Fig. 8.1), δ is pictured as the transmembrane protein anchor and α_2 as entirely extracellular (Jay et al., 1991; Hofmann et al., 1994). In contrast, the β-subunit (55 kDa) is entirely intracellular in location. A fifth subunit, known as γ (25 kDa), has long been known to be part of Ca^{2+} channels in skeletal muscle (Bosse et al., 1990; Jay et al., 1990), and the neuronal homologue of this subunit has recently been found in mouse neurons where it increases steady-state inactivation of voltage-gated Ca^{2+} channels (Letts et al., 1998).

With this brief introduction to the general properties of Ca^{2+} channels, we will proceed to an overview of the multiple types of Ca^{2+} channels, their molecular composition, and their adaptation for specialized roles in cellular Ca^{2+} signaling. In this chapter, we will not touch on other fundamental Ca^{2+} entry mechanisms, such as Ca^{2+} channels not gated by depolarization (Lewis and Cahalan, 1995), nor will we discuss in detail the modulation of Ca^{2+} channels by G-proteins (Hille, 1994).

Ca²⁺ Channel Subtypes Differ in Biophysical and Pharmacological Properties

Multiple types of voltage-gated Ca^{2+} channels were first distinguished on the basis of their voltage- and time-dependence, single-channel conductance, and

pharmacology (e.g., Carbone and Lux, 1984; Nowycky et al., 1985). These criteria have led to a widely accepted classification of Ca^{2+} channels as T-, L-, N-, P-, Q-, and R-type (Tsien et al., 1987a; Llinás et al., 1992; Randall and Tsien, 1995). While these categories make good sense in view of the varied functional roles of the channel types in different organ systems, their relationship to the detailed subunit composition of Ca^{2+} channels in native tissue remains incompletely understood.

One physiologically relevant characteristic which varies considerably among the different Ca^{2+} channel types is the degree of depolarization required to cause significant activation. Based on such requirements, voltage-gated Ca^{2+} channels are sometimes divided into two groups: low-voltage-activated (LVA) and high-voltage-activated (HVA).

T-type Ca²⁺ Channels

The LVA Ca^{2+} channels are exemplified by T-type channels, so-named because they carry *t*iny unitary Ba^{2+} currents (6–8 pS) that occur soon after the depolarizing step, giving rise to a *t*ransient average current (Carbone and Lux, 1984; Nilius et al., 1985; Nowycky et al., 1985). Another defining characteristic of classical T-type channels is their slow deactivation following a sudden repolarization (Matteson and Armstrong, 1986). T-type channel current records also exhibit a distinctive kinetic fingerprint: the superimposed current responses cross over each other in a pattern not found with other rapidly inactivating Ca^{2+} channels, such as the R-type (Randall and Tsien, 1997). The kinetic properties are dominated by a strikingly voltage-dependent delay between the depolarizing step and the channel's first opening (Droogmans and Nilius, 1989). In addition to these properties, T-type channels have a unique pharmacological profile, characterized by mild sensitivity to 1,4-dihydropyridines (DHPs), such as nifedipine or nimodipine (Cohen and McCarthy, 1987), but acute sensitivity to mibefradil and newer experimental drugs (Ertel and Ertel, 1997). Within the overall category of T-type Ca^{2+} channel, further diversity has been found, particularly with respect to kinetic characteristics and pharmacology (Akaike et al., 1989; Kostyuk and Shirokov, 1989; Huguenard and Prince, 1992). Various subtypes of T-type Ca^{2+} channel may coexist in the same cell type and show rates of inactivation differing by as much as 5-fold, while sharing similar voltage-dependence of inactivation (Huguenard and Prince, 1992). Within a given neuron T-type Ca^{2+} channels have been found to contribute a greater percentage of the Ca^{2+} current measured in dendrites than in somata, suggesting that these channels support specialized functional roles in different parts of the cell (Kavalali et al., 1997).

L-type Ca^{2+} Channels

The L-type channels are generally categorized with the HVA group of channels, along with N-, P-, Q- and R-type channels. However, it is important to note that L-type channels may exhibit LVA properties under certain circumstances (Avery and Johnston, 1996). L-type channels in vertebrate sensory neurons and heart cells were initially labeled as *l*arge Ba^{2+} conductance contributing to a *l*ong-lasting current, with characteristic sensitivity to DHPs such as nifedipine or Bay K 8444 (Bean, 1985; Nilius et al., 1985; Nowycky et al., 1985). Members of this group were subsequently identified in other excitable cells, such as vascular smooth muscle, uterus, and pancreatic β-cells. Later, the designation of L-type was extended to refer to all channels with strong sensitivity to DHPs, including those found in skeletal muscle (Hofmann et al., 1988), even though clear-cut biophysical distinctions between skeletal and cardiac L-type channels were already known (Rosenberg et al., 1986). Thus, the category of L-type channels contains individual subtypes of considerable diversity. For example, three subtypes of L-type channel appear to coexist in cerebellar granule neurons: two subtypes that resemble those found in heart and a third that shows prominent voltage-dependent potentiation (Forti and Pietrobon, 1993).

N-type Ca^{2+} Channels

The most extensively characterized non-L-type Ca^{2+} channel was named N-type since it appeared to be largely specific to *n*eurons as opposed to muscle cells and was clearly *n*either T- or L-type (Nowycky et al., 1985). It requires relatively negative resting potentials to be available for opening, somewhat like T-type, but is high-voltage-activated, like L-type. This Ca^{2+} channel is insensitive to DHPs, but is potently and specifically blocked by a peptide toxin derived from the venom of the marine snail, *Conus geographus*, ω-conotoxin GVIA (ω-CTx-GVIA). The N-type channel is found primarily in presynaptic nerve terminals and neuronal dendrites, in addition to cell bodies (Westenbroek et al., 1992).

P-type Ca^{2+} Channels

Currents carried by P-type channels were originally recorded from cell bodies of cerebellar *P*urkinje cells, hence, the label "P-type" (Llinás et al., 1989, 1992). These channels are not blocked by DHPs or ω-CTx-GIVA, but are exquisitely sensitive to block by ω-Aga-IVA, a component of the venom of the funnel-web spider, *Agelenopsis aperta* (Mintz et al., 1992b), with an CI_{50} of ~ 1 nM (Mintz and Bean, 1993). P-type channels support a current that hardly inactivates during depolarizations lasting for several seconds. They are seen in virtual isolation in cerebellar Purkinje neuron cell bodies, but also contribute substantially to somatic currents in many other central neurons (Mintz et al., 1992a).

Q-type Ca^{2+} Channels

Currents supported by Q-type channels were initially characterized in cell bodies of cerebellar granule neurons (Zhang et al., 1993; Randall and Tsien, 1995). This component of current displays prominent inactivation during a ~ 0.1 s depolarization and is ~ 40–100-fold less sensitive to ω-Aga-IVA that the P-type current. Thus, Q-type current differs from P-type current as classically defined in Purkinje neurons and found to coexist with Q-type current in granule neurons. Confidence in the distinct nature of Q-type current is based, in part, on its resemblance to currents generated by expression of cloned α_{1A} subunits (Sather et al., 1993; Stea et al., 1993), as discussed further below. In some circumstances, the designation P/Q-type may serve to indicate ω-Aga-IVA-sensitive current without further distinction between P- and Q-type Ca^{2+} channels.

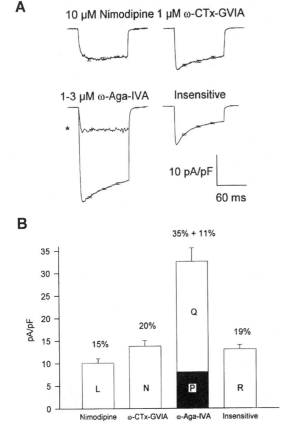

Figure 8.2 Pharmacological dissection of five distinct Ca^{2+} channel currents in cultured rat cerebellar granule neurons. (A) Time course of the Ca^{2+} channel current (5 mM Ba^{2+}) that was blocked by each drug. The asterisk indicates the P-type component of the ω-Aga-IVA-sensitive current. (B) The relative mean current density for each of the five Ca^{2+} channel types expressed as a percentage of the global Ca^{2+} channel current. (Reproduced with permission from Randall and Tsien, 1995.)

R-type Ca^{2+} Channels

The R-type Ca^{2+} channel currents were identified in cerebellar granule cells as a current that remained in the presence of nimodipine, ω-CTx-GIVA, and ω-Aga-IVA, inhibitors of the L-, N-, and P/Q-type channels, respectively (Ellinor et al., 1993; Zhang et al., 1993; Randall and Tsien, 1995). Figure 8.2 illustrates the pharmacological dissection of the various HVA Ca^{2+} currents, including R-type. This HVA current decays rapidly, and is unusually sensitive to block by Ni^{2+}. The biophysical and pharmacological properties of the macroscopic R-type current seem homogeneous enough to allow it to be treated as a single current component (Randall and Tsien, 1997), although it seems likely that further subdivision of this category will be appropriate (Tottene et al., 1996; Newcombe et al., 1998).

Molecular Basis of Ca^{2+} Channel Diversity

In recent years, findings from molecular cloning of Ca^{2+} channels have greatly increased our under-

standing of Ca^{2+} channel diversity. This has allowed (1) a new perspective on the familiar relationships between various channel types, (2) the discovery of Ca^{2+} channels beyond those types uncovered by earlier biophysical and pharmacological analysis, and (3) a more precise description of the pharmacological properties of individual channel types.

Multiple Forms of α_1-Subunit

Much of the diversity of Ca^{2+} channel types seems to arise from the expression of multiple forms of the α_1-subunit, isolated by molecular cloning (e.g., Tanabe et al., 1987; Mikami et al., 1989; Mori et al., 1991; Starr et al., 1991; Dubel et al., 1992; Williams et al., 1992a, 1992b; Soong et al., 1993). Eight different Ca^{2+} channel α_1-subunit genes have been distinguished in mammalian brain (Perez-Reyes et al., 1990; Snutch et al., 1990) and have been labeled as classes A, B, C, D, E (Snutch et al., 1990; Snutch and Reiner, 1992) G, H and I (Perez-Reyes et al., 1998; Cribbs et al., 1998; Talley et al., 1998). In a generally accepted nomenclature (Birmbaumer et al., 1994),

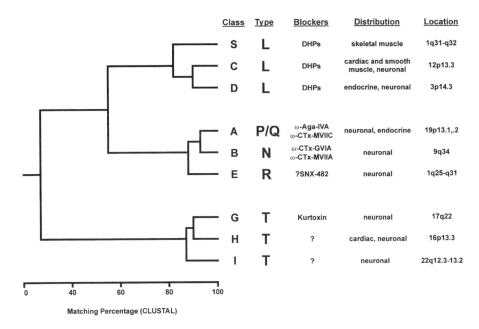

Class	Type	Blockers	Distribution	Location
S	L	DHPs	skeletal muscle	1q31-q32
C	L	DHPs	cardiac and smooth muscle, neuronal	12p13.3
D	L	DHPs	endocrine, neuronal	3p14.3
A	P/Q	ω-Aga-IVA ω-CTx-MVIIC	neuronal, endocrine	19p13.1,.2
B	N	ω-CTx-GVIA ω-CTx-MVIIA	neuronal	9q34
E	R	?SNX-482	neuronal	1q25-q31
G	T	Kurtoxin	neuronal	17q22
H	T	?	cardiac, neuronal	16p13.3
I	T	?	neuronal	22q12.3-13.2

Matching Percentage (CLUSTAL)

Figure 8.3 Phylogenetic relationships, based on amino acid sequence homology of membrane spanning regions, amongst α_1-subunits of voltage-gated Ca^{2+} channels. Dendrogram adapted from Zhang et al., (1993) Tsien (1998) and personal communication from E. Perez-Reyes and L. Cribbs. Chromosome locations according to Diriong et al. (1995), Perez-Reyes et al. (1998), and Cribbs et al. (1998).

the different α_1 isoforms are designated as $\alpha_{1A}, \alpha_{1B}, \alpha_{1C}, \alpha_{1D}, \alpha_{1E}, \alpha_{1G}, \alpha_{1H}, \alpha_{1I}$ and α_{1S}, where α_{1S} refers to the original Ca^{2+} channel clone from skeletal muscle, first isolated by the group of the late Shosaku Numa (Tanabe et al., 1987). Based on sequence homology, the nine α_1-subunits can be assigned to various branches of a family tree as reviewed in Fig. 8.3. As indicated in this diagram, the nine α_1 genes have been localized to at least seven different chromosomes (Diriong et al., 1995).

Three major subfamilies of α_1 subunits clearly emerge on the basis of sequence homology. The first subfamily consists of α_1-subunits of classes S, C, and D. Along with the α_{1S}-subunit from skeletal muscle, these include subunits first derived from heart muscle (α_{1C}) and neuroendocrine tissues (α_{1D}) but found in a variety of other organs, including brain and lung (Mikami et al., 1989; Williams et al., 1992b). These cDNAs encode channels classified as "L-type" because they are responsive to DHPs. The second α_1 subfamily consists thus far of cDNAs derived from nervous tissue, including mammalian brain (A, B, and E) and marine ray electric lobe (doe-1 and doe-4, Ellinor et al., 1993). When expressed, they lack the characteristic DHP-response of L-type channels. Individual genes within this subfamily show > 60% identity with each other but only

~ 45% or less with members of the L-type subfamily. The third subfamily consists so far of α_1 subunits supporting low voltage activated (T-type) channels, and were originally derived from brain (G, I) and heart (H).

The evolutionary divergence of the three channel subfamilies occurred relatively early, as would be expected from the fairly low sequence homology between various subfamily members (Fig. 8.3). This deduction can be corroborated by an examination of the distribution of Ca^{2+} channel types in organisms spread across many phyla, particularly L-type and N, P/Q, R-type channels. Both channel subfamilies are represented in species ranging from marine rays (Horne et al., 1993) through humans (Williams et al., 1992a. 1992b), and in many cases both families of channels are expressed within the same cells (e.g., Randall and Tsien, 1995). Amongst invertebrates, both channel subfamilies have been observed in molluscs (Edmonds et al., 1990), insects (Grabner et al., 1994; Smith et al., 1996), and nematodes (Schafer and Kenyon, 1995). Nematodes also express a relative of α_{1G}. Given the widespread distribution of the three Ca^{2+} channel subfamilies across the animal kingdom, their bifurcation must have occurred quite early during the speciation of Animalia.

Even greater diversity arises from alternative splicing of transcripts from the multiple Ca^{2+} channel genes (Tsien et al., 1991; Snutch and Reiner, 1992), as reviewed below.

Limited Variability of the $\alpha_2\delta$ Complex

Similar or identical forms of the $\alpha_2\delta$ complex are found in a variety of tissues, including skeletal muscle, heart, vascular and intestinal smooth muscle, and brain (Ellis et al., 1988; Biel et al., 1991). The limited variation that has been found has been attributed to alternative splicing of primary transcripts of a single gene (Kim et al., 1992; Williams et al., 1992b).

Multiple Forms of β-Subunit

All Ca^{2+} channels in their native state appear to contain β-subunits. The β-subunit was first identified in voltage-sensitive Ca^{2+} channels purified from skeletal muscle as an integral component with an apparent molecular weight of $\sim 60\,kDa$ (Glossman et al., 1987; Takahashi et al., 1987). Four different types of β-subunit are known to exist in mammals and are now known as $\beta_1-\beta_4$ (Birnbaumer et al., 1994) although they were called CaB1–CaB4 when first isolated (Hofmann et al., 1994). Diversity of these proteins is greatly increased by alternative splicing (designated by lower case letters, $\beta_{2a}, \beta_{2b}, \ldots$). This family of subunits has been extensively reviewed (Hofmann et al., 1994; Isom et al., 1994; De Waard et al., 1996).

cDNAs for the β_{1a}-subunit (L-type Ca^{2+} channel, rabbit skeletal muscle) were first isolated on the basis of peptide sequence and encoded a protein (524 amino acids), lacking homology to other known protein sequences (Ruth et al., 1989). Soon thereafter, cDNAs for two additional β-subunits, β_2 and β_3, were isolated by homology from rabbit heart (Hullin et al., 1992). A fourth β-subunit, β_4 has been cloned from rat brain (Castellano et al., 1993). In general, β-subunits are not found in one organ or tissue exclusively. Whereas β_1 transcripts are expressed primarily in skeletal muscle, they also appear in brain. β_2 is predominantly expressed in heart, aorta, and brain, while β_3 is most abundant in brain but also present in aorta, trachea, lung, heart, and skeletal muscle, β_4 mRNA is expressed almost exclusively in neuronal tissues, with the highest levels being found in the cerebellum.

The sequence of the various β-subunits supports biochemical evidence that they are peripheral membrane proteins associated with the cytoplasmic aspect of the surface membrane. Thus, the β-subunit of Ca^{2+} channel is not homologous to β_1- and β_2-subunits of Na^+ channels, which contain putative transmembrane spanning domains and are significantly glycosylated (Isom et al., 1994). Not only is the β-subunit not an integral membrane protein, but it is also "auxiliary" in the sense that it is not absolutely required for gating or permeation. In most but not all cases, Ca^{2+} channel function can be obtained with α_1, alone, even in the apparent absence of endogenous β-subunits. Nevertheless, β-subunits serve several important and intriguing functions: (1) they play a key role in the appropriate targeting of the complex of Ca^{2+}-channel subunits; (2) they act as modulators of the gating properties of α_1-subunits, thereby contributing greatly to the functional diversity of Ca^{2+} channels; and (3) they are targets of regulation by protein kinases.

All of the β-subunits show the striking ability to increase functional α_1 activity. This has been measured in several ways, for example as increased DHP binding to membranes of mammalian cells transfected with appropriate α_1 subunits, or as increased Ca^{2+} channel current in *Xenopus* oocytes (e.g., Lacerda et al., 1991). Indeed, modulation of Ca^{2+} channel activity by the β-subunit has been observed on coexpression of all six α_1-subunits with four β-subunits in all $\alpha_1-\beta$ combinations tested (De Waard et al., 1995). It has been found by many groups that β-subunits tend to shift the voltage-dependence of activation and, if anything, to accelerate channel opening; inactivation rate was also increased (Lacerda et al., 1991; Singer et al., 1991; Varadi et al., 1991). β-subunits differ strikingly in their ability to speed inactivation, following the order $\beta_3 > \beta_1 \approx \beta_4 > \beta_2$ (Hullin et al., 1992; Ellinor et al., 1993; Sather et al., 1993; De Waard et al., 1995). In addition to the multiplicity of β-subunit genes, alternative splicing of primary transcripts further increases the diversity of the β-subunit isoforms. Because each of the β-subunits appears able to partner with each of the α_1-subunits, β-subunits heterogeneity may contribute to the diversity of Ca^{2+} channels in a multiplicative manner.

Analysis of the family of β-subunit sequences suggests an overall domain structure of **VCVCV**, where **V** denotes highly variable regions located at the N- and C-termini and in a central region, and **C** indicates highly conserved regions (De Waard et al., 1996). Interestingly, the **V** regions are prone to splice variations, while the **C** regions contain the consensus sites for phosphorylation by protein kinases. The second **C** region also incorporates a stretch of amino acids critical for interaction with the α_1-subunit (Witcher et al., 1995). The β-subunit binds to the cytoplasmic linker between repeats I and II of all the Ca^{2+} channel α_1 subunits tested. This linker contains a conserved motif (QQ-E--L-GY--WI--E), positioned 24 amino acids from he IS6 transmembrane domain in each α_1-subunit (De Waard et al., 1995). Mutations within this motif reduce the stimulation of peak currents by the β-subunit and alter inactivation kinetics and voltage-dependence of activation (Pragnell et al., 1994). Glutathione S-transferase fusion proteins containing this motif interact with ^{35}S-labeled β-subunits. The binding is fast and almost irreversible over an 8 h period.

Gene targeting has been used to knock out the β_1 gene (Gregg et al., 1996). The homozygous mice die at birth from asphyxia. In cultured myotubes, caffeine contractures are intact; action potentials are also normal, but fail to induce intracellular Ca^{2+} transients, as expected from a defect in excitation–contraction coupling. Most interesting is the finding that deletion of the β_1-subunit reduces the level of α_{1S} in the membrane to undetectable levels. This may be contrasted with the normal surface membrane localization of β_1 in α_{1S}-null animals (mdg/mdg, Knudson et al., 1989).

The function of the β_3 subunit in neurons has also been investigated using targeted gene disruption (Namkung et al., 1998). Mice deficient in the β_3 subunit were indistinguishable from their wild type brethren by anatomical and gross behavioral criteria. However, examination of whole-cell Ca^{2+} currents in sympathetic neurons revealed a relative decrease in the fraction of L- and N-type current in the absence of β_3 subunit expression. Furthermore, without β_3 subunits, the voltage-dependence of P/Q-type Ca channel activation was altered, facilitating the opening of these channels with weaker depolarizations. This study emphasizes the diverse effects that β subunits may exert on various Ca^{2+} channel subtypes.

Relationship between Structural and Functional Diversity

Clearly, there is an abundance of isoforms of α_1- and β-subunits, each with many possible splice variants, that contribute to the molecular palette for generation of Ca^{2+}-channel diversity. As we learn more about the details of the component subunits of Ca^{2+} channels, lists of the isotype names and splice variations may eventually supplement or supplant the designation of channel types in terms of biophysical and pharmacological properties. However, only partial information is available about how functionally distinct types of Ca^{2+} channels (T-, L-, N-type, etc.) arise from the multiplicity of α_1- and β-subunits, with the L-type Ca^{2+} channel of skeletal muscle ($\alpha_{1S}\beta_{1a}\gamma\alpha_2\delta_a$) as the most notable exception. Intensive efforts are underway to understand the relationship between observed electrophysiological properties and the underlying composition of Ca^{2+} channel subunits. Current approaches include single-cell PCR (Bargas et al., 1994), antisense oligonucleo-tide treatment (Berrow et al., 1995), and gene knock-out strategies (Gregg et al., 1996; Namkung et al., 1998). In the following section, we briefly summarize present-day knowledge about the basis of the individual channel types.

New insights into the Molecular Basis of T-type Ca^{2+} Channels

An understanding of the molecular basis of T-type channels has long been sought. Perez-Reyes and colleagues have recently succeeded in cloning two α_1 subunits—α_{1G}, α_{1H}—each capable of supporting classical T-type calcium channel activity (Cribbs et al., 1998; Perez-Reyes et al., 1998). This goes a long way to closing the circle on Ca^{2+} channel diversity, which began with separation of high and low voltage-activated Ca^{2+} channels (Carbone and Lux, 1984), quickly developed into distinctions between T-, N-, and L-type channels (Nowycky et al., 1985), and eventually P-, Q- and R-type channels (Randall and Tsien, 1995; Llinás et al., 1992). Interestingly, the T-type channels provide the first example in which a brand new Ca^{2+} channel subunit has been isolated with the help of sequence information from the human genome. The evidence is convincing that α_{1G} and α_{1H} are both authentic T-type channel subunits (Cribbs et al., 1998; Perez-Reyes et al., 1998). Both subunits have the signature motifs one would expect for a Ca^{2+} channel, including pseudotetrameric transmembrane repeats, key acidic residues needed to support divalent cation selectivity and permeation, as well as positively charged residues in the S4 transmembrane region, key to voltage-dependent gating. Expression of these subunits in *Xenopus* oocytes or mammalian cell lines yielded currents with key earmarks of classical T-type channels, including a single-channel conductance of 6–8 pS, slow deactivation (shutting-off of conductance following a sudden repolarization), and a characteristic crossing pattern of current families evoked by a series of depolarizing voltage steps. Isolation of a T-type channel α_1-subunits opens up many promising avenues for future work. Localization of T-type channels and understanding their functional roles will be greatly facilitated by development of specific antibodies, knockout mice, and channel-Ca^{2+}-sensor hybrids. Renewed effort will be invested in finding agents to inhibit T-type channels selectively, following up present efforts with mibefradil, a T-type blocker, effective in the treatment of hypertension (Clozel et al., 1997), and kurtoxin, a peptide toxin that blocks T-type channels but spares other voltage-gated Ca^{2+} channels (Chuang et al., 1998).

Many Possible Sources of Diversity of L-type Ca^{2+} Channels

The existence of three α_1-subunits, classes S, C, and D, each capable of supporting L-type channel activity, provides an obvious starting point for attempts at understanding how L-type Ca^{2+} channel diversity might be generated from specific molecular structures. However, little information is yet available to link individual α_1 isoforms to functionally distinct forms of L-type channel activity (e.g., Forti and Pietrobon, 1993; Kavalali and Plummer, 1994). While α_{1S} subunit appears to be largely excluded from neurons according to Northern analysis and electrophysiological criteria, no sharp distinction has been made between currents generated by α_{1C} (Mikami et al., 1989) and α_{1D} (Williams et al., 1992b). Single-channel recordings of expressed α_{1D} channels are lacking and analysis of the functional impact of various β-subunits on α_{1C} and α_{1D} is not extensive.

Most of the attention to date has been focused on splice variations of α_{1C}. These have a marked impact on channel behavior in several cases, producing (1) differences in sensitivity to DHPs in α_{1C} variants found in cardiac or smooth muscle (Welling et al., 1993), (2) differences in the voltage-dependence of DHP binding (Soldatov et al., 1995), and (3) differences in susceptibility to cyclic AMP-dependent phosphorylation (Hell et al., 1993b). Further analysis will be greatly facilitated by knowledge of the genomic structure of the human α_{1C} gene, which spans an estimated 150 kb of the human genome and is composed of 44 invariant and 6 alternative exons (Soldatov, 1994). The L-type channel in chick hair cells incorporates an α_{1D}-subunit that differs from the α_{1D}-subunit in brain due to expression of distinct exons at three locations (Kollmar et al. 1997b). It will be interesting to see if additional splice variations can account for L-type channel activity found at the resting potential of hippocampal neurons, possibly important for setting the resting $[Ca^{2+}]_i$ (Avery and Johnson, 1996).

Advances in Understanding Diversity of N-type Ca^{2+} Channels

As discussed earlier, an important source of channel heterogeneity is the association of α_1-subunits with different ancillary subunits. A good example of this is provided by the N-type Ca^{2+} channel in brain. Biochemical analysis has shown that the α_{1B}-subunit associates with three different isoforms of β-subunit in rabbit brain (Scott et al., 1996). Antibodies against individual β-subunits were each able to immunoprecipitate ω-CTx-GIVA binding activity (a marker of α_{1B}), while immunoprecipitation of α_{1B} showed its association with β_{1b}, β_3, and β_4.

Different isoforms of the N-type Ca^{2+}-channel subunit α_{1B} have been isolated from rat sympathetic ganglia and brain by Lin et al. (1997). Alternative splicing determines the presence or absence of small inserts in the S3–S4 regions of domains III and IV (SFMG and ET, respectively). Different combinations of inserts in these putative extracellular loop regions are dominant in brain ($+$SFMG, ΔET) and ganglia (ΔSFMG, $+$ET). Most interestingly, the gating kinetics of the brain-dominant form are 2–4 fold faster than the ganglia-dominant form. This work provides one of the clearest examples of how alternative splicing contributes to a wide range of functional properties.

What Type(s) of Ca^{2+} Channel Currents Are Generated by α_{1A}?

Initial observations about the class A α_1-subunit suggested that it corresponds to the P-type channel (Mori et al., 1991; Llinás et al., 1992). Closer comparison of the properties of α_{1A}-subunits expressed in *Xenopus* oocytes and those of P-type channels in cerebellar Purkinje cells, however, revealed clear differences. P-type channels activate at relatively negative potentials and support a sustained, non-inactivating current during depolarizing pulses longer than 1 s (Llinás et al., 1992; Usowicz et al., 1992), whereas α_{1A}-subunits expressed in *Xenopus* oocytes activate at less negative potentials and exhibit marked inactivation within 100 ms (Sather et al., 1993). Furthermore, P-type current in cerebellar Purkinje neurons is half-blocked at ~ 1 nM ω-Aga-IVA, while the IC_{50} for blockade of class A channels expressed in oocytes is 100–200 nM (Sather et al., 1993; Stea et al., 1994). These biophysical and pharmacological differences cannot be attributed to a general peculiarity of the oocyte expression system since similar properties of expressed α_{1A}-subunits have been observed in a study of α_{1A} stably expressed in baby hamster kidney cells (Niidome et al., 1994). While it is possible that the kinetic differences between P-type channels and exogenously expressed Ca^{2+} channels might be attributed to alternative splicing of α_{1A} (Stea et al., 1994; Sakurai et al., 1995) or its association with different β-subunits (Liu et al., 1996a; Moreno et al., 1997) there is no explanation yet for the difference in ω-Aga-IVA sensitivity. An intriguing but speculative hypothesis is that the P-type channel arises from the 95 kDa protein often observed in the course of biochemical purification of non-L-type Ca^{2+} channels. Originally interpreted as a subunit of the N-type Ca^{2+} channel, the 95 kDa protein has now been shown to be a truncated version of α_{1A} containing repeats I and II and the II–III loop (Scott et al., 1998). It is conceivable that this "hemi-α_1-subunit" dimerizes to form a channel (cf., dimeric K^+ channels, Ketchum et al., 1995). Systematic studies of the electrophysiological properties of the 95 kDa form of α_{1A} are a logical next step.

While characteristics of α_{1A}-subunits expressed in oocytes of mammalian cells do not match P-type, they do correspond well to a component of Ca^{2+} channel current in cultured cerebellar granule cells, designated Q-type to distinguish it from P-type current (Zhang et al., 1993; Randall and Tsien, 1995). In the granule neurons, about half of the total Ca^{2+} channel current was blocked by high concentrations of ω-Aga-IVA. Analysis of the overall ω-Aga-IVA-responsive current revealed two distinct components: a non-inactivating P-type current, blocked with an IC_{50} of about 1 nM, and an inactivating Q-type current that was half-blocked by 90 nM ω-Aga-IVA. These components were present in a ~1:3 ratio, where the Q-type current represented the largest individual component in the cerebellar granule neurons. In all respects, Q-type current displayed a strong qualitative resemblance to α_{1A} channels heterologously expressed in oocytes (Sather et al., 1993; Stea et al., 1994) or baby hamster kidney cells (Niidome et al., 1994).

Functional Roles of Ca^{2+} Channels

The molecular diversity of voltage-gated Ca^{2+} channels may reflect the variety of functional roles that they are called upon to serve. With the exception of α_{1S}, which appears highly localized to skeletal muscle, α_1-subunits are broadly distributed across the spectrum of exocytotic cells. At the level of individual cells, however, the different channel types often show distinct patterns of localization to different parts of the cell. Some channel types are mainly found at presynaptic release sites where they allow Ca^{2+} entry that triggers neurotransmitter release, while other tend to be found primarily on cell bodies where they may help to shape the action potentials or regulate excitability (Lemos and Nowycky, 1989; Fisher and Bourque, 1996; Reuter, 1996). Here, we provide an overview of the physiological actions of the voltage-gated Ca^{2+} channels.

T-type Ca^{2+} Channels

The T-type Ca^{2+} channels are of considerable importance for many different organ systems. In cardiac cells, T-type Ca^{2+} channels are generally present at much lower density than L-type channels, if at all. Consequently, L-type channels represent the predominant pathway for Ca^{2+} entry and excitation–contraction coupling. However, T-type channels supply a major fraction of the current recorded in cells from the sinoatrial node, the natural source of cardiac rhythms, and thus provide a significant contribution to the inward current that drives the last stages of the pacemaker depolarization (Lei et al., 1998; Hagiwara et al., 1988). T-type channels also support oscillatory activity and repetitive activity in the thalamus (Jahnsen and Llinás, 1984), particularly in the nucleus reticularis, which acts as a major rhythm generator of cortical electrical potentiations. T-type Ca^{2+} channels play a prominent role in dendritic Ca^{2+} signaling in hippocampal and cortical neurons (Magee et al., 1995). Because of their ability to open at relatively negative membrane potentials, they can provide dendritic Ca^{2+} entry during synaptic depolarizations too weak to trigger a regenerative Na^{2+}-dependent action potential. Interestingly, expression of T-type channels in smooth muscle fluctuates in synchrony with the cell cycle (Kuga et al., 1996), and may be associated with cell proliferation (Schmitt et al., 1995).

L-type Ca^{2+} Channels

The L-type Ca^{2+} channels are widely distributed in muscle, nerve, and endocrine cells. Their unique biophysical properties and their subcellular localization put them in a good position to act as transducers that link membrane depolarization to intracellular signaling. In the brain, for example, L-type Ca^{2+} channels are found in the cell bodies and proximal dendrites of hippocampal pyramidal cells, as visualized with a monoclonal antibody (Westenbroek et al., 1990). Class C calcium channels were concentrated in clusters at the base of major dendrites, while class D calcium channels were most generally distributed across cell surface membrane of cell bodies and proximal dendrites (Hell et al., 1993a). The most prominent roles of L-type channels include initiation of skeletal muscle contraction, hormone or neurotransmitter release, and provision of a Ca^{2+} signal involved in the regulation of gene expression.

Excitation–Contraction Coupling

The L-type Ca^{2+} channels play a central role in excitation–contraction coupling in skeletal, cardiac, and smooth muscle. In skeletal muscle. L-type Ca^{2+} channels contain the α_{1S}-subunit and are largely localized to the transverse tubule system. Ca^{2+} through the L-type channel is not required for skeletal muscle contraction (reviewed in Miller and Freedman, 1984), in contrast to cardiac muscle, where Ca^{2+} entry through these channels is essential for contractility (Näbauer et al., 1989). Interestingly, blockade of L-type channels in skeletal muscle by organic Ca^{2+} antagonists completely inhibits contraction

(Eisenberg et al., 1983). The explanation of these findings centers around gating charge movement in the T-tubule membrane, which was known to be essential for intracellular Ca^{2+} release (Schneider and Chandler, 1973). The DHPs eliminate charge movement, thereby blocking skeletal muscle contraction (Ríos and Brum, 1987). The implication of these findings was the DHP-sensitive L-type Ca^{2+} channels act as voltage sensors to link T-tubule depolarization to intracellular Ca^{2+} release.

This hypothesis was tested in elegant experiments by Tanabe, Numa, Beam, and their colleagues. The cloning of the DHP-receptor protein from skeletal muscle led immediately to its identification as a voltage-gated channel (Tanabe et al., 1987). Later, expression of the cloned DHP receptor in dysgenic skeletal muscle myotubes showed that it could restore electrically evoked contractility in these formerly nonresponsive cells (Tanabe et al., 1988), along with L-type Ca^{2+} current (Tanabe et al., 1988; Garcia et al., 1994) and gating charge movement (Adams et al., 1990). While the skeletal DHP receptor allowed contraction even in the absence of extracellular Ca^{2+}, the cardiac L-type Ca^{2+} channel restored contractility only if Ca^{2+} entry occurred (Tanabe et al., 1990). The structural basis of the skeletal-type excitation–contraction coupling was investigated with molecular chimeras. By inserting pieces of the α_{1S} gene into an α_{1C} background, Tanabe et al. (1990) showed that the key domain was the intracellular loop joining repeats II and III of α_{1S} (Fig. 8.4). More recently, other groups have shown that purified II–III loop fragments can directly activate the ryanodine receptor (Lu et al., 1994; el-Hayek et al., 1995) and that this region may contain phosphorylation sites for the regulation of excitation–contraction coupling (Lu et al., 1995).

Excitation–Secretion Coupling

While the vast majority of studies of neurotransmitter release have failed to identify a role for L-type Ca^{2+} channels, this subtype has been implicated in a few specialized forms of exocytosis. For example, activation of L-type channels is required for zona pellucida-induced exocytosis from the acrosome of mammalian sperm (Florman et al., 1992). L-type channels also seem to play an important role in mediating hormone release from endocrine cells. Insulin secretion from pancreatic β-cells (Ashcroft et al., 1994; Bokvist et al., 1995), luteinizing hormone–releasing hormone release from the bovine infundibulum (Dippel et al., 1995), and catecholamine release from adrenal chromaffin cells (Lopez et al., 1994) are all reduced by inhibition of L-type Ca^{2+} channels. L-type channels also seem to play an important role in supporting release of GABA from

retinal bipolar cells (Maguire et al., 1989; Duarte et al., 1992), glutamate release from cochlear hair cells (Kollmar et al., 1997a), as well as dynorphin release from dendritic domains of hippocampal neurons (Simmons et al., 1995).

Excitation–Transcription Coupling

A number of extracellular factors that influence cell growth and activity depolarize the membranes of their target cells (Hill and Triesman, 1995). Membrane depolarization open voltage-gated Ca^{2+} channels and the resulting influx of Ca^{2+} can trigger gene transcription (for a review, see Morgan and Curran, 1989). L-type Ca^{2+} channels are thought to play a role in this cascade because agonists of these channels can induce expression of several protooncogenes in the absence of other stimuli (Morgan and Curran, 1988). Ca^{2+} entry through L-type channels has also been shown to induce expression of acetylcholine receptors in skeletal muscle (Huang et al., 1994). Interestingly, it is only the Ca^{2+} that enters the cell via the L-type channel, and not that released from the sarcoplasmic reticulum, that can trigger this response. These findings suggest that the mode and location of Ca^{2+} entry may be important to how the Ca^{2+} signal is interpreted by the cell (Ghosh et al., 1994; Rosen and Greenberg, 1994). Some recent studies have shed light on the cascade of events that follows influx of Ca^{2+} through L-type channels.

An example of a signal-transduction cascade where Ca^{2+} entry is important involves the cAMP and Ca^{2+} response element (CRE), and its nuclear binding protein (CREB) (Montminy and Bilezikjian, 1987; Hoeffler et al., 1988). The interaction of CREB with the CRE is facilitated when CREB is phosphorylated on serine-133 (Gonzalez and Montminy, 1989). The phosphorylation of CREB is catalyzed by several kinases, including Ca^{2+}-calmodulin kinases II and IV, and cAMP-dependent protein kinase (Greenberg et al., 1992). Thus, rises in $[Ca^{2+}]_i$ can act either directly, via Ca^{2+}-calmodulin and its dependent kinases, or indirectly, by stimulating Ca^{2+}-calmodulin-sensitive adenylate cyclase which leads to increased cAMP levels. Recent work has shown that Ca^{2+} entry through L-type channels can trigger CREB phosphorylation (Yoshida et al., 1995; Deisseroth et al., 1996), but that Ca^{2+} entry through N- and P/Q-type channels is surprisingly ineffective (Deisseroth et al., 1998). Localized actions of Ca^{2+} within close proximity of the mouths of the effective Ca^{2+} channels (Deisseroth et al., 1996) may help explain the channel specificity.

Among the other divalent ions, Zn^{2+} is particularly interesting because it regulates a wide variety of enzymes and DNA-binding proteins, provides an important developmental signal, and may be

Expressed channel structure   ## Force of contraction

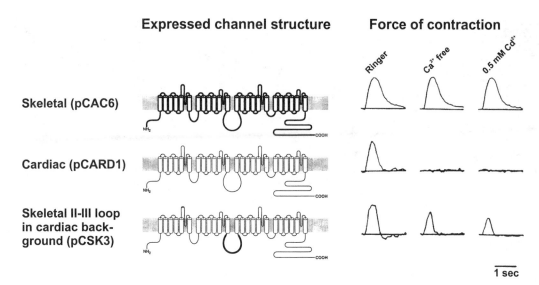

Figure 8.4 Injection of L-type cDNA into mouse dysgenic myotubes rescues contractility. Expression of α_{1S} allows electrically evoked contractions even in the absence of Ca^{2+} influx. In contrast, expression of α_{1C} cannot support contractions without the influx of extracellular Ca^{2+}. Transfer of the cytoplasmic loop joining domains II and III of the skeletal Ca^{2+} channel into the cardiac isoform re-establishes Ca^{2+}-independent contractility. (Adapted from Tanabe et al. 1990, copyright Macmillan Magazines Limited.)

involved in excitotoxity and responses to trauma (for a review, see Smart et al., 1994). Interestingly, L-type Ca^{2+} channels can support Zn^{2+} influx into heart cells, where it can induce transcription of genes driven by a metallothionein promoter (Atar et al., 1995). Morphological studies have revealed that Zn^{2+} is highly enriched in a number of nerve fiber pathways, especially in boutons where it appears to be contained within vesicles (Smart et al., 1994). Furthermore, Zn^{2+} can be released from brain tissue during electrical or chemical stimulation (Assaf and Chung, 1984; Howell et al., 1984; Charton et al., 1985). Given that Zn^{2+} can be released by synaptic activity, and can enter cells via voltage-dependent Ca^{2+} channels, it seems likely that Zn^{2+} may play an important role in excitation–transcription coupling.

N-, P/Q-, and R-type Ca^{2+} Channels

The non-L-type HVA Ca^{2+} channels are widely distributed both pre- and postsynaptically in the central and peripheral nervous systems (Fig. 8.5). In the brain, antibodies to the N-type Ca^{2+} channel α_1-subunit bind primarily on dendrites and nerve terminals in most regions (Westenbroek et al., 1992). P/Q-type channels are concentrated in presynaptic terminals, making synapses on cell bodies and on dendritic shafts and spines of many classes of neurons, and

are present at lower density in the surface membrane of dendrites of most major classes of neurons (Westenbroek et al., 1995). Ca^{2+} channels containing the α_{1E}-subunit, presumably encoding the R-type channel, are found mostly on cell bodies, and in some cases in dendrites, of a broad range of central neurons (Yokoyama et al., 1995). Thus, these classes of Ca^{2+} channels are ideally positioned to allow the Ca^{2+} influx that triggers neurotransmitter release and shapes the postsynaptic response to that release.

Generic Properties of Excitation–Secretion Coupling

The most commonly studied role of Ca^{2+} is its ability to trigger neurotransmitter release. The importance of Ca^{2+} ions in the release of neurotransmitter has been appreciated for more than 60 years (Feng, 1936). Seminal work by Katz (1969) and his colleagues demonstrated that Ca^{2+} ions exert their influence at the nerve terminal where they control the amount of neurotransmitter that is released. The action of Ca^{2+} ions in the regulation of neurotransmission was shown to be cooperative, requiring about four Ca^{2+} ions to bind to their receptor in order to trigger release (Dodge and Rahamimoff, 1967). The centrality of Ca^{2+} action in the nerve terminal was further supported by the observation that injection of Ca^{2+} into the terminal triggered

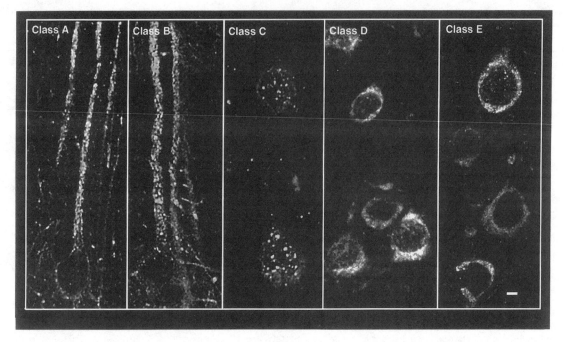

Figure 8.5 Differential subcellular localization of class A–E Ca^{2+} channel subtypes in cortical neurons. Saggital sections through the dorsal cerebral cortex of adult rats were incubated with affinity-purified antipeptide antibodies against unique amino acid sequences in the II–III intracellular loop of the indicatd Ca^{2+} channel α_1-subunits. The various antibodies were detected using fluorescein-tagged secondary antibodies and then visualized using a confocal microscope as described previously (Westenbroek et al., 1995). Reproduced with permission from Catterall et al. (1998).

the release of transmitter at the squid giant synapse (Miledi, 1973). Subsequently, the Ca^{2+}-sensitive protein, aequorin, was used to show that presynaptic $[Ca^{2+}]_i$ increases during neurotransmission (Llinás and Nicholson, 1975).

Studies using simultaneous voltage-clamp of the presynaptic terminal and postsynaptic axon of the squid giant synapse provided direct measurements of the Ca^{2+} currents in the presynaptic membrane that trigger the release of neurotransmitter (Llinás et al., 1981; Augustine et al., 1985). Ongoing issues include the identification of presynaptic Ca^{2+} channels and clarification of the functional consequences of their diversity (for other recent reviews, see Olivera et al., 1994; Dunlap et al., 1995; Reuter, 1996).

Transmitter Release at Peripheral Synapses

At the neuromuscular junction, the release of neurotransmitter is generally mediated by a single Ca^{2+} channel type, although which type predominates varies from species to species. Invertebrate motor end plates utilize primarily P/Q-type channels. In crayfish, for example, inhibitory and excitatory transmit-

ter release onto the claw-opener muscle was completely abolished by ω-Aga-IVA, while ω-Ctx-GVIA and nifedipine were both ineffective (Araque et al., 1994). In locusts and houseflies, motor end plate potentials are blocked by type I and II agatoxins, which inhibit P/Q-type channels, but not by type III agatoxins,which potently block both L- an N-type channels (Bindokas et al., 1991). In nonmammalian vertebrates, unlike invertebrates, neurotransmitter release at the neuromuscular junction is completely blocked by ω-CTx-GIVA. This is true for frogs (Kerr and Yoshikami, 1984; Katz et al., 1995), lizards (Lindgren and Moore, 1989), and chicks (De Luca et al., 1991; Gray et al., 1992). In mammals, on the other hand, ω-CTx-GIVA does not seem to have any effect on the evoked release of acetylcholine at the neuromuscular junction (Sano et al., 1987; Wessler et al., 1990; De Luca et al., 1991; Protti et al., 1991; Bowersox et al., 1995). In contrast, blockade of P/Q-type Ca^{2+} channels by ω-Ctx-MVIIC, ω-Aga-IVA or FTx completely abolishes transmission in mice (Protti and Uchitel, 1993; Bowersox et al., 1995; Hong and Chang, 1995) and humans (Protti et al., 1996). In all of these species, neuromusculer transmission seems to rely on a single type of channel, one or another of those not sensitive to DHPs.

Sympathetic neurons contain both L- and N-type Ca^{2+} channels but not P/Q-type channels (Hirning et al., 1988; Mintz et al., 1992b; Zhu and Ikeda, 1993). However, only N-type Ca^{2+} channels seem to be important for the release of norepinephrine, inasmuch as ω-CTx-GVIA blocks NE secretion (Hirning et al., 1988; Fabi et al., 1993) but DHPs do not (Perney et al., 1986; Hirning et al., 1988; Koh and Hille, 1996). Along similar lines, N- but not L-type Ca^{2+} channels in sympathetic nerve terminals are susceptible to modulation of Ca^{2+} current via autoreceptors for NE or neuropeptide Y (Toth et al., 1993). Thus, sympathetic nerve endings are like motor nerve terminals in relying on a single predominant type of Ca^{2+} channel, in this case N-type, despite the sizeable contribution of L-type channels to the global Ca^{2+} current. Reliance on N-type channels cannot be generalized to all autonomic terminals since P/Q-type channels have been found to play a prominent role in transmitter release in rat urinary bladder (Frew and Lundy, 1995).

Transmitter Release at Central Synapses

At central synapses, unlike synapses in the periphery, neurotransmitter release often involves more than one Ca^{2+} channel type. Central neurons appear to be richly endowed with Ca^{2+} channels, with as many as five or six different types of channels in an individual nerve cell (Mintz et al., 1992a; Randall and Tsien, 1995). Several more recent papers have reported that neurotransmission at specific synapses in the CNS depends upon the concerted actions of more than one type of Ca^{2+} channel (Luebke et al., 1993; Takahashi and Momiyama, 1993; Castillo et al., 1994; Regehr and Mintz, 1994; Wheeler et al., 1994; Mintz et al., 1995; Wu et al., 1998). In general, L-type channels have not been found to play a significant role in the regulation of neurotransmitter release in the brain (Dunlap et al., 1995), although at some synapses they may contribute when potentiated by the DHP agonist Bay K 8644 (e.g., see Sabria et al., 1995). The relative importance of N-type as opposed to P/Q-type Ca^{2+} channels can also vary from one synapse to another. Studies of synapses in hippocampal and cerebellar slices suggest that a large majority of single-release sites are in close proximity to a mixed population of Ca^{2+} channels that jointly contribute to the local Ca^{2+} transient that triggers vesicular fusion (e.g., Mintz et al., 1995; Reid et al., 1998; but see also Reuter, 1995). The synergistic effect of multiple Ca^{2+} channels arises because of limitations on the Ca^{2+} flux through individual channels under physiological conditions. Indeed, the reliance on multiple types of Ca^{2+} chan-

nels was not absolute but was relieved by increasing the Ca^{2+} influx per channel, either by prolonging the presynaptic action potential or increasing $[Ca^{2+}]_o$ (Wheeler et al., 1996).

Interactions between Ca^{2+} Channels and Components of the Release Machinery

The close functional relationship between specific types of Ca^{2+} channels and transmitter release raises interesting questions about a possible molecular basis. Over the last few years, several groups have provided evidence for structural interactions between N- and P/Q-type Ca^{2+} channels and molecular components of transmitter-release machinery. The synaptic membrane protein syntaxin 1 (p35), a key player in the fusion process, binds strongly to N-type Ca^{2+} channels. This was revealed by experiments where antibodies against syntaxin immunoprecipitated ω-CTx-GVIA binding activity from synaptosomal membranes (for review, see Bennett and Scheller, 1994). The syntaxin-binding site on N-type Ca^{2+} channels has been localized to the intracellular loop between repeats II and III, more specifically to an 87 amino acid segment (residues 773–859) which includes a helix–loop–helix–loop–helix structure (Sheng et al., 1994). The corresponding 87 amino acid syntaxin-binding peptide was capable of interfering with the interaction between syntaxin and the solubilized α_{1B}-subunit. The interaction of syntaxin 1A with loop II–III was not seen with α_{1S} (skeletal L-type Ca^{2+} channel), although it has recently been found with α_{1A} (Martin-Moutot et al., 1996; Rettig et al., 1996). The participation of the II-III loop is particularly intriguing because of the involvement of the corresponding region in excitation–contraction coupling.

A different approach to the interaction between syntaxin and various types of voltage-gated Ca^{2+} channels has been provided by coexpression experiments in *Xenopus* oocytes. Syntaxin 1A sharply decreased the availability of N-type channels (Bezprozvanny et al., 1995). This functional effect was due to stabilization of channel inactivation rather than simple block or lack of channel expression, as it was overcome by strong hyperpolarization. It was also found with Q-type but not L-type Ca^{2+} channels. Thus, the syntaxin effect is specific for Ca^{2+} channel types that participate in fast transmitter release in the mammalian CNS. The shift in inactivation was abolished by deletion of syntaxin's carboxy-terminal transmembrane domain in accord with what would be expected from binding studies (Sheng et al., 1994). These results raise the possibility that, in addition to acting as a vesicle docking site, syntaxin may influence presynaptic Ca^{2+} channels, promoting their inactivation in the aftermath of activity-dependent vesicular turnover.

Postsynaptic Ca^{2+} Influx

Much of the electrical and biochemical signal processing in central neurons takes place within their dendritic trees. Ca^{2+} entry through voltage-gated channels is critical for many of these events. The idea that voltage-gated Ca^{2+} channels may contribute to electrogenesis in dendrites first arose in the interpretation of intracellular recordings from hippocampal pyramidal neurons (Spencer and Kandel, 1961). Initial intradendritic voltage recordings were conducted on the dendritic arbors of cerebellar Purkinke neurons (Llinás and Nicholson, 1971; Llinás and Hess, 1976; Llinás and Sugimori, 1980) and apical dendrites of hippocampal pyramidal neurons (Wong et al., 1979). The ability of dendrites to support Ca^{2+}-dependent action potential firing was reinforced by experiments where apical dendrites of pyramidal neurons were surgically isolated from their cell bodies in a hippocampal slice preparation (Benardo et al., 1982; Masukawa and Prince, 1984). These experiments revealed a variety of Ca^{2+}-dependent active responses in the dendrites of central neurons that could be elicited by excitatory postsynaptic potentials or injection of depolarizing current pulses. Postsynaptic Ca^{2+} channels, including N- and L-type, have also been implicated in the release of neuropeptides from dendritic domains (Simmons et al., 1995).

More recent studies of the electrical properties of dendrites have been facilitated by the ability to visualize dendrites in brain slices, thus rendering dendrites accessible to patch electrodes (Stuart et al., 1993). These studies revealed that back-propagating Na^{2+}-dependent action potentials can activate dendritic Ca^{2+} channels, thereby causing substantial increases in intradendritic free Ca^{2+} (Jaffe et al., 1992; Stuart and Sakmann, 1994; Markram et al., 1995; Schiller et al., 1995; Spruston et al., 1995). Subthreshold excitatory postsynaptic potentials can also open Ca^{2+} channels and result in more localized changes in intradendritic Ca^{2+} concentration (Markram and Sakmann, 1994; Yuste et al., 1994; Magee et al., 1995). The presence of multiple types of voltage-gated Ca^{2+} channels on dendrites has been demonstrated by several techniques, including Ca^{2+} imaging (Markram et al., 1995), dendrite-attached patch-clamp recordings (Usowicz et al., 1992; Magee and Johnston, 1995), and immunocytochemistry (Westenbroek et al., 1990, 1992, 1995; Hell et al., 1993a; Yokoyama et al., 1995). In recent results from our group, recordings were made from isolated dendritic segments obtained by acute hippocampal dissociation, and indicated that N-, P/Q-, and R-type channels all contribute to the overall HVA current in dendrites (Kavalali et al., 1997).

Key Functions of Calcium Channels and Their Molecular Basis

There is great interest in understanding the molecular determinants of key properties of voltage-gated Ca^{2+} channels, including generic characteristics common to the entire family of channels, as well as those properties that give individual types of Ca^{2+} channels their distinct character. Structure–function studies have advanced considerably through the analysis of recombinant channels.

Selectivity, Permeation, and Block

As mentioned in the introduction, Ca^{2+} channels operate with great efficiency in allowing Ca^{2+} permeation with rapid turnover rates while also showing exquisite selectivity for Ca^{2+} over other more abundant extracellular ions like Na^{2+} (Tsien et al., 1987b). Permeation through open Ca^{2+} channels has been studied with patch-clamp techniques, often with ~ 100 mM external Ba^{2+} to increase the unitary current size. Under these conditions, the Ca^{2+} channel type with the largest Ba^{2+} conductance (L-type, ~ 25 pS) shows a unitary current amplitude of approximately −1.6 pA at 0 mV, corresponding to a transfer rate of 5 million Ba^{2+} ions/s (Hess et al., 1986). Recordings from L-type channels at physiological levels of external Ca^{2+} (2 mM) yield a much smaller unitary conductance, 2.4 pS (Church and Stanley, 1996). Selectivity of voltage-gated Ca^{2+} channels for Ca^{2+} ions over monovalent cations is > 1000-fold (Hess et al., 1986), so that Ca^{2+} is the main charge carrier even when it is greatly outnumbered by other ions, as under normal physiological conditions. Since the Ca^{2+} channel pore is relatively large (~ 6 Å diameter) (McCleskey and Almers, 1985), the selectivity cannot be explained by molecular sieving (Tsien et al., 1987b).

Recent mutagenesis studies have revealed that the Ca^{2+} channel pore contains a single locus of high-affinity binding within the pore that can either bind a single Ca^{2+} ion with high affinity ($K_d \sim 1 \mu$M) or multiple divalent cations with lower affinity (Ellinor et al., 1995). This locus comprises four highly conserved glutamate residues, one from each of the pore lining H5 regions of the α_1 subunit repeats (Kim et al., 1993; Tang et al., 1993; Yang et al., 1993; Ellinor et al., 1995) (Fig. 8.6). The localization of the Ca^{2+} interaction to a specific cluster of carboxylates fits in with earlier hypotheses about the mechanism of selectivity and permeation (Almers and McCleskey, 1984; Hess and Tsien, 1984; Armstrong and Neyton, 1992;

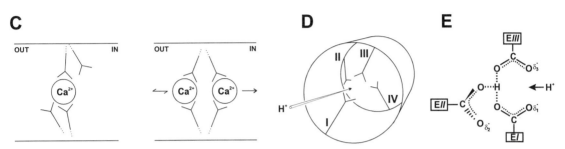

Figure 8.6 Structural models of the locus of Ca^{2+} selectivity in voltage-gated Ca^{2+} channels. (A) Amino acid sequence in P regions. (B) Glutamates in P regions of all four repeats gives rise to a locus of selectivity. (Reprinted from Melzer et al., 1995, with kind permission from Elsevier Science — NL, Sara Burgerhartstraat 25, 1055 KV Amsterdam, The Netherlands.) (C) Carboxylate side chains act to provide a high-affinity interaction with a single Ca^{2+} ion or a lower affinity interaction with two Ca^{2+} ions. (From Yang et al., 1993, copyright Macmillan Magazines Limited.) (D, E) Cartoon representation of how incoordinated action of the same set of carboxylate side chains supports a single protonation site with a pK_a several log units more alkaline than that of a single glutamate carboxylate. (From Chen et al. 1997, by permission of the Rockefeller University Press.)

Kuo and Hess, 1993). Smaller cations such as Na^+ and K^+ are thought to bind weakly and permeate rapidly in the absence of Ca^{2+}, but are rejected when Ca^{2+} occupies the high-affinity site. Ca^{2+} fluxes become appreciable at millimolar levels of $[Ca^{2+}]_o$, when mass action drives more than one Ca^{2+} ion into the pore. The flux rate for Ca^{2+} is then dependent on negative interactions between individual divalent cations with the pore (either through electrostatic repulsion or competition for a limited number of negatively charged oxygen groups). In the absence of Ca^{2+}, larger divalent cations like Ba^{2+} and Sr^{2+} permeate the channel better than Ca^{2+}, since they do not bind as tightly to the high-affinity sites. Ca^{2+} channels are generally impermeable to Mg^{2+}, probably because of its slow rate of dehydration. At supraphysiological concentrations, Mg^{2+} is capable of blocking Ca^{2+} flow through the channel.

Blockade of Ca^{2+} channels can also be demonstrated with larger divalent cations, such as Cd^{2+} and Co^{2+}, which potently inhibit Ca^{2+} influx by binding more tightly than Ca^{2+} to the high-affinity site (Lansman et al., 1986). The order of potency of the blockade depends somewhat on the subtype of Ca^{2+} channel. As with the binding of Ca^{2+}, the blockade of L-type channels by other divalent cations is also strongly reduced in channels with changes in the conserved glutamates (Kim et al., 1993; Ellinor et al., 1995). Open Ca^{2+} channels are also blocked by H^+ ions, which titrate the glutamates that support Ca^{2+} selectivity. Mutagenesis experiments have demonstrated that the carboxylate side chains from repeats I, II, and III act together to form the proton-binding site (Chen et al., 1997). The coordinated action of the multiple oxygen groups creates a site with much higher affinity for protons than a single carboxylate alone. Acidification also decreases the degree of Ca^{2+} channel opening by shifting the voltage-dependence of gating toward more depolarized potentials (Klockner and Isenberg, 1994).

Activation

Membrane depolarization causes many kinds of ion channels to open, a process termed activation. As in the case of other voltage-gated channels, activation of Ca^{2+} channels occurs more quickly and is more complete with larger depolarizations. A positively charged transmembrane segment (S4) has been found in each of the four homologous repeats of Ca^{2+} channels (Tanabe et al., 1987), and is very similar to S4 segments in Na^+ and K^+ channels (Jan and Jan, 1989) where S4 has been firmly established as part of the voltage-sensing mechanism (e.g., Stuhmer et al., 1989; Papazian et al., 1991; Yang and Horn, 1995; Larsson et al., 1996). Analysis of Ca^{2+} channel gating is less extensive than for the other ion channels, but Ca^{2+} channel chimeras have provided insights into specific contributions of individual repeats and motifs within them. The determinants of the rate of activation—slow, skeletal muscle-like activation as opposed to rapid, cardiac-like activation—were first traced to repeat I (Tanabe et al., 1991), then further localized to the membrane-spanning segment IS3 and the external linker between IS3 and IS4 (Nakai et al., 1994). Mutant DHP receptors with α_{1S} sequence in this region activate relatively slowly ($\tau_{act} > 5\,ms$), whereas mutants that have the α_{1C} sequence in the same region activate relatively rapidly ($\tau_{act} < 5\,ms$). More recent analysis has focused on how charge movements associated with Ca^{2+} channel gating are affected by mutations in α_1-subunits.

Inactivation

Inactivation refers to the closing of channels during maintained depolarization, another fairly general property of Ca^{2+} channels. The speed of inactivation varies widely, ranging from very slow (hardly visible during second-long depolarizations), as in the inner segments of photoreceptors, to relatively rapid (complete within tens of milliseconds), as in the case of certain Ca^{2+} channels found on nerve terminals (Lemos and Nowycky, 1989). In most cases, the underlying mechanism of inactivation is dependent on depolarization per se. The rate of this voltage-dependent inactivation is strongly dependent on a region in the first repeat, including IS6 and extracellular and cytoplasmic residues on either side of it (Zhang et al., 1994) (Figure 8.6). Residues in the S6 transmembrane segments of other repeats may also be influential (Hering et al., 1996). The mechanism of channel closing may resemble a kind of inactivation in K^+ channels known as "C-type" (Hoshi et al., 1991; Liu et al., 1996b).

Some Ca^{2+} channels, such as L-type channels, are subject to Ca^{2+}-dependent inactivation along with voltage-dependent inactivation. The Ca^{2+}-dependence is an important negative feedback property that helps limit voltage-gated Ca^{2+} entry as intracellular $[Ca^{2+}]_i$ increases beyond a critical level (Chad and Eckert, 1986). Ca^{2+} influx through one channel can promote the inactivation of another adjacent channel, without a generalized elevation of bulk intracellular Ca^{2+} concentration, a specific example of localized Ca^{2+} signaling (Imredy and Yue, 1992). Intracellular application of the Ca^{2+} chelator BAPTA greatly diminishes such negative interactions within Ca^{2+} channel pairs. The Ca^{2+}-dependent inactivation transpires by a Ca^{2+}-induced shift of channel gating to a low open probability mode, distinguished by a more than 100-fold reduction of entry rate to the open state (Imredy and Yue, 1994). Both calmodulin activation and channel (de)phosphorylation were excluded as significant signaling events underlying Ca^{2+}-induced mode shifts, leaving direct binding of Ca^{2+} to the channel as the most likely initiating event for inactivation. Indeed, the L-type channel contains a consensus EF-hand Ca^{2+}-binding motif in the carboxy-terminal region downstream of repeat IV (Babitch, 1990). Involvement of this motif in Ca^{2+}-dependent inactivation was revealed by analysis of chimeric constructs between α_{1C} and α_{1E}, which lack Ca^{2+}-dependent inactivation (de Leon et al., 1995). Donation of the α_{1C} EF-hand region to the α_{1E} channel conferred the phenotype of Ca^{2+}-dependent inactivation. These results strongly suggest that Ca^{2+}-dependent inactivation is initiated directly by Ca^{2+} binding to the α_{1C}-subunit. Additional aspects of the C-terminal tail beyond the EF-hand region may also participate (Soldatov et al., 1995; Zuhlke and Reuter, 1998). Swapping of an ~ 80 amino acid segment speeds inactivation by 8- to 10-fold in one particular splice variant, and also eliminates the Ca^{2+}-dependence. Figure 8.7 provides a summary of the structural components of the α_1-subunit that are responsible for regulation of Ca^{2+} channel function as discussed above.

Responsiveness to Drugs and Toxins

Another interesting characteristic of voltage-gated Ca^{2+} channels is their ability to respond to drugs and toxins. Sensitivity to neurotoxins, many derived from the venoms of spiders and marine snails, is an important earmark of individual types of voltage-gated Ca^{2+} channels (Olivera et al., 1994). In the case of ω-CTx-GVIA, the potent and selective blocker of N-type channels, high-affinity binding seems to involve extracellular channel domains located in all four repeats of the α_{1B}-subunit, but particularly in repeats I and III (Ellinor et al., 1994). Specific differences between α_{1B} and α_{1A} have been localized to individual amino acids in loops near H5. Thus, the available evidence is consistent with a scenario in which the toxin straddles the mouth of the pore. The precise mechanism of blockade is not known, but observations of competition with permeant cations (Boland et al., 1994;

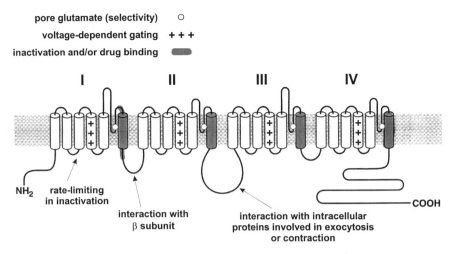

Figure 8.7 Overview of specific domains in the α_1-subunits of voltage-gated Ca^{2+} channels that give rise to key functional aspects.

McDonoguh et al., 1996) suggest the possibility of direct occlusion of the permeation pathway. Further analysis may allow the rigid toxin to be used as a kind of molecular caliper to gain information about the outer aspects of the channel. In this context, it is noteworthy that blockade of N-type channels by ω-CTx-GVIA, ω-CTx-MVIIA, and related toxins is strongly voltage-dependent as a result of enhancement of block by channel inactivation (Stocker et al., 1996). The dependence of toxin blockade on channel gating is a familiar phenomenon in Ca^{2+} channels; for example, ω-Aga-IVA block of P/Q-type channels is strongly antagonized by channel activation (Mintz et al., 1992a).

The pore region has also been implicated in the blockade of L-type Ca^{2+} channels by DHPs and other small organic agents. This was suggested by early photoaffinity-labeling experiments in which photoreactive DHPs labeled a putative extracellular loop between segments IIIS5 and IIIS6 and transmembrane segment IVS6 (Nakayama et al., 1991). General agreement has now been reached that the IVS6 segment is critical for the action of DHPs and phenylalkylamines (Schuster et al., 1996). Some of the most compelling evidence comes from experiments where DHP sensitivity is transferred to chimeras based on the α_{1A} subunit (Grabner et al., 1996). These reinforce the idea that the minimum sequence for DHP sensitivity includes segments IIIS5 and IIIS6 and the connecting linker, as well as the IVS5–IVS6 linker plus segment IVS6. Interestingly, the DHP-responsive α_{1A} chimera still retains sensitivity to ω-Aga-IVA and ω-CTx-MVIIC.

The idea that organic agents interact at or near the pore has been further reinforced by consideration of the Ca^{2+}-dependence of drug binding (Mitterdorfer et al., 1995; Peterson and Catterall, 1995). The high-affinity Ca^{2+} site that regulates DHP binding can be eliminated by replacing the conserved pore glutamate in repeat III with a lysine residue. This finding and other data indicate that high-affinity DHP binding is dependent on Ca^{2+} coordination by the same glutamate residues which form the locus of selectivity (Yang et al., 1993). A three-state model has been proposed whereby affinity for DHPs is promoted by the presence of one bound Ca^{2+} ion within the pore as opposed to zero or two (Peterson and Catterall, 1995).

Ca^{2+} Channels and Disease States

Because voltage-gated Ca^{2+} channels play so many critical roles in cells, acting as initiators of regulated exocytosis and shaping the postsynaptic responsiveness to the secreted transmitters, it is not surprising that a number of pathological states can be attributed to a malfunction these proteins. Here, we provide a few examples of disorders involving Ca^{2+} channels that provide interesting perspectives on the relationship between Ca^{2+} channel structure and function.

Lambert–Eaton Syndrome as an Autoimmune Disease of Ca^{2+} Channels

Lambert–Eaton myasthenic syndrome (LEMS) is an acquired autoimmune disease associated with a decreased probability of neurotransmitter release at the neuromuscular junction. There is substantial evidence that LEMS and related disorders arise from immunologic reactivity against neuronal counterparts of components of neoplastic tumors (for a review, see Lennon, 1994). Mice injected with IgG from patients with LEMS display physiological and morphological features of the syndrome, including fatiguable muscle weakness and selective depletion of particles from presynaptic active zones (Fukunaga et al., 1983). Later studies showed that K^+-induced $^{45}Ca^{2+}$ flux in human small-cell lung carcinoma (SCC) cells was significantly reduced by LEMS IgG, indicating the formation of autoantibodies to tumor cell Ca^{2+} channels (Roberts et al., 1985).

The SCC cells contain mRNAs encoding α_{1A}-, α_{1B}-, and α_{1D}-subunits, suggesting that P/Q-, N-, and L-type Ca^{2+} channels might be expressed (Oguro-Okano et al., 1992; Codignola et al., 1993). However, blockers of P/Q-type Ca^{2+} channels are much more effective in inhibiting $^{45}Ca^{2+}$ flux in SCC cells than inhibitors of N- or L-type Ca^{2+} channels (Lennon et al., 1995). Furthermore, patients with LEMS are much more likely to display immunoreactivity to P/Q-type channels than to N-type channels (Lennon et al., 1995). P/Q-type Ca^{2+} channels are the dominant Ca^{2+} entry pathway at the human neuromuscular junction (Protti et al., 1996) and IgG from LEMS patients has been shown to selectively inhibit Q-like Ca^{2+} channels in a rat insulinoma cells (Magnelli et al., 1996). Thus, it seems likely that the pathogenesis of LEMS involves the formation of autoantibodies to P/Q-type Ca^{2+} channels on paraneoplastic tumor cells, which, in turn, inhibit the corresponding channels on presynaptic terminals that support neuromuscular transmission.

Hypokalemic Periodic Paralysis

Hypokalemic periodic paralysis (HypoPP) is an autosomal dominant muscle disease long believed to arise from an ion channel dysfunction. Expression of the disease often follows increased carbohydrate intake or unaccustomed exercise, and is characterized by transient attacks of muscle weakness of varying duration and severity, accompanied by a drop in serum K^+ concentration. Linkage analysis in three affected families localized the HypoPP locus to chromosome 1q31–32 (Fontaine et al., 1994), the same region as the gene encoding the α_{1S}-subunit (Gregg et al., 1993). The α_{1S}-subunit cosegregated with the disease gene without recombination (Fontaine et al., 1994). Subsequent analysis demonstrated a guanine-to-adenine mutation caused an arginine-to-histidine substitution at residue 528 in transmembrane segment IIS4 of α_{1S} (Boerman et al., 1995). Examination of myotubes from individuals with HypoPP revealed transcription of both the normal and mutant genes but the maximum L-type current density was smaller than in myotubes from normal controls (Sipos et al., 1995). Furthermore, while the voltage-dependence of L-type channel activation appeared normal, the mutant myotubes showed a −40 mV shift in the voltage-dependence inactivation. Given the central role of L-type Ca^{2+} channels in skeletal muscle excitation–contraction coupling, it is not surprising that a mutation that facilitates inactivation should contribute to a paralytic disorder.

Mutational Defects in Migraine and Movement Disorders

Deleterious mutations in the α_{1A}-subunit have been found in individuals who display certain forms of migraine and ataxia (Ophoff et al., 1996) (Fig. 8.8). Familial hemiplegic migraine (FHM) is a rare autosomal dominant form of migraine with aura that is associated with ictal hemiparesis. In FHM families, four different missense mutations were found, including mutations in the S4 membrane-spanning helix of repeat I (RR192Q), in the P region of repeat II (T66M), and in the S6 regions of repeat I (V714A) and of repeat IV (I1811L). Each of these mutations would be expected to affect a key functional property, be it permeation, gating, or both, but this needs to be tested directly by expression of appropriately mutated α_{1A}-subunits.

Episodic ataxia type 2 (EA-2) is another autosomal dominant disease, characterized by attacks of cerebellar ataxia and migraine-like symptoms. The Leiden group (Ophoff et al., 1996) found two mutations in individuals afflicted with EA-2: (1) a single nucleotide deletion in IIIS1 (ΔC_{4073} in codon 1266), which caused a frame shift and a premature stop further downstream (at codon 1333), and (2) a mutation which was predicted to cause aberrant splicing at a splice site in IIIS2. It is possible that both defects cause transcript instability, thereby sharply reducing the levels of α_{1A} in heterozygotes.

A mutation in mice, at the *tottering* locus on chromosome 8, causes ataxia, seizures, and behavioral absence similar to petit mal epilepsy in humans (for a review, see Kostopoulos, 1992). Positional cloning has been used to demonstrate that the *tottering* locus lies within the mouse gene encoding the α_{1A} subunit (Fletcher et al., 1996). The tottering mutants have a single cytosine-to-thymine change, leading to proline-to-leucine substitution in the pore-forming region of repeat II. A more severe of this allele, called *leaner*, was found to be the result of a mutation in a splice donor consensus sequence, leading to truncation of the C-terminus. The mechanisms by which these channel mutations could cause movement disorders

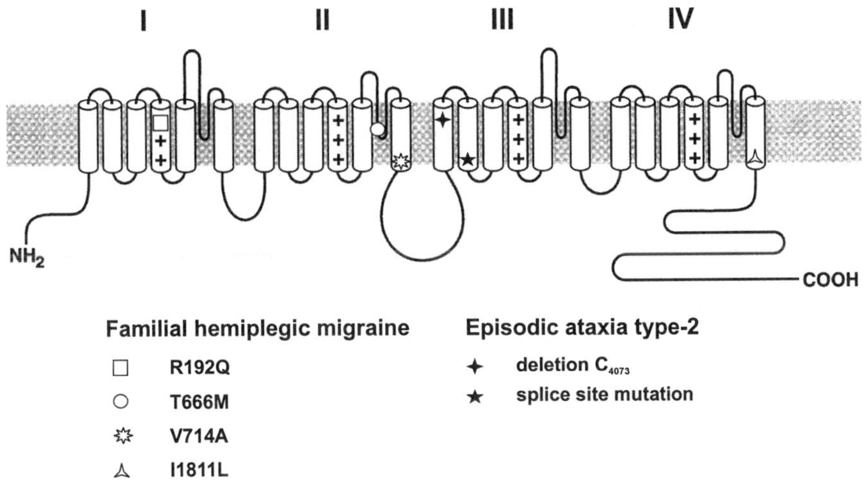

Figure 8.8 Map of the α_{1A}-subunit showing genetic defects underlying some forms of migraine and movement disorders. The location of mutations found in patients with familial hemiplegic migraine are shown by the open symbols and derangements underlying episodic ataxia type 2 are represented by the filled symbols. (Adapted from Ophoff et al., 1996, copyright Cell Press.)

or epileptiform activity remain unclear. However, all of these recent studies raise interesting questions for future investigations into the functional roles of P/Q-type Ca^{2+} channels in the central nervous system.

Concluding Remarks

Understanding of the workings of voltage-gated Ca^{2+} channels has benefited from many approaches over the last decade or so. The identification of multiple types of Ca^{2+} channels on the basis of biophysical and pharmacological criteria has been complemented by major advances in the biochemistry and molecular biology of their underlying subunit components. Considerable progress has been made in clarifying molecular mechanisms of generic properties such as selectivity, permeation, and gating, as well as structural features that allow individual types of Ca^{2+} channels to serve specialized functional roles or to respond to type-selective drugs. It is particularly intriguing to learn that molecular defects in diseases such as migraine and ataxia can not only be traced to a particular Ca^{2+} channel subunit, but also to the very domains that were previously shown to be important for key channel functions. While there have been major developments in understanding the basis of the functional channel classes, the Ca^{2+} channel's three-dimensional structure is completely unknown.

References

Adams, B. A., T. Tanabe, A. Milami, S. Numa, and K. G. Beam (1990) Intramembrane charge movement restored in dysgenic skeletal muscle by injection of dihydropyridine receptor cDNAs. *Nature* 346: 569–72.

Akaike, N., H. Kanaide, T. Kuga, M. Nakamura, J. Sadoshima, and H. Tomoike (1989) Low-voltage-activated calcium current in rat aorta smooth muscle cells in primary culture. *J. Physiol. (London)* 416: 141–60.

Almers, W. and E. W. McCleskey (1984) Non-selective conductance in calcium channels in frog muscle: calcium selectivity in a single-file pore. *J. Physiol.* 353: 565–83.

Araque, A, F. Clarac, and W. Buno (1994) P-type Ca^{2+} channels mediate excitatory and inhibitory synaptic transmitter release in crayfish muscle. *Proc. Natl. Acad. Sci. USA* 91: 4224–8.

Armstrong, C. M. and J. Neyton (1992) Ion permeation through calcium channels: a one-site model. *Ann. N.Y. Acad. Sci.* 635: 18–25.

Ashcroft, F. M., P. Proks, P. A. Smith, C. Ammala, K. Bokvist, and P. Rorsman (1994) Stimulus-secretion coupling in pancreatic β cells. *J. Cell Biochem*, 55(Suppl.): 54–65.

Assaf, S. Y. and S. H. Chung (1984) Release of endogenous Zn^{2+} from brain tissue during activity. *Nature* 308: 734–6.

Atar, D., P. H. Backx, M. M. Appel, W. D. Gao, and E. Marban (1995) Excitation-transcription coupling mediated by zinc influx through voltage-dependent calcium channels. *J. Biol. Chem.* 270:2473–7.

Augustine, G. J., M. P. Charlton, and S. J. Smith (1985) Calcium entry and transmitter release at voltage-clamped nerve terminals of squid. *J. Physiol.* 367: 163–81.

Avery, R. A. and D. Johnston (1996) Multiple channel types contribute to the low-voltage-activated calcium current in hippocampal CA3 pyramidal neurons. *J. Neurosci.* 16: 5567–82.

Babitch, J. (1990) Channel hands [letter]. *Nature* 346: 321–2.

Bargas, J., A. Howe, J. Eberwine, Y. Cao, and D. J. Surmeier (1994) Cellular and molecular characterization of Ca^{2+} currents in acutely isolated, adult rat neostriatal neurons. *J. Neurosci.* 14: 6667–86.

Bean, B. P. (1985) Two kinds of calcium channels in canine atrial cells. Differences in kinetics, selectivity, and pharmacology. *J. Gen. Physiol.* 86: 1–30.

Benardo, L. S., L. M. Masukawa, and D. A. Prince (1982) Electrophysiology of isolated hippocampal pyramidal dendrites. *J. Neurosci.* 2: 1614–22.

Bennett, M. K. and R. H. Scheller (1994) Molecular correlates of synaptic vesicle docking and fusion. *Curr. Opin. Neurobiol.* 4: 324–9.

Berrow, N. S., V. Campbell, E. M. Fitzgerald, K. Brickley, and A. C. Dolphin (1995) Antisense depletion of β-subunits modulates the biophysical and pharmacological properties of neuronal calcium channels. *J. Physiol.* 482: 481–91.

Bezprozvanny, I., R. H. Scheller, and R. W. Tsien (1995) Functional impact of syntaxin on gating of N-type and Q-type calcium channels. *Nature* 378: 623–6.

Biel, M., R. Hullin, S. Freundner, D. Singer, N. Dascal, V. Flockerzi, and F. Hofmann (1991) Tissue-specific expression of high-voltage-activated dihydropyridine-sensitive L-type calcium channels. *Eur. J. Biochem.* 200: 81–8.

Bindokas, V. P., V. J. Venema, and M. E. Adams (1991) Differential antagonism of transmitter release by subtypes of ω-agatoxins. *J. Neurophysiol.* 66: 590–601.

Birnbaumer, L., K. P. Campbell, W. A. Catterall, M. M. Harpold, F. Hofmann, W. A. Horne, Y. Mori, A. Schwartz, T. P. Snutch, T. Tanabe, and R. W. Tsien (1994) The naming of voltage-gated calcium channels. *Neuron* 13: 505–6.

Boerman, R. H., R. A. Ophoff, T. P. Links, R. van Eijk, L. A. Sandkuijl, A. Elbaz, J. E. Vale-Santos, A. R. Wintzen, J. C. van Deutekom, D. E. Isles et al. (1995) Mutation in DHP receptor α_1 subunit (CACLN1A3) gene in a Dutch family with hypokalemic periodic paralysis. *J. Med. Genet.* 32: 44–7.

Bokvist, K., L. Eliasson, C. Ammala, E. Renstrom, and P. Rorsman (1995) Co-localization of L-type Ca^{2+} channels and insulin-containing secretory granules and its significance for the initiation of exocytosis in mouse pancreatic B-cells. *EMBO J.* 14: 50–7.

Boland, L. M. J. A. Morril, and B. P. Bean (1994) ω-Conotoxin block of N-type calcium channels in frog and rat sympathetic neurons. *J. Neurosci.* 14: 5011–27.

Bosse, E., S. Regulla, M. Biel, P. Ruth, H. E. Meyer, V. Flockerzi, and F. Hofmann (1990) The cDNA and deduced amino acid sequence of the gamma subunit of the L-type calcium channel from rabbit skeletal muscle. *FEBS Lett.* 267: 153–6.

Bourinet, E., G. W. Zamponi, A. Stea, T. W. Soong, B. A. Lewis, L. P. Jones, D. Yue, and T. P. Snutch (1996) The α_{1E} calcium channel exhibits permeation properties similar to low-voltage-activated calcium channels. *J. Neurosci.* 16: 4983–93.

Bowersox, S. S., G. P. Miljanich, Y. Sugiura, C. Li, L. Nadasdi, B. B. Hoffman, J. Ramachandran, and C. P. Ko (1995) Differential blockade of voltage-sensitive calcium channels at the mouse neuromuscular junction by novel ω-conopeptides and ω-agatoxin-IVA. *J. Pharmacol. Exp. Ther.* 273: 248–56.

Campbell, K. P., A. T. Leung, and A. H. Sharp (1988) The biochemistry and molecular biology of the dihydropyridine-sensitive calcium channel. *Trends Neurosci.* 11: 425–30.

Carbone, E. and H. D. Lux (1984) A low voltage-activated, fully inactivating Ca channel in vertebrate sensory neurones. *Nature* 310: 501–2.

Castellano, A., X. Wei, L. Birnbaumer, and E. Perez-Reyes (1993) Cloning and expression of a neuronal calcium channel β subunit. *J. Biol. Chem.* 268: 12359–66.

Castillo, P. E., M. G. Weisskopf, and R. A. Nicoll, (1994) The role of Ca^{2+} channels in hippocampal mossy fiber synaptic transmission and long-term potentiation. *Neuron* 12: 261–9.

Catterall, W. A. and B. M. Curtis (1987) Molecular properties of voltage-sensitive calcium channels. *Soc. Gen. Physiol. Ser.* 41: 201–13.

Catterall, W. A., M. J. Seagar, and M. Takahashi (1988) Molecular properties of dihydropyridine-sensitive calcium channels in skeletal muscle. *J. Biol. Chem.* 263: 3535–8.

Catterall, W. A., R. E. Westenbroek, S. Herlitze, and C. T. Yokoyama (1998) Localisation and function of brain calcium channels. In: *Low-Voltage-Activated T-type Calcium Channels* (Tsien, R. W., Clozel, J. P., and Nargeot, J. eds.). Aidis Press, Chester, UK, pp. 207–17.

Chad, J. E. and R. Eckert (1986) An enzymatic mechanism for calcium current inactivation in dialysed Helix neurones. *J. Physiol. (London)* 378: 31–51.

Charton, G., C. Rovira, Y. Ben-Ari, and V. Leviel (1985) Spontaneous and evoked release of endogeneous Zn^{2+} in the hippocampal mossy fiber zone of the rat in situ. *Exp. Brain Res.* 58: 202–5.

Chen, X.-H., I. Bezprozvanny, and R. W. Tsien (1997) Molecular basis of proton block of L-type Ca^{2+} channels. *J. Gen. Physiol.* 108: 363–74.

Chuang, R. S.-I., H. Jaffe, L. Cribbs, E. Perez-Reyes, and K. J. Swartz (1998) Identification of an inhibitor of T-type calcium channels. *Soc. Neurosci. Abstr.* 24(1): 22.

Church, P. J. and E. F. Stanley (1996) Conductance and pharmacological properties of 2 calcium channels in physiological levels of external calcium. *Soc. Neurosci. Abstr.* 22: 713.

Clozel, J. P., E. A. Ertel, and S. I. Ertel (1997) Discovery and main pharmacological properties of mibefradil

(Ro 40-5967), the first selective T-type calcium channel blocker. *J. Hypertension.* Supplement 15: S17–25.

Codignola, A., P. Tarroni, F. Clementi, A. Pollo, M. Lovallo, E. Carbone, and E. Sher (1993) Calcium channel subtypes controlling serotonin release from human small cell lung carcinoma cell lines. *J. Biol. Chem.* 268: 26240–7.

Cohen, C. J. and R. T. McCarthy (1987) Nimodipine block of calcium channels in rat anterior pituitary cells. *J. Physiol.* 387: 195–225.

Cribbs, L. L., J. H. Lee, J. Yang, J. Satin, Y. Zhang, A. Daud, J. Barclay, M. P. Williamson, M. Fox, M. Rees, and E. Perez-Reyes (1998) Cloning and characterization of alpha1H from human heart, a member of the T-type Ca^{2+} channel gene family. *Circ. Res.* 83: 103–9.

De Jongh, K. S., C. Warner, and W. A. Catterall (1990) Subunits of purified calcium channels. α_2 and δ are encoded by the same gene. *J. Biol. Chem.* 265: 14738–41.

de Leon, M., Y. Wang, L. Jones, E. Perez-Reyes, X. Wei, T. W. Soong, T. P. Snutch, and D. T. Yue (1995) Essential Ca^{2+}-binding motif for Ca^{2+}-sensitive inactivation of L-type Ca^{2+} channels. *Science* 270: 1502–6.

De Luca, A., M. J. Rand, J. J. Reid, and D. F. Story (1991) Differential sensitivities of avian and mammalian neuromuscular junctions to inhibition of cholinergic transmission by ω-conotoxin GVIA. *Toxicon* 29: 311–20.

De Waard, M., C. A. Gurnett, and K. P. Campbell (1996) Structural and functional diversity of voltage-activated calcium channels. In: *Ion Channels* (Narahashi, T, ed.) Plenum Press, New York, pp. 41–87.

De Waard, M., D. R. Witcher, M. Pragnell, H. Liu, and K. P. Campbell (1995) Properties of the α_1–β anchoring site in voltage-dependent Ca^{2+} channels. *J. Biol. Chem.* 270: 12056–64.

Deisseroth, K., H. Bito, and R. W. Tsien (1996) Signaling from synapse to nucleus: postsynaptic CREB phosphorylation during multiple forms of hippocampal synaptic plasticity. *Neuron* 16: 89–101.

Deisseroth, K., E. K. Heist, and R. W. Tsien (1998) Translocation of calmodulin to the nucleus supports CREB phosphorylation in hippocampal neurons. *Nature* 392: 198–202.

Dippel, W. W., P. L. Chen, N. H. McArthur, and P. G. Harms (1995) Calcium involvement in luteinizing hormone–releasing hormone release from the bovine infundibulum. *Domest. Anim. Endocrinol.* 12: 349–54.

Diriong, S., P. Lory, M. E. Williams, S. B. Ellis, M. M. Harpold, and S. Taviaux (1995) Chromosomal localization of the human genes for α_{1A}, α_{1B} and α_{1E} voltage-dependent Ca^{2+} channel subunits. *Genomics* 30: 605–9.

Dodge, F. Jr. and R. Rahamimoff (1967) Co-operative action of calcium ions in transmitter release at the neuromuscular junction. *J. Physiol.* 193: 419–32.

Droogmans, G. and B. Nilius (1989) Kinetic properties of the cardiac T-type calcium channel in the guinea-pig. *J. Physiol.* 419: 627–50.

Duarte, C. B., I. L. Ferreira, P. F. Santos, C. R. Oliveira, and A. P. Carvalho, (1992) Ca^{2+}-dependent release of [^{3}H]GABA in cultured chick retina cells. *Brain Res.* 591: 27–32.

Dubel, S. J., T. V. Starr, J. Hell, M. K. Ahlijanian, J. J. Enyeart, W. A. Catterall, and T. P. Snutch (1992) Molecular cloning of the α_1 subunit of an ω-conotoxin-sensitive calcium channel. *Proc. Natl. Acad. Sci. USA* 89: 5058–62.

Dunlap, K., J. I. Luebke, and T. J. Turner (1995) Exocytotic Ca^{2+} channels in mammalian central neurons. *Trends Neurosci.* 18: 89–98.

Edmonds, B., M. Klein, N. Dale, and E. R. Kandel (1990) Contributions of two types of calcium channels to synaptic transmission and plasticity. *Science* 250: 1142–7.

Eisenberg, R. S., R. T. McCarthy, and R. L. Milton (1983) Paralysis of frog skeletal muscle fibres by the calcium antagonist D-600. *J. Physiol. (London)* 341: 495–505.

el-Hayek, R., B. Antoniu, J. Wang, S. L. Hamilton, and N. Ikemoto, (1995) Identification of calcium release-triggering and blocking regions of the II–III loop of the skeletal muscle dihydropyridine receptor. *J. Biol. Chem.* 270: 22116–8.

Ellinor, P. T., J. Yang, W. A. Sather, J.-F. Zhang, and R. W. Tsien (1995) Ca^{2+} channel selectivity at a single locus for high-affinity Ca^{2+} interactions. *Neuron* 15: 1121–32.

Ellinor, P. T., J.-F. Zhang, W. A. Horne, and R. W. Tsien (1994) Structural determinants of the blockade of N-type calcium channels by a peptide neurotoxin. *Nature* 372: 272–5.

Ellinor, P. T., J.-F. Zhang, A. D. Randall, M. Zhou, T. L. Schwartz, R. W. Tsien, and W. A. Horne (1993) Functional expression of a rapidly inactivating neuronal calcium channel. *Nature.* 363: 455–8.

Ellis, S. B., M. E. Williams, N. R. Ways, R. Brenner, A. H. Sharp, A. T. Leung, K. P. Campbell, E. McKenna, W. J. Koch, and A. Hui et al. (1988) Sequence and expression of mRNAs encoding the α_1 and α_2 subunits of a DHP-sensitive calcium channel. *Science* 241: 1661–4.

Ertel, S. I. and E. A. Ertel (1997) Low-voltage-activated T-type Ca^{2+} channels. *Trends Pharmacol. Sci.* 18: 37–42.

Fabi, F., M. Chiavarelli, L. Argiolas, R. Chiavarelli, and B. P. del Basso (1993) Evidence for sympathetic neurotransmission through presynaptic N-type calcium channels in human saphenous vein. *Br. J. Pharmacol.* 110: 338–42.

Feng, T. P. (1936) Studies on the neuromuscular junction II. The universal antagonism between calcium and curarizing agencies. *Chin. J. Physiol.* 10: 513–28.

Fisher, T. E. and C. W. Bourque (1996) Calcium-channel subtypes in the somata and axon terminals of magnocellular neurosecretory cells. *Trends Neurosci.* 19: 440–4.

Fletcher, C. F., C. M. Lutz, T. N. O'Sullivan, J. D. Shaughnessy Jr., R. Hawkes, W. N. Frankel, N. G. Copeland, and N. A. Jenkins (1996) Absence epilepsy in tottering mutant mice is associated with calcium channel defects. *Cell* 87: 607–17.

Florman, H. M., M. E. Corron, T. D. Kim, and D. F. Babcock (1992) Activation of voltage-dependent calcium channels of mammalian sperm is required for zona pellucida-induced acrosomal exocytosis. *Dev. Biol.* 152: 304–14.

Fontaine, B., J. Vale-Santos, K. Jurkat-Rott, J. Reboul, E. Plassart, C. S. Rime, A. Elbaz, R. Heine, J. Guimaraes, J. Weissenbach et al. (1994) Mapping of the hypokalemic periodic paralysis (HypoPP) locus to chromosome 1q31–32 in three European families. *Nat. Genet.* 6: 267–72.

Forti, L. and D. Pietrobon (1993) Functional diversity of L-type calcium channels in rat cerebellar neurons. *Neuron* 10: 437–50.

Frew, R. and P. M. Lundy (1995) A role for Q type Ca^{2+} channels in neurotransmission in the rat urinary bladder. *Br. J. Pharmacol.* 116: 1595–8.

Fukunaga, H., A. G. Engel, B. Lang, J. Newsom-Davis, and A. Vincent (1983) Passive transfer of Lambert-Eaton myasthenic syndrome with IgG from man to mouse depletes the presynaptic membrane active zones. *Proc. Natl. Acad. Sci. USA* 80: 7636–40.

Garcia, J., T. Tanabe, and K. G. Beam (1994) Relationship of calcium transients to calcium currents and charge movements in myotubes expressing skeletal and cardiac dihydropyridine receptors. *J. Gen. Physiol.* 103: 125–47.

Ghosh, A., D. D. Ginty, H. Bading, and M. E. Greenberg (1994) Calcium regulation of gene expression in neuronal cells. *J. Neurobiol.* 25: 294–303.

Glossman, H. and J. Striessnig (1990) Molecular properties of calcium channels. *Rev. Physiol. Biochem. Pharmacol.* 114: 1–105.

Glossmann, H., J. Striessnig, L. Hymel, and H. Schindler (1987) Purified L-type calcium channels: only one single polypeptide (α_1-subunit) carries the drug receptor domains and is regulated by protein kinases. *Biomed. Biochim. Acta* 46: S351–6.

Gonzalez, G. A. and M. R. Montminy (1989) Cyclic AMP stimulates somatostatin gene transcription by phosphorylation of CREB at serine 133. *Cell* 59: 675–80.

Grabner, M., A. Bachmann, F. Rosenthal, J. Striessnig, C. Schultz, D. Tautz, and H. Glossmann (1994) Insect calcium channels. Molecular cloning of an α_1-subunit from housefly (*Musca domestica*) muscle. *FEBS Lett.* 339: 189–94.

Grabner, M., Z. Wang, S. Hering, J. Striessnig, and H. Glossman (1996) Transfer of 1,4-dihydropyridine sensitivity from L-type to class A (BI) calcium channels. *Neuron* 16: 207–18.

Gray, D. B., J. L. Bruses, and G. R. Pilar (1992) Developmental switch in the pharmacology of Ca^{2+} channels coupled to acetylcholine release. *Neuron* 8: 715–24.

Greenberg, M. E., M. A. Thompson, and N. Sheng (1992) Calcium regulation of immediate early gene transcription. *J. Physiol. (Paris)* 86: 99–108.

Gregg, R. G., F. Couch, K. Hogan, and P. A. Powers (1993) Assignment of the human gene for the α_1 subunit of the skeletal muscle DHP-sensitive Ca^{2+} channel (CACNL1A3) to chromosome 1q31–q32. *Genomics* 15: 107–12.

Gregg, R. G., A. Messing, C., Strube, M. Beurg, R. Moss, M. Behan, M. Sukhareva, S. Haynes, J. A. Powell, R. Coronado, and P. A. Powers (1996) Absence of the β subunit (CCHB1) of the skeletal muscle dihydropyridine receptor alters expression of the α_1 subunit and eliminates excitation-contraction coupling. *Proc. Natl. Acad. Sci. USA* 93: 13961–6.

Gurnett, C. A., M. De Waard, and K. P. Campbell (1996) Dual function of the voltage dependent Ca^{2+} channel $\alpha_2\delta$ subunit in current stimulation and subunit interaction. *Neuron* 16: 431–40.

Hagiwara, N., H. Irisawa, and M. Kameyama (1988) Contribution of two types of calcium currents to the pacemaker potentials of rabbit sino-atrial node cells. *J. Physiol.* 395: 233–53.

Hell, J. W., R. E. Westenbroek, C. Warner, M. K. Ahlijanian, W. Prystay, M. M. Gilbert, T. P. Snutch, and W. A. Catterall (1993a) Identification and differential subcellular localization of the neuronal class C and class D L-type calcium channel α_1 subunits. *J. Cell Biol.* 123: 949–62.

Hell, J. W., C. T. Yokoyama, S. T. Wong, C. Warner, T. P. Snutch, and W. A. Catterall, (1993b) Differential phosphorylation of two size forms of the neuronal class C L-type calcium channel α_1 subunit. *J. Biol. Chem.* 268: 19451–7.

Hering, S., S. Aczel, M. Grabner, F. Doring, S. Berjukow, J. Mitterdorfer, M. J. Sinnegger, J. Striessnig, V. E. Degtiar, Z. Wang, and H. Glossmann (1996) Transfer of high sensitivity for benzothiazepines from L-type to class A (BI) calcium channels. *J. Biol. Chem.* 271: 24471–5.

Hess, P. and R. W.Tsien (1984) Mechanism of ion permeation through calcium channels. *Nature* 309: 453–6.

Hess, P., J. B. Lansman, and R. W. Tsien (1986) Calcium channel selectivity for divalent and monovalent cations. Voltage and concentration dependence of single channel current in ventricular heart cells. *J. Gen. Physiol.* 88: 293–319.

Hill, C. S. and R. Treisman (1995) Transcriptional regulation by extracellular signals: mechanisms and specificity. *Cell* 80: 199–211.

Hille, B. (1992) *Ionic Channels of Excitable Membranes* Sinaeur Associates, Sunderland, MA.

Hille, B. (1994) Modulation of ion-channel function by G-protein-coupled receptors. *Trends Neurosci.* 17: 531–6.

Hirning, L. D., A. P. Fox, E. W. McCleskey, B. M. Olivera, S. A. Thayer, R. J. Miller, and R. W. Tsien (1988) Dominant role of N-type Ca^{2+} channels in evoked release of norepinephrine from sympathetic neurons. *Science* 239: 57–61.

Hoeffler, J. P., T. E. Meyer, Y. Yun, J. L. Jameson, and J. F. Habener (1988) Cyclic AMP-responsive DNA-binding protein: structure based on a cloned placental cDNA. *Science* 242: 1430–3.

Hofmann, F., M. Biel, and V. Flockerzi (1994) Molecular basis for Ca^{2+} channel diversity. *Annu. Rev. Neurosci.* 17: 399–418.

Hofmann, F., H. J. Oeken, T. Schneider, and M. Sieber (1988) The biochemical properties of L-type calcium channels. *J. Cardiovasc. Pharmacol.* 12: S25–30.

Hong, S. J., and C. C. Chang (1995) Inhibition of acetylcholine release from mouse motor nerve by a P-type calcium channel blocker, ω-agatoxin IVA. *J. Physiol. (London)* 482: 283–90.

Horne, W. A., P. T. Ellinor, I. Inman, M. Zhou, R. W. Tsien, and T. L. Schwarz (1993) Molecular diversity of Ca^{2+} channel α_1 subunits from the marine ray *Discopyge ommata. Proc. Natl. Acad. Sci. USA* 90: 3787–91.

Hoshi, T., W. N. Zagotta, and R. W. Aldrich (1991) Two types of inactivation in Shaker K$^+$ channels: effects of alterations in the carboxy-terminal region. *Neuron* 7: 547–56.

Howell, G. A., M. G. Welch, and C. J. Frederickson (1984) Stimulation-induced uptake and release of zinc in hippocampal slices. *Nature* 308: 736–8.

Huang, C. F., B. E. Flucher, M. M. Schmidt, S. K. Stroud, and J. Schmidt (1994) Depolarization-transcription signals in skeletal muscle use calcium flux through L channels, but bypass the sarcoplasmic reticulum. *Neuron* 13: 167–77.

Huguenard, J. R. and D. A. Prince (1992) A novel T-type current underlies prolonged Ca^{2+}-dependent burst firing in GABAergic neurons of rat thalamic reticular nucleus. *J. Neurosci.* 12: 3804–17.

Hullin, R., D. Singer-Lahat, M. Freichel, M. Biel, N. Dascal, F. Hofmann, and V. Flockerzi (1992) Calcium channel β subunit heterogeneity: functional expression of cloned cDNA from heart, aorta and brain, *EMBO J.* 11: 885–90.

Imredy, J. P. and D. T. Yue (1992) Submicroscopic Ca^{2+} diffusion mediates inhibitory coupling between individual Ca^{2+} channels. *Neuron* 9: 197–207.

Imredy, J. P. and D. T. Yue (1994) Mechanism of Ca^{2+}-sensitive inactivation of L-type Ca^{2+} channels. *Neuron* 12: 1301–18.

Isom, L. L., K. S. De Jongh, and W. A. Catterall (1994) Auxiliary subunits of voltage-gated ion channels. *Neuron* 12: 1183–94.

Jaffe, D. B., D. Johnston, N. Lasser-Ross, J. E. Lisman, H. Miyakawa, and W. N. Ross (1992) The spread of Na$^+$ spikes determines the pattern of dendritic Ca^{2+} entry into hippocampal neurons. *Nature* 357: 244–6.

Jahnsen, H. and R. Llinás (1984) Ionic basis for the electroresponsiveness and oscillatory properties of guinea-pig thalamic neurones in vitro. *J. Physiol.* 349: 227–47.

Jan, L. Y. and Y. N. Jan (1989) Voltage-sensitive ion channels. *Cell* 56: 13–25.

Jay, S. D., S. B. Ellis, A. F. McCue, M. E. Williams, T. S. Vedvick, M. M. Harpold, and K. P. Campbell (1990) Primary structure of the gamma subunit of the DHP-sensitive calcium channel from skeletal muscle. *Science* 248: 490–2.

Jay, S. D., A. H. Sharp, S. D. Kahl, T. S. Vedvick, M. M. Harpold, and K. P. Campbell (1991) Structural characterization of the dihydropyridine-sensitive calcium channel α_2-subunit and the associated δ peptides. *J. Biol. Chem.* 266: 3287–93.

Katz, B. (1969) *The Release of Neural Transmitter Substances.* Liverpool University Press, Liverpool.

Katz, E., P. A. Ferro, B. D. Cherksey, M. Sugimori, R. Llinás, and O. D. Uchitel (1995) Effects of Ca^{2+} channel blockers on transmitter release and presynaptic currents at the frog neuromuscular junction. *J. Physiol. (London)* 486: 695–706.

Kavalali, E. T. and M. R. Plummer (1994) Selective potentiation of a novel calcium channel in rat hippocampal neurones. *J. Physiol.* 480: 475–84.

Kavalali, E. T., M. Zhuo, H. Bito, and R. W. Tsien (1997) Dendritic Ca^{2+} channels characterized by recordings from isolated hippocampal dendritic segments. *Neuron* 18: 651–63.

Kerr, L. M. and D. Yoshikami (1984) A venom peptide with a novel presynaptic blocking action. *Nature* 308: 282–4.

Ketchum, K. A., W. J. Joiner, A. J. Sellers, L. K. Kaczmarek, and S. A. Goldstein (1995) A new family of outwardly rectifying potassium channel proteins with two pore domains in tandem. *Nature* 376: 690–5.

Kim, H. L., H. Kim, P. Lee, R. G. King, and H. Chin (1992) Rat brain expresses an alternatively spliced form of the dihydropyridine-sensitive L-type calcium channel α_2 subunit. *Proc. Natl. Acad. Sci. USA* 89: 3251–5. Erratum, *Proc. Natl. Acad. Sci. USA* (1992) 89: 5699.

Kim, M. S., T. Morii, L. X. Sun, K. Imoto, and Y. Mori (1993) Structural determinants of ion selectivity in brain calcium channel. *FEBS Lett.* 318: 145–8.

Klockner, U. and G. Isenberg (1994) Calcium channel current of vascular smooth muscle cells: extracellular protons modulate gating and single channel conductance. *J. Gen. Physiol.* 103: 665–78.

Knudson, C. M., N. Chaudhari, A. H. Sharp, J. A. Powell, K. G. Beam, and K. P. Campbell, (1989) Specific absence of the α_1 subunit of the dihydropyridine receptor in mice with muscular dysgenesis. *J. Biol. Chem.* 264: 1345–8.

Koh, D. S. and B. Hille (1996) Modulation by neurotransmitters of norepinephrine secretion from sympathetic ganglion neurons detected by amperometry. *Soc. Neurosci. Abstr.* 22: 507.

Kollmar, R., L. G. Montgomery, J. Fak, L. J. Henry, and A. J. Hudspeth (1997a) Predominance of the alpha 1D subunit in L-type voltage-gated Ca^{2+} channels of hair cells in the chicken's cochlea. *Proc. Natl. Acad. Sci. USA* 94: 14883–8.

Kollmar, R., J. Fak, L. G. Montgomery, and A. J. Hudspeth (1997b) Hair cell-specific splicing of mRNA for the alpha 1D subunit of voltage-gated Ca^{2+} channels in the chicken's cochlea. *Proc. Natl. Acad. Sci. USA* 94: 14889–93.

Kostopoulos, G. K. (1992) The tottering mouse: a critical review of its usefulness in the study of the neuronal mechanisms underlying epilepsy. *J. Neural Transmit. Suppl.* 35: 21–36.

Kostyuk, P. G. and R. E. Shirokov (1989) Deactivation kinetics of different components of calcium inward current in the membrane of mice sensory neurones. *J. Physiol.* 409: 343–55.

Kuga, T., S. Kobayashi, Y. Hirakawa, H. Kanaide, and A. Takeshita (1996) Cell cycle-dependent expression of L- and T-type Ca^{2+} currents in rat aortic smooth muscle cells in primary culture. *Circ. Res.* 79: 14–19.

Kuo, C. C. and P. Hess (1993) Ion permeation through the L-type Ca^{2+} channel in rat phaeochromocytoma cells: two sets of ion binding sites in the pore. *J. Physiol.* 466: 629–55.

Lacerda, A. E., H. S. Kim, P. Ruth, E. Perez-Reyes, V. Flockerzi, F. Hofmann, L. Birnbaumer, and A. M. Brown (1991) Normalization of current kinetics by interaction between the α_1 and β subunits of the skeletal muscle dihydropyridine-sensitive Ca^{2+} channel. *Nature* 352: 527–30.

Lansman, J. B., P. Hess, and R. W. Tsien (1986) Blockade of current through single calcium channels by Cd^{2+}, Mg^{2+}, and Ca^{2+}. Voltage and concentration dependence of calcium entry into the pore. *J. Gen. Physiol.* 88: 321–47.

Larsson, H. P., O. S. Baker, D. S. Dhillon, and E. Y. Isacoff (1996) Transmembrane movement of the shaker K^+ channel S4. *Neuron* 16: 387–97.

Lei, M., H. Brown, and D. Noble (1998) What role do T-type calcium channels play in cardiac pacemaker activity? In: *Low Voltage-Activated T-type Calcium Channels* (Tsien, R. W., Clozel, J. P., and Nargeot, J., eds.). Adis Intl. Ltd, Tattenhall, UK, pp. 103–9.

Lemos, J. R. and M. C. Nowycky (1989) Two types of calcium channels coexist in peptide-releasing vertebrate nerve terminals. *Neuron* 2: 1419–26.

Lennon, V. A. (1994) Paraneoplastic autoantibodies: the case for a descriptive generic nomenclature [see comments]. *Neurology* 44: 2236–40.

Lennon, V. A., T. J. Kryzer, G. E. Griesmann, P. E. O'Suilleabhain, A. J. Windebank, A. Woppmann, G. P. Miljanich, and E. H. Lambert (1995) Calcium-channel antibodies in the Lambert-Eaton syndrome and other paraneoplastic syndromes. *N. Engl. J. Med.* 332: 1467–74.

Letts, V. A., R. Felix, G. H. Biddlecome, J. Arikkath, C. L. Mahaffey, A. Valenzuela, F. S. Bartlett, Y. Mori, K. P. Campbell and W. N. Frankel (1998) The mouse stargazer gene encodes a neuronal Ca^{2+}-channel γ subunit. *Nature Genetics* 19: 340–47.

Lewis, R. S. and M. D. Cahalan (1995) Potassium and calcium channels in lymphocytes. *Annu. Rev. Immunol.* 13: 623–53.

Lin, Z., S. Haus, J. Edgerton, and D. Lipscombe, (1997) Identification of functionally distinct isoforms of the N-type Ca channel in rat sympathetic ganglia and brain. *Neuron* 18: 153–66.

Lindgren, C. A. and J. W. Moore (1989) Identification of ionic currents at presynaptic nerve endings of the lizard. *J. Physiol.* 414: 201–22.

Liu, H., M. De Waard, V. E. S. Scott, C. A. Gurnet, V. A. Lennon, and K. P. Campbell (1996a) Indentification of three subunits of the high affinity ω-conotoxin MVIIC-sensitive Ca^{2+} channel. *J. Biol. Chem.* 271: 13804–10.

Liu, Y., M. E. Jurman, and G. Yellen (1996b) Dynamic rearrangement of the outer mouth of a K^+ channel during gating. *Neuron* 16: 859–67.

Llinás, R. and R. Hess (1976) Tetrodotoxin-resistant dendritic spikes in avian Purkinje cells. *Proc. Natl. Acad. Sci. USA* 73: 2520–3.

Llinás, R. and C. Nicholson (1971) Electrophysiological properties of dendrites and somata in alligator Purkinje cells. *J. Neurophysiol.* 34: 532–51.

Llinás, R. and C. Nicholson (1975) Calcium role in depolarization-secretion coupling: an aequorin study in squid giant synapse. *Proc. Natl. Acad. Sci. USA* 72: 187–90.

Llinás, R. and M. Sugimori (1980) Electrophysiological properties of in vitro Purkinje cell dendrites in mammalian cerebellar slices. *J. Physiol.* 305: 197–213.

Llinás, R., I. Z. Steinberg, and K. Walton (1981) Relationship between presynaptic calcium current and postsynaptic potential in squid giant synapse. *Biophys. J.* 33: 323–51.

Llinás, R., M. Sugimori, D. E. Hillman, and B. Cherksey (1992) Distribution and functional significance of the P-type, voltage-dependent Ca^{2+} channels in the mammalian central nervous system. *Trends Neurosci,* 15: 351–5.

Llinás, R., M. Sugimori, and B. Cherksey (1989) Voltage-dependent calcium conductances in mammalian neurons: the P channel. *Ann. N.Y. Acad. Sci.* 560: 103–11.

Lopez, M. G., A. Albillos, M. T. de la Fuente, R. Borges, L. Gandia, E. Carbone, A. G. Garcia, and A. R. Artalejo (1994) Localized L-type calcium channels control exocytosis in cat chromaffin cells. *Pflugers Arch.* 427: 348–54.

Lu, X., L. Xu, and G. Meissner (1994) Activation of the skeletal muscle calcium release channel by a cytoplasmic loop of the dihydropyridine receptor. *J. Biol. Chem.* 269: 6511–16.

Lu, X., L. Xu, and G. Meissner, (1995) Phosphorylation of dihydropyridine receptor II-III loop peptide regulates skeletal muscle calcium release channel function. Evidence for an essential role of the β-OH group of Ser687. *J. Biol. Chem.* 270: 18459–64.

Luebke, J. I., K. Dunlap,and T. J. Turner (1993) Multiple calcium channel types control glutamatergic synaptic transmission in the hippocampus. *Neuron* 11: 895–902.

Magee, J. C. and D. Johnston (1995) Characterization of single voltage-gated Na^+ and Ca^{2+} channels in apical dendrites of rat CA1 pyramidal neurons. *J. Physiol.* 487: 67–90.

Magee, J. C., G. Christofi, H. Miyakawa, B. Christie, N. Lasser-Ross, and D. Johnston (1995) Subthreshold synaptic activation of voltage-gated Ca^{2+} channels mediates a localized Ca^{2+} influx into the dendrites of hippocampal pyramidal neurons. *J. Neurophysiol.* 74: 1335–42.

Magnelli, V., C. Grassi, E. Parlatore, E. Sher, and E. Carbone, (1996) Down-regulation of non-L-, non-N-type (Q-like) Ca^{2+} channels by Lambert-Eaton myasthenic syndrome (LEMS) antibodies in rat insulinoma RINm5F cells. *FEBS Lett.* 387: 47–52.

Maguire, G., B. Maple, P. Lukasiewicz, and F. Werblin (1989) Gamma-aminobutyrate type B receptor modulation of L-type calcium channel current at bipolar cell terminals in the retina of the tiger salamander. *Proc. Natl. Acad. Sci. USA* 86: 10144–7.

Markram, H. and B. Sakmann (1994) Calcium transients in dendrites of neocortical neurons evoked by single subthreshold excitatory postsynaptic potentials via low-voltage-activated calcium chanels. *Proc. Natl. Acad. Sci. USA* 91: 5207–11.

Markram, H., P. J. Helm, and B. Sakmann (1995) Dendritic calcium transients evoked by single back-propagating action potentials in rat neocortical pyramidal neurons. *J. Physiol.* 485: 1–20.

Martin-Moutot, N., N. Charvin, C. Leveque, K. Sato, T. I. Nishiki, S. Kozaki, M. Takahashi, and M. Seagar (1996) Interaction of snare complexes with p–q-type calcium channels in rat cerebellar synaptosomes. *J. Biol. Chem.* 271: 6567–70.

Masukawa, L. M. and D. A. Prince (1984) Synaptic control of excitability in isolated dendrites of hippocampal neurons. *J. Neurosci.* 4: 217–27.

Matteson, D. R. and C. M. Armstrong (1986) Properties of two types of calcium channels in clonal pituitary cells. *J. Gen. Physiol.* 87: 161–82.

McCleskey, E. W. and W. Almers (1985) The Ca channel in skeletal muscle is a large pore. *Proc. Natl. Acad. Sci. USA* 82: 7149–53.

McDonough, S. I., K. J. Swartz, I. M. Mintx, L. M. Boland, and B. P. Bean (1996) Inhibition of calcium channels in rat central and peripheral neurons by ω-conotoxin MVIIC. *J. Neurosci.* 16: 2612–23.

Melzer, W., A. Herrmann-Frank, and H. C. Lüttgau (1995) The role of Ca^{2+} ions in excitation–contraction coupling of skeletal muscle fibres. *Biochim. Biophys. Acta* 1241: 59–116.

Mikami, A., K. Imoto, T. Tanabe, T. Niidome, Y. Mori, H. Takeshima, S. Narumiya, and S. Numa (1989) Primary structure and functional expression of the cardiac dihydropyridine-sensitive calcium channel. *Nature* 340: 230–3.

Miledi, R. (1973) Transmitter release induced by injection of calcium ions into nerve terminals. *Proc. R. Soc. London [Biol.]* 183: 421–5.

Miller, R. J. and S. B. Freedman (1984) Are dihydropyridine binding sites voltage sensitive calcium channels? *Life Sci.* 34: 1205–21.

Mintz, I. M. and B. P. Bean (1993) Block of calcium channels in rat neurons by synthetic ω-Aga-IVA. *Neuropharmacology* 32: 1161–9.

Mintz, I., M. E. Adams, and B. P. Bean (1992a) P-type calcium channels in rat central and peripheral neurons. *Neuron* 9: 85–95.

Mintz, I. M., B. L. Sabatini, and W. G. Regehr (1995) Calcium control of transmitter release at a cerebellar synapse. *Neuron* 15: 675–88.

Mintz, I. M., V. J. Venema, K. M. Swiderek, T. D. Lee, B. P. Bean, and M. E. Adams (1992b) P-type calcium channels blocked by the spider toxin ω-Aga-IVA. *Nature* 355: 827–9.

Mitterdorfer, J., M. J. Sinnegger, M. Grabner, J. Striessnig, and H. Glossmann (1995) Coordination of Ca^{2+} by the pore region glutamates is essential for high-affinity dihydropyridine binding to the cardiac Ca^{2+} channel α_1 subunit. *Biochemistry* 34: 9350–5.

Montminy, M. R. and L. M. Bilezikjian (1987) Binding of a nuclear protein to the cyclic-AMP response element of the somatostatin gene. *Nature* 328: 175–8.

Moreno, H., B. Rudy, and R. Llinás (1997) Beta subunit influence the biophysical and pharmacological differences between P- and Q-type calcium currents expressed in a mammalian cell line. *Proc. Natl. Acad. Sci. USA* 94(25): 14042–7.

Morgan, J. I. and T. Curran (1988) Calcium as a modulator of the immediate-early gene cascade in neurons. *Cell Calcium* 9: 303–11.

Morgan, J. I. and T. Curran(1989) Stimulus-transcription coupling in neurons: role of cellular immediate-early genes. *Trends Neurosci.* 12: 459–62.

Mori, Y., T. Friedrich, M. S. Kim, A. Mikami, J. Nakai, P. Ruth, E. Bosse, F. Hofmann, V. Flockerzi, T. Furuichi, K. Mikoshiba, K. Imoto, T. Tanabe, and S. Numa (1991) Primary structure and functional expression from complmentary DNA of a brain calcium channel. *Nature* 350: 398–402.

Näbauer, M., G. Callewaert, L. Cleeman, and M. Morad (1989) Regulation of calcium release is gated by calcium current, not gating charge, in cardiac myocytes [see comments]. *Science* 244: 800–3.

Nakai, J., B. A. Adams, K. Imoto, and K. G. Beam (1994) Critical roles of the S3 segment and S3–S4 linker of repeat I in activation of L-type calcium channels. *Proc. Natl. Acad. Sci. USA* 91: 1014–18.

Nakayama, H., M. Taki, J. Striessnig, H. Glossmann, W. A. Catterall, and Y. Kanaoka (1991) Identification of 1,4-dihydropyridine binding regions within the α_1 subunit of skeletal muscle Ca^{2+} channels by photoaffinity labeling with diazipine. *Proc. Natl. Acad. Sci. USA* 88: 9203–7.

Namkung, Y., S. M. Smith, S. B. Lee, N. V. Skrypnyk, H.-L. Kim, H. Chin, R. H. Scheller, R. W. Tsien, and H.-S. Shin (1998) Targeted disruption of the Ca^{2+} channel β_3 subunit reduces N- and L-type Ca^{2+} channel activity and alters voltage-dependent activation of P/Q-type Ca^{2+} chanels in neurons. *Proc. Natl. Acad. Sci. USA* 95: 12010–15.

Newcomb, R., B. Szoke, A. Palma, G. Wang, X.-H. Chen, W. Hopkins, R. Cong, J. Miller, K. Tarczy-Hornoch,

J. A. Loo, D. J. Dooley, L. Nadasdi, R. W. Tsien, J. Lemos, and G. M. Miljanich (1998) Diversity of neuronal "R"-type calcium currents revealed by a selective peptide antagonist of the class E calcium channel from the venom of the tarantula. *Hysterocrates gigas. Biochemistry*, in press.

Niidome, T., T. Teramoto, Y. Murata, I. Tanaka, T. Seto, K. Sawada, Y. Mori, Y., and K. Katayama (1994) Stable expression of the neuronal BI (class A) calcium channel in baby hamster kidney cells. *Biochem. Biophys. Res. Commun.* 203: 1821–7.

Nilius, B., P. Hess, J. B. Lansman, and R. W. Tsien (1985) A novel type of cardiac calcium channel in ventricular cells. *Nature* 316: 443–6.

Nowycky, M. C., A. P. Fox, and R. W. Tsien (1985) Three types of neuronal calcium channel with different calcium agonist sensitivity. *Nature* 316: 440–3.

Oguro-Okano, M., G. E. Griesmann, E. D. Wieben, S. J. Slaymaker, T. P. Snutch, and V. A. Lennon (1992) Molecular diversity of neuronal-type calcium channels identified in small cell lung carcinoma. *Mayo Clinic Proc.* 67: 1150–9.

Olivera, B. M., G. P. Miljanich, J. Ramachandran, and M. E. Adams (1994) Calcium channel diversity and neurotransmitter release: the ω-conotoxins and ω-agatoxins. *Annu. Rev. Biochem.* 63: 823–67.

Ophoff, R. A., G. M. Terwindt, M. N. Vergouwe, R. van Eijk, P. J. Oefner, S. M. G. Hoffman, J. E. Lamerdin, H. W. Mohrenweiser, D. E. Bulman, M. Ferrari, J. Haan, D. Lindhout, G.-J. B. van Ommen, M. H. Hofker, M. D. Ferrari, and R. R. Frants (1996) Familial hemiplegic migraine and episodic ataxia type-2 are caused by mutations in the Ca^{2+} channel gene CACNL1A4. *Cell* 87: 543–52.

Papazian, D. M., L. C. Timpe, Y. N. Jan, and L. Y. Jan (1991) Alteration of voltage-dependence of Shaker potassium channel by mutations in the S4 sequence. *Nature* 349: 305–10.

Perez-Reyes, E., L. L. Cribbs, A. Daud, A. E. Lacerda, J. Barclay, M. P. Williamson, M. Fox, M. Rees, and J. H. Lee (1998) Molecular characterization of a neuronal low-voltage-activated T-type calcium channel. *Nature* 391: 896–900.

Perez-Reyes, E., X. Y. Wei, A. Castellano, and L. Birnbaumer (1990) Molecular diversity of L-type calcium channels. Evidence for alternative splicing of the transcripts of three non-allelic genes. *J. Biol. Chem.* 265: 20430–6.

Perney, T. M., L. D. Hirning, S. E. Leeman, and R. J. Miller (1986) Multiple calcium channels mediate neurotransmitter release from peripheral neurons. *Proc. Natl. Acad. Sci. USA* 83: 6656–9.

Peterson, B. Z. and W. A. Catterall (1995) Calcium binding in the pore of L-type calcium channels modulates high affinity dihydropyridine binding. *J. Biol. Chem.* 270: 18201–4.

Pragnell. M., M. De Waard, Y. Mori, T. Tanabe, T. P. Snutch, and K. P. Campbell (1994) Calcium channel β-subunit binds to a conserved motif in the I–II cytoplasmic linker of the α_1-subunit [see comments]. *Nature* 368: 67–70.

Protti, D. A. and O. D. Uchitel (1993) Transmitter release and presynaptic Ca^{2+} currents blocked by the spider toxin ω-Aga-IVA. *NeuroReport* 5: 333–6.

Protti, D. A., R. Reisen, T. A. Mackinley, and O. D. Uchitel (1996) Calcium channel blockers and transmitter release at the normal human neuromuscular junction. *Neurology* 46: 1391–6.

Protti, D. A., L. Szczupak, F. S. Scornik, and O. D. Uchitel (1991) Effect of ω-conotoxin GVIA on neurotransmitter release at the mouse neuromuscular junction. *Brain Res.* 557: 336–9.

Randall, A. and R. W. Tsien (1995) Pharmacological dissection of multiple types of Ca^{2+} channel currents in rat cerebellar granule neurons. *J. Neurosci.* 15: 2995–3012.

Randall, A. D. and R. W. Tsien (1997) Contrasting biophysical and pharmacological properties of T-type and R-type calcium channels. *Neuropharmacology* 36: 879–93.

Regehr, W. G. and I. M. Mintz (1994) Participation of multiple calcium channel types in transmisssion at single climbing fiber to Purkinje cell synapses. *Neuron* 12: 605–13.

Reid, C. A., J. M. Bekkers, and J. D. Clements (1998) N- and P/Q-type Ca^{2+} channels mediate transmitter release with a similar cooperativity at rat hippocampal autapses. *J. Neurosci.* 18: 2849–55.

Rettig, J., Z. H. Sheng, D. K. Kim, C. D. Hodson, T. P. Snutch, and W. A. Catterall (1996) Isoform-specific interaction of the α_{1A} subunits of brain Ca^{2+} channels with the presynaptic proteins syntaxin and snap-25. *Proc. Natl. Acad. Sci. USA* 93: 7363–8.

Reuter, H. (1995) Measurements of exocytosis from single presynaptic nerve terminals reveal heterogeneous inhibition by Ca^{2+}-channel blockers. *Neuron* 14: 773–9.

Reuter, H. (1996) Diversity and function of presynaptic calcium channels in the brain. *Curr. Opin. Neurobiol.* 6: 331–7.

Rios, E. and G. Brum (1987) Involvement of dihydropyridine receptors in excitation-contraction coupling in skeletal muscle. *Nature* 325: 717–20.

Roberts, A., S. Perera, B. Lang, A. Vincent, and J. Newsom-Davis (1985) Paraneoplastic myasthenic syndrome IgG inhibis $^{45}Ca^{2+}$ flux in a human small cell carcinoma line. *Nature* 317: 737–9.

Rock, D. M., W. A. Horne, S. J. Stoehr, C. Hashimoto, M. Zhou, R. Cong, A. Palma, D. Hidayetoglu, and J. Offord (1998) Does α_{1E} code for T-type calcium channels? A comparison of recombinant α_{1E} calcium channels with GH3 pituitry T-type and recombinant α_{1E} calcium channels. In: *Low-Voltage-Activated T-type Calcium Channels* (Tsien, R. W., Clozel, J. P., and Nargeot, J., eds.). Aidis Press, Chester, UK, pp. 279–89.

Rosen, L. B. and M. E. Grenberg (1994) Regulation of c-fos and other immediate-early genes in PC12 cells as a model for studying specificity in neuronal signaling. *Molec. Neurobiol.* 7: 203–16.

Rosenberg, R. L., P. Hess, J. P. Reeves, H. Smilowitz, and R. W. Tsien (1986) Calcium channels in planar lipid bilayers: insights into mechanisms of ion permeation and gating. *Science* 231: 1564–6.

Ruth, P., A. Rohrkasten, M. Biel, E. Bosse, S. Regulla, H. E. Meyer, V. Flockerzi, and F. Hofmann (1989) Primary structure of the β subunit of the DHP-sensitive calcium channel from skeletal muscle. *Science* 245: 1115–18.

Sabria, J., C. Pastor, M. V. Clos, A. Garcia, and A. Badia (1995) Involvement of different types of voltage-sensitive calcium channels in the presynaptic regulation of noradrenaline release in rat brain cortex and hippocampus. *J. Neurochem.* 64: 2567–71.

Sakurai, T., J. W. Hell, A. Woppmann, G. P. Miljanich, and W. A. Catterall (1995) Immunochemical identification and differential phosphorylation of alternatively spliced forms of the α_{1A} subunit of brain calcium channels. *J. Biol. Chem.* 270: 21234–42.

Sano, K., K. Enomoto, and T. Maeno (1987) Effects of synthetic ω-conotoxin, a new type Ca^{2+} antagonist, on frog and mouse neuromuscular transmission. *Eur. J. Pharmacol.* 141: 235–41.

Sather, W. A., T. Tanabe, J.-F. Zhang, Y. Mori, M. E. Adams, and R. W. Tsien (1993) Distinctive biophysical and pharmacological properties of class A (BI) calcium channel α_1 subunits. *Neuron* 11: 291–303.

Schafer, W. R. and C. J. Kenyon (1995) A calcium-channel homologue required for adaptation to dopamine and serotonin in *Caenorhabditis elegans*. *Nature* 375: 73–8.

Schiller, J., F. Helmchen, and B. Sakmann (1995) Spatial profile of dendritic calcium transients evoked by action potentials in rat neocortical pyramidal neurones. *J. Physiol.* 487: 583–600.

Schmitt, R., J. P. Clozel, N. Iberg, and F. R. Buhler (1995) Mibefradil prevents neointima formation after vascular injury in rats. Possible role of the blockade of the T-type voltage-operated calcium channel. *Arterioscler. Thromb. Vasc. Biol.* 15: 1161–5.

Schneider, M. F. and W. K. Chandler (1973) Voltage dependent charge movement of skeletal muscle: a possible step in excitation-contraction coupling. *Nature* 242: 244–6.

Schuster, A.,. L. Lacinova, N. Klugbauer, H. Ito, L. Birnbaumer, and F. Hofmann (1996) The IVS6 segment of the L-type calcium channel is critical for the action of dihydropyridines and phenylalkylamines. *EMBO J.* 15: 2365–70.

Scott, V. E., M. De Waard, H. Liu, C. A. Gurnett, D. P. Venzke, V. A. Lennon, and K. P. Campbell (1996) B subunit heterogeneity in N-type Ca^{2+} channels. *J. Biol. Chem.* 271: 3207–12.

Scott, V. E., R. Felix, J. Arikkath, and K. P. Campbell (1998) Evidence for a 95 kDa short form of the α_{1A} subunit associated with the ω-CTx-MVIIC receptor of the P/Q-type Ca^{2+} channels. *J. Neurosci.* 18: 641–7.

Sheng, Z. H., J. Rettig, M. Takahashi, and W. A. Catterall (1994) Identification of a syntaxin-binding site on N-type calcium channels. *Neuron*, 13: 1303–13.

Simmons, M. L., G. W. Terman, S. M. Gibbs, and C. Chavkin (1995) L-type calcium channels mediate dynorphin neuropeptide release from dendrites but not axons of hippocampal granule cells. *Neuron* 14: 1265–72.

Singer, D., M. Biel, I. Lotan, V. Flockerzi, F. Hofmann, and N. Dascal (1991) The roles of the subunits in the function of the calcium channel. *Science* 253: 1553–7.

Sipos, I., K. Jurkat-Rott, C. Harasztosi, B. Fontaine, L. Kovacs, W. Melzer, and F. Lehmann-Horn (1995) Skeletal muscle DHP receptor mutations alter calcium currents in human hypokalemic periodic paralysis myotubes. *J. Physiol.* 483: 299–306.

Smart, T. G., X. Xie, and B. J. Krishek (1994) Modulation of inhibitory and excitatory amino acid receptor ion channels by zinc. *Prog. Neurobiol.* 42: 393–41.

Smith, L. A., X. J. Wang, A. A. Peixoto, E. K. Neumann, L. M. Hall, and J. C. Hall (1996) A drosophila calcium channel α_1 subunit gene maps to a genetic locus associated with behavioral and visual defects. *J. Neurosci*, 16: 7868–79.

Snutch, T. P. and P. B. Reiner (1992) Ca^{2+} channels: diversity of form and function. *Curr. Opin. Neurobiol.* 2: 247–53.

Snutch, T. P., J. P. Leonard, M. M. Gilbert, H. A. Lester, and N. Davidson (1990) Rat brain expresses a heterogeneous family of calcium channels. *Proc. Natl. Acad. Sci. USA* 87: 3391–5.

Soldatov, N. M. (1994) Genomic structure of human L-type Ca^{2+} channel. *Genomics* 22: 77–87.

Soldatov, N. M., A. Bouron, and H. Reuter (1995) Different voltage-dependent inhibition by dihydropyridines of human Ca^{2+} channel splice variants. *J. Biol. Chem.* 270: 10540–3.

Soong, T. W., A. Stea, C. D. Hodson, S. J. Dubel, S. R. Vincent, and T. P. Snutch (1993) Structure and functional expression of a member of the low voltage-activated calcium channel family. *Science* 260: 1133–6.

Spencer, W. A. and E. R. Kandel (1961) Electrophysiology of hippocampal neurons IV: fast potentials. *J. Neurophysiol.* 24: 272–85.

Spruston, N., Y. Schiller, G. Stuart, and B. Sakmann (1995) Activity-dependent action potential invasion and calcium influx into hippocampal CA1 dendrites [see comments]. *Science* 268: 297–300.

Starr, T. V., W. Prystay, and T. P. Snutch (1991) Primary structure of a calcium channel that is highly expressed in the rat cerebellum. *Proc. Natl. Acad. Sci. USA* 88: 5621–5.

Stea, A., S. J. Dubel, M. Pragnell. J. P. Leonard, K. P. Campbell, and T. P. Snutch (1993) A β-subunit normalizes the electrophysiological properties of a cloned N-type Ca^{2+} channel α_1-subunit. *Neuropharmacology* 32: 1103–16.

Stea, A., W. J. Tomlinson, T. W. Soong, E. Bourinet, S. J. Dubel, S. R. Vincent, and T. P. Snutch (1994)

Localization and functional properties of a rat brain α_{1A} calcium channel reflect similarities to neuronal Q- and P-type channels. *Proc. Natl. Acad. Sci. USA* 91: 10576–80.

Stocker, J. W., L. Nadasdi, D. Silva, R. W. Aldrich, and R. W. Tsien (1996) Interaction between N-type Ca^{2+} channel inactivation gating and binding of an ω-conotoxin. *Biophys. J.* 70: A238.

Stuart, G. J. and B. Sakmann (1994) Active propagation of somatic action potentials into neocortical pyramidal cell dendrites. *Nature* 367: 69–72.

Stuart, G. J., H. U. Dodt, and B. Sakmann (1993) Patch-clamp recordings from the soma and dendrites of neurons in brain slices using infrared video microscopy. *Pflugers Arch*, 423: 511-18.

Stuhmer, W., F. Conti, H. Suzuki, X. D. Wang, M. Noda, N. Yahagi, H. Kubo, and S. Numa (1989) Structural parts involved in activation and inactivation of the sodium channel. *Nature* 339: 597–603.

Takahashi, M., M. J. Seagar, J. F. Jones, B. F. Reber, and W. A. Catterall (1987) Subunit structure of dihydropyridine-sensitive calcium channels from skeletal muscle. *Proc. Natl. Acad. Sci. USA* 84: 5478–82.

Takahashi, T. and A. Momiyama (1993) Different types of calcium channels mediate central synaptic transmission. *Nature* 366: 156–8.

Talley, E. M., L. L. Cribbs, J.-H. Lee, A. Daud, E. Perez-Reyes, and D. A. Bayliss (1998) CNS distribution of three members of a novel gene family encoding low voltage-activated (T-type) calcium channels. *Soc. Neurosci. Abs.* 24: 1823.

Tanabe, T., B. A. Adams, S. Numa, and K. G. Beam (1991) Repeat I of the dihydropyridine receptor is critical in determining calcium channel activation kinetics. *Nature* 352: 800–3.

Tanabe, T., K. G. Beam, B. A. Adams, T. Niidome, and S. Numa (1990) Regions of the skeletal muscle dihydropyridine receptor critical for excitation-contraction coupling. *Nature* 346: 567–9.

Tanabe, T., K. G. Beam, J. A. Powell, and S. Numa (1988) Restoration of excitation-contraction coupling and slow calcium current in dysgenic muscle by dihydropyridine receptor complementary DNA. *Nature* 336: 134–9.

Tanabe, T., H. Takeshima, A. Mikami, V. Flockerzi, H. Takahashi, K. Kangawa, M. Kojima, H. Matsuo, T. Hirose, and S. Numa (1987) Primary structure of the receptor for calcium channel blockers from skeletal muscle. *Nature* 328: 313–18.

Tang, S., G. Mikala, A. Bahinski, A. Yatani, G. Varadi, and A. Schwartz (1993) Molecular localization of ion selectivity sites within the pore of a human L-type cardiac calcium channel. *J. Biol. Chem.* 268: 13026–9.

Toth, P. T., V. P. Bindokas, D. Bleakman, W. F. Colmers, and R. J. Miller (1993) Mechanism of presynaptic inhibition by neuropeptide Y at sympathetic nerve terminals. *Nature* 364: 635–9.

Tottene, A., A. Moretti, and D. Pietrobon (1996) Functional diversity of P-type and R-type calcium channels in rat cerebellar neurons. *J. Neurosci.* 16: 6353–63.

Tsien, R. W. and R. Y. Tsien (1990) Calcium channels, stores, and oscillations. *Annu. Rev. Cell Biol.* 6: 715–60.

Tsien, R. W. (1998) Key clockwork component cloned. *Nature* 391: 839–41.

Tsien, R. W., P. T. Ellinor, and W. A. Horne (1991) Molecular diversity of voltage-dependent Ca^{2+} channels. *Trends Pharmacol. Sci.* 12: 349–54.

Tsien, R. W., A. P. Fox, P. Hess E. W. McCleskey, B. Nilius, M. C. Nowycky, and R. L. Rosenberg (1987a) Multiple types of calcium channel in excitable cells. *Soc. Gen. Physiol. Ser.* 41: 167–87.

Tsien, R. W., P. Hess, E. W. McCleskey, and R. L. Rosenberg (1987b) Calcium channels: mechanisms of selectivity, permeation, and block. *Annu. Rev. Biophys. Biophys. Chem.* 16: 265–90.

Usowicz, M. M., M. Sugimori, B. Cherksey, and R. Llinás (1992) P-type calcium channels in the somata and dendrites of adult cerebellular Purkinje cells. *Neuron* 9: 1185–99.

Varadi, G., P. Lory, D. Schultz, M. Varadi, and A. Schwartz (1991) Acceleration of activation and inactivation by the β subunit of the skeletal muscle calcium channel. *Nature* 352: 159–62.

Vassort, G. and J. Alvarez (1994) Cardiac T-type calcium current: pharmacology and roles in cardiac tissues. *J. Cardiovasc. Electrophysiol.* 5: 376–93.

Wakamori, M., T. Niidome, D. Furutama, T. Furuichi, K. Mikoshiba, Y. Fujita, I. Tanaka, K. Katayama, A. Yatani, A. Schwartz et al. (1994) Distinctive functional properties of the neuronal BII (class E) calcium channel *Receptors Channels* 2: 303–14.

Welling, A., Y. W. Kwan, E. Bosse, V. Flockerzi, F. Hofmann, and R. S. Kass (1993) Subunit-dependent modulation of recombinant L-type calcium channels. Molecular basis for dihydropyridine tissue selectivity. *Circ. Res.* 73: 974–80.

Wessler, I., D. J. Dooley, H. Osswald, and F. Schlemmer (1990) Differential blockade by nifedipine and ω-conotoxin GVIA of α_1-and β_1-adrenoceptor-controlled calcium channels on motor nerve terminals of the rat. *Neurosci. Lett.* 108: 173–8.

Westernbroek, R. E., M. K. Ahlijanian, and W. A. Catterall (1990) Clustering of L-type Ca^{2+} channels at the base of major dendrites in hippocampal pyramidal neurons. *Nature* 347: 281–4.

Westenbroek, R. E., J. W. Hell, C. Warner, S. J. Dubel, T. P. Snutch, and W. A. Catterall (1992) Biochemical properties and subcellular distribution of an N-type calcium channel α_1 subunit. *Neuron* 9: 1099–115.

Westenbroek, R. E., T. Sakurai, E. M. Elliott, J. W. Hell, T. V. Starr, T. P. Snutch, and W. A. Catterall (1995) Immunochemical identification and subcellular distribution of the α_{1A} subunits of brain calcium channels. *J. Neurosci.* 15: 6403–18.

Wheeler, D. B., A. Randall, and R. W. Tsien (1994) Roles of N-type and Q-type Ca^{2+} channels in supporting hippocampal synaptic transmission. *Science* 264: 107–11.

Wheeler, D. B., A. Randall, and R. W. Tsien (1996) Changes in action potential duration alter reliance of excitatory synaptic transmission on multiple types of Ca^{2+} channels in rat hippocampus. *J. Neurosci.* 16: 2226–37.

Williams, M. E., P. F. Brust, D. H. Feldman, S. Patthi, S. Simerson, A. Maroufi, A. F. McCue, G. Veliçelebi, S. B. Ellis, and M. M. Harpold (1992a) Structure and functional expression of an ω-conotoxin-sensitive human N-type calcium channel. *Science* 257: 389–95.

Williams, M. E., D. H. Feldman, A. F. McCue, R. Brenner, G. Veliçelebi, S. B. Ellis, and M. M. Harpold (1992b) Structure and functional expression of α_1, α_2, and β subunits of a novel human neuronal calcium channel subtype. *Neuron* 8: 71–84.

Williams, M. E., L. M. Marubio, C. R. Deal, M. Hans, P. F. Brust, L. H. Philipson, R. J., Miller, E. C. Johnson, M. M. Harpold, and S. B. Ellis (1994) Structure and functional characterization of neuronal α 1E calcium channel subtypes. *J. Biol. Chem.* 269: 22347–57.

Witcher, D. R., M. De Waard, H. Liu, M. Pragnell, and K. P. Campbell (1995) Association of native Ca^{2+} channel β subunits with the α1 subunit interaction domain. *J. Biol. Chem.* 270: 18088–93.

Wong, R. K., D. A. Prince, and A. I. Basbaum (1979) Intradendritic recordings from hippocampal neurons. *Proc. Natl. Acad. Sci. USA* 76: 986–90.

Wu, L. G., J. G. Borst, and B. Sakmann (1998) R-type Ca^{2+} currents evoke transmitter release at a rat central synapse. *Proc. Natl. Acad. Sci. USA* 95: 4720–5.

Yang, J., P. T. Ellinor, W. A. Sather, J.-F. Zhang, and R. W. Tsien (1993) Molecular determinants of Ca^{2+} selectivity and ion permeation in L-type Ca^{2+} channels [see comments]. *Nature* 366: 158–61.

Yang, N. and R. Horn (1995) Evidence for voltage-dependent S4 movement in sodium channels. *Neuron* 15: 213–18.

Yokoyama, C. T., R. E. Westenbroek, J. W. Hell, T. W. Soong, T. P. Snutch, and W. A. Catterall (1995) Biochemical properties and subcellular distribution of the neuronal class E calcium channel α1 subunit. *J. Neurosci.* 15: 6419–32.

Yoshida, K., J. Imaki, H. Matsuda, and M. Hagiwara (1995) Light-induced CREB phosphorylation and gene expression in rat retinal cells. *J. Neurochem.* 65: 1499–504.

Yuste, R., M. J. Gutnick, D. Saar, K. R. Delaney, and D. W. Tank (1994) Ca^{2+} accumulations in dendrites of neocortical pyramidal neurons: an apical band and evidence for two functional compartments. *Neuron* 13: 23–43.

Zhang, J. F., P. T. Ellinor, R. W. Aldrich, and R. W. Tsien (1994) Molcular determinants of voltage-dependent inactivation in calcium channels. *Nature* 372: 97–100.

Zhang, J.-F., A. D. Randall, P. T. Ellinor, W. A. Horne, W. A. Sather, T. Tanabe, T. L. Schwarz, and R. W. Tsien (1993) Distinctive pharmacology and kinetics of cloned neuronal Ca^{2+} channels and their possible counterparts in mammalian CNS neurons. *Neuropharmacology* 32: 1075–88.

Zhu, Y. and S. R. Ikeda (1993) Adenosine modulates voltage-gated Ca^{2+} channels in adult rat sympathetic neurons. *J. Neurophysiol.* 70: 610–20.

Zühlke, R. D. and H. Reuter (1998) Ca^{2+}-sensitive inactivation of L-type Ca^{2+} channels depends on multiple cytoplasmic amino acid sequences of the α_{1C} subunit. *Proc. Natl. Acad. Sci. USA* 95: 3287–94.

9

Intracellular Calcium Channels

Teiichi Furuichi

Takayuki Michikawa

Katsuhiko Mikoshiba

The calcium ion (Ca^{2+}) plays an important role in physiological function in the cell. Concentration of Ca^{2+} outside the cell is on the order of 10^{-3} M, while that inside the cell is 10^{-7} M, resulting in the great difference in the concentration of Ca^{2+} across the plasma membrane. There are various mechanisms that regulate the intracellular Ca^{2+} concentration. The molecular machines that regulate the cytosolic Ca^{2+} concentrations across the plasma membrane are the Ca^{2+}-entry channels, such as voltage-operated Ca^{2+} channel (VOCC) and receptor-operated Ca^{2+} channel (ROCC), and the Ca^{2+}-excluding molecules, such as Ca^{2+}-ATPase working as Ca^{2+} pump and Na^+–Ca^{2+} exchanger. Gap junction also appears to be important for communication between cells, through which Ca^{2+} moves. In addition, intracellular Ca^{2+} is sequestered into Ca^{2+}-storing organelles, such as the endoplasmic reticulum, by action of a specific Ca^{2+} pump, which also contributes to lowering the cytosolic Ca^{2+} concentrations.

In this intracellular milieu of extremely low Ca^{2+}, which is brought by the strict control mechanisms described above, the intracellular Ca^{2+} signaling is well-documented to be an indispensable cellular signaling system in a variety of organisms, and is triggered by a transient rise of cytosolic Ca^{2+} via either the entry across the plasma membrane or the release from the intracellular store sites. There are pharmacologically two types of Ca^{2+} release store sites inside cells; inositol 1,4,5-trisphosphate (IP_3)-induced Ca^{2+} release (IICR) pools and Ca^{2+}-induced Ca^{2+} release (CICR) pools. The IP_3 receptor (IP_3R) is a key molecule for IICR and the ryanodine receptor (RyR) for CICR. Both receptor channels are localized intracellularly and release Ca^{2+} from the Ca^{2+} pools, and thus are called *intra-cellular Ca^{2+}-release channels*. IP_3 is a second messenger produced through phosphoinositide hydrolysis by phospholipase C (PLC) which is activated through G-protein-coupled or tyrosine kinase-coupled signaling pathways stimulated by various extracellular signals, such as hormones, growth factors, neurotransmitters, etc. (Fig. 9.1). IP_3 (Fig. 9.2) binds to and activates IP_3R, and induces release of Ca^{2+} from the pools, resulting in signal conversion from IP_3 to Ca^{2+}. RyR is activated by increase in cytosolic Ca^{2+} (or mechanical interaction with the dihydropiridine [DHP]-sensitive Ca^{2+} channel), leading to amplification of Ca^{2+} signals.

The genes that encode these receptors have recently been cloned. It was found that these receptors themselves are Ca^{2+}-release channels and their primary sequences are very different from the Ca^{2+} channel on the plasma membrane. However, IP_3R and RyR show some homologies in sequences of a part of their putative channel structures, as well as in properties of the channels, and they are localized on analogous Ca^{2+}-storing organelles, the smooth endoplasmic reticulum and sarcoplasmic reticulum, respectively. Importance of the Ca^{2+} release from these stores has been presented recently. These Ca^{2+} stores participate in the intracellular dynamics of Ca^{2+} signaling, including complex signaling behaviors such as Ca^{2+} waves and Ca^{2+} oscillations, resulting in various physiological functions such as fertilization, cell proliferation and differentiation, muscle contraction, secretion, immuno-response, and memory mechanisms like long-term depression and long-term potentiation.

We describe here the structural and functional aspects of these intracellular Ca^{2+}-release channels, mainly IP_3Rs.

IP$_3$/Ca^{2+} Signaling Pathway

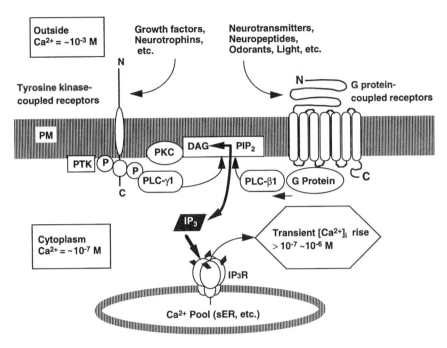

Figure 9.1 IP$_3$/Ca^{2+} signaling pathway. Many extracellular signals stimulate phosphoinositide turnover through the activation of G protein- or tyrosine kinase-coupled receptors on the plasma membrane (PM) followed by the activation of phospholipase C (PLC). As a result, two second messengers, inositol 1,4,5-trisphosphate (IP$_3$), which induces Ca^{2+} release via IP$_3$ receptor (IP$_3$R)-Ca^{2+} release channel from intracellular stores, and diacylglycerol (DAG), which activates protein kinase C (PKC), are produced. PTK, protein tyrosine kinase; PLC-γ1, phospholipase C-γ1; PIP$_2$, phosphatidylinositol 4,5-bisphosphate; PLC-β1, phospholipase C-β1; G-protein, heteromeric GTP-binding protein; sER, smooth endoplasmic reticulum.

IP$_3$R Family

Until the late 1980s, knowledge of the basic features of IP$_3$-induced Ca^{2+} release, as well as [^{3}H]-IP$_3$ binding, had already accumulated enough to start biochemical, molecular biological, and pharmacological studies of IP$_3$R molecules (for a review, see Berridge and Irvine, 1989). To get a clue to uncovering the substance of a vague IP$_3$R molecule at that time, important studies revealed that (1) high [^{3}H]-IP$_3$ binding sites were found in cerebellum (Worley et al., 1987a, 1987b, 1989); (2) the IP$_3$-binding protein was highly purified from cerebellum (Supattapone et al., 1988b) by utilizing [^{3}H]-IP$_3$ binding activity as a guidepost; (3) at single-channel levels, IP$_3$-induced Ca^{2+} currents were recorded from the cerebellar endoplasmic reticulum (ER), which was thought to contain IP$_3$-sensitive Ca^{2+} pools, reconstituted into planar lipid bilayers (Ehrlich and Watras, 1988);

and (4) another intracellular Ca^{2+} release channel, ryanodine receptor (RyR), present in the sarcoplasmic reticulum of muscles, was molecularly cloned in 1989 (Takeshima et al., 1989). Independently of these studies, IP$_3$R protein was originally characterized, by several groups, as a membrane glyco- and phosphoprotein, plentiful in the cerebellum, and variously termed P$_{400}$ (Mallet et al., 1976; Mikoshiba et al., 1979; 1985), PCPP-260 (Walaas et al., 1986), or GP-A (Groswald and Kelly, 1984).

Direct clues to molecularly clone IP$_3$R were gotten from the immunological studies of P$_{400}$ (Maeda et al., 1988, 1989) and the differential subtraction study of PCD6 gene (Nordquist et al., 1988), in both of which interestingly hereditary cerebellar ataxic mutant mice specifically degenerating Purkinje cells (e.g., *Purkinje-cell-degeneration* [*pcd*], *staggerer* [*sg*], *nervous* [*nr*]) or granule cells (e.g., *weaver* [*wv*]) were utilized as a model system. Thus, the complementary

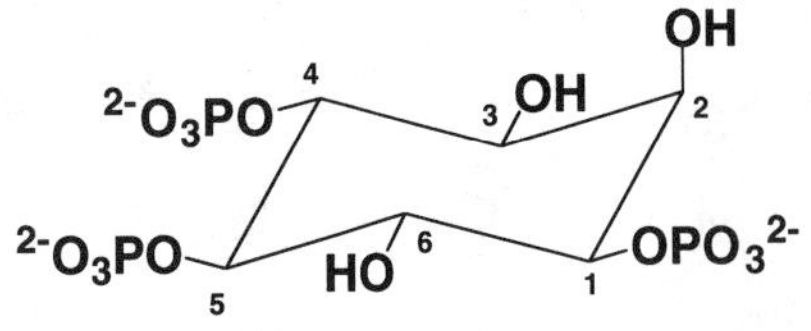

D-*myo*-inositol 1,4,5-trisphosphate
(IP$_3$, or InsP$_3$)

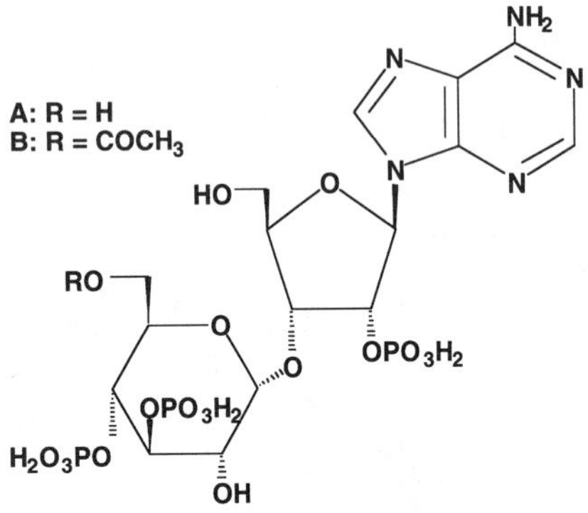

A: R = H
B: R = COCH$_3$

Adenophostin A, B

Figure 9.2 Structure of IP$_3$ and adenophostin. Both D-myo-inositol 1,4,5-trisphosphate (IP$_3$) and adenophostin (A and B) act on the same site of IP$_3$R, leading to activation of its channel. IP$_3$ is a second messenger produced through signaling pathway of phosphoinositide turnover, while adenophostin is a fungal metabolite and is more potent in IICR than IP$_3$.

DNA (cDNA) of cerebellar IP$_3$R was cloned not only by immunoscreening with anti-P$_{400}$ (IP$_3$R) monoclonal antibodies (Furuichi et al., 1989a, 1989b), but also by differential subtraction analysis of cerebellar cDNAs between wild-type and *pcd* mutant mice (Nordquist et al., 1988; Mignery et al., 1989, 1990). A notable feature of the molecular cloning studies was that a part of the putative channel domain of the cloned IP$_3$R shares a significant similarity to that of RyR1 (Furuichi et al., 1989b; Mignery et al., 1989; Furuichi and Mikoshiba, 1995), suggesting that both of the intracellular Ca^{2+} release channels, IP$_3$Rs and RyRs, evolutionarily belong to the same gene family, even though their channels are activated in different manners.

IP$_3$R proteins were purified from rat (Supattapone et al., 1988b), mouse (Maeda et al., 1988, 1989, 1990, 1991), and bovine (Hingorani and Agnew, 1991, 1992) cerebellum by affinities for lectin and heparin. Similarly, IP$_3$Rs were also purified from other tissues: bovine aorta smooth muscles (Chadwick et al., 1990), rat vas deferens (Mourey et al., 1990), and *Xenopus laevis* oocytes (Parys et al., 1992). There are a few differences in the properties of the purified IP$_3$Rs; e.g., while specific binding to IP$_3$ (K_d) was ≈ 100 nM in the cerebellar IP$_3$R (Supattapone et al., 1988b; Maeda et al., 1991), it was ≈ 2.4 nM in aorta smooth muscles (Chadwick et al., 1990). These data suggested the presence of multiple IP$_3$R isoforms having different affinities

for IP$_3$. The PCR study also confirmed the existence of additional types (type 2, 3, and 4) in mouse tissues (Ross et al., 1992).

By means of the sequence homology to the cerebellar IP$_3$R cDNA, the complete sequences of three distinct receptor types (types 1, 2, and 3) have been cloned in mammals, indicating the presence of an IP$_3$R gene family (Furuichi et al., 1994b) (Fig. 9.3). Within this family, the type 1 receptor (IP$_3$R1), first cloned, is the best characterized (Ferris and Snyder, 1992; Furuichi et al., 1992; Mikoshiba, 1993, 1997), and is the predominant type in brain. The complete sequences of IP$_3$R1 cDNAs have been cloned from mouse (Furuichi et al., 1989b), rat (Mignery et al., 1990), human (Yamada et al., 1994; Harnick et al., 1995), and *Xenopus laevis* (Kume et al., 1993), so far. The type 2 IP$_3$R (IP$_3$R2) has been cloned from rat (Südhof et al., 1991), human (Yamamoto-Hino et al., 1994), and *Drosophila melanogaster* (probably a homolog of mammalian type 2) (Hasan and Rosbash, 1992; Yoshikawa et al., 1992). The type 3 IP$_3$R (IP$_3$R3) has been cloned from rat (Blondel et al., 1993) and human (Maranto, 1994; Yamamoto-Hino et al., 1994). Two more types, type 4 (IP$_3$R4) (Ross et al., 1992) and type 5 (IP$_3$R5) (De Smedt et al., 1994), have been found by PCR, although their complete sequences are as yet unknown. It is still unclear whether IP$_3$R4 and IP$_3$R5 are independent isoforms, since the cloned sequences of both are extremely homologous with IP$_3$R2.

IP3R FAMILY

IP3R1: mouse and rat (SI+/SII+), 2749 a.a.; human (SI-/SII-) 2695 a.a.; *Xenopus* (SI-/SII-), 2693 a.a.

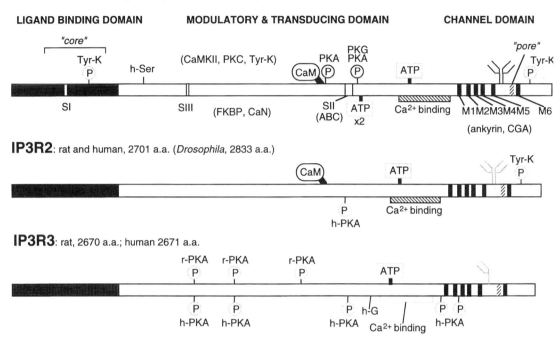

Figure 9.3 Structure of the IP$_3$R family. Functional domains and sites are depicted. The defined sites and putative sites are identified by solid and light symbols, respectively. Putative membrane-spanning domains (MSDs: M1, M2, M3, M4, M5, and M6) and putative "pore" segments are shown by solid and shaded vertical bars, respectively. Branched bars in the channel domain represent *N*-glycosylation sites. Alternative splicing regions are shown, SI, SII, and SIII. h-Ser, human-specific insertion of Ser residue; h-G, human-specific insertion of Gly residue. Phosphorylation sites: PKA, cAMP-dependent protein kinase; PKG, cGMP-dependent protein kinase; PKC, protein kinase C; CaMKII, calmodulin-dependent protein kinase II; Tyr-K, tyrosine kinase, r-PKA and h-PKA represent PKA sites in rat and human receptors, respectively. CaM, calmodulin-binding site; ATP, ATP-binding site; FKBP, FK506-binding protein. CaN, calcineurin; CGA, chromogranin A.

These multiple IP$_3$R types may be involved in differential IP$_3$/Ca^{2+} signaling in various cell types. The molecular cloning of these IP$_3$Rs has contributed to shed light on the structure and function of IP$_3$-induced Ca^{2+}-release channels, most of which constitute the framework for the contents in this chapter; namely, we will try to describe all aspects of IP$_3$Rs, such as biochemical, pharmacological, cytological properties, etc., through the structure–function relationships. Aside from IP$_3$R1, however, little is known about the structures and functions of IP$_3$R2 and IP$_3$R3. In addition, some of the physiological aspects which are still unclear from this viewpoint are outside the scope of this chapter.

IP$_3$R1

The IP$_3$R1 cDNAs were cloned from mouse (Furuichi et al., 1989a, 1989b), rat (Mignery et al., 1989, 1990), human (Yamada et al., 1994; Harnick et al., 1995), and frog (*Xenopus laevis*) (Kume et al., 1993). Cloned IP$_3$R1 from both mouse cerebellum and rat brain is 2749 amino acids in size; however, it comprises a heterogeneous group of receptor subtypes arising from alternative splicing (for details, see "Alternative Splicing Subtypes of IP$_3$R1").

Structure and Function of IP$_3$R1

IP$_3$R1 is a 2749 amino acid polypeptide (calculated $M_r = 313$ kDa) and is structurally divided into three parts: the large N-terminal cytoplasmic arm (-83% of the receptor molecule), the putative six membrane-spanning domains (MSDs) clustered near the C-terminus, and the short C-terminal cytoplasmic tail (5.3%) (Furuichi et al., 1989b; Mignery et al., 1990). Functionally, IP_3R1 is composed of three domains: ligand-binding (or IP_3-binding) domain, modulatory (or transducing) domain, and channel domain. There are subforms of IP_3R1 due to alternative splicing in the SI, SII, and SIII regions (see "Alternative Splicing Subtypes of IP_3R1"). In this chapter, unless otherwise noted, "IP_3R1" means a typical subtype carrying the SI and SII regions but not the SIII region, named $IP_3R1SI+/SII+/SIII-$, or simply IP_3R1SI/SII.

Ligand-Binding Domain   The *ligand-binding domain* is also called the *IP_3-binding domain*. The affinity for IP_3 (K_d) is $\approx 80-100$ nM in purified cerebellar IP_3R1 (Suppattapone et al., 1988b; Maeda et al., 1990), while the expressed full-length (Miyawaki et al., 1990) and the truncated (Südhof et al., 1991) IP_3R1 exhibited $K_d \approx 20$ nM and $K_d \approx 90$ nM, respectively (the difference in this range of K_d values seemed to be dependent upon the assay conditions used, and relative values—as compared to the K_d of cerebellar microsomes—were always constant [Miyawaki et al., 1990]). Since IP_3 has three phosphate moieties in the equatorial positions 1, 4, and 5 of the inositol ring (Fig. 9.2), it was envisaged that the IP_3-binding site has a pocket of positive charges to facilitate ionic interaction with the negative charges on the three phosphate groups (Nahorski and Potter, 1989; Wilcox et al., 1994). In addition, the specific Arg-modifying reagent irreversibly blocked IP_3 binding, suggesting the presence of Arg at the recognition site for IP_3 binding (O'Rourke and Feinstein, 1990). By deletion mutation analyses, the N-terminal tip (650 amino acids) was shown to possess IP_3-binding activity (Mignery and Südhof, 1990; Miyawaki et al., 1991). This region certainly contains many positively charged basic amino acids conserved in the family, and binds heparin (Miyawaki et al., 1991), which is a competitive inhibitor for IP_3 binding (Worley et al., 1987a) (see "IP_3 Binding"). Within this region, residues 476–501 of rat IP_3R1 were specifically labeled by photoaffinity ligands (Mourey et al., 1993). The N-terminal 576 amino acids, when fused to glutathione S-transferase (GST), were found to bind IP_3 whereas further N- or C-terminal deletions abolished all specific binding (Newton et al., 1994). More detailed mutation analyses at the amino acid levels have shown that the residues 225–578 of mouse IP_3R1 are sufficient and close enough to the minimum region for specific IP_3 binding, forming an *IP_3-binding core*, and that the three basic residues, Arg-265, Lys-508, and Arg-511, are critical for the

specific binding (Yoshikawa et al., 1996). Within the core region, the SI splicing region is unlikely to be part of the binding site, because the SI+ and SI− isoforms display the same binding properties (Newton et al., 1994; Liévremont et al., 1996). Newly found IP_3R agonists, referred to as *adenophostin A and B,* (Fig. 9.2) also have three phosphate groups, and are 100-fold more potent than IP_3 in terms of Ca^{2+} release (Takahashi et al., 1993) (see "IP_3 Binding").

Modulatory Domain   The middle portion of the receptor possesses various elements for modulation of IP_3R1 functions, and thus is termed the *modulatory domain*. This portion is often called the *transducing domain* (or "coupling domain") since it is presumed to be involved in the transduction (or coupling) of IP_3-binding information to the channel opening (Mignery and Sudhof, 1990). The modulatory domain contains binding sites for various modulators, such as Ca^{2+} ([Mignery et al., 1992]; residues 2124–2146 [Siehnaert et al., 1996]); residues 2124–2146 [Sienaert et al., 1996]), calmodulin (CaM) (Maeda et al., 1991) (1564–1585 of mouse IP_3R1 [Yamada et al., 1995]), and ATP (Maeda et al., 1991; Ferris et al., 1990a) (1773–1778, 1775–1780, and 2016–2021 of mouse $IP_3R1SII+$) and, by alternative splicing, the creation of an additional possible site (1687–1691/1732 in the SII− [Ferris and Snyder, 1992]) (see "Alternative Splicing Subtypes of IR_3R1") , two sites phosphorylated by cAMP-dependent protein kinase (PKA) ([Supattapone et al., 1988a; Yamamoto et al., 1989]; Ser-1588 and Ser-1755 of mouse and rat IP_3R1 [Danoff et al., 1991; Ferris et al., 1991]), one site phosphorylated by cGMP-dependent protein kinase (PKG) (Ser-1755 of mouse and rat IP_3R1 [Koga et al., 1994; Rooney et al., 1996]; Komalavilas and Lincoln, 1994; 1996]), and potential sites phosphorylated by Ca^{2+}/CaM-dependent protein kinase II (CaMKII) (Yamamoto et al., 1989; Ferris et al., 1990b), protein kinase C (PKC) (Ferris et al., 1990b), and tyrosine kinase (Harnick et al., 1995). Thus, in addition to phosphorylation by CaMKII which is a typical downstream target activated through an $IP_3 \rightarrow Ca^{2+} \rightarrow CaM$ signaling flow, other second-messenger transduction cascades that are linked to protein phosphorylation, i.e., cAMP (via PKA), DAG (via PKC), and cGMP (via PKG), also converge on this modulatory domain, resulting in a fine-tuning of IICR signaling. Autophosphorylation of IP_3R1 has been reported (Ferris et al., 1992a), but little is known about this activity. It has been shown that immunosuppressant FK506-binding protein (FKBP) binds IP_3R1 at residues 1400–1401, a Leu–Pro dipeptide (Cameron et al., 1995a, 1995b, 1997), and then interacts with calcineurin (Cameron et al., 1995b) leading to

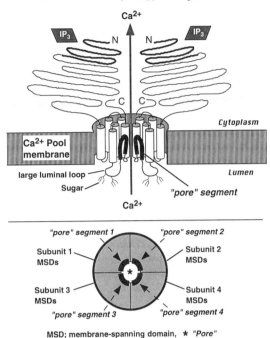

Transmembrane topology of IP₃R-channel

Figure 9.4 Transmembrane topology of IP$_3$R channel and tetrameric structure of channel domain. IP$_3$R has been proposed to traverse six times the membrane of intracellular Ca^{2+} pools such as the endoplasmic reticulum. The large N-terminal cytoplasmic region contains the ligand (IP$_3$)-binding domain, and the modulatory and transducing domain where many modulatory signals, such as phosphorylation, Ca^{2+}, calmodulin, ATP, FKBP, etc., converge. Putative six membrane-spanning domains (MSDs: M1, M2, M3, M4, M5, and M6) cluster near the C-terminus. Putative "pore"-forming segment is located between M5 and M6. The large luminal loop between the M5 and "pore" segment contains the *N*-glycosylation sites (two sites in IP$_3$R1 and IP$_3$R2, one site in IP$_3$R3). The short C-terminal region is also faced into the cytoplasm. An IP$_3$R channel is constituted by an assembly of four subunits, in either homotetrameric or heterotetrameric fashions. Like other ion channels, it has been proposed that this tetrameric assembly results in formation of an ion conduction pathway encircled by four "pore"-forming segments.

Tetrameric structure of channel domain

modification of IP$_3$R functions. In addition, NADH that stimulates IP$_3$R activity seems to act at the same site as ATP (Kaplin et al., 1996).

Channel Domain Six membrane-spanning domains (MSDs [M1–M6], residues 2276–2589 of mouse IP$_3$R1; M1, 2276–2294; M2, 2308–2326; M3, 2352–2372; M4, 2391–2407; M5, 2440–2462; M6, 2570–2589 [Yoshikawa et al., 1992]) clustered near the C-terminus are thought to constitute the channel domain. Between the M5 and M6, there is a large luminal loop (residues 2463–2528 of mouse IP$_3$R1) containing two N-linked glycosylation sites (Asn-2475 and Asn-2503 of mouse IP$_3$R1 [Michikawa et al., 1994]), one putative *pore*-forming sequence (21 residues: 2529–2549 of mouse IP$_3$R1 [Michikawa et al., 1994]) which is thought to form an ion conduction pathway, and a small luminal loop (residues 2550–2569 of mouse IP$_3$R1) (Fig. 9.4). This proposed pore-forming segment of IP$_3$R1 is likely to be compatible with the H5 or SS1–SS2 in the ion channel superfamily which includes the voltage-sensitive Ca^{2+}, Na$^+$, and K$^+$ channels and the nucleotide-gated ion channels (Jan and Jan, 1992). Interestingly, the last two MSDs (M5 and M6), the putative pore segment, and the C-terminal cytoplasmic tail following M6 are homologous with those of

RyRs (Furuichi et al., 1989b; Mignery et al., 1989; Furuichi and Mikoshiba, 1995). This suggests that these homologous regions between IP$_3$R and RyR play an indispensable role in common functions as intracellular Ca^{2+}-release channels. A putative large luminal loop between the M5 and the pore is often called the *variable region* in the family but commonly has a concentration of acidic amino acids, Glu and Asp. Thus, this acidic luminal loop was hypothesized to have an affinity for luminal Ca^{2+} and somehow to concentrate releasable Ca^{2+} near the pore (Yoshikawa et al., 1992). In fact, residues 2463–2528 in this luminal loop have turned out to be capable of binding to ^{45}Ca^{2+} (Sienaert et al., 1996).

It was shown that increased Ca^{2+} induced selective degradation of IP$_3$R1 in cerebellar microsomes with Ca^{2+}-dependent protease calpain, producing a C-terminal 95 kDa polypeptide (Magnusson et al., 1993). Similarly, limited digestion of cerebellar microsomes with trypsin produced a 94 kDa polypeptide that retained the MSDs, glycosylation sites, and immunoreactivity to anti-C-terminus antibody, but not the Ser-1755 for PKA phosphorylation, thereby corresponding to a polypeptide derived from a proteolytic cleavage between Ser-1755 and the M1 to the C-terminus (Joseph et al., 1995a). Therefore, the channel domain repeatedly traversing the membranes appears to form a specific folding structure relatively resistant to degradation. Moreover, this C-terminal polypeptide was shown to interact with the N-terminal polypeptide (68 kDa) produced by limited digestion with trypsin, suggesting that the *head-to-tail* interaction may couple the ligand binding with the channel gating (Joseph et al., 1995a). The monoclonal antibody mAb18A10, which recognizes the C-terminus of IP₃R1 including at least the residues 2736–2747 (Nakade et al., 1991), blocked IICR activity but not IP₃binding (Nakade et al., 1991, 1994; Miyazaki et al., 1992), thereby suggesting an important role of the C-terminal region in the coupling activity. In addition, the mutant IP₃R1 with an N-terminal deletion of 418 amino acids, which lost a part of the ligand-binding domain and thus lost the putative head-to-tail interaction, was capable of assembly with other wild-type subunits to yield a tetrameric channel, and exhibited increase in the sensitivity of IICR channel activity (Fischer et al., 1994). It was, thus, hypothesized that the increased sensitivity was due to the N-terminal deletion which caused a permanent activation of the mutant subunit in a tetramer, independently of IP₃ binding (Fischer et al., 1994).

Tetramer Complex The functional IP₃R/Ca²⁺-release channel is thought to be a complex consisting of four IP₃R subunits, a *tetramer* (Chadwick et al., 1990; Mignery and Südhof, 1990; Maeda et al., 1991) (Fig. 9.4), which under the EM has been observed to exist in the form of a square, ~ 25 nm on each side (Maeda et al., 1990), having a pinwheel appearance (possibly a channel pore) with 4-fold symmetry (Chadwick et al., 1990). The MSDs are apparently required for this intermolecular association (Miyawaki et al., 1991; Joseph et al., 1997; Sayers et al., 1997). In vivo ³⁵S-labeling experiments using WB rat liver epithelial cells showed rapid tetramerization of newly synthesized IP₃R proteins during even a short (10 min) pulse labeling period, suggesting that the tetramerization process is very rapid (Joseph, 1994). It has been shown that IP₃R1 can form heterotetramers with the other types, IP₃R2 and IP₃R3, suggesting great diversities of IICR channels being composed of four subunits, each of which may have distinct properties in terms of IP₃ binding, channel opening, and their modulation (Joseph et al., 1995b, Monkawa et al., 1995; Wojcikiewicz and He, 1995).

The function of the C-terminal cytoplasmic tail (159 amino acids) remains elusive, although the antibody that recognizes the tip of the tail blocked IP₃-induced Ca²⁺-release activity (Nakade et al., 1991; Miyazaki et al., 1992), and the tail appears to have something to do with functional channel formation (Miyawaki et al., 1991).

Alternative Splicing Subtypes of IP₃R1

IP₃R1 has three distinct sites for alternative splicing. Alternative splicing results in variations of the IP₃R1 form in the SI (Mignery et al., 1990; Nakagawa et al., 1991a), SII (Danoff et al., 1991; Nakagawa et al., 1991a) and SIII regions (Nucifora et al., 1995). In all, there are eight alternative splicing patterns: two in the SI (SI+ and SI− subtypes), four in the SII (SIIABC+ [simply expressed as SII+], SIIABC− [simply termed SII−], SIIBC−, and SIIB− subtypes), and two in the SIII (SIII+ and SIII−), which probably occur in combination, resulting in 16 possible variations in IP₃R1 splicing subtypes. In vivo roles of these alternative variations are largely unknown as yet, except that it was shown that the SII splicing affected phosphorylation of IP₃R1 by the catalytic subunit of PKA in vitro (Danoff et al., 1991).

SI The SI region (15 amino acids, 318-VDPDFEEECLEFQPS-332 of mouse and rat IP₃R1 [Mignery et al., 1990; Nakagawa et al., 1991a]) locates in the middle of the ligand-binding domain. Thus, the SI splicing (SI+, long type; SI−, short type) might alter the IP₃ binding property, although no significant difference in ordinary [³H]-IP₃ binding assay has been observed so far (Newton et al., 1994; Lièvremont et al., 1996). However, under in vivo cytosolic conditions it may influence a sensitivity of IP₃, e.g., transducing efficiency. The SI splicing occurs unequally not only in various mouse tissues but also in developmental stages of various brain parts (Nakagawa et al., 1991a, 1991b).

SII The SII region (40 amino acids, 1693-QISIDESENAELPQAPEAENSTEQELEPSPPLR-QLEDHKR-1731 of mouse and rat IP₃R1 [Danoff et al., 1991; Nakagawa et al., 1991a]) is subdivided into three splicing subregions, A (1692-QISIDESENAELPQAPEAENSTE-1714), B (Q-1715), and C (1716-ELEPSPPLRQLEDHKR-1731), and locates between the two PKA phosphorylation consensus sites (Ser-1589 and Ser-1756). *Neuronal IP₃R1* are composed of all SII subtypes, i.e., SII+ (no SII deletion, long type), SII− (a complete SII deletion, short type), SIIB− (a deletion of one residue, Q-1715), and SIIBC− (a deletion of 12 residues, 1715–1731) (Nakagawa et al., 1991a, 1991b), but their substantial expression ratio is controversial (Danoff et al., 1991; Nakagawa et al., 1991a, 1991b; Schell et al., 1993). Contrarily, all IP₃R1 in nonneuronal tissues tested so far (Danoff et al., 1991; Nakagawa et al., 1991a, 1991b;) (*nonneuronal IP₃R1*) is the SII− subtype, except that mouse BW5147 T-lymphoma cell line, mouse NIH3T3 fibroblast cell line, and mouse splenic T-lymphocyte express a novel subtype, SIIA− (Iida and Bourguignon, 1994). The ratio of SII−/SIIA− expressed in mouse splenic T-lymphocytes is 1:1 and increases approximately 1.5-fold during mitogenic stimulation by concanavalin A (Iida and Bourguignon, 1994). The expression of these SII subtypes in various brain areas was also altered during the postnatal brain development (Nakagawa et al., 1991a, 1991b). It is noteworthy that the SII splicing affected the PKA phosphorylation kinetics and sites; SII− subtype from vas deferens was PKA phosphorylated at Ser-1589 with 5-fold higher affinity for PKA (Michaelis constant, $K_m = 3.1\,\text{nM}$) than cerebellar SII+ subtype which was preferentially phosphorylated at Ser-1755 ($K_m = 17.5\,\text{nM}$) (Danoff et al., 1991; Ferris et al., 1991). The SII splicing was also proposed to relate to the creation of an ATP-binding site (G-x-G-x-x-G), i.e., the complete splicing out of the SII region in the short type, SII−, gives rise to an additional putative ATP-binding site (1687-GYGEK-1691/G-1732) just in the spliced junction (Ferris and Snyder, 1992).

SIII The third alternative splicing region, SIII, has been found in human IP$_3$R1 (Nucifora et al., 1995). Note that all the sequences of IP$_3$R1 reported before this finding had spliced out the SIII sequence. The SIII region (9 amino acids, 903-NNDVEKLKS-912 of human IP$_3$R1SI− [Nucifora et al., 1995], corresponding to an insertion between the residues 917 and 918 of mouse IP$_3$R1SI+) is located in the modulatory region. Both the long (SIII+) and short (SIII−) subtypes are expressed in all rat tissues tested, and the SIII+ predominates in most brain regions except the cerebellum while the SIII− predominates in peripheral tissues. A functional diversity in the SIII splicing was proposed; differential creation of a site for protein kinase C (PKC) by which the SIII+ acquires a new PKC potential site (**KLKS**) in the SIII insert but the SIII− has a potential site (**NKGS**) just at the joining point after deleting the SIII (Nucifora et al., 1995).

Expression of IP$_3$R1

IP$_3$R1 is the predominant member of the family that is expressed in many tissues (Furuichi et al., 1990, 1993; Ross et al., 1992; Blondel et al., 1993; De Smedt et al., 1994; Newton et al., 1994; Wojcikiewicz, 1995; Nakanishi et al., 1996).

Widespread Expression of IP$_3$R1 Throughout a Whole Body IP$_3$R1 is prominently expressed in brain and smooth muscle-enriched tissues, and is also present in pituitary, spinal cord, heart, lung, liver, gastrointestinal tract, kidney, adrenal gland, spleen, thymus, uterus, ovary, placenta, vas deferens,

testes, etc. In addition, the alternative splicing isoforms of IP$_3$R1, which differ in the SI, SII, and SIII regions, are differentially expressed in these tissues of adults and even throughout development (Danoff et al., 1991; Nakagawa et al., 1991a, 1991b; Nucifora et al., 1995), and it is a typical example that the SI+ isoforms are exclusively expressed in neuronal cells, whereas non-neuronal cells express only the SII− isoforms (see "Alternative Splicing Subtypes of IP$_3$R1"). These differential expression patterns of IP$_3$R1 isoforms might be responsible for diversity in IP$_3$R1-mediated IICR signaling.

IP$_3$R1 is also expressed in many cell lines (De Smedt et al., 1994; Sugiyama et al., 1994b; Monkawa et al., 1995; Wojcikiewicz, 1995); in comparison with other types, almost exclusively (with high estimated proportions of ≳90%) in neuronal cell lines NG108-15 (mouse neuroblastoma/rat glioma hybrid) and SH-SY5Y (human neuroblastoma); predominantly (≳70%) in HL-60 (human leukemia cell line), GH$_3$ (rat pituitary tumor cell), and A7r5 (rat embryonic aorta); and also present in CHO-K1 (Chinese hamster ovary cell line), OK (American opossum kidney cell line), Swiss 3T3 (mouse embryonic fibroblast), BC$_3$H1 (mouse embryonic myoblast), RINm5F (rat insulinoma), C6 (rat glioma), L-cell (mouse fibroblast), PC12 (rat pheochromocytoma), and human hematopoietic cell lines (Jurkut, Raji, K562, HEL, CMK, ML-1, and THP-1).

Expression in Nervous Systems IP$_3$R1 is expressed at a high level in various types of neurons, but contrarily not in glial cells, throughout the nervous system (Nakanishi et al., 1991; Furuichi et al., 1993; Sharp et al., 1993a, 1993b), and thus is often referred to as *neuronal IP$_3$R*. Among these neurons, the highest expression level of IP$_3$R1 is present in Purkinje neurons of cerebellum (Worley et al., 1987, 1987b, 1989; Maeda et al., 1988, 1989, 1990; Furuichi et al., 1989b, 1993; Ross et al. 1989). The [^{3}H]-IP$_3$ binding activity of the cerebellum is ≈ 4 times and ≈ 8 times higher than that of the cerebrum in rat (Worley et al., 1987a) and mouse (Maeda et al., 1990), respectively, and ≈ 20 times higher than that of brain parts excluding cerebellum in mouse (Furuichi et al., 1993). The amounts of IP$_3$R1 mRNA in the cerebellum are dozens of times greater than that in other brain parts in mouse (Furuichi et al., 1990, 1993). In comparison with expression levels of other types, rat IP$_3$R1 predominates in the cerebellum with estimated high proportions of ≈ 94% for the mRNA (De Smedt et al., 1994) and ≈ 99% for the protein (Wojcikiewicz, 1995). The highest level in the cerebellum is due to extraordinary expression of IP$_3$R1 in Purkinje neurons in spite of the small proportion of their cell number in the cerebellum.

It is worthwhile to note that IP$_3$R1 mRNA in Purkinje cells is localized in the dendrites as well as the soma, suggesting that dendritic IP$_3$R1 protein may be locally synthesized from the mRNA transported from the soma to the dendrites (Furuichi et al., 1993). This dendritic IP$_3$R1 mRNA transport in Purkinje cells has been presumed to be due to involvement of ordered localization machinery (Bian et al., 1996). IP$_3$R1 mRNA is also present in moderate amounts in the olfactory tubercle, cerebral cortex, CA1 region of the hippocampus, caudate-putamen, and choroid plexus (Furuichi et al., 1993). Messenger RNAs for the IP$_3$R1 SI and SII splicing subtypes are also differentially expressed not only throughout cerebellar development but in various brain areas (Nakagawa et al., 1991a, 1991b).

The results of immunohistochemical studies also indicated that IP$_3$R1 protein is predominantly localized in Purkinje cells (Maeda et al., 1988, 1989; Mignery et al., 1989; Ross et al., 1989; Otsu et al., 1990a; Satoh et al., 1990; Nakanishi et al., 1991; Yamamoto et al., 1991; Takei et al., 1992, 1994). The IP$_3$R1 protein is present at low to moderate levels in neurons in the anterior olfactory nucleus, olfactory tubercle, striatum, nucleus accumbens septi, globus pallidus, cerebral cortex, precommissural hippocampus, hippocampus, substantia nigra, pons, and certain hypothalamic nuclei (Nakanishi et al., 1991). IP$_3$R1-immunoreactivity was also found in the axonal pathways of the limbic-hypothalamic pathways, strionigral projection, and part of the corpus callosum (Nakanishi et al., 1991). Colocalization of IP$_3$R1 with RyR in hippocampal neurons (Seymour-Laurent and Barish, 1995), with calsequestrin (luminal Ca^{2+}-binding protein) in Purkinje cells (Volpe et al., 1991), with metabotropic glutamate receptor 1α (mGluR1α) (phosphoinositide turnover-linked neurotransmitter receptor) and ionotropic glutamate receptor 1 (GluR1) in mouse brain (Ryo et al., 1993), and with mGluR1 in rat brain (Fotuhi et al., 1993) were reported. In the retina, IP$_3$R1 has been localized to the presynaptic terminals of photoreceptors and bipolar cells, as well as to the synaptic processes of amacrine cells (Peng et al., 1991). Expression of IP$_3$R1 during embryonic development of nervous systems was also observed (Nakanishi et al., 1991; Dent et al., 1996). Recently, a novel transgenic mouse line carrying the transgene which is composed of the IP$_3$R1 promoter/β-galactosidase (β-gal) fusion gene has been established, and thus the gene expression of IP$_3$R1 can now be visualized by β-gal activity using chromogenic substrates such as X-gal (Furutama et al., 1996).

Intracellular Localization   Subcellular localization of IP$_3$R1 has been well documented in Purkinje cells. In this cell, a principal neuron in the cerebellar cortex, IP$_3$R1 is present in the cell body, axon, axon terminals, dendrites, and dendritic spines (Nakanishi et al., 1991), and is localized in the ER (Mignery et al., 1989; Maeda et al., 1989; Ross et al., 1989; Otsu et al., 1990a; Satoh et al., 1990; Yamamoto et al., 1991; Takei et al., 1992, 1994), which is generally the largest membrane in eukaryotic cells, of all the aforementioned cellular compartments. The major subcellular localization of IP$_3$R1 proteins in Purkinje cells is the smooth ER (in particular, in *stacks* of flattened cisternae and subplasmalemmal cisternae), and a little IP$_3$R1 protein is also localized in the rough ER or the outer nuclear membrane. Similar subcellular localizations were observed in preterminal fibers and terminal boutons in the nuclei of the vestibular complex and in neurons of the dorsal cochlear nucleus (Rodrigo et al., 1994). Thus, the differential localizations in these subcellular compartments probably enhance a tendency toward microheterogeneity of IP$_3$R-mediated Ca^{2+} signaling within the cell.

In Purkinje cells, the ER cisternae, highly enriched IP$_3$R1, are often closely apposed to the mitochondria (Mignery et al., 1989; Otsu et al., 1990a; Satoh et al., 1990). Of the known second messengers, only the Ca^{2+} signal appears to be transferred into the mitochondrial matrix, and it probably modulates certain key enzymes in ATP synthesis (Denton and McCormack, 1990). In HeLa cells, Ca^{2+} microdomains mobilized from IP$_3$R-Ca^{2+} pools are promptly sensed by neighboring mitochondria (Rizzuto et al., 1993), suggesting the involvement of IP$_3$R-mediated Ca^{2+} signaling in ATP homeostasis. From this standpoint, it is noteworthy that mitochondria are abundant in both pre- and postsynapses. Thus, a class of IP$_3$R-mediated Ca^{2+} signals may play an important role in modulating the energy metabolism of synaptic transmission.

Dynamics of IP$_3$R-Ca^{2+} Pools   The *stack formation* of the ER can be induced not only by overexpression of full-length IP$_3$R1 in fibroblastic COS cells, but also by perfusion of blood vessels of rats with saline for $\gtrsim 5$ min (Takei et al., 1994). The overexpression of deletion mutant IP$_3$R1s did not cause any stacking of the ER. Takei et al. (1994) thus suggested that the massive formation of ER stacks in Purkinje cells is due to "head-to-head" interaction of homotypic IP$_3$R1 present on neighboring ER membranes, is not a permanent structure, and represents an adaptive response to nonphysiological conditions, such as anoxia, where a block of Ca^{2+} leakage from the ER may minimize massive rise in cytosolic Ca^{2+} due to the depletion of ATP.

The IP$_3$R-Ca^{2+} pools appears to interact with the cytoskeleton (Rossier and Putney, 1991; Bourguignon et al., 1993; Joseph and Samanta,

1993; Bourguignon and Jin, 1995). This interaction may be required for subcellular anchoring or for dynamics of the pools. Certainly, structural plasticity of the ER (Terasaki and Jaffe, 1991) and $IP_3R1\text{-}Ca^{2+}$ pools (Kume et al., 1993) have been observed in *Xenopus* eggs before and after fertilization.

These intracellular dynamics of $IP_3R\text{-}Ca^{2+}$ pools may be attributed to regulation of selective availability and recruitment of subcellular Ca^{2+} microdomains.

Gene Expression and Protein Stability Concomitant with retinoic acid (RA)- or dimethyl sulfoxide (DMSO)-induced differentiation of human leukemic HL-60 cells, $[^3H]\text{-}IP_3$ binding sites in membranes increased 3–4-fold (Bradford and Autieri, 1991). Upon differentiation of HL-60 cells either into macrophage-like cells by 12-*o*-tetradecanoylphorbol-13-acetate (TPA) or into granulocyte-like cells by DMSO or RA, substantial increases in expression of IP_3R1 mRNAs were observed (Sugiyama et al., 1994a; Yamada et al., 1994), thereby suggesting a possible involvement of IP_3R1 in regulation of cell differentiation. The DNA sequence of the IP_3R1 promoter has been determined and the presence has been found of several consensus sites for transcriptional regulation, including a TPA-responsible element (TRE) which may provide TPA-inducible expression of the IP_3R1 gene (Furutama et al., 1996; Konishi et al., 1997; Kirkwood et al., 1997).

Down-regulation of IP_3R1 was induced by brain ischemia. Alteration of $[^3H]\text{-}IP_3$-binding sites in hippocampal CA1 region was observed after 6 h of hemispheric ischemia in gerbil brain (Nagata et al., 1995). After 45 min of ischemic insult, IP_3R1 mRNA levels in the ischemic cortex decreased to 52% compared with that in the controlateral side at 4 h, and practically no IP_3R mRNA levels could be detected by 16 h (Zhang et al., 1995).

Increased Ca^{2+}-induced degradation of IP_3R1 in cerebellar microsomes with Ca^{2+}-dependent protease calpain, producing a C-terminal 95 kDa polypeptide (Magnusson et al., 1993). In vivo ^{35}S-labeling experiments using WB rat liver epithelial cells showed that half-life of IP_3R proteins was 11 h in comparison with 20 h for total liver microsomal protein (Joseph, 1994). Chronic stimulation of phosphoinositide hydrolysis was shown to cause down-regulation of IP_3R1 at the protein levels in human neuroblastoma SH-SY5Y cells due to calpain-mediated degradation, thereby reducing the sensitivity of IICR to excess IP_3 (Wojcikiewicz et al., 1994; Wojcikiewicz and Oberdorf, 1996), whereas the IP_3R2 protein in AR4-2J cells and cerebellar granule cells was more resistant to this degradative process than the IP_3R1 and IP_3R3 proteins (Wojcikiewicz, 1995). Therefore, the down-regulation of IICR concomitant with the chronic stimulation of phosphoinositide hydrolysis appears to be an adaptive response to excess stimuli and seems to vary from cell type to cell type.

These regulations of IP_3Rs at the gene expression and protein levels, together with the stack formation of $IP_3R\text{-}Ca^{2+}$ pools described above, in combination, might be involved in adaptive Ca^{2+} signaling or Ca^{2+} homeostasis under physiological and pathophysiological states.

IP_3R2

Structure and Function of IP_3R2

The IP_3R2 cDNA has been cloned from rat brain (Südhof et al., 1991) and human Namalwa cells (Burkitt lymphoma cell line) (Yamamoto-Hino et al., 1994). IP_3R2 is composed of 2701 amino acids (calculated $M_r \approx 308\,kDa$) and shares 68–69% sequence identity with IP_3R1. The data obtained in cDNA-transfection experiments have not shown any significant difference in binding specificity but a little in affinity for IP_3 in the family; IP_3R2 was suggested as a *high-affinity IP_3R* because of the relative order of affinity for truncated receptors, $IP_3R2 > IP_3R1 > IP_3R3$, i.e., K_d of IP_3R2 is $\approx$ one fourth that of IP_3R1, and K_d of IP_3R1 is $\approx$ one tenth that of IP_3R3 ($IP_3R2 \approx 27\,nM$ versus $IP_3R1 \approx 90\,nM$ [Südhof et al., 1991]; IP_3R2 15 nM versus IP_3R1 55–65 nM, and $IP_3R1 \approx 6\,nM$ versus $IP_3R3 \approx 66\,nM$ [Newton et al., 1994]). Recently, IICR activity defined as the high-affinity IP_3R2 channel in ventricular cardiac myocytes has been reported by Perez et al. (1997). In comparison with other members of the family, the modulatory and transducing domains are often interspersed with long stretches of diversified amino acids (56–65% identity). Accordingly, most of the putative modulator-binding sites and phosphorylation sites are diversified among the family. Therefore, these differences in the modulatory systems of the individual receptor types may cause more complicated Ca^{2+} signaling. In contrast, some modulatory sites are well conserved; the putative ATP-binding site on the C-terminal side (amino acid positions 2016–2021 of IP_3R1) is conserved (Furuichi et al., 1994b); expressed residues of 1915–2175 of rat IP_3R2 in *Escherichia coli* cells displayed $^{45}Ca^{2+}$ binding activity as did the corresponding region of IP_3R1 (Mignery et al., 1992); expressed residues 1558–1596 of mouse IP_3R2 in *E. coli* cells bound to CaM Sepharose column and, of these, the residues 1565–1587 correspond to the CaM binding site (1564–1585) of mouse IP_3R1 (Yamada et al., 1995). The Ser-1687 of human IP_3R2 is a putative site for PKA phosphorylation. Recently, IP_3R2 has indeed been shown to be phosphorylated by PKA in intact cells with low efficiency (Wojcikiewicz and

Luo, 1998). The channel domain contains six MSDs, M1-M6 (2230-2541 of human IP3R2). In the putative large luminal loop, there are two potential sites for N-glycosylation (Asn-2430 and Asn-2456 of human IP_3R2), which are compatible with the counterparts of mouse IP_3R1.

Expression of IP_3R2

Ross et al. (1992), using RT-PCR and in situ hybridization, showed that the expression levels of IP_3R2 were very low, and that the mRNA was observed at high levels in brain, lung, and placenta, and at low levels in liver, kidney, testes, and spinal cord. Newton et al. (1994), by means of Northern blot analysis, reported that expression of the rat IP_3R2 mRNA was highest in spinal cord and low in lung and testis, and that it was also observed in adrenal gland, intestine, liver, and spleen by using RT-PCR.

IP_3R2 is expressed in inflammatory and hematopoietic cells. Sugiyama et al. (1994a, 1994b) reported that IP_3R2 proteins were detected at high levels in thymocytes and mast cells, and at low levels in macrophages, splenocytes, polymorphonuclear cells, and eosinophils, and that IP_3R2 mRNA was observed in megakaryoblastic cell lines (HEL and CMK), a basophilic cell line (KU812F), T-cell lines (HPB-ALL and Jurkat), and B-cell lines (Namalwa and Jijoye). It was also shown that induction of differentiation of promyelocytic cell line HL-60 by treating with TPA (inducing macrophage-like cells) and with DMSO and retinoic acid (inducing granulocyte cells) resulted in the stimulation of IP_3R2 mRNA expression (Sugiyama et al., 1994a).

By Western blotting analysis Wojcikiewicz (1995) detected the expression of IP_3R2 in several cell lines (high levels, AR4-2J [rat pancreatoma cell line], HL-60 [human leukemic cell line], HEK293 [human embryonic kidney cell line], and COS-1 [African green monkey kidney cell line]; low levels, RINm5F [rat insulinoma]; no expression, SH-SY5Y [human neuroblastoma]), and in rat tissues (brain, cerebellum, pancreas, pituitary, lung, vas deferens, testes, heart, and liver). Quantitative analysis of these expression levels showed that IP_3R2 predominated in AR4-2J cells and liver; $IP_3R1:IP_3R2:IP_3R3 =$ 12:86:2%, and 19:81:0%, respectively. Chronic stimulation of phosphoinositide hydrolysis was shown to cause down-regulation of IP_3R1 at the protein levels in SH-SY5Y cells, thereby reducing the sensitivity of IICR to sustained excess IP_3 (Wojcikiewicz et al., 1994), whereas the IP_3R2 protein in AR4-2J cells and cerebellar granule cells was more resistant to this degradative process than the IP_3R1 and IP_3R3 proteins (Wojcikiewicz, 1995). Therefore, the down-regulation of IICR by the chronic stimulation of phosphoinositide hydrolysis seems to vary from cell type to cell type where each type is differentially expressed. Recently, Wilson et al. (1998) have shown that stimulation of Ca^{2+} increase also causes subcellular redistribution of IP_3R2 in RBL-2H3 (rat basophilic leukemia cell line) and AR4-2 cells, suggesting a possible adaptive mechanism of IP_3/Ca^{2+} signaling at IP_3R protein and/or IP_3R-Ca^{2+} pool levels.

Fujino et al. (1995) reported that IP_3R2 mRNA was detected in the intratubular duct cells of submandibular gland, urinary tubule cells of kidney, epithelial cells of epididymal ducts, and follicular granulosa cells of ovary (by in situ hybridization). Characteristic immunological localizations of IP_3R2 protein in these tissues have also been reported; in kidney, IP_3R2 is expressed in intercalated cells, whereas IP_3R1 and IP_3R3 are in vascular smooth muscle cells and glomerular mesangial cells, and IP_3R3 is also in principal cells of cortical collecting ducts (Monkawa et al., 1998); in ductal epithelial cells of the submandibular gland, IP_3R2 is highly expressed, IP_3R3 is weak, but IP_3R1 is not expressed (Yamamoto-Hino et al., 1998). Lee et al. (1997), however, reported a controversial result that IP_3R1 was localizaed in ductal cells. This discrepancy may be due to differences in the antibodies used. Li et al. (1996) showed that in rat gastrointestinal tract, IP_3R2 was immunohistochemically localized exclusively in goblet cells, especially of duodenum, but not in goblet cells of colon, while IP_3R1 and IP_3R3 were localized predominantly in gastric parietal cells. IP_3R2 may be responsible for secretion of acid glycoprotein in goblet cells. In heart, the myocardium expresses IP_3Rs (Moschella and Marks, 1993); IP_3R1 is predominantly expressed in Purkinje myocytes of the conduction system (Gorza et al., 1993), whereas IP_3R2 is expressed in ventricular myocytes (Perez et al., 1997). The structure of the IP_3R2 gene promoter has been determined and the presence of some tissue-specific transcriptional regulation sites has been deduced (Morikawa et al., 1997).

Although IP_3R2 cDNA was first cloned from a brain cDNA library (Südhof et al., 1991) and the RT-PCR, Northern blot, and Western blot analyses revealed the presence of IP_3R2 mRNA in the brain (Ross et al., 1992; Newton et al., 1994; Wojcikiewicz, 1995), expression sites for IP_3R2 in the brain remain elusive. Interestingly, it has been shown that primarily cultured cerebellar granule cells express both IP_3R1 and IP_3R2 to almost the same extent (Wojcikiewicz, 1995), in marked contrast to the predominant expression of IP_3R1 in Purkinje cells (Furuichi et al., 1993) with high proportion of the mRNA ($\approx$ 94%, De Smedt et al., 1994) and protein ($\approx$ 99%, Wojcikiewicz, 1995). In cultured cortical astrocytes, subcellular localization of IP_3R2 coincides with specialized regions of norepinephrine-induced high Ca^{2+} release (Sheppard et al., 1997).

Monkawa et al. (1995) demonstrated immunologically that IP_3R2 proteins were expressed in MDCK (Madin-Darby canine kidney cell), LLC-PK1 (pig kidney), CHO-K1 (renal epithelial cell lines), RINm5F (rat insulinoma), C6 (rat glioma), NG108-15 (mouse neuroblastoma/rat glioma hybrid cell line), Jurkat (human T-lymphoma), Raji (human pre-B-cell line), and K562 (erythroblastic cell line), in which other IP_3R types were differentially coexpressed and thus physically interacted with IP_3R2 to form a heterotetramer. These data suggest a considerable diversity in actual IICR among cell types.

IP_3R3

Structure and Function of IP_3R3

The IP_3R3 cDNA has been cloned from a rat insulinoma cell line, RIN5mF (Blondel et al., 1993), from a human colon adenocarciroma line cell, HT29 (Maranto, 1994), and from a human umbilical cord vein endothelial cell line, HUVEC (Yamamoto-Hino et al., 1994). IP_3R3 contains 2670 amino acids in rat and 2671 amino acids in human (calculated $M_r \approx 304\,kDa$), and has 62% and 65% identity with mammalian IP_3R1 and IP_3R2, respectively. The ligand-binding domain (72–77%) and channel domain (66–71%) are relatively conserved among the receptor types in the IP_3R family, while the middle modulatory domain comprises a relatively divergent sequence. The cDNA-transfection experiments displayed no significant difference in binding specificity for inositol phosphate. The values of affinity for IP_3 reported ranged from 28.8 to 151 nM; K_d of expressed N-terminal 750 amino acids of human $IP_3R3 = 151$ nM (Maranto, 1994), K_d of expressed full-length of human $IP_3R3 = 28.8$ nM (Yamamoto-Hino et al., 1994), and K_d of expressed N-terminal 576 amino acids of rat IP_3R3 fused with GST = 66 nM (Newton et al., 1994). Compared with the affinities obtained from the IP_3R type, a relative order of affinity for IP_3 was suggested as $IP_3R2 > IP_3R1 > IP_3R3$, and thus IP_3R3 was classified as low-affinity IP_3R (Newton et al., 1994). The modulatory and transducing domains are often interspersed with long stretches of diversified amino acids (56–65% identity compared with other types). Most of the putative modulator-binding sites and phosphorylation sites are diversified among the family. Putative PKA phosphorylation sites differ between rat and human IP3R (three sites [Ser-934, Ser-1133, and Ser-1457] for rat versus five sites [Ser-934, Ser-1133, Thr-1701, Thr-2202, and Ser-2260] for human). Of these, Thr-2202 and Ser-2260 of human IP_3R3 are located in the vicinity of the N-terminal cytoplasmic

end of M1 and M3, respectively, suggesting possible roles in regulation of the channel (Maranto, 1994). Despite in vitro phosphorylation of cell extracts from WB rat epithelial cells by the exogenous PKA, no phosphorylation of IP_3R3 was detected in marked contrast to ready phosphorylation of IP_3R1 (Joseph et al., 1995b). In fact, a more detailed study has recently indicated that PKA phosphorylates IP_3R3 in intact cells with low efficiency (Wojcikiewicz and Luo, 1998). Eighteen potential phosphorylation sites for PKC (S/T-x-R/K) and 16 possible sites for CaMKII (R-x-x-S/T) are present in human IP_3R3 (Maranto, 1994). A putative ATP-binding site (1921–1926 of rat and human IP_3R3) corresponds to the putative ATP-binding site on the C-terminal side of IP_3R1 (amino acid position 2016-2021 of mouse IP_3R1). Human IP_3R3 expressed in NG108-15 cells did not bind to CaM-Sepharose column, whereas both IP_3R1 and IP_3R2 bound to it under the same conditions (Yamada et al., 1995). The channel domain is composed of six MSDs, M1–M6 (2205–2517 of human IP_3R3). IP_3R3 has only one N-glycosylation site (Asn-2405 of human IP_3R3; Asn-2404 of rat IP_3R3) in the large luminal loop, in contrast to two sites in both IP_3R1 and IP_3R2.

Expression of IP_3R3

IP_3R3 displays a wide distribution. Ross et al. (1992) found, by RT-PCR, that IP_3R3 was expressed at highest levels in brain and gastrointestinal tract, and also in lung, liver, kidney, testes, thymus, spleen, and placenta. Blondel et al. (1993) showed that rat IP_3R3 proteins are expressed at high levels in insulin-secreting cell lines RINm5F and HIT-T15, and rat pancreatic islets, and also expressed in COS-7, and that the mRNAs are expressed at highest levels in rat kidney, small intestine, and pancreatic islets, and at low levels in brain, spleen, lung, and heart. Newton et al. (1994) reported that expression of the rat IP_3R3 mRNA was at its highest level in intestine, and was also observed in adrenal gland, kidney, lung, spinal cord, spleen, and testis. Fujino et al. (1995) reported that IP_3R3 mRNA was detected in gastric cells, salivary and pancreatic acinar cells, and epithelial cells of small intestine (by in situ hybridization).

IP_3R3 is expressed in many inflammatory and hematopoietic cells. IP_3R3 proteins were detected at high levels in thymocytes and splenocytes, and at a low level in mast cells (Sugiyama et al., 1994a). IP_3R3 mRNAs were also observed in human hematopoietic cell lines such as erythroblastic cell line (K562), myeloblastic cell line (ML-1), monoblastic cell line (U937), T-cell lines (HPB-ALL, Jurkat, HUT-78), and B-cell lines (Raji, Namalwa, RPMI788, and Jijoye) (Sugiyama et al., 1994a; Yamamoto-Hino et al., 1994).

Wojcikiewicz (1995) reported that, using Western blotting, the expression of IP_3R2 was observed in several cell lines (predominantly in a rat insulinoma cell line RINm5F, at intermediate levels in a human embryonic kidney cell line HEK293 and an African green monkey kidney cell line COS-1, and at a low level in a rat pancreatoma cell line AR4-2J, but not in a human leukemic cell line HL-60 or a human neuroblastoma cell line SH-SY5Y) and in rat tissues (pancreas, pituitary, lung, vas deferens, testes, and heart), and by quantitative analysis of these it was found that IP_3R3 predominated in RINm5F cells; IP_3R1: IP_3R2: $IP_3R3 \approx 4\% : \approx 0\%; \approx 96\%$. The paper also indicated that IP_3R1, IP_3R2, and IP_3R3 proteins were coexpressed in many cell lines, but were expressed predominantly in a neuronal cell line SH-SY5Y ($\approx 99\%$ predominancy), a rat pancreatoma cell line AR4-2J (acinar cell-like; $\approx 86\%$); and a rat insulinoma cell line RINm5F (β-cell-like; $\approx 96\%$), respectively (Wojcikiewicz, 1995). But, De Smedt et al. (1994) estimates were $\approx 54\%$ for proportion of IP_3R2 mRNA in AR4-2J and $\approx 60\%$ for that of IP_3R3 mRNA in RINm5F (by RT-PCR).

In general, cellular localizations associate with cellular functions. Interestingly, IP_3R3 cDNAs have originally been cloned from secretory cell types, and several cell types characteristic of secretion have been reported as the dominant sites for IP_3R3 expression, as described above. In pancreatic islets, IP_3R3 was shown to be localized to the secretory granules of insulin-secreting β-cells and somatostatin-secreting δ-cells (Blondel et al., 1994a) (also see Secretory Granular Membrane IP_3R), and the overexpression of IP_3R3 in insulin-secreting βTC-3 cells, which express the endogenous IP_3R3 at a low level, by cDNA transfection exhibited an enhanced increase in cytosolic Ca^{2+} in response to a muscarinic agonist carbamyl choline (CCh) (Blondel et al., 1994b). Thus, IP_3R3 appears to be involved in endocrine regulation in the pancreatic islets. In pancreas, however, IP_3R3 mRNA, was mainly observed in acinar cells rather than in cells of Langerhans islets (Fujino et al., 1995), suggesting an exocrine Ca^{2+} signaling role. IP_3R3 proteins were localized in the extreme apical region of pancreatic acinar cells, suggesting that IP_3R3 in this trigger zone is responsible for the generation of apical-to-basal Ca^{2+} waves and may be able to regulate apical exocytosis (Nathanson et al., 1994). In rat gastrointestinal tract, IP_3R3 proteins were immunohistochemically localized in the epithelial layer of adult rat jejunum, i.e., throughout cytoplasm (probably the ER) of crypt, villus (which is prominent site and is highly stained in its cytoplasm adjacent to the apical brush border), and enterocytes, but not the smooth muscle layers (Maranto, 1994). IP_3R3, together with IP_3R1, was localized predominantly in gastric parietal cells which secrete hydrochloric

acid (HCl) (Li et al., 1996). In kidney, it has been shown that IP_3R3 is localized in vascular smooth muscle cells, glomerular mesangial cells, and principal cells of cortical collecting ducts (Monkawa et al., 1998). In principal cells, IP_3R3 is subcellularly located to the basolateral side of the cytoplasm, suggesting polarized Ca^{2+} signaling in the cells.

In the brain, Yamamoto-Hino et al. (1995) demonstrated that IP_3R3 proteins were immunohistochemically detected in astrocytes of hippocampus and cerebellum and in Bergman glial cells of cerebellum, in marked contrast to *neuronal IP_3R1*. Thus, *glial IP_3R3* may be involved in glial IICR which appears to be required for some aspects of the complex spatiotemporal dynamics (Ca^{2+} waves and oscillations) characteristic of many glial cells (Finkbeiner, 1993).

Monkawa et al. (1995) demonstrated immunologically that IP_3R3 proteins were expressed in MDCK (Madin-Darby canine kidney cell), LLC-PK1 (renal epithelial cell lines), OK (American opossum kidney cell line), CHO-K1 (Chinese hamster ovary cell line), RINm5F (rat insulinoma), PC12A (rat pheochromocytoma), C6 (rat glioma), NG108-15 (mouse neuroblastoma/rat glioma hybrid cell line), Jurkat (human T-lymphoma), Raji (human pre-B-cell line), and K562 (erythroblastic cell line), in which other IP_3R types were differentially coexpressed and thus physically interacted with IP_3R3 to form a heterotetramer. Joseph et al. (1995b) showed that a rat epithelial cell line, WB, expressed not only homotetrameric IP_3R3 and IP_3R1 but also heterotetramers between them, and in vitro phosphorylation of WB cell extracts by exogenous PKA displayed no phosphorylation of IP_3R3, even having five putative PKA sites (Maranto, 1994), in marked contrast to ready phosphorylation of IP_3R1. These data suggest that composition in a heteromeric IP_3R channel, as well as modification state of each subunit in the tetramer complex, may also contribute toward a diversity in actual IICR.

Recently, the involvement of IP_3R3 in Ca^{2+} entry across the PM has been reported (DeLisle et al., 1996; Khan et al., 1996), suggesting the distinct role of IP_3R3 in Ca^{2+} signaling (see, "IP_3R Present Outside the ER Membranes").

Other Types

IP_3R4 and IP_3R5

By PCR approaches, cDNA sequences homologous to IP_3R were cloned from mouse, and were termed type IV (Ross et al., 1992) and type V (De Smedt et al., 1994) (called IP_3R4 and IP_3R5, respectively, here). On the other hand, IP_3R4 was suggested to be alternatively spliced forms of IP_3R2, because of close similarities between them (Ross et al., 1992). IP$_3$R5 was also shown to have a high sequence identity with IP$_3$R2 (De Smedt et al., 1994). From these data, it was proposed that IP$_3$R2, IP$_3$R4, and IP$_3$R5 form a subfamily (De Smedt et al., 1994). However, to categorize IP$_3$R4 and IP$_3$R5 into independent isoforms we need more clarification of this point, since we know only partial sequences of both IP$_3$R4 and IP$_3$R5, 528 bp and 1269 bp, respectively (Ross et al., 1992; De Smedt et al., 1994).

IP$_3$R4 The 176 amino acids around M5 and M6 were determined, and the wide tissue distribution of IP$_3$R4 was studied, e.g., in brain, spinal cord, testes, lung, kidney, gastrointestinal tract, placenta, etc. (Ross et al., 1992). Rat basophilic leukemia RBL-2H3 cells express IP$_3$R4 predominantly (De Smedt et al., 1994) and are characterized by a much higher IP$_3$-binding affinity than cerebellar microsomes K_d 3.8 nM versus 135 nM; Parys et al., 1995).

IP$_3$R5 The cloned cDNA sequence (1269 bp) has 94.5% sequence identity with the nucleotides 197–1465 of mouse IP$_3$R2 cDNA reported by Ross et al. (1992) (De Smedt et al., 1994). Interestingly, expression of IP$_3$R5 mRNA is species-specific, since it is significant in many mouse cells tested but not in rat cells (De Smedt et al., 1994).

IP$_3$R Present Outside the ER Membranes

Plasma Membrane IP$_3$R Several lines of evidence have indicated the presence of plasma membrane IP$_3$R (*PM-IP$_3$R*), which probably functions as an IP$_3$-activated Ca^{2+}-permeant channel on the plasma membranes. The activities of IP$_3$-activated Ca^{2+} influx across the plasma membranes have been observed in several cell types; e.g., human T-lymphocytes (Kuno and Gardner, 1987; McDonald et al., 1993), rat mast cells (Penner et al., 1988), rat hepatocytes (Guillemette et al., 1988), and Madin-Darby canine kidney epithelial cell line MDCK (Bush et al., 1994). IP$_3$R-like immunoreactivities have also been observed in hepatocytes (Sharp et al., 1992), in endothelium, smooth muscle cells, and keratinocytes (Fujimoto et al., 1992), in T-lymphocytes (Khan et al., 1992a, 1992b), and in olfactory receptor neurons of the lobster (Fadool and Ache, 1992). In the nervous system, there are many studies that show that IP$_3$ elicits Ca^{2+} influx in olfactory neurons; e.g., catfish (Restrepo et al., 1990; Miyamoto et al., 1992), lobster (Fadool and Ache, 1992; Hatt and Ache, 1994), and rat (Okada et al., 1994). IP$_3$-activated inward Ba^{2+} currents have been recorded in excised inside-out patches of primary cultured Purkinje cells (Kuno et al., 1994). However, no information on the

molecular aspects of PM-IP$_3$R is yet available, nor do we know whether IP$_3$R present in the ER (*ER-IP$_3$R*) can be sorted into the PM. Recently, a possibility of the latter case has been reported by two groups (DeLisle et al., 1996; Khan et al., 1996). Khan et al (1996) reported that upon induction of apoptosis with dexamethasone, T-cells increasingly come to express IP$_3$R3 but not IP$_3$R1, and the newly expressed IP$_3$R3 localizes to the PM and is involved in Ca^{2+} entry. DeLisle et al. (1996) showed that the overexpressed IP$_3$R3 but not IP$_3$R1 in *Xenopus* oocytes localizes near the PM (in the subsurface cisternae or the PM) and enhances the IP$_3$-induced Ca^{2+} entry across the PM (capacitative Ca^{2+} influx). Thus, IP$_3$R3 seems to have distinct Ca^{2+} signaling function. However, more convincing data must be provided to certify the localization in the PM. As a matter of fact, if IP$_3$R3 is selectively localized in the subsurface cisternae and is involved in the capacitative Ca^{2+} influx, it would be interesting to know how the IP$_3$R3-Ca^{2+} pools near the PM are coupled with Ca^{2+} entry channels in the PM.

Nuclear Membrane IP$_3$R There are many reports that IP$_3$-binding sites and IICR activities have been detected in liver nuclei (Nicotera et al., 1989, 1990; Malviya et al., 1990; Matter et al., 1993; Gerasimenko et al., 1995). There are no definitive data regarding IP$_3$R types present on the nuclear membranes. *Nuclear IP$_3$R* in rat livers may be IP$_3$R2, since not only does rat liver possess IP$_3$R proteins in a relative order of IP$_3$R2 > IP$_3$R1 $\gg$ IP$_3$R3 (Monkawa et al., 1995; Wojcikiewicz, 1995), but also liver nuclear membranes exhibit relatively higher affinity for IP$_3$, like that of IP$_3$R2. However, there remains a possibility that nuclear IP$_3$R is a distinct type. Interestingly, in cerebellar Purkinje cells, IP$_3$R1-immunoreactivities were substantially localized in the outer nuclear membrane (ONM) (Mignery et al., 1989; Ross et al., 1989; Maeda et al., 1990; Otsu et al., 1990a; Satoh et al., 1990). The ONM is in luminal continuity with the ER, and thus the space between the ONM and the inner nuclear membrane (INM) is an extension of the ER lumen. These data thus suggest that, in response to increased cytosolic IP$_3$, outer nuclear membrane IP3R (*ONM-IP3R*) releases Ca^{2+} into the cytoplasm from the space between the ONM and the INM. It was shown that cytosolic Ca^{2+} increased by Ca^{2+} influx was rapidly transmitted into the nuclei in neuroblastoma cells (Al-Mohanna et al., 1994) and sympathetic neurons (O'Malley, 1994). Considered as a whole, Ca^{2+}-release signals mediated by ONM-IP$_3$R may be more effectively sensed by the nucleus than Ca^{2+}-influx signals, since Ca^{2+} release by ONM-IP$_3$R occurs fairly near the nuclear pores compared with Ca^{2+} influx across the PM. Treatment of rat liver nuclei with TPA displayed PKC phosphorylation of ONM-IP$_3$R that resulted in no change in IP$_3$-binding activity but enhancement of IICR activity (Matter et al., 1993). Patch-clamp electrophysiological techniques have successfully been applied directly to nuclei of *Xenopus* oocytes and have shown that ONM-IP$_3$R displays similar conductance properties of IICR channel as ER-IP$_3$R (Mak and Foskett, 1994; Stehno-Bittel et al., 1995).

There seems to be an inositide cycle in nuclei (see Divecha et al., 1993), and thus it is an intriguing issue to see if IP$_3$R is present inside the INM, namely inner nuclear membrane IP$_3$R (*INM-IP$_3$R*). It has been shown that IP$_3$ injected into the nuclei of *Xenopus* oocytes induces increase in nucleoplasmic Ca^{2+} (Hennager et al., 1995), and that highly purified INMs from rat liver nuclei display high affinity [^{3}H]-IP$_3$ binding, immunoreactivity to anticerebellar IP$_3$R antibody, and IICR from the nuclear envelope to the nucleoplasm (Humbert et al., 1996). It still remains elusive to determine if INM-IP$_3$R is distinct from ONM-IP$_3$R and/or ER-IP$_3$R.

What are the roles of IICR signaling in the nuclei? It was suggested that nuclear IP$_3$R-mediated Ca^{2+} signals may be involved in the fusion of postmitotic nuclear membranes (Sullivan et al., 1993). Berridge (1986) hypothesized that early *ionic events* stimulated by IP$_3$/Ca^{2+} signaling may be causally related to the onset of the *genomic events* which trigger Ca^{2+}-activated transcription (Ghosh and Greenberg, 1995) and protein synthesis (Brostrom and Brostrom, 1990) required for the stable alteration in neural circuitry (such the acquisition of long-term memory) and/or for the cell proliferation.

Secretory Granular Membrane IP$_3$R It is apparent that IP$_3$3 is predominantly expressed in pancreatic islets. However, there are a few discrepancies among the reports (Blondel et al., 1995; Meldolesi et al., 1995; Ravazzola et al., 1996). By immunogold electron microscopy (EM), Blondel et al. (1994a) showed that IP$_3$3 was localized to the secretory granules of insulin-secreting β-cells and somatostatin-secreting δ-cells in rat pancreatic islets, and that the expression increased in RIN5mF cells treated with high concentrations of glucose, and in islets of diabetic rats and rats being refed after a period of fasting, suggesting that the secretory granules themselves act as IP$_3$3-mediated Ca^{2+} signaling pools which regulate secretory response. The overexpression of IP$_3$3 in insulin-secreting βTC-3 cells that expressed the endogenous IP$_3$3 at a low level showed that the overexpressed IP$_3$3 was localized to secretory granules (observed by immunogold EM), and that increase in cytosolic Ca^{2+} in response to a muscarinic agonist carbamyl choline (CCh) was enhanced in the cells (Blondel et al., 1994b). Thus, it has been sug-

gested that IP$_3$R3-mediated release of granule Ca^{2+} stores might facilitate the secretory process (Blondel et al., 1995). However, it is debatable whether secretory granules have an active Ca^{2+}-uptake ability to serve as IP$_3$-sensitive Ca^{2+} pools (Meldolesi et al., 1995), and whether the antibody used for the immunogold EM observation (Blondel et al., 1994a, 1994b) cross-reacts with insulin itself in the granules (Ravazzola et al., 1996).

Properties of IP$_3$R

IP$_3$ Binding

The [^{32}P]-IP$_3$ binding activity was detected in bovine adrenal cortex (Baukal et al., 1985) and in guinea pig hepatocytes and rabbit neutrophils (Spät et al., 1986). Worley et al. (1987b, 1989) then found that rat cerebellum has 100-300 times higher [^{3}H]- and [^{32}P]-IP$_3$ binding activity than those observed in peripheral tissues. These findings accelerated the purification of the IP$_3$-binding sites, i.e., IP$_3$Rs, from the cerebellum of rat (Supattapone et al., 1988b) and mouse (Maeda et al., 1990, 1991) by utilizing [^{3}H]-IP$_3$ binding activity as a guidepost.

The specific IP$_3$ binding to cerebellar IP$_3$Rs displays the K_d value of $\approx 80-100$ nM, and is stoichiometric (Hill coefficient $n_H \approx 1.0$) (Supattapone et al., 1988b; Maeda et al., 1990), while that from aorta smooth muscle displays much higher affinity ($K_d \approx 2.4$ nM) (Chadwick et al. 1990), indicating the presence of a heterogeneity in IP$_3$R isoforms. Expressed IP$_3$-binding sites of distinct IP$_3$R cDNAs have displayed affinity for IP$_3$ (K_d) ranging from 14.8 nM to 151 nM (Furuichi et al., 1989b; Südhof et al., 1991; Miyawaki et al., 1990; Mignery et al., 1990; Maranto, 1994; Newton et al., 1994; Yamamoto-Hino et al., 1994). Although these binding experiments cannot simply be comparable with one another, a potent order of the affinity has been classified as IP$_3$R2 (*high-affinity IP$_3$R*) > IP$_3$R1 > IP$_3$R3 (*low-affinity IP$_3$R*) (Newton et al., 1994). The specificity for inositol phosphates is of the order of (1,4,5)-IP$_3$ > (2,4,5)-IP$_3$ > (1,3,4,5)-IP$_4$ > (1,4)-IP$_2$ > IP$_6$ > (1)-IP$_1$ in cerebellar IP$_3$R1 (Worley et al. 1987a; Miyawaki et al., 1990), and there seems to be no significant difference in the specificity among one another in the family (Furuichi et al., 1989b; Mignery et al., 1990; Miyawaki et al., 1990; Südhof et al., 1991; Yoshikawa et al., 1992; Kume et al., 1993; Maranto, 1994; Newton et al., 1994; Yamamoto-Hino et al., 1994). The stereospecific recognition of (1,4,5)-IP$_3$ was shown in the ligand binding, as well as the Ca^{2+}-releasing activity, of IP$_3$R: D-(1,4,5)-IP$_3$ > the racemate DL-(1,4,5)-IP$_3$

≫ the synthetic enantiomer L-(1,4,5)-IP$_3$ (Willcocks et al., 1987; Strupish et al., 1988).

The IP$_3$-binding activity is increased by alkaline pH between 7.0 and 9.0 (Worley et al., 1987a; White et al., 1991), and is inhibited by Mg^{2+} (Worley et al., 1987a; White et al., 1991; Van Delden et al., 1993), polyamines (Sayers and Michelangeli, 1993), and decavanadate (Föhr et al., 1991; Oldershaw et al., 1992). High Ca^{2+} inhibited the IP$_3$ binding to IP$_3$Rs in cerebellum (Worley et al., 1987a) and myeloid cells (Van Delden et al., 1993), but the effects of cytosolic Ca^{2+} appear to differ from cell type to cell type (see below). Heparin is the most well-known inhibitor of all IP$_3$R types and binds to the IP$_3$-binding site in a competitive manner (Worley et al., 1987a; Ghosh et al., 1988). Other polymeric sulfates also inhibit it (Worley et al., 1987a; O'Rourke and Feinstein, 1990). The sulfhydryl oxidizing reagent thimerosal (Bootman et al., 1992; Poitras et al., 1993; Sayers et al., 1993; Kaplin et al., 1994), the organomercurial thiol-reactive agent mersalyl (Joseph et al., 1995c), and the oxidized glutathione (GSSG) (Renard-Rooney et al., 1995) were shown to stimulate IP$_3$ binding as well as Ca^{2+} release, suggesting regulation of IP$_3$R by intracellular redox state. The calculated association rate constant for IP$_3$ binding to cerebellar microsomes under all tested conditions fell in the $2-5 \times 10 \, M^{-1} \, S^{-1}$, while the half-time ($t_{1/2}$) for IP$_3$ dissociation was affected by temperature, ions (KCl, NaCl, Mg^{2+}, and Ca^{2+}), and pH (≈ 400 ms at 20°C in the presence of 100 mM KCl and 20 mM NaCl at pH 7.4, and reduced to ≈ 125 ms by the addition of Mg^{2+}; concerning Ca^{2+} effect, see "Ca^{2+}" below), suggesting that modulation of IP$_3$ dissociation plays a prominent role in these affinity changes (Hannaert-Merah et al., 1994).

Adenophostin It appears to be a landmark for the study of IP$_3$ binding that new IP$_3$R agonists, referred to as "adenophostin A and B," have been found from the culture broth of *Penicillium brevicompactum* by Takahashi et al. (1993). A particularly striking point is that adenophostins inhibited [^{3}H]-IP$_3$ binding to cerebellar IP$_3$Rs with a low K_i value of 0.18 nM compared with that of IP$_3$ ($K_i = 15$ nM), and were 100-fold more potent than IP$_3$ in terms of Ca^{2+} release (Takahashi et al., 1993). Displacement of [^{3}H]-IP$_3$ binding to highly purified IP$_3$R1 by adenophostin exhibited a positive cooperativity ($n_H \approx 1.9$, $K_i = 10$ nM), whereas IP$_3$ did not ($n_H \approx 1.1$, $K_i = 41$ nM) (Hirota et al., 1995b). Adenophostin was capable of producing the quantal release in the purified IP$_3$R1 in reconstituted lipid vesicles, as does IP$_3$ (Hirota et al., 1995b). The chemical structure of adenophostin is apparently distinct from that of IP$_3$ except for the presence of three phosphate groups, and could not simply be superim-

posed on the IP$_3$ structure in the energy-minimized way (Takahashi et al., 1993). However, the 3″- and 4″-equatorial diphosphate on the glucose ring of adenophostin can be superimposed on the 5′- and 4′-equatorial diphosphate of IP$_3$, respectively. This seems to be consistent with our knowledge about the IP$_3$ molecule as an agonist, i.e., that the vicinal phosphate groups at positions 4′ and 5′ of IP$_3$ are critical for IICR activity while the 1′-phosphate increases affinity for IP$_3$R (Nahorski and Potter, 1989; Wilcox et al., 1994). The 2′-phosphate on the ribose ring of adenophostin is thus thought to be placed to fit IP$_3$R more effectively than the 1′-phosphate of IP$_3$ (Takahashi et al., 1993). Adenophostin A analog 2-hydroxyethyl-α-D-glucopyranoside-2,3′,4′-trisphosphate {Glu(2,3′,4′)P$_3$} has been synthesized (Wilcox et al., 1995; Marchant et al., 1997). Glu(2,3′,4′)P$_3$ is a truncated version of adenophostin A, in which the 2′- and 3′-carbons of the ribose ring, with their terminal phosphate groups, are retained and the remainder of the adenosine residues are excised. Glu(2,3′,4′)P$_3$ displayed [^{3}H]-IP$_3$ from porcine cerebellar microsomes with an affinity (IC$_{50}$ = 130 nM) 5-fold weaker than that of IP$_3$ (IC$_{50}$ = 27 nM), and also release Ca^{2+}, being only 10–12-fold less potent than IP$_3$ in permeabilized neuroblastoma SH-SY5Y cells (EC$_{50}$ = 647 nM versus IP$_3$ EC$_{50}$ = 247 nM). Since Glu(2,3′,4′)P$_3$ was a considerably weaker ligand (≈ 500-fold) and agonist (≈ 1000-fold) than adenophostin A, it was suggested that partial excision of the adenosine residue compromised structural motifs that have favorable interactions with IP$_3$R (Wilcox et al., 1995). An agonistic action of adenophostin has also been shown for IP$_3$R2 in hepatocytes (Marchant et al., 1997) and for IP$_3$R3 in bronchial mucosa 16HBE14o- cells (Missiaen et al., 1998).

Ca^{2+} Cytosolic Ca^{2+} appears to affect IP$_3$ binding differently in cell types expressing certain types of IP$_3$Rs. It was reported that Ca^{2+} inhibited [^{3}H]-IP$_3$ binding to crude cerebellar membranes, predominantly including IP$_3$R1, with an IC$_{50}$ of 300 nM, but not to purified cerebellar IP$_3$R (Supattapone et al., 1988b). Subsequently, it was suggested that "*calmedin*," a membrane protein with an estimated M_r of 300 kDa, was present in the crude cerebellar membrane fraction (but little in peripheral tissues) and conferred a negative feedback regulation of IP$_3$ binding by increased Ca^{2+} (Danoff et al., 1988). It was also proposed that a positive feedback loop operated by reactivation of phospholipase C (PLC), with released Ca^{2+} leading to regeneration of IP$_3$ followed by the triggering of a second cycle of IICR (Harootunian et al., 1991), since PLC is known to be dependent on Ca^{2+} for the catalytic activities (Rhee and Choi, 1992). Based on this Ca^{2+} dependency of PLCs, it was supposed that inhibition of IP_3 binding to crude cerebellar membranes by Ca^{2+} concentrations over $10\mu M$ could be attributed to Ca^{2+}-dependent activation of PLC being present in the crude membranes, resulting in inhibition of $[^3H]$-labeled IP_3 binding by production of IP_3 due to a simple competition mechanism (Mignery et al., 1992). Membranes of human leukemic HL-60 cells containing IP_3R1 proteins (Yamada et al., 1994) with the high proportion of $\approx 71\%$ (Wojcikiewicz, 1995) exhibited inhibition of IP_3 binding by high Ca^{2+} with IC_{50} of $2.5\mu M$ only in the presence of factor(s) from the flow-through of a heparin–Sepharose column, and in this case the inhibition within submicromolar range of Ca^{2+} was revealed not to be due to the activation of PLC (Van Delden, et al., 1993). The data showing that in the presence of excess Ca^{2+} ($100\mu M$ free Ca^{2+}), the half-time ($t_{1/2}$) for $[^3H]$-IP_3 dissociation was accelerated also suggested that inhibition of IP_3 binding to cerebellar membranes by Ca^{2+} may be due to the accelerated dissociation rate but not due to the activation of PLC (Hannaert-Merah et al., 1994).

In comparison with cerebellar IP_3Rs, most of which are IP_3R1 (Furuichi et al., 1993) in high proportion ($\approx 94\%$ in mRNA [De Smedt et al., 1994] and $\approx 99\%$ in protein [Wojcikiewicz, 1995]), IP_3Rs in peripheral tissues display different Ca^{2+} effects on IP_3 binding, e.g., high Ca^{2+} transforms liver IP_3Rs into high-affinity forms inactive for IICR (Pietri et al., 1990; Pietri Rouxel et al., 1992; Watras et al., 1994). Liver expresses prominently IP_3R1 and IP_3R2 (Ross et al., 1992; Monkawa et al., 1995) with a high proportion of IP_3R2 mRNA ($\approx 61\%$, De Smedt et al., 1994) and protein ($\approx 81\%$, Wojcikiewicz, 1995). Liver membranes exhibited the Ca^{2+}-mediated interconversion between two IP_3-binding states with EC_{50} of $140\,nM$ Ca^{2+}: the low-affinity state ($K_d = 65\,nM$) at low Ca^{2+} and the high-affinity state ($K_d = 2.8\,nM$) at high Ca^{2+} (Pietri et al., 1990; Pietri Rouxel et al., 1992). Interestingly, IICR appears to be coupled to the low-affinity state rather than the high-affinity state, suggesting a negative feedback regulation through the interconversion of two affinity sites by released Ca^{2+}, from the active low-affinity state to the inactive high-affinity state (Pietri et al., 1990; Pietri Rouxel et al., 1992; Watras et al., 1994). Although it is as yet unknown whether Ca^{2+} sensitivity is an intrinsic property of the IP_3-binding site itself, the reversion from the high-affinity state to the low-affinity state occurred at $37°C$ but not at $4°C$, suggesting a possibility that action of temperature-sensitive physiological factor(s) is a prerequisite for the reversion (Pietri Rouxel et al., 1992). Some typical Ca^{2+}-dependent regulatory molecules, such as calmodulin, Ca^{2+}-dependent protein kinases (CaMKII and PKC), and

protein phosphatases (calcineurin), appear to have no significant influence upon the representative properties of IP_3 binding in vitro (see "IP_3-Induced Ca^{2+} Release"). These data suggest that the difference in Ca^{2+} regulation of IP_3 binding between liver and cerebellar membranes may be due to a diversity of the IP_3R isoforms being expressed.

Another difference has been reported in the inactivation phenomenon of IP_3 binding by IP_3 itself. Although it was thought that prolonged exposure to IP_3 did not desensitize IP_3R (Oldershaw et al., 1992), by means of an IP_3-induced Mn^{2+} quenching technique over an extended period, inactivation by IP_3 binding itself has been found in permeabilized hepatocytes (Hajnóczky and Thomas, 1994), expressing IP_3Rs in an estimated order of IP_3R2 (≈ 61–81% of total IP_3Rs) > $IP_3R1 \gg IP_3R3$ (De Smedt et al., 1994; Wojcikiewicz, 1995). The inactivation rate was enhanced by increases over the physiological range of cytosolic Ca^{2+} from $30\,nM$ to $1\mu M$. Inactivation of IP_3R by IP_3 binding itself has been reported in permeabilized hepatocytes (Hajnóczky and Thomas, 1994). It has also been reported that prolonged exposure of IP_3 to cerebellar microsomes, containing exclusively IP_3R1, converts IP_3Rs to a state exhibiting higher affinity (Coquil et al., 1996). Independently of Ca^{2+}, which is in marked contrast to Ca^{2+}-induced interconversion in hepatic IP_3Rs, the conversion of cerebellar IP_3R1 to high-affinity site (1.5–2.5-fold) was induced by pre-exposure to only IP_3 (0.01–$1\mu M$) over a period of $2\,s$ to $2\,min$, with the EC_{50} of $\approx 60\,nM$ close to the K_d for IP_3 binding ($\approx 66\,nM$). After preincubation by IP_3 of only 2–$3\,s$, the affinity increased about 2-fold (K_d $107\,nM \rightarrow 53\,nM$), the number of binding sites (B_{max}) was almost unchanged, and no cooperativity was observed. On the other hand, the activation after longer ($10\,min$) preincubation caused a similar change, except for expressing positive cooperativity (apparent Hill coefficient $n_H \approx 1.6$), which was assumed to result from slow conformational transition upon prolonged occupancy of the IP_3-binding site. These data suggested that upon binding, IP_3 not only opens the IP_3R channel (a rapid process) but also initiates a slower regulation of IP_3R1 (Coquil et al., 1996). By using recombinant human IP_3R1 and IP_3R3 protein expressed in insect Sf9 cells, Yoneshima et al. (1997) have shown for the type-specific interconversion by Ca^{2+} that with increasing free Ca^{2+}, IP_3 binding to IP_3R1 decreased, whereas that to IP_3R3 increased. This diverse regulation of IP_3R types by intracellular Ca^{2+} may be one of the molecular bases underlying the complex spatio-temporal dynamics of intracellular Ca^{2+} in certain cell types.

The Ca^{2+}-mediated regulation of IP_3 binding has prompted us to elucidate Ca^{2+}-sensitized IICR that is now considered to be a plausible factor involved in intracellular Ca^{2+} dynamics, such as waves and oscillations. None of the IP_3R types cloned thus far have an authentic Ca^{2+}-binding site such as an *E-F hand* motif. Although the residues 1961–2220 of IP_3R1 and residues 1915–2175 of IP_3R2, both of which are far away from the ligand-binding domain (residues 225–578 [Yoshikawa et al., 1996]) located near the N-terminus, could directly bind $^{45}Ca^{2+}$ on blots (Mignery et al., 1992), there is no information about correlation between the IP_3 binding and Ca^{2+} binding. It should be clarified whether the differential regulation of IP_3 binding by Ca^{2+} is an intrinsic property of IP_3Rs or is mediated by other factor(s) capable of sensing Ca^{2+}, such as calmedin.

IP_3-Induced Ca^{2+} Release

It was shown that IP_3 induced the release of Ca^{2+} from a nonmitochondrial intracellular Ca^{2+} store (Streb et al., 1983). Many reports about IICR using permeabilized cells or microsomes which contain IP_3-sensitive Ca^{2+} pools have been accumulated ever since (see below and Berridge, 1993). Due to its intracellular localization, basic natures of the IICR channels have also been analyzed by utilizing planar lipid bilayers fused with smooth muscle (Ehrlich and Watras, 1988), cerebellar (Watras et al., 1991; Bezprozvanny et al., 1991, 1994; Bezprozvanny and Ehrlich, 1993, 1994; Katfan et al., 1997) ER, planar lipid bilayers incorporating IP_3R purified from cerebellum (Maeda et al., 1991), from smooth muscle (Meyrleitner et al., 1991), or from ventricular cardiac myocytes (Ramos-France et al., 1998), purified cerebellar IP_3R1 in reconstituted lipid vesicles (Ferris et al., 1990a, 1990b; Nakade et al., 1991, 1994; Hirota et al., 1995a, 1995b), molecularly expressed IP_3R1 (Miyawaki et al., 1990; Kaznacheyeva et al., 1998), and patch-clamping of nuclei of *Xenopus* oocytes (Mak and Foskett, 1994, 1997; Stehno-Bittel et al., 1995). The basic conduction properties of the cerebellar IP_3R1 channel have been extensively documented by Ehrlich and Bezprozvanny and their colleagues (Ehrlich et al., 1994; Bezprozvanny and Ehrlich, 1994, 1995; Parys and Bezprozvanny, 1995).

Cerebellar IP_3Rs, in which IP_3R1 is predominantly expressed (Furuichi et al., 1993) with the high proportion of ≈ 94–99% (De Smedt et al., 1994; Wojcikiewicz, 1995), were shown to act as a rather *nonspecific cation-selective channel* permeable to K^+ and Na^+, as well as to Ca^{2+} like RyR. Half-maximal IICR (EC_{50}) from IP_3Rs in cerebellar microsomes or in planar lipid bilayers was observed at $\approx 0.2\,\mu M$ IP_3 with Hill coefficient (n_H) of ≈ 1, and maximal IICR was observed above $\approx 1\,\mu M$ of IP_3 (Watras et al., 1991). Ramos-Franco et al. (1998) have shown that IP_3 sensitivity of the IP_3R2 channel

from ventricular cardiac myocytes is higher than that of the IP_3R1 (EC_{50}; 58 nM for IP_3R2 versus 196 nM for IP_3R1). The vicinal phosphates at positions 4 and 5 are essential for Ca^{2+} release while the position-1 phosphate increases affinity for IP_3R, and a rank order of potency to release Ca^{2+} is (1,4,5)-IP_3 > glycero-1-(4,5)-IP_2 > (2,4,5)-IP_3 > (4,5)-IP_2 > > > (1,4)-IP_2 (see Nahorski and Potter, 1989). Heparin of 0.5–10 $\mu g/ml$ completely inhibits IICR (see Bezprozvanny and Ehrlich, 1994; Ehrlich et al., 1994). Xestospondin (Xe), a class of macrocyclic 1-oxa-quinolizidine isolated from Australian sponge, has been shown to be a potent membrane-permeable blocker of IICR ($IC_{50} = 358\,nM$) without inhibition of IP_3 binding to IP_3R (Gafni et al., 1997). Recently, Lupu et al. (1998) have found that exogenously added PIP_2, a membrane component that is a substrate for PLC, binds to IP_3R, probably to the IP_3 binding site, thereby resulting in inhibition of IICR, and thus have suggested a possible functional link between IP_3R and PIP_2. In addition, the IICR activity of IP_3R1 is specifically inhibited by binding of anti-IP_3R1 monoclonal antibody (mAb) 18A10 which recognizes the C-terminus (Nakade et al., 1991, 1994; Miyazaki et al., 1992). How many IP_3s does each IP_3R complex (being composed of four subunits) bind to open its channel? The data for this issue are not definitive yet, but it was hypothesized that it requires one IP_3 molecule in planar lipid bilayers containing cerebellar IP_3R1 (Watras et al., 1991), and three or four in permeabilized hepatocytes expressing prominently IP_3R2 as well as IP_3R1 (Meyer et al., 1988; Watras et al., 1994), to open a channel. Cerebellar IP_3R incorporated into lipid bilayers exhibited multiple subconductance states occurring in the main substate at $\approx 90\%$ (Watras et al., 1991). At a transmembrane potential of 0 mV and with 55 mM divalent cation in the intraluminal (*trans*) chamber, the current amplitudes and single-channel slope conductances for all four alkaline earth cations fell in the sequence of Ba^{2+} (2.2 pA, 85 pS) > Sr^{2+} (2.0 pA, 77 pS) > Ca^{2+} (1.4 pA, 53 pS) > Mg^{2+} (1.1 pA, 42 pS), and the mean open time was significantly smaller with Ca^{2+} as a current carrier (2.9 ms) than with other alkaline earths (≈ 5.5 ms) (Bezprozvanny and Ehrlich, 1994). The same group also reported that Mn^{2+} was able to pass through the IP_3R channel (0.8 pA, 17 pS) (Striggow and Ehrlich, 1996). A selectivity for divalent cations of IP_3R is considerably lower; e.g., permeability ratio of Ba^{2+} and K^+ (P_{Ba2+}/P_{K+}) of cerebellar $IP_3R = 6.3$ (Bezprozvanny and Ehrlich, 1994), P_{Ca2+}/P_{K+} of smooth muscle $IP_3R \approx 2$ (Meyrleitner et al., 1991) in comparison with the value of $P_{Divalent}/P_{Monovalent} > 1000$ in the L-type Ca^{2+} channel (Lee and Tsien, 1984). Since hydrated Mg^{2+}, which has a water substitution rate more than

3 orders slower than the other alkaline earth cations, Na^+ and K^+ (Hille, 1992), can be permeable via IP_3R differently from its practical impermeability via the L-type Ca^{2+} channels ($Ba^{2+} > Sr^{2+} = Ca^{2+} >> Mg^{2+}$ [Hess et al., 1986]), the narrowest portion of the channel pore of IP_3R is likely to be at least $10\,\text{Å}^2$. In RyR, even large organic cations like Tris can permeate through its channel, implying a fairly wide pore size as estimated for RyR ($\approx 40\,\text{Å}^2$) (Linsay and Williams, 1991). Even under optimal conditions (in the presence of $2\,\mu M$ IP_3, 0.2–$0.5\,\mu M$ Ca^{2+}, and $0.5\,mM$ ATP) the maximal open probability (P_o) of cerebellar IP_3R1 in planar lipid bilayers was at most $\approx 10\%$ which is a marked contrast to the P_o of 100% in RyR, suggesting a low intrinsic efficacy of IP_3 as an agonist (Bezprozvanny et al., 1991; Watras et al., 1991; Bezprozvanny and Ehrlich, 1993, 1994). Under physiological ionic milieu (symmetrical $110\,mM$ K^+ and $2.5\,mM$ luminal free Ca^{2+}) and a transmembrane potential of $0\,mV$, the size of unitary Ca^{2+} current and the mean open time were estimated to be $\approx 0.47\,pA$ and $3.7\,ms$, respectively, suggesting release of 5400 Ca^{2+} per channel opening event, which is ≈ 20-fold fewer than that via RyR (the unitary Ca^{2+} current of $2.0\,pA$ and the mean open time of $\approx 20\,ms$). From these data, Bezprozvanny and Ehrlich (1994) have speculated that there seem to be two kinds of Ca^{2+} release in a cell; slow leakage release through IP_3R or rapid damping release through RyR. The basic IICR properties of *Xenopus* nuclear IP_3R are very similar to those of mammalian IP_3Rs described above (Mak and Foskett, 1994, 1997; Parys and Bezprozpanny, 1995; Stehno-Bittel et al., 1995;), except for slightly larger single-channel conductance for Ca^{2+} ($\approx 85\,pS$) (Mak and Foskett, 1994).

Although it was reported that prolonged exposure to IP_3 did not desensitize hepatic IP_3Rs (Oldershaw et al., 1992), being composed of prominently IP_3R2 and IP_3R1 (De Smedt et al., 1994; Monkawa et al., 1995; Wojcikiewicz, 1995), use of IP_3-induced Mn^{2+} quenching technique over an extended period, has recently shown that the IP_3 binding itself inactivates IP_3R-channel activity in permeabilized hepatocytes in an incremental manner, namely submaximal IP_3 doses inactivate only the sub-population of channels responding to that IP_3 dose (Hajnóczky and Thomas, 1994). The inactivation rate was enhanced by increases over the physiological range of cytosolic Ca^{2+} from $30\,nM$ to $1\,\mu M$, suggesting that activation of IP_3R by IP_3 binding is followed by inactivation dependent on the duration of exposure to IP_3, which is further stimulated by increased Ca^{2+}. Recovery of IP_3R from inactivation of IICR by a few-second prepulse of high Ca^{2+} was measured as only a few seconds in *Xenopus* oocytes (Parker and Ivorra, 1991), rat brain synaptosomes (Finch et al.,

1991), and cerebellar ER fused to planar lipid bilayers (Bezprozvanny and Ehrlich, 1994), in all of which IP_3R1 is prominently expressed (Ross et al., 1992; Furuichi et al., 1993; Kume et al., 1993; De Smedt et al., 1994; Wojcikiewicz, 1995). Recovery from IP_3-induced inactivation in permeabilized hepatocytes predominantly expressing IP_3R2 as well as IP_3R1 also occurred as a result of decreased levels of either IP_3 or Ca^{2+} in a time-dependent manner, e.g., inactivation by a 60-s prepulse of IP_3 gave recovery over a period of $\approx 60\,s$ (Hajnóczky and Thomas, 1994).

It has been shown that IICR form purified cerebellar IP_3R1 in reconstituted lipid vesicles is biexponential, with the fast and slow rate constants ($k_{fast} = 0.3$–$0.7\,s^{-1}$, $k_{slow} = 0.03$–$0.07\,s^{-1}$) indicating the presence of two states for the IICR (Hirota et al., 1995a 1995,b). The former was kinetically the low-affinity state for IP_3 and high permeability of Ca^{2+}, and the latter was constant over the range of 10–$5000\,nM$ IP_3. In addition, the IICR exhibited a moderate positive cooperativity ($n_H \approx 1.8$, $EC_{50} = 100\,nM$), whereas adenophostin-induced Ca^{2+} release displayed a high positive cooperativity ($n_H \approx 3.9$, $EC_{50} = 11\,nM$) (Hirota et al., 1995a, 1995b). Cooperativity, however, remains controversial: no cooperativity (Finch et al., 1991; Watras et al., 1991) versus positive cooperativity (Champeil et al., 1989; Meyer et al., 1990; Somlyo et al., 1992; Hirota et al., 1995a).

ATP   ATP plays a central role in energy metabolism. ATP binds to each subunit of purified cerebellar IP_3R1 with an affinity (K_d) of $2\,\mu M$ (Ferris et al., 1990a) or $17\,\mu M$ (Maeda et al., 1991). All IP_3R members certainly possess consensus sites for the ATP binding (Gly-x-Gly-x-x-Gly) as described above (e.g., 1773–1778, 1775–1780, and 2016–2021 of mouse IP_3R1SI+), but no data concerning the ATP binding has been reported in the other IP_3R types except for IP_3R1. In the IICR pharmacology, this ATP binding gives one of the remarkable features to IP_3R channels (Bezprozvanny and Ehrlich, 1995). In response to the direct binding to ATP, IICR activity significantly potentiates in IP_3Rs from cerebellum (Ehrlich and Watras, 1988; Ferris et al., 1990a; Maeda et al., 1991; Bezprozvanny and Ehrlich, 1993) and smooth muscles (Iino, 1991; Missiaen et al., 1997). Although IP_3R itself was reported to have a Ser kinase activity (Ferris et al., 1992a), this potentiation of IICR by ATP does not require hydrolysis of ATP because of the effectiveness of nonhydrolyzable ATP analogs. ATP does not act as an agonist by itself, in marked contrast with the agonistic action in RyR. Thus, ATP is called an *allosteric modulator of* IICR, AMP, GTP, and adenosine cannot substitute for ATP in this stimulatory effect. It has been shown that $3'$-phosphoadenosine $5'$-phosphosulfate (PAPS), di(adenosine-$5'$)tetraphosphate (Ap4A), and di(adenosine-$5'$pentaphosphate (Ap5A) are more effective than ATP (Missiaen et al., 1997).

There is a slight difference in effective doses among the experiments using different IP_3Rs (native vs. purified) and different assay systems ($^{45}Ca^{2+}$ flux in lipid vesicles vs. single-channel currents in lipid bilayers vs. Ca^{2+} release in permeabilized cells); e.g., IP_3-induced $^{45}Ca^{2+}$ flux by purified cerebellar IP_3R reconstituted into lipid vesicles was stimulated by ATP at $1–10\,\mu M$ (Ferris et al., 1990a), whereas at single-channel levels, submillimolar concentrations of ATP act as an allosteric modulator of cerebellar IP_3R in planar lipid bilayers in a biphase manner (effective at $0.1–1$ mM in purified IP_3R, Maeda et al., 1991; at $0.5–4$ mM in native IP_3R, Bezprozvanny and Ehrlich, 1993). The difference might be due to heterogeneity of the IP_3Rs assayed, although the ATP effects on IICR in these experiments could be attributed to a property of IP_3R1 in consideration of its preferential tissue distribution. The potentiation of IICR by ATP seems to be due to the significant increase in the mean open time and the open probability of the channels, and the ATP binding is thought to facilitate the coupling between the IP_3 binding and the channel gating (Bezprozvanny and Ehrlich, 1993). On the other hand, high concentrations of ATP inhibit IICR, but there are also differences in inhibitory concentrations among the reports; no augmentation of IP_3-induced $^{45}Ca^{2+}$ flux by the reconstituted proteoliposomes at 1 mM ATP (Ferris et al., 1990a), and half-maximal inhibition of IICR-channel currents by the purified IP_3R and cerebellar ER in reconstituted planar lipid bilayers at 2 mM ATP (Maeda et al., 1991) and 10.6 mM ATP (Bezprozvanny and Ehrlich, 1993), respectively. The inhibition of IICR by high concentrations of ATP was indicated to occur at the IP_3-binding site by a simple competitive mechanism, probably through phosphate moieties.

A physiological role of the allosteric modulation by ATP in IICR is as yet unclear, but in functions in cells the ATP dose-dependent regulation of IICR may be influenced by intracellular energy status. For example, ischemia, decreasing ATP levels from the resting level of about 1 mM to below 0.1 mM, was suggested to alter the IP_3R-gating property for changes in intracellular Ca^{2+} signaling (Bezprozvanny and Ehrlich, 1993). In any case, to understand the regulation of IICR signaling dependent upon energy status, it should be clarified whether the resting cytosolic levels of ATP concentrations are stimulatory or inhibitory for IICR in vivo, or how wide the in vivo ranges of fluctuation of ATP concentrations are.

As described previously, all members of the IP_3R family have the putative ATP-binding sites. Particularly striking is that of these ATP sites, one corresponding to the residues 2016–2021 of mouse IP_3R1 is well conserved among all the members reported so far. Therefore, the allosteric action of ATP appears to be employed generally in the regulation of all the family members. However, it has been shown that IICR in bronchial mucosal 16HBE140-cells, predominantly expressing IP_3R3, is 11 times less sensitive to ATP than that in A7r5 cells, predominantly expressing IP_3R1 (Missiaen et al., 1998).

NADH Reduced nicotinamide adenine dinucleotide (NADH) is a product of glycolysis and the citric acid cycle, and together with ATP plays an important role in energy metabolism. Like ATP, NADH was also shown to enhance the IICR activity. NADH increased IP_3-induced Ca^{2+} flux in purified cerebellar IP_3R1 reconstituted in lipid vesicles (maximal effects at $300\,\mu M$ and half-maximal stimulation at $30\,\mu M$) (Kaplin et al., 1996). NADH does not change IP_3 binding, and seems to act at the same site as ATP that also has adenine nucleotide moiety, since both stimulatory effects are not additive. This regulation of IP_3R by NADH is strikingly selective. NADH is four times more potent than NAD and produces 5-fold maximal effect. Moreover, NADPH and NADP that have a $2'$-phosphate on the ring of their adenine nucleotide moiety contrarily showed inhibitory effects on IP_3-mediated Ca^{2+} flux. This up-regulation of IP_3R by NADH was suggested to be responsible for the augmentation of cytosolic Ca^{2+} induced by hypoxia that leads to enhanced glycolytic formation of NADH (Kaplin et al., 1996). In such energy emergencies, NADH may participate to settle a proper level of intracellular Ca^{2+} concentration in concert with ATP in vivo. If so, NADH and ATP are likely to be *dual modulators for emergency IICR*, synergistically binding the same site on IP_3R.

Ca^{2+} As described above (see "Structure and Function of IP_3R1"), IP_3Rs have a long N-terminal arm ($\approx 80\%$ of the molecule) and a short C-terminal tail ($\approx 5\%$) in the cytoplastic side, clustered MSDs and a channel pore embedded in the membranes of Ca^{2+} stores, and a relatively large loop (≈ 100 amino acids) containing the N-glycosylation sites in the luminal side of the pool. In this transmembrane topology, most of the IP_3R part is, thus, exposed to Ca^{2+} in either cytosolic or luminal sides, and some parts likely sense changing Ca^{2+} in either side. The cytosolic Ca^{2+} concentration in the resting cell is controlled at the level of $50–100$ nM in many cell types (Amundson and Clapham, 1993). The effects of cytosolic Ca^{2+} on IICR are well documented: a small increase in cytosolic Ca^{2+} acts as a *coagonist* for

IICR, possibly leading to Ca^{2+}-sensitized (or -activated) IICR, whereas a large increase in Ca^{2+} inhibits it. Whenever a channel opens and Ca^{2+} passes through the channel pore, the area around the vicinity of the pore region on the cytoplasmic side is expected to be exposed to the higher Ca^{2+} released. Thus, there seem to be *positive and negative feedback loops* of IICR regulated by released Ca^{2+}. Recently, using chimeric aequorin with reduced affinity for Ca^{2+} and the ER targeting signal, as well as SR^{2+} (a Ca^{2+} surrogate), luminal Ca^{2+} in intact HeLa cells was estimated to be > 1–2 mM (Montero et al., 1995), but there is much discrepancy among reports (Pozzan et al., 1994; Combettes et al., 1996). The effects of intraluminal Ca^{2+} on IICR are still controversial, although there are some reports about the influence of luminal Ca^{2+} on the IP_3 sensitivity of IICR.

Cytosolic Ca^{2+}   IICR activity was shown to be regulated by cytosolic Ca^{2+} in a *biphasic* dose-dependent manner ($< \approx 300$ nM stimulatory, $> \approx 300$ nM inhibitory) (Iino, 1990; Bezprozvanny et al., 1991; Finch et al., 1991). Thus, IP_3R would be regulated by cytosolic Ca^{2+} at two distinct sites, a stimulatory site and an inhibitory site whose affinity is high and low, respectivly (see Dufour et al., 1997; Hajnóczky and Thomas, 1997; Thrower et al., 1998). However, it is still unclear how changing cytosolic Ca^{2+} concentrations are sensed by these two sites (see below). In skinned smooth muscle, cytosolic Ca^{2+} formed a *positive feedback loop* in IICR below 300 nM, and a negative feedback loop above 300 nM (Iino, 1990). IP_3-induced $^{45}Ca^{2+}$ release from synaptosome-derived microsomal vesicles showed that cytosolic Ca^{2+} acts as a *coagonist* of IICR, and sequential positive and negative feedback regulation by release Ca^{2+} was suggested to contribute to Ca^{2+} oscillations (Finch et al., 1991). Single-channel recording from the cerebellar ER fused with planar lipid bilayers showed the *bell-shaped Ca^{2+} dependence curve* of IICR with maximum probability of opening at $0.2\,\mu M$, which is a striking contrast to the range between 1 and $100\,\mu M$ for maximum activity of RyR (Bezprozvanny et al., 1991). Striggow and Ehrlich (1996) showed that cytosolic Mn^{2+} exhibited a similar bell-shaped modulation of cerebellar IP_3R1, and that the affinity of the stimulatory site is similar for both Ca^{2+} and Mn^{2+}, but that of the inhibitory site is about 15 times lower for Mn^{2+} than for Ca^{2+}. Using photolysis of caged Ca^{2+} (DM-nitrophen), it was shown that a focal rise in Ca^{2+} immediately triggered Ca^{2+}-dependent feedback control of IICR in a permeabilized smooth muscle (Iino and Endo, 1992). Using IP_3R1-enriched cerebellar microsomes, Kaftan et al. (1997) have shown that the Ca^{2+} dose dependency is affected by IP_3 concentrations; at lower IP_3,

the peak of the Ca^{2+}-dependence curve shifts to lower Ca^{2+}, but at higher IP_3 ($189\,\mu M$), IICR activiy persists at Ca^{2+} as high as $30\,\mu M$, suggesting dual regulation by both agonist IP_3 and coagonist Ca^{2+}. In contrast, Ramos-Franco et al. (1998) have shown that IP_3R2 purified from ventricular cardiac myocytes is more sensitive to cytosolic Ca^{2+} (activated at about 25 nM) than the cerebellar IP_3R1, but is not inhibited at high Ca^{2+} (even at millimolar Ca^{2+}). Therefore, type-specific Ca^{2+} dependency implies a complexity of Ca^{2+} regulation probably consisting of Ca^{2+}-dependent alterations of the IP_3 binding, transducing, as well as channel-gating properties.

It is still unclear how IP_3R is regulated by changes in cytosolic Ca^{2+}, and whether Ca^{2+} acts directly or indirectly on IP_3R. For example, at the IP_3-binding level, it was proposed that Ca^{2+}-mediated changes in the binding kinetics lead to the Ca^{2+} dependency of IICR, i.e., increasing Ca^{2+} induces the interconversion of the IP_3-binding affinity from the sensitized low-affinity state (active for IICR) to the desensitized high-affinity state (inactive for IICR), leading to a negative feedback regulation of IICR (Pietri Rouxel et al., 1992; Watras et al., 1994), although it is as yet unclear whether the negative feedback regulation by the Ca^{2+}-mediated interconversion of the IP_3-binding site is equivalent to the biphase regulation of IICR by Ca^{2+}. Since phosphorylation is generally a reversible modification system of protein functions and IP_3Rs have several consensus sites for phosphorylation in the transducing domain, as described above, it is reasonable that a feedback regulation of IP_3Rs via *kinase/phosphatase cycle* dependent upon changing Ca^{2+} is considered for the molecular basis underlying the biphasic regulation of IICR by Ca^{2+}; due to different Ca^{2+} affinities, Ca^{2+}-dependent kinases and Ca^{2+}-dependent phosphatases might opposingly control IICR (see "Phosphorylation"). For example, it was suggested that at cytosolic low Ca^{2+} (between 30 and 100 nM), Ca^{2+}/calmodulin-dependent protein kinase II (CaMKII) activates IICR by phosphorylation of IP_3R, whereas at high cytosolic Ca^{2+} (above 400 nM), Ca^{2+}-dependent phosphatase 2B (PP2B = calcineurin) reverses the kinase action by dephosphorylation (Zhang et al., 1993). Cytosolic Ca^{2+} levels which activate either CaMKII or PP2B, however, seem to remain elusive, since bidirectional synaptic plasticity dependent on cytosolic Ca^{2+} levels, long-term potentiation (LTP) and long-term depression (LTD), are proposed to be due to the activation of CaMKII at high Ca^{2+} levels and of PP2B at moderate Ca^{2+} levels, respectively (Lisman, 1994). Another Ca^{2+}-dependent kinase/phosphatase cycle (protein kinase C/calcineurin) was also suggested (Cameron et al., 1995a). It was shown that increased Ca^{2+}-induced degradation of

IP$_3$R1 with Ca^{2+}-activated neutral thiol-dependent protease calpain in vitro, producing a C-terminal channel polypeptide of 95 kDa, which may be one of the aspects of Ca^{2+}-regulated IICR (Magnusson et al., 1993). It is also possible that Ca^{2+}-dependent and IP$_3$R-associated molecules, such as calmedin (inhibitory effect on the IP$_3$ binding at high Ca^{2+} [Danoff et al., 1988], see "IP$_3$ Binding") and calmodulin (as yet controversial effect in IICR, see "Calmodulin") are involved in the biphasic regulation of IICR by Ca^{2+}.

Besides the effects of Ca^{2+} on IP$_3$R functions, changing Ca^{2+} might act on other molecules that influence IICR signaling. For example, increased Ca^{2+} was proposed to activate Ca^{2+}-sensitive phospholipase C, resulting in the production of IP$_3$ ligand, i.e., release (or increased) Ca^{2+} via IICR triggers the reactivation of IP$_3$/Ca^{2+} signal cascade (reactivation of Ca^{2+}-sensitive phospholipase C [PLC]→regeneration of IP$_3$ →reactivation of IP$_3$R→IICR) (e.g., see Harootunian et al., 1991; Mignery et al., 1992). This positive feedback loop retroactive to PLC activation is still controversial, since it is unclear whether PLC is actually stimulated by released Ca^{2+} in vivo.

It is noteworthy that the putative Ca^{2+}-binding site of IP$_3$R1 and IP$_3$R2 (residues 1961–2220 and 1915–2175, respectively) was mapped near the channel domain (Mignery et al., 1992), i.e., only about 56 amino acides upstream of the first putative membrane-spanning domain, M1. Seinaert et al. (1996) subsequently mapped it to 23 residues between 2124 and 2146. From a similarity in the structural arrangements between IP$_3$R and RyR (Furuichi and Mikoshiba, 1995), this putative Ca^{2+} site of IP$_3$R appears homologous to the high-affinity sites for Ca^{2+} near the channel region of RyR1, in which the Ca^{2+} sites themselves are proposed to trigger the gating in response to the binding to Ca^{2+} (Chen et al., 1992, 1993b). Thus, released Ca^{2+} may act on the putative Ca^{2+} site near the channel, in either a stimulatory or an inhibitory way, or both.

Changing cytosolic Ca^{2+} concentrations might affect a variety of target sites (or molecules) which synergistically leads to complex IICR signaling in cells. The biphase regulation of IICR by Ca^{2+} is now considered to be a plausible factor for the generation of spatiotemporal dynamics of cytosolic Ca^{2+}, such as Ca^{2+} waves or oscillations (for details, see Parker and Ivorra, 1991; DeLisle and Welsh, 1992; Leichleiter and Clapham, 1992; Miyazaki et al., 1992; Amundson and Clapham, 1993; Berridge, 1993; Bezprozvanny 1994; Bootman and Berridge, 1995; Clapham, 1995).

Luminal Ca^{2+}   There are many reports regarding possible involvements of luminal Ca^{2+} in IICR, although the effects are still debatable. It is known that submaximal IP$_3$ concentrations release only a fraction of Ca^{2+} from the IP$_3$-sensitive Ca^{2+} pools (partial release) without affecting their releasability in response to further increases in IP$_3$ concentrations, and in this way IICR has been described as *quantal* in nature (Muallem et al., 1989). Molecular mechanisms underlying the quantal release are as yet unclear, although two models have been proposed: the *steady-state release (or partial emptying)* model in which IP$_3$-sensitive Ca^{2+} pools are uniformly sensitive to IP$_3$, and the luminal levels of Ca^{2+} that decrease due to Ca^{2+} release by submaximal IP$_3$ concentrations somehow limit further release, resulting in partial release (Missiaen et al., 1992a); and the *all-or-none release (or all-or-nothing emptying)* model in which IP$_3$-sensitive Ca^{2+} pools are heterogeneous in sensitivity to IP$_3$, and submaximal IP$_3$ concentrations release the entire contents from each pool sensitive to a particular IP$_3$ concentration (Oldershaw et al., 1991). To account for this *quantal release*, an interaction between luminal Ca^{2+} and IP$_3$ was proposed (Irvine, 1990).

There are some reports that luminal Ca^{2+} increased the sensitivity to IP$_3$ (Missiaen et al., 1992a, 1992b, 1994; Nunn and Taylor, 1992; Oldershaw and Taylor, 1993; Horne and Meyer, 1995). Negative data were also reported (Shuttleworth, 1992; van de Put et al., 1994). The inhibitory action of elevated luminal Ca^{2+} of ≥ 1 mM was observed in single-channel levels in planar lipid bilayer experiments (Bezprozvanny and Ehrlich, 1994). The stimulatory effect of luminal Ca^{2+}, however, was shown to be pronounced only at low luminal Ca^{2+} levels, i.e., below 30% of maximal store filling in a rat embryonic aorta cell line, A7r5 (Parys et al., 1993) or decreased by $\approx 95\%$ in hepatocytes (decreased from 200–300 μM of the total luminal Ca^{2+} to $< 10 \mu$M) (Combettes et al., 1996), thereby suggesting that with discharge (or lowering) of the luminal Ca^{2+} only below certain threshold levels, the IP$_3$ sensitivity can decrease to cause the quantal behavior. In this way, IICR was also shown to be coactivated by both cytosolic and luminal Ca^{2+} (Missiaen et al., 1994). The quantal release was observed even in a reconstituted system using purified IP$_3$R (Ferris et al., 1992b, Hirota et al., 1995a). Using a highly purified IP$_3$R1-isoform in reconstituted lipid vesicles, the IICR kinetics were depicted as a composition of two rate constants (fast and slow), and the quantal release was attributed to the fast component with low affinity for IP$_3$ (Hirota et al., 1995a). In addition, adenophostin, a known IP$_3$ surrogate, also displayed a similar quantal nature of IICR in these reconstituted IP$_3$R1-containing vesicles (Hirota et al., 1995b). These data appear to indicate that a heterogeneity in IP$_3$R types, on which the all-or-none model is grounded as one of the plausible factors, is not indispensable for the quantal release and that the quantal release is an intrinsic nature of IP$_3$R itself, although it cannot be ruled out that differences in modification state of single types, such as phosphorylation, are involved in a quantal behavior. However, besides luminal Ca^{2+}, several factors responsible for the quantal behavior have been proposed: degree of luminal continuity between IP$_3$-sensitive Ca^{2+} pools (Renard-Rooney et al., 1993), desensitization of IP$_3$R by IP$_3$ binding (Hajnóczky and Thomas, 1994), cooperatively in opening of multiple channels with different affinities (Meyer et al., 1990; Kindman and Meyer, 1993), conversion between the low- and high-affinity IP$_3$-binding sites in cytosolic Ca^{2+}- and temperature-dependent manners (Watras et al., 1994), and heterogeneity of IP$_3$R density in IP$_3$-sensitive Ca^{2+} pools (Hirose and Iino, 1994; also see Missiaen et al., 1995).

In the case of the changing luminal Ca^{2+} contents contributing toward the quantal release, how does IP$_3$R sense them? In regard to a possible direct action, the luminal loop, enriched acidic amino acids Glu and Asn, between the M5 and the pore region was proposed to serve as a luminal Ca^{2+}-binding site (Yoshikawa et al., 1992). In fact, residues 2463–2528 in the luminal loop have subsequently turned out to bind ^{45}Ca^{2+} (Sienaert et al., 1996). IP$_3$R may associate with accessory protein(s) that can sense luminal Ca^{2+}. Chromogranin A (CGA), a high-capacity and low-affinity Ca^{2+}-binding protein present abundantly in chromaffin granules and also in IP$_3$-sensitive Ca^{2+} pools, was shown to bind a short luminal loop between the pore region and M6 directly in a pH-dependent manner (Yoo and Lewis, 1994, 1995). Thus luminal Ca^{2+}-binding proteins (e.g., calreticulin and CGA) can sense changing luminal Ca^{2+} and transduce it to IP$_3$R.

The loading state of IP$_3$-sensitive Ca^{2+} pools likely influences the Ca^{2+} entry, as well as Ca^{2+} release. It has been documented that the depletion of IP$_3$-sensitive Ca^{2+} pools by IP$_3$ or thapsigargin, an inhibitor of the Ca^{2+} pump in the ER, is somehow coupled with Ca^{2+} entry across the plasma membrane; i.e., *capacitative Ca^{2+} entry* (for reviews, see Putney, 1990, 1993; Penner et al., 1993; Putney and Bird, 1993; Fasolato et al., 1994; Berridge, 1995).

Caffeine   The xanthine drug caffeine is a well-known agonist of CICR channels, RyRs. In contrast, an inhibitory effect of this drug on IICR channels was demonstrated in *Xenopus* oocytes (Parker and Ivorra, 1991), cerebellar microsomes (Brown et al., 1992), permeabilized smooth muscle cells (Hirose et al., 1993), and cerebellar microsomes fused with planar lipid bilayers (Bezprozvanny et al., 1994). In the permeabilized cells, caffeine inhibits Ca^{2+}-sensitized

IICR with half-maximal inhibition of 2 mM (Hirose et al., 1993). At the single-channel level, caffeine decreases the frequency of channel opening in a non-cooperative fashion with half-inhibition at 1.64 mM (Bezprozvanny et al., 1994). Increased IP$_3$ concentration overcame the inhibitory effect by caffeine, and caffeine did not compete with IP$_3$ binding. How does caffeine act on IP$_3$R? Interestingly, IP$_3$R and RyR share some sequence similarities present mostly in the last two membrane-spanning domains (M5 and M6) and the successive C-terminus, and a little in the ligand-binding domain (Furuichi et al., 1989b; Migner et al., 1989; Furuichi and Mikoshiba, 1995). Although caffeine action is different between IP$_3$R and RyR, these patchy homologous regions may act as a functional site for this drug.

Phosphorylation   Cerebellar IP$_3$R1 has been characterized as a good substrate for cAMP-dependent protein kinase (PKA) by several groups since the early stages of its biochemical study (Groswald and Kelly, 1984; Mikoshiba et al., 1985; Walaas et al., 1986). This PKA phosphorylation was clarified by using purified IP$_3$R (Supattapone et al., 1988a; Yamamoto et al., 1989). Certainly, the cloned members of the IP$_3$R family have several potential sites for phosphorylation by a variety of kinases, such as PKA, protein kinase C (PKC), and Ca^{2+}/calmodulin-dependent protein kinase II (CaMKII), in the modulatory and transducing domains (Furuichi and Mikoshiba, 1995), and in vitro phosphorylation of purified cerebellar IP$_3$R1 reconstituted into lipid vesicles by these three kinases was additive suggesting the existence of different sites for each kinase (Ferris et all, 1990b). These data indicate that subtle regulation of the transducing action by "cross-talk" among these diverse phosphorylation signaling pathways appears to cause differential IP$_3$/Ca^{2+} signaling in various cell types. Note, however, that most of these studies were carried out in vitro, and thus little is known about in vivo phosphorylation. Therefore, the in vivo phosphorylation of IP$_3$R and its effects on in vivo IICR should be studied, since, in most cases, in vitro phosphorylation is suggestive but not always substantial in vivo.

PKA   All mammalian types have the consensus sites for phosphorylation by cAMP-dependent protein kinase (PKA) in the modulatory and transducing domain. As described above, cerebellar IP$_3$R1 is a good substrate for PKA in vitro (Supattapone et al, 1988a; Yamamoto et al., 1989). Two PKA consensus sites, Ser-1589 and Ser-1756, of IP$_3$R1 are separated by the alternative splicing region SII, which is present in the neuronal type (SII +) but absent in the non-neuronal type (SII −) (see "Alternative Splicing Subtypes of IP$_3$R1") (Danoff et al., 1991;

Nakagawa et al., 1991a). Interestingly, this splicing was shown to affect the phosphorylation site and kinetics, i.e., the cerebellar IP_3R1 (neuronal type, SII+) was preferentially phosphorylated at the Ser-1756 with an affinity (K_m) of 17.5 nM, whereas the vas deferens IP_3R1 (nonneuronal type, SII−) was phosphorylated at the Ser-1589 with K_m of 3 nM (Danoff et al., 1991; Ferris et al., 1991). In vivo phosphorylation of IP_3R1 in intact cells was studied using primary cultures of mouse cerebellar neurons (Yamamoto et al., 1989) and rat liver hepatocytes stimulated by glucagon or dibutyryl-cAMP (Joseph and Ryan, 1993). In vitro phosphorylation of cell extracts of WB rat epithelial cells by the exogenous PKA displayed no phosphorylation of IP_3R3 in marked contrast to ready phosphorylation of IP_3R1, though between the two, some populations form a heterotetramer (Joseph et al., 1995b). Wojcikiewicz and Luo (1998) have recently reported that both IP_3R2 and IP_3R3 are phosphorylated by PKA in intact cells, but that the phosphorylation efficiencies of both are not as good as that of IP_3R1.

PKA phosphorylation appears to have no effect on the IP_3 binding in most of studies reported so far, e.g., it exhibited no significant change in $[^3H]$-IP_3 binding affinity of cerebellar microsomes (Supattapone et al., 1988a; Volpe and Alderson-Lang, 1990). In crude IICR assay systems using cerebellar microsomes (Supattapone et al., 1988a; Volpe and Alderson-Lane, 1990), containing predominantly IP_3R1, and of permeabilized hepatocytes (Burgess et al., 1991; Hajnóczky et al., 1993), expressing prominently IP_3R2 as well as IP_3R1, PKA phosphorylation increased in the IP_3-releasable Ca^{2+} amounts, which was proposed to be due to indirect influences, e.g., an augmentation of Ca^{2+}-loading activity of PKA-phosphorylated Ca^{2+} pump (Supattapone et al., 1988a) and a recuitment of additional stores into IP_3-sensitive pools (Hajnóczky et al., 1993), and due to direct changes of IP_3R, e.g., a recruitment of additional IP_3R sensitized by phosphorylation and an increase in the conductance and/or the open probability of the IP_3R channel (Volpe and Alderson-Lang, 1990). Effects of PKA phosphorylation on IICR-channel functions are still controversial, and/or may be dependent on cell types in which IP_3R types are not only differentially expressed but also form heterotetrmers being in different subunit compositions. Phosphorylation of cerebellar microsomes, predominantly IP_3R1, decreased IP_3 sensitivity of IICR by 10% (Supattapone et al., 1988a), or increased the apparent Michaelis constant (K_m) of IICR for IP_3 by 2-fold, from 0.3 to 0.7 μM (Volpe and Alderson-Lang, 1990). Phosphorylation of platelet internal membranes, containing at least IP3R1, by endogenous internal membrane-associated kinase, unlikely to be PKA, increased 2-fold the rate of IICR, but fol-

lowing exogenous phosphorylation with the catalytic subunit of PKA, the enhancement was reversed (Quinton et al., 1996). On the other hand, PKA phosphorylation in permeabilized hepatocytes showed increases in IP_3 sensitivity of IICR (Burgess et al., 1991; Hajnóczky et al., 1993). Stimulation of hepatocytes with either glucagon or dibutyryl-cAMP markedly increased phosphorylation of hepatic IP_3R in vivo, and this phosphorylation enhanced Ca^{2+} sensitivity of IP_3-binding sites (Joseph and Ryan, 1993). Many substrates for PKA could be present in these crude assay systems. More sophisticated experiments using purified cerebellar IP_3R1 reconstituted in lipid vesicles indicated that the IP_3-induced $^{45}Ca^{2+}$ flux was increased after the PKA phosphorylation (Nakade et al., 1994). Therefore, in this debatable issue, the data obtained with the purified IP_3R1 system seem to be conclusive. However, there seems to be heterogeneity in the regulation of IICR by PKA in vivo, since all mammalian IP_3R types have putative PKA phosphorylation sites and are differentially distributed as described above. What is the molecular mechanism underlying PKA phosphorylation? It may cause a conformational change that increases the efficacy of the transducing or gating activities.

In conclusion, the effects of PKA phosphorylation on in vivo IICR are not clear yet and may vary in different cell types that have different IP_3R types. These data indicate that two second messengers, IP_3 and cAMP, produced through the different G-protein pathways (e.g. $G_{q/11} \rightarrow$ phospholipase C versus $G_s \rightarrow$ adenylylcyclase, respectively), likely regulate IICR synergistically in vivo (Mauger et al., 1989; Burgess et al., 1991; Bird et al., 1993), and, thus, actual IICR signaling is dependent not only upon the phosphorylation state of IP_3R but also upon other PKA phosphoproteins, such as the Ca^{2+} pump whose phosphorylation might increase in IP_3-sensitive Ca^{2+} pool sizes.

PKC Through hydrolysis of phosphatidylinositol bisphosphate (PIP_2), two second messengers, diacylglycerol (DAG) and IP_3 are produced. Therefore, when the IP_3/Ca^{2+} signaling pathway is stimulated, some of the protein kinase C (PKC) family are simultaneously activated by DAG and Ca^{2+} released via IICR (Nishizuka, 1992). There are several PKC consensus sites in all IP_3R types. It was shown that PKC exogenously added phosphorylated purified cerebellar IP_3R1 reconstituted into lipid vesicles at three different Ser residues: one major and two minor sites (one minor site is identical to the PKA site) in vitro (Ferris et al., 1990b). With 12-*O*-tetradecanoylphorbol-13-acetate (TPA) treatment, nuclear IP_3R from rat liver was shown to be phosphorylated (Matter et al., 1993). The phosphorylated liver nuclei showed an accelerated and enhanced $^{45}Ca^{2+}$ release by IP_3 without alteration of $[^3H]$-IP_3 binding properties (K_d and B_{max}) (Matter et al., 1993).

Recently, it was shown that calcineurin, a Ca^{2+}/CaM-dependent protein phosphatase, is associated with purified IP_3R-FKBP12 complex (see "IP_3R-binding proteins"). The inclusion of a specific calcineurin inhibitor (cyclosporin A or FK506) in preparation of the complex drastically increased phosphorylation by exogenous PKC more than 10-fold but not significantly by PKA and CaMKII, suggesting that PKC and the associated calcineurin act at the same sites on the cerebellar IP_3R (Cameron et al., 1995a). Interestingly, in the presence of calcineurin inhibitor, cerebellar microsomes phosphorylated by PKC displayed increase in IP_3 potency in augmenting $^{45}Ca^{2+}$ flux, with the IC_{50} for IP_3 reduced about 20-fold, and thus it was suggested that the activation of calcineurin by released Ca^{2+} in turn cancels the potentiation of IICR by PKC phosphorylation (Cameron et al., 1995a). Therefore, IP_3R may be regulated by *differential actions of PKC and calcineurin at the same site.* Since the activation of phospholipase C liberates both DAG and IP_3 triggering the activation of PKC and IP_3R, respectively, in vivo IICR signaling may be subtly regulated by PKC phosphorylation state.

Recently, a novel alternative splicing region, SIII (9 amino acids), has been found between residues 903 and 904 of human IP_3R1SI-(between residues 917 and 918 of mouse IP_3R1), and has been suggested as a differential creation of a site for PKC; i.e., the SIII + acquires a new PKC potential site (**KLKS**) in the SIII insert, but the SIII− has a potential site (**NKGS**) just at the joining point after deleting the SIII (Nucifora et al., 1995) (see "Alternative Splicing Subtypes of IP_3R1").

CaMKII Although there is no information available about downstream regulation of the IICR signaling, we, will first consider Ca^{2+}/calmodulin (CaM)-dependent signaling pathways, since CaM has a wide variety of functions as well as a widespread distribution. Among these functions, Ca^{2+}/CaM-dependent protein kinase II (CaMKII) seems to be an important *CaM mediator* to transmit the Ca^{2+} signal further downstream. Interestingly, purified cerebellar IP_3R1 was shown to be phosphorylated by exogenous CaMKII in vitro (Yamamoto et al., 1989; Ferris et al., 1990b), suggesting a feedback loop of $IP_3R1 \rightarrow Ca^{2+} \rightarrow Ca^{2+}$/CaM→CaMKII→$IP_3T1$. The CaMKII phosphorylated at three diffeent sites (one major site and two minor sites, and mostly at Ser and less than 5% at Thr), all of which differed from the PKA and PKC sites (Ferris et al., 1990b).

Cerebellar membranes phosphorylated by CaMKII displayed a potentiated IP_3-induced $^{45}Ca^{2+}$ flux (Cameron et al., 1995a). A possible feedback regulation of IICR by a *kinase/phosphatase cycle* by cytosolic Ca^{2+} was suggested; Ca^{2+}-dependent kinase and Ca^{2+}-dependent phosphatase, having different Ca^{2+} affinities from one another, opposingly control IICR in a Ca^{2+}-dependent manner: CaMKII activates a channel by phosphorylation, whereas Ca^{2+}-dependent phosphatase 2B (calcineurin) attenuates the channel activation by dephosphorylation (Zhang et al., 1993).

It has also been shown that CaMKII phosphorylation/dephosphorylation cycles of CaMKII sites in HeLa cells are required for histamine-induced Ca^{2+} oscillations, and CaMKII phosphorylation of IP_3R is stimulated by in vivo ^{32}P-labeling of HeLa cells stimulated by histamine (Zhu et al., 1996). In this esperiment, phosphorylated IP_3R was detected only by immunoprecipitation with anti-IP_3R1 antibody, and thus was not defined as a phosphorylated IP_3R type (as IP_3R1) since the proportion of IP_3R types expressed in HeLa cells is as yet unclear and, if multiple types are expressed, other types that form heterotetramers with IP_3R1 might be co-immunoprecipitated by the anti-IP_3R1 antibody.

PKG cGMP-dependent protein kinase (PKG) phosphorylated purified IP_3R1 from rat cerebellum at Ser-1755, identical to the major site for PKA, but not at Ser-1589, the minor site for PKA (Komalavilas and Lincoln, 1994). Purified IP_3R from vascular smooth muscle in which IP_3R1 is prominent was also phosphorylated by PKG (Koga et al., 1994). Although the relaxation of vascular smooth muscle induced by a number of compounds, such as nitric oxide (NO), atrial natriuretic peptides, etc., is likely to be induced by increasing cGMP-mediated reduction in cytosolic Ca^{2+}, the Ca^{2+}-lowering mechanism is yet debatable. By treatment with atrial natriuretic peptide or sodium nitroprusside (NO production), cGMP levels were elevated and simultaneously phosphorylated IP_3R increased by 30–40% in primary culture of rat aortic smooth muscle cells, suggesting that PKG phosphorylation of IP_3R may cause down-regulation of IICR (Komalavilas and Lincoln, 1994). Contrarily, PKG was shown to inhibit IP_3 production, leading to suppression of cytosolic Ca^{2+} transients (Ruth et al., 1993).

It is quite interesting that the Ser-1755 of IP_3R1 serves as the same substrate for both PKG and PKA, with similar high affinities in vitro (Komalavilas and Lincoln, 1994), suggesting a fine-tuning of the IICR property by a *collaborative phosphorylation* with two kinds of cyclic nucleotide-dependent kinases, i.e., PKG and PKA.

Tyrosine Kinase Tyrosine phosphorylation and activation of phospholipase C (PLC)-γl isoform can be achieved by the action of receptor tyrosine kinases (RTKs, such as the receptors for growth factors EGF, PDGF, etc., and neurotrophins NGF, BDNF, etc.) or nonreceptor protein tyrosine kinases (PTKs, such as *fyn* and *lck*) (Rhee and Choi, 1992). Tyrosine phosphorylation signaling thus attracted considerable attention to its involvement in a chain of the IP_3/Ca^{2+} signaling. It was proposed that IP_3R1 has two sites for tyrosine kinases in the ligand-binding domain (EDLvY-482 of human IP_3R1 and EDLvY-482 of mouse IP_3R1) and in the C-terminal cytoplasmic tail (DsTEY-2617 of human $IP_3R1SII-$ and DsTEY-2655 of mouse IP_3R1), and that IP_3R2 has only a site in the tail (DpTEY-2607 of human IP_3R2), whereas IP_3R3 does not have both sites (Harnick et al., 1995). If so, these two sites may be involved in the ligand binding and channel functions, respectively.

In human T-cells (Jurkat), which express all three IP_3R types (Monkawa et al., 1995), activation with the anti-CD3 antibody resulted in IP_3R being tyrosine phosphorylated (Harnick et al., 1995). Activation of T-cell receptor (TCR)-CD3 complex results in recruitment of the *src* tyrosine kinase family (e.g. *fyn* and *lck*) to a cap complex, which is tightly associated with signal transduction during T-cell activation. Upon antigen-stimulation of T-cells (Jurkat) expressing (all) types IP_3R1, IP_3R2, and IP_3R3 (Sugiyama et al., 1994a, 1994b; Monkawa et al., 1995), IP_3R1-immunoreactivity was shown to be co-capped with the TCR-CD3 complex (Khan et al., 1992b), and inhibition of IP_3R1 expression by antisense experiment displayed the requirement of at least functional IP_3R1 for T-cell activation (Jayaraman et al., 1995). From these data, Marks and his colleagues suggested that during T-cell activation, activated tyrosine kinases may act on both the IP_3 signaling step and the IICR signaling step; i.e., activation of tyrosine-phosphorylated PLC-γl produces IP_3 that triggers IICR, and tyrosine-phosphorylated IP_3R may alter IICR properties (Harnick et al., 1995; Jayaraman et al., 1995). These workers have reported that IP_3R1 is phosphorylated by *fyn* in vitro, the phosphorylation is reduced in thymocytes from *fyn*$-/-$ mice, and the phosphorylated IP_3R1 in a planar lipid bilayer displays increase in the P_o with no changes in the conductance and the mean open time (Jayaraman et al., 1996). Recently, Hirota et al. (1998) have demonstrated that IP_3R1-deficient T-cells, expressing both IP_3R2 and IP_3R3 to almost the same extent as wild-type T-cells, display Ca^{2+} mobilization and proliferation properties in response to stimuli. Therefore, at least the notion that only IP_3R1 is critical for T-cell signaling can be ruled out.

Autophosphorylation Rat cerebellar IP_3Rs purified by affinity for both heparin and concanavalin A, followed by reconstitution into liposomes, were shown to be autophosphorylated at Ser residues, and thus were suggested to have serine protein kinase activity (Ferris et al., 1992a). However, its autophosphorylation was not confirmed using mouse cerebellar IP_3Rs purified by immunoaffinity for anti-IP_3R1 antibody, followed by reconstitution into liposomes (Nakade et al., 1994). Therefore, this enzymatic activity is debatable and should be re-examined in detail.

IP_3R-Binding Proteins Recent studies showed the presence of proteins that associate with IP_3R directly. Most of these *IP_3R-binding proteins* (or *IP_3R-accessory proteins*) likely contribute to IP_3R functions in different aspects.

Calmodulin In the intracellular Ca^{2+} signaling flows, we must take into account the possibility of activation of calmodulin (CaM), which acts as a Ca^{2+} *signal transducer* that regulates a wide variety of Ca^{2+}/CaM-dependent target proteins by directly binding, and thereby inducing, their conformational change for functional regulation. CaM binding to IP_3R1 are still controversial. IP_3R1 purified from rat cerebellum failed to adhere to a CaM–agarose column in either the presence or absence of 1 mM $CaCl_2$ (Supattapone et al., 1988b), whereas IP_3R1 purified from mouse cerebellum bound to CaM–Sepharose column in a Ca^{2+}-dependent manner (Maeda et al., 1991). In addition, it has recently been shown that IP_3R purified from porcine aorta binds to CaM–Sepharose column in a Ca^{2+}-dependent manner, but IP_3R from porcine and rat cerebellum do not (Islam et al., 1996).

CaM exerted no effect on [^{3}H]-IP_3 binding to IP_3R1 purified from mouse cerebellum (Maeda et al., 1991). Effects of CaM and CaM antagonists on IICR are still controversial. A CaM antagonist, W7, potentiated IP_3-induced $^{45}Ca^{2+}$ flux in permeabilized pancreatic cells and isolated microsomes (Wolf et al., 1986), whereas other CaM antagonists, W13 and CGS9343B, in addition to W7, inhibited IICR in permeabilized and fura-2-loaded rat liver epithelial (261B) cells (Hill et al., 1988). Rat hepatocytes loaded with fura-2/AM exhibit Ca^{2+} oscillations when they are stimulated by αl-adrenergic agonist phenylphrine, leading to the phosphoinositide breakdown followed by Ca^{2+} release. In this system, CaM antagonists calmidazolium (R24571) and CGS9343B could block or reduce the frequency of Ca^{2+} oscillations, or restore them following overstimulation by the hormone that resulted in a nonoscillatory Ca^{2+} elevation (Somogyi and Stucki, 1991). IICR from mouse cerebellar microsomes was not significantly affected in the presence of CaM or CaM antagonist W7 (Maeda et al., 1991).

However, more careful studies should be done, because much endogenous CaM could contaminate crude brain membranes. In all these studies, it seems that CaM likely acted not only directly on IP_3R but also on CaM mediators such as CaMKII and calcineurin, and thus it is not conclusive whether IP_3R was a single target for CaM action.

The recombinant mouse IP_3R1 (residues 1564–1585) and IP_3R2 (residues 1558–1596) expressed in *E.coli* cells exhibited CaM-binding activity in a Ca^{2+}-dependent manner, whereas the human IP_3R3 (full-length) expressed in NG108-15 (neuroblastoma/glioma hybrid cell line) showed no such activity (Yamada et al., 1995). The estimated K_d of the synthetic peptide corresponding to the CaM-binding site of mouse IP_3R1 (residues 1564–1585) for CaM was $0.7 \mu M$ (Yamada et al., 1995). In contrast, it has recently been proposed that the putative CaM-binding site extends over the joining point after splicing out the SII region in porcine aorta $IP_3R1SII-$ that has the binding activity, but is disrupted by an insertion of the SII region in porcine cerebellar IP_3RSII+ that has no CaM-binding activity (Islam et al., 1996). However, this study in porcine IP_3R1 has not been certified by substantial experimentation, and this needs to be done.

It serves as a reference that all members of the RyR family also have several CaM-binding sites with different affinities (Guerrini et al., 1995). CaM binding itself, without the involvement of CaM-dependent kinases, seems to inhibit Ca^{2+} release from the sarcoplasmic reticulum and single-channel currents from RyR reconstituted in planar lipid bilayers (for a review, Meissner, 1994). Direct observation of CaM/RyR complex, by electron microscopy indicated that CaM binding causes no apparent change in the quaternary structure, and that the CaM-binding site is most likely to be on the cytoplasmic surface of each subunit, at least 10 nm from the central channel-forming region, suggesting that CaM binding induces allosteric, long-range structural changes that influence channel gating (Wagenknecht et al., 1994).

FKBP The immunophilin protein family mediates the actions of immunosuppressant drugs such as cyclosporin A (CsA), FK506, and rapamycin. Within this family, members of the cyclophilin class bind to CsA, while members of the FK506 binding-protein (FKBP) class bind to FK506 and rapamycin. The cyclophilins and FKBPs have peptidylprolyl-*cis-trans* isomerase (PPIase or rotamase) activity which is thought to be associated with protein folding and is inhibited by binding to CsA, and FK506 and rapamycin, respectively. There is a little information about molecular mechanisms underlying immunoreactions through the immunophilin family, e.g.,

inhibition of calcineurin by FK506/FKBP and CsA/cyclophilin complexes (Liu et al., 1991). Recently, it has been shown that FKBP12 is stoichiometrically associated with, and modulates, RyR1 by increasing the channel to full conductance level and by decreasing the open probability (Jayaraman et al., 1992; Timerman et al., 1993, 1995; Brillantes et al., 1994). Recently, it has also been found that IP_3R1 copurifies with FKBP12 from rat cerebellum, and the copurification is inhibited in the presence of FK506 or rapamycin but not in the presence of CsA (Cameron et al., 1995b; Snyder et al., 1998). In rat cerebellar microsomes, FK506 reduced by 50% the $^{45}Ca^{2+}$ uptake into cells but increased by about 10-fold the IP_3 sensitivity of $^{45}Ca^{2+}$ flux (Cameron et al., 1995b). The FKBP12 binding to RyR appears to stabilize the channel, e.g., it likely improves cooperation among the four RyR subunits of the channel to prevent aberrant channel activity such as channel flickering and partial opening (Brillantes et al., 1994). The modulation of the RyR channel by FKBP12 was independent of the rotamase activity (Timerman et al., 1995). Similarly, it has been presumed that FKBP12-bound IP_3R1 is somehow stabilized without rotamase activity, and that when FKBP12 is stripped from the IP_3R1–FKBP12 complex by adding FK506, the channel becomes leaky (Cameron et al., 1995b). The FKBP12/IP_3R1 complex has been shown to associate with, and be modulated by, calcineurin, which likely controls the phosphorylation status of IP_3R1 (Cameron et al., 1995a; Snyder et al., 1998) (see "Calcineurin"). Recently, it has been reported that FKBP12 binds IP_3R1 at residues 1400–1401, a leucylprolyl dipeptide that structurally resembles FK506 (Cameron et al., 1997).

Calcineurin Calcineurin, Ca^{2+}/CaM-dependent Ser/Thr protein phosphatase (PP2B), binds to FK506/FKBP or CsA/cyclophilin complexes in a Ca^{2+}-dependent fashion, and its phosphatase activity is inhibited as a result of the binding, suggesting that calcineurin is the relevant target of these immunosuppressants in vivo and a key molecule in the Ca^{2+}-sensitive and immunosuppressant-sensitive signaling pathways involved in the T-cell receptor (TCR)-mediated transcription in T-lymphocytes and the IgE receptor-mediated exocytosis in mast cells (Liu et al., 1991). As described above, it has been shown that IP_3R1 binds FKBP12 (Cameron et al., 1995b), and calcineurin is copurified with the FKBP12/IP_3R1 complex from rat cerebellum (Cameron et al., 1995a). The data of $^{45}Ca^{2+}$ flux in cerebellar microsomes phosphorylated by various kinases showed that the PKC site was much more sensitive to the associated calcineurin than the CaMKII site, and that FK506 blocked dephosphorylation of a PKC site, thereby suggesting that calcineurin activated by released

Ca^{2+} abolishes the potential activation of IICR by PKC phosphorylation in a feedback regulation (Cameron et al., 1995a). The authors of the report proposed that calcineurin is associated with rat cerebellar IP_3R1 via an FKBP12 anchor in a Ca^{2+}-sensitive manner.

Cytoskeletal proteins The IP_3R-Ca^{2+} pools appear to interact with the cytoskeleton (Rossier and Putney, 1991; Bourguignon et al., 1993; Joseph and Samanta, 1993; Bourguignon and Jin, 1995), which may be required for subcellular anchoring and/or for the dynamics of the pools (Terasaki and Jaffe, 1991). Certainly, the immunostaining pattern with the anti-IP_3R antibody changed in *Xenopus* eggs before and after fertilization, suggesting structural plasticity of the IP_3R-Ca^{2+} pools (Kume et al., 1993). IP_3R was shown to bind ankyrin, which is known to link various transmembrane proteins to the actin network through its interaction with spectrin or fodrin (Bourguignon et al., 1993; Joseph and Samanta, 1993; Bourguignon and Jin, 1995). Recently, RyR was also shown to bind ankyrin (Bourguignon et al., 1995). Although ankyrin was shown to bind to the synthetic peptide 2546-GGVGDVLRKPS-2556 of mouse IP_3R1 (Bourguignon and Jin, 1995), this region has been proposed to be in the luminal side (Yoshikawa et al., 1992; Michikawa et al., 1994) and differs in sequence from the corresponding region of RyR1 (identical in only 5 amino acids out of 11) (Furuichi et al., 1989b).

Chromogranin A Chromogranin A (CAG) is a high-capacity and low-affinity Ca^{2+}-binding protein present in chromaffin granules, and it has been shown to interact with IP_3R in a pH-dependent manner (bound at acidic pH of 5.5 but not near physiological pH of 7.5) (Yoo, 1994; Yoo and Lewis, 1994, 1995). CGA bound to the synthetic peptide 2502-DVLRRPSKDEPLFAARVVYD-2521, corresponding to the short luminal loop just between the putative pore and M6 of rat IP_3R2 (corresponding to the 2550-DVLRKPSKEEPLFAARVIYD-2569 of mouse IP_3R1), at pH 5.5 but not at pH 7.0 (Yoo and Lewis, 1994); the interaction occurred between tetrameric CGA and four molecules of the peptide and was stabilized in the presence of 35 mM Ca^{2+} rather than in the absence of Ca^{2+} (Yoo and Lewis, 1995). Following the passage from *cis*- to *trans*-Golgi network, the acidification of the vesicular milieu appears to occur from $\approx$pH 7.5 to $\approx$pH 5.5, respectively. Thus, it has been proposed that CAG forms tetramers at acidic pH of 5.5 in the *trans*-Golgi and then interacts with the four-luminal loop of tetrameric IP_3R to form a well-ordered structural organization beneath the membrane, and this interaction facili-

tates the cosegregation of CGA and IP_3R into the secretory granules from the *trans*-Golgi network in chromaffin cells (Blondel et al., 1995; Yoo and Lewis, 1995). It has been hypothesized that the molecular interaction between CGA and IP_3R is changed as a result of binding to IP_3 and then Ca^{2+} binding to CGA (buffered Ca^{2+}) near the channel pore becomes Ca^{2+}-releasable by IP_3R (free Ca^{2+}?) in the lumen of secretory granules (Yoo and Lewis, 1995).

Mutation and Disruption of IP₃R Genes

The *opisthotonos (opt)* mouse is a naturally occurring autosomal recessive mutant that exhibits a severe ataxic and convulsive phenotype. Since the IP_3R1 gene was mapped to the vicinity of this *opt* mutation locus on the mouse chromosome 6, a linkage between IP_3R1 and *opt* was suggested (Furuichi et al., 1993). The *opt* mutant has subsequently turned out to have a genomic deletion in the IP_3R1 gene which is a deletion of two exons following the SII region, resulting in the interruption of the translational reading frame thereafter (Street et al., 1997).

Before this finding by Street et al. (1997), IP_3R1 was gene targeted and homozygous mutant mice ($IP_3R1-/-$) were produced (Matsumoto et al., 1996). Most of the $IP_3R1-/-$ mice die in utero, and at postnatal day 10 the $IP_3R1-/-$ frequency is about 5.5%. It is intriguing that $IP_3R1-/-$ pups are born exhibiting similar phenotypes as reported in the *opt* mutant.

IP_3-binding activity of the cerebellar microsome fraction of the $IP_3R1-/-$ mice exhibited a significant reduction in [^{3}H]-IP_3-binding activity (B_{max} value 2.0 pmol per mg protein) compared to the wild-type control (22.8 pmol per mg protein). Corresponding to the lowered binding activity, IP_3-induced Ca^{2+}-release activity is decreased. ATP-dependent Ca^{2+} uptake to the Ca^{2+} store is normal compared to the wild mice. The $IP_3R1-/-$ mice clearly show loss of function of IP_3R1.

Body mass and overall brain sizes of $IP_3R1-/-$ mice are decreased about 50% and 60%, respectively. The gross brain anatomy of the $IP_3R1-/-$ mice appeared normal. Haematoxylin–eosin staining and immunohistochemistry against antimicrotuble-associated protein-2 (MAP-2) antibody showed no significant difference between the $IP_3R1-/-$ and $IP_3R1+/+$ mice. Gross anatomy and light-microscopic observation of paraffin-embedded sections (stained with haematoxylin-eosin) of various tissues of $IP_3R1-/-$ mice showed no difference from the wild-type mice.

Since IP_3R1 is predominantly expressed in cerebellar Purkinje cells, electrophysiological analysis was carried out in Purkinje cells. A combination of high-frequency Na^+ spikes and oscillatory Ca^{2+} spikes, a unique feature of the Purkinje cell, is observed in both types of mice. The persistent and tetrodotoxin-sensitive Na^+ current and the inward rectifier current are also observed in both types of mice. No qualitative difference was observed between passive and active membrane properties of Purkinje cells from the $IP_3R1-/-$ and $IP_3R1+/+$ mice. The time constant of PF (parallel fiber)-mediated excitatory postsynaptic current (e.p.s.c.) decay is significantly shorter in $IP_3R1-/-$ mice. The all-or-none character of CF (climbing fiber) mediated excitatory postsynaptic potentials (e.p.s.p.) suggests that the Purkinje cell is monoinnervated by CF. CF-mediated e.p.s.c. in $IP_3R1-/-$ mice show a significantly stronger depression feature at pulse intervals from 20 ms to 2 s. Therefore, $IP_3R1-/-$ Purkinje cells are not severely impaired but have some quantitative differences. Inoue et al. (1998) have recently shown using cerebellar slices, that $IP_3R1-/-$ mice lack long-term depression (LTD), a model of synaptic plasticity in the cerebellum.

$IP_3R1-/-$ mice show ataxia at about postnatal day 9. This is characterized by loss of balance while standing or walking. Repetitive tonic or tonic–clonic seizures prevailed by postnatal day 20–23, and opisthotonus-like postures are often observed. Electroencephalogram shows clear paroxysmal polyspoke activities, whereas only stable background activities are observed. It is, therefore, clear that IP_3R1 plays a physiologically important function, but also works to suppress, not to cause, epileptic seizures. Intraperitoneal injections of anticonvulsants, pentobarbital or diazepam, suppress the seizures and cerebellar ataxic movements become prominent. After introduction of flumazenil (an antidiazepam drug) intraperitoneally, epileptic seizures suppressed by the diazepam reappeared.

Recently further interesting data on the cellular functions of IP_3Rs have been accumulated using mutation and gene disruption approaches. In the chick DT40 B cell line, disruption of all IP_3R family members caused loss of B-cell receptor-induced response, i.e., Ca^{2+} signaling and apoptosis (Sugawara et al., 1997). In *Drosophila*, disruption of the IP_3R gene resulted in embryonic death *Drosophila* IP_3R has thus been suggested to play an essential role in embryonic and larval development (Acharya et al., 1997; Venkatesh and Hasan, 1997) and in ecdysone release (Venkatesh and Hasan, 1997). In *C. elegans*, five suppressor mutations for fertility defects due to the LET23 (a homolog of the mammalian EGF receptor) mutation have been isolated and mapped to the IP_3R gene, suggesting that IP_3R is involved in normal hermaphrodite fertility (Clandinin et al., 1998).

RyR Family

RyR was originally described in the sarcoplasmic reticulum (SR) of skeletal and cardiac muscle, and was thought to be a component of excitation–contraction (E-C) coupling. (For reviews, see Endo, 1977; Schneider, 1994; Franzini-Armstrong and Jorgensen, 1994; Meissner, 1994.) As a result of molecular cloning, we now know that mammals have at least three distinct types of RyR (Figs. 9.5 and 9.6) (for reviews see McPherson and Campbell, 1993a; Sorrentino and Volpe, 1993), i.e., two muscular RyRs, the skeletal type or type 1 (sRyR or RyR1) (Takeshima et al., 1989, Zorzato et al., 1990) and the cardiac type or type 2 (cRyR or RyR2) (Nakai et al., 1990; Otsu et al., 1990b), and a novel brain type RyR or type 3 (bRyR or RyR3) (Giannini et al., 1992; Hakamata et al., 1992; Nakanishi et al., 1997). The type names are based on their dominant tissue distribution (sRyR and cRyR) or the source of molecular cloning (bRyR). Recently, the RyR family has been shown to be widely localized in nonmuscle cells as well.

RyR1 (sRyR)

The RyR1 cDNAs were cloned from rabbit (Takeshima et al., 1989; Zorzato et al., 1990), and human (Zorzato et al., 1990).

Structure and Function of RyR1

The complete sequence of RyR1 has been reported by two groups: 5037 amino acids long (565 kDa) in the rabbit (Takeshima et al., 1989) and its short-type (5032 amino acids, 564 kDa) lacking five amino acids (residues 3481–3485) in both the rabbit and human (Zorzato et al., 1990, 1994). Recently, an additional alternative splicing region (residues 3865–3870) was detected in mouse RyR1 (Futatsugi et al., 1995). RyR1 has been postulated to have four (M1–M4) (Takeshima et al., 1989) or twelve (M′, M″, and M1–M10) (Zorzato et al., 1990) MSDs clustered near the C-terminus. Grunwald and Meissner (1994) made four site-directed antibodies against RyR1 and showed evidence that suports the transmembrane topology model with four membrane-spanning segments. In this model, the final two MSDs, the most highly conserved sequences between RyR and IP_3R, reveal the same topology as that predicted for the M5 and M6 segments of the IP_3R family. It was shown that the C-terminal 656 amino acids of RyR1 (13% of the receptor protein) are expressed in the brain (Takeshima et al., 1993). This short brain-specific receptor contains a part of the putative modulatory domain (two Ca^{2+}-binding sites and one ATP-binding site) and a complete set of

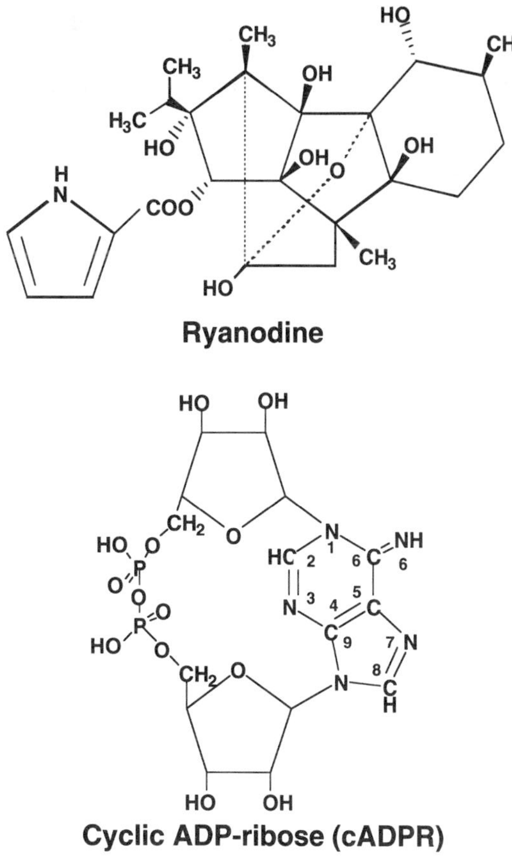

Ryanodine

Cyclic ADP-ribose (cADPR)

Figure 9.5 Structure of ryanodine and cyclic ADP-ribose. Both ryanodine (Ry) and cyclic ADP-ribose (cADPR) activate RyR channels. Ry is a natural plant alkaloid, and acts as an activator at $< 10\mu$M but as an inhibitor at $> 10\mu$M. cADPR is thought to be synthesized from reduced nicotinamide adenine dinucleotide (NAD^+) by ADP-ribosyl cyclase or CD38. 8-NH_2-cADPR is a potent competitive antagonist.

putative channel domains of RyR1. By deletion of the C-terminal residues 183–4004 from RyR1, about 20% of RyR1 has been found to contain structures sufficient to form a functional CICR channel (Bhat et al., 1997a,b). Therefore, the brain-specific truncated RyR1 may function as a CICR channel which is more simply regulated than authentic RyR1, without forming a foot structure. There are the characteristic repeated segments occurring four times in two doublets (R1–R2 and R3–R4) in this large cytoplasmic region, which may form a foot structure (Zorzato et al., 1989). An RyR1 complex observed with an electron microscope (EM) was seen to have a four-leaf clover structure (quatrefoil, @ 27×27 nm square) with a central hole of 1–2 nm diameter (Radermacher et al., 1992).

In vertebrate skeletal muscle, E-C coupling seems to occur by a mechanical coupling mechanism involving protein–protein interactions between the dihydropyridine receptor (DHPR) of the transverse tubule membrane and the RyR of the SR membrane. The cytoplasmic II–III loop peptides of skeletal and cardiac muscle DHPR α1 subunits activate the RyR1

(Lu et al., 1994). PKA-mediated phosphorylation of the cytoplasmic II–III loop peptides of the DHPR α1 subunits fails to activate the RyR (Lu et al., 1995). Recent findings demonstrate that residues 1303–1406 of RyR1, a region that is highly divergent among RyR family members, is critical for E-C coupling &Yamazawa et al., 1997), and that there is a region for retrograde communication from RyR1 to DHPR (Nakai et al., 1996, 1997). Triadin is an intrinsic membrane protein first identified in the skeletal muscle junctional sarcoplasmic reticulum. Triadin was previously proposed to bind RyR and DHPR and to serve as the "linking protein" that mediates the signal transduction process between these two Ca^{2+} channels in skeletal muscle. Luminal domain of triadin is able to bind both RyR and calsequestrin in both skeletal (Caswell et al., 1991) and cardiac muscle SR (Guo et al., 1996).

RyR1 channel activities are regulated by various modulators, including Ca^{2+}, Mg^{2+}, ATP, CaM, caffeine, ryanodine, ruthenium red, FK506-binding protein (FKBP12), procaine, dantrolene, spermidine, and polyanion, and the functional sites of some of

RyR FAMILY

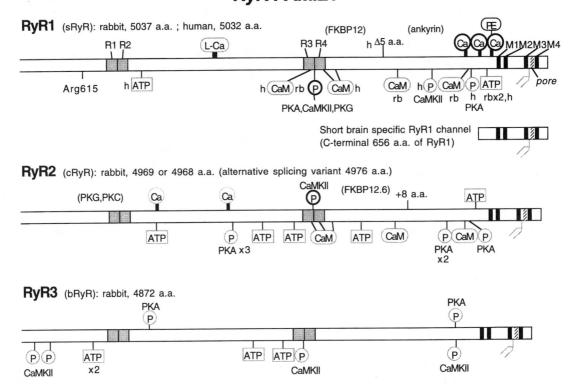

Figure 9.6 Structure of the RyR family. Functional domains and sites are depicted. The defined sites and putative sites are clarified by solid and light symbols, respectively. Putative membrane-spanning domains (MSDs: M1, M2, M3, and M4) and putative "pore" segments are shown by solid and shaded vertical bars, respectively. Branched bars in the channel domain represent N-glycosylation sites. Four long repeats in two doublets (R1–R2 and R3–R4) are shown by shaded boxes in pairs. Phosphorylation sites: PKA, cAMP-dependent protein kinase; PKG, cGMP-dependent protein kinase; PKC, protein kinase C; CaMKII, calmodulin-dependent protein kinase II; Tyr-K, tyrosine kinase. Ca, Ca^{2+}-binding site. PE, Pro-Glu repeat in Ca^{2+}-binding site. CaM, calmodulin-binding site; ATP, ATP-binding site; FKBP, FK506-binding protein.

these factors have been proposed (Takeshima et al., 1989; Zorzato et al., 1990). RyR1 expression in cDNA-transfected cells showed elevation of [^{3}H]ryanodine-binding activity with a K_d of 19 nM (Takeshima et al., 1989), and a large conductance channel activated by ATP and Ca^{2+} and inhibited by Mg^{2+} and inhibited by Mg^{2+} and ruthenium red (Chen et al., 1993a). Three Ca^{2+}-binding sites were found in the N-terminal vicinity of the channel domain in the rabbit RyR1 (Zorzato et al., 1990; Chen et al., 1992). The antibody against residues 4478–4512 containing the third Ca^{2+}-binding site increased the Ca^{2+}-sensitivity of RyR1 channels (Chen et al., 1992). It was suggested that the Pro-Glu (PE) repeat, which matches the third high-affinity Ca^{2+}-binding site (Zorzato et al., 1990), forms a

site involved in the Ca^{2+}-activation mechanism (Chen et al., 1993b). In addition, the antibody against residues 4425–4621 of human RyR1 decreased Ca^{2+}-induced $^{45}Ca^{2+}$ efflux from isolated terminal cisternae, suggesting the existence of a Ca^{2+}-dependent gating domain lying near this epitope region (Treves et al., 1993). The large (2200 kDa) Ca^{2+}-release channel complex of RyR binds with high-affinity ($K_d = 5$–25 nM) 16 CaMs at $\leq 0.1\,\mu M\ Ca^{2+}$ and 4 CaMs at $100\,\mu M\ Ca^{2+}$ (Tripathy et al., 1995). At $\leq 0.2\,\mu M\ Ca^{2+}$, CaM activates the RyR several fold and at micromolar Ca^{2+} concentrations, CaM inhibits the Ca^{2+}-activated channel several fold (Tripathy et al., 1995). These results suggest a role for CaM in modulating RyR1 in skeletal muscle at both resting and elevated Ca^{2+} concentrations. ATP is known to activate the RyR1 channel. Photoaffinity labeling by Bz$_2$ATP showed the presence of a single ATP-binding site per RyR tetramer, but two or more sites with different affinity were also observed (Zarka and Shoshan-Barmatz, 1993). It was shown that one serine residue (Ser-2843) of rabbit RyR1 is phosphorylated by PKA, CaMKII, and cGMP-dependent protein kinase (PKG), and one threonine residue by CaMKII (Suko et al., 1993). Phosphorylation of RyR1 by membrane-associated protein kinases enhances the open probability by increasing the sensitivity toward micromolar Ca^{2+} and ATP (Herrmann-Frank and Varsanyi, 1993). Recordings from excised patches of SR membrane have revealed that Ca^{2+}-release channel activity of RyR1 is inactivated by CaMKII phosphorylation (Wang and Best, 1992). FKBP12 modulates channel gating by increasing channels to full conductance levels, decreasing open probability after caffeine activation, and increasing mean open time (Brillantes et al., 1994). FK506 or rapamycin, an inhibitor of FKBP12 isomerase activity, reverses these stabilizing effects. Thimerosal, a thiol-oxidizing reagent, inhibits ryanodine binding of skeletal RyR by decreasing the binding capacity, but does not affect the binding affinity or the dissociation rate of bound ryanodine (Abramson et al., 1995). Thimerosal (100–200 μM) stimulates single-channel activity without modifying channel conductance. A Ca^{2+}-stimulated channel is first activated and then inhibited in a time-dependent fashion by high concentration of thimerosal (1 mM). Once inactivated, the channel cannot be reactivated by addition of either Ca^{2+} or ATP (Abramson et al., 1995). There is a functional interaction between RyR1 and triadin which involves redox cycling of hyperreactive sulfhydryls in response to channel activation and inactivation (Liu and Pessah, 1994). Ruthenium red block of the RyR is due to binding to multiple sites located in the conduction pore of the channel (Ma, 1993). Both high- and low-affinity binding sites for ryanodine exist in SR membranes derived from skeletal muscle. Carboxyl-terminal tryptic fragment (76 kDa) of RyR1 contains the high-affinity binding site for ryanodine (Callaway et al., 1994; Witcher et al., 1994). Two ryanodine-binding sites are different but are either allosterically or sterically coupled (Wang et al., 1993).

A mutation was identified in the RyR1 of porcine malignant hyperthermia (MH) in which Arg-615 is replaced by Cys (for a review see MacLennan and Phillips, 1992). The porcine MH RyR1-channels were shown to be hypersensitive to various modulators. The region around Arg-615 of RyR1 has fragmentary sequence homology with the corresponding region of IP$_3$R1 (Furuichi and Mikoshiba, 1995). Since this region of IP$_3$R1 is necessary for ligand

binding, Arg-615 of RyR1 may be involved in the binding of regulators of Ca^{2+}-channel gating.

Expression of RyR1

Although RyR1 was shown to be expressed predominantly in fast- and slow-twitch skeletal muscle, expression was also detected in the esophagus and brain. RyR1 is almost exclusively localized in the cerebellar Purkinje cells of the chick (Walton et al., 1991; Ouyang et al., 1993) and the mouse (Kuwajima et al., 1992), but widely distributed throughout the brain of weakly electric gymnotiform fish (Zupanc et al., 1992). By using chimeric genes composed of the upsteam region of the RyR1 gene and the bacterial chloramphenicol acetyltransferase gene, Schmoelzl et al. (1996) showed that the RyR1 gene expression is regulated by at least two novel transcription factors and that tissue specificity results from a transcriptional repression in nonmuscle cells mediated by the first intron.

RyR2 (cRyR)

The RyR2 cDNAs were cloned from rabbit cardiac muscle (Nakai et al., 1990; Otsu et al., 1990b).

Structure and Function of RyR2

RyR2 consists of 4969 amino acids (Otsu et al., 1990b), and 4968 amino acids (or 4976 amino acids with an insertion of 8 amino acids between residues 3715 and 3716) (Nakai et al., 1990). The amino acid sequence of RyR2 was 66% identical to that of RyR1. The functional unit of the RyR2 channel contains only one high-affinity ryanodine-binding site/tetramer (with a four-leaf clover appearance on negative-stain EM) (Anderson et al., 1989).

RyR2 is thought to function as a CICR channel, thereby amplifying and/or regenerating Ca^{2+} that is locally increased by Ca^{2+} influx through cardiac DHPR activated by depolarization. The molecular basis of regulation of the SR Ca^{2+}-release channel in cardiac and skeletal muscle is different and the cardiac SR channel isoform lacks a Ca^{2+}-inactivated site (Chu et al., 1993). The RyR2-channel is activated by micromolar Ca^{2+} and millimolar ATP and inhibited by millimolar Mg^{2+} and CaM (Witcher et al., 1991; McPherson and Campbell, 1993a). CaM inhibits RyR2-channel activity, reducing the open probability, shortening the mean open time, and producing prolonged closures (Witcher et al., 1991). RyR2 is phosphorylated by PKA, PKG, PKC, and CaMKII (Takasago et al., 1989, 1991; Hohenegger and Suko, 1993). The phosphorylation site for CaMKII was determined at Ser-2809 by sequencing phosphopeptides, and this CaMKII phosphorylation of RyR2 reverses the inhibitory effect of CaM on the open probability and restores prolonged channel openings (Witcher et al., 1991). Recent findings indicate that a novel FK506-binding protein (FKBP12.6), which mediates the immunosuppressive effects of FK506, is associated with RyR2 (Lam et al., 1995). It is interesting that FKBP12.6 is not associated with RyR1. Sorcin, a 22 kDa calcium-binding protein initially identified in multidrug-resistant cells, is associated with RyR2 (Meyers et al., 1995). The role of sorcin has been postulated as stabilizer of the RyR2 channel protein. H_2O_2 increases open probability of the cardiac RyR and, therefore, it is probable that H_2O_2 activates the SR Ca^{2+}-release channel via an oxidation of cysteine thiol groups in the channel protein (Boraso and Williams, 1994). Negative surface charges which exist near the luminal mouth of the cardiac RyR may potentiate conduction by increasing the local Ca^{2+} concentration and thus act as a preselection filter for the channel (Tu et al., 1994).

Expression of RyR2

RyR2 is expressed both in cardiac muscle and throughout the brain (particularly cerebellum) but is notably absent from the pituitary (Nakai et al., 1990; Otsu et al., 1990b; Hakamata et al., 1992; Lai et al., 1992). Among the three types, RyR2 is the predominant type in the brain. Expression of RyR2 in the brain has been shown using specific antibodies (Kuwajima et al., 1992; McPherson and Campbell, 1993b; Sharp et al, 1993b) or nucleic acid probes (Furuichi et al., 1994a) and by analyzing phosphorylation of brain RyR (Witcher et al., 1992; Yoshida et al., 1992; McPherson and Campbell, 1993b). RyR2 is expressed in widespread areas of the brain, e.g., cerebral cortex, hippocampus, amygdala, caudate/putamen, olfactory bulb, anterior olfactory nucleus, thalamus, hypothalamus, cerebellum, brain stem, and spinal cord (Kuwajima et al., 1992; Sharp et al., 1993b; Furuichi et al., 1994a).

RyR3 (bRyR)

Hakamota et al. (1992) cloned RyR3 cDNA from a rabbit brain cDNA library by screening with RyR1 cDNA probes. Giannini et al. (1992) independently isolated a partial cDNA clone of RyR3 as a gene $\beta4$ induced by transforming growth factor-$\beta1$ (TGF-$\beta1$) in mink lung epithelial cells (MvlLu).

Structure and Function of RyR3

RyR3 consists of 4872 amino acids (Hakamata et al., 1992). The amino acid sequence of RyR3 is 67% and 70% identical to that of RyR1 and RyR2, respectively. RyR3 also forms a homotetramer (Murayama and Ogawa, 1996). RyR3 demonstrates Ca^{2+}-dependent ryanodine binding and caffeine increase its Ca^{2+}-sensitivity (Murayama and Ogawa, 1996). It has been well documented that the opening of both RyR1 and RyR2 channels is stimulated by either ryanodine or caffeine. In contrast, TGF-$\beta1$-treated MvlLu cells increase $[Ca^{2+}]_i$ in response to ryanodine but not caffeine (Giannini et al., 1992). Takeshima et al. (1995) used myocytes from mice homozygous for a targeted mutation in the RyR1 gene (dyspedic mice) for a study on the function of RyR3, which is predominantly expressed in these cells, and demonstrated caffeine-, ryanodine-, and adenine nucleotide-sensitive CICR in these cells with $\sim$ 10 times lower sensitivity to Ca^{2+} than that of RyR1 (Takeshima et al., 1995). They proposed that RyR3 may induce intracellular Ca^{2+} release in response to a Ca^{2+} rise with a high threshold (Takeshima et al., 1995). Chen et al. (1997a) have shown that recombinant rabbit RyR3 expressed in HEK293 cells displays similar single-channel properties to those of RyR1, except for gating kinetics (longer mean open time), extent of maximal activation by Ca^{2+}, and about 10 times less sensitivity in inhibition by high Ca^{2+} concentrations. Chen et al. (1997b) have also shown that substituting Glu-3885 for Ala (E3885A) causes a reduction in Ca^{2+} sensitivity, without apparent changes in channel conductance and in modulation by ATP and ryanodine.

Expression of RyR3

RyR3 is predominantly present in the hippocampal CA1 region, striatum, and dorsal thalamus of rabbit brain (Hakamata et al., 1992; Furuichi et al., 1994b). RyR3 is abundantly detected immunohistochemically in hippocampus, corpus striatum, and diencephalon (Murayama and Ogawa, 1996). RyR3 is expressed in human Jurkat T-lymphocyte cells (Hakamata et al., 1994). In the murine uterus and vas deferens, expression of RyR3 is localized to the smooth muscle component of these organs (Giannini et al., 1995). In the testis, expressions of RyR1 and RyR3 is detected in germ cells (Giannini et al., 1995).

Disruption of RyR Genes

Among the three ryanodine receptors, gene targeting of RyR1, which is mostly expressed in the skeletal muscle, is reported. RyR1$-/-$ in mice is lethal perinatally and they are dead as neonates. The mice remain purple due to respiratory failure because they cannot breathe. They do not even move after birth. The mutant neonates display an abnormal curvature of the spine, thin limbs and a thick neck area, and abnormal rib cage and arched vertebral column.

These features are similar to mice deficient in myogenin (Hasty et al., 1993; Nabeshima et al. 1993) and muscular dysgenic (*mdg*) mice. The *mdg* mutation is expressed in skeletal muscle as a failure of E-C coupling due to the absence of the DHPR. These three gene-deficient mice, which have mutation of skeletal muscle, have similar abnormal features.

It is considered that DHPR and RyR1 receptors interact at junctions between the SR and surface membrane in developing muscle and between SR and T tubules in adult muscle. Most of the muscle fibers of the wild mice are fully striated, filled with myofibrils and have peripherally located nuclei. However, the skeletal muscle from the RyR1−/− mice is characterized by degeneration (Takeshima et al., 1994). The fibers are small and fragmented by loose and amorphous tissues and the nuclei are located centrally. Electron microscopic study indicates that the myofibrils in muscle fibers of neonates are variable in size and mostly smaller and fewer than those of the controls. The findings obtained from the analysis of RyR1−/− mice and dysgenic mice offered information for understanding the structural component of the junctional gap between the surface membrane and SR. The "foot" structure is composed of cytoplamic domains of RyRs anchored in the SR membrane (Inui et al., 1987; Block et al., 1988; Lai et al., 1988). Tetrads are clusters of four proteins in surface membrane that are located in correspondence to the "feet." In dysgenic mice, myotubes are missing both DHPR protein and tetrads. Tetrads are restored by transfection of dysgenic myotubes with cDNA for DHPRs (Takekura et al., 1994), suggesting that DHPRs are components of the tetrads. But the junctional gap between the SR and T tubule is observed.

In the RyR1−/− mice, the "foot" structure is absent (Takekura et al., 1995). Even though RyR1 is absent, the junction which is formed between SR and surface memebrane is formed in RyR1−/− mice. However, the junctional gap between the SR and T tubule is narrow, probably because the foot structure (which is the N-terminal of RyR1) is missing. Since the junction is formed in the absence of DHPR in the case of dysgenesis mutation, and the absence of RyRs in RyR1−/− mice, coupling the SR and surface membrane can be carried out by some other additional proteins.

The contractile response to electrical stimulation under phsyiological conditions is abolished in the skeletal muscle of RyR1−/− mice. Electrical stimulation of the skeletal muscle induced twitch contractions in skeletal muscle from wild mice neonates and the twitches are evoked by sodium-channel-dependent action potentials. Electrophysiological studies make it clear that the action potentials are generated in the RyR1−/− muscle but that Ca^{2+} release after depolarization of sarcolemma is impaired in the mutant muscle. It is therefore clear that RyR1 functions as the Ca^{2+}-release channel in the skeletal muscle E-C coupling.

The possibility of a retrograde signal from RyR1 to the DHPR using myotubes from RyR1−/− mice has been analyzed. There is roughly 30-fold reduction in L-type Ca^{2+} current density (Nakai et al., 1996). Injection of RyR1−/− myotubes with RyR1 cDNA restores excitation–contraction coupling, and also causes the density of L-type Ca^{2+} current to rise toward normal level.

In the cultured muscle cells from RyR1−/− mice, RT-PCR analysis shows that neither RyR1 nor RyR2 is present. Only RyR3 is predominantly expressed (Takeshima et al., 1995). The skeletal muscle cells is used to characterize the property of CICR of RyR3. In the cultured myocytes, caffeine-, ryanodine-, and adenine nucleotide-sensitive CICR has about 10 times lower sensitivity to Ca^{2+} than that of RyR1. RyR3 does not mediate excitation–contraction coupling of the skeletal muscle type and RyR3 may work as a channel for CICR in response to a Ca^{2+} rise with a high threshold.

Takeshima et al. (1996) showed that RyR3−/− mice have no gross abnormalities, but Bertocchini et al. (1997) reported that skeletal muscle of neonatal RyR3−/− shows impaired contraction. It is interesting that mice lacking both RyR1 and RyR3 (RyR1−/− and RyR3−/− double knockout) have more severe muscular degeneration than RyR1−/− mice and completely lose CICR activity in skeletal muscle (Ikemoto et al., 1997; Barone et al., 1998).

References

Abramson, J. J., Zable, A. C., Favero, T. G., and Salama, G. (1995) Thimerosal interacts with the Ca^{2+} release channel ryanodine receptor from skeletal muscle sarcoplasmic reticulum. *J. Biol. Chem.* 270: 29644–29647.

Acharya, J. K., Jalink, K., Hardy, R. W., Hartenstein, V., and Zuker, C. S. (1997) InsP$_3$ receptor is essential for growth and differentiation but not for vision in *Drosophila. Neuron* 18: 881–887.

Anderson, K., Lai, F. A., Liu, Q. Y., Rousseau, E., Erickson, H. P., and Meissner, G. (1989) Structural and functional characterization of the purified cardiac ryanodine receptor-Ca^{2+} release channel complex. *J. Biol. Chem.* 264: 1329–1315.

Al-Mohanna, F. A., Caddy, K. W. T. and Bolsover, S. R. (1994) The nucleus is insulated from large cytosolic calcium ion changes. *Nature* 367: 745–750.

Amundson, J. and Clapham, D. (1993) Ca waves. *Curr. Opin. Neurobiol.* 3: 375–382.

Barone, V., Bertocchini, F., Bottinelli, R., Protasi, F., Allen, P. O., Armstrong, C. F., Reggiani, C., and Sorrentino, V. (1998) Contractile impairment and structural alternations of skeletal muscles from knockout mice lack type 1 and type 3 ryanodine receptors. *FEBS Lett.* 422: 160–164.

Baukal, A. J., Guillemette, G., Spät A., and Catt, K. J. (1985) Binding sites for inositol trisphosphate in the bovine adrenal cortex. *Biochem. Biophys. Res. Commun.* 133: 532–538.

Berridge, M. J. (1986) Second messenger dualism in neuromodulation and memory. *Nature* 323: 294–295.

Berridge, M. J. (1993) Inositol trisphosphate and calcium signaling. *Nature* 361: 315–325.

Berridge, M. J. (1995) Capacitative calcium entry. *Biochem J.* 312: 1–11.

Berridge, M. J. and Irvine, R. F. (1989) Inositol phosphates and cell signalling. *Nature* 341: 197–205.

Bertocchini, F., Ovitt, C. E., Conti, A., Barone, V., Schöler, H. R., Bottinelli, R., Reggiani, C., and Sorrentino, V. (1997) Requirement for the ryanodine receptor type 3 for efficient contraction in neonatal skeletal muscles. *EMBO J.* 16: 6956–6963.

Bezprozvanny, I. (1994) Theoretical analysis of calcium wave propagation based on inositol (1,4,5)-trisphosphate (InsP$_3$) receptor functional properties. *Cell Calcium* 16: 151–166.

Bezprozvanny, I. and Ehrlich, B. E. (1993) ATP modulates the function of inositol 1,4,5-trisphosphate-gated channels at two sites. *Neuron* 10: 1175–1184.

Bezprozvanny, I. and Ehrlich, B. E. (1994) Inositol (1,4,5)-trisphosphate (InsP$_3$)-gated Ca channels from cerebellum: conduction properties for divalent cations and regulation by intraluminal calcium. *J. Gen. Physiol.* 104: 821–856.

Bezprozvanny, I. and Ehrlich, B. E. (1995) The inositol 1,4,5-trisphosphate (InsP$_3$) receptor. *J. Memb. Biol.* 145: 205–216.

Bezprozvanny, I., Watras, J. and Ehrlich, B. E. (1991) Bell-shaped calcium-response curves of Ins (1,4,5)P$_3$- and calcium-gated channels from endoplasmic reticulum of cerebellum. *Nature* 351: 751–754.

Bezprozvanny, I., Bezprozvannaya, S., and Ehrlich, B. E. (1994) Caffeine-induced inhibition of inositol (1,4,5)-trisphosphate-gated calcium channels from cerebellum. *Mol. Biol. Cell* 5: 97–103.

Bhat, M. B., Ma, J., Zhao, J. Y., Hayek, S., Freeman, E. C., and Takeshima, H. (1997a). Deletion of amino acids 1614–2437 from the foot region of skeletal muscle ryanodine receptor alters the conduction properties of the Ca release channel. *Biophys. J.* 73: 1320–1328.

Bhat, M. B., Zhao, J. Y., Takeshima, H., and Ma, J. (1997b) Functional calcium release channel formed by the carboxyl-terminal portion of ryanodine receptor. *Biophys. J.* 73: 1329–1336.

Bian, F., Chu, T., Schilling, K., and Oberdick, J. (1996) Differential mRNA transport and the regulation of protein synthesis: selective sensitivity of Purkinje cell dendritic mRNAs to translational inhibition. *Mol. Cell. Neurosci.* 7: 116–133.

Bird, G. S. J., Burgess, G. M., and Putney, J. W. Jr. (1993) Sulfhydryl reagents and cAMP-dependent kinase

increase the sensitivity of the inosital 1,4,5-trisphosphate receptor in hepatocytes. *J. Biol. Chem.* 268: 17917–17923.

Block, B. A., Imagawa, T., Campbell, K. P., and Franzini-Armstrong, C. (1988) Structural evidence for direct interaction between the molecular components of the transverse tubule/sarcoplasmic reticulum junction in skeletal muscle. *J. Cell Biol.* 107: 2587–2600.

Blondel, O., Takeda, J., Janssen, H., Seino, S., and Bell, G. I. (1993) Sequence and functional characterization of a third inositol trisphosphate receptor subtype, IP$_3$R-3 expressed in pancreatic islets, kidney, gastrointestinal tract, and other tissues. *J. Biol. Chem.* 268: 11356–11363.

Blondel, O., Moody, M. M., Depaoli, A. M., Sharp, A. H., Ross, C. A., Swift, H., and Bell, G. I. (1994a). Localization of inositol trisphosphate receptor subtype 3 to insulin and somatostatin secretory granules and regulation of expression in islets and insulinoma cells. *Proc. Natl. Acad. Sci. USA* 91: 7777–7781.

Blondel, O., Bell, G. I., Moody, M., Miller, R. J., and Gibbons, S. J. (1994b) Creation of an inositol 1,4,5-trisphosphate-sensitive Ca^{2+} store in secretory granules of insulin-producing cells. *J. Biol. Chem.* 269: 27167–27170.

Blondel, O., Bell, G. I., and Seino, S. (1995) Inositol 1, 4, 5-trisphosphate receptors, secretory granules and secretion in endocrine and neuroendocrine cells. *Trends Neurosci.* 18: 157–161.

Bootman, M. D. and Berridge, M. J. (1995) The elemental principles of calcium signaling. *Cell* 83: 675–678.

Bootman, M. D., Tayler, C. W., and Berridge, M. J. (1992) The thiol reagent, thimerosal, evokes Ca^{2+} spikes in HeLa cells by sensitizing the inositol 1,4,5-trisphosphate receptor. *J. Biol. Chem.* 267: 25113–25119.

Boraso, A. and Williams, A. J. (1994) Modification of the gating of the cardiac sarcoplasmic reticulum Ca(2+)-release channel by H$_2$O$_2$ and dithiothreitol. *Am. J. Physiol.* 267: H1010–H1016.

Bourguignon, L. Y. W. and Jin, H. (1995) Identification of the ankyrin-binding domain of the mouse T-lymphoma cell inositol 1,4,5-trisphosphate (IP$_3$) receptor and its role in the regulation of IP$_3$-mediated internal Ca^{2+} release. *J. Biol. Chem.* 270: 7257–7260.

Bourguignon, L. Y. W., Jin, H., Iida, N., Brandt, N. R., and Zhang, S. H. (1993) The involvement of ankyrin in the regulation of inositol 1,4,5-trisphosphate receptor-mediated internal Ca^{2+} release from Ca^{2+} storage vesicles in mouse T-lymphoma cells. *J. Biol. Chem.* 268: 7290–7297.

Bourguignon, L. Y. W., Chu, A., Jin, H., and Brandt, N. R. (1995) Ryanodine receptor–ankyrin interaction regulates internal Ca^{2+} release in mouse T-lymphoma cells. *J. Biol. Chem.* 270: 17917–17922.

Bradford, P. G. and Autieri, M. (1991) Increased expression of the inositol 1,4,5-trisphosphate receptor in human leukaemic (HL-60) cells differentiated with retinoic acid or dimethyl sulphoxide. *Biochem. J.* 280: 205–210.

Brillantes, A.-M., Ondrias, K., Scott, A., Kobrinsky, E., Ondriasova, E., Moschella, M. C., Jayaraman, T., Landers, M., Ehrlich, B. E., and Marks, A. R. (1994) Stabilization of calcium release channel (ryanodine receptor) function by FK506-binding protein. *Cell.* 77: 513–523.

Brostrom, C. O. and Brostrom, M. A. (1990) Calcium-dependent regulation of protein synthesis in intact mammalian cells. *Annu. Rev. Physiol.* 52: 577–590.

Brown, G. R., Sayers, L. G., Kirk, C. J., Michell, R. H., and Michelangeli, F. (1992) The opening of the inositol 1,4,5-trisphosphate-sensitive Ca^{2+} channel in rat cerebelum is inhibited by caffeine. *Biochem. J.* 282: 309–312.

Burgess, G. M., Bird, G. S. J., Obie, J. F., and Putney, J. W. Jr. (1991) The mechanism for synergism between phospholipase C- and adenylylcyclase-linked hormones in liver. *J. Biol. Chem.* 266: 4772–4781.

Bush, K. T., Stuart, R. O., Li, S.-H., Moura, L. A., Sharp, A. H., Ross, C. A., and Nigam, S. K. (1994) Epithelial inositol 1,4,5-trisphosphate receptors: multiplicity of localization, solubility, and isoforms. *J. Biol. Chem.* 269: 23694–23699.

Callaway, C., Seryshev, A., Wang, J. P., Slavik, K. J., Needleman, D. H., Cantu, C., Wu, Y., Jayaraman, T., Marks, A. R., and Hamilton, S. L. (1994) Localization of the high and low affinity [^{3}H]ryanodine binding sites on the skeletal muscle Ca^{2+} release channel. *J. Biol. Chem.* 269: 15876–15884.

Cameron, A. M., Nucifora, F. C. Jr., Fung, E. T., Livingston, D. J., Aldape, R. A., Ross, C. A., and Snyder, S. H. (1997) FKBP12 binds the inositol 1,4,5-trisphosphate receptor at leucine–proline (1400–1401) and anchors calcineurin to the FK506-link domain. *J. Biol. Chem.* 272: 27582–27588.

Caswell, A. H., Brandt, N. R. Brunschweig, J. P., and Purkeson, S. (1991) Localization and partial characterization of the oligomeric disulfide-linked molecular weight 95,000 protein (triadin) which binds the ryanodine and dihydropyridine receptors in skeletal muscle triadic vesicles. *Biochemistry* 30: 7507–7513.

Chadwick, C. C., Saito, A., and Fleischer, S. (1990) Isolation and characterization of the inositol trisphosphate receptor from smooth muscle. *Proc. Natl. Acad. Sci. USA* 87: 2132–2136.

Champeil, P., Combettes, L., Berthon, B., Doucet, E., Orlowski, S., and Claret, M. (1989) Fast kinetics of calcium release induced by myo-inositol trisphosphate in permeabilized rat hepatocytes. *J. Biol. Chem.* 264: 17665–17673.

Chen, S. R. W., Zhang, L., and MacLennan, D. H. (1992) Characterization of a Ca^{2+} binding and regulatory site in the Ca^{2+} release channel (ryanodine receptor) of rabbit skeletal muscle sarcoplasmic reticulum. *J. Biol. Chem.* 267: 23318–23326.

Chen, S. R., Vaughan, D. M., Airey, J. A., Coronado, R., and MacLennan, D. H. (1993a). Functional expression of cDNA encoding the Ca^{2+} release channel (ryanodine receptor) of rabbit skeletal muscle sarco-plasmic reticulum in COS-1 cells. *Biochemistry* 32: 3743–3753.

Chen, S. R. W., Zhang, L., and MacLennan, D. H. (1993b) Antibodies as probes for Ca^{2+} activation sites in the Ca^{2+} release channel (ryanodine receptor) of rabbit skeletal muscle sarcoplasmic reticulum. *J. Biol. Chem.* 268: 13414–13421.

Chen, S. R. W., Li, X., Ebisawa, K., and Zhang, L. (1997a) Functional characterization of the ryanodine type 3 Ca^{2+} release channel (ryanodine receptor) expressed in HEK293 cells. *J. Biol. Chem.* 272: 24234–24246.

Chen, S. R. W., Ebisawa, K., Li, X., and Zhang, L. (1997b) Molecular identification of the ryanodine receptor Ca^{2+} sensor. *J. Biol. Chem.* 273: 14675–14678.

Chu, A., Fill, M., Stefani, E., and Entman, M. L. (1993) Cytoplasmic Ca^{2+} does not inhibit the cardiac muscle sarcoplasmic reticulum ryanodine receptor Ca^{2+} channel, although Ca^{2+}-induced Ca^{2+} inactivation of Ca^{2+} release is observed in native vesicles. *J. Membr. Biol.* 135: 49–59.

Clandinin, T. R., DeModena, J. A., and Sternberg, P. W. (1998) Inositol trisphosphate mediates a RAS-independent response to LET-23 receptor tyrosine kinase activation in *C. elegans. Cell* 92: 523–533.

Clapham, D. E. (1995) Calcium signaling. *Cell* 80: 259–268.

Combettes, L., Cheek, T. R., and Taylor, C. W. (1996) Regulation of inositol trisphosphate receptor by luminal Ca^{2+} contributes to quantal Ca^{2+} mobilization. *EMBO J.* 15: 2086–2093.

Coquil, J.-F., Mauger, J.-P., and Claret, M. (1996) Inositol 1,4,5-trisphosphate slowly converts its receptor to a state of higher affinity in sheep cerebellar membranes. *J. Biol. Chem.* 271: 3568–3574.

Danoff, S. K., Supattapone, S., and Snyder, S. H. (1988) Characterization of a membrane protein from brain mediating the inhibition of inositol 1,4,5-trisphosphate receptor binding by calcium. *Biochem. J.* 254: 701–705.

Danoff, S. K., Ferris, C. D., Donath, C., Fischer, G. A., Munemitsu, S., Ullrich, A., Snyder, S. H., and Ross, C. A. (1991) Inositol 1,4,5-trisphosphate receptors: distinct neuronal and nonneuronal forms derived by alterntive splicing differ in phosphorylation. *Proc. Natl. Acad. Sci. USA* 88: 2951–2955.

DeLisle, S. and Welsh, M. J. (1992) Inositol trisphosphate is required for the propagation of calcium waves in *Xenopus* oocytes. *J. Biol. Chem.* 267: 7963–7966.

DeLisle, S., Blondel, O., Longo, F. J., Schenabel, W. E., Bell, G. I., and Welsh, M. J. (1996) Expression of inositol 1,4,5-trisphosphate receptors changes the Ca^{2+} signal of *Xenopus* oocytes. *Am. J. Physiol.* 270: C1255–C1261.

Dent, M. A. R., Raisman, G., and Lai, A. (1996) Expression of type 1 inositol 1,4,5-trisphosphate receptor during axogenesis and synaptic contact in the central and peripheral nervous system of developing rat. *Development* 122: 1029–1039.

Denton, R. M. and McCormack, J. G. (1990) Ca^{2+} as a second messenger within mitochondria of the heart and other tissues. *Annu. Rev. Physiol.* 52: 451–466.

De Smedt, H., Missieaen, L., Parys, J. B., Bootman, M. D., Mertens, L., Van Den Bosch, L., and Casteels, R. (1994) Determination of relative amounts of inositol trisphosphate receptor mRNA isoforms by ratio polymerase chain reaction. *J. Biol. Chem.* 269: 21691–21698.

Divecha, N., Banfic, H., and Irvine, R. F. (1993) Inositides and the nucleus and inositides in the nucleus. *Cell* 74: 405–407.

Dufour, J.-F., Arias, I. M., and Turner, T. J. (1997) Inositol 1,4,5-trisphosphate and calcium regulate the calcium channel function of the hepatic inositol 1,4,5-trisphosphate receptor. *J. Biol. Chem.* 272: 2675–2681.

Ehrlich, B. E. and Watras, J. (1988) Inositol 1,4,5-trisphosphate activates a channel from smooth muscle sarcoplasmic reticulum. *Nature* 336: 583–586.

Ehrlich, B. E., Kaftan, E., Bezprozvannaya, S., and Bezprozvanny, I. (1994) The pharmacology of intracellular Ca^{2+}-release channels. *Trends Pharmacol. Sci.* 15: 145–149.

Endo, M. (1977) Calcium release from the sarcoplasmic reticulum. *Physiol. Rev.* 57: 71–108.

Fadool, A. and Ache, B. W. (1992) Plasma membrane inositol 1,4,5-trisphosphate-activated channels mediate signal transduction in lobster olfactory receptor neurons. *Neuron* 9: 907–918.

Fasolato, C., Innocenti, B., and Pozzan, T. (1994) Receptor-activated Ca^{2+} influx: how many mechanisms for how many channels? *Trends Pharmacol. Sci.* 15: 77–83.

Ferris, C. D., and Snyder, S. H. (1992) Inositol 1,4,5-trisphosphate-activated calcium channels. *Annu. Rev. Physiol.* 54: 469–488.

Ferris, C. D., Huganir, R. L., and Snyder, S. H. (1990a) Calcium flux mediated by purified inositol 1,4,5-trisphosphate receptor in reconstituted lipid vesicles is allosterically regulated by adenine nucleotides. *Proc. Natl. Acad. Sci. USA* 87: 2147–2151.

Ferris, C. D., Huganir, R. L., Bredt, D. S., Cameron, A. M., and Snyder, S. H. (1990b) Inositol trisphosphate receptor: phosphorylation by protein kinase C and calcium-calmodulin dependent protein kinases in reconstituted lipid vesicles. *Proc. Natl. Acad. Sci. USA* 88: 2232–2235.

Ferris, C. D., Cameron, A. M., Bredt, D. S., Huganir, R. L., and Snyder, S. H. (1991) Inositol 1,4,5-trisphosphate receptor is phosphorylated by cyclic AMP-dependent protein kinase at serines 1755 and 1589. *Biochem. Biophys. Res. Commun.* 175: 192–198.

Ferris, C. D., Cameron, A. M., Bredt, D. S., Huganir, R. L., and Snyder, S. H. (1992a) Autophosphorylation of inositol 1,4,5-trisphosphate receptors. *J. Biol. Chem.* 267: 7036–7041.

Ferris, C. D., Cameron, A. M., Bredt, D. S., Huganir, R. L., and Snyder, S. H. (1992b) Quantal calcium release by purified reconstituted inositol 1,4,5-trisphosphate receptors. *Nature* 356: 350–352.

Finch, E. A., Turner, T. J., and Goldin, S. M. (1991) Calcium as a coagonist of inositol 1,4,5-tris-phosphate-induced calcium release. *Science* 252: 443–446.

Finkbeiner, S. M. (1993) Glial calcium. *Glia* 9: 83–104.

Fischer, G. A., Clementi, E., Raichman, M., Südhof, T., Ullrich, A., and Meldolesi, J. (1994) Stable expression of truncated inositol 1,4,5-trisphosphate receptor subunits in 3T3 fibroblasts. *J. Biol. Chem.* 269: 19216–19224.

Föhr, K. J., Wahl, Y., Engling, R., Kemmer, T. P., and Gratzl, M. (1991) Decavanadate displaces inositol 1,4,5-trisphosphate (IP_3) from its receptor and inhibits IP_3 induced Ca^{2+} release in permeabilized pancreatic acinar cells. *Cell Calcium* 12: 735–742.

Fotuhi, M., Sharp, A. H., Glatt, C. E., Hwang, P. M., Von Krosigk, M., and Snyder, S. H. (1993) Differential localization of phosphoinositide-linked metabotropic glutamate receptor (mGluR1) and the inositol 1,4,5-trisphosphate receptor in rat brain. *J. Neurosci.* 13: 2001–2012.

Franzini-Armstrong, C. and Jorgensen, A. D. (1994) *Annu. Rev. Physiol.* 56: 509–534.

Fujimoto, T., Nakade, S., Miyawaki, A., Mikoshiba, K., and Ogawa, K. (1992) Localization of inositol 1,4,5-trisphosphate receptor-like protein in plasmalemmal caveolae. *J. Cell Biol.* 119: 1507–1513.

Fujino, I., Yamada, N., Miyawaki, A., Hasegawa, M., Furuichi, T., and Mikoshiba, K. (1995) Differential expression of type 2 and type 3 inositol 1,4,5-trisphosphate receptor mRNAs in various mouse tissues: in situ hybridization study. *Cell Tissue Res.* 280: 201–210.

Furuichi, T. and Mikoshiba, K. (1995) Inositol 1,4,5-trisphosphate receptor-mediated Ca^{2+} signaling in the brain. *J. Neurochem.* 64: 953–960.

Furuichi, T., Yoshikawa, S., and Mikoshiba, K. (1989a) Nucleotide sequence of cDNA encoding P_{400} protein in the mouse cerebellum *Nucleic Acids Res.* 17: 5385–5386.

Furuichi, T., Yoshikawa, S., Miyawaki, A., Wada, K., Maeda, N., and Mikoshiba, K. (1989b) Primary structure and functional expression of the inositol 1,4,5-trisphosphate-binding protein P_{400}. *Nature* 342: 32–38.

Furuichi, T., Shiota, C., and Mikoshiba, K. (1990) Distribution of inositol 1,4,5-trisphosphate receptor mRNA in mouse tissues. *FEBS Lett.* 267: 85–88.

Furuichi, T., Miyawaki, A., Maeda, N., Nakanishi, S., Nakagawa, T., Yoshikawa, S., and Mikoshiba, K. (1992) Structure and function of the inositol 1,4,5-trisphosphate receptor. In *Neuroreceptors, Ion Channels and the Brain* (Kawai, N., Nakajima, T., and Barnard, E., eds.). Elsevier, Amsterdam, pp. 103–111.

Furuichi, T., Simon-Chazottes, D., Fujino, I., Yamada, N., Hasegawa, M., Miyawaki, A., Yoshikawa, S., Guénet, J.-L, and Mikoshiba, K. (1993) Widespread expression of inositol 1,4,5-trisphosphate receptor type 1 gene (*Insp3rl*) in the mouse central nervous system. *Receptor Channels* 1: 11–24.

Furuichi, T., Furutama, D., Hakamata, Y., Nakai, J., Takeshima, H., and Mikoshiba, K. (1994a) Multiple types of ryanodine receptor/Ca^{2+} release channels are differentially expressed in rabbit brain. *J. Neurosci.* 14: 4794–4805.

Furuichi, T., Kohda, K., Miyawaki, A., and Mikoshiba, K. (1994b) Intracellular channels. *Curr. Opin. Neurobiol.* 4: 294–303.

Furutama, D., Shimoda, K., Yoshikawa, S., Miyawaki, A., Furuichi, T., and Mikoshiba, K. (1996) Functional expression of the type 1 inositol 1,4,5-trisphosphate receptor-lacZ fusion genes in transgenic mice. *J. Neurochem.* 66: 1793–1801.

Futatsugi, A., Kuwajima, G., and Mikoshiba, K. (1995) Tissue-specific and developmentally regulated alternative splicing in mouse skeletal muscle ryanodine receptor mRNA. *Biochem. J.* 305: 373–378.

Gafni, J., Munsch, J. A., Lam, T. H., Catlin, M. C., Costa, L. G., Molinski, T. F., and Pessah, I. N. (1997) Xestospongins: potent membrane permeable blockers of the inositol 1,4,5-trisphosphate receptor. *Neuron* 19: 723–733.

Gerasimenko, O. V., Gerasimenko, J. V., Tepikin, A. V., and Petersen, O. H. (1995) ATP-dependent accumulation and inositol trisphosphate- or cyclic ADP-ribose-mediated release of Ca^{2+} from the nuclear envelope. *Cell* 80: 439–444.

Ghosh, A. and Greenberg, M. E. (1995) Calcium signaling in neurons: molecular mechanisms and cellular consequences. *Science* 268: 239–247.

Ghosh, T. K., Eis, P. S., Mullaney, J. M., Ebert, C. L., and Gill, D. L. (1988) Competitive reversible, and potent antagonism of inositol 1,4,5-trisphosphate-activated calcium release by heparin. *J. Biol. Chem.* 263: 11075–11079.

Giannini, G., Clementi, E., Ceci, R., Marziali, G., and Sorrentino, V. (1992) Expression of a ryanodine receptor-Ca^{2+} channel that is regulated by TGF-beta. *Science* 257: 91–94.

Giannini, G., Conti, A., Mammeralla, S., Scrobogna, M., and Sorrentino, V. (1995) The ryanodine receptor/calcium channel genes are widely and differentially expressed in murine brain and peripheral tissues. *J. Cell. Biol.* 128: 893–904.

Gorza, L., Sdiaffino, S., and Volpe, P. (1993) Inositol 1,4,5-trisphosphate receptor in heart evidence for its concentration in Purkinje myocytes of the conduction system. *J. Biol. Chem.* 121: 345–353.

Groswald, D. E., and Kelly, P. T. (1984) Evidence that a cerebellum-enriched, synaptic junction glycoprotein is related to fodrin and resists extraction with triton in a calcium-dependent manner. *J. Neurochem.* 42: 534–546.

Grunwald, R. and Meissner, G. (1995). Luminal sites and C terminus accessibility of the skeletal muscle calcium release channel (ryanodine receptor). *J. Biol. Chem.* 270: 11338–11347.

Guerrini, R., Menegazzi, P., Anacardio, R., Marastoni, M., Tomatis, R., Zorzato, F., and Treves, S. (1995) Calmodulin binding sites of the skeletal, cardiac, and brain ryanodine receptor Ca^{2+} channels: modulation

by the catalytic subunit of cAMP-dependent protein kinase? *Biochemistry* 34: 5120–5129.

Guillemette, G., Balla, T., Baukal, A. J., and Catt, K. J. (1988) Characterization of inositol 1,4,5-trisphosphate receptors and calcium mobilization in a hepatic plasma membrane fraction. *J. Biol. Chem.* 263: 4541–4548.

Guo, W., Jorgensen, A. O., Jones, L. R., and Campbell, K. P. (1996) Biochemical characterization and molecular cloning of cardiac triadin. *J. Biol. Chem.* 271: 458–465.

Hajnóczky, G. and Thomas, A. P. (1994) The inositol trisphosphate calcium channel is inactivated by inositol trisphosphate. *Nature* 370: 474–477.

Hajnóczky, G., Gao, E., Nomura, T., Hoek, J. B., and Thomas, A. P. (1993) Multiple mechanisms by which protein kinase A potentiates inositol 1,4,5-trisphosphate-induced Ca^{2+} mobilization in permealized hepatocytes. *Biochem. J.* 293: 413–422.

Hajnóczky, G. and Thomas, A. P. (1997) Minimal requirements for calcium oscillation driven by the IP$_3$ receptor. *EMBO J.* 16: 3533–3543.

Hakamata, Y., Nakai, J., Takeshima, H., and Imoto, K. (1992) Primary structure and distribution of a novel ryanodine receptor/calcium release channel from rabbit brain. *FEBS Lett.* 312: 229–235.

Hakamata, Y., Nishimura, S., Nakai, J., Nakashima, Y., Kita, T., and Imoto, K. (1994) Involvement of the brain type of ryanodine receptor in T-cell proliferation. *FEBS Lett.* 352: 206–210.

Hannaert-Merah, Z., Coquil, J.-F., Combettes, L., Claret, M., Mauger, J. P., and Champeil, P. (1994) Rapid kinetics of myo-inositol trisphosphate binding and dissociation in cerebellar microsomes. *J. Biol. Chem.* 269: 29642–29649.

Harnick, D. J., Jayaraman, T., Ma, Y., Mulieri, P., Go, L.O., and Marks, A. R. (1995) The human type 1 inositol 1,4,5-trisphosphate receptor from T lymphocytes–structure, localization, and tyrosine phosphorylation. *J. Biol. Chem.* 270: 2833–2840.

Harootunian, A. T., Hao, J. P. Y., Paranjape, S., and Tsien, R. Y. (1991) Generation of calcium oscillations in fibroblasts by positive feedback between calcium and IP$_3$. *Science* 251: 75–78.

Hasan, G. and Rosbash, M. (1992) Drosophila homologues of two mammalian intracellular Ca^{2+} release channels: identification and expression patterns of the inositol 1,4,5-trisphosphate and the ryanodine receptor genes. *Development* 116: 967–975.

Hasty, P., Bradley, A., Morris, J. H., Edmonson, D. G., Venuti, J. M., Olson, E. N., and Klein, W. H. (1993) Muscle deficiency and neonatal death in mice with a targeted mutation in the myogenin gene [see comments]. *Nature* 364: 501–506.

Hatt, H. and Ache, W. (1994) Cyclic nucleotide- and inositol phosphate-gated ion channels in lobster olfactory receptor neurons. *Proc. Natl. Acad. Sci. USA* 91: 6264–6268.

Hennager, D. J., Welsh, M. J., and DeLisle, S. (1995) Changes in either cytosolic or nucleoplasmic inositol

1,4,5-trisphosphate levels can control nuclear Ca^{2+} concentration. *J. Biol. Chem.* 270: 4959–4962.

Herrmann-Frank, A. and Varsanyi, M. (1993) Enhancement of Ca^{2+} release channel activity by phosphorylation of the skeletal muscle ryanodine receptor. *FEBS Lett.* 332: 237–242.

Hess, P., Lansman, J. B., and Tsien, R. W. (1986) Calcium channel selectivity for divalent and monovalent cations. *J. Gen. Physiol.* 88: 293–319.

Hill, T. D., Campos-Gonzales, R., Kindmark, H., and Boynton, A. L. (1988) Inhibition of inositol trisphosphate-stimulated calcium mobilization by calmodulin antagonists in rat liver epithelial cells. *J. Biol. Chem.* 263: 16479–16484.

Hille, B. (1992) *Ionic Channels of Excitable Membranes*, 2nd ed. Sinauer Associates, Sunderland, MA.

Hingorani, S. R. and Agnew, W. S. (1991) A rapid ion-exchange assay for detergent-solubilized inositol 1,4,5-trisphosphate receptors. *Anal. Biochem.* 194: 204–213.

Hingorani, S. R. and Agnew, W. S. (1992) Assay and purification of neuronal receptors for inositol 1,4,5-trisphosphate. *Methods Enzymol.* 207: 573–591.

Hirose, K. and Iino, M. (1994) Heterogeneity of channel density in inositol-1,4,5-trisphosphate-sensitive Ca^{2+} stores. *Nature* 372: 791–794.

Hirose, K., Iino, M., and Endo, M. (1993) Caffeine inhibits Ca^{2+}-mediated potentiation of inositol 1,4,5-trisphosphate-induced Ca^{2+} release in permeabilized vascular smooth muscle cells. *Biochem. Biophys. Res. Commun.* 194: 726–732.

Hirota, J. Michikawa, T., Miyawaki, A., Furuichi, T., Okura, I., and Mikoshiba, K. (1995a) Kinetics of calcium release by immunoaffinity-purified inositol 1,4,5-trisphosphate receptor in reconstituted lipid vesicles. *J. Biol. Chem.* 270: 19046–19051.

Hirota, J., Michikawa, T., Miyawaki, A., Takahashi, M., Tanzawa, K., Okura, I., Furuichi, T., and Mikoshiba, K. (1995b) Adenophostin-mediated quantal Ca^{2+} release in the purified and reconstituted inositol 1,4,5-trisphosphate receptor type 1. *FEBS. Lett.* 368: 248–252.

Hirota, J., Baba, M., Matsumoto, M., Furuichi, T., Takatsu, K., and Mikoshiba, K. (1998) T-cell-receptor signalling in inositol 1,4,5-trisphosphate receptor (IP_3R) type-1-deficient mice: is IP_3R type 1 essential for T-cell-receptor signalling? *Biochem. J.* 333: 615–619.

Hohenegger, M. and Suko, J. (1993). Phosphorylation of the purified cardiac ryanodine receptor by exogenous and endogenous protein kinases. *Biochem. J.* 296: 303–308.

Horne, J. H., and Meyer, T. (1995) Luminal calcium regulates the inositol trisphosphate receptor of rat basophilic leukemia cells at a cytosolic site. *Biochemistry* 34: 12738–12746.

Humbert, J.-P., Matter, N., Artault, J.-C., Koppler, P., and Malviya, A. N. (1996) Inositol 1,4,5-trisphosphate receptor is located to the inner nuclear membrane indicating regulation of nuclear calcium signaling by inositol 1,4,5-trisphosphate. *J. Biol. Chem.* 271: 478–485.

Iida, N. and Bourguignon, L. Y. W. (1994) A new splice variant of inositol 1,4,5-trisphosphate (IP_3) receptor. *Cell. Signalling* 6: 449–455.

Iino, M. (1990) Biphasic Ca^{2+} dependence of inositol 1,4,5-trisphosphate-induced Ca^{2+} release in smooth muscle cells of the guinea pig taenia caeci. *J. Gen. Physiol.* 95: 1103–1122.

Iino, M. (1991) Effects of adenine nucleotides on inositol 1,4,5-trisphosphate-induced calcium release in vascular smooth muscle cells. *J. Gen. Physiol.* 98: 681–698.

Iino, M. and Endo, M. (1992) Calcium-dependent immediate feedback control of inositol 1,4,5-trisphosphate-induced Ca^{2+} release. *Nature* 360: 76–78.

Ikemoto, T., Komazaki, S., Takeshima, H., Nishi, M., Noda, T., Iino, M., and Endo, M. (1997) Functional and morphological features of skeletal muscle from mutant mice lacking both type 1 and type 3 ryanodine receptors. *J. Physiol.* 501.2: 305–312.

Inoue, T., Kato, K., Kohda, K., and Mikoshiba, K. (1998) Type 1 inositol 1,4,5-trisphosphate receptor is required for induction of long-term depression in cerebellar Purkinje neurons. *J. Neurosci.* 18: 5366–5373.

Inui, M., Saito, A., and Fleischer, S. (1987) Purification of the ryanodine receptor and identity with feet structures of junctional terminal cisternae of sarcoplasmic reticulum from fast skeletal muscle. *J. Biol. Chem.* 262: 1740–1747.

Irvine, R. F. (1990) "Quantal" Ca^{2+} release and the control of Ca^{2+} entry by inositol phosphates: a possible mechanism. *FEBS Lett.* 263: 5–9.

Islam, A. D., Yoshida, Y., Koga, T., Kojima, M., Kangawa, K., and Imai, S. (1996) Isolation and characterization of vascular muscle inositol 1,4,5-trisphosphate receptor. *Biochem. J.* 316: 295–302.

Jan, L. Y. and Jan, Y. N. (1992) Tracing the roots of ion channels. *Cell* 69: 715–718.

Jayaraman, T., Brillantes, A.-M. B., Timerman, A. P., Erdjument-Bromage, H., Fleischer, S., Tempst, P., and Marks, A. R. (1992) FK506-binding protein associated with the calcium release channel (ryanodine receptor). *J. Biol. Chem.* 267: 9474–9477.

Jayaraman, T., Ondriasova, E., Ondrias, K., Harnick, D. J., and Marks, A. R. (1995) The inositol 1,4,5-trisphosphate receptor is essential for T-cell receptor signaling. *Proc. Natl. Acad. Sci. USA* 92: 6007–6011.

Jayaraman, T., Ondrias, K., Ondriasova, E., and Marks, A. R. (1996) Regulation of the inositol 1,4,5-trisphosphate receptor by tyrosine phosphorylation. *Science* 272: 1492–1494.

Joseph, S. K. (1994) Biosynthesis of the inositol trisphosphate receptor in WB rat liver epithelial cells. *J. Biol. Chem.* 269: 5673–5679.

Joseph, S. K., and Ryan, S. V. (1993) Phosphorylation of the inositol trisphosphate receptor in isolated rat hepatocytes. *J. Biol. Chem.* 268: 23059–23065.

Joseph, S. K. and Samanta, S. (1993) Detergent solubility of the inositol trisphosphate receptor in rat brain membranes. *J. Biol. Chem.* 268: 6477–6486.

Joseph, S. K., Pierson, S., and Samanta, S. (1995a) Trypsin digestion of the inositol trisphosphate receptor: implications for the conformation and domain organization of the protein. *Biochem. J.* 307: 859–865.

Joseph, S. K., Lin, C., Pierson, S., Thomas, A. P., and Maranto, A. R. (1995b) Heteroligomers of type-I and type-III inositol trisphosphate receptors in WB rat liver epithelial cells. *J. Biol. Chem.* 270: 23310–23316.

Joseph, S. K., Ryan, S. V., Pierson, S., Renard-Rooney, D., and Thomas, A. P. (1995c) The effect of mersalyl on inositol trisphosphate receptor binding and ion channel function. *J. Biol. Chem.* 270: 3588–3593.

Joseph, S. K., Boehning, D., Pierson, S., and Nucchitta, C. V. (1997) Membrane insertion, glycosylation, and oligomerization of inositol trisphosphate receptors in a cell-free translation system. *J. Biol. Chem.* 272: 1579–1588.

Kaftan, E. J., Ehrlich, B. E., and Watras, J. (1997) Inositol 1,4,5-trisphosphate (InsP$_3$) and calcium interact to increase the dynamic range of InsP$_3$ receptor-dependent calcium signaling. *J. Gen. Physiol.* 110: 529–538.

Kaplin, A. I., Ferris, C. D., Voglmaier, S. M., and Snyder, S. H. (1994) Purified reconstituted inositol 1,4,5-trisphosphate receptors: thiol reagents act directly on receptor protein. *J. Biol. Chem.* 269: 28972–28978.

Kaplin, A. I., Snyder, S. H., and Linden, D. T. (1996) Reduced nicotinamide adenine dinucleotide-selective stimulation of inositol 1,4,5-trisphosphate receptors mediates hypoxic mobilization of calcium. *J. Neurosci.* 16: 2002–2011.

Kaznacheyeva, E., Lupu, V. D., and Bezprozvanny, I. (1998) Single-channel properties of inositol (1,4,5)-trisphosphate receptor heterogously expressed in HEK-293 cells. *J. Gen. Physiol.* 111: 847–856.

Khan, A. A., Steiner, J. P., and Snyder, S. H. (1992a) Plasma membrane inositol 1,4,5-trisphosphate receptor of lymphocytes: selective enrichment in sialic acid and unique binding specificity. *Proc. Natl. Acad. Sci. USA* 89: 2849–2853.

Khan, A. A., Steiner, J. P., Klein, M. G., Schneider, M. F., and Snyder, S. H. (1992b) IP$_3$ receptor: localization to plasma membrane of T cells and cocapping with the T cell receptor. *Science* 257: 815–818.

Khan, A. A., Soloski, M. J., Sharp, A. H., Schilling, G., Sabatini, D. M., Li, S.-H., Ross, C. A. and Snyder, S. H. (1996) Lymphocyte apoptosis: mediation by increased type 3 inositol 1,4,5-trisphosphate receptor. *Science* 273: 503–507.

Kindman, L. A., and Meyer, T. (1993) Use of intracellular Ca^{2+} stores from rat basophilic leukemia cells to study the molecular mechanism leading to quantal Ca^{2+} release by inositol 1,4,5-trisphosphate. *Biochemistry* 32: 1270–1277.

Kirkwood, K. L., Homick, K., Dragon, M. B., and Bradford, P. G. (1997) Cloning and characterization of the type I inositol 1,4,5-trisphosphate receptor gene promoter: regulation by 17β-estradiol in osteoblasts. *J. Biol. Chem.* 272: 22425–22431.

Koga, T., Yoshida, Y., Cai, J.-Q., Islam, M. O., and Imai, S. (1994) Purification and characterization of 240-kDa cGMP-dependent protein kinase substrate of vascular smooth muscle. *J. Biol. Chem.* 269: 11640–11647.

Komalavilas, P. and Lincoln, T.M. (1994) Phosphorylation of the inositol 1,4,5-trisphosphate receptor by cyclic GMP-dependent protein kinase. *J. Biol. Chem.* 269: 8701–8707.

Komalavilas, P. and Lincoln, T. M. (1996) Phosphorylation of the inositol 1,4,5-trisphosphate receptor: cyclic GMP-dependent protein kinase mediates cAMP and cGMP dependent phosphorylation in the intact rat aorta. *J. Biol. Chem.* 271: 21933–21938.

Konishi, Y., Kobayashi, Y., Kishimoto, T., Makino, Y., Miyawaki, A., Furuichi, T., Okano, H., Mikoshiba, K., and Tamura, T.-A. (1997) Demonstration of an E-box and its CNS-related binding factors for transcriptional regulation of the mouse type 1 inositol 1,4,5-trisphosphate receptor gene. *J. Neurochem.* 69: 476–484.

Kume, S., Muto, A., Aruga, J. Nakagawa, T., Michikawa, T., Furuichi, T., Nakade, S., Okano, H., and Mikoshiba, K. (1993) The *Xenopus* IP$_3$ receptor: structure, function, and localization in oocytes and eggs. *Cell* 73: 555–570.

Kuno, M. and Gardner, P. (1987) Ion channel activated by inositol 1,4,5-trisphosphate in plasma membrane of human T-lymphocytes. *Nature* 326: 301–304.

Kuno, M., Maeda, N., and Mikoshiba, K. (1994) IP$_3$-activated calcium-permeable channels in the inside-out patches of cultured cerebellar Purkinje cells. *Biochem. Biophys. Res. Commun.* 199: 1128–1135.

Kuwajima, G., Futatsugi, A., Niinobe, M., Nakanishi, S., and Mikoshiba, K. (1992). Two types of ryanodine receptors in mouse brain: skeletal muscle type exclusively in Purkinje cells and cardiac muscle type in various neurons. *Neuron* 9: 1133–1142.

Lai, F. A., Anderson, K., Rousseau, E., Liu, Q. Y., and Meissner, G. (1988) Evidence for a Ca^{2+} channel within the ryanodine receptor complex from cardiac sarcoplasmic reticulum. *Biochem. Biophys. Res. Commun.* 151: 441–449.

Lai, F. A., Dent, M., Wickenden, C., Xu, L., Kumari, G., Misra, M., Lee, H. B., Sar, M., and Meissner, G. (1992) Expression of a cardiac Ca^{2+}-release channel isoform in mammalian brain. *Biochem. J.* 288: 553–564.

Lam, E., Martin, M. M., Timerman, A. P., Sabers, C., Fleischer, S., Lukas, T., Abraham, R. T., O'Keefe, S. J., O'Neill, E. A., and Widerrecht, G. J. (1995) A novel FK506 binding protein can mediate the immunosuppressive effects of FK506 and is associated with the cardiac ryanodine receptor. *J. Biol. Chem.* 270: 26511–26522.

Lee, K. S. and Tsien, R. W. (1984) High selectivity of calcium channels in single dialyzed heart cells of the guinea-pig. *J. Physiol.* 354: 253–272.

Lee, M. G., Xu, X., Zeng, W., Diaz, J., Wojcikiewicz, R. J. H., Kuo, T. H., Wuytack, F., Racymaekers, L., and Muallem, S. (1997) Polarized expression of Ca^{2+} channels in pancreatic and salivary gland cells. *J. Biol. Chem.* 272: 15765–15770.

Leichleiter, J. D., and Clapham, D. E. (1992) Molecular mechanisms of intracellular calcium excitability in *X. laevis* oocytes. *Cell* 69: 283–294.

Li, M., Miyawaki, A., Yamamoto-Hino, M., Yasutomi, D., Furuichi, T., Hasegawa, M., and Mikoshiba, K. (1996) Differential cellular expression of three types of inositol 1,4,5-trisphosphate receptor in rat gastrointestinal epithelium. *Biomed. Res.* 17: 45–51.

Liévremont, J.-P., Lancien, H., Hilly, M., and Mauger, J.-P. (1996) The properties of a subtype of the inositol 1,4,5-trisphosphate receptor resulting from alternative splicing of the mRNA in the ligand-binding domain. *Biochim. J.* 317: 755–762.

Lindsay, A. R. G. and Williams, A. J. (1991) Functional characterization of the ryanodine receptor purified from sheep cardiac muscle sarcoplasmic reticulum. *Biochim. Biophys. Acta* 1064: 89–102.

Lisman, J. (1994) The CaM kinase II hypothesis for the storage of synaptic memory. *Trends Neurosci.* 17: 406–412.

Liu, G. and Pessah, I. N. (1994) Molecular interaction between ryanodine receptor and glycoprotein triadin involves redox cycling of functionally important hyperreactive sulfhydryls. *J. Biol. Chem.* 269: 33028–33034.

Liu, J., Framer, J., Lane, W., Friedman, J., Weissman, I., and Schreiber, S. (1991) Calcineurin is a common target of cyclophilin–cyclosporin A and FKBP–FK506 complexes. *Cell* 66: 807–815.

Lu, X., Xu, L., and Meissner, G. (1994) Activation of the skeletal muscle calcium release channel by a cytoplasmic loop of the dihydropyridine receptor. *J. Biol. Chem.* 269: 6511–6516.

Lu, X., Xu, L., and Meissner, G. (1995) Phosphorylation of dihydropyridine receptor II–III loop peptide regulates skeletal muscle calcium release channel function. Evidence for an essential role of the beta-OH group of Ser687. *J. Biol. Chem.* 270: 18459–18464.

Lupu, V. D., Kaznacheyeva, E., Krishna, U. M., Kalck, J. R., and Bezprozvanny, I. (1998) Functional coupling of phosphatidylinositol 4,5-trisphosphate to inositol 1,4,5-trisphosphate receptor. *J. Biol. Chem.* 273: 14067–14070.

Ma, J. (1993) Block by ruthenium red of the ryanodine-activated calcium release channel of skeletal muscle. *J. Gen. Physiol.* 102: 1031–1056.

MacLennan, D. H. and Phillips, M. S. (1992) Malignant hyperthermia. [Review] *Science* 256: 789–794.

Maeda, N., Niinobe, M., Nakahira, K., and Mikoshiba, K. (1988) Purification and characterization of P_{400} protein, a glycoprotein characteristic of Purkinje cells, from mouse cerebellum. *J. Neurochem.* 51: 1724–1730.

Maeda, N., Niiobe, M., Inoue, Y., and Mikoshiba, K. (1989) Developmental expression and intracellular location of P_{400} protein characteristic of Purkinje cells in the mouse cerebellum. *Dev. Biol.* 133: 67–76.

Maeda, N., Niinobe, M., and Mikoshiba, K. (1990) A cerebellar Purkinje cell marker P_{400} protein is an inositol 1,4,5-trisphosphate ($InsP_3$) receptor protein. Purification and characterization of $InsP_3$ receptor complex. *EMBO J.* 9: 61–67.

Maeda, N., Kawasaki, T., Nakade, S., Yokota, N., Taniguchi, T., Kasai, M., and Mikoshiba, K. (1991) Structural and functional characterization of inositol 1,4,5-trisphosphate receptor channel from mouse cerebellum. *J. Biol. Chem.* 266: 1109–1116.

Magnusson, A., Haung, L. S., Wallas, S. L., and Ostvold, A. C. (1993) Calcium-induced degradation of the inositol (1,4,5)-trisphosphate receptor/Ca^{2+}-channel. *FEBS Lett.* 323: 229–232.

Mak, D.-O. D. and Foskett, J. K. (1994) Single-channel inositol 1,4,5-trisphosphate receptor currents revealed by patch clamp of isolated *Xenopus* oocyte nuclei. *J. Biol. Chem.* 269: 29375–29378.

Mak, D.-O.D. and Foskett, J. K. (1997) Single-channel kinetics, inactivation, and spatial distribution of inositol trisphosphate (IP_3) receptors in *Xenopus* oocyte nucleus. *J. Gen. Physiol.* 109: 571–587.

Mallet, J., Huchet, M., Pougeois, R., and Changeux, J.-P. (1976) Anatomical, physiological and biochemical studies on the cerebellum from mutant mice. III. Protein differences associated with the weaver, staggerer and nervous mutations. *Brain Res.* 103: 291–312.

Malviya, A. N., Rogue, P., and Vincendon, G. (1990) Stereospecific inositol 1,4,5-[^{32}P]trisphosphate binding to isolated rat liver nuclei: evidence for inositol trisphosphate receptor-mediated calcium release from the nucleus. *Proc. Natl. Acad. Sci. USA* 87: 9270–9274.

Maranto, A. R. (1994) Primary structure, ligand binding, and localization of the human type 3 inositol 1,4,5-trisphosphate receptor expressed in intestinal epithelium. *J. Biol. Chem.* 269: 1222–1230.

Marchant, J. S., Beecroft, M. D., Riley, A. M., Jenkins, D. J., Marwood, R. D., Taylor, C. W., and Potter, B. W. (1997) Disaccharide polyphosphates based upon adenophostin A activate hepatic D-*myo* inositol 1,4,5-trisphosphate receptors. *Biochemistry* 36: 12780–12790.

Matsumoto, M., Nakagawa, T., Inoue, T., Nagata, E., Tanaka, K., Takano, H., Minowa, O., Kuno, J., Sakakibara, S., Yamada, M., Yoneshima, H., Miyawaki, A., Fukuuchi, Y., Furuichi, T., Okano, H., Mikoshiba, K., and Noda, T. (1996) Ataxia and epileptic seizures in mice lacking type 1 inositol 1,4,5-trisphosphate receptor. *Nature* 379: 168–171.

Matter, N., Ritz, M.-F., Freyernuth, S., Rogue, P., and Malviya, A. N. (1993) Stimulation of nuclear protein kinase C leads to phosphorylation of nuclear inositol 1,4,5-trisphosphate receptor and accelerated calcium release by inositol 1,4,5-trisphosphate from isolated rat liver nuclei. *J. Biol. Chem.* 268: 732–736.

Mauger, J.-P., Claret, M., Pietri, F., and Hilly, M. (1989) Hormonal regulation of inositol 1,4,5-trisphosphate receptor in rat liver. *J. Biol. Chem.* 264: 8821–8826.

McDonald, T. V., Premack, B. A., and Gardner, P. (1993) Flash photolysis of caged inositol 1,4,5-trisphosphate activates plasma membrane calcium current in human T cells. *J. Biol. Chem.* 268: 3889–3896.

McPherson, P. S. and Campbell, K. P. (1993a) The ryanodine/Ca^{2+} release channel. *J. Biol. Chem.* 268: 13765–13768.

McPherson, P. S., and Campbell, K. P. (1993b) Characterization of the major brain form of the ryanodine receptor/Ca^{2+} release channel. *J. Biol. Chem.* 268: 19785–19790.

Meissner, G. (1994) Ryanodine receptor/Ca^{2+} release channels and their regulation by endogenous effectors. *Annu. Rev. Physiol.* 56: 485–508.

Meldolesi, J., Pozzan, T., and Blondel, O. (1995) IP$_3$ receptors and secretory granules. *Trends Neurosci.* 18: 340–341.

Meyer, T., Holowka, D., and Stryer, L. (1988) Highly cooperative opening of calcium channels by inositol 1,4,5-trisphosphate. *Science* 240: 653–655.

Meyer, T., Wensel, T., and Stryer, L. (1990) Kinetics of calcium channel opening by inositol 1,4,5-trisphosphate. *Biochemistry* 29: 32–37.

Meyers, M. B., Pickel, V. M., Sheu, S. S., Sharma, V. K., Scotto, K. W., and Fishman, G. I. (1995) Association of sorcin with the cardiac ryanodine receptor. *J. Biol. Chem.* 270: 26411–26418.

Meyrleitner, M., Chadwick, C. C., Timerman, A. P., Fleisher, S., and Schindler, H. (1991) Purified IP$_3$ receptor from smooth muscle forms an IP$_3$ gated and heparin sensitive Ca^{2+} channel in planar bilayers. *Cell Calcium* 12: 505–514.

Michikawa, T., Hamanaka, H., Otsu, H., Yamamoto, A., Miyawaki, A., Furuichi, T., Tashiro, Y., and Mikoshiba, K. (1994) Transmembrane topology and sites of N-glycosylation of inositol 1,4,5-trisphosphate receptor. *J. Biol. Chem.* 269: 9184–9189.

Mignery, G. A. and Südhof, T. C., (1990) The ligand binding site and transduction mechanism in the inositol 1,4,5-trisphosphate receptor. *EMBO J.* 9: 3893–3898.

Mignery, G. A., Südhof, T. C., Takei, K. and De Camilli, P. (1989) Putative receptor for inositol 1,4,5-trisphosphate similar to ryanodine receptor. *Nature* 342: 192–195.

Mignery, G. A., Newton, C. L., Archer, B. T. III, and Südhof, T. C. (1990) Structure and expression of the rat inositol 1,4,5-trisphosphate receptor. *J. Biol. Chem.* 265: 12679–12685.

Mignery, G. A., Johnson, P. A., and Südhof, T. C. (1992) Mechanism of Ca^{2+} inhibition of inositol 1,4,5-trisphosphate (InsP$_3$) binding to the cerebellar InsP$_3$ receptor. *J. Biol. Chem.* 267: 7450–7455.

Mikoshiba, K. (1993) Inositol 1,4,5-trisphosphate receptor. *Trends Pharmacol. Sci.* 14: 86–89.

Mikoshiba, K. (1997) The InsP$_3$ receptor and intracellular Ca^{2+} signaling. *Curr. Opin. Neurobiol.* 7: 339–345.

Mikoshiba, K., Huchet, M., and Changeux, J.-P. (1979) Biochemical and immunological studies on the P$_{400}$ protein, a protein characteristic of the Purkinje cell

from mouse and rat cerebellum. *Dev. Neurosci.* 2: 254–275.

Mikoshiba, K., Okano, H., and Tsukada, Y. (1985) P$_{400}$ protein characteristic to Purkinje cells and related proteins in cerebella from neuropathological mutant mice: autoradiographic study by ^{14}C-leucine and phosphorylation. *Dev. Neurosci.* 7: 179–187.

Miller, R. J. (1992) Neuronal Ca^{2+}: getting it up and keeping it up. *Trends Neurosci.* 15: 317–319.

Missiaen, L., De Smedt, H., Droogmans, G., and Casteels, R. (1992a) Ca^{2+} release induced by inositol 1,4,5-trisphosphate is a steady-state phenomenon controlled by luminal Ca^{2+} in permeabilized cells. *Nature* 357: 599–602.

Missiaen, L., De Smedt, H., Droogman, G., and Casteels, R. (1992b) Luminal Ca^{2+} controls the activation of the inositol 1,4,5-trisphosphate receptor by cytosolic Ca^{2+}. *J. Biol. Chem.* 267: 22961–22966.

Missiaen, L., De Smedt, H., Parys, J. B., and Casteels, R. (1994) Co-activation of inositol trisphosphate-induced Ca^{2+} release by cytosolic Ca^{2+} is loading-dependent. *J. Biol. Chem.* 269: 7238–7242.

Missiaen, L., Parys, J. B., De Smedt, H., Sienaert, I., Henning, R. H., Casteels, R., Mezna, M., Michelangeli, F., Hirose, K., and Iino, M. (1995) Opening up Ca^{2+} stores with InsP$_3$. *Nature* 376: 299–301.

Missiaen, L., De Smedt, H. Parys, J. B., Sienaert, I., Vanlingen, S., and Casteels, R. (1996) Threshold for inositol 1,4,5-trisphosphate action. *J. Biol. Chem.* 271: 12287–12293.

Missiaen, L., Parys, J. B., DeSmedt, H., Sienaert, I., Sipama, H., Vanlingen, S., Maes, K., and Casteels, R. (1997) Effect of adenine nucleotides on *myo*-inositol-1,4,5-trisphosphate-induced calcium release. *Biochem. J.* 325: 661–666.

Missiaen, L., Parys, J. B., Sienaert, I., Maes, K., Kunzelmann, K., Takahashi, M., Tanzawa, K., and DeSmedt, H. (1998) Functional properties of the type-3 InsP$_3$ receptor in 16HBE14o- bronchial mucosal cells. *J. Biol. Chem.* 273: 8983–8986.

Miyamoto, T., Restrepo, D., Cragoe, E. J. Jr., and Teeter, J. H. (1992) IP$_3$- and cAMP-induced responses in isolated olfactory receptor neurons from the channel catfish. *J. Membr. Biol.* 127: 173–183.

Miyawaki, A., Furuichi, T., Maeda, N., and Mikoshiba, K. (1990) Expressed cerebellar-type inositol 1,4,5-trisphosphate receptor, P$_{400}$, has calcium release activity in a fibroblast L cell line. *Neuron* 5: 11–18.

Miyawaki, A., Furuichi, T., Ryou, Y., Yoshikawa, S., Nakagawa, T., Saitoh, T., and Mikoshiba, K. (1991) Structure–function relationships of the mouse inositol 1,4,5-trisphosphate receptor. *Proc. Natl. Acad. Sci. USA* 88: 4911–4915.

Miyazaki, S.-I., Yuzaki, M., Nakade, S., Shirakawa, H., Nakanishi, S., Nakade, S., and Mikoshiba, K. (1992) Block of Ca^{2+} wave and Ca^{2+} oscillation by antibody to the inositol 1,4,5-trisphosphate receptor in fertilized hamster eggs. *Science* 257: 251–255.

Monkawa, T., Miyawaki, A., Sugiyama, T., Yoneshima, H., Yamamoto-Hino, M., Furuichi, T., Saruta, T., Hasegawa, M., and Mikoshiba, K. (1995) Heterotetrameric complex formation of inositol 1,4,5-trisphosphate receptor subunits. *J. Biol. Chem.* 270: 14700–14704.

Monkawa, T., Hayashi, M., Miyawaki, A., Sugiyama, T., Yamamoto-Hino, M., Hasegawa, M., Furuichi, T., Mikoshiba, K., and Saruta, T. (1998) Localization of inositol 1,4,5-trisphosphate receptors in the rat kidney. *Kidney Int.* 53: 296–301.

Montero, M., Brini, M., Marsault, R., Alvarez, J., Sitia, R., Pozzan, T., and Rizzuto, R. (1995) Monitoring dynamic changes in free Ca^{2+} concentrations in the endoplasmic reticulum of intact cells. *EMBO J.* 14: 5467–5475.

Morikawa, K., Ohbayashi, T., Nakagawa, M., Konishi, Y., Makino, Y., Yamada, M., Miyawaki, A., Furuichi, T., Mikoshiba, K., and Tamura, T.-A. (1997) Transcription initiation sites and promoter structure of the mouse type 2 inositol 1,4,5-trisphosphate receptor gene. *Gene* 196: 181–185.

Moschella, M. C., and Marks, A. R. (1993) Inositol 1,4,5-trisphosphate receptor expression in cardiac myocytes. *J. Cell Biol.* 120: 1137–1146.

Mourey, R. J., Verma, A., Supattapone, S., and Snyder, S. H. (1990) Purification and characterization of the inositol 1,4,5-trisphosphate receptor protein from rat vas deferens. *Biochem. J.* 272: 383–389.

Mourey, R. J., Estevez, V. A., Marecek, J. F., Barrow, R. K., Prestwich, G. D., and Snyder, S. H. (1993) Inositol 1,4,5-trisphosphate receptors: labeling the inositol 1,4,5-trisphosphate binding site with photoaffinity ligands. *Biochemistry* 32: 1719–1726.

Muallem, S. (1989) Calcium transport pathways of pancreatic acinar cells. *Annu. Rev. Physiol.* 51: 83–105.

Muallem, S., Pandol, S., and Beeker, T. G. (1989) Hormone-evoked calcium release from intracellular stores is a quantal process. *J. Biol. Chem.* 264: 205–212.

Murayama, T. and Ogawa, Y. (1996) Properties of RyR3 ryanodine receptor isoform in mammalian brain. *J. Biol. Chem.* 271: 5079–5084.

Nabeshima, Y., Hanaoka, K., Hayasaka, M., Esumi, E., Li, S., Nonaka, I., and Nabeshima, Y. (1993) Myogenin gene disruption results in perinatal lethality because of severe muscle defect [see comments]. *Nature* 364: 532–535.

Nagata, E., Tanaka, K., Gomi, S., Mihara, B., Shirai, T., Nogawa, H., Mikoshiba, K., and Fukuuchi, Y. (1995) Alteration of inositol 1,4,5-trisphosphate receptor after six-hour hemispheric ischemia. *Neuroscience* 61: 983–990.

Nahorski, S. R. and Potter, B. L. (1989) Molecular recognition of inositol polyphosphates by intracellular receptors and metabolic enzymes. *Trends Pharmacol. Sci.* 10: 139–144.

Nakade, S., Maeda, N., and Mikoshiba, K. (1991) Involvement of the C-terminus of the inositol 1,4,5-trisphosphate receptor in Ca^{2+} release anlayzed using region-specific monoclonal antibodies. *Biochem. J.* 277: 125–131.

Nakade, S., Rhee, S. K. Hamanaka, H., and Mikoshiba, K. (1994) Cyclic AMP-dependent phosphorylation of an immunoaffinity-purified homotetrameric inositol 1,4,5-trisphosphate receptor (type 1) increases Ca^{2+} flux in reconstituted lipid vesicles. *J. Biol. Chem.* 269: 6735–6742.

Nakagawa, T., Okano, H., Furuichi, T., Aruga, J., and Mikoshiba, K. (1991a) The subtypes of the inositol 1,4,5-trisphosphate receptor are expressed in a tissue-specific and developmentally specific manner. *Proc. Natl. Acad. Sci. USA* 88: 6244–6248.

Nakagawa, T., Shiota, C., Okano, H., and Mikoshiba, K. (1991b) Differential localization of alternative spliced transcripts encoding inositol 1,4,5-trisphosphate receptors in mouse cerebellum and hippocampus: in situ hybridization study. *J. Neurochem.* 57: 1807–1810.

Nakai, J., Ogura, T., Protasi, F., Franzini-Armstrong, C., Allen, P. D., and Beam, K. G. (1997) Functional non-equality of the cardiac and skeletal ryanodine receptors. *Proc. Natl. Acad. Sci., USA* 94: 1019–1022.

Nakai, J., Imagawa, T., Hakamat, Y., Shigekawa, M., Takeshima, H., and Numa, S. (1990) Primary structure and functional expression from cDNA of the cardiac ryanodine receptor/calcium release channel. *FEBS Lett.* 271: 169–177.

Nakai, J. Dirksen, R. T., Nguyen, H. T., Pessah, I. N., Beam, K. G., and Allen, P. D. (1996) Enhanced dihydropyridine receptor channel activity in the presence of ryanodine receptor. *Nature* 380: 72–75.

Nakanishi, S., Maeda, N., and Mikoshiba, K. (1991) Immunohistochemical localization of an inositol 1,4,5-trisphosphate receptor, P_{400}, in neural tissue: studies in developing and adult mouse brain. *J. Neurosci.* 11: 2075–2086.

Nakanishi, S., Maeda, N., and Mikoshiba, K. (1991) Immunohistochemical localization of an inositol 1,4,5-trisphosphate receptor, P_{400}, in neural tissue: studies in developing and adult mouse brain. *J. Neurosci.* 11: 2075–2086.

Nakanishi, S., Fujii, A., and Mikoshiba, K. (1996) Immunohistochemical localization of inositol 1,4,5-trisphosphate receptors in non-neuronal tissues, with special reference to epithelia, the reproductive system, and muscular tissues. *Cell Tissue Res.* 285: 235–251.

Nakashima, Y., Nishimura, S., Maeda, A., Barsoumian, E. L., Kakamata, Y., Nakai, J., Allen, P. D., Imoto, K., and Kita, T. (1997) Molecular cloning and characterization of a human ryanodine receptor. *FEBS Lett.* 417: 157–162.

Nathanson, M. H., Fallon, M. B., Padfield, P. J., and Maranto, A. R. (1994) Localization of the type 3 inositol 1,4,5-trisphosphate receptor in the Ca^{2+} wave trigger zone of pancreatic acinar cells. *J. Biol. Chem.* 269: 4693–4696.

Newton, C. L., Mignery, G. A., and Südhof, T. C. (1994) Co-expression in vertebrate tissues and cell lines of multiple inositol 1,4,5-trisphosphate (InsP₃) receptors with distinct affinities for InsP₃. *J. Biol. Chem.* 269: 28618–28619.

Nicotera, P., McConkey, D. J., Jones, D. P., and Orrenius, S. (1989) ATP stimulates Ca^{2+} uptake and increases the free Ca^{2+} concentration in isolated rat liver nuclei. *Proc. Natl. Acad. Sci. USA* 86: 453–457.

Nicotera, P., Orrenius, S., Nilsson, T., and Berggren, P.-O. (1990) An inositol 1,4,5-trisphosphate-sensitive Ca^{2+} pool in liver nuclei. *Proc. Natl. Acad. Sci. USA* 87: 6858–6862.

Nishizuka, Y. (1992) Intracellular signaling by hydrolysis of phospholipids and activation of protein kinase C. *Science* 258: 607–614.

Nordquist, D. T., Kozak, C. A., and Orr, H. T. (1988) cDNA cloning and characterization of three genes uniquely expressed in cerebellum by Purkinje neurons. *J. Neurosci.* 8: 4780–4789.

Nucifora, F. C. Jr., Li, S.-H., Danoff, S., Ullrich, A., and Ross, C. A. (1995) Molecular cloning of cDNA for the human inositol 1,4,5-trisphosphate receptor type 1, and the identification of a third alternatively spliced variant. *Mol. Brain Res.* 32: 291–296.

Nunn, D. L. and Taylor, C. W. (1992) Luminal Ca^{2+} increases the sensitivity of Ca^{2+} stores to inositol 1,4,5-trisphosphate. *Mol. Pharmacol.* 41: 115–119.

Okada, Y., Teeter, J. H., and Restrepo, D. (1994) Inositol 1,4,5-trisphosphate-gated conductance in isolated rat olfactory neurons. *J. Neurophysiol.* 71: 595–602.

Oldershaw, K. A. and Taylor, C. W. (1993) Luminal Ca^{2+} increases the affinity of inositol 1,4,5-trisphosphate for its receptor. *Biochem. J.* 292: 631–633.

Oldershaw, K. A., Nunn, D. L., and Taylor, C. W. (1991) Quantal Ca^{2+} mobilization stimulated by inositol 1,4,5-trisphosphate in permeabilized hepatocytes. *Biochem. J.* 278: 705–708.

Oldershaw, K. A., Richardson, A., and Taylor, C. W. (1992) Prolonged exposure to inositol 1,4,5-trisphosphate does not cause intrinsic desensitization of the intracellular Ca^{2+}-mobilizing receptor. *J. Biol. Chem.* 267: 16312–16316.

O'Malley, D. M. (1994) Calcium permeability of the neuronal nuclear envelope: evaluation using confocal volumes and intracellular perfusion. *J. Neurosci.* 14: 5741–5758.

O'Rouke, F. and Feinstein, M. B. (1990) The inositol 1,4,5-trisphosphate receptor binding sites of platelet membranes. *Biochem. J.* 267: 297–302.

Otsu, H., Yamamoto, A., Maeda, N., Mikoshiba, K., and Tashiro, Y. (1990a) Immunogold localization of inositol 1,4,5-trisphosphate (InsP₃) receptor in mouse cerebellar Purkinje cells using three monoclonal antibodies. *Cell Struct. Func.* 15: 163–173.

Otsu, K., Willard, H. F., Khanna, V. K., Zorzato, F., Green, N. M., and Maclennan, D. H. (1990b). Molecular cloning of cDNA encoding the Ca^{2+} release channel (ryanodine receptor) of rabbit cardiac muscle sarcoplasmic reticulum. *J. Biol. Chem.* 265: 13472–13483.

Ouyang, Y., Deerinck, T. J., Walton, P. D., Airey, J. A., Sutko, J. L., and Ellisman, M. H. (1993) Distribution of ryanodine receptors in the chicken central nervous system. *Brain Res.* 620: 269–280.

Parker, I. and Ivorra, I. (1991) Caffeine inhibits inositol trisphophate-mediated liberation of intracellular calcium in *Xenopus* oocytes. *J. Physiol.* 433: 229–240.

Parys, J. B., and Bezprozvanny, I. (1995) The inositol trisphosphate receptor of *Xenopus* oocytes. *Cell Calcium* 18: 353–363.

Parys, J. B., Sernett, S. W., DeLisle, S., Snyder, P. M. Welsh, M. J., and Campbell, K. P. (1992) Isolation, characterization, and localization of the inositol 1,4,5-trisphosphate receptor protein in *Xenopus laevis* oocytes. *J. Biol. Chem.* 267: 18776–18782.

Parys, J. B., Missiaen, L., De Smedt, H., and Casteels, R. (1993) Loading dependence of inositol 1,4,5-trisphosphate-induced Ca^{2+} release in the clonal cell line A7r5. Implications for the mechanism of quantal Ca^{2+} release. *J. Biol. Chem.* 268: 25206–25212.

Parys, J. B., De Smedt, H., Missiaen, L., Bootman, M. D., Sienaert, I., and Casteels, R. (1995) Rat basophilic leukemia cells as model system for inositol 1,4,5-trisphosphate receptor IV, a receptor of the type II family: functional comparison and immunological detection. *Cell Calcium* 17: 239–249.

Peng, Y.-W., Sharp, A. H., Snyder, S. H., and Yau, K.-W. (1991) Localization of the inositol 1,4,5-trisphosphate receptor in synaptic terminals in the vertebrate retina. *Neuron* 6: 525–531.

Penner, R., Matthews, G., and Neher, E. (1988) Regulation of calcium influx by second messengers in rat mast cells. *Nature* 334: 499–504.

Penner, R., Fasolato, C., and Hoth, M. (1993) Calcium influx and its control by calcium release. *Curr. Opin. Neurobiol.* 3: 368–374.

Perez, P. J., Ramos-Franco, J., Fill, M., and Mignery, G. A. (1997) Identification and functional reconstitution of the type 2 inositol 1,4,5-trisphosphate receptor from ventricular cardiac myocytes. *J. Biol. Chem.* 272: 23961–23969.

Pietri, F., Hilly, M., and Mauger, J.-P. (1990) Calcium mediates the interconversion between two states of the liver inositol 1,4,5-trisphosphate receptor. *J. Biol. Chem.* 265: 17478–17485.

Pietri Rouxel, F., Hilly, M., and Maugeer, J.-P. (1992) Characterization of a rapidly dissociating inositol 1,4,5-trisphosphate-binding site in liver membranes. *J. Biol. Chem.* 267: 20017–20023.

Poitras, M., Bernier, S., Servant, M., Richard, D. E., Boulay, G., and Guillemette, G. (1993) The high affinity state of inositol 1,4,5-trisphosphate receptor is a functional state. *J. Biol. Chem.* 268: 24078–24082.

Pozzan, T., Rizzuto, R., Volpe, P., and Meldolesi, J. (1994) Molecular and cellular physiology of intracellular calcium stores. *Physiol. Rev.* 74: 595–636.

Putney, J. W. Jr. (1990) Capacitative calcium entry revisited. *Cell Calcium* 11: 611–624.

Putney, J. W. Jr. (1993) Excitement about calcium signaling in inexcitable cells. *Science* 262: 676–678.

Putney, J. W. Jr. and Bird, G. S. J. (1993) The signal for capacitative calcium entry. *Cell* 75: 199–201.

Quinton, T. M., Brom, K. D., and Dean, W. L. (1996) Inositol 1,4,5-trisphosphate-mediated Ca^{2+} release from platelet internal membrane is regulated by differential phosphorylation. *Biochemistry* 35: 6865–6871.

Radermacher, M., Wagenknecht, T., Grassucci, R., Frank, J., Inui, M., Chadwick, C., and Fleischer, S. (1992) Cryo-EM of the native structure of the calcium release channel/ryanodine receptor from sarcoplasmic reticulum. *Biophys. J.* 61: 936–940.

Ramos-Franco, J., Fill, M., and Mignery, G. A. (1998) Isoform-specific function of single inositol 1,4,5-trisphosphate receptor channels. *Biophys. J.* 75: 834–839.

Ravazzola, M., Halban, P. A., and Orci, L. (1996) Inositol 1,4,5-trisphosphate receptor subtype 3 in pancreatic islet cell secretory granules revisited. *Proc. Natl. Acad. Sci. USA* 93: 2745–2748.

Renard-Rooney, D. C., Hajnóczky, G., Seitz, M. B., Schneider, T. G., and Thomas, A. P. (1993) Imaging of inositol 1,4,5-trisphosphate-induced Ca^{2+} fluxes in single permeabilized hepatocytes. *J. Biol. Chem.* 268: 23601–23610.

Renard-Rooney, D. C., Joseph, S. K., Seitz, M. B., and Thomas, A. P. (1995) Effect of oxidized glutathione and temperature on inositol 1,4,5-trisphosphate binding in permeabilized hepatocytes. *Biochem. J.* 310: 185–192.

Restrepo, D., Miyamoto, T., Bryant, B. P., and Teeter, J. H. (1990) Odor stimuli trigger influx of calcium into olfactory neurons of the channel catfish. *Science* 249: 1166–1168.

Rhee, S. G. and Choi, K. D. (1992) Regulation of inositol phospholipid-specific phospholipase C isozymes. *J. Biol. Chem.* 267: 12393–12396.

Rizzuto, R., Brini, M., Murgia, M., and Pozzan, T. (1993) Microdomains with high Ca^{2+} close to IP_3-sensitive channels that are sensed by neighboring mitochondria. *Science* 262: 744–747.

Rodrigo, J., Uttenthal, O., Bentura, M. L., Maeda, N., Mikoshiba, K., Martinez-Murillo, R., and Polak, J. M. (1994) Subcellular localization of the inositol 1,4,5-trisphosphate receptor. P_{400}, in the vestibular complex and dorsal cochlear nucleus of the rat. *Brain. Res.* 634: 191–202.

Rooney, T. A., Joseph, S. K., Queen, C., and Thomas, A. P. (1996) Cyclic GMP induces oscillatory calcium signals in rat hepatocytes. *J. Biol. Chem.* 271: 19817–19825.

Ross, C. A., Meldolesi, J., Milner, T. A., Satoh, T., Supattopone, S., and Snyder, S. H. (1989) Inositol 1,4,5-trisphosphate receptor localized to endoplasmic reticulum in cerebellar Purkinje neurons. *Nature* 339: 468–470.

Ross, C. A., Danoff, S. K., Schell, M. J., and Snyder, S. H. (1992) Three additional inositol 1,4,5-trisphosphate receptors: molecular cloning and differential localization in brain and peripheral tissues. *Proc. Natl. Acad. Sci. USA* 89: 4265–4269.

Rossier, M. F. and Putney, J. W. Jr. (1991) The identity of the calcium-storing, inositol 1,4,5-trisphosphate-sensitive organelle in non-muscle cells: calciosome, endoplasmic reticulum or both? *Trends Neurosci.* 14: 310–314.

Ruth, P., Wang, G.-X., Boekhoff, I., May, B., Pfeifer, A., Penner, R., Korth, M., Breer, H., and Hofmann, F. (1993) Transfected cGMP-dependent protein kinase suppresses calcium transients by inhibition of inositol 1,4,5-trisphosphate production. *Proc. Natl. Acad. Sci. USA* 90: 2623–2627.

Ryo, Y., Miyawaki, A., Furuichi, T., and Mikoshiba, K. (1993) Expresion of the metabotropic glutamate receptor mGluR1α and the ionotropic glutamate receptor GluR1 in the brain during the postnatal development of normal mouse and in the cerebellum from mutant mice. *J. Neurosci. Res.* 36: 19–32.

Satoh, T., Ross, C. A., Villa, A., Supattapone, S., Pozzan, T., Snyder, S. H., and Meldolesi, J. (1990) The inositol 1,4,5-trisphosphate receptor in cerebellar Purkinje cells. Quantitative immunogold labeling reveals concentration in an endoplasmic reticulum subcompartment. *J. Cell Biol.* 111: 615–624.

Sayers, L. G. and Michelangeli, F. (1993) The inhibition of the inositol 1,4,5-trisphosphate receptor from rat cerebellum by spermine and other polyamines. *Biochem. Biophys. Res. Commun.* 197: 1203–1208.

Sayers, L. G., Brown, G. R., Michell, B. H., and Michelangeli, F. (1993) The effects of thimerosal on calcium uptake and inositol 1,4,5-trisphosphate-induced calcium release in cerebellar microsomes. *Biochem. J.* 289: 883–887.

Sayers, L. G., Miyawaki, A., Muto, A., Takeshita, H., Yamamoto, A., Michikawa, T., Furuichi, T., and Mikoshiba, K. (1997) Intracellular targeting and homotetramer formation of a truncated inositol 1,4,5-trisphosphate receptor-green fluorescent protein chimera in *Xenopus laevis* oocytes: evidence for the involvement of the transmembrane spanning domain in endoplasmic reticulum targeting and homotetramer complex formation. *Biochem. J.* 323: 273–280.

Schell, M. J., Danoff, S. K., and Ross, C. A. (1993) Inositol (1,4,5)-trisphosphate receptor: characterization of neuron-specific alternative splicing in rat brain and peripheral tissues. *Mol. Brain Res.* 17: 212–216.

Schmoelzl, S., Leeb, T., Brinkmeier, H., Brem, G., and Brenig, B. (1996) Regulation of tissue-specific expression of skeletal muscle ryanodine receptor gene. *J. Biol. Chem.* 271: 4763–4769.

Schneider, M. F., (1994) Control of calcium release in functioning skeletal muscle fibers. *Annu. Rev. Physiol.* 56: 463–483.

Seymour-Laurent, K. J. and Barish, M. E. (1995) Inositol 1,4,5-trisphosphate and ryanodine receptor distributions and patterns of acetylcholine- and caffeine-induced calcium release in cultured mouse hippocampal neurons. *J. Neurosci.* 15: 2592–2608.

Sharp, A. H., Snyder, S. H., and Nigam, S. K. (1992) Inositol 1,4,5-trisphosphate receptors: localization in epithelial tissue. *J. Biol. Chem.* 267: 7444–7449.

Sharp, A. H., Dawson, T. M., Ross, C. A., Fotuhi, M., Mourey, R. J., and Snyder, S. H. (1993a) Inositol 1,4,5-trisphosphate receptors immunohistochemical localization to discrete areas of rat central nervous system. *Neuroscience* 53: 927–942.

Sharp, A. H., McPherson, P. S., Dawson, T. M., Aoki, C., Campbell, K. P., and Snyder, S. H. (1993b). Differential immunohistochemical localization of inositol 1,4,5-trisphosphate- and ryanodine-sensitive Ca^{2+} release channels in rat brain. *J. Neurosci.* 13: 3051–3063.

Sheppard, C. A., Simpson, P. B., Sharp, A. H., Nucifora, F. C., Ross, C. A., Lange, G. D., and Russell, J. T. (1997) Comparison of type 2 inositol 1,4,5-trisphosphate receptor distribution and subcellular Ca^{2+} release sites that support Ca^{2+} waves in cultured astrocytes. *J. Neurochem.* 68: 2317–2327.

Shuttleworth, T. J. (1992) Ca^{2+} release from inositol trisphosphate-sensitive stores is not modulated by intraluminal [Ca^{2+}]. *J. Biol. Chem.* 267: 3573–3576.

Sienaert, I., DeSmedt, H., Parys, J. B., Missiaen, L., Vanlington, S., Sipma, H., and Casteels, R. (1996) Characterization of a cytosolic and a luminal Ca^{2+} binding site in the type I inositol 1,4,5-trisphosphate receptor. *J. Biol. Chem.* 271: 27005–27012.

Snyder, S. H., Sabatini, D. M., Lai, M. M., Steiner, J. P., Hamilton, G. S., and Suzdak, P. D. (1998) Neural actions of immunophilin ligands. *Trends Pharmacol. Sci.* 19: 21–26.

Somlyo, A. V., Horiuti, K., Trentham, D. R., Kitazawa, T., and Somylo, A. P. (1992) Kinetics of Ca^{2+} release and contraction induced by photolysis of caged D-myo-inositol 1,4,5-trisphosphate in smooth muscle. The effects of heparin, procaine, and adenine nucleotides. *J. Biol. Chem.* 257: 22316–22322.

Somogyi, R. and Stucki, J. W. (1991) Hormone-induced calcium oscillations in liver cells can be explained by a simple one pool model. *J. Biol. Chem.* 266: 11068–11077.

Sorrentino, V. and Volpe, P. (1993) Ryanodine receptors: how many, where and why? [Review] *Trends Pharmacol. Sci.* 14: 98–103.

Spät, A., Bradford, P. G., McKinney, J. S., Rubin, R. P., and Putney, J. W. Jr. (1986) A saturable receptor for ^{32}P-inositol-1,4,5-trisphosphate in hepatocytes and neutrophils. *Nature* 319: 514–516.

Stehno-Bittel, L., Lückhoff, A., and Clapham, D. E. (1995) Calcium release from the nucleus by InsP$_3$ receptor channels. *Neuron* 14: 163–167.

Streb, H., Irvine, R. F., Berridge, M. J., and Schulz, I. (1983) Release of Ca^{2+} from a nonmitochondrial intracellular store in pancreatic acinar cells by inositol-1,4,5-trisphosphate. *Nature* 306: 67–96.

Street, V. A., Bosma, M. M., Demas, V. P., Regan, M. R., Lin, D. D., Robinson, L. C., Agnew, W. S., and Tempel, B. L. (1997) The type 1 inositol 1,4,5-trisphosphate receptor gene is altered in the *opisthotonos* mouse. *J. Neurosci.* 17: 635–645.

Striggow, F. and Ehrlich, B. E. (1996) The inositol 1,4,5-trisphosphate receptor of cerebellum. Mn^{2+} permeability and regulation by cytosolic Mn^{2+}. *J. Gen. Physiol.* 108: 115–124.

Strupish, J., Cooke, A. M., Potter, B. V. L., Gigg, R., and Nahorski, S. R. (1988) Stereospecific mobilization of intracellular Ca^{2+} by inositol 1,4,5-trisphosphate. *Biochem. J.* 253: 901–905.

Südhof, T. C., Newton, C. L., Acher, B. T. III, Ushkaryov, Y. A., and Mignery, G. A. (1991) Structure of a novel InsP$_3$ receptor. *EMBO J.* 10: 3199–3206.

Sugawara, H., Kurosaki, M., Takata, M., and Kurosaki, T. (1997) Genetic evidence for involvement of type 1, type 2 and type 3 inositol 1,4,5-trisphosphate receptors in signal transduction through the B-cell antigen receptor. *EMBO J.* 16: 3078–3088.

Sugiyama, T., Yamamoto-Hino, M., Miyawaki, A., Furuichi, T., Mikoshiba, K., and Hasegawa, M. (1994a) Subtypes of inositol 1,4,5-trisphosphate receptor in human hematopoietic cell lines: dynamic aspects of their cell-type specific expression. *FEBS Lett.* 349: 191–196.

Sugiyama, T., Furuya, A., Monkawa, T., Yamamoto-Hino, M., Satoh, S., Ohmori, K., Miyawaki, A., Hanai, N., Mikoshiba, K., and Hasegawa, M. (1994b) Monoclonal antibodies distinctively recognizing the subtypes of inositol 1,4,5-trisphosphate receptor: application to the studies on inflammatory cells. *FEBS Lett.* 354: 149–154.

Suko, J., Maurer-Fogy, I., Plank, B., Bertel, O., Wyskovsky, W., Hohenegger, M., and Hellmenn, G. (1993) Phosphorylation of serine 2843 in ryanodine receptor-calcium release channel of skeletal muscle by cAMP-, cGMP- and CaM-dependent protein kinase. *Biochim. Biophys. Acta.* 1175: 193–206.

Sullivan, K. M., Busa, W. B., and Wilson, K. L. (1993) Calcium mobilization is required for nuclear vesicle fusion in vitro: implication for membrane traffic and IP$_3$ receptor function. *Cell* 73: 1411–1422.

Supattapone, S., Danoff, S. K., Theibert, A., Joseph, S. K., Steiner, J., and Snyder, S. H. (1988a) Cyclic AMP-dependent phosphorylation of a brain inositol trisphosphate receptor decreases its release of calcium. *Proc. Natl. Acad. Sci. USA* 85: 8747–8750.

Supattapone, S., Worley, P. F., Baraban, J. M., and Snyder, S. H. (1988b) Solubilization, purification, and characterization of an inositol trisphosphate receptor. *J. Biol. Chem.* 263: 1530–1534.

Takahashi, M., Tanzawa, K., and Takahashi, S. (1993) Adenophostins, newly discovered metabolites of *Penicillium brevicompactum*, act as potent agonists of the inositol 1,4,5-trisphosphate receptor. *J. Biol. Chem.* 269: 369–372.

Takasago, T., Imagawa, T., and Shigekawa, M. (1989). Phosphorylation of the cardiac ryanodine receptor by cAMP-dependent protein kinase. *J. Biochem.* 106: 872–877.

Takasago, T., Imagawa, T., Furukawa, K., Ogurusu, T., and Shigekawa, M. (1991) Regulation of the cardiac ryanodine receptor by protein kinase-dependent phosphorylation. *J. Biochem.* 109: 163–170.

Takei, K., Stukenbrok, H., Metcalf, A., Mignery, G. A., Südhof, T. C., Volpe, P., and De Camilli, P. (1992) Ca^{2+} stores in Purkinje neurons: endoplasmic reticulum subcompartments demonstrated by the heterogenous distribution of the InsP3 receptor Ca^{2+}-ATPase, and calsequestrin. *J. Neurosci.* 12: 489–505.

Takei, K., Mignery, G. A., Mugnaini, E., Südhof, T. C., and De Camilli, P. (1994) Inositol 1,4,5-trisphosphate receptor causes formation of ER cisternal stacks in transfected fibroblasts and in cerebellar Purkinje cells. *Neuron* 12: 327–342.

Takekura, H., Bennett, L., Tanabe, T., Beam, K. G., and Franzini-Armstrong, C. (1994) Restoration of junctional tetrads in dysgenic myotubes by dihydropyridine receptor cDNA. *Biophys. J.* 67: 793–803.

Takekura, H., Nishi, M., Noda, T., Takeshima, H., and Franzini-Armstrong, C. (1995) Abnormal junctions between surface membrane and sarcoplasmic reticulum in skeletal muscle with a mutation targeted to the ryanodine receptor. *Proc. Natl. Acad. Sci. USA* 92: 3381–3385.

Takeshima, H., Nishimura, S., Matsumoto, T., Ishida, H., Kangawa, K., Minomino, N., Matsuo, H., Ueda, M., Hanaoka, M., Hirose, T., and Numa, S. (1989) Primary structure and expression from complementary DNA of skeletal muscle ryanodine receptor. *Nature* 339: 439–445.

Takeshima, H., Nishimura, S., Nishi, M., Ikeda, M., and Sugimoto, T. (1993) A brain-specific transcript from the 3′-terminal region of the skeletal muscle ryanodine receptor gene. *FEBS Lett.* 322: 105–110.

Takeshima, H., Iino, M., Takekura, H., Nishi, M., Kuno, J., Minowa, O., Takano, H., and Noda, T. (1994) Excitation–contraction uncoupling and muscular degeneration in mice lacking functional skeletal muscle ryanodine-receptor gene. *Nature* 369: 556–559.

Takeshima, H., Yamazawa, T., Ikemoto, T., Takekura, H., Nishi, M., Noda, T., and Iino, M. (1995) Ca^{2+}-induced Ca^{2+} release in myocytes from dyspedic mice lacking the type-1 ryanodine receptor. *EMBO J.* 14: 2999–3006.

Takeshima, H., Ikemoto, T., Nishi, M., Nishiyama, N., Shinuta, M., Sugitani, Y., Kuno, J., Saito, I., Saito, H., Endo, M., Iino, M., and Noda, T. (1996) Generation and characterization of mutant mice lacking ryanodine receptor type 3. *EMBO J.* 271: 19649–19652.

Terasaki, M. and Jaffe, L. A. (1991) Organization of the sea urchin egg endoplasmic reticulum and its reorganization at fertilization. *J. Cell Biol.* 114: 929–940.

Thrower, E. C., Lea, E. J. A., and Dawson, A. P. (1998) The effects of free [Ca^{2+}] on the cytosolic face of the inositol (1,4,5)-trisphosphate receptor at the single channel level. *Biochem. J.* 330: 559–564.

Timerman, A. P., Ogrunbunmi, E., Freund, E., Wiederrecht, G., Marks, A., and Fleischer, S. (1993) The calcium release channel of sarcoplasmic reticulum by FK-506 binding protein. *J. Biol. Chem.* 268: 22992–22999.

Timerman, A. P., Wiedernecht, G., Marcy, A., and Fleischer, S. (1995) Characterization of an exchange reaction between soluble FKBP-12 and FKBP/ryanodine receptor complex. *J. Biol. Chem.* 270: 2451–2459.

Treves, S., Chiozzi, P., and Zorzato, F. (1993) Identification of the domain recognized by anti-(ryanodine receptor) antibodies which affect Ca^{2+}-induced Ca^{2+} release. *Biochem. J.* 291: 757–763.

Tripathy, A., Xu, L., Mann, G., and Meissner, G. (1995) Calmodulin activation and inhibition of skeletal muscle Ca^{2+} release channel (ryanodine receptor). *Biophys. J.* 69: 106–119.

Tu, Q., Velez, P., Cortes-Gutierrez, M., and Fill, M. (1994) Surface charge potentiates conduction through the cardiac ryanodine receptor channel. *J. Gen. Physiol.* 103: 853–867.

Van Delden, C., Foti, M., Lew, D. P., and Krause, K.-H. (1993) Ca^{2+} and Mg^{2+} regulation of inositol 1,4,5-trisphosphate binding in myeloid cells. *J. Biol. Chem.* 268: 12443–12448.

van de Put, F. H., De Pont, J. J., and Willems, P. H. (1994) Heterogeneity between intracellular Ca^{2+} stores as the underlying principle of quantal Ca^{2+} release by inositol 1,4,5-trisphosphate in permeabilized pancreatic acinar cells. *J. Biol. Chem.* 269: 12438–12443.

Venkatesh, K. and Hasan, G. (1997) Disruption of the IP$_3$ receptor gene of *Drosophila* affects larval metamorphosis and ecdysone release. *Curr. Biol.* 7: 500–509.

Volpe, P. and Alderson-Lang, B. H. (1990) Regulation of inositol 1,4,5-trisphosphate-induced Ca^{2+} release II. Effect of cAMP-dependent protein kinase. *Am. J. Physiol.* 258: C1086–C1091.

Volpe, P., Villa, A., Damiani, E., Sharp, A. H., Podini, P., Snyder, S. H., and Meldolesi, J. (1991) Heterogeneity of microsomal Ca^{2+} stores in chicken Purkinje neurons. *EMBO J.* 10: 3183–3189.

Wagenknecht, T., Berkowitz, J., Grassucci, R., Timerman, A. P., and Fleischer, S. (1994) Localization of calmodulin binding sites on the ryanodine receptor from skeletal muscle by electron microscopy. *Biophys. J.* 67: 2286–2295.

Walaas, S. I., Nairn, A. C., and Greengard, P. (1986) PCPP-260, a Purkinje cell-specific cyclic AMP-regulated membrane phosphoprotein of M$_r$ 260,000. *J. Neurosci.* 6: 954–961.

Walton, P. D., Airey, J. A., Sutko, J. L., Beck, C. F., Mignery, G. A., Sudhof, T. C., Derrinck, T. J., and Ellisman, M. H. (1991) Ryanodine and inositol trisphosphate receptors coexist in avian cerebellar Purkinje neurons. *J. Cell Biol.* 113: 1145–1157.

Wang, J. and Best, P. M. (1992) Inactivation of the sarcoplasmic reticulum calcium channel by protein kinase. *Nature* 359: 739–741.

Wang, J. P., Needleman, D. H., and Hamilton, S. L. (1993) Relationship of low affinity [^{3}H]ryanodine binding sites to high affinity sites on the skeletal muscle Ca^{2+} release channel. *J. Biol. Chem.* 268: 20974–20982.

Watras, J., Bezprozvanny, I., and Ehrlich, B. E. (1991) Inositol 1,4,5-trisphosphate-gated channels in cerebellum presence of multiple conductance states. *J. Neurosci.* 11: 3239–3245.

Watras, J., Moraru, I., Costa, D. J., and Kindman, L. A. (1994) Two inositol 1,4,5-trisphosphate binding sites in rat basophilic leukemia cells: relationship between receptor occupancy and calcium release. *Biochemistry* 33: 14359–14367.

White, A. M., Varney, M. A., Watson, S. P., Rigby, S., Changsheng, L., Ward, J. G., Reese, C. B., Graham, H. C., and Williams, R. J. P. (1991) Influence of Mg^{2+} and pH on n.m.r. spectra and radioligand binding of inositol 1,4,5-trisphosphate. *Biochem. J.* 278: 759–764.

Wilcox, R. A., Challiss, R. A. J., Traynor, J. R., Fauq, A. H., Ognayanov, V. I., Kozikowski, A. P., and Nahorski, S. R. (1994) Molecular recognition at the myo-inositol 1,4,5-trisphosphate receptor. *J. Biol. Chem.* 269: 16815–26821.

Wilcox, R. A., Erneux, C., Primrose, W. U., Gigg, R., and Nahorski, S. R. (1995) 2-Hydroxyethyl-α-D-glucopyranoside-2,3′,4′-trisphosphate, a novel metabolically resistant, adenophostin A and myo-inositol-1,4,5-trisphosphate receptor. *Mol. Pharmacol.* 47: 1204–1211.

Willcocks, A. L., Cooke, A. M., Potter, B. V. L., and Nahorski, S. R. (1987) Stereospecific recognition sites for [³H]inositol (1,4,5)-trisphosphate in particulate preparations of rat cerebellum. *Biochem. Biophys. Res. Commun.* 146: 1071–1078.

Wilson, B. S., Pfeiffer, J. R., Smith, A. J., Oliver, J. M., Oberdorf, J. A., and Wojcikiewicz, R. J. H. (1998) Calcium-dependent clustering of inositol 1,4,5-trisphosphate receptors. *Mol. Biol. Cell* 9: 1465–1478.

Witcher, D. R., Kovacs, R. J., Schulman, H., Cefali, D. C., and Jones, L. R. (1991) Unique phosphorylation site on the cardiac ryanodine receptor regulates calcium channel activity. *J. Biol. Chem.* 266: 11144–11152.

Witcher, D. R., Strifler, B. A., and Jones, L. R. (1992) Cardiac-specific phosphorylation site for multifunctional Ca^{2+} calmodulin-dependent protein kinase is conserved in the brain ryanodine receptor. *J. Biol. Chem.* 267: 4963–4967.

Witcher, D. R., McPherson, P. S., Kahl, S. D., Lewis, T., Bentley, P., Mullinnix, M. J., Windass, J. D., and Campbell, K. P. (1994) Photoaffinity labeling of the ryanodine receptor/Ca^{2+} release channel with an azido derivative of ryanodine. *J. Biol. Chem.* 269: 13076–13079.

Wojcikiewicz, R. J. H., (1995) Type I, II, and III inositol 1,4,5-trisphosphate receptors are unequally susceptible to down-regulation and are expressed in markedly different proportions in different cell types. *J. Biol. Chem.* 270: 11678–11683.

Wojcikiewicz, R. J. H. and He, Y. (1995) Type I, II and III inositol 1,4,5-trisphosphate receptor co-immunoprecipitation as evidence for the existence of heterotetrameric receptor complexes. *Biochem. Biophys. Res. Commun.* 213: 334–341.

Wojcikiewicz, R. J. H. and Luo, S. G. (1998) Phosphorylation of inositol 1,4,5-trisphosphate receptors by cAMP-dependent protein kinase. *J. Biol. Chem.* 273: 5670–5677.

Wojcikiewicz, R. J. H. and Oberdorf, J. A. (1996) Regulation of inositol 1,4,5-trisphosphate receptors during cell stimulation is a specific process mediated by cysteine protease activity. *J. Biol. Chem* 271: 16652–16655.

Wojcikiewicz, R. J. H., Furuichi, T., Nakade, S., Mikoshiba, K., and Nahorski, S. R. (1994) Muscarinic receptor activation down-regulates the type I inositol 1,4,5-trisphosphate receptor by accelerating its degradation. *J. Biol. Chem.* 269: 7963–7969.

Wolf, B. A., Colca, J. R., and McDaniel, M. L. (1986) Calmodulin inhibits inositol trisphosphate-induced Ca^{2+} mobilization from the endoplasmic reticulum of islets. *Biochem. Biophys. Res. Commun.* 141: 418–425.

Worley, P. F., Baraban, J. M., Supattaprone, S., Wilson, V. S., and Snyder, S. H. (1987a) Characterization of inositol trisphosphate receptor binding in brain. Regulation by pH and calcium. *J. Biol. Chem.* 262: 12132–12136.

Worley, P. F., Baraban, J. M., Colvin, J. S., and Snyder, S. H. (1987b) Inositol trisphosphate receptor localization in brain: variable stoichiometry with protein kinase C. *Nature* 325: 159–161.

Worley, P. F., Baraban, J. M., and Snyder, S. H. (1989) Inositol 1,4,5-trisphosphate receptor binding: autoradiographic localization in rat brain. *J. Neurosci.* 9: 339–346.

Yamada, N., Makino, Y., Clark, R. A., Pearson, D. W., Mattei, M.-G., Guénet, J.-L., Ohama, E., Fujino, I., Miyawaki, A., Furuichi, T., and Mikoshiba, K. (1994) Human inositol 1,4,5-trisphosphate type 1 receptor, InsP3R1: structure, function, regulation of expression and chromosomal localization. *Biochem. J.* 302: 781–790.

Yamada, M., Miyawaki, A., Saito, K., Nakajima, T., Yamamoto-Hino, M., Ryo, Y., Furuichi, T., and Mikoshiba, K. (1995) The calmodulin-binding domain in the mouse type 1 inositol 1,4,5-trisphosphate receptor. *Biochem. J.* 308: 83–88.

Yamamoto, A., Otsu, H., Yoshimori, T., Maeda, N., Mikoshiba, K., and Tashiro, Y. (1991) Stacks of flattened smooth endoplasmic reticulum highly enriched in inositol 1,4,5-trisphosphate ($InsP_3$) receptor in mouse cerebellar Purkinje cells. *Cell Struct. Func.* 16: 419–432.

Yamamoto, H., Maeda, N., Ninobe, M., Miyamoto, E., and Mikoshiba, K. (1989) Phosphorylation of P_{400} protein by cyclic AMP-dependent protein kinase and Ca^{2+}/calmodulin-dependent protein kinase II. *J. Neurochem.* 53: 917–923.

Yamamoto-Hino, M., Miyawaki, A., Segawa, A., Adachi, E., Yamashina, S., Fujimoto, T., Sugiyama, T., Furuichi, T., Hasegawa, M., and Mikoshiba, K. (1998) Apical vesicles bearing inositol 1,4,5-trisphosphate receptors in the Ca^{2+} initiation site of ductal epithelium of submandibular gland. *J. Cell. Biol.* 141: 135–142.

Yamamoto-Hino, M., Sugiyama, T., Hikichi, K., Mattei, M. G., Hasegawa, K., Sekine, S., Sakurada, K., Miyawaki, A., Furuichi, T., Hasegawa, M., and Mikoshiba, K. (1994) Cloning and characterization of human type 2 and type 3 inositol 1,4,5-trisphosphate receptors. *Receptor Channels* 2: 9–22.

Yamamoto-Hino, M., Miyawaki, A., Kawano, H., Sugiyama, T., Furuichi, T., Hasegawa, M., and Mikoshiba, K. (1995) Immunohistochemical study of inositol 1,4,5-trisphosphate receptor type 3 in rat central nervous system. *NeuroReport* 6: 273–276.

Yamazawa, T., Takeshima, H., Shimuta, M., and Iino, M. (1997) A region of the ryanodine receptor critical for excitation-contraction coupling in skeletal muscle. *J. Biol. Chem.* 272: 8161–8164.

Yoneshima, H., Miyawaki, A., Michikawa, T., Furuichi, T., and Mikoshiba, K. (1997) Ca^{2+} differentially regulates the ligand-affinity states of type 1 and type 3 inositol 1,4,5-trisphosphate receptors. *Biochem J.* 322: 591–596.

Yoo, S. H. (1994) pH-dependent interaction of chromogranin A with integral membrane proteins of secretory vesicle including 260-kDa protein reactive to inositol 1,4,5-trisphosphate receptor anitbody. *J. Biol. Chem.* 269: 12001–12006.

Yoo, S. H. and Lewis, M. S. (1994) pH-dependent interaction of an intracellular loop of inositol 1,4,5-trisphosphate receptor with chromogranin A. *FEBS Lett.* 341: 28–32.

Yoo, S. H. and Lewis, M. S. (1995) Thermodynamic study of the pH-dependent interaction of chromogranin A with an intraluminal loop peptide of the inositol 1,4,5-trisphosphate receptor. *Biochemistry* 34: 632–638.

Yoshida, A., Ogura, A. Imagawa, T., Shigekawa, M., and Takahashi, M. (1992) Cyclic AMP-dependent phosphorylation of the rat brain ryanodine receptor. *J. Neurosci.* 12: 1094–1100.

Yoshikawa, F., Morita, M., Monkawa, T., Michikawa, T., Furuichi, T., and Mikoshiba, K. (1996) Mutational analysis of the ligand binding site of the inositol 1,4,5-trisphosphate receptor. *J. Biol. Chem.* 271: 18277–18284.

Yoshikawa, S., Tanimura, T., Miyawaki, A., Nakamura, M., Yuzaki, M., Furuichi, T., and Mikoshiba, K. (1992) Molecular cloning and characterization of the inositol 1,4,5-trisphosphate receptor in *Drosophila melanogaster. J. Biol. Chem.* 267: 16613–16619.

Zarka, A. and Shoshan-Barmatz, V. (1993) Characterization and photoaffinity labeling of the ATP binding site of the ryanodine receptor from skeletal muscle. *Eur. J. Biochem.* 213: 147–154.

Zhang, B.-X., Zhao, H., and Muallem, S. (1993) Ca^{2+}-dependent kinase and phosphatase control inositol 1,4,5-trisphosphate-mediated Ca^{2+} release. *J. Biol. Chem.* 268: 10997–11001.

Zhang, S. X., Zhang, J.-P., Fletcher, D. L., Zoeller, R. T., and Sun, G. Y. (1995) In situ hybridization of mRNA expression for IP$_3$-3-kinase in rat brain after transient focal cerebral ischemia. *Mol. Brain Res.* 32: 252–260.

Zhu, D. M., Tekle, E., Chock, P. B., and Huang, C. Y. (1996) Reversible phosphorylation as a controlling factor for sustaining calcium oscillations in HeLa cells: involvement of calmodulin-dependent kinase II and a calyculin A-inhibitable phosphatase. *Biochemistry* 35: 7214–7223.

Zorzato, F., Volpe, P., Damiani, E., Quaglino, D. Jr., and Margreth, A. (1989) Terminal cisternae of denervated rabbit skeletal muscle: alterations of functional properties of Ca^{2+} release channels. *Am. J. Physiol.* 257: C504–C511.

Zorzato, F., Fujii, J., Otsu, K., Phillips, M., Green, N. M., Lai, F. A., Meissner, G., MacLennan, D. H. (1990) Molecular cloning of cDNA encoding human and rabbit forms of the Ca^{2+} release channel (ryanodine receptor) of skeletal muscle sarcoplasmic reticulum. *J. Biol. Chem.* 265: 2244–2256.

Zorzato, F., Sacchetto, R., and Margreth, A. (1994) Identification of two ryanodine receptor transcripts in neonatal, slow-, and fast-twitch rabbit skeletal muscles. *Biochem. Biophys. Res. Commun.* 203: 1725–1730.

Zupanc, G. K., Airey, J. A., Maler, L., Sutko, J. L., and Ellisman, M. H. (1992) Immunohistochemical localization of ryanodine binding proteins in the central nervous system of gymnotiform fish. *J. Comp. Neurol.* 325: 135–151.

10

The Calcium Pumps

Danilo Guerini
Ernesto Carafoli

The maintenance of a low free Ca^{2+} concentration is vital to the correct functioning and, eventually, to the survival of the cells (Carafoli, 1987). The flow of information between the extra- and intracellular environment in higher eukaryotes is, indeed, very frequently mediated by fast and transient changes in the free cytosolic Ca^{2+} concentration. The opening of specific channels in the plasma membrane allows Ca^{2+} to enter cells down a 10,000-fold concentration gradient: in the cytosol, free Ca^{2+} oscillates around values of 0.2 μM or less. It is important that changes (increases) in Ca^{2+} concentration are transient, i.e. that they do not last more than a few minutes: sustained increases of cytosolic Ca^{2+} would lead to mitochondrial overloading, activation of proteases, and activation of DNA-fragmenting enzymes. To this aim, efficient systems have evolved to transport the Ca^{2+} out of the cytosol across the plasma membrane and into subcellular organelles (Carafoli, 1987). Two major proteins perform this function in the plasma membrane: the Na^+/Ca^{2+} exchanger and the Ca^{2+}-ATPase (pump). The exchanger couples the transport of Ca^{2+} to the "downhill" cotransport of Na^+ (Chapter 11; Philipson and Nicoll, 1992). The Ca^{2+} pump uses ATP energy to extrude the Ca^{2+} from the cells (Carafoli, 1994). The main Ca^{2+} controlling organelles within the cell are the endo(sarco)-plasmic reticulum: this takes up Ca^{2+} from the cytosol using another Ca^{2+} pump and returns it to the cytosol using ligand-activated channels.

The P-type Pumps

Ion motive ATPases (pumps) are divided into F-, V-, and P-types (Pedersen and Carafoli, 1987a, 1987b). Unlike the F- and V-types, which are multi-subunit hetero-oligomeric complexes, the P-type pumps are normally composed of one or, at most, two subunits, their catalytic function being associated with a large polypeptide chain of 100–130,000 kDa (the catalytic subunit of a K^+-transporting bacterial P-type is significantly smaller). The P-type pumps transport ions as different as K^+, Na^+, Ca^{2+}, H^+ (Pedersen and Carafoli, 1987a), Cd^{2+} (Nucifora et al., 1989), and, possibly, Mg^{2+} (Maguire et al., 1992). Recently, the genetic analysis has led to the suggestion of the existence of Cu^{2+} pumps in higher eukaryotes (Solioz et al., 1994). The distinctive property of P-type pumps is the formation of an energized enzyme intermediate from ATP that is coupled to the translocation of the ion: the γ-phosphate of ATP is transferred to an invariant Asp in the active site, resulting in the formation of a high-energy acyl-phosphate. The reaction is coupled to the transition of the enzyme between two conformational states, called E1 and E2 (Jencks, 1992). In a very simplified description of the reaction mechanism, the E1 conformation would correspond to a state in which the enzyme has high affinity for the ion and is opened to the cytosolic site, and the E2 to a state in which the enzyme acquires low affinity for the ion and is opened to the external (lumenal) site. All P-type pumps have regions of high-sequence homology among them: the site of formation of the phosphorylated intermediate, the ATP-binding pocket, and the region termed the phosphatase domain. The degree of conservation drops considerably in the remainder of the sequences, although some of the residues in the transmembrane domains are also highly conserved.

The Ca^{2+} Pumps in Bacteria and Yeast

A low level of intracellular Ca^{2+} became a necessity as soon as primordial cells began to use ATP as the universal energetic currency: high concentrations of this nucleotide (and of its cleavage product, phosphate) in the cytosol are evidently only possible if the concentration of Ca^{2+} in the environment is low. Although information on the Ca^{2+} homeostasis in prokaryotes is still fragmentary, systems have been described which extrude Ca^{2+} from their cytosol. In most bacteria, Ca^{2+} is exported by Ca^{2+}/H^+ or Ca^{2+}/Na^+ antiporters, which are still poorly characterized (Rosen, 1987). ATP-driven Ca^{2+}-transporting systems have also been identified in some bacteria: in one, *Flavobacterium odoratum*, the activity was found to be associated with a 60,000 Da protein which was purified to apparent homogeneity (Desrosiers et al., 1996). The protein was able to form an aspartyl-phosphate-type phosphorylated intermediate. The cDNA corresponding to this protein has been cloned: surprisingly, none of the sequence features typical of P-type ATPases were found (Pfeiffer et al., 1996). Since even the putative phosphorylation site was very weakly conserved, it appears possible that the Ca^{2+} pump of *Flavobacterium odoratum* is an atypical P-type pump or that the isolated 60,000 Da protein is a subunit of a multimeric pump complex. Although both formation of the phosphorylated intermediate from ATP, and Ca^{2+}-dependent ATP hydrolysis were demonstrated, no direct demonstration that they were coupled to the transport of Ca^{2+} has been possible so far.

The gene of a putative Ca^{2+}-transporting ATPase has been identified in the cyanobacterium *Synechocystis* sp. *PCC 6803* (Geisler et al., 1993). Its sequence had up to 30% identity with that of the SERCA3 cDNA, had a canonical phosphorylation site, and a reasonably high degree of conservation in a domain corresponding to the sequence known to form the ATP-binding site in the SERCA pump. No biochemical or functional information on the product of this gene has been provided.

In the yeast *Saccharomyces cerevisiae*, two genes having sequence homology to those corresponding to Ca^{2+} pumps of mammalian cells have been identified: they have been termed *PMR1* and *PMC1* (Rudolph et al., 1989; Cunningham and Fink, 1994a, 1994b) The first (*PMR1*) encodes a protein that is 30% similar to the SERCA pump but 50% similar to a P-type pump of unknown function for which transcripts have been detected in rat (Gunteski-Hamblin et al., 1992). The protein encoded by the second gene (*PMC1*) has a 40% identity to the PMCA pump (Cunningham and Fink, 1994b), but lacks the calmodulin-binding domain of the latter. The two pumps play important roles in the extrusion of Ca^{2+} from the yeast cytosol, but the deletion of either gene failed to prevent the growth of the yeast under normal conditions. When both genes were deleted, however, the viability of the cells was lost, indicating that at least one functional Ca^{2+} pump was required for yeast survival. The PMC1 gene product has been localized in vacuoles (Cunningham and Fink, 1994b) and that

of the *PMR1* in the Golgi complex (Antebi and Fink, 1992) (Fig. 10.1). The second messenger role of Ca^{2+} in yeast is still poorly investigated, but the maintenance of a low Ca^{2+} concentration in the cytosol is crucial also to the yeast cells to prevent the aberrant activation of Ca^{2+}-dependent enzymes (Cunningham and Fink, 1994a). The role of Ca^{2+} in the control of the functions of protein/enzymes is probably more important in yeast than in prokaryotes, where the extrusion of Ca^{2+} is essentially only needed to prevent cellular Ca^{2+} overload, but probably less sophisticated than in higher eukaryotes where the fluctuations of Ca^{2+} control most of the metabolic processes. Very likely, in the course of evolution, multicellular organisms took advantage of the necessity to keep the cytosolic Ca^{2+} concentrations low and learned to use transient changes of Ca^{2+} concentration as a mean to transmit information. The homeostasis of Ca^{2+} in yeast cells may represent an intermediate evolutionary state: although systems have been developed to control the free cytosolic Ca^{2+} concentration, a probable limited number of yeast enzymes are controlled by Ca^{2+} (Cunningham and Fink, 1994a). Interestingly, although the principal decoder of the Ca^{2+} signal in higher eukaryotes, calmodulin, is essential for the proliferation of yeast, it may not need Ca^{2+} for its function (Geiser et al., 1991).

The SERCA Pump

The Discovery of Sarcoplasmic Reticulum and the SERCA Pump

Early work on muscle has identified the sarcoplasmic reticulum (SR) as a crucial organelle in the control of contraction–relaxation and, in particular, of the Ca^{2+} concentration in the cell. A reticulum Ca^{2+}-dependent ATPase which mediated the transport of Ca^{2+} was discovered well over 30 years ago (Hasselbach and Makinose, 1961; Ebashi and Lipmann, 1962). The enzyme became very popular and its detailed characterization followed rapidly. Relatively recent experiments with the specific inhibitor of the SERCA pump, thapsigargin (Thastrup et al., 1990), and studies on phospholamban (see later) knock-out mice (Luo et al., 1994) have clearly shown that the SERCA pump takes primacy over the two Ca^{2+}-extruding systems of the plasma membrane in the control of cardiac muscle relaxation. This is very likely to be so in the majority of skeletal muscles also.

Most of the biochemical work on the pump has been performed on skeletal muscle, where it represents about 70% of the proteins of the reticulum membrane and could be isolated from it (MacLennan, 1970). The pump is a 100 kDa protein

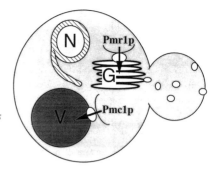

Figure 10.1 Location of the Ca^{2+} pumps in the yeast *Saccharomyces cerevisiae*. The pmr1p and pmc1p pumps found in *Saccharomyces cerevisiae* cells are located on the Golgi complex (G) and on the vacuolar (V) membrane. N, nucleus.

which was initially found in the membranes of sarcoplasmic reticulum but was later detected in the endoplasmic reticulum of most of the eukaryotic cells. Cloning of its cDNAs from muscle and nonmuscle cells confirmed that all intracellular Ca^{2+} pumps belonged to the same gene family, i.e., they have a degree of sequence identity averaging 80%. The pump transports Ca^{2+} ions from the cytosol to the lumen of the SR (or the ER), using an enzymatic cycle (Fig. 10.2) proposed more than 20 years ago (Makinose, 1973), which has received substantial experimental support and is still largely valid today (Jencks, 1992, 1995). The protein binds Ca^{2+} with high affinity at the cytosolic site, a step which is essential for the reaction to progress (Jencks, 1995) and which does not require ATP; the binding of Ca^{2+} can be measured in its the absence (Inesi et al., 1992). The Ca^{2+}-loaded enzyme is phosphorylated by ATP. The phosphate of ATP is transferred to an Asp residue (Bastide et al., 1973; Degani and Boyer, 1973), resulting in the formation of a high-energy acyl phosphate. Asp351 was identified as the sole site of phosphorylation (Allen and Green, 1976; MacLennan et al., 1985), consistent with the ratio of one ATP

hydrolyzed per enzyme cycle. The formation of the phosphorylated intermediate is one of the rate-limiting steps of the reaction cycle (Fig. 10.2), and under favourable conditions the intermediate can be isolated. After phosphorylation, the enzyme undergoes a conformational change from the so-called E1 state to the E2 state, which is a typical feature of P-type pumps (Pedersen and Carafoli, 1987a, 1987b). The E1 state can be operationally viewed as the enzyme state in which the (high-affinity) Ca^{2+}-binding sites are available to the cytosol, whereas in the E2 state the (low-affinity) Ca^{2+}-binding sites are accessible from the lumen of the reticulum. After release of ADP and Ca^{2+} to the lumenal site, the enzyme become slowly dephosphorylated and moves back to the E1 state. The process is fully reversible, i.e. the pump can synthesize ATP from phosphate in the absence of Ca^{2+} (reviewed in Carafoli and Chiesi 1992). Different experiments have established that the SERCA pump transports two Ca^{2+} ions per hydrolyzed ATP (Fig. 10.2) (Inesi, 1987; Inesi et al., 1992; Vilsen, 1995) and it couples the transport of Ca^{2+} to the release of protons to the cytosol. Recent experiments have indicated an electrogenic

Figure 10.2 The catalytic cycle of the SERCA pump. The signs "∼" and "−" indicate the high and low energy content of the complex between the enzyme and the phosphate atom. For simplicity, only the two major conformational states of the pump (E1 and E2) are indicated.

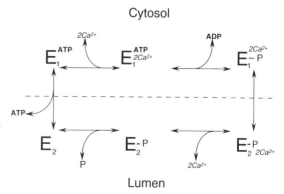

mechanism, in which two H^+ ions per cycle are transported, an operation which could lead to the alkalization of the lumenal site of the SR (Yu et al., 1993).

The treatment of the purified SERCA pump with fluorescein isothiocyanate (FITC), a known competitor of ATP binding, and with different ATP derivatives, has revealed three Lys residues in the ATP-binding domain in the main cytosolic loop of the pump. They are Lys515, which is the FITC-binding site; Lys 684, which is derivatized by pyridoxal-ATP; and Lys492, which becomes photolabelled by TNP-8N3-AMP (Fig. 10.3).

A long-debated question is whether the SERCA pump is active as an oligomer or as a monomer, i.e. whether its activity is influenced by an oligomerization process. The point is important since the concentration of the SERCA pump in the membranes of the SR is very high, so high, in fact, that these membranes have been used to prepare two-dimensional crystals (Toyoshima et al., 1993; Martonosi, 1995). If the high pump concentration in the membrane environment would induce activity changes, it should be expected that the pump of cardiac and skeletal muscle reticulum behaves differently from that in nonmuscle cells, where its concentration is far lower. However, experiments have indicated that the SERCA pump is active as a monomer and that oligomeric forms, although detectable in native membranes or even in solubilized preparations, do not behave differently from monomers (Andersen, 1989). Even cooperative effects observed in the SERCA pump can be accounted for by the monomer state.

The Structure of the SERCA Pump

The high concentration of the SERCA pump in the membranes of sarcoplasmic reticulum in skeletal and cardiac muscles were favourable to the preparation of two-dimensional and even three-dimensional crystals. Electron microscopic observations of SR vesicles has indeed revealed structures that were interpreted as SERCA pump oligomers. These early observations relied on relatively simple contrast staining techniques (Martonosi, 1995). Improvements in the methodology improved the resolution (Martonosi, 1995), and permitted the detection of a large cytosolic structure, a relatively large transmembrane domain, and a very small lumenal domain. These results were used for structure predictions as soon as the sequence of the pump became available (MacLennan et al., 1985). A structural model of the pump is presented in the Fig. 10.3 and will be briefly discussed. The portion anchoring the protein to the membrane consists of 10 transmembrane domains, which are assumed to have α-helical conformation. The first large cytosolic loop protruding between transmembrane domains 2 and 3

contains an extended β-sheet region (Fig. 10.3). Two α-helices following the second and preceding the third transmembrane domain (S1 and S2) are part of the so-called stalk region (see S1–S5 in Fig. 10.3). The second large cytosolic loop between transmembrane domains 4 and 5 (Fig. 10.3) is the site of ATPase activity (MacLennan et al., 1985): sequences in this region were fundamental for the isolation of SERCA cDNAs (MacLennan et al., 1985). This loop contains the aspartic acid, Asp351, which forms the phosphoenzyme intermediate, and the residues that form the ATP-binding site (Lys515, 492, and 684). A more refined tertiary structure prediction of a portion of this region was based on its similarity to the ATP-binding pockets of other proteins (Taylor and Green, 1989): the region between Gly509 and Phe642 in the ATP-binding domain, contains five β-strands that have been proposed to form the inner nucleotide-binding core, surrounded by α-helices, in agreement with the nucleotide-binding domain of the adenylate cyclase (Fig. 10.3) (Taylor and Green, 1989). The model predicted that Gly595, 618, 626, Asp600, 627, and Pro518, 601, and 602 would be critical to the binding of the nucleotide, as it was partially confirmed by site-directed mutagenesis experiments (see below). The "stalk" formed by α-helices above transmembrane domains 1–5 (S1–S5 in Fig. 10.3) was originally suggested to form the high-affinity Ca^{2+}-binding site, but extensive mutation analysis failed to demonstrate its critical role in this (Clarke et al., 1989b).

Crystals of the solubilized SERCA pump that diffracted to 6 Å have been produced (Stokes and Green, 1990), but only very low-resolution maps could so far be generated (Stokes, 1991). A better structure of the pump has been obtained using vanadate as a crystallization inducer and electron microscopy as an observational tool. Although the resolution of the crystals was 14 Å and secondary structures of the pump could not be distinguished, the domains were clearly visible (Toyoshima et al., 1993). Based on this and other available information, a detailed structural model of the SERCA pump could thus be constructed: it confirmed most of the predictions and contained some new features: e.g., a long tilted transmembrane domain, transmembrane 7, which had been predicted to be the longest of the 10 present in the SERCA pump; it contained up to 29 amino acids, in contrast to the 20–23 in the other transmembrane domains (Fig. 10.3). A lumenal domain, mostly made up by the loop between transmembrane domains 7 and 8 (Fig. 10.3), was also discernible, consistent with work with antibodies that had placed the relatively large sequence between the transmembrane domains 7 and 8 between amino acids 870 and 890 in the lumen of the SR (Clarke et al., 1990c; Matthews et al., 1990).

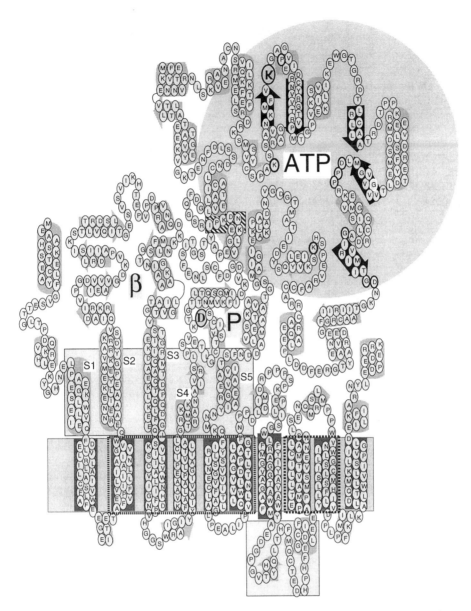

Figure 10.3 A model of the membrane topology of the SERCA pump. The model is based on predictions from the sequence of the SERCA1a protein (the fast-twitch SERCA pump [Brandl et al., 1986]). Also shown are the structural domains described by Toyoshima et al. (1993). α-helices are represented by cylinders, β-strand by large arrows, transmembrane helices are shaded in darker grey. The aspartate (D351) in the catalytic centre (which becomes phosphorylated during the reaction cycle) and the lysine (K515) labelled by FITC are in bold. Two other lysines, which are involved in the binding of ATP (K502 and K684), are encircled in bold.

Three major domains are enclosed in grey shadowed boxes: the transmembrane region, the stalk region (S1, S2, S3, S4, and S5), and the loop between transmembrane helices 7 and 8. The two boxes in the transmembrane region enclosed in rectangles identified by the bold dashed contours identify the transmembrane helices that were predicted to form compact structural units (Toyoshima et al., 1993).

Three other domains involved in the catalysis, the β-sector (β), the phosphorylation (P), and the ATP-binding (ATP) domain, are also indicted. The black β-strands in the ATP-binding domain are those predicted by Taylor and Green (1989) to form a conserved β-fold in the nucleotide-binding domain of P-type pumps. Highly conserved amino acids in the ATP-binding domain of the P-type pump (Taylor and Green, 1989) are encircled in bold.

The striped box indicates the amino acids involved in the binding to phospholamban.

Transmembrane domains 2–5, 6, and 8 form a compact transmembrane structure that is likely to contain the Ca^{2+} channel of the pump, consistent with the results of mutagenesis analysis (MacLennan, 1990). Transmembrane domain 1 probably forms another structural domain which, as was the case of transmembrane domain 10, could not be distinguished in the 14 Å map. The stalk region was clearly visible as a domain that separates the cytosolic domain from the membrane environment. The cytosolic head on the top of the stalk had an asymmetric conformation, but none of its domains (the ATP-binding and the phosphorylation sites) could be assigned. One could mention at this point the recent convincing evidence, based on the isolation and sequencing of membrane-associated peptides after tryptic removal of the cytosolic portion of the pump, for the existence of at least eight transmembrane domains (Shin et al., 1994). A further improvement of this structure has recently been published (Zhang et al., 1998). The resolution of 8 Å permitted the localization of all 10 transmembrane domains. It also revealed the presence of a cavity which has been proposed to represent the Ca^{2+}-binding site and probably the "Ca^{2+} channel".

The use of time resolved x-ray diffraction coupled to the use of caged compounds has allowed the detection of conformational changes in the pump, which in the case of phosphorylation were distributed over all the protein (Blaise et al., 1992). More recently, the same technique has made it possible to demonstrate three independent Ca^{2+}-binding sites in the SERCA pump whose occupancy induced a major conformational change (DeLong and Blaise, 1993). This indicates that the first two steps of the catalytic cycle of the SERCA pump (Fig. 10.2) are associated with long-range conformational changes. This was expected for the phosphorylation step but it was somewhat surprising for the Ca^{2+} binding.

Structure-Function Relationships in the SERCA Pump

Although important information was obtained on preparations of the pump purified from sarcoplasmic reticulum, the cloning of the pump's cDNA (MacLennan et al., 1985; Brandl et al., 1986; Burk et al., 1989; Lytton et al., 1989) and the development of an efficient expression system (Maruyama and MacLennan, 1988) opened the possibility of studying the pump manipulated by site-directed mutagenesis, and thus to test the suggestions based on previous work, on the isolated pump or on the pump embedded in the membrane (MacLennan, 1990; Andersen, 1995a; Andersen and Vilsen, 1995).

The transfection of wild-type SERCA pump's cDNA in COS cells has led to the isolation of microsomal membranes which are 15–20-fold enriched in the pump with respect to control cells (Maruyama and MacLennan, 1988). Since COS cells contain an active endogenous SERCA-type pump, an efficient overexpression of the exogenous pump was necessary to characterize the mutants. The formation of the phosphoenzyme intermediate from phosphate and from ATP and the ATP-dependent Ca^{2+} uptake were easily demonstrated in the microsomes of transfected COS cells (Maruyama and MacLennan, 1988), but the measurement of a Ca^{2+}-dependent ATPase activity and the measurement of the direct binding of Ca^{2+} to the recombinant pump were more difficult. They required expression of the pump in insect Sf9 cells using recombinant baculoviruses (Skerjanc et al., 1993).

Based on the structural model derived from the primary structure and other available information (MacLennan, 1970; Brandl et al., 1986), a critical role in the binding of Ca^{2+} was originally predicted for the charged (negative) amino acid residues which are very abundant in the stalk region (S1–S5, Fig. 10.3). But mutations of 24 of these residues, alone or pairwise, did not significantly influence the activity of the pump (Fig. 10.4) (Clarke et al., 1989b). A more dramatic effect followed the mutation of polar amino acids in the transmembrane domains: mutation of Glu309, Glu771, Asn769, Thr799, Asp800, and Glu908 abolished the ability of the pump to transport Ca^{2+} and to form the phosphoenzyme from ATP (Fig. 10.4) (Clarke et al., 1989a, 1990a). The mutants, however, retained the ability to form the phosphoenzyme from phosphate, a reaction of the catalytic cycle that does not require high-affinity binding of Ca^{2+} (Clarke et al., 1989a) (Fig. 10.2). This compellingly suggests the involvement of these six residues in the formation of high-affinity binding sites for Ca^{2+}, i.e. it was proposed that they would line the "channel" through which Ca^{2+} ions were transported. Mutations of other polar amino acids in the transmembrane domains (see Fig. 10.4) failed to influence the activity of the pump (Clarke et al., 1990a), which was perhaps not surprising since most of the polar amino acids chosen were at the periphery of the transmembrane sectors. By contrast, the six residues critical to the Ca^{2+} transport were located in the middle of transmembrane domains 4, 5, 6, and 8 (Fig. 10.4). However, the activity of one of the mutants that had lost high-affinity Ca^{2+}-binding sites (Glu908Ala, 8th transmembrane domain) was partially rescued at millimolar concentrations of Ca^{2+}, indicating that Glu908 is probably not directly involved in the binding of the Ca^{2+} (Anderson and Vilsen, 1994). This observation was consistent with the finding that this mutant, unlike the other five, could still occlude Ca^{2+} when incubated in the presence of Cr-ATP (Vilsen and Andersen, 1992a;

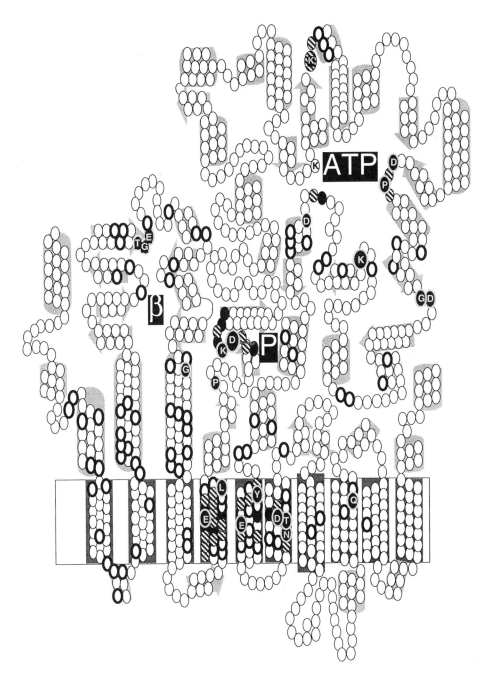

Figure 10.4 A summary of the amino acid mutations that have been performed in the SERCA pump. The model of Fig. 10.3 has been used without identification of all the amino acids. The bold circles show the amino acids whose mutation failed to affect the activity of the pump; the striped and the black circles identify amino acids whose mutation caused a decrease, or the complete loss of the activity of the pump, respectively. The amino acids shown by the single letter code are those whose mutation has been studied in more detail. They are the amino acids involved in the transfer of Ca^{2+} across the protein (those in the transmembrane region), the amino acids involved in the E1 $\rightarrow$ E2 transition (β-sector), and the amino acids directly involved in the catalysis (ATP binding and phosphorylation).

Anderson and Vilsen, 1994, 1995; Andersen 1995a). The Cr-ATP-mediated Ca^{2+} occlusion in this mutant revealed two nonequivalent Ca^{2+}-occluding sites (Vilsen and Andersen, 1992a, 1992b): Glu309 and Asn796 have been proposed to be involved in one, and Glu771 and Thr799 in the other (Andersen, 1995a). Experiments on Ca^{2+} binding to the Glu309Ala mutant expressed in Sf9 cells have shown that a high-affinity Ca^{2+}-binding site was still present (Skerjanc et al, 1993), showing conclusively that two separate Ca^{2+}-binding sites indeed existed and that both were essential for the activity of the pump.

Mutational analysis was also applied to the highly conserved phosphorylation site. As expected, the substitution of Asp351, but also of the conserved Lys352 (Fig. 10.4), abolished the activity of the pump (Maruyama and MacLennan, 1988; Maruyama et al., 1989). Systematic mutations of other conserved amino acids in the region led to decreased activity in all the mutants, although some residues tolerated conservative changes (Maruyama et al., 1989).

Based on the refined structure predictions mentioned above, the nucleotide-binding domain of the SERCA pump was placed between Gly509 and Phe642 (Fig. 10.3) (Taylor and Green, 1989). Mutation of Gly626, Pro602, and Asp601, 627, 703, and 707 abolished the activity of the pump (Clarke et al., 1990b), whereas mutation of Gly510 and 704, and Pro601 resulted in only partial loss of activity. It is likely that all these residues, placed in the loops or turns between predicted α-helices or β-sheets, are critical to the correct folding of the domain and thus to the correct structure of the nucleotide-binding pocket. One of these mutants (Asp703Ala), which was unable to form the phosphoenzyme intermediate from phosphate and from ATP, still occluded Ca^{2+} like the wild-type (Vilsen and Andersen, 1992a): i.e. the mutant had lost the ability to bind ATP, but the Ca^{2+}-binding sites were preserved.

Another region of the SERCA pump in which numerous amino acids are conserved with respect to other P-type pumps is the β-strand sector in the first large cytoplasmic loop (Fig. 10.3). Mutations in this domain produced pumps that were unable to accumulate Ca^{2+}, but still formed the phosphorylated intermediate from both phosphate and ATP (Andersen et al., 1989; Clarke et al., 1990b): mutants Gly233Val, Thr181Ala, Gly182Ala, and Glu183Ala all showed impaired E1P–E2P interconversion, i.e., they were unable to complete the reaction cycle (see Fig. 10.2). Interestingly, mutant Gly233Val also had a lower affinity for phosphate in experiments of phosphoenzyme formation from phosphate (Andersen et al., 1989).

Conserved Gly and Pro residues, some located in the transmembrane domains, were also systematically mutated (Vilsen et al., 1989; Andersen et al., 1992), resulting in variable damage to the pump. While mutations of Pro160, 195, and 812 were well tolerated, those of Pro312, 308, 803, and Gly310, 770, and 801 were not. The latter six residues are located in the transmembrane portion of the pump in the proximity of residues which have been proposed to be involved in the calcium-binding site. These Pro and Gly residues may thus be important to the proper folding of the Ca^{2+}-binding domain. Pro337 could also play a critical role in the conformation changes during the catalytic cycle (Sumbilla et al., 1993). Systematic mutations of residues in transmembrane domain 4 (Clarke et al, 1993; Chen et al., 1996) and of amino acids between this transmembrane domain and the stalk domain (S4, Fig. 10.4) (Vilsen et al., 1991; Chen et al., 1996) confirmed the hypothesis. The tolerance to mutations (Leu319 is one exception) increased with the distance of the residues from the centre of the transmembrane domain, where the Ca^{2+}-binding residues are located. However, the effect of the mutations varied with the residue that was used for the substitution: in the case of Leu319, some mutations were lethal whereas others had only marginal effect (Vilsen et al., 1991; Chen et al., 1996).

A particularly interesting mutant was Tyr763Gly (Andersen 1995b). It was unable to transport Ca^{2+}, but still had normal Ca^{2+}-dependent ATPase activity which was not stimulated by the addition of Ca^{2+} ionophores to the preparation of microsomes: analysis of the phosphoenzyme intermediate revealed that the partial reactions of the catalytic cycle were normal (Fig. 10.2); thus, this mutant was an uncoupled SERCA pump, i.e. a pump that could hydrolyze ATP in a Ca^{2+}-dependent way but was unable to transport Ca^{2+} across the membrane.

Regulation of the Activity of the SERCA Pump

Among the SERCA pumps isoforms, only the cardiac isoform (SERCA2, which is also expressed in slow skeletal and smooth muscles) is regulated by phosphorylation. The regulation has been traditionally considered to be indirect, i.e., due to the small hydrophobic protein phospholamban. Direct phosphorylation of the pump by CaM-kinase (Ser38) has also been described (Xu et al., 1993), but its physiological importance has recently been questioned (Odermatt et al., 1996a; Reddy et al., 1996). The involvement of phospholamban in the regulation of the pump by phosphorylation was described for the heart sarcoplasmic reticulum more than 20 years ago. Dephosphorylated phospholamban binds to the SERCA pump, shifting its K_d for Ca^{2+} to a lower affinity value and thus resulting in the inhibition of the pump activity. Phosphorylation of phospholamban by the cAMP and the Ca^{2+}/calmodulin-dependent protein kinases dissociates it from the SERCA pump, relieving the inhibition. The generation of phospholamban knock-out mice has demonstrated that the regulation of the pump by phospholamban is critical to the positive inotropic effect of catecholamines, i.e. it showed that the increase of the heart-beat rate was mediated by the kinase-promoted dissociation of phospholamban from the pump (Luo et al., 1994).

The site of interaction of phospholamban in the SERCA pump was characterized by cross-linking a derivatized phopholamban to the pump, prepared from cardiac muscle: a short stretch of amino acids close to the phosphorylation site of the pump became labelled (James et al., 1989a). This issue was further explored by cotransfecting the pump cDNA with cDNA for phospholamban in COS cells; as expected, inhibition of the SERCA pump activity was observed (Fujii et al., 1990). Site-directed mutagenesis was then used to define more precisely the site of the interaction of the two proteins. The experiments showed that other amino acids are also involved in the interaction of the phospholamban with the pump (Toyofuku et al., 1993), but confirmed that one of the pump sequences responsible for the interaction with phospholamban, was a six amino acid stretch (Fig. 10.3) located, in agreement with the cross-linking experiments, between Lys397 and Val402 of the SERCA2 isoform (IIe402 in the SERCA1 isoform shown in Fig. 10.3) (Toyofuku et al., 1994b). The importance of the region near the phosphorylation site in the interaction was further supported by the finding that the interaction of the two proteins was inhibited by peptides corresponding to the same region (Vorherr et al., 1992). The SERCA1 pump, which is the isoform of fast-twitch muscles (but not the SERCA3 pump) could also be regulated by phospholamban (Toyofuku et al., 1993), except that the latter is not expressed in fast-twitch muscles. The construction of chimeras of SERCA3 (phospholamban-insensitive) and SERCA2 pumps led to the discovery that another domain is also important in the interaction of the two proteins. The domain is located in the nucleotide binding-hinge region (amino acids 467–762) (Toyofuku et al., 1993) and has not been fully characterized as yet. Mutations performed on phospholamban indicated that residues surrounding the Ser and Thr residues, which are phosphorylated by the two protein kinases, are critical for the interaction with the pump (Toyofuku et al., 1994a).

Inhibitors of the SERCA Pump

The study of the function of the SERCA pump has been greatly helped by the availability of four highly specific inhibitors (Inesi and Sagara, 1994; Rogers et al., 1995). The most specific are thapsigargin and thapsigarcin, which were isolated from the roots of *Thapsia garganica* (Rasmussen et al., 1978). Cyclopiazonic acid and 2,5-di(tert-butyl)hydroquinone (Inesi and Sagara, 1994) have lower affinity for the SERCA pump (K_d in the low micromolar to high nanomolar range), and less is known about their mechanisms of action, but the indications have been that the inhibition of the pump by cyclopiazonic acid could be competed by ATP (Sagara et al., 1992a).

More is known on the mechanism of inhibition by thapsigargin (and on that of thapsigarcin, which is predicted to be very similar) (Inesi and Sagara, 1994). Originally described in rat hepatocytes (Thastrup et al., 1990), the inhibition of thapsigargin was then demonstrated on all SERCA pump isoforms (Lytton et al., 1991). Its affinity for the pump is very high: its K_d was estimated to be in the subnanomolar range (Sagara and Inesi, 1991). The inhibitor could be removed from the pump only after extensive digestion by trypsin, followed by denaturation (Sagara et al., 1992a). Recent measurements of the off-rate of thapsigargin have permitted a more precise estimate of its binding constant. In the absence of detergent, the dissociation rate was $0.0052 \text{ s} - 1$ and the affinity was extrapolated to be 2.2 pM (Davidson and Varhol, 1995). Most of the available information indicates that thapsigargin forms a 1:1 molar complex with the pump (Sagara and Inesi, 1991; Sagara et al., 1992b), although in one case a 2:1 ratio was calculated (Lytton et al., 1991).

The mechanism of the inhibition is now well characterized. The inhibitor reacts with a specific state of the pump (Sagara et al., 1992a, 1992b) forming a dead-end complex with it, i.e. the pump becomes trapped in a conformation from which it can no longer exit and thus cannot perform the catalytic cycle (Sagara et al., 1992b). Thapsigargin reacts preferentially with the Ca^{2+}-free pump, after the pump has performed one run of the catalytic cycle (see Fig. 10.2), i.e. with the E2 conformation (Sagara et al., 1992a). Although the affinity of the thapsigargin pump complex for ATP is 20–70 times lower, the nucleotide can still become bound (De Jesus et al., 1993). SERCA-Na^+K^+ pump chimeras have been used to map the thapsigargin-binding site (Norregard et al., 1993; Sumbilla et al., 1993; Andersen, 1995a): the inhibitor binds close to transmembrane domain 3 (Norregard et al., 1993). The finding that the Gly310Pro mutant, which does not transport Ca^{2+} but is still capable of forming the phosphoenzyme intermediate (Fig. 10.4), is insensitive to thapsigargin is consistent with the proposal that transmembrane domains 3 and 4 form the thapsigargin-binding pocket (Andersen, 1995a). The importance of this region in the catalytic/transport

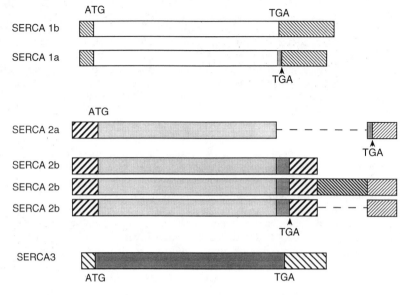

Figure 10.5 A scheme of the SERCA cDNAs. The start (ATG) and the end of the coding sequence (TGA) are indicated. The noncoding sequence is indicated by a striped box, the coding sequences by different shades of grey.

cycle is well documented (see above), and thus it is not surprising that thapsigargin locks the pump in an inactive conformational state (the dead-end complex). Thapsigargin has now also become a useful tool in the study of the function of the SERCA pump in vivo (Wrosek et al., 1992; Rogers et al., 1995). Its use has permitted the analysis of intracellular Ca^{2+} under conditions in which the SERCA pump is inhibited or in which internal Ca^{2+} stores had been emptied.

Genes, Gene Structure, and Isoforms of the SERCA Pump

The SERCA pump is encoded by three genes (Grover and Khan, 1992), termed SERCA1, SERCA2, and SERCA3 (Fig. 10.5). The SERCA1 pump isoform is expressed in high amounts in fast-twitch skeletal muscles, from where it was originally cloned (Brandl et al., 1986), and in lower amounts in slow-twitch muscles (Brandl et al., 1987). Two alternatively spliced isoforms have been described: SERCA1a, which is present predominantly in adult muscles, and SERCA1b, for which high amounts of transcripts have been detected in neonatal tissues (Brandl et al., 1987). SERCA1b lacks a 42 bp exon and corresponds to a 1001 amino acid molecule, 7 amino acids longer than the corresponding adult form (Fig. 10.4). Transcripts of the SERCA2 pump, originally cloned from rabbit skeletal muscle

(MacLennan et al., 1985), have been found in slow-twitch, cardiac, and smooth muscles, in nonmuscle tissues, and to a much lesser extent in fast-twitch skeletal muscles (Brandl et al., 1987). Alternatively spliced products have been described for the SERCA2 pump also. Three encoded an identical protein, the SERCA2b isoform, but different in the 3′ untranslated region (Fig. 10.5) (Gunteski-Hamblin et al., 1988; Lytton and MacLennan, 1988; Eggermont et al., 1989; Lytton et al., 1989). The last four amino acids of the SERCA2a protein are replaced by a stretch of 49 amino acids in the SERCA2b protein (Lytton and MacLennan, 1988). The insertion of these amino acids has no influence on the activity of the pump expressed in COS cells (Campbell et al., 1991), but the C-terminus of the isoform is in the lumen of the ER: the 49 extreme amino acids thus contain an additional (the 11th) transmembrane domain (Campbell et al., 1992). SERCA2b transcripts have been detected in nonmuscle tissues and in smooth muscle, but those for the SERCA2a isoform (Lytton and MacLennan, 1988; Eggermont et al., 1989) have been found only in muscle tissues. Different studies have shown that the expression of the SERCA2 gene is up-regulated during muscle differentiation (Zarain-Herzberg et al., 1990) and that the two alternatively spliced SERCA2 isoforms are differentially regulated: in particular, the SERCA2a isoform is transcribed only when muscle cells differentiate to myotubes (De Smedt et al., 1991; Grover and Khan, 1992). The third isoform, SERCA3, had been cloned from a rat kidney library (Burk et al., 1989) and has been demonstrated to be functional only after expression in COS cells. The corresponding protein is the major SERCA isoform present in platelets (Bobe et al., 1994); Wuytack et al., 1994), but very little information is available on this isoform. A possible up-regulation of this isoform in spontaneously hyperactive rats was also described (Bobe et al., 1994).

The structure of the gene of the SERCA1 pump, which is relatively small, was solved very early (Korczak et al., 1988). Yet very little is known about the regulation of the expression of the isoform. In contrast, even if only a partial sequence of the SERCA2 gene is available (Zarain-Herzberg et al., 1990), more is known about its regulation. These studies were stimulated by the observation that the expression of the gene was strongly influenced by thyroid hormones and by pressure overload in cardiac cells, a behaviour which is similar to that of phospholamban (Nagai et al., 1989). Similar observations were made for the SERCA1 and SERCA2 transcripts in rat soleus muscle (Simonides et al., 1990). Further studies have concentrated on the definition of the *cis* elements that were involved in the regulation by thyroid hormones of the SERCA2 isoform in cardiac cells (Rohrer et al., 1991; Zarain-Herzberg et al., 1994). At least three genuine TREs (thyroid hormone response elements) were identified, and were demonstrated to bind the thyroid hormone receptor (TR) and the retinoic X receptor (RXR) (Hartong et al., 1994). Using a combination of nonmuscle cells and muscle cells that can differentiate to myotubes, it was demonstrated that two distinct regions containing repeats of Sp1-like elements induced high levels of expression of the SERCA2a transcripts in muscle cells (Baker et al., 1996). This may help us to understand why high levels of SERCA-specific mRNA are transcribed in muscle, but not in other cell types. The experiments may also help to clarify the discrepancy between the amount of SERCA2 mRNA and that of the corresponding protein observed in smooth and in cardiac muscles. The SERCA protein is 60–80-fold more abundant in cardiac than in stomach smooth muscles, whereas transcripts are only 6–8-fold more abundant in cardiac than in smooth muscle (Grover and Khan, 1992).

The PMCA Pump

The Plasma Membrane and the Ca^{2+} Pump

The plasma membrane calcium ATPase is essential to the control of the cytosolic Ca^{2+} concentration in nonmuscle cells: its role in muscle cells, particularly heart cells, is likely to be minor compared with that of the much more abundant SERCA pump or with that of the high transport capacity sodium/calcium exchanger. In the plasma membrane of most eukaryotic cells, the amount of PMCA pump probably never exceeds the level of 0.1% of the total membrane protein. This value may perhaps be surpassed in nerve cells, where the level of expression of PMCA is apparently higher than in all other cells examined (Stauffer et al., 1995).

Excitable cells, e.g., muscle cells, rely largely on Ca^{2+} mobilized from internal stores (Carafoli, 1987). However, a minor amount of Ca^{2+} must enter through the plasma membrane (via specific voltage-sensitive and/or receptor-operated channels) to promote its mobilization. It must then be necessarily re-exported via the Na^+Ca^{2+} exchanger and the Ca^{2+} pump. As mentioned, the exchanger predominates in excitable cells, particularly those of the heart, but most other cells depend on the pump: its high calcium affinity qualifies it as a candidate for the fine-tuning of calcium in the cytosdol (Carafoli, 1991). Although most of the initial physiological work on the PMCA pump was on erythrocytes (Schatzmann, 1966, Schatzmann and Buergin, 1978), the introduction of fluorescent Ca^{2+} indicators and refinement of protocols has made it possible to extend the measurement the Ca^{2+} extrusion to other cells (Tepikin et al., 1992; Zhang et al., 1992). Experiments in pancreatic acinar cells, for instance, showed that extrusion of Ca^{2+} mediated by the plasma membrane pump modulated the kinetics of the Ca^{2+} fluctuations in the cytosol. It could also be shown that the amount of Ca^{2+} extruded from the cell by the pump was a significant fraction of that released from the internal stores. The role of the pump was also investigated in stable cell lines that overexpressed it (Guerini et al., 1995; Liu et al., 1996): the extrusion of Ca^{2+} to the extracellular space upon stimulation by agents that mobilized Ca^{2+} from internal stores was greatly increased with respect to control cells. Interestingly, these studies showed that both the SERCA pump (Guerini et al., 1995; Liu et al., 1996) and IP_3 receptor (Liu et al., 1996) were down-regulated in these cells, indicating that cells may be able to adjust the amounts of their Ca^{2+}-transporting systems to fit the requirements of cytosolic Ca^{2+} homeostasis.

General Properties of the PMCA Calcium Pump

The reaction mechanism of the pump is similar to that of the SERCA pump and, indeed, to that of all other P-type pumps (see Fig. 10.6, [Carafoli and Guerini, 1993]). The aspartyl-phosphate intermediate is sufficiently stable and can be detected and characterized (although the PMCA pump forms it with

Cytosol

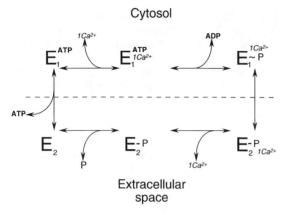

Extracellular
space

Figure 10.6 The catalytic cycle of the PMCA pump. The signs "∼" and "−" indicate the high and low energy content of the complex between the enzyme and the phosphate atom. As in the case of the SERCA pump model (Fig. 10.2), only the two major conformational states of the catalytic cycle of the pump (E1 and E2) are indicated.

lower efficiency than the SERCA pump). Two conformational forms of the phosphorylated state ($E_1 \sim P$ and E_2–P) have been demonstrated also for the PMCA pump (Krebs et al., 1987). The step responsible for the translocation of the bound Ca^{2+} across the protein has not been conclusively identified, but it is likely to correspond to the $E_1 \sim P$ and E_2–P transition. This is consistent with the observation that mutants that cannot form the phosphorylated intermediate from ATP (corresponding to the E1 state) are unable to transport Ca^{2+}, despite their ability to form the phosphointermediate from phosphate (Guerini et al., 1996).

As with all other P-type pumps, the PMCA pump is inhibited by micromolar concentrations of the phosphate analogue orthovanadate $[VO_3(OH)]^{2-}$. The other general inhibitor of P-type ATPases, La^{3+}, acts on the PMCA pump in a peculiar way, i.e. it increases the steady-state level of the phosphorylated intermediate, rather than dissipating it. This allows identification of the plasma membrane Ca^{2+} pump in preparations that contain other P-type pumps as well: La^{3+} enhances significantly the amount of the phosphorylated intermediate of the plasma membrane Ca^{2+} pump, while reducing that of the other pumps. At variance with the SERCA pump, the PMCA pump has a 1:1 molar Ca^{2+}/ATP transport stoichiometry (Niggli et al., 1982b; Hao et al., 1994). The pump exchanges Ca^{2+} for H^+ (Niggli et al., 1982b): early reports on the reconstituted pump indicated an electroneutral Ca^{2+}/H^+ exchanger (Niggli et al., 1982b), but more recent experiments have instead shown partial electrogenicity, i.e. one H^+ is exchanged for one Ca^{2+} (Hao et al., 1994).

The unstimulated pump has low Ca^{2+} affinity ($K_m > 10 \mu M$) and would be inactive at physiological cytosolic Ca^{2+} concentrations. A number of regula-tors, however, increase its Ca^{2+} affinity to a K_m which could be as low as $0.2 \mu M$. The most important is probably calmodulin (Carafoli, 1992), which interacts with the pump with a K_d in the low nanomolar range. Acidic phospholipids have also been shown to activate the pump at a concentration range similar to that found in plasma membranes (Niggli et al., 1981). It has been calculated that their concentration around the pump in vivo may be sufficient for 50% of maximal activation (Niggli et al., 1981). The most active of the phospholipids is the doubly phosphorylated derivative of phosphatidyl inositol (PIP2) (Carafoli and Zurini, 1982; Choquette et al., 1984). Since PIP2 (and PIP1) is the only acidic phospholipid whose concentration in the plasma membrane is rapidly modulated in response to external stimuli, and since the product of its hydrolysis (inositol-trisphosphate) is inactive on the pump (Carafoli, 1991), an activation/deactivation regulation cycle of the pump in vivo has been proposed (Penniston, 1982). One problem with the proposal is that the product of the hydrolysis of PIP2, diacylglycerol, activates protein kinase C, which, in turn, phosphorylates (and activates) the PMCA pump (see above). Another possible problem is the low amounts of phosphoinositides that are metabolized in vivo and which would probably be too low to activate significantly the pump dispersed in the plasma membrane. However, recent experiments indicate that the PMCA pump may be concentrated in portions of the plasma membrane: the pump is enriched in the caveolae—small invaginations present on the plasma membrane (Fujimoto, 1993; Schnitzer et al., 1995). Caveolae are believed to be the site of intense exchange of incoming and outgoing messengers, where various signal transduction proteins have been localized (Anderson, 1993). It will be interesting to quantify the fraction of the pump located in the caveolae as compared with the fraction distributed on the remainder of the plasma membrane surface.

The PMCA pump can be phosphorylated by the cAMP-dependent protein kinase (Caroni and Carafoli, 1981b; Neyses et al., 1985; James et al., 1989b) and by protein kinase C (Wang et al., 1991). The former activates the pump by lowering its K_m for Ca^{2+} to about 1 μM (James et al., 1989b), the second has been claimed to activate the pump, although the magnitude of the effect has varied in different reports (Smallwood et al., 1988). Experiments with a synthetic peptide corresponding to the calmodulin-binding domain of the pump phsophorylated on Thr1102 have shown that the phosphorylation weakens the interaction of the calmodulin-binding domain with the binding ("receptor") site in the pump (Hofmann et al., 1994). The activation by protein kinase A was first observed in heart sarcolemma membranes, where this kinase also activates the L-type Ca^{2+} channels. The overall effect of protein kinase A would thus be to increase the cycling of Ca^{2+} across the sarcolemma, rather than to stimulate the undirectional Ca^{2+} flux.

The phosphorylation effects by both kinases have been studied in reasonable detail in vitro, but are less clear in vivo. This may be due, at least in part, to the presence of different isoforms of the PMCA pump protein in different preparations, but it may be also related to complexity of the tissue-specific mechanism for regulation of the pump (Monteith and Roufogalis, 1995).

The pump may also be activated in vitro by a self-association process (Kosk-Kosicka and Bzdega, 1988) mediated by the calmodulin-binding domain (Vorherr et al., 1991). Self-association occurs at high concentrations of the pump (in excess of 10–20 μM), and transforms it into an activated state that is no longer sensitive to calmodulin (Sackett and Kosk-Kosicka, 1996). Ultracentrifugation experiments have shown that under these conditions the pump distributes between a calmodulin-independent dimer and calmodulin-stimulated monomer, both of which are maximally activated (Sackett and Kosk-Kosicka, 1996). This is at variance with the SERCA pump where dimerization has no effect on activity (Andersen, 1989). Whether dimerization of the PMCA pump also occurs in vivo is an open question, since the protein only represents 0.1–0.01% of the total membrane protein However, the plasma membrane pump could become concentrated in the caveolae (Fujimoto, 1993; Schnitzer et al., 1995), where conditions could thus be more favourable to dimerization. Possibly, dimerization, calmodulin, and acidic phospholipids could all be physiologically significant; the first, however, only for the pump concentrated on specific regions of the plasma membrane (caveolae).

Functional Domains in the PMCA Pump

The purification of the pump on calmodulin columns, first from erythrocytes (Niggli et al., 1979) and then from other plasma membranes, e.g. those of heart (Caroni and Carafoli, 1981a), skeletal muscle (Michalak et al., 1984), and smooth muscle cells (Wuytack et al., 1981), was critical to the identification of functionally important residues and domains.

Experiments on partial trypsin proteolysis of the purified enzyme provided initial information on the domain structure of the pump. Trypsin activates the enzyme by removing portions from both its C- and N-termini (Carafoli, 1991): but the cleaved N-terminal portions, which are predominantly hydrophobic, are likely to remain associated with the remainder of the pump after cleavage. The C-terminally truncated pump gradually loses the ability to respond to calmodulin, becoming irreversibly activated but still retaining the capability to respond to acidic phospholipids (Carafoli, 1994). The loss of the lipid-response after more prolonged trypsinization correlates with the proteolytic attack on a segment of the pump which later work has shown to consist of about 40 predominantly basic amino acids (Zvaritch et al., 1990).

The intracellular Ca^{2+}-dependent neutral protease calpain also attacks the pump, activates it, and makes it calmodulin-insensitive (James et al., 1989c). Calpain generates a fully active 124 kDa peptide, which was later shown to become truncated just upsteam of the calmodulin-binding domain. The fragment has been extensively used to study the role of the calmodulin-binding domain in the auto-inhibition of the pump (James et al., 1989c; Vorherr et al., 1990) and to determine the sequence of the "receptor" site for the domain in the pump (Falchetto et al., 1991, 1992). The irreversible activation of the pump by calpain—which is due to the removal of the autoinhibitory region, including the calmodulin-binding domain (Carafoli, 1992, 1994)— is though to have physiological significance in pathological states of the cell, when the resting level of Ca^{2+} remains elevated for long periods (James et al., 1989c; Salamino et al., 1994).

The calmodulin binding-domain has been identified on the purified human erythrocyte pump using a photoactivatable derivative of calmodulin (James et al., 1988). The domain, whose length is about 30 amino acids, has the propensity to form a basic amphiphylic helix and has a tryptophan close to its N-terminus (Fig. 10.7). It thus reflects the properties of most other calmodulin-binding domains. Extensive work with synthetic versions of the domain showed its tight interaction with calmodulin (Vorherr

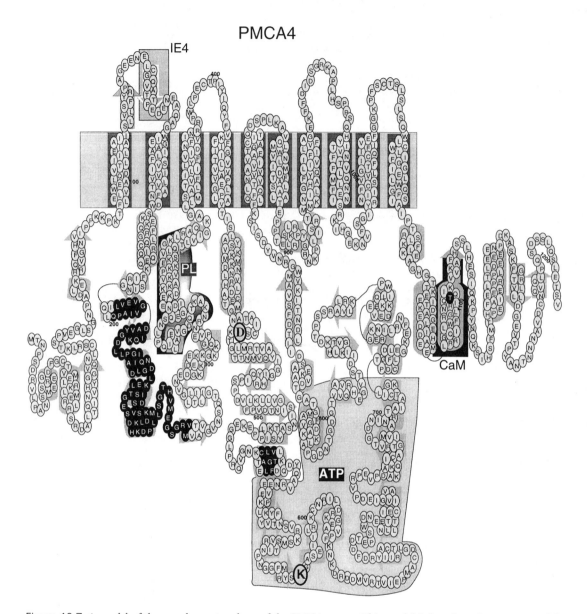

Figure 10.7 A model of the membrane topology of the PMCA pump. This model is based on the sequence of the PMCA4CI protein (Strehler et al., 1990). α-helices are represented by cylinders, β-strand by large arrows, transmembrane helices are shaded in darker grey. D465, the residues phosphorylated during the catalytic cyle and K591, the residue labeled by FITC in the ATP-binding domain are in bold. T1102, the residue located in the calmodulin-binding domain, which is phosphorylated by protein kinase C, is in black.

Five domains are indicated by shadowed or black boxes: the transmembrane region, one of the two phospholipid-binding domains (PL) (the other phospholipid-binding domain is the C-terminal calmodulin-binding site), the loop between transmembrane helices 1 and 2 (IE4), the calmodulin-binding domain (CaM), and the ATP-binding domain (ATP).

The amino acids that form the receptor site for the calmodulin-binding domain are in black.

et al., 1990), which led, as in other calmodulin-binding domains, to the collapse of the elongated structure of calmodulin (Kataoka et al., 1991).

Partial primary sequences of the pump became available before the complete sequence was deduced by cloning the corresponding cDNA. Thus the sequences of two catalytically important regions were determined: that involved in the formation of the phosphoenzyme intermediate (Asp465) (James et al., 1987) and that around the FITC (fluorescein isothiocyanate)-binding site (Lys591), which is a component of the ATP-binding domain (Fig. 10.7) (Filoteo et al., 1987). Hydropathy analysis of the pump and a comparison with the model proposed for the SERCA pump (Verma et al., 1988) led to the proposal of 10 transmembrane domains connected on the external side by five short loops (Fig. 10.7). On the inner side, the pump protrudes into the cytoplasm with four main units: the second one containing the phospholipid-interacting site (Zvaritch et al., 1990); the third, which is the largest, the catalytic site; and the C-terminal one a number of regulatory sites, among them the calmodulin-binding domain (James et al., 1988) and the substrate domains for protein kinases (James et al., 1989b; Wang et al., 1991) (Fig. 10.7). The calmodulin-binding domain also interacts with acidic phospholipids (Brodin et al., 1992), which may thus compete with calmodulin for the activation of the pump.

The model has been supported by work on antibody binding coupled with proteolysis and with the analysis of peptide fragments (Carafoli, 1992). An antibody which recognized the plasma membrane in intact human red blood cells (Feschenko et al., 1992) was shown to be specific for the external sequence between the first and the second transmembrane domains. Experiments with numerous antibodies against the N- and C-terminal portions of the protein, and against its central part, confirmed that most of the pump is located in the cytosol (Foletti et al, 1995; Zvaritch et al., 1995).

The "receptor" sites for the autoinhibitory calmodulin-binding domain (Falchetto et al., 1991, 1992) have been mentioned repeatedly. The 124 kDa calpain-truncated product, which is fully active in the absence of calmodulin, can be restored back to its "inhibited" state by the binding of a synthetic version of the calmodulin-binding domain. The 28 amino acid peptide (Vorherr et al., 1990) was made photoactivatable by coupling a carbene-generating diazirine derivative to Phe9 or Phe25. Upon covalently binding the peptide to the pump, the binding sites were identified. One site (residues 537–544) (Falchetto et al., 1991) is located in the third cytoplasmic unit of the pump between the site of aspartic acid phosphorylation and the site of ATP binding. The other site (residues 206–271) (Falchetto et al.,

1992) is located in the second cytoplasmic protruding unit. The autoinhibitory calmodulin-binding domain would thus "lock" together the second and third cytoplasmic loops of the pump, keeping the pump in an inhibited state (Fig. 10.7).

The C-terminal portion of the pump, encompassing the calmodulin-binding domain, has been expressed in bacteria (Kessler et al., 1992). Also expressed were variants of the C-terminus and portion in which inserts were present due to alternative splicing. The spliced-in sequences contain 2 or 3 histidines, and confer pH-dependency to the binding of calmodulin (Kessler et al., 1992). Since these spliced-in isoforms have been detected in muscle (Carafoli and Guerini, 1993), it could suggest that pH changes during heavy muscular work may influence the abiliyt of calmodulin to interact with the pump.

Also of interest is the detection of three high-affinity Ca^{2+}-binding sites in a large peptide encompassing the calmodulin-binding domain and its acidic flanking regions expressed in *Escherichia coli* (Hofmann et al., 1993). The Ca^{2+} ions bound in this region probably only have an allosteric role, since truncated pumps (in which the region is absent) still transport Ca^{2+}.

The transient expression of the PMCA pump in COS cels and the development of a protocol to measure its activity (Heim et al., 1992b; Enyedi et al., 1993) has permitted testing of the validity of some of the conclusions on domains by site-directed mutagenesis. Mutations were inserted in transmembrane domains (TM) 4, 6, and 8 (Fig. 10.8), since some conserved amino acids in these transmembrane domains (and in transmembrane domain 5) had been shown to be critical to the transport of Ca^{2+} in the SERCA pump (see above) (Clarke et al., 1989a). No charged residues equivalent to Glu771, however, are present in transmembrane domain 5 of the PMCA pump. Mutations of Asp879 and Asp883 (both in TM6) (Adebayo et al., 1995; Guerini et al., 1996), Glu423 (TM4), and Gln971 (TM8) (Guerini et al., 1996), indeed abolished the transport of Ca^{2+} and the formation of the phosphoenzyme intermediate from ATP. However, two of the mutants, Glu423 (TM4) nd Asp883 (TM6), still formed large amounts of phosphointermediate from phosphate (Guerini et al., 1996). Surprisingly, the two other mutants, Asn879 (TM6) and Gln971 (TM8), were retained in the endoplasmic reticulum of the COS cells (Guerini et al., 1996); evidently the mutation has somehow interfered with the conformation of the pump, not just with its Ca^{2+}-transporting activity. Whether the inhibition of the latter had been due to the fact that the two mutated residues are components of the Ca^{2+} path or to a nonspecific conformational disturbance, is an open question. These mutation experiments thus leave only two

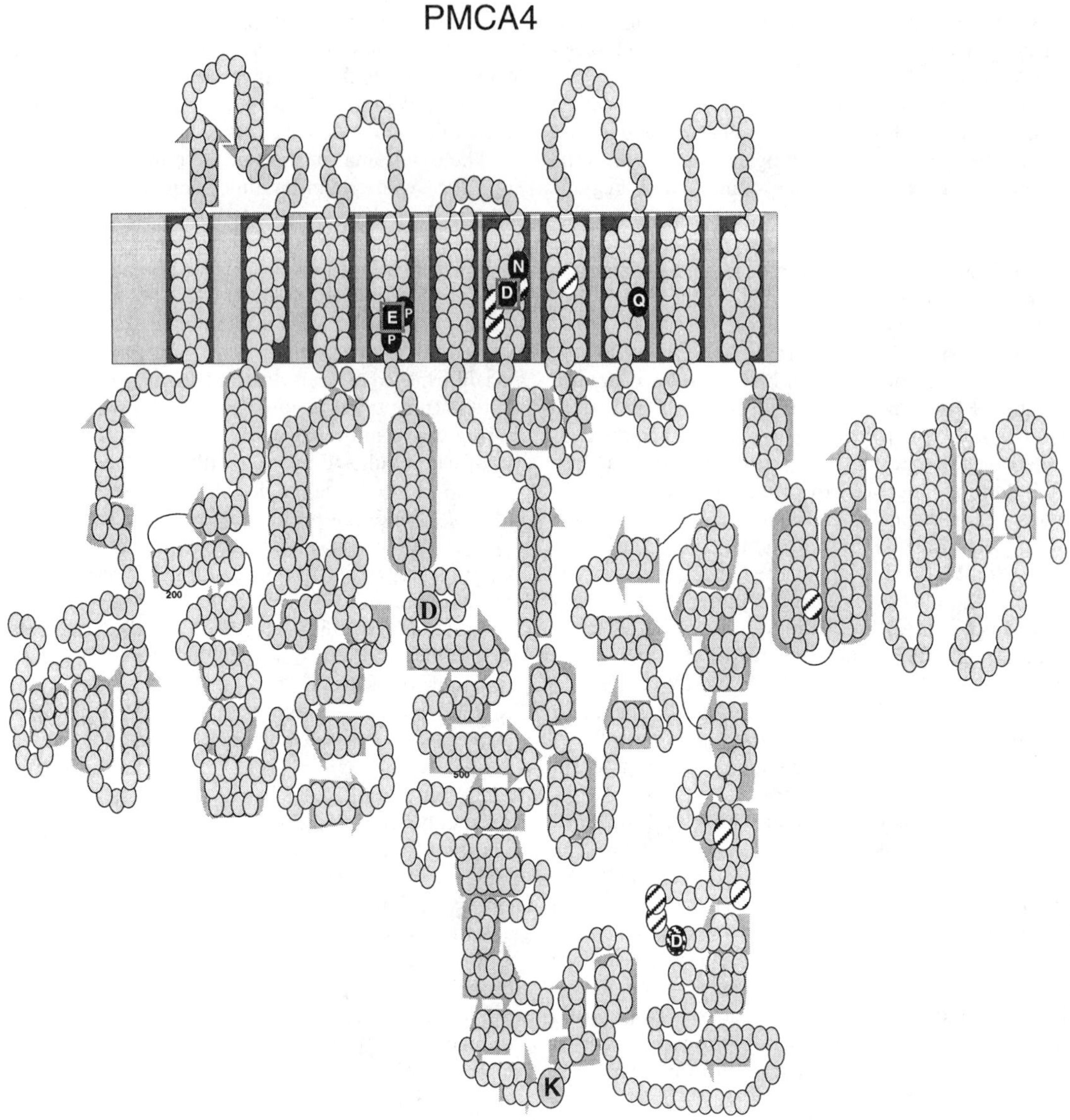

PMCA4

Figure 10.8 Overview of the amino acids whose mutation eliminates the transfer of calcium across the PMCA pump. The folding pattern of Fig. 10.7 has been used. The striped circles represent the amino acids whose mutation failed to cause loss of the activity of the pump or caused only partial inactivation (D672). The mutation of the amino acids in black circles in the transmembrane domains caused complete loss of the pump activity. Two of these amino acids are likely to be involved in the formation of the wall of the transprotein Ca^{2+} path (black circle enclosed in a striped box).

unequivocal ligands for Ca^{2+}, Glu423 (TM4) and Asp883 (TM6), in the PMCA pump. This would be in accordance with the observation that the PMCA pump transports only one Ca^{2+}, instead of two, per reaction cycle (Fig. 10.6). No residues are present in the PMCA that would correspond to the two other amino acids proposed to be components of the high-affinity Ca^{2+}-binding site of the SERCA pump—Glu771 (TM5) and Thr799 (TM6)—and donate part of the oxygen atoms for the second Ca^{2+}-binding site (Andersen, 1995a; Chen et al., 1996). Mutations of met882 (TM6), corresponding to the position of Thr799 in the SERCA pump, and mutations of Ser887 (TM6), had no obvious effects on the activity of the PMCA pump (Adebayo et al., 1995). In contrast, mutations of Pro422 and Pro426 (both in TM4) caused complete loss of activity, an effect that was even more striking than the mutations of the corresponding Pros in the SERCA pump (Guerini et al., 1996). Amino acids in the putative ATP-binding region (see Fig. 10.8), which are predicted to be crucial for the correct folding of the ATP-binding domain of P-type pumps (Taylor and Green, 1989), were also mutated. Only the mutation of Asp672Glu caused a major loss of activity, which resulted from the slower EP1–EP2 conformational transition (Adamo et al., 1995).

Isoforms of the PMCA Pump

Soon after cloning work on the PMCA pump was initiated (Shull and Greb, 1988; Verma et al., 1988), it became clear that the pump was encoded by more than one gene. High-sequence homology (more than 95%) was detected among the same gene products in different mammalian organisms, but significant differences (sequence homology between 70 and 80%) was detected among the products of the different PMCA genes (Fig. 10.9). Four genes for the PMCA pump are now recognized in mammals, but the situation is less well defined in lower organisms. The complete sequences for the rat PMCA1, PMCA2 (Shull and Greb, 1988), PMCA3 (Greb and Shull, 1989), and PMCA4 products (Keeton and Shull, 1995), the human PMCA1 (Verma et al., 1988; Kumar et al., 1993), PMCA2 (Brandt et al., 1992b; Heim et al., 1992a; Latif et al., 1993), and PMCA4 products (Strehler et al., 1990), the pig PMCA1 (De Jaegere et al, 1990) and rabbit PMCA1 products (Kahn and Grover, 1991) are now available. The complete sequence of the human PMCA3 product has just became available (Brown et al., 1996). So far, no additional genes coding for the PMCA pumps have been detected, although a Ca^{2+}-ATPase, which may be a hybrid of the SERCA and PMCA pumps, may exist (Gunteski-Hamblin et al., 1992). This novel pump type, however, has only been detected by cloning, and shows homology to yeast PMR1 gene product (Schlesser et al., 1988; Rudolph et al. 1989), which has been located in the Golgi complex of the yeast *Saccharomyces cerevisiae* (Antebi and Fink, 1992). Other PMCA-type pumps belonging to a different gene family have been described (Pavoine et al., 1987). Another putative Ca^{2+}-ATPase, called PMC1, whose sequence had higher sequence homology to the PMCA than to SERCA protein but does not contain a canonical calmodulin-binding domain

(Cunningham and Fink, 1994b), has been detected in *Saccharomyces cerevisiae* (see above).

Alternative Splicing

As indicated in Fig. 10.9, sequence differences are found mostly in the N-terminal and C-terminal regions of the protein. Other regions where considerable differences have been detected corresponded to sites of alternative splicing of primary transcripts. Alternative spicing is theoretically responsible for the generation of more than 30 PMCA pump isoforms (Carafoli and Guerini, 1993). It was originally detected at four different locations, termed sites A–D (Strehler, 1991), but more detailed analyses led to the conclusion that only the splicings occuring at sites A and C were real (Fig. 10.10) (Keeton et al., 1993; Stauffer et al., 1995). An exception is the intestine, where significant amounts of transcripts for PMCA1 and PMCA4 splicing forms that lack the region corresponding to the 10th transmembrane domain were detected (Howard et al., 1993). Overexpression of the protein corresponding to this isoform, however, showed that it was inactive and was not properly targeted to the plasma membrane: i.e. the translation of this mRNA would not generate an active PMCA pump (Seiz-Preianò et al., 1996). A scheme for the alternative splicing isoforms generated at sites A and C is represented in Fig. 10.10. A maximum of three different introns are inserted or omitted at site A (Adamo and Penniston, 1992; Heim et al., 1992a; Keeton et al., 1993; Stauffer et al., 1993). In the case of PMCA2, up to three (possibly four in the rat) different isoforms may be generated by the alternative splicing process, whereas for PMCA3 and PMCA4 only two splicing products are generated (Fig. 10.10). In the case of PMCA1, no alternative splicing at site A has been detected, although an exon corresponding to the alternatively spliced sequence is present in the PMCA1 (Hilfiker et al., 1993; Stauffer et al., 1993). Alternative transcript splicing is normally assumed to generate products that have differences in function: since the region of alternative splicing site A lays in the vicinity of one of the two phospholipid-binding domains of the pump, differences in the response of the pump to phospholipids could be expected. However, expression of the three variants of the PMCA2 isoform at A site failed to reveal differences in activity or in the stimulation by phospholipids (Hilfiker et al., 1994). In the case of site C, alternative splicing could generate up to seven different isoforms, as shown for rat PMCA3 (Keeton et a., 1993). The insertion or omission of the exons that can be spliced in at site C involves the calmodulin-binding domain (Carafoli and Guerini, 1993) (Fig. 10.10), modifying it substantially, since the insertions occur after its first 18 amino acids

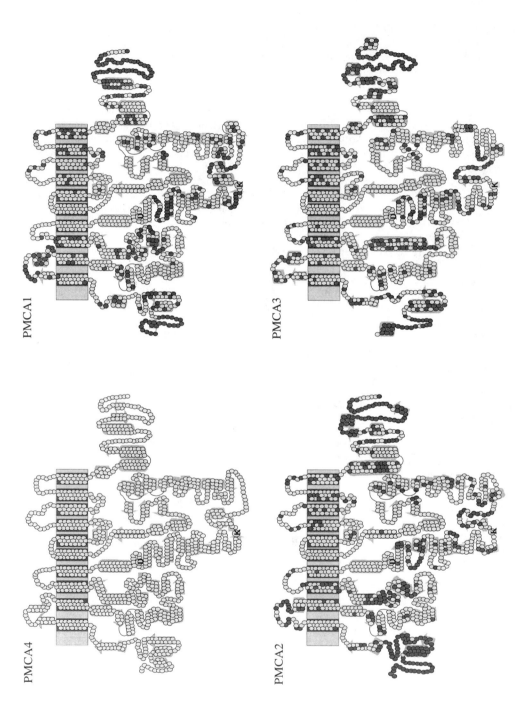

Figure 10.9 Comparison of the amino acid composition of the four PMCA isoforms. The amino acids of the PMCA1, PMCA2, and PMCA3 isoforms were aligned to those of the PMCA4 isoform. The dark dots indicate amino acids that differ from those of the PMCA4 isoform. For the comparison, the sequences of the human PMCA1 (Verma et al., 1988), PMCA2 (Heim et al., 1992a), PMCA4 (Strehler et al., 1990), and of the rat PMCA3 (Greeb and Shull, 1989) have been used. α-helices are represented by cylinders, β-strand by large arrows, transmembrane helices are dark grey.

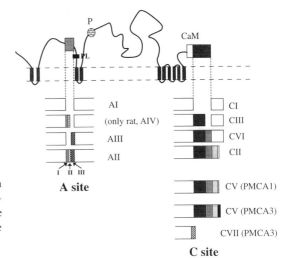

Figure 10.10 Generation of PMCA isoforms by alternative splicing. All alternative splicing variants so far found to be generated by the insertion or the omission of exons at sites A or C are shown. The domain distribution of the PMCAs is shown in the upper part of the figure. PL, acidic phospholipid-binding domain; P, site of the formation of the phosphoenzyme intermediate; CaM, calmodulin-binding domain.

(Q1103, Figs. 10.7 and 10.10). The PMCA isoforms that can potentially be generated by alternative splicing at site C are summarized in Fig. 10.10. In the case of the PMCA1 and PMCA4, only one exon is involved, whereas in that of PMCA2 and PMCA3, two different exons participate. The finding that more than two different products were detected for PMCA1 (Strehler et al., 1989; Stauffer et al., 1993) is due to internal donor splicing sites. A similar process is apparently not used for the PMCA4 (Keeton et al., 1993; Stauffer et al., 1993).

Expression of the Pump Isoforms

The complete PMCA4CI isoform has been expressed in a functionally active state in COS cells and Sf9 cells infected with the recombinant baculovirus (Heim et al., 1992b). The 10–20-fold overexpression afforded by the baculovirus system has permitted the purification of the pump in an active state (Heim et al., 1992b). This system has been successfully used for the measurement of the Ca^{2+}-dependent ATPase activity and the formation of the phosphoenzyme intermediate.

Experiments on the binding of calmodulin to the expressed C-terminal peptides of the PMCA1 pump have indicated that the splicing process modifies the affinity of the pump for calmodulin (Kessler et al., 1992). This was confirmed by the expression of the full-length PMCA4 protein isoforms in COS or Sf9 insect cells (Enyedi et al., 1994; Seiz-Preianò et al., 1996). In both cases, a dramatic reduction of the affinity of the pump for calmodulin (up to 20-fold) was observed. In addition, the PMCA4CII pump, which contained the product of the additional exons (Fig. 10.10), had reproducibly higher basal

activity (Seiz-Preianò et al., 1996), consistent with the concept that the C-terminus of the pump contains autoinhibitory sequences (Benaim et al., 1984; Verma et al., 1988). Full-length PMCA pump isoforms 1, 2, and 4 have now been expressed and their functional properties can thus be compared. The PMCA2 pump had the highest affinity for calmodulin (Hilfiker et al., 1994) and also had different affinity for ATP when the phosphoenzyme was formed from it (Hilfiker et al., 1994). The calmodulin-binding domains of pumps 2 and 4 are essentially identical but differences in the sequences of the pump binding site for the calmodulin-binding domain and in the region C-terminal to the calmodulin-binding sequence are evident (see Fig. 10.9); this would indicate that the affinity for calmodulin is not determined solely by the calmodulin-binding domain itself. Preliminary results have indicated that the PMCA1 pump is much more similar to PMCA4 than PMCA2 (Pan, 1995). Nevertheless, PMCA1 is the only pump isoform for which regulation by protein kinase A has been documented (Neyses et al., 1985; James et al., 1989b).

The Structure of the PMCA Pump Gene

Some of the genes encoding the PMCA pump have been characterized. The analysis was difficult, since the genes are large (Burk et al., 1989; Hilfiker et al., 1993), i.e. they contain 24–26 exons, but the structures of two genes, that of human PMCA1 (Hilfiker et al., 1993) and rat RMCA3 (Burk and Shull, 1992), are now available. As shown in Fig. 10.11, the structure of the exon-intron boundary is conserved in the genes. Only minor differences have been found in three of the 21–23 exons, a remarkable conservation considering that the homology of the two proteins

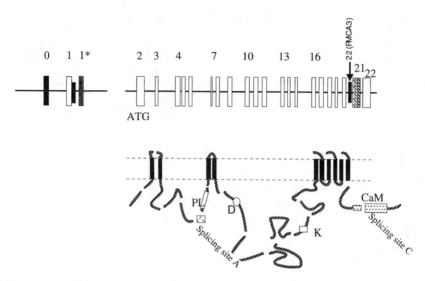

Figure 10.11 Structure of the PMCA genes. The structure, i.e., the distribution of the exons and introns, has been derived from the structure of the human PMCA1 gene (Hilfiker et al., 1993) and from that of the rat PMCA3 gene (Burk and Shull, 1992). The structure of the PMCA protein is shown at the bottom. A break represents the end or the beginning of an exon. The location of the C and A splicing sites is given. The exons present in the rat PMCA3 gene, but absent in the human PMCA gene, are in black; the exon containing internal alternative splicing signals is striped; putative exon 1*, which is present in human and in pig, but that is only transcribed in pig, is in grey. The following domains are shown in the model at the bottom: PL, acidic phospholipid-binding domain; D (D465), site of the formation of the phosphorylated intermediate; K (K591), the FITC-binding site; CaM, calmodulin-binding domain.

only amounts to 70–80%. Major differences have been found in the exons involved in the alternative splicing process, e.g. exon 23 in rat PMCA3 and exon 21 in human PMCA1 (Fig. 10.11). The regions encompassing the $5'$ untranslated sequence, the start of the transcription, and the promoter have been identified for the human (Hilfiker et al., 1993) and mouse PMCA1 gene (Du et al., 1995). No comparison is possible with the corresponding region of the rat PMCA3 gene, since the latter has not been identified as yet. One may expect significant differences here since the expression of PMCA isoform 3 is strikingly tissue-specific.

The PMCA3 gene contains 24 exons; the 22nd and 23rd can be alternatively incorporated in the mRNA to produce isoforms with different C-termini. These properties are similar to those of the human PMCA1 gene: all splicing sites coincide, the only difference being the presence of an additional exon (the 22nd exon) in the rat PMCA3 gene (Fig. 10.11). The length of the exons is the same in the two genes, but that of the intronic sequence differs substantially, mainly at the $5'$ end. The intron between exons 1 and 2 of the human PMCA1 gene is more than 40,000 bp in length, whereas that of the rat PMCA3 gene is only 13,500 bp. The human

PMCA1 and PMCA4 genes have been located to chomosomes 12 (q21–q23) and 1 (q25–q37), respectively (Olson et al., 1991). Since these two genes are thought to produce the housekeeping Ca^{2+} pumps, it is not surprising that their location did not cosegregate with any known genetic defect. Recent work has shown that the human PMCA2 gene is located on chromosome 3 (3p26–p25) (Brandt et al., 1992a; Latif et al., 1993; Wang et al., 1994), and that of human PMCA3 on chromosome X (Xq28) (Wang et al., 1994). Recent observations have suggested that two different exons code for the PMCA1 mRNA start (H. Hilfiker, unpublished). The suggestion is supported by the finding that the $5'$ untranslated region of pig PMCA1 cDNA (De Jaegere et al., 1990) is homologous to genomic sequences preceding exon 1 in the human PMCA1 gene, which is thought to represent the start of the human mRNA. A more detailed analysis of the mouse PMCA1 promoter has shown that the transcription of the mRNA for this isoform may be enhanced by phosphorylation mediated by protein kinases A and C. A cyclic AMP-responsive element (CRE) and different AP1 response elements have been found in the promoter region and shown to be at least partially active (Du et al., 1995).

Tissue Distribution of PMCA Isoforms

The mRNAs of rat and human PMCA1 and PMCA4 pumps have broad tissue distribution, whereas those of PMCA2 and especially PMCA3 pumps have been detected only in a limited number of tissues, e.g. brain and heart (Greeb and Shull, 1989; Stahl et al. 1992; Stauffer et al., 1993; Carafoli and Stauffer, 1994; Keeton and Shull, 1995). Interestingly, the alternatively spliced isoforms of the four PMCA genes also have tissue-specific distribution (Strehler et al., 1989; Carafoli and Stauffer, 1994) and have been shown to be developmentally regulated in the cases of muscle and neuronal cells (Hammes et al., 1994). These findings on transcripts have been validated, at least for the four basic isoforms, with the help of isoform-specific antibodies (Stauffer et al., 1995).

The most commonly detected isoforms in human cells are the PMCA1CI and PMCA4cI (Carafoli and Guerini, 1993; Carafoli and Stauffer, 1994). This led to the proposal that these were the housekeeping isoforms of the PMCA pump. Work with isoform-specific antibodies (Stauffer et al., 1995) largely confirmed this hypothesis for the PMCA1, but showed that the PMCA4 protein was generally present in lower amounts than predicted from the mRNA level. The recent cloning of rat PMCA4 cDNA and studies on the distribution of the corresponding transcripts have shown that the situation in rat is different. For example, no PMCA4 transcripts were ever detected in rat liver (Howard et al., 1994; Keeton and Shull, 1995), and in rat brain the distribution of the PMCA4 transcripts differed markedly from that of the other three isoforms, e.g. its levels were generally much lower (Stahl et al., 1994). A semiquantitative analysis on mRNA had indicated that the PMCA3 and PMCA2 transcripts are probably only present in significant amounts in brain (more than 5% of total mRNA), with exceptions being human liver (Howard et al., 1994), where low amounts of PMCA2 transcripts have been detected, and pancreas, where PMCA2 transcripts were found in α- and β-cells (Varadi et al., 1996). The most unexpected finding was that there is a specific tissue distribution of the spliced isoforms, implying that there must be factors controlling the tissue-specific alternative splicing (Hammes et al., 1994). The products of splicing at site C have been detected in muscle and neuronal tissue also at the protein level (Stauffer et al., 1995).

In situ hybridization with rat brain slices has shown substantial differences in the regional distribution of the mRNAs for the isoforms 1, 2, 3, and 4 (Stahl et al., 1992; Stahl et al., 1994). In particular, high PMCA1 signals have been detected in the hippocampus, PMCA2 signals in Purkinje cells of the cerebellum, and PMCA3 signals in the habenula and the choroid plexuses. Although signals for all three isoforms were generally detected, albeit at low intensity, in all regions of the brain, this was not true of the PMCA4 transcripts, which were very abundant in piriform cortex, but absent in Purkinje cells and the habenula (Stahl et al., 1994). The regional distribution of the isoforms in rat brain is somewhat different from that in human brain (Stauffer et al, 1995), where a PCR-based analysis on samples from the same individual have basically confirmed all the observations at the mRNA level summarized above (Zacharias et al., 1995).

The Membrane Targeting of SERCA and PMCA Pumps

Five P-type pumps are known in eukaryotic cells: the H^+/K^+-ATPase (in the gastric mucosa), the H^+-ATPase (in yeast cells), the Na^+/K^+-ATPase, and, as discussed above, the two Ca^{2+}-ATPases—that of the plasma membrane and that of the sarco/endoplasmic reticulum (Pedersen and Carafoli 1987a, 1987b). The last three pumps are apparently distributed ubiquitously. The two Ca^{2+} pumps have only an overall homology of 32% but share most basic features of functional and membrane architecture. Nevertheless, they have strictly distinct cellular locations, i.e. the sarco(endo)plasmic reticulum or the plasma membrane. Evidently, molecular signals responsible for the proper cellular targeting of the two pump must exist.

The expression of pumps that lack portions of the C-terminal domains have led to the discovery of a hidden retention–degradation signal for ER in the PMCA molecule (Zvaritch et al., 1995): truncated PMCA mutants terminating at Arg1087, the second residue of the calmodulin-binding domain, were retained in the ER and degraded to a fragment of about 80 kDa. The truncation, which under in vivo conditions or in experiments on the purified pump, is accomplished by calpain, exposes a very acidic 18-residue C-terminal domain. As mentioned above, the truncation product is fully active in the absence of calmodulin. Removal of this acidic sequence (Glu1067 and Arg1087), or mutations neutralizing one or more of its negatively charged amino acids, promptly redirected the expressed pump to the plasma membrane and prevented its proteolytic degradation. The hidden signal sequence was equally well recognized by insect, human, and monkey cells, but was ineffective when attached to the C-terminus of unrelated plasma membrane proteins (Vellani, 1995). The truncated PMCA pump and the products carrying mutations in the

C-terminal acidic sequence were active, indicating that the retention of these mutants in the ER was (probably) not the consequence of the gross misfolding of the protein. ER retention and degradation of the truncated PMCA pumps may represent a way to prevent prematurely terminated versions of the PMCA protein during undesirable functional properties from reaching the plasma membrane (Zvaritch et al., 1995).

N-terminal transmembrane domains are generally considered important in determining the topology of membrane proteins, e.g. in the case of their targeting to the Golgi complex (Machamer, 1993). Since the SERCA pump is retained in the ER, signal sequences for its retention could be present in this organelle. Chimeric molecules of the PMCA and SERCA pumps, in which the N-terminal regions were interchanged, have thus been constructed (Foletti et al., 1995). Immuno-histochemistry combined with surface labelling in COS cells expressing the chimeras has revealed that the N-terminal portion of the SERCA protein that encompasses the first transmembrane domain evidently contained a strong ER-retention signal, since constructs in which this portion was followed by the remainder of the PMCA4 pump were retained in the ER (Foletti et al., 1995). Replacement of the region that encompasses the first two transmembrane domains of the SERCA pump with the corresponding PMCA sequence resulted, on the other hand, in the partial (about 10%) appearance of the mutant protein on the surface of the cell, indicating the possible presence of a plasma membrane targeting signal in the N-terminal portion of the PMCA molecule outweighed by additional ER signals elsewhere in the SERCA pump molecule.

Also relevant to the issue of the plasma membrane targeting of the PMCA pump are experiments on an alternatively spliced version of the PMCA4 pump (Seiz-Preianò et al., 1996), which is found almost exclusively in the intestine (Howard et al., 1993). The transcript for this isoform lacks the exon coding for the 10th transmembrane domain of the protein and produces a mature protein that lacks this domain. The protein was still capable of forming the phosphorylated intermediate from phosphate, but not from ATP, and had no detectable Ca^{2+}-dependent ATPase activity: remarkably, it was retained in the ER (Seiz-Preianò et al., 1996). The observation is reminiscent of that on the calpain-truncated product: the elimination of the 10th transmembrane domain probably leads to the exposure of the pump sequences which are normally hidden in the molecule, and which may contain signals for its retention in the ER.

The Calcium Pump in Disease

Although alterations in the pumping functions of the ATPases have been described in a number of disease states, in the majority of cases the effects are likely to be due to secondary mechanisms, i.e. alterations of the membrane environment of the two pumps. This is particularly so for the PMCA pump, whose activity is modulated by a number of factors that could be influenced by diseases, e.g. the lipid environment. Another important factor in the case of the PMCA pump is the variability of the activity in various preparations from human sources: a study on 30 subjects (Reinila et al., 1982) has shown greater variability in the measured activity of the erythrocyte Ca^{2+} pump than in those of the Na^+/K^+ pump; these aspects of the pathology of the pump have been discussed in a recent review (Carafoli, 1991). A limited number of disease conditions can nevertheless be traced back to a primary pump alteration: a conservative estimate includes uncontrolled diabetes (Schaefer et al., 1987; Gonzalez-Flecha et al., 1990; Reddi et al., 1992), and sickle cell disease, for the PMCA pump (Bookchin and Lew, 1980; Litosch and Lee, 1980; Dixon and Winslow, 1981; Gopinath and Vincenzi, 1981; Niggli et al., 1982a), and Brody disease, for the SERCA pump (Brody, 1969; Karpati et al., 1986; Danon et al., 1988; Taylor et al., 1988; Benders et al., 1994). A decreased V_{max} of the PMCA pump in sickle cell anaemia has been repeatedly reported: the decrease would be the result, not the cause, of the increase in erythrocyte Ca^{2+} induced by the sickling pulse (Bookchin and Lew, 1980). The interaction of the pump with calmodulin is disturbed in sickled erythrocytes according to some studies (Dixon and Winslow, 1981; Gopinath and Vincenzi, 1981), but not according to others (Litosch and Lee, 1980; Niggli et al., 1982a). In one of the studies, which was unfortunately performed before the pump was cloned, the enzyme was purified from sickled erythrocytes, and was found to behave normally (Niggli et al., 1982a).

The case of uncontrolled diabetes is relatively clear-cut: decreased Ca^{2+}-ATPase activity has been reported in the erythrocytes of uncontrolled diabetics (Schaefer et al., 1987; Gonzalez-Flecha et al., 1990) and experimentally diabetic rats (Reddi et al., 1992). Indications for the mechanism of the decrease came from the observation that glycosylation of erythrocyte membranes decreased the Ca^{2+}-ATPase activity (Davis et al., 1985; Gonzalez-Flecha et al., 1990) and from the finding that nonenzymatic glycosylation of the purified pump leads to its inhibition (Gonzalez-Flecha et al., 1993), i.e. the high blood glucose level typical of uncontrolled diabetes would promote the inhibitory glycosylation of the erythrocyte PMCA pump. The inhibition of the pump in erythrocytes is unlikely to have consequences on the diabetic disease; however, should the inhibitory glycosylation also occur in other cell types, negative effects that possibly complicate the diabetic condition cannot be ruled out.

Brody's disease is a rare muscular disorder characterized by painless muscle contractures and impairment of muscle relaxation after exercise; a study of a muscle biopsy showed that sarcoplasmic reticulum had decreased Ca^{2+}-ATPase and Ca^{2+} pumping (Karpati et al., 1986). The biochemistry and genetics of the disease will be described in detail elsewhere in this book (Chapter 27): only the essential information will thus be given here. Following the initial biopsy study, other reports (Karpati et al., 1986); Danon et al., 1988; Taylor et al., 1988; Benders et al., 1994) have confirmed the deficiency of sarcoplasmic reticulum Ca^{2+}-ATPase in Brody patient muscles: studies with isoform-specific antibodies have led to the conclusion that the affected isoform was SERCA 1 (Karpati et al., 1986; Danon et al., 1988; Benders et al., 1994). Recent studies have cloned and characterized the cDNA and genomic DNA encoding the SERCA1 pump from Brody patients who had Ca^{2+}-ATPase defects, and later from two Brody disease families (Zhang et al., 1995; Odermatt et al., 1996b). The first report failed to detect mutations in either the coding sequence or the splice juncture sequence (Zhang et al., 1995), but the second (Odermatt et al., 1996b) identified three mutations in the ATP2A1 gene on chromosome 16. One mutation occurred at the splice donor site of intron 3, the other two produced premature stop codons that deleted essential functional domains from the resulting SERCA1 pump.

Acknowledgment This work has been made possible by the financial contribution of the Swiss National Science Foundation (Grant No 31-30858.91).

References

Adamo, H. P. and J. T. Penniston (1992) New Ca^{2+} pump isoforms generated by alternative splicing of rPMCA2 mRNA. *Biochem. J.* 283: 355–359.

Adamo, H. P., A. G. Filoteo, et al. (1995) Mutants in the putative nucleotide-binding region of the plasma membrane Ca^{2+}-pump. *J. Biol. Chem.* 270: 30111–30114.

Adebayo, A. O., A. Enyedi, et al. (1995) Two residues that may ligate Ca^{2+} in transmembrane domain six of the plasma membrane Ca^{2+}-ATPase. *J. Biol. Chem.* 270: 27812–27816.

Allen, G. and M. N. Green (1976) A 31-residue tryptic fragment from the active site of the Ca^{2+}-transporting adenosine triphosphatase of rabbit sarcoplasmic reticulum. *FEBS Lett.* 63: 188–192.

Andersen, J. P. (1989) Monomeric-oligomer equilibrium of sarcoplasmic reticulum Ca-ATPase and the role of subunit interaction in the Ca^{2+} pump mechanism. *Biochim. Biophys. Acta* 988: 47–72.

Andersen, J. P. (1995a) Dissection of functional domains of the sarcoplasmic reticulum Ca^{2+} ATPase by site-directed mutagenesis *Biosci. Rep.* 15: 243–261.

Andersen, J. P. (1995b) Functional consequences of alteration to amino acids at the M5S5 boundary of the Ca^{2+}-ATPase of sarcoplasmic reticulum. Mutation of Tyr763Gly uncouples ATP hydrolysis from Ca^{2+} transport. *J. Biol. Chem.* 270: 908–914.

Andersen, J. P. and B. Vilsen (1994) Amino acids Asn796 and Thr799 of the Ca^{2+}-ATPase of sarcoplasmic reticulum bind Ca^{2+} at different sites *J. Biol. Chem.* 269: 15931–15936.

Andersen, J. P. and B. Vilsen (1995) Structure–function relationships of cation translocation by Ca^{2+}-and Na^+, K^+-ATPases studied by site-directed mutagenesis. *FEBS Lett.* 359: 101–106.

Andersen, J. P., B. Vilsen, et al. (1989) Functional consequences of mutations in the β-strand sector of the Ca^{2+}-ATPase of sarcoplasmic reticulum. *J. Biol. Chem.* 264: 21018–21023.

Andersen, J. P., B. Vilsen, et al. (1992) Functional consequences of alterations to Gly310, Gly770, and Gly801 located in the transmembrane domain of the Ca^{2+}-ATPase of sarcoplasmic reticulum. *J. Biol. Chem.* 267: 2767–2774.

Andersen, R. G. W. (1993) Caveolae: where incoming and outgoing messengers meet. *Proc. Natl. Acad. Sci. USA* 90: 10909–10913.

Antebi, A. and G. R. Fink (1992) The yeast Ca^{2+} ATPase homologue, PMR1, is required for normal Golgi function and localizes in a novel Golgi-like distribution. *Mol. Biol. Cell* 3: 633–654.

Baker, B. L., V. Dave, et al. (1996) Multiple Sp1 binding sites in the cardiac/slow twitch muscle sarcoplasmic reticulum Ca^{2+} ATPase gene promoter are required for expression in So18 muscle cells. *J. Biol. Chem.* 271: 5921–5928.

Bastide, F., G. Meissner, et al. (1973) Similarity of the active site of phosphorylation of the ATPase for transport of sodium and potassium ions in kidney to that for the transport of calcium ion in sarcoplasmic reticulum of muscle. *J. Biol. Chem.* 248: 8385–8391.

Benaim, G., M. Zurini, et al., (1984) Different conformational states of the purified Ca^{2+}-ATPase of erythrocyte plasma membrane revealed by controlled trypsin proteolysis *J. Biol. Chem.* 259: 8471–8477.

Benders, A. A. G. M., J. H. Veerkamp, et al. (1994) Ca^{2+} homeostasis in Brody's disease. A study in skeletal muscle and cultured muscle cells and the effects of dantrolene and verapamil. *J. Clin. Invest.* 94: 741–748.

Blaise, J. K., F. J. Asturias, et al. (1992) Time-resolved X-ray diffraction studies on the mechanism of active Ca^{2+} transport by the sarcoplasmic reticulum Ca^{2+} ATPase. *Ann. N.Y. Acad. Sci.* 671, 11–18.

Bobe, R., R. Bredoux, et al. (1994) The rat platelet 97-kDa Ca^{2+}-ATPase isoform is the sarcoplasmic reticulum Ca^{2+} ATPase 3 protein. *J. Biol. Chem.* 269: 1417–1424.

Bookchin, R. M. and V. L. Lew (1980) Progressive inhibition of the Ca^{2+} pump and Ca^{2+}-Ca^{2+} exchange in sickle red cells *Nature* 284: 561–563.

Brandl, C., N. Green, et al. (1986) Two Ca^{2+}-ATPase genes: homologies and mechanistic implications of deduced amino acid sequences. *Cell* 44: 597–607.

Brandl, C. J., S. deLeon, et al. (1987) Adult forms of the Ca^{2+} ATPase of sarcoplasmic reticulum. *J. Biol. Chem.* 262: 3768–3774.

Brandt, P., E. Ibrahim, et al. (1992a) Determination of the nucleotide sequence and chromosomal localization of the ATP2B2 gene encoding the human Ca^{2+}-pumping ATPase isoform PMCA2. *Genomics* 14: 484–487.

Brandt, P., R. L. Neve, et al. (1992b) Analysis of the tissue-specific distribution of mRNAs encoding the plasma membrane calcium-pumping ATPases and characterisation of an alternately spliced form of PMCA4 at the cDNA and genomic levels. *J. Biol. Chem.* 267: 4376–4385.

Brodin, P., R. Falchetto, et al. (1992) Identification of two domains which mediate the binding of activating phospholipids to the plasma membrane Ca^{2+} pump. *Eur. J. Biochem.* 204: 939–946.

Brody, I. A. (1969) Muscle contracture induced by exercise. A syndrome attributable to decreased relaxing factor. *N. Engl. J. Med.* 281: 187–192.

Brown, B. J., H. Hilfiker, et al. (1996) Primary structure of the plasma membrane Ca^{2+}-ATPase isoform 3. *Biochim. Biophys. Acta* 1283: 10–13.

Burk, S. E. and G. E. Shull (1992) Structure of the rat membrane Ca^{2+}-ATPase isoform 3 gene and characterisation of alternative splicing and transcription products *J. Biol. Chem.* 267: 19683–19690.

Burk, S. E., J. Lytton, et al. (1989) cDNA cloning, functional expression, and mRNA tissue distribution of a third organellar Ca^{2+} pump *J. Biol. Chem.* 264: 18561–18568.

Campbell, M. A., P. D. Kessler, et al. (1991) Nucleotide sequences of avian cardiac and brain SR/ER Ca^{2+} ATPases and functional comparisons with fast twitch Ca^{2+} ATPase. *J. Biol. Chem.* 266: 16050–16055.

Campbell, M. A., P. D. Kessler, et al. (1992) The alternative carboxyl termini of avian cardiac and brain sarcoplasmic/reticulum Ca^{2+} ATPases are on opposite sides of the membranes. *J. Biol. Chem.* 267: 9321–9325.

Carafoli, E. (1987) Intracellular calcium homeostasis *Ann. Rev. Biochem.* 56: 395–433.

Carafoli, E. (1991) Calcium pump of the plasma membrane. *Physiol. Rev.* 71: 129–153.

Carafoli, E. (1992) The Ca^{2+} pump of the plasma membrane *J. Biol. Chem.* 267: 2115–2118.

Carafoli, E. (1994) Biogenesis: plasma membrane calcium ATPase: 15 years of work on the purified enzyme. *FASEB J.* 8: 993–1002.

Carafoli, E. and M. Chiesi (1992) Calcium pumps in the plasma and intracellular membranes. *Curr. Top. Cellular Reg.* 32: 209–241.

Carafoli, E. and D. Guerini (1993) Molecular and cellular biology of plasma membrane calcium ATPase. *Trends Cardiovasc. Med.* 3: 177–184.

Carafoli, E. and T. Stauffer (1994) The plasma membrane calcium pump: functional domains, regulation of the activity, and tissue specificity of isoform expression. *J. Neurobiol.* 25: 312–324.

Carafoli, E. and M. Zurini (1982) The calcium pumping ATPase of plasma membranes. Purification, reconstitution, and properties. *Biochim. Biophys. Acta.* 683: 279–301.

Caroni, P. and E. Carafoli (1981a) The Ca^{2+} pumping ATPase of heart sarcolemma. Characterization, calmodulin dependence, and partial purification. *J. Biol. Chem.* 256: 3263–3270.

Caroni, P. and E. Carafoli (1981b) Regulation of Ca^{2+}-ATPase of heart sarcolemma by phosphorylation/dephosphorylation process. *J. Biol. Chem.* 256: 9371–9373.

Chen, L., C. Sumbilla, et al. (1996) Short and long range functions of amino acids in the transmembrane region of the sarcoplasmic reticulum ATPase. *J. Biol. Chem.* 271: 10745–10752.

Choquette, D., G. Hakim, et al. (1984) Regulation of the plasma membrane Ca^{2+}-ATPase by lipids of the phosphatidylinositol cycle. *Biochim. Biophys. Res. Commun.* 125: 908–915.

Clarke, D. M., T. W. Loo, et al. (1989a) Location of high affinity Ca^{2+}-binding sites within the predicted transmembrane domain of the sarcoplasmic reticulum Ca^{2+}-ATPase. *Nature* 339: 476–478.

Clarke, D. M., T. W. Loo, et al. (1990a) Functional consequences of alterations to polar amino acids located in the transmembrane domain of the Ca^{2+}-ATPase of the sarcoplasmic reticulum. *J. Biol. Chem.* 265: 6262–6267.

Clarke, D. M., T. W. Loo, et al. (1990b) Functional consequences of mutations of conserved amino acids in the β-strand domain of the Ca^{2+}-ATPase of sarcoplasmic reticulum. *J. Biol. Chem.* 265: 14088–14092.

Clarke, D. M., T. W. Loo, et al. (1990c) The epitope for monoclonal antibody A20 (amino acids 870–890) is located on the lumenal surface of the Ca^{2+}-ATPase of sarcoplasmic reticulum. *J. Biol. Chem.* 265: 17405–17408.

Clarke, D. M., T. W. Loo, et al. (1990d) Functional consequences of alterations to amino acids located in the nucleotide binding domain of the Ca^{2+}-ATPase of sarcoplasmic reticulum. *J. Biol. Chem.* 265: 22223–22227.

Clarke, D. M., T. W. Loo, et al. (1993) Functional characterization of alterations to amino acids located in the M4 transmembrane sector of the Ca^{2+}-ATPase of sarcoplasmic reticulum. *J. Biol. Chem.* 268: 18359–18362.

Clarke, D. M., K. Maruyama, et al. (1989b) Functional consequences of glutamate, aspartate, glutamine, and asparagine mutations in the stalk sector of the Ca^{2+}-ATPase of sarcoplasmic reticulum. *J. Biol. Chem.* 264: 11246–11251.

Cunningham, K. W. and G. R. Fink (1994a) Ca^{2+} transport in *Saccharomyces cerevisiae. J. Exp. Biol.* 196: 157–166.

Cunningham, K. W. and G. R. Fink (1994b) Calcineurin-dependent growth control in *Saccharomyces* mutants lacking PMC1, a homologue of plasma membrane Ca^{2+} ATPases. *J. Cell. Biol.* 124: 351–363.

Danon, M. J., G. Karpati, et al. (1988) Sarcoplasmic reticulum adenosine triphosphatase deficiency with probable autosomal dominant inheritance. *Neurology* 38: 812–815.

Davidson, G. A. and R. Varhol (1995) Kinetics of thapsigargin-Ca^{2+}-ATPase (sarcoplasmic reticulum) interaction reveals a two-step binding mechanism and picomolar inhibition. *J. Biol. Chem.* 270: 11731–11734.

Davis, F. B., P. J. Davis, et al. (1985) The effect of in vivo glucose administration on human erythrocyte Ca^{2+} ATPase activity and on enzyme responsiveness in vitro to thyroid hormone and calmodulin. *Diabetes* 34: 639–646.

De Jaegere, S., F. Wuytack, et al. (1990) Molecular cloning and sequencing of the plasma-membrane Ca^{2+} pump of pig smooth muscle. *Biochem. J.* 271: 655–660.

De Jesus, F., J.-L. Girardet, et al. (1993) Characterization of ATP binding inhibition to the sarcoplasmic reticulum Ca^{2+}-ATPase by thapsigargin. *FEBS Lett.* 332: 229–232.

De Smedt, H., J. A. Eggermont, et al. (1991) Isoform switching of the sarco(endo)plasmic reticulum Ca^{2+} pump during differentiation of BC3H1 myoblasts. *J. Biol. Chem.* 266: 7092–7095.

Degani, C., and P. D. Boyer (1973) A borohydride reduction method for characterization of the acyl phosphate linkage in proteins and its application to sarcoplasmic reticulum adenosine triphosphatase. *J. Biol. Chem.* 248: 8222–8226.

DeLong, L. J. and J. K. Blaise (1993) Effect of Ca^{2+} binding on the profile structure of the sarcoplasmic reticulum membrane using time-resolved X-ray diffraction. *Biophys. J.* 64: 1750–1759.

Desrosiers, M. G., L. J. Gately, et al. (1996) Purification and characterization of the Ca^{2+}-ATPase of *Flavobacterium odoratum. J. Biol. Chem.* 271: 3945–3951.

Dixon, E. and R. M. Winslow (1981) The interaction between (Ca^{2+}–Mg^{2+}) ATPase and the soluble activator (calmodulin) in erythrocytes containing hemoglobins. *Br. J. Haematol.* 47: 391–397.

Du, Y., L. Carlock, et al. (1995) The mouse plasma membrane Ca^{2+} pump isoform 1 promoter: cloning and characterization. *Arch. Biochem. Biophys.* 316: 302–310.

Ebashi, S. and F. Lipmann (1962) Adenosine-triphosphate-linked concentration of calcium ions in a particular fraction of rabbit muscle. *J. Cell Biol.* 14: 389–400.

Eggemont, J., F. Wuytack, et al. (1989) Evidences for two isoforms of the ER Ca-pump in pig smooth muscle. *Biochem. J.* 260: 757–761.

Enyedi, A., A. K. Verma, et al. (1993) A highly active 120-kDa truncated mutant of the plasma membrane Ca^{2+} pump. *J. Biol. Chem.* 268: 10621–10626.

Enyedi, A., A. K. Verma, et al. (1994) The Ca^{2+} affinity of the plasma membrane Ca^{2+} pump is controlled by alternative splicing. *J. Biol. Chem.* 269: 41–43.

Falchetto, R., T. Vorherr, et al. (1991) The plasma membrane Ca^{2+} pump contains a site that interacts with its calmodulin-binding domain. *J. Biol. Chem.* 266: 2930–2936.

Falchetto, R., T. Vorherr, et al. (1992) The calmodulin binding site of the plasma membrane Ca^{2+} pump interacts with the transduction domain of the enzyme. *Protein Sci.* 1: 1613–1621.

Feschenko, M. S., E. Zvaritch, et al. (1992) A monoclonal antibody recognizes an epitope in the first extracellular loop of the plasma membrane Ca^{2+} pump. *J. Biol. Chem.* 267: 4097–4104.

Filoteo, A. G., J. P. Gorski, et al. (1987) The ATP-binding site of the enterocyte membrane Ca^{2+} pump. *J. Biol. Chem.* 267: 6526–6530.

Foletti, D., D. Guerini, et al. (1995) Subcellular targeting of the endoplasmic reticulum and plasma membrane Ca^{2+} pumps: a study using recombinant chimeras. *FASEB J.* 9: 670–680.

Fujii, J., K. Maruyama, et al. (1990) Co-expression of slow-twitch cardiac muscle Ca^{2+} ATPase (SERCA2) and phospholamban *FEBS Lett.* 273: 232–234.

Fujimoto, T. (1993) Calcium pump of the plasma membrane is localized in caveolae. *J. Cell. Biol.* 120: 1147–1157.

Geiser, J. R., D. van Tuinen, et al. (1991) Can calmodulin function without binding of calcium? *Cell* 65: 949–969.

Geisler, M., J. Richter, et al. (1993) Molecular cloning of a P-type ATPase gene from the *Cyanobacterium synechocystis* sp. PCC 6803. *J. Mol. Biol.* 234: 1284–1289.

Gonzalez-Flecha, F. L., M. Bermunez, et al. (1990) Decreased Ca^{2+}-ATPase activity after glycosylation of enterocyte membranes in vivo and in vitro. *Diabetes* 39: 707–711.

Gonzalez-Flecha, F. L. G., P. R. Castello, et. al. (1993) The red cell calcium pump is inhibited by nonenzymatic glycation. Studies with the "in situ" and the purified enzyme. *Biochem. J.* 293: 369–375.

Gopinath, R. M. and F. F. Vincenzi (1981) (Ca^{2+} + Mg^{2+}) ATPase activity of sickle cell membranes decreased activation by red blood cells cytoplasmic factor. *Am. J. Hematol.* 7: 303–312.

Greeb, J. and G. E. Shull (1989) Molecular cloning of a third isoform of the calmodulin-sensitive plasma membrane Ca^{2+} transporting ATPase that is expressed predominantly in brain and skeletal muscle. *J. Biol. Chem.* 264: 18569–18576.

Grover, A. K. and I. Khan (1992) Calcium pump isoforms: diversity, selectivity and plasticity. *Cell Calcium* 13: 9–17.

Guerini, D., D. Foletti, et al. (1996) Mutation of conserved residues in transmembrane domains 4, 6 and 8 causes loss of Ca^{2+} transport by the plasma membrane Ca^{2+} pump. *Biochemistry* 35: 3290–3296.

Guerini, D., S. Schröder, et al. (1995) Isolation and characterization of a stable Chinese ovary cell line overexpressing the plasma membrane Ca^{2+}-ATPase. *J. Biol. Chem.* 270. 14643–14650.

Gunteski-Hamblin, A.-M., D. M. Clark, et al. (1992) Molecular cloning and tissue distribution of alternatively spliced mRNAs encoding possible mammalian homologues of the yeast secretory pathway calcium pump. *Biochemistry* 31: 7600–7608.

Gunteski-Hamblin, A.-M., J. Greeb, et al. (1988) A novel Ca^{2+} pump expressed in brain, kidney, and stomach is encoded by an alternative transcript of the slow-twitch muscle sarcoplasmic reticulum Ca-ATPase gene. *J. Biol. Chem.* 263: 15032–15040.

Hammes, A., S. Oberdorf, et al. (1994) Differentiation-specific isoform mRNA expression of the calmodulin-dependent plasma membrane Ca^{2+}-ATPase. *FASEB J.* 8: 428–435.

Hao, L., J.-L. Rigaud, et al. (1994) Ca^{2+}/H^+ countertransport and electrogenicity in proteoliposomes containing erythrocyte plasma membrane Ca-ATPase and exogenous lipids. *J. Biol. Chem.* 269: 14268–14275.

Hartong, R., N. Wang, et al. (1994) Delineation of three different thyroid hormone-responsive elements in promoter of rat sarcoplasmic reticulum Ca^{2+} ATPase gene. *J. Biol. Chem.* 269: 13021–13029.

Hasselbach, W. and M. Makinose (1961) Die Calcium Pumpe der "Erschlaffungsgrana" des Muskels und ihre Abhängigkeit von der ATP-Spaltung. *Biochem. Z.* 333: 518–528.

Heim, R., M. Hug, et al. (1992a) Microdiversity of human-plasma-membrane calcium-pump isoform 2 generated by alternative RNA splicing in the N-terminal coding region. *Eur. J. Biochem.* 205: 333–340.

Heim, R., T. Iwata, et al. (1992b) Expression, purification, and properties of the plasma membrane Ca^{2+} pump and its N-terminally truncated 105-kDa fragment. *J. Biol. Chem.* 267: 24476–24484.

Hilfiker, E., M.-A. Strehler, et al. (1993) Structure of the gene encoding the human plasma membrane calcium isoform 1. *J. Biol. Chem.* 268: 19717–19725.

Hilfiker, H., D. Guerini, et al. (1994) Cloning and expression of isoform 2 of the human plasma membrane Ca^{2+} ATPase. *J. Biol. Chem.* 269: 26178–26183.

Hofmann, F., J. Anagli, et al. (1994) Phosphorylation of the calmodulin binding domain of the plasma membrane Ca^{2+} pump by protein kinase C reduces its interaction with calmodulin and with its pump receptor site. *J. Biol. Chem.* 269: 24298–24303.

Hofmann, F., P. James, et al. (1993) The C-terminal domain of the plasma membrane Ca^{2+} pump contains three high affinity Ca^{2+} binding sites. *J. Biol. Chem.* 268: 10252–10259.

Howard, A., N. F. Barley, et al. (1994) Plasma-membrane calcium pump isoforms in human and rat liver. *Biochem. J.* 303: 275–279.

Howard, A., S. Legon, et al. (1993) Human and rat intestinal plasma membrane pump isoforms. *Am. J. Physiol.* 265: G917–G925.

Inesi, G. (1987) Sequential mechanism of calcium binding and translocation in sarcoplasmic reticulum adenosine triphosphate. *J. Biol. Chem.* 262: 16338–16342.

Inesi, G. and Y. Sagara (1994) Specific inhibitors of intracellular Ca^{2+} transport ATPases. *J. Membr. Biol.* 141: 1–6.

Inesi, G., T. Cantilina, et al. (1992) Long-range intramolecular linked functions in activation and inhibition of SERCA ATPases. *Ann. N.Y. Acad. Sci.*, 671: 32–48.

James, P., M. Inui, et al. (1989a) Nature and site of Phospholamban regulation of the Ca^{2+} pump of sarcoplasmic reticulum. *Nature* 342: 90–92.

James, P., M. Maeda, et al. (1988) Identification and primary structure of a calmodulin binding domain of the Ca^{2+} pump of human erythrocytes. *J. Biol. Chem.* 263: 2905–2910.

James, P., M. Pruschy, et al. (1989b) Primary structure of the cAMP-dependent phosphorylation site of the plasma membrane calcium pump. *Biochemistry* 28: 4253–4258.

James, P., T. Vorherr, et al. (1989c) Modulation of the erythrocyte Ca^{2+}-ATPase by selective calpain cleavage of the calmodulin binding domain. *J. Biol. Chem.* 264: 8289–8296.

James, P., E. Zvaritch, et al. (1987) The amino acid sequence of the phosphorylation domain of the erythrocyte Ca^{2+} ATPase. *Biochem. Biophys. Res. Commun.* 149: 7–12.

Jencks, W. P. (1992) On the mechanism of the ATP-driven Ca^{2+} transport by the calcium ATPase of sarcoplasmic reticulum. *Ann. N.Y. Acad. Sci.* 671: 49–57.

Jencks, W. P. (1995) The mechanism of coupling chemical and physical reactions by the calcium ATPase of sarcoplasmic reticulum and other coupled vectorial systems. *Biosci. Rep.* 15: 283–287.

Kahn, I. and A. K. Grover (1991) Expression of cyclic-nucleotide-sensitive and -insensitive isoforms of the plasma membrane Ca^{2+} pump in smooth muscle and other tissues. *Biochem. J.* 277: 345–349.

Karpati, G., J. Charuk, et al. (1986) Myopathy caused by a deficiency of Ca^{2+}-adenosine triphosphatase in sarcoplasmic reticulum (Brody's disease). *Ann. Neurol.* 20: 38–49.

Kataoka, M., J. F. Head, et al. (1991) Small-angle X-ray scattering study of calmodulin bound to two peptides corresponding to parts of the calmodulin-binding domain of the plasma membrane Ca^{2+} pump. *Biochemistry* 30: 6247–6251.

Keeton, T. P. and G. E. Shull (1995) Primary structure of rat plasma membrane Ca^{2+}-ATPase isoform 4 and analysis of alternative splicing patterns at splice site A. *Biochem. J.* 306: 779–785.

Keeton, T. P., S. E. Burk, et al. (1993) Alternative splicing of exons encoding the calmodulin-binding domains and C-termini of plasma membrane Ca^{2+}-ATPase isoforms 1, 2, 3 and 4. *J. Biol. Chem.* 268: 2740–2748.

Kessler, F., R. Falchetto, et al. (1992) Study of calmodulin binding to the alternative spliced C-terminal domain of the plasma membrane Ca^{2+} pump. *Biochemistry* 31: 11785–11792.

Korczak, B., A. Zarain-Herzberg, et al. (1988) Structure of the rabbit fast-twitch skeletal muscle Ca-ATPase gene. *J. Biol. Chem.* 263: 4813–4819.

Kosk-Kosicka, D. and T. Bzdega (1988) Activation of the erythrocyte Ca^{2+}-ATPase by either self-association or interaction with calmodulin. *J. Biol. Chem.* 263: 18184–18189.

Krebs, J., M. Vasak, et al. (1987) Conformational differences between the E1 and E2 states of the calcium adenosine triphosphatase of the erythrocyte plasma membrane as revealed by circular dichroism and fluorescence spectroscopy. *Biochemistry* 26: 3921–3926.

Kumar, R., J. D. Haugen, et al. (1993) Molecular cloning of the plasma membrane calcium pump from normal human osteoblasts. *J. Bone Miner. Res.* 8: 505–513.

Latif, F., F.-M. Duh, et al. (1993) Von Hippel-Lindau syndrome: cloning and identification of the plasma membrane Ca^{2+}-transporting ATPase isoform 2 gene that resides in the van Hippel-Lindau gene region. *Cancer Res.* 53: 861–867.

Litosch, I. and K. S. Lee (1980) Sickle red cell Ca^{2+} metabolism: studies on Ca^{2+}–Mg^{2+} ATPase and Ca^{2+} binding properties of sickle red cell membranes. *Am. J. Hematol.* 8: 377–387.

Liu, B.-F., X. Xu, et al. (1996) Consequences of functional expression of the plasma membrane Ca^{2+} pump isoform 1a. *J. Biol. Chem.* 271: 5536–5544.

Luo, W., I. L. Grupp, et al. (1994) Targeted ablation of the phospholamban gene is associated with markedly enhanced myocardial contractility and loss of β-agonist stimulation. *Circ. Res.* 75: 401–409.

Lytton, J. and D. H. MacLennan (1988) Molecular cloning of cDNAs from human kidney coding for two alternatively spliced products of the cardiac Ca^{2+}-ATPase gene. *J. Biol. Chem.* 263: 15024–15031.

Lytton, J., M. Westlin, et al. (1991) Thapsigargin inhibits the sarcoplasmic and endoplasmic reticulum Ca-ATPase family of calcium pumps. *J. Biol. Chem.* 266: 17067–17071.

Lytton, J., A. Zarain-Herzberg, et al. (1989) Molecular cloning of the mammalian smooth muscle sarco(endo)plasmic reticulum Ca^{2+}-ATPase. *J. Biol. Chem.* 264: 7059–7065.

Machamer, C. E. (1993) Targeting and retention of Golgi membrane proteins. *Curr. Opin. Cell Biol.* 5: 606–612.

MacLennan, D. H. (1970) Purification and properties of an adenosine triphosphatase from sarcoplasmic reticulum. *J. Biol. Chem.* 245: 4508–4518.

MacLennan, D. H. (1990) Molecular tools to elucidate problems in excitation–contraction coupling. *Biophys. J.* 58: 1355–1365.

MacLennan, D. H., C. Brandl, et al. (1985) Amino-acid sequence of a Ca^{2+}-Mg^{2+}-dependent ATPase from rabbit muscle SR, deduced from its complementary DNA sequence. *Nature* 316: 696–700.

Maguire, M. E., M. D. Snavely, et al. (1992) Mg^{2+} *Transporting P-type ATPases of Salmonella typhimurium*. The New York Academy of Sciences, New York.

Makinose, M. (1973) Possible functional states of the enzyme of the sarcoplasmic calcium pump. *FEBS Lett.* 37: 140–143.

Martonosi, A. N. (1995) The structure and interaction of Ca^{2+}-ATPase. *Biosci. Rep.* 15: 262–281.

Maruyama, K. and D. H. MacLennan (1988) Mutation of aspartic acid-351, lysine-352, and lysine-515 alters the Ca^{2+} transport activity of the Ca^{2+}-ATPase expressed in COS-1 cells. *Proc. Natl. Acad. Sci. USA* 85: 3314–3318.

Maruyama, K., D. M. Clarke, et al. (1989) Functional consequences of alterations to amino acids located in the catalytic center (isoleucine 348 to threonine 357) and nucleotide binding domain of the Ca^{2+}-ATPase of sarcoplasmic reticulum. *J. Biol. Chem.* 264: 13038–13042.

Matthews, I., R. P. Sharma, et al. (1990) Transmembraneous organization of $(Ca^{2+}$–$Mg^{2+})$-ATPase from sarcoplasmic reticulum: evidence for lumenal location of residues 877–888. *J. Biol. Chem.* 265: 18737–18740.

Michalak, M., R. Famulski, et al. (1984) The calcium pumping ATPase in skeletal muscle sarcolemma. Calmodulin dependence, regulation by cAMP dependent phosphorylation, and purification. *J. Biol. Chem.* 259: 15540–15547.

Monteith, G. R. and B. D. Roufogalis (1995) The plasma membrane calcium pump—a physiological perspective on its regulation. *Cell Calcium* 18: 459–470.

Nagai, R., A. Zarain-Herzberg, et al. (1989) Regulation of myocardial Ca^{2+}-ATPase and phospholamban mRNA expression in response to pressure overload and thyroid hormones. *Proc. Natl. Acad. Sci. USA* 86: 2966–2970.

Neyses, L., L. Reinlieb, et al. (1985) Phosphorylation of the Ca^{2+}-pumping ATPase of heart sarcolemma and erythrocyte plasma membrane by the cAMP dependent protein kinase. *J. Biol. Chem.* 260: 10283–10287.

Niggli, V., E. S. Adunyah, et al. (1981) Acidic phospholipids, unsaturated fatty acids and limited proteolysis mimic the effect of calmodulin on the purified erythrocyte Ca^{2+}-ATPase. *J. Biol. Chem.* 256: 8588–8592.

Niggli, V., E. S. Adunyah, et al. (1982a) The Ca^{2+} pump of sickle cell plasma membranes. Purification and reconstitution of the ATPase enzyme. *Cell Calcium* 3: 131–151.

Niggli, V., J. T. Penniston, et al. (1979) Purification of the $(Ca^{2+}$–$Mg^{2+})$-ATPase from human erythrocytes membranes using a calmodulin affinity column. *J. Biol. Chem.* 254: 9955–9958.

Niggli, V., E. Sigel, et al. (1982b) The purified Ca^{2+} pump of human erythrocyte membranes catalyzes electroneutral Ca^{2+}:H^+ exchange in reconstituted liposomal systems. *J. Biol. Chem.* 257: 420–424.

Norregard, A., B. Vilsen, et al. (1993) Chimeric Ca^{2+}-ATPase/Na^+,K^+-ATPase molecules. Their phosphoenzyme intermediates and sensitivity to Ca^{2+} and thapsigargin. *FEBS Lett.* 336: 248–254.

Nucifora, G., L. Chu, et al. (1989) Cadmium resistance from *Staphylococcus aureus* plasmid pl258 cadA gene results from a cadmium-efflux ATPase. *Proc. Natl. Acad. Sci. USA* 86: 3544–3548.

Odermatt, A., K. Kurzydlowski, et al. (1996a) The Vmax of the Ca^{2+}-ATPase of cardiac sarcoplasmic reticulum (SERCA2a) is not altered by Ca^{2+}/calmodulin-dependent phosphorylation or by interaction with phospholamban. *J. Biol. Chem.* 271: 14206–14213.

Odermatt, A., P. E. M. Taschner, et al. (1996b) Mutations in the gene encoding SERCA1, the human fast-twitch muscle sarcoplasmic reticulum Ca^{2+} ATPase, are associated with Brody disease. *Nat. Genet.* 14: 191–194.

Olson, S., M. G. Wang, et al. (1991) Localisation of two genes encoding plasma membrane Ca^{2+}-transporting ATPases to human chromosomes 1q25–32 and 12q21–23. *Genomics* 9: 629–641.

Pan, B. (1995) Overexpression and functional characterization of three isoforms of plasma membrane Ca^{2+}-ATPase. Thesis No. 11310, Swiss Federal Institute of Technology (ETH), Zurich, Switzerland.

Pavoine, C., S. Lotersztjan, et al. (1987) The high affinity (Ca^{2+}–Mg^{2+})–ATPase in liver plasma membranes is a Ca^{2+} pump. *J. Biol. Chem.* 262: 5113–5177.

Pedersen, P. L. and E. Carafoli (1987a) Ion motive ATPases. I. Ubiquity, properties, and significance for cell function. *Trends Biochem. Sci.* 12: 146–150.

Pedersen, P. L. and E. Carafoli (1987b) Ion motive ATPases. II. Energy coupling and work output. *Trends Biochem. Sci.* 12: 186–189.

Penniston, J. T. (1982) Plasma membrane Ca^{2+}-pumping ATPase. *Ann. N.Y. Acad. Sci.* 402: 296–303.

Pfeiffer, W. E., M. G. Desrosiers, et al. (1996) Cloning and expression of the unique Ca^{2+}-ATPase for *Flavobacterium odoratum*. *J. Biol. Chem.* 271: 5095–5100.

Philipson, K. D. and D. A. Nicoli (1992) Sodium–calcium exchange. *Curr. Opin. Cell Biol.* 4: 678–683.

Rasmussen, U., S. B. Christensen, et al. (1978) Thapsigargin and thapsigarcin, two new histamine liberators from *Thapsia garganica*. *Acta Pharmaceut. Sued.* 267: 133–140.

Reddi, A. S., A. Dasmahapatra, et al. (1992) Erythrocyte Ca, Na/K-ATPase in long-term streptozotocin diabetic rats: effect of good glycemic control and Ca antagonist. *Am. J. Hypertens.* 5: 863–868.

Reddy, L. G., L. R. Jones, et al. (1996) Purified, reconstituted cardiac Ca^{2+}-ATPase is regulated by phospholamban but not direct phosphorylation with Ca^{2+}/

calmodulin dependent protein kinase. *J. Biol. Chem.* 271: 14964–14970.

Reinila, M., E. MacDonald, et al. (1982) Standardized method for determination of human erythrocyte membrane adenosine triphosphatases. *Anal. Biochem.* 124: 19–26.

Rogers, T. B., G. Inesi, et al. (1995) Use of thapsigargin to study Ca^{2+} homeostasis in cardiac cells. *Biosci. Rep.* 15: 341–349.

Rohrer, D. K., R. Hartong, et al. (1991) Influence of thyroid hormone and retinoic acid on slow sarcoplasmic reticulum Ca^{2+} ATPase and myosin heavy chain alpha gene expression in cardiac myocites. *J. Biol. Chem.* 266: 8638–8646.

Rosen, B. P. (1987) Bacterial calcium transport. *Biochim. Biophys. Acta* 906: 101–110.

Rudolph, H. K., A. Antebi, et al. (1989) The yeast secretory pathway is perturbed by mutations in PMR1, a member of a Ca^{2+} ATPase family. *Cell* 58: 133–145.

Sackett, D. J. and D. Kosk-Kosicka (1996) The active species of plasma membrane Ca^{2+}-ATPase are a dimer and a monomer–calmodulin complex. *J. Biol. Chem.* 271: 9987–9991.

Sagara, Y. and G. Inesi (1991) Inhibition of the sarcoplasmic reticulum Ca^{2+} transport ATPase by thapsigargin at subnanomolar concentrations. *J. Biol. Chem.* 266: 13503–13506.

Sagara, Y., F. Fernandez-Belda, et al. (1992a) Characterization of the inhibition of intracellular Ca^{2+} transport ATPases by thapsigargin. *J. Biol. Chem.* 267: 12606–12613.

Sagara, Y., J. B. Wade, et al. (1992b) A conformational mechanism for formation of a dead-end complex by the sarcoplasmic reticulum ATPase with thapsigargin. *J. Biol. Chem.* 267: 1286–1292.

Salamino, F., B. Sparatore, et al. (1994) The plasma membrane calcium pump is the preferred calpain substrate within the erythrocyte. *Cell Calcium* 15: 28–35.

Schaefer, W., J. Priessen, et al. (1987) Ca^{2+}–Mg^{2+}-ATPase activity of human red blood cells in healthy and diabetic volunteers. *Klin. Wochenschr* 65: 17–21.

Schatzmann, H. J. (1966) ATP-dependent Ca^{2+} extrusion from human red cells. *Experientia Basel* 22: 364–368.

Schatzmann, H. J. and H. Buergin (1978) Calcium in human blood red cells. *Ann. N.Y. Acad. Sci.* 307: 125–147.

Schlesser, R. B., S. Ulaszewski, et al. (1988) A second transport ATPase gene in *Saccharomyces cerevisiae*. *J. Biol. Chem.* 263: 19480–19487.

Schnitzer, J. E., P. Oh, et al. (1995) Caveolae from luminal plasmalemma of rat lung endothelium: microdomains enriched in caveolin, Ca^{2+}-ATPase, and inositol triphosphate receptor. *Proc. Natl. Acad. Sci. USA* 92: 1759–1763.

Seiz-Preianò, B., D. Guerini, et al. (1996) Expression and functional characterization of isoforms 4 of the plasma membrane calcium pump. *Biochemistry* 35: 7946–7953.

Shin, J. M., M. Kajimura, et al. (1994) Biochemical identification of transmembrane segments of the Ca^{2+}-

ATPase of sarcoplasmic reticulum. *J. Biol. Chem.* 269: 22533–22537.

Shull, G. E. and J. Greeb (1988) Molecular cloning of two isoforms of the plasma membrane Ca^{2+}-transporting ATPase from rat brain. *J. Biol. Chem.* 263: 8646–8657.

Simonides, W. S., G. C. van der Linde, et al. (1990) Thyroid hormone differentially affects mRNA levels of Ca-ATPase isoenzymes of sarcoplasmic reticulum in fast and slow skeletal muscle. *FEBS Lett.* 274: 73–76.

Skerjanc, I. S., T. Toyofuku, et al. (1993) Mutation of glutamate 309 to glutamine alters one Ca^{2+}-binding site in the Ca^{2+}-ATPase of sarcoplasmic reticulum expressed in Sf9 cells. *J. Biol. Chem.* 268: 15944–15950.

Smallwood, J. I., B. Gügi, et al. (1988) Regulation of the Ca^{2+} pump activity by protein kinase C. *J. Biol. Chem.* 263: 2195–2205.

Solioz, M., A. Odermatt, et al. (1994) Copper pumping ATPases: common concepts in bacteria and man. *FEBS Lett.* 346: 44–47.

Stahl, W. L., T. J. Eakin, et al. (1992) Plasma membrane Ca^{2+}-ATPase isoforms: distribution of mRNAs in rat brain by in situ hybridisation. *Mol. Brain Res.* 16: 223–231.

Stahl, W. L., T. P. Keeton, et al. (1994) The plasma membrane Ca^{2+}-ATPase mRNA isoform PMCA4 is expressed at high levels in neurons of rat piriform cortex and neocortex. *Neurosci. Lett.* 178: 267–270.

Stauffer, T., D. Guerini, et al. (1995) Tissue distribution of the four gene products of the plasma membrane Ca^{2+} pump. *J. Biol. Chem.* 270: 12184–12190.

Stauffer, T., H. Hilfiker, et al. (1993) Quantitative analysis of alternative splicing options of human plasma membrane calcium pump. *J. Biol. Chem.* 268: 25993–26003.

Stokes, D. L. (1991) P-type pumps: structure determination may soon catch up with structure predictions *Curr. Opin. Struct. Biol.* 1: 555–561.

Stokes, D. L. and M. N. Green (1990) Structure of Ca-ATPase: electron microscopy of frozen-hydrated crystals at 6 Å resolution in projection *J. Mol. Biol.* 213: 529–538.

Strehler, E. E. (1991) Recent advances in the molecular characterisation of plasma membrane Ca^{2+} pumps. *J. Membr. Biol.* 120: 1–15.

Strehler, E. E., P. James, et al. (1990) Peptide sequence analysis and molecular cloning reveal two calcium isoforms in the human erythrocyte membrane. *J. Biol. Chem.* 265: 2835–2842.

Strehler, E. E., M.-A. Strehler-Page, et al. (1989) mRNAs for plasma membrane calcium pump isoforms differing in their regulatory domain are generated by alternative splicing that involves two internal donor sites in a single exon. *Proc. Natl. Acad. Sci. USA* 86: 6908–6912.

Sumbilla, C., L. Lu, et al. (1993) Ca^{2+}-dependent and thapsigargin-inhibited phosphorylation of Na^+, K^+-ATPase catalytic domain following chimeric recombination with Ca^{2+}-ATPase. *J. Biol. Chem.* 268: 21185–21192.

Taylor, D. J., M. J. Brosnan, et al. (1988) Ca^{2+}-ATPase deficiency in a patient with an exertional muscle pain syndrome. *J. Neurol. Neurosurg. Psych.* 51: 1425–1433.

Taylor, W. P. and M. N. Green (1989) The predicted secondary structures of the nucleotide-binding sites of six cation-transporting ATPases lead to a probable tertiary fold. *Eur. J. Biochem.* 179: 241–248.

Tepikin, A. V., S. G. Voronina, et al. (1992) Pulsatile Ca^{2+} extrusion from single pancreatic acinar cells during receptor-activated cytosolic Ca^{2+} spiking. *J. Biol. Chem.* 267: 14073–14076.

Thastrup, O., P. J. Cullen, et al. (1990) Thapsigargin, a tumor promoter, discharges intracellular Ca^{2+} stores by specific inhibition of the endoplasmic reticulum. *Proc. Natl. Acad. Sci. USA* 87: 2466–2470.

Toyofuku, T., K. Kurzydlowsky, et al. (1993) Identification of regions in the Ca^{2+}-ATPase of sarcoplasmic reticulum that affect functional association with phospholamban. *J. Biol. Chem.* 268: 2809–2815.

Toyofuku, T., K. Kurzydlowski, et al. (1994a) Amino acids Glu2 to Ile18 in the cytoplasmic domain of phospholamban are essential for functional association with the Ca^{2+}-ATPase of sarcoplasmic reticulum. *J. Biol. Chem.* 269: 3088–3094.

Toyofuku, T., K. Kurzydlowski, et al. (1994b) Amino acids Lys-Asp-Asp-Lys-Pro-Val402 in the Ca^{2+}-ATPase of cardiac sarcoplasmic reticulum are critical for functional association with phospholamban. *J. Biol. Chem.* 269: 22929–22932.

Toyoshima, C., H. Sasabe, et al. (1993) Three-dimensional cryo-electron microscopy of the calcium ion pump in the sarcoplasmic reticulum membrane. *Nature* 362: 469–471.

Varadi, A., E. Molnar, et al. (1996) A unique combination of plasma membrane Ca^{2+}-ATPase isoforms is expressed in islets of Langerhans and pancreatic β-cell lines. *Biochem. J.* 314: 663–669.

Vellani, F. (1995) P-type ATPases from yeast to man. Thesis No. 11254, Swiss Federal Institute of Technology (ETH), Zurich, Switzerland.

Verma, A. K., A. G. Filoteo, et al. (1988) Complete primary structure of a human plasma membrane Ca^{2+} pump. *J. Biol. Chem.* 263: 14152–14159.

Vilsen, B. (1995) Structure–function relationships in the Ca^{2+}-ATPase of sarcoplasmic reticulum studied by use of the substrate analogue CrATP and by side-directed mutagenesis. Comparison with the Na^+, K^+-ATPase. *Acta Physiol. Scand.* 154: 1–146.

Vilsen, B. and J. P. Andersen (1992a) CrATP-induced Ca^{2+} occlusion in mutants of the Ca^{2+}-ATPase of sarcoplasmic reticulum. *J. Biol. Chem.* 267: 25739–25743.

Vilsen, B. and J. P. Andersen (1992b) Interdependence of Ca^{2+} occlusion sites in the unphosphorylated sarcoplasmic reticulum Ca^{2+} ATPase complex with CrATP. *J. Biol. Chem.* 267: 3539–3550.

Vilsen, B., J. P. Andersen, et al. (1989) Functional consequences of proline mutations in the cytoplasmic and transmembrane sectors of the Ca^{2+}-ATPase of sarcoplasmic reticulum. *J. Biol. Chem.* 264: 21024–21030.

Vilsen, B., J. P. Andersen, et al. (1991) Functional consequences of alterations to hydrophobic amino acids located at the M_4S_4 boundary of the Ca^{2+}-ATPase of sarcoplasmic reticulum. *J. Biol. Chem.* 266: 18839–18845.

Vorherr, T., M. Chiesi, et al. (1992) Regulation of the calcium ion pump of sarcoplasmic reticulum: reversible inhibition by phospholamban and by calmodulin binding domain of the plasma membrane calcium ion pump. *Biochemistry* 31: 371–376.

Vorherr, T., P. James, et al. (1990) Interaction of calmodulin with the calmodulin binding domain of the plasma membrane Ca^{2+} pump. *Biochemistry* 29: 355–365.

Vorherr, T., T. Kessler, et al. (1991) The calmodulin-binding domain mediates the self-association of the plasma membrane Ca^{2+} pump. *J. Biol. Chem.* 266: 22–27.

Wang, G. M., Y. Huafang, et al. (1994) Localization of two genes encoding plasma membrane Ca^{2+} ATPases isoforms 2 (ATP2B2) and 3 (ATP2B3) to human chromosomes 3p23–p25 and Xq28. *Cytogenet. Cell Genet.* 67: 41–45.

Wang, K. K. W., L. C. Wright, et al. (1991) Protein kinase C phosphorylates the carboxyl terminus of the plasma membrane Ca^{2+}-ATPase from human erythrocytes. *J. Biol. Chem.* 266: 9078–9085.

Wrosek, A., H. Schneider, et al. (1992) Effect of thapsigargin on cardiac muscle cells. *Cell Calcium* 13: 281–292.

Wuytack, F., G. DeSchutter, et al. (1981) Partial purification of $(Ca^{2+}–Mg^{2+})$ ATPase from pig smooth muscle and reconstitution of an ATP-dependent Ca^{2+}-transport system. *Biochem. J.* 198: 265–271.

Wuytack, F., B. Papp, et al. (1994) A sarco/endoplasmic reticulum Ca^{2+}-ATPase 3-type Ca^{2+} pump is expressed in platelets, in lymphoid cells, and in mast cells. *J. Biol. Chem.* 269: 1410–1416.

Xu, A., C. Hawkins, et al. (1993) Phosphorylation and activation of the Ca^{2+}-pumping ATPase of cardiac sarcoplasmic reticulum by Ca^{2+}/calmodulin-dependent protein kinase. *J. Biol. Chem.* 268: 8394–8397.

Yu, X., S. Carrol, et al. (1993) H^+ countertransport and electrogenicity of the sarcoplasmic reticulum Ca^{2+} pump in reconstituted proteoliposomes. *Biophys. J.* 64: 1232–1242.

Zacharias, D. A., S. J. Dalrymple, et al. (1995) Transcripts distribution of plasma membrane Ca^{2+} pump isoforms and splice variants in the human brain. *Mol. Brain Res.* 28: 263–272.

Zarain-Herzberg, A., D. MacLennan, et al. (1990) Characterization of rabbit cardiac sarco(endo)plasmic reticulum Ca^{2+}-ATPase gene. *J. Biol. Chem.* 265: 4670–4677.

Zarain-Herzberg, A., J. Marques, et al. (1994) Thyroid hormone receptor modulates the expression of the rabbit cardiac sarco(endo)plasmic reticulum Ca^{2+} ATPase gene. *J. Biol. Chem.* 269: 1460–1467

Zhang, B.-X., H. Zhao, et al. (1992) Activation of the plasma membrane Ca^{2+} pump during agonist stimulation of pancreatic acini. *J. Biol. Chem.* 267: 15419–15425.

Zhang, P., C. Toyoshima, et al. (1998) Structure of the calcium pumps from sarcoplasmic reticulum at 8-Å resolution. *Nature* 392: 835–839.

Zhang, Y., J. Fujii, et al. (1995) Characterization of cDNA and genomic DNA encoding SERCA1, the Ca^{2+}-ATPase of human fast-twitch skeletal muscle sarcoplasmic reticulum, and its elimination as a candidate gene for Brody disease in three patients. *Genomics* 30: 415–424.

Zvaritch, E., P. James, et al. (1990) Mapping of functional domains in the plasma membrane Ca^{2+} pump using trypsin proteolysis. *Biochemistry* 29: 8070–8076.

Zvaritch, E., F. Vellani, et al. (1995) A signal for endoplasmic retention at the carboxyl terminus of the plasma membrane Ca^{2+}-ATPase isoform 4CI. *J. Biol. Chem.* 270: 2679–2688.

11

Sodium–Calcium Exchange

Kenneth D. Philipson

The plasma membrane Na^+–Ca^{2+} exchanger catalyzes the countertransport of three Na^+ ions for one Ca^{2+} ion. In many cells, this exchange mechanism is sufficiently active to mediate rapid and substantial fluxes of Ca^{2+}. The first demonstrations of Na^+/Ca^{2+} countertransport were made using isotope fluxes in guinea pig atria (Reuter and Seitz, 1968) and the squid giant axon (Baker et al., 1969). The potential importance of this transport pathway was immediately appreciated. Indeed, Baker et al. (1969) correctly proposed, based on their initial observations in squid axon, that Na^+–Ca^{2+} exchange could account for the cardiotonic action of cardiac glycosides. In the ensuing years, Na^+–Ca^{2+} exchange has been a topic of intense interest. This chapter will review some of the physiological and biochemical background on Na^+–Ca^{2+} exchange research and will then focus on recent molecular biological advances. The emphasis is to illustrate advances and issues in Na^+–Ca^{2+} exchange research and not to review the literature exhaustively. Other literature focusing on various aspects of Na^+–Ca^{2+} exchange which may interest the reader include Bers (1991), Philipson and Nicoll (1993), Reeves (1995) and Philipson (1996). Also, three volumes devoted to Na^+–Ca^{2+} exchange have been published. These books result from three international conferences on Na^+–Ca^{2+} exchange (Allen et al., 1989; Blaustein et al., 1991; Hilgemann et al., 1995).

Energetics

In general, Na^+–Ca^{2+} exchange is a Ca^{2+}-efflux mechanism. The exchanger essentially uses the inwardly directed electrochemical gradient of Na^+ (produced by the ATP-dependent Na^+ pump) to extrude Ca^{2+} from a cell. However, the net direction of exchange depends on three factors: the Na^+ gradient, the Ca^{2+} gradient, and the membrane potential. Membrane potential is a factor because of the stoichiometry of exchange (three Na^+ for one Ca^{2+}). With each reaction cycle, there is a net movement of one positive charge in the same direction as the three Na^+ ions. The thermodynamics of Na^+–Ca^{2+} exchange can be summarized in the following equation:

$$[Ca^{2+}]_i = [Ca^{2+}]_o \left(\frac{[Na^+]_i}{[Na^+]_o} \right)^3 \exp\left(\frac{E_m F}{RT} \right)$$

where E_M is the membrane potential and R, T, and F have the usual meanings. The equation defines the equilibrium intracellular Ca^{2+} level when Na^+–Ca^{2+} exchange is the dominant Ca^{2+}-transport mechanism. Although this condition is often not met, the equation has useful applications for understanding the role of exchange under different conditions. Under typical cellular conditions, the reversal potential for Na^+–Ca^{2+} exchange is about −40 mV. For polarized excitable cells such as myocytes or neurons at −80 mV, Ca^{2+} efflux prevails. A rise in intracellular Na^+ or membrane depolarization tends to reverse the exchanger and bring about Ca^{2+} influx. Theoretical predictions of the net direction of Ca^{2+} flux, however, are often not easily made. Ca^{2+} flux is critically dependent on internal Na^+ (because of the cube dependence) and accurate knowledge of this value is often difficult to obtain.

A further complication is possible nonhomogeneities of ion levels in the cytoplasm. This has been considered in detail for cardiac muscle (see below). Thus, the concentrations of Na^+ and Ca^{2+} to which the exchanger is exposed may differ substantially from levels in the bulk cytoplasm due to possible proximity of ion channels and areas of restricted diffusion. The increased uncertainty of internal ion levels at the exchanger complicates modeling of exchanger function.

Physiology

The Na^+-Ca^{2+} exchange across the plasma membrane can be detected in almost all tissues though the level of activity varies considerably. Cardiac myocytes have especially high activity and the physiological consequences of exchange have been studied most extensively for cardiac muscle. I will present some detail on the role of Na^+–Ca^{2+} exchange in heart as an example of the importance of this transporter and then more briefly review the importance of exchange in other tissues.

Heart

Substantial transmembrane fluxes of Ca^{2+} occur with each contraction of cardiac muscle. The action potential initiates contraction by opening voltage-dependent Ca^{2+} channels. The Ca^{2+} influx can directly activate myofibrils but also triggers a release of a larger amount of Ca^{2+} from the sarcoplasmic reticulum by a "Ca^{2+}-induced Ca^{2+}-release mechanism." Contraction is terminated by a decline in myoplasmic Ca^{2+}. Ca^{2+} is either resequestered by the sarcoplasmic reticulum or extruded from the cell. The sarcolemma of cardiac myocytes has two mechanisms for Ca^{2+} extrusion: an ATP-dependent Ca^{2+} pump and Na^+–Ca^{2+} exchange. In cardiac muscle, Na^+-Ca^{2+} exchange is the dominant Ca^{2+} efflux mechanism and it is difficult to demonstrate any role for the plasma membrane Ca^{2+} pump. Thus, the quantity of Ca^{2+} which had entered the myocytes through the L-type Ca^{2+} channels must be extruded by the Na^+–Ca^{2+} exchanger to maintain cellular Ca^{2+} homeostasis and to bring about muscle relaxation. It is estimated that 10–20% of the rise in intracellular Ca^{2+} is due to Ca^{2+} influx and the remaining Ca^{2+} rise is due to release from the sarcoplasmic reticulum. Thus, the exchanger is responsible for removing up to 20% of the Ca^{2+} from the myoplasm. The remaining Ca^{2+} is transported back into the sarcoplasmic reticulum by an ATP-dependent Ca^{2+} pump. Because cardiac muscle undergoes continuous stimulation, any modification of exchange activity rapidly leads to an alteration in myoplasmic Ca^{2+} levels and sarcoplasmic reticular Ca^{2+} loading. Such events change the contractility of cardiac muscle. The best known modulator of exchange activity, digitalis, acts indirectly on the exchanger by inhibiting the Na^+ pump and causing a rise in internal Na^+. Other factors, such as stimulation frequency or pathological insults, can also lead to altered exchange activity.

The scheme just presented is generally accepted by researchers studying cardiac excitation–contraction coupling. However, certain aspects involving Na^+–Ca^{2+} exchange are controversial and the topics of much current research. For example, it is agreed that Na^+–Ca^{2+} exchange is the major cardiac Ca^{2+}-efflux pathway, but it is contentious as to whether the exchanger also has a role in Ca^{2+} influx. In one scenario, as initially proposed by Leblanc and Hume (1990), the channel-mediated influx of Na^+ which accompanies the action potential is sufficient to cause a rise in Na^+ in a subsarcolemmal space. The elevated Na^+ then induces Ca^{2+} influx through the exchanger, which contributes to the contractile event. This proposal has led to much discussion (Lederer et al., 1990; Langer and Peskoff, 1996) on regions of restricted diffusion ("fuzzy space") and to experiments both for and against the hypothesis (Sham et al., 1992; Lipp and Niggli, 1994).

It has also been proposed that depolarization of the sarcolemmal membranes by itself, such as occurs with a normal action potential, can induce Ca^{2+} influx through the exchanger and trigger sarcoplasmic reticular Ca^{2+} release. This undoubtedly occurs under some experimental conditions. The question is whether "reverse" exchange is of a sufficient magnitude and rapidity vis-à-vis Ca^{2+} influx through channels to have a significant effect under physiological conditions. Again, articles on both sides of this issue have been published (Levi et al., 1994; Sham et al., 1995).

Detailed modeling of myocardial Ca^{2+} fluxes during excitation-contraction coupling predicts an important role for Na^+–Ca^{2+} exchange in the specialized region of the sarcolemma (the transverse tubule) which is close to the Ca^{2+}-release sites on the junctional sarcoplasmic reticulum (Langer and Peskoff, 1996). Large transient Ca^{2+} levels in the space between these membranes will greatly increase the effectiveness of Ca^{2+} efflux through the exchanger following Ca^{2+} release. Resolving the exact role of restricted diffusion spaces in both exchanger-mediated Ca^{2+} influx and efflux is likely to remain a central issue in exchanger research.

Levels of cardiac Na^+–Ca^{2+} exchanger protein and mRNA decrease substantially during development (Artman, 1992; Boerth et al., 1994). The exchange activity of the fetal or neonatal heart is a few fold higher than that in the adult heart. An increasing level of sarcoplasmic reticular Ca^{2+} transport correlates with the decline in Na^+–Ca^{2+} exchange activity. Properties of cardiac excitation–contraction coupling change in parallel with the changes in Ca^{2+} transport pathways. Cardiac exchange activity may also be upregulated in some pathophysiologic situations (e.g., Kent et al., 1993).

Brain

Neurotransmitter release at the synapse is triggered by an influx of Ca^{2+} through Ca^{2+} channels. This critical signaling event has been the subject of intense pharmacological and biochemical investigation. Imaging techniques have also been used to study the temporospatial characteristics of the Ca^{2+} signal. Nevertheless, the mechanism extruding Ca^{2+} out of the presynaptic terminal has received relatively little attention. Ca^{2+} efflux is certainly an important event in regulating neuronal Ca^{2+} and synaptic transmission. As in heart, the neuronal plasma membrane possesses both an ATP-dependent Ca^{2+} pump and a Na^+–Ca^{2+} exchanger. Technical difficulties have precluded a clear definition of the relative importance of each pathway. Studies from the Blaustein laboratory have indicated that the exchanger is concentrated in the presynaptic membrane at neuromuscular junctions (Luther et al., 1992) and that the exchanger is much more potent than the Ca^{2+} pump in extruding Ca^{2+} from presynaptic terminals (Sanchez-Armass and Blaustein, 1987).

Additionally, a favorable system for the study of Na^+–Ca^{2+} exchange has been the squid giant axon. Much detailed kinetic and regulatory information has derived from this preparation (DiPolo and Beaugé, 1991).

Other Tissues

The Na^+–Ca^{2+} exchange has been measured and characterized in many tissues, though in many cases clear physiologic importance is difficult to prove. In kidney, for example, exchange activity is low in membrane vesicle preparations. However, recent studies have indicated that exchange activity is concentrated in only certain portions of the nephron which may have high activity. Both Bourdeau et al. (1993) and Reilly et al. (1993) used immunocytochemistry to demonstrate that the exchanger is most prominent in the basolateral membrane of a majority of the cells in connecting tubules. It is a reasonable hypothesis that the Na^+–Ca^{2+} exchanger is involved in the regulated reabsorption of Ca^{2+} in this region of the nephron.

Na^+–Ca^{2+} exchange has also been implicated in the regulation of vascular tone by arterial smooth muscle cells. Exchange activity is quite low in these cells though experiments with muscle strips strongly suggest that modulation of exchange activity can affect contraction (Ashida and Blaustein, 1987). The importance of exchange for in vivo regulation of vascular tone, however, is still controversial.

Photoreceptors

The outer segments of rod photoreceptors (ROS) have extremely high Na^+–Ca^{2+} exchange activity. The exchanger extrudes Ca^{2+} brought in by cGMP-gated channels and is important in visual adaptation.

ROS exchange, however, is mediated by a protein with little sequence identity to the·exchangers discussed so far (see below). Whereas Na^+–Ca^{2+} exchange is the countertransport of three Na^+ for one Ca^{2+}, the exchange activity of ROS is the countertransport of four Na^+ for one Ca^{2+} plus one K^+. Isolated rod outer segments are an excellent system for studying exchange kinetics but have attracted relatively few investigators. The detailed kinetics of the ROS exchanger have been considered elsewhere (e.g., Schnetkamp and Szerencsei, 1993; Schnetkamp, 1995).

Mitochondria

The mitochondrial inner membrane also contains a Na^+–Ca^{2+} exchange mechanism, as first noted by Carafoli and colleagues. This Na^+–Ca^{2+} exchanger may be very important in regulating metabolism. There has been relatively little progress on molecular characterization of this transporter though isolation of the protein has been reported (Li et al., 1992). The mitochondrial exchanger has some characteristics which clearly distinguish it from the plasma membrane Na^+–Ca^{2+} exchanger. For example, Li^+ can substitute for Na^+ on the mitochondrial exchanger but not on the plasma membrane exchanger. The mitochondrial exchanger had been thought to be electroneutral but some recent experiments suggest otherwise (Jung et al., 1995). Mitochondrial Na^+–Ca^{2+} exchange has been reviewed by Crompton (1990).

Membrane localization

Immunolocalization studies of the Na^+–Ca^{2+} exchanger have now been carried out in several tissues. Frank et al. (1992) reported that the exchanger was present at highest abundance in the portion of the sarcolemma which forms the transverse tubules in cardiac myocytes from rat and guinea pig. (Curiously, this nonhomogenous distribution of exchangers appears to be species-dependent and is not seen in rabbit myocytes [Frank et al., 1992; Chen et al., 1995]). Such a localization may have significance in understanding the Ca^{2+} fluxes which occur during excitation–contraction coupling. The presence of exchangers in the T-tubules places them in proximity to the Ca^{2+}-release sites on the sarcoplasmic reticulum. In contrast, Kieval et al. (1992) found a homogenous distribution of exchangers in the cardiac sarcolemma of rats and guinea pigs. The reason for this discrepancy is unclear.

A domain localization of the Na^+–Ca^{2+} exchanger has now also been described in other tissues. As mentioned above, this is true in kidney (Bourdeau et al., 1993; Reilly et al., 1993). Additionally, the exchanger is localized in the plasma membrane of smooth muscle cells in regions near the underlying sarcoplasmic reticulum (Moore et al., 1993; Juhaszova et al., 1994) and the exchanger is present in high concentrations at presynaptic nerve terminals in neuromuscular preparations (Luther et al., 1992).

If the Na^+–Ca^{2+} exchanger is localized to particular domains of the plasma membrane, some mechanism must constrain the exchanger in the plane of the membrane. Li et al. (1993) found that the exchanger bound to the cytoskeletal element ankyrin. This may be the mechanism responsible for specialized localization. Condrescu et al. (1995) report that disruption of the actin cytoskeleton inhibited exchange activity in CHO cells transfected with the Na^+–Ca^{2+} exchanger. Perhaps, multiple proteins and mechanisms are involved in exchanger membrane localization.

Vesicles

A key advance in Na^+–Ca^{2+} exchange research was the development of techniques for studying exchange in isolated plasma membrane vesicles. This was first accomplished by Reeves and Sutko (1979) using cardiac sarcolemmal vesicles. Their elegant study made unequivocal measurements of Na^+–Ca^{2+} exchange available at the biochemical level. Membrane vesicles from several tissues have now been used for Na^+–Ca^{2+} exchange measurements but the most detailed studies were accomplished with cardiac sarcolemmal vesicles due to their high exchange activity. With the advent of cloning and molecular studies, vesicles have now become less popular for exchange studies. Nevertheless, vesicles are still useful for many applications. Cardiac sarcolemmal vesicles have been used to measure many basic properties of Na^+-Ca^{2+} exchange, such as stoichiometry, electrogenicity, and ion interactions (for a review see Reeves and Philipson, 1989).

In addition, the cardiac exchanger is readily solubilized and reconstituted. This has allowed the dependence of exchange activity on membrane lipid environment to be explored in detail. Most striking is the requirement for anionic phospholipids, most notably phosphatidyl serine, in the intracellular leaflet of the lipid bilayer for optimal exchange activity. These studies were initiated using reconstituted proteoliposomes (reviewed in Philipson, 1990) but have subsequently been extended in the Hilgemann laboratory (Hilgemann and Collins, 1992) using the giant excised patch technique (see below).

The Na^+–Ca^{2+} exchanger can be stimulated by several interventions. These include protease treatment, high pH, redox modification, altered lipid environment, and Ca^{2+}-EGTA (reviewed in Reeves and Philipson, 1989). Most of these phenomena were first noted using sarcolemmal vesicles. These modulations can all be speculated to have possible physiological consequences. Recently, the probable site where proteases cleave the cardiac Na^+–Ca^{2+} exchanger has been identified (Iwata et al., 1995).

Inhibitors

High-affinity, specific inhibitors of the Na^+–Ca^{2+} exchanger have not yet been developed. Amiloride derivatives such as dichlorobenzamil have received attention (Kaczorowski et al., 1989) but have almost no specificity for the Na^+–Ca^{2+} exchanger. Khananshvili et al. (1993, 1995) have described the use of small, charged peptides to inhibit exchange. The specificity of these peptides has not been assessed but possibly the peptides could form the basis for development of more potent inhibitors. A peptide inhibitor, XIP (exchanger inhibitory peptide), has been developed which acts at the intracellular surface of the exchanger (Li et al., 1991; He et al., 1997). XIP has the same amino acid sequence as a putative auto-inhibitory region of the exchanger and will be discussed below. More recently, Iwamoto et al. (1996b) have described an isothiourea derivative (compound number 7943) as a selective inhibitor of Na^+–Ca^{2+} exchange, though another group (Watano et al., 1996) has found that 7943 also blocks Na^+, Ca^{2+}, and K^+ currents.

Regulation

Evidence for physiological regulation of the Na^+–Ca^{2+} exchanger has been somewhat elusive. Until recently, most data have been obtained with the squid giant axon, where two types of regulation have been clearly identified. First, the exchanger is activated by intracellular Ca^{2+} at a high-affinity binding site which is distinct from the Ca^{2+} transport site. (As discussed below, molecular information on the regulatory Ca^{2+}-binding site is now available.) Second, the squid exchanger is activated by ATP. The effect of ATP appears to be due to a phosphorylation reaction, though the nature of the kinase which is involved has not yet been elucidated. Stimulation of the squid axon exchanger by ATP or Ca^{2+} affects several of the kinetic properties of exchange. ATP, for example, increases the affinity of the exchanger for regulatory Ca^{2+}. The apparent affinity of the Ca^{2+}-regulatory site increases from about $4\,\mu M$ to about $0.4\,\mu M$ upon addition of ATP. Regulation of the Na^+–Ca^{2+} exchanger of

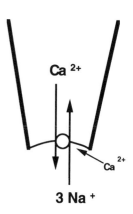

Figure 11.1 Schematic of a giant excised patch. The patch is inside-out and can be about 30 μm in diameter. With the ion gradients as shown, outward Na$^+$–Ca^{2+} exchange currents ("reverse exchange") will be generated. Typically, exchange activity is initiated by the rapid application of Na$^+$ to the bath (intracellular surface), as in Figs 11.2 and 11.7. The presence of Na$^+$ and Ca^{2+} on opposite sides of the membrane is not sufficient, by itself, to initiate exchange currents: micromolar levels of non-transported, regulatory Ca^{2+} must also be at the intracellular surface. Usually, the patches are excised from cardiac myocytes or from oocytes expressing cloned exchangers.

the squid giant axon is reviewed in DiPolo and Beaugé, 1991.

The effects of phosphorylation on Na$^+$–Ca^{2+} exchangers in other tissues have not been demonstrated as convincingly. Conflicting reports appear in the literature. Recently, Iwamoto et al. (1995, 1996a) have demonstrated, for the first time, direct phosphorylation of the Na$^+$-Ca^{2+} exchanger protein. Their data suggest that the Na$^+$–Ca^{2+} exchanger of cardiac and vascular smooth muscle cells is activated by protein kinase C-dependent phosphorylation in response to growth factors. Other laboratories are investigating exchanger phosphorylation and more information may be available shortly. Also, multiple exchanger gene products and splice variants are being described (see below) and some of these exchangers may prove to be substrates for different kinases.

A major technological advance for studying Na$^+$–Ca^{2+} exchange, especially regulation, was the giant excised patch system developed by Hilgemann (1989, 1990). The giant excised patches (about 30 μm in diameter) have allowed the small currents associated with transporters to be measured in inside-out excised patches for the first time (Fig. 11.1). This patch system permits free access and rapid solution changes at the intracellular surface of the exchanger where regulatory reactions are occurring (Fig. 11.2). This has greatly facilitated the study of Ca^{2+} regulation. Also, a new form of intrinsic regulation of the exchanger, called Na$^+$-dependent inactivation (Fig. 11.2), was discovered. When three Na$^+$ ions bind to the transport sites of the exchanger at the intracellular surface, the protein can undergo either of two reactions. Either the exchanger can undergo a conformational change translocating the Na$^+$ ions to the extracellular surface or the exchanger can enter an inactivated state (I_1, in the terminology of Hilgemann). The inactivation is manifested in the transient component of outward exchange current observed upon application of intracellular Na$^+$ to

initiate the exchange reaction (Fig. 11.2). The kinetic properties of Ca^{2+} regulation and Na$^+$-dependent inactivation have been modeled in detail by Hilgemann et al. (1992a, 1992b). Some molecular information on the parts of the exchanger molecule involved in these forms of regulation has been deduced (see below).

The giant excised patch has also been useful in investigating the effects of lipid environment and ATP on the cardiac exchanger. Hilgemann and Collins (1992) found that the application of anionic phospholipids to the inner leaflet of excised sarcolemmal patches stimulated Na$^+$–Ca^{2+} exchange currents. A similar effect was induced by the addition of ATP. It was initially hypothesized that ATP stimulated the exchanger by modulating the bilayer distribution of anionic phospholipid. More recently, however, Hilgemann and Ball (1996) have provided elegant evidence that ATP stimulates exchange by increasing the sarcolemmal level of phosphatidyl inositol-4,5-bisphosphate (PIP$_2$). Regulation of the exchanger by PIP$_2$ may be important in cellular signaling events.

Reaction Mechanism

A transport cycle in which three Na$^+$ ions are exchanged for one Ca^{2+} ion across a biological membrane will, by necessity, be complicated. Multiple electrogenic steps and conformational changes are certain to be involved. Nevertheless, some advances have been made and techniques have been developed for further studies. The exchange reaction is best modeled as being consecutive (Fig. 11.3). That is, for example, Na$^+$ ions are first bound to the exchanger on one surface, translocated across the membrane, and released. Then, Ca^{2+} binds and is translocated across the membrane in the opposite direction (Khananshvili, 1990; Hilgemann et al., 1991; Kappl and Hartung, 1996). With a consecutive

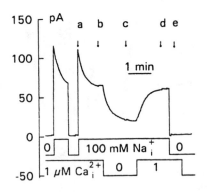

Figure 11.2 Outward Na^+–Ca^{2+} exchange current from the wild-type exchanger in a giant inside-out patch of oocyte membrane. Ca^{2+} is present within the pipette at the extracellular membrane surface at all times. An outward exchange current is initiated by the application of Na^+ to the bath at arrow a. Regulatory Ca^{2+} is also present in the bath. Note that the current reaches a peak and then partially decays to a new steady-state value. The decay is due to the Na^+-dependent inactivation process. At arrow b, regulatory Ca^{2+} is removed from the bath and current slowly decays. This demonstrates the Ca^{2+} regulatory process. (Reprinted with permission from Matsuoka et al., 1993.)

reaction, conditions can be devised in which partial reactions of the exchange cycle can be detected under appropriate conditions. For example, rapid application of Na^+ or Ca^{2+} by solution switch or by photo-release can induce transient currents which have been interpreted as being due to a half-cycle of the exchanger (Hilgemann et al., 1991; Niggli and Lederer, 1991). Overall, the data support a model in which the major electrogenic step involves Na^+ movement near the extracellular surface of the exchanger (Hilgemann et al., 1991). The turnover rate for the exchanger is quite high for a transporter (about $5000\,s^{-1}$), as determined by analysis of noise and of voltage-induced charge movements (Hilgemann, 1996).

Isolation

The development of the vesicle assay system and solubilization–reconstitution techniques was the *sine qua non* for attempts to isolate the exchanger protein. Progress was initially slow for two reasons. First, the Na^+–Ca^{2+} exchanger is a low-abundance protein and somewhat labile in the solubilized state. Second, the only marker for the exchanger is transport. No high-affinity, labeled ligands were available to follow the exchanger during protein purification. Thus, to isolate the exchanger protein, purified plasma membranes must be solubilized and the membrane proteins fractionated. Each fraction must then be reconstituted into liposomes to assay for exchange activity. The first isolation of the mature Na^+–Ca^{2+} exchanger protein (120 kDa) was described by Philipson et al. (1988). Isolation of the ROS Na^+–Ca^{2+}, K^+ exchanger (220 kDa) has also been achieved (Cook and Kaupp, 1988; Nicoll and Applebury, 1989).

Cloning

Molecular research on the Na^+–Ca^{2+} exchanger was initiated after cloning was accomplished. For this step, isolated Na^+–Ca^{2+} exchanger protein was first used as antigen to produce a specific polyclonal antibody (Philipson et al., 1988). Screening of a dog heart cDNA expression library with the antibody identified

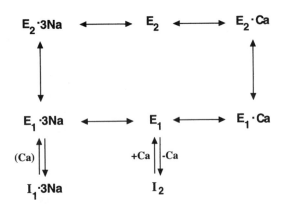

Figure 11.3 Transport cycle and regulation of the Na^+–Ca^{2+} exchanger. The sites for ion binding alternately face the intracellular or extracellular medium in the E_1 and E_2 states, respectively. A consecutive reaction mechanism is shown with the Na^+ and Ca^{2+} ions being transported in separate steps. I_1 and I_2 represent two inactivated states. When three Na^+ ions bind at the intracellular surface, the $E_1.3Na^+$ state is formed from which a fraction of the exchangers enters the I_1 inactivated state. Ca^{2+} modulates this reaction. The I_2 (Na^+-independent) inactivation is directly controlled by Ca^{2+} binding at a regulatory site. (Reprinted with permission from Matsuoka et al., 1995.)

a clone which coded for a peptide with strong immunoreactivity. Further screening with this initial partial clone led to the isolation of a cDNA encoding the full-length exchanger protein. Identity of the clone was proven when RNA synthesized from the cDNA induced activity in *Xenopus* oocytes (Nicoll et al., 1990). Thus, the dog heart Na^+–Ca^{2+} exchanger, NCX1, was the first exchanger cloned, but sequences of NCX1s from several other species and tissues have now been reported. In general, the NCX1 amino acid sequence is highly conserved across different species. Identities of 95% or greater are typical. Subsequently, the bovine ROS Na^+–Ca^{2+}, K^+ exchanger was also cloned (Reiländer et al., 1992).

Sequence Analysis

The predicted primary amino acid sequence of the cardiac Na^+–Ca^{2+} exchanger has some features of interest. Figure 11.4 summarizes some of our current knowledge.

Processing

The cDNA codes for a protein with 970 amino acids. However, the first 32 amino acids represent a signal sequence which is cleaved from the protein during biosynthesis in the endoplasmic reticulum (Hryshko et al., 1993). Cleaved signal sequences are often found on secreted proteins but are somewhat unusual for integral membrane proteins. The role of the cleavable signal sequence is unknown. The exchanger can be mutated to prevent signal peptide cleavage or to delete the first 32 amino acids comprising the signal peptide. Surprisingly, neither mutation prevents targeting or expression of exchange activity (Furman et al., 1995; Loo et al., 1995; Sahin-Toth et al., 1995). Curiously, the cleaved signal sequence is the region of

the Na^+–Ca^{2+} exchangers which shows least sequence conservation (Tsuruya et al., 1994).

Glycosylation

The Na^+–Ca^{2+} exchanger has N-linked glycosylation at Asn 9 (Fig. 11.4). (Amino acids are numbered from the site of cleavage of the signal sequence.) Mutation of Asn 9 to prevent glycosylation apparently does not alter Na^+–Ca^{2+} exchange activity (Hryshko et al., 1993). Thus, the role of glycosylation, if any, is unknown. Glycosylation adds about 10 kDa to the apparent molecular weight of the protein.

Topology

The amino acid sequence of the Na^+–Ca^{2+} exchanger has several stretches of hydrophobic residues which are modeled to be transmembrane α-helices (Fig. 11.4). The protein has 11 such segments (a 12th hydrophobic segment is part of the signal sequence removed during processing). Five putative transmembrane segments are in the amino terminal portion of the protein and are separated from six carboxyl-terminal transmembrane segments by a large hydrophilic intracellular loop. The intracellular loop, by itself, is 520 amino acids in length and comprises more than half of the exchanger protein. Several of the transmembrane segments form amphipathic helices (Nicoll and Philipson, 1991). The hydrophilic faces of these helices may form the ion translocation pathway.

Only a few features of the topological model have experimental confirmation. The first hydrophilic segment is glycosylated and is thus extracellular as shown. The large hydrophilic loop in the middle of the protein is assigned to be intracellular for two reasons: First, monoclonal antibodies with epitopes on the large loop bind to cells only after permeabili-

Figure 11.4 Secondary structure model of the Na^+–Ca^{2+} exchanger. The exchanger is modeled to have 11 transmembrane segments. A large intracellular loop comprising more than half the protein separates transmembrane segments 5 and 6. A single site of glycosylation is present on the first extracellular segment. The α-repeats are shown and represent regions of inter- and intramolecular homology among exchangers. The endogenous XIP region is marked. A peptide with the same sequence as the endogenous XIP region inhibits exchange activity. Mutations of the endogenous XIP region eliminate the Na^+-dependent inactivation process. The site of binding for regulatory Ca^{2+} is also indicated.

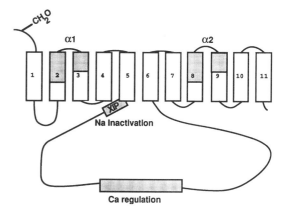

zation (Porzig et al.,1993). Second, mutations of the hydrophilic loop affect regulatory properties associated with the intracellular surface of the exchanger (Matsuoka et al., 1993). The loop between putative transmembrane segments 4 and 5 is likely to be extracellular, as modeled, as ascertained by the glycosylation of a site engineered at that location (Sahin-Toth et al., 1995). More attention to determining the correct topology is needed.

Alpha and Beta Repeats

Schwarz and Benzer (1997) have noted that there are two regions of the exchanger sequence which have similarity to other portions of the exchanger sequence. The first has been termed the α-repeat, and describes a region (α-1) in the N-terminal transmembrane segments which is homologous to a region (α-2) in the C-terminal transmembrane segments. Specifically, the amino acid sequence of the α-1 region spanning transmembrane segments 2 and 3 shows homology to the α-2 region spanning transmembrane segments 8 and 9 (Fig. 11.4). This intramolecular homology indicates that evolution of the Na^+–Ca^{2+} exchanger involved a gene-duplication event. The α-repeat will be discussed in more detail below.

The β-repeat sequences are present in the large intracellular loop of the Na^+–Ca^{2+} exchanger. These are stretches of about 60 amino acids with similarity to each other and also to the integrin β4-subunit. The functional significance of these homologies is unknown.

The Exchanger Superfamily

The NCX Family

The molecular evidence for the existence of many cation exchangers is growing. First, there are mammalian exchangers closely related to the initially cloned exchanger, NCX1. Two of these additional NCX-type clones have been isolated from rat brain. These are NCX2 (Li et al., 1994) and NCX3 (Nicoll et al., 1996b). All three of these gene products have been expressed and are Na^+–Ca^{2+} exchangers. NCX1 is expressed in high levels in cardiac muscle but significant levels are present in most tissues. mRNA for both NCX2 and NCX3, in contrast, is readily detectable in only two tissues: brain and skeletal muscle. The importance of this very specific tissue distribution is unclear. The chromosomal locations of the mammalian NCX genes have been mapped (Nicoll, et al., 1996b).

NCX-type exchangers have also been cloned from two more primitive species: the *Drosophila* Na^+–Ca^{2+} exchanger (Schwarz and Benzer, 1997; Ruknudin et al., 1997) and the squid brain Na^+–Ca^{2+} exchanger (He et al., 1998). Analysis of the amino acid sequences and functional properties of these exchangers should be productive. The *Drosophila* exchanger exhibits "anomalous" Ca^{2+} regulation, as described below. In addition, the partial sequence of an NCX from *Caenorhabditis elegans* has appeared in the databases as part of a large-scale sequencing project.

The intron–exon organization of the human NCX1 gene and of the related gene NCX3 has been described (Kraev et al., 1996; this paper mistakenly refers to NCX3 as NCX2). The NCX1 gene consists of 12 exons spread over 200 kb.

Others

The Na^+–Ca^{2+}, K^+ exchanger of ROS discussed above is not included as a member of the NCX family. This exchanger has a different ion specificity and stoichiometry and shows only a quite small amount of sequence similarity to the NCX exchangers (Reiländer et al., 1992). Strikingly, the regions of similarity are confined to the α-repeat regions discussed above (Fig. 11.3). That is, homology exists only in proposed transmembrane segments 2 and 3 (α-1) and transmembrane segments 8 and 9 (α-2). Thus, the α-repeats are regions conserved both intramolecularly and intermolecularly. Despite the lack of sequence similarity in regions outside the α-repeats, the proposed topology of the ROS exchanger is similar to that of the NCX exchangers.

Sequences are available for other genes with low sequence similarity to the NCX family. These genes are found in yeast, *C. elegans*, bacteria, and plants and are assigned to be members of the exchanger superfamily by certain criteria. All are predicted by hydropathy analysis to have multiple transmembrane segments and similar topologies. All have homology to other members of the exchanger superfamily only in the α-repeat regions. All members of this "other" family have relatively little similarity to one another outside the α-repeat regions. The conservation of the α-repeats strongly suggests that these α-repeat regions have essential structural or functional significance.

The yeast (Cunningham and Fink, 1996; Pozos et al., 1996) and plant (*Arabidopsis*; Hirschi et al., 1996) genes of the exchanger superfamily code for vacuolar Ca^{2+}–H^+ exchangers. The sequences for the putative exchangers of bacteria (*Escherichia coli, Methanococcus jannaschii, Methanobacterium thermoautotrophicum, Synechocystis*) and *C. elegans* appeared in the databases as a result of genome-sequencing projects. (This *C. elegans* gene is distinct from the *C. elegans* NCX mentioned above. That is, the *C. elegans* genome contains sequences for both "NCX-" and "other-" type members of the exchanger superfamily.) No information is available on the function or expression of these genes. We consider them to be candidate cation exchangers.

Alternative Splicing

NCX1 is the most widely distributed vertebrate exchanger and has been analyzed in the most detail. It has become apparent that there are several splice variants of NCX1 which are expressed in a tissue-specific manner. The amino acid sequences of the splicing isoforms differ in only a small portion of the large intracellular loop. The exchanger gene contains six small exons coding for this variable portion

of the protein. The exons are spliced together in different combinations specific for different tissues (Fig. 11.5). Alternative splicing has been observed in several laboratories (e.g., Reilly and Shugrue, 1992; Furman et al., 1993) and detailed analyses have recently been described in four reports (Nakasaki et al., 1993; Kofuji et al., 1994; Lee et al., 1994; Quednau et al., 1997). NCX3 also has at least three splice variants (Quednau et al., 1997).

Much detail on the tissue-specific distribution of the splice variants has been described, but little information is available on the functional relevance of alternative splicing. Twelve splice variants of NCX1 have been reported, increasing the mystery of their significance. It is tempting to speculate that different combinations of exons may affect exchanger kinetics, regulation, processing, or membrane targeting, but

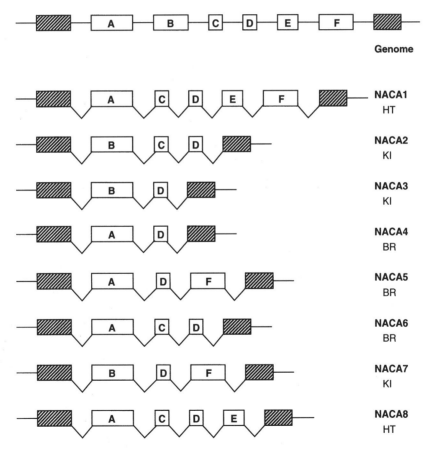

Figure 11.5 Schematic representation of the splice variants of NCX1. Six different exons, A–F, are present in the gene and code for a small portion of the large intracellular loop. These small exons are spliced together in different combinations, as indicated, and are denoted NACA1–8 using the terminology of Kofuji et al. (1994) and Lee et al. (1994). NACA8 was initially described by Reilly and Lattanzi (1995) as NACA7. Most isoforms are predominantly found in a single tissue. HT, heart; KI, kidney; BR, brain.

data are lacking. The only exchanger splice variant which has been functionally characterized is that found in cardiac muscle. Expression of the cardiac isoform of NCX1 in *Xenopus* oocytes has allowed this analysis by use of the giant excised patch technique. Analysis of other splice variants by this approach should be fruitful.

Matsuoka et al. (1993) created a chimeric exchanger clone in which the intracellular loop of a kidney splice variant replaced the loop of the cardiac isoform of NCX1. This cardiac/kidney chimera, containing the kidney sequence in the variable region, retained regulatory mechanisms characteristic of the cardiac clone. Specifically, the chimeric protein displayed Ca^{2+} regulation and Na^+-dependent inactivation. More detailed analyses were not carried out but, at least in this case, different combinations of exons did not appear to alter function drastically.

Alternative splicing also occurs in the $5'$-untranslated region of NCX1 in a tissue-specific manner. Lee et al. (1994) found different $5'$-untranslated regions in brain, kidney, and cardiac muscle. Apparently, different transcription start sites are used in these tissues. The three different initial exons are then spliced to a common core at nucleotide 34 in the $5'$-untranslated region of NCX1. The presence of different transcription start sites allows the regulation of tissue-specific expression by different promoters. The different promoters could respond in different ways to various environmental stimuli. It has also been reported that alternative splicing can modify the carboxyl-terminal transmembrane segments of the Na^+–Ca^{2+} exchanger (Gabellini et al., 1995). A stretch of nucleotides in the $3'$-untranslated region was spliced on the ends of various truncated forms of the exchanger. In this case, the splicing originated from a 6 kb exchanger cDNA in transfected cells and not from the exchanger gene. Low-level expression of one of these truncated exchangers has been reported (Gabellini et al., 1996). Expression of this "half" exchanger may have important implications in understanding exchanger function.

Structure–Function Studies

Mutational analyses of the Na^+–Ca^{2+} exchanger have centered on two different regions of the exchanger molecule. First, mutations have been made in putative transmembrane segments. The goal of these experiments is to begin to understand the parts of the molecule involved in ion translocation. Second, mutation of the large intracellular loop of the exchanger affects regulation of transport activity. Another useful approach has been to use synthetic peptides as inhibitors to provide insight into exchanger function. This section will focus on research from the laboratory of the author.

Transmembrane Segment Mutations

An initial problem is to determine appropriate candidate amino acid residues for mutation. For the Na^+–Ca^{2+} exchanger, it is reasonable to guess that hydrophilic residues, particularly anionic residues, modeled to be in transmembrane segments would be important in the translocation of cations. Also, portions of the transmembrane segments which are highly conserved in other exchangers are likely to be significant in exchange function. The most striking regions of conservation are the α-repeats discussed above (Fig. 11.4). As described, portions of transmembrane segments 2, 3, 8, and 9 are similar in all members of the exchanger superfamily. In addition to this intermolecular conservation, there is intramolecular conservation: the sequence spanning transmembrane segments 2 and 3 is similar to the sequence spanning transmembrane segments 8 and 9. Thus, sequence analysis strikingly points to an important role for the α-repeats in exchanger function. Indeed, mutagenesis work (Fig. 11.6) demonstrates that exchanger function is highly sensitive to mutations within the α-repeat regions (Nicoll et al., 1995, 1996a). As shown, 18 amino acids within the α-repeat regions have been mutated. In 17 cases, the mutations reduced or abolished exchange activity. More detailed analyses are in progress.

XIP

The Na^+–Ca^{2+} exchanger has a segment of amino acids which resembles a calmodulin-binding site. This segment is located at the very beginning of the large intracellular loop (Fig. 11.4). Although the exchanger does not appear to bind calmodulin, such segments on proteins often have autoregulatory roles. A synthetic peptide, XIP, with the same amino acid sequence as the putative autoregulatory region of the exchanger, is a relatively specific and potent inhibitor of exchange activity (Li et al., 1991). The peptide acts at the intracellular surface of the exchanger. Perhaps the synthetic, exogenously added, XIP binds to the same site on the exchanger with which the "endogenous XIP" site interacts. Inhibition of exchanger function by XIP is consistent with the hypothesis that the region of the exchanger from which XIP was designed (the endogenous XIP region) has some important regulatory role. Mutagenesis (see below) bears out this contention. It has been suggested that the endogenous XIP region of the exchanger modulates exchange activity by interacting with the membrane phospholipids (Shannon et al., 1994).

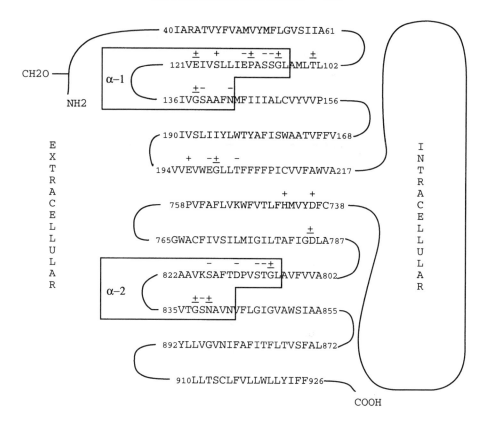

Figure 11.6 Mutational analysis of transmembrane segment amino acid residues. The amino acid sequence of each of the 11 transmembrane segments of the canine Na^+–Ca^{2+} exchanger is shown, with the first transmembrane segment at the top. The α-repeat regions are boxed. Residues which have been mutated are indicated with a "–" for inactive mutants, "±" for mutants with reduced activity, and "+" for mutants with wild-type levels of activity. Activity is highly sensitive to mutations within the α-repeat regions.

Mutations Affecting Regulation

As shown in Fig. 11.2, the cardiac exchanger NCX1 displays two forms of intrinsic regulation: (1) Na^+-dependent inactivation manifested as the slow partial decline in outward exchange current upon application of Na^+ to the intracellular surface, (2) secondary Ca^{2+} regulation seen as the requirement for low micromolar levels of nontransported Ca^{2+} at the intracellular surface for activation of exchange current. Mutations can affect each of these properties giving insight into the regions of the molecule involved in regulation.

The large intracellular loop of the exchanger was first implicated in regulation when a large deletion was made in the molecule (Matsuoka et al., 1993). The loop is modeled to comprise 520 amino acids, and an exchanger $\Delta240$–679) was constructed in which 440 of these residues were deleted. Surprisingly, mutant $\Delta240$–679 still demonstrated

transport function. Clearly, the large loop is not essential for exchange activity. However, Na^+-dependent inactivation and Ca^{2+} regulation were all lost after loop deletion. This suggested that the loop was involved in exchanger regulation.

Subsequently, a less drastic deletion of 124 amino acids was introduced into the exchanger ($\Delta562$–685). Interestingly, this deletion mutant retained Na^+-dependent inactivation kinetics though Ca^{2+} regulation was lost (Matsuoka et al., 1993). A small deletion ($\Delta680$–685) within this region (Fig. 11.7) was also Ca^{2+}-insensitive (L. V. Hryshko, D. A. Nicoll, and K. D. Philipson, unpublished observations). Although it turned out that the Ca^{2+}-binding site is not within amino acids 562–685, these studies provided the impetus which led to the identification of the Ca^{2+}-regulatory site.

The regulatory Ca^{2+}-binding site on the exchanger was localized by Levitsky et al. (1994) using the

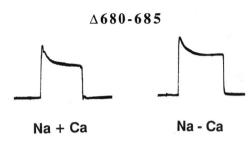

Δ680-685

Na + Ca   **Na - Ca**

Figure 11.7 Outward Na^+–Ca^{2+} exchange currents generated by deletion mutant Δ680-685. The deletion of six amino acids in mutant Δ680–685 eliminates Ca^{2+} regulation. The current on the left was initiated and then terminated by application of bath Na^+ in the continuous presence of regulatory Ca^{2+}. An identical current is obtained if Na^+ is added and then removed in the absence of regulatory Ca^{2+}. Note that, although Ca^{2+} regulation is eliminated, the Na^+-dependent inactivation process is still present.

$^{45}Ca^{2+}$ overlay technique. A series of fusion proteins was constructed comprising different regions of the intracellular loop. The fusion proteins were analyzed for Ca^{2+} binding and it was determined that the Ca^{2+}-binding site resided within amino acids 371–525. Mutations of aspartate residues within two acidic clusters of this region decreased the Ca^{2+} affinity of this binding site. The study identified a Ca^{2+}-binding site on the large intracellular loop but did not indicate whether this site was the functionally relevant regulatory site.

Subsequently Matsuoka et al. (1995) examined whether mutations which affected $^{45}Ca^{2+}$ binding also affected Ca^{2+} regulation. Any mutation which had decreased the affinity of the binding site for $^{45}Ca^{2+}$ also decreased the apparent Ca^{2+} affinity for regulation. These data demonstrated that the Ca^{2+}-regulatory site had indeed been identified. Interestingly, mutants with altered binding of regulatory Ca^{2+} also had somewhat decreased Na^+-dependent inactivation, indicating that these two regulatory processes are not independent phenomena.

An "anomalous" Ca^{2+}-regulatory mechanism (Hryshko et al., 1996) is present in a Na^+–Ca^2 exchanger cloned from *Drosophila* (Schwarz and Benzer, 1997; Ruknudin et al., 1997). In this case, exchange activity is inhibited by regulatory Ca^{2+} instead of being stimulated. In excised patches, full activity is present for outward exchange currents (reverse exchange) with EGTA at the intracellular surface, but activity is almost abolished when the regulatory Ca^{2+} is raised to $1 \mu M$. This is the exact opposite of the response of NCX1 to regulatory Ca^{2+}. The physiological significance of this mechanism is unclear though the contrast in the regulatory responses of different exchangers is striking.

Mutational analysis of the endogenous XIP region (Fig. 11.4) of the exchanger has implicated this region in the Na^+-dependent inactivation process (Matsuoka et al., 1997). Although it had been speculated that the endogenous XIP region has an autoregulatory role (Li et al., 1991), direct support for this hypothesis had previously been lacking. The proper-

ties of Na^+-dependent inactivation are sensitive to any alteration of the XIP region and in some cases the inactivation process is completely eliminated. Elimination of the inactivation alters some of the kinetics of the secondary Ca^{2+} regulation, again demonstrating that the two processes are related. Thus, it appears that the endogenous XIP region is directly involved in exchanger regulatory properties.

Perspectives

Interest in Na^+–Ca^{2+} exchange has been growing. Many questions about Na^+–Ca^{2+} exchange are unresolved but progress should now be rapid and new issues will arise. Some molecular advances can now be applied to address physiological questions. Use of antisense technology (Lipp et al., 1995) and transgenic mice (Adachi-Akahane et al., 1997) has begun. Unresolved issues include the physiological significance of different exchanger genes and splice variants. The topology of the exchanger needs to be determined as a first step towards structural models. The difficulties of unraveling the workings of a membrane transporter are a challenge for the future.

Acknowledgments I thank Drs. A. Doering, Z. He, D. Nicoll, B. Quednau, L. Santacruz-Toloza, and J. Weiss for helpful comments on this review.

References

Adachi-Akahane, S., Lu, L., Li, Z., Frank, J. S., Philipson, K. D., and Morad, M. (1997) Calcium signaling in transgenic mice overexpressing cardiac Na^+–Ca^{2+} exchanger. *J. Gen. Physiol.* 109: 717–729.

Allen, T. J. A., Noble, D., and Reuter, H. (1989) *Sodium–Calcium Exchange*. Oxford University Press, Oxford.

Artman, M. (1992) Sarcolemmal Na^+–Ca^{2+} exchange activity and exchanger immunoreactivity in developing rabbit hearts. *Am. J. Physiol.* 263: H1506–H1513.

Ashida, T. and Blaustein, M. P. (1987) Regulation of cell calcium and contractility in mammalian arterial smooth muscle: the role of sodium–calcium exchange. *J. Physiol.* 392: 617–635.

Baker, P. F., Blaustein, M. P., Hodgkin, A. L., and Steinhardt, R. A. (1969) The influence of calcium on sodium efflux in squid axons. *J. Physiol.* 200: 431–458,.

Bers, D. M. (1991) *Excitation–Contraction Coupling and Cardiac Contractile Force.* Kluwer, Boston.

Blaustein, M. P., DiPolo, R., and Reeves, J. P., eds (1991) *Sodium–Calcium Exchange: Proceedings of the Second International Conference.* New York Academy of Sciences, New York.

Boerth, S. R., Zimmer, D. B., and Artman, M. (1994) Steady-state mRNA levels of the sarcolemmal Na^+–Ca^{2+} exchanger peak near birth in developing rabbit and rat hearts. *Circ. Res.* 74: 354–359.

Bourdeau, J. E., Taylor, A.N., and Iacopino, A. M. (1993) Immunocytochemical localization of sodium–calcium exchanger in canine nephron. *J. Am. Soc. Nephrol.* 4: 105–110.

Chen, F. H., Mottino, G., Klitzner, T. S., Philipson, K. D., and Frank, J. S. (1995) Distribution of the Na^+/Ca^{2+} exchange protein in developing rabbit myocytes. *Am. J. Physiol.* 37: C1126–C1132.

Condrescu, M., Gardner, J. P., Chernaya, G., Aceto, J. F., Kroupis, C., and Reeves, J. P. (1995) ATP-dependent regulation of sodium–calcium exchange in Chinese hamster ovary. *J. Biol. Chem.* 270: 9137–9146.

Cook, N. J. and Kaupp, U. B. (1988) Solubilization, purification, and reconstitution of the sodium–calcium exchanger from bovine retinal rod outer segments. *J. Biol. Chem.* 263: 11382–11388.

Crompton, M. (1990) The role of Ca^{2+} in the function and dysfunction of heart mitochondria. In: Langer, G. A. ed.). *Calcium and the Heart.* Raven, New York. pp. 167–198.

Cunningham, K. W. and Fink, G. R. (1996) Calcineurin inhibits *VCX1*-dependent H^+/Ca^{2+} exchange and induces Ca^{2+} ATPases in *Saccharomyces cerevisiae*. *Mol. Cell. Biol.* 2226–2237.

DiPolo, R., and Beaugé, L. (1991) Regulation of Na-Ca exchange: an overview. In: *Sodium–Calcium Exchange: Proceedings of the Second International Conference* (Blaustein, M. P., DiPolo, R., and Reeves, J. P., eds.). New York Academy of Sciences, New York, pp. 100–111.

Frank, J. S., Mottino, G., Reid, D., Molday, R. S., and Philipson, K. D. (1992) Distribution of the Na^+–Ca^{2+} exchange protein in mammalian cardiac myocytes — an immunofluorescence and immunocolloidal gold labeling study. *J. Cell Biol.* 117: 337–345.

Furman, I., Cook, O., Kasir, J., and Rahamimoff, H. (1993) Cloning of two isoforms of the rat brain Na^+–Ca^{2+} exchanger gene and their functional expression in HeLa cells. *FEBS Lett.* 319: 105–109.

Furman, I., Cook, O., Kasir, J., Low, W., and Rahamimoff, H. (1995) The putative amino-terminal signal peptide of the cloned rat brain Na^+–Ca^{2+} exchanger gene

(Rbe-1) is not mandatory for functional expression. *J. Biol. Chem.* 270: 19120–19127.

Gabellini, N., Iwata, T., and Carafoli, E. (1995) Alternative splicing site modifies the carboxyl-terminal trans-membrane domains of the Na^+/Ca^{2+} exchanger. *J. Biol. Chem.* 270: 6917–6924.

Gabellini, N., Zatti, A., Rispoli, G., Navangione, A., and Carafoli, E. (1996) Expression of an active Na^+–Ca^{2+} exchanger isoform lacking the six C-terminal transmembrane segments. *Eur. J. Biochem.* 239: 897–904.

He, Z., Tong, Q., Quednau, B. D., Philipson, K. D., and Hilgemann, D. W. (1998) Cloning, expression, and characterization of the squid Na^+–Ca^{2+} exchanger (NCX-5Q1). *J. Gen. Physiol.* 111: 857–863.

He, Z., Petesch, N., Voges, K.-P., Röben, W., and Philipson, K. D. (1997) Identification of important amino acid residues of the Na^+–Ca^{2+} exchanger inhibitory peptide, XIP. *J. Memb. Biol.* 156: 149–156.

Hilgemann, D. W. (1989) Giant excised cardiac sarcolemmal membrane patches — sodium and sodium–calcium exchange currents. *Pflug. Arch.* 415: 247–249.

Hilgemann, D. W. (1990) Regulation and deregulation of cardiac Na^+–Ca^{2+} exchange in giant excised sarcolemmal membrane patches. *Nature* 344: 242–245.

Hilgemann, D. W. (1996) Unitary cardiac Na^+, Ca^{2+} exchange current magnitudes determined from channel-like noise and charge movements of ion transport. *Biophys. J.* 71: 759–768.

Hilgemann, D. W. and Ball, R. (1996) Regulation of cardiac Na^+, Ca^{2+} exchange and K_{ATP} potassium channels by PIP_2. *Science* 273: 956–960.

Hilgemann, D. W. and Collins, A. (1992) Mechanism of cardiac Na^+–Ca^{2+} exchange current stimulation by MgATP — possible involvement of aminophospholipid translocase. *J. Physiol.* 454: 59–82.

Hilgemann, D.W., Nicoll. D. A., and Philipson, K. D. (1991) Charge movement during Na^+ translocation by native and cloned cardiac Na^+/Ca^{2+} exchanger. *Nature* 352: 715–718.

Hilgemann, D.W., Collins, A., and Matsuoka, S. (1992a) Steady-state and dynamic properties of cardiac sodium-calcium exchange — secondary modulation by cytoplasmic calcium and ATP. *J. Gen. Physiol.* 100: 933–961.

Hilgemann, D.W., Matsuoka, S., Nagel, G. A., and Collins, A. (1992b) Steady-state and dynamic properties of cardiac sodium–calcium exchange — sodium-dependent inactivation. *J. Gen. Physiol.* 100: 905–932.

Hilgemann, D.W., Philipson, K. D., and Vassort, G. eds. (1995) *Sodium-Calcium Exchange: Proceedings of the Third International Conference.* New York Academy of Sciences, New York.

Hirschi, K.D., Zhen, R.-G., Cunningham, K. W., Rea, P. A., and Fink, G. R. (1996) CAX1, an H^+/Ca^{2+} antiporter from *Arabidopsis. Proc. Natl. Acad. Sci. USA* 93: 8782–8786.

Hryshko, L.V., Nicoll, D. A., Weiss, J. N., and Philipson, K. D. (1993) Biosynthesis and initial processing of the cardiac sarcolemmal Na^+–Ca^{2+} exchanger. *Biochim. Biophys. Acta* 1151: 35–42.

Hryshko, L.V., Matsuoka, S., Nicoll, D. A., Weiss, J. N., Schwarz, E. M., Benzer, S., and Philipson, K. D. (1996) Anomalous regulation of the *Drosophila* Na^+–Ca^{2+} exchanger by Ca^{2+}. *J. Gen. Physiol.* 108: 67–74.

Iwamoto, T., Wakabayashi, S., and Shigekawa, M. (1995) Growth factor-induced phosphorylation and activation of aortic smooth muscle Na^+–Ca^{2+} exchanger. *J. Biol. Chem.* 270: 8996–9001.

Iwamoto, T., Pan, Y., Wakabayashi, S., Imagawa, T., Yamanaka, H. I., and Shigekawa, M. (1996a) Phosphorylation-dependent regulation of cardiac Na^+/Ca^{2+} exchanger via protein kinase C. *J. Biol. Chem.* 271: 13609–13615.

Iwamoto, T., Watano, T., and Shigekawa, M. (1996b) A novel isothiourea derivative selectively inhibits the reverse mode of Na^+/Ca^{2+} exchange in cells expressing NCX1. *J. Biol. Chem.* 271: 22391–22397.

Iwata, T., Galli, C., Dainese, P., Guerini, D., and Carafoli, E. (1995) The 70 kD component of the heart sarcolemmal Na^+/Ca^{2+}-exchanger preparation is the C-terminal portion of the protein. *Cell Calcium* 17: 263–269.

Juhaszova, M., Ambesi, A., Lindenmayer, G. E., Bloch, R. J., and Blaustein, M. P. (1994) Na^+–Ca^{2+} exchanger in arteries — identification by immunoblotting and immunofluorescence microscopy. *Am. J. Physiol.* 266: C234–C242.

Jung, D. W., Baysal, K., and Brierley, G. P. (1995) The sodium-calcium antiport of heart mitochondria is not electroneutral. *J. Biol. Chem.* 270: 672–678.

Kaczorowski, G.J., Slaughter, R. S., King, V. F., and Garcia, M. L. (1989) Inhibitors of sodium–calcium exchange: identification and development of probes of transport activity. *Biochim. Biophys. Acta* 988: 287–302.

Kappl, M. and Hartung, K. (1996) Rapid charge translocation by the cardiac Na^+–Ca^{2+} exchanger after a Ca^{2+} concentration jump." *Biophys. J.* 71: 2473–2485.

Kent, R.L., Rozich, J. D., McCollam, P. L., McDermott, D. E., Thacker, U. F., Menick, D. R., McDermott, P. J., and Cooper, G. (1993) Rapid expression of the Na^+–Ca^{2+} exchanger in response to cardiac pressure overload. *Am. J. Physiol.* 265: H1024–H1029.

Khananshvili, D. (1990) Distinction between the 2 basic mechanisms of cation transport in the cardiac Na^+–Ca^{2+} exchange system. *Biochemistry* 29: 2437–2442.

Khananshvili, D., Price, D. C., Greenberg, M. J., and Sarne, Y. (1993) Phe-Met-Arg-Phe-NH_2 (FMRFa)-related peptides inhibit Na^+–Ca^{2+} exchange in cardiac sarcolemma vesicles. *J. Biol. Chem.* 268: 200–205.

Khananshvili, D., Shaulov, G., Weilmaslansky, E., and Baazov, D. (1995) Positively charged cyclic hexapeptides, novel blockers for the cardiac sarcolemma Na^+–Ca^{2+} exchanger. *J. Biol. Chem.* 270: 16182–16188.

Kieval, R. S., Block, R. J., Lindenmayer, G. E., Ambesi, A., and Lederer, W. J. (1992) Immunofluorescence localization of the Na^+–Ca^{2+} exchanger in heart cells. *Am. J. Physiol.* 263: C545–C550.

Kofuji, P., Lederer, W. J.., and Schulze, D. H. (1994) Mutually exclusive and cassette exons underlie alternatively spliced isoforms of the Na/Ca exchanger. *J. Biol. Chem.* 269: 5145–5149.

Kraev, A., Chumakov, I., and Carafoli. E. (1996) The organization of the human gene NCX1 encoding the sodium–calcium exchanger." *Genomics* 37:105–112.

Langer, G. A. and Peskoff, A. (1996) Calcium concentration and movement in the diadic cleft space of the cardiac ventricular cell. *Biophys. J.* 70: 1169–1182.

Leblanc, N. and Hume, J. R. (1990) Sodium current induced release of calcium from cardiac sarcoplasmic reticulum. *Science* 248: 372–376.

Lederer, W. J., Niggli, E., and Hadley, R. W. (1990) Sodium-calcium exchange in excitable cells — fuzzy space. *Science* 248: 283.

Lee, S. L., Yu, A. S. L., and Lytton, J. (1994) Tissue-specific expression of Na^+–Ca^{2+} exchanger isoforms. *J. Biol. Chem.* 269: 14849–14852.

Levi, A. J., Spitzer, K. W., Kohmoto, O., and Bridge, J. H. B. (1994) Depolarization-induced Ca entry via Na–Ca exchange triggers SR release in guinea pig cardiac myocytes. *Am. J. Physiol.* 226: H1422–H1433.

Levitsky, D. O., Nicoll, D. A., and Philipson, K. D. (1994) Identification of the high affinity Ca^{2+}-binding domain of the cardiac Na^+–Ca^{2+} exchanger. *J. Biol. Chem.* 269: 22847–22852.

Li, Z., Nicoll, D. A., Collins, A., Hilgemann, D. W., Filoteo, A. G., Penniston, J. T., Weiss, J. N., Tomich, J. M., and Philipson, K. D. (1991) Identification of a peptide inhibitor of the cardiac sarcolemmal Na^+–Ca^{2+} exchanger. *J. Biol. Chem.* 266: 1014–1020.

Li, W. H., Shariatmadar, Z., Powers, M., Sun, X. C., Lane, R. D., and Garlid, K. D. (1992) Reconstitution, identification, purification, and immunological characterization of the 110-kDa Na^+/Ca^{2+} antiporter from beef heart mitochondria. *J. Biol. Chem.* 267: 17983–17989.

Li, Z., Burke, E. P., Frank, J. S., Bennett, V., and Philipson, K. D. (1993) The cardiac Na^+-Ca^{2+} exchanger binds to the cytoskeletal protein ankyrin. *J. Biol. Chem.* 268: 11489–11491.

Li, Z., Matsuoka, S., Hryshko, L. V., Nicoll, D. A., Bersohn, M. M., Burke, E. P., Lifton, R. P., and Philipson, K. D. (1994) Cloning of the NCX2 isoform of the plasma membrane Na^+–Ca^{2+} exchanger. *J. Biol. Chem.* 269: 17434–17439.

Lipp, P. and Niggli, E. (1994) Sodium current-induced calcium signals in isolated guinea-pig ventricular myocytes. *J. Physiol.* 474: 439–446.

Lipp, P., Schwaller, B., and Niggli, E. (1995) Specific inhibition of Na–Ca exchange function by antisense oligodeoxynucleotides. *FEBS Lett.* 364: 198–202.

Loo, T.W., Ho, C., and Clarke, D. M. (1995) Expression of a functionally active human renal sodium–calcium exchanger lacking a signal sequence. *J. Biol. Chem.* 270: 19345–19350.

Luther, P.W., Yip, R. K., Block, R. J., Ambesi, A., Lindenmayer, G. E., and Blaustein, M. P. (1992) Presynaptic localization of sodium calcium exchangers in neuromuscular preparations. *J. Neurosci.* 12: 4898–4904.

Matsuoka, S., Nicoll, D. A., Reilly, R. F., Hilgemann, D. W., and Philipson, K. D. Initial localization of regulatory regions of the cardiac sarcolemmal Na^+–Ca^{2+} exchanger. *Proc. Natl. Acad. Sci. USA* 90: 3870–3874.

Matsuoka, S., Nicoll, D. A., Hryshko, L. V., Levitsky, D. O., Weiss, J. N., and Philipson, K. D. (1995) Regulation of the cardiac Na^+-Ca^{2+} exchanger by Ca^{2+}-mutational analysis of the Ca^{2+}-binding domain. *J. Gen. Physiol.* 105: 403–420.

Matsuoka, S., Nicoll, D. A., He, Z., and Philipson, K. D. (1997) Regulation of the cardiac Na^+–Ca^{2+} exchanger by the endogenous XIP region. *J. Gen. Physiol.* 109: 273–286.

Moore, E. D. W., Etter, E. F., Philipson, K. D., Carrington, W. A., Fogarty, K. E., Lifshitz, L. M., and Fay, F. S. (1993) Coupling of the Na^+/Ca^{2+} exchanger, Na^+/K^+ pump and sarcoplasmic reticulum in smooth muscle. *Nature* 365: 657–660.

Nakasaki, Y., Iwamoto, T., Hanada, H., Imagawa, T., and Shigekawa, M. (1993) Cloning of the rat aortic smooth muscle Na^+/Ca^{2+} exchanger and tissue-specific expression of isoforms. *J. Biochem.* 114: 528–534.

Nicoll, D. A. and Applebury, M. L. (1989) Purification of the bovine rod outer segment Na^+/Ca^{2+} exchanger. *J. Biol. Chem.* 264: 16207–16213.

Nicoll, D. A. and Philipson, K. D. (1991) Molecular studies of the cardiac sarcolemmal sodium–calcium exchanger. In *Sodium–Calcium Exchange: Proceedings of the Second International Conference.* (Blaustein, M. P., DiPolo, R., and Reeves, J. P., eds.). New York Academy of Sciences, New York. pp. 181–188.

Nicoll, D. A., Longoni, S., and Philipson, K. D. (1990) Molecular cloning and functional expression of the cardiac sarcolemmal Na^+–Ca^{2+} exchanger. *Science* 250: 562–565.

Nicoll, D. A., Hryshko, L. V., Matsuoka, S., Frank, J. S., and Philipson, K. D. (1995) Mutagenesis studies of the cardiac Na^+–Ca^{2+} exchanger. In: *Sodium–Calcium Exchange: Proceedings of the Third International Conference.* (Hilgemann, D. W., Philipson, K. D., and Vassort, G., eds.). New York Academy of Sciences, New York.

Nicoll, D. A., Hryshko, L. V., Matsuoka, S., Frank, J. S., and Philipson, K. D. (1996a) Mutation of amino acid residues in the putative transmembrane segments of the cardiac sarcolemmal Na^+–Ca^{2+} exchanger. *J. Biol. Chem.* 271: 13385–13391.

Nicoll, D. A., Quednau, B., Qin, Z., Xia, Y.-R., Lusis, A. J., and Philipson, K. D. (1996b) Cloning of a third mammalian Na^+–Ca^{2+} exchanger, NCX3. *J. Biol. Chem.* 271: 24914–24921.

Niggli, E. and Lederer, W. J. (1991) Molecular operations of the sodium calcium exchanger revealed by conformation currents. *Nature* 349: 621–624.

Philipson, K. D. (1990) The cardiac Na^+–Ca^{2+} exchanger —dependence on membrane environment. *Cell Biol. Intl. Rep.* 14: 305–309.

Philipson, K.D. (1996) The Na^+–Ca^{2+} exchanger: molecular aspects. In *Molecular Physiology and Pharmacology of Cardiac Ion Channels and Transporters.* (Morad, M., ed.). Kluwer, Boston.

Philipson, K. D. and Nicoll, D. A. (1993) Molecular and kinetic aspects of sodium–calcium exchange. In *Molecular Biology of Receptors and Transporters: Pumps, Transporters, and Channels.* (Friedlander, M. and Mueckler, M. eds.). Academic Press, New York. pp. 199–227.

Philipson, K. D., Longoni, S., and Ward, R. (1988) Purification of the cardiac Na^+-Ca^{2+} exchange protein. *Biochim. Biophys. Acta* 945: 298-306.

Porzig, H., Li, Z., Nicoll, D. A., and Philipson, K. D. (1993) Mapping of the cardiac sodium–calcium exchanger with monoclonal antibodies. *Am. J. Physiol.* 265: C748–C756.

Pozos, T.C., Sekler, I., and Cyert, M. S. (1996) The product of *HUM1*, a novel yeast gene, is required for vacuolar Ca^{2+}/H^+ exchange and is related to mammalian Na^+/Ca^{2+} exchangers. *Mol. Cell. Biol.* 16: 3730–3741.

Quednau, B. D., Nicoll, D. A., and Philipson, K. D. (1997) Tissue specificity and alternative splicing of the Na^+/Ca^{2+} exchanger isoforms NCX1, NCX2, and NCX3 in rat. *Am. J. Physiol.* 272: C1250–C1261.

Reeves, J. P. (1995) Cardiac sodium–calcium exchange system. In *Physiology and Pathophysiology of the Heart*, 3rd Ed. Kluwer, Boston, pp. 309–318.

Reeves, J. P. and Philipson, K. D. (1989) Sodium–calcium exchanger activity in plasma membrane vesicles. In *Sodium–Calcium Exchange* (Noble, D., Reuter, H., and Allen, T. J. A., eds.). Oxford University Press, Oxford, pp. 27–33.

Reeves, J. P. and Sutko, J. L. (1979) Sodium–calcium ion exchange in cardiac membrane vesicles. *Proc. Natl. Acad. Sci. USA* 76: 590–594.

Reiländer, H., Achilles, A., Friedel, U., Maul, G., Lottspeich, F., and Cook, N. J. (1992) Primary structure and functional expression of the Na/Ca, K-exchanger from bovine rod photoreceptors. *EMBO J.* 11: 1689–1695.

Reilly, R. F. and Lattanzi, D. (1995) Identification of a novel alternatively-spliced isoform of the Na–Ca exchanger (NACA7) in heart. In *Sodium-Calcium Exchange: Proceedings of the Third International Conference.* (Hilgemann, D. W., Philipson, K. D., and Vassort, G., eds.). New York Academy of Sciences, New York.

Reilly, R. F. and Shugrue, C. A. (1992) cDNA cloning of a renal Na^+–Ca^{2+} exchanger. *Am. J. Physiol.* 262: F1105–F1109.

Reilly, R. F., Shugrue, C. A., Lattanzi, D., and Biemesderfer, D. (1993) Immunolocalization of the Na^+/Ca^{2+} exchanger in rabbit kidney. *Am. J. Physiol.* 265: F327–F332.

Reuter, H. and Seitz, N. (1968) The dependence of calcium efflux from cardiac muscle on temperature and external ion composition. *J. Physiol.* 195: 451–470.

Ruknudin, A., Valdivia, C., Kofuji, P., Lederer, W. J., and Schulze, D. H. (1997) Na^+/Ca^{2+} exchanger in *Drosophila*: cloning expression, and transport differences. *Am. J. Physiol.* 273: C257–C265.

Sahin-Toth, M., Nicoll, D. A., Frank, J. S., Philipson, K. D., and Friedlander, M. (1995) The cleaved N-terminal signal sequence of the cardiac Na^+–Ca^{2+} exchanger is not required for functional membrane integration. *Biochem. Biophys. Res. Commun.* 212: 968–974.

Sanchez-Armass, S. and Blaustein, M. P. (1987) Role of sodium-calcium exchange in regulation of intracellular calcium in nerve terminals. *Am. J. Physiol.* 252: C595–C603.

Schnetkamp, P. P. M. (1995) How does the retinal rod Na-Ca, K exchanger regulate cytosolic free Ca^{2+}. *J. Biol. Chem.* 270: 13231–13239.

Schnetkamp, P. P. M. and Szerencsei, R. T. (1993) Intracellular Ca^{2+} sequestration and release in intact bovine retinal rod outer segments—role in inactivation of Na-Ca, K exchange. *J. Biol. Chem.* 268: 12449–12457.

Schwarz, E. M. and Benzer, S. (1997) Calx, a Na–Ca exchanger gene of *Drosophila melanogaster. Proc. Natl. Acad. Sci. USA* 94: 10249–10254.

Sham, J. S. K., Cleemann, L., and Morad, M. (1992) Gating of the cardiac Ca^{2+} release channel—the role of Na^+ current and Na^+–Ca^{2+} exchange. *Science* 255: 850–853.

Sham, J. S. K., Cleemann, L., and Morad, M. (1995) Functional coupling of Ca^{2+} channels and ryanodine receptors in cardiac myocytes. *Proc. Natl. Acad. Sci. USA* 92: 121-125.

Shannon, T. R., Hale, C. C., and Milanick, M. A. (1994) Interaction of cardiac Na–Ca exchanger and exchange inhibitory peptide with membrane phospholipids. *Am. J. Physiol.* 266: C1350–C1356.

Tsuruya, Y., Bersohn, M. M., Li, Z., Nicoll, D. A., and Philipson, K. D. (1994) Molecular cloning and functional expression of the guinea pig cardiac Na^+–Ca^{2+} exchanger. *Biochim. Biophys. Acta* 1196: 97–99.

Watano, T., Kimura, J,. Morita, T., and Nakanishi, H., (1996) A novel antagonist, No. 7943, of the Na^+/Ca^{2+} exchange current in guinea-pig cardiac ventricular cells. *Brit. J. Pharmacol.* 119: 555–563.

12

The Plasma Membrane Calcium Sensor

Edward M. Brown
Stephen M. Quinn
Peter M. Vassilev

Calcium ions (Ca^{2+}) perform numerous essential functions both extra- and intracellularly. Cytosolic calcium (Ca^{2+}_i) serves as a key intracellular second messenger and as a cofactor for a variety of enzymes and other processes (Berridge and Irvine, 1984; Pietrobon et al., 1990), as discussed in detail elsewhere in this volume. In this capacity, Ca^{2+}_i regulates a myriad of processes, such as cellular proliferation, differentiation, and motility, as well as muscular contraction, glycogen metabolism, and hormonal secretion (Berridge and Irvine, 1984; Pietrobon et al., 1990). The resting level of Ca^{2+}_i is generally ~ 100 nanomolar (nM), 10,000-fold below the ionized calcium concentration (Ca^{2+}_o) (~ 1 mM) in the extracellular fluid (ECF). Moreover, it undergoes large and rapid fluctuations in the course of carrying out its intracellular signaling functions, owing to the release of calcium ions from intracellular stores and/or influx through the plasma membrane (Berridge and Irvine, 1984). Ca^{2+}_o, on the other hand, remains remarkably constant, varying by only a few percent over the course of a day or even much of a lifetime (Aurbach et al., 1985; Brown, 1991; Stewart and Broadus, 1987).

The stability of Ca^{2+}_o is essential because, like Ca^{2+}_i, it participates in a large variety of vitally important functions, including regulation of neuromuscular excitability, promotion of blood clotting, and maintenance of skeletal integrity (Brown, 1991; Stewart and Broadus, 1987). Because of its near constancy, Ca^{2+}_o has not generally been thought to serve as an extracellular messenger. The maintenance of mineral ion homeostasis, however, requires that parathyroid, kidney, and other cells involved in regulating Ca^{2+}_o be capable of coordinating their activities by directly sensing (i.e., recognizing and responding to) Ca^{2+} as an extracellular signal (Brown, 1991). In addition, while the serum or

blood ionized calcium concentration is nearly constant, several other compartments of the extracellular fluid, such as the urine or the brain ECF, can undergo considerably larger changes in their levels of Ca^{2+}_o (Brown, 1991), as noted in more detail later. The cloning of a G-protein-coupled, Ca^{2+}_o-sensing receptor (CaR) (Brown, et al., 1993), as well as the identification of inherited diseases of Ca^{2+}_o-sensing resulting from mutations in the CaR (Brown, et al., 1995), have provided strong support for the concept that Ca^{2+}_o has important functions in extracellular signaling as a first messenger. This chapter will summarize our current state of knowledge concerning the role of the CaR as a Ca^{2+}_o sensor in various extracellular compartments.

Ca^{2+}_o Sensing and the Maintenance of Extracellular Calcium Homeostasis

The near invariance of Ca^{2+}_o is the result of a sophisticated homeostatic system comprising two key components (Fig. 12.1) (Aurbach et al., 1985; Brown, 1991; Stewart and Broadus, 1987). The first are Ca^{2+}_o-sensing cells, such as those in the parathyroid gland, thyroid gland (e.g., its calcitonin-secreting C-cells) (Brown, 1991), and renal proximal tubule, that synthesize and secrete calcitropic hormones. In response to perturbations in Ca^{2+}_o, these cells modify their release of parathyroid hormone (PTH), calcitonin (CT), and 1,25-dihydroxyvitamin D [1,25(OH)$_2$D], respectively, in a manner designed to normalize Ca^{2+}_o. They do so by acting on the second important element of the homeostatic system, the effector tissues (i.e., bone, intestine, and kidney) that modify their transport of calcium and phosphate ions into or out of the ECF (Fig. 12.1).

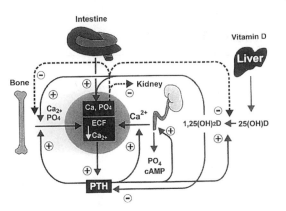

Figure 12.1 Diagram showing the operation of the system maintaining mineral ion homeostasis, as well as the manner in which calcium acts as an extracellular messenger (dashed lines), similar to the more classical calciotropic hormones, PTH and 1,25-dihydroxyvitamin D (solid lines). Abbreviations: Ca^{2+}, calcium; PO_4, phosphate; ECF, extracellular fluid; PTH, parathyroid hormone; $1,25(OH)_2D$, 1,25-dihydroxyvitamin D; $25(OH)D$, 25-hydroxyvitamin D; the plus signs represent positive actions and the minus signs show inhibitory effects. [Modified with permission from Brown, E. M., Pollak, M., and Hebert, S. C. (1994) Cloning and characterization of extracellular Ca^{2+}-sensing receptors from parathyroid and kidney: molecular physiology and pathophysiology of Ca^{2+}-sensing. *The Endocrinologist* 4: 419-426.]

The calciotropic hormones secreted by Ca^{2+}_o-sensing cells regulate their target tissues in a receptor-dependent fashion that has been studied intensely over the past two decades (Aurbach et al., 1985; Stewart and Broadus, 1987). Indeed, the receptors for all three hormones have been cloned within the past decade (Bringhurst et al., 1993; Gorn et al., 1992; Whitfield et al., 1995). The mechanistic basis for Ca^{2+}_o-sensing, on the other hand, has been poorly understood until much more recently. The actions of Ca^{2+}_o on these cells in earlier, indirect studies had suggested that Ca^{2+}_o-sensing might involve a cell surface receptor that functioned similarly to the G-protein-coupled receptors (GPCRs) for the so-called "Ca^{2+}-mobilizing" hormones (Brown, 1991; Juhlin et al., 1990; Nemeth and Scarpa, 1986; Shoback et al., 1988). For instance, raising the level of Ca^{2+}_o activates phospholipase C (PLC) (Kifor et al., 1992), with accompanying transient followed by sustained increases in Ca^{2+}_i (Nemeth and Scarpa, 1986), in bovine parathyroid cells as well as a pertussis toxin-sensitive inhibition of adenylate cyclase (Chen et al., 1989), much like the actions of angiotensin II that are mediated via the AT-1 receptor in its target tissues (Griendling and Alexander, 1993). Ca^{2+}_o also directly modulates the functions of a number of additional cell types, particularly those within the kidney (for review, see Brown, 1991). Ca^{2+}_o decreases glomerular filtration rate (Humes et al., 1978), reduces renal blood flow (Edvall, 1958), and inhibits the actions of vasopressin on both the medullary thick ascending limb of Henle's loop (MTAL) and the collecting duct (Jones et al., 1988; Suki et al., 1969). In addition, elevated levels of peritubular but not luminal Ca^{2+}_o or Mg^{2+}_o reduce the reabsorption of both of these divalent cations in the thick ascending limb (TAL) (Quamme, 1989). The actions of Ca^{2+}_o on renal function are, in some cases, reminiscent of those of Ca^{2+}_o on parathyroid cells, suggesting a similar mechanism of action. For example, high Ca^{2+}_o produces a pertussis toxin-sensitive inhibition of vasopressin-elicited cAMP accumulation in the TAL (Takaichi and Kurokawa, 1986, 1988). Thus, it appeared that certain cells in the kidney might also express Ca^{2+}_o-sensing receptors which could regulate renal function.

Isolation of cDNAs Encoding Parathyroid and Kidney Ca^{2+}_o-Sensing Receptors

We utilized *Xenopus laevis* oocytes as a means for screening a cDNA library constructed from bovine parathyroid mRNA (Brown et al., 1993), since expression of G-protein-coupled, Ca^{2+}-mobilizing receptors in *X. laevis* oocytes produces agonist-dependent stimulation of a Ca^{2+}-activated chloride channel that provides a convenient readout of the biologically active receptor (Chen et al., 1994; Racke et al., 1993). The full-length cDNA clone of the Ca^{2+}_o-sensing receptor, BoPCaR (Bovine Parathyroid Ca^{2+}_o-sensing Receptor) (Brown et al., 1993), isolated by this approach enabled subsequent cloning by hybridization-based techniques of additional CaRs from human parathyroid (Garrett et al., 1995a) and kidney (Aida et al., 1995b) as well as from rat brain (Ruat et al., 1995) and kidney (Riccardi et al., 1995) and a rat C-cell tumor line (Garrett et al., 1995c). All share greater than 90% identity at the amino acid level and represent the various species homologs of a common ancestral gene.

The CaRs that have been studied to date show similar pharmacological profiles, with an order of potency for CaR agonists (Gd^{3+} > neomycin $\gg$ Ca^{2+} > Mg^{2+}) that is the same as that for the

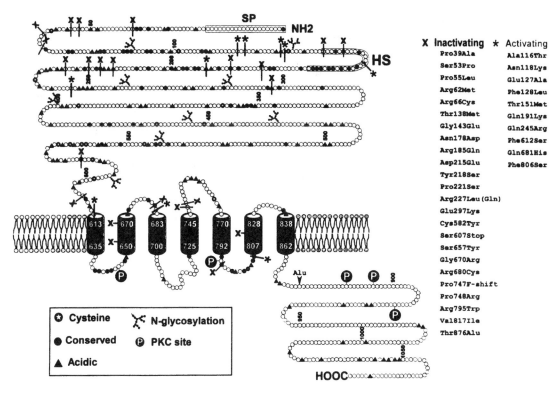

Figure 12.2 Schematic diagram showing the major structural features of the predicted human CaR protein. Symbols are given in the key. Abbreviations: SP, predicted signal peptide; HS, hydrophobic segment; NH2, amino-terminus; HOOC, carboxy-terminus; PKC, protein kinase C. The diagram also illustrates the mutations that have been described in the disorders familial hypocalciuric hypercalcemia (FHH) and autosomal dominant hypocalcemia (ADH) using the three-letter code for amino acid residues (see text for details). [Reproduced with permission from Brown, E. M., Bai, M., and Pollak, M. R. (1997) Familial benign hypocalciuric hypercalcemia and other syndromes of altered responsiveness to extracellular calcium. In *Metabolic Bone Diseases and Clinically Related Disorders*, 3rd ed. (Krane, S. and Avioli, L. eds.). Academic Press, San Diego, 479–499.]

responses of bovine parathyroid cells to the same polycationic agonists (Brown et al., 1990; Ridefelt et al., 1992). The cloned CaRs activate PLC (Brown et al., 1993) and PLA$_2$ (although it is not clear whether the latter action is a direct or indirect one) (Ruat et al., 1996) and probably also inhibit adenylate cyclase (Brown et al., 1993; Chang et al., 1998). Therefore, a single isoform of the receptor may mediate the actions of Ca^{2+}_o on both PI turnover and cAMP metabolism in parathyroid and kidney cells, similar to the actions of several other GPCRs, such as the AT-1 (Griendling and Alexander, 1993) and thrombin receptors (Coughlin, 1994). The activation of phospholipase C in parathyroid cells is pertussis toxin-insensitive (Hawkins et al., 1989) and likely involves a member of the G_q/G_{11} family of heterotrimeric G-proteins (Varrault et al., 1995), while the inhibition of adeny-

late cyclase (Chen et al., 1989) is probably mediated by one of the isoforms of G_i.

Deduced Structure and Receptor–Effector Coupling of the CaR.

The predicted amino acid sequence of the CaR has three principal domains (Fig. 12.2) (Brown et al., 1993): a large (>600 amino acid) amino-terminal extracellular domain (ECD) is followed by 250 amino acids containing seven putative, membrane-spanning helices, indicating that the CaR is a member of the superfamily of GPCRs. The third domain is a ~200 amino acid cytoplasmic, carboxy-terminal tail. Included within the intracellular loops and carboxy-terminal tail of the human (Garrett et al., 1995a) and bovine receptors (Brown et al., 1993; Chang et al., 1998) are five and four predicted protein kinase C

(PKC) phosphorylation sites, respectively. It is likely that these contribute to the apparent uncoupling of the CaR from activation of PLC following treatment of bovine parathyroid cells with activators of protein kinase C (Kifor et al., 1990; Nygren et al., 1988; Racke and Nemeth, 1993). Indeed, Bai et al. (1998) have recently shown that the PKC phosphorylation site at amino acid 888 in the human CaR is responsible for much of the uncoupling of this receptor from PLC induced by activators of PKC when expressed in human embryonic kidney (HEK293) cells.

Preliminary studies using chimeric receptors indicate that polycationic agonists likely bind to the ECD of the CaR (Hammerland et al., 1995). Among the GPCRs, the various species homologs of the CaR share significant homology in their amino acid sequences only with the metabotropic glutamate receptors (mGluRs) (Brown et al., 1993; Garrett et al., 1995a; Nakanishi 1992), which are receptors for glutamate, the major excitatory neurotransmitter in the central nervous system (Nakanishi, 1992). This has made it feasible to create chimeric receptors in which the extracellular domains of the CaRs or mGluRs are swapped (Hammerland et al., 1995). When a receptor is constructed containing the ECD of the CaR and the transmembrane and cytoplasmic domains of an mGluR, for example, the resultant chimeric receptor responds to CaR agonists and not to glutamate, formally demonstrating that the principal determinants of ligand specificity reside within the ECD. Although the identities of the amino acid residues within the extracellular domain to which calcium and other CaR ligands bind are not yet known, there are several highly acidic regions that could potentially represent binding sites. These acidic regions are reminiscent of those in other low-affinity, Ca^{2+}-binding proteins (e.g., calsequestrin) (Brown et al., 1993). Recent studies have shown that some isoforms of the mGluRs can also sense Ca^{2+}_o in addition to glutamate (Kubo et al., 1998).

Conklin and Bourne (1994) have suggested, based on putative homologies between the mGluRs and the bacterial periplasmic binding proteins (PBPs) pointed out by O'Hara et al., (1993), that the CaR is a hybrid molecule in which an extracellular "sensing" domain has been fused to a "serpentine" (i.e., seven membrane-spanning) signal-transducing motif. The PBPs sense and/or transport into bacteria extracellular ligands such as nutrients, polyvalent cations and anions (including phosphate and sulfate) (Adams and Oxender, 1989; Sharff et al., 1992). Thus, the CaR potentially may have evolved from an evolutionarily ancient family of cell surface receptors that play a key role in the sensing of a variety of ligands in the extracellular environment.

The CaR Gene and the Regulation of CaR Gene Expression

The CaR gene has at least seven exons, five of which code for the extracellular domain of the receptor (Pearce et al., 1995), while the most 3′ of the exons encode the remainder of the receptor (Pearce et al., 1995; Pollak et al., 1993), including all of the membrane-spanning helices as well as the carboxy-terminal tail. There is currently little known about the upstream regulatory regions of the gene, although several factors have been found to alter the level of CaR transcripts, potentially through changes in gene transcription. When bovine parathyroid cells are placed in culture, there is a precipitous drop in the level of CaR mRNA, with a half-time of less than 3 hours, while the level of the CaR protein falls more slowly ($t_{1/2} =\sim 30–36$ hours) (Mithal et al., 1995). Exposure of ACTH-secreting AtT-20 cells to high Ca^{2+}_o, in contrast, increases CaR mRNA levels about 2-fold (Emanuel et al., 1996). Similarly, administration of $1,25(OH)_2D$ to rats in vivo raised the level of CaR transcripts in parathyroid and kidney in one study (Brown et al., 1996), although the same vitamin D metabolite did not have this effect when given in a somewhat different manner in another study (Rogers et al., 1995). Additional work will be necessary to characterize in more detail the factors regulating the expression of the CaR gene and the mechanisms by which they do so.

Tissue Distribution of the CaR

Transcripts for the CaR are abundant in tissues that sense Ca^{2+}_o, including parathyroid, kidney, and C-cells (Brown et al., 1993; Garrett et al., 1995a; Riccardi et al., 1995, 1996; Ruat et al., 1995). Transcripts are also present in several other tissues, specifically lung, intestine, pituitary, and various regions of the brain [particularly hippocampus, cerebellum, prefrontal cortex, hypothalamus, and subfornical organ (SFO)] (Rogers et al., 1997; Ruat et al., 1995). Some of these tissues (i.e., the brain) are not traditionally thought to play any role in mineral ion metabolism. Rats, however, exhibit a specific "calcium appetite" (Tordoff, 1994). In addition, hypercalcemic individuals not infrequently exhibit excessive thirst (Aurbach et al., 1985; Stewart and Broadus, 1987), a symptom that could potentially be mediated by CaRs in the SFO, an important thirst center. As discussed below, ensuring adequate intake of water may be important to replace the loss of water resulting from hypercalcemia-evoked impairment of urinary concentrating ability. Other hypercalcemic symptoms, such as obtundation, poor memory, and reduced attention span (Aurbach et al., 1985), could likewise be the result of direct actions of Ca^{2+}_o on CaRs in various regions of the brain. Thus, there is much to be learned about the role of the CaR in the overall regulation of mineral ion metabolism, as well as the integration of this system with other homeostatic mechanisms (e.g., the control of water homeostasis), including possible behavioral aspects of these processes.

Intrarenal Localization of the CaR

The application of immunohistochemistry with specific, anti-CaR antibodies, in situ hybridization, and reverse transcription–polymerase chain reaction (RT-PCR) on identified, microdissected tubular segments has defined the intrarenal distribution of the CaR in some detail (Riccardi et al., 1995, 1996; 1998). Receptor transcripts are nearly ubiquitous, being located within the glomerulus, proximal tubule, TAL, distal convoluted tubule (DCT), cortical collecting duct (CCD), and inner medullary (IMCD) and papillary collecting ducts (Riccardi et al., 1996). The receptor is most heavily expressed in the CTAL, where it is localized principally on the basolateral surface of the cell (Riccardi et al., 1998). In the IMCD, on the other hand, the receptor protein is expressed on the apical surface of the epithelial cells (Sands et al., 1997), suggesting that it is positioned to respond to changes in Ca^{2+}_o within the tubular fluid rather than in the blood or systemic ECF, as in the CTAL, C-cell, and parathyroid. The CaR is co-expressed with the PTH receptor in CTAL and DCT (Riccardi et al., 1996), raising the possibility of mutually antagonistic interactions between the two receptors in the regulation of renal calcium handling, as described in more detail later.

Inherited Diseases of Ca^{2+} Homeostasis as Experiments in Nature Elucidating the Role of the CaR in Regulating its Target Tissues

The identification of inherited human diseases resulting from inactivating or activating mutations in the CaR has provided strong evidence in favor of a key role for the receptor in the regulation of Ca^{2+}_o (Brown et al., 1995). The clinical characteristics of these patients have also enabled further understanding of the manner in which the CaR controls key aspects of the function of several tissues involved in mineral ion homeostasis. Indeed, the clinical features of one such disorder, the hypercalcemic syndrome, familial hypocalciuric hypercalcemia (FHH), provided important clues to the possible existence of a shared Ca^{2+}_o-sensing mechanism in parathyroid gland and kidney a decade before the actual cloning of the CaR (Attie et al., 1983; Davies et al., 1984; Law and Heath, 1985; Marx et al., 1981a).

Individuals with FHH have mild to moderate hypercalcemia as a result of their failure to show appropriate suppression of circulating PTH levels within the upper part of the normal range of blood calcium concentration that is a crucial part of the negative feedback loop maintaining the normality of Ca^{2+}_o (Law and Heath, 1985; Marx et al., 1981a). This elevated set-point of the parathyroid gland (e.g., the level of Ca^{2+}_o half-maximally suppressing PTH) is accompanied by a failure of the kidneys to increase their excretion of calcium normally in the face of hypercalcemia. The abnormality in renal divalent cation handling persists, even following total parathyroidectomy, indicating that it is PTH-independent (Attie et al., 1983; Davies et al., 1984). That is, it is not solely the result of the increase in the set-point of the parathyroid gland for Ca^{2+}_o that is present in FHH combined with the stimulatory effect of PTH on renal tubular Ca^{2+} reabsorption. Thus, these clinical studies provided evidence that the FHH gene was involved in the sensing and/or handling of Ca^{2+}_o by both parathyroid and kidney.

Following the cloning of the CaR, it was possible to show that FHH is the consequence of heterozygous mutations in this receptor that reduce its activity (Aida et al., 1995a; Chou et al., 1995; Heath et al., 1996; Janicic et al., 1995; Pearce et al., 1995; Pollak et al., 1993). These mutations are most commonly missense mutations (e.g., they substitute a new amino acid for the one normally coded for). The majority are present within the first half of the extracellular domain or within the transmembrane helices, and intra- or extracellular loops (Fig. 12.2). In a few cases, nonsense mutations (i.e., those inserting a stop codon) (Pearce et al., 1995b), frameshift mutations (Pearce et al., 1995), or the insertion of an alu sequence within the carboxyterminal tail of the receptor (Janicic et al., 1995) have also been described. When expressed in *X. laevis* oocytes (Heath et al., 1996) or in mammalian cells [i.e., human embryonic kidney (HEK293) cells] (Bai et al., 1996), receptors engineered to contain FHH mutations exhibit a range of reductions in function, varying from complete loss of biological activity to modest ($\sim 20\%$) increases in the level of Ca^{2+}_o required to achieve a given increase in Ca^{2+}_i (e.g., the EC_{50}) as a result of CaR-mediated activation of PLC.

In some cases, consanguineous marriages within FHH families have resulted in individuals homozygous for such mutations (Aida et al., 1995a; Janicic et al., 1995; Marx et al., 1985; Pearce et al., 1995; Pollak et al., 1993, 1994b). These homozygous patients have much more severe hypercalcemia than observed in heterozygotes, which is usually accompanied by markedly elevated PTH levels and skeletal demineralization as a result of their severe hyperparathyroidism. In such cases, the disorder can be fatal if parathyroidectomy is not carried out early in life. The homozygous form of FHH is termed neonatal severe hyperparathyroidism (NSHPT) and represents, in effect, the human "knockout" of the CaR, thereby further supporting the central, nonredundant role of the CaR in maintaining Ca^{2+}_o at its normal level. Subsequent to the recognition that inactivating mutations in the CaR cause FHH and NSHPT [some cases of the latter can also result from either inherited or de novo heterozygous mutations that interfere with the function of the normal allele of the CaR (Pearce et al., 1995; Bai et al., 1997), mice have been developed with targeted disruption of the CaR gene (Ho et al., 1995). Like their human counterparts, mice heterozygous for inactivation of the CaR gene are phenotypically normal and have mild hypercalcemia with slight elevations in PTH, while homozygotes have severe hypercalcemia and elevations in PTH and die within the first few weeks of life.

Recently, several families have been identified in which activating mutations of the CaR cause a form of autosomal dominant hypocalcemia (ADH) (Baron et al., 1996; De Luca et al., 1997; Pearce et al., 1996; Perry et al., 1994; Pollak et al., 1994a). Affected members of these families show stable hypocalcemia, with values of Ca^{2+}_o that are usually 10–20%, but sometimes as much as 40%, below the lower limit of normal, and occasional sporadic cases with de novo heterozygous activating mutations of the CaR (Baron et al., 1996; De Luca et al., 1997). They also exhibit higher than expected levels of urinary calcium excretion, likely due to the effects of overactive renal CaRs on tubular handling of calcium. Because of the latter abnormality in renal function, it is important not to treat such patients overly aggressively with vitamin D in attempts to normalize their levels of Ca^{2+}_o, as this can produce marked hypercalciuria and resultant renal damage. Of the activating mutations in the CaR that have been described to date, most reside within the ECD, confirming its importance in the activation of the receptor (Pearce et al., 1996; Perry et al., 1994; Pollak et al., 1994a), although several are also present in transmembrane domains (Fig. 12.2) (Baron et al., 1996; De Luca et al., 1997), similar to the activating mutations that cause other human diseases. Mutations in the luteinizing hormone (LH) receptor produce premature puberty (Shenker et al., 1993), for example, while those in the thyrotropin-stimulating hormone receptor produce hyperthyroidism (Tonacchera et al., 1996). The mutations described in ADH produce an increase in the apparent affinity of the CaR when it is expressed in HEK293 cells, probably as a result of increased affinity of the ECD for Ca^{2+}_o and/or more efficient signal transduction (Bai et al., 1996; De Luca et al., 1997; Pearce et al., 1996).

Acquired Disorders of Ca^{2+}_o-Sensing

The cloning of the CaR and the availability of specific anti-CaR antibodies have made it possible to examine the possible involvement of the Ca^{2+}_o-sensing receptor in hypercalcemic disorders other than FHH and NSHPT, such as various forms of primary and secondary hyperparathyroidism. Hosokawa et al. (1995) sought mutations in the CaR gene similar to those causing FHH and NSHPT in some 40 parathyroid tumors from individuals with parathyroid adenoma, carcinoma, and various types of primary and uremic, secondary/tertiary hyperparathyroidism. None harbored CaR mutations, suggesting that somatic mutations that produce abnormal sensing of Ca^{2+}_o limited to the parathyroid must be an uncommon cause of hyperparathyroidism. Kifor et al. (1996) and Goqusev et al. (1997) subsequently employed immunohistochemistry with anti-CaR antibodies to document that there is a substantial reduction (50–60% on average) in CaR immunoreactivity in parathyroid adenomas compared with that present in normal parathyroid glands. Eight hyperplastic glands from two patients with severe uremic secondary/tertiary hyperparathyroidism showed a similar decrease in CaR immunoreactivity (Kifor et al., 1996). The reason(s) underlying this diminution in CaR expression in hyperparathyroidism is/are unknown. It has recently been demonstrated, however, that a few parathyroid tumors exhibit loss of heterozygosity in the portion of chromosome 3 in which the CaR gene is localized (Thompson et al., 1995). While loss of one allele of the CaR gene has not been formally proven in such cases, it would presumably reduce total CaR expression, similar to the decrease in CaR immunoreactivity documented in parathyroid and kidney in mice heterozygous for targeted disruption of the CaR gene (Ho et al., 1995).

The Functions of the CaR in Parathyroid Cells

The Ca^{2+}_o-sensing receptor inhibits PTH release rather than stimulating hormonal secretion (as is the case with most GPCRs activating PLC) through mechanisms whose elucidation remains an important unresolved issue in parathyroid physiology (Brown, 1991). As noted above, the documentation of markedly blunted, high Ca^{2+}_o-mediated inhibition of PTH secretion in individuals with homozygous inactivating mutations in the CaR (Cooper et al., 1986; Marx et al., 1986) or in mice homozygous for targeted disruption of the CaR gene (Ho et al., 1995) provides strong evidence that the CaR plays a key role in Ca^{2+}_o-regulated PTH secretion. Conversely, there is a reduction in the set-point for Ca^{2+}_o-controlled PTH release in ADH due to increased responsiveness of mutant CaRs that bear activating mutations to

Ca^{2+}_o (Brown et al., 1995). The CaR also appears to control other aspects of parathyroid function. The use of specific "calcimimetic" CaR agonists, which activate the CaR in the presence but not in the absence of Ca^{2+}_o (suggesting that they function as allosteric modifiers of the CaR) (Nemeth, 1995; Silverberg et al., 1997), has documented that the receptor also mediates high Ca^{2+}_o-induced reduction in the level of prepro-PTH mRNA (Garrett et al., 1995b). Finally, the chief cell hyperplasia observed in parathyroid glands in individuals with NSHPT (Cooper et al., 1986), as well as in mice homozygous for CaR knockout (Ho et al., 1995), suggest a direct or indirect involvement of the receptor in suppressing parathyroid cellular proliferation.

Role of the CaR in the C-cell

C-cells also express a CaR that is identical to that present in parathyroid and kidney, (Freichel et al., 1996; Garrett et al., 1995c) as documented by direct sequencing of PCR-amplified, reverse-transcribed mRNA from the C-cell-derived, rMTC44-2 cell line (Garrett et al., 1995c). In contrast, the TT C-cell line, which is not responsive to alterations in Ca^{2+}_o, does not express transcripts for the CaR (Garrett et al., 1995c). Depending on the cellular context in which it is expressed, therefore, the same receptor that suppresses PTH secretion can apparently stimulate CT secretion (McGehee et al., 1997). This stimulation of secretion is not unique to C-cells, as high Ca^{2+}_o also enhances ACTH release from AtT-20 cells, a murine, pituitary-derived cell line that expresses the mouse homolog of the CaR (Emanuel et al., 1996).

The Role of the CaR in Controlling Kidney Cells

Of the known actions of Ca^{2+}_o on the function of the kidney, there are several that may be mediated by the CaR. High Ca^{2+}_o decreases the absorption of sodium chloride (Hebert et al., 1996), calcium, and magnesium (Quamme, 1989) in the thick ascending limb, probably through activation of CaRs present on the basolateral surface of the tubular epithelial cells (Brown et al., 1995) (Fig. 12.3). The CaR may also regulate Ca^{2+} reabsorption in the distal convoluted tubule, an important site for stimulation of calcium reabsorption by PTH, although the precise location and mechanism(s) through which the CaR regulates this segment of the nephron remain to be established (Riccardi et al., 1996). The CaR-mediated reduction in NaCl transport in the medullary thick limb would be expected to reduce vasopressin (AVP)-stimulated generation of the countercurrent gradient in the renal medulla, thereby contributing to the decreased maximal urinary concentrating ability observed in some hypercalcemic individuals (Suki et al., 1969). High

luminal Ca^{2+}_o also directly inhibits transepithelial water flow in the collecting duct, separate from its effects on the countercurrent gradient (Sands et al., 1997). We recently showed that the CaR is present predominantly on the apical membrane of the cells of the inner medullary and papillary collecting ducts (Sands et al., 1997), where it could modulate water permeability by modulating the activity and/or availability of aquaporin water channel(s).

Several of the clinical features described in patients with FHH or ADH support the involvement of the CaR in mediating the effects of high Ca^{2+}_o on several of these physiological processes in the kidney (Brown et al., 1995). The excessive renal tubular reabsorption of calcium and magnesium in FHH, even following parathyroidectomy (Attie et al., 1983; Davies et al., 1984), likely reflects an impaired ability to reduce divalent cation transport because of the reduced responsiveness of the mutant CaRs to Ca^{2+}_o. In a reciprocal fashion, the hypercalciuria in patients with ADH is presumably the result of overactive CaRs that decrease tubular Ca^{2+} and Mg^{2+} reabsorption, even in the face of hypocalcemia (Baron et al., 1996; Pearce et al., 1996). Patients with FHH and ADH also show alterations in renal water handling compatible with the postulated role of the CaR in MTAL and or collecting duct in reducing generation of the countercurrent gradient and/or water transport, respectively. Individuals with FHH can concentrate their urine normally despite being hypercalcemic (Marx et al., 1981b). Those with ADH, in contrast, can develop symptoms of polyuria during therapy with vitamin D, even when their serum calcium concentrations are normal or still overtly hypocalcemic (Pearce et al., 1996). They presumably develop impaired urinary concentrating ability at inappropriately low levels of Ca^{2+}_o because they harbor CaRs that are excessively responsive to extracellular calcium.

The intracellular mechanism(s) through which the CaR regulates renal function remain to be fully elucidated. The PTH-stimulated accumulation of cAMP in the CTAL is thought to stimulate NaCl transport via the apical Na–K–2Cl cotransporter, thereby producing a lumen-positive, transepithelial potential gradient driving passive, paracellular reabsorption of NaCl, Ca^{2+}, and Mg^{2+} (Fig. 12.3) (Brown and Hebert, 1995). Thus, the suppression of PTH-stimulated cAMP accumulation by high Ca^{2+}_o in this segment of the nephron (Firsov et al., 1995; Takaichi and Kurokawa, 1986, 1988), which is presumably mediated via the CaR, could be an important element in the previously described decrease in renal tubular Ca^{2+} and Mg^{2+} reabsorption with high peritubular but not luminal levels of Ca^{2+}_o. Additional CaR-regulated second messenger pathways may also participate in Ca^{2+}_o-induced modulation of ion trans-

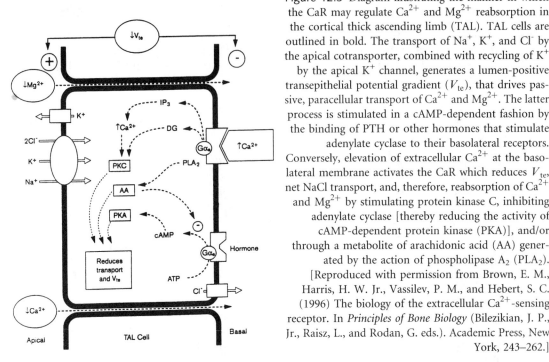

Figure 12.3 Diagram illustrating the manner in which the CaR may regulate Ca^{2+} and Mg^{2+} reabsorption in the cortical thick ascending limb (TAL). TAL cells are outlined in bold. The transport of Na^+, K^+, and Cl^- by the apical cotransporter, combined with recycling of K^+ by the apical K^+ channel, generates a lumen-positive transepithelial potential gradient (V_{te}), that drives passive, paracellular transport of Ca^{2+} and Mg^{2+}. The latter process is stimulated in a cAMP-dependent fashion by the binding of PTH or other hormones that stimulate adenylate cyclase to their basolateral receptors. Conversely, elevation of extracellular Ca^{2+} at the basolateral membrane activates the CaR which reduces V_{te}, net NaCl transport, and, therefore, reabsorption of Ca^{2+} and Mg^{2+} by stimulating protein kinase C, inhibiting adenylate cyclase [thereby reducing the activity of cAMP-dependent protein kinase (PKA)], and/or through a metabolite of arachidonic acid (AA) generated by the action of phospholipase A_2 (PLA$_2$). [Reproduced with permission from Brown, E. M., Harris, H. W. Jr., Vassilev, P. M., and Hebert, S. C. (1996) The biology of the extracellular Ca^{2+}-sensing receptor. In *Principles of Bone Biology* (Bilezikian, J. P., Jr., Raisz, L., and Rodan, G. eds.). Academic Press, New York, 243–262.]

port in the CTAL. Wang et al. 1997 have demonstrated that an arachidonic acid metabolite, probably 20-HETE, may be an important mediator of the high Ca^{2+}_o-elicited inhibition of an apical K^+ channel. By blocking recycling into the tubular lumen of intracellular potassium ions taken up by the apical Na–K–2Cl cotransporter, the inhibition of the K^+ channel substantially reduces NaCl transport by limiting the availability of K^+ in the lumen for subsequent transport along with sodium and chloride via the cotransporter (Fig. 12.3) (Brown and Hebert, 1995). CaR-induced activation of PLC, and, in turn, protein kinase C, may also play some role in mediating the various effects of high Ca^{2+}_o on renal ion and water transport. Further studies are needed to determine whether the CaR plays any role in the other known actions of high Ca^{2+}_o on kidney function, including inhibition of the 1-hydroxylation of vitamin D in the proximal tubule (Weisinger et al., 1989) as well as reductions in glomerular filtration rate (Humes et al., 1978) and renal blood flow (Edvall, 1958).

Role of the CaR in Integrating Renal Handling of Water and Ca^{2+}

When the calcium concentration in the blood rises, there is a homeostatically appropriate reduction in tubular reabsorption of calcium in the CTAL and a corresponding increase in the urinary concentration

of calcium ions (Quamme, 1989). The rise in urinary calcium concentration, however, entails some risk of precipitation of calcium-containing salts, since even normal individuals are often metastably supersaturated with respect to the concentration of calcium oxalate within the urine (Insogna and Broadus, 1987). Moreover, if the urinary concentrating mechanism continued unabated in this circumstance, there would be a further increase in the risk of stone formation. It is in this setting that CaRs on the apical surface of the inner medullary and papillary collecting duct may become active and inhibit water channel availability and/or function, thereby decreasing urinary concentrating ability and, in turn, the risk of nephrolithiasis (Fig. 12.4) (Sands et al., 1997). Persistent reduction in maximal urinary concentration, however, could eventually lead to a substantial water deficit. CaRs in the subfornical organ may reduce the risk of clinically significant dehydration by stimulating thirst and drinking behavior. Indeed, it has been demonstrated that rats made hypercalcemic by administration of pharmacological doses of vitamin D increase their intake of water before they develop a defect in maximal urinary concentrating ability (Peterson, 1990).

In addition to the rapid effect of elevated tubular calcium concentrations on water transport in the collecting duct, hypercalcemia may also exert a more slowly developing effect on urinary concentration

Figure 12.4 Possible mechanisms through which calcium and water homeostasis are integrated by the CaR in various locations. With a calcium load leading to an increase in Ca^{2+}_o and an accompanying increase in urinary calcium concentration, the resultant activation of apical CaRs in the IMCD, coupled with reduced generation of the medullary gradient through CaR-evoked inhibition of NaCl transport in the MTAL, will reduce urinary concentration and, therefore, the risk of urinary stone formation. The risk of an ensuing net loss of free water in the kidney is mitigated by a hypercalcemia-mediated increase in thirst and resultant drinking, perhaps coupled with CaR-induced slowing of GI motility, allowing greater time for absorption of water in the GI lumen. [Reproduced with permission from Brown, E. M., Harris, H. W. Jr., Vassilev, P. M., and Hebert, S. C. (1996) The biology of the extracellular Ca^{2+}-sensing receptor. In *Principles of Bone Biology* (Bilezikian, J.P., Jr., Raisz, L., and Rodan, G. eds.). Academic Press, New York, 243–262.]

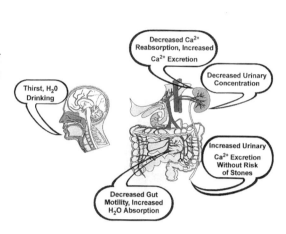

through its presence in the MTAL. In this nephron segment, the receptor is also on the basolateral side of the tubular cells but appears mainly to be involved in regulating NaCl transport independent of Ca^{2+} and Mg^{2+} transport because of differences in the permeability of the paracellular pathways in the MTAL and CTAL (Brown and Hebert, 1995). A high Ca^{2+}_o-elicited decrease in NaCl transport in MTAL would reduce the medullary countercurrent gradient, further reducing maximal urinary concentrating ability. Thus, in humans and other free-living terrestrial organisms, it is possible that the ability to coordinate renal handling of salt, divalent cations, and water may be an important element in our overall capacity to adapt to a variety of conditions where the availability of these constituents varies.

Is the CaR also a Mg^{2+}_o-Sensing Receptor?

Several lines of evidence suggest that the CaR also serves as a physiologically relevant, Mg^{2+}_o-sensing receptor. Elevated levels of Mg^{2+}_o activate the CaR when it is expressed in *X. laevis* oocytes (Brown et al., 1993) or in mammalian cell lines (Ruat et al., 1996), although somewhat higher concentrations of Mg^{2+}_o than Ca^{2+}_o are required to achieve a given degree of CaR activation. Moreover, patients with FHH have levels of serum magnesium that are in the upper part of the normal range or even frankly elevated (Law and Heath, 1985; Marx et al., 1981a) and show a positive correlation between serum calcium and mag-

nesium in this condition (Marx et al., 1981a) that contrasts with the inverse relationship between these parameters in primary hyperparathyroidism where CaR mutations are not present (Hosokawa et al., 1995). There can be more marked elevations in Mg^{2+}_o in individuals homozygous for CaR mutations (Aida et al., 1995a) further suggesting that the CaR contributes to "setting" Mg^{2+}_o. Finally, in ADH there is frequently a decrease not only in Ca^{2+}_o but also in Mg^{2+}_o (Pearce et al., 1996), providing additional indirect evidence that the CaR functions as a Mg^{2+}_o-sensing receptor. It is likely that Mg^{2+}_o acts on CaRs in the kidney to regulate reabsorption of Mg^{2+} in the CTAL (Quamme, 1989) and, perhaps, in the parathyroid to control PTH secretion.

Is Ca^{2+}_o a Systemic Hormone?

The cloning of the Ca^{2+}_o-sensing receptor provides strong evidence that extracellular calcium ions can act in an informational capacity, being utilized by distant tissues to communicate with one another. The concept that extracellular calcium ions serve as a calciotropic "hormone" may be unfamiliar, since Ca^{2+}_o, unlike Ca^{2+}_i, is usually thought of as serving a "metabolic" role. That is, analogous to glucose, it is important that the level of Ca^{2+}_o be maintained at a nearly constant level to provide a reliable supply of calcium ions for their numerous intracellular roles. Ca^{2+}_o (again, similar to glucose) is usually thought of as the *regulated* variable, while PTH, calcitonin, and $1,25(OH)_2D$ are the *regulating* factors. It is useful, however, to consider the thyroid hormones, T3 and T4, as well as thyroid-stimulating hormone (TSH). All three are considered to be hormones and are integral parts of a negative feedback loop, similar to the one incorporating Ca^{2+}_o, PTH, and $1,25(OH)_2D$ in the mineral ion homeostatic system. In other words, the secretion of TSH is adjusted via "sensing" of T4 and T3 by the pituitary thyrotropes in order to ensure relatively constant circulating levels of thyroid hormones for their intracellular roles. Thus, while Ca^{2+}_o is crucial for numerous aspects of the function of essentially all cells, many of which do not express CaRs, the ability of specialized cell types to sense Ca^{2+}_o is crucial to the integrity of the multihormonal, negative feedback loop that regulates mineral ion metabolism. In addition, the presence of the Ca^{2+}_o-sensing receptor within effector elements of the homeostatic system (e.g., the kidney) adds additional sophistication to this mechanism, whereby an increase in Ca^{2+}_o not only reduces renal tubular calcium reabsorption indirectly by suppressing PTH secretion, but also exerts direct, local actions on the kidney that enhance calcium excretion. Thus, the CaR enables Ca^{2+}_o to act both as a systemic and as a paracrine hormone in the control of mineral ion homeostasis.

Possible Functions of the CaR in Local Ca^{2+}_o Signaling in the Brain

The Ca^{2+}_o Signal in Brain

Experimental evidence strongly suggests that the level of Ca^{2+}_o within the brain is much less constant than that measured in blood. The use of Ca^{2+}-sensitive electrodes has shown that Ca^{2+}_o within the narrow intercellular spaces of the brain [which comprise only 15% of the total cellular plus extracellular volume in some regions of the brain (McBain et al., 1990)] can change substantially with alterations in neuronal activity. For example, strong electrical stimulation of the cerebellum of an anesthetized rat (Nicholson et al., 1977) can reduce Ca^{2+}_o by as much as 90%. An even more physiologically relevant stimulus, such as repetitive stroking of a cat's paw with a brush, elicits readily detectable, several percent decreases in Ca^{2+}_o in the ECF of the primary somatosensory cortex that receives impulses from the stimulated region (Heinemann et al., 1977).

Such neuronal activity-dependent reductions in Ca^{2+}_o are caused by activation of Ca^{2+}-permeable plasma membrane channels, with resultant Ca^{2+} influx as shown by the use of specific channel agonists. Iontophoretic application of excitatory amino acids to hippocampal slices reduces Ca^{2+}_o by 20–

30% (Lucke et al., 1995), an effect that can be blocked by specific antagonists of these channels (Arens et al., 1992; Lucke, et al., 1995). Addition of blockers of voltage-dependent Ca^{2+} channels (VDCC) likewise mitigates the NMDA agonist-evoked reduction in Ca^{2+}_o, presumably reflecting activation of VDCC as a result of cellular depolarization due to activation of NMDA channels.

While such studies have so far only measured Ca^{2+}_o in perineuronal ECF accessible to Ca^{2+}-sensitive microelectrodes, it is likely that comparable or even larger changes in Ca^{2+}_o could occur in less accessible but biologically critical compartments of the nervous system, such as the synaptic cleft. We have recently used computer modeling to estimate the magnitude of the changes in Ca^{2+}_o that might be expected at various stimulation frequencies as a result of calcium influx into the postsynaptic dendritic spines via glutamate-evoked activation of NMDA receptor channels with subsequent Ca^{2+} efflux through the Ca^{2+} pump and Na^+–Ca^{2+}-exchanger (Vassilev et al., 1997). Stimulation-elicited release of glutamate from the presynaptic terminal activates the Ca^{2+}-permeable NMDA channels in the postsynaptic membrane, which would be predicted to result in an initial depletion of Ca^{2+}_o within the synaptic cleft. At low stimulation frequencies (e.g., < 10 Hz), the ensuing diffusion of Ca^{2+} from the periphery of the cleft and extrusion of Ca^{2+} from the postsynaptic spine by efflux pathways would restore Ca^{2+}_o to a normal level before the next stimulation. With higher frequency stimulation, however, such as that used to induce long-term potentiation (LTP), the model predicts a summation of neuromediator-elicited reductions in Ca^{2+}_o within the synaptic cleft, with reductions in Ca^{2+}_o averaging from 10% to as much as 30–40%, depending on the parameter values used to estimate inward and outward fluxes of Ca^{2+} from the postsynaptic neuron (Vassilev et al., 1997). These results raise the possibility that activity-dependent reductions in Ca^{2+}_o within the synaptic cleft could potentially be of sufficient magnitude to modulate the activity of CaRs recently shown to be present pre- and/or postsynaptically (Ruat et al., 1995), as described in the next section.

The Regulation of Neuronal Function by the CaR

What would be the impact of putative, CaR-mediated detection of changes in Ca^{2+}_o during neuronal excitation? Much remains to be learned in this regard, but the location of the receptor and recent studies on how it regulates brain cell function are providing some clues in this regard (Fig. 12.5). In the brain, CaR mRNA and protein are present in a variety of regions throughout the central nervous sys-

Figure 12.5 Possible roles of pre- and postsynaptic CaRs (denoted schematically as shown in Figure 12.2) in the brain. Presynaptic CaRs could regulate exocytosis of synaptic vesicles directly or indirectly through regulation of ion channels, such as voltage-dependent calcium channels (VDCC). Postsynaptic CaRs could regulate excitability and neurotransmission through effects of VDCC or other key ion channels such as glutamate- or other receptor-operated channels (ROC). [Reproduced with permission from Brown, E. M., Vassilev, P. M. and Hebert, S. C. (1995) Calcium as an extracellular messenger. *Cell* 83: 679–682.

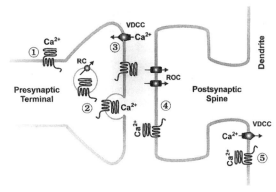

tem (CNS), with particularly high levels of expression in cerebellum, hippocampus, thalamus, hypothalamus, olfactory bulbs, basal ganglia, ependymal zones of the cerebral ventricles, and some cerebral arteries (Rogers et al., 1997; Ruat et al., 1995). In hippocampal pyramidal neurons, the receptor protein is present on the soma of the neurons and also in synaptic areas of the dendrites (Ruat et al., 1995). It remains to establish definitively the predominant localization of the CaR in this region.

We have recently found that pyramidal neurons, cultured from rat hippocampi and studied using the cell-attached mode of the patch-clamp technique, show CaR agonist-dependent activation of a nonselective cation channel (NCC) that exhibits substantial permeability to Ca^{2+}, Na^+, and K^+ (Ye et al., 1996a). High levels of Ca^{2+}_o and neomycin (a potent activator of the cloned CaR) both produce 3- 4-fold increases in the open-state probability of the channel. The same two agonists activate a similar channel in HEK293 cells stably transfected with the human CaR (but not in nontransfected HEK cells which do not express the CaR) (Ye et al., 1996b). Because of the permeability of this NCC to Ca^{2+}, CaR-dependent activation of this NCC at a postsynaptic level could potentially have important actions on the excitability or other properties of pyramidal neurons known to be modulated by increases in Ca^{2+}_i. This possibility, as well as potential effects of the CaR at a presynaptic level that might modulate neurotransmission, requires additional investigation in future studies. It will also be of considerable interest to determine the role of CaRs in other regions of the brain on the functions of the cells expressing them. For example, could the CaRs expressed by ependymal cells in the cerebral ventricles be involved in local Ca^{2+}_o homeostasis, regulating the fluxes of calcium and perhaps other ions into and out of the cerebrospinal fluid, thereby modulating its ionic composition?

Summary and Future Prospects

The cloning of a G-protein-linked, Ca^{2+}_o-sensing receptor directly documents that parathyroid, kidney, and various other cell types sense small, physiologically relevant changes in Ca^{2+}_o by a receptor-mediated mechanism similar to that through which a wide variety of cells respond to numerous hormones, neurotransmitters, and other extracellular messengers. Therefore, Ca^{2+}_o serves not only as a nearly ubiquitous, intracellular second messenger but also in a "hormone-like" capacity as an extracellular first messenger. In addition to being present in tissues that are key elements in mineral ion metabolism and have been known to sense Ca^{2+}_o for many years (e.g., parathyroid and C-cells), the CaR is also expressed in several regions of the kidney and may mediate well-recognized, direct effects of Ca^{2+}_o on renal function. The latter include the increase in the urinary excretion of calcium and magnesium ions in response to hypercalcemia, which augments the reduction in renal tubular calcium reabsorption that results from high Ca^{2+}_o-induced suppression of PTH release. The decrease in maximal renal concentrating ability observed in some hypercalcemic patients likely reflects a functionally important coordination of the homeostatic systems regulating renal calcium and water handling, whose purpose may be to mitigate the risk of pathological renal deposition of calcium salts at times when calcium loads must be disposed of in the urine. Further evidence in favor of a role for the CaR in regulating renal function and for coordinating renal calcium and water handling has come from the investigation of human syndromes of Ca^{2+}_o "overresponsivenss" or "resistance" due to gain-of-function or loss-of-function mutations in the CaR, respectively. Much remains to be learned, however, about the function of the CaR in locations such as in the brain, where it probably responds to changes in local rather than systemic levels of Ca^{2+}_o.

Moreover, the CaR likely also plays a role in mediating longer term effects of Ca^{2+}_o on cellular function, such as the regulation of gene expression (Emanuel et al., 1996), cellular proliferation, or cellular differentiation (for review, see Brown, 1991). Finally, it remains to be determined whether there are additional receptors/sensors for Ca^{2+}_o (Quarles, 1997; Malgaroli et al., 1989; Zaidi et al., 1989) or other ions (Smith et al., 1989a, 1989b, 1989c). Such ion receptor/sensors could malfunction, in turn, in certain disease states and might be amenable to pharmacological manipulation with appropriate therapeutics. Indeed, clinical trials are currently under way to test the efficacy of a "calcimimetic" CaR agonist in states where the CaR is inappropriately underactive, namely in primary and uremic secondary hyperparathyroidism (Heath et al., 1995; Nemeth, 1995; Silverberg et al., 1997).

Acknowledgments The authors have been supported by generous grants from the USPHS (DK41415, DK44588, DK46422, DK48330 and 52005 to E. M. B.), from NPS Pharmaceuticals, Inc. (to E. M. B.), the Theodore and Vada Stanley Foundation (to E. M. B. and P. M. V.) and the St. Giles Foundation (to E. M. B. and P. M. V.).

References

Adams, M. and Oxender, D. (1989) Bacterial periplasmic binding protein tertiary structures. *J. Biol. Chem.* 264: 15739–15742.

Aida, K., Koishi, S., Inoue, M., Nakazato, M., Tawata, M., and Onaya, T. (1995a) Familial hypocalciuric hypercalcemia associated with mutation in the human Ca^{2+}-sensing receptor gene. *J. Clin. Endocrinol. Metab.* 80: 2594–2598.

Aida, K., Koishi, S., Tawata, M., and Onaya, T. (1995b) Molecular cloning of a putative Ca^{2+}-sensing receptor cDNA from human kidney. *Biochem. Biophys. Res. Commun.* 214: 524–529.

Arens, J., Stabel, J., and Heinemann, U. (1992) Pharmacological properties of excitatory amino acid induced changes in extracellular calcium concentration in rat hippocampal slices. *Can. J. Physiol. Pharmacol.* 70: S194–S205.

Attie, M., Gill, J. J., Stock, J., Spiegel, A., Downs, R. J., Levine, M., and Marx, S. (1983) Urinary calcium excretion in familial hypocalciuric hypercalcemia. *J. Clin. Invest.* 72: 667–676.

Aurbach, G., Marx, S., and Spiegel, A. (1985) Parathyroid hormone, calcitonin, and the calciferols. In *Textbook of Endocrinology* 7th ed. (Wilson, J. D. and Foster, D. W., eds.). W.B. Saunders, Philadelphia, PA, pp. 1137–1217.

Bai, M., Quinn, S., Trivedi, S., Kifor, O., Pearce, S., Pollak, M., Krapcho, K., Hebert, S., and Brown, E. (1996) Expression and characterization of inactivating and activating mutations of the human Ca^{2+}_o-sensing receptor. *J .Biol. Chem.* 271: 19537–19545.

Bai, M., Pearce, S. H. S., Kifor, O., Trivedi, S., Stauffer, U. G., Thakker, R. V., Brown, E. M., and Steinmann, B. (1997) In vivo and in vitro characterization of neonatal hyperparathyroidism resulting from a de novo, heterozygous mutation in the Ca^{2+}-sensing receptor gene: normal maternal calcium homeostasis as a cause of secondary hyperparathyroidism in familial benign hypocalciuric hypercalcemia. *J. Clin. Invest.* 99: 88–96.

Bai, M., Trivedi, S., Lane, C. R., Yang, Y., Quinn, S. J., and Brown, E. M. (1998) Protien kinase C phosphorylation of Ca^{2+}_o-sensing receptor (CaR) inhibits its coupling to Ca^{2+} store release. *J. Biol. Chem.* (in press).

Baron, J., Winer, K., Yanovski, J., Cunningham, A., Laue, L., Zimmerman, D., and Cutler G. Jr., (1996) Mutations in the Ca^{2+}-sensing receptor gene cause autosomal dominant and sporadic hypoparathyroidism. *Hum. Mol. Genet.* 5: 601–606.

Berridge, M. and Irvine, R. (1984) Inositol trisphosphate, a novel second messenger in cellular signal transduction. *Nature* 312: 315–321.

Bringhurst, F., Juppner, H., Guo, J., Urena, P., Potts, J. J., Kronenberg, H., Abou-Samra, A., and Segre, G. (1993) Cloned, stably expressed parathyroid hormone (PTH)/PTH-related peptide receptors activate multiple messenger signals and biological responses. *Endocrinology* 132: 2090–2098.

Brown, E. (1991) Extracellular Ca^{2+} sensing, regulation of parathyroid cell function, and role of Ca^{2+} and other ions as extracellular (first) messengers. *Physiol. Rev.* 71: 371–411.

Brown, E. and Hebert, S. (1995) A cloned Ca^{2+}-sensing receptor: a mediator of direct effects of extracellular Ca^{2+} on renal function? *J. Am. Soc. Nephrol.* 6: 1530–1540.

Brown, E., Fuleihan, G. E.-H., Chen, C., and Kifor, O. (1990) A comparison of the effects of divalent and trivalent cations on parathyroid hormone release, $3',5'$-cyclic-adenosine monophosphate accumulation, and the levels of inositol phosphates and bovine parathyroid cells. *Endocrinology* 127: 1064–1071.

Brown, E., Gamba, G., Riccardi, D., Lombardi, D., Butters, R., Kifor, O., Sun, A., Hediger, M., Lytton, J., and Hebert, S. (1993) Cloning and characterization of an extracellular Ca^{2+}-sensing receptor from bovine parathyroid. *Nature* 366: 575–580.

Brown, E., Pollak, M., Seidman, C., Seidman, J., Chou, Y.-H., Riccardi, D., and Hebert, S. (1995) Calcium-ion-sensing cell-surface receptors. *New Engl. J. Med.* 333: 234–240.

Brown, A. J., Zhong, M., Finch, J., Ritter, C., McCracken, R., Morrissey, J., and Slatopolsky, E. (1996) Rat calcium-sensing receptor is regulated by vitamin D but not by calcium. *Am. J. Physiol.* 270: F454–F460.

Chang, W., Pratt, S., Chen, T.-S., Nemeth, E., Huang, Z., and Shoback, D. (1998) Coupling of calcium receptors to inositol phosphate and cAMP generation in mammalian cells and *Xenopus* oocytes and immunodetection of receptor protein by region-specific antipeptide antisera. *J. Bone Miner. Res.* 13: 570–580.

Chen, C., Barnett, J., Congo, D., and Brown, E. (1989) Divalent cations suppress $3',5'$-adenosine monophosphate accumulation by stimulating a pertussis toxin-sensitive guanine nucleotide-binding protein in cultured bovine parathyroid cells. *Endocrinology* 124: 233–239.

Chen, T., Pratt, S., and Shoback, D. (1994) Injection of parathyroid poly(A)+ RNA into *Xenopus* oocytes confers sensitivity to high extracellular calcium. *J. Bone Miner. Res.* 9: 293–300.

Chou, Y.-H., Pollak, M., Brandi, M., Toss, G., Arnqvist, H., Atkinson, A., Papapoulos, S., Marx, S., Brown, E., Seidman, J., and Seidman, C. (1995) Mutations in the human calcium-sensing receptor gene. *Am. J. Hum. Genet.* 56: 1075–1079.

Conklin, B. and Bourne, H. (1994) Marriage of the flytrap and the serpent. *Nature* 367: 22.

Cooper, L., Wertheimer, J., Levey, R., Brown, E., LeBoff, M., Wilkinson, R., and Anast, C. (1986) Severe primary hyperparathyroidism in a neonate with two hypercalcemic parents: management with parathyroidectomy and heterotopic autotransplantation. *Pediatrics* 78: 263–268.

Coughlin, S. (1994) Expanding horizons for receptors coupled to G-proteins-diversity and disease. *Curr. Opin. Cell. Biol.* 6: 191–197.

Davies, M., Adams, P., and Lumb, G. (1984) Familial hypocalciuric hypercalcemia: evidence for continued enhanced renal tubular reabsorption of calcium following total parathyroidectomy. *Acta Endocrinol.* 106: 499–504.

De Luca, F., Ray, K.,Mancilla, E. E., Fan, G.-F., Winer, K. K., Gore, P., Spiegel, A. M., and Baron, J. (1997) Sporadic hypoparathyroidism caused by *de novo* gain-of-function mutations in the Ca^{2+}-sensing receptor gene. *J. Clin. Endocrinol. Metab.* 82: 2710–2715.

Edvall, C. (1958) Renal function in hyperparathyroidism: a clinical study of 30 cases with special reference to selective renal clearance and renal vein catheterization. *Acta Chir. Scand.* 229: 1–54.

Emanuel, R., Adler, G., Kifor, O., Quinn, S., Fuller, F., Krapcho, K., and Brown, E. (1996) Ca^{2+}-sensing receptor expression and regulation by extracellular calcium in the AtT-20 pituitary cell line. *Mol. Endocrinol.* 10: 555–565.

Firsov, D., Aarab, L., Mandon, B., Siaume-Perez, S., Rouffignac, S. D., and Chabardes, D. (1995) Arachidonic acid inhibits hormone-stimulated cAMP accumulation in the medullary thick ascending limb of the rat kidney by a mechanism sensitive to pertussis toxin. *Pfleugers Arch.* 429: 636–646.

Freichel, M., Zinc-Lorenz, A., Hollishi, A., Hafner, M., Flockerzi, V., and Raue, F. (1996) Expression of a calcium-sensing receptor in a human medullary thyroid carcinoma cell line and its contribution to calcitonin secretion. *Endocrinology* 137: 3842–3848.

Garrett, J., Capuano, I., Hammerland, L., Hung, B., Brown, E., Hebert, S., Nemeth, E., and Fuller, F. (1995a) Molecular cloning and characterization of the human parathyroid calcium receptor. *J. Biol. Chem.* 270: 12919–12925.

Garrett, J. E., Steffey, M. E., and Nemeth, E. F. (1995b) The calcium receptor agonist R-568 suppresses PTH mRNA levels in cultured bovine parathyroid cells. *J. Bone Miner. Res.* 10: S387(Abstract).

Garrett, J. E., Tamir, H., Kifor, O., Simin, R. T., Rogers, K. V., Mithal, A., Gagel, R. F., and Brown, E. M. (1995c) Calcitonin-secreting cells of the thyroid gland express an extracellular calcium-sensing receptor gene. *Endocrinology* 136: 5202–5211.

Gogusev, J., Duchambon, P., Hory, B., Giovannini, M., Goureau, Y., Sarfati, E., and Drueke, T. (1997) Depressed expression of calcium receptor in parathyroid gland tissue of patients with hyperparathyroidism. *Kidney Int.* 51: 328–336.

Gorn, A., Lin, H. Y., Auron, P., Flannery, M., Tapp, D., Manning, C., Lodish, H., Krane, S., and Goldring, S. (1992) Cloning and characterization of a human calcitonin receptor from an ovarian carcinoma cell line. *J Clin Invest* 90: 1726–1735.

Griendling, K. K. and Alexander, R. W. (1993) The angiotensin (AT1) receptor. *Semin. Nephrol.* 13: 559–566.

Hammerland, L. G., Krapcho, K. J., Alasti, N., Garrett, J. E., Capuano, I. V., Hung, B. C. P., and Fuller, F. (1995) Cation binding determinants of the calcium receptor revealed by functional analysis of chimeric receptors and a deletion mutant. *J. Bone Miner. Res.* 10: S156(Abstract).

Hawkins, D., Enyedi, P., and Brown, E. M. (1989) The effects of high extracellular Ca^{2+} and Mg^{2+} concentrations on the levels of inositol 1,3,4,5-tetrakisphosphate in bovine parathyroid cells. *Endocrinology* 124: 838–844.

Heath, H. III., Odelberg, S., Jackson, C., Teh, B., Hayward, N., Larsson, C., Buist, N., Krapcho, K., Hung, B., Capuano, I., Garrett, J., and Leppert, M. (1996) Clustered inactivating mutations and benign polymorphisms of the calcium receptor gene in familial benign hypocalciuric hypercalcemia suggest receptor functional domains. *J. Clin. Endocrinol. Metab.* 81: 1312–1317.

Heath, H. III., Sanguinetti, E., Oglesby, S., and Marriot, T.B. (1995) Inhibition of human parathyroid hormone secretion in vivo by NPS R-568, a calcimimetic drug that targets the parathyroid-cell surface calcium receptor. *Bone* 16(Suppl. 1): 85S.

Hebert, S., Pollak, M., Riccardi, D., and Brown, E. (1996) A Ca^{2+}-sensing receptor: from physiology to inherited disorders of calcium homeostasis. *Adv. Nephrol.* 25: 245–255.

Heinemann, U., Lux, H. D., and Gutnick, M. J. (1977) Extracellular free calcium and potassium during paroxysmal activity in cerebral cortex of the rat. *Expt. Brain Res.* 27: 237–243.

Ho, C., Conner, D. A., Pollak, M., Ladd, D. J., Kifor, O., Warren, H., Brown, E. M., Seidman, C. E., and Seidman, J. G. (1995) A mouse model for familial hypocalciuric hypercalcemia and neonatal severe hyperparathyroidism. *Nature Genet.* 11: 389–394.

Hosokawa, Y., Pollak, M., Brown, E., and Arnold, A. (1995) Mutational analysis of the extracellular Ca^{2+}-sensing receptor gene in human parathyroid tumors. *J. Clin. Endocrinol. Metab.* 80: 3107–3110.

Humes, H. D., Ichikawa, I., Troy, J. L., and Brenner, B. M. (1978) Evidence for a parathyroid hormone-dependent influence of calcium on the glomerular ultrafiltration coefficient. *J. Clin. Invest.* 61: 32–40.

Insogna, K. L. and Broadus, A. E. (1987) Nephrolithiasis. In *Endocrinology and Metabolism* 2nd ed., (Broadus, A. E., Frohman L. A., Felig, P., and Baxter, J. D. eds.). McGraw-Hill, New York, pp. 1500–1577.

Janicic, N., Pausova, Z., Cole, D. E. C., and Hendy, G. N. (1995) Insertion of an alu sequence in the Ca^{2+}-sensing receptor gene in familial hypocalciuric hypercalcemia and neonatal severe hyperparathyroidism. *Am. J. Hum. Genet.* 56: 880–886.

Jones, S., Frindt, G. and Windhager, E. (1988) Effect of peritubular [Ca] or ionomycin on hydrosmotic response of CCD to ADH or cAMP. *Am. J. Physiol.* 254: F240–F253.

Juhlin, C., Lundgren, S., Johansson, H., Rastad, J., Akerstrom, G., and Klareskog, L. (1990) 500 Kilodalton calcium sensor regulating cytoplasmic Ca^{2+} in cytotrophoblast cells of human placenta. *J. Biol. Chem.* 265: 8275–8280.

Kifor, O., Congo, D., and Brown, E. M. (1990) Phorbol esters modulate the high Ca^{2+}-stimulated accumulation of inositol phosphates in bovine parathyroid cells. *J. Bone Miner. Res.* 5: 1003–1011.

Kifor, O., Kifor, I., and Brown, E. M. (1992) Effects of high extracellular calcium concentrations on phosphoinositide turnover and inositol phosphate metabolism in dispersed bovine parathyroid cells. *J. Bone Miner. Res.* 7: 1327–1335.

Kifor, O., Moore, F. J., Wang, P., Goldstein, M., Vassilev, P., Kifor, I., Hebert, S., and Brown, E. (1996) Reduced immunostaining for the extracellular Ca^{2+}-sensing receptor in primary and uremic secondary hyperparathyroidism. *J. Clin. Endocrinol. Metab.* 81: 1598–1606.

Kubo, Y., Miyashita, T., and Murata, Y. (1998) Structural basis for a Ca^{2+}-sensing function of the metabotropic glutamate receptors. *Science* 279: 1722–1725.

Law, W. J. and Heath, H. III. (1985) Familial benign hypercalcemia (hypocalciuric hypercalcemia). Clinical and pathogenetic studies in 21 families. *Ann. Int. Med.* 105: 511–519.

Lucke, A., Kohling, R., Straub, H., Moskopp, D., Wassman, H., and Speckmann, E.-J. (1995) Changes of extracellular calcium concentration induced by application of excitatory amino acids in the human neocortex in vitro. *Brain Res.* 671: 222–226.

Malgaroli, A., Meldolesi, J., Zambone-Zallone, A., and Teti, A. (1989) Control of cytosolic free calcium in rat and chicken osteoclasts. The role of extracellular calcium and calcitonin *J. Biol. Chem.* 264: 14342–14349.

Marx, S., Fraser, D., and Rapoport, A. (1985) Familial hypocalciuric hypercalcemia. Mild expression of the disease in heterozygotes and severe expression in homozygotes. *Am. J. Med.* 78: 15–22.

Marx, S., Lasker, R., Brown, E., Fitzpatrick, L., Sweezey, N., Goldbloom, R., Gillis, D., and Cole, D. (1986) Secretory dysfunction in parathyroid cells from a neonate with severe primary hyperparathyroidism. *J. Clin. Endocrinol Metab.* 62: 445–449.

Marx, S. J., Attie, M. F., Levine, M. A., Spiegel, A. M., Downs, R. W. J., and Lasker, R. D. (1981a) The hypocalciuric or benign variant of familial hypercalcemia: clinical and biochemical features in fifteen kindreds. *Medicine (Baltimore)* 60: 397–412.

Marx, S. J., Attie, M. F., Stock, J. L., Spiegel, A. M., and Levine, M. A. (1981b) Maximal urine-concentrating ability: familial hypocalciuric hypercalcemia versus typical primary hyperparathyroidism. *J. Clin. Endocrinol. Metab.* 52: 736–740.

McBain, C. J., Traynelis, S. F., Dingledine, R. (1990) Regional variation of extracellular space in the hippocampus. *Science* 249: 674–677.

McGehee, D. S., Aldersberg, M., Liu, J.-P., Hsuing, S.-C., Heath, M. J. S., and Tamir, H. (1997) Mechanism of extracellular Ca^{2+}-receptor-stimulated hormone release from sheep thyroid parafollicular cells. *J. Physiol.* 502: 31–44.

Mithal, A., Kifor, O., Vassilev, P., Krapcho, K., Simin, R., Fuller, F., Hebert, S. C., and Brown, E. M. (1995) The reduced responsiveness of cultured bovine parathyroid cells to extracellular Ca^{2+} is associated with marked reduction in the expression of extracellular Ca^{2+}-sensing receptor mRNA and protein. *Endocrinology* 136: 3087–3092.

Nakanishi, S. (1992) Molecular diversity of glutamate receptors and implications for brain function. *Science* 258: 597–603.

Nemeth, E. (1995) Ca^{2+} receptor-dependent regulation of cellular function. *News Physiol. Sci.* 10: 1–5.

Nemeth, E. F. and Scarpa, A. (1986) Cytosolic Ca^{++} and the regulation of secretion in parathyroid cells. *FEBS Lett.* 203: 15–19.

Nicholson, C., Ten Bruggencate, G., Steinberg, R., and Strokle, H. (1997) Calcium modulation in brain extracellular microenvironment demonstrated with ion-selective microelectrode. *Proc. Natl. Acad. Sci. USA* 74: 1287–1290.

Nygren, P., Gylfe, E., Larsson, R., Johansson, H., Juhlin, C., Klareskog, L., Akerstrom, G., and Rastad, J. (1988) Modulation of the Ca^{2+}-sensing function of parathyroid cells in vitro and in hyperparathyroidism. *Biochim. Biophys. Acta* 968: 253–260.

O'Hara, P., Sheppard, P., Thogersen, H., Venezia, D., Haldeman, B., McGrane, V., Houamed, K.,

Thomsen, C., Gilbert, T., and Mulvihill, E. (1993) The ligand binding domain in metabotropic glutamate receptors is related to bacterial periplasmic binding proteins. *Neuron* 11: 41–52.

Pearce, S. H. S., Williamson, C., Kifor, O., Bai, M., Coulthard, M. G., Davies, M., Lewis-Barned, N., McCredie, D., Powell, H., Kendall-Taylor, P., Brown, E. M., and Thakker, R. V. (1996) A familial syndrome of hypocalcemia with hypercalciuria due to mutations in the calcium-sensing receptor. *N. Engl. J. Med.* 335: 1115–1122.

Pearce, S., Trump, D., Wooding, C., Besser, G., Chew, S., Heath, D., Hughes, I., and Thakker, R. (1995) Calcium-sensing receptor mutations in familial benign hypercalcaemia and neonatal hyperparathyroidism. *J. Clin. Invest.* 96: 2683–2692.

Perry, Y., Finegold, D., Armitage, M., and Ferell, R. (1994) A missense mutation in the Ca-sensing receptor causes familial autosomal dominant hypoparathyroidism. *Am. J. Human Genet.* 55(Suppl.): A17(Abstract).

Peterson, L. (1990) Vitamin D-induced chronic hypercalcemia inhibits thick ascending limb NaCl reabsorption in vivo. *Am. J .Physiol.* 259: F122–F129.

Pietrobon, D., Virgilio, F. D., and Pozzan, T. (1990) Structural and functional aspects of calcium homeostasis in eukaryotic cells. *Eur. J. Biochem.* 120: 599–622.

Pollak, M., Brown, E., Chou, Y.-H., Hebert, S., Marx, S., Steinmann, B., Levi, T., Seidman, C., and Seidman, J. (1993) Mutations in the human Ca^{2+}-sensing receptor gene cause familial hypocalciuric hypercalcemia and neonatal severe hyperparathyroidism. *Cell* 75: 1297–1303.

Pollak, M., Brown, E., Estep, H., McLaine, P., Kifor, O., Park, J., Hebert, S., Seidman, C., and Seidman, J. (1994a) An autosomal dominant form of hypocalcemia caused by a mutation in the human Ca^{2+}-sensing receptor gene. *Nature Genet.* 8: 303–308.

Pollak, M., Chou, Y.-H., Marx, S., Steinmann, B., Cole, D., Brandi, M., Papapoulos, S., Menko, F., Hendy, G., Brown, E., Seidman, C., and Seidman, J. (1994b) Familial hypocalciuric hypercalcemia and neonatal severe hypercalcemia: the effects of mutant gene dosage on phenotype. *J. Clin. Invest.* 93: 1108–1112.

Quamme, G. (1989) Control of magnesium transport in the thick ascending limb. *Am. J. Physiol. (Renal Fluid Electrolyte Physiol.)* 256: F197–F210.

Quarles, L. D. (1997) Cation-sensing receptors in bone: a novel paradigm for regulating bone remodeling? *J. Bone Miner. Res.* 12: 1971–1974.

Racke, F., and Nemeth, E. (1993) Cytosolic calcium homeostasis in bovine parathyroid cells and its modulation by protein kinase C. *J. Physiol.* 468: 141–162.

Racke, F., Hammerland, L., Dubyak, G., and Nemeth, E. (1993) Functional expression of the parathyroid calcium receptor in *Xenopus* oocytes. *FEBS Lett.* 333: 132–136.

Riccardi, D., Park, J., Lee, W.-S., Gamba, G., Brown, E., and Hebert, S. (1995) Cloning and functional expression of a rat kidney extracellular calcium-sensing receptor. *Proc. Natl. Acad. Sci. USA* 92: 131–135.

Riccardi, D., Lee, W.-S., Lee, K., Segre, G. V., Brown, E. M., and Hebert, S. C. (1996) Localization of the extracellular Ca^{2+}-sensing receptor and parathyroid hormone-related protein in rat kidney. *Am. J. Physiol.* 271: F951–956.

Riccardi, D., Hall, A. E., Chattopadhay, N., Xu, J., Brown, E. M., and Hebert, S. C. (1998) Localization of the extracellular Ca^{2+}/(polyvalent cation)-sensing receptor protein in rat kidney. *Am. J. Physiol.* 274: F611–F622.

Ridefelt, T., Hellman, P., Wallfelt, C., Akerstrom, G., Rastad, J., and Gylfe, E. (1992) Neomycin interacts with Ca^{2+}-sensing of normal and adenomatous parathyroid cells. *Mol. Cell Endocrinol.* 83: 211–218.

Rogers, K. V., Dunn, C. E., Brown, E. M., and Hebert, S. C. (1997) Localization of calcium receptor mRNA in the adult rat central nervous system by in situ hybridization. *Brain Res.* 744: 47–56.

Rogers, K., Dunn, C., Conklin, R., Hadfield, S., Petty, B., Brown, E., Hebert, S., Nemeth, E., and Fox, J. (1995) Calcium receptor mRNA levels in the parathyroid glands and kidney of vitamin D deficient rats are not regulated by plasma calcium or 1,25-dihydroxyvitamin D_3. *Endocrinology* 136: 499–504.

Ruat, M., Molliver, M., Snowman, A., and Snyder, S. (1995) Calcium sensing receptor: molecular cloning in rat and localization to nerve terminals. *Proc. Natl. Acad. Sci. USA* 92: 3161–3165.

Ruat, M., Snowman, A. M., Hester, L. D., and Snyder, S. H. (1996) Cloned and expressed rat Ca^{2+}-sensing receptor. *J. Biol. Chem.* 271: 5972–5976.

Sands, J. M., Naruse, M., Baum, M., Jo, I., Hebert, S. C., Brown, E. M., and Harris, W. H. (1997) Apical extracellular calcium/polyvalent cation-sensing receptor regulates vasopressin-elicited water permeability in rat kidney inner medullary collecting duct. *J. Clin. Invest.* 99: 1399–1405.

Sharff, A., Rodseth, L., Spurlino, J., and Quiocho, F. (1992) Crystallographic evidence for a large ligand-induced hinge-twist motion between the two domains of the maltodextrin binding protein involved in active transport and chemotaxis. *Biochemistry* 31: 10657–10663.

Shenker, A., Laue, L., Kosugi, S., Merendino, J. J., Mineyshi, T., and Cutler, G. J. (1993) A constitutively activating mutation of the luteinizing hormone receptor in familial male precocious puberty. *Nature* 65: 652–654.

Shoback, D., Membreno, L., and McGhee, J. (1988) High calcium and other divalent cations in increased inositol trisphosphate in bovine parathyroid cells. *Endocrinology* 123: 382–389.

Silverberg, S. J., Bone, H. G. III, Marriott, T. B., Locker, F. G., Thys-Jacobs, S., Dziem, G., Kaatz, S., Sanguinetti, E. L., and Bilezikian, J. P. (1997) Short-term inhibition of parathyroid hormone secretion by a calcium-receptor agonist in patients with primary hyperparathyroidism. *N. Engl. J. Med.* 337: 1506–1510.

Smith, J., Dwyer, S., and Smith, L. (1989a) Cadmium evokes inositol polyphosphate formation and calcium mobilization. *J. Biol. Chem.* 264: 7115–7118.

Smith, J., Dwyer, S., and Smith, L. (1989b) Decreasing extracellular Na$^+$ concentration triggers inositol polyphosphate production and Ca^{2+} mobilization. *J. Biol. Chem.* 264: 831–837.

Smith, J., Dwyer, S., and Smith, L. (1989c) Lowering extracellular pH evokes inositol polyphosphate formation and calcium mobilization. *J. Biol. Chem.* 264: 8723–8728.

Stewart, A. and Broadus, A. (1987) Mineral metabolism. In *Endocrinology and Metabolism* 2nd ed. (Broadus, A. E., Frohman L.A., Felig, P., and Baxter, J. D. eds.). McGraw-Hill, New York, pp. 1317–1453.

Suki, W., Eknoyan, G., and Rector, F. (1969) The renal diluting and concentrating mechanism in hypercalcemia. *Nephron* 6: 50–61.

Takaichi, K. and Kurokawa, K. (1986) High Ca^{2+} inhibits peptide hormone-dependent cAMP production specifically in thick ascending limbs of Henle. *Miner. Electrolyte Metab.* 12: 342–346.

Takaichi, K. and Kurokawa, K. (1988) Inhibitory guanosine triphosphate-binding protein-mediated regulation of vasopressin action in isolated single medullary tubules of mouse kidney. *J. Clin. Invest.* 82: 1437–1444.

Thompson, D., Samowitz, W., Odelberg, S., Szabo, J., and Heath, H. III. (1995) Genetic abnormalities in sporadic parathyroid adenomas: loss of heterozygosity for chromosome 3q markers flanking the calcium receptor locus. *J. Clin. Endocrinol. Metab.* 80: 3377–3380.

Tonacchera, M., Van Sande, J., Cetani, F., Swillens, S., Schvartz, C., Winiszewski, P., Portmann, L., Dumont, J., Vassart, G., and Parma, J. (1996) Functional characterization of three new germline mutations of the thyrotropin receptor gene causing autosomal dominant toxic nodular thyroid hyperplasia. *J. Clin. Endocrinol. Metab.* 81: 547–554.

Tordoff, M. G. (1994) Voluntary intake of calcium and other minerals by rats. *Am. J. Physiol.* 167: R470–R475.

Varrault, A., Rodriguez-Pena, M., Goldsmith, P., Mithal, A., Brown, E., and Spiegel, A. (1995) Expression of G-protein alpha-subunits in bovine parathyroid. *Endocrinology* 136: 4390–4396.

Vassilev, P. M., Mitchel, J., Vassilev, M., Kanazirska, M., and Brown, E. M. (1997) Assessment of frequency-dependent alterations in the level of extracellular Ca^{2+} in the synaptic cleft. *Biophys J.* 72: 1–14.

Wang, W.-H., Lu, M., Balazy, M., and Hebert, S. C. (1997) Phospholipase A$_2$ is involved in mediating the effect of extracellular Ca^{2+} on apical K$^+$ channels in rat TAL. *Am. J. Physiol.* 273: F421–F429.

Weisinger, J., Favus, M., Langman, C., and Bushinsky, D. (1989) Regulation of 1,25-dihydroxyvitamin D$_3$ by calcium in the parathyroidectomized, parathyroid hormone replete rat. *J. Bone Miner. Res.* 4: 929–935.

Whitfield, G., Hsieh, J., Juretka, D., Selznick, S., Haussler, C., MacDonald, P., and Haussler, M. (1995) Genomic actions of 1,25-dihydroxyvitamin D$_3$. *J. Nutrition* 125 (6 Suppl.): 1690S–1694S.

Ye, C.-P., Kanazirska, M., Quinn, S., Brown, E., and Vassilev, P. (1996a) Modulation by polycationic Ca^{2+}-sensing receptor agonists of nonselective cation channels in rat hippocampal neurons. *Biochem. Biophys. Res. Commun.*, 224: 271–280.

Ye, C., Rogers, K., Bai, M., Quinn, S. J., Brown, E. M., and Vassilev, P. M. (1996b) Agonists of the Ca^{2+}-sensing receptor (CaR) activate nonselective cation channels in HEK293 cells stably transfected with the human CaR. *Biochem. Biophys. Res. Commun.* 226: 572–579.

Zaidi, M., Datta, H., Patchell, A., Moonga, B., and MacIntyre, I. (1989) "Calcium-activated" intracellular calcium elevation: a novel mechanism of osteoclast regulation. *Biochem. Biophys. Res. Commun.* 183: 1461–1465.

13

Calcium/Calmodulin-Dependent Protein Kinases

Howard Schulman
Andrew Braun

Protein phosphorylation is a ubiquitous mechanism for controlling cellular function and it is therefore not surprising that Ca^{2+} can regulate protein phosphorylation and dephosphorylation. What is unusual about the Ca^{2+} signaling system is the number of distinct protein kinases that are primarily regulated by Ca^{2+}. Regulatory phosphorylation offers the advantage of a ready source of activated substrate in the form of ATP and several possible amino acid targets on proteins, including serine, threonine, and tyrosine. All of the characterized Ca^{2+}-dependent kinases are serine/threonine kinases, although there is no reason to exclude direct regulation of tyrosine kinases by Ca^{2+}. Furthermore, Ca^{2+} activates the known Ca^{2+}-dependent kinases via calcium-binding domains present on either a distance protein, such as calmodulin, or as part of a built-in calmodulin-like domain.

Protein kinases, including Ca^{2+}/calmodulin-dependent protein kinases (CaM kinases) are either dedicated to the phosphorylation of one or a small number of related substrates or are multifunctional, with a broad substrate specificity (Fig. 13.1). Very precise and focused response specificity can be obtained by activation of dedicated protein kinases. These include myosin light chain kinase (MLCK), phosphorylase kinase, and CaM kinase III (elongation factor-2 kinase) which regulate the myosin light chain (MLC), glycogen phosphorylase, and elongation factor-2, respectively. Coordination of many cellular processes can be achieved by multifunctional protein kinases which can regulate different subsets of their many protein targets in different cells. The multifunctional CaM kinases family includes CaM kinases I (and V), II, and IV. The activity of each kinase is highly dependent on Ca^{2+}/calmodulin. CaM kinase II, MLCK, and phosphorylase kinase were the first protein kinases found to be regulated by Ca^{2+}/calmodulin and extended the functional role of cal-

modulin beyond an involvement in cyclic nucleotide synthesis and degradation (Cohen et al., 1978; Dabrowska et al., 1978; Schulman and Greengard, 1987a,b; Yagi et al., 1978). In addition to the direct requirement for calmodulin, CaM kinases I and IV also require phosphorylation of their "activation segment" by CaM kinases for maximal activity.

The aim of this chapter is to review recent studies on the structure/function of this class of protein kinase. These enzymes share several regulatory features and it is most instructive to combine the description of their regulation by Ca^{2+}/calmodulin. Additional focus will be placed on CaM kinase II and MLCK as examples of a multifunctional and dedicated CaM kinase, respectively. Some selective functions of the CaM kinase II will be described in order to highlight approaches used in studying protein kinases. A number of earlier reviews will provide the reader with more detailed analysis of each of the kinases, including the more historical literature which cannot be given adequate space in this chapter (Dunkley, 1991; Heilmeyer, 1991; Means et al., 1991a; Colbran, 1992; Hanson and Schulman, 1992b; Rostas and Dunkley, 1992; Kemp et al., 1994, 1996; Nairn and Picciotto, 1994; Braun and Schulman, 1995b; Soderling, 1995). General properties of CaM kinases I, II, IV, MLCK, and phosphorylase kinase will first be summarized before a detailed treatment of their regulation is given.

CaM Kinase II CaM kinase II is a ubiquitous kinase that was first detected in membranes from nerve terminal preparations (Schulman and Greengard, 1978a, 1978b) and in soluble extracts from rat brain (Yamauchi and Fujisawa, 1980; Kennedy and Greengard, 1981), and was subsequently purified from a variety of neuronal and non-neuronal tissues (reviewed in Hanson and Schulman, 1992b). This kinase is highly abundant in brain,

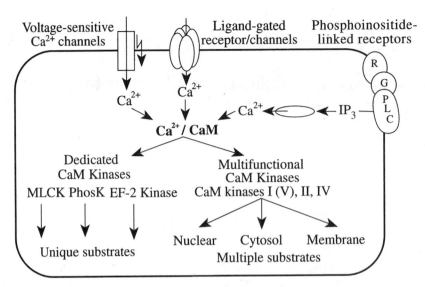

Figure 13.1 Regulation of protein phosphorylation by Ca^{2+}. A variety of extracellular signals increase Ca^{2+} influx or redistribution of intracellular Ca^{2+} stores leading to activation of both dedicated and multifunctional CaM kinases.

accounting for as much as 2% of hippocampal protein and approximately 0.25% of protein in total brain (Erondu and Kennedy, 1985). Although its level in nonneuronal tissues is 1/50th that of brain, it is found in all tissues and has been purified and characterized from many of these, including skeletal muscle, liver, and heart. The kinase is encoded by four genes (α, β, γ, and δ) and each can give rise to three or more isoforms by alternative splicing in the association domain (reviewed in Hanson and Schulman, 1992b). The subunits range in size from 54 to 72 kDa. The enzyme is multimeric, containing 6–12 subunits per holenzyme which can be either homomultimers or heteromultimers. The kinase is a member of a larger family of Ca^{2+}/calmodulin-dependent protein kinases that includes CaM kinases I, II, III, and IV (Hanks and Quinn, 1991). The arrangement of the catalytic, regulatory, and association domains which are responsible for substrate binding and catalysis, autoinhibition and calmodulin binding, and for association between subunits and with cellular targeting proteins, respectively, is shown in Fig. 13.2 in comparison to other protein kinases.

CaM kinase II is considered to be multifunctional because it phosphorylates numerous substrates in vitro and in situ. The processes and structures regulated by CaM kinase via its target substrates include modulation of synaptic release (synapsin I), neurotransmiter synthesis (tyrosine hydroxylase and tryptophan hydroxylase), cytoskeleton (tau and microtubule-associated protein 2), membrane current

(Ca^{2+}, Cl^-, K^+ channels, and ligand-gated channels), carbohydrate metabolism (glycogen synthetase and pyruvate kinase), and has been implicated in long-term potentiation and neuronal memory (reviewed in Hanson and Schulman, 1992b; Soderline, 1995).

CaM Kinases I and IV   CaM kinase I was first identified in soluble extracts of rat brain as a synapsin I kinase that phosphorylated a distinct site on synapsin I to that phosphorylated by CaM kinase II (Huttner et al., 1981; Kennedy and Greengard, 1981). The enzyme was first purified and characterized as a 42 kDa monomeric CaM kinase from bovine brain (Nairn and Greengard, 1987). The heterogeneous preparation was resolved as apparently distinct proteins termed CaM kinase Ia and Ib when purified from rat brain (DeRemer et al., 1992a, 1992b). The precise relationship between these proteins, as well as between these and a kinase purified from rat brain that is termed CaM kinase V, is not clear. The properties described for CaM kinase I, Ia, and V are very similar and all three are activated by an activator protein that was subsequently identified as a CaM kinase kinase (see below). CaM kinase Ib may be the phosphorylated form of CaM kinase Ia since phosphatase treatment decreases the activity of CaM kinase Ib (DeRemer et al., 1992a, 1992b). Alternatively, these may be related to the protein products of two new recently cloned isoforms, CaM kinase Iβ1 and Iβ2, that differ by alternative splicing and CaM kinase Iγ (Naito et al.,

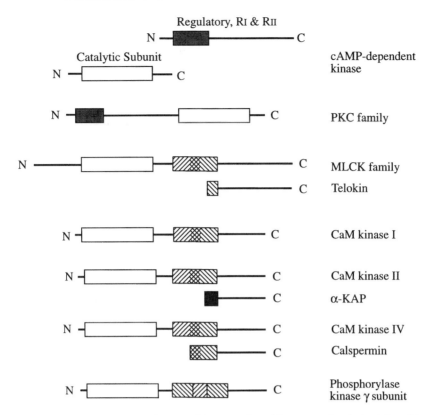

Figure 13.2 Structure/function domains of protein kinases and nonkinase proteins derived from kinase genes. The relative position of the catalytic and regulatory (shaded) domains of several protein kinases is shown schematically. For each of the CaM kinases, the autoinhibitory and calmodulin-binding domains are shown on the N-terminal and C-terminal ends of the regulatory domain, respectively.

1997). We will use the name CaM kinase I when generally referring to either CaM kinase I, Ia, Ib or V. Molecular cloning has revealed a similar domain structure to CaM kinase II but the absence of an association domain (Fig. 13.2) (Picciotto et al., 1993; Cho et al., 1994).

Consistent with a role as a multifunctional kinase, CaM kinase I phosphorylates diverse substrates, including synapsins I and II at the sites previously shown for cyclic AMP-dependent protein kinase (PKA) (Nairn and Greengard, 1987), the cAMP response element-binding protein (CREB) (Sheng et al., 1991), and the cystic fibrosis transmembrane conductance regulator (Picciotto et al., 1992). In addition, it has a broad tissue distribution (Nairn and Greengard, 1987; Picciotto et al., 1993, 1995; Ito et al., 1994).

CaM kinase IV was first purified from rat cerebellum, where it is enriched in cerebellar granule cells and was thus named CaM kinase Gr (Ohmstede et al., 1989). Cloning of the complete cDNA for the

kinase from rat and mouse revealed homology with CaM kinase II (Jones et al., 1991; Means et al., 1991b; Ohmstede et al., 1991). The kinase is not restricted to cerebellar granule cells nor to brain. In cerebellum, it appears transiently and decreases markedly by 14 days after birth as its level in granule cells increases from a negligible level at birth (Jensen et al., 1991a). It is broadly distributed in brain, including the hippocampus, cerebral cortex, striatum, cerebellum, and other regions (Frangakis et al., 1991; Jensen et al., 1991b). Its developmental induction is under the control of thyroid hormone T_3 (Krebs et al., 1996). It has a more restrictive distribution than the other CaM kinases; in the periphery it is found in meiotic male germ cells, thymus, and T-lymphocytes (Frangakis et al., 1991; Hanissian et al., 1993; Means et al., 1991b).

Cerebellum contains two monomeric species of CaM kinase IV with differing mobilities on sodium dodecyl sulfate–polyacrylamide gels (SDS-Page) and differential immunoreactivity (Ohmstede et al.,

1989). The different mobilities are due to two alternatively spliced forms, now termed CaM kinase IVα and CaM kinase IVβ. CaM kinase IVβ contains a 28-amino acid insert at its N-terminus and is confined to cerebellar granule cells whereas the α isoform is more widely distributed (Ohmstede et al., 1989; Jones et al., 1991; Means et al., 1991b; Sakagami and Kondo, 1993; Sun et al., 1995). The kinase is activated by the Ca^{2+} rise that accompanies stimulation of T-lymphocytes (Hanissian et al., 1993).

The substrate specificity of CaM kinase IV has not been extensively characterized. CaM kinase IV phosphorylates synapsin I at both the PKA (and CaM kinase I) and the CaM kinase II site (Ohmstede et al., 1989). PKA and CaM kinase IV both phosphorylate CREB at Ser133 (Enslen et al., 1994; Matthews et al., 1994), serum response factor (Miranti et al., 1995), and may also share the same phosphorylation site on Rap-1b, a ras-related GTP-binding protein enriched in brain (Sahyoun et al., 1991). Phosphorylation of these substrates, as well as of MAP-2, myelin basic protein, and numerous unidentified cerebellar proteins suggest that CaM kinase IV is a multifunctional CaM kinase (Miyano et al., 1992; Cruzalegui and Means, 1993). To date, there are few demonstrations of in situ substrates. However, transcription of several immediate–early genes in PC12 cells is largely blocked by KN-62, probably acting via CaM kinase IV or I (Enslen et al., 1994). Transcription of a reporter gene under a GAL4 promoter is enhanced by cotransfection of cells with a constitutive CaM kinase IV construct and a CREB/GAL4 fusion construct (Sun et al., 1994).

It is not possible at this stage to properly assess the relative roles of CaM kinase I, CaM kinase II, and CaM kinase IV in Ca^{2+} signaling. Functional studies are needed to determine whether each Ca^{2+}-linked signal activates all three when present in the same cell, whether they show distinct temporal or spatial patterns of activation, or whether there is selective stimulation of one or another of these kinases by certain signals. The three kinases share a strict dependence on Ca^{2+}/calmodulin and a relatively low K_{act} for calmodulin (20–100 nM), but differ in their mode of regulation by phosphorylation. CaM kinase II exhibits maximal activity in the presence of Ca^{2+}/calmodulin and is made partially Ca^{2+}-independent after autophosphorylation. Maximal activity of CaM kinases I and IV is quite low without prior phosphorylation by CaM kinase kinases. In addition, the Ca^{2+}-independent activity of CaM kinase IV, but not of CaM kinase I, is significantly enhanced by this phosphorylation.

Myosin Light Chain Kinase (MLCK) MLCK catalyzes the Ca^{2+}-dependent phosphorylation of Ser19 in the 20 kDa regulatory MLC. MLCK is a dedicated protein kinase; the myosin light chain is its only established physiological substrate. Calmodulin was shown to mediate the Ca^{2+}-dependence of smooth muscle MLCK (smMLCK) and skeletal muscle MLCK (skMLCK) at a time when it was primarily considered an activator of phosphodiesterase (Dabrowska et al., 1978; Yagi et al., 1978). Phosphorylation of the MLC by MLCK directly triggers the Ca^{2+}-dependent contraction of smooth muscle removing the block on the myosin ATPase and is thus the key regulatory step in contraction (Walsh, 1994). By contrast, in skeletal muscle, regulation of the acto-myosin ATPase, and thus cross-bridging cycling, is triggered by Ca^{2+} binding to the troponin complex. The apparent function of MLC phosphorylation by MLCK in skeletal muscle is a modulatory enhancement of the rate of development and maximal extent of isometric contraction (Sweeney et al., 1993). In nonmuscle cells, MLCK appears to regulate the cytoskeleton, suggesting that it may play a role in cell shape and motility (Shoemaker et al., 1990).

MLCK has been isolated and cloned from a number of tissues, including smooth muscle (Olson et al., 1990; Gallagher et al. 1991), chicken embryo fibroblasts (Shoemaker et al., 1990), and skeletal muscle (Takio et al., 1986; Roush et al., 1988). The various isozymes fall into one of two major categories: striated muscle and smooth muscle/nonmuscle. The smooth muscle and nonmuscle forms (M_r 107–140 kDa) are alternative splice variants, arising from the same gene, whereas the skeletal muscle MLCK ($M_r \sim 66$ kDa) is the product of a separate gene. MLCK consists of a single polypeptide, in which the variable N- and C-termini are separated by a strongly conserved catalytic core/calmodulin-binding regulatory domain (Fig. 13.3).

Phosphorylase Kinase The rapid production of energy in many tissues, such as muscle, often involves the enzymatic breakdown of stored glycogen to glucose-1-phosphate, leading to the synthesis of ATP via the Krebs cycle. Glycogen breakdown is stimulated by conversion of glycogen phosphorylase b to the active form, phosphorylase a, by the key regulatory enzyme phosphorylase kinase. Phosphorylase kinase is a large complex involving association of four tetrameric units, $(\alpha\beta\delta\gamma)_4$, with a combined relative molecular mass of $\sim 1.3 \times 10^3$ kDa (Heilmeyer, 1991). The kinase activity is contained within the 45 kDa γ-subunit, whereas the 17 kDa δ-subunit is identical to calmodulin (Cohen et al., 1978). The δ-subunit is very tightly associated with the complex in the absence of Ca^{2+} and therefore functions as a subunit of the enzyme. This tight binding is likely due to the presence of two putative, closely spaced CaM-binding sequences in the γ-subunit, at positions 302–326 and 342–366 which flank an autoinhibitory domain

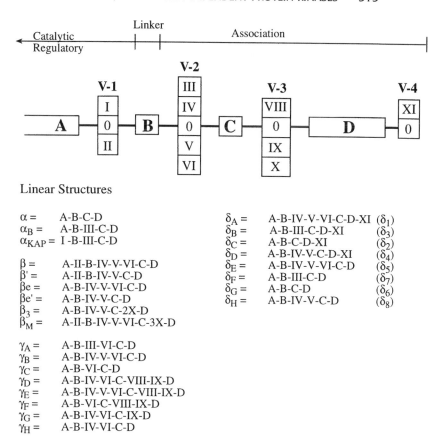

Linear Structures

α =   A-B-C-D
α_B =   A-B-III-C-D
α_KAP = I -B-III-C-D

β =   A-II-B-IV-V-VI-C-D
β' =   A-II-B-IV-V-C-D
βe =   A-B-IV-V-VI-C-D
βe' =   A-B-IV-V-C-D
β_3 =   A-B-IV-V-C-2X-D
β_M =   A-II-B-IV-V-VI-C-3X-D

γ_A =   A-B-III-VI-C-D
γ_B =   A-B-IV-V-VI-C-D
γ_C =   A-B-VI-C-D
γ_D =   A-B-IV-VI-C-VIII-IX-D
γ_E =   A-B-IV-V-VI-C-VIII-IX-D
γ_F =   A-B-VI-C-VIII-IX-D
γ_G =   A-B-IV-VI-C-IX-D
γ_H =   A-B-IV-VI-C-D

δ_A =   A-B-IV-V-VI-C-D-XI (δ_1)
δ_B =   A-B-III-C-D-XI   (δ_3)
δ_C =   A-B-C-D-XI   (δ_2)
δ_D =   A-B-IV-V-C-D-XI   (δ_4)
δ_E =   A-B-IV-V-VI-C-D   (δ_5)
δ_F =   A-B-III-C-D   (δ_7)
δ_G =   A-B-C-D   (δ_6)
δ_H =   A-B-IV-V-C-D   (δ_8)

Fiture 13.3 Diversity of CaM kinase II isoforms. The core structure of a composite CaM kinase II consists of the catalytic domain (A), a conserved linker (B), and segments C and D that constitute the association domain. Inserts I–X are introduced into one or more isoforms in four positions (V1–V4) that contain these variable inserts. A shorthand notation is used to describe the arrangement and content of core and variable segments of each of the currently known isoforms.

(Fig. 13.2) (Dasgupta and Blumenthal, 1995; Huang et al., 1995). These interactions also have the effect of significantly enhancing the affinity of the δ-subunit for Ca^{2+}, making the kinase highly Ca^{2+}-sensitive. Exogenous calmodulin (or troponin) can also interact with the complex, in the conventional Ca^{2+}-dependent manner, and further activate the kinase. This interaction occurs via the γ-subunit. The α- and β-subunits ($M_r \sim 138$ and $125 \, kDa$, respectively) appear to function as regulatory sites governing the complex regulation of phosphorylase kinase activity by magnesium and adenine nucleotide, in addition to multiple signaling pathways (Heilmeyer, 1991).

Phosphorylase kinase is known to be activated by both Ca^{2+} and cyclic AMP second messenger cascades. Ca^{2+} activates phosphorylase kinase by binding to the δ-subunit (calmodulin), which, in the absence of Ca^{2+}, is bound to the γ-subunit at a region C-terminal to the catalytic domain. Cyclic AMP stimulates PKA-mediated phosphorylation of at least two critical residues in the β-subunit (Ser26 and Ser700) (Yeaman and Cohen, 1975). Phosphorylation of the β-subunit appears to be essential for the full expression of phosphorylase kinase activity through a complex mechanism involving enhanced Mg^{2+} affinity of the β subunit and, in turn, increased affinity of the δ-subunit for calcium (Heilmeyer, 1991). In addition, phosphorylation of the α-subunit may also play a role in the activation of the holoenzyme (Pickett-Gies and Walsh, 1985).

Structure/Function Domains

CaM kinases consist of a catalytic domain, a regulatory domain that suppresses kinase activity until

Ca^{2+}/calmodulin binds, and may contain an association domain that is involved in binding to other subunits of the kinase or to intracellular targeting proteins. These functional domains consist of structural units along the polypeptide chain that are ordered differently in the various kinases and can be either contiguous on one polypeptide or reside on separate polypeptides. Unlike PKA, however, all of the CaM kinases contain both catalytic and autoregulatory domains on one polypeptide. The regulatory domain of the CaM kinases is on the C-terminal side of the catalytic domain, the opposite orientation of the protein kinase C family (Fig. 13.2). The association domain, when present, is on the C-terminal side of the catalytic domain.

Catalytic Domain

The catalytic domain is responsible for binding ATP, recognition of the protein substrate, and catalysis of the phosphotransferase reaction. A segment of each kinase consisting of approximately 260 amino acids constitutes the catalytic domain. The catalytic domains of the CaM kinases show significant homology to each other and other protein kinases (Hanks and Quinn, 1991). Evolutionary conserved regions include residues involved in binding ATP, coordinating with Mg^{2+}, and mediating the catalytic reaction. The crystal structure of PKA, CaM kinase I, and twitchin kinase show extensive similarities in their catalytic domains and this will likely extend to the other CaM kinases.

Based on the crystal structures of PKA and CaM kinase I (Knighton et al., 1991; Goldberg et al., 1996), the overall structure of the catalytic domain of all CaM kinases is expected to be bilobal, with a small N-terminal lobe formed primarily of an antiparallel β-sheet responsible for binding ATP and a larger C-terminal lobe formed primarily of α-helices containing the protein-binding/catalytic segment. All of the CaM kinases utilize ATP as the phosphate donor. Catalysis occurs in a cleft between the two lobes and largely involves amino acid residues of the larger lobe. All of the CaM kinases discussed in this chapter prefer substrates with one or more basic residues near the Ser/Thr that is phosphorylated, with the corresponding Glu and Asp residues that interact with these basic residues likely to be on the surface of the large lobe.

The CaM kinases contain all of the boxes of conserved residues characteristic of protein kinases (Lin et al., 1987; Hanks and Quinn, 1991; Means et al., 1991b; Picciotto et al., 1993). Each kinase contains a glycine-rich loop that is followed by a key lysine residue, each of which is necessary for binding of ATP by the small lobe. Mutation of the Lys to a Met or Arg produces a nearly inactive kinase in all kinases that have been examined and this is often used as an inactive control. The large lobe contains a conserved Asp-Phe-Gly sequence that is involved in binding Mg^{2+} and a contiguous stretch of amino acids, largely charged residues, referred to as a "catalytic loop" because of its involvement in catalysis. In CaM kinases I and IV, phosphorylation of a Thr in an "activation loop," corresponding to Thr197 of PKA, is required to make these enzymes competent kinases (see below).

Substrate Consensus

The core consensus sequences for the CaM kinases are shown in Table 13.1. The substrate recognition determinants are based on analysis of sites phosphorylated within various substrates of these kinases, as well as detailed analysis of synthetic peptides (Kennelly and Krebs, 1991). As expected, there are fewer determinants for substrate recognition by the multifunctional kinases than by MLCK, enabling them to phosphorylate a broader range of substrates. All of the kinases require a basic residue, usually an Arg, 3 amino acids N-terminal (p−3) of the phosphorylated residue at P0, but differ in other aspects of their consensus sequences.

CaM Kinase II   CaM kinase II requires only one basic residue N-terminal of the phosphorylatable residue, rather than two as preferred by PKA, and can phosphorylate either a Ser or Thr (Pearson et al., 1985; White et al., 1998). An Arg is only 2–3-fold better than a Lys at this position (Pearson et al., 1985; Stokoe et al., 1993). A hydrophobic amino acid is preferred at P−5 (Stokoe et al., 1993), as found for CaM kinase I (Lee et al. 1994). A hydrophobic residue at P+1 is found in several CaM kinase II substrates whereas a basic residue at this position is not tolerated (Stokoe et al., 1993). In addition, several "anomalous" phosphorylation sites, e.g., a complete lack of basic residue at P−3, have been found. An acidic residue at P+2 that is commonly found in CaM kinase II substrates plays an important role in the phosphorylation of vimentin, one of the substrates lacking a basic residue at P−3 (Ando et al., 1991).

CaM Kinase I   The critical determinants of its substrate specificity have been identified by analysis of synthetic peptides (Lee et al., 1994; Dale et al., 1995). These studies have indicated the importance of hydrophobic residues, in addition to the basic residue for substrate recognition. CaM kinase I prefers a hydrophobic residue at both P−5 and at P+4. Comparison of the consensus determinants for CaM kinases I and II suggests why they phosphorylate synapsin I at distinct sites. The presence of an

Table 13.1 Preferred or Consensus Sequences for Some Protein Kinases

Protien Kinase	Consensus Sequence
CaM kinase I	Φ-X-R-X_2-S/T-X_3-Φ
CaM kinase II	Φ-X-R-X_2-S/T-Φ
CaM kinase IV	Φ-X-R-X_2-S/T
smMLCK	$(K/R_2,X)$-X_{1-2}-K/R_3-X_{2-3}-R-X_2-S-N-V-F
PKA	R-R/K-X-S/T

Single-letter designations for amino acids are used. X and Φ are polar and hydrophobic amino acids, respectively.

Arg at P−2, in addition to P−3 in the CaM kinase I site, is a positive determinant for CaM kinase I and a negative determinant for CaM kinase II (Lee et al., 1994; Stokoe et al., 1993). In addition, an Asp at P+1 has a greater negative effect on CaM kinase II than on CaM kinase I. The two homologous sites in synapsin I phosphorylated by CaM kinase II may be poor substrates for CaM kinase I because both have a Pro instead of a hydrophobic residue at P+4 and only one has a hydrophobic residue at P−5.

CaM Kinase IV A core consensus of R-X-X-S/T has been proposed for substrates of CaM kinase IV based on reduced phosphorylation of peptides whose Arg at P−3 is replaced by a neutral amino acid (Cruzalegui and Means, 1993; White et al., 1998). The core consensus for CaM kinase IV and that for CaM kinase II are quite similar, resulting in some overlap in substrate specificity. However, CaM kinase II does not tolerate a basic residue at P−2 as does CaM kinase IV, whereas CaM kinase IV has no preference for a hydrophobic residue at P+1 as does CaM kinase II (White et al., 1998).

MLCK MLCK is a prototype for dedicated protein kinases in that it has a very strict substrate specificity, with only MCL as substrate, and only one site (Ser19) on MLC that is phosphorylated. Both smooth and skeletal muscle MLC are phosphorylated by skMLCK whereas smMLCK only phosphorylates smooth muscle MLC. Substrate recognition is largely dependent on the primary sequence, as short peptides derived from MLC are phosphorylated with affinities equal to the entire protein. Unlike the multifunctional CaM kinases for which a single basic residue is sufficient, smMLCK utilizes at least four basic residues over a more extended region on the N-terminal side of the phosphorylated Ser as part of the consensus recognition sequence. In the core sequence of smooth muscle MLC(11–22), (K-K-R-P-Q-R-A-T-S-V-N-F), Lys11 (P−8), Lys12 (P−7), Arg13 (P−6), and Arg16 (P−3) are primary determinants of specificity, with the first

three having the largest impact on kinetics of peptide phosphorylation (Kemp and Pearson, 1985). The apparent K_m increased approximately 200-fold and 1000-fold in peptides in which the sequence up to and including either Lys11 or Lys12 was truncated, respectively. The strict substrate specificity is also based on the strict requirement that the three basic residues be spaced either two or three amino acids from Arg at P−3 (Kemp and Pearson, 1985). The K_m increases markedly and relative phosphorylation of Thr at P−1 versus the Ser increases at this spacing is either larger or smaller. The presence of Glu at P−3 in skeletal muscle MLC makes it a poor substrate for smMLCK. Although MLCK phosphorylates only MLC, other protein kinases that require only a single basic residue, such as CaM kinase II, can phosphorylate MLC (Fukunaga et al., 1982).

Phosphorylase Kinase Although phosphorylase kinase has been shown to phosphorylate several proteins in vitro, it is likely dedicated to the phosphorylation of only a single substrate, phosphorylase b, in vivo. Its phosphorylation site in phosphorylase from diverse tissues and species conforms to the pattern K/R-R-K/R-Q-I-S-V/I-R-G-L. Analysis of peptide substrates suggests features of the primary sequence that contribute to affinity and efficacy of phosphorylation (reviewed in Graves, 1983). Lys at P−5, Lys at P−3 and Arg at P+2 are important for substrate recognition (Tabatabai and Graves, 1978). It would not be unexpected, however, to find that dedicated kinases, such as phosphorylase kinase, utilize additional features of the conformation of phosphorylase to improve specificity. Indeed, peptides corresponding to the primary sequence around the phosphorylation site are considerably poorer substrates than the intact protein and, thus, no simple consensus sequence can be described for this kinase.

Calmodulin-Binding Domain

The sequence of the autoinhibitory and calmodulin-binding domains for several CaM kinases is shown in

Table 13.2. Ca^{2+}/calmodulin activates its target enzymes by direct binding to a distinct site with certain common features but without a clear consensus sequence (James et al., 1995). The calmodulin-binding domain of the CaM kinases (with the exception of the γ-subunit of phosphorylase kinase) are bracketed by Arg residues. The hydrophobic residues N-terminal of the calmodulin-binding domain (Table 13.2; Phe or Tyr in bold) may serve to anchor this segment of the kinase at the active site. When arranged in an α-helical wheel pattern, these domains have hydrophobic residues on one side and basic or polar residues on the other side. The specific interactions between calmodulin and its target sequences on skMLCK, smMLCK, and CaM kinase II have been determined based on x-ray and NMR structures of calmodulin bound to peptides derived from these protein kinases (Ikura et al., 1992; Meador et al., 1992, 1993). All three structures show calmodulin collapsed around its peptide, making numerous hydrophobic and electrostatic interactions. The target peptide assumes an α-helical conformation along most of its length (amino acids 294–210 of α-CaM kinase II and 799–813 of smMLCK) (Meador et al., 1993). Two hydrophobic residues (Trp800 and Leu813 of chicken smMLCK and Leu299 and Leu308 of α-CaM kinase II) were found to be critical for anchoring of calmodulin to the peptides. The distance between these anchors is 13 amino acids for MLCK and CaM kinase I and 9 for CaM kinase II. The central helix of calmodulin unwinds in order to maximize interactions by accommodating to different arrangements of basic and hydrophobic residues in its targets. There are differences in calmodulin affinity of these enzymes, with CaM kinase II having a relatively low affinity (K_{act} of 20–100 nM) while others, such as MLCK, show half-maximal activation at 1 nM. Differences in interaction between calmodulin and CaM kinase II are also suggested by the finding that CLP, a calmodulin-like protein isolated from human mammary epithelium, is equipotent to calmodulin at activation of CaM kinase II but does not activate many other calmodulin-dependent enzymes, including smMLCK; it weakly binds to smMLCK but does not activate it (Edman et al., 1994).

Domain Structure: Multifunctional CaM Kinase II

Mammalian CaM kinase II comprises a family of closely related isoforms, currently numbering 24, that are derived from four genes: α (Hanley et al., 1987; Lin et al., 1987; Srinivasan et al., 1994; Bayer et al., 1996) β (Bennett and Kennedy, 1987; Bulleit et al., 1988; Brocke et al., 1995; Urquidi and Ashcroft, 1995; Bayer et al., 1998), γ (Tobimatsu et al., 1988;

Tobimatsu and Fujisawa, 1989; Nghiem et al., 1993; Kwiatkowski and McGill, 1995; Tombes and Krystal, 1997), and δ (Schworer et al., 1993; Edman and Schulman, 1994; Mayer et al., 1994). Kinase isoforms originating from any of the four genes are highly homologous throughout the catalytic domain and regions of the regulatory and association domains that form the core kinase structure. There are 11 known types of sequences, ranging from 9–127 amino acids, that are inserted into one or more isoforms (Fig. 13.3; designate as inserts I–XI). It is assumed that diversity is generated by alternative splicing, although direct evidence for this is only available for the β-CaM kinases (Bulleit et al., 1988; Karls et al., 1992).

The core sequence of all CaM kinases includes the catalytic/regulatory domains (A), a segment that may link this domain to the association domain (B), a segment that follows the region with the largest number of variable inserts (C), and the core association domain (D) (Fig. 13.3). Inserted sequences are found at four positions (VI–V4) in mammalian CaM kinase II—between A and B, B and C, C and D, and after D at the very C-terminus of the kinase. Two isoforms, α- and δ_G-CaM kinase II (δ_6) contain none of the inserts and are therefore the smallest of the kinase isoforms with a structure that can be depicted as A-B-C-D.

Are the isoforms redundant or does alternative splicing make some or all of them distinct? Possible insight into the function of the inserts can be gained from an examination of their location in the holoenzyme. Based on electron micrographs of the multimeric kinase, it appears that the association domains self-assemble into a large globular structure from which the catalytic/regulatory domains extend radially as smaller particles attached via a short linker (segment B), like petals of a flower (Kanaseki et al., 1991). If the assignment of domains is correct, V1 (N-terminal of the linker) would be part of the catalytic/regulatory domain and affect some aspect of catalysis and calmodulin binding, whereas the remaining inserts would be part of the central globular mass and may affect self-assembly or intracellular targeting of the holoenzyme. The effects of inserts on CaM kinase II would be less predictable if the catalytic/regulatory domains and central globular domain interact in solution. In fact, inserts within segment D of the *Drosophila* CaM kinase II (Cho et al., 1991; Griffith, 1993; Ohsako et al., 1993) are at some distance from the catalytic/regulatory domains based on the electron micrographic data, yet they exhibit differences in calmodulin affinity and efficacy of activation by various calmodulin mutants, suggesting interactions between these domains (GuptaRoy and Griffith, 1996; GuptaRoy et al., 1996).

Table 13.2 Regulatory Domains

CaMK-II	HPW	^{273}HRSTVASCMHRQETVDCLKK**F**NARR**RK**LKGAILTTMLATRNFSSR**S**MITKKGEGSQVKE^{329}
CaMK-I	HPWIA	^{278}GDTALDKNIHQSVSEQIKKN**F**AKS**K**WKQAFNRTA VVRH**MRK**LQLGHQPGGTGTDS^{332}
CaMK-IV	HPWVT	^{252}GKAANFVHMDTAQKKLQE**F**NARR**KL**KAAVKAV VASS**RL**GSASSSHTNIQESNKASS^{307}
skMLCK	HPWLNN	^{554}LAEKAKRCNRRLKSQILLKK**Y**LMK**RR**WKKNFIAV SAAN**RF**KKISSSGALMALGV^{607}
smMLCK	HPWLQ	^{776}KDTKNMEAKKLSKDRMKK**Y**MARR**KW**QKTGHAV RAIG**RL**SSMAMISG^{821}
Phosphorylase kinase γ	HPFF	^{288}QQYVVEEVRHFSPRGK**F**KVICLTVLASVRIY YQYR**RV**KPVTREIVIRDPYALRPLRRLIDA^{348}

Intracellular targeting is clearly regulated by Insert III (Fig. 13.3). This insert targets the large CaM kinase II holoenzyme to the nucleus (Srinivasan et al., 1994; Brocke et al., 1995). CaM kinases α_B, γ_A, and δ_B contain a homologous 11-amino acid insert III (KRKSSSSVQLM in α_B-CaM kinase II) as does the nonkinase protein α-KAP (see below). Alternative splicing introduces this sequence after a Lys in the previous exon, resulting in the sequence KKRK, one of the canonical nuclear localization sequences (NLS) also found in SV40 T-antigen. Indeed, cells transfected with either α_B- or δ_B-CaM kinase show immunostaining that is largely restricted to the nucleus, whereas isoforms lacking this inert are excluded from the nucleus (Srinivasan et al., 1994). Phosphorylation of a serine residue immediately adjacent to the NLS by either CaM kinase I or CaM kinase IV specifically blocks this nuclear translocation of α_B-CaM kinase II (Heist et al., 1988). α_B-CaM kinase is found primarily in the midbrain/diencephalon (Brocke et al., 1995). This is the only area of brain where α-CaM kinase immunostain is found in the nucleus, consistent with targeting of α_B-CaM kinase in transfected cells. Nuclear and nonnuclear isoforms can coassemble; the ratio of the two types of subunits in a given holoenzyme will determine whether it distributes to the nucleus, cytoplasm, or both.

A nonkinase product of the α-CaM kinase gene, termed α-KAP, has been identified (Bayer et al., 1996). α-KAP contains the association domain but lacks the entire catalytic domain and calmodulin-binding domain, and in their place contains a 25-amino acid insert I followed by the linker B, the NLS insert III, and the remainder of the association domain segment D (I-B-III-D). The transcription of α-KAP occurs within an intron of the CaM kinase II gene (Bayer et al., 1996). The short hydrophobic segment (insert I) is unlikely to encode a catalytic function. This insert is dominant to the targeting information provided by the NLS, since α-KAP is found in sarcomeres in skeletal muscle, where it is most abundant, and not in muscle nuclei. α-KAP serves as a targeting and anchoring protein for CaM kinase II; it assembles with the catalytically active subunits and targets them to the sarcoplasmic reticulum via the hydrophobic segment (Bayer et al., 1998).

There is evidence for additional targeting of the kinase. β_3-CaM kinase II (found in pancreas and testes) and β_M-CaM kinase II (found in skeletal muscle) have inserts suggestive of a targeting domain. The β_3 (Urquidi and Ashcroft, 1995) and β_M (Bayer et al., 1998) isoforms contains an 86-amino acid sequence with tandem repeat of a sequence rich in proline residues and characteristic of the binding sites for SH3 domains (Fig. 13.3; insert X). This novel insert may be involved in targeting the kinase to adapter molecules or enzymes that contain SH3 domains. CaM kinase II is also targeted to synaptic vesicles. When present on vesicles, α-CaM kinase II interacts with the C-terminal end of synapsin I via the autoinhibitory domain of the kinase (Benfenati et al., 1992). Upon autophosphorylation at Thr286 the kinase selectively translocates to the postsynaptic density (PSD), a postsynaptic thickening that may organize synaptic signaling, because of an increase in affinity to the C-terminal end of the NR2B subunit of the NMDA receptor (Yoshimura and Yamauchi, 1997; Strack and Colbran, 1998). Such anchoring may serve to associate the kinase with the PSD under physiological conditions where it can phosphorylate the NMDA (Omkumar et al., 1996) and AMPA (Barria et al., 1997) type glutamate receptors. Although the kinase appears to be the major protein of rat brain PSD, much of the kinase may associate with the PSD only after decapitation (Suzuki et al., 1994), perhaps resulting from physical effects of autophosphorylation (Dunkley, 1991; Hudmon et al., 1996).

Some unique features of other isoforms are as follows: β-CaM kinase II shows distinct developmental regulation in brain. The earliest CaM kinase isoforms to appear during brain development, termed embryonic isoforms β_e-CaM kinase and β_e'-CaM kinase, differ from the adult isoforms in that they lack insert II (Brocke et al., 1995). At embryonic day 10–13, the level of these isoforms declines as the adult isoforms increase in prominence. The gene for δ-CaM kinase II gives rise to the largest number of isoforms, four with the C-terminal insert XI and four without it for a total of eight isoforms (Mayer et al., 1994).

Domain Structure: Multifunctional CaM Kinase I and IV and CaM Kinase Kinase

The predicted amino acid sequence of CaM kinases I and IV indicates a similar domain structure to the other members of the CaM kinase II family (Fig. 13.2). We know most about the catalytic/regulatory domains of CaM kinase I from its crystal structure, which will be described in the next section. The full-length cloning of CaM kinase IV from rat brain (Means et al., 1991b) and mouse brain (Jones et al., 1991) showed it to encode a protein of 53 kDa, now referred to as CaM kinase IVα. A second clone, CaM kinase IVβ, encoding a 56 kDa protein with a 28-amino acid insert at its N-terminus was subsequently obtained (Sakagami and Kondo, 1993). Analysis of the gene organization of CaM kinase IV and cloning of its cDNAs have clarified the presence of multiple bands in some preparations of the kinase (Sakagami and Kondo, 1993; Sun et al., 1995). The larger, CaM kinase IVβ, begins with exon I, whereas CaM kinase IVα begins with exon II. CaM kinase IVβ expression in adult rat was confined to cerebellar granule cells, consistent with the initial description of CaM kinase Gr (Ohmstede et al., 1989). The catalytic domain which is 58% homologous to CaM kinase II is at the N-terminal end of the protein. This is followed by a regulatory domain containing an autoinhibitory segment overlapping with a calmodulin-binding domain. In fact, eight residues (Phe293 to Lys300) in the autoinhibitory domain of CaM kinase II are found in a comparable position (Phe316 to Lys323) in CaM kinase IV. The C-terminal end of CaM kinase IV is distinct from the association domain of CaM kinase II. This region is not involved in self-assembly, as CaM kinase IV is monomeric (Ohmstede et al., 1989; Miyano et al., 1992; Cruzalegui and Means, 1993). The C-terminal domain is enriched in acidic residues, including stretches of polyglutamate residues, and it has been suggested that this may target it to subcellular sites such as the nucleus or axons (Jensen et al., 1991b).

The gene encoding CaM kinase IV also encodes a nonkinase protein called calspermin which is found in testis (Jones et al., 1991; Means et al., 1991b; Ohmstede et al., 1991) (Fig. 13.2). Calspermin is a 169-amino acid protein containing the calmodulin-binding and C-terminal domains of CaM kinase IV but lacking the catalytic domain of the kinase (Ono et al., 1989; Means et al., 1991b). Genomic cloning indicates that the use of three alternative promoters produces distinct transcripts for α- and β-CaM kinase IV and calspermin from a single gene (Sun et al., 1995). The transcript for calspermin is initiated at a testes-specific promoter in an intron of the CaM kinase IV gene.

The kinase phosphorylating CaM kinase IV (CaMK IV kinase) is itself a Ca^{2+}/calmodulin-dependent enzyme. CaMK IV kinase cloned from rat brain cDNA is comprised of 505 amino acids with a molecular mass of 56 kDA (Tokumitsu et al., 1995). It is related to the CaM kinase II family of kinases, with 30–40% sequence identity in its catalytic domain and no homology outside this domain. Like CaM kinase IV, it has a highly polar C-terminus with acidic residues accounting for 24% of the amino acids from residues 374–505 (Tokumitsu et al., 1995). A novel segment for CaM kinases is a 22-amino acid insert rich in Pro, Arg, and Gly in the catalytic domain, prior to the DGF kinase motif. Although the expressed protein binds calmodulin, there is no calmodulin-binding domain resembling those seen in CaM kinases I, II, and IV, although a segment (amino acids 444–457) with both basic and hydrophobic residues may comprise such a domain.

Domain Structure: MLCK

The MLCK family consists of monomeric proteins with similar catalytic cores and calmodulin-binding regulatory domains flanked by variable domains on both the N- and C-terminal sides. The long N-terminal end of smMLCK contains two repeats of unc1 (fibronectin type III) and one copy of unc2 (immunoglobulin C2), whereas one copy of unc1 follows the regulatory domain (Kemp et al., 1996) (Fig. 13.4). These flanking domains were originally found in the giant members of the MLCK family, twitchin kinase, the *unc22* gene product of *Caenorhabditis elegans*, and titin kinase, a large structural protein of muscle. The presence of these domains may thus provide a structural role for MLCK, or aid in the subcellular localization of MLCK to cytoskeletal elements. In addition, the extreme amino-terminus of smooth muscle MLCK contains an actin-binding region. Interestingly, the C-terminal portion alone of MLCK can be isolated from smooth muscle, and this protein, called telokin (Ito et al., 1989), appears to be an alternative product derived from the *MLCK* gene as a result of a second transcriptional initiation site within the gene. All of these N- and C-terminal structural features are lacking in skeletal muscle isoforms. It is interesting that alternative splicing of genes for CaM kinase II, CaM kinase IV, and MLCK each generate a nonkinase protein — α-KAP, calspermin, and telokin, respectively, containing the C-terminal portion of the kinase (Fig. 13.2).

Domain Structure: Phosphorylase Kinase

The γ-subunit of phosphorylase kinase follows a structural organization similar to other members of the CaM kinases family (Fig. 13.2). An autoregulatory domain (residues 302–367) lies just C-terminal to the catalytic domain (1–300) (reviewed in Heilmeyer, 1991). Unlike the other CaM kinases, there are two discontinuous calmodulin-binding domains (one spanning residues 287–331 and the other, residues 332–371) (Dasgupta et al., 1989). These segments of the γ-subunit likely interact with the δ-subunit (calmodulin) in the complex similar to the mechanism proposed for maintaining troponin C bound to troponin I in striated muscle. Synthetic peptides from these domains can interact with a single calmodulin molecule in solution and it is likely that they cooperate in binding calmodulin in the native complex.

Auto-inhibitory Gate Blocks the Active Site

Second messengers activate their cognate protein kinases by disrupting an autoinhibitory segment and thus deinhibiting them. The autoinhibitory

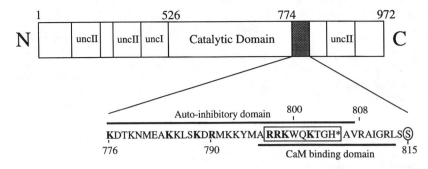

Figure 13.4 Proposed linear domain structure of chicken gizzard smooth muscle MLCK. An expanded view of the regulatory domain (shaded) indicates the positions of (1) the overlapping autoinhibitory and calmodulin-binding regions, (2) the pseudosubstrate sequence (boxed) with the phosphorylation site residue (H) denoted by an asterisk, (3) Ser815 (circled) which is phosphorylated in vivo by CaM kinase, resulting in decreased calmodulin-binding affinity, and (4) critical basic residues (bold) that are hypothesized to form charge–charge interactions with acidic residues in the catalytic domain, and thereby contribute to autoinhibition. (Figure adapted from Walsh, 1994; Kemp et al., 1996.)

segment serves as a gate of the catalytic site. The binding site for second messengers, e.g., Ca^{2+}/calmodulin, is adjacent to or overlapping with the autoinhibitory domain, and binding of the second messenger likely activates the kinases by displacing this inhibitory gate. Since calmodulin wraps around its target sequence and this sequence overlaps with the autoinhibitory domain of CaM kinases, the autoinhibitory domain can either maintain intrasteric inhibition of the active site or participate in intermolecular interactions with calmodulin, but not both (Kemp et al., 1994; James et al., 1995).

The autoinhibitory domain may be positioned in the active site, like a substrate, and mimic interactions with the catalytic site, thus serving as a pseudosubstrate. The concept of a pseudosubstrate was initially formulated in suggesting that PKI competitively blocked substrate binding to PKA by virtue of an Arg that would mimic the Arg (P−3) in its substrates (Demaille et al., 1977). The presence of such a sequence in the regulatory subunit of PKA and a substrate-like sequence in MLCK (which may inhibit the enzyme) were suggested to serve as pseudosubstrates (reviewed in Kemp et al., 1994; Taylor et al., 1990). We prefer the more general term, autoinhibitory segment, since such intrasteric regulation will not fully mimic interactions with a kinase having multiple substrates.

CaM Kinase II

The autoinhibitory domain of CaM kinase II has been examined with synthetic peptides, by truncation of recombinant enzyme or proteolyis of native enzyme, and by site-directed mutagenesis (reviewed

in Hanson and Schulman, 1992b; Kemp et al., 1996) (Fig. 13.5). As for other calmodulin-dependent enzymes, partial proteolysis of CaM kinase II generates a 30 kDa fragment with Ca^{2+}-independent activity, suggesting that the enzyme was under inhibitory control (LeVine and Sahyoun, 1987). The general location of an autoinhibitory domain was identified by showing that a proteolytic fragment ending at amino acid 293 is inactive, whereas further digestion to amino acid 271 produces an active enzyme (Yamagata et al., 1991). Recombinant techniques refined the location of the autoinhibitory domain by demonstrating that a construct truncated at amino acid 294 is inactive, whereas a construct truncated at amino acid 290 is constitutively active (Cruzalegui et al., 1992).

Peptides corresponding to amino acids 273–302 or 281–309 (of α-CaM kinase II) are effective inhibitors of CaM kinase ($IC_{50} = 1$–$2\,\mu M$). Peptide 281–309 can inhibit a catalytic fragment of CaM kinase and bind calmodulin (Colbran et al., 1988; Kelly et al., 1988). Binding of calmodulin to this peptide blocks its inhibitory effect on a catalytic fragment, suggesting that the calmodulin-binding segment extends into the inhibitory domain. Peptides (273–302 or 281–302) lacking the amino acids critical for calmodulin binding inhibit the kinase with or without calmodulin present (Malinow et al., 1989). The autoinhibitory domain in intact CaM kinase II differs from that of MLCK and others in that it is bifunctional, blocking both peptide substrate- and ATP-binding sites (Smith et al., 1992). Activation of CaM kinase II is accompanied by an increase in affinity for ATP of approximately 10-fold (Colbran, 1993) and increased incorporation of a photoaffinity label at the ATP-

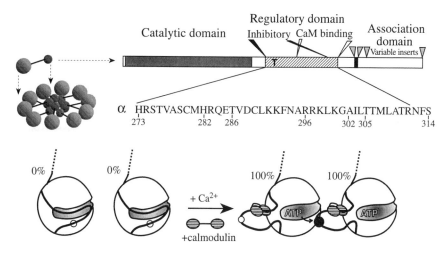

Figure 13.5 An expanded view of the domain structure of CaM kinase II showing the sequence of α-CaM kinase II. Displacement of the autoinhibitory domain activates two neighbouring subunits in a holoenzyme. Binding of Ca^{2+}/ calmodulin is also necessary to expose the autonomy site of one subunit for phosphorylation by an active neighbouring subunit in an inter-subunit reaction.

binding site (Shields et al., 1984). His282 has been suggested to be in the ATP-binding site because peptides containing a substitution of His282 with an Ala are no longer competitive with ATP (Smith et al., 1992).

A molecular understanding of how the autoinhibitory segment of CaM kinase II competes with ATP and peptide substrates must await the crystal structure, but some insights can be gained from biochemical analysis and comparison with the structures of PKA and CaM kinase I. Since the autoinhibitory segment contains the primary autophosphorylation site (Thr286) within an appropriate consensus of R-X-X-T, it might have been natural to expect that this region lies in the substrate-binding site. However, even when high ATP is used, to overcome the low affinity of the kinase for ATP in the basal state, there is no autophosphorylation of Thr286; rather, there is a slow autophosphorylation of Thr306, suggesting that it is closer to the P0 site than Thr286 in the basal state (Hanson and Schulman 1992a; Colbran, 1993) It is not known whether such basal autophosphorylation occurs in vivo to regulate the level of activatable kinase. This phosphorylation of Thr306 is an intra-subunit reaction (Mukherji and Soderling, 1994). Although this seemed paradoxical, subsequent studies described below demonstrate that Thr286 is not phosphorylated by the catalytic site on the subunit that it is inhibiting but by the catalytic site on a proximate neighboring subunit in the holoenzyme (Fig. 13.5; see below). Site-directed mutagenesis suggests that Arg297 is positioned at P−3 and Lys 300 is

at P0 (Mukherji and Soderling, 1995). Replacement of Lys300 with a Ser leads to phosphorylation of the Ser in the basal state. Computer modeling can accommodate such an assignment of Lys300 near the P0 site (Brickley et al., 1994).

The catalytic domain of CaM kinase (amino acids 1–270) can be modeled based on the homologous sequences in the crystal structure of PKA, but the positioning of the critical autoinhibitory segment is less clear (Cruzalegui et al., 1992; Brickey et al., 1994). In one model (Brickey et al., 1994), the region preceding the autoinhibitory segment is looped over the large lobe in order to position His282 in the adenosine-binding pocket to directly block ATP binding. This is followed by an α-helical region containing Thr286 near the active site, but with the side-chain hydroxyl facing away from it, and the sequence mimicking substrate binding by placement of Arg297 at P−3 and Lys300 at P0. Another model (Cruzalegui et al., 1992) uses the same orientation of the autoinhibitory domain seen in PKI and loops the segment preceding the autoinhibitor domain under the large lobe, positions Thr286 far from the active site, and anchors the inhibitory domain by placing Phe293 of CaM kinase II in a position of Phe (P−11) of PKI. This general arrangement is closer to the orientation of the autoinhibitory domain of twitchin kinase (Hu et al., 1994) and CaM kinase I (Goldberg et al., 1996), whose structures were recently determined. The data support an important role of the region near His282 and Thr286 in restricting binding of peptide substrate and ATP to CaM kinase II in the basal state. However, this region may do so indirectly by serving as an anchor to position sequences on its C-terminal side at the peptide- and ATP-binding sites.

CaM Kinases I and IV

Regulation of CaM kinase I by an autoinhibitory sequence can be inferred from analysis of synthetic peptides, mutagenesis of recombinant enzyme, and the crystal structure of the kinase. C-terminal truncation of rat CaM kinase I was used to test for an autoinhibitory segment at a position corresponding to that found in CaM kinase II (Haribabu et al., 1995; Yokokura et al., 1995). Indeed, constructs reduced to 1–321 of the human enzyme retained Ca^{2+}/calmodulin-stimulated activity, whereas 1–294 of the rat enzyme and 1–299 of the human enzyme neither bound calmodulin nor had significant Ca^{2+}-independent activity. Further truncation to 1–293/1–294 produced a constitutively active enzyme with activity comparable to the full-length enzyme maximally stimulated by Ca^{2+}/calmodulin (Haribabu et al., 1995; Yokokura et al., 1995). CaM kinase (294–321) inhibits the constitutively active fragment competitively with peptide substrate and noncompetitively with ATP. CaM kinases I and II therefore have an autoinhibitory domain in the same position relative to the catalytic domain and overlapping with the calmodulin-binding domain.

CaM kinase I is the only kinase with a Ca^{2+}/calmodulin-binding regulatory domain whose crystal structure has been solved (Goldberg et al., 1996). The small ATP-binding lobe comprises residues 10–100; the larger lobe is contained in residues 101–275, followed by the regulatory domain (autoinhibitory and calmodulin-binding domains) in residues 276–316. Comparison of the active form of PKA and the inactive form of CaM kinase I suggests that activation of CaM kinase I involves rotation of the two lobes into a "closed" conformation that helps to form the active site. The inactive conformation of CaM kinase I may be stabilized by interactions between residues 305–316 at the C-terminal end of the autoinhibitory domain and the ATP-binding domain. Activation of CaM kinase I by CaM kinase kinase involves the phosphorylation of a site (Thr177) in the activation loop that is comparable to activation of PKA and other kinases (Haribabu et al., 1995; Johnson et al., 1996). In the inactive state, the autoinhibitory domain may also serve to disorder the activation loop (Goldberg et al., 1996).

The regulatory domain forms a novel helix–loop–helix structure that is involved in hydrophobic interactions over an extensive area on the surface of both N- and C-terminal lobes of the kinase (Goldberg et al., 1996). The autoinhibitory domain likely maintains a low basal activity by both pseudosubstrate and other interactions. The consensus sequence for CaM kinase I includes a hydrophobic residue at P−5 and an Arg at P−3 (Table 13.1) which are likely simulated by Phe298 and Lys300 in the autoinhibitory domain, respectively. This feature is consistent with the pseudosubstrate hypothesis, although it is uncertain whether the conformation of the peptide-binding region in the autoinhibited kinase and in an activated kinase with substrate bound are the same. The remaining portion of the autoinhibitory domain does not mimic peptide substrate or ATP, as it loops away from the site of phosphate transfer and then wedges residues 305–316 outside the ATP-binding domain and distorts it. This leads to the positioning of a Phe from the ATP-binding loop at a site that would be occupied by the ribose ring of ATP. Thus, the autoinhibitory domain does not literally occupy the ATP-binding site as it does in twitchin kinase and in PKA complexed with PKI.

The autoinhibitory domain of CaM kinase IV has been roughly mapped based on homology with CaM kinase II. In fact, one of the autoinhibitory peptides used for CaM kinase II (peptide 281–302) also inhibits CaM kinase IV, although with lower potency (Tokumitsu et al., 1994). When CaM kinase II is truncated down to residues 1–290, the kinase becomes constitutively active (Cruzalegui et al., 1992). CaM kinase IV (1–293) would be predicted to lack calmodulin binding and to be similarly constitutively active based on sequence alignment and, indeed, this was found to be the case (Cruzalegui and Means, 1993). The autoinhibitory domain may include amino acids between residues 305 and 316 since recombinant constructs with acidic amino substitutions in this region are almost fully constitutively active (Tokumitsu et al., 1994).

MLCK

The pseudosubstrate hypothesis has been most thoroughly studied for MLCK (Kemp et al., 1994, 1996). The hypothesis was given impetus by the finding that the arrangement of basic residues in the calmodulin-binding domain of smMLCK and the arrangement of basic residues serving as substrate consensus determinants in MLC are similar (Kemp et al., 1987). Furthermore, peptides corresponding to this region in smMLCK (Kemp et al., 1987) and a similar region of skMLCK (Kennelly et al., 1987) inhibited constitutively active fragments of the respective kinases (Tanaka et al., 1980), providing direct support for an autoinhibitory domain in MLCK. Tryptic digestion of gizzard smMLCK yielded in inactive 64 kDa fragment (Thr283–Arg808) which could not be activated by Ca^{2+}/calmodulin, although it retained the catalytic core of the kinase (Pearson et al., 1988) (Fig.

13.4). Additional proteolysis at the C-terminus generated a 61 kDa fragment (Thr283–Lys779) that was constitutively active in the absence of Ca^{2+}/calmodulin because it removed the autoinhibitory domain. The peptide smMLCK (787–807) is autoinhibitory and peptide smMLCK (796–813) constitutes the core calmodulin-binding domain.

In its strictest sense, pseudosubstrate would be positioned in the peptide-binding site and would be autophosphorylated if it had a Ser or Thr in the P0 position. Furthermore, contacts between basic and hydrophobic residues that are important determinants of substrate consensus and the active site should be emulated by corresponding residues in the pseudosubstrate in the inactive form of the kinase. Indeed, recombinant smMLCK with the phosphorylation site of MLC introduced into the putative pseudosubstrate site as the sequence $RATSNV^{807}$ resulted in Ca^{2+}-independent autophosphorylation of the Ser (Bagchi et al., 1991). The crystal structure of MLCK has not yet been obtained. However, modeling of MLCK based on the crystal structure of a complex of PKI bound to PKA (Knighton et al., 1992) indicates that it is possible to arrange the binding of the autoinhibitory region (Ser787 to Val807) to the substrate-binding site in a pseudosubstrate fashion. The crystal structures of twitchin kinase and CaM kinase I show that the catalytic site is occupied by a segment of the kinase, consistent with the pseudosubstrate hypothesis (Hu et al., 1994).

Other data indicate that the core pseudosubstrate segment is not sufficient to explain inhibition of MLCK in the basal state. For example, substitution of the critical basic residues in the putative pseudosubstrate segment of nonmuscle MLCK with Ala did not produce a constitutively active kinase, indicating that autoinhibition exists even without substrate-like interactions (Shoemaker et al., 1990) Tanaka et al. (1995) have recently shown that a recombinant MLCK mutant truncated to produce a C-terminus ending at Tyr794, and thus lacking key basic residues (Fig. 13.4; Arg797, Arg798, Lys799, and Lys802) proposed to contribute to the pseudosubstrate sequence, is inactive. Autoinhibition of MLCK may depend upon multiple basic residues, extending from Arg776 to Lys802 (or chicken smMLCK). The basic residues listed above likely mimic those contained within the consensus phosphorylation sequence of the substrate. However, additional amino acids N-terminal of this region appear to interact with acidic residues contained in the substrate-binding/active site that are not expected to interact with the substrate (Knighton et al., 1992; Krueger et al., 1995). These and other residues in the truncated smMLCK above may be sufficient to provide autoinhibition in the complete absence of the proposed pseudosubstrate

sequence. The involvement of multiple contacts are also consistent with full autoinhibition in recombinant constructs with mutated pseudosubstrate sequences and with the crystal structure of the related twitchin kinase (Hu et al., 1994) and of CaM kinase I (Goldberg et al., 1996). Amino acids outside the pseudosubstrate segment may also contribute to autoinhibition by favoring an inactive conformation of the small and large lobes, as suggested for CaM kinase I (Goldberg et al., 1996).

Phosphorylase Kinase

Although phosphorylase kinase has been extensively studied, its autoregulatory mechanisms are less well understood than some of the other CaM kinases because of its more complex multi-subunit structure. Like other CaM kinases, the presence of an autoinhibitory domain was suggested by the ability to produce proteolytically active fragments of the enzyme. Proteolytic fragments or truncation mutants of the γ-subunit that lack its carboxyl-terminus display constitutive kinase activity (Harris et al., 1990; Huang et al., 1993). Recent findings indicate that synthetic peptides derived from the autoregulatory domain act as competitive inhibitors of kinase activity versus substrate. A peptide derived from the first calmodulin-binding region of the γ-subunit, peptide 303–327, was found to inhibit phosphorylation of phosphorylase b with K_i values of 0.3–2 μM, depending on whether the truncated γ-subunit or the phosphorylase kinase complex was used (Dasgupta and Blumenthal, 1995; Huang et al., 1995). This inhibition could be relieved by exogenous calmodulin, suggesting that the calmodulin-bound peptide 303–327 could no longer interact with the active site of the kinase. Within the autoregulatory domain are two calmodulin-binding regions, which may keep the δ-subunit (calmodulin) bound to γ-subunit in the absence of calcium (Dasgupta et al., 1989). This peptide is within the first calmodulin-binding region. Evidence for a role for the β-subunit in autoinhibition has also been presented (Sanchez and Carlson, 1993), and further study will be needed to clarify how and whether two distinct domains can have overlapping functions.

A distinct pseudosubstrate sequence (326–334) lying between the two calmodulin-binding regions in the autoregulatory domain was identified by use of a weighted matrix algorithm to score residues resembling the substrate consensus sequence of phosphorylase kinase (Lanciotti and Bender, 1995). A synthetic peptide corresponding to this sequence acts as a competitive inhibitor of the truncated, constitutively active form (1–300) of the γ-subunit. Consistent with the positioning of this segment in the catalytic site, replacement of Val332 by Ser332 in the full-length γ-subunit leads to autophosphorylation at the substituted Ser residue. The mechanism of this autophosphorylation is hypothesized to be intramolecular. These observations are thus consistent with a mechanism of intrasteric autoinhibition, contained within the γ-subunit, which maintains the phosphorylase kinase holoenzyme in its inactive state. This conclusion is further supported by the crystal structure of the catalytic domain (1–297) of the γ-subunit, which reveals a very similar structure to that of PKA and CaM kinase I (Owen et al., 1995).

Regulation of Multifunctional CaM Kinases by Phosphorylation

Covalent modification by phosphorylation of either the autoinhibitory gate, the calmodulin-binding domain, or the activation loop also regulates the state of activity of several of these enzymes. Maximal activity of CaM kinases I and IV is under control of CaM kinase kinases which generate a considerably more activatable form of the kinase. One feature that distinguishes CaM kinase II from the others is the ability of autophosphorylation to disrupt its autoinhibitory gate directly and thereby to generate a partially Ca^{2+}-independent species.

CaM Kinase II Activity is Potentiated by Autophosphorylation of the Inhibitory Gate

The regulatory domain of CaM kinase II is modified by two sequential steps of autophosphorylation which will be summarized here and elaborated below. The first involves Thr286 (Autonomy site) in the autoinhibitory domain and the second involves primarily Thr305Thr306 (inhibitory sites) in the calmodulin-binding domain. Ca^{2+}/calmodulin promotes an inter-subunit phosphorylation of the autonomy site while protecting the inhibitory sites from being phosphorylated. Modulation of the autonomy site traps bound calmodulin by greatly reducing its rate of dissociation from the kinase, as well as disabling the autoinhibitory domain so that the kinase remains partially active or autonomous after calmodulin dissociates. Phosphorylation of the inhibitory sites blocks rebinding of calmodulin; it is initiated by dissociation of calmodulin from an autonomous kinase and likely occurs by an intra-subunit reaction. Modulation of the autonomy site potentiates brief Ca^{2+} signals and may enable high-frequency Ca^{2+} spikes to be particularly effective in activating the kinase. No physiological role has been ascribed to the inhibitory phosphorylation.

Ca^{2+}/calmodulin-stimulated autophosphorylation of Thr286 converts CaM kinase II from an enzyme with one of the weakest affinities for calmodulin (45 nM) to an enzyme with one of the highest affinities (60 pM) (Meyer et al., 1992). This is primarily due to a reduced rate of dissociation of calmodulin at either high or low Ca^{2+}. Fluorescently labeled calmodulin dissociates at high Ca^{2+} in 0.4 second from an unphosphorylated kinase and in several hundred seconds after autophosphorylation, a 1000-fold change. Even when free Ca^{2+} is reduced below 100 nM, it takes approximately 3 seconds for calmodulin to dissociate from an autophosphorylated subunit, effectively trapping calmodulin and keeping that subunit fully active until calmodulin dissociates.

Autophosphorylated subunits do not fully deactivate after calmodulin dissociates and retain partial Ca^{2+}-independent or autonomous activity that can be as high as 70–80% of maximal Ca^{2+}/calmodulin-stimulated activity for some substrates. This initial finding for the *Aplysia* enzyme by Saitoh and Schwartz (1985) was followed by more extensive analysis for the purified enzyme from rat and identification of the autonomy site as Thr286 (reviewed in Hanson and Schulman, 1992b). Autophosphorylation of Thr286, like Ca^{2+}/calmodulin, may disrupt the autoinhibitory domain since this autophosphorylation also reduced the K_m for protein substrates (Ikeda et al., 1991). The region around Thr286 may be critical in appropriate folding or positioning of the autoinhibitory domain. Substitution of Asp for Thr286, and even protonation of His282, disrupts this region and produces an autonomous kinase, whereas substitution of Leu or Ala for Thr286 produces a kinase that neither traps calmodulin nor becomes autonomous (Fong et al. 1989; Hanson et al., 1989; Fong and Soderling, 1990; Waldmann et al., 1990; Waxham et al., 1990; Meyer et al., 1992; Smith et al., 1992). Autophosphorylation of Thr286 is therefore both necessary and sufficient for autonomy and trapping.

An explanation for the critical role of Thr286 has been suggested from the crystal structure of CaM kinase I (Goldberg et al., 1996). Thr286 in CaM kinase II and the corresponding amino acid (Val290) in CaM kinase I are flanked by nearby hydrophobic amino acids. In the crystal structure of CaM kinase I, these amino acids interact extensively with a hydrophobic channel on the surface of the catalytic core and are presumably displaced by binding of Ca^{2+}/calmodulin. Relaxation of the structure back to the inhibited state after dissociation of calmodulin requires re-establishment of these hydrophobic interactions, but this would be destabilized by the large negatively charged phosphate on Thr286 (Goldberg et al., 1996).

The mechanism of autophosphorylation of CaM kinase II may have important consequences for its biological function. Autophosphorylation occurs within each holoenzyme (Kuret and Schulman, 1985) but is it intra-subunit or inter-subunit? This has been addressed by use of monomeric constructs and modified heteromultimers. With monomers [e.g., truncated constructed CaM kinase (1–326)], this autophosphorylation is an intermolecular reaction, suggesting that the reaction in the holenzyme would be inter-subunit (Hanson et al., 1994). Autophosphorylation in the holoenzyme was examined by using two distinguishable types of subunits—one, an active kinase and the other, an inactive kinase that can only serve as an obligate substrate for autophosphorylation. The data show that Thr286 on the inactive kinase subunits becomes phosphorylated, indicating that the autophosphorylation in the holoenzyme is inter-subunit (Fig. 13.5) (Hanson et al., 1994; Mukherji and Soderling, 1994).

Calmodulin has a dual role in the autophosphorylation of Thr286. In a given reaction, one calmodulin functions to activate a subunit which serves as kinase (enzyme-directed effect) while another calmodulin must be bound to a neighbouring subunit that serves as substrate (substrate-directed effect) (Hanson et al., 1994). This was determined by testing monomeric CaM kinase II autophosphorylated at Thr286 with a disabled calmodulin-binding domain for its ability to phosphorylate two substrates—a standard peptide substrate and unphosphorylated monomeric CaM kinase II. While the kinase did not require Ca^{2+}/calmodulin to phosphorylate its standard substrate, it was a poor kinase for the added monomers until calmodulin was added. It is likely that displacement of the autoinhibitory domain by the binding of Ca^{2+}/calmodulin also exposes Thr286 for autophosphorylation by the neighbouring subunit of the holoenzyme. Calmodulin must therefore be simultaneously bound to both the subunit that serves as kinase and subunit that serves as substrate in each autophosphorylation reaction as depicted in Fig. 13.5. This has also been shown in holoenzymes engineered to have distinct obligate "kinase" subunits (made constitutively active but unable to bind calmodulin) and obligate "substrate" subunits (made catalytically inactive). Binding of calmodulin to the "substrate" subunits was needed for their phosphorylation by the constitutively active "kinase" subunits in the same holoenzyme (Rich and Schulman, 1998). A dual requirement for calmodulin would suggest that autophosphorylation of CaM kinase II and resultant trapping of calmodulin is a cooperative process. Indeed, the rate of calmodulin trapping and autophosphorylation were found to be cooperative with increasing calmodulin concentration (Hanson et al., 1994; De Koninck and Schulman, 1998). By contrast, activation of each subunit in the holoenzyme is independent of any other subunit, i.e., each is activated upon binding one molecule of calmodulin. This would suggest that activation occurs more readily

than autophosphorylation and autonomy, particularly with stimuli that only slightly increase Ca^{2+}/calmodulin.

What might be the consequences of autonomy and calmodulin trapping? One certain effect is the potentian of Ca^{2+} signals which can often be rather brief. This has been demonstrated in numerous systems: CaM kinase II is activated and converted to an autonomous species in response to influx of Ca^{2+} via ligand-gated channels and voltage-sensitive channels, as well as in response to stimulation of the phosphoinositol signaling pathway in response to IP_3 and release of intracellular Ca^{2+} (Fig. 13.1) (reviewed in Schulman, 1993). The autonomous activity is also increased by stimulation of Jurkat T cells (Nghiem et al., 1994) and the increase following induction of LTP is sustained for at least 1 hour (Fukunaga et al., 1993). These studies document that the in vitro characterized autophosphorylation of Thr286 also occurs in situ, presumably potentiating CaM kinase action, and indicates that CaM kinase II responds to a multitude of signal transduction pathways that elevate Ca^{2+}.

An equally important effect, Ca^{2+} spike frequency-dependent activation, would arise under conditions in which individual Ca^{2+} spikes do not maximally activate CaM kinase II. Submaximal activation is likely because of (1) the relative low affinity of the kinase for calmodulin (2) the very high concentration of the kinase in some tissues, (3) the sequestration of calmodulin in cells that reduces the effective free calmodulin, and (4) subsecond rises in Ca^{2+} in many excitable systems (Hanson and Schulman, 1992b; Schulman et al., 1992). The role of cooperative trapping of calmodulin in activation of a CaM kinase holoenzyme during repetitive Ca^{2+} spikes is shown in Fig. 13.6. At high frequency of stimulation, the interspike interval is too brief for significant dissociation of trapped calmodulin and binding of calmodulin would occur onto holoenzymes still retaining calmodulin trapped from a previous stimulus. The efficiency of autophosphorylation and calmodulin trapping increases during a series of rapid spikes because the probability of proximate neighbors with bound calmodulin increases with calmodulin occupancy (Hanson et al., 1994). Successive stimulation will involve an accumulation of calmodulin on CaM kinase II and an increase in kinase activity with each stimulus. At low frequency of stimulation, the long interspike interval would enable any trapped calmodulin to dissociate and the autophosphorylated subunits to be dephosphorylated so that there would be no accumulation of calmodulin with successive spikes. The cooperativity of trapping may allow the kinase to function as a frequency detector with a threshold frequency beyond which CaM kinase II becomes

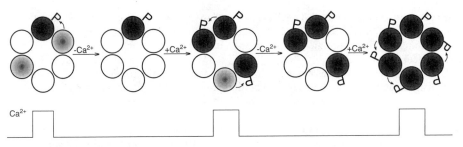

Figure 13.6 Trapping of calmodulin by a CaM kinase II holoenzyme during successive Ca^{2+} spikes. For simplicity, the holoenzyme is depicted as a hexamer and only half of the subunits bind calmodulin when first exposed to a brief Ca^{2+} spike (shaded subunits). Only one subunit is autophosphorylated (darker shading with a "P") after the first Ca^{2+} spike because there was only one pair of proximate neighbors with bound calmodulin. During the subsequent short interspike intervals, this subunit traps calmodulin and is active. The second spike again leads to addition of three calmodulins, but now a total of four subunits have bound calmodulin and are active. Furthermore, there is higher probability of autophosphorylation with four than with three subunits with bound calmodulin so that the second Ca^{2+} spike leads to an addition of two autophosphorylated and calmodulin-trapping subunits. By the third Ca^{2+} spike, the kinase is fully saturated and maximally active.

increasingly saturated with calmodulin (Hanson et al., 1994).

Indeed, activation and autophosphorylation of the kinase in vitro is sensitive to the frequency of Ca^{2+} oscillations (De Koninck et al., 1989). Kinase immobilized in plastic tubing and subjected to rapid superfusion with high and low Ca^{2+} and calmodulin indicated that it can decode the frequency of Ca^{2+} spikes into distinct amounts of kinase activity. The frequency response was modulated by the amplitude and duration of individual spikes, as well as the subunit composition and previous state of activation of the kinase (De Koninck et al., 1998).

Phosphorylation of the Calmodulin-Binding Domain Blocks Stimulation by Calmodulin

CaM Kinase II Calmodulin-binding domains of numerous proteins, including MLCK and CaM kinase II, are regulated by phosphorylation. For MLCK, this phosphorylation is catalyzed by other kinases whereas for CaM kinase II this occurs via autophosphorylation that is distinct from phosphorylation of the autonomy site described above. When calmodulin is removed from CaM kinase whose autonomy site is phosphorylated, it undergoes a "burst" of autophosphorylation at sites distinct from Thr286 (Miller and Kennedy, 1986; Hashimoto et al., 1987; Lou and Schulman, 1989). This second phase of autophosphorylation is inhibitory in that it blocks the ability of Ca^{2+}/calmodulin to activate the kinase (Hashimoto et al., 1987; Lou and Schulman, 1989). Thus, the activity of a kinase that undergoes complete sequential autophosphory-

lation at the autonomy site, followed by autophosphorylation of the inhibitory sites, is capped at the level of the Ca^{2+}-independent activity; the residual Ca^{2+}/calmodulin stimulatable activity is blocked. The Ca^{2+}-independent autophosphorylation occurs on Thr305, within the calmodulin-binding domain, and Ser314, just C-terminal of that domain; inhibition correlates with phosphorylation of Thr305 (Patton et al., 1990). Site-directed mutagenesis was used to show that Thr305, Thr306, and Ser314 are phosphorylated during the burst of Ca^{2+}-independent autophosphorylation and that phosphorylation of either Thr305 or Thr306 is sufficient to block subsequent binding and activation of the kinase by Ca^{2+}/camodulin (Hanson and Schulman, 1992a) (Fig. 13.5). Ca^{2+}-independent autophosphorylation of the PSD CaM kinase II leads to phosphorylation of Thr253, although a functional effect of this phosphorylation has not been described (Dosemeci et al., 1994). Phosphorylation within the calmodulin-binding domain is likely to be direct, as phosphorylation of the corresponding calmodulin-binding peptide blocks its interaction with calmodulin (Colbran and Soderling, 1990) and positioning of a phosphate moiety on Thr305 and/or Thr306 would be expected to interfere with the large area of interaction between calmodulin and this region (Meador et al., 1993).

CaM kinase IV Binding of calmodulin to CaM kinase IV can also be regulated by autophosphorylation. If Ca^{2+}/calmodulin is removed subsequent to phosphorylation of CaM kinase IV by CaM kinase IV kinase (see below) and autophosphorylation, the enzyme undergoes additional phosphorylation on a cluster of serine residues (Watanabe et al., 1996). The critical inhibitory phosphorylation is Ser332 in the calmodulin-binding domain, whose phosphorylation blocks Ca^{2+}/calmodulin-stimulated activity without affecting its Ca^{2+}-independent activity. Interestingly, the activity of CaM kinase IV kinase is itself inhibited by phosphorylation of its calmodulin-binding domain by PKA (Wayman et al., 1997; Tokumitsu et al., 1997).

MLCK The calmodulin-binding domain of MLCK is regulated by other kinases, as well as by autophosphorylation. MLCK is stoichiometrically phosphorylated in vitro at Ser815 and Ser828 by other protein kinases (i.e., PKA, PKC, and CaM kinase) (Stull et al., 1993). Phosphorylation of Ser815 just C-terminal of the calmodulin-binding domain (Fig. 13.4) decreases the calmodulin-binding affinity of MLCK, whereas phosphorylation of Ser828 does not affect enzymatic properties. Physiologically, phosphorylation of Ser815 is proposed to contribute to the "desensitization" of smooth muscle contraction that can be observed in response to calcium-elevating agents in situ (Stull et al., 1993).

In tracheal smooth muscle, phosphorylation of Ser815 was observed in response to elevation of intracellular Ca^{2+} by carbachol or high extracellular KCl, but not following elevation of cyclic AMP by isoproterenol, or stimulation of PKC by phorbol dibutyrate. This phosphorylation of ser815 could further be blocked by KN-62, an inhibitor of CaM kinase (Tansey et al., 1992). These findings support a role for CaM kinase in mediating the phosphorylation of smooth muscle MLCK at Ser815 in vivo. Since phosphorylation of Ser815 is prevented by the presence of bound calmodulin and MLCK has a higher affinity for calmodulin than CaM kinase II (~ 1 nM versus $200\text{--}100$ nM), it is unclear how Ser815 would be phosphorylated during a rise in Ca^{2+}. It is possible that this is achieved after Ca^{2+} levels decline to baseline by the autonomous form, CaM kinase II, whose affinity for calmodulin (~ 100 pM) is higher than that of MLCK. Phosphorylation of MLCK by CaM kinase II may thus be an important regulatory mechanism contributing to the desensitization of agonist-dependent regulation of smooth muscle contraction.

MLCK is also regulated by an intramolecular autophosphorylation at Thr803 and Ser815, within the autoregulatory domain, and Ser823 (Fig. 13.4) (Tokui et al., 1995). Autophosphorylation of Ser815 decreases the affinity of MLCK for calmodulin and can be blocked in the presence of Ca^{2+}/calmodulin, whereas phosphorylation of the other sites produces modest or no effects. It is unknown whether MLCK undergoes autophosphorylation in vivo; however,

this possibility may help to understand some aspects of the regulation of smooth muscle contraction in the presence of low intracellular calcium.

CaM Kinases I and IV are Activated by a Kinase Cascade

Each of the three multifunctional CaM kinases is subject to Ca^{2+}/calmodulin-stimulated phosphorylation. For CaM kinase II, this is its autophosphorylation; each subunit serves as a CaM kinase II kinase in phosphorylating a proximate neighbor in the holoenzyme. For CaM kinases I and IV, a distinct Ca^{2+}/calmodulin-dependent kinase termed CaM kinase kinase (CaMK-I kinase and/or CaMK-IV kinase) is responsible for much of the activation previously ascribed to autophosphorylation. The sites and consequences of phosphorylation of the three multifunctional kinases are distinct. CaM kinase II is regulated by phosphorylation of its autoinhibitory domain and leads to an increase in autonomous activity without a change in maximal activity. Both CaM kinase I and IV are phosphorylated in their activation loops at the active site, leading to an increase in maximal Ca^{2+}/calmodulin-stimulated activity. Autonomous activity of CaM kinase IV, but not of CaM kinase I, is also increased by phosphorylation which may involve additional sites.

The involvement of an activator protein, now known to be CaMK-I kinase, was first determined for CaM kinase Ia (DeRemer et al., 1992a, 1992b). Up to the last step of purification, rat brain CaM kinase Ia appeared to undergo activation by an autophosphorylation which increased maximal activity 50-fold, as previously found for bovine brain CaM kinase I (Nairn and Greengard, 1987). The final purification step, a calmodulin-affinity column, resolved CaM kinase Ia from its activator protein leading to an apparent loss of activity. Addition of fractions containing the activator reconstituted the apparent autophosphorylation and consequent activation of the kinase (DeRemer et al., 1992a, 1992b). A similar separation of activator and kinase was found for CaM kinase V (Mochizuki et al., 1993) a variant of CaM kinase I. A partially purified fraction was found to stimulate phosphorylation and activation of CaM kinase V (Mochizuki et al., 1993) and subsequently of CaM kinase IV as well (Sugita et al., 1994). Involvement of a CaM kinase kinase was also reinforced by parallel studies on CaM kinase IV described below.

The CaM kinase Ia activator was purified from porcine brain and shown to be a monomeric protein of 52 kDa capable of binding Ca^{2+}/calmodulin (Lee and Edelman, 1994). Although it was suspected to be a CaM-I kinase, it was only shown to be so by the demonstration that irreversible inhibition of the

ATP-binding site on the activator blocked the phosphorylation and a similar treatment of CaM kinase Ia did not (Lee and Edelman, 1995). Bacterially expressed CaM kinase I is much less active than the native enzyme and exhibits very slow autophosphorylation, suggesting that some of the autophosphorylation reported earlier for the native kinase was due to contamination by CaMK-I kinase (Picciotto et al., 1993). Addition of CaMK-Ia kinase to human CaM kinase I expressed using baculovirus led to a phosphorylation and a 25-fold increase of maximal CaM kinase I activity with only a small increase in autonomous activity (Haribabu et al., 1995). Activation of CaM kinase I (1–294) requires Ca^{2+}/calmodulin, even though this construct lacks a calmodulin-binding domain, suggesting that CaMK-Ia kinase is itself a calmodulin-dependent enzyme. CaMK-I kinase from rat brain (64 kDa) is larger than the porcine brain enzyme but also activates CaM kinase I by decreasing its K_m for substrate (Inoue et al., 1995). The activity of CaM kinase I alone is less than 10% of the activity of many other kinases and CaMK-I kinase enables it to achieve a level of activity that is comparable to the other kinases. Interestingly, CaM kinase Iβ can activate CaM kinase Iα by phosphorylation of Thr177, the site of CaM kinase I kinase activation (Naito et al., 1997).

The activation site on human CaM kinase I was identified by site-directed mutagenesis as Thr177 (Haribabu et al., 1995). A comparable site was earlier found to be the site of apparent autophosphorylation of the rat brain enzyme (Picciotto et al., 1993). Based on the crystal structure, this phosphorylation occurs in the activation loop at one end of the catalytic cleft (Goldberg et al., 1996), as found earlier for enzymes such as MAP kinase and PKA (Taylor and Radzio-Andzelm, 1994). Depolarization of PC12 cells followed by Ca^{2+} entry led to the phosphorylation of CaM kinase I, likely by CaM kinase I kinase, leading to its activation (Aletta et al., 1996).

The first indication that CaM kinase IV was similarly regulated in a kinase cascade was the finding that the enzyme expressed in either baculovirus (Cruzalegui and Means, 1993) or *Escherichia coli* (Okuno and Fukisawa, 1993) had very low activity and little autophosphorylation when compared with the native enzyme purified from brain. Some of the apparent autophosphorylation seen in initial studies (Ohmstede et al., 1989; McDonald et al., 1993) is likely due to a contaminating CaMK-IV kinase (Okuno et al., 1995). This was confirmed by the demonstration that addition of brain extract to the bacterially expressed enzyme led to its phosphorylation and marked activation (Okuno and Fujisawa, 1993). The tissue distribution of CaMK-IV kinase was similar to that found for CaM kinase IV, with brain being enriched with both kinases relative to nonneuronal tissue (Okuno and Fujisawa, 1993). The enzyme was subsequently purified from rat brain and found to be a Ca^{2+}/calmodulin-dependent protein kinase of 66–68 kDa (Okuno et al., 1994; Tokumitsu et al., 1994).

CaM kinase IV is phosphorylated by native CaMK-Ia kinase or recombinant CaMK-IV kinase on Thr196 situated in the activation loop in its catalytic site (Selbert et al., 1995; Tokumitsu and Soderling, 1996). Thr196 is homologous to Thr177 in CaM kinase I, suggesting a common mechanism for their activation. Activation increased maximal activity at least 10-fold and made the enzymes 38% autonomous of Ca^{2+}/calmodulin (Selbert et al., 1995). Activation is due to a decrease in K_m for substrate. The phosphorylation of CaM kinase IV may be more complex, however, and may involve multiple sites of phosphorylation (McDonald et al., 1993; Selbert et al., 1995; Okuno et al., 1995). This may result from a two-step process in which its initial phosphorylation by CaMK-IV kinase is followed by enhanced autophosphorylation (Okuno et al., 1995). This is consistent with the finding that replacement of Thr196 with an Ala blocks essentially all phosphorylation in the presence of CaM kinase Ia kinase, even though phosphorylation of Thr196 is only one of approximately five sites of phosphorylation (Selbert et al., 1995). Unresolved is the report that phosphorylation of Ser437 at the C-terminal end of the kinase correlates with activation (Kameshita and Fukisawa, 1993) since activation of CaM kinase IV (1–373), a truncated construct lacking this site, has been demonstrated (Tokumitsu et al., 1994). Analysis of the human CaM kinase IV indicates both in vitro and in situ a CaM kinase phosphorylated at a site that is homologous to Thr196 of the rat enzyme and that this is followed by autophosphorylation of serine residues at its N-terminal end (Ser12 and Ser13) (Chatila et al., 1996). Binding of Ca^{2+}/calmodulin, in the absence of a CaM kinase IV kinase, can stimulate the autophosphorylation of the N-terminal serines, relieving the enzyme of intrasteric inhibition and increasing its basal activity. CaM kinase IV becomes autonomous following activation and autophosphorylation, but this is not due to phosphorylation of the N-terminal serine residues and the mechanism for autonomous activity is not understood. Deactivation of CaM kinase IV is mediated by protein phosphatase 2A which is complexed with the kinase in both Jurkat T cells and brain (Westphal et al., 1998).

Can a single CaM kinase regulate both CaM kinase I and IV? The purified CaM kinase Ia kinase phosphorylates and activates both CaM kinase I (Haribabu et al., 1995) and CaM kinase IV (Selbert et al., 1995). Similarly, the kinase that phosphorylates CaM kinase V also activates CaM kinase IV (Sugita et al., 1994). CaMK-IV kinase has been cloned from rat brain and characterized (Tokumitsu et al., 1995). The recominant protein requires Ca^{2+}/calmodulin for activation and phosphorylates and activates CaM kinase IV with functional effects similar to those reported earlier for the purified enzyme (Tokumitsu et al., 1994, 1995). The enzyme has a similar domain structure as CaM kinase II, with partially overlapping autoinhibitory and calmodulin-binding domains (Tokumitsu et al., 1997). Northern blot analysis shows mRNA for this kinase to be highly expressed in brain, thymus, and spleen, as earlier found for the CaMK-IV kinase activity (Okuno and Fujisawa, 1993). It does not appear to correlate with the tissue distribution of mRNA of CaM kinase I, suggesting that a distinct CaMK-I kinase may be present in some tissues (Picciotto et al., 1993). If distinct, they are likely to be similar in specificity, since recombinant CaMK-IV kinase was clearly demonstrated to activate CaM kinase I but not CaM kinase II (Tokumitsu et al., 1995). Two species of CaMK-I have been purified from rat brain and are referred to as CaMk-I kinase-α and CaMK-I kinase-β (Edelman et al., 1996). Partial amino acid sequencing indicates that they are unlikely to be alternatively spliced variants and are therefore likely to be derived from two genes. The α isoform corresponds to the one previously cloned and characterized (Tokumitsu et al., 1995).

Calmodulin has both a kinase-directed and a substrate-directed role in the phosphorylation of CaM kinase I and IV (Hawley et al., 1995; Tokumitsu and Soderling, 1996), as previously found for autophosphorylation of CaM kinase II (Hanson et al., 1994). The requirement that calmodulin be bound to CaM kinase I in order for it to be phosphorylated on Thr177 was shown by the use of AMP-activated protein kinase kinase (AMPKK), which can weakly substitute for CaMK-I kinase (Hawley et al., 1995). Similarly, CaMK-IV kinase (1–434), a constitutively active form of the kinase, requires Ca^{2+}/calmodulin for the phosphorylation of intact CaM kinase IV but not for a fragment containing Thr196 (Tokumitsu and Soderling, 1996). Activation of CaM kinase IV must be highly dependent on calmodulin, since both CaM kinase kinase and CaM kinase IV are calmodulin-dependent enzymes. Basal activity may even be kept low by sequestering the enzyme in the nucleus away from calmodulin. It has been shown, for example, that activation of CaM kinase IV in hippocampal neurons by synaptic activity involves translocation of calmodulin to the nucleus where it likely activates CaM kinase IV kinase, which in turn activates CaM kinase IV (Deisseroth et al., 1998).

There is a striking mechanistic parallel in the activation of CaM kinase I and IV and AMPK in their respective cascades and in the potentiation of CaM kinase II by autophosphorylation. Each of the CaM kinases is phosphorylated by a CaMK kinase that is also activated by Ca^{2+}/calmodulin. CaM kinase II can be considered its own kinase kinase, utilizing an inter-subunit autophosphorylation in the holoenzyme rather than intermolecular phosphorylation by a specialized kinase kinase. The AMPK cascade, which regulates cellular responses to ATP depletion, also has a dual requirement for its second messenger (AMP). AMP must be bound to AMPK (Substrate-directed effect) in order for it to be phosphorylated by AMPKK which requires AMP for activity (kinase-directed effect) (Hawley et al., 1995). A dual requirement for second messengers is expected to result in cooperative activation as shown for CaM kinase II (Meyer et al., 1992; Hanson, et al., 1994). This would produce a steeper dependence on the second messenger and a greater signal-to-noise ratio for its activation in response to stimuli that elevate the second messenger or activator.

Approaches to Identifying CaM Kinase II Functions

Several criteria (Krebs, 1973; Schulman, 1988) can be used to identify in situ functions of a CaM kinase. (1) Physiological or pharmacological elevation of intracellular Ca^{2+} should modify the in situ function or lead to phosphorylation of the substrate in question. (2) There should be a correspondence of sites phosphorylated by Ca^{2+} elevation in situ and direct phosphorylation with purified kinase and substrate in vitro. (3) Inhibition of the kinase by a selective inhibitor should block the effect of elevated Ca^{2+} on the function or substrate. (4) Activators of the kinase should simulate elevated Ca^{2+}. (5) Introduction of a Ca^{2+}-independent form of the kinase should modify a subset of Ca^{2+}-stimulated cellular changes that are subserved by that particular kinase. (6) Reduction in the cellular content of the kinase by knockout or antisense should produce a corresponding reduction in Ca^{2+} signals mediated by the kinase.

A number of approaches have been used to fulfill these criteria for a variety of substrates and cellular processes (Table 13.3). Many of the examples concern various aspects of synaptic transmission since they exemplify the wide range of experimental approaches and their limitations, some of which are detailed below. Neurons use Ca^{2+} as a key index of synaptic activity and respond with increased release of neurotransmitters, resynthesis of neurotransmitters, modulation of synaptic strength, and a variety of other cellular changes. CaM kinase II appears to coordinate some of these cellular responses to elevated Ca^{2+}. For example, it phosphorylates tyrosine

Table 13.3 Selected In Situ Substrates and Functional Effects of CaM Kinase II in Mammalian Systems

Substrates	Effect	Approach
Synapsin I	Neurotransmitter release	^{32}P, I, CaMK*
Tyrosine hydroxylase	Catecholamine synthesis	^{32}P, I
MLCK	Smooth muscle contraction	^{32}P, I
MAP-2	Microtubule dynamics	^{32}P
Calcium channels[a]	Facilitation	I
c/EBPβ	Gene expression	CaMK*, ^{32}P
AMPA receptor	LTP	^{32}P
?	LTP; Learning and memory	I, KO, T
?	Block of IL-2 induction[b]	CaMK*
?	Chloride conductance	I, CaMK*

[a]L-type calcium channels (McCarron et al., 1992; Anderson et al., 1994), [b](Nghiem et al., 1994). All other studies are cited in Hanson and Schulman (1992b) and/or are detailed in the text. The approaches used are ^{32}P (correlation of phosphorylation sites in situ and in vitro), I (peptide inhibitors or KN-62), CaMK* (constitutively active CaM kinase), KO (knockout mice), and T (transgenic animals).

hydroxylase, the critical enzyme in catecholamine synthesis, at the same site in vitro that is known to be regulated by depolarization and Ca^{2+} influx in situ (Campbell et al., 1986; Griffith and Schulman, 1988; Waymire et al., 1988). CaM kinase II may adjust the release process to enable release to keep up with high-frequency stimulation by phosphorylation of synapsin I. This synaptic vesicle protein can cross-link synaptic vesicles with actin filaments away from release sites, thereby limiting release (Valtorta et al., 1992; Ceccaldi et al., 1995). Phosphorylation of synapsin I at two homologous C-terminal sites by CaM kinase II promotes release by decreasing its interaction with both synaptic vesicles and with actin filaments. Indeed, evidence for a facilitatory role of CaM kinase II was obtained by introduction of an autoinhibitory peptide or of an autonomous CaM kinase II into nerve terminal preparations (synaptosomes) by use of a free-thaw technique (Nichols et al., 1990). Introduction of the autophosphorylated autonomous kinase enhanced the rates of stimulated catecholamine and glutamate release.

The strength of neurotransmission at some synapses can be increased or decreased for hours or days based on the history of synaptic activity and CaM kinase II has been implicated in this activity-dependent plasticity (for reviews, see Lisman, 1994; Soderling, 1995). This process is often studied in the hippocampus where high-frequency stimulation leads to a long-term potentiation (LTP) and low-frequency stimulation leads to long-term depression (LTD), two processes that may underlie spatial learning (Malenka, 1994; Bliss and Collingridge, 1993). Interestingly, both processes are induced by a rise

in intracellular Ca^{2+} in the postsynaptic neuron. Brief repetitive stimulation of a hippocampal circuit increases Ca^{2+} for short periods but increases the autonomous species of CaM kinase for at least an hour (Fukunaga et al., 1993), leading to phosphorylation of substrates such as synapsin I and MAP-2 (Fukunaga et al., 1995). Induction of LTP also increases transcription of α- but not of β-CaM kinase mRNA and may be important for late changes in synaptic plasticity which are known to require transcription (Mackler et al., 1992). Within 30 minutes of tetanic stimulation of hippocampal CA1 neurons, there is an increase in both the level of CaM kinase II protein, as well as the level of phospho-Thr286 in apical dendrites, by a process that is likely to involve increased translation of dendritic mRNA for α-CaM kinase II (Ouyang et al., 1997).

Although the process of LTP is complex, a pharmacological approach using membrane-impermeant autoinhibitory peptides from CaM kinase II (273–302 and 281–302) can yield useful information in a system where the peptides can be microinjected while monitoring induction of LTP by elecrophysiological measurements. Indeed, these peptides, but not control peptides, blocked LTP in the hippocampus (Malinow et al., 1989), although the specificity of the peptides is in question (Smith et al., 1990; Hvalby et al., 1994). More selective inhibitors have been designed based on autocamtide-2 and autocamtide-3, two selective peptide substrates. Such inhibitors offer greater promise for use in microinjection experiments where they selectively block CaM kinase II and have little or no effect on PKC or on CaM kinases I and IV (Braun and Schulman 1995a; Ishida et al., 1995).

LTP may involve the regulation of AMPA receptors for glutamate (Tan et al., 1994), since they are responsible for both basal and potentiated synaptic currents in the hippocampus (Bliss and Collingridge, 1993). Introduction of an autonomous form of CaM kinase (autothiophosphorylated) into cultured hippocampal neurons increased the amplitude of the AMPA current 3-fold (McGlade-McCulloh et al., 1993) , and phosphorylation of the AMPA receptor by endogenous CaM kinase II was demonstrated in synaptosomes and PSD preparations. In culture, glutamate activates CaM kinase II and increases phosphorylation of AMPA receptors (Tan et al., 1994). Mutational analysis of the expressed AMPA receptor (GluR1) suggested that phosphorylation of Ser627, which is conserved in many non-NMDA glutamate receptors, accounts for the observed phosphorylation and increased current (Yakel et al., 1995). However, new analysis of the topology of the AMPA receptor suggests that this site is likely to be extracellular. Phosphoselective antibodies and phosphopeptide maps have been used to demonstrate that the key phosphorylation of GluR1 in transfected cells and in hippocampal slices is Ser831 rather than Ser624 (Mammen et al., 1997). Induction of LTP in hippocampal slices has been shown to increase autophosphorylation of CaM kinase II at Thr286, detected by use of phosphoselective antibody (Barria et al., 1997; Ouyang et al., 1997).

One of the more powerful techniques in the study of protein function is gene-targeted knockouts and this has been applied to α-CaM kinase (Silva et al., 1992a, 1992b). The knockout produced a null mutation in α-CaM kinase II with no change in the level of β-CaM kinase II, although some aspects of the activity and localization of the β isoform may be altered in the mutant since there is no α isoform for coassembly into α/β heteromultimers. The mice developed normally, since much of α-CaM kinase II develops postnatally, but LTP in hippocampal slices from the mutant animals was blocked. Furthermore, α-CaM kinase II mutants were learning-impaired in a Morris hidden-platform task which monitors their ability to learn spatial cues (Silva et al., 1992a). This supports a role of α-CaM kinase II in LTP, an important model of neuronal plasticity, and links LTP with certain forms of learning.

The critical role of autophosphorylation of CaM kinase II was demonstrated by generation of mice in which Thr286 was substituted with Ala286 in the endogenous α-CaM kinase II gene (Giese et al., 1998). Despite that the mutation did not alter the distribution of the kinase or its maximal stimulated activity, the mutant mice were as defective for induction of hippocampal lTP as the null mutants and displayed a similar inability to perform spatial learning tasks. This learning disability is likely due to the improper function of hippocampal place cells which are believed to be involved in developing cognitive maps of space (Cho et al., 1998).

A different approach to reduce CaM kinase II activity in vivo, expression of a minigene that encodes an autoinhibitory peptide, has been used to study the role of this kinase in behavioural conditioning in *Drosophila* (Griffith et al., 1993). Transgenic strains were generated with an inhibitory minigene under control of the heat-shock promoter. The flies were deficient in courtship conditioning, an associative learning assay. After exposure to a mated unresponsive female fly, male flies have a reduced courtship behavior when presented with a virgin female. This associative phenomenon is reduced in mutants; males that express the inhibitory minigene take less time to initiate a new courtship after exposure to a mated female and do it for an extended time when compared with controls.

Transgenic *Drosophila*, particularly after heat shock to induce even higher levels of the inhibitory minigene, show hyperexcitability at the neuromuscular junction that may relate to modulation of potassium channels (Griffith et al., 1994). Interestingly, α-CaM kinase II knockout mice also exhibited hyperexcitability (Butler et al., 1995). Stimuli that are normally subconvulsive induced prolonged limbic seizure activity in the mutants.

Introduction of a constitutively active form of a given CaM kinase should selectively simulate the actions of that kinase in situ. This can be achieved either by microinjecting CaM kinase II made Ca^{2+}-independent by prior autophosphorylation (or autothiophosphorylation to reduce dephosphorylation) or by partial proteolysis. Alternatively, a modified form of the kinase engineered to be constitutively active can be introduced in the form of nucleic acid into cells. Is the introduction of a Ca^{2+}-independent CaM kinase II sufficient to induce LTP? This has been examined by microinjection of a catalytic fragment of the kinase into CA1 hippocampal neurons (Lledo et al., 1995). Indeed, the kinase produced a time-dependent increase in postsynaptic currents with electrophysiological characteristics of authentic LTP. Introduction of Ca^{2+}/calmodulin into CA1 neurons is also sufficient to induce a time-dependent potentiation of synaptic responses (Wang and Kelly, 1995). These findings are consistent with an earlier study in which a vaccinia virus encoding CaM kinase II truncated down to its catalytic domain [CaM kinase (1–290)] was used to infect postmitotic CA1 neurons and potentiate synaptic responses to minimal or near-threshold stimuli within 4–8 hours after infection (Pettit et al., 1994). The augmentation of synaptic strength by LTP and CaM kinase II (or Ca^{2+}/calmodulin) appears to share a common mechanism since a prior induction of LTP attenuated potentiation by introduction of the kinase and a prior potentiation by the kinase (or Ca^{2+}/calmodulin) blocks LTP (Pettit et al., 1994; Lledo et al., 1995; Wang and Kelly, 1995).

Changes in AMPA receptor function occurs during maturation of synapses, as well as by induction of LTP in mature neurons, and both phenomena may involve the appearance of AMPA receptors on the surface of postsynaptic sites that previously only exhibited functional NMDA receptors. Postsynaptic expression of a constitutively active form of CaM kinase II in *Xenopus* tadpoles mimicked such maturation of synapses seen during development (Wu et al., 1996). CaM kinase II may act as an activity-dependent mediator of morphological maturation of neurons, as well as of AMPA receptors. Increasing CaM kinase II activity in young neurons by viral expression of a constitutively active form of the enzyme slowed dendritic growth to a level seen in mature neurons, whereas inhibition of the kinase increased dendritic growth (Wu and Cline, 1998).

Transgenic expression of a catalytically active form of CaM kinase II (with Asp in place of Thr286) implicates the kinase in a critical regulation of synaptic strength in hippocampus (Mayford et al., 1995). Placing the transgenic construct under its own promoter leads to selective expression of the mRNA in forebrain, including the dendritic layers of the hippocampus, similar to the targeting of the endogenous mRNA. Both LTD and LTP could be induced in these mice but there was a rightward shift in the frequency response that favored LTD (Mayford et al., 1995) and a disruption of spatial learning (Bach et al., 1995). Learning spatial cues involves hippocampal place cells, which are selectively activated when the animal is at a specific location relative to spatial cues. The hippocampus of mice expressing the constitutively active form of the kinase exhibited fewer place cells and the cells which were present were less precise and less stable than in wild-type animals (Rotenberg et al., 1996). The autonomous kinase appears to regulate this threshold frequency, suggesting that CaM kinase II may change the rules that govern the relationship between synaptic activity and the direction in change of synaptic strength (Bear, 1995; Deisseroth et al., 1995; Mayford et al., 1995). New technical advances enable regulated expression of the constitutively active CaM kinase II (Mayford et al., 1996). Similar deficits in the induction of LTP at low frequency and in spatial memory were found in hippocampal CA1 neuron (postsynaptic) expression of the activated kinase under a tetracycline transactivator control as found in animals with unregulated transgene expression. Suppression of transgene expression reversed both the physiological and memory deficits, suggesting that the deficits are not due to effects of the activated kinase during development.

Conclusions and Future Directions

CaM kinases mediate many cellular responses to elevated Ca^{2+}. The response to Ca^{2+} includes both a focused phosphorylation of select proteins by dedicated CaM kinases, such as MLCK and phosphorylase kinase, and a coordinated phosphorylation of a host of substrate proteins by the CaM kinase II family of multifunctional CaM kinases. As new CaM kinases are studied and their regulation is understood, common regulatory themes have emerged. Ca^{2+}/calmodulin serves to displace an auto-inhibitory domain from a catalytic domain that would otherwise be active. Activity of the multifunctional CaM kinases is increased or potentiated by phosphorylation. It is probably more than fortuitous that this phosphorylation requires that calmodulin be bound to the protein being phosphorylated as well as to the protein serving as kinase in the reaction. There are a number of intriguing consequences of the CaM kinase cascade involved in the regulation of CaM kinases I and IV and in the autophosphorylation of CaM kinase II. The cooperative requirement for calmodulin provides for a narrow range of Ca^{2+}/calmodulin concentrations that trigger the cascade or activation. For CaM kinase II, this may also enable it to function as a detector of the frequency by which Ca^{2+} levels oscillate or spike.

Important issues remaining for the CaM kinase field include (1) a molecular understanding for regulation of all of these kinases by the autoinhibitory domain and mechanism for activation by calmodulin (2) an appreciation for the functions regulated by each of the kinases. Identification of new substrates will require more selective inhibitors and a fuller application of mouse knockouts, dominant-negative constructs, and antisense technology; (3) an understanding of response-specificity by these kinases. Do all substrates of a given kinase in a cell get phosphorylated in response to elevated Ca^{2+} regardless of the stimulus or does a circumscribed localization of kinase, substrate, and elevated Ca^{2+} enable a more focused response-specificity? This will require additional analysis of Ca^{2+} signaling, as well as identification of anchoring proteins for specific kinase isoforms followed by in situ analysis of specific signal transduction pathways; and finally (4) knowledge about the network of modulatory enzymes, including CaM kinases, other kinases, and phosphoprotein phosphatases that enable cells to change their functional states rapidly in response to extracellular signals.

References

Aletta, J. M., Selbert, M. A., Nairn, A. C., and Edelman, A. M. (1996) Activation of a calcium-calmodulin-dependent protein kinase I cascade in PC12 cells. *J. Biol. Chem.* 271: 20930–20934.

Anderson, M. E., Braun, A. P. Schulman, H., and Premack, B. A. (1994) Multifunctional Ca^{2+}/calmodulin-dependent protein kinase mediates Ca^{2+}-induced enhancement of the L-type Ca^{2+} current in rabbit ventricular myocytes. *Circ. Res.* 75: 854–861.

Ando, S., Tokui, T., Yamauchi, T., Sugiura, H., Tanabe, K., and Inagaki, M. (1991) Evidence that Ser-82 is a unique phosphorylation site on vimentin for Ca^{2+}/calmodulin-dependent protein kinase II. *Biochem. Biophys. Res. Commun.* 175: 955–962.

Bach, M. E., Hawkins, R. D., Osman, M., Kandel, E. R., and Mayford, M. (1995) Impairment of spatial but not contextual memory in CaMKII mutant mice with a selelctive loss of hippocampal LTP in the range of the theta frequency. *Cell* 81: 905–915.

Bagchi, I. C., Kemp, B. E., and Means, A. R. (1991) Intrasteric regulation of myosin light chain kinase: the pseudosubstrate prototope binding to the active site. *Mol. Endocrinol.* 6: 621–626.

Barria, A., Muller, D., Derkach, V., Griffith, L. C., and Soderling, T. R. (1997) Regulatory phosphorylation of AMPA-type glutamate receptors by CaM-KII during long-term potentiation. *Science* 267: 2042–2045.

Bayer, K. U., Harbers, K., and Schulman, H. (1998) αKAP is an anchoring protein for a novel CaM kinase II isoform in skeletal muscle. *EMBO. J.* 17: 5598–5605.

Bayer, K. U., Lohler, J., and Harbers, K., (1996) An alternative, nonkinase product of the brain-specifically expressed Ca^{2+}/calmodulin-dependent kinase II alpha isoform gene in skeletal muscle. *Mol. Cell. Biol.* 16: 29–36.

Bear, M. F. (1995) Mechanism for a sliding synaptic modification threshold. *Neuron* 15: 1–4.

Benefenati, F., Valtorta, F., Rubenstein, J. L., Gorelick, F. S., Greengard, P., and Czernik, A. J. (1992) Synaptic vesicle-associated Ca^{2+}/calmodulin-dependent protein kinase II is a binding protein for synapsin I. *Nature* 359: 417–420.

Bennett, M. K. and Kennedy, M. B. (1987) Deduced primary structure of the β subunit of brain type II Ca^{2+}/calmodulin-dependent protein kinase determined by molecular cloning, *Proc. Natl. Acad. Sci. USA* 84: 1794–1798.

Bliss, T. V. P. and Collingridge, G. L. (1993) A synaptic model of memory: long-term potentiation in the hippocampus. *Nature* 361: 31–39.

Braun, A. and Schulman, H. (1995a) A non-selective cation current activated via the multifunctional Ca^{2+}/calmodulin-dependent protein kinase in human epithelial cells. *J. Physiol* 488: 37–55.

Braun, A. P. and Schulman, H. (1995b) The multifunctional calcium/calmodulin-dependent protein kinase: from form to function. *Annu. Rev. Physiol.* 57: 417–445.

Brickey, D. A., Bann, J. G., Fong, Y.-L., Perrino, L., Brennan, R. G., and Soderling, T. R. (1994) Mutational analysis of the autoinhibitory domain of calmodulin kinase II. *J. Biol. Chem.* 269: 29047—29054.

Brocke, L., Srinivasan, M., and Schulman, H. (1995) Developmental and regional expression of multifunctional Ca^{2+}/calmodulin-dependent protein kinase isoforms in rat brain. *J. Neurosci.* 15: 6797–6808.

Bulleit, R. F., Bennett, M. K., Molloy, S. S., Hurley, J. B., and Kennedy, M. B. (1988) Conserved and variable regions in the subunits of brain type II Ca^{2+}/calmodulin-dependent protein kinase. *Neuron* 1: 63–72.

Butler, L. S., Silva, A. J., Abeliovich, A., Watanabe, Y., Tonegawa, S., and McNamara, J. O. (1995) Limbic epilepsy in transgenic mice carrying a Ca^{2+}/calmodulin-dependent kinase II α-subunit mutation. *Proc. Natl. Acad. Sci. USA* 92: 6852–6855.

Campbell, D. G., Hardie, D. G., and Vulliet, P. R. (1986) Identification of four phosphorylation sites in the N-terminal region of tyrosine hydroxylase. *J. Biol. Chem.* 261: 10489–10492.

Ceccaldi, P.-E., Grohovaz, F., Benefanti, F., Chieregatti, E., Greengard, P., and Valtorta, F. (1995) Dephosphorylated synapsin I anchors synaptic vesicles to actin cytoskeleton: an analysis by videomicroscopy. *J. Cell. Biol.* 128: 905–912.

Chatila, T., Anderson, K. A., Ho, N., and Means, A. R. (1996) A unique phosphorylation-dependent mechanism for the activation of Ca^{2+}/calmodulin-dependent protein kinase type IV/GR. *J. Biol. Chem.* 271: 21542–21548.

Cho, F. S., Phillips, K. S., Bogucki, B., and Weaver, T. E. (1994) Characterization of a rat cDNA clone encoding calcium/calmodulin-dependent protein kinase I. *Biochim. Biophys. Acta* 1224: 156–160.

Cho, K., Wall. J. B., Pugh, P. C., Ito, M., Mueller, S. A., and Kennedy, M. B. (1991) The α subunit of type II Ca^{2+}/calmodulin-dependent protein kinase is highly conserved in *Drosophila. Neuron* 7: 439–450.

Cho, Y. H., Giese, K. P., Tanila, H., Silva, A. J., and Eichenbaum, H. (1998) Abnormal hippocampal spatial representations in αCaMKIIT^{286A} and CREB$^{\alpha\Delta-}$mice. *Science* 279: 867–869.

Cohen, P., Burchell, A., Foulkes, J. G., Cohen, P. T. W., Vanaman, T. C., and Nairn, A. C. (1978) Identification of the Ca^{2+}-dependent modulator protein as the fourth subunit of rabbit skeletal muscle phosphorylase kinase. *FEBS Lett.* 92: 287–293.

Colbran, R. J. (1992) Regulation and role of brain calcium/calmodulin-dependent protein kinase II. *Neurochem. Int.* 21: 469–497.

Colbran, R. J. (1993) Inactivation of Ca^{2+}/calmodulin-dependent protein kinase II by basal autophosphorylation. *J. Biol. Chem.* 268: 7163–7170.

Colbran, R. J. and Soderling, T. R. (1990) Calcium/calmodulin-independent autophosphorylation sites of calcium/calmodulin-dependent protein kinase II. Studies on the effect of phosphorylation of threonine 305/306 and serine 314 on calmodulin binding using synthetic peptides. *J. Biol. Chem.* 265: 11213–11219.

Colbran, R. J., Fong, Y. L., Schworer, C. M., and Soderling, T. R. (1988) Regulatory interactions of the calmodulin-binding, inhibitory, and autophosphorylation domains of the Ca^{2+}/calmodulin-dependent protein kinase II. *J. Biol. Chem.* 263: 18145–18151.

Cruzalegui, F. H. and Means, A. R. (1993) Biochemical characterization of the multifunctional Ca^{2+}/calmodulin-dependent protein kinase type IV expressed in insect cells. *J. Biol. Chem.* 268: 26171–26178.

Cruzalegui, F. H., Kapiloff, M. S., Morfin, J. P., Kemp, B. E., Rosenfeld, M. G., and Means, A. R. (1992) Regulation of intrasteric inhibition of the multifunctional calcium-calmodulin-dependent protein kinase. *Proc. Natl. Acad. Sci. USA* 89: 12127–12131.

Dabrowska, R., Aromatorio, D., Sherry, J. M. F., and Hartshorne, D. J. (1978) Modulator protein as a component of the myosin light chain kinase from chicken gizzard. *Biochemistry* 17: 253–258.

Dale, S., Wilson, W. A., Edelman, A. M., and Hardie, D. G. (1995) Similar substrate recognition motifs for mammalian AMP-activated protein kinase, higher plant HMG-CoA reductase kinase-A, yeast SNF1, and mammalian calmodulin-dependent protein kinase I. *FEBS Lett.* 361: 191–195.

Dasgupta, M. and Blumenthal, D. K. (1995) Characterization of the regulatory domain of the gamma-subunit of phosphorylase kinase. *J. Biol. Chem.* 270: 22283–22289.

Dasgupta, M., Honeycutt, T., and Blumenthal, D. K. (1989) The gamma-subunit of skeletal muscle phosphorylase kinase contains two noncontiguous domains that act in concert to bind calmodulin. *J. Biol. Chem.* 264: 17156–17163.

Deisseroth, K., Bito, H., Schulman, H., and Tsien, R. W. (1995) A molecular mechanism for metaplasticity. *Curr. Biol.* 5: 1334–1338.

Deisseroth, K., Heist, E. K., and Tsien, R. W. (1988) Translocation of calmodulin to the nucleus supports CREB phosphorylation in hippocampal neurons. *Nature* 392: 198–202.

De Koninck, P. and Schulman, H. (1998) Sensitivity of Ca^{2+}/calmodulin-dependent protein kinase II to the frequency of Ca^{2+} oscillations. *Science* 279: 227–230.

Demaille, J. G., Peters, K. A., and Fischer, E. H. (1977) Isolation and properties of the rabbit skeletal muscle protein inhibitor of $3',5'$-monophosphate-dependent protein kinases. *Biochemistry* 16: 3080–3086.

DeRemer, M. F., Saeli, R. J., Brautigan, D. L., and Edelman, A. M. (1992a) Ca^{2+}-calmodulin-dependent protein kinases Ia and Ib from rat brain. II. Enzymatic characteristics and regulation of activities by phosphorylation and dephosphorylation. *J. Biol. Chem.* 267: 13466–13471.

DeRemer, M. F., Saeli, R. J., and Edelman, A. M. (1992b) Ca^{2+}-calmoduli-dependent protein kinases Ia and Ib from rat brain. I. Identification, purification, and structural comparisons. *J. Biol. Chem.* 267: 13460–13465.

Dosemeci, A., Gollop, N., and Jaffe, H. (1994) Identification of a major autophosphorylation site on postsynaptic density-associated Ca^{2+}/calmodulin-dependent protein kinase. *J. Biol. Chem.* 269: 31330–31333.

Dunkley, P. R. (1991) Autophosphorylation of neuronal calcium/calmodulin-stimulated protein kinase II. *Mol. Neurobiol.* 5: 179–202.

Edelman, A. M., Mitchelhill, K., Selbert, M. A., Anderson, K. A., Hook, S. S., Stapleton, D., Goldstein, E. G., Means, A. R., and Kemp, B. E. (1996) Multiple Ca^{2+}-calmodulin-dependent protein kinase kinases from rat brain. Purification, regulation by Ca^{2+}-calmodulin, and partial amino acid sequence. *J. Biol. Chem.* 271: 10806–10810.

Edman, C. F. and Schulman, H. (1994) Identification and characterization of δ_B-CaM kinase and δ_C-CaM kinase from rat heart: two new multifunctional Ca^{2+}/calmodulin-dependent protein kinase isoforms. *Biochim. Biophys. Acta* 1221: 90–102.

Edman, C. F., George, S. E., Means, A. R., Schulman, H., and Yaswen, P. (1994) Selective activation and inhibition of calmodulin dependent enzymes by a calmodulin-like protein found in human epithelial cells. *Eur. J. Biochem.* 226: 725–730.

Enslen, H., Sun, P., Brickey, D., Soderling, S. H., Klamo, E., and Soderling, T. R. (1994) Characterization of Ca^{2+}/calmodulin-dependent protein kinase IV: role in transcriptional regulation. *J. Biol. Chem* 269: 15520–15527.

Erondu, N. E. and Kennedy, M. B. (1985) Regional distribution of type II Ca^{2+}/calmodulin-dependent protein kinase in rat brain. *J. Neurosci.* 5: 3270–3277.

Fong, Y.-L. and Soderling, T. R. (1990) Studies on the regulatory domain of Ca^{2+}/calmodulin-dependent protein kinase. II. Functional analyses of arginine 283 using synthetic inhibitory peptides and site-directed mutagenesis of the α subunit. *J. Biol. Chem.* 265: 11091–11097.

Fong. Y.-L., Taylor, W. L., Means, A. R., Soderling, T. R. (1989) Studies of the regulatory mechanism of Ca^{2+}/calmodulin-dependent protein kinase II. Mutation of threonine 286 to alanine and aspartate. *J. Biol. Chem.* 264: 16759–16763.

Frangakis, M. V., Chatila, T., Wood, E. R., and Sahyoun, N. (1991) Expression of a neuronal Ca^{2+}/calmodulin-dependent protein kinase, CaM kinase-Gr, in rat thymus. *J. Biol.Chem.* 266: 17592–17596.

Fukunaga, K., Muller, D., and Miyamoto, E. (1995) Increased phosphorylation of Ca^{2+}/calmodulin-dependent protein kinase II and its endogenous substrates in the induction of long-term potetiation. *J. Biol. Chem.* 270: 6119–6124.

Fukunaga, K., Stoppini, L., Miyamoto, E., and Muller, D. (1993) Long-term potentiation is associated with an increased activity of Ca^{2+}/calmodulin-dependent protein kinase II. *J. Biol. Chem.* 268: 7863–7867.

Fukunaga, K., Yamamoto, H., Matsui, K., Higashi, K.., and Miyamoto, E. (1982) Purification and characterization of a Ca^{2+}-and calmodulin-dependent protein kinase from rat brain. *J.Neurosci.* 39: 1607–1617.

Gallagher, P. J., Herring, B. P., Griffin, S. A., and Stull, J. T. (1991) Molecular characterization of a mammalian smooth muscle myosin light chain kinase. *J. Biol. Chem.* 266: 23936–23944.

Giese, K. P., Fedorov, N. B., Filiplowski, R. K., and Silva, A. J. (1998) Autophosphorylation at Thr^{286} of the α calcium-calmodulin kinase II in LTP and learning. *Science* 279: 870–873.

Goldberg, J., Nairn, A. C., and Kuriyan, J. (1996) Structural basis for the auto-inhibition of calcium/calmodulin-dependent protein kinase-I. *Cell* 84: 875–887.

Graves, D. J. (1983) Use of peptide substrates to study the specificity of phosphorylase kinase phosphorylation. *Methods Enzymol.* 99: 268–278.

Griffith, L. C. (1993) The diversity of calcium/calmodulin-dependent protein kinse II isoforms in *Drosophila* is generated by alternative splicing of a single gene. *J. Neurochem.* 61: 1534–1537.

Griffith, L. C., and Schulman, H. (1988) The multifunctional Ca^{2+}/calmodulin-dependent protein kinase mediates Ca^{2+}-dependent phosphorylation of tyrosine hydroxylase. *J. Biol. Chem.* 263: 9542–9549.

Griffith, L. C., Verselis, L. M., Aitken, K. M., Kyriacou, C. P., Danho W., and Greenspan R. J. (1993) Inhibition of calcium/calmodulin-dependent protein kinase in *Drosophila* disrupts behavioral plasticity. *Neuron* 10: 501–509.

Griffith, L. C., Wang, J., Zhong, Y, Wu, C. F., and Greenspan, R. J. (1994) Calcium/calmodulin-dependent protein kinase II and potassium channel subunit eag similarly affect plasticity in *Drosophila*. *Proc. Natl. Acad. Sci. USA* 91: 10044–10048.

GuptaRoy, B., and Griffith, L. C. (1996) Functional heterogeneity of alternatively spliced isoforms of *Drosophila* Ca^{2+}/calmodulin-dependent protein kinase II. *J. Neurochem.* 66: 1282–1288.

GuptaRoy, B., Beckingham, K., and Griffith, L. C. (1996) Functional diversity of alternatively spliced isoforms of *Drosophila* Ca^{2+}/calmodulin-dependent protein kinase II: a role for the variable domain in activation. J. Biol. Chem. 271: 19846–19851.

Hanissian, S. H., Frangakis, M., Bland, M. M., Jawahar, S., and Chatila, T. A. (1993) Expression of a Ca^{2+}/calmodulin-dependent protein kinase, CaM kinase-Gr, in human T lymphocytes. Regulation of kinase activity by T cell receptor signaling. *J. Biol. Chem.* 268: 20055–20063.

Hanks, S. K. and Quinn, A. M. (1991) Protein kinase catalytic domain sequence database: identification of conserved features of primary structure and classification of family members. *Methods Enzymol.* 200: 38–62.

Hanley, R. M., Means, A. R., Ono, T., Kemp, B. E., Burgin, K. E., Waxham, N., and Kelly, P. T. (1987) Functional analysis of a complementary DNA for the 50-kilodalton subunit of calmodulin kinase II. *Science* 237: 293–297.

Hanson, P. I. and Schulman, H. (1992a) Inhibitory autophosphorylation of multifunctional Ca^{2+}/calmodulin-dependent protein kinase analyzed by site-directed mutagenesis. *J. Biol. Chem.* 267: 17216–17224.

Hanson, P. I. and Schulman H. (1992b) Neuronal Ca^{2+}/calmodulin-dependent protein kinases. *Annu. Rev. Biochem.* 61: 559–601.

Hanson, P. I., Kapiloff, M. S., Lou, L. L., Rosenfeld, M. G., and Schulman, H. (1989) Expression of a multifunctional Ca^{2+}/calmodulin-dependent protein kinase and mutational analysis of its autoregulation. *Neuron* 3: 59–70.

Hanson, P. I., Meyer, T., Stryer, L., and Schulman, H. (1994) Dual role of calmodulin in autophosphorylation of multifunctional CaM kinase may underlie decoding of calcium signals. *Neuron* 12: 943–956.

Hariabu, B., Hook, S. S., Selbert, M. A., Goldstein, E. G., Tomhave, E. D., Edelman, A. M., Snyderman, R., and Means, A. R. (1995) Human calcium-calmodulin dependent protein kinase I: cDNA cloning, domain structure and activation by phosphorylation at threonine-177 by calcium-calmodulin dependent protein kinase I kinase. *EMBO J.* 14: 3679–3686.

Harris, W. R., Malencik, D. A., Johnson, C. M., Carr, S. A., Roberts, G. D., Byles, C. A., Anderson, S. R., Heilmeyer, L. M., Fischer, E. H., and Crabb, J. W. (1990) Purification and characterization of catalytic fragments of phosphorylase kinase gamma subunit missing a calmodulin-binding domain. *J. Biol. Chem.* 265: 11740–11745.

Hashimoto, Y., Schworer, C. M., Colbran, R. J., and Soderling, T. R. (1987) Autophosphorylation of Ca^{2+}/calmodulin-dependent protein kinase II. Effects on total and Ca^{2+}-independent activities and kinetic parameters. *J. Biol. Chem.* 262: 8051–8055.

Hawley, S. A., Selbert, M. A., Goldstein, E. G., Edelman, A. M., Carling, D., and Hardie, D. G. (1995) 5′-AMP activates the AMP-activated protein kinase cascade, and Ca^{2+}/calmodulin activates the calmodulin-dependent protein kinase I cascade, via three independent mechanisms. *J. Biol. Chem.* 270: 27186–27191.

Heilmeyer, L. M. G. (1991) Molecular basis of signal integration in phosphorylase kinase. *Biochim. Biophys. Acta* 1094: 168–174.

Heist, K. E., Srinivasan, M., and Schulman, H. (1998) Phosphorylation at the nuclear localization signal of Ca^{2+}/calmodulin-dependent protein kinase II blocks its nuclear targeting. *J. Biol. Chem.* 273: 19763–19771.

Hu, S.-H., Parker, M. W., Lei, J. Y., Wilce, M. C. J., Benian, G. M., and Kemp, B. E. (1994) Insights into autoregulation from the crystal structure of twitchin kinase. *Nature* 369: 581–584.

Huang, C. F., Yuan, C.-J., Blumenthal, D. K., and Graves, D. J. (1995) Identification of the substrate and pseudosubstrate binding sites of phosphorylase kinase gamma-subunit. *J. Biol. Chem.* 270: 7183–7188.

Huang, C. F., Yuan, C.-J., Livanova, N. B., and Graves, D. J. (1993) Expression, purification, characterization, and deletion mutations of phosphorylase kinase gamma subunit: identification of an inhibitory domain in the gamma subunit. *Mol. Cell. Biochem.* 127/128: 7–18.

Hudmon, A., Aronowski, J., Kolb, S. J., and Waxham, M. N. (1996) Inactivation and self-association of $Ca^{2+}/$ calmodulin-dependent protein kinase II during autophosphorylation. *J. Biol. Chem.* 271: 8800–8808.

Huttner, W. B., DeGennaro, L. J., and Greengard, P. (1981) Differential phosphorylation of multiple sites in purified protein I by cyclic AMP-dependent protein kinases. *J. Biol. Chem.* 256: 1482–1488.

Hvalby, O., Hemmings, H. C. Jr., Paulsen, O., Czernik, A. J., Nairn, A. C., Godfraind, J. M., Jensen, V., Raastad, M., Storm, J. F., Andersen, P., and Greengard, P. (1994) Specificity of protein kinase inhibitor peptides and induction of long-term potentiation. *Proc. Natl. Acad. Sci. USA* 91: 4761–4765.

Ikeda, A., Okuno, S., and Fujisawa, H. (1991) Studies on the generation of $Ca^{2+}/$calmodulin-independent activity of calmodulin-dependent protein kinase II by autophosphorylation. Autothiophosphorylation of the enzyme. *J. Biol. Chem.* 266: 11582–11588.

Ikura, M., Clore, G. M., Gronenborn, A. M., Zhu, G, Klee, C. B., and Bax, A. (1992) Solution structure of a calmodulin-target peptide complex by multidimensional NMR. *Science* 256: 632–638.

Inoue, S.,Mizutani, A., Sugita, R., Sugita, K., and Hidaka, H. (1995) Purification and characterization of a novel protein activator of $Ca^{2+}/$calmodulin-dependent protein kinase I. *Biochem. Biophys. Res. Commun.* 215: 861–867.

Ishida, A., Kameshita, I., Okuno, S., Kitani, T., and Fujisawa, H. (1995) A novel highly specific and potent inhibitor of calmodulin-dependent protein kinase II. *Biochem. Biophys. Res. Commun.* 212: 806–812.

Ito, M., Dabrowska, R., Guerriero, V., and Hartshorne, D. J. (1989) Identification in turkey gizzard of an acidic protein related to the C-terminal portion of smooth muscle myosin light chain kinase. *J. Biol. Chem.* 264: 13971–13974.

Ito, T., Mochizuki, H., Kato, M., Ninura, Y., Hani, T., Usuda, N., and Hidaka, H. (1994) $Ca^{2+}/$calmodulin-dependent protein kinase V: tissue distribution and immunohistochemical localization in rat brain. *Arch. Biochem. Biophys.* 31: 278–284.

James, P., Vorherr, T., and Carafoli, E. (1995) Calmodulin-binding domains: just two faced or multi-faceted? *Trends Biochem. Sci.* 20: 38–42.

Jensen, K. F., Ohmstede, C. A., Fisher, R. S., Olin, J. K., and Sahyoun, N. (1991a) Acquisition and loss of a neuronal $Ca^{2+}/$calmodulin-dependent protein kinase during neuronal differentiation. *Proc. Natl. Acad. Sci. USA* 88: 4050–4053.

Jensen, K. F., Ohmstede, C.-A., Fisher, R. S., and Sahyoun, N. (1991b) Nuclear and axonal localization of $Ca^{2+}/$calmodulin-dependent protein kinase type Gr in rat

cerebellar cortex. *Proc. Natl. Acad. Sci. USA* 88: 2850–2853.

Johnson, L. N., Noble, M. E. M., and Owen, D. J. (1996) Active and inactive protein kinases: structural basis for regulation. *Cell* 85: 149–158.

Jones, D. A., Glod, J., Wilson-Shaw, D., Hahn, W. E., and Sikela, J. M. (1991) cDNA sequence and differential expression of the mouse $Ca^{2+}/$calmodulin-dependent protein kinase IV genes. *FEBS Lett.* 289: 105–109.

Kameshita, I. and Fukisawa, H. (1993) Autophosphorylation of calmodulin-dependent protein kinase IV from rat cerebral cortex. *J. Biochem. (Tokyo)* 113: 583–590.

Kanaseki, T., Ikeuchi, Y., Sugiura, H., and Yamauchi, T. (1991) Structural features of $Ca^{2+}/$calmodulin-dependent protein kinase II revealed by electron microscopy. *J. Cell Biol.* 115: 1049–1060.

Karls, U., Muller, U., Gilbert, D. J., Copeland, N. G., Jenkins, N. A., and Harbers, K. (1992) Structure, expression, and chromosome location of the gene for the β subunit of brain-specific $Ca^{2+}/$calmodulin-dependent protein kinae II identified by transgene integration in an embryonic lethal mouse mutant. *Mol. Cell. Biol.* 12: 3644–3652.

Kelly, P. T., Weinberger, R. P., and Waxham, M. N. (1988) Active site-directed inhibition of $Ca^{2+}/$calmodulin-dependent protein kinase II by a bifunctional calmodulin-binding peptide. *Proc. Natl. Acad. Sci. USA* 85: 4991–4995.

Kemp, B. E. and Pearson, R. B. (1985) Spatial requirements for location of basic residues in peptide substrates for smooth muscle myosin light chain kinase. *J. Biol. Chem.* 260: 3355–3359.

Kemp, B. E., Barden, J. A., Kobe, B., House, C., and Parker, M. W. (1996) Intrasteric regulation of calmodulin-dependent protein kinases. *Adv. Pharmacol.* (Hidaka, H. and Nairn, A. C., eds.). 36: 221–249.

Kemp, B. E., Faux, M. C., Means, A. R., House, C., Tiganis, T., Hu, S.-H., and Mitchelhill, K. I. (1994) Structural aspects: pseudosubstrate and substrate interactions. In *Protein Kinases* (Woodgett, J. R., ed). Oxford University Press, New York, pp. 30–67.

Kemp, B. E., Pearson, R. B., Guerriero, V., Bagchi, I. C., and Means, A. R. (1987) The calmodulin binding domain of chicken smooth muscle myosin light chain kinase contains a pseudosubstrate sequence. *J. Biol. Chem.* 262: 2542–2548.

Kennedy, M. B. and Greengard, P. (1981) Two calcium/calmodulin-dependent protein kinases, which are highly concentrated in brain, phosphorylate protein I at distinct sites. *Proc. Natl. Acad. Sci. USA* 78: 1293–1297.

Kennelly, P. J. and Krebs, E. G. (1991) Consensus sequences as substrate specificity determinants for protein kinases and protein phosphatases. *J. Biol. Chem.* 266: 15555–15558.

Kennelly, P. J., Edelman, A. M., Blumenthal, D. K., and Krebs, E. G. (1987) Rabbit skeletal muscle myosin light chain kinase. The calmodulin binding domain as a potential active site-directed inhibitory domain. *J. Biol. Chem.* 262: 11958–11963.

Kitani, T., Okuno, S., and Fujisawa, H. (1997) Studies on the site of phosphorylation of Ca^{2+}/calmodulin-dependent protein kinase (CaM-kinase) IV by CaM-kinase kinase. *J. Biochem. (Tokyo)* 121: 804–810.

Knighton, D. R., Pearson, R. B., Sowadski, J. M., Means, A. R., Eyck, L. F. T., Taylor, S. S., and Kemp, B. E. (1992) Structural basis of the intrasteric regulation of myosin light chain kinases. *Science* 258: 130–135.

Knighton, D. R., Zheng, J., Eyck, L. F. T., Ashford, V. A., Xuong, N., Taylor, S. S., and Sowadski, J. M. (1991) Crystal structure of the catalytic subunit of cyclic adenosine monophosphate-dependent protein kinase. *Science* 253: 407–414.

Krebs, E. G. (1973) The mechanism of hormonal regulation by cyclic AMP. In *Endocrinology, Proceedings of the 4th International Congress.* Excerpta Medica, Amsterdam, pp. 17–29.

Krebs, J., Means, R. L., and Honegger, P. (1996) Induction of calmodulin kinase IV by the thyroid hormone during rat brain development. *J. Biol. Chem.,* 271: 11055–11058.

Krueger, J. K., Padre, R. C., and Stull, J. T. (1995) Intrasteric regulation of myosin light chain kinase. *J. Biol. Chem.* 270: 16848–16853.

Kuret, J. and Schulman, H. (1985) Mechanism of autophosphorylation of the multifunctional Ca^{2+}/calmodulin-dependent protein kinase. *J. Biol. Chem.* 260: 6427–6433.

Kwiatkowski, A. P. and McGill, J. M. (1995) Human biliary epithelial cell line Mz-ChA-1 expresses new isoforms of calmodulin-dependent kinase II. *Gastroenterology* 109: 1316–1323.

Lanciotti, R. A. and Bender, P. K. (1995) The gamma subunit of phosphorylase kinase contains a pseudosubstrate sequence. *Eur. J. Biochem.* 230: 139–145.

Lee, J. C. and Edelman, A. M. (1994) A protein activator of Ca^{2+}-calmodulin-dependent protein kinase Ia. *J. Biol. Chem.* 269: 2158–2164.

Lee, J. C. and Edelman, A. M. (1995) Activation of Ca^{2+}-calmodulin-dependent protein kinase Ia is due to direct phosphorylation by its activator. *Biochem. Biophys. Res. Commun.* 210: 631–637.

Lee, J. C., Kwon, Y. G., Lawrence, D. S., and Edelman, A. M. (1994) A requirement of hydrophobic and basic amino acid residues for substrate recognition by Ca^{2+}/calmodulin-dependent protein kinase Ia. *Proc. Natl. Acad. Sci. USA* 91: 6413–6417.

LeVine, H. and Sahyoun, N. E. (1987) Characterization of a soluble M_r-30000 catalytic fragment of the neuronal calmodulin-dependent protein kinase II. *Eur. J. Biochem.* 168: 481–486.

Lin, C. R., Kapiloff, M. S., Durgerian, S., Tatemoto, K., Russo, A. F., Hanson, P., Schulman, H., and Rosenfeld, M. G. (1987) Molecular cloning of a brain-specific calcium/calmodulin-dependent protein kinase. *Proc. Natl. Acad. Sci. USA* 84: 5962–5966.

Lisman, J. (1994) The CaM kinase II hypothesis for the storage of synaptic memory. *Trends Neurosci.* 17: 406–412.

Lledo, P.-M., Hjelmstad, G. O., Mukherji, S., Soderling, T. R., Malenka, R. C., and Nicoll, R. A. (1995) Calcium/calmodulin-dependent kinase II and long-term potentiation enhance synaptic transmission by the same mechanism. *Proc. Natl. Acad. Sci. USA* 92: 11175–11179.

Lou L. L. and Schulman, H. (1989) Distinct autophosphorylation sites sequentially produce autonomy and inhibition of the multifunctional Ca^{2+}/calmodulin-dependent protein kinase. *J. Neurosci.* 9: 2020–2032.

Mackler, S. A., Brooks, B. P., and Eberwine, J. H. (1992) Stimulus-induced coordinate changes in mRNA abundance in single postsynaptic hippocampal CA1 neurons. *Neuron* 9: 539–548.

Malenka, R. C. (1994) Synaptic plasticity in the hippocampus: LTP and LTD. *Cell* 78: 535–538.

Malinow, R., Schulman, H., and Tsien, R. W. (1989) Inhibition of postsynaptic PKC or CaMKII blocks induction but not expression of LTP. *Science* 245: 862–866.

Mammen, A. L., Kameyama, K., Roche, K. W., and Huganir, R. L. (1997) Phosphorylation of the alpha-amino-3-hydroxy-5-methylisoxazole-4-propionic acid receptor GluR1 subunit by calcium/calmodulin-dependent protein kinase II. *J. Biol. Chem.* 272: 32528–32533.

Matthews, R. P., Guthrie, C. R., Wailes, L. M., Zhao, X., Means, A. R., and McKnight, G. S. (1994) Calcium/calmodulin-dependent protein kinase types II and IV differentially regulate CREB-dependent gene expression. *Mol. Cell. Biol.* 14: 6107–6116.

Mayer, P., Mohlig, M., Schatz, H., and Pfeiffer, A. (1994) Additional isoforms of multifunctional calcium/calmodulin-dependent protein kinase II in rat heart tissue. *Biochem. J.* 298: 757–758.

Mayford, M., Bach, M. E., Huang, Y. Y., Wang, L., Hawkins, R. D., and Kandel, E. R. (1996) Control of memory formation through regulated expression of a CaMKII transgene. *Science* 274: 1678–1683.

Mayford, M., Wang, J., Kandel, E. R., and O'Dell, T. J. (1995) CaMKII regulates the frequency response function of hippocampal synapses for the production of both LTD and LTP. *Cell.* 81: 891–904.

McCarron, J. G., McGeown, J. G., Reardon, S, Ikebe, M., Fay, F. S., and Walsh, J. V. Jr. (1992) Calcium-dependent enhancement of calcium current in smooth muscle by calmodulin-dependent protein kinase II. *Nature* 357: 74–77.

McDonald. O. B., Merrill, B. M., Bland, M. M., Taylor, L. C., and Sahyoun, N. (1993) Site and consequences of the autophosphorylation of Ca^{2+}/calmodulin-dependent protein kinase type "Gr". *J. Biol. Chem.* 268: 10054–10059.

McGlade-McCulloh, E., Yamamoto, H., Tan, S.-E., Brickley, D. A., and Soderling, T. R. (1993) Phosphorylation and regulation of glutamate receptors by calcium/calmodulin-dependent protein kinase II. *Nature* 362: 640–642.

Meador, W. E., Means, A. R., Quiocho, F. A. (1992) Target enzyme recognition by calmodulin: 2.4 Å structure of a calmodulin-peptide complex. *Science* 257: 1251–1255.

Meador, W. E., Means, A. R., and Quiocho, F. A. (1993) Modulaiton of calmodulin plasticity in molecular recognition on the basis of X-ray structures. *Science* 262: 1718–1721.

Means, A. R., Bagachi, I. C., VanBerkum, M. F., and Kemp, B. E. (1991a) Regulation of smooth muscle myosin light chain kinase by calmodulin. *Adv. Exp. Med. Biol.* 304: 11–24.

Means, A. R., Cruzalegui, F., LeMagueresse, B., Needleman, D. S., Slaughter, G. R., and Ono, T. (1991b) A novel Ca^{2+}/calmodulin-dependent protein kinase and a male germ cell-specific calmodulin-binding protein are derived from the same gene. *Mol. Cell. Biol.* 11: 3960–3971.

Meyer, T., Hanson, P. I., Stryer, L., and Schulman, H. (1992) Calmodulin trapping by calcium-calmodulin-dependent protein kinase. *Science.* 256: 1199–1201.

Miller, S. G. and Kennedy, M. B. (1986) Regulation of brain type II Ca^{2+}/calmodulin-dependent protein kinase by autophosphorylation: a Ca^{2+}-triggered switch. *Cell* 44: 861–870.

Miranti, C. K., Ginty, D. D., Huang, G., Chatila, T., and Greenberg, M. E. (1995) Calcium activates serum response factor-dependent transcription by a Ras- and Elk-1-independent mechanism that involves a Ca^{2+}/calmodulin-dependent kinase. *Mol. Cell. Biol.* 15: 3672–3684.

Miyano, O., Kameshita, I., and Fujisawa, H. (1992) Purification and characterization of a brain-specific multifunctional calmodulin-dependent protein kinase from rat cerebellum. *J. Biol. Chem.* 267: 1198–1203.

Mochizuki, H., Sugita, R., Ito, T., and Hidaka, H. (1993) Phosphorylation of Ca^{2+}/calmodulin-dependent protein kinase V and regulation of its activity. *Biochem. Biophys. Res. Comm.* 197: 1595–1600.

Mukherji, S. and Soderling, T. R. (1994) Regulation of Ca^{2+}/calmodulin-dependent protein kinase II by inter- and intrasubunit-catalyzed autophosphorylations. *J. Biol. Chem.* 269: 13744–13747.

Mukherji, S. and Soderling, T. R. (1995) Mutational analysis of Ca^{2+}-independent autophosphorylation of calcium/calmodulin-dependent protein kinase II. *J. Biol. Chem.* 270: 14062–14067.

Nairn, A. C. and Greengard, P. (1987) Purification and characterization of Ca^{2+}/calmodulin-dependent protien kinase I from bovine brain. *J. Biol. Chem.* 2621: 7273–7281.

Nairn, A. C. and Picciotto, M. R. (1994) Calcium/calmodulin-dependent protein kinases. *Sem. Cancer Biol.* 5: 295–303.

Naito, Y., Watanabe, Y., Yokokura, H., Sugita, R., Nishio, M., and Hidaka, H. (1997) Isoform-specific activation and structural diversity of calmodulin kinase I. *J. Biol. Chem.* 272: 32704–32708.

Nghiem, P., Ollick, T., Gardner, P., and Schulman, H. (1994) Interleukin-2 transcriptional block by multifunctional Ca^{2+}/calmodulin kinase. *Nature* 371: 347–350.

Nghiem, P., Saati, S. M., Martens, C. L., Gardner, P., and Schulman, H. (1993) Cloning and analysis of two new isoforms of multifunctional Ca^{2+}/calmodulin-dependent protein kinase. Expression in multiple human tissues. *J. Biol. Chem.* 268: 5471–5479.

Nichols, R. A., Sihra, T. S., Czernik, A. J., Nairn, A. C., and Greengard, P. (1990) Calcium/calmodulin-dependent protein kinase II increases glutamate and noradrenaline release from synaptosomes. *Nature* 343: 647–651.

Ohmstede, C. A., Bland, M. M., Merrill, B. M., and Sahyoun, N. (1991) Relationship of genes encoding Ca^{2+}/calmodulin-dependent protein kinase Gr and calspermin: a gene within a gene. *Proc. Natl. Acad. Sci. USA* 88: 5784–5788.

Ohmstede, C. A., Jensen, K. F., and Sahyoun, N. E. (1989) Ca^{2+}/calmodulin-dependent protein kinase enriched in cerebellar granule cells. Identification of a novel neuronal calmodulin-dependent protein kinase. *J. Biol. Chem.* 264: 5866–5875.

Ohsako, S., Nishida, Y., Ryo, H., and Yamauchi, T. (1993) Molecular characterization and expression of the *Drosophila* Ca^{2+}/calmodulin-dependent protein kinase II gene. Identification of four forms of the enzyme generated from a single gene by alternative splicing. *J. Biol. Chem.* 268: 2052–2062.

Okuno, S. and Fujisawa, H. (1993) Requirement of brain extract for the activity of brain calmodulin-dependent protein kinase IV expressed in *Escherichia coli. J. Biochem. (Tokyo)* 114: 167–170.

Okuno, S., Kitani, T., and Fujisawa, H. (1994) Purification and characterization of Ca^{2+}/calmodulin-dependent protein kinase IV kinase from rat brain. *J. Biochem. (Tokyo)* 116: 923–930.

Okuno, S., Kitani, T., and Fujisawa, H. (1995) Full activation of brain calmodulin-dependent protein kinase IV requires phosphorylation of the amino-terminal serine-rich region by calmodulin-dependent protein kinase IV kinase. *J. Biochem. (Tokyo)* 117: 686–690.

Olson, N. J., Pearson, R. B., Needleman, D. S., Hurwitz, M. Y., Kemp, B. E., and Means, A. R. (1990) Regulatory and structural motifs of chicken gizzard myosin light chain kinase. *Proc. Natl. Acad. Sci. USA* 87: 2284–2288.

Omkumar, R. V., Kiely, M. J., Rosenstein, A. J., Min, K. T., and Kennedy, M. B. (1996) Identification of a phosphorylation site for calcium/calmodulin-dependent protein kinase II in the NR2B subunit of the N-methyl-D-aspartate receptor. *J. Biol. Chem.* 271: 31670–31678.

Ono, T., Slaughter, G. R., Cook, R. G., and Means, A. R. (1989) Molecular cloning sequence and distribution of rat calspermin, a high affinity calmodulin-binding protein. *J. Biol. Chem.* 264: 2081–2087.

Ouyang, Y., Kantor, D., Harris, K. M., Schuman, E. M., and Kennedy, M. B. (1997) Visualization of the distribution of autophosphorylated calcium/calmodulin-dependent protein kinase II after tetanic stimulation in the CA1 area of the hippocampus. *J. Neurosci.* 17: 5416–5427.

Owen, D. J., Noble. M. E., Garman, E. F., Papagerorgiou, A. C., and Johnson, L. N. (1995) Two structures of the catalytic domain of phosphorylase kinase: an active protein kinase complexed with substrate analogue and product. *Structure* 3: 467–482.

Patton, B. L., Miller, S. G., and Kennedy, M. B. (1990) Activation of type II calcium/calmodulin-dependent protein kinase by Ca^{2+}/calmodulin is inhibited by autophosphorylation of threonine within the calmodulin-binding domain. *J. Biol. Chem.* 265: 11204–11212.

Pearson, R. B., Wettenhall, R. E., Means, A. R., Hartshorne, D. J., and Kemp, B. E. (1988) Autoregulation of enzymes by pseudosubstrate prototopes: myosin light chain kinase. *Science* 241: 970–973.

Pearson, R. B., Woodgett, J. R., Cohen, P., and Kemp, B. E. (1985) Substrate specificity of a multifunctional calmodulin-dependent protein kinase. *J. Biol. Chem.* 260: 14471–14476.

Pettit, D. L., Perlman, S., and Malinow, R. (1994) Potentiated transmission and prevention of further LTP by increased CaMKII activity in postsynaptic hippocampal slice neurons. *Science* 266: 1881–1885.

Picciotto, M. R., Cohn, J. A., Bertuzzi, G., Greengard, P., and Nairns, A. C. (1992) Phosphorylation of the cystic fibrosis transmembrane conductance regulator. *J. Biol. Chem.* 267: 12742–12752.

Picciotto, M. R., Czernik, A. J., and Nairn, A. C. (1993) Calcium/calmodulin-dependent protein kinase I. cDNA cloning and identification of autophosphorylation site. *J. Biol. Chem.* 268: 26512–26521.

Piccioto, M. R., Nastiuk, K. L., and Nairn, A. C. (1996) Structure, regulation, and function of calcium/calmodulin-dependent protein kinase. I. *Adv. Pharmacol.* 36: 251–275.

Picciotto, M. R., Zoli, M., Bertuzzi, G., and Nairn, A. C. (1995) Immunochemical localization of calcium/calmodulin-dependent protein kinase I. *Synapse* 20: 75–84.

Pickett-Gies, C. A. and Walsh, D. A. (1985) Subunit phosphorylation and activation of skeletal muscle phosphorylase kinase by the cAMP-dependent protein kinase. Divalent metal ion, ATP, and protein concentration dependence. *J. Biol. Chem.* 260: 2046–2056.

Rich, R. C. and Schulman, H. (1998) Substrate-directed function of calmodulin in autophosphorylation of Ca^{2+}/calmodulin-dependent protein kinase II. *J. Biol. Chem.* 273: 28424–28429.

Rostas, J. A. P. and Dunkley, P. R. (1992) Multiple forms and distribution of calcium/calmodulin-stimulated protein kinase II in brain. *J. Neurochem,* 59: 1191–1202.

Rotenberg, A., Mayford, M., Hawkins, R. D., Kandel, E. R., and Muller, R. U. (1996) Mice expressing activated CaMKII lack low frequency LTP and do not form stable place cells in CA1 region of the hippocampus. *Cell.* 87: 1351–1361.

Roush, C. L., Kennelly, P. J., Glaccum, M. B., Helfman, D. M., Scott, J. D., and Krebs, E. G. (1988) Isolation of the cDNA encoding rat skeletal muscle myosin light chain kinase. *J. Biol. Chem.* 263: 10510–10516.

Sahyoun, N., McDonald, O. B., Farrell, F., and Lapetina, E. G. (1991) Phosphorylation of a ras-related GTP-binding protein, rap-1b, by a neuronal Ca^{2+}/calmodulin-dependent protein kinase, CaM kinase Gr. *Proc. Natl. Acad. Sci. USA* 88: 2643–2647.

Saitoh, T. and Schwartz, J. H. (1985) Phosphorylation-dependent subcellular translocation of a Ca^{2+}/calmodulin-dependent protein kinase produces an autonomous enzyme in *Aplysia* neurons. *J. Cell Biol.* 100: 835–842.

Sakagami, H. and Kondon, H. (1993) Cloning and sequencing of a gene encoding the beta polypeptide of Ca^{2+}/calmodulin-dependent protein kinase IV and its expression confined to the mature cerebellar granule cells. *Brain Res. Mol. Brain. Res.* 19: 215–218.

Sanchez, V. E. and Carlson, G. M. (1993) Isolation of an autoinhibitory region from the regulatory beta-subunit of phosphorylase kinase. *J. Biol. Chem.* 268: 17889–17895.

Schulman, H. (1988) The multifunctional Ca^{2+}/calmodulin-dependent protein kinase. *Adv. Second Mess. Phosphoprotein Res.* 22: 39–112.

Schulman, H. (1993) The multifunctional Ca^{2+}/calmodulin-dependent kinases. *Curr. Opin. Cell Biol.* 5: 247–253.

Schulman, H. and Greengard, P. (1978a) Ca^{2+}-dependent protein phosphorylation system in membranes from various tissues and its activation by "calcium-dependent regulator." *Proc. Natl. Acad. Sci. USA* 75: 5432–5436.

Schulman, H. and Greengard, P (1978b) Stimulation of brain membrane protein phosphorylation by calcium and an endogenous heat-stable protein. *Nature* 271: 478–479.

Schulman, H., Hanson, P. I., and Meyer, T. (1992) Decoding calcium signals by multifunctional CaM kinase. *Cell Calcium* 13: 401–411.

Schworer, C. M., Rothblum, L. I., Thekkumkara, T. J., and Singer, H. A. (1993) Identification of novel isoforms of the δ isoform of Ca^{2+}/calmodulin-dependent protein kinase II. Differential expression in rat brain and aorta. *J. Biol. Chem.* 268: 14443–14449.

Selbert, M. A., Anderson, K. A., Huang, Q.-H., Goldstein, E. G., Means, A. R., and Edelman, A. M. (1995) Phosphorylation and activation of Ca^{2+}-calmodulin-dependent protein kinase IV by Ca^{2+}-calmodulin-dependent protein kinase IA kinase. *J. Biol. Chem.* 270: 17616–17621.

Sheng, M., Thompson, M. A., and Greenberg, M. E. (1991) CREB: a Ca^{2+}-regulated transcription factor phosphorylated by calmodulin-dependent kinases. *Science* 252: 1427–1430.

Shields, S. M., Vernon, P. J., and Kelly, P. T. (1984) Autophosphorylation of calmodulin-kinase II in synaptic junctions modulates endogenous kinase activity. *J. Neurochem.* 43: 1599–1609.

Shoemaker, M. O., Lau, W., Shattuck, R. L., Kwiatkowski, A. P., Matrisian, P. E., Guerra-Santos, L., Wilson, E., Lukas, T. J., VanEldik, L. J., and Watterson, D. M. (1990) Use of DNA sequence and mutant analyses and antisense oligodeoxynucleotides to examine the molecular basis of nonmuscle myosin light chain kinase autoinhibition, calmodulin regulation, and activity. *J. Cell. Biol.* 111: 1107–1125.

Silva, A. J., Paylor, R., Wehner, J. M., and Tonegawa, S. (1992a) Impaired spatial learning in α-calcium-calmodulin kinase II mutant mice. *Science* 257: 206–211.

Silva, A. J., Stevens, C. F., Tonegawa, S., and Wang, Y. (1992b) Deficient hippocampal long-term potentiation in α-calcium-calmodulin kinase II mutant mice. *Science* 257: 201–206.

Smith, M. K., Colbran, R. J., Brickley D. A., and Soderling, T. R., (1992) Functional determinants in the autoinhibitory domain of calcium/calmodulin-dependent protein kinase II. Role of His282 and multiple basic residues. *J. Biol. Chem.* 267: 1761–1768.

Smith, M. K., Colbran, R. J., and Soderling, T. R. (1990) Specificiities of autoinhibitory domain peptides for four protein kinases. Implications for intact cell studies of protein kinase function. *J. Biol. Chem.* 265: 1837–1840.

Soderling, T. R. (1995) Calcium-dependent protein kinases in learning and memory. *Adv. Second Mess. Phosphoprotein Res.* 30: 175–189.

Soderling, T. R. (1996) Structure and regulation of calcium/calmodulin-dependent protein kinases II and IV. *Biochim. Biophys. Acta* 1297: 131–138.

Srinivasan, M., Edman, C., and Schulman, H. (1994) Alternative splicing introduces a nuclear localization signal that targets multifunctional CaM kinase to the nucleus. *J. Cell Biol.* 126: 839–852.

Stokoe, D., Caudwell, B., Cohen, P. T., and Cohen, P. (1993) The substrate specificity and structure of mitogen-activated protein (MAP) kinase-activated protein kinase-2. *Biochem. J.* 296: 843–849.

Strack, S. and Colbran, R. J. (1998) Autophosphorylation-dependent targeting of calcium/calmodulin-dependent protein kinase II by the NR2B subunit of the N-methyl-D-aspartate receptor. *J. Biol. Chem.* 273: 20689–20692.

Stull, J. T., Tansey, M. G., Tang, D.-C., Word, R. A., and Kamm, K. E. (1993) Phosphorylation of myosin light chain kinase: a cellular mechanism for Ca^{2+} desensitization. *Mol. Cell. Biochem.* 127–128: 229–237.

Sugita, R., Mochizuki, H., Ito, T., Yokokura, H., Kobayashi, R., and Hidaka, H. (1994) Ca^{2+}/calmodulin-dependent protein kinase kinase cascade. *Biochem. Biophys. Res. Commun.* 203: 694–701.

Sun, P., Enslen, H., Myung, P. S., and Maurer, R. A. (1994) Differential activation of CREB by Ca^{2+}/calmodulin-dependent kinases type II and type IV involves phosphorylation of a site that negatively regulates activity. *Genes Dev.* 8: 2527–2539.

Sun, Z., Means, R. L., LeMagueresse, B., and Means, A. R. (1995) Organization and analysis of the complete rat calmodulin-dependent protein kinase IV gene. *J. Biol. Chem.* 270: 29507–29514.

Suzuki, T., Okumara-Noji, K., Tanaka, R., and Tada, T. (1994) Rapid translocation of cytosolic Ca^{2+}/calmodulin-dependent protein kinase II into postsynaptic density after decapitation. *J. Neurochem.* 63: 1529–1537.

Sweeney, H. L., Bowman, B. F., and Stull. J. T., (1993) Myosin light chain phosphorylation in vertebrate striated muscle: regulation and function. *Am. J. Physiol.* 264: C1085–C1095.

Tabatabai, L. B. and Graves, D. J. (1978) Kinetic mechanism and specificity of the phosphorylase kinase reaction. *J. Biol. Chem.* 253: 2196–2202.

Takio, K., Blumenthal, D. K., Walsh, K. A., Titani, K., and Krebs, E. G. (1986) Amino acid sequence of rabbit skeletal muscle myosin light chain kinase. *Biochemistry* 25: 8049–8057.

Tan, S.-E., Wenthold, R. J., and Soderling, T. R. (1994) Phosphorylation of AMPA-type glutamate receptors by calcium/calmodulin-dependent protein kinase II and protein kinase C in cultured hippocampal neurons. *J. Neurosci.* 14: 1123–1129.

Tanaka, M., Ikebe, R., Matsuura, M., and Ikebe, M. (1995) Pseudosubstrate sequence may not be critical for autoinhibition of smooth muscle myosin light chain kinase. *EMBO J.* 14: 2839–2846.

Tanaka, T., Naka, M., and Hidaka, H. (1980) Activation of myosin light chain kinase by trypsin. *Biochem. Biophys. Res. Commun.* 92: 313–318.

Tansey, M. G., Word, R. A., Hidaka, H., Singer, H. A., Schworer, C. M., Kamm, K. E., and Stull, J. T. (1992) Phosphorylation of myosin light chain kinase by the multifunctional calmodulin-dependent protein kinase II in smooth muscle cells. *J. Biol. Chem.* 267: 12511–12516.

Taylor, S. S. and Radzio-Andzelm, E. (1994) Three protein kinase structures define a common motif. *Structure* 2: 345–355.

Taylor, S. S., Buechler, J. A., and Yonemoto, W. (1990) cAMP-dependent protein kinase: framework for a diverse family of regulatory enzymes. *Annu. Rev. Biochem.* 59: 971–1005.

Tobimatsu, T. and Fujisawa, H. (1989) Tissue-specific expression of four types of rat calmodulin-dependent protein kinase II mRNAs. *J. Biol. Chem.* 264: 17907–17912.

Tobimatsu, T., Kameshita, I., and Fujisawa, H. (1988) Molecular cloning of the cDNA encoding the third polypeptide (γ) of brain calmodulin-dependent protein kinase. II. *J. Biol. Chem.* 263: 16082–16086.

Tokui, T., Ando, S., and Ikebe, M. (1995) Autophosphorylation of smooth muscle myosin light chain kinase at its regulatory domain. *Biochemistry* 34: 5173–5179.

Tokumitsu, H. and Soderling, T. R. (1996) Requirements for calcium and calmodulin in the calmodulin kinase activation cascade. *J. Biol. Chem.* 271: 5617–5622.

Tokumitsu, H., Brickley, D. A., Glod, J., Hidaka, H., Sikela, J., and Soderling, T. R. (1994) Activation mechanisms for Ca^{2+}/calmodulin-dependent protein kinase IV. Identification of a brain CaM-kinase IV kinase. *J. Biol. Chem.* 269: 28640–28647.

Tokumitsu, H., Enslen, H., Soderling, T. R. (1995) Characterization of a Ca^{2+}/calmodulin-dependent protein kinase cascade. Molecular cloning and expression of calcium/calmodulin-dependent protein kinase kinase. *J. Biol. Chem.* 270: 19320–19324.

Tokumitsu, H., Wayman, G. A., Muramatsu, M., and Soderling, T. R. (1997) Calcium/calmodulin-dependent protein kinase kinase: identification of regulatory domains. *Biochemistry* 36: 12823–12827.

Tombes, R. M. and Krystal, G. W. (1997) Identification of novel human tumor cell-specific CaMK-II variants. *Biochim. Biophys. Acta* 1355: 281–292.

Urquidi, V. and Ashcroft, S. J. H. (1995) A novel pancreatic β-cell isoform of calcium/calmodulin-dependent protein kinase II ($\beta3$ isoform) contains a proline-rich tandem repeat in the association domain. *FEBS Lett.* 358: 23–26.

Valtorta, F., Benfenati, F., and Greengard, P. (1992) Structure and function of the synapsins. *J. Biol. Chem.* 267: 7195–7198.

Waldmann, R., Hanson, P. I., and Schulman, H. (1990) Multifunctional Ca^{2+}/calmodulin-dependent protein kinase made Ca^{2+}-independent for functional studies. *Biochemistry* 29: 1679–1684.

Walsh, M. P. (1994) Calmodulin and the regulation of smooth muscle contraction. *Mol. Cell. Biochem.* 135: 21–41.

Wang, J. H. and Kelly, P. T. (1995) Postsynaptic injection of Ca^{2+}/CaM induces synaptic potentiation requiring CaMKII and PKC activity. *Neuron* 15: 443–452.

Watanabe, S., Okuno, S., Kitani, T., and Fujisawa, H. (1996) Inactivation of calmodulin-dependent protein kinase IV by autophosphorylation of serine 332 within the putative calmodulin-binding domain. *J. Biol. Chem.* 271: 6903–6910.

Waxham, M. N., Aronowski, J., Westgate, S. A., and Kelly, P. R. (1990) Mutagenesis of Thr-286 in monomeric Ca^{2+}/calmodulin-dependent protein kinase II eliminates Ca^{2+}/calmodulin-independent activity. *Proc. Natl. Acad. Sci. USA* 87: 1273–1277.

Wayman, G. A., Tokumitsu, H., and Soderling, T. R. (1997) Inhibitory cross-talk by cAMP kinase on the calmodulin-dependent protein kinase cascade. *J. Biol. Chem.* 272: 16073–16076.

Waymire, J. C., Johnston, J. P.., Hummer-Lickteig, K., Lloyd, A., Vigny, A., and Craviso, G. L. (1988) Phosphorylation of bovine adrenal chromaffin cell tyrosine hydroxylase. Temporal correlation of acetylcholine's effect on site phosphorylation, enzyme activation, and catecholamine synthesis. *J. Biol. Chem.* 263: 12439–12447.

Westphal, R. S., Anderson, K. A., Means, A. R., and Wadzinski, B. E. (1998) A signaling complex of Ca^{2+}-calmodulin-dependent protein kinase IV and protein phosphatase 2A. *Science* 280: 1258–1261.

White, R. R., Kwon, Y.-G., Taing, M., Lawrence, D. S., and Edelman, A. M. (1998) Definition of optimal substrate recognition motifs of Ca^{2+}-calmodulin-dependent protein kinases IV and II reveals sharp and distinctive features. *J. Biol. Chem.* 273: 3166–3172.

Wu, G., Manilow, R., and Cline, H. T. (1996) Maturation of a central glutamatergic synapse. *Science* 274: 972–976.

Wu, G. Y. and Cline, H. T. (1998) Stabilization of dendritic arbor structure in vivo by CaMKII. *Science* 279: 222–226.

Yagi, K., Yazawa, M., Kakiuchi, S., Ohshima, M., and Uenishi, K. (1978) Identification of an activator protein for myosin light chain kinase as the Ca^{2+}-dependent modulator protein. *J. Biol. Chem.* 253: 1338–1340.

Yakel, J. L., Vissavajjhala, P., Derkach, V. A., Brickey, D. A., and Soderling, T. R. (1995) Identification of a Ca^{2+}/calmodulin-dependent protein kinase II regulatory phosphorylation site in non-*N*-methyl-D-aspartate glutamate receptors. *Proc. Natl. Acad. Sci. USA* 92: 1376–1380.

Yamagata, Y., Czernik, A. J., and Greengard, P. (1991) Active catalytic fragment of Ca^{2+}/calmodulin-dependent protein kinase II. Purification, characterization, and structural analysis. *J. Biol. Chem.* 266: 15391–15397.

Yamauchi, T. and Fujisawa, H. (1980) Evidence for three distinct forms of calmodulin-dependent protein kinases from rat brain. *FEBS Lett.* 116: 141–144.

Yeaman, S. J. and Cohen, P. (1975) The hormonal control of activity of skeletal phosphorylase kinase. Phosphorylation of the enzyme at two sites *in vivo* in response to adrenalin. *Eur. J. Biochem.* 51: 93–104.

Yokokura, H., Picciotto, M. R., Nairn, A. C., and Hidaka, H. (1995) The regulatory region of calcium/calmodulin-dependent protein kinase I contains closely associated autoinhibitory and calmodulin-binding domains. *J. Biol. Chem.* 270: 23851–23859.

Yoshimura, Y. and Yamauchi, T. (1997) Phosphorylation-dependent reversible association of Ca^{2+}/calmodulin-dependent protein kinase II with the postsynaptic densities. *J. Biol. Chem.* 272: 26354–26359.

14

Calcium-Regulated Protein Dephosphorylation

Claude Klee
Xutong Wang
Hao Ren

The regulation of cell function by protein phosphorylation is controlled by an equilibrium between protein kinases and phosphatases described more than 30 years ago for the hormonal control of glycogen metabolism (Krebs and Fischer, 1956). The role of protein kinase cascades in the regulation of cellular processes has been extensively studied over the past 10 years. It is only recently, with the identification and isolation of an increasing number of protein phosphatases and the availability of specific inhibitors, that the role of phosphatases in cellular regulation has attracted more attention (Cohen, 1996; Cohen et al., 1996). There is good evidence that their role is not limited to maintaining proteins in a dephosphorylated state to allow regulation by protein kinases. They can themselves trigger cellular responses to external signals. In contrast to the many protein kinases under the control of Ca^{2+} acting as a second messenger, only two Ca^{2+}-regulated protein phosphatases have been isolated so far: the insulin-sensitive pyruvate dehydrogenase of mitochondria (Chapter 22, this volume) and the calmodulin-regulated protein phosphatase, calcineurin. A third Ca^{2+}-regulated protein phosphatase, a product of the Drosophila *retinal degeneration* C *(rdgC)* gene, has been postulated on the basis of the sequence of a cDNA clone showing a domain which has 30% identity with the catalytic domain of protein phosphatases 1, 2A, and 2B in addition to a calmodulin-like regulatory domain (Steele et al., 1992). Human homologues of RDGC have recently been identified (Sherman et al., 1997). Although neither the RDGC protein nor its human homologues have yet been isolated the hyperphosphorylation of rhodopsin which accompanies the loss of RDGC supports the identification of RDGC as a protein phosphatase (Vinos et al., 1997). This chapter is devoted to calcineurin, the only known

protein phosphatase under the control of both Ca^{2+} and calmodulin and, thus, ideally suited to play a critical role in the coupling of Ca^{2+} signals to cellular responses. Its stimulation by the multifunctional protein calmodulin ensures the coordinated regulation of its protein phosphatase activity with that of the many enzymes, including a large number of protein kinases, under Ca^{2+} and calmodulin control.

General Properties

Calcineurin, first discovered as a major calmodulin-binding protein in brain extracts, was later found to be a calmodulin-stimulated serine/threonine protein phosphatase (Klee et al., 1980; Stewart et al., 1982).

On the basis of its Ca^{2+} dependence, resistance to endogenous protein phosphatase 1 inhibitors, and its high activity on the α-subunit of phosphorylase kinase, it was classifed as a protein phosphatase 2B by Ingebritsen and Cohen (1983a). Many different calcineurin isoforms have been identified by the isolation of calcineurin cDNAs from many eukaryotes, including the budding yeast, and many tissues. Nevertheless, it remains the sole member of this class of protein phosphatases, characterized by its dependence on Ca^{2+} and calmodulin, unique subunit structure, resistance to a family of potent inhibitors of protein phosphatase 1 and 2A, and its predominance in neural tissues (see Klee and Cohen, 1988; Klee et al., 1988; Giri et al., 1992 for reviews of the early literature). It became clear during the past years that these perfectly conserved structural features are responsible for the unique ability of calcineurin to interact with two classes of immunosuppressive drugs, cyclosporin A and FK506 (also called tacrolimus) complexed with their respective binding-proteins, cyclophilin A and the FK506-binding protein,

(FKBP) (as reviewed by Schreiber, 1992; Schreiber et al., 1992; Stoddard and Flick, 1996).

Catalytic Properties

Substrate Specificity

The latent serine/threonine phosphatase activity is specifically activated upon binding of the Ca^{2+}/calmodulin complex. The activated enzyme dephosphorylates a large number of substrates but with widely different efficiencies. The accurate assessment of the substrate specificity of calcineurin is impaired by the variability of enzyme preparations (Blumenthal et al., 1985a), and the existence of many isoforms with different kinetic properties and different catalytic properties depending on the metal ions used to promote activity (Seki et al., 1995). There is general agreement that the two protein phosphatase 1 inhibitors, inhibitor 1 and DARPP-32 (dopamine- and cAMP-regulated protein M_r = 32,000), the type II regulatory subunit of cAMP-dependent protein kinase, G-protein, and neurogranin are among the best substrates with K_m values of 2–10 μM and V_{max} of 1.4–0.2 μmol/min mg (King et al., 1984; Blumenthal et al., 1985b; Klee et al., 1988; Seki et al., 1995). Neuromodulin (also called Gap-43) and MARCKs (myristoylated alanine-rich C-kinase substrates) are also dephosphorylated at significant rates (Liu and Storm, 1989; Seki et al., 1995) while myelin basic protein, synapsin 1, casein, and histones are poor substrates. It is noteworthy that calmodulin kinase II, whose calmodulin activation depends on autophosphorylation, is not dephosphorylated by calcineurin (Klee et al., 1988). Other substrates of physiological interest, but which have not yet been studied in detail, include NO synthase, a GTPase involved in endocytosis (dynamin, previously called dephosphin), vimentin, and the heat-shock protein, hsp25, (Evans, 1989; Gaestel et al., 1992; Dawson et al., 1993; Liu et al., 1994; Nichols et al., 1994). Recently, it was demonstrated that the transcription factor NF-ATp (Nuclear Factor-Activated T-cells) , responsible for the activation of the interleukin-2 gene during T-cell activation, is specifically dephosphorylayed in vivo by calcineurin, despite its very low concentration (Jain et al., 1993; McCaffrey et al., 1993; Rao, 1994). Thus, the specificity of calcineurin in cell physiology may be even greater than proposed originally on the basis of the kinetic analysis of its dephosphorylation of readily available substrates (King et al., 1984).

Phosphorylated peptides, also dephosphorylated by calcineurin although with K_m values 2–3 orders of magnitude larger than their protein counterparts (Blumenthal et al., 1985a), are powerful tools to study calcineurin activity in vitro. The most commonly used is a 23-residue peptide which includes the phosphorylation site of the type II regulatory subunit of cAMP-dependent protein kinase (Chan et al., 1986; Hubbard and Klee, 1991). The specificity of calcineurin is not due to a consensus sequence but rather is determined by both primary and higher order structural features in contrast to protein kinases or to other phosphatases (Blumenthal et al., 1985a; Donella-Deana et al., 1994). An extended amino-terminal stretch appears to be necessary, and a basic residue at position −3 relative to the phosphorylated residue plays a relevant role, whereas acidic residues on the carboxyl-terminal side are powerful negative determinants (Donella-Deana et al., 1994). In vitro calcineurin also dephosphorylates phosphotyrosyl peptides and tyrosyl phosphates (Chan et al., 1986; Martin and Graves, 1986). The physiological relevance of this activity remains to be demonstrated but these low-molecular-weight substrates, including the conveniently measurable p-nitrophenylphosphatase, have been extensively used to elucidate the mechanism of calcineurin catalysis.

Mechanism of Action

Failure to detect a common phosphoenzyme intermediate or transphosphorylation activity, and inhibition by the two products of the reaction indicated that the dephosphorylation reaction catalyzed by calcineurin, unlike that of tyrosine and alkaline and acid phosphatases, does not involve the formation of a covalent intermediate (Martin and Graves, 1986). The similarity between the calcineurin-catalyzed dephosphorylation of phosphite and phosphate esters of tyrosine with their acid hydrolysis led Graves and his colleagues (Wang and Graves, 1991; Martin and Graves, 1994) to propose the acid-catalyzed cleavage of the phosphoester bond as illustrated in Fig. 14.1. This metal-assisted catalysis occurs in two steps: the first step affects the V_{max}/K_m term and is a protonation of the phosphoester bond by a metal-activated water molecule (Fig. 14.1A); the second step affects only V_{max} and is the cleavage of the bond by a second metal-activated water molecule (Fig. 14.1B). This mechanism, similar to the one proposed for the hydrolytic reaction catalyzed by bovine spleen purple acid phosphatase (Mueller et al., 1993) is consistent with the requirement of the calcineurin-catalyzed reaction for a divalent metal. Ions, such as Mn^{2+}, Ni^{2+}, and Mg^{2+}, greatly increase the V_{max} with a small effect on the K_m value of the enzyme for phosphopeptides or p-nitrophenylphosphate (reviewed by Pallen and Wang, 1985; Klee at al., 1988). Since calcineurin in crude tissue extracts does not require added metal ions for activity and is 10–20 times more active (spe-

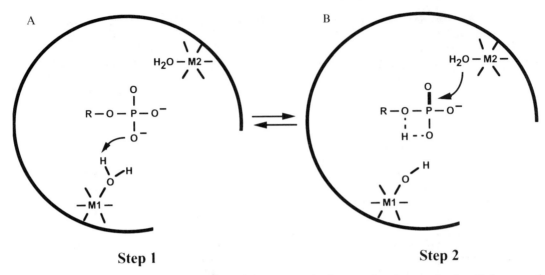

A   B

Step 1   **Step 2**

Figure 14.1 Mechanism of calcineurin catalysis. (A) In step 1, the first metal, acting as a Lewis acid, increases the nucleophilicity of water to donate a proton to the phosphate ester, which is subsequently transferred to the leaving group. (B) In step 2, the cleavage reaction is initiated by addition of a second metal-bound activated water molecule. (Reproduced with permission from Martin and Graves, 1994.)

cific activity as high as 20–50 μmol/min mg at saturating concentrations of the peptide substrate) than the purified enzyme, it is likely that these particular metal ions are not the natural cofactors bound to the native enzyme (Stemmer et al., 1995). Definitive identification of the native metal cofactors was impaired by their partial dissociation upon prolonged exposure to Ca^{2+} and calmodulin during the affinity chromatography step, which could foster substitution by contaminating metals during purification (Stewart et al., 1983) . The identification of calcineurin as an iron–zinc enzyme by King and Huang (1984) has gained strong support during the past years. In the recently published crystal structures of calcineurin (Griffith et al., 1995; Kissinger et al., 1995; and reviewed by Barford,1996), these two metal ions are included, modeled on the structure of the [Fe(III)–Zn(II)] kidney bean purple acid phosphatase (Strater et al. 1995). The oxidation state of iron required for calcineurin activity has not yet been definitively established but the inactivation of calcineurin by the superoxide anion and its protection and reactivation by ascorbate strongly suggests that reduced iron is required for activity (Wang et al., 1996b).

Calcineurin Inhibitors

As with other protein phosphatases, calcineurin is inhibited by nonspecific phosphatase inhibitors such as inorganic phosphate, vanadate, and sodium fluor-

ide. An irreversible "suicide" inhibitor, whose specificity has not been tested, has also been reported (Born et al., 1995). Unlike protein phosphatase 1 and 2A, which are specifically inhibited by nanomolar concentrations of okadaic acid, caliculin, or microcystin, inhibition of calcineurin requires greater than micromolar concentrations of these inhibitors (Cohen et al., 1990). The Ca^{2+} chelator EGTA is a good but not specific inhibitor since it cannot distinguish calcineurin from other metal-dependent phosphatases or unrelated enzymes. Calmodulin inhibitors [phenothiazine derivatives, calmidazolium, and the many calmodulin-binding peptides such as mastoparans, melittin, and the calmodulin-binding domains of skeletal muscle myosin light chain kinase (M13) and calcineurin] are routinely used to inhibit calcineurin in vitro (Klee et al., 1988). Their usefulness in vivo is limited because of their ability to inhibit the entire family of calmodulin-regulated enzymes. A more specific inhibitor is a peptide corresponding to the autoinhibitory domain of calcineurin, but it has only a moderate affinity ($K_i = 5$–20 μM) for the enzyme (Hashimoto et al., 1990; Parsons et al., 1994).

This lack of specific inhibitors prevented the quantitification of calcineurin in crude extracts and delayed the identification of its physiological roles until the discovery that calcineurin was the target of the immunosuppressive drugs FK506 and cyclosporin complexed with their respective immunophilins (Liu et al., 1991a). These drugs, in the form of complexes with their binding proteins (the most commonly used are FKBP-12 and cyclophilin A), act as potent inhibitors of all the isoforms of calcineurin regardless of source, including a constitutively active truncated derivative (refered to as calcineurin-45, according to the molecular weight of its catalytic subunit or Ca^{2+}-independent form of calcineurin). They have no inhibitory effect on protein phosphatases 1, 2A, and 2C even at micromolar concentrations (Liu et al., 1992; Mukai et al., 1993). Both FK506/FKBP and cyclosporin A/cyclophilin act as noncompetitive, slowly reversible inhibitors of the protein phosphatase and activators of the *p*-nitrophenylphosphatase activities of calcineurin with K_i values of between 10 and 100 nM (Liu et al., 1992; Etzkorn et al., 1994). Because of their ability to cross membranes easily, these inhibitors are powerful tools to identify roles of calcineurin in vivo (Clipstone and Crabtree, 1992; Fruman et al., 1992a; O'Keefe et al., 1992). The extent of inhibition depends on the intracellular concentration of endogenous binding proteins which is usually sufficient to ensure the formation of the inhibitory complex. The effective concentration of drug in vivo (1/10 to 1/100 of that needed to inhibit calcineurin in vitro) may depend on the ability of the cells to accumulate the drug and may vary with the intracellular concentrations of calcineurin and binding proteins. The immunosuppressive drugs could inhibit the prolyl *cis-trans* isomerase activities of their binding proteins or may also prevent their interaction with target proteins such as ryanodine, inositol 1,4,5-trisphosphate (IP_3), and the TGF-β receptors, and exert some effects independently of calcineurin inhibition (Brillantes et al., 1994; Cameron et al., 1994). This uncertainty is minimized by testing the effect of other inhibitors, such as EGTA, and the calmodulin-binding and autoinhibitory peptides. Another powerful tool is the immunosuppressant, rapamycin, that interacts with, and inhibits the isomerase activity of, FKBP but does not promote its interaction with calcineurin and is routinely used to identify calcineurin definitively as the target responsible for the effect under study (Liu et al., 1991a; Schreiber et al., 1992).

Calcineurin Assays

A standardized calcineurin assay using phosphorylated peptides has been described in detail by Hubbard and Klee (1991). This assay has been modified to measure calcineurin in crude tissue extracts or cell lysates where the FK506/FKBP- and M13-sensitive, Ca^{2+}-dependent phosphatase activity measured in the presence of 0.5 μM okadaic acid is representative of calcineurin activity (Stemmer et al. 1995; Fruman et al., 1996). A nonradioactive assay procedure has also been described (Enz et al., 1994). Calcineurin itself can be quantified by immunologic techniques using specific anticalcineurin antibodies (Kuno et al., 1992) or by a simple binding assay based on its ability to bind the FKBP/FK506 complex (Asami et al., 1993). These methods together with readily available synthetic peptide substrates greatly facilitate the characterization of calcineurin in crude extracts and allow its purification (see Klee et al., 1988, for a review of the purification procedures).

Structure

Subunit Composition and Isoforms

Calcineurin is a heterodimer of a 58,000–64,000 MW catalytic and calmodulin-binding subunit, calcineurin A, tightly bound, even in the presence of only low (nanomolar) concentrations of Ca^{2+}, to a regulatory 19,000 MW Ca^{2+}-binding regulatory subunit, calcineurin B (reviewed by Klee et al., 1988). Thus, calcineurin is not only a Ca^{2+}/calmodulin-regulated enzyme but is itself a Ca^{2+}-binding protein. This two-subunit structure, unique among the protein phosphatases, is conserved from yeast to man and is essential for activity. Calcineurin A, dissociated from calcineurin B under denaturing conditions, is dependent on calcineurin B for activity (Merat et al., 1985) and more recent evidence has confirmed that, with the exception of *Neurospora crassa* calcineurin A, calcineurin B is absolutely required to reconstitute a fully active enzyme from its recombinant subunits (see below).

Two mammalian isoforms of calcineurin B have been detected at the mRNA level and isolated as proteins. The major protein is a highly conserved isoform, calcineurin B1. Tightly bound to calcineurin A, it was originally identified as an "EF-hand" Ca^{2+}-binding protein of the calmodulin family on the basis of its amino acid sequence (Aitken et al., 1984). Its structure as determined by multidimensional NMR is similar to that of the calmodulin dumbbell with two lobes, each composed of two adjacent "EF-hand" Ca^{2+}-binding loops connected by a flexible helix linker (Anglister et al., 1993, 1994; Grzesiek and Bax, 1993). As predicted from its sequence, it binds 4 moles of Ca^{2+}, one with high affinity ($K_d < 10^{-7}$ M) and three with moderate affinities in the micromolar range (Kakalis et al., 1995). Two high-affinity sites for the lanthanide Eu^{3+} have been localized to the C-terminal domain and two with moderate affinities to the N-terminal domain (Burroughs et al., 1994). Thus, although the overall structure of calcineurin B is similar to that of calmodulin, its metal-binding properties are more

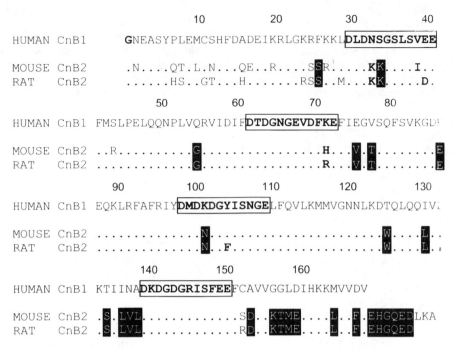

Figure 14.2 Comparison of calcineurin B1 and calcineurin B2 sequences. The deduced amino acid sequences of mouse and rat testis-specific calcineurin B2 (Mukai et al., 1991; Ueki et al., 1992) are aligned with the sequence of human calcineurin B1 (Guerini et al. 1989). Identities with calcineurin B1 are indicated by a dot; conserved amino acid substitutions in rat and mouse calcineurin B2 are represented by white letters on black. Amino acids forming the predicted "EF-hand" Ca^{2+} loops and the conserved amino-terminal glycine that is myristoylated in calcineurin B1 are indicated by boxed bold letters.

like those of another four "EF-hand" Ca^{2+}-regulated protein, troponin C.

The amino acid sequences of mouse and human calcineurin B1 deduced from their cDNA sequences (Guerini et al., 1989) and that of bovine calcineurin B1 determined by protein sequencing (Aitken et al., 1984) are identical with the exception of one amino acid substitution in the mouse protein and a loss of the carboxyl-terminal valine in the bovine protein[1] (Fig. 14.2). Equally conserved from yeast to man is the myristoylation of the amino-terminal glycine (Aitken et al., 1982; Cyert and Thorner, 1992). The isolation from testis cDNA libraries of cDNA clones encoding a 176 amino acid calcineurin B-like protein revealed the presence of another isoform, calcineurin B2 (Mukai et al., 1991; Sugimoto et al., 1991). This protein, 82% identical to calcineurin B1, also contains four amino acid sequences characteristic of the "EF-hand" calcium-binding loops and an additional carboxyl-terminal hydrophilic tail (Fig. 14.2).

This protein has only been isolated from testis which contains both calcineurin B1 and calcineurin B2 (Nishio et al., 1992a). In contrast to the almost perfect conservation of calcineurin B1, there is only 88% identity between rat and mouse calcineurin B2 (Ueki et al., 1992). It is not known if the amino-terminus of calcineurin B2 is myristolated.

Three different mammalian isoforms of calcineurin A (α, β, and γ) have been identified on the basis of their cDNA sequences and detected at the protein level (Guerini and Klee, 1989; Ito et al., 1989; Kuno et al., 1989; Kincaid et al., 1990; Muramatsu et al., 1992). As illustrated in Fig. 14.3, with the exception of the amino- and carboxyl-terminal tails, the three isoforms exhibit a remarkable degree of identity (77–81% identity among their entire sequences). An amino-terminal polyproline motif is a conserved feature of the β isoform (Guerini and Klee, 1989) whereas several additional basic residues in the variable carboxyl-terminal tail are responsible for the high pI (7.1) of the γ isoform as opposed to pIs of 5.6 and 5.8 for the α and β isoforms (Muramatsu et al., 1992). The α and β isoforms are highly conserved

[1]Cys-11 and Cys-153 were erroneously identified as Met-11 and Ser-153.

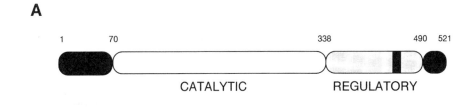

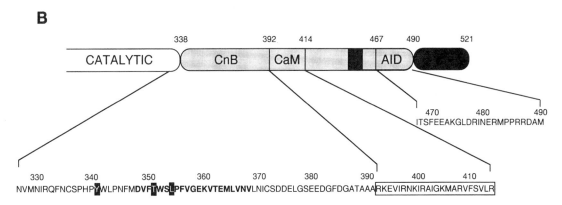

Figure 14.3 Functional domain organization of calcineurin A. (A) All eukaryotic calcineurins A are composed of a catalytic and a regulatory domain flanked by variable amino- and carboxyl-terminal tails. The variable regions and the 10-amino acid insert, resulting from alternative splicing, in the α and β isoforms of mammalian calcineurin A are shown by a black area. (B) Extended representation of the regulatory domain with the amino acid sequences of the calcineurin B, calmodulin-binding, and autoinhibitory domains shown at the bottom (the numbering of the amino acids is that of calcineurin Aα). Residues forming the calcineurin B-binding helix (Griffith et al., 1995) are indicated in bold letters; residues critical for interaction with cyclophilin (Tyr-341 and Thr-351) and FKBP (Thr-351 and Leu-354) are represented by white letters on black.

across species with 99% identity among the rat, mouse, and human proteins. Additional isoforms of calcineurin Aα and calcineurin Aβ, detected only at the cDNA level, are derived from alternative splicing: a 10-residue deletion between the calmodulin-binding and the autoinhibitory domain of the γ isoform has been reported for the spliced α and β isoforms (Kincaid et al., 1990; McPartlin et al., 1991; Muramatsu and Kincaid 1992). This is the location where the two isoforms of calcineurin Aβ (βI and βII) show divergent carboxyl-termini (Guerini and Klee, 1989).

Expression of Recombinant Calcineurin Subunits

The combination of site-directed mutagenesis and assays based on the reconstitution of the active enzyme from recombinant subunits expressed in heterologous systems has been widely used to study the interaction of calcineurin A and calcineurin B at the molecular level. The most commonly used expression system is the translation/transcription of the calcineurin subunits in reticulocyte lysates as this does not require purification of the recombinant proteins. The *Escherichia coli* expression system is the system of choice for the large-scale expression of soluble myristoylated or not myristoylated calcineurin B (Anglister et al., 1994; Rokosz et al., 1995; Sikkink et al., 1995; Kennedy et al., 1996). In SF9 cells, recombinant calcineurin B is usually partially myristoylated (Perrino et al., 1995). The α and β isoforms of calcineurin A and selectively modified derivatives have been expressed in *E. coli* and SF9 cells (Haddy et al., 1992; Perrino et al., 1995; Rokosz et al., 1995; Sikkink et al., 1995). The insoluble β isoform expressed in *E. coli* requires solubilization in denaturating solvents prior to reconstitution (Rokosz et al., 1995; H. Ren and C. Klee, in preparation). Regardless of the expression systems, the two calcineurin A isoforms are inactive and require calcineurin B and Mn^{2+} for activity (Haddy et al., 1992;

Perrino et al., 1995). It is only by coexpression of the two subunits that a fully active enzyme, with stoichiometric amounts of zinc and iron, was expressed in *E. coli* and used to determine the crystal structure of calcineurin described below (Kissinger et al., 1995).

Functional Domain Organization of Calcineurin A

The functional domain organization of calcineurin A was revealed by limited proteolysis experiments (Hubbard and Klee, 1989; Wang et al., 1989). As illustrated in Fig. 14.3, the amino-terminal two-thirds of calcineurin A contains a catalytic domain resistant to proteolysis. This active domain, whose enzymatic activity is repressed in the native protein, becomes fully active when severed by proteases from a regulatory domain that encompasses the carboxyl-terminal third of the molecule. This active domain, still associated with calcineurin B, is often refered to as the Ca^{2+}-independent form of calcineurin. The regulatory domain, readily susceptible to proteolysis, is divided into two subdomains: a carboxyl-terminal autoinhibitory subdomain and a calmodulin-binding subdomain located between the autoinhibitory subdomain and the catalytic domain. The mapping of these functional domains was facilitated by the elucidation of the amino acid sequences of calcineurin A isoforms deduced from their cDNA sequences. A strong homology between calcineurin A and the known sequences of the catalytic subunits of protein phosphatase-1 and -2A suggests that the catalytic domain spans residues 70–328 of calcineurin Aα. A 23-residue peptide (Lys-392 to Arg-414) was proposed by Kincaid et al. (1988) to be the calmodulin-binding domain on the basis of its sequence similarity with known calmodulin-binding peptides that all share a high content of hydrophobic and basic amino acids (Chapter 7, this volume). This peptide, located within a 9000-M_r fragment (residues 332–423) isolated by limited proteolysis of calcineurin in the presence of calmodulin, is a potent inhibitor of the calmodulin stimulation of the enzyme, with a K_d of 1.5 nM (Hubbard and Klee, 1987). Inhibition of calmodulin-stimulated activity of calcineurin by synthetic peptides with sequences overlapping the carboxyl-terminal sequence of calcineurin A localized the inhibitory domain between Ile-467 and Ser-492 (Hashimoto et al., 1990). A highly conserved sequence rich in hydrophobic and acidic amino acids (residues 328–390), between the postulated catalytic and calmodulin-binding domains, was a highly likely candidate for the calcineurin B-binding domain. A peptide corresponding to Gln-333 to Lys-360 binds calcineurin B, but its moderate affinity for calcineurin B ($K_d = 0.5 \times 10^{-6}$ M as opposed to less than 10^{-13}

M for calcineurin A) suggested that this peptide was only part of the calcineurin B-binding domain of calcineurin A (Anglister et al., 1995; H. Ren and C. Klee, unpublished observation). The binding to recombinant calcineurin A of larger peptides expressed in *E. coli* and their ability to inhibit the reconstitution of an active enzyme from its subunits identified a larger peptide (Pro-328 to Ala-390) with a slightly higher affinity (Sikkink et al., 1995; Watanabe et al., 1995). Serial deletions of the carboxyl-terminus of calcineurin A that prevent interaction of the truncated derivatives with calcineurin B were used to identify the carboxyl-terminus of the calcineurin B-binding domain as Leu-377 (Clipstone et al., 1994). Several residues within this domain were shown by site-directed mutagenesis to be critical for the calcineurin B/calcineurin A interaction (Cardenas et al., 1995; Kawamura and Su, 1995; Watanabe et al., 1995). Thus, the calcineurin B-binding domain of calcineurin A spans a large peptide rich in hydrophobic and acidic amino acids as opposed to the calmodulin-binding domain which is rich in basic amino acids (Chapter 7, this volume). This difference may, at least in part, explain the different binding characteristics and different roles that these two Ca^{2+}-binding proteins play in the regulation of calcineurin (Stemmer and Klee, 1994).

Different regions of calcineurin B are involved in the interaction with calcineurin A. In agreement with the study of the calcineurin B/calcineurin A peptide by NMR spectroscopy (Anglister et al., 1995), calcineurins B with single amino acid substitutions in the linker between helix 1 and the Ca^{2+}-binding loop 1, the central helix linker (Phe-82), the linker between Ca^{2+}-binding loops 3 and 4 (Leu-115 and Val-116), and the carboxyl-terminal tail (Phe-153 and Met-166) fail to form heterodimers with calcineurin A (Milan et al., 1994; Watanabe et al., 1996).

Immunophilin-Binding Domains

The identification of calcineurin as the target of the immunosuppressive drugs FK506 and cyclosporins complexed with FKBP and cyclophilins prompted the search for another family of interaction sites on the calcineurin subunits. A role for calcineurin B in these interactions was first suggested by the calcineurin B requirement for inhibition of the protein phosphatase and stimulation of the *p*-nitrophenyl-phosphatase activities of calcineurin by the FKBP/FK506 complex (Haddy et al., 1992). The selective Ca^{2+}- and cyclosporin A-dependent cross-linking of cyclophilin A and calcineurin B demonstrated this essential role of calcineurin B and provided a molecular basis for the exquisite specificity of the immunosuppressive agents for calcineurin (Li and Handschumaker, 1993). While calcineurin B interacts with cyclophilin, calcineurin A is selectively labeled by an affinity-labeling derivative of cyclosporin A (Ryffel et al., 1993), providing a possible explanation for the dependence on calcineurin A for the cross-linking of calcineurin B with cyclophilin. Using the two-hybrid system, Clipstone et al. (1994) confirmed the calcineurin B requirement for interaction, and by serial deletions of the carboxyl-terminal regions of calcineurin A identified the calcineurin B-binding domain as a region of calcineurin A that is also required for the interaction with either one of the two immunophilins. It was shown earlier that a synthetic peptide corresponding to part of the calcineurin B-binding domain is sufficient to promote the interaction of calcineurin B with the FKBP/FK506 complex (Husi et al., 1994). Protein modeling and site-directed mutagenesis of calcineurin B allowed Milan et al. (1994) to identify a "latch region" formed by the peptide connecting the third and fourth Ca^{2+}-binding domains of calcineurin B when it binds calcineurin A as the calcineurin B interaction site with FKBP. The short linker peptide between the two carboxyl-terminal Ca^{2+}-binding sites in calmodulin may prevent the formation of the "latch" region and thus explain the failure of calmodulin to substitute for calcineurin B (Anglister et al., 1995).

The apparent competitive binding of two structurally different immunophilins, FKBP and cyclophilin, to the same site on calcineurin remained a puzzle until the isolation of yeast calcineurin A mutants specifically resistant to FK506 or cyclosporin suggested that the interaction sites may not, in fact, be identical. Mutations of residues in yeast calcineurin A corresponding to Val-314 in the carboxyl-end of the catalytic domain and Tyr-341 and Thr-351 in the calcineurin B-binding domain were identified in cyclosporin-resistant mutants while a mutation of the equivalent of Trp-52 was identified in a FK506-resistant mutant (Cardenas et al., 1995). Site-directed mutagenesis also suggested that the recognition sites for cyclosporin and FK506 are not identical. Substitution of either Thr-351, Leu-354, or Lys-360, with alanine does not interfer with the calcineurin B/calcineurin A interaction and the activation of NFκB in vivo but renders the cells more resistant to FK506 without affecting their sensitivity to cyclosporin A (Kawamura and Su, 1995). Thus, despite their similar immunosuppressive effects and calcineurin inhibition, cyclophilin and FKBP interact with distinct, but possibly overlapping, sites on calcineurin A. Residues of FKBP and cyclophilin involved in interaction with calcineurin are residues exposed to solvent on the surface of the molecule (Futer et al., 1995). These residues identify sites distinct from the isomerase catalytic and drug-binding sites. Mutations which decrease the affinity of FKBP

and cyclophilin for calcineurin do not affect their isomerase activities or their affinity for FK506 and cyclosporin A (Zydowsky et al., 1992; Yang et al., 1993; Etzkorn et al., 1994).

Crystal Structure of Calcineurin and Calcineurin-FKBP12-FK506 Complex

The crystal structures of the recombinant α isoform of human calcineurin and that of its complex with FKBP12–F506, as well as that of the complex of the proteolytic fragment of bovine calcineurin lacking the regulatory domain and the amino-terminal 16 residues with FKBP12–FK506, (shown in Fig. 14.4) have been determined at 2.1, 3.5, and 2.5 Å, respectively (Griffith et al. 1995; Kissinger et al., 1995). With the exception of the amino- and carboxyl-terminal fragments of calcineurin A, missing in the bovine protein, the crystal structure of the Ca^{2+}-saturated form of the full-length human calcineurin is similar to the structure of the bovine protein. The catalytic domain, similar to that of protein phosphatase-1, consists of a sandwich of a sheet of six β-strands covered by three α-helices and three β-strands and a sheet of five β-strands covered by an all-helical structure (reviewed by Barford, 1996). The two divalent metal ions, iron and zinc, bound to residues provided by the two faces of the β-sandwich, define the catalytic center.

The catalytic center is blocked in the full-length protein by two short α-helices and five residues in an extended conformation with Glu-481 hydrogen-bonded to water molecules bound to the two active-site metal ions. This 18-residue segment (Ser-469 to Arg-486) corresponds to the previously identified inhibitory domain. Segments of calcineurin A not visible on the electron density map of the full length protein (residues 1–13, 374–468, and 487–521) include the amino- and carboxyl-termini and part of the regulatory domain, including the calmodulin-binding domain. Their disordered structure is consistent with their extreme sensitivity to proteolytic attack (Hubbard and Klee, 1989).

The last β-sheet extends into a five-turn amphipathic α-helix (residues 350–370) whose top face, completely nonpolar, is covered by a 33 Å groove formed by the amino- and carboxyl-terminal lobes of calcineurin B together with the carboxyl-terminal strand of calcineurin B. Interaction of residues 14 to 23 of calcineurin A with the carboxyl-terminal lobe of calcineurin B in the full length protein may provide the additional binding energy to account for the very high affinity of calcineurin B for calcineurin A ($K_d < 10^{-13}$ M) as opposed to the relatively low affinity of the calcineurin B-binding peptides of calcineurin A for calcineurin B (Anglister et al., 1995; Watanabe et al., 1995). In bovine calcineurin B, the

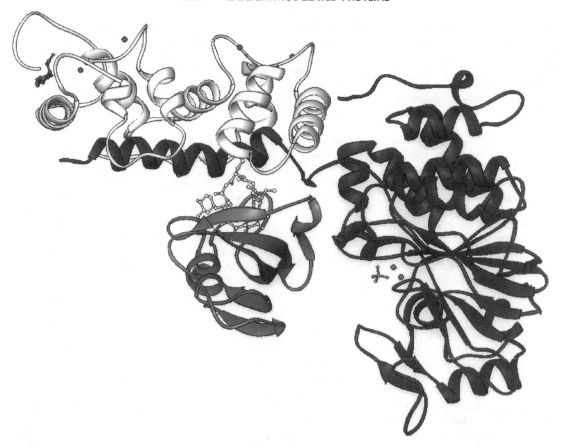

Figure 14.4 Ribbon representation of the crystal structure of truncated calcineurin complexed with FKBP–FK506. Calcineurin Aα is shown in black, calcineurin B in white, and FKBP in grey. Iron and zinc in the active site and the four Ca^{2+} in the calcineurin B sites are shown as grey spheres. The bound phosphate in the active site of calcineurin A, myristic acid covalently linked to the N-terminal glycine and running parallel to the N-terminal helix of calcineurin B, and FK506 are shown in ball-and-stick representation. (PDB code 1TCO, Griffith et al., 1995.)

myristic acid covalently linked to the amino-terminal glycine and located close to the carboxyl-terminal end of the calcineurin B-binding domain lays parallel to the hydrophobic face of the amino-terminal helix of calcineurin B, whereas the nonmyristoylated amino-terminus of the recombinant protein is disordered. This highly conserved post-translational modification of calcineurin B, regardless of its subcellular distribution, is not required for activity or interaction with the immunosuppressive drugs in yeast (Zhu et al., 1995; Kennedy et al., 1996). As was described for another Ca^{2+}-binding protein, recoverin, and for cAMP-dependent protein kinase (Zheng et al., 1993), it may serve as a stabilizing structural element (Ames et al., 1995).

Unlike calmodulin, calcineurin B interacts with calcineurin A in an extended conformation. The more polar bottom face of the calcineurin B-binding helix, together with calcineurin B and the phospha-

tase domain of calcineurin A exposed to solvent, make contact with three distinct regions of FKBP12 while the interface of the calcineurin B-binding domain and calcineurin A forms the binding site of FK506 (Fig. 14.4). Two-thirds of the surface contact between FKBP12–FK506 and calcineurin B comes from the latch region identified by Milan et al. (1994) as the major site of interaction of calcineurin B with the cyclophilin–cyclosporin A. This latch region formed by calcineurin B upon binding to calcineurin A may be recognized by each of the two immunosuppressive complexes, thus explaining their competitive binding to calcineurin (Liu et al. 1991a).

Regulation

The phosphatase activity of calcineurin is under the tight control of Ca^{2+}, mediated by two structurally similar but functionally different Ca^{2+}-binding proteins: calmodulin and calcineurin B.

Activation of Calcineurin by Calmodulin

The protein phosphatase activity of the Ca^{2+}-liganded form of the full length holenzyme, is very low. Full activation requires, in addition to the concentration of Ca^{2+} reported in stimulated cells (between 0.5 and 1 μM), an equimolar amount of calmodulin (Klee et al., 1988). Calcineurin affinity for calmodulin ($K_d < 10^{-9}$ M) is one of the highest reported for calmodulin-regulated enzymes (Hubbard and Klee, 1987). The low basal activity is stimulated more than 10-fold on the combined addition of Ca^{2+} and calmodulin. The calmodulin stimulation is strictly the result of an increased V_{max}. The Ca^{2+}-dependence of this calmodulin stimulation is highly cooperative, with a Hill coefficient of 2.5–3 which indicates that it requires occupancy of at least three of the calmodulin Ca^{2+} sites. Consistent with the fact that activation is the result of the Ca^{2+}-dependent binding of calmodulin to calcineurin, the concentration of Ca^{2+} needed for activation decreases with increasing concentrations of calmodulin. Conversely, the presence of calcineurin should increase the affinity of calmodulin for Ca^{2+}. Because calcineurin is itself a Ca^{2+}-binding protein, only the effect of its calmodulin-binding domain on the affinity of calmodulin could be tested and this was shown to dramatically increase the affinity of calmodulin for Ca^{2+} (Stemmer and Klee, 1994). The cooperative Ca^{2+}-dependent activation of calcineurin by calmodulin allows the enzyme to respond to narrow Ca^{2+} thresholds following cell stimulation.

As for most calmodulin-regulated enzymes, the mechanism of calmodulin activation of calcineurin involves the displacement of an autoinhibitory domain. Removal of this domain by limited proteolysis results in activation of the enzyme and loss of calmodulin dependence (Hubbard and Klee, 1989). A general mechanism for the stimulation of calmodulin-stimulated enzymes based on the elucidation of the structure of calmodulin complexed with the calmodulin-binding peptide of myosin light chain kinase is described by Tjandra et al. (Chapter 7, this volume). Although this simple mechanism applies to many calmodulin-regulated enzymes, different enzymes interact with, and are activated by, calmodulin differently: they exhibit different affinities for Ca^{2+} and calmodulin, and in some cases the activation process involves autophosphorylation of the enzyme (Chapter 13, this volume). These differences, which may ensure the temporal and topological coordination of calmodulin-regulated cellular processes, may require additional sites of interaction between calmodulin and its targets. The crystal structure of the calcineurin/calmodulin complex may reveal such additional sites and help to elucidate how some calmodulin variants only partially activate calmodulin (Newton and Klee, 1984; Putkey et al., 1986; Hurwitz et al., 1988).

Ca^{2+}/Calmodulin-Dependent Inactivation of Calcineurin

The specific inhibition of calcineurin by immunosuppressive drugs has also been used to demonstrate that, in crude tissue extracts, calcineurin is 10–20 times more active than purified calcineurin. The crude enzyme is almost completely dependent on calmodulin, and does not depend on added metals for activity, but is subject to a time- and Ca^{2+}/calmodulin-dependent inactivation facilitated by small heat-stable inactivator(s) (Stemmer et al., 1995). The search for factors responsible for the high phosphatase activity and instability of crude calcineurin led to the finding that, in crude extracts, calcineurin is protected against inactivation by superoxide dismutase as illustrated in Fig. 14.5. It was proposed that the displacement of the autoinhibitory domain upon binding of Ca^{2+}/calmodulin exposes the metal cofactors in the active site of the enzyme to the damaging effects of superoxide anion (Wang et al., 1996b). The reversibility of the calmodulin-dependent inactivation of calcineurin by ascorbate illustrated in Fig. 14.5 suggests that the inactivation of calcineurin is the result of the oxidation of Fe^{2+} at the active site. The protective effect of superoxide dismutase on calcineurin activity has also been observed in vivo, in yeast cells lacking the gene for the cytoplasmic superoxide dismutase and in hippocampal neurons after prolonged Ca^{2+} stimulation (Wang et al., 1996b, Bito et al., 1997a). The redox state of iron provides a reversible mechanism to regulate calcineurin activity by desensitizing the enzyme and serves to couple Ca^{2+}-dependent protein dephosphorylation to the redox state of the cells.

Role of Calcineurin B

The elucidation of the role of calcineurin B in the Ca^{2+} regulation of calcineurin has been somewhat elusive. Exposure of calcineurin to very low concentration of Ca^{2+} neither decalcifies the protein nor dissociates the two subunits. Complete decalcification, however, results in irreversible inactivation of the enzyme and indicates that the high-affinity sites of calcineurin B are always occupied and thus cannot play a regulatory role, but may be involved in the correct folding and stabilization of calcineurin B (Stemmer and Klee, 1994). The small Ca^{2+}-dependent increase of the phosphatase activity, in the absence of calmodulin, indicates that Ca^{2+} binding

Repressed   **Activated**   **Inactivated**

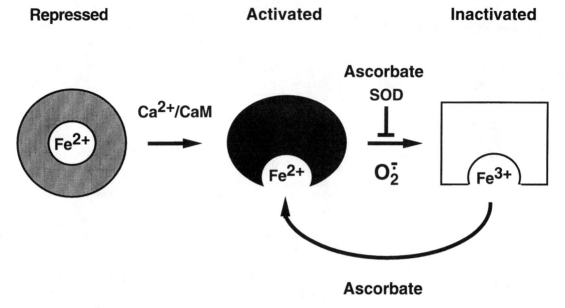

Figure 14.5 Ca^{2+}/calmodulin-dependent inactivation of calcineurin. Activation of crude calcineurin by calmodulin is accompanied by a conformational change, displacement of the autoinhibitory domain, and exposure of the reduced iron (Fe^{2+}) in the active site. Oxidation of Fe^{2+} by the superoxide anion (O_2^-) inactivates calcineurin. The enzyme is protected against inactivation by superoxide dismutase, and reduction of Fe^{3+} by ascorbate reactivates the enzyme (Wang et al., 1996b).

to moderate-affinity sites ($K_d = 0.2–0.5\ \mu M$) plays a regulatory role (Stemmer and Klee, 1994; Perrino et al., 1995). Unlike Ca^{2+} binding to calmodulin, Ca^{2+} binding to the low-affinity Ca^{2+} sites of calcineurin B results in a decrease of the K_m value with a small increase of V_{max}, and may therefore not fully displace the inhibitory domain but transform this pseudoirreversible inhibitor into a competitive one by weakening the interaction between the inhibitory domain and the catalytic site (Stemmer and Klee, 1994; Perrino et al., 1995). The truncated form of calcineurin, calcineurin-45, which still binds calcineurin B but lacks the entire regulatory domain (residues 392–521), has an affinity for Ca^{2+} 10 times that of the native enzyme and still depends on a very low concentration of Ca^{2+} to decrease the K_m value of the enzyme (Stemmer and Klee, 1994). Serial deletions of the carboxyl-terminus of calcineurin A have now revealed that this increased affinity of calcineurin B for Ca^{2+} requires not only the removal of the inhibitory domain (residues 458–524) but also that of residues 420–457 (Perrino et al., 1995). Perhaps this segment of calcineurin A, linking the calmodulin-binding and the autoinhibitory domains, interacts with calcineurin B when the low-affinity sites are not occupied. Occupancy of the four Ca^{2+} sites may promote dissociation of this domain and result in the disordered structure of the regulatory

domain observed in the crystal structure of the full-length enzyme. Preferential binding of the linker to the amino-terminal lobe of calcineurin B depleted of Ca^{2+} should decrease the affinity of this lobe for Ca^{2+} and maintain the enzyme in the inactive state by strengthening the interaction between the inhibitory domain and the catalytic site. Removal of residues 420–457 would increase the affinity of calcineurin B for Ca^{2+} and partially activate the enzyme.

The third, and perhaps most important, function of Ca^{2+} binding to calcineurin B is to allow the Ca^{2+}-dependent and calmodulin-facilitated interaction of calcineurin with the immunophilins bound to endogenous ligands or other target proteins. It has been reported that calcineurin binding to the ryanodine and IP$_3$ receptors is mediated by FKBP. The release of calcineurin from these receptors upon exposure to FK506 led to the proposal that a structural motif on the receptors may act as an endogenous FK506 analog (Cameron et al., 1995; 1997).

Another potentially important partner of calcineurin is AKAP79 (A-kinase anchoring protein), a member of the family of proteins whose function is to bring kinases or phosphatases close to their substrates (Coghlan et al., 1995; Faux and Scott, 1996; Klauck et al., 1996; Dell'Acqua et al. 1998). The function of AKAP79 is apparently to bring into close proximity calcineurin and the two other partners of AKAP79: the type II regulatory subunit of cAMP-dependent protein kinase and protein kinase $C\beta$ (known substrates of calcineurin) and their substrates. It is not clear, however, how calcineurin will be stimulated in response to a Ca^{2+} signal since AKAP79 is a noncompetitive inhibitor of calcineurin whose interaction with calcineurin is Ca^{2+}-independent.

The tight binding of the transcription factor NF-AT is another example of calcineurin interaction that is possibly mediated by calcineurin B (Wesselborg et al., 1996). This interaction is prevented by exposure of calcineurin to the FK506/FKBP complex. It does not require phosphorylation of NF-AT and allows the cotranslocation of calcineurin and NF-AT to the nucleus where calcineurin may be needed to maintain NF-AT dephosphorylated. It may also be required to ensure the dephosphorylation of substrates, which, like NF-AT, are present at low concentrations inside the cells.

Phosphorylation

The report that purified brain calcineurin contains substoichiometric amounts of covalently bound phosphate suggests that it may be regulated by phosphorylation in vivo (King and Huang, 1984). Calcineurin is readily phosphorylated in vitro at Ser-411 in the calmodulin-binding domain by protein kinase C and the autoactivated form of calmodulin-kinase II (Tung, 1986; Hashimoto et al., 1988; Hashimoto and Soderling, 1989; Martensen et al., 1989). Although this phosphorylation is inhibited by calmodulin, it decreases only slightly the affinity of calcineurin for calmodulin (Calalb et al., 1990). In the presence of calmodulin, calcineurin is preferentially phosphorylated at two unidentified sites by casein kinase I, and to a lesser extent by casein kinase II and phosphorylase kinase, whereas no phosphorylation is observed with cAMP-dependent kinase, tyrosine kinase, and calmodulin kinase II and myosin light chain kinase in presence of calmodulin (Singh and Wang, 1987). The physiological significance of these observations needs to be ascertained by the identification of the sites phosphorylated in vivo and a better characterization of the effect of phosphorylation on enzyme activity or regulation.

Chromosome Localization and Gene Structure

Calcineurin A

The α, β, and γ isoforms of human calcineurin A are the products of three different genes located on chromosomes 4, 10, and 8, respectively (Giri et al., 1991; Muramatsu and Kincaid, 1992; Wang et al., 1996a). The rat homologs are each located on chromosome 15 (Yamada et al., 1994). Two calcineurin A genes have been identified in *Drosophila melanogaster*. One gene (*DRO1*), located at position 21EF, encodes a 64.5 kDa protein. Transcripts of this gene have not yet been detected (Guerini et al., 1992). A second gene, *DRO2*, located at 14D1-4, encodes a 62 kDa calcineurin A isoform (Brown et al., 1994). A single gene has been reported in *Neurospora crassa*. It encodes a 59,580 Da protein, 75% identical to mammalian calcineurin A (Higuchi et al., 1991). Two isoforms of calcineurin A (MW 63,000 and 69,000) are encoded by two intronless genes (*CMP1* and *CMP2*) located on chromosome 4 in *Saccharomyces cerevisiae* (Cyert et al., 1991; Liu et al., 1991b; Ye and Bretscher, 1992). A single calcineurin A gene has been isolated in *Aspergillus nidulans* (Rasmussen et al., 1994) and the fission yeast, *Schizosaccharomyces pombe* (Yoshida et al., 1994; Plochocka-Zulinska et al., 1995).

Although the full structure of the mammalian calcineurin genes has not yet been determined, a comparison of the partial sequences of the calcineurin A genes from different species helped to interpret the splicing events in the calcineurin genes. The conserved intron/exon boundaries from flies to mammals revealed a good correlation between the intron/exon boundaries and the boundaries of the functional domains. The calcineurin $A\beta$ gene is relatively large with many introns that span at least 50 kb. A 1.8 kb intron, located at the position where the calcineurin $A\beta I$ and βII cDNAs become divergent has confirmed that calcineurin $A\beta II$ is generated by an alternative splicing which involves the loss of the exon containing the 3' end of calcineurin A βI cDNA and the next intron (Wang et al., 1996a). The 3' end of this intron, which is conserved in the human calcineurin $A\alpha$ genes, is the site of the 30 base pair deletion in calcineurin $A\alpha$, βII, and γ cDNAs (Kincaid et al., 1990; McPartlin et al., 1991; Muramatsu et al., 1992). The site of the 54 base pair insert in calcineurin $A\beta I$ corresponds to an intron also found in the *Drosophila* (Guerini and Klee, 1989; Guerini et al., 1992) and *Aspergillus nidulans* (Rasmussen et al., 1994) genes. Two highly conserved introns, one immediately after or within the variable amino-terminal region and the other within the coding sequence for the catalytic domain, are present in all organisms examined except for the budding and fission yeasts (Higuchi et al., 1991; Guerini et al., 1992; Rasmussen et al., 1994). The *Drosophila* and human genes share two more introns, one located in the catalytic domain and another at the carboxyl end of the calcineurin B-binding domain (Guerini et al., 1992). An intron within the 3' end of the coding sequence has been found in *Neurospora* and *Aspergillus* (Higuchi et al., 1991; Rasmussen et al., 1994), and another found within the aminoterminal region of the catalytic domain is apparently unique to the fission yeast (Yoshida et al., 1994; Plochocka-Zulinska et al., 1995). Thus, the calmodulin-binding domain, the autoinhibitory domain, and the amino-terminal variable region are each encoded by individual exons while the calcineurin B-binding domain of *Drosophila* calcineurin A is encoded by two exons (one corresponding to the cyclophilin-binding region and the other to the FKBP-binding region of this domain) and the catalytic domain by five exons. Less is known about the 5' untranslated region. The core promoter region of the calcineurin Aα gene, as well as a sequence critical for its cell type-specific expression, is located -107 to $+157$ nt with respect to the major transcription initiation site (Chang et al., 1992).

Calcineurin B

Human calcineurin B is the product of a 10 kb gene that contains four introns and is located on human chromosome 2 (Wang et al., 1996a). In testis, the alternative utilization of two promoters yields another isoform of calcineurin B1 (CNBα2) with 46 additional amino acids at the amino-terminus. Developmentally regulated transcripts for this variant are only detected in testis (Chang et al., 1994). *Drosophila* calcineurin B is the product of a single intronless gene located at position 4F on the X chromosome (Guerini et al., 1992). In the budding yeast, calcineurin B is the product of a single gene, with a single intron following the start codon, located on chromosome 4 (Kuno et al., 1991; Cyert and Thorner, 1992).

Distribution

Tissue Distribution

Isolation of calcineurin from different tissues, immunostaining of crude tissue extracts with anticalcineurin antibodies, and measurement of enzymatic activity firmly established the wide distribution and coordinated expression of the two subunits of calcineurin in mammalian tissues (for a review and recent references, see Klee et al., 1988; Papadopoulos et al., 1989; Gagliardino et al., 1991; Natarajan et al., 1991; Ellis and Edwards, 1994; Hubbard, 1995). Calcineurin is particularly abundant in brain (0.8–1% of total protein) where its concentration is 6–7 times that of skeletal muscle and 10–20 times that of other tissues and a number of cell lines. It is almost undetectable in liver and smooth muscle (Dawson et

al., 1994; Tumlin et al., 1995). A similar differential expression of the calcineurin subunits was observed at the mRNA level (Guerini and Klee, 1989; Guerini et al., 1989; Kincaid et al., 1990; Mukai et al., 1991; Ueki et al., 1992; Chang et al., 1994).

Regional Distribution in Brain

The high concentration of calcineurin in brain has facilitated a detailed analysis of its cellular and regional distribution. Contrary to the Ca^{2+}-binding protein, S100, which is specifically expressed in glial cells, calcineurin is predominantly expressed in neurons and is particularly enriched in the striatum, the dentate gyrus, and the hippocampus (Goto et al., 1986, 1988; Matsui et al., 1987; Klee et al., 1990; Kuno et al., 1992; Dawson et al., 1994). In the striatum, very high levels are found in the medium spiny neurons of the caudate-putamen which constitute 90% of the striatal neurons and are distinct from the NADPH-, somatostatin-, and acetylcholine-expressing neurons (Goto et al., 1987). The calcineurin-containing neurons project to the globus pallidus and the substantia nigra. Accordingly, immunoreactivity is primarily found in the perikarya, dendrites, postsynaptic densities and spines of the caudate-putamen, and in the axons and axon terminals of the globus pallidus and substantia nigra (Klee et al., 1990). Lesions of the nigrostriatal dopaminergic pathway do not affect calcineurin activity (Chung et al., 1989; Goto et al., 1989). Kainic acid, which destroys GABAnergic striatal neurons, affects calcineurin-rich neurons to a lesser extent (Chung et al., 1987). In the hippocampus, a high level of immunoreactivity is observed in the pyramidal cells of the CA1–CA2 regions and in the granule cells of the dentate gyrus (Matsui et al., 1987). Low levels of calcineurin are detected in the cortex and the cerebellum and very low levels in white matter and the hypothalamus (Goto et al., 1986). Moderate immunoreactivity is visible in some Purkinje cells and in the granular cells and their projections to the molecular layer of the cerebellum (Goto et al. 1986; Klee et al., 1990). The abundance of calcineurin in brain parallels equally high levels of its binding protein, the immunophilin FKBP (Dawson et al., 1994). The similar regional localization of the two proteins and the increased phosphorylation of endogenous substrates upon exposure to FK506 or cyclosporin A suggests a physiological link between calcineurin and the immunophilins (Steiner et al., 1992).

Isoform Specific Expression

Recent studies, using specific riboprobes and antibodies, deal with the differential tissue distribution of the calcineurin isoforms at both the mRNA and protein levels. Similar levels of the 4.1–4.2 kb calcineurin Aα mRNA and the 3.1–3.2 kb calcineurin Aβ mRNA are observed in brain (Takaishi et al., 1991; A. Stump et al., in preparation). Very low levels of calcineurin AβI mRNA are detected only by PCR (D. Guerini, unpublished observations). Whereas similar levels of calcineurin Aα and β transcripts are detected in brain, skeletal muscle, and kidney, only β transcripts are found in the thymus and γ transcripts in the testis (Takaishi et al., 1991; Buttini et al., 1995; Tumlin et al., 1995; A. Stump et al. unpublished observations). At the protein level, calcineurin Aα (80% of total calcineurin) is the predominant isoform in brain (Kuno et al., 1989; Takaishi et al., 1991). A post-transcriptional regulation of the isoforms in brain may be responsible for the discrepancy between the mRNA and protein levels. The 2.8–2.9 kb calcineurin B1 mRNA is detected in all tissues examined whereas the 4.2 kb calcineurin B2 mRNA and that of the high-molecular variant of calcineurin B1 are found only in testis (Guerini et al., 1989; Mukai et al., 1991; Ueki et al., 1992; Chang et al., 1994). This tissue-specific expression of calcineurin B2 was confirmed at the protein level (Nishio et al., 1992a)

The regional distribution of α and β isoforms of calcineurin A mRNAs in brain by in situ hybridization showed that the α isoform is predominant in the caudate putamen and substantia nigra, with intense signals in the pyramidal cell layer of the hippocampus, and dentate gyrus. The β isoform is more widely distributed, being predominant in the thalamus and the olfactory bulb, with moderate levels in the granular layer of the cerebellum (Takaishi et al., 1991). Specific antibodies to the α and β isoforms showed that calcineurin Aα is enriched in the cortex, the striatum, and the hippocampus while the β isoform is specifically expressed in Purkinje cells (Kuno et al., 1992) although a specific nuclear localization of calcineurin Aα has been reported in these cells (Usuda et al., 1996).

The localization of calcineurin has also been extensively studied in the testis, where a 2.8 kb mRNA, encoding the γ isoform of calcineurin A, and two transcripts of calcineurin B2 are specifically and coordinately expressed in spermatocytes during late stages of spermatogenesis (Muramatsu et al., 1992; Nishio et al., 1992b; Miyamoto et al., 1994). CnB2 accumulates in spermatid nuclei in scallop testis at the beginning of nuclear elongation (Moriya et al., 1995). In the kidney, both calcineurin Aα and β, and spliced variants, are differentially expressed in specific segments of the nephron with relatively high calcineurin activity in the proximal tubules (Buttini et al., 1995; Tumlin et al., 1995).

Species Distribution

Calcineurin has also been detected at the protein and mRNA levels in invertebrates and low eukaryotes. In *Drosophila melanogaster*, the transcript of the *DRO1* gene has not yet been detected (Guerini et al., 1992). Two transcripts of the *DRO2* gene (3 and 3.5 kb) are highly expressed in the early embryo while the 3.5 kb transcript is also detectable in adult females (Brown et al., 1994). The two isoforms of calcineurin A (MW 63,000 and 69,000) have been identified in yeast with specific antibodies, with the *CMP1* gene product being the major isoform (Liu et al 1991b; Ye and Bretscher, 1992). Only one calcineurin B isoform has been detected at the protein and mRNA level in flies and yeast.

Functions

T-cell Activation and Regulation of Gene Expression

The identification of calcineurin as the target of immunosuppressive drugs revealed an essential role for this enzyme in T-cell activation (Liu et al., 1991a), while the identification of FK506 and cyclosporin as specific inhibitors of calcineurin provided powerful tools to reveal the many roles of this enzyme in the transduction of Ca^{2+} signals (Clipstone and Crabtree, 1992; Fruman et al., 1992a; Liu et al., 1992; O'Keefe et al., 1992). During the past 7 years, studies of calcineurin have been dominated by the search for its physiological roles and attempts to understand the mechanisms of the Ca^{2+}-dependent, calcineurin-mediated regulation of biological processes. None is better understood than the Ca^{2+}-dependent calcineurin-mediated transcription of the T-cell growth factor, interleukin-2, in response to T-cell activation as illustrated in Fig. 14.6 (for recent reviews see Lane et al., 1993; Crabtree and Clipstone, 1994; Heitman et al., 1994; Rao, 1994; Cardenas and Heitman, 1995; Kincaid, 1995). The translocation of the transcription factor, NF-ATp, in response to an increase of intracellular Ca^{2+} induced by the occupancy of the T-cell receptor, dependent upon its dephosphorylation by calcineurin, is the first well-documented example of the transduction of a signal at the plasma membrane to the nucleus. A prolonged Ca^{2+} signal and the cotranslocation of calcineurin and NF-ATp to the nucleus ensure the sustained activation of gene expression (Shaw et al. 1995; Luo et al., 1996; Shibasaki et al., 1996; Timmerman et al., 1996). Glycogen synthase kinase-3 and casein kinase I together with MEKK1 have recently been implicated in the export of NF-AT to the cytoplasm and thus, terminating the signal (Beals et al., 1997; Zhu et

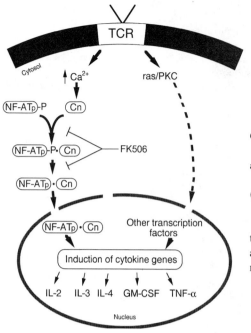

Figure 14.6 Calcineurin-mediated T-cell activation. Calcineurin-mediated signal transduction from the cytoplasm to the nucleus leading to induction of cytokine genes according to Crabtree and Clipstone (1994), Rao (1994), and Shibasaki et al. (1996). Stimulation of the T-cell receptor (TCR) is accompanied by a sustained increase of intracellular Ca^{2+}, activation of calcineurin (Cn), interaction of calcineurin with the cytoplasmic phosphorylated form of the transcription factor NF-ATp, dephosphorylation of NF-ATp, and translocation of the calcineurin–NF-ATp complex to the nucleus for induction of the cytokine genes [interleukin (IL)-2, -3, -4, granulocyte-macrophage colony-stimulating factor (GM-CFS), and tumor necrosis factor α (TNF-α)]. Gene activation may require the concomitant activation of the protein kinase C (PKC) or the ras21 pathways.

al., 1998). Regulation of cell growth involves the complex interplay of an array of growth factors, whose expression or activity is often modulated by calcineurin solely in response to an increase in intracellular Ca^{2+} concentration or sometimes in concert with other signaling pathways (as reviewed by Crabtree and Clipstone (1994) and Rao (1994). Calcineurin is necessary but not sufficient alone for IL-2 gene expression that depends also on the activation of the p21ras pathway or protein kinase C activation (Woodrow et al., 1993). Dephosphorylation of the transcription factor κ3 by calcineurin is necessary and sufficient for the induction of the tumor necrosis factor, TNF-α (Goldfeld et al., 1994; McCaffrey et al., 1994). Calcineurin-mediated dephosphorylation and inactivation of the inhibitor of the transcription factor, NFkB, is responsible, along with the activation of protein kinase C, for the nuclear translocation of NFκB in a wide variety of cell types (Frantz et al., 1994). Among other transcription factors whose activity is controlled by calcineurin are NF(P), required for the induction of interleukin-4 (Kubo et al., 1994), and the granulocyte-macrophage colony-stimulating factor gene (Tsuboi et al., 1994).

The recent recognition of the broad tissue distribution of NF-AT isoforms suggests that many other functions of calcineurin mediated by this transcription factor remain to be identified. NF-AT-mediated cardiac hypertrophy in transgenic mice overexpressing the constitutive form of calcineurin is such an example (Molkentin et al., 1998). The protective effect of FK506 against congestive heart failure in this mice model illustrates the broad therapeutic potential of specific calcineurin inhibitors.

Programmed Cell Death

The excellent correlation between the inhibition of calcineurin phosphatase activity and the loss of the ability of lymphoid cells to undergo apoptosis upon treatment with cyclosporin A suggested that calcineurin plays a role in T-cell receptor signalling that leads to programmed cell death (Fruman et al., 1992b; Bonnefoy-Berard et al., 1994). Conversely, overexpression of calcineurin facilitates Ca^{2+}-activated cell death (Shibasaki and McKeon, 1995). The mechanism of action of calcineurin involves the dephosphorylation of a yet unknown substrate and results in the activation of the DNA-binding activity of the orphan steroid receptor, Nur77, whose induction is necessary to induce apoptosis in T-cell hybridomas (Yazdanbakhsh et al., 1995). In contrast, calcineurin inhibits the early stages of glucocorticoid-induced apoptosis (Zhao et al., 1995).

Neuronal Function and Regulation of Receptor Function

The diverse roles of calcineurin in the control of neuronal function that are covered in several recent com-

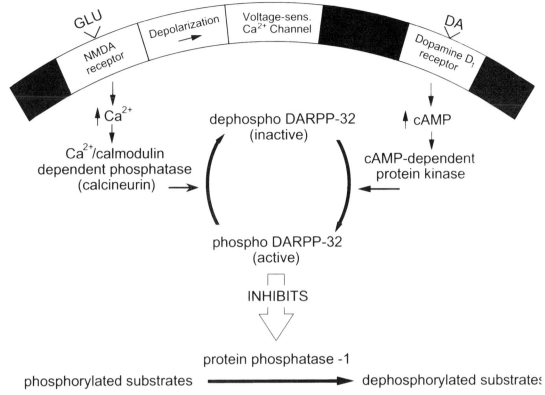

Figure 14.7 Protein phosphatase cascade triggered by stimulation of NMDA receptors of striatal neurons. Illustration of the mechanism proposed by Halpain et al. (1990) to explain the opposite effects of dopamine and glutamate on the excitability of striatal neurons. In response to dopamine (DA) binding to the dopamine D_1 subclass of dopamine receptors, adenylate cyclase is activated and the increase in cAMP results in activation of cAMP-dependent protein kinase, and phosphorylation of DARPP-32 which is converted into a potent inhibitor of protein phosphatase-1, thus favoring protein phosphorylation. Activation of the NMDA receptors by glutamate is accompanied by an increase of intracellular Ca^{2+}, activation of calcineurin, dephosphorylation of DARPP-32, and activation of protein phosphatase-1, thus favoring protein dephosphorylation. Thus, calcineurin-mediated protein dephosphorylation involving a protein phosphatase cascade is inhibited not only by calcineurin-specific inhibitors but also by inhibitors of protein phosphatase-1, such as okadaic acid. (Reproduced with permission from Halpain et al., 1990.)

prehensive reviews (Bear and Malenka, 1994; Bear and Abraham, 1996; Bito et al., 1997b; Yakel, 1997; Antoni et al., 1998) are only briefly reviewed here. The very high concentration of calcineurin in the striatal and hippocampal neurons suggests essential roles for the enzyme in these areas of the brain. The specific dephosphorylation of the dopamine and cAMP-regulated phosphoprotein, DARPP-32, an inhibitor of protein phosphatase 1, by calcineurin provided an explanation for the antagonistic effects of glutamate and dopamine on the excitability of striatal neurons (Halpain et al., 1990). The *N*-methyl-D-aspartate (NMDA) receptor-mediated inhibition of the stimulatory effect of dopamine in striatal neu-

rons, illustrated in Fig. 14.7, was the first evidence for the protein phosphatase cascade postulated by Ingebritsen and Cohen (1983b). In contrast, the Ca^{2+}-dependent and calcineurin-mediated desensitization of NMDA receptor-coupled Ca^{2+} channels in hippocampal neurons is insensitive to inhibitors of protein phosphatase 1 and 2A, implying a direct and specific modulation of these Ca^{2+} channels by calcineurin. The inhibitory effects of calcineurin blockers on long-term depression (LTD), induced by low-frequency stimulation, may, in part, be explained by a decreased desensitization of the NMDA receptors and increased postsynaptic Ca^{2+} concentrations (Mulkey et al., 1993, 1994; Bear and

Malenka, 1994; Lieberman and Mody, 1994; Tong et al., 1995). The large increase of intracellular Ca^{2+} that follows high-frequency stimulation which is needed for the activation of Ca^{2+}/calmodulin-stimulated protein kinase II favors long-term potentiation (LTP) as described by Shulman and Braun (Chapter 13, this volume). Thus, inhibitors of calcineurin not only inhibit LTD but also facilitate LTP in the visual cortex (Funauchi et al., 1994; Torii et al., 1995).

A series of elegant experiments, using transgenic mice overexpressing calcineurin, have now clarified the role of calcineurin in the development of long lasting LTP and in the conversion of short- to long-term memory (Winder et al., 1998; Mansuy et al., 1998). These experiments provide another example of the involvement of calcineurin in a protein phosphatase cascade explaining the antagonistic effects of Ca^{2+} and cAMP.

Calcineurin-mediated activation of nitric oxide synthase induced by the massive release of glutamate and stimulation of NMDA receptors following vascular strokes has been proposed to explain the protective effect of FK506 against glutamate neurotoxicity (Dawson et al., 1993; Snyder and Sabatini, 1995). The Ca^{2+}-mediated destabilization of the GABA receptor in hippocampal neurons is another example of a Ca^{2+}-induced negative feedback mechanism (Stelzer, 1992) that may be related to calcineurin.

Secretion, Endocytosis, and Neurotransmitter Release

Inhibition of the specific dephosphorylation of dynamin (a GTPase involved in endocytosis) by cyclosporin A and FK520 suggested a role for calcineurin in the Ca^{2+}-dependent release of neurotransmitters. The inhibition of calcineurin by these agents alone, or in combination, had no effect on Ca^{2+}-induced release of glutamate but completely blocked the release of glutamate evoked by K^+-depolarization by 4-aminopyridine (Nichols et al., 1994). This observation is consistent with the previous report that calcineurin "serves as a Ca^{2+} switch for depolarization-evoked synaptic vesicle recycling" (Liu et al., 1994).

Cell Motility

The involvement of calcineurin in cell motility was revealed by its regulation of organelle transport in melanophores (Thaler and Haimo, 1990), the correlation between the inhibition of calcineurin by immunosuppressive drugs and calcineurin inhibitory peptide and the loss of "catch" contraction in molluscan smooth muscle (Castellani and Cohen, 1992), and inhibition of the migration of neutrophils on

vibronectin in response to Ca^{2+} transients generated by exposure to a chemoattractant (Hendey et al., 1992). More recent studies have been devoted to the elucidation of the mechanism of action of calcineurin in cell motility. The Ca^{2+}-dependent release of cell adhesion mediated by calcineurin is the result in the polarized redistribution of $\alpha v\beta 3$ integrins that is responsible for neutrophil migration (Lawson and Maxfield, 1995; Pomies et al., 1995). The spatial redistribution of calcineurin has also been proposed as a potential mechanism for the regulation of motility and neurite outgrowth (Ferreira et al., 1993; Chang et al., 1995). Potential substrates involved in this process include GAP-43, a regulator of cone morphology, the microtubule-associated protein, tau, and dynamin.

Modulation of Ca^{2+} Fluxes

In addition to regulating the Ca^{2+} channels coupled to the NMDA receptors, recent studies indicate that calcineurin may play a role in the regulation of the two intracellular Ca^{2+} channels: the ryanodine receptor involved in excitation-contraction coupling in skeletal muscle and other excitable cells, and the IP_3 receptor involved in Ca^{2+} release by hormones and neurotransmitters (Cameron et al., 1995). Together, these two classes of receptors can potentially affect all cellular processes under Ca^{2+} control. Like calcineurin, the two receptors interact with FKBP12, but this interaction does not require FK506, and the modulation of the conductance of these channels does not require the isomerase activity of FKBP12 (Jaramayan et al., 1992; Brillantes et al., 1994; Timerman et al., 1995). The demonstration that FKBP-mediated anchoring of calcineurin to these two receptors is accompanied by a modulation of Ca^{2+} release from internal stores suggests that an additional role of FKBP is to ensure the proximity of calcineurin to its substrates. The dissociation of FKBP and calcineurin from the channel by FK506 led Cameron et al. (1995; 1997) to propose the attractive hypothesis that the receptors may be endogenous analogs of the immunosuppressive drugs.

Another important physiological role of calcineurin is to regulate the activity of the Na^+/K^+ ATPase (Aperia et al., 1992; Lea et al., 1994). The inhibition of this enzyme by the immunosuppressive drugs may be responsible for their nephrotoxicity. In cerebellar neurons, the calcineurin activation of the Na^+, K^+ ATPase is required to prevent neurotoxicity due to the excess of Na^+ entry induced by glutamate binding to NMDA receptors (Marcaida et al., 1996).

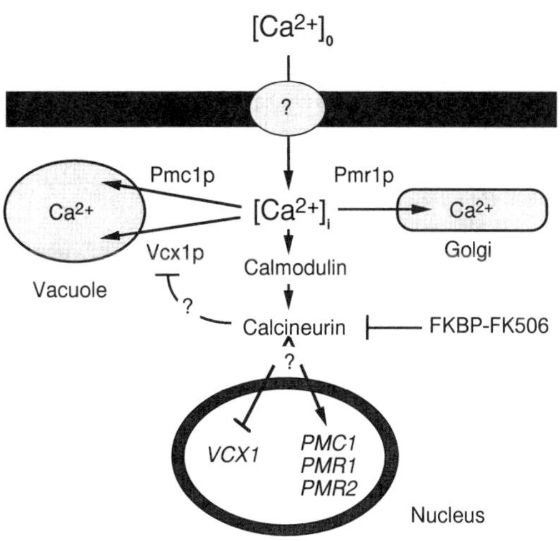

Figure 14.8 Calcineurin-regulated Ca^{2+} homeostasis in yeast. Proposed model to explain the role of calcineurin in the regulation of ion transport in the budding yeast *S. cerevisiae* (Cunningham and Fink, 1996). Lines with arrowheads indicate positive functions or activating interactions, and lines with bars indicate inhibitory interactions. $[Ca^{2+}]_o$, extracellular Ca^{2+}; $[Ca^{2+}]_i$, intracellular Ca^{2+}. (Reproduced with permission from Cunningham and Fink, 1996.)

Role of Calcineurin in Yeast and Fungi

The identification of the physiological roles of calcineurin in yeast was greatly facilitated by the combined use of the immunosuppressive drugs and yeast genetics (for a review see Heitman et al., 1993; Ohya et al., Chapter 19, this volume). A role for calcineurin in the pheromone response pathway was first revealed by the demonstration that yeast cells which lack the regulatory subunit of calcineurin or both isoforms of the catalytic subunit do not recover from G1 arrest after pheromone treatment (Cyert and Thorner, 1992; Nakamura et al., 1993). The recovery from α-factor arrest is inhibited by treatment with FK506, and the inhibition is prevented by disruption of the *FKBP12* gene (Foor et al., 1992). The induction of one calcineurin isoform by α-factor (Ye and Bretscher, 1992) and the subsequent Ca^{2+}-dependent dephosphorylation of substrate(s) that remained to be identified may explain the previously reported Ca^{2+} requirement for viability of yeast cells after exposure to pheromone (Iida et al., 1990).

In contrast to *A. nidulans*, where calcineurin is required for growth (Rasmussen et al., 1994), calcineurin is not essential for growth of yeast cells but its deletion becomes lethal under specific growth conditions or when associated with other mutations (Liu et al., 1991b, Cyert and Thorner, 1992; Nakamura et al., 1993; Parent et al., 1993; Breuder et al., 1994). Calcineurin was shown to play a role in adaptation to high-salt stress, alkaline pH, and treatment with vanadate (Liu et al., 1991b). Failure of calcineurin B null mutants to grow in the presence of high concentrations of NaCl or LiCl and the rescue of NaCl-sensitive mutants by the calcineurin B gene suggested a role for calcineurin in the export of these cations (Nakamura et al., 1993). This transport defect is due,

at least in part, to a decreased expression of the product of the *ENA1/PMR2* gene, a P-type Ca^{2+}-ATPase, involved in Na^+ and Li^+ efflux (Mendoza et al., 1994; Hirata et al., 1995).The identification of a cAMP phosphodiesterase gene as a suppressor of the calcineurin null mutant, and a decrease of the ATPase expression by disruption of a gene that leads to the production of constitutively active cAMP-dependent protein kinase, indicated that the activity or expression of the ATPase is subject to negative control by cAMP-dependent phosphorylation and positive control by a Ca^{2+}-dependent dephosphorylation (Wilson et al., 1991; Hirata et al., 1995). Calcineurin is apparently also involved in the regulation of K^+ uptake in response to Na^+ stress (Mendoza et al., 1994; Nakamura et al., 1993). It was proposed that in *CNB1* null cells, the K^+-uptake system is not converted to the high-affinity state needed to allow the discrimination of K^+ over Na^+ that is required for the maintenance of high internal K^+. Calcineurin-mediated protein dephosphorylation may be involved directly or indirectly in this conversion, believed to be the result of a post-translational modification (Mendoza et al., 1994). Thus, adaptation to high-salt stress involves a Ca^{2+} signal, mediated by calcineurin, that regulates the expression and activity of proteins involved in Na^+, Li^+, and K^+ transport.

The complex role of calcineurin in the regulation of Ca^{2+} homeostasis involves the interplay of at least three Ca^{2+}-transporting enzymes, as well as the vacuolar H^+-ATPase (Fig. 14.8). Deletion of the calcineurin genes, or inhibition of calcineurin by FK506, in yeast mutants deficient in a vacuolar Ca^{2+}-ATPase (product of the *PCM1* gene) restores the ability of these mutants to grow in the presence of high concentrations of Ca^{2+} (Cunningham and Fink,

1994, 1996). Another set of mutants are deficient in the vacuolar H^+-ATPase (*vma* mutants), or fail to assemble this ATPase (*vph6* mutants), and do not sequester Ca^{2+} in the vacuole and require calcineurin for vegetative growth (Garrett-Engele et al., 1995; Hemenway et al., 1995; Tanida et al., 1995, 1996). These opposite effects of calcineurin on the Ca^{2+} modulation of cell growth can now be explained by the recent identification of the product of the *VCX1* (*HUM1*) gene as a H^+/Ca^{2+} exchanger and the involvement of this exchanger in the regulation of Ca^{2+} sequestration (Cunningham and Fink, 1996; Pozos et al., 1996). In response to increased intracellular Ca^{2+} concentration, calcineurin can decrease Ca^{2+} tolerance by a post-translational modification of the transporter that inhibits Ca^{2+} sequestration in the *pcm1* mutants, while in *vma* and *vph6* mutants calcineurin can increase Ca^{2+} tolerance by stimulating the expression of the *PMC1* gene product up to 500-fold (Cunningham and Fink, 1996). A calcineurin-mediated activation or induction of the P-type ion pumps (products of the *PMR1* and *PMR2A* genes) in response to high concentrations of Na^+ may be responsible for the Mn^{2+} and Na^+ tolerance of *pcm1* cells (Cunningham and Fink, 1996). An alternative mechanism for Mn^{2+} tolerance is to prevent Mn^{2+} entry into the cells by a calcineurin-mediated mechanism (Farcasanu et al., 1995).

Calcineurin may also be involved in the regulation of the plasma membrane H^+-ATPase and the H^+ transport. Calcineurin is required for growth at high pH and extracellular acidification in response to glucose (Hemenway et al., 1995). Three other genes whose deletion renders the cells hypersensitive to FK506 and requiring calcineurin for growth have been cloned but their function has not yet been characterized: the *FKS1* gene which encodes a transmembrane protein, and the genes encoding a protein kinase (*pkc1*) and a MAP kinase (*mpk1/slt2*). These three genes are required for cell membrane integrity and cell wall morphogenesis (Garrett-Engele et al., 1995). In these mutants, the calcineurin requirement may be due to altered ion homeostasis since it is corrected by the Ca^{2+} or the Ca^{2+}-independent form of calcineurin (Eng et al., 1994; Garrett-Engele et al., 1995). Recent evidence indicates that calcineurin together with MPK1p are also involved in the regulation of the onset of mitosis in the budding yeast (Mizunuma et al., 1998).

The transcription factor TCN1p (for target of calcineurin, also called Crz1p) has now been identified. This factor is responsible for the calcineurin-dependent differential induction of the *PCM1*, *PMR1*, *PMR2A* and *FKS2* genes involved in tolerance to high Ca^{2+}, Na^+, Mn^{2+} and cell wall damage but not in the inhibition of the vacuolar H^+/Ca^{2+} exchanger or the inhibition of a pheromone-stimu-

lated Ca^{2+} uptake (Stathopoulos and Cyert, 1997; Matheos et al., 1997).

In the fission yeast, *Schizosaccharomyces pombe*, calcineurin is also not essential for viability but is involved in the regulation of many cellular processes, including cytokinesis, mating, cation and drug transport, nuclear and spindle pole body positioning, cell shape and Cl^- homeostasis (Yoshida et al., 1994; Sugiura et al., 1998). Disruption of the calcineurin A gene or exposure to cyclosporin A results in slow growth at low temperature, delayed cytokinesis and appearance of a large number of multiseptate and branched cells. The impaired polarity control and the cold sensitivity are reminiscent of cold-sensitive β-tubulin mutants. Additive effects on cation transport of the deletions of calcineurin and the *STS1* genes, whose functions are interrelated, led to the suggestion that activation of calcineurin by *STS1* gene products may be involved in the regulation of cation transport across membranes (Yoshida et al., 1994). A role for calcineurin in the gene expression cascade essential for mating and sporulation is also suggested by the induction of calcineurin A by nitrogen starvation that favors mating and is mediated by the *STE11* gene product in this organism (Plochocka-Zulinska et al., 1995).

Concluding Remarks

The immunosuppressive drugs and yeast genetics have opened the way to a better understanding of the role of Ca^{2+}-dependent protein dephosphorylation and have demonstrated the essential role of the calmodulin-stimulated protein phosphatase, calcineurin, in the transduction of Ca^{2+} signals. The elucidation of the multidomain structure of calcineurin and that of its complex with the immunophilin FKBP bound to FK506 have established the important role of the regulatory Ca^{2+}-binding subunit of calcineurin in the interaction of calcineurin with the immunosuppressive drugs and possibly other partners such as AKAP79 and the ryanodine and IP_3 receptors. The challenge now is to determine how Ca^{2+} binding to calcineurin B and calmodulin modulate the interplay of these multiple regulatory proteins in the fine-tuning of enzyme activity. What is the nature of the conformational change of calcineurin A induced by Ca^{2+} binding to calcineurin B that allows activation of calcineurin by calmodulin and its interaction with its target proteins? How does interaction of calcineurin with proteins such as FKBP, cyclophilin, AKAP79, and the ryanodine and IP_3 receptors modulate Ca^{2+} signaling? Is the prolyl *cis-trans*-isomerase activity of the immunophilins playing a role in the calcineurin-mediated transduction of Ca^{2+}-signaling pathways? How can dephosphorylation of a transcription factor promote its translocation to the nucleus? The answers to these questions may help us to design better immunosuppressants and to develop therapeutic approaches for the increasing number of diseases where an alteration of calcineurin activity is implicated.

References

Aitken, A., P. Cohen, et al. (1982) Identification of the NH_2-terminal blocking group of calcineurin B as myristic acid. *FEBS Lett.* 150: 314–317.

Aitken, A., C. B. Klee, and P. Cohen (1984) The structure of the B subunit of calcineurin. *Eur. J. Biochem.* 139: 663–671.

Ames, J. B., T. Porumb, et al. (1995) Amino-terminal myristoylation induces cooperative calcium binding to recoverin. *J. Biol. Chem.* 270: 4526–4533.

Anglister, J., S. Grzesiek, et al. (1993) Isotope-edited multidimensional NMR of calcineurin B in the presence of the non-deuterated detergent CHAPS. *J. Biomol. NMR* 3: 121–126.

Anglister, J., S. Grzesiek et al. (1994) 1H, 13C, 15N nuclear magnetic resonance backbone assignments and secondary structure of human calcineurin B. *Biochemistry* 33: 3540–3547.

Anglister, J., H. Ren, et al. (1995) NMR identification of calcineurin B residues affected by binding of a calcineurin A peptide. *FEBS Lett* 375: 108–112.

Antoni, F. A., S. M. Smith, et al. (1998) Calcium control of adenylyl cyclase: the calcineurin connection. *Adv. Second Messenger Phosphoprotein Res.* 32: 153–172.

Aperia, A., F. Ibarra, et al. (1992) Calcineurin mediates alpha-adrenergic stimulation of Na + ,K + -ATPase activity in renal tubule cells. *Proc. Natl. Acad. Sci. USA* 89: 7394–7397.

Asami, M., T. Kuno, et al. (1993) Detection of the FK506-FKBP-calcineurin complex by a simple binding assay. *Biochem. Biophys. Res. Commun.* 192: 1388–1394.

Barford, D. (1996) Molecular mechanisms of the protein serine/threonine phosphatases. *Trends Biochem. Sci.* 21: 407–412.

Beals, C. R., C. M. Sheridan, et al. (1997) Nuclear export of NF-ATc enhanced by glycogen synthase kinase-3. *Science* 275: 1930–1934.

Bear, M. F. and W. C. Abraham (1996) Long-term depression in hippocampus. *Annu. Rev. Neurosci.* 19: 437–462.

Bear, M. F. and R. C. Malenka (1994) Synaptic Plasticity: LTP and LTD. *Curr. Opin. Neurobiol.* 4: 389–399.

Bito, H., K. Deisseroth, and R. W. Tsien (1997a) CREB phosphorylation and dephosphorylation: a Ca^{2+}- and stimulus duration-dependent switch for hippocampal gene expression. *Cell* 87: 1203–1214.

Bito, H., K. Deisseroth, and R. W. Tsien (1997b) Ca^{2+}-dependent regulation in neuronal gene expression. *Curr. Opin. Neurobiol.* 7: 419–429.

Blumenthal, D. K., C. P. Chan et al. (1985a) Substrate specificity of calmodulin-dependent protein phosphatases. *Adv. Protein Phosphatases* 1: 163–174.

Blumenthal, D. K., K. Takio, et al. (1985b) Dephosphorylation of cAMP-dependent protein kinase regulatory subunit (type II) by calmodulin-dependent protein phosphatase. *J. Biol. Chem.* 261: 8140–8145.

Bonnefoy-Berard, N., L. Genestier, et al. (1994) The phosphoprotein phosphatase calcineurin controls calcium-dependent apoptosis in B cell lines. *Eur. J. Immunol.* 24: 325–329.

Born, T. L., J. K. Myers, et al. (1995) Fluoro(methyl)phenyl phosphate acts as a mechanism-based inhibitor of calcineurin. *J. Biol. Chem.* 270: 25651–25655.

Breuder, T., C. S. Hemenway, et al. (1994) Calcineurin is essential in cyclosporin A- and FK506-sensitive yeast strains. *Proc. Natl. Acad. Sci. USA* 91: 5372–5386.

Brillantes, A.-M. B., K. Ondrias, et al. (1994) Stabilization of calcium release channel (ryanodine receptor) function by FK506-binding protein. *Cell* 77: 513–523.

Brown, L., M. X. Chen, and P. T. W. Cohen (1994) Identification of a cDNA encoding a *Drosophila* calcium/camodulin regulated protein phosphatase which has its most abundant expression in early embryos. *FEBS Lett.* 339: 124–128.

Burroughs, S. E., W. D. Horrocks Jr., et al. (1994) Characterization of the lanthanide ion-binding properties of calcineurin-B using laser-induced luminescence spectroscopy. *Biochemistry* 33: 10428–10436.

Buttini, M., S. Limonta, et al. (1995) Distribution of calcineurin A isoenzyme mRNAs in rat thymus and kidney. *Histochem. J.* 27: 291–299.

Calalb, M. B., R. L. Kincaid, and T.R. Soderling (1990) Phosphorylation of calcineurin: effect on calmodulin binding. *Biochem. Biophys. Res. Commun.* 172: 551–556.

Cameron, A. M., J. P. Steiner, et al. (1994) Immunophilin FK506-binding protein associated with inositol 1,4,5,-trisphosphate receptor modulates calcium flux. *Proc. Natl. Acad. Sci. USA* 92: 1784–1788.

Cameron, A. M., J. P. Steiner, et al. (1995) Calcineurin associated with the inositol 1,4,5-trisphosphate receptor-FKBP12 complex modulates Ca^{2+} flux. *Cell* 83: 463–472.

Cameron, A. M., F. C. Nucifora, et al. (1997) FKBP12 binds the inositol 1,4,5-trisphosphate receptor at leucine–proline (1400–1401) and anchors calcineurin to this FK506-like domain *J. Biol. Chem.* 272: 27582–27588.

Cardenas, M. E. and J. Heitman (1995) Role of calcium in T-lymphocyte activation. *Adv. in Second Mess. and Phosphoprotein Res.* 30: 281–298.

Cardenas, M. E., R. S. Muir, et al. (1995) Targets of immunophilin-immunosuppressant complexes are distinct highly conserved regions of calcineurin A. *EMBO J.* 14: 2772–2783.

Castellani, L. and C. Cohen (1992) A calcineurin-like phosphatase is required for catch contraction. *FEBS Lett.* 309: 321–326.

Chan, C. P., B. Gallis et al. (1986) Characterization of phosphotyrosyl protein phosphatase activity of calmodulin-dependent protein phosphatase. *J. Biol. Chem.* 261: 9890–9895.

Chang, C. D., T. Takeda, et al. (1992) Molecular cloning and characterization of the promoter region of the calcineurin A alpha gene. *Biochem. J.* 288: 801–805. [Erratum, *Biochem. J.* (1993) 291: 951.]

Chang, C. D., H. Mukai, et al. (1994) cDNA cloning of an alternatively spliced isoform of the regulatory subunit of Ca^{2+}/calmodulin-dependent protein phosphatase (calcineurin B alpha 2). *Biochim. Biophys. Acta* 1217: 174–180.

Chang, H. Y., K. Takei, et al. (1995) Asymmetric retraction of growth cone filopodia following focal inactivation of calcineurin. *Nature* 376: 686–690.

Chung, E., H. C. Li, et al. (1987) Ca^{2+}/calmodulin-dependent phosphoprotein phosphatase activity of calcineurin in rat striatum: effect of kainic acid lesions. *Neuropharmacology* 26: 633–636.

Chung, E., M. T. Dvorozniak, et al. (1989) Regional distribution of calcium/calmodulin-dependent phosphatase activity of calcineurin in rat brain. *Res. Commun. Chem. Pathol. Pharmacol.* 64: 357–371.

Clipstone, N. A. and G. R. Crabtree (1992) Identification of calcineurin as a key signalling enzyme in T-lymphocyte activation. *Nature* 357: 695–697.

Clipstone, N. A., D. F. Fiorentino, and G. R. Crabtree (1994) Molecular analysis of the interaction of calcineurin with drug-immunophilin complexes. *J. Biol. Chem.* 269: 26431–26437.

Coghlan, V. M., B. A. Perrino, et al. (1995) Association of protein kinase A and protein phosphatase 2B with a common anchoring protein. *Science* 267: 108–111.

Cohen, P. (1996) Dissection of protein kinase cascades that mediate cellular response to cytokines and cellular stress. *Adv. Pharmacol.* 36: 15–27.

Cohen, P., C. F. B. Holmes, and Y. Tsukitani (1990) Okadaic acid: a new probe for the study of cellular regulation. *Trends Biochem. Sci.* 15: 98–102.

Cohen, P. T., M. X. Chen, and C. G. Armstrong (1996) Novel protein phosphatases that may participate in cell signaling. *Adv. Pharmacol.* 36: 67–89.

Crabtree, G. R. and N. A. Clipstone (1994) Signal transmission between the plasma membrane and the nucleus of T lymphocytes. *Ann. Rev. Biochem.* 63: 1045–1083.

Cunningham, K. W. and G. R. Fink (1994) Calcineurin-dependent growth control in *Saccharomyces cerevisiae* mutants lacking *PMC1*, a homolog of plasma membrane Ca^{2+} ATPases. *J. Cell Biol.* 124: 351–363.

Cunningham, K. W. and G. R. Fink (1996) Calcineurin inhibits *VCX1*-dependent H^+/Ca^{2+} exchange and induces Ca^{2+} ATPases in *Saccharomyces cerevisiae*. *Mol. Cell. Biol.* 16: 2226–2237.

Cyert, M.S. and J. Thorner (1992) Regulatory subunit (*CNB1* gene product) of yeast Ca^{2+}/calmodulin-dependent phosphoprotein phosphatases is required for adaptation to pheromone. *Mol. Cell. Biol.* 12: 3460–3469.

Cyert, M. S., R. Kunisawa, et al. (1991) Yeast has homologs (*CNA1* and *CNA2* gene products) of mammalian calcineurin, a calmodulin-regulated phosphoprotein phosphatase. *Proc. Natl. Acad. Sci. USA* 88: 7376–7380.

Dawson, T. M., et al. (1993) Immunosuppressant FK506 enhances phosphorylation of nitric oxide synthase and protects against glutamate neurotoxicity. *Proc. Natl. Acad. Sci. USA* 90: 9808–9812.

Dawson, T. M., J. P. Steiner, et al. (1994) The immunophilins, FK506 binding protein and cyclophilin, are discretely localized in the brain: relationship to calcineurin. *Neuroscience* 62: 569–580.

Dell'Acqua, M. L., M. C. Faux, et al. (1998) Membrane-targeting sequences on AKAP79 bind phosphatidylinositol-4,5-bisphosphate. *EMBO J.* 17: 2246–2260.

Donella-Deana, A., M. H. Krinks, et al. (1994) Dephosphorylation of phosphopeptides by calcineurin (protein phosphatase 2B). *Eur. J. Biochem.* 219: 109–117.

Ellis, D. Z. and S. C. Edwards (1994) Characterization of a calcium/calmodulin-dependent protein phosphatase in the *Limulus* nervous tissue and its light regulation in the lateral eye. *Visual Neurosci.* 11: 851–860.

Eng, W. K., L. Faucette, et al. (1994) The yeast FKS1 gene encodes a novel membrane protein, mutations in which confer FK506 and cyclosporin A hypersensitivity and calcineurin-dependent growth. *Gene* 151: 61–71.

Enz, A., G. Shapiro, et al. (1994) Nonradioactive assay for protein phosphatase 2B (calcineurin) activity using a partial sequence of the subunit of cAMP-dependent protein kinase as substrate. *Anal. Biochem.* 216: 147–153.

Etzkorn, F. A., Z. Y. Chang, et al. (1994) Cyclophilin residues that affect noncompetitive inhibition of the protein serine phosphatase activity of calcineurin by the cyclophilin.cyclosporin A complex. *Biochemistry* 33: 2380–2388.

Evans, R. M. (1989) Phosphorylation of vimentin in mitotically selected cells. In vitro cyclic AMP-independent kinase and calcium-stimulated phosphatase activities. *J. Cell Biol.* 108: 67–78.

Farcasanu, I. C., D. Hirata, et al. (1995) Protein phosphatase 2B of *Saccharomyces cerevisiae* is required for tolerance to manganese, in blocking the entry of ions into the cells. *Eur. J. Biochem.* 232: 712–717.

Faux, M. C. and J. D. Scott (1996) Molecular glue: kinase anchoring and scaffold proteins. *Cell* 85: 9–12.

Ferreira, A., R. Kincaid, and K. S. Kosik (1993) Calcineurin is associated with the cytoskeleton of cultured neurons and has a role in the acquisition of polarity. *Mol. Biol. Cell* 4: 1225–1238.

Foor, F., S. A. Parent, et al. (1992) Calcineurin mediates inhibition by FK506 and cyclosporin of recovery from alpha-factor arrest in yeast. *Nature* 360: 682–684.

Frantz, B., E. C. Nordby, et al. (1994) Calcineurin acts in synergy with PMA to inactivate I kappa B/MAD3, an inhibitor of NF-kappa B. *EMBO J.* 13: 861–870.

Fruman, D. A., C. B. Klee, et al. (1992a) Calcineurin phosphatase activity in T lymphocytes is inhibited by FK 506 and cyclosporin A. *Proc. Natl. Acad. Sci. USA* 89: 3686–3690.

Fruman, D. A., P. E. Mather, et al. (1992b) Correlation of calcineurin phosphatase activity and programmed cell death in murine T cell hybridomas. *Eur. J. Immunol.* 22: 2513–2517.

Fruman, D. A., S.-Y. Pai, et al. (1996) Measurement of calcineurin phosphatase activity in cell extracts. *A Companion to Methods Enzymal.* 9: 146–154.

Funauchi, M., H. Haruta, and T. Tsumoto (1994) Effects of an inhibitor for calcium/calmodulin-dependent protein phosphatase, calcineurin, on induction of long-term potentiation in rat visual cortex. *Neurosci. Res.* 19: 269–278.

Futer, O., M. T. DeCenzo, et al. (1995) FK506 binding protein mutational analysis. Defining the surface residue contributions to stability of the calcineurin co-complex. *J. Biol. Chem.* 270: 18935–18940.

Gaestel, M., R. Benndorf, et al. (1992) Dephosphorylation of the small heat shock protein hsp25 by calcium/calmodulin-dependent (type 2B) protein phopshatase. *J. Biol. Chem.* 267: 21607–21611.

Gagliardino, J. J., M. H. Krinks, and E. E. Gagliardino (1991) Identification of the calmodulin-regulated protein phosphatase, calcineurin, in rat pancreatic islets. *Biochim. Biophys. Acta.* 1091: 370–373.

Garrett-Engele, P., B. Moilanen, and M. S. Cyert (1995) Calcineurin, the Ca^{2+}/calmodulin-dependent protein phosphatase, is essential in yeast mutants with cell integrity defects and in mutants that lack a functional vacuolar H^+-ATPase. *Mol. Cell. Biol.* 15: 4103–4114.

Giri, P. R., S. Higuchi, and R. L. Kincaid (1991) Chromosomal mapping of the human genes for the calmodulin-dependent protein phosphatase (calcineurin) catalytic subunit. *Biochem. Biophys. Res. Commun.* 181: 252–258.

Giri, P. R., C. A. Marietta, et al. (1992) Molecular and phylogenetic analysis of calmodulin-dependent protein phosphatase (calcineurin) catalytic subunit genes. *DNA Cell Biol.* 11: 415–424.

Goldfeld, A. E., E. Tsai, et al. (1994) Calcineurin mediates human tumor necrosis factor alpha gene induction in stimulated T and B cells. *J. Exp. Med.* 180: 763–768.

Goto, S., Y. Matsukado, et al. (1986) The distribution of calcineurin in rat brain by light and electron microscopic immunohistochemistry and enzyme-immunoassay. *Brain Res.* 397: 161–172.

Goto, S., Y. Matsukado, et al. (1987) Morphological characterization of the rat striatal neurons expressing calcineurin immunoreactivity. *Neuroscience* 22: 189–201.

Goto, S., Y. Matsukado, et al. (1988) A comparative immunohistochemical study of calcineurin and S-100 protein in mammalian and avian brains. *Exp. Brain Res.* 69: 645–650.

Goto, S., A. Hirano, and R. R. Rojas-Corona (1989) Calcineurin immunoreactivity in striatonigral degeneration. *Acta Neuropathol. (Berlin)* 78: 65–71.

Griffith J. P., J. L. Kim, et al. (1995) X-ray structure of calcineurin inhibited by the immunophilin-immunosuppressant FKBP12-FK506 complex. *Cell* 82: 507–522.

Grzesiek, S. and A. Bax (1993) Measurement of amide proton exchange rates and NOEs with water in $^{13}C/^{15}N$-enriched calcineurin B. *J. Biomol. NMR* 3: 627–638.

Guerini, D. and C. B. Klee (1989) Cloning of human calcineurin A: evidence for two isozymes and identification of a polyproline structural domain. *Proc. Natl. Acad. Sci. USA* 86: 9183–9187.

Guerini, D., M. H. Krinks, et al. (1989) Isolation and sequence of a cDNA clone for human calcineurin B, the Ca^{2+}-binding subunit of the Ca^{2+}/calmodulin-stimulated protein phosphatase. *DNA* 8: 675–682.

Guerini, D., G. Montell, and C. B. Klee (1992) Molecular cloning and characterization of the genes encoding the two subunits of *Drosophila melanogaster* calcineurin. *J. Biol. Chem.* 267: 22542–22549.

Haddy, A., S. K. Swanson, et al. (1992) Inhibition of calcineurin by cyclosporin A-cyclophilin requires calcineurin B. *FEBS Lett.* 314: 37–40.

Halpain, S., J.-A. Girault, and P. Greengard (1990) Activationof NMDA receptors induces dephosphorylation of DARPP-32 in rat striatal slices. *Nature* 343: 369–372.

Hashimoto, Y. and T.R. Soderling (1989) Regulation of calcineurin by phosphorylation. Identification of the regulatory site phosphorylated by Ca^{2+}/calmodulin-dependent protein kinase II and protein kinase C. *J. Biol. Chem.* 264: 16524–16529.

Hashimoto, Y., M. M. King, and T. R. Soderling (1988) Regulatory interactions of calmodulin-binding proteins: phosphorylation of calcineurin by autophosphorylated Ca^{2+}/calmodulin-dependent protein kinase II. *Proc. Natl. Acad. Sci. USA* 85: 7001–7005.

Hashimoto, Y., B. A. Perrino, and T. R. Soderling (1990) Identification of an autoinhibitory domain in calcineurin. *J. Biol. Chem.* 265: 1924–1927.

Heitman, J., A. Koller, et al. (1993) Identification of immunosuppressive drug targets in yeast. *A Companion to Methods Enzymol.* 5: 176–187.

Heitman, J., M. E. Cardenas, et al. (1994) Antifungal effects of cyclosporin and FK 506 are mediated via immunophilin-dependent calcineurin inhibition. *Transplant. Proc.* 26: 2833–2834.

Hemenway, C. S., K. Dolinski, et al. (1995) *vph6* Mutants of *Saccharomyces cerevisiae* require calcineurin for growth and are defective in vacuolar H^+-ATPase assembly. *Genetics* 141: 833–844.

Hendey, B., C. B. Klee, and F. R. Maxfield (1992) Inhibition of neutrophil chemokinesis on vitronectin by inhibitors of calcineurin. *Science* 258: 296–299.

Higuchi, S., J'i Tamura, et al. (1991) Calmodulin-dependent protein phosphatase from *Neurospora crassa*. *J. Biol. Chem.* 266: 18104–18112.

Hirata, D., S. Harada, et al. (1995) Adaptation to high-salt stress in *Saccharomyces cerevisiae* is regulated by Ca^{2+}/calmodulin-dependent phosphoprotein phosphatase (calcineurin) and cAMP-dependent protein kinase. *Mol. Gen. Genet.* 249: 257–264.

Hubbard, M. J. (1995) Calbindin 28 kDa and calmodulin are hyperabundant in rat dental enamel cells. Identification of the protein phosphatase calcineurin as a principal calmodulin target and of a secretion-related role for calbindin 28 kDa. *Eur. J. Biochem.* 230: 68–79.

Hubbard M. J. and C. B. Klee (1987) Calmodulin binding by calcineurin. Ligand-induced renaturation of protein immobilized on nitrocellulose. *J. Biol. Chem.* 262: 15062–15070.

Hubbard M. J. and C. B. Klee (1989) Functional domain structure of calcineurin A: mapping by limited proteolysis. *Biochemistry* 28: 1868–1874.

Hubbard, M. J. and C. B. Klee (1991) Exogenous kinases and phosphatases as probes of intracellular modulation. In *Molecular Neurobiology, A Pratical Approach* (Chad, J. and Wheal, H., eds.) IRL Press, Oxford, pp. 135–157.

Hurwitz, M. Y., J. A. Putkey, et al. (1988) Domain II of calmodulin is involved in activation of calcineurin. *FEBS Lett.* 238: 82–86.

Husi, H., M. A. Luyten, and M. G. Zurini (1994) Mapping of the immunophilin-immunosuppressant site of interaction on calcineurin. *J. Biol. Chem.* 269: 14199–14204.

Iida, H., Y. Yagawa, and Y. Anraku (1990) Essential role for induced Ca^{2+} influx followed by $[Ca^{2+}]_i$ rise in maintaining viability of yeast cells late in the mating pheromone response pathway. *J. Biol. Chem.* 265: 13391–13399.

Ingebritsen, T. S. and P. Cohen (1983a) The protein phosphatases involved in cellular regulation. 1. Classification and substrate specificities. *Eur. J. Biochem.* 132: 255–261.

Ingebritsen, T. S. and P. Cohen (1983b) Protein phosphatases: properties and role in cellular regulation. *Science* 221: 331–338.

Ito, A., T. Hashimoto, et al. (1989) The complete primary structure of calcineurin A, a calmodulin binding protein homologous with protein phosphatases 1 and 2A. *Biochem. Biophys. Res. Commun.* 163: 1492–1497.

Jain, J., P. G. McCaffrey, et al. (1993) The T-cell transcription factor NFATp is a substrate for calcineurin and interacts with Fos and Jun. *Nature* 365: 352–355.

Jarayaman, T., A.-M. Brillantes, et al. (1992) FK506-binding protein associated with the calcium release channel (ryanodine receptor). *J. Biol. Chem.* 267: 9474–9477.

Kakalis, L. T., M. Kennedy, et al. (1995) Characterization of the calcium-binding sites of calcineurin B. *FEBS Lett.* 362: 55–58.

Kawamura, A. and M. S. Su (1995) Interaction of FKBP12-FK506 with calcineurin A at the B subunit-binding domain. *J. Biol. Chem.* 270: 15463–15466.

Kennedy, M. T., H. Brockman, and F. Rusnak (1996) Contributions of myristoylation to calcineurin structure/function. *J. Biol. Chem.* 271: 26517–26521.

Kincaid, R. L. (1995) The role of calcineurin in immune system responses. *J. Allergy Clin. Immunol.* 96: 1170–1177.

Kincaid, R. L., M. S. Nightingale, and B. M. Martin (1988) Characterization of a cDNA clone encoding the calmodulin-binding domain of mouse brain calcineurin. *Proc. Natl. Acad. Sci. USA* 85: 8983–8987.

Kincaid, R. L., P. R. Giri, et al. (1990) Cloning and characterization of molecular isoforms of the catalytic subunit of calcineurin using nonisotopic methods. *J. Biol. Chem.* 265: 11312–11319.

King, M. M. and C. Y. Huang (1984) The calmodulin-dependent activation and deactivation of the phosphoprotein phosphatase, calcineurin, and the effect of nucleotides, pyrophosphate, and divalent metal ions. *J. Biol. Chem.* 259: 8847–8856.

King, M. M., C. Y. Huang, et al. (1984) Mammalian brain phosphoproteins as substrates for calcineurin. *J. Biol. Chem.* 259: 8080–8083.

Kissinger, C. R., H. E. Parge, et al. (1995) Crystal structures of human calcineurin and the human FKBP12–FK506–calcineurin complex. *Nature* 378: 641–644.

Klauck, T. M., M. C. Faux, et al. (1996) Coordination of three signaling enzymes by AKAP79, a mammalian scaffold protein. *Science* 271: 1589–1592.

Klee, C. B. and P. Cohen (1988) The calmodulin-regulated protein phosphatase. In *Molecular Aspects of Cellular Regulation*, Vol. 5 (Cohen, P., and Klee, C. B., eds.). Elsevier, Amsterdam, pp. 225–248 .

Klee, C. B., T. H. Crouch, and M. H. Krinks (1980) Calcineurin: a calcium- and calmodulin-binding protein of the nervous system. *Proc. Natl. Acad. Sci.* 76: 6270–6273.

Klee, C. B., Draetta, G. F., and Hubbard, M. J. (1988) Calcineurin. In *Advances Enzymology*, Vol. 61 (Meister, A., ed.). Interscience, New York, pp. 149–200.

Klee, C. B., D. Guerini, et al. (1990) Calcineurin: a major Ca^{2+}/calmodulin-regulated protein phosphatase in brain. In *Neurotoxicity of Excitory Amino Acids* (Guidotti, A., ed.). Raven Press, New York, pp. 95–108.

Krebs, E. G. and E. H. Fischer (1956) The phosphorylase *b* to *a* converting enzyme of rabbit skeletal muscle. *Biochim. Biophys. Acta* 20: 150–157.

Kubo, M., R. L. Kincaid, and J. T. Ransom (1994) Activation of the interleukin-4 gene is controlled by the unique calcineurin-dependent transcriptional factor NF(P). *J. Biol. Chem.* 269: 19441–19446.

Kuno, T., T. Takeda, et al. (1989) Evidence for a second isoform of the catalytic subunit of calmodulin-dependent protein phosphatase (calcineurin A). *Biochem. Biophys. Res. Commun.* 165: 1352–1358.

Kuno, T., H. Tanaka, et al. (1991) cDNA cloning of a calcineurin B homolog in *Saccharomyces cerevisiae*. *Biochem. Biophys. Res. Commun.* 180: 1159–1163.

Kuno, T., H. Mukai, et al. (1992) Distinct cellular expression of calcineurin A alpha and A beta in rat brain. *J. Neurochem.* 58: 1643–1651.

Lane, B. C., L. N., Miller, et al. (1993) Evaluation of calcineurin's role in the immunosuppressive activity of FK506, related macrolactams, and cyclosporine. *Transplant Proc.* 25: 644–646.

Lawson, M. A. and F. R. Maxfield (1995) Ca^{2+}- and calcineurin-dependent recycling of an integrin to the front of migrating neutrophils. *Nature* 377: 75–79.

Lea, J. P., J. M. Sands, et al. (1994) Evidence that the inhibition of Na^+/K^+-ATPase activity by FK506 involves calcineurin. *Kidney Int.* 46: 647–652.

Li, W. and R. E. Handschumacher (1993) Specific interaction of the cyclophilin–cyclosporin complex with the B subunit of calcineurin. *J. Biol. Chem.* 268: 14040–14044.

Lieberman, D. N. and I. Mody (1994) Regulation of NMDA channel function by endogenous Ca^{2+}-dependent phosphatase. *Nature* 369: 235–239.

Liu, J., J. D. Farmer, et al. (1991a) Calcineurin is a common target of cyclophilin-cyclosporin A and FKBP-FK506 complexes. *Cell* 66: 807–815.

Liu, J., M. W. Albers, et al. (1992) Inhibition of T cell signaling by immunophilin–ligand complexes correlates with loss of calcineurin phosphatase activity. *Biochemistry* 31: 3896–3901.

Liu, J. P., A. T. Sim, and P. J. Robinson (1994) Calcineurin inhibition of dynamin I GTPase activity coupled to nerve terminal depolarization. *Science* 265: 970–973.

Liu, Y. and D. R. Storm (1989) Dephosphorylation of neuromodulin by calcineurin. *J. Biol. Chem.* 264: 12800–12804.

Liu, Y., S. Ishii, et al. (1991b) The *Saccharomyces cerevisiae* genes (CMP1 and CMP2) encoding calmodulin-binding proteins homologous to the catalytic subunit of mammalian protein phosphatase 2B. *Mol. Gen. Genet.* 227: 52–59.

Luo, C., K. T.-Y. Shaw, et al. (1996) Interaction of calcineurin with a domain of the transcription factor NFAT1 that controls nuclear imports. *Proc. Natl. Acad. Sci. USA* 93: 8907–8912.

Mansuy, I. M., M. Mayford, et al. (1998) Restricted and regulated overexpression reveals calcineurin as a key component in the transition from short-term to long-term memory. *Cell* 92: 39–49.

Marcaida, G., E. Kosenko, et al. (1996) Glutamate induces a calcineurin-mediated dephosphorylation of Na^+, K^+-ATPase that results in its activation in cerebellar neurons in culture. *J. Neurochem.* 66: 99–104.

Martensen, T. M., B. M. Martin, and R. L. Kincaid (1989) Identification of the site on calcineurin phos-

phorylated by Ca^{2+}/CaM-dependent kinase II: modification of the CaM-binding domain. *Biochemistry* 28: 9243–9247.

Martin, B. L. and D. J. Graves (1986) Mechanistic aspects of the low-molecular-weight phosphatase activity of the calmodulin-activated phosphatase, calcineurin. *J. Biol. Chem.* 261: 14545–14550.

Martin, B. L. and D. J. Graves (1994) Isotope effects on the mechanism of calcineurin catalysis: kinetic solvent isotope and isotope exchange studies. *Biochim. Biophys. Acta* 1206: 136–142.

Matheos, D. P., T. J. Kingbury, et al. (1997) Tcn1p/Crz1p, a calcineurin-dependent transcription factor that differentially regulates gene expression in *Saccharomyces cerevisiae*. *Genes Dev.* 11: 3445–3458.

Matsui, H., A. Doi, et al. (1987) Immunohistochemical localization of calcineurin, calmodulin-stimulated phosphatase, in the rat hippocampus using a monoclonal antibody. *Brain Res.* 402: 193–196.

McCaffrey, P. G., B. A. Perrino, et al. (1993) NF-ATp, a T lymphocyte DNA-binding protein that is a target for calcineurin and immunosuppressive drugs. *J. Biol. Chem.* 268: 3747–3752.

McCaffrey, P. G., A. E. Goldfeld, and A. Rao (1994) The role of NFATp in cyclosporin A-sensitive tumor necrosis factor-alpha gene transcription. *J. Biol. Chem.* 269: 30445–30450.

McPartlin, A. E., H. M. Barker, and P. T. W. Cohen (1991) Identification of a third alternatively spliced cDNA encoding the catalytic subunit of protein phosphatase 2Bbeta. *Biochim. Biophys. Acta* 1088: 308–310.

Mendoza, I., F. Rubio, et al. (1994) The protein phosphatase calcineurin is essential for NaCl tolerance of *Saccharomyces cerevisiae*. *J. Biol. Chem.* 269: 8792–8796.

Merat, D. L., Z. Y. Hu, et al. (1985) Bovine brain calmodulin-dependent protein phosphatase: regulation of subunit A activity by calmodulin and subunit B. *J. Biol. Chem.* 260: 11053–11059.

Milan, D., J. Griffith, et al. (1994) The latch region of calcineurin B is involved in both immunosuppressant–immunophilin complex docking and phosphatase activation. *Cell* 79: 437–447.

Miyamoto, K., H. Matsui, et al. (1994) In situ localization of rat testis-specific calcineurin B subunit isoform beta 1 in the developing rat testis. *Biochem. Biophys. Res. Commun.* 203: 1275–1283.

Mizunuma, M., D. Hirata, et al. (1998) Role of calcineurin and Mpk1 in regulating the onset of mitosis in budding yeast. *Nature* 392: 303–306.

Molkentin, J. D., J. R. Lu, et al. (1998) A calcineurin-dependent transcriptional pathway for cardiac hypertrophy. *Cell* 93: 215–228.

Moriya, M., K. Fujinaga, et al. (1995) Immunohistochemical localization of the calcium/calmodulin-dependent protein phosphatase, calcineurin, in the mouse testis: its unique accumulation in spermatid nuclei. *Cell Tissue Res.* 281: 273–281.

Mueller, E. G., M. W. Crowder, et al. (1993) Purple acid phosphatase: an enzyme that catalyzes a direct phospho group transfer to water. *J. Am. Chem. Soc.* 115: 2974–2975.

Mukai, H., C. D. Chang, et al. (1991) cDNA cloning of a novel testis-specific calcineurin B-like protein. *Biochem. Biophys. Res. Commun.* 179: 1325–1330.

Mukai, H., T. Kuno, et al. (1993) FKBP12–FK506 complex inhibits phosphatase activity of two mammalian isoforms of calcineurin irrespective of their substrates or activation mechanisms. *J. Biochem. (Tokyo)* 113: 292–298.

Mulkey, R. M., C. E. Herron, and R. C. Malenka (1993) An essential role for protein phosphatases in hippocampal long term potentiation. *Science* 261: 1051–1055.

Mulkey, R. M., S. Endo, et al. (1994) Involvement of a calcineurin/ inhibitor-1 phosphatase cascade in hippocampal long-term depression. *Nature* 369: 486–488.

Muramatsu, T. and R. L. Kincaid (1992) Molecular cloning and chromosomal mapping of the human gene for the testis-specific catalytic subunit of calmodulin-dependent protein phosphatase (calcineurin A). *Biochem. Biophys. Res. Commun.* 188: 265–271.

Muramatsu, T., P. R. Giri, et al. (1992) Molecular cloning of a calmodulin-dependent phosphatase from murine testis: identification of a developmentally expressed non neural isozyme. *Proc. Natl. Acad. Sci. USA* 89: 529–533.

Nakamura, T., Y. Liu, et al. (1993) Protein phosphatase type 2B (calcineurin)-mediated, FK506-sensitive regulation of intracellular ions in yeast is an important determinant for adaptation to high salt stress conditions. *EMBO J.* 12: 4063–4071.

Natarajan, K., J. Ness, et al. (1991) Specific identification and subcellular localization of three calmodulin-binding proteins in the rat gonadotrope: spectrin, caldesmon, and calcineurin. *Biol. Reprod.* 44: 43–52.

Newton, D. and C. B. Klee (1984) CAPP-calmodulin: a potent competitive inhibitor of calmodulin actions. *FEBS Lett.* 165: 269–272.

Nichols, R. A., G. R. Suplick, and J. M. Brown (1994) Calcineurin-mediated protein dephosphorylation in brain nerve terminals regulates the release of glutamate. *J. Biol. Chem.* 269: 23817–23823.

Nishio, H., Matsui H, et al. (1992a) Identification of testis specific calcineurin beta subunit isoform by a monoclonal antibody and detection of a specific six amino acid sequence. *Biochem. Biophys. Res. Commun.* 182: 34–38.

Nishio, H., H. Matsui, et al. (1992b) The evidence for post-meiotic expression of a testis-specific isoform of a regulatory subunit of calcineurin using a monoclonal antibody. *Biochem. Biophys. Res. Commun.* 187: 828–831.

O'Keefe, S. J., J. Tamura, et al. (1992) FK-506- and CsA-sensitive activation of the interleukin-2 promoter by calcineurin. *Nature* 357: 692–694.

Pallen, C. J. and J. H. Wang (1985) A multifunctional calmodulin-stimulated protein phosphatase. *Arch. Biochem. Biophys.* 237: 282–291.

Papadopoulos, V., A. S. Brown, and P. F. Hall (1989) Isolation and characterisation of calcineurin from adrenal cell cytoskeleton: identification of substrates for Ca^{2+}-calmodulin-dependent phosphatase activity. *Mol. Cell. Endocrinol.* 63: 23–38.

Parent, S. A., J. B. Nielsen, et al. (1993) Calcineurin-dependent growth of an FK506- and CsA-hypersensitive mutant of *Saccharomyces cerevisiae*. *Gen. Microbiol.* 39: 2973–2984.

Parsons, J. N., G. J. Wiederrecht, et al. (1994) Regulation of calcineurin phosphatase activity and interaction with the FK-506.FK-506 binding protein complex. *J. Biol. Chem.* 269: 19610–19616.

Perrino, B. A., L. Y. Ng, and T. R. Soderling (1995) Calcium regulation of calcineurin phosphatase activity by its B subunit and calmodulin. Role of the autoinhibitory domain. *J. Biol. Chem.* 270: 340–346.

Plochocka-Zulinska, D., G. Rasmussen, and C. Rasmussen (1995) Regulation of calcineurin gene expression in *Schizosaccharomyces pombe*. Dependence on the ste11 transcription factor. *J. Biol. Chem.* 270: 24794–24799.

Pomies, P., P. Frachet, and M. R. Block (1995) Control of the alpha 5 beta 1 integrin/fibronectin interaction in vitro by the serine/threonine protein phosphatase calcineurin. *Biochemistry* 34: 5104–5112.

Pozos, T. C., I. Seckler, and M. S. Cyert (1996) The product of *HUM1*, a novel yeast gene, is required for vacuolar Ca^{2+}/H^+ exchange and is related to mammalian Na^+/Ca^{2+} exchangers. *Mol. Cell. Biol.* 16: 3730–3741.

Putkey, J. A., G. F. Draetta, et al. (1986) Genetically engineered calmodulins differentially activate target enzymes. *J. Biol. Chem.* 261: 9896–9903.

Rao, A. (1994) NF-ATp: a transcription factor required for the coordinate induction of several cytokine genes. *Immunol. Today* 15: 274–281.

Rasmussen, C., C. Garen, et al. (1994) The calmodulin-dependent protein phosphatase catalytic subunit (calcineurin A) is an essential gene in *Aspergillus nidulans*. *EMBO J.* 13: 3917–3924.

Rokosz, L. L., S. J. O'Keefe, et al. (1995) Reconstitution of active human calcineurin from recombinant subunits expressed in bacteria. *Protein Expr. Purif.* 6: 655–664.

Ryffel B., G. Woerly, et al. (1993) Binding of active cyclosporins to cyclophilin A and B, complex formation with calcineurin A. *Biochem. Biophys. Res. Commun.* 94: 1074–1083.

Schreiber, S. L. (1992) Immunophilin-sensitive protein phosphatase action in cell signaling pathways. *Cell* 70: 365–368.

Schreiber, S. L., J. Liu, et al. (1992) Molecular recognition of immunophilins and immunophilin–ligand complexes. *Tetrahedron* 48: 2545–2558.

Seki, K., H. C. Chen, et al. (1995) Dephosphorylation of protein kinase C substrates, neurogranin, neuromodulin, and MARCKS, by calcineurin and protein phosphatases 1 and 2A. *Arch. Biochem. Biophys.* 316: 673–679.

Shaw, K.-Y., A. M. Ho, et al. (1995) Immunosuppressive drugs prevent a rapid dephosphorylation of transcription factor NFAT1 in stimulated immune cells. *Proc. Natl. Acad. Sci. USA* 92: 11205–11209.

Sherman, P. M., H. Sun, et al. (1997) Identification and characterization of a conserved family of protein serine/threonine phosphatases homologous to *Drosophilia* retinal degeneration C (rdgC). *Proc. Natl. Acad. Sci. USA* 94: 11639–11644.

Shibasaki, F. and F. McKeon (1995) Calcineurin functions in Ca^{2+}-activated cell death in mammalian cells. *J. Cell. Biol.* 131: 735–743.

Shibasaki, F., E. Roydon Price, et al. (1996) Role of kinases and the phosphatase calcineurin in the nuclear shuttling of transcription factor NF-AT. *Nature* 382: 370–373.

Sikkink, R., A. Haddy, et al. (1995) Calcineurin subunit interactions: mapping the calcineurin B binding domain on calcineurin A. *Biochemistry* 34: 8348–8356.

Singh, T. J. and J. H. Wang (1987) Phosphorylation of calcineurin by glycogen synthase (casein) kinase-1. *Biochem. Cell. Biol.* 65: 917–921.

Snyder, S. H. and D. M. Sabatini (1995) Immunophilins and the nervous system. *Nature Med.* 1: 32–37.

Stathopoulos, A. M. and M. S. Cyert (1997) Calcineurin acts through the *CRZ1/TCN1*-encoded transcription factor to regulate gene expression in yeast. *Genes Dev.* 11: 3432–3444.

Steele, F. R., T. Washburn, et al. (1992) Drosophila *retinal degeneration C* (rdgC) encodes a novel serine/threonine protein phosphatase. *Cell* 69: 669–676.

Steiner, J. P., T. M. Dawson, et al. (1992) High brain densities of the immunophilin FKBP colocalized with calcineurin. *Nature* 358: 584–587.

Stelzer, A. (1992) GABA receptors control the excitability of neuronal populations. *Int. Rev. Neurobiol.* 33: 195–287.

Stemmer, P. M. and C. B. Klee (1994) Dual calcium ion regulation of calcineurin by calmodulin and calcineurin B. *Biochemistry* 33: 6859–6866.

Stemmer, P. M., X. Wang, et al. (1995) Factors responsible for the Ca^{2+}-dependent inactivation of calcineurin in brain. *FEBS Lett.* 374: 237–240.

Stewart, A. A., T. S. Ingebritsen, et al. (1982) Discovery of a calcium ion and calmodulin-dependent protein phosphatase: probable identity with calcineurin (CaM-BP80). *FEBS Lett.* 137: 80–84.

Stewart, A. A., T. S. Ingebritsen, and P. Cohen (1983) The protein phosphatases involved in cellular regulation: purification and properties of a Ca^{2+}/calmodulin-dependent protein phosphatase (2B) from rabbit skeletal muscle. *Eur. J. Biochem.* 132: 289–295.

Stoddard, B. L. and K. E. Flick (1996) Calcineurin-immunosuppressor complexes. *Curr. Opin. Struct. Biol.* 6: 770–775.

Strater, N., T. Klabunde, et al. (1995) Crystal structure of a purple acid phosphatase containing a dinuclear Fe(III)-Zn(II) active site. *Science* 268: 1489–1492.

Sugiura, R., T. Toda, et al. (1998) *pmp1$^+$*, a suppressor of calcineurin deficiency, encodes a novel MAP kinase phosphatase in fission yeast. *EMBO J.* 177: 140–148.

Sugimoto, M., H. Matsui, et al. (1991) Isolation and sequence of rat testis cDNA for a calcium-binding polypeptide similar to the regulatory subunit of calcineurin. *Biochem. Biophys. Res. Commun.* 180: 1476–1482.

Takaishi, T., N. Saito, et al. (1991) Differential distribution of the mRNA encoding two isoforms of the catalytic subunit of calcineurin in the rat brain. *Biochem. Biophys. Res. Commun.* 174: 393–398.

Tanida, I., A. Hasegawa, et al. (1995) Cooperation of calcineurin and vacuolar H^+-ATPase in intracellular Ca^{2+} homeostasis of yeast cells. *J. Biol. Chem.* 270: 10113–10119.

Tanida, I., Y. Takita, et al. (1996) Yeast Cls2p localized in the endoplasmic reticulum membrane regulates a nonexchangeable intracellular Ca^{2+} pool cooperatively with calcineurin. *FEBS Lett.* 379: 38–42.

Thaler, C. D. and L. T. Haimo (1990) Regulation of organelle transport in melanophores by calcineurin. *J. Cell Biol.* 111: 1939–1948.

Timerman, A. P., G. Weiderrecht, et al. (1995) Characterization of an exchange reaction between soluble FKBP12 and the FKBP-ryanodine receptor complex. *J. Biol. Chem.* 270: 2451–2459.

Timmerman, L. A., N. A. Clipstone, et al. (1996) Rapid shuttling of NF-AT in discrimination of Ca^{2+} signals and immunosupression. *Nature* 383: 837–840.

Tong, G., D. Shepherd, and C. E. Jahr (1995) Synaptic desensitization of NMDA receptors by calcineurin. *Science* 267: 1510–1512.

Torii, N., T. Kamishita, et al. (1995) An inhibitor for calcineurin, FK506, blocks induction of long-term depression in rat visual cortex. *Neurosci. Lett.* 185: 1–4

Tsuboi, A., E. S. Masuda, et al. (1994) Calcineurin potentiates activation of the granulocyte-macrophage colony-stimulating factor gene in T cells: involvement of the conserved lymphokine element 0. *Mol. Biol. Cell.* 5: 119–128.

Tumlin, J. A., J. T. Someren, et al. (1995) Expression of calcineurin activity and alpha-subunit isoforms in specific segments of the rat nephron. *Am. J. Physiol.* 269: 558–563.

Tung, H. Y. L. (1986) Phosphorylation of the calmodulin-dependent protein phosphatase by protein kinase C. *Biochem. Biophys. Res. Commun.* 138: 783–788.

Ueki, K., T. Muramatsu, and R. L. Kincaid (1992) Structure and expression of two isoforms of the murine calmodulin-dependent protein phosphatase regulatory subunit (calcineurin B). *Biochem. Biophys. Res. Commun.* 187: 537–543.

Usuda, N., H. Arai, et al. (1996) Differential subcellular localization of neural isoforms of the catalytic subunit of calmodulin-dependent protein phosphatase (calcineurin) in central nervous system neurons: immunohistochemistry on formalin-fixed paraffin sections employing antigen retrieval by microwave irradiation. *J. Histochem. Cytochem.* 44: 13–18.

Vinos, J., K. Jalink, et al. (1997) A G protein-coupled receptor phosphatase required for rhodopsin function. *Science* 277: 687–690.

Wang, H. and D. J. Graves (1991) Calcineurin-catalyzed reaction with phosphite and phosphate esters of tyrosine. *Biochemistry* 30: 3019–3024.

Wang, K. K., B. D. Roufogalis, and A. Villalobo (1989) Characterization of the fragmented forms of calcineurin produced by calpain I. *Biochem. Cell. Biol.* 67: 703–711.

Wang, M. G., H. Yi, et al. (1996a) Calcineurin A alpha (PPP3CA), calcineurin A beta (PPP3CB) and calcineurin B (PPP3R1) are located on human chromosomes 4,10q21→q22 and 2p16→p15 respectively. *Cytogenet. Cell Genet.* 72: 236–241.

Wang, X., V. C. Culotta, and C. B. Klee (1996b) Superoxide dismutase protects calcineurin from inactivation. *Nature* 383: 434–437.

Watanabe, Y., B. A. Perrino, and T. R. Soderling (1995) Identification in the calcineurin A subunit of the domain that binds the regulatory B subunit. *J. Biol. Chem.* 270: 456–460.

Watanabe Y., B. A. Perrino, and T. R. Soderling (1996) Activation of calcineurin A subunit phosphatase activity by its calcium-binding B subunit. *Biochemistry* 35: 562–566.

Wesselborg, S., D. A. Fruman, et al. (1996) Identification of a physical interaction between calcineurin and nuclear factor of activated T cells (NFATp). *J. Biol. Chem.* 271: 1274–1277.

Wilson, R. B., A. A. Brenner, et al. (1991) The *Saccharomyces cerevisiae SRK1* gene, a suppressor of *bcy1* and *ins1*, may be involved in protein phosphatase function. *Mol. Cell. Biol.* 11: 3369–3373.

Winder, D. G., I. M. Mansuy, et al. (1998) Genetic and pharmacological evidence for a novel, intermediate phase of long-term potentiation suppressed by calcineurin. *Cell* 92: 25–37.

Woodrow, M., N. A. Clipstone, and D. Cantrell (1993) p21ras and calcineurin synergize to regulate the nuclear factor of activated T cells. *J. Exp. Med.* 178: 1517–1522.

Yakel, J. L. (1997) Calcineurin regulation of synaptic function: from ion channels to transmitter release and gene transcription. *Trends Pharmacol. Sci.* 18: 124–134.

Yamada, T., J. K. Kim, et al. (1994) Chromosomal assignments of the genes for the calcineurin A alpha (Calna1) and A beta subunits (Calna2) in the rat. *Cytogenet. Cell. Genet.* 67: 55–57.

Yang, D., M. K. Rosen, and S. L. Schreiber (1993) A composite FKBP-FK506 contacts calcineurin. *J. Am. Chem. Soc.* 115: 819–820.

Yazdanbakhsh, K., J.-W. Choi, et al. (1995) Cyclosporin A blocks apoptosis by inhibiting the DNA binding activity of the transcription factor Nur77. *Proc. Natl. Acad. Sci. USA* 92: 437–441.

Ye, R. R. and A. Bretscher (1992) Identification and molecular characterization of the calmodulin-binding subunit gene (*CMP1*) of protein phosphatase 2B from *Saccharomyces cerevisiae. Eur. J. Biochem.* 204: 713–723.

Yoshida, T., T. Toda, and M. Yanagida (1994) A calcineurin-like gene ppb1+ in fission yeast: mutant defects in cytokinesis, cell polarity, mating and spindle pole body positioning. *J. Cell. Sci.* 107: 1725–1735.

Zhao, Y., Y. Tozawa, et al. (1995) Calcineurin activation protects T cells from glucocorticoid-induced apoptosis. *J. Immunol.* 154: 6346–6354.

Zheng, J., D. R. Knighton, N.-H. Xuong, S. S. Taylor, J. M. Sowadski, and L. F. Ten Eyck (1993) Crystal structures of the myristylated catalytic subunit of cAMP-dependent protein kinase reveal open and closed conformations. *Protein Sci.* 2: 1559–1573.

Zhu, D., M. E. Cardenas, and J. Heitman (1995) Myristoylation of calcineurin B is not required for function or interaction with immunophilin–immunosuppressant complexes in the yeast *Saccharomyces cerevisiae. J. Biol. Chem.* 270: 24831-24838.

Zhu, J., F. Shibasaki, et al. (1998) Intramolecular masking of nuclear import signal on NF-AT4 by casein kinase I and MEKK1. *Cell* 93: 851–861.

Zydowsky, L. D., F. A. Etzkorn, et al. (1992) Active site mutants of human cyclophilin A separate peptidyl-prolyl isomerase activity from cyclosporin A binding and calcineurin inhibition. *Protein Sci.* 1: 1092–1099.

15

Calpain: From Structure to Biological Function

Sandro Pontremoli
Edon Melloni
Franca Salamino

Calpain, a term originally proposed for defining a neutral calcium-dependent cysteine endopeptidase, today indicates a family of several genetically distinct isozymes, widely distributed in all mammalian cells (Suzuki, 1987; Murachi, 1989; Croall and De Martino, 1991; Melloni et al., 1992; Sorimachi et al., 1994;). However, for the purposes of this chapter, only two of the calpain isoforms will be taken into consideration, since from the original discovery of this protease, most of the research has involved the characterization of the molecular structure and of the catalytic properties of the two ubiquitous isozymes currently termed calpain I and II. The first one has also been named μ-calpain, the second, m-calpain, as they express catalytic activity at micromolar or milli-molar calcium concentration, respectively (Mellgren, 1980; Goll et al., 1990).

With respect to the understanding of the physio-logical role of calpains, considerable progress has been made in recent years, also due to several reports that have suggested a role of calpain in pathological conditions involving alteration in cell structure and function (Nilsson et al., 1990; Pontremoli and Melloni, 1990; Karlsson et al., 1992; Saito et al., 1993, Saido et al., 1994a; Richard et al., 1995; Spencer et al., 1995).

In order to provide a brief review of the present knowledge concerning this family of proteases, we will try to answer to a number of basic crucial ques-tions.

Since calpains are soluble neutral proteinases, the first point concerns the nature of the molecular mechanism(s) that maintain(s) the proteinase in an inactive state and the kind of signal that induces transition from the inactive to the active form. A second point is related to the nature of positive and negative effectors responsible for the modulation and for the site-directed specificity of calpains. Finally, it is important also to clarify the physiological role(s) of calpain to establish more precisely if these proteases are involved in certain pathological conditions; this is an obligatory condition for the definition of new and more efficient therapeutic devices for the control of still unsolved human disorders.

Molecular Structure of μ-Calpain and m-Calpain

As previously indicated, the most investigated cal-pain isoforms, present in all animal tissues, are the μ-calpain and m-calpain isoforms. Both proteinases are present in a heterodimeric form consisting of a large subunit (80 kDa), containing the catalytic site, and of a small regulatory subunit (30 kDa). In both calpain isoforms, the small subunit is identical, whereas the 80 kDa subunit occurs as one of two closely related but distinct polypeptides (Table 15.1) (Suzuki, 1990).

In both types of calpain, the large catalytic sub-unit contains four domains (Suzuki, 1987, 1990). Domain I, rich in basic amino acid residues, is prob-ably involved in the maintenance of the native conformation, as well as in providing membrane-associating capacities. Domain II, containing the catalytic site, is characterized by an amino acid sequence highly homologous to that of other cysteine proteases, including papain. The role of domain III still remains unclear. Its participation in the conformational transition induced by the binding of different effectors has been suggested. Domain IV, bearing EF-hand structures, contains the Ca^{2+} binding sites and shows a close homology with calmodulin.

In the small subunit, two domains have been identified (Sakihama et al., 1985; Emori et al.,

Table 15.1 Molecular Properties of μ-Calpain and m-Calpain

Subunit	Domain	Homology	Intramolecular Function
80 kDa	I	—	Involved in binding to membranes, modulates the accessibility to the active site
80 kDa	II	Thiol proteinase (papain, cathepsins)	Contains the essential residues for catalysis
80 kDa	III	—	Participates in the transduction of conformational signals
80 kDa	IV	Calmodulin	Contains binding sites for Ca^{2+}
30 kDa	V	—	Involved in binding to membranes, modulates the accessibility to the active site
30 kDa	VI	Calmodulin	Contains binding sites for Ca^{2+}

For experimental details, see Saido et al. (1994a).

1986a): domain V, with a high content in glycine residues, could be functionally related to domain I; domain VI is homologous to domain IV and thus provides additional Ca^{2+}-binding sites.

Comparative analyses have established significant differences between the amino acid sequence of the 80 kDa subunits present in μ-calpain and in m-calpain, respectively (Sakihama et al., 1985; Aoki, 1986; Emori et al., 1986b; Imajoh et al., 1988; Saido et al., 1994a). This indicates the existence of separate genes that encode for each of the two polypeptides (Ohno et al., 1990).

Recently, three additional calpain isoforms, termed n-calpains, have been discovered and characterized by cDNA cloning (Sorimachi et al., 1994). Their main feature seems to be tissue-specificity as one isoform (n-1) is expressed exclusively in skeletal muscle (Sorimachi et al., 1993b) and the other two found in cDNA libraries from the rat stomach (Sorimachi et al., 1993a). In these calpain isozymes, the sequence of the single domains, as well as the global structure, is significantly different from that present in μ- and m-calpain (Fig. 15.1).

All together, these findings open interesting considerations concerning the existence of many more isomeric isoforms of calpain than those actually identified, and suggest more strict structural relationships between a specialized cell function and the presence of a specific type of calpain.

Native Conformation of Calpains

As far as μ-calpain and m-calpain are concerned, on the basis of their structural characteristics discussed above and additional indirect experimental evidence, the following conclusions can be drawn. First, it is conceivable that, as a result of a genetic fusion (Ohno et al., 1984), the protease papain-like domain is expressed in a linear amino acid sequence that contains three other polypeptide segments.

Second, at a conformational level, the resulting calpain molecule folds into a very tight conformation in which the active site is masked; this results in the presence of an inactive protease in the intracellular cytoplasmic space (Fig. 15.2), further stabilized by the small 30 kDa subunit.

Activation of Calpains: Multiple roles of Ca^{2+}

In both μ- and m-calpain, the binding of calcium ions to domains IV and VI induces a transition from an inactive to an active molecular form (Suzuki, 1987; Melloni and Pontremoli, 1989; Murachi, 1989). Since calpain activation is the result of several sequential steps in which Ca^{2+} plays an essential role, it is appropriate to attribute to this metal ion multiple effects, such as conformational changes, and expres-

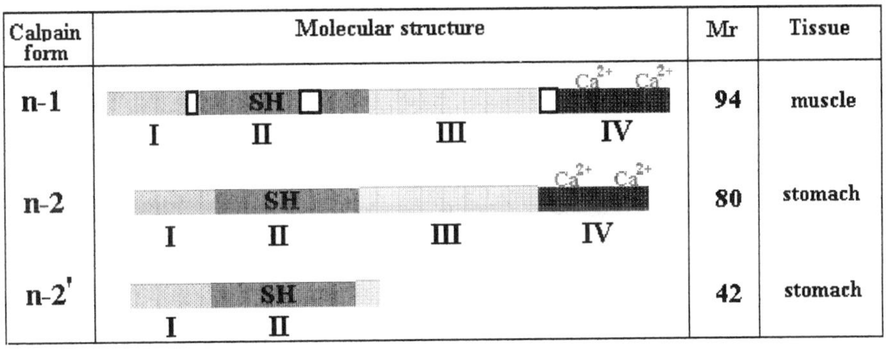

Calpain form	Molecular structure				Mr	Tissue
n-1	I	SH II	III	Ca²⁺ Ca²⁺ IV	94	muscle
n-2	I	SH II	III	Ca²⁺ Ca²⁺ IV	80	stomach
n-2'	I	SH II			42	stomach

Figure 15.1 Molecular properties and tissue distribution of newly discovered calpain isozymes. For experimental details, see Sorimachi et al. (1994).

General Mechanism of Calpain Activation

It is generally accepted that purified preparations of μ- or m-calpain isoforms require for full activity approximately $10–50\,\mu M$ and $500\,mM$ Ca^{2+}, respectively (Mellgren, 1980; Pontremoli and Melloni, 1986; Suzuki, 1987; Melloni and Pontremoli, 1989; Murachi, 1989; Goll et al., 1990; Melloni et al., 1992; Sorimachi et al., 1994).

In the sequence of molecular events initiated by the binding of Ca^{2+}, it has been suggested that a first step may involve the dissociation of calpain into the two constituent subunits (Fig. 15.3). This dissociating effect of Ca^{2+} has been shown to occur in the case of the red cell calpain isoform (Pontremoli and Melloni, 1986) and more recently confirmed also for the enzyme from rabbit and chicken skeletal muscle

sion of catalytic activity, including autoproteolytic degradation of the protease. To try to illustrate these effects, it is important to distinguish those obtained in enzymological studies in vitro using the purified protease, from those in which the physiological conditions were reproduced as much as possible.

(Yoshizawa et al., 1995a, 1995b). Dissociation could be visualized as an intermediate step that facilitates the transition from a "tight" to a "relaxed" conformation of the 80 kDa subunit, a condition that must be reached in order to unmask the active site and thereby express the proteolytic activity.

If the effects of Ca^{2+} were limited to these modifications of the enzyme conformation, then the removal of Ca^{2+} should restore the tight native inactive conformation as well as the heterodimeric structure of calpain. The occurrence of these events is, however, still controversial and difficult to establish since both μ-calpain and m-calpain, in the presence of optimal concentrations of Ca^{2+}, not only undergo a conformational change but also an autolytic degradation that removes small peptides from the N-terminus of both subunits (Coolican et al., 1986; De Martino et al., 1986; Cong et al., 1989; Brown and Crawford, 1993). As a result, the large subunit is converted into a 76 kDa polypeptide, the smaller one into a 18 kDa fragment (Fig. 15.4), and the resulting protease becomes active at much lower concentrations of Ca^{2+}.

Thus, this intramolecular autoproteolytic conversion, of the type inactive proenzyme/active enzyme, does not represent a mechanism that promotes the

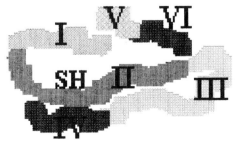

Figure 15.2 Speculative representation of the inactive conformation of calpain. The roman numbers indicate the distinct domains (see Table 15.1). SH indicates the approximate location of the essential Cys residue. Domains I and V are located in proximity of the active site for their suggested interference with the binding of substrates.

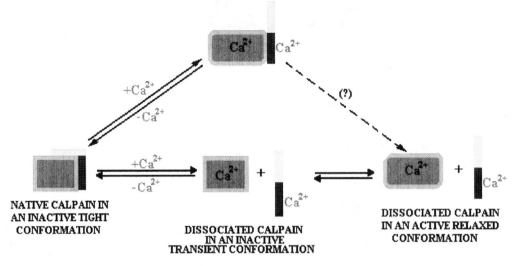

Figure 15.3 Proposed model for the reversible initial step involved in the activation of calpain. Binding of Ca^{2+} can produce two alternative effects on the calpain molecule. The first one can promote conformational changes on both calpain subunits, the second one can induce the dissociation of the heterodimeric structure, followed by conformational changes.

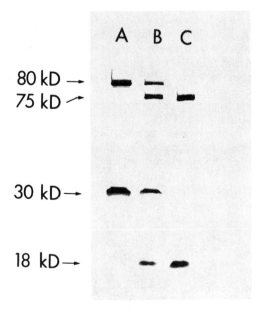

Figure 15.4 Calcium-induced autoproteolysis of calpain subunits. Degradation as revealed by the appearance in SDS-PAGE of new protein bares having M_r 75 kDa and 18 kDa, respectively. The figure shows a time course of the autodigestion of human erythrocyte calpain. A, native enzyme; B, after 30 seconds of incubation in 100 μM Ca^{2+}; C, after 1 min of incubation in 100 μM Ca^{2+}. For experimental details, see Melloni et al. (1992).

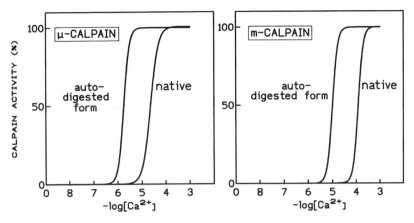

Figure 15.5 Ca^{2+} requirement of native and autoproteolysed calpain isoforms. Native and autodigested μ and m-calpain (see Fig. 15.4) were assayed in the presence of the indicated concentrations of Ca^{2+}. For experimental details, see Melloni et al. (1992).

formation of a constitutively active protease, since the presence of Ca^{2+} still remains essential for the expression of the proteolytic activity of both auto-lysed forms of μ- and m-calpains (Fig. 15.5). Apart from this consideration, autoproteolysis of calpain poses a number of questions both at the structural and at the functional level. It is, in fact, still contro-versial whether the expression of calpain activity must be preceded by an autolytic modification.

In a recent report by Saido et al. (1994b), it has been suggested that this occurs in the case of μ-cal-pain but not in the case of m-calpain, implying the presence in the μ-isoform of a highly susceptible pep-tide bond, specifically exposed to the active site. On the contrary, results from Molinari et al. (1994) have reported that erythrocyte calpain can express cataly-tic activity without undergoing autoproteolysis, indi-cating that the conformational changes induced by the binding of Ca^{2+} are sufficient to promote activa-tion of the erythrocyte proteinase. In any case, since in both calpain isoforms, the autolytic degradation results in the removal of a small fragment from domain I, it is tempting to speculate that the con-straints that maintain the tight inactive conformation are largely imposed by the folding of this terminal protein segment.

Furthermore, experimental evidence (Pontremoli et al., 1985a) has indicated that, upon or following autolysis, calpain loses its anchoring capacity to cell membranes, an observation that may suggest an additional role for the N-terminal of domain I. A final consideration is that autolysis represents also a way for inactivation of calpain accomplished through extensive degradation to inactive fragments.

In conclusion, enzymological studies have pro-vided basic information relative to the molecular mechanisms underlying the activation of calpain and to the essential role played by Ca^{2+} ions within the sequential events that promote expression of pro-teolytic activity. These can be summarized as follows (Fig. 15.6): upon binding of Ca^{2+}, a series of still-reversible conformational events are promoted which lead to the unmasking of the catalytic site, possibly through an initial dissociation of the dimeric structure. Once the active conformation is reached, an autolytic digestion of both subunits can also occur; this promotes a new irreversible conformation that confers to the protease a much higher sensitivity to Ca^{2+} ions.

Mechanism of Activation of Calpain In Vivo

The general mechanisms of activation of calpain, reported before, have been constructed on the basis of experiments performed in vitro in the presence of a concentration of Ca^{2+} too high (largely above $6 - 10\,mM$) with respect to the physiological condi-tions of the cell. Thus, for the same mechanism to operate in vivo, additional devices are required in order to reduce to a significant extent, the overall requirement of Ca^{2+} ions. This is, in fact, what has been shown to occur in a variety of cells, essentially by means of a translocation of both μ- and m-calpain to the membranes induced by concentrations of Ca^{2+} insufficient, in vitro, to promote activation and activ-ity of the isolated enzyme (Fig. 15.7).

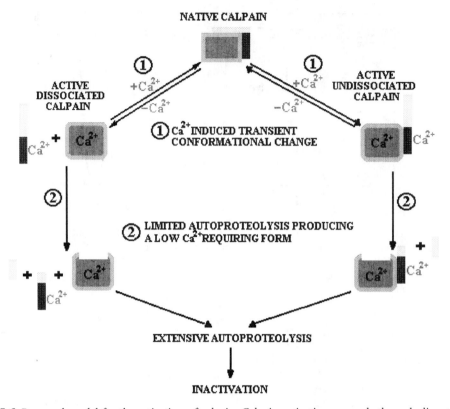

Figure 15.6 Proposed model for the activation of calpain. Calpain activation proceeds through discrete steps. In the first one, calpain undergoes transient conformational changes (see also Fig. 15.3), without (right side) or with (left side) the dissociation of the heterodimer. In the following step, calpain acquires a permanent active conformation by the removal of both N-terminal regions in domains I and V. A further extensive autoproteolysis causes complete loss of enzyme activity.

In their associated form, sustained by the anchoring regions of the proteinases (i.e. domain I of the catalytic subunit and domain V of the regulatory subunit) (Pontremoli et al., 1985b, 1989a; Saido et al., 1993), both calpains acquire a new conformation that corresponds, on a catalytic basis, to that promoted by autodigestion. In fact, in this associated form, calpain, at the same concentrations of Ca^{2+} that promote translocation, can express proteolytic activity and thereby undergo to autolytic degradation. Thus, in vivo, this overall process involves an additional role for Ca^{2+} which includes a site-directed activation mechanism that locates active calpain in close proximity of target proteins constitutively present on membranes (Wang et al., 1989; Schwarz-Ben Meier et al., 1991; Salamino et al., 1992, 1994a) or associated with the cell cytoskeleton (Tashiro and Ishizaki, 1982; Shoeman and Traub, 1990; Fox et al., 1991; Yang and Ksiezak-Reding, 1995). Alternatively, cytoplasmic proteins that become co-translocated to the cell membrane can become suitable substrates (Kishimoto et al., 1983; Melloni et al., 1985).

The autoproteolytic degradation, as indicated earlier, removes the membrane-anchoring regions from calpain and thereby promotes its release in a form that still requires Ca^{2+}, at much lower concentration, however.

In conclusion, an intracellular increase in Ca^{2+} concentration could promote in vivo a sequential multistep process that involves translocation of calpain to the activation sites coincident with the major specific site of action of the protease (Pontremoli et al., 1989a; Goll et al., 1992; Saito et al., 1993; Mellgren and Qin, 1994).

Expression of Calpains in Bacterial and Eukaryotic Cells

Although both calpain subunits were initially cloned in the 1980s (Emori et al., 1986a, 1986b; Sakihama

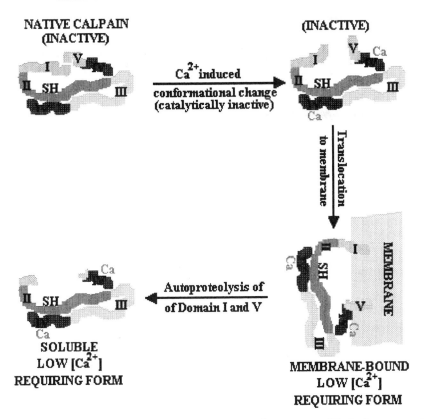

Figure 15.7 Effect of membranes on the activation of calpain. In the presence of limited amounts of Ca^{2+}, native calpain associates with membranes acquiring a conformation in which the active site is accessible, this resulting in an active form of the proteinase. The subsequent autodigestion removes the membrane-anchoring regions of the proteinase and promotes its release from the membranes. For experimental details see Pontremoli et al. (1985a, 1985b).

et al., 1985), only recently has the production of recombinant calpains been obtained. Human μ-calpain cDNA has been expressed using the baculovirus system and the properties of the recombinant enzyme are identical to those of native μ-calpain (Meyer et al., 1996). Both calpain subunits can be expressed separately. In the absence of the 30 kDa subunit, the catalytic 80 kDa subunit is still expressed, although in lower amount. In addition, the isolated recombinant 80 kDa subunit has a substantial, but lower, catalytic activity with respect to that of the native heterodimer. These observations could be interpreted on the basis of the different conformations that the catalytic subunit can achieve in the absence or presence of the 30 kDa small subunit.

m-Calpain has been expressed in bacterial cells, first as an 80 kDa inactive form, then as a complete enzyme retaining full catalytic activity (De Luca et al., 1993; Graham-Siegenthaler et al., 1994).

Positive Modulators of Calpain Activation

It has already been pointed out that activation of calpain in vivo is the result of a concerted action promoted by Ca^{2+} and membranes. Phospholipids have been identified as the membrane components that are involved in calpain activation, the most potent effect being expressed by phosphatidylinositol phosphate (PIP), followed, with lower efficiency, by phosphatidylinositol (PI) and phosphatidylserine (PS) (Pontremoli et al., 1985c; Imajoh et al., 1986).

These conclusions were obtained from studies in which the interaction between calpain and membranes was explored using phospholipid vesicles (Fig. 15.8). In these conditions, an almost complete association was observed only with μ-calpain, whereas complete activation of the m-isoform was reached following interaction with a natural activator protein still located at the membrane (Takeyama et al., 1986; Michetti et al., 1991; Salamino et al., 1993).

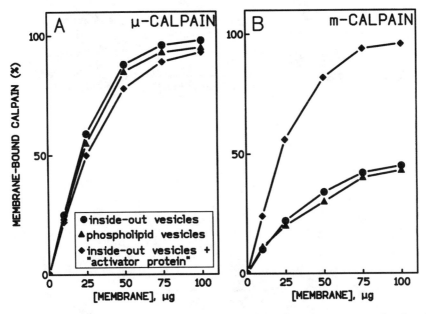

Figure 15.8 Binding of (A) μ-calpain and (B) m-calpain to vesicles. Calpain isozymes were incubated with inside-out vesicles (●), artificial phospholipid vesicles (▲), and inside-out vesicles enriched with the calpain protein activator (◆) in the presence of micromolar Ca^{2+}. The amount of associated calpain was determined as described in Pontremoli et al. (1985a, 1985b, 1985c) and Salamino et al. (1993), following sedimentation of the vesicles.

Thus, within "physiological" fluctuations in intracellular $[Ca^{2+}]$, both calpains can become active following translocation and association to phospholipids (μ-calpain) or with a peripheral membrane protein (m-calpain).

In all, these results are indicative of the site of action of calpain (at or near the cell membrane) and of its potential role in the down-regulation processes involving receptors and cotranscriptional factors.

Negative Modulator of Calpain Activity

We have so far analysed those sequential events that lead to an active calpain form. We have now to consider the role of a natural protein inhibitor, defined calpastatin (Murachi, 1989; Adachi et al., 1991; Melloni et al., 1992), also present in the cytoplasm and characterized by an unique molecular structure, together with multiple interacting capacities and specificities.

Calpastatin is present in all mammalian cells that contain the two calpain forms. In order to understand better its inhibitory mechanism, the following general properties are outlined:

1. Calpastatin is a specific competitive inhibitor of μ-calpain and m-calpain (Nishimura and Goll, 1991; Ma et al., 1993; Croall and McGrody, 1994).
2. At a molecular level, calpastatin is composed by a linear arrangement of repeated consensus regions (inhibitory domains) separated by short sequences (Emori et al., 1987; Maki et al., 1988; Kawasaki et al., 1989).
3. Calpastatin inhibition is expressed only in the presence of Ca^{2+} (Kapprell and Goll, 1989), which also promotes its binding to the cell membrane (Mellgren, 1988).
4. Calpastatin inhibits the binding of calpain to membranes either by a competitive mechanism or by promoting dissociation of the membrane-bound calpain (Salamino et al., 1994b).
5. Calpastatin is a substrate of calpain (Nakamura et al., 1989; Pontremoli et al., 1991).
6. The amount of calpastatin in relation to that of calpain varies from cell to cell and generally is higher than the amount of the protease (Salamino et al., 1994c).

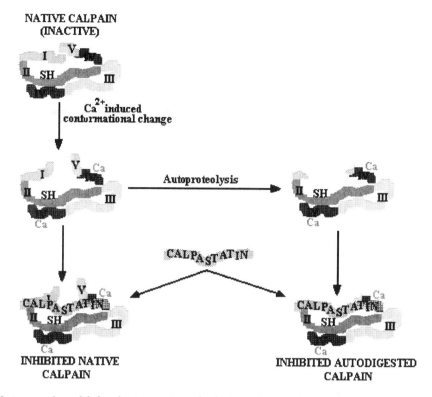

Figure 15.9 Proposed model for the interaction of calpain with calpastatin. As shown in Figs. 15.3 and 15.6, following binding of Ca^{2+}, calpain undergoes modifications which reversibly (transient conformational changes, on the left) or irreversibly (autodigestion, on the right) remove the N-terminal end of domain I and V. Calpastatin interacts with both calpain forms, covering the active site. The calpastatin molecule, bound to calpain, could replace those molecular constraints removed by interaction with Ca^{2+}.

On the basis of these properties, it can be assumed, even in conditions in which all molecular constraints have been removed in native calpain through a concerted action of Ca^{2+} and activators, that calpastatin is still capable to repress the expression of the proteolytic activity by directly interacting at the active site (Fig. 15.9). This could prevent unnecessary activation of calpain following a transient increase in Ca^{2+}. Activation, on the contrary, could take place if the Ca^{2+} concentration is sufficient to promote a significant proteolytic degradation of calpastatin. This has been experimentally demonstrated in rat red cells deprived of calpastatin (Fig. 15.10A) (Salamino et al., 1992) and evaluated also by an increase in the degradation of the specific substrate Ca^{2+}-ATPase (Fig. 15.10B). In accordance with these observations, in cells almost completely lacking calpastatin, a small increase in intracellular Ca^{2+} induces large degradation of target proteins (Fig. 15.10C).

Post-Translational Modifications of Calpastatin

Although most cells contain two types of calpains (μ- and m-calpain), only a single calpastatin form has been found.

We have recently demonstrated that in rat skeletal muscle cells the calpastatin form recovered can be resolved, by ion exchange chromatography, into two different molecular species of the inhibitor (Pontremoli et al., 1991). If the bulk of calpastatin activity had been previously submitted to treatment with a protein phosphatase (Pontremoli et al., 1992), all calpastatin activity is recovered in a single peak in the same chromatographic conditions. Based on these observations, further experiments were performed (Pontremoli et al., 1992) which revealed that the two peaks of calpastatin corresponded to a phosphorylated and a dephosphorylated form of the inhibitor.

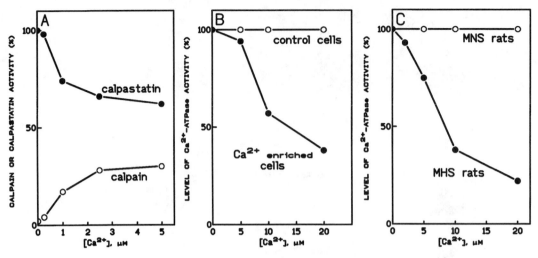

Figure 15.10 (A) Modification of the level of calpastatin and of calpain activity in rat erythrocytes loaded with Ca^{2+}. Rat red cells were incubated with $1\,\mu M$ A23187 in the presence of the indicated $[Ca^{2+}]$. After 1 hour at 37°C, the cells were collected, lysed, and calpain and calpastatin were directly assayed (Salamino et al., 1992). (B) Digestibility of plasma membrane Ca^{2+}ATPase in rat erythrocytes carrying a reduced level of calpastatin. Red cells, treated as in part (A), were then collected, washed, and exposed to the indicated $[Ca^{2+}]$ in the presence of $1\,\mu M$ A23187. Control cells were treated in the same conditions, without exposure to the ionophore. After 30 min at 37°C, the cells were recovered, lysed, and the plasma membranes were collected by centrifugation. The levels of Ca^{2+}-ATPase were determined as described by Salamino et al. (1992). (C) Digestibility of the Ca^{2+}-ATPase in rat erythrocytes of MNS and MHS rats loaded with Ca^{2+}. MHS rats are characterized by a genetically determined essential hypertension. It has been established that red cells from this rat strain contain normal levels of calpain and a 90–95% reduced amount of calpastatin as compared with normal MNS rats. Freshly collected erythrocytes were treated with the indicated amount of Ca^{2+}, in the presence of $1\,\mu M$ A23187. The level of plasma membrane Ca^{2+}-ATPase was determined as in part (B).

The dephosphorylated form was found to be maximally efficient toward μ-calpain (Fig 15.11A), and the phosphorylated one against m-calpain (Fig. 15.11B). Upon examination of different cell types, it has been demonstrated (Salamino et al., 1994c) that the levels of the phosphorylated and dephosphorylated calpastatin correlated with those of m-calpain present in the same cells. Accordingly, in kidney, in which m-calpain is highly represented, phosphorylated calpastatin is also present at the higher levels (Table 15.2).

Preliminary results using perfused rat heart (Salamino et al., 1994b) indicated that an elevation in intracellular cAMP is followed by a 2-fold increase in the phosphorylated calpastatin, whereas an increase in intracellular $[Ca^{2+}]$ induces a rapid decrease in the amount of this calpastatin form.

In conclusion, the flexibility in the inhibitory activity of calpastatin is apparently increased by the presence of an interconvertible system capable of regulating the amount of phosphorylated or dephosphorylated calpastatin in relation to the amount of m- or μ-calpain, respectively. Finally, the importance of this regulatory process is strongly supported by the observation that in those cells in which a single calpain (usually μ-calpain) is present, the single calpastatin does not undergo conversion to the dephosphorylated form.

General Consideration on the Physiological Role of Calpain

The presence of calpain in all mammalian cells, its localization, the nature of activation signal, the activation mechanism (similar to that of transductional proteins), the site of action (at the membrane or at the membrane cytoskeleton array), the substrate specificity (receptors, transcription factors, transductional proteins, cell-matrix associated proteins), and the "limited proteolysis" catalysed by the proteinase (Suzuki, 1987; Melloni and Pontremoli, 1989; Murachi, 1989; Nixon, 1989; Wang et al., 1989; Goll et al., 1990; Croall and De Martino, 1991;

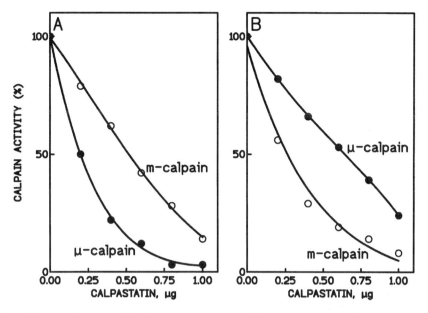

Figure 15.11 Inhibitory efficiency of (A) dephosphorylated and (B) phosphorylated calpastatin against μ- and m-calpain. Rat skeletal muscle μ- and m-calpain isoforms were incubated with the indicated amounts of purified dephosphorylated and phosphorylated calpastatin. The residual activity was expressed as a percentual of the total (Pontremoli et al., 1991).

Melloni et al., 1992) are all supporting evidences in favour of a specific role of calpain in the transduction of extracellular signals, mediated by changes in permeability of membranes to Ca^{2+} or by mobilization of this ion from internal stores.

Substrate Specificity of μ-Calpain and m-Calpain

To understand the physiological role of calpain, it is important to establish first the significance of the presence in a single cell of two distinct calpain iso-forms. In this respect, pertinent data are those indicating a different substrate specificity of the two isoenzymes.

This is the case with the troponins, two of which, troponin I and T (Fig. 15.12), are rapidly degraded by μ-calpain, but not by m-calpain (Di Lisa et al., 1995).

Similarly, a different susceptibility to degradation of calpastatin (Fig. 15.13) is promoted by μ- and m-calpain (Pontremoli et al., 1991). Digestion of calpastatin by μ-calpain results in the formation of small

Table 15.2 Correlation between the Levels of m-Calpain and of Phosphorylated Calpastatin in Rat Tissues

Rat Tissue	Calpain Activity (% of total)		Calpastatin Activity (% of total)	
	μ	m	Dephos.[a]	Phos.[a]
Skeletal muscle	65	35	82	18
Heart	54	46	73	27
Kidney	40	60	63	37
Erythrocytes	100	—	100	—

[a]Dephos. = dephosphorylated form of calpastatin; Phos. = phosphorylated form of calpastatin. For experimental details, see Pontremoli et al. (1992).

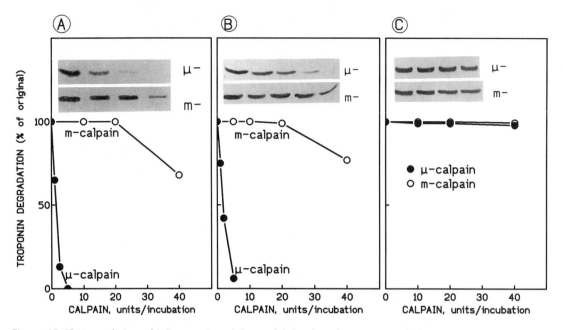

Figure 15.12 Degradation of (A) troponin I, (B) T and (C) C by calpain isozymes. The troponin complexes were isolated from rat heart and submitted to digestion with the indicated amounts of homologous μ- and m-calpain. The digestion was determined following the disappearance of the troponin bands in Western blotting, using specific monoclonal antibodies (Di Lisa et al., 1995), shown in the inserts.

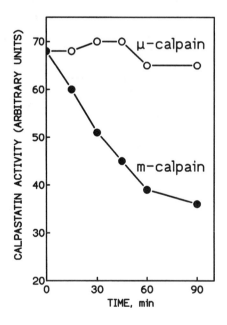

Figure 15.13 Digestion of calpastatin by μ- and m-calpain. Calpastatin was incubated with homologous μ- and m-calpain in the presence of 100 μM [Ca^{2+}]. At the times indicated, aliquots of the incubation mixture were collected for the assay of the residual calpastatin activity (Melloni et al., 1992).

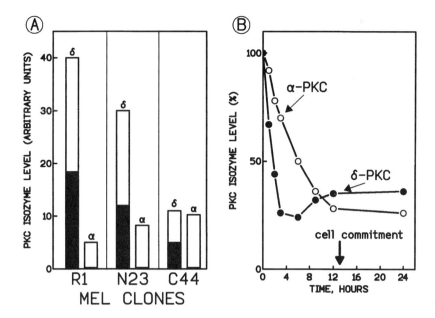

Figure 15.14 (A) Levels of δ- and α-PKC in MEL clones with different sensitivity to the chemical inducer. The level of δ- and α-PKC in MEL clone resistant to differentiation (R1), intermediate sensitivity (N23), and highly sensitive to differentiation (C44) were determined by Western blotting using specific monoclonal antibodies (Patrone et al., 1994). The filled part of the bars indicates the amount of each δ and α- isozyme-PKC recovered associated with the plasma membranes. (B) Changes in levels of δ- and α-PKC in MEL cells exposed to the inducer hexamethylene bisacetamide (HMBA). MEL cells (N23) were incubated at 37°C with 5 mM HMBA in α-MEM tissue culture, supplemented with 10% fetal calf serum. At the times indicated, aliquots of the suspensions were removed and the cells were collected for the determination of levels of δ- and α-PKC as above (Patrone et al., 1994).

amounts of lower M_r fragments, still capable of expressing full inhibitory activity; digestion of calpastatin with m-calpain proceeds to complete inactivation.

Recent observations (Cressman et al., 1995) have indicated that protein kinase C (PKC) is converted by μ-calpain into the catalytically active fragment (PKM) and into multiple inactive products by m-calpain.

If observations of this kind are further extended, the concept of a different specificity and thereby of different function of the two calpain isoforms will be definitively and more precisely established.

The Biological Role of Calpain in the Modulation of PKC Activity

The biological system that, at present, provides direct evidence(s) for a specific biochemical function of calpain is that concerning the degradation and inactivation of Ca^{2+}-dependent protein kinase (PKC)

activity (Melloni et al., 1985). Degradation can initially proceed also as an activation mechanism generating the constitutively active free catalytic domain, called PKM (Melloni et al., 1985; Cressman et al., 1995).

The physiological significance of this calpain-mediated irreversible modification and inactivation of almost all the multiple members of the PKC family has been recently clarified in the course of studies that aimed to identify the biochemical events occurring in the early stages of MEL cell differentiation (Sparatore et al., 1994). These cells contain five PKC isozymes (Pessino et al., 1994) and many observations have indicated that the δ- and α-PKC play divergent effects on the differentiation process (Sparatore et al., 1993). The δ-PKC isoform is highly expressed in cells characterized by a low responsiveness to chemical inducer, whereas α-PKC isoform is particularly represented in MEL clones with high sensitivity to differentiation (Fig. 15.14A). Both isoforms undergo down-regulation with different kinetics (Fig. 15.14B) when cells are exposed to the chemical inducer (Patrone et al., 1994). The rapid degradation of the δ-PKC mediated by calpain corresponds to the removal of a negative signal on cell commitment, delivered by this kinase form, a statement supported by the observation that the changes in the level of δ-PKC correlate directly with the sensitivity of cells to the chemical inducer.

Down-regulation of α-PKC appears to be considerably delayed and, unlike that of δ-PKC, directly correlated to the onset of the irreversible cell commitment and to the appearance of benzidine-positive cells.

Thus, by means of a comparable mechanism of degradation of δ- and α-PKC, calpain removes a negative signal and enhances the delivery of a positive one, respectively. The extent and the rate at which calpain promotes degradation of PKC(s) therefore represents an efficient mechanism for the modulation of the differentiation process of murine erythroleukaemic cells (Sparatore et al., 1994).

From Physiology to Pathology

In recent years, a number of reports have suggested that calpain is responsible for the degradation of proteins that are crucially involved in maintaining the integrity of cell structure and function (Saido et al., 1994a; Squier et al., 1994; Martin and Green, 1995; Spencer et al., 1995; Trump and Berezesky, 1995; Yang and Ksiezak-Reding, 1995). These events cannot be related to a physiological role of calpain but rather to a primary defect in a cell function that implies intracellular increase of Ca^{2+}.

A pertinent example is the modification of the calpain/calpastatin system that occurs in rats of Milan hypertensive strain (MHS). In red blood cells, as well as in leukocytes and kidney of these rats (Pontremoli and Melloni, 1990), a calpain/calpastatin ratio of 1 occurs as compared with a value of 0.5 in cells from control animals. A similar change in the calpain/calpastatin ratio has been found in red cells of humans affected by essential hypertension (Pontremoli et al., 1989b).

Thus, whereas in normal conditions the amount of calpastatin largely exceeds that of calpain, in hypertensives, calpain is present in larger amounts with respect to its natural inhibitor protein and thereby becomes much more susceptible to activation. This mechanism could explain those alterations (reduced cellular volume, altered cellular shape, defects in sodium and potassium transport, increased intracellular calcium) that have been recognised as characteristic of the hypertensive red cell (Table 15.3). This hypothesis has been supported on an experimental basis since it has been shown that small variations in the intracellular concentration of Ca^{2+} are sufficient to induce, in red cells only from hypertensives, an extensive degradation of the transmembrane band 3 protein (anion exchange protein) as well as of Ca^{2+}-ATPase (Salamino et al., 1994a).

Preliminary analysis in humans with essential hypertension has provided indications that the extent in the defect in calpastatin is significantly correlated to the severity of the disease and to the effectiveness of the therapeutic treatment (Salamino et al., 1991).

Other suggestions supporting an implication of calpain in the development of pathological conditions have recently been reviewed by Saido et al. (1994a) and by Wang and Yuen (1994). Among these, particular interest has been focused on the proteolytic damage occurring in the post-ischaemic period in both cardiac muscle and brain tissue with emphasis on the potential therapeutic effects of specific calpain inhibitors. However, it still remains to be established whether the degenerative lesions occurring in these acute pathological conditions are a direct or indirect effect of calpain-mediated degradation of specific target proteins.

Experimental evidence collected in recent years shows that the muscle-specific calpain form (Sorimachi et al., 1993b) is implicated in the aetiology of limb-girdle muscular dystrophy type 2A (Richard et al., 1995b). Very little information is provided on the properties of this tissue-specific calpain due to its rapid autolysis; however, due to its association with connectin, a gigantic filamentous protein of muscle sarcomere, a new regulatory mechanism for this proteinase is now proposed (Sorimachi et al., 1995).

In addition, immunohistochemical studies have suggested that calpains are involved in the early stage of degradation of muscle fibres that occurrs in Duchenne and Becker muscular dystrophy (Kumamoto et al., 1995).

Conclusive Remarks

Calpain is a product of a gene fusion that occurred in the course of evolution to provide, within the molecular conformation, appropriate capability either to maintain in an inactive state an intracellular soluble proteinase or to provide very sensitive mechanism(s) for the modulation of its proteolytic activity. As a result, the activation of the enzyme takes place through the transduction of an activating signal that corresponds to the binding of calcium to the calmodulin-like domain, followed by a sequential progressive unmasking of the catalytic site (domain II), expression of the proteolytic activity of which also includes intramolecular autodigestion and generation of a constitutive "active" proteinase form.

Table 15.3 Possible Relationship between Structural and Functional Alterations in Red Cells from Hypertensives and Types of Membrane Proteins Highly Susceptible to Digestion by Calpain

Cell Alteration	Calpain Target Proteins
Increased cell volume and changes in cell shape	Cytoskeletal proteins (spectrin, ankirin, adducin, band 4.1)
Changes in cell permeability	Transmembrane proteins, ion pumps
Deficiency in calpastatin	
Rapid activation of calpain	

A positive modulation of calpain is accomplished by a protein molecule which increases Ca^{2+} sensitivity, thus favouring transition to the active conformation at a physiological concentration of Ca^{2+}. All the experimental evidence available so far suggests that calpain catalyses one of the crucial steps that are part of different signal transduction pathways. In this respect, activation and down-regulation of PKC by calpain represents a striking example. A negative modulation is promoted by a natural inhibitor protein, calpastatin, generally present in the same cell in large excess with respect to calpain, and thereby representing an additional mechanism for repressing uncontrolled activation of the proteinase.

A single calpastatin form can efficiently modulate the activation-activity of both calpain isoforms (μ- and m-calpain) normally present in all mammalian cells, through a post-translational reversible modification involving a phosphorylation–dephosphorylation process. The phosphorylated and dephosphorylated calpastatin forms express a higher inhibitory capacity with respect to one or other calpain isoform.

Furthermore, an increasing amount of evidence supports the direct or indirect involvement of calpain in different pathological conditions. These hypotheses, however, still require a direct demonstration that can only be obtained by the availability of specific calpain inhibitors, also an important goal to be achieved in consideration of their potential therapeutic use.

For the definition of a precise role of calpain, further investigations are required to establish (1) the precise concentration of Ca^{2+}, for the intracellular activation of each calpain isozyme, (2) if activation directed to different cell sites, and thus to different target protein substrates, confers, in vivo, further specificity to that directly resulting from the general properties of the enzyme, (3) if inhibition by different forms of calpastatin, produced by a post-translational interconvertible modification of a single protein molecule, is an important device for an increased efficiency of the natural protein inhibitor in vivo.

Acknowledgments This work was supported, in part, by grants from the National Research Council (CNR), Target Project "Prevention and Control of Disease Factors" (SP8), Control of Cardiovascular Diseases (grant n. 91.000.237; PF41), Target Project "Ingenneria Genetica" (SP4), Malattie ereditarie, and by the Ministero della Ricerca Scientifica e Tecnologica.

References

Adachi, Y., Ishida-Takahashi, A., Takahashi, C., Takano, E., Murachi, T., and Hatanaka, M. (1991) Phosphorylation and subcellular distribution of calpastatin in human hemopoietic system cells. J. Biol. Chem. 266: 3968–3972.

Aoki, K., Imajoh, S., Ohno, S., Emori, Y., Koike, M., Kosaki, G., and Suzuki, K. (1986) Complete amino acid sequence of the large subunit of a low-Ca^{2+}-requirement form of human Ca^{2+}-activated neutral protease (CANP) deduced from its cDNA sequence. FEBS Lett. 205: 313–317.

Brown, N. and Crawford, C. (1993) Structural modifications associated with the change in Ca^{2+} sensitivity on activation of m-calpain. FEBS Lett 322: 65–68.

Cong, J. Y., Goll, D. E., Peterson, A. M., and Kapprell, H. P. (1989) The role of autolysis in activity of the Ca^{2+} dependent proteinases (μ- and m-calpain). J. Biol. Chem. 264: 10096–10103.

Coolican, S. A., Haiech, J., and Hathana, D. R. (1986) The role of subunit autolysis in activation of smooth muscle Ca^{2+}-dependent proteases. J. Biol. Chem. 261: 4170–4178.

Cressman, C. M., Mohan P. S., Nixon R. A., and Shea T. B. (1995) Proteolysis of protein kinase C by mM and μM calcium-requirement calpains have different abilities to generate and degrade the free catalytic subunit, protein kinase M. FEBS Lett 367: 223–227.

Croall, D. E. and De Martino, G.N. (1991) Calcium-activated neutral protease (calpain) system: structure, function and regulation. Physiol. Rev. 71: 813–847.

Croall, D. E. and Mc Grody, K. S. (1994) Domain structure of calpain: mapping the binding site for calpastatin. *Biochemistry* 33: 13223–13230.

De Luca, C. I., Davies, P. L., Samis, J. A., and Else, J. S. (1993) Molecular cloning and bacterial expression of cDNA for rat calpain II 80 kDa subunit. *Biochim. Biophys. Acta* 1216: 81–93.

De Martino, G. N., Huff, C. A., and Croall, D. E. (1986) Autoproteolysis of the small subunit of calcium-dependent protease II activates and regulates protease activity. *J. Biol. Chem.* 261: 12047–12052.

Di Lisa, F., De Tullio, R., Salamino, F., Barbato, R., Melloni, E., Siliprandi, N., Schiaffino, S., and Pontremoli, S. (1995) Specific degradation of troponin T and I by μ-calpain and its modulation by substrate phosphorylation. *Biochemi. J.* 308: 57–61.

Emori, Y., Kawasaki, H., Imajoi, S., Kawashima, S., and Suzuki, K. (1986a) Isolation and sequence analysis of cDNA clones for the small subunit of rabbit calcium-dependent protease. *J. Biol. Chem.* 261: 9472–9476.

Emori, Y., Kawasaki, H., Sugihara, H., Imajoi, S., Kawashima, S., and Suzuki, K. (1986b) Isolation and sequence analysis of cDNA clones for the large subunits of two isozymes of rabbit calcium-dependent protease. *J. Biol. Chem.* 261: 9465–9471.

Emori, Y., Kawasaki, H., Imahori, K., and Suzuki, K. (1987) Endogenous inhibition for calcium-dependent cysteine protease contains four internal repeats that could be responsible for its multiple reactive sites. *Proc. Natl. Acad. Sci. USA* 84: 3590–3594.

Fox, J. E. B., Austin, C. B., Reynolds, C. C,. and Steffen, P. K. (1991) Evidence that agonist-induced activation of calpain causes the shedding of procoagulant containing microvesicles from the membrane of aggregating platelets. *J. Biol. Chem.* 266: 13289–13295.

Goll, D. E., Kleese, W. C., Okitani, A., Kumamoto, T., Cong, J., and Kapprell, H. P. (1990) Historical background and current status of the Ca^{2+}-dependent proteinase system. In *Intracellular Calcium-Dependent Proteolysis* (Mellgren, R. L. and Murachi, T. eds.). CRC Press, Boca Raton, FL, pp. 3–24.

Goll, D. E., Thompson, V. F., Taylor, R. G., and Zolewska, T. (1992) Is calpain activity regulated by membranes and autoproteolysis or by calcium and calpastatin? *BioEssays* 14, 549–556.

Graham-Seigenthaler, K., Gauthier, S., Davies, P. L., and Elce, J. S. (1994) Active recombinant rat calpain II. *J. Biol. Chem.* 269: 30457–30460.

Imajoh, S. Kawasaki, H., and Suzuki, K (1986) The amino terminal hydrophobic region of the small subunit of calcium-activated neutral protease (CANP) is essential for its activation by phosphatidylinositol. *J. Biochem.* 99, 1281–1284.

Imajoh, S., Aoki, K., Ohno, S., Kawasaki, H., Sugihara, H., and Suzuki, K (1988) Molecular cloning of the cDNA for the large subunit of the Ca^{2+} requiring form of human Ca^{2+}-activated neutral protease. *Biochemistry* 27: 8122–8128.

Kapprell, H. P. and Goll, D. E. (1989) Effect of Ca^{2+} on binding of calpains to calpastatin. *J. Biol. Chem.* 264: 17888–17896.

Karlsson, J. O., Blennow, K., Gottfries, C. G. (1992) Increased proteolytic activity in erythrocytes from patients with Alzheimer's disease *Dementia* 3: 200–204.

Kawasaki, H., Emori, Y., Inajoh, O. S., Minami, Y., and Suzuki, K. (1989) Identification and characterization of inhibitory sequences in four repeating domains of the endogenous inhibitor for calcium dependent protease. *J. Biol. Chem.* 106: 274–281.

Kishimoto, A., Kajikawa, N., Shiota, M., Nishizuka, Y. (1983) Proteolytic activation of calcium-activated phospholipid-dependent protein kinase by calcium-dependent neutral protease. *J Biol. Chem.* 258: 1156–1164.

Kumamoto, T., Ueyama, H., Watanabe, S., Yoshioka, K., Miike, T., Goll. D. E., Ando, M., and Tsuda, T. (1995) Immunohistochemical study of calpain and its endogenous inhibitor in the skeletal muscle of muscolar dystrophy. *Acta Neuropathol.* 89: 399–403.

Ma H., Yang Q., Takano E., Lee W. J., Hatanaka M., and Maki M. (1993) Requirement of different subdomains of calpastatin for calpain inhibition and for binding to calmodulin-like domains. *J. Biochem.* 113: 591–599.

Maki, M., Takano, E., Osawa, T., Ooi, T., Murachi, T., and Hatanaka, M. (1988) Analysis of structure–function relationship of pig calpastatin by expression of mutated cDNAs in *Escherichia coli*. *J. Biol. Chem.* 263: 10254–10261.

Martin, S. and Green, D. R. (1995) Protease activation during apoptosis: death by a thousand cuts? *Cell* 62: 348–352.

Mellgren, R. L. (1980) Canine cardiac calcium-dependent proteases: resolution of two forms with different requirement for calcium. *FEBS Lett.* 109:129–133.

Mellgren, R. L. (1988) On the mechanism of binding of calpastatin, the protein inhibitor of calpains to biological membranes. *Biochem. Biophys. Res. Commun.* 150: 170–176.

Mellgren, R. L. and Lu Q. (1994) Selective nuclear transport of μ-calpain. *Biochem. Biophys. Res. Commun.* 204: 544–550.

Melloni, E. and Pontremoli, S. (1989) The calpains. *Trends Neurosci.* 12: 438–444.

Melloni, E., Salamino, F., and Sparatore, B. (1992) The calpain–calpastatin system in mammalian cells: properties and possible functions. *Biochimie* 74:217–223.

Melloni, E., Pontremoli, S., Michetti, M., Sacco, O., Sparatore, B., Salamino, F., and Horecker, B. L. H. (1985) Binding of protein kinase C to neutrophil membranes in the presence of Ca^{2+} and its activation by a Ca^{2+}-requirement proteinase. *Proc. Natl. Acad. Sci. USA* 82: 6435–6439.

Meyer, S. L., Bozyczko-Coyne, D., Mallya, S. K., Spais, C. M., Bihovsky, R., Kaywooya, J. K., Lang D. M., Scott, R. W., and Siman, R. (1996) Biologically active monomeric and heterodimeric recombinant human calapin I produced using the baculovirus expression system. *Biochem. J.* 314: 511–519.

Michetti, M., Viotti, P. L., Melloni, E., and Pontremoli, S. (1991) Mechanism of action of the calpain activator protein in rat skeletal muscle. *Eur. J. Biochem.* 202: 1177–1180.

Molinari, M., Anagli, J., and Carafoli, E. (1994) Ca^{2+}-activated neutral protease is active in the erythrocyte membrane in its non autolyzed 80 kDa form. *J. Biol. Chem.* 269, pp. 27992–27995.

Murachi, T. (1989) Intracellular regulatory system involving calpain and calpastatin. *Biochem. Int.* 18: 263–294.

Nakamura, M., Inomata, M., Imajoh, S., Suzuki, K., and Kawashima S. (1989) Fragmentation of an endogenous inhibitor upon complex formation with high and low-Ca^{2+} requirement forms of calcium-activated neutral proteases. *Biochemistry* 28: 449–455.

Nilsson, E., Alafuzoff, I., Blennow, K., Blomgren, K., Hall, C. M., Janson, I., Karlsson, I., Wallin, A., Gottfries, C. G., and Karlsson, J. O. (1990) Calpain and calpastatin in normal and Alzheimer-degenerated human brain tissues. *Neurobiol. Aging*: 11, 425–431.

Nishimura, T. and Goll, D. E. (1991) Binding of calpain fragments to calpastatin. *J. Biol. Chem.* 266, 11842–11850.

Nixon, R. A. (1989) Calcium activated neutral proteinases as regulators of cellular function. *Ann. of N.Y. Acad. Sci.* 568: 198–208.

Ohno, S., Emori, Y., Imajoh, S., Kawasaki, H., Kisaregi, M., and Suzuki, K. (1984) Evolutionary origin of a calcium-dependent protease by fusion of genes for a thiol protease and a calcium binding protein? *Nature* 312: 566–570.

Ohno, S., Minoshima, S., Kudoh, J., Fukuyama, R., Shinuzu, Y., Ohmi-Imajoh, S., Shimazu, N., and Suzuki, K. (1990) Four genes for the calpain family locate on four distinct human chromosomes. *Citogenet. Cell Genet.* 53, 225–229.

Patrone, M., Pessino, A., Passalacqua, M., Sparatore. B., Melloni, E., and Pontremoli, S. (1994) Protein kinase C isoforms in murine erythroleukemia cells and their involvement in the differentiation process. *FEBS Lett.* 344: 91–95.

Pessino, A., Sparatore, B., Patrone, M., Melloni, E., and Pontremoli, S. (1994) Differential expression of protein kinase C isoform genes in three murine erythroleukemia cell variants: implication for chemical induced differentiation. *Biochem. Biophys. Res. Commun.* 204: 461–467.

Pontremoli, S. and Melloni, E. (1986) Regulation of Ca^{2+}-dependent proteinase of human erythrocytes. In *Calcium and Cell Function* Vol. VI. (Wai Yiu Cheung, ed.). Academic Press, New York, pp. 159–181.

Pontremoli, S. and Melloni, E. (1990) Erythrocyte calpain in health and disease. In *Intracellular Calcium-Dependent Proteolysis* (Mellgren, R. L. and Murachi, T. eds.). CRC Press, Boca Raton, FL, pp. 225–239.

Pontremoli, S., Salamino, F., Sparatore, B., Michetti, M., Sacco, O., and Melloni, E. (1985a) Following associa-

tion to membrane, human erythrocyte procalpain is converted and released as fully active calpain. *Biochim. Biophys. Acta* 831: 335–339.

Pontremoli. S., Melloni, E., Sparatore, B., Salamino, F., Michetti, M., Sacco, O., and Horecker, B. L. (1985b) Binding to erythrocyte membrane is the physiological mechanism for activation of Ca^{2+}-dependent neutral proteinase. *Biochem. Biophys. Res. Commun.* 128: 331–338.

Pontremoli, S., Melloni, E., Sparatore, B., Salamino, F., Michetti, M., Sacco, O., and Horecker, B. L. (1985c) Role of phospholipids in the activation of the Ca^{2+}-dependent neutral proteinase of human erythrocytes. *Biochem. Biophys. Res. Commun.* 129: 389–395.

Pontremoli, S., Melloni, E., Salamino, F., Patrone, M., Michetti, M., and Horecker, B. L. (1989a) Activation of neutrophyl calpain following its translocation to the plasma membrane induced by phorbol ester or fMet-Leu-Phe. *Biochem. Biophys. Res. Commun.* 160: 737–743.

Pontremoli, S., Melloni, E., Sparatore, B., Salamino, F., Pontremoli, R., Tizianello, A., Barlassina, C., Cusi, D., Colombo, R. and Bianchi, G. (1989b) Erythrocytes deficiency in calpain inhibitor activity in essential hypertension. *Hypertension* 12: 474–480.

Pontremoli, S., Melloni, E., Viotti, P. L., Michetti, M., Salamino, F., and Horecker, B. L. (1991) Identification of two calpastatin forms in rat skeletal muscle and their susceptibility to digestion by homologous calpains. *Arch. Biochem. Biophys.* 288: 646–652.

Pontremoli, S., Viotti, P. L., Michetti, M., Salamino, F:, Sparatore, B., and Melloni, E. (1992) Modulation of inhibitory efficiency of rat skeletal muscle calpastatin by phosphorylation. *Biochem. Biophys. Res. Commun.* 187: 751–759.

Richard, I., Broux, O., Allamond, V., Fougerousse, F., Chiannilkulchai, N., Bourg, N., Brenguier, L., Devaud, C., Pasturaud, P., Boudant, C., Hillaire, D., Passos-Bueno, M. R., Zatz, M., Tischfield, J. A., Fardeau, M., Jackson, C., Cohen, D., and Beckmann, J. S. (1995) Mutations in the proteolytic enzyme calpain 3 cause limb-girdle muscular dystrophy type 2A. *Cell* 81: 27–40.

Saido, T. C., Suzuki, H., Yamazaki, H., Tanoue, K., and Suzuki, K. (1993) In situ capture of μ-calpain activation in platelets. *J. Biol. Chem.* 268: 7422–7426.

Saido, T. C., Sorimachi, H., and Suzuki, K. (1994a) Calpain: new perspectives in molecular diversity and physiological–pathological involvement. *FASEB J.* 8: 814–822.

Saido, T. C., Nagao, S., Shiramine, M., Tsukagushi, M., Yoshizawa, T., Sorimachi, H., Ito, H., Tsuchiya, T., Kawashima, S., and Suzuki, K. (1994b) Distinct kinetics of subunit autolysis in mammalian m-calpain activation. *FEBS Lett.* 346: 263–267.

Saito, K., Elce, J. S., Hannos, J. E., and Nixon, R. A. (1993) Widespread activation of calcium-activated neutral proteinase (calpain) in the brain in Alzheimer disease:

a potential molecular basis for neuronal degeneration. *Proc. Natl. Acad. Sci. USA* 90: 2628–2632.

Sakihama, T., Kakidani, S., Zenita, K., Yumoto, N., Kikuchi, T:, Sasaki, T., Kannagi, R., Nakanishi, S., Ohmori, M., Takio, K , Titani, K., and Murachi, T. (1985) A putative Ca^{2+}-binding protein: structure of the light subunit of porcine calpain elucidated by molecular cloning and protein sequence analysis. *Proc. Natl. Acad. Sci.* 82: 6075–6079.

Salamino, F., Sparatore, B., De Tullio, R., Pontremoli, R., Melloni, E., and Pontremoli, S. (1991) The calpastatin defect in hypertension is possibly due to a specific degradation by calpain. *Biochim. Biophys. Acta* 1096: 265–269.

Salamino, F., De Tullio, R., Mengotti, P., Melloni, E., and Pontremoli, S. (1992) Different susceptibility of red cell membrane proteins to calpain degradation. *Arch. Biochem. Biophys.* 298: 287–292.

Salamino, F., De Tullio, R:, Mengotti, P., Viotti, P. L., Melloni, E., and Pontremoli, S. (1993) Site-directed activation of calpain is promoted by a cytoskeletal associated natural activator protein. *Biochem. J.* 290: 191–197.

Salamino, F., Sparatore, B., Melloni, E., Michetti, M., Viotti, P. L., and Pontremoli, S. (1994a) The plasma membrane calcium pump is the preferred calpain substrate within the erythrocytes. *Cell Calcium* 15: 28–35.

Salamino, F., De Tullio, R., Mengotti, P., Melloni, E., and Pontremoli, S. (1994b) Differential regulation of μ-calpain and m-calpain in rat hearts perfused with Ca^{2+} and cAMP. *Biochem. Biophys. Res. Commun.* 202: 1197–1203.

Salamino, F., De Tullio, R., Michetti, M., Mengotti, P., Melloni, E., and Pontremoli, S. (1994c) Modulation of calpastatin specificity in rat tissues by reversible phosphorylation and dephosphorylation. *Biochem. Biophys. Res. Commun.* 199: 1326–1332.

Schwarz-Ben Meier, N., Glasez, T., and Kosower, N. (1991) Band 3 protein degradation by calpain is enhanced in erythrocytes of old people. *Biochem. J.* 275: 47–52.

Shoeman, R. L. and Traub, P. (1990) Calpains and the cytoskeleton. *Intracellular Calcium-Dependent Proteolysis* (Mellgren, R. L. and Murachi, T. eds.). CRC Press, Boca Raton, FL, pp. 191–209.

Sorimachi, H., Toyama-Sorimachi, N., Saido, T., Kawasaki, H., Sugita, H., Miyasaka, M., Arahata, K., Ishiura S. and Suzuki, K. (1993b) Muscle-specific calpain, p94, is degraded by autolysis immediately after translation, resulting in disappearance from muscle. *J. Biol. Chem.* 268: 10593–10605.

Sorimachi, H., Ishiura, S., and Suzuki, K. (1993) A novel tissue-specific calpain species expressed predominantly in the stomach comprises two alternative splicing products with and without Ca^{2+}-binding domain. *J. Biol. Chem.* 268: 19476–19482.

Sorimachi, H., Saido, T. C., and Suzuki, K. (1994) New era of calpain research: discovery of tissue-specific calpains. *FEBS Lett.* 343: 1–5.

Sorimachi, H., Kinbara, K., Kimura, S., Takahashi, M., Ishiura, S., Sasagawa, N., Sorimachi, N., Shimada, H., Tagawa, K., Maruyama K., and Suzuki, K. (1995) Muscle specific calpain, p94, responsible for limb girdle muscular dystrophy type 2A, associates with connectin through IS2, a p94-specific sequence. *J. Biol. Chem.* 270, 31158–31162.

Sparatore, B., Pessino, A., Patrone, M., Passalacqua, M., Melloni, M., and Pontremoli, S. (1993) Role of δ-PKC on the differentiation process of murine erythroleukemia cells. *Biochem. Biophys. Res. Commun.* 193: 220–227.

Sparatore, B., Passalacqua, M., Pessino, A., Melloni, E., Patrone, M., and Pontremoli, S. (1994) Modulation of the intracellular Ca^{2+} dependent proteolytic system is critically correlated with the kinetics of differentiation of murine erythrocytes cells. *Eur. J. of Biochem.* 225: 173–178.

Spencer, M. J., Croall, D. E., and Tidball, J. G. (1995) Calpains are activated in necrotic fibers from mdx dystrophic mice. *J. Biol. Chem.* 270: 10909–10914.

Squier, M. K. T. Miller, A. C. K., Malkinson. A. M., and Cohen, J. J. (1994) Calpain activation on apoptosis. *J. Cell. Physiol.* 159: 229–237.

Suzuki, K. (1987) Calcium activated neutral protease: domain structure and activity regulation. *Trends Biol. Sci.* 12: 103–105.

Suzuki, K. (1990) The structure of calpain and the calpain gene. In *Intracellular Calcium-Dependent Proteolysis* (Mellgren, R. T. and Murachi, T. eds.). CRC Press, Boca Raton, FL, pp. 25–35.

Takeyama. Y., Nakanishi, H., Uratsuji, Y., Kishimoto, A., and Nishizuka, Y. (1986) A calcium insoluble activator associated with brain microsomal insoluble elements. *FEBS Lett.* 194, 110–114.

Tashiro, T. and Ishizaki, Y. (1982) A calcium-dependent protease selectively degrading the 160,000 M_r component of neurofilaments is associated with the cytoskeletal preparation of the signal cord and has an endogenous inhibitory factor. *FEBS Lett.* 141: 41–45.

Trump, B. F. and Berezesky, Y. (1995) Calcium-mediated cell injury and cell death. *FASEB J.* 9: 219–228.

Wang, K. and Yuen, P. W. (1994) Calpain inhibition: an overview of its therapeutic potential. *Trends Pharmacol. Sci.* 15: 412–419.

Wang, K. K. W., Villalobo, A., and Roufogalis, B. D. (1989) Calmodulin-binding proteins as calpain substrates. *Biochem. J.* 262: 693–706.

Yang, L. S. and Ksiezak-Reding, H. (1995) Calpain induced proteolysis of normal human tau and tau associated with paired helical filaments. *Eur. J. Biochem.* 233, 9–17.

Yoshizawa, T., Sorimachi, H., Tomiska, S., Ishiura, S., and Suzuki, K. (1995a) Calpain dissociates into subunits in the presence of calcium ions. *Biochem. Biophys. Res. Commun.* 208: 376–383.

Yoshizawa, T., Sorimachi, H., Tomioka, S., Ishiura, S., and Suzuki, K. (1995b) A catalitic subunit of calpain possesses full proteolytic activity. *FEBS Lett.* 358: 101–103.

16

cAMP Phosphodiesterase/Adenylate Cyclase

Chen Yan
Guy C. K. Chan
Daniel R. Storm
Joseph A. Beavo

Calcium and cyclic AMP (cAMP) can both act as second messengers for a number of signaling molecules. In addition, changes in intracellular free Ca^{2+} often alter cAMP and vice versa (Rasmussen and Goodman, 1977; Berridge, 1984). Many of these interactions are mediated via Ca^{2+} binding to calmodulin (CaM). The two most direct mechanisms by which Ca^{2+} modulates cAMP are through $Ca^{2+}/$ CaM regulation of adenylyl cyclases and cyclic nucleotide phosphodiesterases (PDEs). Ca^{2+}-sensitive kinases can also modulate both adenylyl cyclase and phosphodiesterase activities. Conversely, cAMP can also directly and indirectly (e.g., through cAMP-dependent protein kinases and phosphatases) modulate Ca^{2+} channels, Ca^{2+} pumps, and Ca^{2+}-exchange mechanisms. The discovery of multiple genes coding for different adenylyl cyclases and phosphodiesterases has increased interest in the regulatory properties of these enzymes. Presumably, cell type-specific expression of different isozymes provides distinct mechanisms for cross-talk between the cAMP and other signal transduction systems. For example, interactions between cAMP and cGMP have been discussed recently (Beavo, 1995). This chapter will focus on the mechanisms for $Ca^{2+}/$ CaM control of cAMP synthesis and degradation.

Both the amplitude and the duration of cAMP signals are regulated by the activities of adenylyl cyclase(s) and phosphodiesterases present in a particular tissue. In many cases, these signals are transient and, in fact, prolonged cAMP increases can be toxic for some cells. In other systems, longer term changes in the steady-state level of cAMP have been demonstrated. Because of the diverse physiological and biochemical processes regulated by the cAMP signal transduction system, one would anticipate a variety of cAMP signaling patterns. For example, acute regulation of ion channel activity by cAMP may require only brief cAMP increases lasting a few milliseconds whereas stimulation of transcription requires more prolonged increases lasting many minutes. The kinetics for cAMP transients stimulated by various agonists depend upon the cellular and subcellular distribution of specific adenylyl cyclases and phosphodiesterases. Until the regulatory properties of these enzymes and their distributions are completely defined, it will not be possible to predict quantitatively the effect of various hormones and drugs on cAMP levels in specific tissues.

It has become increasingly evident that cross-talk between the cAMP and Ca^{2+} signal transduction systems is important for various physiological processes ranging from neuroplasticity, to olfaction, to muscle contraction. For example, some forms of synaptic plasticity in the hippocampus require coupling of intracellular Ca^{2+} to changes in cAMP. This process seems to have both short- and longer term components. Similarly, olfactory signal transduction may also require rapid Ca^{2+} control of both synthesis and degradation of cAMP. The high expression of specific isozymes of Ca^{2+}-regulated adenylyl cyclases and phosphodiesterases in olfactory sensory neurons provides mechanisms for these cAMP transients. Many of the effects of intracellular Ca^{2+} on cellular function are mediated by the calcium-binding protein, calmodulin, which mediates Ca^{2+} regulation of various enzymes, including adenylyl cyclases, phosphodiesterases, and protein kinases. Specific examples of relationships among these isozymes are discussed in more detail in the following sections.

Ca²⁺-Regulated Adenylyl Cyclases

The adenylyl cyclases can be regulated by stimulatory and inhibitory receptors coupled to their catalytic subunits through the guanyl nucleotide regulatory proteins, G_s and G_i (Xia and Storm, 1996). Several can also be regulated directly by Ca^{2+}/CaM or indirectly by Ca^{2+}-dependent phosphorylation mechanisms. There are at least eight different genes that encode adenylyl cyclases. Of these, only I-AC and VIII-AC are directly stimulated by Ca^{2+}/calmodulin. These Ca^{2+}-sensitive adenylyl cyclases provide one mechanism for "cross-talk" between the Ca^{2+} and cAMP signal transduction systems. Furthermore, the adenylyl cyclases are regulated by other calcium-sensitive enzymes, including protein kinase C (PKC) (Choi et al., 1993; Jacobowitz et al., 1993) and the CaM-kinases (Wayman et al., 1996). Recent evidence also suggests that neurons may express voltage-sensitive adenylyl cyclase activity (Reddy et al., 1995). In some cases, adenylyl cyclases can function as signal integrators and respond synergistically to multiple extracellular and intracellular signals. Although the adenylyl cyclase family is regulated by multiple effector molecules, including Ca^{2+} and hormones, each enzyme has its own unique regulatory properties. The diversity of this enzyme system reflects the need for different mechanisms for regulation of cAMP levels in animal cells, and the variety of physiological processes that are regulated by intracellular cAMP.

The following sections focus on the CaM-regulated enzymes, I-AC and VIII-AC, as well as III-AC, an enzyme that is regulated by CaM-kinase II. In addition, it has been proposed that Ca^{2+} may directly inhibit the activity of V-AC and VI-AC.

The Type I Adenylyl Cyclases

Regulatory Properties of I-AC

The existence of distinct CaM-stimulated and CaM-insensitive adenylyl cyclases in brain was first demonstrated by the separation of two forms of the enzyme using CaM–Sepharose affinity chromatography (Westcott et al., 1979). In addition, polyclonal antibodies were isolated that distinguished between CaM-sensitive and CaM-insensitive adenylyl cyclases in brain (Rosenberg and Storm, 1987). The isolation of cDNA clones for I-AC (CaM-sensitive) and II-AC (CaM-insensitive) confirmed the existence of at least two classes of adenylyl cyclases (Krupinski et al., 1989; Feinstein et al., 1991). I-AC has been of particular interest to neurobiologists because it is neuro-specific and data from invertebrates and mammals suggest that it is important for synaptic plasticity and spatial memory. The regulatory properties of

VIII-AC complement I-AC and it may also be important for modulation of synaptic plasticity. Ca^{2+} inhibition of the type III adenylyl cyclase provides a mechanism for attenuation of hormone-stimulated cAMP increases which may be important for olfactory signal transduction.

Stimulation by Ca²⁺

I-AC is directly stimulated by Ca^{2+} and CaM in vitro (Tang et al., 1991; Choi et al., 1992a) and in vivo (Choi et al., 1992a; Wu et al., 1993); half-maximal stimulation occurs at approximately 150 nM free Ca^{2+} and 20 nM CaM. To characterize Ca^{2+} stimulation of the enzyme in vivo, the enzyme was expressed in HEK-293 cells and its sensitivity to Ca^{2+} was examined using A23187 to increase intracellular Ca^{2+} (Choi et al., 1992a). Although Ca^{2+} did not affect intracellular cAMP in control cells transfected with the expression vector alone, it significantly increased cAMP in cells stably expressing I-AC. The increase in cAMP depended upon the concentration of Ca^{2+} applied and it was detectable within a few minutes after addition of the Ca^{2+} ionophore.

HEK-293 cells express muscarinic acetylcholine receptors that are coupled to activation of phospholipase C and release of intracellular Ca^{2+} from IP_3-sensitive pools. Treatment of these cells with carbachol increased intracellular free Ca^{2+} to approximately 300 μM. Carbachol also increased cAMP levels approximately 3-fold in HEK-293 cells stably transfected with I-AC but was without effect on cAMP in control cells that lacked I-AC (Fig. 16.1). Half-maximal stimulation was at 30 μM carbachol and cAMP increases were inhibited by the muscarinic acetylcholine receptor antagonist, atropine. Carbachol-stimulated increases in intracellular Ca^{2+} and cAMP were completely blocked by BAPTA, an intracellular Ca^{2+} chelator. Furthermore, a CaM-insensitive mutant of I-AC, containing an Arg in place of Phe within its CaM-binding domain, was not stimulated by intracellular free Ca^{2+} (Wu et al., 1993). Therefore, Ca^{2+} stimulation of I-AC in vivo is very likely mediated by binding of CaM to the enzyme.

Synergistic Stimulation of I-AC by Ca²⁺ and G_s-Coupled Receptors In Vivo

To examine the effect of intracellular Ca^{2+} on the coupling of G_s-coupled receptors to I-AC in vivo, stably transfected HEK-293 cell lines expressing the glucagon receptor with I-AC or III-AC were prepared (Wayman et al., 1994). HEK-293 cells contain endogenous β-adrenergic receptors that couple to stimulation of some adenylyl cyclases. For example,

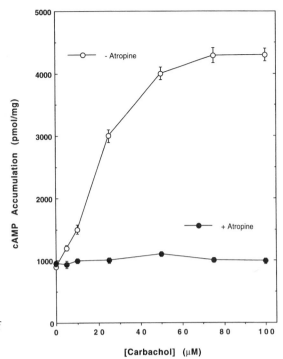

Figure 16.1 Carbachol stimulation of intracellular cAMP levels in HEK-293 cells expressing the I-AC. Cells were treated with varying concentrations of carbachol for 30 minutes in the absence or presence of 1.0 μM atropine and assayed for intracellular cAMP. Data are from Choi et al., 1992a.

activation of either β-adrenergic or glucagon receptors in whole cells stimulated III-AC activity 2.5- and 380-fold, respectively. I-AC, on the other hand, was not stimulated by isoproterenol or glucagon in vivo. Although I-AC was not stimulated by isoproterenol alone, it was synergistically activated by combinations of A23187 with increasing concentrations of isoproterenol (Fig. 16.2). Isoproterenol stimulation was blocked by the β-adrenergic receptor antagonist, propranolol, and stimulation by A23187 and isoproterenol was dependent upon the concentration of extracellular Ca^{2+}. To test the generality of this phenomena, and to ensure that the insensitivity to isoproterenol alone was not due to low levels of endogenous β-adrenergic receptors in HEK-293 cells, stable transformants coexpressing glucagon receptors and I-AC were also analyzed. In the absence of Ca^{2+} increases, the enzyme was not stimulated by glucagon. However, it was stimulated by glucagon when it was also activated by intracellular Ca^{2+}. Intracellular Ca^{2+} signals generated by physiologically relevant agonists also increased the sensitivity of I-AC to G_s-coupled receptors. For example, carbachol stimulated I-AC activity 5-fold and combinations of carbachol with isoproterenol stimulated it 14-fold.

Synergism between intracellular Ca^{2+} and hormones might be attributable to activation of Ca^{2+} or cAMP-stimulated protein kinases which phos-phorylate a component of the adenylyl cyclase system. However, synergism between A23187 and isoproterenol was not blocked by KN 62 (an inhibitor of CaM-kinases), calphostin C (a protein kinase C inhibitor), or the PKA inhibitors H89 or Rp-cAMP. Similarly, carbachol and isoproterenol synergism was not inhibited by these inhibitors, suggesting that activation of a protein kinase, in response to cAMP or Ca^{2+}, was not required for these phenomena.

Synergistic activation of I-AC could be due to conformational changes in the catalytic subunit induced by interactions with CaM or due to other Ca^{2+}-dependent phenomena. This question was addressed using a CaM-insensitive mutant of I-AC containing an Arg in place of Phe-503 within the CaM-binding domain (FR-I-AC). This mutant enzyme was insensitive to CaM but its basal and forskolin-stimulated activities were indistinguishable from wild-type I-AC (Wu et al., 1993). Since FR-I-AC was not synergistically stimulated by isoproterenol and A23187, the increased sensitivity to β-adrenergic receptors caused by Ca^{2+} requires CaM binding to I-AC.

Stimulation of I-AC by PKC

Activation of PKC by phorbol esters stimulates the activity of several adenylyl cyclases in vivo

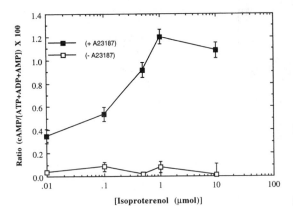

Figure 16.2 Synergistic stimulation of I-AC by isoproterenol and Ca^{2+} ionophore in vivo. HEK-293 cells expressing I-AC were exposed to the β-adrenergic agonist isoproterenol in the absence (□) or presence of (■) 10 μM A23187 + 1.8 mM CaCl$_2$ and assayed for intracellular cAMP. Data are from Wayman et al. (1994).

(Jacobowitz et al., 1993). To determine if PKC modulates the activity of I-AC, the effects of phorbol esters were examined in vivo (Choi et al., 1993). The phorbol ester, TPA, stimulated I-AC 200% within 15 minutes after treatment and half-maximal stimulation was at 100 nM TPA. To evaluate the specificity of the cAMP increase caused by TPA, the effects of two other phorbol esters were examined. TPA and PDBu are potent activators of PKC and both stimulated I-AC. However, the inactive phorbol ester, 4-a-PMA, did not stimulate I-AC at concentrations up to 100 nM.

Since I-AC is stimulated by Ca^{2+} in vivo and activators of PKC may elevate intracellular Ca^{2+} levels in some cells (Hopkins and Johnston, 1988), the increases in cAMP caused by phorbol esters could be due to increased intracellular Ca^{2+}. To evaluate this possibility, the effect of BAPTA on TPA-stimulated cAMP increases was examined. Exposure of cells to BAPTA, prior to TPA treatment, did not block TPA stimulation of I-AC. Furthermore, TPA had no effect on intracellular free Ca^{2+} levels in HEK-293 cells. Therefore, TPA stimulation of I-AC in whole cells was not due to an increase in intracellular free Ca^{2+}. Although the mechanism for PKC regulation of I-AC is not known, it may be due to direct phosphorylation of the enzyme. For example, V-AC is directly phosphorylated by PKC in vitro. (Kawabe et al., 1994).

Stimulation of I-AC and other adenylyl cyclases by PKCs provides an interesting mechanism for cross-talk between the phosphoinositide and cAMP signal transduction systems. This may play an important role in various physiological processes, including modulation of synaptic plasticity in the nervous system. For example, PKC activity is stimulated during some forms of long-term potentiation (LTP) in the hippocampus and this may contribute to increased cAMP associated with LTP.

Mechanisms for Inhibition of I-AC

Although mechanisms for inhibition of I-AC in vivo have not been extensively characterized, CaM stimulation of this enzyme was inhibited by M_4-muscarinic (Dittman et al., 1994), somatostatin, and dopamine D_2 receptor agonists in vivo (Nielsen et al., 1996). Inhibition was blocked by pertussis toxin, indicating that receptor coupling is most probably mediated by G_i. These data are consistent with studies showing that CaM-stimulation of I-AC is directly inhibited by GTP-activated G_{ia} in vitro (Taussig et al., 1993).

It was also recently discovered that the I-AC is inhibited by CaM-kinase IV in vivo (G. A. Wayman and D. R. Storm, unpublished observations). Expression of constitutively active or wild-type CaM-kinase IV inhibited Ca^{2+} stimulation of I-AC without affecting basal or forskolin-stimulated activities. Ca^{2+} stimulation of I-AC was inhibited approximately 50% by CaM-kinase IV, whereas CaM-kinase II had no effect. In addition, CaM-kinase IV catalyzed the phosphorylation of I-AC in vitro. The other CaM-stimulated adenylyl cyclase, VIII-AC, was not inhibited by CaM-kinase II or IV. These data suggest that I-AC may be phosphorylated on a domain that affects stimulation by CaM without affecting basal catalytic activity. I-AC contains five CaM-kinase consensus domains, two of which are close to the CaM-binding domain. Mutagenesis of either of these serines to alanines completely abolished CaM-kinase IV inhibition of I-AC (Wayman et al., 1996). These data suggest that I-AC is directly phosphorylated by CaM-kinase IV in vivo.

CaM-kinase IV (Nakamura et al., 1995) and I-AC (Xia et al., 1991) are both expressed in the neocortex, hippocampus, and cerebellar cortex with high levels of mRNA expression in granule cells. Consequently, Ca^{2+} stimulation of I-AC, in some neurons, may subsequently be inhibited by CaM-kinase IV. In

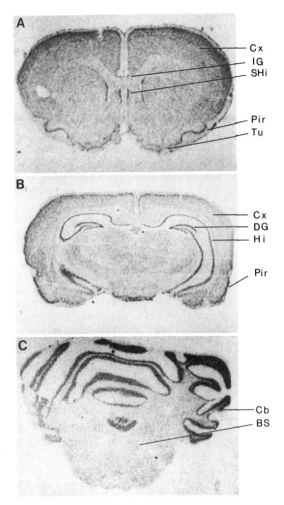

Figure 16.3 In situ hybridization analysis of I-AC mRNA distribution in adult rat brain. Middle rat brain sections were hybridized with an ^{35}S-labeled antisense riboprobe specific for I-AC. Abbreviations: BS, brain stem; Cb, cerebellum; Cx, neocortex; DG, dentate gyrus; Hi, hippocampus; IG, indusium griseum; Pir, piriform cortex; SHi, septohippocampal nucleus; Tu, olfactory tubercle. Data are from Xia et al. (1991).

cells expressing CaM-kinase IV, I-AC, and CaM-stimulated phosphodiesterases, cAMP signals generated by Ca^{2+} stimulation of I-AC may be rapidly attenuated by Ca^{2+} activation of CaM-kinase IV and Ca^{2+}-stimulated phosphodiesterases.

Tissue and Cellular Distribution of I-AC

It is becoming increasingly clear that isozymes of many enzymes have evolved in order for the organism to be able to express forms of the enzyme that have different regulatory and kinetic features in a tissue and cell type-specific manner. Therefore, it is also now clear that in order to determine a physiological role for any given isoenzyme, it is important to determine in which tissues and cell types within a tissue it is expressed. An analysis of the distribution of I-AC mRNA in a large number of bovine and rodent tissues indicates that this enzyme is neurospe-cific (Xia et al., 1993). The only bovine tissues showing a positive signal for I-AC mRNA were brain, retina, and adrenal medulla. Several cultured cell lines, including neuroblastoma cell N1E-115, neuro-glio-hybridoma cell NG-108, rat glioma 36B-10 cell, and PC-12 cells also do not express mRNA for I-AC. The restricted expression of I-AC to neural tissues contrasts sharply with some of the other mammalian enzymes that show fairly broad distribution in both neural and nonneuronal tissues.

The distribution of mRNA encoding the I-AC in rat brain was also examined by in situ hybridization (Xia et al., 1991). In situ hybridizations in adult rat brain revealed high levels of I-AC mRNA in specific areas of brain, including the hippocampal formation, the neocortex, entorhinal cortex, cerebellar cortex, and parts of the olfactory system (Fig. 16.3). The dentate gyrus in the hippocampal formation showed very intense labeling that was associated with the granule cell layer. Moderately strong labeling was also evident in association with the pyramidal cells in CA1, CA2, and CA3 layers of the hippocampus. I-AC mRNA was not detected in the brain stem.

Since I-AC is not expressed throughout the brain, it probably does not play a general regulatory role (e.g., in regulation of cell metabolism), and it may be important for specific neuronal functions. Messenger RNA for I-AC is highly localized to specific regions of brain, including areas that show long-term potentiation and have been implicated in learning and memory. The neurospecific expression of the I-AC and its limited distribution in brain are consistent with the proposal that this enzyme may by important for some forms of synaptic plasticity.

Genetic Studies—The Drosophila Learning Mutant rutabaga

The first evidence that Ca^{2+}-sensitive adenylyl cyclases may be important for synaptic plasticity came from studies of the *Drosophila* learning mutant, *rutabaga*. *Rutabaga* is an X-linked recessive mutant that is deficient in associative learning (Dudai and Zvi, 1984; Livingston et al., 1984). In contrast to wild-type *Drosophila*, the *rutabaga* fly lacks Ca^{2+}/calmodulin-sensitive adenylyl cyclase activity. The gene for an adenylyl cyclase similar to the I-AC maps within a region on the X chromosome that includes the *rut* locus and a single point mutation in this gene is sufficient to destroy all enzyme activity (Levin et al., 1992). Feany proposed that Ca^{2+} responsiveness, rather than the overall cAMP synthesis, may be the crucial component of adenylyl cyclase activity required for associative learning in *Drosophila* (Feany, 1990). In *rutabaga* larvae, voltage-clamp analysis of neuromuscular transmission indicated deficient synaptic facilitation and post-tetanic potentiation (Zhong and Wu, 1991).

Disruption of the Gene for the I-AC Leads to Deficiencies in LTP and Spatial Learning

More recently, the gene for I-AC in mice was disrupted to evaluate the role of the enzyme for synaptic plasticity, as well as learning and memory (Wu et al., 1995). I-AC mutant mice had normal growth, motor coordination, and longevity. They showed no detectable anatomical differences in the hippocampus, neocortex, or cerebellum. Compared with wild-type mice, Ca^{2+}-sensitive adenylyl cyclase activities in the cerebellum, neocortex, and hippocampus of mutant mice were decreased 62%, 38%, and 46%, respectively. Furthermore, the Ca^{2+}-sensitivity of the residual adenylyl cyclase activity in mutant mice was lower than I-AC. There was only a minor reduction in Ca^{2+}-sensitive adenylyl cyclase activity in the brain stem, a region that does not express I-AC.

To determine if Ca^{2+} stimulation of intracellular cAMP levels is depressed in neurons from mutant mice, the effect of A23187 on cAMP was examined. Intracellular cAMP levels in cultured cerebellar neurons from wild-type and mutant mice both increased when neurons were treated with A23187. However, Ca^{2+}-stimulated cAMP increases were depressed 50% in mutant mice relative to wild-type mice, indicating that coupling of Ca^{2+} to cAMP increases is reduced in neurons from mutant mice.

To determine if the I-AC is crucial for synaptic plasticity, several forms of hippocampal LTP were compared in wild-type and mutant mice. Both wild-type and mutant mice exhibited long-lasting CA1 LTP (L-LTP) that persists beyond 3 hours; however, there were several quantitative differences in the response of the mutant mice that are evident during the first hour after stimulation. The rate of increase of the EPSP slope for the mutant mice after tetanic stimulation (from 1 minute to 30 minutes) was half that of wild-type (1.3 ± 0.05 vs. 2.6 ± 0.1; $p < .001$). The maximum field EPSP slope above baseline was also reduced approximately 40% in mutant mice. Facilitation of the synaptic response that occurred after tetanic stimulation developed more slowly and reached a lower level in mutants lacking I-AC. Mossy fiber/CA3 LTP in mutant and wild-type mice was also compared. This form of LTP is presynaptic in origin and maintenance and independent of NMDA receptors. The mutant mice showed greatly depressed mossy fiber/CA3 LTP, indicating that the I-AC is important for this type of LTP (E. C. Villacres and D. R. Storm, unpublished observations).

Mutant and wild-type mice were analyzed for spatial learning by the Morris water task, a set of assays used to examine spatial learning in mice (Morris et al., 1982, 1986; Morris, 1990; Davis et al., 1992). Both sets of animals showed decreased escape latencies with training, and there were no statistically significant differences in the ability of the mutant and wild-type mice to find the visible or hidden platform. However, escape latencies in the hidden-platform task are a poor indicator of spatial learning, and even rodents with hippocampal lesions that affect other forms of spatial learning can learn to find the hidden platform in the Morris water task (Davis et al., 1992). A better indicator of spatial learning is the transfer test in which the animal is trained to find the hidden platform at a specific site in the pool. The platform is then removed, and the number of times that a mouse swims across the target area (a) or the time in the target quadrant (A) is quantitated. There were significant and reproducible differences in transfer test behavior between the mutant and wild-type mice. Wild-type mice crossed the target area 6 ± 0.4

times during a 60 second trial whereas the mutant mice crossed only 4 ± 0.4 times ($p < .002$). The difference in transfer ability was also evident when the time spent in various quadrants was analyzed. Only wild-type mice showed a bias for the target quadrant, A. They spent 42% ± 3.0 of their time in quadrant A searching for the platform. The mutant mice showed no significant preference for quadrant A (27% ± 2.0), indicating an impaired ability in this specific task ($p < .001$). These data illustrate that the mutant mice have a significant and lasting place navigational impairment that is dissociated from visual, motivational, or motor requirements of the test. The relationships between the defects in spatial memory and LTP have not been established.

Molecular Models for the Role of the Calmodulin-Stimulated Adenylyl Cyclases in Synaptic Plasticity

Because various forms of LTP are associated with increased intracellular Ca^{2+}, either presynaptically or postsynaptically, and the I-AC is expressed in the hippocampus, one might expect that activation of NMDA receptors or other Ca^{2+} channels during LTP would elevate cAMP in these regions. Indeed, activation of NMDA receptors or LTP increased cAMP in area CA1 of the hippocampus (Chetkovich et al., 1991) and LTP in the dentate gyrus also increased cAMP (Stanton and Sarvey, 1985). There is also evidence that adenylyl cyclases, cAMP, and cAMP-dependent protein kinases may play an important role in some forms of LTP, particularly in the hippocampus. For example, stimulation of adenylyl cyclase activity in the dentate gyrus by norepinephrine produced LTP (Stanton and Sarvey, 1985; Hopkins and Johnston, 1988). Decremental LTP (D-LTP) in the CA1 persists only 1–2 hours and is initiated by a single train of high-frequency stimulation, whereas L-LTP requires multiple trains of high-frequency stimulation and lasts up to 10 hours. Since D_1 dopamine antagonists blocked L-LTP and D_1 receptors are coupled to stimulation of adenylyl cyclase (Frey et al., 1991), PKA may play a pivotal role in L-LTP. In fact, dibutyryl cAMP induced increases in synaptic efficacy in the CA1 region of the hippocampus (Slack and Pockett, 1991) and L-LTP in the CA1 was blocked by Rp-cAMPS, an inhibitor of PKA (Frey et al., 1993). Furthermore, Sp-cAMPS, which activates PKA, produced L-LTP in the CA1. Finally, mossy fiber/CA3 LTP was inhibited by blockers of the cAMP signal transduction system and cAMP mimicked tetanus-induced mossy fiber LTP.

What is the molecular role of Ca^{2+}-stimulated adenylyl cyclases for various forms of LTP? Some forms of LTP may be due, in part, to cAMP control of transcription through DNA elements such as the cAMP response element (CRE). For example, L-LTP, but not D-LTP, in the CA1 stimulates CRE-mediated transcription (Impey et al., 1996). Synergistic stimulation of adenylyl cyclases by Ca^{2+} and neurotransmitters or PKC may produce exceptionally strong or prolonged cAMP signals required for stimulation of transcription. Indeed, stimulation of transcription by cAMP, which requires the nuclear translocation of cAMP-dependent protein kinase (Nigg et al., 1985; Hagiwara et al., 1993), requires higher or more persistent cAMP signals than other cAMP-regulated events, particularly in neurons. Stimulation of nuclear translocation of PKA in *Aplysia* neurons (Bacskai et al., 1993) and serotonin stimulation of transcription through CRE (Kaang et al., 1993) are relatively slow processes that require multiple doses of serotonin for adenylyl cyclase activation. Robust cAMP signals may be required for transcriptional control in neurons because a significant cAMP gradient must be established from the synapse to the cell body. Elevated cAMP signals arising from synergistic activation of the I-AC by Ca^{2+} and neurotransmitters, or other signals, may play an important role in synaptic plasticity. For example, I-AC coupled Ca^{2+} to CRE-mediated transcription and simultaneous activation by Ca^{2+} and β-adrenergic receptors caused synergistic stimulation of CRE-mediated transcription in HEK-293 cells and cultured neurons (Impey et al., 1994).

The coupling of the Ca^{2+} and cAMP systems also may result in simultaneous or sequentially ordered activation of Ca^{2+} and cAMP-stimulated protein kinases, or provide positive feedback regulation of Ca^{2+} channels by PKA. For example, Nicoll and colleagues have proposed that mossy fiber/CA3 LTP may depend upon Ca^{2+} stimulation of I-AC (Weisskopf et al., 1994). Entry of Ca^{2+} into the presynaptic terminal is hypothesized to activate I-AC and causes a persistent increase of glutamate release through PKA. The fact that I-AC mutant mice show depressed mossy fiber/CA3 LTP is consistent with this hypothesis. All of these mechanisms are dependent upon the unique property of the I-AC to integrate multiple signals for modification of synaptic function.

Developmental Expression of I-AC

Synaptogenesis and the expression of LTP in rodents occur during the first 3 weeks following birth, with a gradual decline at later stages of development (Pokorny and Yamamoto, 1981; Harris and Teyler, 1984; Teyler et al., 1989). Because I-AC may be important for regulation of synaptic plasticity, basal and Ca^{2+}-stimulated adenylyl cyclase activities were measured in rodent brains during the first 3 weeks of postnatal development (Villacres et al., 1995). Ca^{2+}-stimulated adenylyl cyclase activity in membranes from the rat hippocampus increased 5.5-fold between PD1 and PD16 and declined after PD16. Although basal activity also increased during this period, the relative increase in Ca^{2+}-stimulated adenylyl cyclase activity was greater.

The developmental increase in Ca^{2+}-stimulated adenylyl cyclase in rat brain could be due to changes in gene expression or modulation of adenylyl cyclase activity by other proteins. Consequently, I-AC mRNA was quantitated in brains from PD2, PD8, and PD16 rats by Northern analysis. In these experiments, poly(A)$^+$-selected RNA was isolated from the hippocampus, cerebellum, or whole rat brain and the amount of I-AC mRNA was normalized to EF-1a mRNA. In the hippocampus, cerebellum, or whole brain, I-AC mRNA increased during the period from P2 to P16. In the hippocampus, it increased 7-fold between PD2 and PD16. This is consistent with the increase in Ca^{2+}-stimulated adenylyl cyclase activity in this tissue. In contrast, VIII-AC mRNA increased 2-fold between PD2 and PD16, suggesting that the developmental increase in Ca^{2+}-stimulated adenylyl cyclase activity in brain is due primarily to expression of I-AC.

To evaluate the contribution of I-AC to developmental increase in Ca^{2+}-stimulated activity, adenylyl cyclase activity was measured in the cerebellum from developing wild-type and I-AC mutant mice. Like rats, wild-type mice showed a significant increase in basal and Ca^{2+}-stimulated adenylyl cyclase activity during the first 2 weeks of postnatal development. The developmental increase in Ca^{2+}-stimulated activity was greater than basal activity (6.5- vs. 3.0-fold). The mutant mice showed only a 2-fold increase in Ca^{2+}-stimulated adenylyl cyclase activity between P2 and P16 and no significant increase in basal activity. Changes in expression of the I-AC during the period of long-term potentiation development are consistent with the hypothesis that this enzyme is important for neuroplasticity in vertebrates.

The Type VIII Adenylyl Cyclases

The only other known Ca^{2+}/CaM-stimulated adenylyl cyclase besides I-AC is VIII-AC. Since this enzyme was only recently discovered (Cali et al., 1994), there is less known about its regulatory properties and physiological functions. VIII-AC is stimulated by Ca^{2+} and CaM in vitro, but its Ca^{2+}-sensitivity is approximately 4–5 times lower than I-AC (Villacres et al., 1995). It is also stimulated by increases in intracellular Ca^{2+}, presumably through CaM, although the mechanism for Ca^{2+} stimulation in vivo has not been defined. Like I-AC, VIII-AC is not stimulated by G_s-coupled receptors in vivo, even

though it is stimulated by GTP-activated G_{sa} in vitro. In contrast to I-AC, VIII-AC is not synergistically stimulated by Ca^{2+} and G_s-coupled receptors in vivo. The activity of I-AC is attenuated by G_i-coupled receptors and CaM-kinase IV in vivo; VIII-AC activity is not. VIII-AC apparently functions as a pure Ca^{2+} detector and responds to relatively high concentrations of intracellular Ca^{2+} in the micromolar range.

Physiological Role(s) of VIII-AC

Although the physiological function of VIII-AC is not known, it may also contribute to some forms of synaptic plasticity. For example, it may respond to prolonged, high-level Ca^{2+} signals generated by the stimulus paradigm used to generate L-LTP in the hippocampus. In this respect, it is interesting that I-AC mutant mice showed quantitative defects only in the early stage of CA1 L-LTP but not in the later stages of L-LTP which depend upon transcription. The two Ca^{2+}-stimulated adenylyl cyclases may both contribute to CA1 LTP as complimentary activities responding to low and high Ca^{2+} signals. I-AC may be particularly important for amplifying initial Ca^{2+} signals, whereas the VIII-AC may respond to higher Ca^{2+} signals and couple Ca^{2+} to cAMP-stimulated transcription. The relative contribution of the two Ca^{2+}-sensitive adenylyl cyclases for various types of synaptic plasticity will be more clearly defined when VIII-AC mutant mice become available.

Localization of VIII-AC

Since VIII-AC mRNA has been detected only in brain, but not heart, kidney, liver, testes, or skeletal muscle, it also may be a neurospecific adenylyl cyclase (Cali et al., 1994). In situ hybridization data indicate that VIII-AC is expressed in the CA1–3 pyramidal cell layer and the granule cell layer of the dentate gyrus. VIII-AC mRNA is also detected in the neocortex, thalamus, and hypothalamus.

The Type III Adenylyl Cyclases

In contrast to the type I and VIII adenylyl cyclases, which are highly stimulated by Ca^{2+} and CaM in vitro, III-AC shows little stimulation by Ca^{2+} and CaM. To determine if Ca^{2+} and receptor-activated G_s synergistically stimulate III-AC in membranes, the sensitivity of the enzyme to CaM and Ca^{2+} was analyzed in the presence of glucagon. In membrane preparations, III-AC was stimulated by glucagon with an EC_{50} of 7 nM. Glucagon-stimulated III-AC activity was enhanced only 45% by CaM and Ca^{2+} and the EC_{50} for glucagon was not significantly

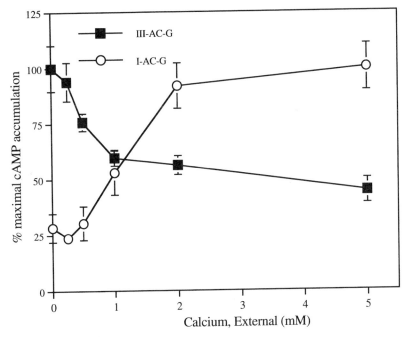

Figure 16.4 Ca^{2+} concentration dependence for inhibition of glucagon-stimulated III-AC activity in vivo. HEK-293 cells stably expressing I-AC (I-AC-G) or III-AC (III-AC-G) were treated with 100 nM glucagon, 10 μM A23187, and increasing concentrations of $CaCl_2$j, and then cAMP accumulation was measured. III-AC and I-AC data are presented as percentage of the ratio (cAMP/[ATP + ADP + AMP]) $\times$ 100 with no added $CaCl_2$ and are the mean $\pm$SD of triplicate assays. Data are from Wayman and Storm (1995).

affected by CaM. Nevertheless, these data suggest that Ca^{2+} and hormones might synergistically activate III-AC in vivo. Therefore, the Ca^{2+}-sensitivity of hormone-stimulated III-AC was evaluated in intact HEK-293 cells. Glucagon stimulated III-AC 222-fold in vivo. However, increases in intracellular Ca^{2+}, generated by A23187 and extracellular Ca^{2+}, inhibited glucagon-stimulated III-AC activity 60% and had no effect on basal activity. Isoproterenol and forskolin-stimulated III-AC activities were also inhibited by increases in intracellular Ca^{2+} caused by A23187 or carbachol. Carbachol inhibition of isoproterenol-stimulated III-AC activity was insensitive to pertussis toxin and therefore not mediated by G_i.

The Ca^{2+} dependencies for inhibition of III-AC and stimulation of I-AC in vivo were compared using A23187 and varying amounts of extracellular Ca^{2+} (Fig. 16.4). Glucagon-stimulated III-AC activity was inhibited by Ca^{2+} concentrations which stimulated I-AC, and the curves were almost mirror images of each other. The concentration of free intracellular Ca^{2+} for half-maximal inhibition of glucagon-stimulated III-AC activity was estimated at 150–200 nM.

Role of Inhibition of III-AC by CaM-kinase II

Ca^{2+} inhibition of III-AC might be due to the action of one of the Ca^{2+}-sensitive protein kinases. This question was addressed by examining the effect of several protein kinase inhibitors on Ca^{2+} inhibition. The PKA inhibitors H89 and Rp-cAMP, and calphostin C (an inhibitor of protein kinase C), did not affect Ca^{2+} inhibition of III-AC. KN-62, a specific inhibitor of CaM-kinases, blocked Ca^{2+} inhibition of III-AC. Calmidazolium, a CaM antagonist, also blocked Ca^{2+} inhibition. These data suggested that Ca^{2+} activation of CaM-kinases may contribute to Ca^{2+} inhibition of III-AC.

To determine if CaM-kinase II inhibits III-AC activity in vivo, stable transfectants that express CaM-kinase II under the control of a metallothionein promoter were made. The CaM-kinase II used (KII290) contained a point mutation that truncates the protein, removes its autoinhibitory domain, and makes it constitutively active (Matthews et al., 1994). These cells were transiently transfected with a construct that encodes III-AC and the sensitivity of the adenylyl cyclase to CaM-kinase II was evaluated by

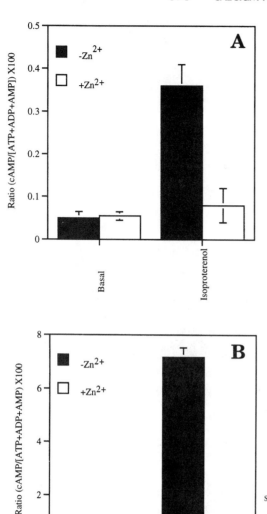

Figure 16.5 Inhibition of isoproterenol and forskolin-stimulated III-AC by calmodulin-dependent protein kinase II-290. HEK-293 cells stably transfected with the Zn^{2+}-inducible, constitutively active CaM-kinase II-290, which were transiently transfected with III-AC, were exposed to either (A) isoproterenol or (B) forskolin ± induction of CaM-kinase II-290 by Zn^{2+}. CaM-kinase II-290 is under control the Zn^{2+}-inducible metallothionein promoter. Data are from Wayman and Storm (1995).

inducing the expression of the kinase with Zn^{2+}. Zn^{2+} treatment of cells that do not express KII290 had no effect on basal, isoproterenol, or forskolin-stimulated activities (Figure 16.5). However, induction of CaM-kinase II activity in KII290 cells completely inhibited isoproterenol and forskolin-stimulated activities. Therefore, Ca^{2+} inhibition of III-AC in vivo may be mediated by CaM-kinase II.

Evidence for phosphorylation of III-AC by CaM-kinase II has come from in vivo phosphorylation studies using an antibody specific for III-AC (Wei et al., 1996). To determine if III-AC is phosphorylated by CaM-kinases in vivo, HEK-293 cells were preloaded with $[^{32}P]P_i$ to label ATP. The cells were treated with A23187 to increase intracellular Ca^{2+}, the membranes were solubilized, the enzyme was immunoprecipitated, and then analyzed for phosphorylation by autoradiography of SDS gels. Treatment with the Ca^{2+} ionophore A23187 resulted in phosphorylation of III-AC which was blocked by

KN-93 and KN-62. Since HEK-293 cells express CaM-kinase II but not CaM-kinase IV, these data strongly suggest that the enzyme is directly phosphorylated by CaM-kinase II in vivo.

The physiological significance of Ca^{2+} inhibition of hormone-stimulated III-AC activity remains to be established. Adenylyl cyclase activity in most tissues is inhibited by millimolar levels of Ca^{2+} and this has been attributed to formation of complexes between ATP and Ca^{2+}, or binding of Ca^{2+} to a Mg^{2+}-regulatory site on adenylyl cyclases (Steer and Levitzki, 1975). Heart muscle expresses adenylyl cyclase activity that is inhibited by sub-micromolar levels of Ca^{2+} and III-AC is expressed in heart (Xia et al., 1992). This may provide a mechanism whereby the positive ionotropic and chronotropic effects of β-adrenergic agonists are attenuated by increased intracellular Ca^{2+}.

Type III Adenylyl Cyclase and Olfactory Transduction

Although III-AC is expressed in several tissues, including heart and brain (Xia et al., 1992), it is most abundant in olfactory tissue (Bakalyar and Reed, 1990). Since III-AC is stimulated by G_s-coupled receptors in vivo and inhibited by Ca^{2+}, it seems likely that its regulatory properties may explain at least part of the kinetics for odorant-induced cAMP changes in olfactory cilia. Various odorants stimulate rapid cAMP increases in olfactory cilia that rise and fall within milliseconds to seconds depending upon the preparation examined and the technique used to measure the kinetics for cAMP changes (Jaworsky et al., 1995). These increases in cAMP are likely to be due to stimulation of the III-AC and other adenylyl cyclases through G_s-coupled olfactory receptors. There are several possible mechanisms for the subsequent decreases in cAMP which include the actions of CaM-dependent cyclic nucleotide phosphodiesterases, which are discussed later in this chapter. Since intracellular Ca^{2+} is elevated during odorant exposure (Restrepo et al., 1990; Hirono et al., 1992), Ca^{2+} inhibition of III-AC activity and stimulation of CaM-sensitive phosphodiesterases may both contribute to the biphasic cAMP response.

Hormone Stimulation of III-AC Induces Ca^{2+} Oscillations

Although intracellular free Ca^{2+} can affect cAMP levels by modulation of adenylyl cyclase or phosphodiesterase activities, cAMP can also affect intracellular Ca^{2+} by regulating Ca^{2+} ion channel activity. Since activation of PKA can increase intracellular free Ca^{2+}, and hormone stimulation of III-AC is inhibited by Ca^{2+}, one might expect Ca^{2+} hormone stimulation of III-AC to generate Ca^{2+} oscillations. This question was addressed using HEK-293 cells expressing the glucagon receptor and III-AC (Wayman et al., 1995a). HEK-293 cells that stably express the glucagon receptor (293-G), the glucagon receptor with I-AC (I-AC-G), or III-AC (III-AC-G) were treated with glucagon and individual cells were Ca^{2+}-imaged using fura-2 (Fig. 16.6). Treatment of 293-G cells with glucagon caused a single spike of intracellular Ca^{2+} (Fig. 16.6A). Cells that express I-AC and the glucagon receptor gave a similar response: a single peak of Ca^{2+} with no additional increase with subsequent exposures to glucagon (Fig. 16.6B). In contrast, Ca^{2+} oscillations were generated when cells that express the glucagon receptor and III-AC were treated with glucagon (Fig. 16.6C). These oscillations were dependent upon the continued presence of glucagon and were not generated by transient exposure to the hormone. Exposure of III-AC-G cells to isoproterenol or forskolin also caused Ca^{2+} oscillations that strongly resembled those induced by glucagon. Ca^{2+} oscillations were dependent upon the activity of PKA and CaM-kinases but not solely due to cAMP increases since dibutyryl-cAMP or Sp-cAMP did not stimulate Ca^{2+} oscillations. Although Ca^{2+} oscillations were not dependent upon extracellular Ca^{2+}, they were blocked when the IP_3-sensitive Ca^{2+} pools were depleted.

What is the mechanism for hormone-stimulated Ca^{2+} oscillations in III-AC-G cells? The data are most consistent with the model schematically depicted in Fig. 16.7. When III-AC is activated by hormones, cAMP stimulates PKA which phosphorylates and activates IP_3 receptors. As intracellular Ca^{2+} rises, III-AC activity is attenuated by CaM-kinase II and intracellular cAMP levels decrease because of cAMP phosphodiesterases. When cAMP levels drop below a threshold point and the IP_3 receptor is dephosphorylated, Ca^{2+} is resequestered and the cycle is repeated if III-AC is chronically exposed to an activator such as forskolin or glucagon. The periodicity of the oscillations may depend upon the concentrations of adenylyl cyclase, phosphodiesterases, and protein kinases found in a particular cell. This model also predicts that other cells expressing III-AC may also show cAMP oscillations.

Ca^{2+}/CaM and Phosphodiesterase Activity

The intracellular cyclic AMP level is controlled not only by its rate of synthesis, catalyzed by adenylyl cyclases, but also by its rate of degradation, catalyzed by one or more cyclic nucleotide phosphodiesterases (PDEs). In most tissues, the maximal capacity for degradation of cyclic nucleotide is much greater

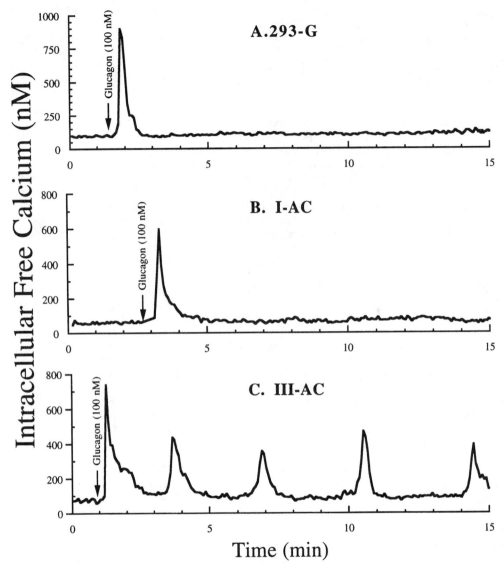

Figure 16.6 Glucagon stimulation of Ca^{2+} oscillations in HEK-293 cells expressing the glucagon receptor and III-AC. Cells expressing the rat glucagon receptor (293-G), or the glucagon receptor with I-AC (I-AC-G), or III-AC (III-AC-G) were treated with 100 nM glucagon and Ca^{2+} imaged using Fura-2. Data are from Wayman et al. (1995a).

than for synthesis. Therefore, most of PDE activity in cells must be controlled at less than maximal capacity. It is increasingly clear that cyclic nucleotide degradation is not a constitutive function of the cell, but it is regulated by different mechanisms in different physiological conditions. At least nine different families of PDE isoenzymes exist in higher eukaryotes. These are the Ca^{2+}/CaM-stimulated PDE (PDE1 family), the cGMP-stimulated PDE (PDE2 family), the cGMP-inhibited PDE (PDE3 family), the high-affinity/rolipram-sensitive cAMP PDE (PDE4 family), the cGMP-specific and cGMP-binding PDE (PDE5 family), the light-stimulated photoreceptor PDE (PDE6 family), the high-affinity/rolipram-insensitive cAMP PDE (PDE7 family, the cAMP-specific PDE8 family, and the

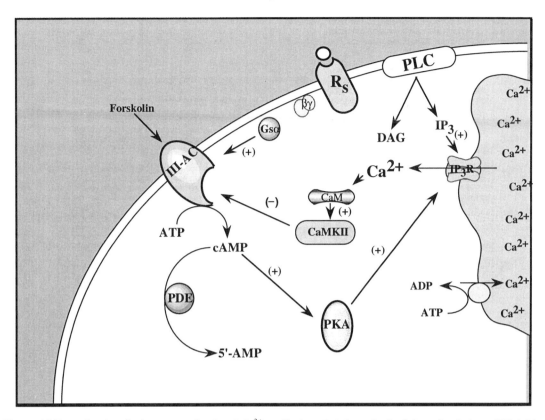

Figure 16.7 Mechanism for hormone-stimulated Ca^{2+} oscillations. It is hypothesized that stimulation of III-AC by hormones or forskolin leads to activation of PKA, stimulation of IP_3 receptors, and increases in intracellular Ca^{2+}. As intracellular Ca^{2+} increases, III-AC activity is inhibited by CaM-kinases and cAMP levels are decreased by cAMP phosphodiesterases. When cAMP drops below a threshold level, Ca^{2+} is resequestered and the cycle is repeated as long as activators of III-AC are present. Abbreviations: R_s, adenylyl cyclase stimulatory receptor; PLC, phospholipase C; DAG, diacylglycerol; IP_3, inositol triphosphate; IP_3R, IP_3 receptor/channel; PDE, cAMP phosphodiesterase. (Figure is from Wayman et al. 1995a with permission.)

high affinity cGMP-specific PDE9 family). Most of these families contain multiple genes and most of the mRNAs from these genes can undergo tissue-specific alternative splicing to yield different translation products. Therefore, there are a number of ways to generate distinct PDE isozymes. More importantly, the individual PDE isozymes often have different kinetic properties and respond to different signals (for a comprehensive review, see Beavo, 1995). Thus, in addition to the selective and differential expression of the receptors and cyclases, to which they are coupled, the great diversity of PDEs leads to distinct regulatory pathways for the precise control of cAMP in each cell type. Moreover, like the cyclases, the PDEs can be regulated by multiple stimuli, thereby providing additional means of integration between the cyclic nucleotide and other signal transduction pathways. For example, regulatory properties of the PDE1 family provide a mechanism by which signal transduction pathways that generate Ca^{2+} can alter cyclic nucleotide metabolism. Similarly, the PDE2 family and PDE3 family provide mechanisms by which cGMP can alter the concentration of cAMP in the cell in either a positive or a negative manner, respectively.

The following sections give a brief review on the control of cyclic nucleotide degradation by Ca^{2+}, with particular emphasis on the role of Ca^{2+}/CaM-stimulated PDEs (CaM-PDEs). The first part deals with some general properties of CaM-PDEs. In the second part, the role of CaM-PDEs in mediating cross-talk between cyclic nucleotide and calcium signal systems is discussed by considering a number of separate examples.

```
MMPDE1A2   ...MGSTDTD IEELENATYK YLIGEQTEKM WQRLKGI... ......LRCL VKQLEKGDVN VVDLKKNIEY AASVLEAVYI   68
MMPDE1C1   ...MESPTKE IEEFESNSLK HLQPEQIEKI WLRLRGLRKY KKTSQRLRSL VKQLERGEAS VVDLKKNLEY AATVLESVYI   77
MMPDE1B1   MELSPRSPPE MLESDCPSPL ELKSAPSKKM WIKLRSL... ......LRYM VKQLENGEVN IEELKKNLEY TASLLEAVYI   71
           - - * - - _       *         * *- * -*--* ** _ ***** *-- _ -******* *--** ***

MMPDE1A2   DETRRLLDTE DELSDIQTDS VPSEVRDWLA STPTRKMGMM KKKPEEKPKF RSIVHAVQAG IFVERMYRKN YHMVGLTYPA   148
MMPDE1C1   DETRRLLDTE DELSDIQSDA VPSEVRDWLA STPTRQMGMM LRRSDEKPRF KSIVHAVQAG IFVERMYRRT SNMVGLSYPP   157
MMPDE1B1   DETRQILDTE DELRELRSDA VPSEVRDWLA STPTQQTRAK GRRAEEKPKF RSIVHAVQAG IFVERMPRRT YTSVGPTYST   151
           **** ***** *** _* _*  ********** ****    _      -- _***_* _********* ******_*_  ** _*

MMPDE1A2   AVIVTLKEVD KWSPDVFALN EASGEHSLKF MIYELPTRYD LINRFKIPVS CLIAPAEALE VGYSKHKNPY HNLVHAADVT   228
MMPDE1C1   AVIDALKDVD TWSPDVFSLN EASGDHALKF IFYELLTRYD LISRFKIPIS ALVSFVEALE VGYSKHKNPY HNLMHAADVT   237
MMPDE1B1   AVHNCLKNLD LWCFDVFSLN RAADDHALRT IVFELLTRHS LISRFKIPTV FLMSFLEALE TGYGKYKNPY HNQIHAADVT   231
           **   ** _*   * ******* *   _* **  - _** ** ** ***   *- *-**** **_* **** ** _******

MMPDE1A2   QTVHYIMLHT GIMHWLTELE ILAMVPAAAI HDYEHTGTTN NPHIQTRSDV AILYNDRSVL ENHHVSAAYR LMQE.EEMNI   307
MMPDE1C1   QTVHYLLYKT GVANWLTELE IPAIIFSAAI HDYEHTGTTN NPHIQTRSDP AILYNDRSVL ENHHLSAAYR LLQEDEEMNI   317
MMPDE1B1   QTVHCFLLRT GMVHCLSEIE VLAIIFAAAI HDYEHTGTTN SPHIQTKSEC AILYNDRSVL ENHHISSVFR MMQD.DEMNI   310
           ****   _   * **_- *_*** - * *__* *** ********** ****_*_  ********** ****_* __* __*_ _****

MMPDE1A2   LVNLSKDDWR DLRNLVIEMV LATDMSGHFQ QIKNIRNSLQ QPEGIDRAKT MSLILHAADI SHPAKTWKLH YRWTMALMEE   387
MMPDE1C1   LVNLSKDDWR EFRTLVIEMV MATDMSCHFQ QIKAMKTALQ QPEAIEKPKA LSLMLMHTADI SHPAKAWDLH HRWTMSLLEE   397
MMPDE1B1   FINLTKDEPA ELRALVIEMV LATDMSCHFQ QVKTMKTALQ QLERIDKSKA LSLLLHAADI SHPTKQWSVH SRWTKALMEE   390
           -**-**__  - * ****** _***** ***  *_* __  ** * * *_* ***-** *** *** * _*  *** *_**

MMPDE1A2   FFLQGDKEAE LGLPFSPLCD RKSTMVAQSQ IGPFIDFIVEP TFSLLTDSTE KIVIPLIEEA SKSQSSNYGA SSSSTMIGFH   467
MMPDE1C1   FFRQGDKEAE LGLPFSPLCD RKSTMVAQSQ VGPFIDFIVEP TFTVLTDMTE KIVSPLIDES SQTGG...TG QRRSSLNSIN   474
MMPDE1B1   FFRQGDKEAE LGLPFSPLCD RTSTLVAQSQ IGPFIDFIVEP TFSVLTDVAE KSVQPLADDD SKPKS..... ..........   455
           ** ***_*** ********** * **_***** _******** **_*** * *  **__**   -

MMPDE1A2   VADSLRRSNT KGSVCDGSYA PDYSLSAVDL KSFKNNLVDI IQQNKERW.. .........K ELAAQGELDL HKNSEELGNT   536
MMPDE1C1   SSDA.KRSGV KSSGSDGSAP INNSVIPVDY KSFKATWTEV VQINRERWRA KVPKEEKAKK EAEEKARLAA EEKQKEMEAK   553
MMPDE1B1   .QPSFQWR.. .....QPSLD VDVGDPNPDV VSFRATWTKY IQENKQKWK. .....ERAAS GITNQMSIDE LSPCEEEAPS   521
                       *      -  *  **-  -* *-_-*       -      -        *

MMPDE1A2   EEKHADTRP* .......... .......... .......... .......... .......... .......... ..........   545
MMPDE1C1   SQAEQGTTSK GEKKTSGEAK SQVNGTRKGD NPRGKNSKGE KAGEKQQNGD LKDGKNKADK KDHSNTGNES KKTDDPEE*   631
MMPDE1B1   SPAEDEHNQN GNLD*..... .......... .......... .......... .......... .......... ..........   535
```

Figure 16.8 Amino acid sequence alignment and comparison. Comparison of deduced amino acid sequences from cDNAs encoding mouse PDE1B1 (L01111695), mouse PDE1A2 (L56649), and mouse PDE1C1 (L76944). Positions which are identical for all three isozymes are indicated by stars. Hyphens represents conservative substitutions. Periods are gaps generated to achieve best sequence alignment. The catalytic domain is enclosed in the box.

General Properties of CaM-PDEs

Multiple CaM-PDE Isozymes

All CaM-PDEs that have been isolated and characterized are activated by the Ca^{2+}/CaM complex in cell-free systems. At present, at least six different CaM-PDE isoforms, 59, 61, 63, 68, and 75 kDa as well as olfactory-enriched forms, have been reported (Beavo, 1995). These isoforms differ in subunit composition, molecular weight, and substrate kinetics (reviewed in Beavo, 1995). Three different genes in the CaM-PDE family have been isolated, which are named as PDE1A (Sonnenburg et al., 1993), PDE1B (Bentley et al., 1992; Polli and Kincaid, 1992; Repaske et al., 1992), and PDE1C (Loughney et al., 1994; Yan et al., 1995). Each of the genes may encode more than one alternative splice variant. Two splice variants of the *PDE1A* gene (PDE1A1 and PDE1A2) and five splice variants of the *PDE1C* gene (PDE1C1, -1C2, -1C3, -1C4, and -1C5) have been cloned (Sonnenburg et al., 1993, 1995; Yan et al., 1995, 1996). To date, only one product from the *PDE1B*

gene has been identified (Bentley and Beavo, 1992). Comparison of the cDNA sequences with the sequences of purified proteins indicated that the 59 kDa bovine heart and 61 kDa bovine brain isozymes, with differences only in their N-terminus, are encoded by PDE1A1 and PDE1A2, respectively (Charbonneau et al., 1991; Novack et al., 1991; Sonnenburg et al., 1993, 1995). The bovine brain 63 kDa CaM-PDE is encoded by the *PDE1B* gene (Bentley et al., 1992; Polli and Kincaid, 1992; Repaske et al., 1992). PDE1C2, which is highly expressed in olfactory sensory neurons, is most likely to encode the high-affinity CaM-PDE form detected in olfactory cilia in early studies (Borisy et al., 1992; Yan et al., 1995). It is possible that PDE1C5 mainly represents the high-affinity CaM-PDE activity in the testis 68–70 kDa CaM-PDE preparations (Purvis and Rui, 1988; Rossi et al., 1988).

Sequence alignment among several representative PDE1 family members indicates that they are very homologous (Fig. 16.8). There are several regions where more than 10 consecutive amino acids are

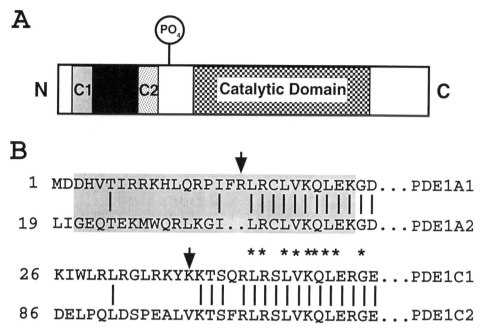

A

B

Figure 16.9 Domain organization. (A) Domain map of PDE1A1 and Y-1A2 isozymes. The boxes labeled C1 and C2 indicate the first putative CaM-binding domain and the second CaM-binding domain, respectively (Charbonneau et al., 1991; Novack et al., 1991; Sonnenburg et al., 1995). The inhibitory domain (Sonnenburg et al., 1995) and the catalytic domain are indicated by the black and stippled boxes, respectively. (B) The N-terminal sequence comparison between PDE1A1 and PDE1A2, or between PDE1C1 and PDE1C2. The amino acids within the gray box are the first CaM-binding domain. The arrows indicate the divergence point in either the PDE1A pair or the PDE1C pair. The stars indicate amino acids that are conserved between PDE1A and PDE1C isozymes. The N-terminal sequences of the PDE1A1 and 1A2 midway through the first CaM-binding domain are divergent. The divergence point between PDE1C1 and Y-1C2 corresponds to that between PDE1A1 and Y-1A2.

identical, suggesting the possibility that cross-immunoreactivities exist among PDE1 family members. This raises questions about many of the studies carried out previously with the antibodies generated with an intact biochemically purified brain CaM-PDE. However, despite the extensive homology between the PDE1 family members, the C-terminal regions are different enough to be used to generate isozyme-specific antibodies.

Structural Organization of CaM-PDEs

As with other PDEs, all CaM-PDEs contain a well-conserved core region of about 250 amino acids which is located toward the C-terminus and functions as a catalytic domain (Charbonneau et al., 1986) (Fig. 16.9A) In addition, all CaM-PDE isozymes contain at least one domain located toward the N-terminus that is involved in their regulation by Ca^{2+}/CaM. Studies on the structure and function of PDE1A isozymes by mutagenesis and peptide com-

petition have characterized two CaM-binding domains (Charbonneau et al., 1991; Novack et al., 1991; Sonnenburg et al., 1995) (Fig. 16.9A). The first CaM-binding domain is located near the N-terminus, while the second one lies approximately 100 residues C-terminal to the first CaM-binding domain. An inhibitory domain which lies between the first and the second CaM-binding domains has also been reported (Sonnenburg et al., 1995). The first binding domain is only partially conserved among different CaM-PDE isozymes and is likely to determine the difference in binding affinities of these enzymes to the Ca^{2+}/CaM complex (Fig. 16.9B). For example, the N-terminal sequences midway through the first CaM-binding domain for the PDE1A1 and Y-1A2, which respond to Ca^{2+}/CaM with different sensitivities, are divergent. Similarly, the divergence point between PDE1C1 and Y-1C2, which also have distinct Ca^{2+} activation constants, corresponds well to the difference between PDE1A1 and Y-1A2. These results further confirm that the first CaM-binding domain determines the binding affinities of CaM-PDEs to the Ca^{2+}/CaM complex. The second CaM-binding domain and the inhibitory domain are highly conserved among all the identified CaM-PDEs, suggesting that they may be involved in a common mechanism for regulating CaM-PDE activity. The molecular events by which the CaM-PDE is activated are not known in any detail. Several hypothetical models have been proposed and a current model suggests that the inhibitory domain maintains the enzymes in a less active state at low Ca^{2+} concentration until binding of a Ca^{2+}/CaM complex to the CaM-PDEs releases the inhibition of the inhibitory domain and thereby activates the enzyme (Sonnenburg et al., 1995).

Kinetic Properties

Substantial variation in the substrate specificity of the previously purified isozymes has been reported. For example, the 59, 61, and 63 kDa CaM-PDEs catalyze cGMP hydrolysis at higher affinities than those for cAMP (Sharma et al., 1984; Sharma and Wang, 1986a, 1986b; Grewal et al., 1989). A 75 kDa isozyme from bovine brain appears to hydrolyze specifically cGMP (Shenolikar et al., 1985). However, the testis 68–70 kDa and olfactory-enriched isoforms have high affinity for both cAMP and cGMP (Rossi et al., 1988; Borisy et al., 1992). Nevertheless, it should be mentioned that many of the kinetic parameters determined using biochemically purified enzymes from various tissues are often complicated by the presence of more than one isozyme in the enzyme preparation because several CaM-PDEs are often expressed in the same tissue. This unquestionably has contributed to the large variation in the kinetic data reported previously from different groups (Wang et al., 1990). Recent molecular cloning and expression of the recombinant enzymes have allowed a more precise determination of the kinetic characteristics of individual isozymes (Yan et al., 1996) (Table 16.1).

In general, for all the CaM-PDEs, calmodulin in the presence of Ca^{2+} increases the maximum velocity (V_{max}) 5–10-fold with little effect on the apparent K_m (Sonnenburg et al., 1995). However, several different CaM-PDEs possess notable differences in Ca^{2+} and calmodulin activation properties which are believed to be determined by the first CaM-binding domain in the N-termini of the CaM-PDEs. For example, the 59 kDa isozyme (PDE1A1) has about a 10-fold higher Ca^{2+} or calmodulin activation constant than that of the 61 kDa isozyme (PDE1A2) (Hansen and Beavo, 1986; Sharma and Wang, 1986b; Sharma, 1991; Sonnenburg et al., 1995; Yan et al., 1995). Similarly, among PDE1C splice variants, PDE1C2 has a higher Ca^{2+}-sensitivity than that of the others

(Yan et al., 1996). In addition, Ca^{2+} or calmodulin activation of some isozymes is attenuated by phosphorylation in vitro (Sharma and Wang, 1985, 1986a; Hashimoto et al., 1989; Sharma, 1991). The 63 kDa isozyme serves as a substrate for CaM-dependent protein kinase II and the 59 kDa and 61 kDa isozymes are substrates for the cAMP-dependent protein kinase. In each case, phosphorylation blunts stimulation of enzyme activity by Ca^{2+}/CaM. Recent site-directed mutagenesis studies indicated that the phosphorylation of serine residue 120 by PKA is responsible for the reduction in the CaM-binding affinity of the 61 kDa isozyme (Florio et al., 1994) (Fig. 16.9A).

Tissue-Specific Distribution

CaM-PDE activities have been either purified or described in many different tissues, including brain, heart, olfactory epithelium, pancreas, and testis. The cellular and subcellular distribution of CaM-PDE in the rodent brain has been studied extensively. Using histochemical methods (Ariano and Adinolfi, 1977; Ariano and Appleman, 1979) and immunocytochemical techniques (Kincaid et al., 1987; Balaban et al., 1989; Billingsley et al., 1990), CaM-PDE has been found in most brain regions, but it is highly enriched in neuronal populations and selectively present in the soma and dendrites of these regional output neurons. Electron microscopic immunocytochemical studies further demonstrate a highly localized distribution of CaM-PDE in the postsynaptic region of specific classes of neurons (Ludvig et al., 1991). However, in none of these early studies has it been entirely clear which isozyme was being detected. For example, the histochemical techniques do not differentiate between isozymes, and it has not been certain whether or not the antibodies used in immunochemical studies would interact with more than one isoform.

Using the technique of in situ hybridization with gene-specific probes, the expression of the different CaM-PDE genes in mouse brain has been more precisely mapped recently (Yan et al., 1994, 1996). The CaM-PDE genes exhibit very different and somewhat overlapping distribution patterns. The PDE1A mRNA has a restricted pattern with a high level in the cerebral cortex and in the pyramidal cells of the hippocampus. The expression of *PDE1B* gene, on the other hand, has a wide but uneven distribution with very high levels in the caudate putamen, nucleus accumbens, olfactory tubercle, and the dentate gyrus of the hippocampus. The expression of PDE1C mRNA, however, is found in granule cells and Purkinje cells of the cerebellum, the central amygdaloid nucleus, the caudate putamen, the olfactory bulb, and some regions in the brain stem. In

Table 16.1 Enzymatic Properties of Recombinant CaM-PDes

Genes	cDNAs	Isozyme (kDa) M.W.[b]	$K_m(\mu M)$[c] cAMP	cGMP	V_{max} Ratio[c] (cAMP/cGMP)	$EC_{50}(\mu M)$[d] (for Ca^{2+})
PDE1A	BTODE1A1	59	87.2 ± 11.4	3.5 ± 0.5	2.2 ± 0.1	0.27 ± 0.01
	BTPDE1A2	61	112.7 ± 7.9	5.1 ± 0.6	2.9 ± 0.1	1.99 ± 0.02
PDE1B	BTPDE1B1	63	24.3 ± 2.9	2.7 ± 0.2	0.9 ± 0.1	1.25 ± 0.06
PDE1C	MMPDE1C1	72	3.5 ± 0.3	2.2 ± 0.1	1.3 ± 0.1	3.01 ± 0.04
	RNPDE1C2	87	1.2 ± 0.1	1.1 ± 0.2	1.2 ± 0.1	0.83 ± 0.01
	HSPDE1C3	81	0.3 ± 0.1	0.6 ± 0.1	0.8 ± 0.1	?
	MMPDE1C4/5[a]	74	1.1 ± 0.0	1.0 ± 0.1	1.0 ± 0.0	2.43 ± 0.03

[a]MMPDE1C4 and MMPDE1C5 cDNAs encode the same protein but have different 3′-noncoding sequences (Yan et al., 1996).

[b]The molecular weights (M.W.) for BTPDE1A1, BTPDE1A2, and BTPDE1B1 are the apparent sizes on SDS gels. The M.W. for PDE1C isozymes are predicted according to their cDNA sequences.

[c]The K_m values and V_{max} ratio for HSPDE1C3 were measured with the HSPDE1C3 protein expressed in yeast cell extracts (Loughney et al., 1994), for BTPDE1A1 in High-Five[TM] insect cell extracts (W. K. Sonnenburg et al., unpublished results; Sonnenburg et al., 1995), and for the rest of the isozymes in COS-7 cell extracts (Yan et al., 1996).

[d]The EC_{50} values of Ca^{2+} activation were measured in COS-7 cells extracts (Yan et al., 1996).

addition, all CaM-PDE mRNAs are localized exclusively in neuronal cells.

In addition to brain, the expression of different CaM-PDE genes has been also examined in other tissues, such as olfactory epithelium and testis. In olfactory epithelium, only the PDE1C gene shows substantial expression (Fig. 16.10). PDE1C2 mRNA is exclusively localized in the olfactory sensory neurons and PDE1C2 protein is highly concentrated in the cilia of these neurons (Yan et al., 1995; and unpublished observation from authors' laboratory). In mouse testis, both PDE1A and PDE1C genes are highly expressed in developing germ cells. The expression of each is tightly controlled to different stages of spermatogenesis (unpublished observations from the authors' laboratory).

We can now begin to appreciate why certain cell types prefer to express one particular CaM-PDE isozyme rather than another. One answer can be drawn from the fact that the kinetic and regulatory properties of a particular CaM-PDE fit better than the others to the particular functional requirements of the cell. For example, a CaM-PDE activity (PDE1C2) that has a particularly high affinity for cAMP and for Ca^{2+} has been proposed to play an important role in the rapid Ca^{2+}-regulated olfactory signal termination process (Yan et al., 1995). The high expression levels and high affinity for cAMP of PDE1C2 are well suited for quickly restoring the sub-micromolar level of cAMP after odorant stimulation.

Physiological Roles of CaM-PDEs

Role of CaM-PDEs in Cross-Talk between Ca^{2+} and cAMP Second Messenger Systems

It is thought that, like the CaM-dependent adenylyl cyclases, CaM-PDEs work as a coordinator between the pathways that increase intracellular Ca^{2+} concentration and the pathways that regulate cyclic nucleotide-dependent processes. In many different cell types and tissues, it is found that agonists that increase intracellular Ca^{2+} concentration stimulate cAMP or cGMP degradation. Early examples of this include the effect of muscarinic agonists to modulate cAMP level through activating CaM-PDE activity in 1321 neuroblastoma cells (Tanner et al., 1986), thyroid cells (Erneux et al., 1985), and WI-38 fibroblasts (Nemecek and Honeyman, 1982), as well as in dog thyroid slices (Miot et al., 1983). In NG108-15 cells, it was found that the effect of opioid agonists on inhibition of cAMP accumulation is through a dual mechanism, via both inhibition of an adenylyl cyclase activity and activation of a CaM-PDE activity (Law and Loh, 1993). A similar dual mechanism probably exists in many different cell types.

Definitive experimental evidence demonstrating physiologically relevant activation of CaM-PDEs in the nervous system is not as well developed. However, the high levels of CaM-PDE activity existing in several regions suggest important physiologi-

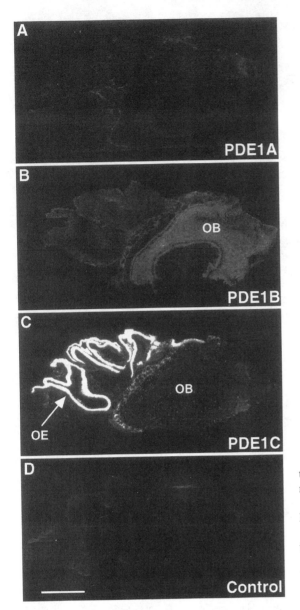

Figure 16.10 Selective expression of PDE1C in olfactory epithelium. Dark-field images of sagittal sections of a mouse nasal cavity and olfactory bulb after hybridization with antisense riboprobes specific to (A) PDE1A, (B) PDE1B, and (C) PDE1C. (D) The section hybridized with a corresponding PDE1C sense probe. The results using the sense probes for PDE1A and Y-1B are similar to those shown in panel (D) (data not shown). Only PDE1C is highly expressed in olfactory epithelium. Abbreviations: OB, olfactory bulb; OE, olfactory epithelium.

cal roles for these enzymes in the regional regulation of cyclic nucleotides through activation of Ca^{2+} pathways. For example, in the striatum, several different types of dopamine receptors are prominently expressed. The high level of CaM-PDE1B mRNA there suggest that this PDE1B isozyme modulates dopamine functions by attenuating the magnitude and duration of dopamine-induced cAMP signals. D_1-like receptors have been shown to couple not only to adenylyl cyclase but also to increased inositol trisphosphate (IP$_3$) and, consequently, Ca^{2+} levels.

As discussed earlier in relation to type III adenylyl cyclase, an important characteristic of olfactory signal transduction is its transient nature. Following stimulation of odorants, cAMP levels decline nearly as rapidly (within about 200 milliseconds) as they rise (within 50–100 milliseconds) (Breer et al., 1990). Shut-off of cAMP synthesis may be achieved by multiple mechanisms, including the phosphorylation of odorant receptors by protein kinase A (Boekhoff and Breer, 1992; Boekhoff et al., 1992), or β-adrenergic receptor kinase (Dawson et al., 1993; Schleicher et al., 1993). In addition, recent studies strongly indi-

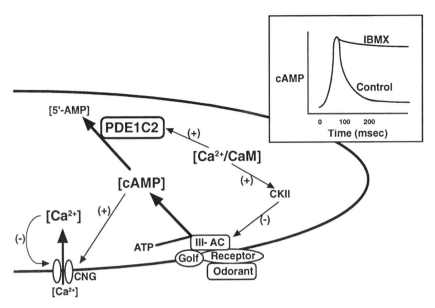

Figure 16.11 Model for the role of PDE1C2 in odorant signal transduction. The olfactory transduction initially involves the binding of an odorant to an odorant receptor. Activation of the receptor is transduced via G_{olf} to III-AC. Activation of the adenylyl cyclase results in an increase in the cAMP concentration in the cilia. Opening of cyclic nucleotide-gated (CNG) channels as a result of elevation of cAMP allows a Ca^{2+} influx into the cilia. The rapid termination of the response to the stimulus is proposed to be partially due to the stimulation of cAMP hydrolysis, which is mediated by the CaM-PDE, PDE1C2. In addition, other mechanisms have also been suggested for the termination of the response, such as the decrease in the binding affinity of CNG channels to cAMP at the high level of Ca^{2+} and probably indirect inhibition of III-AC via activation a CaM-dependent kinase II (CKII). Inset: theoretical time course for cAMP levels in olfactory cilia, which is similar to actual data obtained by Breer et al. (1990).

cate that the rapid decline of odorant-stimulated cAMP also must involve hydrolysis of cAMP by PDE(s) (Firestein et al., 1991; Boekhoff and Breer, 1992). The predominant PDE activity in olfactory cilia is CaM-dependent (Borisy et al., 1992). In addition, the role of Ca^{2+} in control of cAMP signal termination (Kurahashi, 1989; Kurahashi and Shibuya, 1990; Jaworsky et al., 1995) is consistent with the idea that the elevated Ca^{2+} level after odorant stimulation provides an important control for signal termination via stimulation of CaM-PDE activity (Jaworsky et al., 1995; Yan et al., 1995). The major CaM-PDE found in olfactory cilia (PDE1C2) has a high affinity for cAMP and therefore is well suited for restoring the sub-micromolar level of cAMP after odorant stimulation (Yan et al., 1995). Figure 16.11 illustrates a model in which entry of Ca^{2+} through the olfactory cyclic nucleotide-gated (CNG) channel after odorant stimulation provides an important regulatory control for olfactory cAMP termination.

Possible Roles for CaM-PDEs in Synaptic Plasticity

One interesting possible physiological role for CaM-PDE is in synaptic transmission. As discussed above, substantial evidence now exists implicating cAMP and the CaM-sensitive adenylyl cyclases as having an important role in various types of synaptic plasticity. More recently, attention has begun to be focused on a possible complementary role for the CaM-dependent PDEs in this process. Some of the earliest data for a role of cAMP and cAMP-PDE activity came from one of the original *Drosophila* learning mutants, the so-called DUNCE mutant. This genetic defect was eventually traced to a mutation in one of the two major cyclic nucleotide phosphodiesterases present in this organism. Substantial evidence for a role of PDE4 in this process has now been developed.

A role for the CaM-dependent PDE1s in LTP or other forms of synaptic plasticity is much more speculative. However, a very reasonable model based on the localization and phosphorylation data described above can be made. First, PDE1A and PDE1B are highly expressed in neurons of the hippocampus, the cerebellum, and the cortex, all areas where cyclic nucleotide modulation of synaptic efficacy has been noted. Second, as described in more detail below, a biochemical "feed forward" regulation cycle has been noted for both PDE1A and -1B, just as it has for autophosphorylation and activation of CaM-kinase II, another molecule implicated in synaptic function.

Regulation of CaM-PDEs by Phosphorylation/Dephosphorylation

The activation of several of the CaM-PDEs by Ca^{2+} and CaM can be modified by phosphorylation/ dephosphorylation in vitro. The PDE1A1 and Y-1A2 isozymes are phosphorylated by cAMP-dependent protein kinase (PKA) (Sharma and Wang, 1985), and PDE1B1 is phosphorylated by CaM-kinase II (Hashimoto et al., 1989). In all cases, the phosphorylation results in a decrease in the enzyme's affinity for Ca^{2+} and calmodulin. Conversely, the phosphorylation of CaM-PDE is blocked by Ca^{2+} and calmodulin. In addition, the phosphorylated CaM-PDEs can be dephosphorylated by a CaM-dependent phosphatase (Sharma and Wang, 1985). In general, it is thought that this complex regulation of phosphodiesterase allows for a greater versatility in the interaction between Ca^{2+} and cyclic nucleotides.

These in vitro studies suggest that the inhibitory effects of phosphorylation on the CaM-PDE activities may have important physiological roles. Although not yet demonstrated by direct experiments in vivo, the in vitro data would suggest that an initial increase in cAMP concentration during cell activation by neurotransmitters or hormones would cause phosphorylation of the PDE1A enzyme. Phosphorylation of this enzyme should then prevent the enzyme from being activated by low levels of Ca^{2+} at early stages of cell activation. Thus, the inhibition of cAMP hydrolysis, possibly coordinately with the stimulation of cAMP synthesis, ensures a rapid and sharp rise in intracellular cAMP. As cAMP levels continue to increase, the enzyme would become more highly phosphorylated and less active. This is, in effect, a "feed forward" type of regulation that should cause the levels of cAMP in a neuron to remain high enough for a long enough time to allow alterations in cAMP-dependent nuclear transcription and therefore longer term changes in synaptic function. They would further predict that with additional increases in intracellular Ca^{2+} at the later stages of cell activation, the PDE1A activity would finally be activated by the high Ca^{2+}, reversing the change in cAMP and PDE phosphorylation. This then brings about a decline in cAMP concentration. If shown to occur in intact neurons, this type of theoretical mechanism for potentiating intracellular cAMP could have important physiological implications in neuronal plasticity (Beavo, 1995).

Coordinate Regulation of cAMP by CaM-Dependent Adenylyl Cyclase and Phosphodiesterases

Obviously, there is an inverse relationship between adenylyl cyclase activity and phosphodiesterase activity on the amplitude and duration of the cAMP signals. In some tissues, notably brain, CaM-dependent adenylyl cyclase (type I) and CaM-PDE have been found in many of the same cells, such as cerebellar granule cells, and hippocampal granule and pyramidal cells (Yan et al., 1994; Beavo, 1995). To date, with one exception, the evidence for the subcellular colocalization of these two enzymes is lacking. In olfactory cilia, adenylyl cyclase (type III) and PDE1C2 are highly concentrated not only in the same cell but also in the same subcellular region (unpublished observations from authors' laboratory). In other cells, is it likely that both CaM-stimulated adenylyl cyclase and CaM-stimulated PDEs are present in the same compartments. Therefore, it has been considered paradoxical that calmodulin stimulation of both adenylyl cyclase and PDE should occur. Several models to explain the relationship of Ca^{2+} and cAMP are possible.

Spatially Sequential Stimulation Model

One can easily imagine that the influx of Ca^{2+} through the cell membrane or the release of membrane-bond Ca^{2+} in response to stimuli could activate adenylyl cyclase, leading to an increase in intracellular cAMP. Ca^{2+} subsequently arriving in a cytosolic compartment then might activate the Ca^{2+}-dependent PDEs, thus returning the elevated level of cAMP to its steady-state level. This sequential stimulation, initially of adenylyl cyclase and then of phosphodiesterases, could allow a transient stimulation of cAMP under certain conditions. Initial support for this model is based on the early observations that all identified CaM-dependent adenylyl cyclases were membrane-bound and that many CaM-PDEs are cytosolic. It is now known, however, that some CaM-PDEs may associate with cellular membranes, such as olfactory-enriched CaM-PDE (Borisy et al., 1992). Nevertheless, this model may be appropriate for some cell types or even regions of a cell.

Ca²⁺ Concentration-Dependent Stimulation Model

This model depends on having different affinities of the CaM-dependent adenylyl cyclases and phosphodiesterases for Ca^{2+}/CaM. Support for this model comes from data showing that in many cases adenylyl cyclase is activated at lower free Ca^{2+} levels (< 0.1 μM) and is inhibited at slightly high levels (> 0.1 μM) while only at the higher Ca^{2+} concentrations is phosphodiesterase activated. Thus, a model for differential modulation of cAMP levels in the brain by differing Ca^{2+} concentration has been proposed (Piascik et al., 1980). Such coordinated regulation of adenylyl cyclase and phosphodiesterase activities by Ca^{2+} should be able to modulate cyclic nucleotide concentrations within a fairly narrow range. It has been found, more recently, that the sensitivity of the phosphodiesterases to Ca^{2+}/CaM depends on which isozyme is expressed and also on its state of phosphorylation (Sonnenburg et al., 1995). Both modes of regulation give cells a mechanism for regulating the coordination between Ca^{2+}/CaM and cAMP in a cell type-specific manner.

Kinase Regulation Model of Ca²⁺/CaM and cAMP Coordination

One example of coordinate regulation of adenylyl cyclase and CaM-PDE activities by Ca^{2+} appears to be present in olfactory signal transduction (Yan et al., 1995). As stated earlier, a CaM-dependent adenylyl cyclase (III-AC) and a CaM-dependent PDE have been found to be highly concentrated in most olfactory cilia, indicating that they are probably colocalized. This olfactory adenylyl cyclase was slightly activated when the intracellular Ca^{2+} concentration was low and was inhibited when the Ca^{2+} concentration was high (Shirley et al., 1986; Sklar et al., 1986; Anholt and Rivers, 1990; Jaworsky et al., 1995). At the Ca^{2+} concentrations which inhibit the adenylyl cyclase, the olfactory CaM-PDE is likely to be activated (Jaworsky et al., 1995; Yan et al., 1995). In the early stages of odorant stimulation, the initial elevation of cAMP through activation of the G_{olf}-activated type III adenylyl cyclase results in Ca^{2+} entry through cyclic nucleotide-gated channels. At lower concentrations, Ca^{2+} would stimulate the G_{olf}-stimulated adenylyl cyclase, thus leading to formation of more cAMP and rapid amplification of the initial G-protein-induced signal. However, it is likely that the predominant effect at later stages will be for the III-AC to be phosphorylated by CaM-kinase II and thereby inhibited. This inhibition of cyclase coupled to stimulation of the PDE should operate in concert to cause a rapid decline in cAMP. It remains to be shown whether or not high concentrations of

CaM-kinase II are present in olfactory cilia. Physiologically, it seems likely that this rapid resetting of the cAMP signal is necessary for the animal to be able to discriminate rapidly between odorant intensities.

Most studies have concentrated on Ca^{2+} regulation of either the cyclases or the phosphodiesterases, and there have been few attempts to unravel how these two enzyme systems may interact with each other during cell activation. All the models presented above are, at present, rather speculative, but do have some data to support them. It is hoped that they will help to draw attention to the cooperative interactions that undoubtedly exist between cyclases and PDEs and thereby stimulate experiments to determine just how such interactions do occur in a particular cell. While it is clear that the amplitude and duration of the cAMP signal will be directly determined by which specific isozymes of CaM-dependent cyclases and PDEs are expressed, much work remains to be done to elucidate the important details of how they interact in a physiological setting.

References

Anholt, R. R. and A. M. Rivers (1990) Olfactory transduction: cross-talk between second-messenger systems. *Biochemistry* 29: 4049–4054.

Ariano, M. A. and A. M. Adinolfi (1977) Cyclic nucleotide phosphodiesterase: subcellular localization in caudate following selective interruption of striatal afferents. *Exp. Neurol.* 57: 426–433.

Ariano, M. A. and M. M. Appleman (1979) Biochemical characterization of postsynaptically localized cyclic nucleotide phosphodiesterase. *Brain Res.* 177: 301–309.

Bacskai, B. J., B. Hochner, M. Mahaut-Smith, S. R. Adams, B. K. Kaang, E. R. Kandel, and R. Y. Tsien (1993) Spatially resolved dynamics of cAMP and protein kinase A subunits in Aplysia sensory neurons. *Science* 260: 222–226.

Bakalyar, H. A. and R. R. Reed (1990) Identification of a specialized adenylyl cyclase that may mediate odorant detection. *Sience* 250: 1403–1406.

Balaban, C. D., M. L. Billingsley, and R. L. Kincaid (1989) Evidence for transsynaptic regulation of calmodulin-dependent cyclic nucleotide phosphodiesterase in cerebellar Purkinje cells. *J. Neurosci.* 9: 2374–2381.

Beavo, J. A. (1995) Cyclic nucleotide phosphodiesterases—functional implications of multiple isoforms [review]. *Physiol. Rev.* 75: 725–748.

Bentley, J. K. and J. A. Beavo (1992) Regulation and function of cyclic nucleotides. *Curr. Opin. Cell. Biol.* 4: 233–240.

Bentley, J. K., A. Kadlecek, C. H. Sherbert, D. Seger, W. K. Sonnenburg, H. Charbonneau, J. P. Novack, and J. A. Beavo (1992) Molecular cloning of cDNA encoding a

"63"-kDa calmodulin-stimulated phosphodiesterase from bovine brain. *J. Biol. Chem.* 267(26): 18676–18682.

Berridge, M. J. (1984) Cellular control through interactions between cyclic nucleotides and calcium. *Adv. Cyclic Nucleotide Protein Phosphor. Res.* 17: 329–335.

Billingsley, M. L., J. W. Polli, C. D. Balaban, and R. L. Kincaid (1990) Developmental expression of calmodulin-dependent cyclic nucleotide phosphodiesterase in rat brain. *Brain Res. Dev. Brain Res.* 53: 253–63.

Boekhoff, I. and H. Breer (1992) Termination of second messenger signaling in olfaction. *Proc. Natl. Acad. Sci. USA* 89(2): 471–474.

Boekhoff, I., S. Schleicher, J. Strotmann, and H. Breer (1992) Odor-induced phosphorylation of olfactory cilia proteins. *Proc. Natl. Acad. Sci. USA* 89: 11983–11987.

Borisy, F. F., G. V. Ronnett, A. M. Cunningham, D. Juilfs, J. Beavo and S. H. Snyder (1992) Calcium/calmodulin-activated phosphodiesterase expressed in olfactory receptor neurons. *J. Neurosci.* 12: 915–923.

Breer, H., I. Boekhoff, and E. Tareilus (1990) Rapid kinetics of second messenger formation in olfactory transduction. *Nature* 345: 65–68.

Cali, J. J., C. Zwaagstra, N. Mons, D. M. F. Cooper, and J. Krupinski (1994) Type VIII adenylyl cyclase: A Ca^{2+}/calmodulin stimulated enzyme expressed in discrete regions of rat brain. *J. Biol. Chem.* 269: 12190–12196.

Charbonneau, H., N. Beier, K. A. Walsh and J. A. Beavo (1986). Identification of a conserved domain among cyclic nucleotide phosphodiesterases from diverse species. *Proc. Natl. Acad. Sci. USA* 83: 9308–9312.

Charbonneau, H., S. Kumar, J. P. Novack, D. K. Blumenthal, P. R. Griffin, J. Shabanowitz, D. F. Hunt, J. A. Beavo, and K. A. Walsh (1991). Evidence for domain organization within the 61-kDa calmodulin-dependent cyclic nucleotide phosphodiesterase from bovine brain. *Biochemistry* 30: 7931–7940.

Chetkovich, D. M., R. Gray, and J. D. D. Sweatt (1991) N-methyl-D-aspartate receptor activation increases cAMP levels and voltage gated Ca^{2+} channel activity in area CA1 of hippocampus. *Proc. Natl. Acad. Sci. USA* 88: 6467–6471.

Choi, E. J., S. T. Wong, A. H. Dittman and D. R. Storm (1993) Phorbol ester stimulation of the type I and type III adenylyl cyclases in whole cells. *Biochemistry* 32: 1891–1894.

Choi, E. J., S. T. Wong, T. R. Hinds, and D. R. Storm (1992a) Calcium and muscarinic agonist stimulation of type I adenylyl cyclase in whole cells. *J. Biol. Chem.* 267: 12440–12442.

Choi, E. J., Z. Xia, and D. R. Storm (1992b) Stimulation of the type III olfactory adenylyl cyclase by calcium and calmodulin. *Biochemistry* 31: 6492–6498.

Davis, S., S. P. Butcher, and R. G. Morris (1992) The NMDA receptor antagonist D-2-amino-5-phosphono-pentanoate (D-AP5) impairs spatial learning and LTP in vivo at intracerebral concentrations comparable to those that block LTP in vitro. *J. Neurosci.* 12: 21–34.

Dawson, T. M., J. L. Arriza, D. E. Jaworsky, F. F. Borisy, H. Attramadal, R. J. Lefkowitz, and G. V. Ronnett (1993) Beta-adrenergic receptor kinase-2 and beta-arrestin-2 as mediators of odorant-induced desensitization. *Science* 259: 825–829.

Dittman, A. H., J. P. Weber, T. J. Hinds, E. J. Choi, J. C. Migeon, N. M. Nathanson, and D. R. Storm (1994) A novel mechanism for coupling of m4 muscarinic acetylcholine receptors to calmodulin sensitive adenylyl cyclases: cross-over from G protein coupled inhibition to stimulation. *Biochemistry* 33: 943–951.

Dudai, Y. and S. Zvi (1984) Adenylate cyclase in the Drosophila memory mutant rutabaga displays an altered Ca^{2+} sensitivity. *Neurosci. Lett.* 47: 19–24.

Erneux, C., S. J. Van, F. Miot, P. Cochaux, C. Decoster, and J. E. Dumont (1985) A mechanism in the control of intracellular cAMP level: the activation of a calmodulin-sensitive phosphodiesterase by a rise of intracellular free calcium. *Mol. Cell. Endocrinol.* 43: 123–34.

Feany, M. B. (1990) Rescue of the learning defect in dunce, a Drosophila learning mutant, by an allele of rutabaga, a second learning mutant. *Proc. Natl. Acad. Sci. USA* 87: 2795–2799.

Feinstein, P. G., A. Schrader, H. A. Bakalyar, W. J. Tang, J. Krupinski, G. A. Gilman and R. R. Reed (1991) Molecular cloning and characterization of a calcium calmodulin insensitive adenylyl cyclase (type II) from rat brain. *Proc. Natl. Acad. Sci. USA.* 88: 10173-10177.

Firestein, S., B. Darrow, and G. M. Shepherd (1991) Activation of the sensory current in salamander olfactory receptor neurons depends on a G protein-mediated cAMP second messenger system. *Neuron* 6: 825–835.

Florio, V. A., W. K. Sonnenburg, R. Johnson, K. S. Kwak, G. S. Jensen, K. A. Walsh, and J. A. Beavo (1994) Phosphorylation of the 61-kDa calmodulin-stimulated cyclic nucleotide phosphodiesterase at serine 120 reduces its affinity for calmodulin. *Biochemistry* 33: 8948–8954.

Frey, U., Y. Y. Huang and E. R. Kandel (1993) Effects of cAMP stimulate a late stage of LTP in hippocampal CA1 neurons. *Science* 260: 1661–1664.

Frey, U., K. G. Reymann, and H. Matthies (1991) The effect of dopaminergic D1 receptor blockade during tetanization on the expression of long-term potentiation in the rat CA1 region in vitro. *Neurosci. Lett.* 129: 111–114.

Grewal, J., N. Karuppiah, and B. Mutus (1989) A comparative kinetic study of bovine calmodulin-dependent cyclic nucleotide phosphodiesterase isozymes utilizing cAMP, cGMP and their 2'-O-anthraniloyl-, 2'-O-(N-methylanthraniloyl)-derivatives as substrates. *Biochem. Int.* 19: 1287–1295.

Hagiwara, M., P. Brindle, A. Harootunian, R. Armstrong, J. Rivier, J. Vale, R. Tsien, and M. R. Montminy (1993) Coupling of hormonal stimulation and transcription via the cAMP responsive factor CREB is rate limited by nuclear entry of protein kinase A. *Mol. Cell. Biol.* 13: 4852–4859.

Hansen, R. S. and J. A. Beavo (1986) Differential recognition of calmodulin–enzyme complexes by a conformation-specific anti-calmodulin monoclonal antibody. *J. Biol. Chem.* 261: 14636–14645.

Harris, K. M. and T. J. Teyler (1984) Developmental onset of long-term potentiation in area CA1 of the rat hippocampus. *J. Physiol.* (*London*) 346: 27–48.

Hashimoto, Y., R. K. Sharma, and T. R. Soderling (1989) Regulation of Ca^{2+}/calmodulin-dependent cyclic nucleotide phosphodiesterase by the autophosphorylated form of Ca^{2+}/calmodulin-dependent protein kinase II. *J. Biol. Chem.* 264: 10884–10887.

Hirono, J., T. Sato, and M. Tonoike (1992) Simultaneous recording of $[Ca^{2+}]_1$ increases in isolated olfactory receptor neurons retaining their original spatial relationship in intact tissue. *J. Neurosci. Methods* 42: 185–194.

Hopkins, W. F. and D. Johnston (1988) Noradrenergic enhancement of long-term potentiation at mossy fiber synapses in the hippocampus. *J. Neurophysiol.* 59: 667–687.

Impey, S., M. Mark, E. C. Villacres, S. W. Poser, C. Charkin, and D. R. Storm (1996) Induction of CRE-mediated gene expression by stimuli that generate long-lasting LTP in area CA7 of the hippocampus. *Neuron* 16: 973–982.

Impey, S., G. Wayman, Z. Wu, and D. R. Storm (1994) Type I adenylyl cyclase functions as a coincidence detector for control of cyclic AMP response element-mediated transcription: synergistic regulation of transcription by Ca^{2+} and isoproterenol. *Mol. Cell. Biol.* 14: 8272–8281.

Jacobowitz, O., J. Chen, and R. T. Premont (1993) Stimulation of specific types of G_s-stimulated adenylyl cyclases by phorbol ester treatment. *J. Biol. Chem.* 268: 3829–3832.

Jaworsky, D. E., O. Matsuzaki, and G. V. Ronnett (1995) Calcium modulates the rapid kinetics of the odorant-induced cyclic AMP signal in rat olfactory cilia. *J. Neurosci.* 15: 310–318.

Kaang, B. K., E. R. Kandel, and S. G. Grant (1993) Activation of cAMP-responsive genes by stimuli that produce long-term facilitation in Aplysia sensory neurons. *Neuron.* 10: 427–435.

Kawabe, J., G. Iwami, and T. Ebina (1994) Differential activation of adenylyl cyclase by protein kinase C isoenzymes. *J. Biol. Chem.* 269: 6554–6558.

Kincaid, R. L., H. Takayama, M. L. Billingsley, and M. V. Sitkovsky (1987) Differential expression of calmodulin-binding proteins in B, T lymphocytes and thymocytes. *Nature* 330: 176–178.

Krupinski, J., F. Coussen, H. A. Bakalyar, W. J. Tang, P. G. Feinstein, K. Orth, C. Slaughter, R. R. Reed, and A. G. Gilman (1989) Adenylyl cyclase amino acid sequence: possible channel- or transporter-like structure. *Science* 244: 1558–1564.

Kurahashi, T. (1989) Activation by odorants of cation-selective conductance in the olfactory receptor cell isolated from the newt. *J. Physiol.* (*London*) 419: 177–192. [Erratum, *J. Physiol.* (*London*) (1990) 424: 561–562.]

Kurahashi, T. and T. Shibuya (1990) Ca(2+)-dependent adaptive properties in the solitary olfactory receptor cell of the newt. *Brain Res.* 515: 261–268.

Law, P. Y. and H. H. Loh (1993) delta-Opioid receptor activates cAMP phosphodiesterase activities in neuroblastoma × glioma NG108-15 hybrid cells. *Mol. Pharmacol.* 43: 684–693.

Levin, L. R., P. L. Han, P. M. Hwang, P. G. Feinstein, R. L. Davis, and R. R. Reed (1992) The drosophila learning and memory gene rutabaga encodes a Ca^{2+}/calmodulin-responsive adenylyl cyclase. *Cell* 68: 479–489.

Livingston, M. S., P. P. Sziber, and W. G. Quinn (1984) Loss of calcium calmodulin responsiveness in adenylyl cyclase of rutabaga, a Drosophila learning mutant. *Cell* 37: 205–215.

Loughney, K., T. Martins, B. Sonnenburg, J. Beavo, and K. Ferguson (1994) Isolation and characterization of cDNAs corresponding to two human calmodulin stimulated $3',5'$ cyclic nucleotide phosphodiesterase genes. *FASEB J.* 8: A469.

Ludvig, N., V. Burmeister, P. C. Jobe, and R. L. Kincaid (1991) Electron microscopic immunocytochemical evidence that the calmodulin-dependent cyclic nucleotide phosphodiesterase is localized predominantly at postsynaptic sites in the rat brain. *Neuroscience* 44: 491–500.

Matthews, R. P., C. R. Guthrie, L. M. Wailes, X. Zhao, A. R. Means, and G. S. McKnight (1994) Calcium/calmodulin-dependent protein kinase types II and IV differentially regulate CREB-dependent gene expression. *Mol. Cell. Biol.* 14: 6107–6116.

Miot, F., J. E. Dumont, and C. Erneux (1983) The involvement of a calmodulin-dependent phosphodiesterase in the negative control of carbamylcholine on cyclic AMP levels in dog thyroid slices. *FEBS Lett.* 151: 273–276.

Morris, R. G. (1990) Toward a representational hypothesis of the role of hippocampal synaptic plasticity in spatial and other forms of learning. *Cold Spring Harbor Symp. Quant. Biol.* 50: 161–173.

Morris, R. G., E. Anderson, and G. S. Lynch (1986) Selective impairment of learning and blockade of long-term potentiation by an N-methyl-D-aspartate receptor antagonist, AP5. *Nature* 319: 774–776.

Morris, R. G., P. Garrud, and J. N. Rawlins (1982) Place navigation impaired in rats with hippocampal lesions. *Nature* 297: 681–683.

Nakamura, Y., S. Okuno, and F. Sato (1995) An immuno-histochemical study of Ca^{2+}/calmodulin-dependent protein kinase IV in the rat central nervous system: light and electron microscopic observations. *J. Neurosci.* 68: 181–194.

Nemecek, G. M. and T. W. Honeyman (1982) The role of cyclic nucleotide phosphodiesterase in the inhibition of cyclic AMP accumulation by carbachol and phosphatidate. *J. Cyclic Nucleotide Res.* 8: 395–408.

Nielsen, M. D., G. C. K. Chan, S. W. Poser and D. R. Storm (1996) Differential regulation of type I and type VIII Ca^{2+}-stimulated adenylyl cyclases by Gi-coupled receptors in vivo. *J. Biol. Chem.* 271: 33308–33316.

Nigg, E. A., H. Hilz, H. M. Eppenberger, and F. Dutly (1985) Rapid and reversible translocation of the catalytic subunit of cAMP-dependent protein kinase type II from the Golgi complex to the nucleus. *EMBO J.* 4: 2801–2806.

Novack, J. P., H. Charbonneau, J. K. Bentley, K. A. Walsh, and J. A. Beavo (1991) Sequence comparison of the 63-, 61-, and 59-kDa calmodulin-dependent cyclic nucleotide phosphodiesterases. *Biochemistry* 30: 7940–7947.

Piascik, M. T., P. L. Wisler, C. L. Johnson, and J. D. Potter (1980) Ca^{2+}-dependent regulation of guinea pig brain adenylate cyclase. *J. Biol. Chem.* 255(9): 4176–4181.

Pokorny, J. and T. Yamamoto (1981) Postnatal ontogenesis of hippocampal CA1 area in rats. II. Development of ultrastructure in stratum lacunosum and moleculare. *Brain Res. Bull.* 7: 113–120.

Polli, J. W. and R. L. Kincaid (1992) Molecular cloning of DNA encoding a calmodulin-dependent phosphodiesterase enriched in striatum. *Proc. Natl. Acad. Sci. USA* 89: 11079–11083.

Purvis, K. and H. Rui (1988) High-affinity, calmodulin-dependent isoforms of cyclic nucleotide phosphodiesterase in rat testis. *Methods Enzymol.* 159: 675–685.

Rasmussen, H. and D. B. Goodman (1977) Relationships between calcium and cyclic nucleotides in cell activation. *Physiol. Rev.* 57: 421–509.

Reddy, R., D. Smith, G. Wayman, Z. Wu, E. C. Villacres, and D. R. Storm (1995) Voltage-sensitive adenylyl cyclase activity in cultured neurons: a calcium independent phenomenon. *J. Biol. Chem.* 270: 14340–14346.

Repaske, D. R., J. V. Swinnen, S. L. Jin, W. J. J. Van, and M. Conti (1992) A polymerase chain reaction strategy to identify and clone cyclic nucleotide phosphodiesterase cDNAs. Molecular cloning of the cDNA encoding the 63-kDa calmodulin-dependent phosphodiesterase. *J. Biol. Chem.* 267: 18683–18688.

Restrepo, D., T. Miyamoto, and B. P. Bryant (1990) Odor stimuli trigger influx of Ca^{2+} into olfactory neurons of the channel catfish. *Science* 249: 1166–1168.

Rosenberg, G. B. and D. R. Storm (1987) Immunological distinction between calmodulin-sensitive and calmodulin-insensitive adenylyl cyclases. *J. Biol. Chem.* 262: 7623–7628.

Rossi, P., M. Giorgi, R. Geremia, and R. L. Kincaid (1988) Testis-specific calmodulin-dependent phosphodiesterase. A distinct high affinity cAMP isoenzyme immunologically related to brain calmodulin-dependent cGMP phosphodiesterase. *J. Biol. Chem.* 263): 15521–15527.

Schleicher, S., I. Boekhoff, J. Arriza, R. J. Lefkowitz, and H. Breer (1993) A beta-adrenergic receptor kinase-like enzyme is involved in olfactory signal termination. *Proc. Natl. Acad. Sci. USA* 90: 1420–1424.

Sharma, R. K. (1991) Phosphorylation and characterization of bovine heart calmodulin-dependent phosphodiesterase. *Biochemistry* 30: 5963–5968.

Sharma, R. K. and J. H. Wang (1985) Differential regulation of bovine brain calmodulin-dependent cyclic nucleotide phosphodiesterase isoenzymes by cyclic AMP-dependent protein kinase and calmodulin-dependent phosphatase. *Proc. Natl. Acad. Sci. USA* 82: 2603–2607.

Sharma, R. K. and J. H. Wang (1986a) Calmodulin and Ca^{2+}-dependent phosphorylation and dephosphorylation of the 63-kDa subunit-containing bovine brain calmodulin-stimulated cyclic nucleotide phosphodiesterase isozyme. *J. Biol. Chem.* 261: 1322–1328.

Sharma, R. K. and J. H. Wang (1986b) Purification and characterization of bovine lung calmodulin-dependent cyclic nucleotide phosphodiesterase. An enzyme containing calmodulin as a subunit. *J. Biol. Chem.* 261: 14160–14166.

Sharma, R. K., A. M. Adachi, K. Adachi, and J. H. Wang (1984) Demonstration of bovine brain calmodulin-dependent cyclic nucleotide phosphodiesterase isozymes by monoclonal antibodies. *J. Biol. Chem.* 259: 9248–9254.

Shenolikar, S., W. J. Thompson, and S. J. Strada (1985) Characterization of a Ca^{2+}-calmodulin-stimulated cyclic GMP phosphodiesterase from bovine brain. *Biochemistry* 24: 672–678.

Shirley, S. G., C. J. Robinson, K. Dickinson, R. Aujla, and G. H. Dodd (1986) Olfactory adenylate cyclase of the rat. Stimulation by odorants and inhibition by Ca^{2+}. *Biochem. J.* 240: 605–607.

Sklar, P. B., R. R. Anholt, and S. H. Snyder (1986) The odorant-sensitive adenylate cyclase of olfactory receptor cells. Differential stimulation by distinct classes of odorants. *J. Biol. Chem.* 261: 15538–15543.

Slack, J. R. and S. Pockett (1991) Cyclic AMP induces long-term increase in synaptic efficacy in CA1 region of rat hippocampus. *Neurosci. Lett.* 130: 69–72.

Sonnenburg, W. K., D. Seger, and J. A. Beavo (1993) Molecular cloning of a cDNA encoding the "61-kDa" calmodulin-stimulated cyclic nucleotide phosphodiesterase. Tissue-specific expression of structurally related isoforms. *J. Biol. Chem.* 268: 645–652.

Sonnenburg, W. K., D. Seger, K. S. Kwak, J. Huang, H. Charbonneau, and J. A. Beavo (1995) Identification of inhibitory and calmodulin-binding domains of the PDE1A1 and PDE1A2 calmodulin-stimulated cyclic nucleotide phosphodiesterases. *J. Biol. Chem.* 270: 30989–31000.

Stanton, P. K. and J. M. Sarvey (1985) The effect of high-frequency electrical stimulation and norepinephrine on cyclic AMP levels in normal versus norepinephrine-depleted rat hippocampal slices. *Brain Res.* 358: 343–348.

Steer, M. L. and A. Levitzki (1975) The control of adenylate cyclase by calcium in turkey erythrocyte ghosts. *J. Biol. Chem.* 250: 2080–2084.

Tang, W. J., J. Krupinski, and A. G. Gilman (1991) Expression and characterization of calmodulin activated (type I) adenylyl cyclase. *J. Biol. Chem.* 266: 8595–8603.

Tanner, L. I., T. K. Harden, J. N. Wells, and M. W. Martin (1986) Identification of the phosphodiesterase regulated by muscarinic cholinergic receptors of I32INI human astrocytoma cells. *Mol. Pharmacol.* 29: 455–460.

Taussig, R., J. A. Iniguez-Lluhi, and A. G. Gilman (1993) Inhibition of adenylyl cyclase by G_{ia}. *Science* 261: 218–221.

Teyler, T. J., A. T. Perkins, and K. M. Harris (1989) The development of long-term potentiation in hippocampus and neocortex. *Neuropsychologia* 27: 31–39.

Villacres, E. C., Z. Wu, W. Hua, M. D. Nielsen, J. J. Watters, C. Yan, J. Beavo, and D. R. Storm (1995) Developmentally expressed Ca(2 +)-sensitive adenylyl cyclase activity is disrupted in the brains of type I adenylyl cyclase mutant mice. *J. Biol. Chem.* 270: 14352-14357.

Wang, J. H., R. K. Sharma, and M. J. Mooibroek (1990) Calmodulin-stimulated cyclic nucleotide phosphodiesterases. In *Cyclic Nucleotide Phosphodiesterases: Structure, Regulation and Drug Action* (J. A. Beavo and M. D. Houslay). John Wiley & Sons, Chichester, pp. 19–60.

Wayman, G. A. and D. R. Storm (1995) Ca^{2+} inhibition of type III adenylyl cyclase in vivo. *J. Biol. Chem.* 270: 21480–21486.

Wayman, G. A., T. R. Hinds, and D. R. Storm (1995a) Hormone stimulation of type III adenylyl cyclase induces Ca^{2+} oscillations in HEK-293 cells. *J. Biol. Chem.* 270: 24108–24115.

Wayman, G. A., S. Impey, and D. R. Storm (1995b) Ca^{2+} inhibition of type III adenylyl cyclase in vivo. *J. Biol. Chem.* 270: 21480–21486.

Wayman, G. A., S. Impey, Z. Wu, W. Kindsvogel, L. Prichard, and D. R. Storm (1994) Synergistic activation of the type I adenylyl cyclase by Ca^{2+} and Gs-coupled receptors in vivo. *J. Biol. Chem.* 269): 25400–25405.

Wayman, G. A., J. Wei, S. Wong, and D. R. Storm (1996) Regulation of type I adenylyl cyclase by CaM-kinase IV in vivo. *Mol. Cell. Biol.* 17: 6075–6082.

Wei, J., G. A. Wayman, and D. R. Storm (1996) Phosphorylation and inhibition of type III adenylyl cyclase by calmodulin-dependent protein kinase II in vivo. *J. Biol. Chem.* 271: 24231–24235.

Weisskopf, M. G., P. E. Castillo, R. A. Zalutsky, and R. A. Nicoll (1994) Mediation of hippocampal mossy fiber long-term potentiation by cyclic AMP. *Science* 23: 1878–1882.

Westcott, K. R., D. C. LaPorte, and D. R. Storm (1979) Resolution of adenylate cyclase sensitive and insensitive to Ca^{2+} and calmodulin by calmodulin–Sepharose affinity chromatography. *Proc. Natl. Acad. Sci. USA* 76: 204–228.

Wu, Z. L., S. A. Thomas, E. C. Villacres, Z. Xia, M. L. Simmons, C. Chavkin, R. D. Palmiter, and D. R. Storm (1995) Altered behavior and long-term potentiation in type I adenylyl cyclase mutant mice. *Proc. Natl. Acad. Sci. USA* 92: 220–224.

Wu, Z., S. T. Wong, and D. R. Storm (1993) Modification of the calcium and calmodulin sensitivity of the type I adenylyl cyclase by mutagenesis of its calmodulin binding domain. *J. Biol. Chem.* 268: 23766–23768.

Xia, Z. and D. R. Storm (1996) *Regulatory Properties of the Mammalian Adenylyl Cyclases.* R. G. Landes, Austin, TX.

Xia, Z., E. J. Choi, F. Wang, C. Blazynski, and D. R. Storm (1993) Type I calmodulin-sensitive adenylyl cyclase is neural specific. *J. Neurochem.* 60: 305–311.

Xia, Z., E. J. Choi, F. Wang, and D. R. Storm (1992) The type III calcium/calmodulin-sensitive adenylyl cyclase is not specific to olfactory sensory neurons. *Neurosci. Lett.* 144: 169–173.

Xia, Z. G., C. D. Refsdal, K. M. Merchant, D. M. Dorsa, and D. R. Storm (1991) Distribution of mRNA for the calmodulin-sensitive adenylate cyclase in rat brain: expression in areas associated with learning and memory. *Neuron* 6: 431–443.

Yan, C., J. K. Bentley, W. K. Sonnenburg, and J. A. Beavo (1994) Differential expression of the 61 kDa and 63 kDa calmodulin-dependent phosphodiesterases in the mouse brain. *J. Neurosci.* 14: 973–984.

Yan, C., A. Z. Zhao, J. K. Bentley, and J. A. Beavo (1996) The calmodulin-dependent phosphodiesterase gene PDE1C encodes several functionally different splice variants in a tissue specific manner. *J. Biol. Chem.* 271: 25699–25706.

Yan, C., A. Z. Zhao, J. K. Bentley, K. Loughney, K. Ferguson, and J. A. Beavo (1995) Molecular cloning and characterization of a calmodulin-dependent phosphodiesterase enriched in olfactory sensory neurons. *Proc. Natl. Acad. Sci. USA* 92: 9677–9681.

Zhong, Y. and C. F. Wu (1991) Altered synaptic plasticity in Drosophila memory mutants with a defective cyclic AMP cascade. *Science* 251: 198–201.

PART III

Specialized Functions of Calcium

17

Calcium in Plants

David E. Evans

Perhaps one of the most remarkable aspects of calcium physiology described in all organisms, including higher plants, is the fact that an element which is cytotoxic at concentrations much above $1\,\mu M$ is also essential to cell function. Plant and fungal cells, like animal cells, maintain low (submicromolar) cytoplasmic free calcium concentrations (Table 17.1); but they also require calcium for cell signalling, as well as for structural and metabolic functions, and the concentrations of calcium within the plant as a whole are therefore much higher than this.

Calcium was first described as an essential macronutrient element in plants more than a century ago (see Burström, 1968). It is taken up by terrestrial plants from soil solution and conveyed in aqueous solution in the xylem to aerial parts of the plant. All cells and tissues are thereby supplied with and surrounded by calcium-containing sap. Calcium mobility within cells is, however, very low as a consequence of its low free concentration and rapid chelation (see DeMarty et al., 1984). It moves predominantly apoplastically (i.e. in the walls and spaces between cells) rather than symplastically (i.e. in the cytoplasm) and the suberinised casparian strip of the endodermis (which encircles the root vascular tissue) poses a major barrier to calcium movement from the root tissue into the xylem flow. Thus, most of the calcium transported through the plant enters in regions of the root prior to the point of differentiation of the endodermis, near to the root cap. Calcium concentrations in the phloem sap are severely limited by the presence of phosphate and other chelators (Raven, 1977).

Some indications of the function of calcium in plants come from symptoms of its deficiency. Calcium deficiency can occur in circumstances of low soil availability, competition by other cations or of low transpiration where xylem flow is inadequate to supply the calcium requirements of rapidly growing tissue. Calcium deficiency results in stunted root growth and altered leaf appearance (see Marschner, 1986). Commonly observed and more severe symptoms are generally the result of failure of cell membrane integrity and include bitter-pit in apple fruit, blossom end rot in tomato, and tipburn in lettuce (Shear, 1975). Deficiency symptoms develop first in regions of the plant most distal to the regions of calcium uptake and furthest from points of xylem water flow and transpiration.

Plants vary markedly in calcium content and requirements. Calcium content can vary between 0.1 and $> 5.0\%$ dry wt (Marschner 1986) for different plants and organs, while soil calcium also varies widely from $< 0.01\%$ calcium in acid laterites to very high abundance in chalky soils (Burström, 1968). The abilities of different plant species to grow on soils of varying calcium status cannot solely be considered in terms of calcium availability or uptake as calcium-rich soils also tend to be base-rich, while calcium-poor soils are generally acidic. Such natural soils will also vary in respect of the availability of other nutrients and of toxic ions (for instance aluminium, which becomes available at acidic pH). The terms calcicole (to describe species able to tolerate calcium and base-rich soils) and calcifuge (to describe species tolerating acidic, low calcium soils) are in widespread use. However, it is evident that plants vary in their calcium requirements and ability to extract calcium from complex soil environments. In particular, monocots require less calcium for optimal growth than do dicots (Loneragan et al., 1968; Loneragan and Snowball, 1969). Cation exchange (in the root, the xylem, and at cell walls and the plasma membrane [PM]) is central to calcium transport within the plant and calcium competes with other cations both for these sites and for uptake from the soil. The presence of high levels of calcium is known to ameliorate the effects of the uptake of toxic cations (Al^{3+}, Na^+) from the soil while the presence of high levels of other cations

Table 17.1 Cytosolic Free Calcium Concentrations in Plant Tissues

Tissue		Technique	Concentration	Reference(s)[b]
Carrot	Protoplasts	Quin-2	360 nM	1
		Quin-2	120–360 nM	2
Barley	Protoplasts Aleurone	Aequorin	< 200 nM	3
	Protoplasts	Indo-1	250 nM	3
Mung bean	Root	Quin-2	170 nM	4
	Root hairs	Fura-2	30–90 nM	5
Tradescantia	Stamen hairs	Arsenoazo III	< 1 μM	6
Diatom	Protoplasts	Fura-2	110 nM	7
Fucus	Rhizoids (subtip region)	Microelectrode	280 nM	8
		Quin-2	< 1 μM	
Chara	Internode	Aequorin	220 nM	9
Nitella	Internode	Electrode (dark)	250–400 nM	10
		(light)	145 nM	
Riccia	Rhizoid	Electrode	109–187 nM	11
Zea mays	Root hair	Electrode	145–231 nM	
	Coleoptiles	Electrode	84–143 nM	11, 12
	Epidermis	Fluo-3	255 nM	13
	Protoplasts	Indo-1	93 nM	14
Tobacco	Epidermis (stimulated)	Aequorin	10 μM	15

[a]Estimates of cytoplasmic free calcium concentrations in a variety of plant tissues by a variety of techniques.
[b]References: 1, Gilroy et al. (1987); 2, Gilroy et al. (1989); 3, Bush and Jones (1987); 4, Gilroy et al. (1986); 5, Clarkson et al. (1988); 6, Hepler and Callaham (1987); 7, Brownlee et al. (1987); 8, Brownlee and Wood (1986); 9, Williamson and Ashley (1982); 10, Miller and Sanders (1987); 11, Felle (1988a); 12, Felle (1988b); 13, Gehring et al. (1990a); 14, Lynch et al. (1989); 15, Haley et al. (1995).

(K^+, Mg^{2+}) may reduce calcium uptake. Calcium requirements for optimal growth are thus strongly dependent on the presence of other cations (Wyn-Jones and Lunt, 1967; Burström, 1968).

Calcium Homeostasis in Plant Cell Cytoplasm

Plant cells, like animal cells, maintain a low and regulated cytoplasmic free calcium concentration. Measurements of calcium concentration in a variety of unstimulated plant cells are presented in Table 17.1 and range from 30 nM to 400 nM in higher plants. These values were estimated using a variety of measuring techniques; the values obtained are comparable to those of mammalian and other cells and indicate that cytoplasmic calcium concentrations are maintained against a steep concentration gradient across the plasma membrane. This gradient has been estimated to be between 1000:1 and 10,000:1 out:in

(based on an apoplastic calcium concentration of about 1 mM).

Plant Calcium Pumps

Maintaining the low cytosolic calcium concentrations observed in plant cells requires active transport from the cytosol. Active efflux pumping is also a prerequisite for the restoration of cytosolic free calcium levels after signalling events. Calcium-transport in plants serves two functions: preservation (or restoration) of low cytosolic calcium and the maintenance of "pools" of accumulated calcium for signalling transients. Pumping therefore occurs both across the PM (efflux pumping) and into intracellular organelles. Plant cells differ from those of animals both in the classes of transporters used to pump calcium and in the organelles into which calcium is accumulated. They also differ markedly in that plant cells (in common with fungi) operate a proton-based physiology. The major primary ion pumps in plants are proton pumps — located at the plasma membrane (a P-type

Table 17.2 Properties and Locations of Calcium Pumping ATPases in Plant Cells

Type	Plant	Notes	Reference(s)
Calmodulin-stimulated at intracellular membranes	Carrot	ER	1
	Brassica	BCA1 Tonoplast	2, 3, 4
	Barley	ER (aleurone)	5
	Maize	ER	6, 7
	Arabidopsis	ACA2 ER	8
	Bryonia dioica	ER	9
Calmodulin-stimulated at the plasma membrane	Maize		10
	Red beet		11
	Radish		12
	Arabidopsis		13
	Cucumber		14
	Brassica		15
Non calmodulin-stimulated Ca pumps	*Arabidopsis*	ACA 1 Plastid envelope(?)	
		PM-type, lacking CaM binding domain	16
	Tomato	LCA 1 ER-type, tonoplast, PM, NE	17, 18, 19
	Tobacco	ER-type (ER?) predicted from sequence	20
	Arabidopsis	ECA1/ACA 3 (ER-type; ER?)	21

Key calmodulin-stimulated and non-calmodulin-stimulated calcium pumps in higher plants giving evidence from both membrane fractionation and molecular cloning (clone identities presented where known). The list is not exhaustive; several further *Arabidopsis* clones have been identified, of unknown location and function. References: 1, Hsieh et al. (1991); 2, Askerlund and Evans (1992); 3, Askerlund (1997); 4, Malmstron et al. (1997); 5, Gilroy and Jones (1993); 6, Brauer et al. (1990); 7, Theodoulou et al. (1994); 8, Harper et al. (1998); 9, Liss et al. (1998); 10, Robinson et al. (1988); 11, Thomson et al. (1993); 12, Rasi-Caldogno et al. (1995), 13, De Michelis et al. (1993); 14, Erdei and Matsumoto (1991); 15, Askerlund and Evans (1993); 16, Huang et al. (1993); 17, Wimmers et al. (1992); 18, Ferrol and Bennett (1996); 19, Downie et al. (1998); 20, Perez-Prat et al. (1992); 21, Hwang et al. (1997).

pump) and at the vacuolar membrane or tonoplast (a V-type pump; see Sze, 1985). Most plant cells, other than meristematic cells, are highly vacuolate and thus the tonoplast represents a very significant area for ion transport and the vacuole itself a large pool for ion sequestration.

Calcium Efflux Pumps at the Plasma Membrane

Preparations of plant plasma membrane of high purity obtained have been used to identify the calcium pumps of these membranes (see Briskin, 1990; Evans et al., 1991, 1992; Evans and Williams, 1998 for reviews). In all cases, direct calcium pumps have been identified which show the properties of P-type ion translocating ATPases (see Table 17.2). Nucleotide specificity is broad (30–60% activity achieved with GTP and ITP); the pumps are inhibited by erythrosin B ($IC_{50} \leq 1 \mu M$) and estimates of calcium affinity are in the range $0.4 \mu M$–$12 \mu M$. It seems likely that one of the plant plasma membrane calcium pumps (in common with that from mammalian tissues) is stimulated by calmodulin (CaM), although it has not been possible to demonstrate this activity in all the tissues tested. Plant plasma membrane has high concentrations of endogenous calmodulin (Collinge and Trewavas, 1989) and, in consequence, demonstration of calmodulin-sensitivity requires extensive washing of the membrane in calcium-chelating buffers (Williams et al., 1990; Evans et al., 1992). Studies using proteolysis (Rasi-Caldogno et al., 1995) suggest that CaM-sensitivity is the result of the presence of an autoinhibitory calmodulin-binding domain similar to that in the mammalian PM calcium pump. The relative molecular mass of this pump in plant tissues has been estimated to be between 115,000Da and 135,000Da by phosphorylated intermediate formation, radiation inactivation and FITC binding (Askerlund and Evans, 1993; Basu and Briskin, 1995; Rasi-Caldogno et al., 1995). In two cases (*Commelina communis* and *Zea mays*), a calcium pump has been purified and reconstituted from purified plant PM. In neither case was CaM-stimulation of the purified protein identifiable (Gräf and Weiler, 1990; Kasai and Muto, 1991).

In addition to a P-type pump, the presence of a low-affinity Ca^{2+}/H^+ antiport mechanism at the PM has been suggested (Kasai and Muto, 1990). The presence of such an antiport mechanism suggests a similar mechanism to the low-affinity/high-capacity sodium/calcium antiport of mammalian cells. It seems likely that in most plants the role of such a high-capacity, low-affinity calcium transporter is fulfilled by an antiport located at the vacuolar membrane (see below).

Calcium Transport at the Tonoplast and Calcium Sequestration in the Vacuole

Numerous studies of calcium transport using either intact, purified vacuoles or highly purified tonoplast membranes indicate the presence of a low-affinity, high-capacity Ca^{2+}/H^+ antiporter, utilising the H^+ gradient generated by the tonoplast V-type H^+-pump and by a proton-pumping pyrophosphatase to sequester calcium in the vacuole (Schumaker and Sze, 1985; Blumwald and Poole, 1986; Bush and Sze, 1986; Chanson, 1994). Given that in many cells the vacuole can occupy more than 50% of cell volume (in some cases, more than 95%), it is evident that trans-tonoplast calcium transport makes a very significant contribution to the regulation of cytosolic calcium concentrations. Calcium concentrations within the vacuole range from 0.1 to 10 mM (Macklon, 1984). The affinity values reported for the tonoplast Ca^{2+}/H^+ antiport (ca. 1.3 mM; see Evans et al., 1991) are low in comparison with the affinities displayed for the calcium pumps; however, the antiport may well be of importance in controlling transient rises in cytosolic calcium during signalling events. The stoichiometry of the antiporter has been suggested to be $> 3H^+:1Ca^{2+}$ (Blackford et al., 1990). It has been functionally reconstituted after solubilisation from oat root membranes (Schumaker and Sze,

1990) but the molecular structure of the protein has not been elucidated. Several groups have now also identified direct (P-type) calcium pumping ATPase activity in tonoplasts (Fukumoto and Venis, 1986; Gavin et al., 1993; Pfeiffer and Hager, 1993; Askerlund, 1997). An effect of calmodulin on tonoplast calcium uptake was suggested by Andreev et al. (1990). Malmstron et al. (1997) have sequenced such a calmodulin-stimulated calcium pumping ATPase which is localised to the tonoplast in *Brassica oleracea*. This pump is unique in having a CaM binding domain at the N-terminus (in contrast to the C-terminal CaM binding domains previously described).

Endoplasmic Reticulum

In mammalian cells, the endoplasmic reticulum (ER) is an important pool for calcium and cell-signalling. In plants, there is evidence for ER calcium-binding proteins (e.g. calreticulin; Denecke et al., 1995) and for the activity of P-type calcium transport into the ER lumen (reviewed by Evans et al., 1991; Chanson, 1993). The ER calcium pump has been described as a non-calmodulin stimulated pump of M_r ca. 100,000Da (based on phosphorylated intermediate formation); but more recently, a calmodulin-stimulated calcium pump of M_r 110,000–120,000Da has been ascribed an ER location by several groups (Hsieh et al., 1991; Chen, et al., 1993; Gilroy and Jones 1993; Theodoulou et al., 1994; Logan and Venis, 1995). Although this location is tentative in some instances (a vacuolar location being a second possibility; see Askerlund and Evans, 1992; Gavin et al., 1993), it is clear that plant cells differ significantly from other eukaryotes in having a calmodulin-stimulated P-type calcium pump at an intracellular membrane (see Evans, 1994a, 1994b).

A role for a calmodulin-stimulated Ca^{2+}-pumping ATPase at the ER has been identified in germinating barley seeds. Here, hydrolytic enzymes are released from the aleurone layer during germination, resulting in mobilisation of seed storage reserves. Amongst these enzymes, α-amylase, which is secreted, requires calcium for stability. Stimulation of enzyme synthesis and secretion by gibberellic acid results in increased insertion of a calmodulin-stimulated calcium pump into the ER and enhanced calcium transport into the lumen of the organelle (Gilroy and Jones, 1992, 1993; Jones, et al., 1993).

In spite of the evidence for a calmodulin-stimulated calcium pump at plant ER, evidence from molecular cloning suggests that plants have a homologue of the mammalian ER/SR calcium pump, which lacks a CaM-binding domain (Perez-Pratt et al., 1992; Wimmers et al., 1992; Liang et al., 1997). This pump is highly homologous with the mammalian form; it shows $> 50\%$ overall sequence identity,

> 70% in sacroplasmic reticulum (SR)-calcium pump specific regions, including transmembrane domains. The deduced amino acid sequence of the clone from tomato specifies 1048 amino acids that give a 116 kDa protein. All 11 residues essential for calcium transport in the SR pump are present. Immunocytochemical evidence suggests that a SERCA pump in plants may, in common with that in animal tissue, be localised at the nuclear envelope (Downie et al., 1998).

Mitochondria and Plastids

Plant mitochondria show electrophoretic calcium influx (see Evans, 1988, for review) that appears to be a ruthenium-red and mersalyl-sensitive $Ca^{2+}/$inorganic phosphate symporter (Chanson, 1993) and a phosphate-independent calcium-efflux pathway (Silva et al. 1992). The affinity of the plant mitochondrial import system for calcium is, however, very low (10 times that of the P-type pumps) and it therefore is unlikely that mitochondrial calcium uptake will be significant other than after extreme rises in cytosolic calcium (Evans, 1988).

Calcium transport in chloroplasts appears to be a much more important phenomenon. Uptake occurs as a result of a uniport driven by changes in membrane potential driven by photosynthetic electron tranport. Thus chloroplasts take up calcium upon illumination (Muto et al., 1982; Kreimer, et al., 1985, 1988; Miller and Sanders, 1987). Miller and Sanders (1987) observed that cells of a Characean alga showed lower cytosolic free calcium concentrations in the light (204 nM) than in the dark (350 nM) and suggested that this may be involved in regulating the activity of enzymes of metabolism in the cell cytoplasm. Chloroplasts contain large amounts of calcium (4–23 mM; Portis and Heldt, 1976) but the stromal free calcium concentration is much lower (2.4–6.3 μM; Kreimer et al., 1988). It has been suggested that this low concentration is maintained by the action of a P-type efflux pump; such a pump (of 90–95 kDa, showing 40–45% identity with mammalian calcium pumps) has recently been identified by molecular cloning and antibody localisation (Huang et al., 1993), and would be unique in exporting calcium to the cytoplasm. Biochemical evidence for such a calmodulin-stimulated calcium-pumping ATPase has also been obtained (Nguyen and Siegenthaler, 1985).

Summary: Calcium Homeostasis

Plant cells have been shown to have a variety of calcium pumps and antiports at several membranes (see Fig. 17.1). A homologue of the mammalian "PM-type" calcium pump is present (though it has yet to be cloned and sequenced) but, in contrast to mammalian tissue, it appears to be located at an intracellular membrane (ER/tonoplast) as well as at the PM (Evans, 1994a, 1994b). Several isoforms of homologues of the mammalian SERCA calcium pumps have also been revealed by moleclar techniques and have been suggested to be localised at intracellular membranes and possibly the plasma membrane (see Evans and Williams, 1998). Calcium proton antiport activity is found at the tonoplast, where it appears to be a major low-affinity/high-capacity Ca^{2+}-efflux transport. An entirely novel form of P-type calcium pump, not present in animal cells, has also been identified, which has an N-terminal CaM-binding autoinhibitory domain, located at the tonoplast and possibly other membranes (Malmstron et al., 1997; Harper et al., 1998). Chloroplasts also accumulate Ca^{2+} upon illumination by an electrogenic mechanism; however, molecular cloning suggests a P-type Ca^{2+} pump which may act to reduce stromal calcium concentration. All the pumps and antiports (apart from the chloroplast pump) act to efflux calcium from the cytoplasm, thereby maintaining the low concentrations typical of all eukaryotic cells. Some (at the vacuole and ER) also act to generate calcium-signalling pools.

Why is there such complexity, and indeed so much apparent confusion about the location and function of the calcium pumps in plant cells, when the pattern in mammalian cells appears to be so much simpler? At least in part, the answer must lie with the diversity of plant cell types studied. Whereas mammalian cell biologists can isolate — and therefore ask questions of — a single cell type, this is seldom possible for plants, and biochemical studies are undertaken using mixed cell populations, some of which are highly vacuolate, while others are scarcely vacuolated. Even among highly vacuolate cells, vacuole function and content varies — from highly hydrated mesophyll to protein-filled cotyledons. The studies described above have also involved tissue from different species, of different ages, from different locations. Essentially, therefore, it seems likely that the type and location of plant calcium pumps varies depending on cell type, function, and morphology, and it is unlikely that all the types and locations of calcium transporters are present simultaneously in all cell types. This will be highlighted further as the role of calcium in cell signalling is discussed.

Calcium Channels in Plant Membranes

Extensive studies using electrophysiological and related techniques have identified calcium channels located at the plasma membrane and tonoplast of plant cells; these are likely to be involved in calcium release involved in signal transduction. Early studies

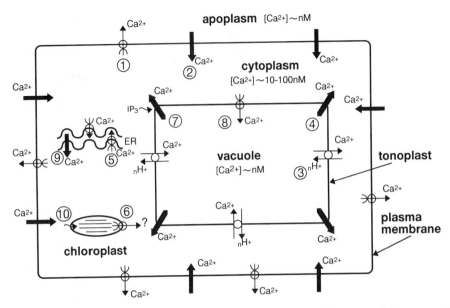

Figure 17.1 Calcium pumps, antiporters, and channels described in plant cells. 1, PM calcium pump (P-type); 2, PM calcium channel; 3, tonoplast calcium/proton antiporter; 4, vacuolar calcium channel; 5, ER calcium pump (P-type); 6, putative plastid calcium pump (P-type); 7, vacuolar calcium channel (ligand gated); 8, vacuolar calcium pump (P-type); 9, ER calcium channel; 10, chloroplast electrogenic calcium influx. All of these channels, pumps, and antiporters are described in detail in the text.

on giant Characean algal cells identified a plasma membrane calcium conductance which was blocked by nifedipine. This channel is believed to be involved both in turgor regulation and in the Characean action potential (Johannes et al., 1991). Plasma membrane calcium channels have also been identified in higher plants; early studies on carrot (Graziana et al., 1988) and maize (Harvey et al., 1989) identified verapamil-binding proteins which, in the case of maize, could be reconstituted by the planar lipid bilayer technique to form a calcium channel (Tester and Harvey, 1989). More recent studies have used electrophysiological techniques to identify calcium channels in the PM of a number of plants. Two calcium channels (which vary in their sensitivities to La^{3+}, and to Gd^{3+}) are present in maize roots (Marshall et al., 1994); a wheat root PM calcium channel analagous to mammalian L-type voltage-gated channels has been described (Huang et al., 1994); and nifedipine-sensitive calcium influx channels activated by cytokinin have been suggested in moss protonemal protoplasts (Schumaker and Gizinski, 1993). Many studies on calcium channels have concentrated on stomatal guard cells, as calcium release into the cytosol has been implicated in stomatal closure. Influx of calcium at the PM of *Commelina communis* guard cells has been demonstrated (MacRobbie and Banfield, 1988). Guard cell PM contains non selective calcium influx conductance (Schroeder, 1992), and

calcium permeation through K^+ channels in *Vicia faba* guard cell PM has been demonstrated (Fairley-Grenot and Assmann, 1992). Schroeder and Hagiwara (1990) demonstrated that repetitive stimulation of a nonselective calcium channel by abscisic acid [ABA] resulted in increases in guard cell cytosolic free calcium, and the same authors had previously observed that alteration in $[Ca^{2+}]_{cyt}$ regulated ion channel activity in the same cells (Schroeder and Hagiwara, 1989).

Much attention has focused on plant vacuolar calcium channels. Perhaps most exciting is an $InsP_3$-gated, voltage-activated channel which releases calcium from the vacuole (Ranjeva et al., 1988; Brosnan and Sanders, 1990; Johannes et al., 1992a, 1992b). This channel is similar in properties to the mammalian equivalent which resides at the ER; both are inhibited by heparin and both show high-affinity $InsP_3$ binding. Release of calcium from intracellular stores induced by photolysis of caged $InsP_3$ in guard cells (Gilroy et al., 1990) and abscisic acid induced rises in cytosolic calcium preceding closure of stomata (McAinsh et al., 1990), suggesting that this vacuolar channel is a key component of signal transduction in plants. A second, $InsP_3$-insensitive channel has also been identified in the tonoplast. This is a voltage-gated channel which is insensitive to heparin (Johannes, et al., 1992a, 1992b). Patch-clamp studies of stomatal guard cells also reveal the presence of voltage-sensitive calcium channels. Two such channels — one nifedipine- and Gd^{3+}-sensitive and of 27 pS conductance, the second a less abundant 14 pS channel — are presented in *Vicia faba* (Allen and Sanders, 1994). These channels would also function as release channels for calcium from the vacuole.

After many years of fruitless attempts to identify an ER calcium release channel in plants, a gadolinium-sensitive, voltage-dependent calcium channel has been identified (Klusener et al., 1995). This channel was detected using reconstituted protein from highly purified ER of the tendrils of *Bryonia dioica*. These tendrils are strongly touch-sensitive and the authors suggest that calcium-released from ER may be directly involved in this touch response — particularly as gadolinium also inhibits this action. Voltage gating is sensitive to the calcium gradient across the membrane and, in consequence, its open state will be dependent on the accumulation of calcium into the ER driven by the ER calcium-ATPase.

Plant Calmodulin

Higher plant calmodulin was first identified as a regulator of the enzyme NAD kinase (Muto and Miyachi, 1977; Anderson and Cormier, 1978). Since that time, involvement of calmodulin in a wide variety of plant cell processes has been demonstrated and its role as a primary response element for calcium signals in plants is now without question (for detailed review, see Zielinski, 1998).

Isoforms and Structure of Plant Calmodulins

Spinach calmodulin was first sequenced in 1984 (Lukas et al., 1984), revealing a very high level of structural and sequence homology with vertebrate calmodulin (ca. 90%). The protein undergoes conformational change upon binding calcium and has four calcium-binding domains of similar properties to those of vertebrates (Dieter et al., 1985; Yoshida et al., 1983). Plant calmodulin will also activate calmodulin-dependent enzymes from other organisms, but the degree of activation varies between calmodulin sources, especially for the activation of plant NAD kinase (Roberts et al., 1986).

Calmodulin sequences from a variety of plants were compared by Roberts and Harmon (1992). High levels of homology (97–99%) were noted with spinach calmodulin (alfalfa, barley calmodulin 1 isoform, wheat, *Arabidopsis*) with slightly less homology (92%) observed for potato calmodulin. Two genes for calmodulin, differing by four conservative amino acid substitutions have been identified in *Arabidopsis* together with evidence for other calmodulin isoforms (Ling et al., 1991). A putative calmo-

dulin homologue from barley has been identified showing some amino acid substitutions characteristic of vertebrate calmodulin; it is, however, only 76% homologous with barley calmodulin 1 and may either not be expressed or expressed only at low levels (Roberts and Harmon, 1992).

Calmodulin and Plant Cell Function

NAD kinase

As already indicated, calmodulin was first identified in plants as heat-stable activator of NAD kinase (Muto and Miyachi, 1977; Anderson and Cormier, 1978). Plant cells contain several forms of this enzyme, with only the cytosolic form being calmodulin-dependent. This enzyme, which appears to be a 55 kDa protein (Dieter and Marmé, 1980), shows almost complete dependence on calmodulin, with a $K_{0.5}$ of 0.1 nM (Harmon et al., 1984). Major changes in NAD(H)/NADP(H) ratio occur between illuminated and nonilluminated green tissue, affecting the utilisation of metabolites in the cell. However, the major change in NAD(H)/NADP(H) ratio in these circumstances is brought about by a chloroplast NAD kinase which is not calmodulin-regulated (Roberts et al., 1986). It has been proposed that calmodulin-regulated NAD kinase may be important in nongreen tissue; Dieter and Marmé (1986) noted that far red illumination of dark-grown maize coleoptiles resulted in a rise in NADP and that this effect could be mimicked by treatment with calcium ionophore A23187 in the presence of calcium. Increases in the level of NAD kinase in response to red light have also been noted, with up to a 17-fold increase following red light treatment (Kansara et al., 1989)

Nuclear NTPase

A nuclear NTPase that is stimulated 3.5-fold by calmodulin has been identified in pea. The enzyme has a relative molecular mass of ca. 47 kDa and has been shown to be regulated by phytochrome; however, its function remains uncertain (Matsumoto et al., 1984; Roux et al., 1986; Chen et al.,1987; Roberts and Harmon, 1992).

Glutamate Decarboxylase

Evidence for a plant calmodulin-binding glutamate decarboxylase has been obtained by molecular cloning using a petunia cDNA library; the translated sequence is of a 58 kDa protein showing 67% amino acid homology with its *Escherichia coli* equivalent (Baum et al., 1993). The calmodulin-binding domain lies near the C-terminal and is unique to the plant enzyme. It is suggested that calcium/calmodulin is involved in regulating γ-aminobutyric acid synthesis in plants. Activity of the enzyme is stimulated 2–8-fold by CaM at pH 7.0, but is not stimulated at pH 5.8; at pH 7, V_{max} is increased 2.4-fold, with a 55% reduction in K_m (Snedden et al., 1995).

Calmodulin-Stimulated Ca^{2+} Pumps

Plant cells, like animal cells, posess P-type calmodulin-stimulated calcium-pumping ATPases involved in the regulation of cytoplasmic free calcium. These pumps bear a calmodulin-binding autoinhibitory domain (Rasi-Caldogno et al., 1995); however, in contrast to mammalian cells there is evidence to suggest that at least, in some tissues, they are located at an intracellular membrane as well as at the plasma membrane (Evans, 1994a,b). and at least in some forms, the calmodulin binding domain is at the N-terminus rather than the C-terminus (Malmstrom et al., 1997; Harper et al., 1998). Further description of these pumps is presented elsewhere in this chapter.

Association with Microtubules and Tubulin

Since the demonstration by Keith et al. (1983) that calmodulin can depolymerise microtubules in vivo and of non-calcium-dependent binding of calmodulin to microtubules (MT) of mammalian cultured cells (Deery et al., 1984), interest has been shown in calmodulin binding to MTs in plants. Initially, calmodulin was colocalised with the spindle apparatus, phragmoplast, and pre–prophase band microtubules in plant cells (Vantard et al., 1985; Wick et al., 1985). Subsequently, immunolocalisation with cortical microtubules has also been shown (Fisher and Cyr, 1993). A homologue of mammalian elongation factor-1α has been recently purified by tubulin-affinity chromatography followed by calmodulin-affinity chromatography of plant cell extracts. This protein bundles plant microtubules in vitro in a calcium- and calmodulin-dependent manner (Durso and Cyr, 1994). Need for caution in interpreting results of experiments on calmodulin location in fixed tissue has, however, been indicated by Vos and Hepler (1998).

Calcium and Calmodulin-Dependent Protein Kinases

Numerous calcium-dependent protein kinases have been identified in plant tissue and several of these show 2–3-fold stimulation by calmodulin. Roberts and Harmon (1992), in a recent review, point out that many such studies, which used membrane preparations to demonstrate activity, are now equivocal in the light of the identification of a family of calmodulin-homologous protein kinases (calcium-dependent, calmodulin-independent protein kinase [CDPK]). However, molecular cloning has resulted in the identification of some plant protein kinases with calmodulin-binding domains. These include an apple calmodulin-binding serine-threonine protein kinase homologous to the mammalian type II protein kinase (Watillon et al., 1993) and a lily anther calmodulin-dependent protein kinase with a neural visinin-like calcium-binding domain (Patil et al., 1995). It seems likely that further calcium- and calmodulin-dependent protein kinases that have roles in signal transduction will be identified.

Calmodulin Localisation, Modulation, and Modification in Plants

In plants, as well as animals, modulation of calmodulin levels and activity is important in signal transduction; thus, cell processes respond not only to the calcium concentration, but also to the amount and localisation of calmodulin within a cell or cellular compartment. Roberts and Harmon (1992) point out that quantifying active calmodulin in plant cells is a difficult task and much of the data presented relies too heavily on techniques of debatable accuracy. However, studies reveal uneven distribution of calmodulin in various plant tissues. Calmodulin appears to be more strongly expressed in meristematic regions of leaves (Zeilinski, 1987), fruit (Ling et al., 1991), shoots (Muto and Miyachi, 1984), and roots (Allen and Trewavas, 1985; Dauwalder et al., 1986; Lin et al., 1986). Other studies suggest that post-translational modification of calmodulin, by methylation of lysine 115, is also of importance; in cell cultures of carrot, methylation is low in the lag and early exponential phases, but rises during exponential growth; unmethylated calmodulin is restored after 12 hours in fresh culture medium (Oh et al., 1992). Studies on whole tissue reveal that levels of methylation are higher in the elongation zone of roots than at the root meristem (Oh and Roberts, 1990). Together, the data suggest developmental regulation of calmodulin levels and of calmodulin methyl transferase. While the regulatory significance of calmodulin methylation in plants is unknown, both NAD kinase activation and ubiquitin binding are affected by methylation levels (see Roberts and Harmon, 1992).

High levels of calmodulin appear to be associated with plant cell plasma membranes (Collinge and Trewavas, 1989); indeed, it is believed that, in many instances, the plant plasma membrane calmodulin-stimulated calcium pump is fully saturated with calmodulin in vivo, making identification of calmodulin activation difficult (Evans et al., 1992).

Calcium-Dependent, Calmodulin-Independent Protein Kinase

Plant cells posess a family of protein kinases which contain a calcium-binding domain homologous to calmodulin. Known as CDPK, they appear to represent a new class of protein kinases (Roberts and Harmon, 1992). CDPK from soybean appears on SDS-PAGE as two bands, of M_r 52 and 54 kDa. Molecular cloning suggests isoenzymes; CDPKα, the first to be fully characterised, being 508 amino acids in length with a predicted M_r of 57,175 Da (Harper, et al., 1991). CDPKα residues 32–295 encode a serine-threonine protein kinase catalytic domain 33–42% identical to the catalytic domains of calcium/calmodulin-dependent protein kinases; residues 301–322 have 32% identity with the autoinhibitory domain of CaM-PKβ from rat brain; and residues 329–473 are 39% identical to spinach calmodulin (Roberts and Harmon, 1992). Three new CDPKs identified in *Arabidopsis* of predicted MW 55,376–59,974 Da were shown to contain a single polypeptide chain with a putative EF-hand structure at the C-terminus and a kinase domain at the N-terminus (Hong et al., 1996). Expression of these isoforms appeared to be developmentally regulated.

CDPK is suggested to be widely distributed in higher plants and algae; forms of the enzyme ranging from 40–90 kDa have been identified. CDPK is activated by micromolar concentrations of free calcium, does not require calmodulin or phosphatidyl serine for activity, shows calcium-dependent binding in hydrophobic interaction chromatography and phosphorylates histone H1 (Harmon et al., 1987; Putnam-Evans et al., 1990). CDPK has been shown to be associated with plasma membrane (Schaller and Sussman, 1988; Klucis and Polya, 1988) and soybean root nodule membranes (Weaver et al., 1991). Soluble, chromatin-associated and cytoskeleton-associated CDPK have also been found (see Roberts and Harmon, 1992).

Investigation of the regulatory functions of plant CDPK is progressing rapidly. In reviewing the field, Roberts and Harmon (1992) reported that CDPK had been shown to phosphorylate the purified plasma membrane proton pump (Schaller and Sussman, 1988; see Roberts and Harmon, 1992) and was localised to plant actin filaments. More recent work has identified several key substrates for CDPK. In Characean algae, CDPK is associated with actin bundles involved in cytoplasmic streaming, colocalising with myosin, and, suggesting a role in the inhibition of streaming by transient rises in $[Ca^{2+}]_{cyt}$, influencing myosin light chain kinase (McCurdy and Harmon, 1992a, 1992b). Plant phospholipid transfer proteins have been observed to be excellent substrates for CDPK (Polya et al. 1992; Neumann et al., 1993, 1994b, 1995). Nodulin 26, a nodule-specific channel protein associated with the symbiosome of soybean root nodules is also a CDPK substrate (Weaver et al., 1991, 1992, 1994; Weaver and Roberts, 1992) as is a plant protease inhibitor (Neumann et al., 1994a). Finally, expression of two *Arabidopsis thaliana* CDPKs have been shown to be induced by drought and salinity (Urao et al., 1994) and *Arabidopsis* CDPK isoform CPK-1 has been shown to be stimulated by members of the 14-3-3 family of regulatory proteins (Camoni et al., 1998).

Phosphoinositides and Calcium

Given the extensive interactions of compounds of phosphoinositide metabolism and calcium-signalling pathways in animal cells, it is inevitable that considerable interest has been expressed in similar interactions in plants; however, it is by no means inevitable that exact parallel systems will exist. Over the last 10 years or so, very considerable evidence has been accumulated to suggest that calcium signalling, in at least some plant cells, is linked with phosphoinositide signals; but it is also clear that some major questions remain. Given the recent publication of a number of excellent reviews of the field (Drøbak, 1991, 1992; Coté and Crain, 1993) and the considerable supporting literature involved, only key elements of our understanding of the systems involved will be presented here.

Perhaps the most convincing evidence for a role for the key polyphosphoinositide metabolite IP$_3$ (Ins [1,4,5]P$_3$) in plants comes from studies where release of calcium from intracellular stores (the vacuole, not the ER as in animal systems) was described (Drøbak and Ferguson, 1985; Schumaker and Sze, 1987). Release was shown to be specific for IP$_3$ and occurs at low ligand concentrations; the presence of IP$_3$-gated calcium channels has been demonstrated and although the nature of the channel/ligand interaction has yet to be fully elucidated, a high-affinity binding site for IP$_3$ has been shown (Brosnan and Sanders, 1993). Other components of an IP$_3$-dependent signalling pathway have also been shown in plants; in reviewing the area, Drøbak (1991) considered that the following were established:

1. Presence of phosphatidyl inositol, phosphatidyl inositol-4-phosphate and phosphatidyl inositol-4,5-bisphosphate in plant membranes (Drøbak et al., 1988; Irvine et al., 1989).
2. Presence of kinases/phosphatases involved in polyphophoinositide turnover (Drøbak et al., 1988; Sommarin and Sandelius, 1988).

3. Presence of phospholipase(s) C capable of hydrolysing phosphatidyl inositol-4,5-bisphosphate (McMurray and Irvine, 1988).
4. Presence of enzymes which, in several respects, resemble mammalian protein kinase C (Drøbak et al., 1988).
5. Presence of enzymes capable of rapidly metabolising Ins(1,4,5)P$_3$ (Joseph et al., 1989; Drøbak et al., 1991).

Others have also obtained evidence for protein kinase C-like enzymes (Schafer et al., 1985; Elliott and Kokke, 1987a, 1987b; Elliott et al., 1988) and molecular cloning has revealed protein kinases that show some homology with the mammalian enzyme (Lawton et al., 1989).

Coté and Crain (1993), however, comment that "the establishment of causal links among signal reception, Ins(1,4,5)P$_3$ production, Ca^{2+} mobilization and cell response has remained elusive". In part, this may be because numerous different cell systems have been studied and the evidence from each is somewhat fragmentary. The role of IP$_3$ in guard cell function is discussed elsewhere in this chapter, as is egg fertilisation in Fucus. Studies on the responses of lucerne and tobacco cells to fungal attack have demonstrated enhanced turnover of inositol phospholipids. In tobacco, treatment with a fungal elicitor resulted in inositol-1,4-bisphosphate levels increasing 15-fold over controls after only 10 minutes, while levels of inositol-1,4,5 trisphosphate increased, to a smaller extent (38%). Phosphatidyl inositol kinase, but not phospholipase C, was increased by treatment (Kamada and Muto, 1994a, 1994b). Lucerne cells showed a 100–160% rise in a compound co-chromatographing with IP$_3$ within 1 minute of stimulation, with levels falling to below those of controls in 3 minutes (Walton et al., 1993).

In summary, it is evident that many of the components of an IP$_3$-based signalling pathway that interact with a Ca^{2+}-signalling pathway are present in plants. However, certain key areas remain to be fully elucidated and there are clear contrasts with the animal cell systems—not least the fact that calcium is released from the plant vacuole not the ER.

Calcium and Mitosis in Plants

The involvement of calcium ions in the regulation and coordination of mitosis in plants has been a subject of research for more than 20 years. Research on plant tissue (for instance, the stamen hairs of Tradescantia and the naked (wall-less) endosperm cells of Haemanthus) have permitted significant advances in understanding mitosis in plant and animal cells.

Key calcium-related features of the higher plant mitotic apparatus are the spindle microtubular arrays, membranes of the endoplasmic reticulum which permeate the spindle, and the presence of calcium-response elements within the mitotic apparatus. The sensitivity of plant microtubules to calcium and the presence of calcium response elements (in particular, calmodulin) associated with plant tubulin have been discussed elsewhere in this chapter. Association of calcium with the membranes of the plant mitotic apparatus [MA] have been indicated by Ca^{2+}-dependent antimonate precipitation and by chlorotetracycline fluorescence (Wick and Hepler, 1980, 1982; Hepler and Wolniak, 1984); recent studies (Theodoulou et al., 1994; Downie et al., 1998) also suggest that the membranes of the plant MA may contain a calcium pump, similar to that shown for nonplant systems (see Hepler, 1992, for review). Calmodulin distribution patterns similar to (but not identical with) the distribution of microtubules in the phragmoplast have been shown in onion (Wick et al., 1985) and Haemanthus (Vantard et al., 1985), though a recent study by Vos and Hepler (1998) of living (unfixed) Tradescantia stamen hairs suggested that microtubule association of CaM may be an artefact of fixation.

It appears that calcium has an important role in progression through metaphase and entry into anaphase. Using Tradescantia stamen hairs, which show marked uniformity in progression through metaphase, Hepler (1985) demonstrated that factors which restrict calcium entry into the cell (external chelation, application of lanthanum, calcium channel blockers, etc.) prolong metaphase transit times, without affecting subsequent rates of chromosome movement, in a calcium-dependent manner (Hepler, 1985; Wolniak and Bart, 1986). Introduction of calcium chelator (Quin-2) into the cells also prolonged metaphase (Wolniak and Bart, 1985). Measurement of calcium transients in mitosis suggested that calcium may be released from spindle-associated ER briefly prior to the onset of chromosome movement in anaphase (Wolniak et al., 1983, 1984); ionic changes (which may include calcium) have also been demonstrated (Wolniak et al., 1983). However, Hepler (1994) comments that while there is "little direct evidence that Ca^{2+}-transients trigger the onset of anaphase, spindle fibres and chromosome movement at anaphase are exquisitely sensitive to the ion at physiological levels".

The evidence for a role for calcium in plant mitosis is therefore compelling, with calcium accumulated in ER closely associated with the kinetochore microtubules during metaphase and released at the onset of anaphase, when it interacts with microtubule-associated response elements involved in chromosome movement. Some caution is needed, however, as much of the data is correlative; in reviewing the field, Hepler and Wayne (1985) and Hepler (1992) point out that we are far from understanding the mechanism of action of calcium in this process.

Calcium and Gravitropisms

Plant tropisms (growth responses, particularly to light and gravity) have fascinated plant scientists for more than a century. At first sight, tropisms provide all the elements for an ideal model system for the study of signal perception, transduction, and response: perception of light or gravity, signalling from zones of perception to regions of growth; and finally a clear response—alterations in elongation rate giving bending of the plant organ to or from the stimulus. Sadly, progress in understanding the signal transduction pathway has been slow, with much apparently conflicting information and debate as to which out of several models is correct. It is into this debate that our understanding of a role for calcium must be placed and, while it is not possible to detail all the relevant literature here, a summary of the alternative signal perception and transduction routes is appropriate.

Gravitropism occurs in many plant organs, which may be either positively or negatively geotropic or diageotropic (i.e. remain at right angles to gravity). In many organs (e.g. rapidly growing roots), the induction time for geotropism is less than a minute (Johnsson and Pickard, 1979). Of several possible perception mechanisms suggested, the sedimentation of amyloplasts (starch-containing organelles of comparatively high density) has remained favoured as one of the few components of the cell which sediments significantly within this time (Pickard, 1985)—although the amyloplast theory is not without problems. Other initial perception mechanisms may be involved, at least in some tissues (see Masson, 1995). In a recent review, Pickard (1994) cites accumulating evidence for a mechanism involving protoplast interaction aginst the cell wall, detected by mechanosensory ion channels in the plasma membrane (see also Pickard and Ding, 1993). In roots, gravity sensing occurs at the root tip, possibly in specialised columella cells which contain large amyloplasts (Masson, 1995) and a signal transduction pathway is then involved in conveying that signal to the elongation zone of the root. Here, changes in cell wall rigidity coupled with the internal pressure of cell turgor influence elongation and induce curvature of the organ until vertical growth is restored. Alterations in growth rate are related (inter alia) to the activity of the plasma membrane proton pump, which acidifies the wall space, resulting in loosening of the wall.

Calcium is believed to be involved directly in the transmission of the signal between root tip and elongation zone; but its consideration is a relatively recent phenomenon when compared with auxin, which dominated research from the 1920s to 1980s. Uneven distribution of auxin in root tips in response to gravistimulation was first suggested by Cholodny (1924) and Went (1926); evidence for the so-called Cholodny/Went hypothesis was collected and then hotly debated in subsequent years (eg Hall, et al., 1980; Trewavas, 1981, 1991, 1992a,b). Recently, Li et al. (1991) used transgenic tobacco seedlings that express a reporter gene (β-glucuronidase) coupled to an auxin-responsive promoter to demonstrate both redistribution of an auxin signal on gravistimulation and also a rapid alteration of auxin-regulated gene expression in the vicinity of the elongation zone. Auxin involvement in gravitropism cannot be denied. However, auxin is not the only signal to redistribute during graviresponse. In 1983, Slocum and Roux demonstrated marked assymetry in calcium across gravistimulated oat coleoptiles within 10 minutes using an antimonate precipitation technique. Calcium redistribution preceded curvature of the organ. This demonstration was rapidly followed by studies suggesting similar redistribution using a variety of other techniques. Goswami and Audus (1976) had already shown redistribution of [45]Ca in gravistimulated roots; roots whose tips were deprived of calcium were shown to lose the ability to respond to gravity (Lee et al., 1983a), and calcium applied to one side of the root resulted in curvature towards the calcium source (Lee et al., 1983b, 1984). Decapped roots and those grown in the presence of auxin transport inhibitors did not redistribute [45]Ca to the same extent as controls (Lee et al, 1983b,1984). Using microelectrodes, Bjorkman and Cleland (1991) have provided further confirmation of redistribution in gravistimulated maize roots and have also shown that calcium redistribution is not just the result of pH changes in the tissue. Legue et al. (1997), however, found no evidence for alternation in cytoplasmic calcium in root cells of *Arabidopsis* in response to gravity.

Perhaps one of the most interesting aspects of the involvement of calcium in gravitropism is the fact that the redistribution of calcium is chiefly apoplastic (i.e. outside the cell) in nature. Transport of [45]Ca from donor blocks to receivers placed at the root surface, for instance, requires a transport pathway across the root, and while intracellular transport may be involved, the net effect is one of the gross movement of the ion outside the cells; this is unusual in calcium signalling, which is normally considered to be an intracellular event. Such transport could result from the unequal activity of channels (on one face of the cell) and pumps (on the opposite face); Sievers et al. (1984), for instance, speculated that the initial trigger for calcium redistribution was a calcium pump on ER impinged upon by amyloplasts. However, while this may be part of an initial trigger, plasma membrane channels and calcium pumps must also be involved. Calmodulin has also been implicated in graviresponse. Study of *Arabidopsis* has revealed that whilst wild-type plants show a three-fold increase in CaM expression in the early stages of graviresponse, CaM expression is decreased when a gravity-insensitive mutant is exposed to a gravitational field (Sinclair et al., 1996).

Several key questions remain: which comes first—redistribution of calcium or of auxin? Second, how is the uneven calcium distribution propagated to the responsive tissue? Third, to what extent are the auxin and calcium signals interactive? Young and Evans (1994) inhibited auxin redistribution in roots by applying calcium chelators and then restored redistribution with calcium; they proposed that calcium redistribution is essential for development of the auxin gradient. Such inhibition studies are at best suggestive, however, as calcium may be required for the signal without being the initial signal itself. It is evident that much effort still remains if we are to be certain about the perception and transmission of stimuli in gravitropism; however, it is without question that calcium signalling is a key component of the system.

Calcium and Photoregulation

A variety of plant cell processes are regulated by light quality. The ability to sense and respond to light is vital for the optimal growth of autotrophic organisms dependent on light for their existence. Higher plants are able to sense and respond to light; they possess pigments sensitive to specific wavelengths of light and signal transduction pathways that induce a range of responses from chloroplast movement and phototropism (growth to or from the light) to seed germination and the induction of gene expression. Key amongst the photoreceptors identified are the phytochromes, which respond to light of red and far-red wavelengths; blue light responses have also been identified. In reviewing the involvement of calcium in photoresponses, Hepler and Wayne (1985) identified a range of processes in which calcium was known to be involved, including chloroplast movement and rotation (see below), spore germination (Raghavan, 1980; Appenroth et al., 1994), and leaflet closure in the sensitive plant *Mimosa* (Toriyama and Jaffé, 1972). Since 1985, perhaps the most significant progress has been made in dissecting signal transduction pathways involving light and calcium; and it is

on the elucidation of these pathways that this section will concentrate.

The phytochrome family of photoreceptors possess an amino-terminus that confers photosensory specificity and a C-terminus that specifies subsequent transmission of the signal (Quail et al., 1995). Phytochrome regulation of gene expression has been observed during greening of etioplasts to form fully functional chloroplasts. Using phytochrome-deficient tomato mutants, Bowler et al., (1994) have discovered the operation of three separate but interactive signalling pathways operating via heterotrimeric G-proteins, cGMP, calcium, and calmodulin. The same authors demonstrated that G-protein activation and a combination of cGMP and calcium were able to induce the development of fully mature chloroplasts; calcium and calmodulin alone failed to induce the formation of fully active chloroplasts or anthocyanin production, and cGMP alone induced anthocyanin production. Similar results were obtained using a GUS reporter system which demonstrated that phytochrome phototransduction first involved the activation of two or more G-proteins which activated separate signal transduction pathways, one requiring calcium and activated calmodulin, the other being calcium-independent (Neuhaus et al., 1993).

A direct link between phytochrome and transient rises in cytosolic free calcium has been demonstrated by Shacklock et al. (1992) using wheat leaf protoplasts. Red light induces swelling of these protoplasts by a mechanism believed to be similar to that in phytochrome-regulated leaf photomorphogenesis. Red-light-induced protoplast swelling could be mimicked by release of caged calcium or IP_3 within the cells and red light induced a transient rise in cytosolic free calcium followed by a decrease to below resting levels. In other instances, phytochrome has been shown to induce an influx of calcium into the cell. Red-light stimulation of the spores of *Onoclea* induces net calcium uptake, while far-red light inhibits (Wayne and Hepler, 1984); in chloroplast rotation in the alga *Mougeotia*, red light has been noted to stimulate calcium uptake transiently and, again, far-red light inhibits (Dreyer and Weisenseel, 1979).

Chloroplast movement and rotation has received considerable attention (see Hepler and Wayne, 1985, for review of early work) as, in *Mougeotia*, a clear link with calcium has been demonstrated. External calcium is essential for chloroplast rotation (Wagner and Klein, 1981), and a small calcium influx occurs, accompanied by a rapid calcium efflux, possibly due to release of calcium from intracellular stores (Hale and Roux, 1980). X-ray microanalysis suggests the presence of calcium-containing vesicles (Wagner and Rossbacher, 1980) in the vicinity of

Mougeotia chloroplasts which release calcium on light stimulation. Further evidence for calcium involvement in chloroplast rotation came from the elegant experiments of Serlin and Roux (1984), who induced chloroplast rotation in the absence of light by positioning the calcium ionophore A23187 locally at the plasma membrane. Chloroplast movement in other species is influenced by blue light, rather than red light and the phytochrome system. In reviewing calcium and chloroplast movement in a variety of species, Wada et al. (1993) comment that chloroplast rotation results from reorganisation of the actomyosin system and that local changes in membrane properties are the earliest stages of the signal transduction pathway. However, they comment that "the role of calcium and calmodulin, although obviously involved in some species, is not clear as yet".

Calcium and Mechanical Signalling

Plants are known to be sensitive to a variety of mechanical signals. In addition to obviously "sensitive" plants like *Mimosa pudica*, which shows progressive leaflet collapse on touch, and the insectivorous plants which show rapid leaf movement on stimulation of trigger hairs, many other plants show both short- and long-term responses to stimulation. Persistent wind movement causes stem thickening and strengthening, and organs like coleoptiles and roots will bend upon unilateral stroking (Trewavas and Knight, 1994). Evidence for an involvement of calcium signals in these responses requires a rapid and sensitive assay method for calcium in tissues as well as cells; something not achievable with fluorescent indicator dyes like quin and fura.

Measurement of short-duration transient rises in $[Ca^{2+}]_{cyt}$ in plant cells has recently been achieved by developing recombinant plants that express the coelenterate calcium indicator aequorin (Knight et al., 1991, 1992, 1993; Knight and Knight, 1995). Apoaequorin (recombinant aequorin) luminesces in the presence of coelenterazine; use of *h*-coelenterazine (a semisynthetic recombinant form of coelenterazine) permits quantification of even very small fluxes in $[Ca^{2+}]_{cyt}$ that are obtained when plants are subjected to mechanical stimuli. This approach has been used very successfully both on whole plants (tissues and organs) and on cell strips, originally those of *Nicotiana plumbaginifolia* and more recently on the moss *Physcomitrella patens*.

Wind stimulation of *Nicotiana* expressing aequorin results in an immediate increase in aequorin luminescence, probably reflecting increased $[Ca^{2+}]_{cyt}$ (Knight et al., 1992). The magnitude of the response increases as the stimulus is increased either in magnitude or duration. Prolonged stimulation results in a temporary loss of the response; however, sensitivity is recovered after 40–60 seconds. Other mechanical stimuli that result in disturbance of the tissue also result in transient rises in $[Ca^{2+}]_{cyt}$ proportional to the magnitude and duration of stimulus. Localised wounding, for instance, induces a rise in $[Ca^{2+}]_{cyt}$ in small isolated groups of cells some distance from the wound as well as at the wound site itself (Knight et al, 1993). Touch also causes a localised rise in luminescence indicative of a rise in $[Ca^{2+}]_{cyt}$ (Knight et al., 1991). When isolated epidermal strips and even mesophyll protoplasts are stimulated (in these cases, by a stream of isotonic medium), a calcium-dependent increase in aequorin luminescence is also observed (Haley et al., 1995). Similar rises in $[Ca^{2+}]_{cyt}$ are also evident in transformed tobacco after cold shock and treatment with fungal elicitors (Knight et al., 1991), as well as when tissues are placed under oxidative stress (Price et al., 1994).

It appears that the calcium signal results from movement of the tissue—possibly as a result of mechanosensory calcium channels that detect flexing and compression of cells. However, while localised deformation of tissue may explain rises in $[Ca^{2+}]_{cyt}$ in intact tissue, it seems less likely for protoplasts. Indeed, as La^{3+}, Gd^{3+}, and EGTA did not inhibit the protoplast signal, release of calcium from an internal store seems likely and the same situation pertains for isolated epidermal strips (Haley et al., 1995). Clearly, significant further effort is required to describe the calcium signal transduction pathway in detail.

These data indicate that transient rises in $[Ca^{2+}]_{cyt}$ are commonplace phenomena in plants and reinforce the concept of calcium as a key component of signalling pathways in plant as well as animal cells. It is evident that if altered $[Ca^{2+}]_{cyt}$ is to result in specific responses to these many stimuli, the calcium signal must contain more information than a rise in concentration alone can confer, and localisation and duration of the signal, as well as abundance and localisation of response elements, are likely to be of key importance. It is interesting to note that alterations in expression of calmodulin and calmodulin-related proteins have been noted in response to mechanical stimulation (Braam and Davies, 1990) This can be very rapid; for instance, transient up-regulation of calmodulin in *Brassica napus* by touch and wind has beennoted in only 30 min (Oh et al., 1996).

Calcium and Stomatal Closure

Stomatal pores are surrounded by two cells, the guard cells; altered turgor within these cells results in changes in cell shape, such that the stomatal pore opens and closes. These turgor changes occur predominantly as a result of cation (predominantly potassium) and anion influx and efflux. It is well established that these turgor changes are regulated by light, carbon dioxide, and the plant growth substance abscisic acid, which is produced in response to wilting. It is now clear that alterations in cell calcium are vital components of the signal transduction involved.

The wealth of information on calcium and stomatal action means that citation of all the relevant literature is impossible within the confines of this chapter; however, several excellent recent reviews are available (e.g. MacRobbie, 1992; Schroeder, 1992; Blatt and Thiel, 1993). Alterations in guard cell turgor involve the activity of a number of ion channels, permitting the rapid movement of both anions and cations. ABA is believed to first activate inward-directed ion currents which depolarise the plasma membrane and initiate K^+-efflux; subsequent stages involve inactivation of inward-rectifying K^+ channels, activation of voltage-gated anion efflux channels, a rise in cytoplasmic pH, and, finally activation of outward-rectifying K^+-efflux channels. All of these stages are complete within about 1 minute (see MacRobbie, 1992; Blatt and Thiel, 1993).

ABA is believed to interact with an (as yet undiscovered) receptor protein which may be cytosolic or at the plasma membrane (Blatt and Thiel, 1993; Anderson et al., 1994). This interaction initiates two components of a signalling cascade, one involving calcium and the other cytosolic pH (Blatt and Armstrong, 1993; reviewed by Blatt and Thiel, 1993). It is established that stomatal closure is frequently associated with a transient rise in cytosolic free calcium concentration, detectable by fluorescent dyes and other means (McAinsh et al., 1990, 1992; Gilroy et al., 1991). This calcium transient can be also be triggered by the photolysis of caged IP_3 in the guard cell (Gilroy et al., 1990), which is able to inactivate $I_{k\ in}$ (Blatt et al., 1990). ABA has been shown to activate nonselective Ca^{2+}-permeable ion channels at the plasma membrane (the first inward-directed ion current) and this, coupled with calcium release from intracellular stores (probably the vacuole, by IP_3), results in the observed rise in cytosolic free calcium, which progresses as a wave across the cell. Addition of IP_3 has been shown to inhibit the inward K^+ channel and to increase background conductance (known as the "leak current"), as well as initiating stomatal closure. Some involvement of calcium influx at the plasma membrane is also suggested by the fact that low external calcium concentrations and the presence of inhibitors of calcium influx partially inhibit closure (Gilroy et al., 1990; McAinsh et al., 1990, 1992).

There are, however, some questions that remain concerning the involvement of calcium transients in stomatal closure. The fact that calcium-channel blockers acting externally, and external calcium chelation, do not inhibit stomatal closure in all cases (Gilroy et al., 1990; MacRobbie, 1990; McAinsh et al., 1990) has been cited as one such problem (Blatt and Thiel, 1993); however, it is possible that calcium from intracellular stores is sufficient to convey the signal and that influx at the plasma membrane is required chiefly to replenish those stores. More problematical is the fact that calcium transients are not observed in all guard cells when stomata close; however, given the short duration of such calcium transients and the technical difficulties involved in measuring intracellular calcium in plant cells, such transients may be occurring but are undetected.

Finally, the importance of interaction between the signalling pathways involved should not be overlooked. While cytoplasmic free calcium seems likely to play a very important role, alterations in cytoplasmic pH (which will indirectly also affect free calcium), together with phosphoinositide-based signalling, must also be taken into account. These are likely to affect both the magnitude and duration of calcium transients, as well as the cells responsiveness to them. We are some way from a fully integrated picture of stomatal action but calcium is clearly identified as a major component of the system.

Calcium and Fertilisation Signals in Plants

Fundamentally, fertilisation in plant cells is identical to that in animals in that it requires the interaction of male and female gametes. However, while lower plants (e.g. the algae) have simple fertilisation mechanisms in which a sessile egg and motile sperm interact in an aqueous environment, higher plants show much more complexity. In the angiosperms, the male gametes reach the ovule (contained within the embryo sac) via a pollen tube which develops from one of two haploid cells within the pollen grain. Fertilisation, in which two haploid male cells simultaneously fuse with two female cells (one destined to form the zygote, embryo, and sporophyte; the other to form the storage reserve tissue of the endosperm), takes place some distance from the point where the pollen grain first makes contact (the stigma). The whole process requires extensive temporal and spatial coordination; and it is in this coordination that involvement of calcium has been suggested or demonstrated. While the events of pollen tube growth have been extensively studied in higher plants, the signalling mechanisms of egg activation and fertilisation are far better understood in

Table 17.3 Events of *Fucus Egg Fertilisation*

Sperm recognition
Rapid membrane depolarisation
Na^+ and Ca^{2+} influx via channels
Release of calcium from sub-PM stores
Localised rise in $[Ca^{2+}]_{cyt}$ subtending PM
Wall deposition
Repolarization phase
K^+-efflux channels active
Depolarisation of PM
Decline in $[Ca^{2+}]_{cyt}$ to resting state

After Brownlee (1994).

the simpler algal cell system and these two systems will be summarised here. For further critical appraisal and general detail, the review of Brownlee (1994) is strongly recommended.

Calcium and *Fucus* Egg Fertilisation

Fertilisation events in *Fucus* eggs have been investigated using electrophysiological techniques and fluorescent indicators by Brownlee and coworkers (see Brownlee, 1994, for a review, and Taylor and Brownlee, 1992, 1993; Roberts et al., 1993, 1994). Unfertilised eggs are wall-less; upon fertilisation, the eggs demonstrate a characteristic membrane depolarisation, the fertilisation potential, which lasts for 2 or more minutes (Brawley, 1991). This is followed by a gradual repolarisation to the zygote resting potential. Meantime, the fertilised egg rapidly lays down a cell wall, develops turgor, and becomes resistant to fertilisation by other sperm.

While *Fucus* eggs present many experimental problems for both electrophysiology and microscopy, it has been possible to measure ion channel activity in unfertilized eggs by patch-clamping (Taylor et al., 1992), as well as monitoring membrane potential through fertilisation. Based on this data, Brownlee (1994) suggests that the fertilisation potential observed upon gamete fusion results from the activity of two voltage-gated ion channels present at the egg plasma membrane: one a calcium influx channel and the second a potassium efflux channel. A rapid initial depolarisation results from sperm recognition and involves the opening of calcium influx channels. Release of Ca^{2+} from intracellular stores (possibly ER subtending the PM) propagates the signal around the egg just below the PM and induces exocytosis of wall materials; finally, a high activity of K^+-efflux channels (typical of the zygote) repolarises the plasma membrane (see Table 17.3).

Rises in $[Ca^{2+}]_{cyt}$ in regions just subtending the PM have been observed following fertilisation by the use of the calcium indicator calcium-green dextran (Roberts et al., 1994), but are not of the magnitude or global nature observed for animal deuterosomes (Brownlee, 1994). Egg activation cannot be inhibited by injection of calcium buffers sufficient to block the internal rise in calcium concentration (although this was achieved at higher buffer concentrations). Roberts et al. (1994) therefore suggest that a large, transient global rise in $[Ca^{2+}]_{cyt}$ is not required for egg activation although localised influx is needed for early fertilisation events.

Subsequent to fertilisation, algal zygotes develop a distinct polarity that results in an unequal cell division to give rise to rhizoid and thallus. The polarisation is determined in response to light, with the rhizoid pole growing away from the light (Jaffé, 1958). Involvement of a calcium-influx current in polarisation seems likely as measurements using Ca^{2+}-specific electrodes (Brownlee and Wood, 1986) and Ca^{2+}-indicators (CTC; Fura-2 and calcium-green; Kropf and Quatrano, 1987; Brownlee and Pulsford, 1988; Brownlee, 1989; Berger and Brownlee, 1993) all demonstrate elevated $[Ca^{2+}]_{cyt}$ at the rhizoid tip of the cell.

Signals that Involve Calcium in Pollination in Higher Plants

Growth of a pollen tube from the stigma down the style to the ovule requires very significant spatial coordination. Pollen tubes are tip-growing organs and a role for $[Ca^{2+}]$ in determining the direction of this growth has been suggested (see Brownlee, 1994). $[Ca^{2+}]_{cyt}$ is elevated near the apex of the growing tube and injection with calcium buffers inhibits tube growth (Obermeyer and Weisenseel, 1991; Rathore et al., 1991; Miller et al., 1992). Pollen tubes also show positive chemotropism toward elevated calcium (Steer and Steer, 1989), although it seems likely that chemotropism toward other substances is also involved (Reger et al., 1992a, 1992b, 1993). Pollen tubes finally grow to the micropyle, through which they enter the embryo sac. Closely associated with the ovule within the embryo sac in many angiosperms are the synergids. These cells have high vacuolar Ca^{2+} concentrations (Chaubal and Reger, 1990, 1992) and may provide the calcium gradient towards which the pollen tube grows having entered the embryo sac. Subsequent events surrounding the fusion of the two male haploid cells with the two female cells are much less clearly understood, not least because achieving fusion in vitro has proved difficult. Apart from electrofusion, fusion has only been achieved under conditions of high Ca^{2+} and pH, suggesting that [Ca] may be involved at this stage in vivo (Faure et al., 1994).

Conclusions

Calcium is firmly established both as an essential plant nutrient and as an intracellular signalling molecule with widespread functions in plants. The molecular dissection of its function in plants has already yielded many insights into its importance in integrating metabolic and structural processes. Calcium in plants clearly remains a major subject of research and advances can be expected in the near future; perhaps most significant of these will be the development of more detailed overall models for the initiation and integration of calcium signals in a diversity of plant tissues and organs.

Acknowledgments The author is grateful to Dr. M. R. Blatt, L. Downie, and I. M. Williamson for their comments on sections of the manuscript. The author is a Royal Society 1983 University Research Fellow.

References

Allan, E. and A. Trewavas (1985) Quantitative changes in calmodulin and NADkinase during early cell-development in the root apex of *Pisum sativum* L. *Planta* 165: 493–501.

Allen, G. J. and D. Sanders (1994) Two voltage-gated, calcium-release channels coreside in the vacuolar membrane of broad bean guard-cells. *Plant Cell* 6: 685–694.

Anderson, B. E., J. M. Ward, and J. I. Schroeder (1994) Evidence for an extracellular reception site for abscisic-acid in *Commelina* guard-cells. *Plant Physiol.* 104:1177–1183.

Anderson, J. M. and M. J. Cormier (1978) Calcium-dependent regulator of NADkinases in higher plants. *Biochem. Biophys. Res. Commun.* 84: 595–602.

Andreev, I. G., V. Koren'Kov, and Y. G. Molotkovsky (1990) Calmodulin stimulation of Ca^{2+}/nH^+ transport across the vacuolar membrane of sugar beet taproot. *J. Plant Physiol.* 136: 3–7.

Appenroth, K. J., S. Durr, H. Gabrys, and R. Scheuerlein (1994) No regulation of [^{45}]-calcium-uptake or release by phytochrome as an essential step in the transduction chain *Plant Physiol. Biochem.* 32: 429–435.

Askerlund, P. (1997) Calmodulin-stimulated Ca^{2+}-ATPases in the vacuolar and plasma membranes in cauliflower. *Plant Physiol.* 114: 999–1007.

Askerlund, P. and D. E. Evans (1992) Reconstitution and characterization of a calmodulin-stimulated Ca^{2+}-pumping ATPase purified from *Brassica oleracea* L. *Plant Physiol.* 100: 1670–1681.

Askerlund, P. and D. E. Evans (1993) Detection of distinct phosphorylated intermediates of Ca^{2+}-ATPase and H^+-ATPase in plasma-membranes from *Brassica-oleracea*. *Plant Physiol. Biochem.* 31: 787–791.

Basu, S. and D. P. Briskin (1995) Target molecular size suggesting a dimeric structure for the red beet plasma membrane Ca^{2+} ATPase. *Phytochemistry* 38: 15–17.

Baum, G., Y. L. Chen,, T. Arazi, H. Takatsuji, and H. Fromm (1993) A plant glutamate-decarboxylase containing a calmodulin-binding domain—cloning, sequence, and functional-analysis. *J. Biol. Chem.* 268: 19610–19617.

Berger, F. and C. Brownlee (1993) Ratio confocal imaging of free cytoplasmic calcium gradients in polarising and polarised *Fucus* zygotes. *Zygote*, 1: 9–15.

Bjorkman, T. and R. E. Cleland (1991) The role of extracellular free-calcium gradients in gravitropic signaling in maize roots, *Planta*, 185: 379–384.

Blackford, S., P. A. Rea, and D. Sanders (1990) Voltage sensitivity of H^++Ca^{2+} antiport in higher-plant tonoplast suggests a role in vacuolar calcium accumulation. *J. Biol. Chem.* 265: 9617–9620.

Blatt, M. R. and F. Armstrong (1993) K^+-channels of stomatal guard cells—abscisic acid-evoked control of the outward rectifier mediated by cytoplasmic pH. *Planta*, 191: 330–341

Blatt, M. R. and G. Thiel (1993) Hormonal-control of ion-channel gating. *Annu. Rev. Plant Physiol. Plant Molec. Biol.* 44: 543–567.

Blatt, M. R., G. Thiel, and D.R. Trentham (1990) Reversible inactivation of K^+-channels of *Vicia* stomatal guard cells following the photolysis of caged inositol 1,4,5-trisphosphate. *Nature*: 346, 766–769.

Blumwald, E. and R. J. Poole (1986) Kinetics of Ca^{2+}/H^+ antiport in isolated tonoplast vesicles from storage tissue of *Beta vulgaris* L. *Plant Physiol.* 80: 727–731.

Bowler, C., H. Yamagata, G. Neuhaus and N. H. Chua (1994) Phytochrome signal-transduction pathways are regulated by reciprocal control mechanisms. *Genes Dev.* 8: 2188–2202.

Braam, J. and R. W. Davies (1990) Rain-, wind-, and touch-induced expression of calmodulin and calmodulin-related genes in *Arabidopsis*. *Cell* 60: 357–364.

Brauer, D., C. Schubert and S. I. Tsu (1990) Characterization of a Ca^{2+}-translocating ATPase from corn root microsomes. *Physiol. Plant.* 78: 335–344.

Brawley, S. H. (1991) The fast block against polyspermy in fucoid algae is an electrical block. *Dev Biol.*, 144: 94–106.

Briskin, D. P. (1990) Ca^{2+}-Translocating ATPase of the plant plasma membrane. *Plant Physiol.* 94: 397–400.

Brosnan, J. M. and D. Sanders (1990) Inositol trisphosphate-mediated Ca^{2+} release in beet microsomes is inhibited by heparin. *FEBS Lett.* 260: 70–72.

Brosnan, J. M. and D. Sanders (1993) Identification and characterization of high-affinity binding-sites for inositol trisphosphate in red beet. *Plant Cell* 5: 931–940.

Brownlee, C. (1989) Visualizing cytoplasmic calcium in polarizing zygotes and growing rhizoids of *Fucus-serratus*. *Biol. Bull.* 176: 14–17.

Brownlee, C. (1994) Signal-transduction during fertilisation in algae and vascular plants. *New Phytol.* 127: 399–423.

Brownlee, C. and A. L. Pulsford (1988) Visualization of the cytoplasmic Ca^{2+} gradient in *Fucus-serratus* rhizoids—correlation with cell ultrastructure and polarity. *J. Cell Sci.*, 91: 249–256.

Brownlee, C. and J. W. Wood (1986) A gradient of cytoplasmic free calcium in growing rhizoid cells of *Fucus-serratus Nature* 320: 624–626.

Brownlee, C., J. W. Wood, and D. Briton (1987) Cytoplasmic free calcium in single cells of centric diatoms—the use of FURA-2. *Protoplasma* 140: 118–122.

Burström, H. G. (1968) Calcium and plant growth. *Biol. Rev.*, 43: 287–316.

Bush, D. S. and R. L. Jones (1987) Measurement of cytoplasmic calcium in aleurone protoplasts using INDO-1 AND FURA-2. *Cell Calcium* 8: 455–472.

Bush, D. S. and H. Sze (1986) Ca^{2+} transport in tonoplast and endoplasmic reticulum vesicles isolated from cultured carrot cells. *Plant Physiol.*, 80: 549–555.

Camoni, L., J. F. Harper, and M. G. Palmgren (1998) 14-3-3 proteins activate a plant calcium-dependent protein kinase (CDPK). *FEBS Lett.* 430: 381–3842.

Chanson, A. (1993) Active-transport of proton and calcium in higher-plant cells. *Plant Physiol. Biochem.* 31: 943-955.

Chanson, A. (1994) Characterization of the tonoplast Ca^{2+}/H^+ antiport system from maize roots. *Plant Physiol. Biochem.* 32, 341–346.

Chaubal, R. and B. J. Reger (1990) Relatively high calcium is localized in synergid cells of wheat ovaries. *Sexual Plant Reprod.*, 3: 98–102.

Chaubal, R. and B. J. Reger (1992) Calcium in the synergid cells and other regions of pearl-millet ovaries. *Sexual Plant Reprod.* 5: 34–46.

Chen, F. H., D. M. Ratterman, and H. Sze (1993) A plasma-membrane type Ca^{2+}-ATPase of 120 kDa on the endoplasmic reticulum from carrot (*Daucus carrota*) cells. *Plant Physiol.* 102: 651–661.

Chen, Y. R., N. Datta, and S. J. Roux (1987) Purification and partial characterization of a calmodulin-stimulated nucleoside triphosphatase from pea nuclei. *J. Biol. Chem.* 262: 10689–10694.

Cholodny, N. (1924) Ueber die hormonale wirkung der organspitze bei der geotropischen krümmung. *Ber. Det. Bot. Ges.* 42: 356–362.

Clarkson, D. T., C. Brownlee, and S. M. Ayling (1988) Cytoplasmic calcium measurements in intact higher-plant cells—results from fluorescence ratio imaging of FURA-2. *J. Cell Sci.* 91: 71–80.

Collinge, M. and A. J. Trewavas (1989) The location of calmodulin in the pea plasma-membrane *J. Biol. Chem.* 264: 8865–8872, 1989.

Coté, G. G. and R. C. Crain (1993) Biochemistry of phosphoinositides. *Annu. Rev. Plant Physiol. Plant Molec. Biol.* 44: 333–356.

Dauwalder, M., S. J. Roux, and L. Hardison (1986) Distribution of calmodulin in pea-seedlings—immunocytochemical localization in plumules and root apices. *Planta* 168: 461–470.

DeMarty, M., C. Morvan, and M. Thellier (1984) Calcium and the cell-wall. *Plant, Cell Environ.* 7: 441–448.

De Michelis, M. I., A. Carnelli and F. Rasi-Caldogno (1993) The Ca^{2+} pump of the plasma membrane of *Arabidopsis thaliana*: characteristics and sensitivity to fluorescein derivatives. *Bot. Acta* 106: 20–25.

Deery, W. J., A. R. Means, and B. R. Brinkley (1984) Calmodulin-microtubule association in cultured mammalian-cells *J. Cell Biol.* 98: 904–910.

Denecke, J., L. E., Carlsson, S. Vidal, A. S. Hoglund, B. Ek, M. J. Vanzeijl, K. M. C. Sinjorgo, and E. T. Palva (1995) The tobacco homolog of mammalian calreticulin in protein complexes in vivo. *Plant Cell* 7: 391–406.

Dieter, P. and D. Marmé (1980) Partial purification of a plant NADkinase by calmodulin-sepharose affinity chromatography. *Cell Calcium* 1: 279–286.

Dieter, P. and D. Marmé (1986) NADkinase in corn—regulation by far red-light is mediated by Ca^{2+} and calmodulin *Plant Cell Physiol.* 27: 1327–1333.

Dieter, P., J. A. Cox, and D. Marmé (1985) Calcium-binding and its effect on circular-dichroism of plant calmodulin. *Planta* 166: 216–218.

Downie, L., J. Priddle, C. Hawes, and D. E. Evans (1998) A calcium pump at the higher plant nuclear envelope? *FEBS Lett.* 429: 44–48.

Drøbak, B. K. (1991) Plant signal perception and transduction—the role of the phosphoinositide system. *Essays Biochem.* 26: 27–37.

Drøbak, B. K. (1992) The plant phosphoinositide system. *Biochem. J.* 288: 697–712.

Drøbak, B. K. and I. B. Ferguson (1985) Release of Ca^{2+} from plant hypocotyl microsomes by inositol- 1,4,5-trisphosphate. *Biochem. Biophys. Res. Commun.* 130: 1241–1246.

Drøbak, B. K., I. B. Ferguson, A. P. Dawson, and R. F. Irvine (1988) Inositol-containing lipids in suspension-cultured plant-cells—an isotopic study. *Plant Physiol.* 87: 217–222.

Drøbak, B. K., P. A. C. Watkins, J. A. Chattaway, K. Roberts, and A. P. Dawson (1991) Metabolism of inositol (1,4,5) trisphosphate by a soluble enzyme fraction from pea (*Pisum-sativum*) roots. *Plant Physiol.* 95: 412–419.

Dreyer, E. M. and M. H. Weisenseel (1979) Phytochrome mediated uptake of calcium in *Mougeotia* cells. *Planta* 146: 31–39

Durso, N. A. and R. J. Cyr (1994) A calmodulin-sensitive interaction between microtubules and a higher-plant homolog of elongation factor-1-alpha. *Plant Cell* 6: 893–905.

Elliott, D. C. and Y. S . Kokke (1987a) Cross-reaction of a plant protein-kinase with antiserum raised against a sequence from bovine brain protein kinase-C regulatory sub-unit. *Biochem. Biophys. Res. Commun.* 145: 1043–1047.

Elliott, D. C. and Y. S. Kokke. (1987b) Partial-purification and properties of a protein kinase-c type enzyme from plants. *Phytochemistry* 26: 2929–2935.

Elliott, D. C., A. Fournier, and Y. S. Kokke (1988) Phosphatidylserine activation of plant protein kinase-C. *Phytochemistry* 27: 3725–3730.

Erdei L. and H. Matsumoto (1991) Activation of the Ca^{2+}-Mg^{2+}-ATPase by Ca-EGTA in plasmalemma from the roots of cucumber. *Biochem. Physiol. Pflanzen* 187: 189–195.

Evans, D. E. (1988) Regulation of cytoplasmic free calcium by plant-cell membranes. *Cell Biol. Int. Rep.* 12: 383–396.

Evans, D. E. (1994a) Calmodulin-stimulated calcium-pumping ATPases located at higher-plant intracellular membranes—a significant divergence from other eukaryotes. *Physiol. Plant.* 90: 420–426.

Evans, D. E. (1994b) PM-type calcium pumps are associated with higher-plant cell intracellular membranes. *Cell Calcium* 15: 241–246.

Evans, D. E. and L. E. Williams (1998) P-type calcium ATPases in higher plants—biochemical, molecular and functional properties. *Biochim. Biophys. Acta–Reviews on Biomembranes* 1376: 1–25.

Evans, D. E., S.-A. Briars, and L. E. Williams (1991) Active calcium transport by plant cell membranes. *J. Exp. Bot.* 42: 285–303.

Evans, D. E., P. Askerlund, J. M. Boyce, S.-A. Briars, D. Coates, J. Coates, D. T. Cooke, and F. L. Theodoulou (1992) Studies on the higher plant calmodulin-stimulated ATPase. in *Transport and receptor proteins of plant cell membranes* (Cooke, D. T., and Clarkson, D. T., eds.). Plenum Press, New York, pp. 39–53.

Fairley-Grenot, K. A. and S. M. Assmann (1992) Permeation of Ca^{2+} through K^+ channels in the plasma membrane of *Vicia-faba* guard cells. *J. Membr. Biol.* 128: 103–113.

Faure, J. E., C. Digonnet and C. Dumas (1994) An *in-vitro* system for adhesion and fusion of maize gametes. *Science* 263: 1598–1600.

Felle, H. (1988a) Auxin causes oscillations of cytosolic free calcium and pH in *Zea mays* coleoptiles. *Planta* 174: 495–499.

Felle, H. (1988b) Cytoplasmic free calcium in *Riccia fluitans* 1 and *Zea mays* L.—interaction of Ca^{2+} and pH. *Planta* 176: 248–255.

Ferrol, N. and A. B. Bennett (1996) A single-gene may encode differentially localized Ca^{2+}-ATPases in tomato. *Plant, Cell Environ.* 8: 1159–1169.

Fisher, D. D. and R. J. Cyr (1993) Calcium levels affect the ability to immunolocalize calmodulin to cortical microtubules. *Plant Physiol.* 103: 543–551.

Fukumoto, M. and M. A. Venis (1986) ATP-dependent Ca^{2+} transport in tonoplast vesicles from apple fruit. *Plant Cell Physiol.* 27: 491–497.

Gavin, O., P. E. Pilet, and A. Chanson (1993) Tonoplast location of a calmodulin-stimulated Ca^{2+} pump from maize roots. *Plant Sci.* 92: 143–150.

Gehring, C. A., H. R. Irving, and R. W. Parish (1990) Effects of auxin and abscisic-acid on cytosolic calcium and pH in plant-cells. *Proc. Natl. Acad. Sci. USA* 87: 9645–9649.

Gilroy, S. and R. L. Jones (1992) Gibberellic acid and abscisic acid coordinately regulate cytoplasmic calcium and secretory activity in barley aleurone protoplasts. *Proc. Natl. Acad. Sci. U.SA* 89: 3591–3595.

Gilroy, S. and R. L. Jones (1993) Calmodulin-stimulation of unidirectional calcium uptake by the endoplasmic reticulum of barley aleurone. *Planta* 190: 289–296.

Gilroy, S., W. A. Hughes and A. J. Trewavas (1986) The measurement of intracellular calcium levels in protoplasts from higher-plant cells. *FEBS Lett.* 199: 217–221.

Gilroy, S., W. A. Hughes, and A. J. Trewavas (1987) Calmodulin antagonists increase free cytosolic calcium levels in plant-protoplasts in vivo. *FEBS Lett.* 212: 133–137.

Gilroy, S., W. A. Hughes, and A. J. Trewavas (1989) A comparison between quin-2 and aequorin as indicators of cytoplasmic calcium levels in higher-plant cell protoplasts. *Plant Physiol.* 90: 482–491.

Gilroy, S., N. D. Read, and A. J. Trewavas (1990) Elevation of cytoplasmic calcium by caged calcium or caged inositol trisphosphate initiates stomatal closure. *Nature* 346: 769–771.

Gilroy, S., M. D. Fricker, N. D. Read, and A. J. Trewayas (1991) Role of calcium in signal transduction of *Commelina* guard-cells. *Plant Cell* 3: 333–44.

Goswami, K. K. A. and L. J. Audus (1976) Distribution of calcium, potassium and phosphorus in *Helianthus anuus* hypocotyls and *Zea mays* coleoptiles in relation to tropic stimuli and curvature. *Ann. Bot.* 40: 49–64.

Gräf, P. and E. W. Weiler (1990) Functional reconstitution of an ATP-driven Ca^{2+} transport system from the plasma membrane of *Commelina communis* L. *Plant Physiol.* 94: 634–640.

Graziana, A., M. Fosset, R. Ranjeva, A. M. Hetherington, and M. Lazdunski (1988) Ca^{2+} channel inhibitors that bind to plant-cell membranes block ca^{2+} entry into protoplasts. *Biochemistry* 27: 764–768.

Hale, C. C. and S. J. Roux (1980) Photoreversible calcium fluxes induced by phytochrome in oat coleoptile cells. *Plant Physiol.* 65: 658–662.

Haley, A., A. Russell, N. Wood, A. C. Allan, M. Knight, A.K. Campbell, and A. J. Trewavas (1995) Effects of mechanical signaling on plant cell cytosolic calcium. *Proc. Natl. Acad. Sci. USA.* 92: 4124–4128.

Hall, A. B., R. D. Firn and J. Digby (1980) Auxins and shoot tropisms—a tenuous connection. *J. Biol. Educ.* 14: 195–199.

Harmon, A. C., H. W. Jarrett, and M. J. Cormier (1984) Any enzymatic assay for calmodulins based on plant NADkinase-activity. *Anal. Biochem.* 141: 168–178.

Harmon, A. C., C. Putnam-Evan, and M. J. Cormier (1987) A calcium-dependent but calmodulin-independent protein-kinase from soybean. *Plant Physiol.* 83: 830–837.

Harper, J. F., M. R. Sussman, G. E. Schaller, C. Putnam-Evans, H. Charbonneau, and A. C. Harmon (1991) A calcium-dependent protein-kinase with a regulatory domain similar to calmodulin. *Science* 252: 951–954.

Harper, J. F., B. M. Hong, I. D. Hwang, H. Q. Guo, R. Stoddard, J. F. Juang, and M. G. Palmgren (1998) A novel calmodulin-regulated Ca^{2+}-ATPase (ACA2) from Arabidopsis with an N-terminal autoinhibitory domain. *J. Biol. Chem.* 273: 1099–1106.

Harvey, H. J., M. A. Venis, and A. J. Trewavas (1989) Partial-purification of a protein from maize (*Zea mays* L.) coleoptile membranes binding the Ca^{2+} channel antagonist verapamil. *Biochem. J.* 257: 95–100.

Hepler, P. K. (1985) Calcium restriction prolongs metaphase in dividing *Tradescantia* stamen hair-cells. *J. Cell Biol.* 100: 1363–1368.

Hepler, P. K. (1992) Calcium and mitosis. *Int. Rev. Cytol.— A Survey of Cell Biol.* 138: 239–268.

Hepler, P. K. (1994) The role of calcium in cell-division. *Cell Calcium* 61: 322–330.

Hepler, P. K. and D. A. Callaham (1987) Free calcium increases during anaphase in stamen hair-cells of *Tradescantia. J. Cell Biol.* 105: 2137–2143.

Hepler, P. K. and R. O. Wayne (1985) Calcium and plant development. *Ann. Rev. Plant Physiol.* 36: 397–439,.

Hepler, P. K. and S. M. Wolniak (1984) Membranes in the mitotic apparatus—their structure and function. *Int. Rev. Cytol.* 90: 169–238.

Hong, Y., M. Takano, C. M. Liu, A. Gasch, M. L. Chye, and N. H. Chua (1996) Expression of 3 members of the calcium-dependent protein-kinase gene family in *Arabidopsis-thaliana. Plant Mol. Biol.* 30: 1259–1275.

Hsieh, W.-L., W. S. Pierce and H. Sze (1991) Calcium pumping ATPase in vesicles from carrot cells. Stimulation by calmodulin or phosphatidylserine and formation of a 120 kilodalton phosphoenzyme. *Plant Physiol.* 97: 1535–1544.

Huang, J. W. W., D. L. Grunes, and L. V. Kochian (1994) Voltage-dependent Ca^{2+} influx into right-side-out plasma-membrane vesicles isolated from wheat roots—characterization of a putative Ca^{2+} channel. *Proc. Natl. Acad. Sci. USA* 91: 3473–3477.

Huang, L. Q., T. Berkelman, A. E. Franklin and N. E. Hoffman (1993) Characterization of a gene encoding a Ca^{2+}-ATPase-like protein in the plastid envelope. *Proc. Natl. Acad. Sci. USA* 90: 10066–10070.

Hwang, I., D. M. ratterman, and H. Sze (1997) Distinction between endoplasmic reticulum-type and plasma membrane-type Ca^{2+} pumps. *Plant Physiology* 113: 535–548.

Irvine, R. F., A. J. Letcher, D. J. Lander, B. K. Drobak, A. P. Dawson, and A. Musgrave (1989) Phosphatidylinositol(4,5)bisphosphate and phosphatidylinositol(4)phosphate in plant-tissues. *Plant Physiol.* 89: 888–892.

Jaffé, L F. (1958) Tropistic responses of zygotes of the fucaceae to polarized light. *Exp. Cell Res.* 15: 282–299.

Johannes, E., J. M. Brosnan, and D. Sanders (1991) Calcium channels and signal transduction in plant-cells. *BioEssays* 13: 331–336.

Johannes, E., J. M. Brosnan, and D. Sanders (1992a) Calcium channels in the vacuolar membrane of plants—multiple pathways for intracellular calcium mobilization. *Phil. Trans. R Soc. Ser. B* 338: 105–112.

Johannes, E., J. M. Brosnan, and D. Sanders (1992b) Parallel pathways for intracellular Ca^{2+} release from the vacuole of higher-plants. *Plant J.* 2: 97–102.

Johnsson, A. and B. G. Pickard (1979) The threshold stimulus for geotropism. *Physiol. Plant.* 45: 315–319.

Jones, R. L., S. Gilroy, and S. Hillmer (1993) The role of calcium in the hormonal regulation of enzyme synthesis and secretion in barley aleurone. *J. Exp. Bot.* 44S: 207–212.

Joseph, S. K., T. Esch, and W. D. Bonner (1989) Hydrolysis of inositol phosphates by plant-cell extracts. *Biochem. J.* 264: 851–856.

Kamada, Y. and S. Muto (1994a) Stimulation by fungal elicitor of inositol phospholipid turnover in tobacco suspension-culture cells. *Plant Cell Physiol.* 35: 397–404.

Kamada, Y. and S. Muto (1994b) Protein-kinase inhibitors inhibit stimulation of inositol phospholipid turnover and induction of phenylalanine ammonia-lyase in fungal elicitor-treated tobacco suspension-culture cells. *Plant Cell Physiol.* 35: 405–409.

Kansara, M. S., J. Ramdas, and S. K. Srivastava (1989) Phytochrome mediated photoregulation of NADkinase in terminal buds of pea-seedlings. *J. Plant Physiol.* 134: 603–607.

Kasai, M. and S. Muto (1990) Ca^{2+} pump and Ca^{2+}/H^{+} antiporter located in plasma membrane vesicles isolated by aqueous two phase partition from corn leaves. *J. Membr. Biol.* 114: 133–142.

Kasai, M. and S. Muto (1991) Solubilisation and reconstitution of Ca^{2+}-pump from corn leaf plasma membrane. *Plant Physiol.* 96: 565–570.

Keith, C., M. Dipaola, F. R. Maxfield, and M. L. Shelanski (1983) Microinjection of Ca^{2+}-calmodulin causes a localized depolymerization of microtubules. *J. Cell Biol.* 97: 1918–1924.

Klucis, E. and G. M. Polya (1988) Localization, solubilization and characterization of plant membrane-associated calcium-dependent protein-kinases. *Plant Physiol.* 88: 164–171.

Klusener, B., G. Boheim, H. Liss, J. Engelberth, and E. W. Weiler (1995) Gadolinium-sensitive, voltage-dependent calcium-release channels in the endoplasmic reticulum of a higher plant mechanoreceptor organ. *EMBO J.* 14: 2708–2714.

Knight, H. and M. R. Knight (1995) Recombinant aequorin methods for intracellular calcium measurements in plants. *Methods Cell Biol.* 49: 199–213.

Knight, M. R., A. K. Campbell, S. M. Smith, and A. J. Trewavas (1991) Transgenic plant aequorin reports the effects of touch and cold-shock and elicitors on cytoplasmic calcium. *Nature* 352: 524–526.

Knight, M. R., S. M. Smith, and A. J. Trewavas. (1992) Wind-induced plant motion immediately increases cytosolic calcium. *Proc. Natl. Acad. Sci. USA* 89: 4967–4971.

Knight, M. R., N. D. Read, A. K. Campbell and A. J. Trewavas (1993) Imaging calcium dynamics in living plant cells using semi-synthetic recombinant aequorins. *J. Cell Biol.* 121: 83–90.

Kreimer, G., M. Melkonian, J. A. M. Holtum, and E. Latzko (1985) Characterization of calcium fluxes across the envelope of intact spinach-chloroplasts. *Planta* 166: 515–523.

Kreimer, G., M. Melkonian, J. A. M. Holtum, and E. Latzko (1988) Stromal free calcium-concentration and light-mediated activation of chloroplast fructose-1,6-bisphosphatase. *Plant Physiol.* 86: 423–428.

Kropf, D. L. and R. S. Quatrano (1987) Localization of membrane-associated calcium during development of fucoid algae using chlorotetracycline. *Planta* 171: 158–170.

Lawton, M. A., R. T. Yamamoto, S. K. Hanks, and C.J . Lamb (1989) Molecular-cloning of plant transcripts encoding protein-kinase homologs. *Proc. Natl. Acad. Sci. USA* 86: 3140–3144.

Lee, J. S., T. J. Mulkey, and M. L. Evans (1983a) Gravity-induced polar transport of calcium across root-tips of maize. *Plant Physiol.* 73: 874–876.

Lee, J. S., T. J. Mulkey, and M. L. Evans (1983b) Reversible loss of gravitropic sensitivity in maize roots after tip application of calcium chelators. *Science* 220: 1375–1376.

Lee, J. S., T. J. Mulkey, and M. L. Evans (1984) Inhibition of polar calcium movement and gravitropism in roots treated with auxin-transport inhibitors. *Planta* 160: 536–543.

Legue, V., E. Blancaflor, C. Wymer, G. Perbal, D. Fantin, and S. Gilroy (1997) Cytoplasmic free Ca^{2+} in *Arabidopsis* roots changes in response to touch but not gravity. *Plant Physiol.* 114: 789–800.

Li, Y., G. Hagen, and T. J. Guilfoyle (1991) An auxin-responsive promoter is differentially induced by auxin gradients during tropisms. *Plant Cell* 3: 1167–1175.

Liang, F., K. W. Cunningham, J. F. Harper, and H. Sze (1997) ECA1 complements yeast mutants defective in Ca^{2+} pumps and encodes an endoplasmic reticulum-type Ca^{2+}-ATPase in *Arabidopsis thaliana*. *Proc. Natl. Acad. Sci. USA* 94: 8579–8584.

Lin, C. T., D. Sun, G.X . Song, and J. Y. Wu (1986) Calmodulin—localization in plant-tissues. *J. Histochem. Cytochem.* 34: 561–567.

Ling, V., I. Perera, and R. E. Zielinski (1991) Primary structures of *Arabidopsis* calmodulin isoforms deduced from the sequences of cDNA clones. *Plant Physiol.* 96: 1196–1202.

Liss, H., C. Bockelmann, N. Werner, H. Fromm, and E. W. Weiler (1998) Identification and purification of the calcium-regulated Ca^{2+}-ATPase from the endoplasmic reticulum of a higher plant mechanoreceptor organ. *Physiol. Plant.* 102: 561–572.

Logan, D. C. and M. A. Venis (1995) Characterisation and immunological identification of a calmodulin-stimulated Ca^{2+}-ATPase from maize shoots. *J. Plant Physiol.* 145: 702–710.

Loneragan, J. F. and K. Snowball (1969) Calcium requirements of plants. *Aust. J. Agric. Res.* 20: 465–478.

Loneragan, J. F., K. Snowball, and W. J. Simmons (1968) Response of plants to calcium concentration in solution culture. *Aust. J. Agric. Res.* 19: 845–857.

Lukas, T. J., D. B. Iverson, M. Schleicher, and D. M. Watterson (1984) Structural characterization of a higher-plant calmodulin—*Spinacia oleracea*. *Plant Physiol.* 75: 788–795.

Lynch, J., V. S. Polito, and A. Lauchli (1989) Salinity stress increases cytoplasmic-Ca activity in maize root protoplasts. *Plant Physiol.* 90: 1271–1274.

Macklon, A. E. S. (1984) Calcium fluxes at plasmalemma and tonoplast. *Plant Cell Environ.* 7: 407–413.

MaCrobbie, E. A. C. (1990) Calcium-dependent and calcium-independent events in the initiation of stomatal closure by abscisic-acid. *Proc. R. Soc. Ser. B* 241: 214–219.

MaCrobbie, E. A. C. (1992) Calcium and ABA-induced stomatal closure. *Phil. Trans. R. Soc. Ser. B* 338: 5–18.

MaCrobbie, E. A. C. and J. Banfield (1988) Calcium influx at the plasmalemma of *Chara-corallina*. *Planta* 176: 98–108.

Malmstron, S., P. Askerlund, and M. G. Palmgren (1997) A calmodulin-stimulated Ca^{2+}-ATPase from plant vacuolar membranes with a putative regulatory domain at its N-terminus. *FEBS Lett.* 400: 324–328.

Marschner, H. (1986) *Mineral Nutrition of Higher Plants*. Academic Press, London, pp. 252–254.

Marshall, J., A. Corzo, R. A. Leigh, and D. Sanders (1994) Membrane potential dependent calcium transport in right-side out plasma membrane vesicles from *Zea mays* L. roots. *Plant J.* 5: 683–694.

Masson, P. H. (1995) Rest gravitropism. *BioEssays* 17: 119–127.

Matsumoto, H., T. Yamaya, and M. Tanigawa (1984) Activation of atpase activity in the chromatin fraction of pea nuclei by calcium and calmodulin. *Plant Cell Physiol.* 25: 191–195.

McAinsh, M. R., C. Brownlee, and A. M. Hetherington (1990) Abscisic acid-induced elevation of guard-cell cytosolic Ca^{2+} precedes stomatal closure. *Nature* 343: 186–188,.

McAinsh, M. R., C. Brownlee, and A. M. Hetherington (1992) Visualizing changes in cytosolic-free Ca^{2+} during the response of stomatal guard-cells to abscisic-acid *Plant Cell* 4: 1113–1122.

McCurdy, D. W. and A. C. Harmon (1992a) Phosphorylation of a putative myosin light chain in *Chara* by calcium-dependent protein-kinase. *Protoplasma* 171: 85–88.

McCurdy, D. W. and A. C. Harmon (1992b) Calcium-dependent protein-kinase in the green-alga *Chara*. *Planta* 188: 54–61.

McMurray, W. C. and R. F. Irvine. Phosphatidylinositol 4,5-bisphosphate phosphodiesterase in higher-plants. *Biochem. J.* 249: 877–881.

Miller, A. J. and D. Sanders (1987) Depletion of cytosolic free calcium induced by photosynthesis. *Nature* 326: 397–400.

Miller, D. D., D. A. Callaham, D. J. Gross, and P. K. Hepler (1992) Free Ca^{2+} gradient in growing pollen tubes of *Lilium*. *J. Cell Sci.* 101: 7–12.

Muto, S.Y . and S. Miyachi (1977) Properties of a protein activator of NADkinase from plants. *Plant Physiol.* 59: 55–60.

Muto, S. and S. Miyachi (1984) Production of antibody against spinach calmodulin and its application to radioimmunoassay for plant calmodulin. *Z. fur Pflanzenphysiol.* 114: 421–431.

Muto, S., S. Izawa, and S. Miyachi (1982) Light-induced Ca^{2+} uptake by intact chloroplasts. *FEBS Lett.* 139: 250–254.

Neuhaus, G., C. Bowler, R. Kern, and N. H. Chua (1993) Calcium/calmodulin-dependent and calcium/calmodulin-independent phytochrome signal-transduction pathways. *Cell* 73: 937–952.

Neumann, G. M., R. Condron, B. Svensson, and G. M. Polya (1993) Phosphorylation of barley and wheat phospholipid transfer proteins by wheat calcium-dependent protein-kinase. *Plant Science* 92: 159–167.

Neumann, G. M., R. Condron, and G. M. Polya (1994a) Phosphorylation of a plant protease inhibitor protein by wheat calcium-dependent protein-kinase. *Plant Sci.* 96: 69–79.

Neumann, G. M., R. Condron, I. Thomas, and G. M. Polya (1994b) Purification and sequencing of a family of wheat lipid transfer protein homologs phosphorylated by plant calcium-dependent protein-kinase. *Biochim. Biophys. Acta* 1209: 183–190.

Neumann, G. M., R. Condron, I. Thomas, and G. M. Polya (1995) Purification, characterization and sequencing of a family of petunia petal lipid transfer proteins phosphorylated by plant calcium-dependent protein-kinase. *Plant Sci.* 107: 129–145.

Nguyen, T. D. and P. A. Siegenthaler (1985) Purification and some properties of an Mg^{2+}-stimulated, Ca^{2+}-stimulated and calmodulin-stimulated ATPase from spinach chloroplast envelope membranes. *Biochim. Biophys. Acta* 840: 99–106.

Obermeyer, G. and M. H. Weisenseel (1991) Calcium-channel blocker and calmodulin antagonists affect the gradient of free calcium-ions in lily pollen tubes. *Eur. J. Cell Biol.* 56: 319–327.

Oh, S. H. and D. M. Roberts (1990) Analysis of the state of posttranslational calmodulin methylation in developing pea-plants. *Plant Physiol.* 93: 880–887.

Oh, S. H., H. Y. Steiner, D. K. Dougall, and D. M. Roberts (1992) Modulation of calmodulin levels, calmodulin methylation, and calmodulin binding-proteins during carrot cell-growth and embryogenesis. *Arch. Biochem. Biohys.* 297: 28–34.

Oh, S. A., J. M. Kwak, I. C. Kwun, and H. G. Nam (1996) Rapid and transient induction of calmodulin-encoding gene(s) of brassica-napus by a touch stimulus. *Plant Cell Reports* 15: 586–590.

Patil, S., D. Takezawa, and B. W. Poovaiah (1995) Chimeric plant calcium/calmodulin-dependent protein-kinase gene with a neural visinin-like calcium-binding domain. *Proc. Natl. Acad. Sci. USA* 92: 4897–4901.

Perez-Prat, E., M. L. Narashimhan, M. L. Binzel, M. A. Botella, Z. Chen, V. Valpuesta, R. A. Bressan and P. M. Hasegawa (1992) Induction of a putative Ca^{2+} ATPase messenger RNA in NaCl-adapted cells. *Plant Physiol.* 100: 1471–1478.

Pfeiffer, W. and A. Hager (1993) A Ca^{2+} ATPase and a Mg^{2+}/H^{+} antiporter are present on tonoplast membranes from roots of *Zea mays* L. *Planta* 191: 377–385.

Pickard, B. G. (1985) Early events in geotropism of seedling shoots. *Annu. Rev. Plant Physiol.* 36: 55–75.

Pickard, B. G. (1984) Contemplating the plasmalemmal control center model. *Protoplasma* 182: 1–9.

Pickard, B. G. and J. P. Ding (1993) The mechanosensory calcium-selective ion-channel—key component of a plasmalemmal control center. *Aust. J. Plant Physiol.* 20: 439–459.

Polya, G. M., S. Chandra, R. Chung, G. M. Neumann and P. B. Hoj (1992) Purification and characterization of wheat and pine small basic-protein substrates for plant calcium-dependent protein-kinase. *Biochim. Biophys. Acta* 1120: 273–280.

Portis, A. R. and H. W. Heldt (1976) Light-dependent changes of the Mg^{2+} concentration in the stroma in relation to the Mg^{2+} dependency of CO_2 fixation in intact chloroplasts. *Biochim. Biophys. Acta* 449: 434–446.

Price A. H., A. Taylor, S. Ripley, A. Griffiths, A. . Trewavas, and M. R. Knight (1994) Oxidative signals in tobacco increase cytosolic calcium. *Plant Cell* 6: 1301–1310,.

Putnam-Evans, C. L., A. C. Harmon, and M. J. Cormier (1990) Purification and characterization of a novel calcium-dependent protein-kinase from soybean. *Biochemistry* 29: 2488–2495.

Quail, P. H., M.T . Boylan, B. M. Parks, T. W. Short, Y. Xu, and D. Wagner (1995) Phytochromes—photosensory perception and signal-transduction. *Science* 268: 675–680.

Raghavan, V. (1980) Cytology, physiology and biochemistry of germination of fern spores. *Int Rev Cytol.* 62: 69–118.

Ranjeva, R., A. Carrasco, and A. M. Boudet (1988) Inositol trisphosphate stimulates the release of calcium from intact vacuoles isolated from acer cells. *FEBS Lett.* 230: 137–141.

Rasi-Caldogno, F., A. Carnelli, and M. I. DeMichelis (1995) Identification of the plasma membrane Ca^{2+} ATPase and of its autoinhibitory domain. *Plant Physiol.* 108: 105–113,.

Rathore, K. S., R. J. Cork, and K. R. Robinson (1991) A cytoplasmic gradient of Ca^{2+} is correlated with the growth of lily pollen tubes. *Dev.Biol.* 148: 612–619.

Raven, J. A. (1977) H^+ and Ca^{2+} in phloem and symplast: relation of relative immobility of the ions to the cytoplasmic nature of the transport path. *New Phytol.* 79: 465–480.

Reger, B. J., R. Chaubal and R. Pressey (1992a) Chemotropic responses by pearl-millet pollen tubes. *Sexual Plant Reprod.* 5: 47–56.

Reger, B. J., R. Pressey, and R. Chaubal (1992b) *In vitro* chemotropism of pearl-millet pollen tubes to stigma tissue—a response to glucose produced in the medium by tissue-bound invertase. *Sexual Plant Reprod.* 5: 201–205.

Reger, B. J., R. Chaubal, and R. Pressey (1993) Chemotropism by pearl-millet pollen tubes. *Plant Physiol.* 102ss: 120.

Roberts, D. M. and A. C. Harmon (1992) Calcium-modulated proteins—targets of intracellular calcium signals in higher-plants. *Annu. Rev. Plant Physiol. Molec. Biol.* 43: 375–414.

Roberts, D. M., T. J. Lukas, and D. M. Watterson (1986) Structure, function, and mechanism of action of calmodulin. *CRC Crit Rev. Plant Sci.* 4: 311–339.

Roberts, S. K., F. Berger, and C. Brownlee (1993) The role of Ca^{2+} in signal-transduction following fertilisation in *Fucus-serratus. J. Exp. Biol.* 184: 197–212,.

Roberts, S. K., I. Gillot, and C. Brownlee (1994) Cytoplasmic calcium and *Fucus* egg activation. *Development* 120: 155–163.

Robinson, C., C. Larsson, and T. J. Buckhout. (1988) Identification of a calmodulin-stimulated (Ca^{2+} + Mg^{2+})-ATPase in a plasma membrane fraction isolated from maize (*Zea mays*) leaves. *Physiol. Plant.* 72: 177–184.

Roux, S. J., N. Datta, Y. R. Chen, and S. H. Kim (1985) Light, calcium and calmodulin regulation of enzyme-activities in isolated-nuclei. *J. Cell Biochem.* S110b: 14, 1986.

Schafer, A., F. Bygrave, S. Matzenauer, and D. Marmé. Identification of a calcium-dependent and phospholipid-dependent protein-kinase in plant-tissue. *FEBS Lett.* 187: 25–28.

Schaller, G. E. and M. R. Sussman (1988) Isolation and sequence of tryptic peptides from the proton-pumping ATPase of the oat plasma-membrane. *Plant Physiol.* 86: 512–516.

Schroeder, J. I. (1992) Plasma-membrane ion channel regulation during abscisic acid-induced closing of stomata. *Phil. Trans. R. Soc. Ser. B* 338: 83–89.

Schroeder, J. I. and S. Hagiwara (1989) Cytosolic calcium regulates ion channels in the plasma-membrane of *Vicia-faba* guard-cells. *Nature* 338: 427–430.

Schroeder, J.I . and S. Hagiwara (1990) Repetitive increases in cytosolic Ca^{2+} of guard-cells by abscisic-acid activation of nonselective Ca^{2+} permeable channels. *Proc. Natl. Acad. Sci. USA* 87: 9305–9309.

Schumaker, K. S. and M. J. Gizinski (1993) Cytokinin stimulates dihydropyridine-sensitive calcium-uptake in moss protoplasts. *Proc. Natl. Acad. Sci. USA* 90: 10937–10941.

Schumaker, K. S. and H. Sze (1985) A Ca^{2+}/H^+ antiporter driven by the proton electrochemical gradient of a tonoplast H^+-ATPase from oat roots. *Plant Physiol.* 79: 1111–1117.

Schumaker, K. S. and H. Sze (1987) Inositol 1,4,5-trisphosphate releases Ca^{2+} from vacuolar membrane-vesicles of oat roots. *J. Biol. Chem.* 262: 3944–3946.

Schumaker, K. S. and H. Sze (1990) Solubilization and reconstitution of the oat root vacuolar H^+/Ca^{2+} exchanger. *Plant Physiol.* 92: 340–345.

Serlin, B. S. and S. J. Roux (1984) Modulation of chloroplast movement in the green-alga mougeotia by the Ca^{2+} ionophore A23187 and by calmodulin antagonists. *Proc. Natl. Acad. Sci. USA* 81: 6368–6372.

Shacklock, P.S ., N. D. Read. and A. J. Trewavas (1992) Cytosolic free calcium mediates red light-induced photomorphogenesis. *Nature* 358: 753–755.

Shear, C. B. (1975) Calcium related disorders of fruit and vegetables. *Hort. Sci.* 10: 361–365.

Sievers, A., H. M. Behrens, T. J. Buckhout, and D. Gradmann (1984) Can a Ca^{2+} pump in the endoplasmic-reticulum of the *Lepidium* root be the trigger for rapid changes in membrane-potential after gravistimulation? *Z. Pflanzenphysiol.* 114: 195–200.

Silva, M. A. P., E. G.S . Carnieri, A. E. Vercesi (1992) Calcium-transport by corn mitochondria—evaluation of the role of phosphate. *Plant Physiol.* 98: 452–457.

Sinclair, W., L. Oliver, P. Maher, and A. Trewavas (1996) The role of calmodulin in the gravitropic response of the *Arabidopsis-thaliana* agr-3 mutant. *Planta* 199: 343–351.

Slocum, R. D. and S. J. Roux (1983) Cellular and sub-cellular-localization of calcium in gravistimulated oat coleoptiles and its possible significance in the establishment of tropic curvature. *Planta* 157: 481–492.

Snedden, W. A., T. Arazi, H. Fromm, and B. J. Shelp (1995) Calcium/calmodulin activation of soybean glutamate decarboxylase. *Plant Physiol.* 108: 543–549.

Sommarin, M. and A. S. Sandelius (1988) Phosphatidylinositol and phosphatidylinositolphosphate kinases in plant plasma-membranes. *Biochim. Biophys. Acta* 958: 268–278.

Steer, M. W. and J. M. Steer (1989) Pollen-tube tip growth. *New Phytol.* 111: 323–358.

Sze, H. (1985) H^+ translocating-ATPases: advances using membrane vesicles. *Annu. Rev. Plant Physiol.* 36: 175–208.

Taylor, A. R. and C. Brownlee (1992) Localized patch clamping of plasma-membrane of a polarized plant-cell—laser microsurgery of the *Fucus-spiralis* rhizoid cell-wall. *Plant Physiol.* 99: 1686–1688.

Taylor, A. and C. Brownlee (1993) Calcium and potassium currents in the *Fucus* egg. *Planta* 189: 109–119.

Taylor, A. R., S. K. Roberts, and C. Brownlee (1992) Calcium and related channels in fertilisation and early development of *Fucus*. *Phil. Trans. R. Soc. Ser. B* 338: 97–104,.

Tester, M. and H. J. Harvey (1989) Verapamil binding fraction forms Ca^{2+} channels in planar lipid bilayers. In *Plant Membrane Transport, the Current Position*, Proceedings of the 8th International Workshop on Plant Membrane Transport (Dainty, J., DeMichelis, M. I., Marré, E., Rasi-Caldogno, F. eds.). Elsevier, Amsterdam, pp. 277–278.

Theodoulou, F. L., F. M. Dewey, and D. E. Evans (1994) Calmodulin-stimulated ATPases of maize cells—functional reconstitution, monoclonal antibodies and subcellular localization. *J. Exp. Bot.* 45: 1553–1564.

Thomson, L.J ., T. Xing, J. L. Hall, and L. E. Williams (1993) Investigation of the calcium-transporting ATPase at the endoplasmic reticulum and plasma membrane of red beet (*Beta vulgaris*). *Plant Physiol.* 102: 553–564.

Toriyama, H. and M. J. Jaffé (1972) Migration of calcium and its role in the regulation of siesmonasty in the motor cell of *Mimosa pudica* L. *Plant Physiol* 49: 72–81.

Trewavas, A. J. (1981) How do plant growth substances work? *Plant Cell Environ.* 4: 203–228.

Trewavas, A. J. (1991) How do plant growth substances work (2)? *Plant Cell Environ.* 14: 1–12.

Trewavas, A. J. (1992a) What remains of the cholodny-went theory—introduction. *Plant Cell Environ.* 15: 761.

Trewavas, A. J . (1992b) What remains of the cholodny-went theory—a summing-up. *Plant Cell Environ.* 15: 793–794

Trewavas, A. J. and M. R. Knight (1994) Mechanical signalling and plant form. *Plant Molec. Biol.* 26: 1329–1341.

Urao, T., T. Katagiri, T. Mizoguchi, K. Yamaguchishinozaki, and N. Hayashida (1994) Genes that encode Ca^{2+}-dependent protein-kinases are induced by drought and high-salt stresses in *Arabidopsis-thaliana*. *Mol. Gen. Genet.* 244: 331–340.

Vantard, M., A. M. Lambert, J. Demey, P. Picquot, and L. J. VanEldik (1985) Characterization and immunocytochemical distribution of calmodulin in higher-plant endosperm cells—localization in the mitotic apparatus. *J. Cell Biol.* 101: 488–499.

Vos, J. W. and P. K. Hepler (1998) Calmodulin is uniformly distributed during cell division in living stamen hair cells of *Tradescantia virginiana*. *Protoplasma* 201: 158–171.

Wada, M., F. Grolig, and W. Haupt (1993) Light-oriented chloroplast positioning—contribution to progress in photobiology. *J. Photochem. Photobiol. B* 17: 3–25.

Wagner, G. and K. Klein (1981) Mechanism of chloroplast movement in *Mougeotia*. *Protoplasma* 109: 169–185.

Wagner, G. and R. Rossbacher (1980) X-ray microanalysis and chlorotetracycline staining of calcium vesicles in the green alga *Mougeotia*. *Planta* 149: 298–305.

Walton, T. J., C. J. Cooke, R. P. Newton and C. J. Smith (1993) Evidence that generation of inositol 1,4,5-trisphosphate and hydrolysis of phosphatidylinositol 4,5-bisphosphate are rapid responses following addition of fungal elicitor which induces phytoalexin synthesis in lucerne (*Medicago-sativa*) suspension-culture cells. *Cell. Signalling* 5: 345–356.

Watillon, B., R. Kettmann, P. Boxus, and A. Burny (1993) A calcium calmodulin-binding serine threonine protein-kinase homologous to the mammalian type-ii calcium calmodulin-dependent protein-kinase is expressed in plant-cells. *Plant Physiol.* 101: 1381–1384.

Wayne, R. and P. K. Hepler (1984) The role of calcium-ions in phytochrome-mediated germination of spores of *Onoclea-sensibilis* L. *Planta* 160: 12–20.

Weaver, C. D. and D. M. Roberts (1992) Determination of the site of phosphorylation of nodulin-26 by the calcium-dependent protein-kinase from soybean nodules. *Biochemistry* 31: 8954–8959.

Weaver, C. D., B. Crombie, G. Stacey, and D. M. Roberts (1991) Calcium-dependent phosphorylation of symbiosome membrane-proteins from nitrogen-fixing soybean nodules—evidence for phosphorylation of nodulin-26. *Plant Physiol.* 95: 222–227.

Weaver, C. D., L. J. Ouyang, D. A. Day, and D. M. Roberts (1992) Structural and functional-characterization of soybean nodulin-26 phosphorylation by the calcium-dependent protein-kinase. *Molec. Biol. Ce..* 3s: 124.

Weaver, C. D., N. H. Shomer, C. F. Louis, and D. M. Roberts (1994) Nodulin-26, a nodule-specific symbiosome membrane-protein from soybean, is an ion-channel. *J. Biol. Chem.* 269: 17858–17862.

Went, F. W. (1926) On growth accelerating substances in the coleoptile of *Avena sativa. Proc. Kon. Acad. Wetensch. Amst.* 30: 10–19.

Wick, S. M. and P. K. Hepler (1980) Localization of Ca^{2+} containing antimonate precipitates during mitosis. *J. Cell Biol.* 86: 500–513.

Wick, S. M. and P. K. Hepler (1982) Selective localization of intracellular Ca^{2+} with potassium antimonate. *J. Histochem Cytochem.* 30: 1190–1204

Wick, S. M., S. Muto, and J. Duniec (1985) Double immu-nofluorescence labeling of calmodulin and tubulin in dividing plant-cells. *Protoplasma* 126: 198–206.

Williams, L. E., S. B. Schueler, and D. P. Briskin (1990b) Further characterization of the red beet plasma membrane Ca^{2+} ATPase using GTP as an alternative substrate. *Plant Physiol.* 92: 747–754.

Williamson, R. E. and C. C. Ashley (1982) Free Ca^{2+} and cytoplasmic streaming in the alga *Chara. Nature* 296: 647–651.

Wimmers, L. E., N. N. Ewing, and A. B. Bennett (1992) Higher plant Ca^{2+} ATPase—primary structure and regulation of messenger RNA abundance by salt. *Proc. Natl. Acad. Sci. USA* 89: 9205–9209.

Wolniak, S. M. and K. M. Bart (1985) The buffering of calcium with quin2 reversibly forestalls anaphase onset in stamen hair-cells of *Tradescantia. Eur.J. Cell Biol.* 39: 33–40.

Wolniak, S. M. and K. M. Bart (1986) Nifedipine reversibly arrests mitosis in stamen hair-cells of *Tradescantia. Eur. J. Cell Biol.* 39: 273–277.

Wolniak, S. M., P. K. Hepler, and W. T. Jackson (1983) Ionic changes in the mitotic apparatus at the metaphase anaphase transition. *J. Cell Biol.* 96: 598–605.

Wolniak, S. M., K. M. Bart, and P. K. Hepler (1984) A change in the intracellular free calcium-concentration accompanies the onset of anaphase. *J. Cell Biol.* 99: A429.

Wyn-Jones, R. G. and O. R. Lunt (1967) The function of calcium in plants. *Bot. Rev.* 33: 407–426.

Yoshida, M., O. Minowa, and K. Yagi (1983) Divalent-cation binding to wheat-germ calmodulin. *J. Biochem.* 94: 1925–1933.

Young, L. M. and M. L. Evans (1994) Calcium-dependent asymmetric movement of ^{3}H-indole-3-acetic-acid across gravistimulated isolated root caps of maize. *Plant Growth Reg.* 14: 235–242.

Zielinski, R.E. (1987) Calmodulin messenger-RNA in barley (*Hordeum-vulgare* L.)—apparent regulation by cell-proliferation and light. *Plant Physiol.* 84: 937–943.

Zielinski, R. E. (1998) Calmodulin and calmodulin-binding proteins in plants. *Ann. Rev. Plant Physiol. Plant Mol. Biol.* 49: 697–725.

18

Calcium in Yeasts and Fungi

Patrice Catty
André Goffeau

The question as to whether calcium is required by yeast has been debated on the basis that this microorganism grows well in the absence of added calcium and even in the presence of divalent cation chelators, such as EGTA or EDTA (Morris, 1958) (see Chapter 19). This unexpected behavior results from a combination of factors, identified by Kovac in 1985, as follows. It is notoriously difficult to work in the complete absence of calcium, which is even present on glassware and in water. In addition, the calcium–EDTA complex enters the yeast cell and releases free calcium intracellularly. Moreover, yeast cells contain very efficient calcium-trapping systems, or "intracellular stores," accounting for the observation that their total calcium content can change 1000-fold without any effect on cell growth or morphology. In fact, calcium is an absolute requirement for yeast growth, as first indicated in 1974 by Duffus and Patterson for *Schizosaccharomyces pombe*, the nuclear division of which is blocked by the calcium ionophore A23187. Recent physiological and genetic works, for the most part carried out in the yeast *Saccharomyces cerevisiae*, have amply confirmed the essential role of calcium, not only in nuclear division, but also in many other cellular functions (see reviews by Cunningham and Fink, 1994b; Davis, 1995).

Inventory of all Calcium-Related Proteins Encoded by the Genome of *Saccharomyces Cerevisiae*

A variety of calcium-related gene products have been identified by the complete systematic sequencing of the *S. cerevisiae* genome. The screening of the 6000 predicted yeast protein sequences (Goffeau et al., 1996) was carried out by search of yeast homologs to mammalian proteins known to be involved in calcium metabolism and its regulation, and by identifi-

cation of yeast proteins containing a calcium-binding motif. Homology searches were performed using the BLAST program (Altschul et al., 1990) and by considering an homology between two proteins to be significant at a BLAST P value lower than 10^{-9}. Tables 18.1 to 18.4 list the 84 detected calcium-related proteins encoded by the *S. cerevisiae* genome and display some of their main characteristics, such as the number of amino acid residues, the number of transmembrane spans predicted by the KKD algorithm (Klein et al., 1985; Nakai and Kanehisa, 1992), the presence or absence of a calcium-binding motif predicted by the PROSITE program, the codon adaptation index (CAI) which predicts the expression level of the gene (Sharp and Li, 1986) and, when available, the presumed physiological or biochemical function. Tables 18.1 and 18.2 pool proteins involved in calcium transport and in calcium signaling, respectively. Tables 18.3 and 18.4 pool membrane and soluble proteins, respectively, with putative calcium-binding site or with enzymatic activity modulated by calcium. The following sections will be devoted to the detailed description of the role of each of these proteins.

Calcium Transport

Early Physiological Data

The apparent K_m for active calcium uptake by both the fission yeast *Schizosaccharomyces pombe* and the budding yeast *Saccharomyces cerevisiae* is lower than $100\,\mu M$ (Fuhrmann and Rothstein, 1968; Boutry et al., 1977; Pena, 1978; Roomans et al., 1979). This high-affinity calcium uptake is markedly stimulated at alkaline external pH and requires a negative-inside transmembrane electric potential (Boutry et al., 1977; Roomans et al., 1979).

Table 18.1 Membrane Proteins Involved in Calcium Transport

ORF Name	Accession Number	Length	CAI	TMS	Protein Name and Presumed Function
YGR217w	Z73002	2039	0.113	16	Putative Ca^{2+}-channel subunit
YGL176c	Z72698	554	0.125	0	Putative Ca^{2+}-channel subunit
YOR088w	Z74995	482	0.155	4	Putative capacitative calcium entry channel
YBR086c	Z35955	946	0.175	7	Putative Ca^{2+}-channel subunit
YDL073w	Z74121	984	0.134	2	Putative Ca^{2+}-channel subunit
YLR220w	U19027	302	0.120	4	Putative Ca^{2+}-channel subunit
YDL128w	U18944	411	0.163	10	**Vxc1p/Hum1p**: vacuolar Ca^{2+}/H^+ antiporter
YNL321w	Z71597	908	0.145	12	Putative H^+/Ca^{2+} antiporter
YDL206w	Z74254	762	0.109	12	Putative Na^{2+}/Ca^{2+} antiporter
YJR106w	Z49606	725	0.104	12	Putative Na^+/Ca^{2+} antiporter
YGL006w	U03060	1173	0.159	10	**Pmc1p**: vacuolar Ca^{2+}-ATPase
YGL167c	X85757	950	0.196	8	**Pmr1p**: Golgi-like Ca^{2+}-ATPase
YOR291w	Z75199	1472	0.135	9	Putative Ca^{2+}-ATPase
YEL031w	U18530	1215	0.257	8	Putative Ca^{2+}-ATPase
YAL026c	U12980	1355	0.199	7	**Drs2p**: putative Ca^{2+}-ATPase or aminophospholipid translocase
YER166w	U18922	1571	0.163	8	Related to Drs2p
YIL048w	Z47047	1151	0.164	10	Related to Drs2p
YMR162c	Z49705	1656	0.140	7	Related to Drs2p
YDR093w	Z47746	1612	0.160	9	Related to Drs2p

ORF, open reading frame; accession numbers are given for the GENBANK database; length is given in number of amino acids; CAI, codon adaptation index; TMS, putative transmembrane segment.

Table 18.2 Proteins Involved in Calcium-Signaling Pathways

ORF Name	Accession Number	Length	CAI	Calcium Binding	Protein Name and Presumed Function
YBR109c	Z35978	146	0.220	+	**Cmd1p**: calmodulin
YLR433c	U21094	553	0.161	−	**Cna1p**: calcineurin catalytic subunit 1
YML057w	Z46729	604	0.151	−	**Cna2p**: calcineurin catalytic subunit 2
YKL190w	Z28190	174	0.152	+	**Cnb1p**: calcineurin regulatory subunit
YNL027w	Z71303	678	0.146	−	**Crz1p**: calcineurin-dependent transcription factor
YFR014c	D50617	446	0.163	−	**Cmk1p**: calcium/calmodulin-dependent kinase
YOL016c	X56961	447	0.165	−	**Cmk2p**: calcium/calmodulin-dependent kinase
YLR248w	U20865	610	0.172	−	**Cmk3p**: calcium/calmodulin-dependent kinase
YGL158w	Z72680	512	0.149	−	**Rck1p**: putative calcium/calmodulin-dependent kinase
YBL105c	Z35866	1151	0.171	+/−	**Pkc1p**: protein kinase C
YPL268w	L13036	869	0.133	+	**Plc1p**: phospholipase C

ORF, open reading frame; accession numbers are given for the GENBANK database; length is given in number of amino acids; CAI, codon adaptation index.

Table 18.3 Membrane Proteins Involved in Calcium-Dependent Pathways

ORF Name	Accession Number	Length	CAI	TMS	Ca²⁺-Binding SQ	Ca²⁺-Binding BE	Protein Name and Presumed Function
YKR038c	Z28263	421	0.116	1	EF	ND	Putative glycoprotease
YEL042w	U18779	518	0.227	1	–	+	**Gda1p**: guanosine diphosphatase
YBL048w	Z35809	103	0.153	3	R	ND	Protein of unknown function
YNL087w	Z71363	1178	0.153	3	C2	–	Protein of unknown function
YLR247c	U20865	1556	0.136	1	EF	ND	Putative DNA repair protein
YNL083w	Z71359	494	0.128	1	EF	ND	Putative mitochondrial carrier protein
YJR066w	Z49566	2470	0.138	1	C2	–	**Tor1p**: putative phosphatidyl inositol kinase
YKL203c	Z28203	2473	0.152	3	C2	–	**Tor2p**: putative phosphatidyl inositol kinase
YDR420w	U33007	1781	0.099	2	EF	ND	**Hkr1p**: involved in β-glucan synthesis
YAR050w	L28920	1513	0.259	1	–	ND	**Flo1p**: involved in flocculation
YLR056w	Z73228	365	0.249	1	–	–	**Syr1p**: C-5 sterol desaturase
YAL058w	U12980	482	0.109	1	–	–	**Cne1p**: homologous to calnexin
YNL291c	Z71567	548	0.133	4	–	ND	**Mid1p**: involved in calcium influx
YLR332w	U20618	356	0.125	1	L	ND	**Mid2p**: required for mating
YBR036c	Z35905	410	0.142	9	EF	ND	**Csg2p**: involved in calcium tolerance
YLR220w	U19027	302	0.120	5	–	ND	**Ccc1p**: involved in calcium regulation
YLL004w	Z73109	616	0.133	1	EF	ND	**Orc3p**: origin recognition complex subunit
YLR153c	Z73325	683	0.371	1	EF	ND	**Acs2p**: acetyl-CoA synthase
YOR018w	U40561	837	0.123	1	–	–	**Rod1p**: involved in resistance to *o*-dinitrobenzene

ORF, open reading frame; accession numbers are given for the GENBANK database; length is given in number of amino acids; CAI, codon adaptation index; TMS, putative transmembrane segment; SQ, predicted from primary sequence; BE, biochemical evidences; EF, EF-hand calcium-binding motif; R, related to EF-hand; L, lactalbumin-like calcium-binding motif; C2, calcium-dependent phospholipid binding domain; ND, not determined.

In addition to the high-affinity calcium-uptake system, the existence of a low-affinity calcium-transport system, with K_m in the millimolar range, has been proposed on the basis of the non-Michaelis-Menten kinetics of energy-dependent calcium uptake (see discussion in Borst-Pauwels, 1981). Using kinetic analysis, two distinct calcium-uptake systems have been demonstrated by Eilam and Chernichovsky (1987), who distinguished rapid (< 20 s) and slow (3 min) energy-dependent cellular calcium influxes.

One of these uptake systems seems to be a channel, which opens when the transmembrane electric potential falls below a threshold value of the order of −70 mV negative inside. This nonspecific channel is driven electrophoretically by the weak hyperpolarization created by the efflux of potassium, which is released outside the cell when the proton motive force is collapsed by the presence of protonophores and ATPase inhibitors (Boutry et al., 1977; Eilam et al.,

1984, 1985b; Eilam and Chernichovsky, 1987). A yet unexplained observation is that this channel-like calcium influx is activated by glucose (Boutry et al., 1977), but is unaffected by intracellular ATP or cAMP levels (Eilam and Othman, 1990; Eilam et al., 1990).

The efflux of intracellular calcium is also energy-dependent in both *S. pombe* (Boutry et al., 1977) and *S. cerevisiae* (Eilam, 1982a). One component involved is stimulated by potassium influx and is believed to act via a Ca^{2+}-efflux/K^+-influx antiporter (Eilam, 1982b), while another component is activated by divalent cations (Eilam, 1982a).

The study of cellular calcium efflux also led to the discovery of at least two types of exchangeable intracellular calcium pools with different turnover rates. The fast turnover pool has now been equated with the cytoplasmic pool, whereas the major component of the slow turnover pool is likely to be vacuolar

Table 18.4 Non-Membrane Proteins Involved in Calcium-Dependent Pathways

ORF Name	Accession Number	Length	CAI	Ca²⁺-Binding		Protein Name and Presumed Function
				SQ	BE	
YNL238w	M22870	700	0.150	–	ND	**Kex2p**: late golgi endoprotease
YLR120c	L31651	548	0.147	–	ND	**Yap3p**: calcium-dependent protease
YNL239w	Z71515	483	0.257	–	+	**Bhl1p**: cysteine proteinase
YMR154c	Z49705	727	0.115	–	ND	Similar to calpain
YPR160w	U28371	901	0.249	–	ND	**Gph1p**: glycogen phosphorylase
YDR001c	Z48008	751	0.168	–	ND	**Nth1p**: neutral trehalase
YBR001c	Z35870	780	0.123	–	ND	**Nth2p**: trehalase
YMR267w	Z49260	280	0.136	–	ND	**Ppa2p**: mitochondrial inorganic pyrophosphatase
YBR053c	Z35922	358	0.167	R	ND	Similar to regucalcin
YDR373w	U28373	190	0.150	EF	ND	Similar to neurocalcin
YGR058w	Z72843	335	0.131	R	ND	Similar to mouse calcium-binding proteins PMP41 and ALG-2
YGL106w	Z72628	149	0.217	–	ND	Similar to calmodulins and myosin light chains
YPR188c	U25841	163	0.128	EF		Similar to calmodulins
YIR010w	Z47047	576	0.113	EF	ND	Protein of unknown function
YER125w	U18916	809	0.186	C2	ND	**Rsp5p**: ubiquitin-protein ligase
YOR086c	Z47994	1186	0.220	C2	ND	Protein of unknown function
YBL061c	Z35823	696	0.113	EF	ND	**Chs4p**: required for chitin synthase III activity
YJL041w	Z49316	823		R	–	**Nsp1p**: nucleoporin
YIR006c	Z47047	1480	0.156	EF	ND	**Pan1p**: involved in organization of the actin cytoskeleton
YKL062w	Z28062	630	0.160	EF	ND	**Msn4p**: transcription activator
YHR065c	U00061	543	0.184	EF	ND	**Rrp3p**: helicase
YPL048w	L01879	415	0.397	–	+	**Tef3p**: translation elongation factor EF-1 gamma
YGR167w	X52272	233	0.228	R	+	**Clc1p**: clathrin light chain involved in endocytosis
YNL084c	X89016	349	0.185	EF	ND	**End3p**: involved in endocytosis
YJL034w	M31006	640	0.440	–	–	**Kar2p**: molecular chaperone
YDR129c	Z48179	642	0.234	R	ND	**Sac6p**: fimbrin
YDL058w	X54378	1790	0.171	R	ND	**Uso1p**: putative integrin
YBL047c	Z35808	1381	0.197	EF	ND	Putative integrin
YHR166c	U00027	626	0.122	R	–	**Cdc23p**: involved in cell cycle
YAL041w	U12980	736	0.129	R	ND	**Cdc24p**: GTP-GDP exchange factor
YOR257w	M14078	161	0.108	EF	ND	**Cdc31p**: centrin involved in spindle pole body duplication
YGL155w	M74109	376	0.111	–	ND	**Cdc43p**: geranylgeranyl transferase β subunit
YJL030w	Z49305	196	0.115	R	ND	**Madwp**: spindle-assembly checkpoint protein
YCR077c	X59720	797	0.180	EF	ND	**Pat1p**: topoisomerase II-associated protein
YDR150w	Z50046	2748	0.162	R	ND	**Num1p**: nuclear migration protein
YJL031c	Z49307	290	0.154	R	ND	**Bet4p**: geranylgeranyl transferase type II α-subunit
YDL020c	Z74068	531	0.140	EF	ND	**Rpn4p**: involved in protein degradation
YPR103w	X68662	212	0.209	–	ND	**Pre2p**: involved in protein degradation

ORF, open reading frame; accession numbers are given for the GENBANK database; length is given in number of amino acids; CAI, codon adaptation index; SQ, predicted from primary sequence; BE, biochemical evidences; EF, EF-hand calcium-binding motif; R, related to EF-hand; C2, calcium-dependent phospholipid binding domain; ND, not determined.

(Eilam et al., 1985a). Indeed, whereas yeast mitochondria have low calcium transport and storage capacity (Carafoli et al., 1970; Uribe et al., 1992), the vacuole calcium content has been estimated to be of the order of 1.3 mM, several thousand-fold higher than that of the cytoplasm which has been estimated, using the calcium-sensitive fluorescence dye, indo-1, to be of the order of 100 to 350 nM (Halachmi and Eilam, 1989, 1993).

The cellular-efflux and vacuole-influx transport system might not be the only components of the remarkable calcium homeostasis machinery of the yeast cytoplasm, in which the calcium level is maintained at a level near 200 nM, even in the presence of up to 200 mM extracellular calcium (Halachmi and Eilam, 1993). Indeed, other subcellular calcium stores exist, some of which exhibit a variety of organelle-specific Ca^{2+}-ATPase activities (Halachmi et al., 1992; Okorokov et al., 1993; Okorokov 1994).

Calcium Channels

In addition to the early physiological data reported above, suggesting the existence of at least two calcium-uptake and two calcium-efflux systems in the yeast plasma membrane, more recent observations on increased calcium influx on exposure to mating pheromone or glucose have been reported (Ohsumi and Anraku, 1985; Anand and Prasad, 1987; Eilam et al., 1990; Iida et al., 1990; Nakajima-Shimada et al., 1991; Prasad and Rosoff, 1992). This calcium-influx system might be channel-like, as suggested by Gustin et al. (1988), who identified, by patch-clamp, a mechano-sensitive cation channel activity in the yeast plasma membrane, permitting passage of calcium ions in a rather nonselective manner.

The existence of a vacuolar channel, activated by cytoplasmic calcium, was first demonstrated by Wada et al. (1987) in electrophysiological studies using artificial planar bilayer membranes. However, the high concentration of calcium needed to trigger channel opening led the authors to wonder if the observation was physiologically relevant. Bertl and Slayman (1990, 1992) reported that reducing agents, such as dithiothreitol or β-mercaptoethanol, decrease the calcium requirement of the calcium-activated channel in the vacuolar membrane (called YVC1). Furthermore, measurements on isolated patches have shown that YVC1 activity may be regulated by calmodulin and cytoplasmic pH (Bertl and Slayman, 1992) and by inositol(1,4,5)trisphosphate (Belde et al., 1993). Although YVC1 also conducts sodium and potassium from the vacuoles to the cytoplasm, it has been suggested that it may regulate cytoplasmic calcium levels (Bertl and Slayman, 1990, 1992).

Table 18.1 shows that three open reading frames, recently identified on chromosome 7 and 15 during systematic yeast genome sequencing, might encode yeast calcium-channel subunits. The first, *YGR217W* (Van Der Aart et al., 1995), encodes a protein that shows significant homology to voltage-dependent calcium channels (BLAST *P* value around 10^{-45}). *YGR217W* was recently identified as a recessive suppressor of the temperature-sensitive and calcium-sensitive *cdc-1* mutant and demonstrated to be together with *MID1*, an essential component of the calcium uptake in yeast (Fischer et al., 1997; Paidhungat and Garrett, 1997). The second, *YGL176C*, enclodes a protein with very weak partial homology with the calcium-channel α-1-subunit of *Discopyge* ommata (BLAST *P* value of 0.42; Bertani et al., 1995). The third, *YOR088W*, encodes a protein with partial homology with the capacitative calcium entry channel of the "trp" family (Birnbaumer et al., 1996). Notably, it displays about 30–35% identity in a region corresponding to the "Pore-S6" domain of the trp channels. Databases suggest weak similarity of Ybr086cp, Yd1073wp and Ylr200wp with components of calcium channels. However, no clear experimental or homology data supports these proposals.

Calcium/Proton Exchanger

Biochemical and genetic studies have suggested the presence of a vacuolar Ca^{2+}/H^+ antiporter. Early experiments showed that calcium transport into vacuole membrane vesicles was blocked by inhibitors of the vacuolar H^+-ATPase (Ohsumi and Anraku, 1983). This observation was backed by the finding that nine calcium-sensitive mutants are mutated in genes that encode structural components of the vacuolar membrane H^+-ATPase (Ohya et al., 1991a). Calcium uptake into purified vacuoles depends on a transmembrane proton gradient, mainly attributed to the activity of the vacuolar H^+-ATPase (Dunn et al., 1994). The *VCX1* gene that encodes the vacuolar Ca^{2+}/H^+ antiporter has been recently identified as a multicopy suppressor of calcium sensitivity of the *pmc1* mutant (Cunningham and Fink, 1996). The same gene, called *HUM1*, was also discovered at the same time as a multicopy suppressor of manganese sensitivity of calcineurin mutants (Pozos et al., 1996). In wild-type strain and in normal growth conditions, Vex1p/Hum1p post-translationally inhibited through a calcineurin-dependent mechanism, does not contribute greatly to calcium tolerance. On the contrary, Vcx1p ensures the essential part of calcium tolerance in *cnb1 pmc1* or *cmd1 pmc1* double mutants (the *CNB1* gene encodes for the calcineurin-regulatory subunit, *CMD1* for the calmodulin and *PMC1* for the vacuolar Ca^{2+}-ATPase). The *VCX1/HUM1* gene product has eleven putative transmembrane domains and shows limited homology to the retinal Na^+/Ca^{2+}, K^+ and cardiac Na^+/Ca^{2+} exchangers. Three other putative Ca^{2+}/H^+ or Ca^{2+}/Na^+ exchangers identified in the yeast genome (Paulsen et al., 1998) are listed in Table 18.1.

Calcium-ATPases

The *S. cerevisiae* genome contains 18 genes encoding for P-type ATPases (Catty et al., 1997), easily recognized by multiple conserved amino acid sequences (Wach et al., 1992; Fagan and Saier 1994). Two of the encoded proteins, Pmr1p and Pmc1p, characterized in terms of subcellular location and physiological function, belong to the subfamily of calcium-transporting ATPases. Another one, Drs2p, previously classified in this subfamily, was recently described as a putative aminophospholipid translocase. None of these proteins have been isolated and studied biochemically. In addition, six genes, encoding putative Ca^{2+}-ATPases, are known only from their DNA sequences recently identified during the systematic sequencing of the yeast genome.

The Vacuolar Calcium-ATPase

In addition to the calcium-channel and the H^+/Ca^{2+} exchanger, yeast vacuoles contain the Pmc1 Ca^{2+}-ATPase (Cunningham and Fink, 1994a). The *PMC1* gene, the expression of which is induced by a calcineurin/calmodulin-dependent process (Cunningham and Fink, 1996; Matheos et al., 1997; Stathopoulos and Cyert, 1997), encodes a protein showing 40% identity with the mammalian plasma membrane Ca^{2+}-ATPases. Using an epitope-tagged *PMC1* gene, expressed either from the chromosomal locus or from a multicopy plasmid, immunofluorescence studies have localized Pmc1p to the vacuolar membranes. The *pmc1* mutant shows a dramatic decrease in the nonexchangeable calcium pool and high sensitivity to exogenous high calcium concentrations, suggesting that Pmc1p regulates the cytosolic calcium concentration by contributing to calcium uptake in the vacuoles.

The Golgi-Like Organelle Calcium-ATPase

The *PMR1* gene (Rudolph et al., 1989) encodes a P-type ATPase found in a Golgi-like organelle (Antebi and Fink, 1992) and showing 50% identity with the rat RS10-31 Ca^{2+}-ATPase (Gunteski-Hamblin et al., 1992). Disruption of the *PMR1* gene results in poor growth on solid medium that contains micromolar calcium and in hypersensitivity to EGTA. The *pmr1* mutants have a severely affected secretory pathway,

as shown by nonglycosylation of secreted proteins, incomplete proteolytic processing by the Kex2 calcium-dependent protease, and oversecretion of heterologously expressed proteins (Rudolph et al., 1989; Antebi and Fink, 1992; Harmsen et al., 1993). Consistent with Pmr1p having an essential role in the yeast secretory pathway is the finding that *PMR1* is identical to the *SSC1* gene, mutation which was shown to affect protein secretion (Smith et al., 1985), and the ability of the *pmr1* mutation to suppress several mutant phenotypes affected in the secretory pathway (Antebi and Fink, 1992).

The *pmr1* null mutation also suppresses oxidative damage caused by loss of functional Cu/Zn superoxide dismutase Sod1p. As this effect depends on extracellular manganese, a possible role for Pmr1p in manganese homeostasis has been suggested (Lapinskas et al., 1995). Recently, *PMR1* transcription was shown to be modulated by manganese through a calcineurin-mediated mechanism (Cunningham and Fink, 1996). A recent study revealed that Pmr1p displayed biochemical properties (sensitivity to inhibitors and affinity for substrates) distinct from SERCA and PMCA ATPases, suggesting that it might belong to a novel Ca^{2+}-ATPases subfamily (Sorin et al., 1997).

The Putative Endoplasmic Reticulum Calcium-ATPase

The *DRS2* gene was isolated by complementation of mutants deficient in the assembly of ribosomal subunits, suggesting that Drs2p may be located in the endoplasmic reticulum membrane. Drs2p was thought to modulate calcium levels at the ER membrane, which, in turn, could control targeting of the translation apparatus through Tef3p, a putative calcium-binding protein homologous to the eukaryotic elongation factor, EF-1γ (Ripmaster et al., 1993). This hypothesis is actually debated since DRs2p exhibits significant sequence homology to the newly discovered bovine ATPase II, a P-type ATPase involved in aminophospholipids translocation. Accordingly, a *drs2* null mutant was shown to be deficient in phosphatidyl serine transport (Tang et al., 1996).

The Other Putative Calcium-ATPase Genes

The *S. cerevisiae* genome might contain six other Ca^{2+}-ATPases. Four of them, Yer166wp, Ymr162cp, Yil048wp, and Ydr093wp, which are very similar to one another, as shown by BLAST *P* values of 10^{-200} to 10^{-110}, are closely related to Drs2p. In this group, Yer166wpo and Ydr093wp, the genes for which are located on chromosomes 5 and 4, respectively, might be isoforms of the same protein. Two other open reading frames, *YEL031w* and *YOR291w*, encode for unusual P-type ATPases that display a low degree of homology with known calcium pumps (André, 1996; Catty et al., 1997).

Biochemical Evidence for Calcium-ATPases

Purification of a putative plasma membrane Ca^{2+}-ATPase, inhibited *in vitro* by the mating pheromone α-factor, has been reported (Hiraga et al., 1991). Moreover, calcium-uptake measurements in the *pmr1* null mutant and subcellular fractionation studies have suggested the existence of several ATP-dependent calcium-pump activities, two possibly located in the Golgi apparatus one in endoplasmic or endoplasmic-like membranes (Okorokov et al., 1993; Okorokov and Lehley, 1997).

Calcium Signaling

As in mammalian cells, calcium signaling functions in yeast are largely carried out by calmodulin and calmodulin-dependent kinase and phosphatase, but also, in a less well-characterized system, by phospholipase C and protein kinase C. Table 18.2 lists the open reading frames presumed to be involved in yeast calcium-signaling.

Calmodulin

Calmodulin is a small, acidic calcium-binding protein, which, in response to intracellular calcium stimuli, regulates a wide variety of cellular functions by modulating the activity of calmodulin/calcium-dependent enzymes (see Chapter 4. In *Saccharomyces cerevisiae*, calmodulin is encoded by *CMD1*, a single-copy gene located on chromosome 2. Disruption or deletion of *CMD1* is lethal, indicating that yeast calmodulin, which shares only 60% identity with other known calmodulins, is an essential protein (Davis et al., 1986). However, yeast calmodulin does not share all of the properties of calmodulins from other species. It is a poor activator of vertebrate phosphodiesterase, myosin light chain (Luan et al., 1987), and pea NAD kinase (Ohya et al., 1987), and an antibody raised against yeast calmodulin does not cross-react with bovine calmodulin (Ohya et al., 1987). Yeast calmodulin, which contains four calcium-binding motifs, only binds three calcium ions compared with four for the vertebrate calmodulins (Luan et al., 1987; Matsuura et al., 1991) and its affinity for calcium, as well as its calcium-exchange rates, differ from those of other calmodulins (Starovasnik et al., 1993).

The study of yeast calmodulin mutants with modifications of the putative calcium-binding loops suggests that high-affinity calcium binding may not be required for protein function (Geiser et al., 1991). Vertebrate calmodulins can, however, functionally replace yeast calmodulin (Davis and Thorner 1989; Ohya and Anraku 1989a). The functional domains of Cmd1p were characterized by structural studies on wild-type and mutant yeast calmodulins expressed in *Escherichia coli* (Brockerhoff et al., 1992), as well as by a variety of recombination DNA studies (Persechini et al., 1991; Sun et al., 1991; Matsuura et al., 1993; Harris et al., 1994).

Calmodulin-dependent lethality was demonstrated by use of episomal *CMD1* expression under the control of the galactose-inducible *GAL1* promoter. Analysis of cells arrested by shift from galactose to glucose medium suggested involvement of calmodulin in nuclear division (Ohya and Anraku, 1989b). A study of the temperature-sensitive mutant, *cmd1-1*, which bears two point mutations in the *CMD1* sequence, showed a complex phenotype of cells arrested in cell cycle (Davis, 1992). A dosage-dependent suppressor of *cmd1-1* has been isolated, which encodes the Hcm1 protein, related to the fork head family of DNA-binding proteins (Zhu et al., 1993). In the temperature-sensitive *cmd1-101* mutant, expressing only the carboxy-terminal half of calmodulin, nuclear integrity and spindle body function are affected (Sun et al., 1991, 1992). Numerous conditional lethal mutants of yeast calmodulin were recovered by site-directed mutagenesis of phenylalanine residues (Ohya and Bostein, 1994a, 1994b). This approach made it possible to define four intragenic complementation groups of temperature-sensitive mutants: *cmd1A* (abnormal actin organization), *cmd1B* (abnormal calmodulin distribution), *cmd1C* (defect in nuclear division), and *cmd1D* (bud emergence affected).

These studies point out the essential role of calmodulin in yeast, especially in the cell cycle (see also Chapters 19, 20 and 21, and Anraku et al., 1991; Ohya and Anraku, 1992).

Calmodulin and the Cytoskeleton

Two studies (Geiser et al., 1993; Stirling et al., 1994) have reported an interaction between calmodulin and Nuf1p/Spc110p, an essential component of the spindle pole body (Kilmartin et al., 1993); the authors have mapped the calmodulin-binding site to the C-terminus of the Nuf1p/Spc110p protein and suggest that the interaction may not require calcium. Thus, one proposed role for calmodulin would be to regulate the correct assembly of Nuf1p/Spc10p into the spindle pole body.

Other studies suggests that calmodulin may also be implicated in actin cytoskeletal organization. Actin accumulates in growing regions, suggesting that it may be involved in the direction of new wall and membrane materials (Reviewed in Welch et al., 1994). Calmodulin has also been immunolocalized at sites of active surface growth during the cell cycle (Brockerhoff and Davis, 1992); this localization, which does not require calcium binding, overlaps with that of actin. Taken together with data obtained from analysis of the *cmd1A* mutant (Ohya and Bostein, 1994a, 1994b), these observations suggest an involvement of calmodulin in polarization of the yeast actin cytoskeleton. Calmodulin action might be mediated via a calcium-independent interaction with Myo2p (Brockerhoff et al., 1994), an unconventional myosin, a member of the actin-binding protein family and essential for secretory vesicles movement and polarized growth in yeast (Johnston et al., 1991). It is not excluded that another unconventional myosin protein, Moy4p, which contains six putative calmodulin-binding sites, also mediates the effect of calmodulin on actin distribution (Haarer et al., 1994).

Calmodulin and Endocytosis

Actin, together with Sac6p, a fimbrin belonging to the family of actin-bundling proteins (Kübler and Riezman, 1993; Otto, 1994; see also Section 5), is involved in the internalization step of endocytosis. Using the temperature-sensitive *cmd1-1* mutant (Davis, 1992), Kübler et al. (1994) have demonstrated an apparently calcium-independent involvement of calmodulin in endocytosis. Myo2p does not seem to be involved, unlike the clathrin light chain, which plays an important role in endocytosis (Tan et al., 1993) and, as suggested by analogy with the vertebrate homologs (Brodsky et al., 1991), might be another potential calmodulin target in yeast (Kübler et al., 1994).

Other Calmodulin-Related Processes

MARs (*m*atrix-*a*ssociation *r*egions) or SARs (*s*caffold-*a*ttached *r*egions) are DNA sequences that interact with nonhistone proteins in the assembly of the yeast nuclear scaffold. In a *cmd1* mutant strain, in which seven amino acids of the carboxy-terminal region of calmodulin are deleted, the interaction between MARs and the nuclear matrix is disrupted, suggesting a role for calmodulin in the chromatin reorganization occurring during each mitotic cycle (Fishel et al., 1993).

As with plant, fungal, and vertebrate calmodulins, yeast calmodulin can be covalently linked to ubiquitin (Ziegenhagen and Jennissen, 1990; Jennissen et al., 1992). This step, a prerequisite for protein degradation and involving at least two proteins (an ubiquitin protein ligase and Ubc4p, an ubiquitin-conjugating enzyme), is calcium-dependent (Parag et al., 1993).

Calmodulin-Binding Proteins

Several other techniques have been used to detect calmodulin target proteins in yeast. By screening a λ gt11 expression library with ^{125}I-labeled calmodulin as probe, Liu et al. (1991) reported the isolation of the two genes, *CMP1* and *CMP2*, which encode two catalytic subunits of calcineurin (see also Ye and Bretscher, 1992). Identification of nuclear calmodulin-binding proteins by the electrophoretic gel overlay method using ^{125}I-calmodulin, followed by the chromatographic isolation of potential calmodulin target proteins (Hiraga et al., 1993a), has resulted in the cloning of the gene for the yeast elongation factor, *EF-1β* (Hiraga et al., 1993b), and the genes *SSE1* and *SSE2*, which belong to the heat-shock protein 70 family (Mukai et al., 1993). Primary sequence analysis has shown that Pmr2p, a putative Na^+-ATPase, also contains a putative calmodulin-binding domain (Rudolph et al., 1989).

Calmodulin-Dependent Protein Kinases

Two early studies have reported the existence of calmodulin-dependent protein kinase activity in *Saccharomyces cerevisiae* (Londesborough and Nuutinen, 1987; Miyakawa et al., 1989). Purification of yeast calmodulin-dependent protein kinases has led to the identification of the *CMK1* and *CMK2* gene products, whose amino acid sequences show 60% homology and whose disruption is not lethal (Ohya et al., 1991c; Paush et al., 1991). Cmk1p and Cmk2p, which belong to the type II calmodulin-dependent protein kinase family, display different responses to calmodulin stimulation, with Cmk1p being more sensitive to bovine calmodulin than to yeast calmodulin, while the converse holds for Cmk2p, with Cmk2p being converted to active protein kinase by autophosphorylation, while Cmk1p is not (Ohya et al., 1991c). Although *CMK1* and *CMK2* are both involved in yeast thermotolerance, individual disruption of *CMK1* or *CMK2* reveals a greater involvement of *CMK1* (Iida et al., 1995). Calmodulin-dependent kinase activity might also be involved in the glucose-induced activation of the plasma membrane PMal H^+-ATPase, as suggested by site-directed mutagenesis at the putative calmodulin-dependent kinase site (Portillo et al., 1991) and by physiological studies using calmidozolium, a calmodulin-dependent kinase inhibitor (Brandao et al., 1994). Van Heusden et al. (1995) have reported the isolation of the essential *BMH1* and *BMH2* yeast genes, which encode 14-3-3-like proteins, one activity of which requires the presence of calcium calmodulin-dependent protein kinase II. The existence of a yeast calmodulin-dependent protein kinase III, known to inactivate elongation factor 2 in mammalian cells (Mitsui et al., 1993), has been suggested by two independent groups (Donovan and Bodley, 1991; Perentesis et al., 1992). Two other genes, *RCK1* and *RCK2* (also called *CMK3*), whose products are more distantly related to known calmodulin-dependent kinases, have also been discovered (Dahlkvist and Sunnerhagen, 1994) and shown to be suppressors of cell cycle checkpoint mutations in *Schizosaccharomyces pombe* (Dahlkvist et al., 1995).

Calcineurin

Calcineurin is the calmodulin-dependent protein phosphatase 2B. In yeast, calcineurin is comprised of two homologous catalytic subunits, Cna1p/Cmp1p and Cna2p/Cmp2p, which interact with calcium/calmodulin (Cyert et al., 1991; Cyert and Thorner, 1992; Liu et al., 1991; Nakamura et al., 1992; Ye and Bretscher, 1992) and one regulatory subunit, Cnb1p (Kuno et al., 1991; Cyert and Thorner, 1992), which is myristoylated and contains four calcium-binding sites. Disruption of either both of the catalytic subunit-encoding genes or the regulatory subunit-encoding gene shows, under normal growth conditions, calcineurin not to be an essential protein.

Calcineurin and the Mating Response

Mating pheromones cause yeast cells to be arrested at the G1 phase of the cell cycle and induce the expression of many genes involved in conjugation. This is transient and cells that have not mated can recover and proliferate, as long as calcium is present in the medium (Iida et al., 1990). A potential role for calcineurin in the mating process was first suggested by observations that the *Mata cna1cna2* or *Mata cnb1* mutants were unable to recover from α-factor-induced growth arrest (Cyert et al., 1991; Cyert and Thorner, 1992). Moreover, *CNA1* expression could be regulated by an α-factor-involving mechanism (Ye and Bretscher, 1992). Indirect proof of a requirement for calcineurin in the mating response comes from studies using cyclosporin A (CsA) and FK506. In mammalian cells, these two compounds are immunosuppressive drugs, known to inhibit a calcium-dependent signal leading to interleukin-2 transcription and T-cell activation, the target of both being calcineurin (reviewed in Kunz and Hall, 1993). In yeast, CsA and FK506 block recovery from mating pheromone-induced G1 arrest (Foor et al., 1992). The effects of CsA and FK506 are mediated by the respective specific receptors, cyclophilin and FKBP-12, which belong to the immunophilin family (Breuder et al., 1994). It appears that, in the absence of exogenous ligand, the immunophilins are able to interact with, and regulate, calcineurin, possibly via their intrinsic peptidlyprolyl isomerase activity, and that CsA and FK506 potentiate this interaction (Cardenas et al., 1994). This would explain how impairment of calcineurin activity, either by mutation or drugs, affecting the mating process.

Calcineurin was recently shown to be involved in cell proliferation, in participating in the control of the expression of Swe1p, a tyrosine kinase that regulates the activity of Cdc28p, a protein required for the G2 to M transition (Mizunuma et al., 1998).

Calcineurin and Calcium Homeostasis

One possible explanation for calcineurin involvement in the mating response could be its ability to regulate calcium homeostasis. Indeed, exposure of *Mata* cells to α-factor induces intracellular calcium accumulation (Ohsumi and Anraku, 1985). Using calcium imaging with fura-2 as probe, Iida et al. (1990) demonstrated that the rise in concentration of intracellular calcium in response to α-factor was generated essentially by increased calcium influx through the plasma membrane.

In yeast, the vacuoles have been demonstrated to be the major calcium storage organelle and Pmc1p, a Ca^{2+}-ATPase, triggers energy-dependent calcium influx in this subcellular compartment (Cunningham and Fink, 1994a). A strain lacking functional Pmc1p fails to grow in medium containing a high concentration of calcium. This calcium-dependent growth inhibition is mediated by calcineurin, since double mutants lacking Pmc1p plus calcineurin protein or Pmc1p plus calmodulin exhibit this phenotype. One possible explanation is that abolition of calcium sequestration in the vacuole by *PCM1* disruption leads to an increase in cytosolic calcium concentration, which then inhibits growth by constant activation of calcineurin (Cunningham and Fink, 1994a). In agreement with this hypothesis, it has been suggested that calcium/calmodulin-activated calcineurin negatively regulates another vacuolar calcium transporter, such as the proton/calcium antiporter (Garrett-Engele et al., 1995). In the *pmc1* null mutant, loss of active calcineurin would then permit this exchanger to partially substitute for the absent Pmc1p and produce partial calcium tolerance. A gene encoding this antiporter, *VCX1/HUM1*, has been isolated and the activity of the gene product is post-translationally inhibited by calcineurin (Cunningham and Fink, 1996; Pozos et al., 1996). Calcineurin also induces *PMC1* expression (Cunningham and Fink 1996; Matheos et al., 1997; Stathopoulos and Cyert, 1997). Another study, using a *vma* mutant of a structural gene of the vacuolar H^+-ATPase, suggests that calcineurin may inhibit calcium influx into an intracellular compartment distinct from the vacuole (Tanida et al., 1995). Indeed, in the *vma* mutant, which displays no detectable calcium uptake into the vacuole, inhibition of calcineurin by FK506 leads to increased calcium sequestration and a parallel decrease in cytosolic calcium concentration. These observations might reflect the existence of other potential calcineurin targets, such as Pmr1p (Rudolph et al., 1989), expression of which was shown to be induced by calcineurin (Cunningham and Fink, 1996; Matheos et al., 1997) or Cls2p/Csg2p, a calcium-regulatory membrane protein in the endoplasmic reticulum (Takita et al., 1995).

Calcineurin and Salt Tolerance

While calcineurin is not required for cell growth under normal conditions, the calcineurin mutants *cna1cna2* or *cnb1* are unable to grow in the presence of high concentrations of NaCl or LiCl (Nakamura et al., 1993; Breuder et al., 1994). Furthermore, in the presence of the calcineurin inhibitor FK506, neither growth sensitivity to high-salt medium nor intracellular levels of Na^+ and K^+ are modified in a calcineurin-deficient strain (Nakamura et al., 1993). An explanation for the sodium or lithium toxicity in strains that lack calcineurin has been provided by Mendoza et al. (1994), who demonstrated that the Cnb1 regulatory subunit of calcineurin regulates the expression of a Na^+-ATPase-encoding gene, *ENA1/PMR2* (Rudolph et al., 1989; Haro et al., 1991). *PMR2* expression was recently found to be induced through a calcineurin mechanism, in response to high sodium concentration (Cunningham and Fink, 1996; Matheos et al., 1997; Stathopoulos and Cyert, 1997). It was also shown that, in addition to this calcineurin-dependent regulation, Ena1p/Pmr2p undergoes a second calcium/calmodulin-dependent (but calcineurin-independent) post-translational regulatory step (Wieland et al., 1995). Accordingly, the PPZ phosphatases, whose disruption results in increased *ENA1/PMR2* expression, also regulate salt tolerance in yeast (Posas et al., 1995).

Calcineurin and Cell Wall Integrity

The *fks1-1* mutant, described in 1993 by Parent et al., shows a 100–1000-fold higher sensitivity to FK506 or CsA than the wild type. It displays reduced growth, partially suppressed by calcium and osmotic stabilizing agents, such as sorbitol, and also has a pronounced tendency to spontaneous lysis (Parent et al., 1993; Garrett-Engele et al., 1995), a phenotype characteristic of strains in which cell wall integrity is affected by mutation of the PKC1/MAP kinase path-

way (reviewed by Herskowitz, 1995). Surprisingly, calcineurin is an essential protein for this strain. The *FKS1/CND1* gene, first isolated by complementation of the drug-hypersensitive phenotype of the *fks1-1* mutant (Eng et al., 1994) and, later, by complementation of the inability of the *cnd1-8* mutant to grow on dextrose (Garrett-Engele et al., 1995), encodes an integral membrane protein, subunit of the β-1,3 glucan synthase (Douglas et al., 1994).

A second subunit of the β-1,3 glucan synthase, Fks2p, highly homologous to Fks1p was identified by Mazur et al. (1995), providing an explanation for the colethality between *fks1/cnd1* and calcineurin mutants. The double null mutant *fks1/fks2* was shown to be lethal, while transcription of *FSK2* was demonstrated as being regulated through a calcineurin-dependent mechanism (Mazur et al., 1995; Stathopoulos and Cyert, 1997).

Phospholipase C and Protein Kinases C

In mammalian cells, phospholipid signaling is one of the major pathways for signal transduction (Divecha and Irvine, 1995). In response to extracellular signals, the plasma membrane phospholipase C hydrolyzes phosphatidyl inositol(4,5)bisphosphate $[(PtdIns(4,5)P_2]$ into inositol(1,4,5)trisphosphate $[Ins(1,4,5)P_3]$ and 1,2-diacylglycerol (DAG). These two molecules act as second messengers, $Ins(1,4,5)P_3$ promoting calcium release from specific organelles, while DAG activates protein kinase C (PKC). In this signaling pathway, calcium is required for phospholipase C activity and for the activity of certain types of PKC.

In yeast, phospholipase C is encoded by the *PLC1* gene (Flick and Thorner, 1993; Payne and Fitzgerald-Hayes, 1993; Yoko-o et al., 1993). Plc1p contains an EF-hand calcium-binding domain and its activity is strictly calcium-dependent (Flick and Thorner, 1993). *plc1* mutants show severe growth defects (Yoko-o et al., 1993, 1995), osmotic sensitivity, defects in nutrient utilization (Flick and Thorner, 1993), and aberrant mitotic chromosome segregation, partially suppressed by the addition of exogenous calcium (Payne and Fitzgerald-Hayes, 1993). Several reports have identified Plc1p as a component of a yeast phospholipid-signaling pathway. The addition of ammonium sulfate to starved cells increases the levels of $Ins(1,4,5)P_3$ and DAG in a Cdc25p-dependent fashion (Schomerus and Kuntzel, 1992). Moreover, $INS(1,4,5)P_3$ causes the release of calcium from vacuolar vesicles, suggesting the existence, in this organelle, of a calcium-transport system distinct from the Ca^{2+}/H^+ antiporter (Belde et al., 1993). It may also be mentioned that the temperature-sensitive phenotype produced by mutation of the Pik1 phosphatidyl inositol 4-kinase, a protein involved in the biosynthesis of PtdIns(4,5)P$_2$ (Flanagan et al., 1993), is suppressed at elevated calcium concentrations (Garcia-Bustos et al., 1994).

The family of mammalian protein kinases C has been divided into three groups according to their dependence upon the combination of calcium and diacylglycerol, diacylglycerol alone, or neither (reviewed by Hug and Sarre, 1993). In yeast, the *PKC1* gene, disruption of which causes bud growth deficiency leading to cell growth arrest, encodes protein kinase C (Levin et al., 1990). Amino acid sequence analysis shows that the C$_2$ region, corresponding to the putative calcium-binding site in other PKCs, is poorly conserved in Pkc1p. Biochemical studies have confirmed that the Pkc1p activity does not require either calcium or diacylglycerol (Antonsson et al., 1994; Watanabe et al., 1994). However, Levin and Bartlett-Heubusch (1992) have reported that the growth arrest phenotype of *pkc1* mutants was suppressed by addition of extracellular calcium. It is now admitted that Pkc1p ensures cell wall integrity through a cascade where it is upstream controlled in a Plc1p-independent fashion and where it downstream regulates a MAP kinase pathway (Kamada et al., 1995; Herskowitz, 1995). A second protein kinase C, related to the first group of PKCs, has been partially purified and biochemically characterized (Ogita et al., 1990).

Calcium-Dependent Enzymatic Activities

Calcium-Dependent Proteases

KEX2

The *kex2* mutants are unable to produce the active α-mating pheromone because of impaired proteolytic conversion of the inactive prohormone into active peptide (Julius et al., 1984). The *KEX2* gene encodes a calcium-dependent subtilisin-like serine protease, specific for pairs of basic residues and containing one transmembrane C-terminus domain (Mizuno et al., 1988, 1989; Fuller et al., 1989; Brenner and Fuller 1992; reviewed by Steiner et al., 1992). As with the eukaryotic endoproteases, Kex2p undergoes several maturation steps by which the inactive proenzyme is converted into an active endopeptidase. in the endoplasmic reticulum, Kex2p cleaves off its own N-terminus (Wilcox and Fuller, 1991; Germain et al., 1992, 1993; Gluschankof and Fuller, 1994), producing a truncated protein, which is, in turn, cleaved by the Ste13p dipeptidyl aminopeptidase (Brenner and Fuller 1992). The resulting enzyme is N- and O-glycosylated in the endoplasmic reticulum and Golgi apparatus; with the late Golgi compartment being the final destination of the mature protein

(Payne and Schekman, 1989; Redding et al., 1991; Wilcox et al., 1992; Wilsbach and Payne, 1993). Kex2p cleaves a wide variety of precursors of eukaryotic proteins expressed in yeast, such as proalbumin (Ledgerwood et al., 1995) and prosomatostatin (Bourbonnais et al., 1993, 1994). In yeast, in addition to its essential role in maturation of pro-α-factor, Kex2p is involved in the processing of killer toxin (Julius et al., 1984) and exoglucanase (Larriba et al., 1993).

YAP3

The *YAP3* gene was first isolated as a partial suppressor of the α-factor-processing defect of the *kex2* null mutant. It encodes an aspartyl protease (Egel-Mitani et al., 1990), which seems to be calcium-dependent (Azaryan et al., 1993). This enzyme, shown to be involved in the processing of heterologously expressed prosomatostatin (Bourbonnais et al., 1993), might be attached to the plasma membrane through a glycophosphatidyl inositol anchor (Ash et al., 1995).

BLH1

Ycp1p, also called Blh1p (Enenkel and Wolf, 1993; Magdolen et al., 1993) or Lap3p (Trumbly and Bradley, 1983) is a cysteine proteinase, homologous to the bleomycin hydrolase. Isolated as a calcium-dependent membrane-binding protein (Kambouris et al., 1992), Ycp1p does not, however, share any calcium-binding motif. Ycp1p activity was shown to be markedly reduced by phosphatidyl serine, an effect reversed by addition of 1 mM free calcium. It was suggested that the calcium could affect not only the enzymatic activity but also the targeting to specific organelles of Ycp1p (Kambouris et al., 1992).

YKR038C

The *YKR038C* open reading frame encodes a 421 amino acids protein that contains a putative EF-hand calcium-binding motif and shows significant homology with the *Pasteurella haemolytica* O-sialoglycoprotein endopeptidase (Abdullah et al., 1991). This newly discovered yeast protein might therefore belong to the glycoprotease family (reviewed by Mellors and Sutherland, 1994).

YMR154C

This open reading frame, identified during the systematic sequencing of the yeast genome, still has no known function. The encoded protein exhibits 27% identity with the calcium-dependent protease calpain of *Caenorhabditis elegans*.

Other Calcium-Dependent Enzymatic Activities

Guanosine Diphosphatase (GDA1 Gene)

The guanosine diphosphatase, encoded by the *GDA1* gene and involved in the processing of glycoproteins, converts GDP into GMP. Enriched in Golgi vesicles, GDPase activity was shown to be dependent on calcium (Yanagisawa et al., 1990).

Glycogen Phosphorylase (GPH1 Gene)

Glycogen phosphorylase activity, which releases α-D-glucose-1-phosphate from glycogen, was shown to be stimulated by micromolar free calcium (François and Hers, 1988).

α-Mannosidase (MNS1 Gene)

The α-mannosidase encoded by the *MNS1* gene is an ER-resident protein that converts $Man_9GlcNac$ into $Man_8GlcNAc$, a key step in glycoprotein biosynthesis. Mns1p has an amino-terminal EF-hand calcium-binding motif (Camirand et al., 1991), probably essential for stabilizing its enzymatic activity (Jelinek-Kelly and Herscovics, 1988).

Trehalase (NTH1 and NTH2 Genes)

Trehalase, which converts α, α-trehalose to glucose, has been shown to require free calcium with an apparent affinity in the micromolar range (Nevés and François, 1992).

Mitochondrial Pyrophosphatase (PPA2 Gene)

The activity of mitochondrial inorganic pyrophosphatase is inhibited by calcium through competition between Ca^{2+}-pyrophosphate and Mg^{2+}-pyrophosphate. The required inhibitory calcium concentrations are compatible with the intramitochondrial calcium content (Uribe et al., 1993).

Putative Calcium-Binding Proteins

YBR053C

The *YBR053C* open reading frame encodes a 358 amino acids protein showing some similarities with the rat calcium-binding protein, regucalcin (Aljinovic and Pohl, 1995). Cloned in 1993(a) by Shimokawa and Yamaguchi, rat regucalcin, which shows no significant homology with other known calcium-binding proteins, contributes, nevertheless, to the liver calcium pathway. The expression of rat regucalcin is stimulated by calcium (Yamaguchi et al., 1995) through a calmodulin- and calcitonin-dependent mechanism (Shimokawa and Yamaguchi, 1993b; Yamaguchi et al., 1994; Yamaguchi and Kanayama, 1995). Moreover, regucalcin may be an activator of the Ca^{2+}/Mg^{2+}-ATPase of the rat liver plasma membrane (Takahashi and Yamaguchi, 1993).

YDR373W

The *YDR373W* open reading frame encodes a small soluble protein showing strong similarities to the calcium-binding proteins frequenin from *Xenopus laevis* (BLAST *P* value 8.4×10^{-81}), rat neurocalcin (BLAST *P* value 1.2×10^{-80}), and rat visinin (BLAST *P* value 1.9×10^{-69}). These proteins belong to a subclass of the EF-hand family of calcium-binding proteins, the best known member of which is recoverin, a calcium sensor of the rod photoreceptor, involved in desensitization of light-activated rhodopsin by direct interaction with rhodopsin kinase (Chen et al., 1995).

YGR058W

The *YGR058W* sequence encodes a protein of unknown function, homologous to the calcium-binding proteins PMP41 (BLAST *P* value of 8.4×10^{-14}) and ALG-2 (BLAST *P* value of 9.9×10^{-14}) recently shown to participate in cell apoptosis (Vito et al., 1996), and, to a lesser extent, to the calcium-dependent protease calpain (BLAST *P* values lower than 10^{-8}).

YGL106W

The Ygl106w protein is of unknown function and does not possess any canonical EF-hand motifs (as predicted by the PROSITE program); it shares striking similarity with plant, mammalian, and fungi calmodulins (BLAST *P* values of 10^{-30}), as well as with myosin light chains (BLAST *P* value 10^{-22}).

YIR010W and YBL048W

The Yir010w and Ybl048w proteins that, respectively, possess an EF-hand and EF-hand-like calcium-binding domain, do not share any significant similarity with known proteins.

YPR188C

Ypr188cp is a protein of unknown function that contains one EF-hand calcium-binding motif and displays weak homology with calmodulins (BLAST *P* value 10^{-10}).

YLR247C

The Ylr247c protein, of unknown function, contains, in addition to an EF-hand, a zinc-finger and two leucine-zipper motifs. It displays partial homology with fungal DNA repair proteins (reviewed in Prakash et al., 1993), from *Schizosaccharomyces pombe* (accession numbers Z99292 and AL023287; BLAST *P* values 10^{-18} and 9×10^{-17} respectively) and from *Saccharomyces cerevisiae* (protein Rad16p; BLAST *P* values 5×10^{-15}).

YNL083W

Ynl083wp exhibits highest homology with the peroxisomal calcium-dependent solute carrier from rabbit (accession number AF004161; BLAST *P* value 2×10^{-23}; Weber et al., 1997). It contains one EF-hand motif and a mitochondrial energy transfer protein signature (PROSITE motif PS00215). The hypothesis of an involvement of Ynl083wp in mitochondrial transport was tested by expression in *Escherichia coli* and transport activity measurements, of a truncated form of the protein (Mayor et al., 1997). However, this work does not provide any relevant informations about Ynl083wp function in yeast.

Putative Calcium-Dependent Phospholipid-Binding Proteins

The CalB domain involved in the calcium-dependent lipid binding of the cytosolic form of phospholipase A2 (Clark et al., 1991) is present in several classes of proteins, including some isoenzymes of protein kinases C, mammalian phospholipases C and A2 (cytosolic form), and the synaptic membrane proteins synaptotagmins (Azzi et al., 1992; Fukada et al., 1996). In *S. cerevisiae*, five proteins exhibit such a domain. In the Rsp5 ubiquitin protein ligase (Huibregtse et al., 1995), this domain is located at the amino-terminal end of the protein, whereas in Ynl087wp and Yor086cp (two proteins of unknown function), two Ca1b domains are located as in the synaptotagmins in the carboxy-terminal moiety of the proteins. Drr1p and Drr2p (also called Tor1p and Tor2p), two homologous proteins that share similarity to the phosphatidyl inositol 3-kinase, were also reported to contain this domain (Cafferkey et al., 1993; Helliwell et al., 1994).

Calcium in Yeast Physiology

Calcium in Cell Wall and Membranes

CHS4

CHS4 encodes a protein that contains an EF-hand calcium-binding domain (Scherens et al., 1993). Chs4p is required for the activity of chitin synthase III (Bulawa, 1992), an enzyme involved in the formation of the chitin ring around the base of emerging buds and in the biosynthesis of cell wall chitin (reviewed by Klis, 1994); its activity might be modulated by Chs4p either by proteolytic activation or by a direct interaction (Choi et al., 1994).

HKR1

The *HKR1* gene was identified as a multicopy suppressor of the sensitivity of *S. cerevisiae* to the *Hansenula mrakii* killer toxin (Kasahara et al., 1994a). *HKR1*, whose disruption is lethal, encodes an EF-hand motif-containing protein, possibly involved in synthesis of the β-glucans, components of the cell wall known to mediate the action of the HM-1 killer toxin (Kasahara et al., 1994b).

FLO1

Saccharomyces cerevisiae undergoes calcium-dependent flocculation (Stratford and Assinder, 1991). The "calcium-bridging hypothesis," in which calcium was thought to link flocculating cells directly, was more recently replaced with the "lectin theory of flocculation," which claims that calcium binding to lectin-like proteins induces them to interact with specific cell wall sugar receptors from coflocculating cells (reviewed by Stratford, 1992, 1993, 1994). The dominant *FLO1* gene, which causes calcium-dependent and mannose-specific flocculation, might encode a lectin-like protein (Teunissen et al., 1993; Watari et al., 1994) or a lectin-activation protein (Stratford and Carter, 1993). The calcium-dependent flocculation seen in the *tup1* mutant has recently been related to Tup1p repression of *FLO1* transcription (Teunissen et al., 1995).

SYR1

The phytotoxin syringomycin strongly affects the growth of several fungi, including *S. cerevisiae*. Associated to growth defect, an increase in calcium influx has been observed in treated cells. The syringomycin-resistant *syr1-1* mutant is unable to grow in 400 mM $CaCl_2$ and exhibits higher calcium uptake and higher intracellular calcium content than the wild-type strain (Takemoto et al., 1991). the *SYR1* gene cloned by complementation of the *syr* mutation (Taguchi et al., 1994) is identical to the previously identified *ERG3* gene which encodes for the C-5 sterol desaturase (Arthington et al., 1991). Changes in membrane permeability or inhibition of some calcium-transport mechanisms like Ca^{2+}-ATPase (Hiraga et al., 1991) or calcium channels (Takemoto et al., 1991) have been proposed to relate the desaturated sterol deficiency and the calcium sensitivity of *syr* mutant.

NSP1

NSP1 is an essential gene encoding for nucleoporin. The carboxy-terminal domain of Nsp1p involved in the interaction of Nsp1p with other nuclear pore components was reported to contain an EF-hand like calcium-binding motif. Although mutations in this region strongly affect Nsp1p function, neither calcium binding to Nsp1p nor calcium dependency of nuclear pore assembly have been reported so far (Nehrbass et al., 1990).

Calcium and Gene Expression

MSN4

The *MSN4* gene, identified as a multicopy suppressor of Snf1p protein kinase mutant (Estruch and Carlson, 1993), encodes for a zinc-finger protein that acts as a positive transcription regulator of some genes that contain a response element (Martinez-Pastor et al., 1996). Although Msn4p contains an EF-hand motif at its amino-terminal end, no calcium regulation of this protein has been reported.

CAM1/TEF3

Cam1p/Tef3p was initially identified by screening for proteins able to bind secretory vesicles or lipids in a calcium-dependent manner (Creutz et al., 1991). The *CAM1* gene was later shown to encode a protein homologous to the elongation factor-1-γ of *Artemia salina* (Kambouris et al., 1993). Since no known calcium-binding motifs were identified in its primary sequence, it was suggested that Cam1p/Tef3p might harbor a novel type of calcium-binding site or that the calcium-dependent membrane binding might be indirect. The *TEF3/CAM1* gene was reisolated as a suppressor of the cold-sensitive phenotype of the *drs2* mutant, mutated in Drs2, a putative Ca^{2+}-ATPase, and found to be affected in ribosome assembly (Ripmaster et al., 1993). It was suggested that changes in calcium concentration induced by Drs2p activity could modulate the targeting of the translational apparatus to the endoplasmic reticulum via membrane binding of Tef3p. Recent data (Tang et al., 1996), however, suggest that Drs2p could modulate this targeting by regulating aminophospholipid distribution at the cytoplasmic layer of the endoplasmic reticulum membrane.

RRP3

The essential *RRP3* gene was found by a PCR-based strategy, developed for the search of yeast proteins that contain the DEAD box, a sequence included in the ATP-binding site of RNA-dependent ATPases. The gene disruption demonstrated the involvement of Rrp3p in the processing of the 35S primary transcript of pre-rRNA and in the maturation of the 18S rRNA. The protein Rrp3p, that contains one EF-hand calcium-binding motif as well as characteristic motifs of RNA helicases, was purified to homogeneity and showed a very weak RNA-dependent ATPase activity (O'Day et al., 1996).

Calcium and the Secretory Pathway

Calcium plays an essential role in the yeast secretory pathway, e.g., in the fusion of endoplasmic reticulum-derived vesicles and Golgi membranes (Rexach and Schekman, 1991) and protein transport (Baker et al., 1990). Along the secretory pathway, each organelle probably possesses its own calcium transporters, ATPases or other transport system, which regulate the intraorganelle calcium concentration, which, in turn, modulates the activity of calcium-dependent secretory pathway proteins. Indeed, loss of active Pmr1p, a Ca^{2+}-ATPase of a Golgi-like organelle, affects the secretory pathway (Rudolph et al., 1989; Antebi and Fink, 1992). Creutz et al. (1991) have identified at least five potential annexin-like proteins which show calcium-dependent binding to secretory vesicles. Heterologous expression of mammalian annexins in yeast secretory mutants suggests that such proteins are liable to participate in yeast intracellular membrane trafficking (Creutz et al., 1992). Using a gel overlay assay, Parlati et al. (1995a) have reported the existence, in the endoplasmic reticulum of *S. cerevisiae*, of six soluble calcium-binding proteins, which might be components of the secretory pathway. Six gene products, with known or putative calcium-dependent functions in folding and sorting, are described below.

CLC1

In yeast, clathrin is involved in the retention of Golgi proteins during secretion (Payne and Schekman, 1989; Graham et al., 1994; Rad et al., 1995; Stepp et al., 1995) and in the endocytic internalization of mating pheromone receptors (Tan et al., 1993). Yeast clathrin light chain is encoded by the *CLC1* gene, whose disruption leads to a slow-growth phenotype (Silveira et al., 1990). Like its mammalian homologs, Clc1p contains an EF-like calcium-binding site (Näthke et al., 1990) and, as in mammalian cells, calcium may influence the correct assembly/disassembly of the yeast clathrin coat (reviewed by Brodsky et al., 1991).

CNE1

The *CNE1* gene, initially isolated during yeast genome systematic sequencing (De Virgilio et al., 1993), encodes a protein showing significant homology with the mammalian calcium-binding proteins, calnexins, considered to be the molecular chaperone of neosynthesized proteins (Wada et al., 1991, reviewed by Williams 1995; Bergeron et al., 1994). In spite of the absence of a C-terminal endoplasmic retention signal, Cne1p is an intrinsic endoplasmic reticulum membrane protein, but does not seem to bind calcium. Study of the *CNE1*-disrupted strain has suggested that Cne1p, like its mammalian homologs, could be a chaperone (Parlati et al., 1995a).

END3

The *END3* gene was isolated by complementation of the temperature-sensitive growth of the *end3* mutant (Benedetti et al., 1994), which is defective in the internalization step of endocytosis (Raths et al., 1993). In addition to its inability to internalize the α-factor receptor and the fluid-phase marker, lucifer yellow, the *ned3* mutant also displays cell wall defects. End3p, which is involved in the endocytosis of uracil permease under stress conditions (Volland et al., 1994), contains an EF-hand calcium-binding site. However, the role of calcium in the activity of the protein has not yet been demonstrated.

KAR2

The *KAR2* gene (Normington et al., 1989; Rose et al., 1989) encodes an endoplasmic reticulum-resident protein, which shows 67% identity with the mammalian BIP molecular chaperone (Munro and Pelham, 1986). In yeast Kar2p is required for protein translocation across the endoplasmic reticulum membrane, for nuclear fusion, and for the correct folding of newly synthesized proteins. Although no evidence exists for a calcium-dependent function of Kar2p in yeast, it is interesting to note that, in mammalian cells, *BIP* transcription is regulated by calcium (Wenfeng et al., 1993) and the BIP protein has been identified as a calcium-binding protein (Macer and Koch, 1988), which interacts with endoplasmic reticulum proteins in a calcium-dependent fashion (Suzuki et al., 1991).

SAC6

The yeast Sac6 fimbrin, which belongs to the family of actin-bundling proteins, was shown to participate to endocytosis. Identified through dominant suppression of a temperature-sensitive actin mutation (Adams et al., 1989), Sac6p, like vertebrate fimbrins, contains two potential EF-hand-like calcium-binding domains. It was suggested, however, that calcium-coordinating residues of these domains were not well enough conserved to ensure an efficient ion binding (Adams et al., 1991).

USO1

Uso1p is a hydrophilic protein involved in vesicular transport from endoplasmic reticulum to Golgi, which shares similarity with mammalian integrins. A component of the cytoskeleton, Uso1p has an EF-hand-like motif that suggests a regulation through calcium binding (Nakajima et al., 1991). Yb1047cp, a Uso1p homolog that like Pan1p and End3p displays an EHeps15 homolog domain, also contains an EF-hand motif.

PAN1

The essential *PAN1* gene, previously thought to encode a PAB (Poly(A) binding protein)-dependent poly(A) ribonuclease (Sachs and Deardorff, 1992), was reisolated by screening for rapid-death mutants, in a Start-deficient *cdc28-4* strain (Tang and Cai, 1996). The *pan1-4* mutant, a multicopy suppressor of which was found to be *END3*, displays abnormal growth bud and abnormal actin distribution. It is therefore defective in receptor-mediated and fluid-phase endocytosis (Tang et al., 1997). The Pan1 protein, whose localization to cortical actin patches might be dependent on End3p, contains an EF-hand calcium-binding motif, a proline-rich carboxy-terminal motif suggested to be a SH3-binding domain as well as two eps 15 homology (EH) domains. Pan1p appears to play an essential role in actin organization through an interaction with End3p and in endocytosis through interactions with the clathrin assembly proteins Yhr161cp and Ygr241cp (Wendland and Emr, 1998).

Calcium and Cell Cycle (see Chapters 19 and 20)

CDC23

Cdc23p belongs to a 20S complex involved in ubiquitin-mediated proteolysis (Hilt and Wolf, 1996). Together with Cdc16p and Cdc27p, Cdc23p promotes the metaphase-to-anaphase transition by acting on DNA replication (Heichman and Roberts,

1996). Although it posseses a potential EF-hand motif at its N-terminus (Doi and Doi, 1990), Cdc23p has not been reported to be submitted to calcium regulation.

CDC24

Early studies of temperature-sensitive cell division cycle (*cdc*) mutants demonstrated the essential role of *CDC24* in morphogenic adaptation to cell cycle progression (Reid and Hartwell, 1977; Sloat and Pringle, 1978; Sloat et al., 1981). *CDC24* was first thought to be involved in chitin organization, a step in the so-called cell polarization prior to bud emergence. Links between calcium pathways and a *CDC24*-dependent mechanism emerged when the *cls4* mutation (calcium-ssensitive) was shown to be allelic to *cdc24* (Ohya et al., 1986b). Disruption of the *CLS4/CDC24* gene is lethal and the amino acid sequence of Cls4p/Cdc24p reveals two putative calcium-binding sites, only one being of the Ef-hand type (Miyamoto et al., 1987, 1991). Cls4p.Cdc24p also contains a domain homologous to the *dbl* oncogene product described as a GDP-dissociation stimulator of the human GTPase Cdc24Hs (Hart et al., 1991). Genetic studies have shown that Cdc24p could interact with at least three proteins involved in bud emergence: Cdc42p, a Rho-like GTPase (Bender and Pringle, 1989; Miller and Johnson, 1994; Ziman and Johnson, 1994), Rsr1p, a Ras-like GTPase (Bender and Pringle, 1989), and Bem1p, a protein of unknown function (Peterson et al., 1994). Cdc24p acts on Cdc42p as a GDP–GTP exchange stimulator (Zheng et al., 1994) and stabilizes the GTP-bound active form of Rsr1p (Zheng et al., 1995). The Cdc24p/Bem1p interaction is inhibited by calcium (Zheng et al., 1995); its functional role is unknown, but could be related to the molecular event impaired in the *cls4* mutant (Ohya et al., 1986b).

CDC31

The *CDC31* gene was cloned by complementation of a temperature-sensitive *cdc* mutant in which nuclear division was affected (Schild et al., 1981; Baum et al., 1986). It encodes a calcium-binding protein very similar to the centrin/caltractin proteins, major components of plant and mammalian spindle poles and centrosomes (reviewed by Salisbury, 1995). Purified Cdc31p has a high affinity for calcium and is located in the half-bridge of the spindle pole body (Spang et al., 1993). Genetic studies suggest that Cdc31p and Kar1p (Vallen et al., 1992a, 1992b) play synergetic roles in nuclear division (Biggins and Rose, 1994; Vallen et al., 1994). A direct *in vitro* interaction between Kar1p and Cdc31p was demonstrated by protein blotting (Biggins and Rose, 1994). Mislocalization of Cdc31p in the *kar1* mutant suggested that the function of kar1p might be to target Cdc31p to the spindle pole body. As proposed for other species, calcium could play a role in centrin distribution, perhaps by modulating the Kar1/Cdc31p interaction (Baron et al., 1994).

CDC43

In addition to *CDC24* and *CDC24*, the essential *CDC43* gene is also involved in cell polarity and budding (Adams et al., 1990). It encodes the β-subunit of the type I geranylgeranyl transferase (Johnson et al., 1990; Mayer et al., 1992), which may, together with other proteins, post-translationally process Cdc42p (Ohya et al., 1991b, 1993). Interestingly, *CDC43* is identical to the *CAL1* gene isolated by complementation of the cal1-1 mutation (Ohya et al., 1991b), characterized by calcium-dependent growth (Ohya et al., 1984).

BET4

Originally misidentified as *MAD2* (Li and Murray, 1991), the essential gene *BET4* encoded for α-subunit of the geranylgeranyl transferase type II, an enzyme directly involved in cell cycle through its action on the small ATPase rab proteins (Brown and Goldstein, 1993). Although Bet4p contains three domains structurally related to EF-hand motif, no calcium regulation of Bet4p function has been reported.

PAT1

The *PAT1* gene, deletion of which leads to a reduced growth rate and a default in chromosome segregation, encodes for a proline- and glutamine-rich protein that interacts with the leucine-rich region of topoisomerase II and possesses one EF-hand calcium-binding motif (Wang et al., 1996).

ORC3

The nuclear DNA replication occurs during the S phase of the cell cycle. In yeast it is initiated by the binding of the origin recognition complex ORC, whose ORc2p is an essential component, to specific DNA sequences called ARS (autonomously replicating sequences) (reviewed in Toyn et al., 1995). The Orc3 protein, that contains an EF-hand motif at its amino-terminal end, was identified by two approaches: identification of mutants that were lethal in combination with *orc2* mutant (synthetic lethal screen) and identification of proteins that interact with orc2p (two-hybrid method). The *ORC3* gene was found to be essential for cell viability, temperature-sensitive mutants exhibiting a so-called "dumbell" morphology, characteristic of cells damaged in DNA replication or nuclear division (Hardy, 1996).

MAD2

The *MAD2* gene (mitotic arrest defect) was identified during the search of benomyl-sensitive yeast mutants, defective in feedback control of mitosis (Li and Murray, 1991). it encodes a protein of 196 amino acids that exhibits three regions structurally related to the EF-hand motif and whose homolog has been found in *Schizosaccharomyces pombe* (BLAST *P* value 3×10^{-26}) Mad2p-like proteins exist also in frog and human where they have been localized at the kinetochore, in prometaphase only, suggesting they might act as monitor of the kinetochore to spindle assembly (Chen et al., 1996; Li and Benezra, 1996).

NUM1

The *NUM1* gene encodes a 313 kDa protein, transiently expressed during the cell cycle at the G2/S (Kormanec et al., 1991) and targeted to the cell cortex thanks to its carboxyterminal PH (pleckstrin homology) domain (Gibson et al., 1994; Farkasovsky and Küntzel, 1995). While growing normally in rich and synthetic media, *num1* null mutants often exhibit default in nuclear migration, leading to the presence of two nuclei in the budded mother cell. Cytological and genetical data have suggested that Num1p might control the interaction between the G2 nucleus and the bud neck cytoskeleton, by affecting astral microtubule functions (Farkasovsky and Küntzel, 1995). Interestingly, Num1p has a potential carboxy-terminal EF-hand calcium-binding domain. The expression of a modified Num1 protein, where this domain has been removed, does not complement the *NUM1* deficiency, suggesting a possible role of calcium in Num1p function (Farkasovsky and Küntzel, 1995).

Calcium and Mating

MID1 and MID2

The *MID1* and *MID2* genes (mating pheromone-induced death), whose mutation cause cell death upon exposure to mating pheromone, encode proteins linked to the calcium pathway. Mid1p contains four putative transmembrane domains with partial homology to voltage-gated and cGMP-gated ion channels. The *mid-1* phenotype is characterized by low calcium-uptake activity, suggesting that Mid1p may be involved in the calcium influx seen in

response to mating pheromone (Iida et al., 1990, 1994). The *MID2* gene, whose expression is enhanced by mating pheromone (Ono et al., 1994), encodes integral calcium-binding protein. The *MID2* gene is homologous to the *SMS1* gene, a suppressor of the *htr1* temperature-sensitive mutant (Takeuchi et al., 1995), characterized by an inability to recover from G1 arrest induced by mating pheromone (Kikuchi et al., 1994).

Calcium and Protein Degradation

RPN4

The *RPN4/SON1/UFD5* gene encodes a protein of 531 amino acids, recently demonstrated to be a component of the regulatory particle of the 26S proteasome (Fujimuro et al., 1998). The Rpn4 protein found to be involved in the degradation of ubiquitinated proteins (Johnson et al., 1995) contains a zinc-finger-like motif, as well as three domains rich in aspartic and glutamic acids, one of them being considered as an EF-hand motif.

Calcium and Metabolism

ACS2

Two acetyl-CoA synthetases exist in *Saccharomyces cerevisiae*. Despite their high degree of identity (61.2%), these proteins ensure different roles in yeast metabolism. The Acs1 protein is required for growth on ethanol, while Acs2p is required for growth on glucose, the first corresponding to the so-called "aerobic", while the second corresponds to the "anaerobic" acetyl-CoA synthetase (Van den Berg and Steensma, 1995; Van den Berg et al., 1996). Although Acs2p displays an EF-hand motif, no involvement of calcium in the enzymatic activity of Acs2p has yet been demonstrated.

Calcium-Dependent and Calcium-Sensitive Mutants (see Chapter 19)

A unique feature of yeast is the case with which it is possible to obtain and analyze mutants modified in specific physiological traits. Several yeast mutants, listed in Table 18.5, exhibit either calcium-dependent or calcium-sensitive cell growth.

The *cal1-1* calcium-dependent mutant (Ohya et al., 1984) was isolated by screening thermosensitive mutants able to grow at 37°C in YPD medium only when supplemented with 100 mM CaCl$_2$. In calcium-poor medium, the growth of the mutant is blocked at the G2 stage of cell division. At 23°C, both the wild-type and *cal1-1* strains show the same sensitivity to

Table 18.5 Main Characteristics of Mutants in Calcium-Dependent Pathways

Mutation	Divalent Cations Sensitivity						TFP	Pet$^-$	Ca Content	Ca Uptake	Gene name
	Ca^{++}	Mg^{++}	Zn^{++}	Mn^{++}	Cu^{++}	Sr^{++}					
cal1-1	−						s				CAL1/CDC43
cls1	+						s	+	I	N	
cls2	+						s	+	I	N	CLS2/CSG2
cls3	+						s	+	I	N	
cls4	+							+	N	N	CDC24
cls5	+						s	+	I	I	
cls6	+						s	+	I	I	
cls7	+		+	+				−			VMA3
cls8	+		+	+				−	N	I	VMA1/TFP1
cls9	+		+	+				−	N	I	VMA11/VMA16/TFP3
cls10	+		+	+				−	N	I	VMA12/VPH2
cls11	+		+	+				−	N	I	VMA13
cls12	+		+	+				−			
cls13	+	+	+			+	s	+	I	I	VMA1/VSP11/PEP5/ END1
cls14	+	+	+				s	+	I	I	VAM6/VSP33/PEP14/ SLP1
cls15	+	+	+	+		+	s	+	I	I	
cls16	+	+			+		s	+	I	I	
cls17	+		+				s	+	I	I	VAM9/VSP16
cls18	+	+	+				s	+	I	I	VAM8/VSP18/PEP3
csg1	+					−			I		CCC1 and CCC2
csg2	+					−			I	I	CSG2/CLS2

The calcium-dependent cal-1 mutant is described in Ohya et al., 1984. The calcium sensitive cls mutants are described in Ohya et al., 1986a. The csg mutants (for calcium sensitive growth) are described in Beeler et al., 1994. Divalent cations sensitivity was measured in YPD medium plus 100 mM CaCl$_2$ or 100 mM MgCl$_2$, or 3 mM ZnCl$_2$, or 3 mM MnCl$_2$, or 10 mM CuCl$_2$ or 50 mM SrCl$_2$. The trifluoroperazine (TFP) sensitivity was measured in YUPD medium plus 40 μM trifluoroperazine. The Pet$^-$ phenotype was tested in 1% Bacto-yeast extract, 2% Polypeptone plus either 2% glycerol or 2% potassium acetate or 2% D(L)-sodium lactate. In cls mutants, the calcium content was measured by atomic absorption, the calcium uptake was measured using 20 or 200 μM of ^{45}Ca in 10 mM glucose. The calcium uptake and calcium content in csg mutants were determined by spectrophotometry using calcium indicator arsenazo III. Abbreviations: s, sensitive; I, increased; N, normal.

trifluoroperazine (TFP), an anticalmodulin drug. At 37°C, unlike the wild-type strain, cal1-1 cannot grow in calcium-poor medium in the presence of 0–40 μM TFP, while neither strain can grow in the presence of 80 μM TFP. The TFP sensitivity of cal1-1 is lost in the TFP-dependent tfr1 pseudorevertant, which grows in calcium-poor medium at 37°C only in the presence of 20–80 μM TFP. This suggests an interaction between the CAL1 and TFR1 gene products and a role for calcium and a calmodulin-like protein in yeast cell division. The CAL1 gene, cloned by com-plementation of the cal1-1 mutation (Ohya et al., 1991b), encodes a 376 amino acids protein identical to Cdc43p, the β-subunit of geranylgeranyl transfer-ase (Johnson et al., 1990), an enzyme which catalyzes the post-translational addition of the geranylgeranyl group to the cysteine residue of proteins bearing the carboxy-terminal sequence CAAX (C = cysteine, A = alanine, X = any amino acid).

In contrast to the scarcity of calcium-dependent mutants, 18 complementation groups of cls (calcium-sensitive) mutants have been isolated and characterized (Ohya et al., 1986a). Unable to grow at 30°C in calcium-rich medium (100 mA CaCl$_2$), they have been classified into four types. The type I mutants, *cls5 cls6, cls13, cls14, cls15, cls16, cls17* and *cls18*, have an elevated calcium content and increased calcium uptake; the type II mutant, *cls4*, has a normal calcium content and normal calcium uptake; the type III mutants, *cls1,* cls2, and *cls3*, have an elevated calcium content, but normal calcium uptake; and the type IV mutants, *cls8, cls9, cls10,* and *cls11*, have a normal calcium content, but increased calcium uptake. The *cls7* and *cls12* mutants do not belong to a specific *cls* type. The type I and III mutants (elevated calcium levels) are TFP-sensitive, suggesting modification of a calmodulin-mediated mechanism for maintaining calcium homeostasis. All type IV mutants are unable to utilize nonfermentable carbon sources, this being usually ascribed to mitochondrial defects. The products of certain of the genes, mutation of which results in the *cls* phenotype, have been identified. *CLS2* (a type III mutant) encodes a membrane protein required for calcium tolerance (Takita et al., 1995). The *CLS2* disruption can be suppressed by expression of the *BCL21* gene (also called *SUR1*), which also suppresses the *rvs 161* mutation, responsible for reduced viability of starved cells and leading to high salt-sensitivity, as well as the inability to grow on nonfermentable sources (Desfarges et al., 1993). *CLS4* is identical to *CDC24* (Ohya et al., 1986b) and encodes the GDP/GTP exchange factor of the Rho-like Cdc42 protein, a member of the small GTPases family, which includes the Ras proteins. All other known *CLS* genes are involved in the structure of the vacuolar H$^+$-ATPase or the vacuolar sorting protein (Ohya et al., 1991a), emphasizing the essential role of vacuoles in the maintenance of calcium-dependent processes.

Another genetic approach to the isolation of calcium-sensitive mutants was reported in 1994 by Dunn and collaborators (Beeler et al., 1994), who isolated the *csg* (calcium sensitive growth) mutants, in which vacuolar function is presumably not affected, as indicated by a normal sensitivity to strontium. Two major complementation groups, *csg1* and *csg2*, have been isolated, both of which are unable to grow in the presence of 100 mM CaCl$_2$ and have an increased intracellular calcium content.

The *CSG2* gene is identical to the *CLS2* gene. It encodes a 410 amino acids protein, containing nine putative transmembrane spans. The *csg2* null mutant fails to grow in the presence of 50 mM CaCl$_2$ and accumulates 22 times more calcium than the wild-type strain, this being stored in an exchangeable pool. Csg2p may be involved in calcium efflux from a specific organelle. Interestingly, it contains an EF-hand calcium-binding motif, which may act as a sensor of the cytosolic calcium concentration (Beeler et al., 1994). Second site suppressors of the *csg2* null mutant were isolated by screening csg2::LEU2 strains able to grow at 37°C in YPD medium, supplemented with 100 mM CaCl$_2$. Out of the seven classes of *scs* mutants (suppressor of calcium sensitivity), the *scs1* and *scs2* mutants were found to require at least 10 mM CaCl$_2$ growth. The *SCS1* gene, also called *LCB2*, was cloned by complementation of the calcium-dependent growth phenotype of the *scs1* mutant and encodes a subunit of serine palmitoyltransferase, suggesting a coupling between sphingolipid metabolism and the calcium pathway (Zhao et al., 1994; Nagiec et al., 1994).

The two genes *CCC1* and *CCC2* were reported to be calcium cross-complementers of the *csg1* mutation. the *CCC1* gene, located on chromosome 12, encodes a membrane protein of 322 amino acids, with five putative membrane-spanning segments and a potential signal peptidase cleavage site (Fu et al., 1994), and was recently isolated as a multicopy suppressor of manganese sensitivity of calcineurin mutants (Pozos et al., 1996). BLAST analysis reveals a very weak homology between the *CCC1* gene product and soybean nodulin-21 protein (*P* value of 4.7×10^{-7}), while the *CCC2* gene encodes a P-type ATPase (Fu et al., 1995), related to the copper-transporting ATPases involved in Menkes' and Wilson's diseases (Bull et al., 1993; Chelly et al., 1993) and to the other putative yeast Cu^{2+}-ATPase, Pca1p (Rad et al., 1994). Ccc2p was recently found to be essential for copper transport from the cytosol to an undefined intracellular compartment (Yuan et al., 1995).

ROD1

Rod1p, found to confer yeast resistance to *o*-dinitrobenzene when overproduced, does not share any similarity with proteins of known function. Together with Yap1p, a transcriptional regulator involved in multidrug resistance in *S. cerevisiae*, Rod1p was shown to be involved in yeast calcium tolerance as judged by the high sensitivity of the *rod1* mutant to millimolar CaCl$_2$. However, the molecular basis of this effect is not yet understood (Wu et al., 1996).

PRE2

The *PRE2/PRG1/DOA3* essential gene encodes the β5-subunit of the 20S proteasome, required for chymotryptic activity and degradation of ubiquitinated proteins (Heinemeyer et al., 1993; Freidman and Snyder, 1994; Chen and Hochstrasser, 1995; Groll et al., 1997). While originally identified as a suppressor of *crc1-1* (calcium regulated component), a mutation that leads to the loss of chromosome in YPD medium containing 4 mM EGTA (Friedman et al.,

1992), the link between calcium and Pre2p activity is not yet understood.

Calcium in Other Yeasts and Fungi

As expected, although less well-characterized than in *Saccharomyces cerevisiae*, similar calcium pathways exist in other yeast types. Experiments using the calcium ionophore A23187 have suggested that there is a calcium requirement for the cell cycle in *Schizosaccharomyces pombe* and *Kluyveromyces fragilis* (Duffus and Patterson, 1974; Penman and Dufus, 1975). *Schizosaccharomyces pombe* shows energy-dependent calcium uptake, with an apparent K_m of $45\,\mu M$ at an external pH of 4.5 (Boutry et al., 1977). In this organism, the *cta3* gene encodes a Ca^{2+}-ATPase (Ghislain et al., 1990). Biochemical studies on nystatin-permeabilized cells have demonstrated that the cta3p-dependent calcium uptake is located on a nonvacuolar organelle and suggested the existence of other organellar pumps (Halachmi et al., 1992). By polymerase chain reaction amplification of genomic *S. pombe* DNA and by using degenerated oligonucleotides corresponding to regions conserved across P-type ATPases, we have identified a new gene that encodes a protein homologous to Yil048wp from *S. cerevisiae* (P. Catty, unpublished results). This gene has also been found during the systematic sequencing of the *S. pombe* genome (accession number Z69731), together with a gene encoding for a protein-sharing similarity with Yer166wp (accession number Q09891). Calmodulin-regulated Ca^{2+}-ATPase activity, which is markedly inhibited *in vitro* by the mating pheromone rhodotorucine A, has been noted in *Rhodopiridium toruloides* membranes (Miyakawa et al., 1987). Kulakovskaya et al. (1993) have suggested the existence of a Ca^{2+}/H^+ antiporter in the vacuolar membranes of *Yarrowia lypolitica.*

The *cam1* and *CMD1* genes from, respectively, *S. pombe* and *Candida albicans* encode calmodulin proteins which, as in *S. cerevisiae*, are essential proteins involved in cell cycle progression and cell morphology (Takeda and Yamamoto, 1987; Takeda et al., 1989; Paranjape et al., 1990; Saporito and Sypherd, 1991). A recent study using nuclear magnetic resonance spectroscopy demonstrated that, in contrast to *S. cerevisiae*, calmodulin function in *S. pombe* requires calcium binding (Moser et al., 1995). In *R. toruloides*, cell differentiation induced by rhodotorucine A was shown to be mediated by a calcium/calmodulin-dependent process (Miyakawa et al., 1985, 1986). In *S. pombe*, the *ppb1* gene encodes a calcineurin-like calmodulin-dependent phosphatase (Yoshida et al., 1994) and a genetic approach has pinpointed the existence of a calmodulin-dependent

protein kinase II (Rasmussen and Rasmusen, 1994). In the *S. pombe* genome two genes, *pkc1* and *pkc2*, encode calcium-independent protein kinases C (Mazzei et al., 1993; Toda et al;., 1993), and the *pip1* gene encodes phospholipase C (Andoh et al., 1995).

The homologs of several *S. cerevisiae* proteins involved in a calcium pathway are found in *S. pombe*, e.g., scd1p, which shows 32% identity with Cdc24p (Chang et al., 1994), and Krp1p, a Kex2-like protease (Davey et al., 1994). Other Kex2-like proteases, Xpr6p and Kex1p, have been found, respectively, in *Yarrowia lipolytica* (Enderlin and Ogrydziak, 1994) and *Kluyveromyces lactis* (Tanguy-Rougeau et al., 1988). The fission yeast also possesses a recently discovered calmodulin-like protein, cdc4p (McCollum et al., 1995) as well as a calnexin-like protein encoded by the cnx1 gene, which, in contrast to Cne1p of *S. cerevisiae*, binds calcium ions (Parlati et al., 1995b).

In fungi, calcium has been shown to play an essential role in proliferative and morphogenetic processes, including the yeast-to-mycelium transition in *Sporothrix schenckii* (Rodriguez-del-Valle and Rodriguez-Medina, 1993), morphological changes in *Wangiella dermatitidis* (Szaniszlo et al., 1993), branching in *Achlya* (Harold and Harold, 1986), branching and apical growth in *Neurospora crassa* (Reissig and Kinney, 1983; Schmid and Harold, 1988) and cell cycle progression in *Aspergillus nidulans* (Lu et al., 1992).

As in yeast, an important calmodulin-dependent pathway exists in fungi. In *Aspergillus nidulans*, calmodulin (Rasmussen et al., 1990) is involved in the cell cycle (Lu et al., 1992, 1993; Rasmussen et al., 1992) and has been shown to modulate protein kinase activity (Bartelt et al., 1988; Kornstein et al., 1992). In *Neurospora crassa*, in addition to calmodulin (Capelli et al., 1993; Melnick et al., 1993), almost all the major components of a calmodulin-dependent pathway have been found, including calmodulin-dependent protein kinase (Ulloa et al., 1991), calmodulin-dependent protein phosphatase (Tropschug et al., 1989; Higuchi et al., 1991), calmodulin-activated phosphodiesterase (Tellez-Inon et al., 1985; Ortega Perez et al., 1983), and calmodulin-activated adenylate cyclase (Reig et al., 1984). Calmodulin has been found in *Achlya klebsiana* (Lejohn, 1989).

Calcium transport in fungi is well-documented in *Neurospora crassa*, which contains a plasma membrane ATP-driven H^+-influx/Ca^{2+}-efflux transport system (Stroobant and Scarborough, 1979; Miller et al., 1990). As in *S. cerevisiae*, vacuoles are the major intracellular calcium store (Miller et al., 1990), the calcium being released via an inositol 1,4,5-trisphosphate-dependent mechanism (Cornelius et al., 1989; Schultz et al., 1990). Calcium transport is stimulated by cytokinins in *Achlya* (Lejohn et al., 1974) and by pyrophosphate in *Phytophthora infestans* (Okorokov et al., 1978).

In fungi, calcium has been shown to regulate the transport of several compounds, such as polyamine in *N. crassa* (Davis et al., 1991) or amino acids in *Achlya* (Cameron and Lejohn, 1972), and to bind to several enzymes, such as α-amylase from *Aspergillus niger* (Boel et al., 1990), thermomycolase from *Malbranchea pulchella* (Voordouw and Roche, 1975) and proteinase K from *Tritirachium album* Limber (Muller et al., 1994).

Acknowledgements   We gratefully acknowledge Dr. Kyle Cunningham and Dr. Hans Rudolph for critical reading of this manuscript. This work was supported by grants from the Services de la Politique Scientifique: Action Science de la Vie and the Fonds National de la Recherche Scientifique (Belgium), and from the Commission of the European Communities as part of the BioTECH Programme No. BIO2-CT930422.

References

Abdullah, K. M., R. Y. Lo, and A. Mellors (1991) Cloning, nucleotide sequence and expression of the *Pasteurella haemolytica* A1 glycoprotease gene. *J. Bacteriol.* 173: 5597–5603.

Adams, A. E. M., D. Botstein, and D. G. Drubin (1989) A yeast actin-binding protein is encoded by SAC6, a gene found by suppression of an actin mutation. *Science* 243: 231–233.

Adams, A. E., D. I. Johnson., R. M. Longnecker, B. F. Sloat, and J. R. Pringle (1990) *CDC42* and *CDC43*, two additional genes involved in budding and the establishment of cell polarity in the yeast. *Saccharomyces cerevisiae. J. Cell. Biol.* 111: 131–142.

Adams, A. E. M., D. Botstein, and D. G. Drubin (1991) Requirement of yeast fimbrin for actin organization and morphogenesis *in vivo. Nature* 354: 404–408.

Aljinovic, G. and T. M. Pohl (1995) Sequence and analysis of 24 kb on the chromosome II of *Saccharomyces cerevisiae. Yeast* 11: 475–479.

Altschul, S. F., W. Gish, W. Miller, E. W. Myers, and D. J. Lipman (1990) Basic local alignment search tool. *J. Mol. Biol.* 215: 403–410.

Anand, S. and R. Prasad (1987) Status and calcium influx in cell cycle of *S. cerevisiae. Biochem. Int.* 14: 963–970.

Andoh, T., T. Yoko, Y. Matsui, and A. Toh (1995) Molecular cloning of the *plc1+* gene of *Schizosaccharomyces pombe*, which encodes a putative phosphoinositide-specific phospholipase C. *Yeast* 11: 179–185.

André, B. (1995) An overview of membrane transport proteins in *Saccharomyces cerevisiae. Yeast* 11: 1575–1611.

Anraku, Y., Y. Ohya, and H. Iida (1991) Cell cycle control by calcium and calmodulin in *Sacchraomyces cerevisiae. Biochim. Biophys. Acta* 1093: 169–177.

Antebi, A. and G. R. Fink (1992) The yeast Ca^{2+}-ATPase homologue *PMR1* is required for normal Golgi function and localizes in a novel Golgi-like distribution. *Mol. Biol. Cell* 3: 633–654.

Antonsson, B., S. Montessuit, L. Friedli, M. A. Payton, and G. Paravicini (1994) Protein kinase C in yeast. *J. Biol. Chem.* 269: 16821–16828.

Arthington, B. A., L. G. Bennet, P. L. Skatrud, C. J., Guynn, R. J. Barbuch, C. E. Ulbright, and M. Bard (1991) Cloning, disruption and sequence of the gene encoding yeast C-5 sterol desaturase. *Gene* 102: 39–44.

Ash, J., M. Dominguez, J. J. Bergeron, D. Y. Thomas, and Y. Bourbonnais (1995) The yeast proprotein convertase encoded by *YAP3* is a glycophosphatidylinositol-anchored protein that localizes to the plasma membrane. *J. Biol. Chem.* 270: 20847–20854.

Azaryan, A. V., M. Wong, T. C. Friedman, N. X. Cawley, F. E. Estivariz, H. Chen, and Y. P. Loh (1993) Purification and characterization of a paired basic residue-specific yeast aspartic protease encoded by the *YAP3* gene. Similarity to the mammalian propiomelanocortin-converting enzyme. *J. Biol. Chem.* 268: 11968–11975.

Azzi, A., D. Boscoboinik and C. Hensey (1992) The protein kinase C family. *Eur. J. Biochem.* 208: 547–577.

Baker, D., L. Wuestehube, R. Schekman, D. Botstein, and N. Segev (1990) GTP-binding Ypt1 protein and Ca^{2+} function independently in a cell-free protein transport reaction. *Proc. Natl. Acad. Sci. USA* 87: 355–359.

Baron, A., V. Sunman, E. Nemeth, and J. L. Salisbury (1994) The pericentriolar lattice of PtK2 cells exhibits temperature and calcium-modulated behavior. *J. Cell Biol.* 107: 2993-3003.

Bartelt, D. C., S. Fidel, L. H. Farber, D. J. Wolff, and R. L. Hammell (1988) Calmodulin-dependent multifunctional protein kinase in *Aspergillus nidulans. Proc. Natl. Acad. Sci. USA* 85: 3279–3283.

Baum, P., C. Furlong, and B. Byers (1986) Yeast gene required for spindle pole body duplication: homology of its product with Ca^{2+}-binding proteins. *Proc. Natl. Acad. Sci. USA* 83: 5512–5516.

Beeler, T., K. Gable, C. Zhao, and T. Dunn (1994) A novel protein, CSG2p, is required for Ca^{2+} regulation in *Saccharomyces cerevisiae. J. Biol. Chem.* 269: 7279–7284.

Belde, P. J., J. H. Vossen, W. Borst-Pauwels, and A. P. Theuvenet (1993) Inositol 1,4,5-trisphosphate releases Ca^{2+} from vacuolar membrane vesicles of *Saccharomyces cerevisiae. FEBS Lett.* 323: 113–118.

Bender, A. and J. R. Pringle (1989) Multicopy suppression of the *cdc24* budding defect in yeast by *CDC42* and three newly identified genes including the ras related gene *RSR1. Proc. Natl. Acad. Sci. USA* 86: 9976–9980.

Benedetti, H., S. Raths, F. Crausaz, and H. Riezman (1994) The *END3* gene encodes a protein that is required for the internalization step of endocytosis and for actin cytoskeleton organization in yeast. *Mol. Biol. Cell.* 5: 1023–1037.

Bergeron, J. J. M., M. B. Brenner, D. Y. Thomas, and D. B. Williams (1994) Calnexin: a membrane-bound chaperone of the endoplasmic reticulum. *Trends Biochem. Sci.* 19: 124–128.

Bertani, I., M. Coglievina, P. Zaccaria, R. Klima, and C. Bruschi (1995) The sequence of an 11.1 kb fragment on the left arm of *Saccharomyces cerevisiae* chromosome VII reveals six open reading frames including *NSP49*, *KEM1* and four putative new genes. *Yeast* 11: 1187–1194.

Bertl, A. and C. L. Slayman (1990) Cation-selective channels in the vacuolar membrane of *Saccharomyces*: dependence on calcium, redox state, and voltage. *Proc. Natl. Acad. Sci. USA* 87: 7824–7828.

Bertl, A. and C. L. Slayman (1992) Complex modulation of cation channels in the tonoplast and the plasma membrane of *Saccharomyces cerevisiae*: single-channel studies. *J. Exp. Biol.* 172: 271–287.

Biggins, S. and M. D. Rose (1994) Direct interaction between yeast spindle pole body components: Kar1p is required for Cdc31p localization to the spindle pole body. *J. Cell. Biol.* 1252: 843–852.

Birnbaumer, L., X. Zhu, M. Jiang, G. Boulay, M. Peyton, B. Vannier, D. Brown, D. Platano, H. Sadeghi, E. Stefani, and M. Birnbaumer (1996) On the molecular basis and regulation of cellular capacitative calcium entry: roles of the Trp proteins. *Proc. Natl. Acad. Sci. USA*, 93: 15195–15202.

Boel, E., L. Brady, A. M. Brzozowski, Z. Derewenda, G. G. Dodson, V. J. Jensen, S. B. Petersen, H. Swift, L. Thim, and H. F. Woldike (1990) Calcium binding in alpha-amylases: an X-ray diffraction study at 2.1-Å resolution of two enzymes from *Aspergillus*. *Biochemistry* 29: 6244–6249.

Borst-Pauwels, G. W. F. H. (1981) Ion transport in yeast. *Biochim. Biophys. Acta* 650: 88–127.

Bourbonnais, Y., J. Ash, M. Daigle, and D. Y. Thomas (1993) Isolation and characterization of *S. cerevisiae* mutants defective in somatostatin expression: cloning and functional role of a yeast gene encoding an aspartyl protease in precursor processing at monobasic cleavage sites. *EMBO J.* 12: 285–294.

Bourbonnais, Y., D. Germain, J. Ash, and D. Y. Thomas (1994) Cleavage of prosomatostatins by the yeast Yap3 and Kex2 endoprotease. *Biochimie* 76: 226–233.

Boutry, M., F. Foury, and A. Goffeau (1977) Energy-dependent uptake of calcium by the yeast *Schizosaccharomyces pombe*. *Biochim. Biophys. Acta* 464: 602–612.

Brandao, R. L., N. M. de Magalhaes-Rocha, R. Alijo, J. Ramos, and J. M. Thevelein (1994) Possible involvement of a phosphatidylinositol-type signaling pathway in glucose-induced activation of plasma membrane $H^{(+)}$-ATPase and cellular proton extrusion in the yeast *Saccharomyces cerevisiae*. *Biochim. Biophys. Acta.* 1223: 117–124.

Brenner, C. and R. S. Fuller (1992) Structural and enzymatic characterization of a purified prohormone-processing enzyme: secreted, soluble Kex2 protease. *Proc. Natl. Acad. Sci. USA* 89: 922–926.

Breuder, T., C. S. Hemenway, N. R. Movva, M. E. Cardenas, and J. Heitman (1994) Calcineurin is essential in cyclosporin A- and FK506-sensitive yeast strains. *Proc. Natl. Acad. Sci. USA* 91: 5372–5376.

Brockerhoff, S. E. and T. N. Davis (1992) Calmodulin concentrates at regions of cell growth in *Saccharomyces cerevisiae*. *J. Cell. Biol.* 118: 619–629.

Brockerhoff, S. E., C. G. Edmonds, and T. N. Davis (1992) Structural analysis of wild-type and mutant yeast calmodulins by limited proteolysis and electrospray ionization mass spectrometry. *Protein Sci.* 1: 504–516.

Brockerhoff, S. E., R. C. Stevens, and T. N. Davis (1994) The unconventional myosin, Myo2p, is a calmodulin target at sites of cell growth in *Saccharomyces cerevisiae*. *J. Cell. Biol.* 124: 315–323.

Brodsky, F. M., L. H. Beth, S. L. Acton, I. Nathke, D. H. Wong, S. Ponnambalam, and P. Parham (1991) Clathrin light chains: arrays of protein motifs that regulate coated-visicle dynamics. *Trends Biochem. Sci.* 16: 208–213.

Brown, M. S. and J. L. Goldstein (1993) Mad bet for Rab. *Nature* 366: 14–15.

Bulawa, C. E. (1992) *CSD2*, *CSD3* and *CSD4*, genes required for chitin synthesis in *Sacchraomyces cerevisiae*: the *CSD2* gene product is related to chitin synthase and to developmentally regulated proteins in *rhizobium* species and *Xenopus laevis*. *Mol. Cell. Biol.* 12: 1764–1776.

Bull, P. C., G. R. Thomas, J. M. Rommens, J. R. Forbes, and D. W. Cox (1993) The Wilson disease gene is a putative copper transporting P-type ATPase similar to the Menkes gene. *Nature Genet.* 5: 327–336.

Cafferkey, R., P. R. Young, M. M. McLaughlin, D. J. Bergsma, Y. Koltin, G. M. Sathe, L. Faucette, W.-K. Eng, R. K. Johnson, and G. P. Livi (1993) Dominant missense mutation in a novel yeast protein related to mammalian phophatidyl 3-kinase and VPS34 abrogate rapamycin cytotoxicity. *Mol. Cell. Biol.* 13: 6012–6023.

Cameron, L. E. and H. B. LeJohn (1972) On the involvement of calcium in amino acid transport and growth of the fungus *Achlya*. *J. Biol. Chem.* 247: 4720–4739.

Camirand, A., A. Heysen, B., Grondin, and A. Herscovics (1991) Glycoprotein biosynthesis in *Saccharomyces cerevisiae*. *J. Biol. Chem.* 266: 15120–15127.

Capelli, N., D. van Tuinen, R. Ortega Perez, J. F. Arrighi, and G. Turian (1993) Molecular cloning of a cDNA encoding calmodulin from *Neurospora crassa*. *FEBS Lett.* 321: 63–68.

Carafoli, E., W. X. Balcavage, A. L. Lehninger, and J. R. Mattoon (1970) Ca^{2+} metabolism in yeast cells and mitochondria. *Biochim. Biophys. Acta* 205: 18–26.

Cardenas, M. E., C. Hemenway, R. S. Muir, R. Ye, D. Fiorentino, and J. Heitman (1994) Immunophilins interact with calcineurin in the absence of exogenous immunosuppressive ligands. *EMBO J.* 13: 5944–5957.

Catty, P., A. de Kerchove d'Exaerde, and A. Goffeau (1997) The complete inventory of the yeast *Saccharomyces cerevisiae* P-type transport ATPase5. *FEBS Lett.* 409: 325–332.

Chang, E. C., M. Barr, Y. Wang, V. Jung, H.-P. Xu, and M. H. Wigler (1994) Cooperative interaction of *S. pombe* proteins required for mating and morphogenesis. *Cell* 79: 131–141.

Chelly, J., Z. Tumer, T. Tonnesen, A. Petterson, Y. Ishikawa-Brush, N. Tommerup, N. Horn, and A. P. Monaco (1993) Isolation of a candidate gene for Menkes disease that encodes a potential heavy metal binding protein. *Nature Genet.* 3: 14–19.

Chen, C. K., J. Inglese, R. J. Lefkowitz, and J. B. Hurley (1995) $Ca^{(2+)}$-dependent interaction of recovering with rhodopsin kinase. *J. Biol. Chem.* 270: 18060–18066.

Chen, P. and M. Hochstrasser (1995) Biogenesis, structure and function of the yeast 20S proteasome. *EMBO J.* 14: 2620–2630.

Chen, R.-H., J. C. Waters, E. D. Salmon, and A. W. Murray (1996) Association of spindle assembly checkpoint component XMAD2 with unattached kinetochores. *Science* 274: 242–246.

Choi, W.-J., A. Sburlati, and E. Cabib (1994) Chitin synthase 3 from yeast has zymogenic properties that depend on both the *CAL1* and *CAL3* genes. *Proc. Natl. Acad. Sci. USA* 91: 4727–4730.

Clark, J. D., L.-L. Lin, R. W. Kriz, C. S. Ramesha, L. A. Sultzman, A. Y. Lin, N. Milona, and J. L. Knopf (1991) A novel arachidonic acid-selective cytosolic PLA2 contains a Ca^{2+}-dependent translocation domain with homology to PKC and GAP. *Cell* 65: 1043–1051.

Cornelius, G., G. Gebauer, and D. Techel (1989) Inositol trisphosphate induces calcium release from *Neurospora crassa* vacuoles. *Biochem. Biophys. Res. Commun.* 162: 852–856.

Creutz, C. E., S. L. Snyder, and N. G. Kambouris (1991) Calcium-dependent secretory vesicles-binding and lipid-binding proteins of *Saccharomyces cerevisiae*. *Yeast* 7: 229–244.

Creutz, C. E., N. G. Kambouris, S. L. Snyder, H. C. Hamman, M. R. Nelson, W. Liu, and P. Rock (1992) Effects of the expression of mammalian annexins in the yeast secretory mutants. *J. Cell. Biol.* 103: 1177–1192.

Cunningham, K. W. and G. R. Fink (1994a) Calcineurin-dependent growth control in *Saccharomyces cerevisiae* mutants lacking *PMC1*, a homolog of plasma membrane Ca^{2+} ATPases. *J. Cell. Biol.* 124: 351–365.

Cunningham, K. and G. R. Fink (1994b) Ca^{2+} transport in *Saccharomyces cerevisiae*. *J. Exp. Biol.* 196: 157–166.

Cunningham, K. and G. R. Fink (1996) Calcineurin inhibits *VCX1*-dependent H^+/Ca^{2+} exchange and induces Ca^{2+} ATPase in *Saccharomyces cerevisiae*. *Mol. Cell. Biol.* 16: 2226–2237.

Cyert, M. S. and J. Thorner (1992) Regulatory subunit (*CNB1* gene product) of yeast Ca^{2+}/calmodulin-dependent phosphoprotein phosphatases is required for adaptation to phermomone. *Mol. Cell. Biol.* 12: 3460–3469.

Cyert, M. S., R. Kunisawa, D. Kaim, and J. Thorner (1991) Yeast has homologs (*CNA1* and *CNA2* gene products) of mammalian calcineurin, a calmodulin-regulated phosphoprotein phosphatase. *Proc. Natl. Acad. Sci. USA* 88: 7376–80 Erratum, *Proc. Natl. Sci. USA* (1992a) 89: 4220.

Dahlkvist, A. and P. Sunnerhagen (1994) Two novel deduced serine/threonine protein kinases from *Saccharomyces cerevisiae*. *Gene* 139: 27–33.

Dahlkvist, A., G. Kanter-Smoler, and P. Sunnerhagen (1995) The *RCK1* and *RCK2* protein kinase genes from *Saccharomyces cerevisiae* suppress cell cycle checkpoint mutations in *Schizosaccharomyces pombe*. *Mole. Gen. Genet.* 246: 316-326.

Davey, J., K. Davis, Y. Imai, M. Yamamoto, and G. Matthews (1994) Isolation and characterization of krp, a dibasic endopeptidase required for cell viability in the fission yeast *Schizosaccharomyces pombe*. *EMBO J.* 13: 5910–5921.

Davis, R. H., J. L. Ristow, A. D. Howard, and G. R. Barnett (1991) Calcium modulation of polyamine transport is lost in a putrescine-sensitive mutant of *Neurospora crassa*. *Arch. Biochem. Biophys.* 285: 297–305.

Davis, T. N. (1992) A temperature-sensitive calmodulin mutant loses viability during mitosis. *J. Cell Biol.* 118: 607–617.

Davis, T. N. (1995) Calcium in *Saccharomyces cerevisiae*. *Adv. Second Mess. Phosphoprotein Res.* 30: 339–358.

Davis, T. N. and J. Thorner (1989) Vertebrate and yeast calmodulin, despite significant sequence divergence, are functionally interchangeable. *Proc. Natl. Acad. Sci. USA* 86: 7909–7913.

Davis, T. N., M. S. Urdea, F. R. Masiarz, and J. Thorner (1986) Isolation of the yeast calmodulin gene: calmodulin is an essential protein. *Cell* 47: 423–431.

De Virgilio, C., N. Bürckert, J.-M. Neuhaus, T. Boller, and A. Wiemken (1993) CNE1, a *Saccharomyces cerevisiae* homologue of the gene encoding mammalian calnexin and calreticulin. *Yeast* 9: 185–188.

Desfarges, L., P. Durrens, H. Juguelin, C. Cassagne, M. Bonneu, and M. Aigle (1993) Yeast mutants affected in viability upon starvation have a modified phospholipid composition. *Yeast* 9: 267–277.

Divecha, N. and R. F. Irvine (1995) Phospholipid signaling. *Cell* 80: 269–278.

Doi, A. and K. Doi (1990) Cloning and nucleotide sequence of the *CDC23* gene of *Saccharomyces cerevisiae*. *Gene* 91: 123–126.

Donovan, M. G. and J. W. Bodley (1991) *Saccharomyces cerevisiae* elongation factor 2 is phophorylated by an endogenous kinase. *FEBS Lett.* 291: 303–306.

Douglas, C. M., F. Foor, J. A. Marrinan, N. Morin, J. B. Nielsen, A. D. Dahl, P. Mazur, W. Baginsky, M. El-Sherbeini, J. A. Clemas, S. A. Mandala, B. R. Frommer, and M. B. Kurtz (1994) The *Saccharomyces cerevisiae FKS1 (EGT1)* gene encodes an integral membrane protein which is a subunit of 1,3-β-D-glucan synthase. *proc. Natl. Acad. Sci. USA* 91: 12907–12911.

Duffus, J. and L. Patterson (1974) Control of cell division in yeast using the ionophore, A23187 with calcium and magnesium. *Nature* 251: 626–627.

Dunn, T., K. Gable, and T. Beeler (1994) Regulation of cellular Ca^{2+} by yeast vacuoles. *J. Biol. Chem.* 269: 7273–7278.

Egel-Mitani, M., H. P. Flygenring, and M. T. Hansen (190) A novel aspartyl protease allowing *KEX2*-independent MFalpha propheromone processing in yeast. *Yeast* 6: 127–137.

Eilam, Y. (1982a) The effect of monovalent cations on calcium efflux in yeasts. *Biochim. Biophys. Acta* 687: 8–16.

Eilam, Y. (1982b) The effect of potassium ionophores and potassium on cellular calcium in the yeast *Saccharomyces cerevisiae*. *J. Gen. Microbiol.* 128: 2611–2614.

Eilam, Y. and D. Chernichovsky (1987) Uptake of Ca^{2+} driven by the membrane potential in energy-depleted yeast cells. *J. Gen Microbiol.* 133: 1641–1649.

Eilam, Y. and M. Othman (1990) Activation of Ca^{2+} influx by metabolic substrates in *Saccharomyces cerevisiae*: role of membrane potential and cellular ATP levels. *J. Gen. Microbiol.* 136: 861–866.

Eilam, Y., H. Lavi, and N. Grossowicz (1984) Effects of inhibitors of plasma-membrane ATPase on potassium and calcium fluxes, membrane potential and proton motive force in the yeast *Saccharomyces cerevisiae*. *Microbios* 41: 177–189.

Eilam, Y. H. Lavi, and N. Grossowicz (1985a) Cytoplasmic Ca^{2+} homeostasis maintained by a vacuolar Ca^{2+} transport system in the yeast *Saccharomyces cerevisiae*. *J. Gen. Microbiol.* 131: 623–629.

Eilam, Y., H. Lavi, and N. Grossowicz (1985b) Mechanism of stimulation of Ca^{2+} uptake by miconazole and ethidium bromide in yeast: role of vacuoles in Ca^{2+} detoxification. *Microbios* 44: 51–66.

Eilam, Y., M. Othman, and D. Halachmi (1990) Transient increase in Ca^{2+} influx in *Saccharomyces cerevisiae* in response to glucose: effects of intracellular acidification and cAMP levels. *J. Gen. Microbiol.* 136: 2537–2543.

Enderlin, C. S. and M. D. Ogrydziak (1994) Cloning, nucleotide sequence and functions of *XPR6*, which codes for a dibasic processing endoprotease from the yeast *Yarrowia lipolytica*. *Yeast* 10: 67–79.

Enenkel, C. and D. H. Wolf (1993) *BHL1* codes for a yeast thiol aminopeptidase, the equivalent of mammalian bleomycin hydrolase. *J. Biol. Chem.* 268: 7036–7043.

Eng, W. K., L. Faucette, M. M. McLaughlin, R. Cafferkey, Y. Koltin, R. A. Morris, P. R. Young, R. K. Johnson, and G. P. Livi (1994) The yeast *FKS1* gene encodes a

novel membrane protein, mutations in which confer FK506 and cyclosporin A hypersensitivity and calcineurin-dependent growth. *Gene* 151: 61–71.

Estruch, F. and M. Carlson (1993) Two homologous zinc finger genes identified by multicopy suppression in a SNF1 protein kinase mutant of *Saccharomyces cerevisiae*. *Mol. Cell. Biol.* 13: 3872–3881.

Fagan, M. J. and M. H. Saier Jr. (1994) P-type ATPases of eukaryotes and bacteria: sequence analyses and construction of phylogenetic trees. *J. Mol. Evol.* 38 57–99.

Farkasovsky, M. and H. Küntzel (1995) Yeast Num1p associates with the mother cell cortex during the S/G2 phase and affects microtubular functions. *J. Cell Biol.* 131: 1003–1014.

Fischer, M. N., Schnell, J. Chattaway, P. Davies, G. Dixon, and D. Sanders (1997) The Saccharomyces cerevisiae CCH1 gene is involved in calcium influx and mating. *FEBS Lett.* 419: 259–262.

Fishel, B. R., A. O. Sperry, and W. T. Garrard (1993) Yeast calmoduloin and a conserved nuclear protein participate in the in vivo binding of a matrix association region. *Proc. Natl. Acad. Sci. USA* 90: 5623–5627.

Flanagan, C. A., E. A. Schnieders, A. W. Emerick, R. Kunisawa, A. Admon, and J. Thorner (1993) Phosphatidylinositol 4-kinase: gene structure and requirement for yeast cell viability. *Science* 262: 1444–1558.

Flick, J. S. and J. Thorner (1993) Genetic and biochemical characterization of a phosphatidylinositol-specific phospholipase C in *Saccharomyces cerevisiae*. *Mol. Cell. Biol.* 13: 5861–5876.

Foor, F., S. A. Parent, N. Morin, A. M. Dahl, N. Ramadan, G. Chrebet, K. A. Bostian, and J. B. Nielsen (1992) Calcineurin mediates inhibition by FK506 and cyclosporin of recovery from alpha-factor arrest in yeast. *Nature* 360: 682–684.

François, J. and H.-G. Hers (1988) The control of glycogen metabolism in yeast. *Eur. J. Biochem.* 174: 561–567.

Friedman, H. and M. Snyder (1994) Mutations in *PRG1*, a yeast proteasome related gene, cause defects in nuclear division and are suppressed by the deletion of a mitotic cyclin gene. *Proc. Natl. Acad. Sci. USA* 81: 2031–2035.

Friedman, H., M. Goebel, and M. Snyder (1992) A homolog of the proteasome-related *RING10* gene is essential for yeast cell growth. *Gene* 122: 203–206.

Fu, D., T. Beeler, and T. Dunn (1994) Sequence, mapping and disruption of *CCC1*, a gene that cross-complements the Ca^{2+}-sensitive phenotype of *csg1* mutants. *Yeast* 10: 515–521.

Fu, D., T. Beeler, and T. Dunn (1995) Sequence, mapping and disruption of *CCC2*, a gene that cross-complements the Ca^{2+}-sensitive phenotype of *csg1* mutants and encodes a P-type ATPase belonging to the Cu^{2+}-ATPase subfamily. *Yeast* 11: 283–292.

Fujimuro, M., K. Tanaka, H. Yokosawa, and A. Toh-e (1998) Son1p is a component of the 26S proteasome of the yeast *Saccharomyces cerevisiae*. *FEBS Lett.* 423: 149–154.

Fukada, M., T. Kojima, and K. Mikoshiba (1996) Phospholipid composition dependence of Ca^{2+}-dependent phospholipid binding to the C2A domain of synaptotagmin IV. *J. Biol. Chem.* 271: 8430–8434.

Fuller, R. S., A. Brake, and J. Thorner (1989) Yeast pro-hormone processing enzyme (*KEX2* gene product) is a Ca^{2+}-dependent serine protease. *proc. Natl. Acad. Sci. USA* 86: 1434–1438.

Furhmann, G. F. and A. Rothstein (1968) The transport of Zn^{2+}, Co^{2+} and Ni^{2+} into yeast cells. *Biochim. Biophys. Acta* 163: 325–330.

Garcia-Bustos, F. J., F. Marini, I. Stevenson, C. Frei, and M. N. Hall (1994) *PIK1*, and essential phosphatidylinositol 4-kinase associated with the yeast nucleus. *EMBO J.* 13: 2352–2361.

Garrett-Engelle, P., B. Moilanen, and M. S. Cyert (1995) Calcineurin, the Ca^{2+}/calmodulin-dependent protein phosphatase, is essential in yeast mutants with cell integrity defects and in mutants that lack a functional vacuolar $H^{(+)}$-ATPase. *Mol. Cell. Biol.* 15: 4103–4114.

Geiser, J. R., D. van Tuinen, S. E. Brockerhoff, M. M. Neff, and T. N. Davis (1991) Can calmodulin function without binding calcium? *Cell* 65: 949–959.

Geiser, J. R., H. A. Sundberg, B. H. Chang, E. G. Muller, and T. N. Davis (1993) The essential mitotic target of calmodulin is the 110-kilodalton component of the spindle pole body in *Saccharomyces cerevisiae*. *Mol. Cell Biol.* 13: 7913–7924.

Germain, D., F. Dumas, T. Vernet, Y. Bourbonnais, D. Y. Thomas, and G. Boileau (1992) The pro-region of the Kex2 endoprotease of *Saccharomyces cerevisiae* is removed by self-processing. *FEBS Lett.* 299; 283–286.

Germain, D., D. Y. Thomas, and G. Boileua (1993) Processing of Kex2 pro-region at two interchangeable cleavage sites. *FEBS Lett.* 323: 129–131.

Ghislain, M., A. Goffeau, D. Halachmi, and Y. Eilam (1990) Calcium homeostasis and transport are affected by disruption of *cta3*, a novel gene encoding $Ca^{2(+)}$-ATPase in *Schizosaccharomyces pombe*. *J. Biol. Chem.* 265: 18400–18407.

Gibson, T. J., M. Hyvönen, A. Musacchio, and M. Saraste (1994) PH domain: the first anniversary. *Trends Biochem. Sci.* 19: 349–353.

Gluschankof, P. and R. S. Fuller (1994) A C-terminal domain conserved in precursor processing proteases is required for intramolecular N-terminal maturation of pro-Kex2 protease. *EMBO J.* 13: 2280–2288.

Goffeau, A., B. G. Barrell, H. Bussey, R. W. Davis, B. Dujon, H. Feldmann, F. Galibert, J. D. Hoheisel, C. Jacq, M. Johnston, E. J. Louis, H. W. Mewes, Y. Murakami, P. Philippsen, H. Tettelin, and S. G. Oliver (1996) Life with 6000 genes. *Science* 274: 546–567.

Graham, T. R., M. Seeger, G. S. Payne, V. L. MacKay, and S. D. Emr (1994) Clathrin-dependent localization of alpha 1,3 mannosyltransferase to the Golgi complex of *Saccharomyces cerevisiae*. *J. Cell Biol.* 127: 667–678.

Groll, M., L. Ditzel, J. Löwe, D. Stock, M. Bochtler, H. D. Bartunik, and R. Huber (1997) Structure of the 20S proteasome from yeast at 2.4Å resolution. *Nature* 386: 463–471.

Gunteski-Hamblin, A. M., D. M. Clarke, and G. E. Shull (1992) Molecular cloning and tissue distribution of alternatively spliced mRNAs encoding possible mammalian homologues of the yeast secretory pathway calcium pump. *Biochemistry*, 31: 7600–7608.

Gustin, M. C., X. L. Zhou, B. Martinac, and C. Kung (1988) A mechanosensitive ion channel in the yeast plasma membrane. *Science* 242: 762–765.

Haarer, B. K., A. Petzold, S. H. Lillie, and S. S. Brown (1994) Identification of *MYO4*, a second class V myosin gene in yeast. *J. Cell Sci.* 107: 1055–1064.

Halachmi, D. and Y. Eilam (1989) Cytosolic and vacuolar Ca^{2+} concentrations in yeast cells measured with the Ca^{2+}-sensitive fluorescence dyle indo-1. *FEBS Lett.* 256: 55–61.

Halachmi, D. and Y. Eilam (1993) Calcium homeostasis in yeast cells exposed to high concentrations of calcium. Roles of vacuolar H^+-ATPase and cellular ATP, *FEBS Lett.* 316: 73–78.

Halachmi, D., M. Ghislain, and Y. Eilam (1992) An intracellular ATP-dependent calcium pump with the yeast. *Schizosaccharomyces pombe*, encoded by the gene *cta3*. Eur. J. Biochem. 207: 1003–1008.

Hardy, C. F. J. (1996) Characterization of an essential Orc2p-associated factor that plays a role in DNA replication. *Mol. Cell. Biol.* 16: 1832–1841.

Harmsen, M. M., A. C. Langedijk, E. van Tuinen, R. H. Geerse, H. A. Raue, and J. Maat (1993) Effect of *pmr1* disruption and different signal sequences on the intracellular processing and secretion of *Cyamopsis tetragonoloba* alpha-galactosidase by *Saccharomyces cerevisiae*. *Gene* 125: 115–123.

Haro, R., B. Garciadeblas, and A. Rodriguez-Navarro (1991) A novel P-type ATPase from yeast involved in sodium transport. *FEBS Lett.* 291: 189–191.

Harold, R. L. and F. M. Harold (1986) Ionophores and cytochalasins modulate branching in *Achlya bisexualis*. *J. Gen Microbiol.* 132: 213–219.

Harris, E., D. M. Watterson, and J. Thorner (1994) Functional consequences in yeast of single-residue alternations in a consensus calmodulin. *J. Cell Sci.* 107: 3235–3249.

Hart, M. J., A. Eva, T. Evans, S. A. Aaronson, and R. Cerione (1991) Catalysis of guanine nucleotide exchange on the CDC42Hs protein by the *dbl* oncogene product. *Nature* 354: 311–314.

Heichman, K. A. and J. M. Roberts (1996) The *CDC16* and *CDC27* genes restrict DNA replication to once per cell cycle. *Cell* 85: 39–48.

Heinemeyer, W., A. Gruhler, V. Möhrle, Y. Mahé, and D. H. Wolf (1993) PRE2, highly homologous to the human major histocompatibility complex-linked *RING10* gene, codes for a yeast proteasome subunit necessary for the chymotruptic activity and degradation of ubiquitinated proteins. *J. Biol. Chem.* 268: 5115–5120.

Helliwell, S. B., P. Wagner, J. Kunz, M. Deuter-Reinhard, R. Henriquez, and M. N. Hall (1994) *TOR1* and *TOR2* are structurally and functionally similar but not identical phosphatidylinositol kinase homologues in yeast. *Mol. Biol. Cell.* 5: 105–118.

Herskowitz, I. (1995) MAP kinase pathways in yeast: for mating and more. *Cell,* 80: 187–197.

Higuchi, S., J. Tamura, P. R. Giri, J. W. Polli, and R. L. Kincaid (1991) Calmodulin-dependent protein phosphatase from *Neurospora crassa.* Molecular cloning and expression of recombinant catalytic subunit. *J. Biol. Chem.* 266: 18104–18112.

Hilt, W. and D. H. Wolf (1996) Proteasomes: destruction as a programme. *Trends Biochem. Sci.* 21: 96–102.

Hiraga, K., H. Tahara, N. Taguchi, E. Tsuchiya, S. Fukui, and T. Miyakawa (1991) Inhibition of membrane Ca^{2+}-ATPase of *Saccharomyces cerevisiae* by mating pheromone alpha-factor in vitro. *J. Gen Microbiol.* 137: 1–4.

Hiraga, K., K. Suzuki, E. Tsuchiya, and T. Miyakawa (1993a) Identification and characterization of nuclear calmodulin-binding proteins of *Saccharomyces cerevisiae. Biochim. Biophys. Acta.* 1177: 25–30.

Hiraga, K., K. Suzuki, E. Tsuchiya, and T. Miyakawa (1993b) Cloning and characterization of the elongation factor EF-1 beta homologue of *Saccharomyces cerevisiae.* EF-1 beta is essential for growth. *FEBS Lett.* 316: 165–169.

Hug, H. and T. F. Sarre (1993) Protein kinase C isoenzymes: divergence in signal transduction? *Biochem. J.* 291: 329–343.

Huibregtse, J. M., M. Scheffner, S. Beaudenon, and P. M. Howley (1995) A family of proteins structurally and functionally related to the E6-AP ubiquitin-protein ligase. *proc. Natl. Acad. Sci. USA* 92: 2563–2567.

Iida, H., Y. Yagawa, and Y. Anraku (1990) Essential role for induced Ca^{2+} influx followed by $[Ca^{2+}]_i$ rise in maintaining viability of yeast cells late in the mating pheromone response pathway. *J. Biol. Chem.* 265: 13391–13399.

Iida, H., H. Nakamura, T. Ono, M. S. Okumura, and Y. Anraku (1994) *MID1* a novel *Saccharomyces cerevisiae* gene encoding a plasma membrane protein is required for Ca^{2+} influx and mating. *Mol. Cell. Biol.* 14: 8259–8271.

Iida, H., Y. Ohya, and Y. Anraku (1995) Calmodulin-dependent protein kinase II and calmodulin are required for induced thermotolerance in *Saccharomyces cerevisiae. Curr. Genet.* 27: 190–193.

Jelinek-Kelly, S. and A. Herscovics (1988) Glycoprotein biosynthesis in *Saccharomyces cerevisiae. J. Biol. Chem.* 263: 14757–14763.

Jennissen, H. P., G. Botzet, M. Majetschak, M. Laub, R. Ziegenhagen, and A. Demiroglou (1992) $Ca^{(2+)}$-dependent ubiquitination of calmodulin in yeast. *FEBS Lett.* 296: 51–56.

Johnson, D. I., J. M. O'Brien, and C. W. Jacobs (1990) Isolation and sequence analysis of *CDC43,* a gene involved in the control of cell polarity in *Saccharomyces cerevisiae. Gene* 90: 93–98. Corrigendum, *Gene* (1991) 98: 149–150.

Johnston, E. S., P. C. M. Ma, I. M. Ota, and A. Varshasky (1995) A proteolytic pathway that recognizes ubiquitin as a degradation signal. *J. Biol. Chem.* 270: 17442–17456.

Johnston, G. C., J. A. Prendergast, and R. A. Singer (1991) The *Saccharomyces cerevisiae MOY2* gene encodes an essential myosin for vectorial transport of vesicles. *J. Cell. Boil.* 113: 539–551.

Julius, D., A. Brake, L. Blair, R. Kunisawa, and J. Thorner (1984) Isolation of putative structural gene for the lysine-arginine-cleaving endopeptidase required for processing of yeast pre-pro-alpha-factor. *Cell* 37: 1075–1089.

Kamada, Y., U. S. Jung, J. Piotrowski, and D. E. Levin (1995) The protein kinase C-activated MAP kinase pathway of *Saccharomyces cerevisiae* mediates a novel aspect of the heat shock response. *Genes Dev.* 9: 1559–1571.

Kambouris, N. G. D., D. J. Burke, and C. E. Creutz (1992) Cloning and characterization of a cysteine proteinase from *Saccharomyces cerevisiae. J. Biol. Chem.* 267: 21570–21576.

Kambouris, N. G., D. J. Burke, and C. E. Creutz (1993) Cloning and genetic characterization of a calcium- and phospholipid-binding protein from *Saccharomyces cerevisiae* that is homologous to translation elongation factor-1 gamma. *Yeast* 9: 151–163.

Kasahara, S., H. Yamada, T. Mio, Y. Shiratori, C. Miyamoto, T. Yabe, T. Nakajima, E. Ichishima, and Y. Furuichi (1994a) Cloning of the *Saccharomyces cerevisiae* gene whose over expression overcomes the effect of HM-1 killer toxin, which inhibits beta-glucan synthesis. *J. Bacteriol.* 176: 1488–1499.

Kasahara, S., S. Ben Inoue, T. Mio, T. Yamada, T. Nakajima, E. Ichishima, Y. Furuichi, and H. Yamada (1994b) Involvement of cell wall beta-glucan in the action of HM-1 killer toxin. *FEBS Lett.* 348: 27–32.

Kikuchi, Y., Y. Oka, M. Kobayashi, Y. Uesono, A. Toh-e, and A. Kikuchi (1994) A new yeast gene, *HTR1,* required for growth at high temperature, is needed for recovery from mating pheromone-induced G1 arrest. *Mol. Gen. Genet.* 245: 107–116.

Kilmartin, J. V., S. L. Dyos, D. Kershaw, and J. T. Finch (1993) A spacer protein in the *Saccharomyces cerevisiae* spindle pole body whose transcript is cell cycle-regulated. *J. Cell. Biol.* 123: 1175–1184.

Klein, P., M. Kanehisa, and C. Delisi (1985) The detection and classification of membrane-spanning proteins. *Biochim. Biophys. Acta* 815: 468–476.

Klis, F. M. (1994) Review: cell wall assembly in yeast. *Yeast* 10: 851–869.

Kormanec, J., I. Schaaff-Gerstenschläger, F. K. Zimmermann, D. Perecko, and H. Küntzel (1991) Nuclear migration in *Saccharomyces cerevisiae* is controlled by the highly repetitive 313 kDa NUM1 protein. *Mol. Gen. Genet.* 230: 277–287.

Kornstein, L. B., M. L. Gaiso, R. L. Hammell, and D. C. Bartelt (1992) Cloning and sequence determination of a cDNA encoding *Aspergillus nidulans* calmodulin-dependent multifunctional protein kinase. *Gene* 113; 75–82.

Kovac, L. (1985) Calcium and *Saccharomyces cerevisiae*. *Biochim. Biophys. Acta* 840: 317–323.

Kübler, E. and H. Riezman (1993) Actin and fimbrin are required for the internalization step of endocytosis in yeast. *EMBO J.* 12: 2855–2862.

Kübler E., F. Schimmoller, and H. Riezman (1994) Calcium-independent calmodulin requirement for endocytosis in yeast. *EMBO J.* 1357: 5539–5546.

Kulakovskaya, T. V., R. N. Matyashova, N. V. Shishkanova, T. V. Finogenova, and L. A. Okorokov (1993) Change in transport activities of vacuoles of the yeast *Yarrowia lipolytica* during its growth on glucose. *Yeast* 9: 121–126.

Kuno, T., H. Tanaka, H. Mukai, C. D. Chang, K. Hiraga, T. Miyakawa, and C. Tanaka (1991) cDNA cloning of a calcineurin B homolog in *Saccharomyces cerevisiae*. *Biochem. Biophys. Res. Commun.* 180: 1159–1163.

Kunz, J. and M. N. Hall (1993) Cyclosporin A, FK506 and rapamycin: more than just immunosuppression. *Trends Biochem. Sci.* 18: 334–338.

Lapinskas, P. J., K. W. Cunningham, X. F. Liu, G. R. Fink, and V. C. Culotta (1995) Mutations in *PMR1* suppress oxidative damage in yeast cells lacking superoxide dismutase. *Mol. Cell. Biol.* 15: 1382–1388.

Larriba, G., R. D. Basco, E. Andaluz, and J. P. Luna-Arias (1993) Yeast exoglucanases. Where redundancy implies necessity. *Arch. Med. Res.* 24: 293–299.

Ledgerwood, E. C., P. M. George, R. J. Peach, and S. O. Brennan (1995) Endoproteolytic processing of recombinant proalbumin variants by the yeast Kex2 protease. *Biochem. J.* 308: 321–325.

LeJohn, H. B. (1989) Structure and expression of fungal calmodulin gene. *J. Biol. Chem.* 264: 19366–19372.

LeJohn, H. B., L. E. Cameron, R. M. Stevenson, and R. U. Meuser (1974) Influence of cytokinins and sulfhydryl group-reacting agents on calcium transport in fungi. *J. Biol. Chem.* 249: 4016–4020.

Levin, D. E. and E. Bartlett-Heubusch (1992) Mutants in the *S. cerevisiae PKC1* gene display a cell cycle-specific osmotic stability defect. *J. Cell Biol.* 116: 1221–1229.

Levin, D. E., F. O. Fields, R. Kunisawa, J. M. Bishop, and J. Thorner (1990) A candidate protein kinase C gene, *PKC1*, is required for the *S. cerevisiae* cell. *Cell* 62: 213–224

Li, R. and A. W. Murray (1991) Feedback control of mitosis in budding yeast. *Cell* 66: 519–531.

Li, Y. and R. Benezra (1996) Identification of a human mitotic checkpoint gene: hsMAD2. *Science* 274: 246–248.

Liu, Y., S. Ishii, M. Tokai, H. Tsutsumi, O. Ohki, R. Akada, K. Tanaka, E. Tsuchiya, S. Fukui, and T. Miyakawa (1991) The *Saccharomyces cerevisiae* genes (*CMP1* and *CMP2*) encoding calmodulin-binding proteins homologous to the catalytic subunit of mammalian protein phosphatase 2B. *Mol. Gen. Genet.* 227: 52–59.

Londesborough J. and M. Nuutinen (1987) Ca^{2+}/calmodulin-dependent protein kinase in *Saccharomyces cerevisiae*. *FEBS Lett.* 219: 249–253.

Lu, K. P., C. D. Rasmussen, G. S. May, and A. R. Means (1992) Cooperative regulation of cell proliferation by calcium and calmodulin in *Aspergillus nidulans*. *Mol. Endocrinol.* 6: 365–374.

Lu, K. P., S. A. Osmani, A. H. Osmani, and A. R. Means, (1993) Essential roles for calcium and calmodulin in G2/M progression in *Aspergillus nidulans*. *J. Cell. Biol.* 121: 621–630.

Luan, Y., I. Matsuura, M. Yazawa, T. Nakamura, and K. Yagi (1987) Yeast calmodulin: structural and functional differences compared with vertebrate calmodulin. *J. Biochem. (Tokyo)* 102: 1531–1537.

Macer, D. R. and G. L. Koch (1988) Identification of a set of calcium-bearing proteins in reticuloplasm, the luminal content of the endoplasmic reticulum. *J. Cell. Sci.* 91: 61–70.

Magdolen, U., G. Müller, V. Magdolen, and W. Bandlow (1993) A yeast gene (*BLH1*) encodes a polypeptide with homology to vertebrate bleomycin hydrolase, a family member of thio proteinase. *Biochim. Biophys. Acta* 1171: 299–303.

Martinez-Pastor, M. T., G. Marchler, C. Schüller, A. Marchler-Bauer, H. Ruis, and F. Estruch (1996) The *Saccharomyces cerevisiae* zinc finger proteins Msn2p and Msn4p are required for transcriptional induction through the stress-response element (STRE). *EMBO J.* 15: 2227–2235.

Matheos, D. P., T. J. Kinsbury, U. S. Ahsan, and K. W. Cunningham (1997) Tcn1p/Crz1p, a calcineurin-dependent transcription factor that differentially regulates gene expression in *Saccharomyces cerevisiae*. *Genes Dev.* 1: 3445–3458.

Matsuura, I., K. Ishihara, Y., Nakai, M. Yazawa, H. Toda, and K. Yagi (1991) A site-directed mutagenesis study of yeast calmodulin. *J. Biochem. (Tokyo)* 109: 190–197.

Matsuura, I., E. Kimura, K. Tai, and M. Yazawa (1993) Mutagenesis of the fourth calcium-binding domain of yeast calmodulin. *J. Biol. Chem.* 268: 13267–13273.

Mayer, M. L., B. E. Caplin, and M. S. Marshall (1992) *CDC43* and *RAM2* encode the polypeptides subunits of a yeast type I protein geranylgeranyltransferase. *J. Biol. Chem.* 267: 20589–20593.

Mayor, J. A., D. Kakhniashvili, D. A. Gremse, C. Campbell, R. Krämer, A. Schroers, and R. S. Kaplan (1997) Bacterial overexpression of putative yeast mitochondrial transport proteins. *J. Bioernerg. Biomemb.* 29: 6541–547.

Mazur, P., N. Morin, W. Baginsky, M. el-Sherbeini, J. A. Clemas, J. B. Nielsen, and F. Foor (1995) Differential expression and function of two homologous subunits of yeast 1,3-β-D-glucan synthase. *Mol. Cell. Biol.* 15: 5671–5681.

Mazzei, G. J., E. M. Schmid, J. K. C. Knowles, M. A. Payton, and K. G. Maundrell (1993) A Ca^{2+}-independent protein kinase C from fission yeast. *J. Biol. Chem.* 268: 7401-7406.

McCollum, D., M. K. Balasubramanian, L. E. Pelcher, S. M. Hemmingsen, and K. L. Gould (1995) *Schizosaccharomyces pombe cdc4+* gene encodes a novel EF-hand protein essential for cytokinesis. *J. Cell. Biol.* 130: 651–660.

Mellors, A. and D. R. Sutherland (1994) Tools to cleave glycoproteins. *Trends Biotechnol.* 12: 15–18.

Melnick, M. B., C. Melnick, M. Lee, and D. O. Woodward (1993) Structure and sequence of the calmodulin gene from *Neurospora crassa. Biochim. Biophys. Acta* 1171: 334–336.

Mendoza, I., F. Rubio, A. Rodriguez-Navarro, and J. M. Pardo (1994) The protein phosphatase calcineurin is essential for NaCl tolerance of *Saccharomyces cerevisiae. J. Biol. Chem.* 269: 8792–8796.

Miller, A. J., G. Vogg, and D. Sanders (1990) Cytosolic calcium homeostasis in fungi: roles of plasma membrane transport and intracellular sequestration of calcium. *Proc. Natl. Acad. Sci. USA* 87: 9348–9352.

Miller, P. J. and D. I. Johnson (1994) Cdc42p GTPase is involved in controlling polarized cell growth in *Schizosaccharomyces pombe. Mol. Cell. Biol.* 14: 1075–1083.

Mitsui, K., M. Brady, H. C. Palfrey, and A. C. Nairn (1993) Purification and characterization of calmodulin-dependent protein kinase III from rabbit reticulocytes and rat pancreas. *J. Biol. chem.* 268: 13422–13433.

Miyakawa, T., T. Tachikawa, Y. E. Jeong, E. Tsuchiya, and S. Fukui (1985) Transient increase of Ca^{2+} uptake as a single for mating pheromone-induced differentiation in the heterobasiomycetous yeast. *Rhosporidium toruloides. J. Bacteriol.* 162: 1304–1306.

Miyakawa, T., T. Tachikawa, R. Akada, E. Tsuchiya, and S. Fukui (1986) Involvement of Ca^{2+}/calmodulin in sexual differentiation induced by mating pheromone rhodotorucine A in *Rhosporidium toruloides. J. Gen Microbiol.* 132: 1453–1457.

Miyakawa, T., T. Tachikawa, Y. K. Jeong, E. Tsuchiya, and S. Fukui (1987) Inhibition of membrane Ca^{2+}-ATPase in vitro by mating pheromone in *Rhodosporidium toruloides*, a heterobasiomycetous yeast. Biochem. Biophys. Res. Commun. 143: 893–900.

Miyakawa, T., Y. Oka, E. Tsuchiya, and S. Fukui (1989) *Saccharomyces cerevisiae* protein kinase dependent on Ca^{2+} and calmodulin. *J. Bacteriol.* 17: 1417–1422.

Miyamoto, S., Y. Ohya, Y. Ohsumi, and Y. Anraku (1987) Nucleotide sequence of the *CLS4* (*CDC24*) gene of *Saccharomyces cerevisiae. Gene* 54: 125–132.

Miyamoto, S., Y. Ohya, Y. Sano, S. Sakaguchi, H. Iida, and Y. Anraku (1991) A DBL-homologous region of the yeast *CLS4/CDC24* gene product is important for $Ca^{(2+)}$-modulated bud assembly. *Biochem. Biophys. Res. Commun.* 181: 604–610.

Mizuno, K., T. Nakamura, T. Ohshima, S. Tanaka, and H. Matsuo (1988) Yeast *KEX2* gene encodes an endopeptidase homologous to subtilisin-like serine proteases. *Biochem. Biophys. Res. Commun.* 156: 246–254.

Mizuno, K., T. Nakamura, T. Ohshima, S. Tanaka, and H. Matsuo (1989) Characterization of *KEX2*-encoded endopeptidase from yeast *Saccharomyces cerevisiae. Biochem. Biophys. Res. Commun.* 159: 305–311.

Minzunuma, M., D. Hirata, K. Miyahara, E. Tsuchiya, and T. Miyakawa (1998) Role of the calcineurin and Mpk1 in regulating the onset of mitosis in budding yeast. *Nature* 392: 303–306.

Morris, E. O. (1958) *The Chemistry and Biology of Yeast* (Cook, A. H., ed.) Academic Press, New York, pp. 251–321.

Moser, M. J., S. Y. Lee, R. E. Klevit, and T. N. Davis (1995) Ca^{2+} binding to calmodulin and its role in *Schizosaccharomyces pombe* as revealed by mutagenesis and NMR spectroscopy. *J. Biol. Chem.* 270: 20643–20652.

Mukai, K., T. Kuno, H. Tanaka, D. Hirata, T. Miyakawa, and C. Tanaka (1993) Isolation and characterization of *SSE1* and *SSE2*, new members of the yeast HSP70 multigene family. *Gene* 132: 57–66.

Muller, A., W. Hinrichs, W. M. Wolf, and W. Saenger (1994) Crystal structure of calcium-free proteinase K at 1.5-Å resolution. *J. Biol. Chem.* 269: 23108–23111.

Munro, S. and H. R. B. Pelham (1986) An hsp70-like protein in the ER: identity with the 78 kd glucose-regulated protein and immunoglobulin heavy chain binding protein. *Cell* 46: 291–300.

Nagiec, M. M., J. A. Baltisberger, G. B. Wells, R. L. Lester, and R. C. Dickson (1994) The *LCB2* gene of *Saccharomyces* and the related *LCB1* gene encode subunits of serine palmitoyltransferase, the initial enzyme in sphingolipid synthesis. *proc. Natl. Acad. Sci USA* 91: 7899–7902.

Nakai, K. and M. Kanehisa (1992) A knowledge for predicting protein localization sites in eukaryotic cells. *Genomics* 14: 897–911.

Nakajima, H., A. Hirata, Y. Ogawa, T. Yonehara, K. Yoda, and M. Yamasaki (1991) A cytoskeleton-related gen, *USO1*, is required for intracellular protein transport in *Saccharomyces cerevisiae. J. Cell Biol.* 113: 245–260.

Nakajima-Shimada, J., H. Iida, F. Tsuji, and Y. Anraku (1991) Monitoring of intracellular calcium in *Saccharomyces cerevisiae* with apoaequorin cDNA expression system. *Proc. Natl. Acad. Sci. USA* 88: 6878–6882.

Nakamura, T., H. Tsutsumi, H. Mukai, T. Kuno, and T. Miyakawa (1992) Ca^{2+}/calmodulin-activated protein phosphatase (PP2B) of *Saccharomyces cerevisiae*. PP2B activity is not essential for growth. *FEBS Lett.* 309: 103–106.

Nakamura, T., Y. Liu, D. Hirata, H. Namba, S. Harada, T. Hirokawa, and T. Miyakawa (1993) Protein phosphatase type 2B (calcineurin)-mediated, FK506-sensitive regulation of intracellular ions in yeast is an important determinant for adaptation to high salt stress conditions. *EMBO J.* 12: 4063–4071.

Näthke, I., B. E. Hill, P. Parham, and F. M. Brodsky (1990) The calcium-binding site of clathrin light chains. *J. Biol. Chem.* 265: 18621–18627.

Nehrbass, U., H. Kern, A. Mutvei, H. Horstmann, B. Marshallsay, and E. C. Hurt (1990) NSP1: a yeast nuclear envelope protein localized at the nuclear pore exerts its essential function by its carboxy-terminal domain. *Cell* 6: 979–989.

Neves, M. J. and J. François (1992) On the mechanism by which a heat shock induces trehalose accumulation in *Saccharomyces cerevisiae*. *Biochem. J.* 288: 859–864.

Normington, K., K. Kohno, Y. Kozutsumi, M.-J. Gething, and J. Sambrook (1989) *S. cerevisiae* encodes an essential protein homologous in sequence and function to mammalian BiP. *Cell* 57: 1223–1236.

O'Day, C. L., F. Chavanikamannil, and J. Abelson (1996) 18S rRNA processing requires the RNA helicase-like protein Rrp3. *Nucleic Acids Res.* 24: 3201–3207.

Ogita, K., S. Miyamoto, H. Koide, T. Iwai, M. Oka, K. Ando, A. Kishimoto, K. Ikeda, Y. Fukami, and Y. Nishizuka (1990) Protein kinase C in *Saccharomyces cerevisiae*: comparison with the mammalian enzyme. *Proc. Natl. Acad. Sci. USA* 87: 5011–5015.

Ohsumi, Y. and Y. Anraku (1983) Calcium transport driven by a proton motive force in vacuolar membrane vesicles of *Saccharomyces cerevisiae*. *J. Biol. Chem.* 258: 5614–5616.

Ohsumi, Y. and Y. Anraku (1985) Specific induction of Ca^{2+} transport activity in *MAT*a cells of *Saccharomyces cerevisiae* by mating pheromone, α factor. *J. Biol. Chem.* 260: 1048–10486.

Ohya, Y. and Y. Anraku (1989a) Functional expression of chicken calmodulin in yeast. *Biochem. Biophys. Res. Commun.* 158: 541–547.

Ohya, Y. and Y. Anraku (1989b) A galactose-dependent *cmd1* mutant of *Saccharomyces cerevisiae*: involvement of calmodulin in nuclear division. *Curr. Genet.* 15; 113–120.

Ohya, Y. and Y. Anraku (1992) Yeast calmodulin: structural and functional elements essential for the cell cycle. *Cell Calcium* 13: 445–455.

Ohya, Y. and D. Botstein (1994a) Structure-based systematic isolation of conditional-lethal mutations in the single yeast calmodulin gene. *Genetics* 138: 1041–1054.

Ohya, Y. and D. Botstein (1994b) Diverse essential functions revealed by complementing yeast calmodulin mutants *Science* 263: 963–966.

Ohya, Y., Y. Ohsumi, and Y. Anraku (1984) Genetic study of the role of calcium ions in the cell division cycle of *Saccharomyces cerevisiae*: a calcium-dependent mutant and its trifluoroperazine-dependent pseudorevertants. *Mol. Gen. Genet.* 193: 386–394.

Ohya, Y., Y. Ohsumi, and Y. Anraku (1986a) Isolation and characterization of Ca^{2+}-sensitive mutants of *Saccharomyces cerevisiae*. *J. Gen Microbiol.* 132: 979–988.

Ohya, Y., S. Miyamoto, Y. Ohsumi, and Y. Anraku (1986b) Calcium-sensitive *cls4* mutant of *Saccharomyces cere-*

visiae with a defect in bud formation. *J. Bacteriol.* 165: 28–33.

Ohya, Y., I. Uno, T. Ishikawa, and Y. Anraku (1987) Purification and biochemical properties of calmodulin from *Saccharomyces cerevisiae*. *Eur. J. Biochem.* 168: 13–19.

Ohya, Y., N. Umemoto, I. Tanida, A. Ohta, H. Iida, and Y. Anraku (1991a) Calcium sensitive *cls* mutants of *Saccharomyces cerevisiae* showing a Pet⁻ phenotype are ascribable to defects of vacuolar membrane H^+-ATPase activity. *J. Biol. Chem.* 266: 13971–13977.

Ohya, Y., M. Goebl, L. E. Goodman, S. Petersen-Bjorn, J. D. Friesen, F. Tamanoi, and Y. Anraku (1991b) Yeast *CAL1* is a structural and functional homologue to the *DPR1* (RAM) gene involved in ras processing. *J. Biol. Chem.* 266: 12356–12360.

Ohya, Y., H. Kawasaki, K. Suzuki, J. Londesborough, and Y. Anraku (1991c) Two yeast genes encoding calmodulin-dependent protein kinases. Isolation, sequencing and bacterial expressions of *CMK1* and *CMK2*. *J. Biol. Chem.* 266: 12784–12794.

Ohya, Y., H. Qadota, Y. Anraku, J. R. Pringle, and D. Bostein (1993) Suppression of geranylgeranyl I defect by alternative prenylation of two target GTPases, Rho1p and Cdc42p. *Mol. Biol. Cell* 4: 1017–1025.

Okorokov, L. (1994) Several compartments of *Saccharomyces cerevisiae* are equipped with Ca^{2+}-ATPase(s). *FEMS Microbiol. Lett.* 117: 311–318.

Okorokov, L. and L. Lehle (1997) Ca^{2+}-ATPases of *Saccharomyces cerevisiae*: diversity and possible role in protein sorting. *FEMs Microbiol. Lett.* 162: 83–91.

Okorokov, L. A., V. A. Sysuev, and I. S. Kulaev (1978) Pyrophosphate-stimulated uptake of calcium into the germlings of *Phytophthora infestans*. *Eur. J. Biochem.* 83: 507–511.

Okorokov, L. A., W. Tanner, and L. Lehle (1993) A novel primary Ca^{2+}-transport system from *Saccharomyces cerevisiae*. *Eur. J. Biochem.* 216: 573–577.

Ono, T., T. Suzuki, Y. Anraku, and H. Iida (1994) The *MID2* gene encodes a putative integral membrane protein with $Ca^{(2+)}$-binding domain and shows mating pheromone-stimulated expression in *Saccharomyces cerevisiae*. *Gene* 151: 203–208.

Ortega Perez, R., D. Van Tuinen, D. Marme, and G. Turian (1983) Calmodulin-stimulated cyclic nucleotide phosphodiesterase from *Neurospora crassa*. *Biochim. Biophys. Acta* 758: 84–87.

Otto, J. J. (1994) Actin-bundling proteins. *Curr. Opin. Cell. Biol.* 6: 105–109.

Paidhungat, M. and S. Garrett (1997) A homolog of mammalian, voltage-gated calcium channels mediates yeast pheromone-stimulated Ca^{2+} uptake and exacerbates the *cdc1* (Ts) growth defect. *Mol. Cell Biol.* 17: 6339–6347.

Parag, H. A., D. Dimitrovsky, B. Raboy, and R. G. Kulka (1993) Selective ubiquitination of calmodulin by UBC4 and a putative ubiquitin protein ligase (E3) from *Saccharomyces cerevisiae*. *FEBS Lett.* 325: 242–246.

Paranjape, V., B. G. Roy, and A. Datta (1990) Involvement of calcium, calmodulin and protein phosphorylation in morphogenesis of *Canadida albicans*. *J. Gen Microbiol.* 136: 2149–2154.

Parent, S. A., J. B. Nielsen, N. Morin, G. Chrebet, N. Ramadan, A. M. Dahjl, M. J. Hsu, K. A. Bostian, and F. Foor (1993) Calcineurin-dependent growth of an FK506- and CsA-hypersensitive mutant of *Saccharomyces cerevisiae*. *J. Gen. Microbiol.* 139: 2973–2984.

Parlati, F., M. Dominguez, J. J. M. Bergeron, and D. Y. Thomas (1995a) *Saccharomyces cerevisiae CNE1* encodes an endoplasmic reticulum (ER) membrane protein with sequence similarity to calnexin and calreticulin and functions as a constituent of the ER quality control apparatus. *J. Biol. Chem.* 270: 244–253.

Parlati, F., D. Dignard, J. J. M. Bergeron, and D. Y. Thomas (1995b) The calnexin homologue *cnx1+* in *Schizosaccharomyces pombe*, in an essential gene which can be complemented by its soluble ER domain. *EMBO J.* 14: 3064–3072.

Paulsen, I. T., M. K. Sliwinski, B. Nelissen, A. Goffeau, and M. H. Saier (1998) Unified inventory of established and putative transporters encoded with the complete genome of *Saccharomyces cerevisiae*. *FEBS Lett* 430: 116–125.

Pausch, M. H., D. Kaim, Kunisaw, A. Admon, and J. Thorner (1991) Multiple Ca^{2+}/calmodulin-dependent protein kinase genes in a unicellular eukaryote. *EMBO J.* 10: 1511–1522.

Payne, G. S. and R. Schekman (1989) Clathrin: a role in the intracellular retention of a Golgi membrane protein. *Science* 245: 1358–1365.

Payne, W. E. and M. Fitzgerald-Hayes (1993) A mutation in *PLC1*, a candidate phosphoinositide-specific phospholipase C gene from *Saccharomyces cerevisiae*, causes aberrant mitotic chromosome segregation. *Mol. Cell. Biol.* 13: 4351–4364.

Pena, A. (1978) Effect of ehtidium bromide on Ca^{2+} uptake by yeast. *J. Membr. Biol.* 42: 199–213.

Penman, C. S. and J. H. Duffus (1975) 2′-deoxyadenosine and A2317 as agents for inducing synchrony in the budding yeast, *Kluyveromyces fragilis*. *J. Gen Microbiol.* 90: 76–80.

Perentesis, J. P., L. D. Phan, W. B. Gleason, D. C. LaPorte, D. M. Livingston, and J. W. Bodley (1992) *Sacchraomyces cerevisiae* elongation factor 2. Genetic cloning, characterization of expression, and G-domain modeling. *J. Biol. Chem.* 267: 1190–1197.

Persechini, A., R. H. Kretsinger, and T. N. Davis (1991) Calmodulins with deletions in the central helix functionally replace the native protein in yeast cells. *Proc. Natl. Acad. Sci. USA* 88: 449–452.

Peterson, J., Y. Zheng, L. Bender, A. Myers, R. Cerione, and A. Bender (1994) Interactions between the bud emergence proteins Bem1p and Bem2p and Rho-type GTPases in yeast. *J. Cell Biol.* 127: 1395–1406.

Portillo, F., P. Eraso, and R. Serrano (1991) Analysis of the regulatory domain of yeast plasma membrane H$^+$-ATPase by directed mutagenesis and intragenic suppression. *FEBS Lett.* 287: 71–74.

Posas, F., M. Camps, and J. Arino (1995) The PPZ protein phosphatases are important determinants of salt tolerance in yeast cells. *J. Biol. Chem.* 270: 13036–13041.

Pozos, T. M., I. Sekler, and M. S. Cyert (1996) The product of *HUM1*, a novel yeast gene, is required for vacuolar Ca^{2+}/H^+ exchange and is related to mammalian Na^+/Ca^{2+} exchangers. *Mol. Cell Biol.* 16: 3730–3741.

Prakash, S., P. Sung, and L. Prakash (1993) DNA repair genes and proteins of *Saccharomyces cerevisiae*. *Annu. Rev. Genet.* 23: 33–70.

Prasad, K. R. and P. M. Rosoff (1992) Characterization of the energy-dependent, mating factor-activated Ca^{2+} influx in *Saccharomyces cerevisiae*. *Cell Calcium* 13: 615–626.

Rad, R. M., L. Kirchrath, and C. P. Hollenberg (1994) A putative P-type Cu^{2+}-transporting ATPase gene on chromosome II of *Saccharomyces cerevisiae*. *Yeast* 10: 1217–1225.

Rad, M. R., H. L. Phan, L. Kirchrath, P. K. Tan, T. Kirchhausen, C. P. Hollenberg, and G. S. Payne (1995) *Saccharomyces cerevisiae* Ap12p, a homologue of the mammalian clathrin AP beta subunit, plays a role in chathrin-dependent Golgi functions. *J. Cell Sci.* 108: 1605–1615.

Rasmussen, C. and G. Rasmussen (1994) Inhibition of G2/M progression in *Schizosaccharomyces pombe* by a mutant calmodulin kinase II with constitutive activity. *Mol. Cell. Biol.* 5 785–795.

Rasmussen, C. D., R. L. Means, K. P. Lu, G. S. May, and A. R. Means (1990) Characterization and expression of the unique calmodulin gene of *Aspergillus nidulans*. *J. Biol. Chem.* 265: 13767–13775.

Rasmussen, C. D., K. P. Lu, R. L. Means, and A. R. Means (1992) Calmodulin and cell cycle control. *J. Physiol. (Paris)* 86: 83–88.

Raths, S., J. Rohrer, F. Crausaz, and H. Riezman (1993) *end3* and *end4*: Two mutants defective in receptor-mediated and fluid-phase endocytosis in *Saccharomyces cerevisiae*. *J. Cell. Biol.* 120: 565–65.

Redding, K., C. Holcomb, and R. S. Fuller (1991) Immunolocalization of Kex2 protease identifies a putative late Golgi compartment in the yeast *Saccharomyces cerevisiae*. *J. Cell Biol.* 13: 527–538.

Reid, B. J. and L. H. Hartwell (1977) Regulation of mating in the cell cycle of *Saccharomyces cerevisiae*. *J. Cell Biol.* 75: 355–365.

Reig, J. A., M. T. Tellez-Inon, M. M. Flawia, and H. N. Torres (1984) Activation of *Neurospora crassa* soluble adenylate cyclase by calmodulin. *Biochem. J.* 221: 541–543.

Reissig, J. L. and S. G. Kinney (1983) Calcium as a branching signal in *Neurospora crassa*. *J. Bacteriol.* 154: 1397–1402.

Rexach, M. F. and R. W. Schekman (1991) Distinct biochemical requirements for the budding, targeting and fusion of ER-derived transport vesicles. *J. Cell Biol.* 114: 219–229.

Ripmaster, T. L., G. P. Vaughn, and J. L. Woolford Jr (1993) *DRS1* to *DRS7*, novel genes required for ribosome assembly and function in *Saccharomyces cerevisiae*. *Mol. Cell. Biol.* 13: 7901–7912.

Rodriguez-del Valle, N. and J. R. Rodriguez-Medina (1993) Calcium stimulates molecular and cellular events during the yeast-to-mycelium transition in *Sporothrix schenckii*. *J. Med. Vet. Mycol.* 31: 43–53.

Roomans, G. M., A. P. Theuvenet, T. P. Van den Berg, and G. W. Borst-Pauwels (1959) Kinetics of Ca^{2+} and Sr^{2+} uptake by yeast. Effects of pH, cations and phosphate. *Biochim. Biophys. Acta* 551: 187–196.

Rose, M. D., L. M. Misra, and J. P. Vogel (1989) *KAR2*, a kayogamy gene is the yeast homolog of the mammalian *BiP/GRP78* gene. *Cell* 57: 1211–1221.

Rudolph, H. K., A. Antebi, G. R. Fink, C. M. Buckley, T. E. Dorman, J. LeVitre, L. S. Davidow, J. I. Mao, and D. T. Moir (1989) The yeast secretory pathway is perturbed by mutations in *PMR1*, a member of a Ca^{2+} ATPase family. *Cell* 58: 133–145.

Sachs, A. B. and J. A. Deardorff (1992) Translation initiation requires the PAB-dependent poly(A) ribonuclease in yeast. *Cell* 70: 961–973.

Salisbury, J. L. (1995) Centrin, centrosomes and mitotic spindle poles. *Curr. Opin. Cell. biol.* 7: 39–45.

Saporito, S. M. and P. S. Sypherd (1991) The isolation and characterization of a calmodulin-encoding gene (*CMD1*) from the dimorphic fungus *Candida albicans*. *Gene* 106: 43–49.

Scherens, B., M. El Bakkoury, F. Vierendeels, E. Dubois, and F. Messenguy (1993) Sequencing and functional analysis of a 32560 bp segment on the left arm of yeast chromosome II. Identification of 26 open reading frames including the KIP1 and SEC17 genes. *Yeast* 9: 1355–1371.

Schild, D., H. N. Ananthaswamy, and R. K. Mortimer (1981) An endomitotic effect of a cell cycle mutation of *Saccharomyces cerevisiae*. *Genetics* 97: 551–562.

Schmid, J. and F. M. Harold (1988) Dual roles for calcium ions in apical growth of *Neurospora crass*. *J. Gen. Microbiol.* 134: 2623–2631.

Schomerus, C. and H. Kuntzel (1992) *CDC25*-dependent induction of inositol 1,4,5-trisphosphate and diacylglycerol in *Saccharomyces cerevisiae* by nitrogen. *FEBS Lett.* 307: 249–252.

Schultz, C., G. Gebauer, T. Metschies, L. Rensing, and B. Jastorff (1990) cis,cis-Cyclohexane 1,3,5-triol polyphosphates release calcium from *Neurospora crass* via an unspecific Ins 1,4,5-P3 receptor. *Biochem. Biophys. Res. Commun.* 166: 1319–1327.

Sharp, P. M. and W. H. Li (1986) The codon adaptation index—a measure of directional synonymous codon usage bias, and its potential application. *Nucleic Acids Res.* 15: 1280–1295.

Shimokawa, N. and M. Yamaguchi (1993a) Molecular cloning and sequencing of the cDNA coding for a calcium-binding protein regucalcin from rat liver. *FEBS Lett.* 327: 251–255.

Shimokawa, N. and M. Yamaguchi (1993b) Expression of hepatic calcium-binding protein regucalcin mRNA is mediated through Ca^{2+}/calmodulin in rat liver. *FEBS Lett.* 316: 79–84.

Silveira, L. A., D. H. Wong, F. R. Masiarz, and R. Schekman (1990) Yeast clathrin has a distinctive light chain that is important for cell growth. *J. Cell Biol.* 111: 1437–1449.

Sloat, B. F. and J. R. Pringle (1978) A mutant of yeast defective in cellular morphogenesis. *Science* 200: 1171–1173.

Sloat, B. F., A. Adams, and J. R. Pringle (1981) Roles of the *CDC24* gene product in cellular morphogenesis during the *Saccharomyces cerevisiae* cell cycle. *J. Cell. Biol.* 89: 395–405.

Smith, R. A., M. J. Duncan, and D. T. Moir (1985) Heterologous protein secretion from yeast. *Science* 229: 1219–1224.

Sorin, A., G. Rosas, and R. Rao (1997) PMR1, a Ca^{2+}-ATPase in yeast Golgi, has properties distinct from sarco/endoplasmic reticulum and plasma membrane calcium pumps. *J. Biol. Chem.* 272: 9895–9901.

Spang, A., I. Courtney, U. Fackler, M. Matzner, and E. Schiebel (1993) The calcium-binding protein cell division cycle 31 of *Saccharomyces cerevisiae* is a component of the half bridge of the spindle pole body. *J. Cell Biol.* 123: 405–416.

Starovasnik, M. A., T. N. Davis, and R. E. Klevit (1993) Similarities and differences between yeast and vertebrate calmodulin: an examination of the calcium-binding and structural properties of calmodulin from the yeast *Saccharomyces cerevisiae*. *Biochemistry* 32: 3261–3270.

Stathopoulos, A. M. and M. S. Cyert (1997) Calcineurin acts through the CRZ1/TCN1-encoded transcription factor to regulate gene expression in yeast. *Genes Dev.* 11: 3432–3444.

Steiner, D. F., S. P. Smeekens, S. Ohagi, and S. J. Chan (1992) The new enzymology of precursor processing endoprotease. *J. Biol. Chem.* 267: 23435–23438.

Stepp, J. D., A. Pellicena-Palle, S. Hamilton, T. Kirchhausen, and S. K. Lemmon (1995). A late Golgi sorting function for *Saccharomyces cerevisiae* Apm1p, but not for Apm2p, a second yeast clathrin AP medium chain-related protein. *Mol. Biol. Cell* 6: 41–58.

Stirling, D. A., K. A. Welch, and M. J. Stark (1994) Interaction with calmodulin is required for the function of Spc110p, an essential component of the yeast spindle pole body. *EMBO J.* 13: 4329–4342.

Stratford, M. (1992) Yeast flocculation: reconciliation of physiological and genetic viewpoints. *Yeast* 8: 25–38.

Stratford, M. (1993) Yeast flocculation: flocculation onset and receptor availability. *Yeast* 9: 85–94.

Stratford, M. (1994) Another brick in the wall? Recent developments concerning the yeast cell envelope. *Yeast* 10: 1741–1752.

Stratford, M. and S. Assinder (1991) Yeast flocculation: Flo1 and Newflo phenotypes and receptor structure. *Yeast* 7: 559–574.

Stratford, M. and A. T. Carter (1993) Yeast flocculation: lectin synthesis and activation. *Yeast* 9: 371–378.

Stroobant, P. and G. A. Scarborough (1979) Active transport of calcium in *Neurospora* plasma membrane vesicles. *proc. Natl. Acad. Sci. USA* 76: 3102–3106.

Sun, G. H., Y. Ohya, and Y. Anraku (1991) Half-calmodulin is sufficient for cell proliferation. Expressions of N- and C-terminal halves of calmodulin in the yeast *Saccharomyces cerevisiae*. *J. Biol. Chem.* 266: 7008–7015.

Sun, G. H., A. Hirata, Y. Ohya, and Y. Anraku (1992) Mutations in yeast calmodulin cause defects in spindle pole body functions and nuclear integrity. *J. Cell Biol.* 119: 1625–1639.

Suzuki, C. K., J. S. Bonifacino, A. Y. Lin, M. M. Davis, and R. D. Klausner (1991) Regulating the retention of T-cell receptor alpha chain variants within the endoplasmic reticulum: Ca^{2+}-dependent association with BiP. *J. Cell Biol.* 114: 189–205.

Szaniszlo, P. J., S. M. Karuppayil, L. Mendoza, and R. J. Rennard (1993) Cell cycle regulation of polymorphism in *Wangiella dermatitidis*. *Arch. Med. Res.* 24: 251–261.

Taguchi, N., Y. Takano, C. Julmanop, Y. Wang, S. Stock, J. Takemoto, and T. Miyakawa (1994) Identification and analysis of the *Saccharomyces cerevisiae SYR1* gene reveals that ergosterol is involved in the action of syringomycin. *Microbiology* 140: 353–359.

Takahashi, H. and M. Yamaguchi (1993) Regulatory effect of regucalcin on $Ca^{(2+)}$-$Mg^{(2+)}$-ATPase in rat liver plasma membranes: comparison with the activation by Mn^{2+} and Co^{2+}. *Mol. Cell. Biochem.* 124: 169–174.

Takeda, T. and M. Yamamoto (1987) Analysis and in vivo disruption of the gene coding for calmodulin in *Schizosaccharomyces pombe*. *Proc. Natl. Acad. Sci. USA* 84: 3580–3584.

Takeda, T., Y. Imai, and M. Yamamoto (1989) Substitution at position 116 of *Schizosaccharomyces pombe* calmodulin decreases its stability under nitrogen starvation and results in a sporulation-deficient phenotype. *Proc. Natl. Acad. Sci. USA* 86: 9737–9741.

Takemoto, J. Y., L. Zhang, N. Taguchi, T. Tachikawa, and T. Miyakawa (1991) Mechanism of action of the phytotoxin syringomycin: a resistant mutant of *Saccharomyces cerevisiae* reveals an involvement of Ca^{2+} transport. *J. Gen. Microbiol.* 137: 653–659.

Takeuchi, J., M. Okada, A. Toh-e, and Y. Kikuchi (1995) The *SMS1* gene encoding a serine-rich transmembrane protein suppresses the temperature sensitivity of the htr1 disruptant in *Saccharomyces cerevisiae*. *Biochim. Biophys. Acta* 1260: 94–96.

Takita, Y., Y. Ohya, and Y. Anraku (1995) The *CLS2* gene encodes a protein with multiple membrane-spanning domains that is important Ca^{2+} tolerance in yeast. *Mol. Gen. Genet.* 246: 269–281.

Tan, P. K., N. G. Davis, G. F. Sprague, and G. S. Payne (1993) Clathrin facilitates the internalization of seven transmembrane segment receptors for mating pheromone in yeast. *J. Cell Biol.* 123: 1707–1716.

Tang, H.-Y. and M. Cai (1996) The EH-domain-containing protein Pan1 is required for the normal organization of the actin cytoskeleton in *Saccharomyces cerevisiae*. *Mol. Cell. Biol.* 16: 4897–4914.

Tang, H.-Y., A. Munn, and M. Cai (1997) EH domain proteins Pan1p and End3p are components of a complex that plays a dual role in organization of the cortical actin cytoskeleton and endocytosis in *Saccharomyces cerevisiae*. *Mol. Cell. Biol.* 17: 4294–4304.

Tang, X., M. S. Halleck, R. A. Schlegel, and P. Williamson (1996) A subfamily of P-type ATPases with aminophospholipid transporting activity. *Science* 272: 1495–1497.

Tanguy-Rougeau, C., M. Wesolowski-Louvel, and H. Fukuhara (1988) *The Kluyveromyces lactis KEX1* gene encodes a subtilisin-type serine proteinase. *FEBS Lett.* 234: 464–470.

Tanida, I., A. Hasegawa, H. Iida, Y. Ohya, and Y. Anraku (1995) Cooperation of calcineurin and vacuolar $H^{(+)}$-ATPase in intracellular Ca^{2+} homeostasis of yeast cells. *J. Biol. Chem.* 270: 10113–10119.

Tellez-Inon, M. T., R. M. Ulloa, G. C. Glikin, and H. N. Torres (1985) Characterization of *Neurospora crassa* cyclic AMP phosphodiesterase activated by calmodulin. *Biochem. J.* 232: 425–430.

Teunissen, A. W. R. H., E. Holub, J. Van Der Hucht, J. A. Van Den Berg, and Y. Steensma (1993) Sequence of the open reading frame of the *FLO1* gene from *Saccharomyces cerevisiae*. *Yeast* 9: 423–427.

Teunissen, A. W. R. H., J. A. Van Den Berg, and Y. Steensma (1995) Transcriptional regulation of flocculation genes in *Saccharomyces cerevisiae*. *Yeast* 111: 435–446.

Toda, T., M. Shimanuki, and M. Yanagida (1993) Two novel protein kinase C-related genes of fission yeast are essential for cell viability and implicated in cell shape control. *EMBO J.* 12: 1987–1995.

Toyn, J. H., W. M. Toone, B. A. Morgan, and L. H. Johnston (1995) The activation of DNA replication in yeast. *Trends Biochem. Sci.* 20: 70–73.

Tropschug, M., I. B. Barthelmess, and W. Neupert (1989) Sensitivity to cyclosporin A is mediated by cyclophilin in *Neurospora crassa* and *Saccharomyces cerevisiae*. *Nature* 342: 953–955.

Trumbly, R. J. and G. Bradley (1983) Isolation and characterization of aminopeptidase mutants of *Saccharomyces cerevisiae*. *J. Bacteriol.* 156: 36–48.

Ulloa, R. M., H. N. Torres, C. M. Ochatt, and M. T. Tellez-Inon (1991) Ca^{2+} calmodulin-dependent protein kinase activity in the ascomycetes *Neurospora crassa*. *Mol. Cell. Biochem.* 102: 155–163.

Uribe, S., P. Rangel, and J. P. Pardo (1992) Interactions of calcium with yeast mitochondria. *Cell Calcium* 13: 211–217.

Uribe, S., P. Rangel, J. P. Pardo, and L. Pereira-Da-Silva (1993) Interactions of calcium and magnesium with the mitochondrial inorganic pyrophosphatase from *Saccharomyces cerevisiae*. *Eur. J. Biochem.* 217: 657–660.

Vallen, E. A., M. A. Hiller, T. Y. Scherson, and M. D. Rose (1992a) Separate domains of *KAR1* mediate distinct functions in mitosis and nuclear fusion. *J. Cell Biol.* 117: 1277–1287.

Vallen, E. A., T. Y. Scherson, T. Roberts, K. van Zee, and M. D. Rose (1992b) Asymmetric mitotic segregation of the yeast spindle pole body. *Cell* 69: 505–515.

Vallen, E. A., W. Ho, M. Winey, and M. D. Rose (1994) Genetic interactions between *CDC31* and *KAR1*, two genes required for duplication of the microtubule organizing center in *Saccharomyces cerevisiae*. *Genetics* 137: 407–422.

Van den Berg, M. A. and H. Y. Steensma (1995) ACS2, a *Saccharomyces cerevisiae* gene encoding acetyl-coenzyme A synthetase essential for growth on glucose. *Eur. J. Biochem.* 231: 704–713.

Van den Berg, M. A., P. de Jong-Gubbels, C. J. Kortland, J. P. van Dijken, J. T. Pronk, and H. Y. Steensma (1996) The two acetyl-coenzyme A synthetases of *Saccharomyces cerevisiae* differ with respect to kinetic properties and transcriptional regulation. *J. Biol. Chem.* 271: 28953–28959.

Van Der Aart, Q. J. M., K. Kleine, and H. Y. Steensma (1995) Sequence analysis of the 43 KB CRM1-YLM9-PET54-SMI1-PHO81-YHB4-PFK1 region from the right arm of *Saccharomyces cerevisiae* chromosome VII. *Yeast* 12: 385–290.

Van Heusden, G. P., D. J. Griffiths, J. C. Ford, T. F. Chin-A-Woeng, P. A. Schrader, A. M. Carr, and H. Y. Steensma (1995) The 14-3-3 proteins encoded by the *BMH1* and *BMH2* genes are essential in the yeast *Saccharomyces cerevisiae* and can be replaced by a plant homologue. *Eur. J. Biochem.* 229: 45–53.

Vito, P., E. Lacana, and L. D'Adamio (1996) Interfering with apoptosis: Ca^{2+}-binding protein ALG-2 and Alzheimer's disease gene *AGL − 3*. *Science* 271: 521–525.

Volland, C. D., Urban-Grimal, G. Geraud, and R. Haguenauer-Tsapis (1994) endocytosis and degradation of the yeast uracil permease under adverse conditions. *J. Biol. Chem.* 269: 9833–9841.

Voordouw, G. and R. S. Roche (1975) The role of bound calcium ions in thermostable, proteolytic enzymes. I. Studies on thermomycolase, the thermostable protease from the fungus *Malbranchea pulchella*. *Biochemistry* 14: 4659–4666.

Wach, A., A. Schlesser, and A. Goffeau (1992) An alignment of 17 deduced protein sequences from plant, fungi and flagellate H^+-ATPase genes. *J. Bioenerg. Biomembr.* 24: 309–317.

Wada, I., D. Rindress, P. H. Cameron, W.-J. Ou, J. J. Doherty II, D. Louvard, A. W. Bell, D. Dignard, D. Y. Thomas, and J. J. M. Bergeron (1991) SSRalpha and associated calnexin are major calcium binding proteins of the endoplasmic reticulum membrane. *J. Biol. Chem.* 266: 19599–19610.

Wada, Y., Y. Ohsumi, M. Tanifuji, M. Kasai, and Y. Anraku (1987) Vacuolar ion channel of the yeast. *Saccharomyces cerevisiae*. *J. Biol. Chem.* 262: 17260–17263.

Wang, X., P. M. Watt, E. J. Louis, R. H. Borts, and I. D. Hickson (1996) Part 1: a topoisomerase II-associated protein required for faithful chromosome transmission in *Saccharomyces cerevisiae*. *Nucleic Acids Res.* 24: 4791–4797.

Watanabe, M., C. Y. Chen, and D. E. Levin (1994) *Saccharomyces cerevisiae PKC1* encodes a protein kinase C (PKC) homolog with a substrate specificity similar to that of mammalian PKC. *J. Biol. Chem.* 269: 16829–16836.

Watari, J., Y. Takata, M. Ogawa, H. Sahara, S. Koshino, M.-L. Onnella, U. Airaksinen, R. Jaatinen, M. Penttilä, and S. Keränen (1994) Molecular cloning and analysis of the yeast flocculation gene *FLO1*. *Yeast* 10: 211–225.

Weber, F. E., G. Menestrini, J. H. Dyer, M. Werder, D. Boffelli, S. Compassi, E. Wehrli, R. M. Thomas, G. Schulthess, and H. Hauser (1997) Molecular cloning of a peroxisomal Ca^{2+}-dependent member of the mitochondrial carrier superfamily. *Proc. Natl. Acad. Sci. USA* 94: 8509–8514.

Welch, M. D., D. A. Holtzman, and D. G. Drubin (1994) The yeast actin cytoskeleton. *Curr. Opin. Cell Biol.* 6: 110–19.

Wendland, B. and S. D. Emr (1998) Pan1p, yeast eps15, functions as a multivalent adaptor that coordinates protein-protein interactions essential for endocytosis. *J. Cell Biol.* 141: 71–84.

Wenfeng, W., S. Alexandre, X. Cao, and A. S. Lee (1993) Transactivation of the grp78 promoter by Ca^{2+} depletion. *J. Biol. Chem.* 268: 12003–12009.

Wieland, J., A. M. Nitsche, J. Strayle, H. Steiner, and H. K. Rudolph (1995) The *PMR2* gene cluster encodes functionally distinct isoforms of a putative Na^+ pump in the yeast plasma membrane. *EMBO J.* 14 3870–3882.

Wilcox, C. A. and R. S. Fuller (1991) Posttranslational processing of the prohormone-cleaving Kex2 protease in the *Saccharomyces cerevisiae* secretory pathway. *J. Cell. Biol.* 115: 297–307.

Wilcox, C. A., K. Redding, R. Wright, and R. S. Fuller (1992) Mutation of a tyrosine localization signal in the cytosolic tail of yeast Kex 2 protease disrupts Golgi retention and results in default transport to the vacuole. *Mol. Biol. Cell* 3: 1353–1371.

Williams, D. B. (1995) Calnexin: a molecular chaperone with a taste for carbohydrate. *Biochem. Cell Biol.* 73: 123–132.

Wilsbach, K. and G. S. Payne (1993) Vps1p, a member of the dynamin GTPase family, is necessary for Golgi membrane protein retention in *Saccharomyces cerevisiae*. *EMBO J.* 12: 3049–3059.

Wu, A.-L., T. C. Hallstrom, and W. S. Moye-Rowley (1996) *ROD1*, a novel gene conferring multiple resistance phenotypes in *Saccharomyces cerevisiae*. *J. Biol. Chem.* 271: 2914–2920.

Yamaguchi, M. and Y. Kanayama (1995) Enhanced expression of calcium-binding protein regucalcin mRNA in regenerating rat liver. *J. Cell. Biochem.* 5705: 185–190.

Yamaguchi, M., Y. Kanayama, and N. Shimokawa (1994) Expression of calcium-binding protein regucalcin in mRNA in rat liver is stimulated by calcitonin: the hormonal effect is mediated through calcium. *Mol. Cell Biochem.* 136: 43–48.

Yamaguchi, M., K. Oishi, and M. Isogai (1995) Expression of hepatic calcium-binding protein regucalcin mRNA is elevated by refeeding of fasted rats: involvement of glucose, insulin and calcium as stimulating factors. *Mol. Cell. Biochem.* 142: 35–41.

Yanagisawa, K., D. Resnick, C. Abeijon, P. W. Robbins, and C. B. Hirschberg (1990) A guanosine diphosphatase enriched in Golgi vesicles of *Saccharomyces cerevisiae*. *J. Biol. Chem.* 265: 19351–19355.

Ye, R. R. and A. Bretscher (1992) Identification and molecular characterization of the calmodulin-binding subunit gene (*CMP1*) of protein phosphatase 2B from *Saccharomyces cerevisiae*. An alpha-factor inducible gene. *Eur. J. Biochem.* 204: 713–723.

Yoko-o, T., Y. Matsui, H. Yagisawa, H. Nojima, I. Uno, and A. Toh0-e (1993) The putative phosphoinositide-specific phospholipase C gene, *PLC1*, of the yeast *Saccharomyces cerevisiae* is important for cell growth. *Proc. Natl. Acad. Sci. USA* 90: 1804–1808.

Yoko-o, T., H. Kato, Y. Matsui, T. Takenawa, and A. Toh-e (1995) Isolation and characterization of temperature-sensitive *plc1* mutants of the yeast *Saccharomyces cerevisiae*. *Mol. Gen Genet.* 247: 148–156.

Yoshida, T., T. Toda, and M. Yanagida (1994) A calcineurin-like gene *ppb1* + in fission yeast: mutant defects in cytokinesis, cell polarity, mating and spindle pole body positioning. *J. Cell Biol.* 107: 1725–1735.

Yuan, D. S., R. Stearman, A. Dancis, T. Dunn, T. Beeler, and R. D. Klausner (1995) The Menkes/Wilson disease gene homologue in yeast provides copper to a ceruloplasmin-like oxidase required for iron uptake. *Proc. Natl. Acad. Sci. USA* 92: 2632–3636.

Zhao, C., T. Beeler, and T. Dunn (1994) Suppressors of the Ca^{2+}-sensitive yeast mutant (*csg2*) identify genes involved in sphingolipid biosynthesis. *J. Biol. Chem.* 269: 21480–21488.

Zheng, Y., R. A. Cerione, and A. Bender (1994) Control of the yeast bud-site assembly CTPase Cdc42. *J. Biol. Chem.* 269: 2369–2372.

Zheng, Y., A. Bender, and R. A. Cerione (1995) Interactions among proteins involved in bud-site selection and bud-site assembly in *Saccharomyces cerevisiae*. *J. Biol. Chem.* 270: 626–630.

Zhu, G., E. G. Muller, S. L. Amacher, J. L. Northrop, and T. N. Davis (1993) A dosage-dependent suppressor of a temperature-sensitive calmodulin mutant encodes a protein related to the fork head family of DNA-binding proteins. *Mol. Cell. Biol.* 13: 1779–1787.

Ziegenhagen, R. and H. P. Jennissen (1990) Plant and fungus calmodulins are polyubiquitinated at a single site in $Ca^{2(+)}$-dependent manner. *FEBS Lett.* 273: 253–256.

Ziman, M. and D. I. Johnson (1994) Genetic evidence for a functional interaction between *Saccharomyces cerevisiae CDC24* and *CDC42*. *Yeast* 10: 463–474.

19

Calcium Genetics: Exploitation and Perspective

Yoshikazu Ohya
Isei Tanida
Yasuhiro Anraku

Calcium (Ca^{2+}) is an essential element for cell growth and formation of multicellular organisms. It is involved in diverse intracellular regulatory processes, including metabolic switching, cell cycle progression, motility, and other cellular mechanodynamics by acting as a second messenger of chemical and light signal transduction. In eukaryotic cells, a number of Ca^{2+}-dependent modulators and Ca^{2+}-requiring enzymes play important roles by receiving the intracellular Ca^{2+}-signals which are generated as a cellular response to maintain life against a large variety of environmental circumstances.

The extracellular concentration of Ca^{2+} affects physiological cellular state, since a moderate concentration of extracellular Ca^{2+} is necessary to drive intracellular Ca^{2+}-signaling efficiently in many cellular processes. We have long since thought that the growth phenotype triggered by an extracellular Ca^{2+} change is genetically available to study Ca^{2+}-regulatory process and/or Ca^{2+} homeostasis. The *calcium genetics* that we have explored was initiated in 1984 by using unicellular eukaryotic organisms, the budding yeast *Saccharomyces cerevisiae*. We have predicted that yeast mutants which impair intracellular Ca^{2+}-regulatory process or Ca^{2+} homeostasis may alter the response to extracellular Ca^{2+}. Our strategy was simply to look for yeast mutants that show Ca^{2+}-dependent on Ca^{2+}-sensitive phenotypes for growth. This chapter presents and summarizes our current accomplishments with the yeast calcium mutants and the genetic concept of Ca^{2+} homeostasis and signaling deduced therefrom.

Calcium Homeostasis in the Cellular Compartments

Ca^{2+} Is Essential for Life

Calcium (Ca^{2+}) is ubiquitously present in living organisms and their surrounding environments. Ca^{2+} ions are incorporated into cellular compartments, sequestered into the organelles, and utilized to support cell growth (Anraku et al., 1991). The cytosolic free Ca^{2+} concentration (referred to hereafter as $[Ca^{2+}]$) has been known to be maintained at homeostatic levels of 100–200 nM despite the differences in organisms and environmental Ca^{2+} concentrations (Anraku et al., 1991). Thus, in mammalian cells, deprivation of extracellular Ca^{2+} causes growth arrest of the cell cycle in G1 due to a break of $[Ca^{2+}]$ (Pardee et al., 1978; Campbell, 1983). Cells of *S. cerevisiae*, when deprived of medium Ca^{2+} and the vacuolar Ca^{2+} pool, undergo transient G1 arrest with a simultaneous decrease in the cAMP level, followed by a block mostly at G2/M (Iida et al., 1990a). Examination of the terminal phenotype suggests that Ca^{2+} is required at all the stages of the cell cycle except for the initiation of DNA synthesis (see Chapter 21).

There has been increasing evidence that $[Ca^{2+}]$ and its transient break play important roles to trigger signal transduction. For instance, during T-cell activation, antigens bind to specific receptors of quiescent T-cells, and Ca^{2+} channels on the endoplasmic reticulum and the plasma membrane open to induce an elevation of $[Ca^{2+}]$ (Gardner, 1989). This rapid rise of $[Ca^{2+}]$ is necessary for expression of many genes, including interleukin-2. Calcineurin, also known as Ca^{2+}/calmodulin-dependent phosphoprotein phosphatase, is a key signaling enzyme in this process (Clipstone and Crabtree, 1992; O'Keefe et al., 1992).

In *S. cerevisiae*, the mating pheromone pathway involves mechanisms similar to that of the T-cell activation. The pheromone-induced influx of medium Ca^{2+} followed by a rise of $[Ca^{2+}]$ is essential for the late stage of this pathway (Iida et al., 1990b). Yeast mutants that lack calcineurin activity have a defect in recovery from α-factor arrest in G1 (Cyert et al., 1991; Cyert and Thorner, 1992), and the recovery from the G1 stage is highly sensitive to the immunosuppressant drugs FK506 and cyclosporin A and their respective intracellular receptors, FKBP-12 and Cyp-18 (Foor et al., 1992). These findings imply that the Ca^{2+}/calcineurin signal transduction is indispensable for the progression of the late stage of the yeast mating pheromone pathway.

Ca^{2+} Homeostasis is a Prerequisite for Responding to Environmental Signals

In most eukaryotic cells, a very low level of $[Ca^{2+}]$ is maintained against a large gradient across the plasma membrane. The basal level of $[Ca^{2+}]$ in yeast cells is maintained homeostatically by active transport from the cytosol in organelles and by pumping out from the cell through the plasma membrane (Anraku, 1987b). In *S. cerevisiae*, the vacuole is a major intracellular Ca^{2+} pool, accumulating over 95% of the total calcium associated with cells (Iida et al., 1990b; Tanida et al., 1995). Ohsumi and Anraku (1983) established a method for measuring in vitro vacuolar Ca^{2+} uptake and indicated that the vacuole takes up Ca^{2+} by a Ca^{2+}/H^+ antiporter system that uses the proton motive force generated by the vacuolar H^+-ATPase (Kakinuma et al., 1981).

We think that these organelle Ca^{2+}-transport systems are only a few among many, the functions of which may all be concerned with maintenance of intracellular Ca^{2+} homeostasis. Calcium genetics has been needed to elaborate the calcium circuit in the cell (Anraku et al., 1991).

Genetic Searches for Yeast Mutants Defective in Response to Ca^{2+} in the Media

Isolation of *cal1* and *cls* Mutants

Ohya et al. (1984, 1986b) initiated systematic genetic searches to screen yeast mutants that are either dependent on Ca^{2+} or sensitive to Ca^{2+} for growth. At the time of their investigations, no information on

$[Ca^{2+}]$ in *S. cerevisiae* was available, so 100 mM $CaCl_2$ was used as the selection condition for mutations, simply due to the fact that yeast wild-type parental strains can grown on a Ca^{2+}-poor YPD medium (1% yeast extract/2% polypepton/2% glucose: calcium content, 0.18 mM) supplemented with 100 mM $CaCl_2$ (Ca^{2+}-rich medium) but not with 400 mM $CaCl_2$ (Ohya et al., 1986b). They isolated one *cal* mutant (for calcium-dependent) (Ohya et al., 1984) and 30 *cls* mutants (for calcium-sensitive) (Ohya et al., 1986b), each of which had a single recessive chromosomal mutation. The *cls* mutants were further divided into 18 complementation groups (Table 19.1).

The *cal-1* mutant can grow well in Ca^{2+}-rich medium but stops growing in Ca^{2+}-poor medium at 37°C (Ohya et al., 1984). Mg^{2+} ions cannot replace Ca^{2+} ions. In Ca^{2+}-poor medium, the mutant cells show a *cdc* phenotype and arrest homogeneously at the G2 stage of the cell cycle with a single tiny bud. Based on the phenotypic analysis of the *cal-1* cells, Ohya et al., (1984) suggested involvement of Ca^{2+} in the progression of the yeast cell cycle.

Among the 18 *cls* mutants, some show a phenotype of sensitivity to only Ca^{2+}, whereas others show phenotypes of sensitivities to several divalent cations (Table 19.1). The *cls* mutants were examined for their calcium contents and activities of Ca^{2+} uptake and were classified into four types (Ohya et al., 1986b). Type I mutants (see Table 19.1) have high (ratio of > 2 in mutant: wild-type cells) calcium content and high Ca^{2+} uptake activities. Type II mutants have a normal calcium content (ratio of 0.8–1.0 in mutant: wild-type cells) and a normal uptake activity. Type III mutants have high calcium content and normal Ca^{2+} uptake activities, whereas Type IV mutants have normal calcium content but high Ca^{2+} uptake activities. Two *cls* mutants in this subgroup, *cls7* and *cls12*, showed intermediary properties. Unlike the *cal1* mutation, most *cls* mutants, except *cls4* and *cls5*, do not show any obvious terminal phenotype and they all arrested at random stages of the cell cycle.

Interestingly, type IV *cls* mutants, though they all possess mitochondrial DNA (ρ^+-factor), cannot grow in YPD medium containing nonfermentable carbon sources, such as glycerol, acetate, and D(L)-lactate, thus showing a typical Pet$^-$ phenotype (Table 19.1). Segregation of the Pet$^-$ phenotype in tetrads of crosses between the type IV *cls* mutants and wild-type strain was 2+: 2− in more than 20 tetrads. This result indicates that the *cls* mutation that confers these two phenotypes originated from the same mutational defect.

Sherman (1963) first described *pet* mutants of *S. cerevisiae*, and, in subsequent genetic studies, a large number of *pet* mutants of 200 complementation groups have been obtained, though they were not

Table 19.1 Growth Phenotypes of *cls* Mutants

Mutation	*cls* Subtype	Pet	Divalent Cation Sensitivity
cls1	III	+	Ca, Mn
cls2	III	+	Ca
cls3	III	+	Ca
cls4	II	+	Ca
cls	I	+	Ca
cls6	I	+	Ca
cls7	IV	−	Ca, Mn, Zn
cls8	IV	−	Ca, Mn, Zn
cls9	IV	−	Ca, Mn, Zn
cls10	IV	−	Ca, Mn, Zn
cls11	IV	−	Ca, Mn, Zn
cls12	IV	−	Ca, Mn, Zn
cls13	I	+	Ca, Mn, Zn, Cu
cls14	I	+	Ca, Mg, Zn
cls15	I	+	Ca, Mg, Zn, Mn, Cu
cls16	I	+	Ca, Mg, Cu
cls17	I	+	Ca, Zn
cls18	I	+	Ca, Mg, Zn

tested for Ca^{2+}-sensitivity (Tzagoloff and Myers, 1986). Tzagoloff and Dieckmann (1990) classified *pet* mutants into six major types with defects in (1) mitochondrial F_1F_0-ATP synthease, (2) ubiquinone: cytochrome C oxidoreductase, (3) ferrocytochrome C: oxygen oxidoreductase, (4) ATP/ADP translocator, (5) splicing and processing for mRNAs of the oxidoreductases, and (6) other mitochondrial functions, including aminoacyl-tRNA synthetase.

Type IV Pet–*cls* mutants were examined for aerobic metabolism and oxygen consumption with glucose, succinate, and glycerol as substrates and the activities of mitochondrial F_1-ATPase and succinate dehydrogenase (Ohya et al., 1991b). Interestingly, there are no obvious biochemical defects and the mutant cells are arrested in the G1 phase of the cell cycle. The only defect that the mutants exhibit is a decreased level (15–50% of the wild-type cells) of activities of phosphatidyl serine decarboxylase, which was not repressed by the addition of inositol. This suggests that the Pet − *cls* mutations belong to the seventh class of *pet* mutants. In collaboration with the Tzagoloff's group, the Anraku laboratory has attempted to identify the Pet − *cls* mutations among the 200 *pet* complementation groups. Two new Pet − *cls* mutants (*VMA14* and *VMA15*) were found from this luxurious collection (unpublished

results of the Y. Anraku and A. Tzagoloff laboratories).

Identification of the *CLS* Genes Responsible for Ca^{2+} Homeostasis

Vacuolar Function and Morphogenesis Meet Jointly to Understand the *CLS* Function

Parallel to the aforementioned studies, and since 1981, the Anraku laboratory has studied physiological functions of yeast vacuoles, including their roles in sequestration of Ca^{2+} ion and basic amino acids (Anraku, 1987a, 1987b). Ohsumi and Anraku (1981, 1983) demonstrated that the purified vacuoles with right-side-out orientation generated a proton motive force inside positive and acidic, depending on ATP hydrolysis and take up Ca^{2+} ions and basic amino acids actively into the lumen, as proven by simultaneous discoveries of the vacuolar H^+-ATPase (Kakinuma et al., 1981), Ca^{2+}/H^+-antiporter (Ohsumi and Anraku, 1983), and basic amino acids/H^+-antiporters (Sato et al., 1984). These independent findings were brought a priori into a crucible of thought to consider if the *cls* phenotype might be derived from a defect either of vacuolar H^+-ATPase activity or of Ca^{2+}/H^+-antiporter. Also, it was highly conceivable that some *cls* mutants had a defect of vacuolar morphogenesis.

Type IV *cls* Mutants Show Vma⁻ defects

The first decisive evidence came when Ohya et al. (1991b) demonstrated that all the null mutants of *VMA1*, *VMA2*, and *VMA3* (for vacuolar membrane ATPase; Anraku, 1996; Anraku et al., 1989, 1992) that encode the 69, 60, and 17 kDa subunit, respectively, of the vacuolar H⁺-ATPase show the same phenotype as Pet⁻ *cls*. Consistent with these results, all type IV *cls* mutants are defective in vacuolar acidification in vivo, like the null *vma1*, *vma2*, and *vma3* mutants, and isolated vacuoles from the *cls* mutants have no vacuolar H⁺-ATPase activity. Genetic complementation tests revealed that *cls7* and *cls8* are not complemented by *vma3* and *vma1*, respectively, and that *vma2* complemented all the *cls* mutants tested. Thus, *CLS7* and *CLS8* are identical with *VMA3* and *VMA1*, respectively, and *CLS9*, *CLS10*, and *CLS11* are referred to as *VMA11*, *VMA12*, and *VMA13* based on their biochemical Vma⁻ defects mentioned above (Ohya et al., 1991b).

The [Ca²⁺] in individual cells of the Pet⁻ *cls* mutants was measured with the Ca²⁺-specific fluorescent dye fura-2 using epifluorescent microscopy combined with a digital calcium-imaging apparatus ARGUS 100 (Iida et al., 1990b). All the five mutant cells contained increased levels of [Ca²⁺], amounting to $900 \pm 100 \, nM$, which is 6-fold higher than in the wild-type cells. This indicates that in the *cls* mutants with Vma⁻ defects, serious breakdown of Ca²⁺ homeostasis is triggered by the defect of vacuolar H⁺-ATPase, which results in an injurious effect for cell growth.

It was shown that the vacuoles from vacuolar H⁺-ATPase-defective *vma* mutants cannot take up Ca²⁺, so the mutants become Ca²⁺-sensitive for growth (Ohya et al., 1991b). Cunningham and Fink (1994) discovered a putative Ca²⁺-ATPase, the *PMC1* gene product, and located it on the vacuolar membrane. The *pmc1* null mutant does not grow in the presence of 200 mM CaCl₂. Another putative Ca²⁺-ATPase, the *PMR1* gene product, is thought to transport Ca²⁺ into the Golgi complex (Rudolph et al., 1989; Antebi and Fink, 1992). The *pmc1 pmr1* double mutant shows synthetic lethality, indicating that the function of *PMC1* is required in the *pmr1* mutants.

Type I *cls* Mutants show the Vam⁻ Phenotype

By 1987, the Anraku group had listed chemiosmotic solute transport systems residing on the vacuolar membrane and characterized their functional properties and mechanisms for vacuolar acidification and ionic homeostasis in the cytosol (Anraku, 1987a;

Anraku et al., 1992; Wada and Anraku, 1994). Similar to Ca²⁺ sequestration into vacuoles, lysine is largely stored in the vacuole via basic amino acids/H⁺ antiport and lysine/H⁺ antiport systems and the cytosolic lysine pool is maintained at low level (Kitamoto et al., 1988a). Kitamoto et al. (1988b) isolated a lysine-sensitive mutant that cannot grow in YPD medium supplemented with 10 mM lysine and showed that it has no central vacuole. In addition, this *slp1* mutant (for small lysine pool) is hypersensitive to Ca²⁺ and cannot grow in YPD medium supplemented with 10 mM CaCl₂, indicating that the large volume of the vacuole is indispensable for vacuolar chemiosmotic work which regulates homeostasis of lysine and Ca²⁺ in the cytosolic pools (Wada et al., 1990). Therefore, it was most likely that any of the mutations of vacuolar morphology and/or assembly would result in the Ca²⁺-sensitive phenotype.

Based on their basic understanding of the mechanisms of vacuolar acidification, Wada et al. (1992), developed logical screenings for *vam* mutations (for vacuolar morphogenesis), in which wild-type yeast cells were treated with chloroquine, an acidophilic antimalarial agent, or wild-type cells with *ade1* background (that form a red colony) were subjected to pigmentation/depigmentation assays. Three stable mutants were further subjected to genetic analyses, which indicated that the Vam⁻ phenotype in each strain was responsible for a single recessive mutation on the chromosome. Complementation tests then established *vam-1*, *vam2-1*, and *vam3-1* mutations. A more direct method for isolating mutants with the Vam⁻ phenotype in the *ade1* background was very successful (Wada et al., 1992). Consequently, a total of 18 alleles of nine complementation groups of *VAM* genes were established. The nine *VAM* genes were classified into two classes according to the mutant phenotypes. The class 1 *vam* mutants (*vam1*, *vam5*, *vam8*, and *vam9*) contain a few small vesicles that are stained with histochemical markers, ade fluorochrome and lucifer yellow CH, for the vacuolar compartment. These mutants also have defects in the maturation of vacuolar marker proteins, and their growth is hypersensitive to high concentrations of CaCl₂ or a temperature of 37°C. The class II *vam* mutants (*vam2*, *vam3*, *vam4*, *vam6*, and *vam7*) contain numerous small vesicles stained with the vacuolar histochemical markers and mature forms of the vacuolar proteins, and do not show any apparent growth defects in the presence of CaCl₂ or at 37°C.

The *vam* mutations were identified by cytological screening for mutants with the Vam⁻ phenotype; however, the class I *vam* mutants show clear defects not only in vacuolar morphology but also in growth under the various conditions mentioned above. Wada et al. (1992) carried out complementation tests and found the genetic overlaps of *VAM* and other genes, including the *CLS* genes and the *SLP* gene. They examined the vacuolar morphology of type I *cls* mutants and demonstrated that the mutants show the Vam⁻ phenotype. Complementation tests by crossing the *cls* and *vam* mutants showed that the Vam⁻ and Cls− phenotypes were not complementary in the following combinations: *cls13/vam1, cls14/vam5, cls17/vam9,* and *cls18/vam8* (see Table 19.2). Complementation tests also revealed that the *slp-1* mutation is allelic to the *vam5-1* mutation, and DNA fragments of the *SLP1* gene complemented the *vam5-1* mutation.

The *CLS2* Gene Product is an ER Membrane Protein That Plays a Role in [Ca²⁺] Homeostasis

Among the three complementation groups of type III *cls* mutants, the *cls2-2* mutant is the only one that is sensitive specifically to Ca²⁺ (Ohya et al., 1986b). Takita et al. (1995) cloned and sequenced the *CLS2* gene that encodes a membrane protein comprising 410 amino acids. Immunofluorescent staining of the yeast cells expressing influenza haemagglutinin-tagged Cls2 protein revealed that the gene is localized to the endoplasmic reticulum (ER) membrane. Several pieces of indirect evidence suggest that Cls2p participates in regulation of Ca²⁺ homeostasis in the lumen of ER (see below for details).

Identification of *CAL* and *CLS* Genes Involved in Cell Cycle Progression and Ca²⁺ Signaling

CAL1 Encodes a β-subunit of Geranylgeranyl Transferase I Which Requires Ca²⁺ for its Activity

Structural analysis of the *CAL1* gene (Ohya et al., 1991a) has revealed that it encodes the β-subunit of a protein prenyltransferase, geranylgeranyl transferase I. The same gene was isolated during the study of the cell cycle gene, *CDC43* (Johnson et al., 1991). The *CAL1* gene shows extensive similarity with *DPR1/RAM1* (Goodman et al., 1988; He et al., 1991) with the gene encoding the catalytic subunit of the other protein prenyltransferase, farnesyltransferase. Consistent with this similarity, multiple copies of *CAL1* suppress the growth defects of a *dpr1* null mutant in a Ca²⁺-dependent manner, though the predicted amino acid sequence of the *cal1* gene product does not have a Ca²⁺-binding EF-hand motif (Ohya et al., 1991a). It is also noted that a double *cal1-1*

Δ*dpr1* mutant (Δ*dpr* = deletion of DPR gene) shows synthetic lethal phenotype.

Geranylgeranyl transferase I has been well characterized biochemically in the budding yeast by using the recombinant prenyltransferase (Mayer et al., 1992; Caplin et al., 1994). Analysis of bacterially expressed Cal1p and Ram2p, the other α-subunit of the enzyme, has revealed that the yeast geranylgeranyl transferase I is a Mg²⁺-requiring, Zn²⁺ metalloenzyme (Mayer et al., 1992). This is consistent with the results from studies of the mammalian farnesyl transferase and geranylgeranyl transferase I which demonstrated a requirement for Mg²⁺ in isoprenoid transfer and for Zn²⁺ in binding of protein substrate. Yeast geranylgeranyl transferase I differs from mammalian farnesyl transferase because geranylgeranyl transferase I can also function with Ca²⁺ as the only divalent cation (Mayer et al., 1992). Ca²⁺ is likely to bind to both Mg²⁺- and Zn²⁺-binding sites of the yeast geranylgeranyl transferase I to mediate both isoprenoid transfer and substrate binding. This in vitro observation is consistent with the in vivo finding that *cal1-1* was originally isolated as a mutant that exhibited the Ca²⁺_-dependent phenotype for growth. All evidence suggests involvement of Ca²⁺ in the regulation of the geranylgeranyl transferase I function.

Identification of *CLS4/CDC24* as the Gene Regulating the Yeast Cell Cycle in a Ca²⁺-Dependent Manner

The *cls4-1* mutant is the only one among the 18 *cls* complementation group mutants that has normal calcium content and normal activity of Ca²⁺ uptake. From the beginning, the mutant attracted our attention since the mutant cells stop dividing in the YPD medium in the presence of 100 mM CaCl₂ with the terminal phenotype of large, round unbudded cell shape(Ohya et al., 1986a). Genetic analysis revealed that *cls4* is identical to *cdc24*, which has defects in bud emergence (Ohya et al., 1986a).

Primary structure of the *CLS4/CDC24* gene predicts putative Ca²⁺-binding sites in the molecules. In addition to the original *cls4-1* mutation with a Gly⁶¹⁵ to Ser alteration, Miyamoto et al. (1991) reported another Ca²⁺-sensitive allele (*cls4-2*) during site-directed mutagenesis of CLS4/CDC24. These results suggest that the Cls4p/Cdc24p activity is somehow regulated by intracellular Ca²⁺ (Ohya and Anraku, 1992).

Evidence has accumulated to suggest that the *CDC24/CLS4* gene product (Cdc24p) interacts with other bud assembly gene products such as *CDC42, RSR1/BUD1,* and *BEM1* to establish cell polarity (Drubin, 191). One of the biochemical activities of Cls4p/Cdc24p is a GDP/GTP exchange activity toward Cdc42p (Zheng et al., 1994). Another in vitro activity of Cls4p/Cdc24p is binding to Rsr1p and Bem1p (Peterson et al., 1994, Zheng et al., 1995). it is interesting to note an in vitro study which indicates that Ca^{2+} affects interaction between Cls4p/Cdc24p and Bem1p (Zheng et al., 1995). Binding between Cls4p/Cdc24p and Bem1p is observed only in the absence of Ca^{2+}. This would explain why a certain cls allele of CLS4/CDC24 results in loss of function in the presence of high concentration of Ca^{2+} in the medium.

CLS5 Is Identical to PFY1

The cls5 mutant belongs to type I cls mutants which have high calcium contents and high Ca^{2+} uptake activities. cls5 has attracted our attention, because of its abnormal cell morphology: cls5 mutants stopped growing with large, round shape of cells in the presence of high concentration of Ca^{2+} in the medium (authors' unpublished results). Even under the permissive condition (YPD), a population of cls5 cells exhibit a round shape. cls5 is specifically sensitive to Ca^{2+} among several divalent cations examined (Ohya et al., 1986b).

Cloning of the CLS5 gene has revealed that CLS5 is identical to PFY1, a gene encoding yeast profilin which regulates actin assembly and function. Immunofluorescent study revealed that cls5-1 mutant cells, in fact, have altered actin morphology. We do not know why a mutant of profilin shows the Ca^{2+}-sensitive phenotype, but the Ca^{2+}-sensitivity likely results from loss of profilin function, because pfy1 deletion mutant also shows the Ca^{2+}-sensitive phenotype. During the cloning study of CLS5, SMY1 was isolated as a multicopy suppressor of the cls5-1 mutation. Overproduction of SMY1 suppressed the Ca^{2+}-sensitivity of cls5-1 and a deletion mutation of profilin, but did not suppress elevated Ca^{2+} uptake due to the cls5 mutation (authors' unpublished results). Since multiple copies of SMY1 also suppress a mutation of MYO2, a class V yeast myosin, SMY1 may have the ability to stabilize actin filaments or to positively regulate actin morphology and its function.

The First Stage of Calcium Genetics: A Milestone for the Future

Our attempt to genetically identify cellular components essential for Ca^{2+} homeostasis and Ca^{2+}-regulatory process has yielded one CAL and 18 CLS genes (Table 19.2). Although screening of cal and cls mutants has not been completed as yet, the current view of the yeast calcium mutants tells us how yeast cells orchestrate the cellular regulatory networks to

tolerate and require extracellular Ca^{2+}. So far, CAL1 and 12 CLS genes have been cloned, characterized at the molecular level, and found to play indispensable roles in the cellular Ca^{2+}-signalling. Half of the CLS genes (CLS7, CLS8, CLS9, CLS10, CLS11, CLS13, CLS14, CLS17, CLS18) are involved in vacuolar homeostasis and its biogenesis. CLS2 functions on the ER membrane to regulate exchangeable Ca^{2+} pools. CAL1, CLS4, and CLS5 gene products have essential cellular functions that are regulated by intracellular Ca^{2+} (Table 19.2).

After our initial exploitation on the calcium genetics had appeared (Ohya et al., 1984), many investigators started to use the YPD medium containing 100–300 mM $CaCl_2$ to study yeast Ca^{2+}-regulatory processes. Levin et al. (1991) has reported that yeast protein kinase C mutants (pkc1) exhibit a Ca^{2+}-dependent phenotype, which is very similar to cal1-1. Some calmodulin mutants exhibit a Ca^{2+}-dependent phenotype (Ohya and Anraku, 1992). Cunningham and Fink (1994) and Beeler et al. (1994) reported genes important for Ca^{2+} homeostasis. These studies have clearly demonstrated that the medium containing a high concentration of Ca^{2+} is useful to genetically analyze yeast Ca^{2+}-regulatory processes. The use of EGTA in the selective media has also been useful for the characterization of mutants in Ca^{2+}-regulatory processes (Shih et al., 1988; Rudolph et al. 1989; Umemoto et al., 1991).

Although further screening of the mutants is necessary to cover all CAL and CLS genes, alternative genetic approaches will serve as a powerful tool for revealing the Ca^{2+}-regulatory network. Classical genetic studies, such as suppression analysis by mutation or high dosage, will promise to define new genes functionally related to Ca^{2+}-signaling. As a matter of fact, we have already observed genetic interaction between cal1-1 and cls. All type IV cls mutations suppression cal1-1 (authors' unpublished results). The studies of this line are presented in the next section.

Suppressor Genetics for Genes Involved in [Ca^{2+}] Homeostasis and Calcium Signal-Dependent Cell Cycle Control

Attempts to Isolate Revertants that Suppress the Vma− Phenotype

We have currently screened suppressor mutations of the yeast vma3 disruptant in the hope of constructing a genetic network relating to the regulation of [Ca^{2+}] homeostasis (authors' unpublished results). Standard genetic procedures were applied to isolate suv mutants (for suppressor of vma mutations) and five

Table 19.2 Summary of *CAL1*, *CLS* and *MID* Genes and their Functions

Gene	Other Name	Function of Gene Product
CLA1	*CDC43*	β-subunit of geranylgeranyl transferase I
CLS1		Unknown
CLS2	*CSG2*	Transmembrane protein in ER
CLS3		Unknown
CLS4	*CDC24*	Bud assembly factor
CLS5	PFY1	Profilin
CLS6		Unknown
CLS7	VMA3	Integral subunit of vacuolar H^+-ATPase
CLS8	*VMA1*	Catalytic subunit of vacuolar H^+-ATPase
CLS9	*VMA11*	Integral subunit of vacuolar H^+-ATPase
CLS10	*VMA12*	Assembly factor of vacuolar H^+-ATPase
CLS11	*VMA13*	Regulatory subunit of vacuolar H^+-ATPase
CLS13	*VAM1, VPS11*	Involved in vacuolar sorting/morphology
CLS14	*VAM5, SLP1, VPS33, PEP4*	Involved in vacuolar sorting/morphology
CLS15		Unknown
CLS16		Unknown
CLS17	*VAM9, VPS16*	Involved in vacuolar sorting/morphology
CLS18	*VAM8, VPS18, PEP3*	Involved in vacuolar sorting/morphology
MID1		Unknown
MID2	*KA11, SMS1*	Unknown
MID5	*BCK1*	Protein kinase (MAP kinase kinase kinase)

suv complementation groups were identified. Interestingly, one *suv* mutation (*suv5*) only suppresses the Pet⁻ phenotype, whereas the remaining four (*suv1, suv2, suv3,* and *suv4*) suppress all of the Pet⁻ *cls* phenotypes to some extent. These *suv* mutations also suppress the phenotypes of the Δvma1, Δvma2, and Δvma*1 1* mutants (Δ*vma* = deletion of VMA gene), suggesting that the *suv* mutations suppress the defects not specific to the *vma3* mutant but due to loss of the vacuolar H^+-ATPase activity.

To investigate whether the *suv* mutations suppress the defects in [Ca^{2+}] homeostasis of the Δ*vma* cells, Tanida et al. (1995) examined the initial Ca^{2+}-uptake activity and cellular nonexchangeable Ca^{2+} pool. As described above, the Δ*vma3* mutants have high Ca^{2+}-uptake activity, probably reflecting the higher [Ca^{2+}] in the Δ*vma3* mutant than in the wild-type (Ohya et al., 1986b, 1991b). The *suv1* and *suv5* mutations were found to suppress the elevated Ca^{2+}-uptake activities to almost the same level as the wild-type, and the *suv3* and *suv4* mutations also exhibited partial suppression. These results suggested that the *suv* mutations may decrease the [Ca^{2+}] of the Δ*vma3* cells (authors' unpublished results).

The yeast vacuole accumulates more than 95% of the total cell-associated calcium (Eilam et al., 1985), the majority of which is nonexchangeable in a pulse-chase experiment (Eilam, 1982). In the *vma3* mutant, the nonexchangeable Ca^{2+} pool decreases to 20% compared with that in the wild-type strain, since the *vma3* mutants lack Ca^{2+}-uptake activity into the vacuole due to loss of the vacuolar H^+-ATPase activity (Ohya et al., 1991b, Tanida et al., 1995). The *suv1* and *suv2* mutations restore the nonexchangeable Ca^{2+} pools to a level 2.5–2.7-fold higher than that in the *vma3* mutant (authors' unpublished results). But no quinacrine is accumulated in the *suv vma3* cells under this condition, indicating that the *suv* mutations do not recover the vacuolar acidification in the cells. These results suggest that cytosolic free Ca^{2+} is probably sequestered into (a) nonvacuolar compartment(s) in the *suv1 vma3* and *vma3* mutants. In contrast, the Ca^{2+} pools in the *suv3 vma3* and *suv5 vma3* mutants do not increase. Further biochemical and molecular biological analysis of the *suv* mutations will lead to novel clues on the regulation of [Ca^{2+}] homeostasis in yeast.

The Vacuolar H^+-ATPase Cooperates with Calcineurin to Maintain [Ca^{2+}] Homeostasis

Recently, Cunningham and Fink (1994) reported that inactivation of calcineurin by *cnb1* mutation or by addition of the immunosuppressant FK506 suppresses the Ca^{2+}-sensitivity of the *pmc1* mutant. This observation caught our attention as the *pmc1* mutant shows a Ca^{2+}-sensitive phenotype similar to *cls/vma* mutants (Ohya et al., 1991b).

Tanida et al. (1995) demonstrated that FK506 makes the Δ *vma3* mutant hypersensitive to Ca^{2+}, namely, the mutant cannot grow in the YPD medium with 5 mM $CaCl_2$ in the presence of 1 μg/ml of FK506. More important, FK506 was found to confer Ca^{2+}-tolerance on the wild-type cells, so that they can grow in UYPD medium supplemented with up to 500 mM $CaCl_2$ in the presence of 1 μg/ml of FK506, while in its absence the growth ceased at Ca^{2+} concentrations up to 300 mM. This FK506 effect on Ca^{2+}-sensitivity for growth occurs in a dose-dependent manner and requires the presence of the FK506-binding protein FKBP-12, the product of the *FKB1* gene, in the cytosol; thus, it is being mediated by the regulation of calcineurin activity. To verify this possibility, Tanida et al. (1995) constructed a *vma3 fkb1* double disruptant and showed that FK506 has no more inhibitory effect on growth of the mutant in YPD medium with 5 mM $CaCl_2$. Genetic evidence also supports this possibility since a mutation of the *CNB1* gene that encodes an essential calcineurin subunit shows synthetic lethal interaction with the *vma3* mutation. Cyclosporin A, the other immunosuppressant drug capable of inhibiting calcineurin activity in the presence of its cytosolic binding protein, Cyp-18, produces effects similar to FK506. In addition, this immunosuppressant effect is not restricted to the *vma3* mutant, and is common to all the *vma* mutants tested (Tanida et al., 1995; Garrett-Engele et al., 1995). Thus, it was concluded that calcineurin regulates $[Ca^{2+}]$ in cooperation with the vacuolar H^+-ATPase (Garrett-Engele et al., 1995; Hemenway et al., 1995; Tanida et al., 1995).

The other important feature of the immunosuppressant effect was that the drugs decrease $[Ca^{2+}]$ in the cytosol of the *vma3* cells. An epifluorescence microscopic method, using fura-2 as a Ca^{2+}-specific indicator, was applied for measurements of $[Ca^{2+}]$ in the *vma3* and *vma3 fkb1* cells in the presence and absence of 2 μg/ml of FK506 for 1 hour. The average values of $[Ca^{2+}]$ (the mean ± SD in nanomoles) of *vma3* cells with or without FK506 were 448 ± 177 and 651 ± 174, respectively, whereas those of *vma3 fkb1* cells were 562 ± 117 and 563 ± 172, respectively (Tanida et al., 1995). Then, using $^{45}Ca^{2+}$, the authors measured cellular exchangeable and nonexchangeable Ca^{2+} pools against extracellular free Ca^{2+} under the experimental conditions given above. In the wild-type cells, a nonexchangeable pool accounted for 88% of total Ca^{2+} pool, whereas that in the *vma3* cells was decreased to 20% due to loss of the vacuolar H^+-ATPase activity. By the addition of 1 μg/ml of FK506 to the *vma3* cells, the nonex-changeable pool increased to a level 8.9-fold higher than that without FK506, suggesting that calcineurin activates (an) internal compartment(s) other than the vacuole, which is able to sequester free Ca^{2+} ions from the cytosol.

Recent studies have revealed that the *csg2* mutant (Beeler et al., 1994), another allele of *cls2-2* (Takita et al., 1995), accumulate much higher amounts of Ca^{2+} presumably in a nonvacuolar compartment, when the cells are cultured in a Ca^{2+}-rich medium. Tanida et al. (1996) demonstrated that the *cls2* null mutation results in a 3.4-fold increase of the nonexchangeable Ca^{2+} pool in the Δ*vma3* cells under normal conditions, and this level of Ca^{2+} accumulation is synergistically enhanced by the addition of FK506. They also showed that the *cls2* mutation in the genetic background Δ*vma3* confers on the mutant hypersensitivity to FK506, namely, a *cls2 Δvma3* mutant cannot grow in the presence of 0.1 μg/ml of FK506, and this expression of hypersensitivity depends on FKBP-12 in the cytosol. Since the Cls2p is largely localized on the ER membrane and the synergistic controlling effect of FK506 on nonvacuolar Ca^{2+} pools is evident, it may be possible that calcineurin and Cls2p regulate a putative Ca^{2+} pool in the ER in a cooperative manner but in an opposite direction.

Essential ρ-GTPase Gene RHO1 was Isolated as a Multiple Copy Suppressor of the cal1-1 Mutation

Qadota et al. (1992) identified two *RHO* genes as multicopy suppressors of *cal1-1*. There are five ρ-type GTPases thus far known: *RHO1* and *CDC42* encode essential small GTPases for growth, and *RHO2*, *RHO3*, and *RHO4* encode a nonessential GTPase. Multiple copies of either *RHO1* or *RHO2* genes suppressed temperature-sensitive growth of the *cal1-1* mutant. Multiple copies of neither *CDC42* nor *RHO3* nor *RHO4* suppressed cal1-1. Rho2p is likely to act as a homolog of Rho1p. The ability of the rho GTPases overproduction to suppress *cal1-1* may be due, in part, to the fact that the ρ-GTPases are substrates of geranylgeranyl transferase I. However, the ρ-GTPases also appear to enhance protein modification of another substrate by geranylgeranyl transferase I in vivo suggesting that ρ-GTPases carry out positive feedback regulation of geranylgeranyl transferase I activity (Qadota et al., 1992).

Although *CAL1* is normally an essential gene for growth, it can be made nonessential when the dosages of the two GTPases, Rho1p and Cdc42p, are simultaneously elevated (Ohya et al., 1993). Overproduction of neither Rho1p nor Cdc42p alone suppressed the *cal1/cdc43* deletion. The lethality of the *CAL1* deletion is most efficiently suppressed by provision of both Rho1p and Cdc42p with altered C-terminal sequences (Cys-Ala-Ala-Met) corresponding to the C-termini of substrates of far-nesyl transferase. Since the yeast geranylgeranyl transferase I prenylates these two GTPases, Cdc42p and Rho1p are implicated genetically as the two most essential substrates of geranylgeranyl transferase I.

Ohya et al. (1996) were puzzled by the failure of a high dosage of *CDC42* to suppress *cal1-1*. Overproduction of *CDC42*, instead is deleterious in *cal1-1*. There are seven temperature-sensitive mutations in the *CAL1/CDC43* gene thus far known (*cal1-1, cdc43-2* to *cdc43-7*). Mutational analysis of the *CAL1/CDC43* gene revealed that the cal1-1 mutation, located most proximal to the C-terminus of the protein, differs from the other *cdc43* mutations in several criteria. Soluble Rho1p was increased in the cal1-1 strain grown at the restrictive temperature, while soluble Cdc42p was increased in *cdc43* strains such as *cdc43-5*. The temperature-sensitive phenotype of *cal1-1* is most efficiently suppressed by overproduction of Rho1p, while *cdc43-5* is suppressed by overproduction of Cdc42p. Overproduction of Cdc42p is deleterious in *cal1-1* cells, but not deleterious in the other *cdc43* mutants Thus, several phenotypic differences were observed among the *cal1/cdc43* mutations, possibly due to the alteration of substrate specificity caused by the mutations (Ohya et al., 1996).

Genes Involved in Ca^{2+} Mobilization and Signaling

The *MID* Gene Family Functions in the Late Stage of the Mating Process

Yeast mating pheromone, the α-factor, induces Cu^{2+} influx into *MATa* cells with a lag of about 30 minutes, followed by a rise in the [Ca^{2+}] (Ohsumi and Anraku, 1985; Iida et al., 1990b). This rise is essential for maintaining the viability of the cells that have differentiated into *shmoos* having a mating projection in the late stage of the peromone response pathway. Iida et al. (1994) devised a new screening method for isolating conditional mutants defective in the α-factor-dependent Ca^{2+} influx and signaling, and identified a new family of *MID* genes (for mating pheromone-induced death). The *Mid* mutants die specifically after differentiating into *shmoos* in a medium containing 0.1 mM CaCl$_2$, but do not die with 1 mM CaCl$_2$, and are classified into five complementation groups. The *mid1* and *mid3* mutants are indeed defective in Ca^{2+} influx, and the remaining *mid* mutants, *mid2, mid4*, and *mid5*, are normal with regard to Ca^{2+} influx. The *MID* genes belonging to the latter *mid* groups function downstream of those

belonging to the first groups, suggesting that the *MID2, MID4*, and *MID5* genes play essential roles in Ca^{2+}-signal transduction.

The *MID1* gene encodes a new 62 kDa N-glycosylated, integral plasma membrane protein and is nonessential for vegetative growth (Iida et al., 1994). The Mid1p is predicted to have four hydrophobic domains, the fourth one being homologous to the membrane-spanning region (S3/H3) that resides in several ion channels, including voltage-gated Ca^{2+} channels (Jan and Jan, 1990). Overexpression of *MID1* on a multicopy vector does not affect Ca^{2+} influx, suggesting that the Mid1p itself may not be a putative Ca^{2+} channel, but rather its subunit or regulatory component.

The *MID2* gene has six potential TATA boxes and two pheromone-response elements in its 5'-upstream region and encodes a new 40 kDa membrane protein (Ono et al., 1994). The Mid2p contains a putative N-terminal signal sequence followed by a long serine-rich region that could be O-glycosylated, a potential transmembrane domain, and a conserved Ca^{2+}-binding domain. In *MATa* cells, the expression of *MID2* is enhanced 3-fold by the addition of α-factor.

Summary and Perspective

We have described our current results and the accomplishments that have derived from our first attempt to explore calcium genetics. It is now understood that since the selection condition for the *cls* mutations was rather severe, major *cls* mutants isolated are defective in vacuolar functions, which participate in the [Ca^{2+}] homeostasis in the cell. Based on the lucky coincidence of these mutational analyses with biochemistry, the definition of type IV *cls/vma* and type I *cls/vam* mutants was clearly obtained at the molecular level. Genetic screening for the *MID* family is another success, which will give useful information relating to the initiation of a pheromone-dependent Ca^{2+}-signaling cascade.

Genetics aiming at the definition of suppressors and synthetic lethal interaction for the *cal, cls*, and *mid* mutants is also a fruitful area for research in the future. Accumulating evidence suggests that the *VMA* genes cross-talk with the *SUV* genes in respect to the [Ca^{2+}] homeostasis and the signal transducing calmodulin-calcineurin pathway. The *CAL1/CDC43-RHO1* and *CLS4/CDC24* pathways are involved in bud emergency and have a close correlation with the cell wall remodeling.

During our genetic and biochemical studies on the type IV *cls/vma* mutants, we have noticed that the Vma$^-$ defects phenotypically relate to pH and osmotic sensitivities (Umemoto et al., 1991; Hill and

Stevens, 1994). Thus, calcium genetics is a means for exploring cellular chemiosmotic function and solute-ion transport, which are now recognized to influence widely the homeostasis of intracellular pH and cell osmolarity, both factors potentially capable of regulating the chemical signal transducing systems in living organisms.

Acknowledgements The authors are particularly grateful to Drs. Y. Ohsumi and H. Iida for their collaboration and to the many colleagues whose names are cited in the references. This work was supported in part by a grant-in-aid for Scientific Research from the Ministry of Education, Science, Sports and Culture of Japan (YO) and by a grant from HFSP (YA).

References

Anraku, Y. (1987a) Unveiling the mechanism of ATP-dependent energization of yeast vacuolar membranes: discovery of a third type of H$^+$-translocating adenosine triphosphatase. In *Structure and Function of Energy Transducing Systems* (Ozawa, T. and Papa, S., eds). Springer-Verlag, New York, pp. 249–262.

Anraku (1987b) Active transport of amino acids and calcium ions in fungal vacuoles. In *Plant Vacuoles* (Marin B. P., ed). Plenum Press, New York, pp. 255–265.

Anranku, Y. (1996) Structure and function of the yeast vacuolar membrane H$^+$-ATPase. In *Handbook of Biological Physics*, Vol. 2 (Konings, W. N., Kaback, H. R. and Lokema, J. S., eds). Elsevier Science, New York.

Anraku, Y., R. Hirata, Y. Wada, and Y. Ohya (1992) Molecular genetics of the yeast vacuolar H$^+$-ATPase. *J. Exp. Biol.* 172: 67–81.

Anraku, Y., Y. Ohya, and H. Iida (1991) Cell cycle control by calcium and calmodulin in *Saccharomyces cerevisiae*. *Biochim. Biophys Acta* 1093: 169–177.

Anraku, Y., N. Umemoto, R. Hirata, and Y. Wada (1989) Structure and function of the yeast vacuolar membrane proton ATPase. *J. Bioenerg. Biomembr.* 21: 589–603.

Antebi, A. and G. R. Fink (1992) The yeast Ca^{2+} ATPase homologue, *PMR1*, is required for normal Golgi function and localizes in a novel Golgi-like distribution. *Mol. Biol. Cell* 3: 633–654.

Beeler, T., K. Gable, C. Zhao, and T. Dunn (1994) A novel protein, Csg2p, is required for Ca^{2+} regulation in *Saccharomyces cerevisiae*. *J. Biol. Chem.* 269: 7279–7284.

Campbell, A. K. (1983) *Intracellular Calcium: Its Universal Role as Regulator*. John Wiley & Sons, Chichester, UK.

Caplin, B., L. A. Hettich, and M. Marshall (1994) Substrate characterization of the *Saccharomyces cerevisiae* protein farnesyl transferase and type I geranylgeranyl transferase. *Biochem. Biophys. Acta* 1205: 39–48.

Clipostone, N. A. and G. R. Crabtree (1992) Identification of calcineurin as a key signalling enzyme in T-lymphocyte activation. *Nature* 357: 695–697.

Cunningham, K. W. and G. R. Fink (1994) Calcineurin-dependent growth control on *Saccharomyces cerevisiae* mutants lacking *PMC1*, a homolog of plasma membrane Ca^{2+} ATPases. *J. Cell Biol.* 124: 351–363.

Cyert, M. S. and J. Thorner (1992) Regulatory subunit (*CNB1* gene product) of yeast Ca^{2+}/calmodulin-dependent phosphoprotein phosphatases is required for adaptation to pheromone. *Mol. Cell Biol.* 12: 3460–3469.

Cyert, M. S., R. Kunisawa, D. Kaim, and J. Thorner (1991) Yeast has homologs (*CNA1* and *CNA2* gene products) of mammalian calcineurin, a calmodulin-regulated phosphoprotein phosphatase. *Proc. Natl. Acad. Sci. USA* 88: 7376–7380.

Drubin, D. (1991) Development of cell polarity in budding yeast. *Cell* 65: 1093–1096.

Eilam, Y. (1982) The effect of monovalent cations on calcium efflux in yeasts. *Biochim. Biophys. Acta* 687: 8–16.

Eilam, Y., H. Lavi, and N. Grossowicz (1985) Cytoplasmic Ca^{2+} homeostasis maintained by a vacuolar Ca^{2+} transport system in the yeast *Saccharomyces cerevisiae*. *J. Gen. Microbiol.* 131: 623–629.

Foor, F., S. A. Parent, N. Morin, A. M. Dahl, N. Ramadan, G. Chrebet, K. A. Bostian, and J. B. Nielsen (1992) Calcineurin mediates inhibition by FK506 and cyclosporin A of recovery from alpha-factor arrest in yeast. *Nature* 360: 682–684.

Gardner, P. (1989) Calcium and T lymphocyte activation. *Cell* 59: 15–20.

Garrett-Engele, P., B. Moilanen, and M. S. Cyert (1995) Calcineurin, the Ca^{2+}/calmodulin-dependent protein phosphatase, is essential in yeast mutants with cell integrity defects and in mutants that lack a functional vacuolar H$^+$-ATPase. *Mol. Cell. Biol.* 15: 4103–4114.

Goodman, L. E., C. M. Perou, A. Fujiyama, and F. Tamanoi (1988) Structure and expression of yeast *DPR1*, a gene essential for the processing and intracellular localization of ras protein. *Yeast* 4: 271–281.

He, B., P. Chen, S.-Y. Chen, K. L. Vancura, S. Michaelis, and S. Powers (1991) *RAM2*, an essential gene of yeast, and *RAM1* encode the two polypeptide components of the farnesyltransferase that prenylates α-factor and Ras protein. *Proc. Natl. Acad. Sci. USA* 88: 1371–11377.

Hemenway, C. S., K. Dolinski, M. E. Cardenas, M. A. Hillier, E. W. Jones, and J. Heitman (1995) *vph6* mutants of *Saccharomyces cerevisiae* require calcineurin for growth and are defective in vacuolar H$^+$-ATPase assembly. *Genetics* 141: 833–844.

Hill, K. J. and T. H. Stevens (1994) Vma21p is a yeast membrane protein retained in the endoplasmic reticulum by a di-lysine motif and is required for the assembly of the vacuolar H⁺-ATPase complex. *Mol. Biol. Cell* 5: 1039–1050.

Iida, H., H. Nakamura, T. Ono, M. K. Okumura, and Y. Anraku (1994) *MID1*, a novel *Saccharomyces cerevisiae* gene encoding a plasma membrane protein, is required for Ca^{2+} influx and mating. *Mol. Cell. Biol.* 14: 8259–8271.

Iida, H., S. Sakaguchi, Y. Yagawa, and Y. Anraku (1990a) Cell cycle control by Ca^{2+} in *Saccharomyces cerevisiae*. *J. Biol. Chem.* 265: 21216–21222.

Iida, H., Y. Yagawa, and Y. Anraku (1990b) Essential role for induced Ca^{2+} influx followed by $[Ca^{2+}]_i$ rise in maintaining viability of yeast cells late in the mating pheromone response pathway: a study of $[Ca^{2+}]_i$ in single *Saccharomyces cerevisiae* cells with imaging of fura-2. *J. Biol. Chem.* 265: 13391–13399.

Jan, L. Y. and Y. N. Jan (1990) A superfamily of ion channels. *Nature* 345: 672.

Johnson, D. I., J. M. O'Brien, and C. W. Jacobs (1991) Isolation and sequence analysis of *CDC43*, a gene involved in the control of cell polarity in *Saccharomyces cerevisiae*. *Gene* 98: 149–150.

Kakinuma, Y., Y. Ohsumi, and Y. Anraku (1981) Properties of H⁺-translocating adenosine triphosphatase in vacuolar membranes of *Saccharomyces cerevisiae*. *J. Biol. Chem.* 256: 10859–10863.

Kitamoto, K., K. Yoshizawa, Y. Ohsumi, and Y. Anraku (1988a) Dynamic aspects of vacuolar and cytosolic amino acid pools of *Saccharomyces cerevisiae*. *J. Bacteriol.* 170: 2683–2686.

Kitamoto, K., K. Yoshizawa, Y. Ohsumi, and Y. Anraku (1988b) Mutants of *Saccharomyces cerevisiae* with defective vacuolar function. *J. Bacteriol.* 170: 2687–2691.

Levin, D. E., F. O. Fields, R. Kunisawa, J. M. Bishop, and J. Thorner (1990) A candidate protein kinase C gene, *PKC1*, is required for the *S. cerevisiae* cell cycle. *Cell* 62: 213–224.

Mayer, M. L., B. E. Caplin, and M. S. Marshall (1992) *CDC43* and *RAM2* encode the polypeptide subunits of a yeast type I protein geranylgeranyltransferase. *J. Biol. Chem.* 267: 20589-20593.

Miyamoto, S., Y. Ohya, Y. Sano, S. Sakaguchi, H. Iida, and Y. Anraku (1991) A *DBL*-homologous region of the yeast *CLS4/CDC24* gene product is important for Ca^{2+}-modulated bud assembly. *Biochem. Biophys. Res. Commun.* 181: 604–610.

Ohsumi, Y. and Y. Anraku (1981) Active transport of basic amino acids driven by a proton motive force in vacuolar membrane vesicles of *Saccharomyces cerevisiae*. *J. Biol. Chem.* 256: 2079–2082.

Ohsumi, Y. and Y. Anraku (1983) Calcium transport driven by a proton motive force in vacuolar membrane vesicles of *Saccharomyces cerevisiae*. *J. Biol. Chem.* 258: 5614–5617.

Ohsumi, Y. and Y. Anraku (1985) Specific induction of Ca^{2+} transport activity in *MAT*a cells of *Saccharomyces cerevisiae* by a mating pheromone, α-factor. *J. Biol. Chem.* 2650: 10482–10486.

Ohya, Y. and Y. Anraku (1992) Yeast calmodulin: Structural and functional elements essential for the cell cycle. *Cell Calcium* 13: 445–455.

Ohya, Y., B. E. Caplin, H. Quadota, M. F. Tibbets, Y. Anraku, J. R. Pringle, and M. S. Marshall (1996) Mutational analysis of the β-subunit of yeast geranylgeranyl transferase I. *Mol. Gen. Genet.* 252: 1–10.

Ohya, Y., M. Goebl, L. E. Goodman, S. Petersen-Bjorn, J. D. Friesen, F. Tamanoi, and Y. Anraku (1991a) Yeast *CAL1* is a structural and functional homologue to the *DPR1/RAM1* gene involved in ras processing. *J. Biol. Chem.* 266: 12356–12360.

Ohya, Y., S. Miyamoto, Y. Ohsumi, and Y. Anraku (1986a) Calcium-sensitive *cls4* mutant of *Saccharomyces cerevisiae* with a defect in bud formation. *J. Bacteriol.* 165: 28–33.

Ohya, Y., Y. Ohsumi, and Y. Anraku (1984) Genetic study of the role of calcium ions in the cell division cycle of *Saccharomyces cerevisiae*: a calcium-dependent mutant and its trifuloperazine-dependent pseudorevertants. *Mol. Gen. Genet.* 193: 389–394.

Ohya, Y., Y. Ohsumi, and Y. Anraku (1986b) Isolation and characterization of Ca^{2+}-sensitive mutants of *Saccharomyces cerevisiae*. *J. Gen. Microbiol.* 132: 979–988.

Ohya, Y., H. Qadota, Y. Anraku, J. R. Pringle, and D. Botstein (1993) Suppression of yeast geranylgeranyl transferase I defect by alternative prenylation of two target GTPases. Rho1p and Cdc42p. *Mol. Biol. Cell* 4: 1017–1025.

Ohya, Y., N. Umemoto, I. Tanida, A. Ohta, H. Iida, and Y. Anraku (1991b) Calcium-sensitive *cls* mutants of *Saccharomyces cerevisiae* showing a Pet⁻ phenotype are described to defects of vacuolar membrane H⁺-ATPase activity. *J. Biol. Chem.* 266: 13971–13977.

O'Keefe, S. J., J. Tamura, R. L. Kincaid, M. J. Tocci, and E. A. O'Neill (1992) FK506- and CsA-sensitive activation of the interleukin-2 promoter by calcineurin. *Nature* 357: 692–694.

Ono, T., T. Suzuki, Y. Anraku, and H. Iida (1994) The *MID2* gene encodes a putative integral membrane protein with a Ca^{2+}-binding domain and shows mating pheromone-stimulated expression in *Saccharomyces cerevisiae*. *Gene* 151: 203–208.

Pardee, A. B., J. Dubrow, J. L. Hamlin, and R. F. Kletzien (1978) Animal cell cycle. *Annu. Rev. Biochem.* 47: 715–750.

Peterson, J., Y. Zheng, L. Bender, A. Myers, R. Cerione, and A. Bender (1994) Interactions between the bud emergence proteins Bem1p and Bem2p and Rho-type GTPases in yeast. *J. Cell Biol.* 127: 1395–1406.

Qadota, H., I. Ishii, A. Fujiyama, Y. Ohya, and Y. Anraku (1992) *RHO* gene products, putative small GTP-binding proteins, are important for activation of the *CAL1/CDC43* gene products, a protein geranylgeranyltransferase in *Saccharomyces cerevisiae*. *Yeast* 8: 735–741.

Rudolph, H. K., A. Antebi, G. R. Fink, C. M. Buckley, T. E. Dorman, J. LeVitre, L. S. Davidow, J. Mao, and D. T. Moir (1989) The yeast secretory pathway is perturbed by mutations in *PMR1*, a member of a Ca^{2+} ATPase family. *Cell* 58: 133–145.

Sato, T., Y. Ohsumik, and Y. Anraku (1984) Substrate specificities of active transport systems for amino acids in vacuolar membrane vesicles of *Saccharomyces cerevisiae*. *J. Biol. Chem.* 259: 11505–11508.

Sherman, F. (1963) Respiratory-deficient mutants of yeast. I. Genetics. *Genetics* 48: 375–385.

Shih, C., K., Wagner, R. Feinstein, S., C. Kanik-Ennulat, and N. Neff (1988) A dominant trifluoperazine resistance gene from *Saccharomyces cerevisiae* has homology with F_oF_1 ATP synthase and confers calcium-sensitive growth. *Mol. Cell. Biol.* 8: 3094–3103.

Takita, Y., Y. Ohya, and Y. Anraku (1995) The *CLS2* gene encodes a protein with multiple membrane-spanning domains that is important for Ca^{2+} tolerance in yeast. *Mol. Gen. Genet.* 246: 269–281.

Tanida, I., A. Hasegawa, H. Iida, Y. Ohya, and Y. Anraku (1995) Cooperation of calcineurin and vacuolar H^+-ATPase in intracellular Ca^{2+} homeostasis of yeast cells. *J. Biol. Chem.* 270: 10113–10119.

Tanida, I., Y. Takita, A. Hasegawa, Y. Ohya, and Y. Anraku (1996) Yeast Cls2p/Csg2p localized on the endoplasmic reticulum membrane regulates a non-exchangeable intracellular Ca^{2+}-pool cooperatively with calcineurin. *FEBS Lett.* 379: 38–42.

Tzagoloff, A. and C. L. Dieckmann (1990) *PET* genes of *Saccharomyces cerevisiae*. *Microbiol. Rev.* 54: 211–225.

Tzagolof, A. and A. M. Myers (1986) Genetics of mitochondrial biogenesis. *Annu. Rev. Biochem.* 55: 249–285.

Umemoto, N. Y. Ohya, and Y. Anraku (1991) *VMA11*, a novel gene that encodes a putative proteolipid, is indispensable for expression of yeast vacuolar membrane H^+-ATPase activity. *J. Biol. Chem.* 266: 24526–24532.

Wada, Y. and Y. Anraku (1994) Chemiosmotic coupling of ion transport in the yeast vacuole: its role in acidification inside organelles. *J. Bioenerg. Biomembr.* 26: 631–637.

Wada, Y., K. Kitamoto, T. Kanbe, K. Tanaka, and Y. Anraku (1990) The *SLP1* gene of *Saccharomyces cerevisiae* is essential for vacuolar morphogenesis and function. *Mol. Cell Biol.* 10: 2214–2223.

Wada, Y., Y. Ohsumi, and Y. Anraku (1992) Genes for directing vacuolar morphogenesis in *Saccharomyces cerevisiae*: I. Isolation and characterization of two classes of *vam* mutants. *J. Biol. Chem.* 267: 18665–18670.

Zheng, Y., A. Bender, and R. Cerione (1995) Interactions among proteins involved in bud-site selection and bud-site assembly in *Saccharomyces cerevisiae*. *J. Biol. Chem.* 270: 626–630.

Zheng, Y., R. Cerione, and A. Bender (1994) Control of the yeast bud-site assembly GTPase Cdc42: catalysis of guanine-nucleotide exchange by Cdc24 and stimulation of GTPase activity by Bem3. *J. Biol. Chem.* 269: 2369–2372.

20

Calcium in the Nucleus

Luigia Santella
Stephen Bolsover

Cell activation by calcium is likely to be a three-phase process. The rise in intracellular calcium concentration [Ca^{2+}] has its first, rapid phase of action on pre-existing proteins, turning on the system for which a particular cell is specialized. In a second phase of activation by Ca^{2+}, gene transcription is turned on so that proteins used in the cell's operation are replaced. In many cases, calcium has a third phase of action in promoting cell division, so that a stimulated organ enlarges. This chapter is concerned with the second and third phases of action which require the participation of the cell nucleus. In the first part of the chapter, we will describe those nuclear processes that have been found to be calcium-sensitive. The second part of the chapter is concerned with the contentious issue of the regulation of nuclear calcium concentration.

Calcium-Sensitive Processes in the Nucleus

That some nuclear enzymes and complex processes are regulated by calcium has been known for a long time; the first to be described was perhaps myosin light chain kinase (MLCK) which was documented in the nucleus well over 10 years ago (Simmen et al., 1984). MLCK is a calmodulin-binding protein, which would imply the existence of calmodulin (CaM) in the nucleus. The existence of a separate nuclear calmodulin pool was frequently questioned, because most of the early work had been performed on isolated nuclei that could have been contaminated by extranuclear calmodulin during the procedure of isolation. However, nuclear calmodulin has now been conclusively documented, e.g. by in situ immunocytochemical work. It seems likely, therefore, to be a major target for calcium in the nucleus. Interestingly, recent work has shown that calmodulin, despite its small size, does not pass through nuclear pores passively but instead moves into the nucleus by a carrier-mediated process (Pruschy et al., 1994). Activated calmodulin then modulates a number of nuclear processes, which will be discussed in this chapter. However, some of the nuclear processes claimed to be under Ca^{2+} control, e.g. the breakdown of DNA, may not involve CaM. The most important processes, which are either nuclear or originate in the nucleus, and that are presently claimed to be controlled by Ca^{2+} are:

1. DNA replication.
2. DNA fragmentation, which is an essential component of the process of programmed cell death.
3. The transcription of some genes.
4. The cell cycle, both mitotic and meiotic.

The exact details (i.e. the molecular mechanisms) by which calcium modulates these processes are not established, but the proposal of Ca^{2+} modulation is now generally accepted. The role of Ca^{2+} in other processes in the nucleus is, on the other hand, based essentially on the presence of nuclear targets of Ca^{2+} (and CaM) known to be involved in Ca^{2+}-related processes elsewhere in the cell: this is the case of a putative motility system, whose presence in the nucleus is indicated by the finding of Ca^{2+}- and CaM-modulated motility proteins, e.g. MLCK and caldesmon. One recent development that deserves to be considered in this chapter is that of nuclear calpain. Calpain is a Ca^{2+}-dependent neutral protease which has frequently been mentioned as a mediator of apoptosis; a novel isoform of calpain has now been documented in the nucleus. Finally, recent work has indicated a possible role for Ca^{2+} and CaM in nuclear processes hitherto not postulated to be influenced by them, i.e. the processing of RNA in the heterogeneous ribonucleoprotein particles (hnRNPs) and in the nucleolus.

DNA Replication

DNA replication, which is the initial step of the cell cycle, occurs in supramolecular structures of the nuclear matrix called replisomes (Berezney, 1991). The effect of CaM antagonists on the triggering of DNA replication has led to the suggestion of a role for CaM in the process. Calcium-binding proteins different from calmodulin, e.g. Mbh1 (Prendergast and Ziff, 1991) and annexin II, which are actin-DNA-binding proteins (Jindal et al., 1991; Boyko et al., 1994), may also influence DNA replication, but the main mediator of the effect of calcium in the process appears, indeed, to be calmodulin. Perhaps more significant than the findings with anti-CaM drugs, whose specificity is questionable, are recent studies in which the synthesis of DNA was blocked by monoclonal antibodies against CaM (Reddy et al., 1992). Parallel studies have described similar effects, using inhibitors of CaM-modulated proteins, e.g. of calmodulin kinase II: the addition of the inhibitor KN-62 to cultured K562 cells blocked them in the S phase (Minami et al., 1994), whereas gene disruption experiments on the catalytic subunit of the calmodulin-modulated protein phosphatase calcineurin blocked *Aspergillus nidulans* cells in the G_1 phase (Rasmussen et al., 1994). Therefore, it appears that CaM-dependent phosphorylation and dephosphorylation events may modulate DNA replication.

However, Ca^{2+}/CaM could also be involved *directly*, i.e. in a manner not linked to the stimulation of protein kinases and/or phosphatases, in the process: immunogold labelling work has shown that CaM associates to the DNA replicative machinery, possibly by binding to proteins like p62, La/SS-B, hnRNP A2, and the CaM-binding protein-68 (Bachs and Agell, 1995), and purified DNA polymerase α has been claimed to be activated by CaM-binding proteins belonging to the family of microtubule-associated proteins (MAPs) (Shioda et al., 1991). A particularly interesting possibility for the intervention of Ca^{2+}/CaM in the replication of DNA involves the retinoblastoma protein (pRB) (Takuwa et al., 1992, 1993). This is a nuclear protein which has tumour suppressor function but which has also been proposed to regulate the progression of the cell cycle, and will thus also be mentioned in the section on the cell cycle (see below). The protein has been suggested to influence the transcription of genes that encode DNA replicative enzymes (DNA polymerases α and δ and thymidine kinase) (Pearson et al., 1991; Dou et al., 1992). It exists in both under- and hyperphosphorylated states depending on the cell cycle stage, its level of phosphorylation being

apparently controlled by Ca^{2+}. In the hypophosphorylated form, pRB binds to nuclear proteins, at least in some cell types, including transcription factors involved in the expression of genes of the DNA synthesis pathway. Phosphorylation then releases it from the bound state. Among the pRB-binding proteins is the transcription factor E2F-1, which complexes DNA at a specific site to activate the transcription of the DNA replication genes mentioned above. The mechanism by which Ca^{2+} controls the phosphorylation state of pRB is debated: some of the published evidence is controversial, but it appears that cyclin-dependent kinases are central to it. Evidence coming essentially from experiments on the effects of extracellular Ca^{2+} and of CaM inhibitors on the level of pRB phosphorylation have indicated that Ca^{2+}/CaM may stimulate the transcription of genes of cyclin-dependent kinases (cdks), which, in turn, promote the hyperphosphorylation of the pRB protein (Colomer et al., 1994). Anti-CaM drugs have also been shown to affect the amount of the proliferating cell nuclear antigen (PCNA), which is a cofactor of DNA polymerase δ, in NRK cells (Wang, 1991). This is probably due to the release of a transcriptional block at the level of the first intron of the PCNA gene. CaM has been found associated with the replicase complexes in CHEF/18 cells during the S phase (Subramanyam et al., 1990).

In addition to CaM, several CaM-binding proteins of unknown function have also been detected in the replication/transcription factories. The 68 kDa CaM-binding protein mentioned above activates DNA synthesis when added to permeabilized cells (Cao et al., 1995).

Ca^{2+} and Calmodulin in the Process of DNA Fragmentation and in Apoptotic Cell Death

DNA fragmentation is a characteristic component of apoptotic cell death and is thought to involve the action of Ca^{2+}-dependent endonucleases (Dowd et al., 1991; Gaido and Cidlowski, 1991). Other Ca^{2+}-dependent mechanisms may be involved in the nuclear alterations of apoptosis: changes in the expression of genes, activation of the protease calpain, activation of protein kinases and phosphatases, and changes in the conformation of chromatin. However, the activation of the endonucleases, which produces the well-known phenomenon of DNA laddering, is one of the distinctive features of most forms of apoptosis. Some forms, however, particularly in neurones, do not exhibit it, and in any case the phenomenon occurs late in the process. More than one endonuclease becomes activated during apoptosis: one is NUC18 (Gaido and Cidlowski, 1991), which is a Ca^{2+}–Mg^{2+}-dependent enzyme active at neutral pH. Another is a lysosomal enzyme

(DNase II) which is active at acidic pH and is not known to require Ca^{2+}: the proposal of an involvement of DNase II in apoptosis has been recently reinforced by the observation of rapid acidification in some cell types during apoptosis. The third is DNase I, which is also activated by Ca^{2+} and Mg^{2+} and is active at neutral pH. However, DNase I does not produce the DNA ladder pattern typical of apoptosis when incubated with isolated nuclei, and it is localized in the endoplasmic reticulum, not in the nucleus. NUC18 cleaves the DNA linker regions between nucleosomes, producing nucleosome-size fragments which can be seen as a DNA ladder in agarose gels (necrosis produces, instead, random DNA fragmentation, not the regular splitting pattern typical of apoptosis). Very interestingly, NUC18 has high sequence homology to cyclophilin, the receptor for the immunosuppressive drug cyclosporin (Montague et al., 1994), and recombinant cyclophilins have nuclease activity. These findings may be relevant to the proposal that the Ca^{2+}/CaM-dependent phosphatase, calcineurin, which is the target of the complex between immunophilins and the immunosuppressive drugs, is a component of the signalling pathway that leads to some forms of apoptosis (see below).

That the elevation of nuclear Ca^{2+} is essential to some forms of apoptosis has been shown by a number of findings: for example, the induction of apoptosis by tumour necrosis factor α (TNF-α) in mammary adenocarcinoma cells produces elevation of nuclear Ca^{2+}, and is followed by the fragmentation of DNA in the ladder-like pattern. Although the Ca^{2+}–Mg^{2+} activated nuclear endonuclease is not known to be calmodulin-sensitive, the nuclear Ca^{2+} increase and the DNA fragmentation seen in apoptosis have been claimed to be inhibited by calmodulin antagonists (Bellomo et al., 1992). The role of elevated nuclear Ca^{2+} in the endonuclease-mediated fragmentation is further supported by studies in isolated liver nuclei, in which an ATP-stimulated nuclear Ca^{2+} uptake results in endonuclease activation with a DNA fragmentation pattern identical to that observed in apoptotic cells (Jones et al., 1989; McConkey et al., 1989; Nicotera et al., 1989, 1994). Surprisingly, a recent study on heart has claimed that the increase in nuclear Ca^{2+} and the fragmentation of DNA correlate with the accumulation of Ca^{2+} in mitochondria, not in the cytosol (Faulk et al., 1995). A number of experiments have shown that the mechanism leading to the activation of the endonuclease involves the action of proteases. A protease cascade leads to the cleavage of the nuclear enzymes poly(ADP-ribose) polymerase (PARP) and DNA-dependent protein kinase. PARP is a key enzyme in DNA repair and is cleaved to an inactive 85 kDa fragment during apoptosis in most systems examined, indicating that its inactivation is crucial to the fragmentation of the DNA by the endonuclease. Apoptosis is under genetic control and a number of apoptosis-regulating genes which are switched on as the cell undergoes apoptosis have been identified, e.g. the c-fos gene. As will be discussed in the next section, the latter has been shown to be linked to the increase of calcium concentration, and to the activation of a calmodulin-dependent kinase, which phosphorylates a transcription factor (the cAMP-response element binding protein, CREB). This transcription factor is responsible for the induction of the gene, which could thus be the first step in a complex cascade that leads to cell death (Smeyne et al., 1993). Overexpression of the gene for Bcl-2, which is the first oncogene shown to have a role in the control of apoptosis, opposes the apoptosis induced by the increase of intracellular Ca^{2+} induced by thapsigargin (Lam et al., 1994). This has led to the suggestion that Bcl-2 regulates intracellular Ca^{2+} fluxes which may initiate the apoptotic signal. Other apoptosis-regulating genes include p53, which is mutated in the majority of human cancers; in normal cells the tumour suppressor protein p53 is located in the nucleus (Perry and Levine, 1993) and stops cells in the G_1 phase to allow the repair of DNA damage. A role for nuclear calcium in regulating p53 activity is supported by the finding that this protein is a substrate of protein kinase C (PKC), which, in turn, may be Ca^{2+}-regulated. The phosphorylation of p53 by PKC stimulates its DNA-binding activity, which has been found to be inhibited by the Ca^{2+}-binding protein S100b (Baudier et al., 1992). Mutations in the p53 gene affect DNA repair processes, resulting in an abnormal progression of the cell cycle and, in the end, the malignant transformation.

The involvement of CaM in the apoptotic process is supported by the finding that the expression of the calmodulin gene increases during glucocorticoid-induced lymphocyte apoptosis: the effect is likely to be at the transcriptional level, since it is blocked by the transcriptional inhibitor actinomycin D (Dowd et al., 1991). In the apoptosis of murine T-cell hybridomas, calmodulin may act by activating the Ca^{2+}/CaM dependent serine/threonine phosphatase calcineurin (protein phosphatase B). The suggestion that calcineurin phosphatase activity may be required for the apoptotic signalling pathway was prompted by the finding that the immunosuppressive agents cyclosporin A and FK506, which inhibit calcineurin phosphatase activity, also inhibit cell death during the Ca^{2+}-dependent process of (TcR)/CD3-mediated apoptosis of T-cell receptors (Fruman et al., 1992) and during activation in the murine WEHI-231 B-cell line (Genestier et al., 1994).

Chromatin conformational changes are required for gene activation, and the induction of DNA cleavage by eukaryotic topoisomerase II is promoted by calcium (Osheroff and Zechiedrich, 1987). Accordingly, endonuclease-mediated DNA fragmentation during apoptosis in thymocytes or isolated liver nuclei is inhibited by polyamines which stabilize chromatin (and nuclear enzymes) (Brüne et al., 1991). Thus, nuclear Ca^{2+} may play a role in modifying the conformation of chromatin regions accessible to enzymes such as DNase I or other endonucleases (Nicotera and Rossi, 1994).

Ca^{2+} and Calmodulin in Gene Transcription

Many genes are regulated in response to signalling cascades linked to the mobilization of intracellular Ca^{2+} which is normally initiated by the interaction of first messengers with plasma membrane receptors (Berridge, 1993). Extracellular Ca^{2+}, however, can also influence the expression of certain genes: a striking example is that of parathyroid cells (Yamamoto et al., 1989), in which the interaction of external Ca^{2+} with a recently cloned plasma membrane sensor (Brown et al., 1993) inhibits the expression of the parathyroid hormone gene (expression of the gene is induced by the lowering of extracellular Ca^{2+}).

The role of intracellular Ca^{2+} (and CaM) in the regulation of gene transcription has now become a booming topic, which was initiated by the pioneering observation that the addition of Ca^{2+} to GH3 cells in culture stimulated by up to 200-fold the transcription of the prolactin gene (White 1985). That CaM was involved in the effect of Ca^{2+} was indicated by the finding that anti-CaM drugs attenuated the Ca^{2+}-dependence of the expression of the gene (White, 1985). Calcium regulation of genes in response to external stimuli affects both immediate-early and -late genes. Regulation of immediate-early genes (which are expressed rapidly and transiently) was first demonstrated for the c-*fos* proto-oncogene in response to the depolarization of the plasma membrane of PC12 cells with either elevated extracellular KCl or with agonists of the nicotinic acetylcholine receptor (Greenberg, et al., 1986). The induction of c-*fos* gene expression was inhibited by Ca^{2+}-channel antagonists, showing that Ca^{2+} entry was essential and, interestingly, also inhibited by the calmodulin inhibitors chlorpromazine, W7, and trifluoperazine. This again suggested a concerted action of Ca^{2+} and calmodulin in gene regulation (Morgan and Curran, 1986). The depletion of the intracellular Ca^{2+} stores by thapsigargin (in the absence of extracellular calcium), induced a rapid expression of c-*fos* and c-*jun* (Schontal et al., 1991), showing that, as an alternative to Ca^{2+} influx, the liberation of Ca^{2+} stored in intracellular deposits was also effective. The response of the c-*fos* gene to the elevation of intracellular Ca^{2+} was very rapid, a 1 minute transient being sufficient

for full stimulation by 30 minutes. The transcription stimulation, however, may be longer lasting and more general, since the c-*fos* gene product can regulate the expression of other genes (Werlen et al., 1993). This is one of the cases of regulation of late-response genes, whose induction occurs with delayed kinetics and depends, albeit not obligatorily, on de novo protein synthesis. Late-response genes that are under Ca^{2+} control include many that code for endocrine factors, for neuropeptides that are involved in the biosynthesis of neurotransmitters, e.g. proenkephalin, for neutrophic factors, for cytokinesis, and for the family of glucose-regulated proteins which are expressed in several cells under conditions of glucose starvation (reviewed by Lee, 1987; Crabtree, 1989; Gardner, 1989; Williamson and Monck, 1989; Rao, 1991). Coming back to the issue of the regulation of immediate-early genes, Ca^{2+} could affect transcription initiation, but also influence the elongation of the nascent transcript, the stability of mRNA, and its translation. In the extensively studied case of the c-*fos* proto-oncogene, Ca^{2+} sensitivity is conferred by a 3′ DNA domain (the calcium-response element, CaRE) located at −60 from the transcription start site (Sheng et al., 1988). The CaRE contains a five-base pair match with the consensus cAMP-response element (CRE), which mediates transcription activation by binding to the CRE-binding protein (CREB) (Montminy and Bilezikjian, 1987; Hoeffler et al., 1988). CREB is a member of a large family of transcription factors which contain leucine-zipper dimerization motifs, and which are phosphorylated by cAMP-dependent protein kinase (PKA). In in vitro experiments, phosphorylation was shown to occur on serine 133 (Gonzales and Montminy, 1989), leading to the transcriptional activation of the CRE-containing gene. However, in the absence of an increase in cAMP level, serine 133 is still phosphorylated by treatments that increase intracellular Ca^{2+}, suggesting a distinct signalling pathway involving CaM-dependent kinases (Sheng et al., 1991). CREB phosphorylation stabilizes the binding of RNA polymerase II to the DNA domain involved in the initiation of transcription (Chrivia et al., 1993). In vitro assays have shown that CaMKs I, II, and IV all phosphorylate CREB on serine 133 (Dash et al., 1991; Sheng et al., 1991), but CaMKII also promotes the phosphorylation of serine 142, which blocks the activating effect of the phosphorylation of serine 133: thus, CaMKIV is more effective than CaMKII. CaMKIV has been detected in the nucleus of different cellular types, e.g. by immunocytochemical and immunoblot experiments in granule cells: they have shown that CaMKIV is largely associated with regions of dispersed chromatin (Jensen et al., 1991). CaMKII has also been detected in the nucleus, but of all CaMKII isoforms only δ-CaMKII is translocated to it. At the moment, then, even if it appears clear that CaMKs mediate the transcriptional effect of Ca^{2+}, no final decision is possible on which of the CaMKs is involved. One particularly important aspect of the CREB phosphorylation matter is that CREB is apparently permanently bound to DNA (Sheng et al., 1990); thus, its phosphorylation by calcium and CaM must, by definition, occur within the nucleus. One last point on the issue of Ca^{2+} and the c-*fos* gene must be mentioned: although the Ca^{2+}/CRE interaction in the transcriptional activation is now documented, other elements in the c-*fos* promoter are Ca^{2+}-responsive (Sheng et al., 1988). A serum-response element (SRE) at -300 relative to the transcription start site is involved in the induction of the gene in response to not only growth factors, phorbol esters, and NGF, but also to treatments increasing cellular Ca^{2+} (Bading et al., 1993).

The finding that a fraction of the cellular pool of the Ca^{2+}/CaM-dependent protein phosphatase calcineurin (phosphatase 2B) is nuclear (Pujol et al., 1993; Bosser et al., 1993; Hiraga et al., 1993) indicates that nuclear CaM, in addition to activating the specific kinases that regulate gene transcription, may also terminate their action by promoting the dephosphorylation of their substrates. Thus, it has been shown that calcineurin is a component of the signalling pathway that regulates the transcription of the *lac-1* gene, which encodes the enzyme laccase in the fungus *Cryphonectria parasitica* (Larson and Nuss, 1994). Calcineurin is an important regulator of transcription factors that stimulate the expression of a number of early genes in T-cells, such as NF-AT and OAP/Oct-1 (Clipstone and Crabtree, 1992; O'Keefe et al., 1992; Clipstone et al., 1994). Since the pool of calcineurin involved in these processes appears to be cytosolic, not nuclear, the Ca^{2+}-sensitive steps are evidently cytosolic. One of the two subunits of NF-AT (nuclear factor-activated T cells), NF-ATc, is rapidly translocated to the nucleus in response to the increase in cytosolic Ca^{2+}, which can be assumed to promote its dephosphorylation. In the nucleus, NF-ATc interacts with the second subunit, NF-ATn, and then binds to DNA to activate transcription. Calcineurin also induces the transcription of the genes for IL-2, IL-4, and the granulocyte-macrophage colony-stimulating factor GM-CSF (Kubo et al., 1994; Tsuboi et al., 1994a, 1994b): also, the up-regulation of these genes is apparently initiated by a Ca^{2+}-dependent event occurring in the cytosol. Interestingly, whereas Ca^{2+} mobilization and calcineurin are sufficient for the activation of the IL-4 promoter, up-regulation of the genes for IL-2 and GM-CSF also requires activation of protein kinase C. The cooperation of calcineurin and protein kinase C is not a unique example of the interplay between the phosphatase

and other agents; thus, overexpression of constitutively activated calcineurin does not lead to the up-regulation of the gene for NF-AT unless the p21 *ras* oncogene is also coexpressed (Woodrow et al., 1993). Similarly, agents that elevate cAMP, or prostaglandin E2, oppose the activating effect of calcineurin on the transcription of the gene for IL-2 (Paliogianni et al., 1993).

Although not explicitly stated, the underlying assumption in the discussion above was that the role of Ca^{2+}/CaM in the regulation of gene transcription was activatory. CaM, however, could also down-regulate genes, as has been proposed for the effect on transcription factors of the helix–loop–helix group. Several of these factors which bind to E-box domains of the promoter region of several genes have been found to bind CaM (Corneliussen et al., 1994). The binding of CaM to these transcription factors impairs their ability to bind to DNA, i.e. in these cases, CaM would act as a negative modulator of transcription. The suggestion has been given some support by in vivo experiments in which cells have been cotransfected with transcription factors of this group and an E-box reporter plasmid from the luciferase gene: the expression of luciferase is inhibited by the Ca^{2+} ionophore ionomycin (luciferase expression is not inhibited by the expression of transcription factors of this group that are unable to bind CaM).

Ca^{2+} and Calmodulin in the Regulation of the Cell Cycle (see Chapter 21)

Evidence from numerous experimental systems supports a role for Ca^{2+} in the regulation of G_2/M progression and in the metaphase/anaphase transition (Means, 1994; reviewed by Hepler, 1992, 1994). Work with fluorescent dyes (Grynciewicz et al., 1985) has revealed Ca^{2+} transients associated to the breakdown of the nuclear envelope (NEB) which initiates mitosis in sea urchin embryos (Steinhardt and Alderton, 1988). More recent work has revealed three separate calcium peaks, preceding the onset of anaphase and during cleavage (Ciapa et al., 1994). Periodic oscillations of calcium correlated with the cell division cycle have been observed in medaka eggs following fertilization and in *Xenopus* embryos (Yoshimoto et al., 1985; Fluck et al., 1991; Grandin and Charbonneau, 1991; Kubota et al., 1993). The finding that the injection of different BAPTA-type Ca^{2+}-buffers retards or inhibits the events of cell division (Kao et al., 1990; Tombes et al., 1992; Snow and Nuccitelli, 1993), and the parallel finding that NEB is restored by the microinjection of calcium (Silver, 1989) further support the cell cycle role of Ca^{2+}. The finding that the mobilization of stored calcium by microinjected inositol 1,4,5-trisphosphate ($InsP_3$) induces rapid chromatin condensation and NEB

(Twigg et al., 1988) establishes, on the other hand, the role of InsP$_3$ signalling in the cell cycle role of Ca^{2+}. In line with this, a recent report using antibodies against different domains of the InsP$_3$ receptor has shown that the fusion of the vesicles derived from the disassembled nuclear envelope of *Xenopus* eggs during mitosis, to reconstitute the nuclear envelope, requires the flux of Ca^{2+} through InsP$_3$ receptors located on their membranes (Sullivan et al., 1993).

In CHO-K1 cells, the cellular calmodulin content reaches a peak during release of CHO-K1 cells from the stationary phase at G$_0$/G$_1$ and, again, with the entry to S phase (Chafouleas et al., 1984). On the basis of these and other results, it has been proposed that the effects of Ca^{2+} on mitosis entry and progression are mediated by Ca^{2+} in conjunction with calmodulin (Rasmussen and Means, 1989; Ohya and Botstein, 1994). Certainly, calmodulin is essential for progression through the cell cycle; for instance, expression of antisense CaM RNA in mouse C127 cells causes a transient cell cycle arrest in G$_1$ (Rasmussen and Means, 1989). Also, consistent with this is the location of CaM in the mitotic apparatus (Welsh et al., 1978; Davis, 1992). In the budding yeast *Saccharomyces cerevisiae*, CaM is required for chromosome segregation, and plays a role in the polarized growth required to form the bud (Davis, 1992; Sun et al., 1992; Geiser et al., 1993); unexpectedly, however, mutant calmodulins completely defective in Ca^{2+} binding can still influence the yeast mitotic function (Geiser et al., 1991). This has led to the proposal that the effects of CaM in the yeast cell nucleus may not be linked to its ability to bind Ca^{2+}.

Very likely, many of the effects of CaM on the cell cycle are mediated by CaM-dependent protein kinases, i.e. CaMKII (Whitaker and Patel, 1990; Planas-Silvas and Means, 1992). As mentioned in the preceding section, CaMKII has been documented in the nuclei of many cells (Sahyoun et al., 1984; Ohta et al., 1990; Yano et al., 1994) and its intranuclear location differs, suggesting that its nuclear targets may be relatively cell type-specific. A nuclear localization, an 11-amino acid sequence (Chelsky et al., 1989) inserted by alternative splicing (Srinivasan et al., 1994) at the beginning of the association domain of the δ-CaMKII isoform, is responsible for the targeting the protein to the nucleus. In line with the idea of a role of CaMKII, microinjection of specific CaMKII peptide inhibitors into sea urchin zygotes after pronuclear fusion delays or even prevents NEB (Baitinger et al., 1990). Experiments in several cell types, e.g. sea urchin eggs, C127 cells, and the fungus *Aspergillus nidulans* (Lu and Means, 1993), strongly indicate that the cell cycle step controlled by CaM is the G$_1$/S transition; this idea comes

from the transient overexpression of CaM antisense RNA (see above), and from the use of CaM mutants: loss of CaM function in all cases leads to arrest in G$_2$ or in G$_1$/S (Rasmussen et al., 1992). As discussed in the section on DNA replication, phosphorylation (and inactivation) of the retinoblastoma protein (pRB) promotes entry into S phase (Riley et al., 1994), due to the dissociation of the heterodimer of the pRB protein with the transcription factor E2F-1; the latter then activates the transcription of genes whose products are required for S-phase progression (Pearson, et al., 1991; Dou et al., 1992). The finding that overexpression of a constitutively active form of that kinase, i.e. a form that is permanently active in the absence of CaM, arrests a mouse cell line and *Aspergillus nidulans* in G$_2$ (Planas-Silva and Means, 1992) is at variance with the suggestion that CaMKII mediates the cell cycle effects of CaM. However, the G$_2$ arrest in this case must be due to some other phosphorylation step promoted by CaMKII (Lu and Means, 1993), since the mitotic kinase is indeed active in these cells. When mitosis is arrested at G$_2$ by the decreased function of calmodulin, two mitotic kinases become inactivated: p34^{cdc2} in the tyrosine phosphorylated state, because a phosphatase which is a target of CaMKII and which should dephosphorylate it, cdc25, is blocked in the inactive dephosphorylated state (Whitaker, 1995); and NIMA, which is a direct substrate of CaMKII, in the dephosphorylated state. CaMKII reactivates both, strongly indicating Ser/Thr phosphorylation as a mechanism for their reactivation (Means, 1994).

Numerous studies suggest a role for calcium- and calmodulin-dependent kinases in meiosis as well. Oocytes of molluscs or bivalves are naturally arrested in the metaphase of the first maturation division. In these oocytes, a large transient increase of Ca^{2+} follows the addition of spermatozoa. A Ca^{2+}/CaM-dependent kinase step, in this case catalysed by another CaM kinase isoform, CaMKIII, is required to initiate the metaphase–anaphase transition: i.e. inactivation of the M-phase promoting factor (MPF), which is composed of the p34^{cdc2} kinase and a cyclin B regulatory subunit, and completion of meiosis (Abdelmajid et al., 1993; Deguchi and Osanai, 1994). Phosphorylation of elongation factor 2 (eEF-2), which is a substrate of CaMKIII (Nairn et al., 1985), results in the decrease of the rate of de novo protein synthesis (Ryazanov and Spirin, 1990) and thus rapidly inactivates MPF, since its cyclin component disappears (Dubé and Dufresne, 1990; Colas et al., 1992). A Ca^{2+} spike is also associated with the progression of meiotic anaphase II in mouse oocytes arrested at metaphase II (Tombes et al., 1992). CaMKII has been claimed to be involved in the progression of the meiotic cycle in *Xenopus* oocytes (Lorca et al., 1993). During fertilization, these oocytes are arrested at the second meiotic metaphase by a cytostatic factor (CSF) which prevents the degradation of cyclins and thus the inactivation of the MPF. The transient increase in cytoplasmic Ca^{2+} at fertilization inactivates both CSF and MPF, releasing the eggs from the meiotic metaphase arrest. Specific inhibitors of CaMKII prevent the Ca^{2+}-promoted cyclin degradation in cell-free extracts, and the direct microinjection of constitutively active CaMKII into unfertilized eggs inactivates MPF and CSF even in the absence of a calcium transient (Lorca et al., 1993).

Starfish oocytes offer advantages to the study of the function of Ca^{2+}/CaM in meiosis, since the hormone 1-methyladenine (Kanatani et al., 1969), which generates a plasma membrane signalling cascade linked to G-proteins (Yoshikuni et al., 1988; Jaffe et al., 1993), reinitiates meiosis after the arrest at the prophase of the first maturation division. Although it has been claimed that the breakdown of the germinal vesicle (GVBD) in these oocytes is not calcium-dependent (Witchel and Steinhardt, 1990; Stricker et al., 1994; Stricker, 1995; Whitaker, 1995), recent work (Santella and Kyozuka, 1994) has shown that a rise in nuclear free calcium is actually required for it: i.e. the intranuclear injection of BAPTA blocks the disassembly of the nuclear envelope. Mobilization of intracellular calcium has also been shown to initiate GVBD in the oocytes of mouse and pig (DeFelici et al., 1991; Kaufman and Homa, 1993). Mouse oocytes exhibit fluorescence oscillations in the nuclear region, supporting the involvement of nuclear calcium in the meiotic process (Lefèvre et al., 1995).

Changes in the Ca^{2+} concentration could also regulate several events involved in spindle formation and function (Hepler, 1994), a process which is not nuclear but which, in a sense, has its origin in the nucleus. Elevated levels of Ca^{2+} are required for microtubule depolymerization and for the movement of the chromosomes to the spindle poles (Ratan et al., 1988; Tombes and Borisy, 1989; Zhang et al., 1992). Direct demonstration that Ca^{2+} participates in cleavage furrow formation comes from work on medaka fish eggs in which a Ca^{2+} wave initiates near the mitotic apparatus and spreads over the cell (Fluck et al., 1991). Experiments in which BAPTA injection inhibits cytokinesis in *Xenopus* embryos and plants (Miller et al., 1993; Jurgens et al., 1994) are in line with this.

Also, these effects of Ca^{2+} may be mediated by Ca^{2+}-binding proteins: the binding of chromosomes to the mitotic spindle is initiated by the phosphorylation of lamin B by the main kinase $p34^{cdc2}$ and protein kinase C. Two other Ca^{2+}-binding proteins which are major substrates for tyrosine and serine kinases, annexin II (calpactin I) and CAP-50, have

also been localized to the nucleus and proposed to play a role in the regulation of the cytoskeletal structure during the cell cycle.

Nuclear Calmodulin in RNA Processing

This topic is still in its infancy. The functional significance of the observations that have been made has far-reaching potential implications, but at the moment they are still in the realm of speculation. The basic finding is the phosphorylation of several proteins of the heterogeneous ribonucleoprotein particles (hnRNPs) by casein kinase II (CKII) in nuclear extract from rat liver cells, and the inhibition of their phosphorylation by CaM (Bosser et al., 1993, 1995). Two of these proteins (A2, also called p36, and C) bind CaM, and the suggestion has been put forward that their association with CaM would block the CK II phosphorylation site, either directly or through other CaM-binding proteins. The interest of these findings is heightened by the demonstration, obtained by immunoelectronmicroscopy experiments, that such interactions also occur in vivo. The work on the hnRNPs has now been extended to the nucleus of starfish oocytes, where these particles are much better defined (Santella, 1996). Figure 20.1 shows the colocalization of one of these proteins, A2, and of CaM in the hnRNPs of a prophase-arrested oocyte (Fig. 20.1A and B); interestingly, when the oocytes are induced to resume meiosis by the hormone 1-methyladenine (see the section on the cell cycle), the particles disintegrate, and the A2 protein and CaM translocate to the cytosol across the still intact nuclear envelope (Fig. 20.1C).

The hnRNP proteins bind to nascent RNA transcripts of RNA polymerase II and influence their structure, facilitating their interaction with other molecules needed for the processing of pre-mRNA (Dreyfuss et al., 1993). The transport of mRNA to the cytoplasm has also been proposed as a role for some of the hnRNP proteins (Pinõl-Roma and Dreyfuss, 1991). The finding that CaM specifically binds to some of these proteins, preventing their phosphorylation, is at the moment only an interesting observation, but it is certainly compatible with the suggestion that nuclear CaM could play a role in mRNA processing (Bosser et al., 1995).

The work on starfish oocytes has also led to one additional observation of high potential interest. In the dormant oocytes, nuclear CaM is weakly associated with the nucleolus. After the addition of 1-methyladenine, at the time when the hnRNP particles disappear from the nucleoplasm, CaM and at least one of the two CaM-binding proteins of the hnRNPs, the A2 protein, become prominently associated with the nucleolus (Fig. 20.4D). The nucleolus is a complex structure consisting of

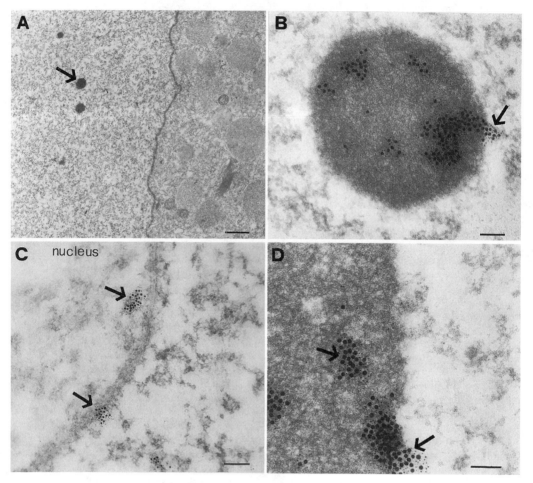

Figure 20.1 Immunogold labelling of starfish oocytes using antibodies against calmodulin and the A2 protein of the hnRNPs. (A) A prophase-arrested oocyte showing electron-dense bodies (hnRNPs) scattered in the nucleoplasm (arrow). Bar = 1 mm. (B) Higher magnification of a hnRNP particle. The antibodies against calmodulin (large gold particles) and against the A2 protein (small gold particles) are clustered on the electron-dense particle (arrow). Bar = 100 nm. (C) 10 minutes after treatment with the hormone the hnRNPs disappear from the nucleoplasm. The two types of gold particles translocate to the matrix milieu and to the cytosol through the still intact nuclear envelope (arrows). Bar = 200 nm. (D) Portion of a nucleolus of an oocyte 10 minutes after treatment with 1-MA, showing labelling with the antibodies against calmodulin and the A2 protein (arrows). Nucleoli of prophase-arrested oocytes not treated with 1-MA were only labelled with antibodies against the A2 protein and against calmodulin to a very minor extent (not shown). The upper portion of the panel shows the chromatin in the nuclear matrix without labelling. Bar = 100 nm. (Modified from Santella and Kyozuka, 1997.)

(1) a dense fibrillar region, (2) fibrillar centres, (3) a granular region, (4) nucleolar vacuoles, and (5) condensed nucleolar chromatin. It presides over ribosomal gene transcription, ribosomal RNA processing, and preribosomal particle formation. The finding of CaM associated with the nucleolus may be taken as an indication of the participation of CaM in these processes.

The Ca^{2+}-Dependent Neutral Protease Calpain in the Nucleus

Calpain has been frequently proposed as one of the mediators of apoptosis (see Squier et al., 1994, for a recent contribution). This is a plausible suggestion, considering its ability to induce membrane and cytoskeletal alterations in a number of cell pathology conditions (see Patel et al., 1996, for a recent review).

Calpain is a heterodimeric cysteine protease consisting of a larger (80 kDa) catalytic subunit and a smaller (30 kDa) regulatory subunit: both subunits contain a domain with strong homology to calmodulin, which is normally assumed to be responsible for the Ca^{2+}-sensitivity of the enzyme (for a recent review, see Saido et al., 1994). Two isoforms of calpain have long been known, the m and the μ isoforms, which are optimally stimulated by mmolar and μmolar Ca^{2+}, respectively. A protein inhibitor of calpain, calpastatin, which probably controls the activity of calpain in vivo, has also been described. A distinctive feature of calpain is that it does not digest protein substrates completely, but degrades them in a limited way, inducing functional and/or structural alterations which, in the latter case, may make them more susceptible to other proteases. Since it contains a CaM-like sequence, calpain preferentially degrades substrates which contain calmodulin-binding domains (Molinari et al., 1995) and, interestingly, a number of proteins that are localized in the nucleus, among them lamins (Traub et al., 1988) and microtubule-associated proteins (Takahashi, 1990). Together with α-spectrin and myosin light chain kinase, which are also preferred calpain substrates, the latter are components of the karyoskeleton, indicating a possible role of calpain in the structural organization of the nucleus. This is also supported by the finding of a Ca^{2+}-dependent protease, which is likely to be calpain, in the nuclear scaffold of hepatocytes (Tökés and Clawson 1990).

One problem with the role of calpains in physiology and pathology is their requirement, particularly in the case of m-calpain, for nonphysiological Ca^{2+} concentrations. Considerable evidence nevertheless indicates that they do act in vivo, particularly in the regulation of the cell cycle. Thus, m-calpain relocates among PtK cell compartments during mitosis, i.e. it moves from the plasma membrane to the mitotic chromosomes and to a location around the reforming nuclei as the cell cycle progresses from the interphase to telophase (Schollmeyer, 1988). These results, however, are at variance with the findings on A431 cells, which do not show binding of m-calpain to chromosomes during mitosis (Lane et al., 1992). m-Calpain promotes the onset of metaphase when microinjected near the nucleus of PtK1 cells, and causes the precocious disassembly of the mitotic spindle at the onset of anaphase. The simultaneous microinjection of calpastatin removes the effect of calpain (Schollmeyer, 1988). A possible way out of the difficulty of reconciling the apparently documented action of calpains in vivo with their requirement for very high Ca^{2+} concentrations is provided by a series of studies on the proteolysis of high-molecular-weight proteins (60–200 kDa) in the matrix of rat liver nuclei by m-calpain (Mellgren, 1991): proteoly-

sis occurs in the presence of as little as 3 mM Ca^{2+}, provided that DNA is present (Mellgren et al., 1993). Treatment of the nuclei with DNase I eliminates the activity at low Ca^{2+} concentrations (RNase has no effect). Since DNA fails to alter the high Ca^{2+} requirement for the proteolysis of casein and other proteins by m-calpain, its effect is evidently specific for the high-molecular-weight matrix protein. Their function is presently unknown, but a number of them are capable of binding DNA, suggesting a complex interaction between m-calpain, DNA, and the high-molecular-weight DNA-binding matrix proteins. This interaction would be responsible for the significant enhancement of the sensitivity of m-calpain to Ca^{2+} (Mellgren et al., 1993).

A recent exciting development in the area of calpain and the nucleus is the discovery that a novel calpain isoform, termed p94, contains a potential nuclear localization sequence (Sorimachi et al., 1989) and is indeed found to be specifically located in the nuclear envelope and in the nucleus (Sorimachi et al., 1993). μ-Calpain had been previously detected in the nucleus of c-33A cervical carcinoma cells by immunocytochemical work (Lane et al., 1992), and μ-calpain had been found to be transported into the nuclei in an ATP-dependent process when added to digitonin-permeabilized A431 cells (Mellgren and Lin, 1994). What makes p94, which is specifically expressed in muscle, particularly interesting is the finding that it becomes apparently active in the cytosol immediately after translation, but then disappears very rapidly through autolysis unless stabilized by connectin (Sorimachi et al., 1993). The latter is a very large filamentous protein of the sarcomere, which apparently replaces the regulatory 30 kDa subunit in the case of p94 (p94 is a member of a new family of atypical calpains, which are monomers of a large catalytic subunit). Connectin binds to the domain of p94 that contains a nuclear localization sequence, which would be inconsistent with the nuclear translocation of the protease, but the suggestion has been put forward that a fragment of p94 moves from connectin to the nucleus after its autolysis (Sorimachi et al., 1993). Although the issue still has unclear aspects, p94 somehow ends up in the nucleus, and is detected there, while disappearing from the cytosol, when expressed in COS cells (Sorimachi et al., 1993).

This relatively long discussion on p94 would have been unjustified had it not been for a very exciting recent finding by Richard et al. (1995). Mutations in the p94 gene that presumably destroy the proteolytic activity of the protein cause limb-girdle muscular dystrophy type 2A. Since the work described above strongly indicates that the natural site of action of p94 is the nuclear compartment, it is tempting to suggest that the dystrophy results from the inability of p94 to proteolyse one or more nuclear proteins. Thus, p94 would be a regulatory rather than a purely degradative enzyme, participating in the proteolytic activation of (nuclear) proteins essential for muscle function. This is reminiscent of the proposed regulation of transcription by restricted proteolysis of c-*jun* and c-*fos* (Hirai et al., 1991), and of the IκB-α inhibitor of NF-κB activities (Miyamoto et al., 1994) by m and/or μ-calpains.

A Putative Ca^{2+}-Regulated Motility System in the Nucleus

The first indication for the existence of a Ca^{2+}/CaM-modulated motility system in the nucleus was the discovery that MLCK, an enzyme that regulates the contraction of actin–myosin filaments in nonmuscle cells, was present in the nuclei of different cell types (Simmen et al., 1984; Bachs et al., 1990; Pujol et al., 1993) MLCK is a calmodulin-modulated kinase which phosphorylates the light chains of myosin, relieving the inhibition of the actin-stimulated ATPase of myosin by the regulatory light chains, allowing the sliding of the actin filaments along the myosin filaments, i.e. contraction, to occur; myosin and actin (Bremer et al., 1981; De Boni, 1994;) have been documented in the nucleus, and the suggestion has been made that nuclear actin, which is organized in filaments, could perform intranuclear contractile movements (Kumar et al., 1984). A second CaM-binding protein which is related to the actin–myosin motility system, caldesmon, has also been identified in the nucleus (Bachs et al., 1990). Caldesmon binds to actin filaments in the absence of Ca^{2+} and CaM, hindering the binding of actin to myosin. CaM binding to caldesmon releases it from actin, permitting myosin to bind to it and allowing the contraction of the filament. The components of a potential actin–myosin contractile system are thus all present in the nucleus: what could the function of such a system be? Although no conclusive data are available, suggestions have been made. For example, an actin structural scaffold in the nucleoplasm has been proposed (Clark and Rosenbaum, 1979). Or, it has been suggested that actin could promote chromosome condensation in mitosis and meiosis, and the formation of heterochromatin in interphase (Scheer et al., 1984). Possibly, actin could control nucleocytoplasmic transport through the pores of the envelope (Schindler and Jiang, 1990), act as a transcription initiation factor (Egly et al., 1984), and play a role in the processing of RNA (Sahlas et al., 1993).

α-Spectrin, a CaM-binding protein which is the anchor for actin in the plasma membrane, has also been surprisingly detected in the nuclear matrix and the envelope (Bachs et al., 1990) of several cell types. By analogy with its function in the plasma membrane, it could act as an anchor for actin to the inner side of the envelope.

The Ca^{2+}-related nuclear processes discussed in the previous sections are those on which most of the interest of the Ca^{2+}/CaM field is presently concentrated. They are also the processes for which either reasonably detailed information and/or plausible functional proposal have been made. Still, a selection was made and it could be argued that the nucleus contains additional Ca^{2+}/CaM targets of potential interest, which would have also deserved specific discussion. This may well be true. The area of nuclear calcium is expanding very rapidly, and an update of this chapter a few years from now will certainly demand the inclusion of topics which have been omitted in the present version: among them, the numerous Ca^{2+}-binding and/or Ca^{2+}-triggered proteins different from calmodulin which are being increasingly described in the nucleus, e.g. S100 proteins, annexins, calreticulin, gelsolin/severin-related proteins, and calnexin; or, the nuclear targets of calmodulin which are still unidentified and/or orphan of a function; or, finally, protein kinase C, whose function in the nucleus is still mysterious, but which is known to translocate to it. Protein kinase C has been mentioned a couple of times in this chapter, and it was felt that this is sufficient at the present state of knowledge. It can be safely predicted, however, that protein kinase C will soon become an important actor in nuclear Ca^{2+} research.

The Regulation of Nuclear Calcium Concentration

The Nucleoplasm Is Separated from the Cytosol by Two Membranes

The nucleus is surrounded by the two membranes of the nuclear envelope, whose lipid composition is very similar to that of endoplasmic reticulum (Fig. 20.2) (James et al., 1981). The membrane is continuous with those membranes of the endoplasmic reticulum (ER) and shares many characteristics with that structure: for instance, it bears ribosomes (Kirschner et al., 1977). Between the inner and outer nuclear membranes is the intermembrane space, which is of the order of 40 nm across and is continuous with the lumen of the ER. Inner and outer nuclear membranes come together at the nuclear pores, where macromolecules enter and leave the nucleus. In electron microscopy preparations stained for lipid, the nuclear pore appears as a channel almost 100 nm in diameter, linking the nucleoplasm and cytosol. In 1975, Paine et al. studied the ease with which dextrans of different molecular weight passed from the cytosol to nucleoplasm of amphibian oocytes and concluded that the

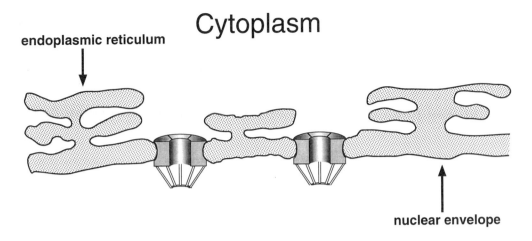

Cytoplasm

endoplasmic reticulum

nuclear envelope

Nucleoplasm

Figure 20.2 A scheme of the nuclear envelope, showing its derivation from the endoplasmic reticulum and the nuclear pores, depicted in the fully opened state.

nuclear pore acted as if it were a hole 9 nm in diameter and 15 nm long. Clearly, something was occluding much of the pore, and this has been revealed to be the nuclear pore complex, a multicomponent structure of more than a hundred different proteins with, possibly, some RNA. Although most macromolecules of MW > 50,000 Da cannot pass through nuclear pores, RNAs and proteins that bear a nuclear targeting signal are recognised by the pore machinery and transported through in an ATP-dependent manner. In sharp contrast to the system that allows passage of protein through the double-membrane envelope of the mitochondrion, proteins do not unfold before they traverse the nuclear pore but pass through in their normal three-dimensional folded state, implying that the pore has an extraordinary ability to expand during active transport.

When not transporting macromolecules, the nuclear pore complex resembles a wagon wheel. The central hole, where the axle would go through the wheel, is thought to represent the 9 nm diameter hole deduced by Paine et al. (1975). Eight spokes lead to the rim which lies against and within the lipid membrane of the pore. Recently, electron microscopy has revealed a second component of the nuclear pore complex that lies on the nuclear side of the wagon wheel. Again, this "pepperpot" has 8-fold symmetry and comprises a central cap or plug connected to the rim of the wagon wheel by eight struts (Davis, 1995).

If this organization persists when nuclei are isolated from the cell and suspended in artificial media,

then the nucleoplasm of isolated nuclei should freely exchange solutes of less than 50,000 Da with the bathing solution. Only the intermembrane space would form an independent, membrane-bound compartment. Gerasimenko et al. (1995) obtained exactly this result. Bathing isolated liver nuclei in the membrane-permeable form of fura-2, fura-2 AM, caused the dye to accumulate in the intermembrane space, while fura-2 dextran in the bathing medium moved into the nucleoplasm. However, a number of laboratories have reported results that are more consistent with the hypothesis that the pores shut when nuclei are isolated. Nicotera and Rossi (1994) bathed isolated liver nuclei in the fura-2 AM and saw a uniform fluorescence across the nucleus, implying that the dye was contained within, and could not leave, the nucleoplasm. The groups of both DeFelice and Bustamante have sealed glass micropipettes to the nuclear surface. For significant stretches of the record, no current flows (Mazzanti et al., 1990; Bustamante, 1992; Dale et al., 1994). DeFelice and Bustamante both argue that the density of nuclear pores is so high that several pores must be present in the area of membrane sampled, and therefore in these isolated nuclei the pores must, for a significant fraction of time, exist in a form that is impermeable to ions.

At other times, typical single-channel currents are recorded from patches of nuclear envelope. While many scientists assume that the channels recorded lie in the outer nuclear membrane, and allow ion flow between cytosol and intermembrane space (e.g.

Stehno-Bittel et al., 1995), Bustamante (1992) believes that these represent the opening of the nuclear pore, allowing ion flow between cytosol and nucleoplasm. He found that these channels close in the presence of a peptide inhibitor of protein kinase A; however, introduction of this inhibitor at higher concentration does not slow the passage of calcium ions across the nuclear envelope of intact cells, inconsistent with the identification of the channels as nuclear pores (Parkinson et al., 1996).

Can the nuclear pores shut under physiological conditions, when the nuclei are in situ and bathed in cytosol? Oocytes, large cells which arrest in a dormant state to await hormonal activation, have a very high density of nuclear pores but carry out little protein synthesis. Hydrophilic dyes injected into the nucleus of intact oocytes of starfish or *Xenopus* do not pass out into the cytosol, indicating that in these cells the nuclear pores are shut (Santella and Kyozuka, 1994; Hennager et al., 1995). Closure of the nuclear pore is not a phenomenon restricted to dormant cells. In kidney and HeLa cell lines, depletion of the calcium content of the endoplasmic reticulum (and of the intermembrane space, which is continuous with it) stops movement through the nuclear pore. Both the ATP-dependent active transport of proteins bearing a nuclear localization signal, and the passive diffusion of a 10 kDa dextran, are reversibly blocked by ER calcium depletion. The appearance of the pore viewed in the electron microscope does not change when it shuts; in particular, fusion between the inner and outer nuclear membranes is maintained. One of the transmembrane proteins that contributes to the nuclear pore complex, gp210, contains potential EF-hand calcium-binding domains, suggesting that these domains must bind calcium in order for the pore to be open (Greber and Gerace, 1995). In a similar study, Stehno-Bittel et al. (1995), instead, depleted the Ca^{2+} stores of isolated nuclei reaching similar conclusions. More recently, Perez-Terzic et al. (1997), using the atomic force microscope, observed a "plug" in the nuclear pores when their permeability became restricted by the emptying of the nuclear Ca^{2+} stores.

Methodologies and Problems in Nuclear Calcium Measurement

Each of the laboratories working in the field of nuclear calcium uses different cells, indicators, and measurement techniques, so it is perhaps not surprising that the results obtained are often markedly different. However, different workers often also differ greatly in the intellectual framework within which they analyse their data. We cannot hope to resolve these differences in this chapter, but we do hope to

lay down a common framework for assessing different ent laboratories' data.

Confocal Versus Conventional Microscopy

Many measurements of nucleoplasmic calcium concentration ($[Ca^{2+}]_N$) have used a fluorescent indicator dye that is present in both cytoplasm and nucleus. Conventional fluorescent microscopes collect information from the entire thickness of the preparation, so that where the nucleus is entirely surrounded by fluorescent cytoplasm, the measurement of uncontaminated nuclear signal is impossible (Silver et al., 1992). Some cells have such a thin layer of cytoplasm over the nucleus that a conventional fluorescence microscope can give reasonable estimates of $[Ca^{2+}]_N$: osteoclasts and hepatocytes have been studied in this manner (Shankar et al., 1993; Lin et al., 1994). For most other cells, calcium indicators that equilibrate between the cytosol and the nucleoplasm can only measure $[Ca^{2+}]_N$ accurately if a confocal microscope is used to reject light emitted from cytosolic dye above and below the nucleus. Clearly, this problem does not arise if the dye or other calcium indicator is present only in the nucleus. As with any complex piece of equipment, one should never accept the manufacturer's claims about resolution, so it is necessary to test that signals generated in the cytosol do not contaminate the nuclear signal. Where papers do show confocal images of fluorescence that should be confined to the cytoplasm, the nuclei appear reassuringly dark (Gillot and Whitaker, 1994; Al-Mohanna et al., 1994).

Acetoxymethyl Ester Loading

The explosion of knowledge about the role of calcium as an intracellular messenger has, in large part, been due to the ease with which cells can be loaded with fluorescent indicators by bathing them in the membrane-permeable acetoxymethyl (AM) ester of the dye. Once the dye has crossed the membranes, it is hydrolysed by endogenous esterases to release the hydrophilic, membrane-impermeant, calcium-sensitive molecule (Tsien, 1981). The AM ester can cross internal membranes as well as the plasmalemma, and so the loaded cell will contain dye in membrane-bound internal spaces, such as the endoplasmic reticulum and nuclear intermembrane space, as well as in the cytosol. Serious misinterpretation can arise if the signal from the cytoplasm—that is, the entire volume outside the nucleus—is assumed, wrongly, to come entirely from cytosolic dye (Connor, 1993). The fraction of dye in membrane-bound organelles can be estimated by sequentially lysing different cell membranes with increasing concentrations of detergent and measuring the fraction of dye that is lost. When done properly, this approach requires biochemical analysis of released proteins as markers for the lysis of specific organelles—and therefore can become more time-consuming than the calcium measurements themselves.

Isolated nuclei can be loaded with dye by bathing in AM ester. Esterases are likely to be active in both the nucleoplasm and the intermembrane space; however, if the nuclear pores are open, nucleoplasmic dye will be lost when the nuclei are rinsed with dye-free medium, leaving only the dye in the intermembrane space. Nicotera and Rossi (1994) and Gerasimenko et al. (1995) used this technique on liver nuclei, using confocal microscopy to examine the loaded nuclei. While Gerasimenko et al. found that the trapped dye was indeed located to the intermembrane space, Nicotera and Rossi saw a uniform fluorescence across the nucleus, implying that the dye was also contained within the nucleoplasm and that the nuclear pores were shut. These different conclusions as to the location of the dye lead the two groups to different interpretations of the calcium fluxes induced by intracellular messengers in these nuclei.

Injection of Untargeted Dye

Some of the problems associated with AM loading can be avoided by injecting hydrophilic dye directly into the cell from a sharp micropipette or patch pipette. Low-molecular-weight anionic dyes can be taken up into organelles by carrier-mediated transport, leading to problems of the same type as those seen after AM loading, but these can be avoided by the use of pharmacological inhibitors of anion transport or by using dyes conjugated to high-molecular-weight dextrans (DiVirgilio et al., 1990; Al-Mohanna et al., 1994). In all cells studied, the nuclear region fluoresces brightly after introduction of dye into the cytosol, implying that the dye moves through the nuclear pores into the nucleoplasm. However, results from starfish and *Xenopus* oocytes are inconsistent with this simple interpretation. When fura-2 is injected directly into the nucleus, then it does not leave but remains in the nucleus (Santella and Kyozuka, 1994; Hennager et al., 1995). If an injected starfish nucleus is subsequently removed from the cell, it continues to trap the dye. If, on the other hand, fura-2 is injected into the cytosol, the nuclear region glows brightly, but when the nucleus is removed from the cell it is seen to be dark; that is, it retains no dye. There are two possible explanations of these data, which lead to very different interpretations of subsequent $[Ca^{2+}]$ measurements. In the first explanation, the nuclear pores are impermeable to dye in these dormant cells. When dye is injected into the nucleus, it is trapped, and remains trapped when the nucleus is removed from the cell. When dye

is injected into the cytosol, it cannot enter the nucleus, so that when the nucleus is removed from the cell it is seen to be dark. In this model, the bright glow of the nuclear region upon injection of dye into the cytosol is not produced by nucleoplasmic dye but by dye in the perinuclear space, which in these cells is free of organelles that exclude the dye. In this model, which may apply to other cells too, fluorescence from the nuclear region of cells in which dye has been introduced into the cytosol is not, as is usually assumed, reporting $[Ca^{2+}]_N$ but instead reports $[Ca^{2+}]$ in the cytosol immediately outside the nucleus. In the second explanation, the nuclear pores are normally permeable to dye. When dye is injected into the cytosol, it enters the nucleus, and this is the source of the bright signal from the nuclear region. Upon removal of the nucleus from the cell, the dye leaves through the open pores, and the nucleus appears dark. In this model, the act of injecting dye into the nucleus somehow causes the nuclear pores to become impermeant to dye, so that when the nucleus is removed from the cytosol it glows brightly. To further complicate matters, impermeability to dye need not imply impermeability to calcium. Injected nuclei of *Xenopus* oocytes retain dye, but calcium ions can cross from cytosol to nucleoplasm within seconds (Hennager et al., 1995). Resolving these questions clearly requires further work, but whatever the explanation of these results it is clear that the nuclear pore is a much more plastic, adaptable structure than is normally assumed.

Targeted Indicators

Molecules bearing a nuclear localization signal are recognized by the nuclear pore complex and transported into the nucleoplasm. This approach has been used with the calcium-sensitive photoprotein aequorin by the groups of Rizzuto and Campbell (Brini et al., 1993; Badminton et al., 1995). Cells transfected with recombinant DNA express a chimeric protein comprising aequorin plus the localization signal, which therefore accumulates in the nucleus. Addition of the cofactor coelentrazine generates the active photoprotein, which reports $[Ca^{2+}]_N$. Allbritton et al. (1994) have, in contrast, attached peptides that contain a nuclear localization signal to calcium green dextran. Upon injection into the cytosol, the dextran accumulates in the nucleoplasm and reports $[Ca^{2+}]_N$. These time-consuming approaches are carried out in only a few specialist laboratories, but give results that are enormously useful in confirming or throwing into doubt results obtained using commercially available indicators.

Calcium Movement Into and Out of the Intermembrane Space

Calcium-ATPase

The outer nuclear membrane is continuous with the membrane of the endoplasmic reticulum and bears a calcium-ATPase that appears to be identical to the pump on the endoplasmic reticulum (Lanini et al., 1992; Humbert et al, 1996). This pump will take up calcium from a bathing solution with 100 nM free calcium when given ATP (Nicotera and Rossi, 1994; Gerasimenko et al., 1995; Humbert et al, 1996). In Fig. 20.3, we show the ATPase on both the inner and outer nuclear membranes, but recent work by Humbert et al. (1996) indicates that, in fact, it may be absent from the inner nuclear membrane of rat liver cells.

InsP$_3$ Receptors

The outer nuclear membrane of some cells, like the endoplasmic reticulum membrane, bears InsP$_3$ receptors that can release calcium from the intermembrane space into the cytosol (Fig. 20.3) (Allbritton et al., 1994; Stehno-Bittel et al., 1995).

More interesting is the question of InsP$_3$ receptors in the inner nuclear membrane. A number of laboratories have reported InsP$_3$ receptors on the inner nuclear membrane, where they might function to release calcium from the intermembrane space to the nucleoplasm. In particular, Humbert et al. (1996) found that the vesicles of inner, but not of outer, nuclear membrane from liver immunostained for InsP$_3$ receptor, and released calcium upon InsP$_3$ application. There are therefore two pathways by which InsP$_3$ can increase the calcium concentration of the nucleoplasm. In the first pathway, InsP$_3$ in the cytosol causes calcium release from the endoplasmic reticulum (and the outer nuclear membrane if this bears InsP$_3$ receptors), and some of this calcium passes through the nuclear pores into the nucleoplasm. In the second pathway, InsP$_3$ in the nucleoplasm causes calcium release from the inner nuclear membrane directly into the nucleoplasm. Allbritton et al. (1994) have shown clearly that the $[Ca^{2+}]_N$ increase that results when antigen causes InsP$_3$ to be generated in the cytosol of leukocytes is entirely the result of calcium released into the cytosol, some of which then finds its way to the nucleus. A similar result was obtained by Hennager et al. (1995) in *Xenopus* oocytes: injection of InsP$_3$ into the cytosol caused a $[Ca^{2+}]_N$ increase, and the presence of the InsP$_3$ receptor antagonist heparin in the cytosol entirely blocked this increase. However, Hennager et al. (1995) also showed that injection of InsP$_3$ into the nucleus caused a $[Ca^{2+}]_N$ increase that was entirely blocked by the presence of heparin in the nucleus, and starfish oocytes behave similarly. Isolated liver nuclei bathed in calcium buffer to remove any calcium ions released outward through the outer nuclear membrane show a transient $[Ca^{2+}]_N$ increase upon stimulation with InsP$_3$, again implying that there are functional InsP$_3$-receptors on the inner nuclear membrane (Gerasimenko et al., 1995). Thus, InsP$_3$ receptors are present on the inner nuclear membrane, and will respond when InsP$_3$ is supplied to the nucleoplasm, but seem to play no role in the response of the cell to the well-characterized signalling pathways that hydrolyse PIP$_2$ at the plasmalemma to release InsP$_3$ into the cytosol.

Inositol Tetrakisphosphate Receptors

The role and molecular biology of InsP$_4$ receptors in cell biology is unclear, so there is little context in which to place results obtained on nuclei. Humbert et al. (1996) found InsP$_4$-binding sites on both the inner and outer nuclear membranes of liver cells. The inner membrane sites were of low affinity ($K_{1/2} = 61$ nM) and there is no evidence that they are involved in calcium transport. The InsP$_4$-binding sites on the outer nuclear membrane had high affinity ($K_{1/2} = 7$ nM) and, on binding InsP$_4$, allowed calcium ions to cross the membrane.

Ryanodine Receptors

The ryanodine receptor, like the InsP$_3$ receptor, is a calcium channel of the sarco/endoplasmic reticulum. Like the InsP$_3$ receptor, its open probability increases as $[Ca^{2+}]_C$ increases, and it mediates calcium-induced calcium release in a number of cells (Furuichi et al., 1994). Recent work has suggested that in at least some cells, cyclic ADP-ribose (cADPR) may be a natural agonist at the ryanodine receptor, filling a role homologous to InsP$_3$ at the InsP$_3$R (Galione et al., 1991). Both the inner and outer nuclear membranes of starfish oocytes label when incubated with antibodies to the ryanodine receptor (Fig. 20.4) and photorelease of cADPR in the nucleus increases $[Ca^{2+}]_N$, implying that ryanodine receptors on the inner nuclear membrane were activated, allowing calcium to leave the intermembrane space to pass into the nucleoplasm. Isolated liver nuclei show a $[Ca^{2+}]_N$ increase in response to both ryanodine and cADPR, once again implying that the channels on the inner nuclear membrane are functional (Fig. 20.3) (Gerasimenko et al., 1995). The biological function of these channels, including whether the physiological trigger to opening is a rise of $[Ca^{2+}]_N$ or a rise of cADPR, is at present unclear.

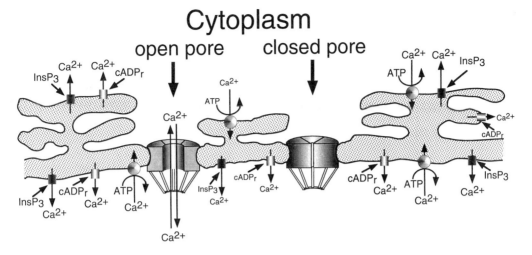

Cytoplasm

open pore   closed pore

Nucleoplasm

Figure 20.3 The nuclear envelope in the regulation of nuclear calcium. Pore complexes are represented in both the calcium-permeable and calcium-sealed states. The latter state is indicated graphically by the adjoining of the pore subunits. The model shows the calcium-transporting systems of the envelope: the Ca^{2+}-ATPase, the $InsP_3$-sensitive calcium-release channels, and the cyclic ADP-ribose (cADPR)-sensitive channel (ryanodine receptor), which are visualized on both the inner and outer membranes. The $InsP_4$-sensitive channel, which has been detected in the envelope (Humbert et al., 1996), has been omitted since its function is still obscure. We show the Ca^{2+}-ATPase as present on the inner nuclear membrane, but recent work by Humbert et al., (1996) suggests that, in fact, it is absent from the inner nuclear membrane of rat liver cells.

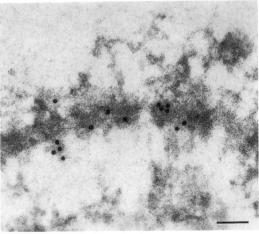

Figure 20.4 Immunogold labelling of the cADPR-sensitive Ca^{2+} channel with polyclonal antibodies against the skeletal muscle ryanodine receptor in both the inner and outer membrane of the nuclear envelope of prophase-arrested oocytes. Bar = 100 nm. (From Santella and Kyozuka, 1997.)

Modification of Cytosolic Calcium Transients at the Nuclear Envelope

Calcium Permeability of the Nuclear Pores

Given that under most circumstances the nuclear pore is freely permeable to solutes of up to 50 kDa, one would expect calcium ions to permeate the nuclear pore easily. Most experiments in which calcium transients have been initiated on one side of the nuclear envelope and calcium concentration monitored on the other side have shown a delay of less than a second in transmission of the calcium diffusion wave (Allbritton et al., 1994; Al-Mohanna et al., 1994; O'Malley, 1994). There is a possible problem in the interpretation of dye data. Introduction of calcium indicator dyes into the cytosol will speed the diffusion of calcium (Nowycky and Pinter, 1993). Kasai and Petersen (1994) have argued that calcium will diffuse a distance of only $4\,\mu$m in cells whose calcium-handling systems have not been perturbed by the experimenter, implying that the nucleus will be effectively isolated from the cytosol during calcium-signalling episodes of a physiological time course. In contradiction of this prediction, the few studies in which calcium indicator has been targeted to the nucleus, so that it is absent from the cytosol and will not affect cytosolic calcium handling, largely confirm the results obtained using untargeted dyes. Brini et al. (1993), using targeted aequorin in HeLa cells, and Allbritton et al. (1994) using targeted calcium green dextran in leukocytes, found that cytosolic calcium transients propagated rapidly to the nucleus.

In oocytes, calcium traffic across the nuclear envelope seems to be a little more sluggish. In sea urchin oocytes, the cytosolic $[Ca^{2+}]$ wave sweeps past the nucleus and $[Ca^{2+}]_N$ equilibrates within a few seconds (Gillot and Whitaker, 1994). In starfish oocytes, the hormone 1-methyladenine induces a $[Ca^{2+}]_C$ increase; $[Ca^{2+}]_N$ increases too, but with a lag of about 20 seconds (Santella and Kyozuka, 1994). In *Xenopus* oocytes, release of calcium from the endoplasmic reticulum into the cytosol causes $[Ca^{2+}]_N$ to rise over about 5 seconds (Hennager et al., 1995). Even this relatively slow transmission of the calcium signal across the nuclear envelope of starfish and *Xenopus* is perhaps surprising, since these nuclei retain the injected indicator dye for many minutes, implying that the nuclear pores are essentially impermeable to the dye. Indeed, the results of Hennager et al. (1995) imply that the nuclear envelope of *Xenopus* oocytes is also impermeable to InsP₃. Taken at face value, these findings imply that in these oocytes, although not in others (e.g. sea urchin; Gillot and Whitaker 1994), the nuclear pores are permeable to low-molecular-weight cations but block the passage of low-molecular-weight anions such as InsP₃ and calcium dyes.

A large number of laboratories have reported more or less stable calcium gradients across the nuclear envelope of nondormant cells: that is, in an unstimulated cell $[Ca^{2+}]_N$ is reported by an intracellular dye to be different in nucleoplasm and cytosol (Williams et al., 1985; Ishida et al., 1991; Himpens et al., 1992, 1994; Wahl et al., 1992). If these results are correct, then they imply either that there is a constant diffusion of calcium from a source in one compartment to a sink in the other, or that in these cells the nuclear envelope is a considerable barrier to the movement of calcium ions. However, these measurements are subject to many errors, including the effect of the very different chemical environments on dye behaviour (Konishi et al., 1988; Al-Mohanna et al., 1994), intracellular gradients of autofluorescence intensity (O'Malley, 1994), dye loading into organelles (Connor, 1993), and active dye uptake by organelles (DiVirgilio et al., 1990; Al-Mohanna et al., 1994). Until measurements of stable calcium gradients are confirmed using dextran-conjugated dyes that are injected into the cytosol or nucleoplasm and then calibrated in situ, they must be regarded as suggestive only.

Nuclear Calcium Amplification

Although many investigators have seen $[Ca^{2+}]_N$ to follow $[Ca^{2+}]_C$ with a delay of a few seconds or less (Gillot and Whitaker, 1993; Al-Mohanna et al., 1994; Allbritton et al., 1994; O'Malley, 1994), a number of laboratories have reported deviations from this rule. The deviations fall into two classes, which we shall call nuclear calcium amplification and nuclear calcium insulation. In 1990, Adams' laboratory reported the first measurements of transient $[Ca^{2+}]$ gradients using a confocal microscope (Hernandez-Cruz et al., 1990). When bullfrog sympathetic neurones were loaded with calcium indicator through a whole-cell patch pipette and then depolarised to cause a calcium influx across the plasmalemma, a calcium wave propagated radially inward and, upon reaching the nucleus, produced a larger fractional fluorescence change in the nucleus than in the cytoplasm, implying that the calcium change was greater in the nucleoplasm than in the cytosol. O'Malley (1994), working in Adams' laboratory, has shown that this result was an artifact that had its origin in the very different autofluorescence intensities of cytoplasm and nucleoplasm. After correction for local autofluorescence, fluo-3 indicated that $[Ca^{2+}]_N$ equilibrated with $[Ca^{2+}]_C$ after a delay of about 200 milliseconds; that is, the data were entirely consistent with passive calcium movements through the nuclear pore. In general, an apparent and artifactual nuclear calcium amplification will be produced whenever the fluorescence signal from cytosolic calcium indicator is contaminated by an unsuspected or uncorrected signal that is insensitive to $[Ca^{2+}]_C$. The dye itself can be the source of this $[Ca^{2+}]_C$-insensitive signal. Connor (1993) showed that fibroblasts loaded with fura-2 by the acetoxymethylester (AM) method (Tsien, 1981) contain much dye in the organelles. This dye does not report $[Ca^{2+}]_C$, and if its signal is not subtracted from that of cytosolic dye then ratio images give the erroneous impression that $[Ca^{2+}]$ changes are amplified in the nucleus. Fluo-3 injected into sea urchin oocytes and neurones is actively taken up by the organelles, once again causing the appearance of nuclear calcium amplification (Gillot and Whitaker, 1993; Al-Mohanna et al., 1994). In no case has the phenomenon of nuclear calcium amplification been observed when autofluorescence is independently measured in cytosol and nucleus and where dextran-conjugated dye, which cannot cross cell membranes either passively or actively, is used. We take this to be strong evidence that nuclear calcium amplification does not, in fact, occur in any cell. Rand et al. (1994), however, argue that the act of microinjecting dextran-conjugated dye damages cellular calcium handling, so that the absence of nuclear calcium amplification in injected cells is not proof that the phenomenon does not exist in unperturbed cells. This question will eventually be resolved by determining whether microinjection eliminates nuclear calcium amplification in cells preloaded with dye by the AM technique.

Nuclear Calcium Insulation

Al-Mohanna et al. (1994) loaded rat sensory neurones with fluo-3 through a whole-cell patch pipette and then depolarised the plasmalemma to cause a calcium influx across it. Transient increases of $[Ca^{2+}]_C$ of less than 300 nM amplitude propagated rapidly to the nucleoplasm, but $[Ca^{2+}]_C$ increases of greater than 300 nM amplitude were delayed for a few seconds at the nuclear envelope, so that by the time $[Ca^{2+}]_N$ and $[Ca^{2+}]_C$ equilibrated the calcium concentration in the cytosol had already fallen from its peak. More extreme examples of nuclear calcium insulation have been seen in COS and smooth muscle cells, although in each case the results are less clear cut. Badminton et al. (1995) used aequorin to study the $[Ca^{2+}]$ changes induced by ionophore and complement in COS cells. Both agonists raised $[Ca^{2+}]_C$, which remained high for about a minute. In different cells expressing nuclear targeted aequorin, the same agonists caused $[Ca^{2+}]_N$ changes with a similar time course but significantly smaller amplitude. Neylon et al. (1990) studied the $[Ca^{2+}]$ changes evoked in smooth muscle cells by thrombin. Waves of elevated

$[Ca^{2+}]$ propagated through the cytoplasm but did not enter the nuclei, even when the calcium concentration immediately outside the nuclei remained high for a minute. This result was obtained using AM-loaded dye, and must be treated with caution until repeated using dextran-conjugated dyes. Badminton et al. (1996) have measured cytoplasmic and nuclear $[Ca^{2+}]$ in HeLa cells using aequorin targeted to either the cytoplasm or the nucleus, and found that ATP, histamine, and ionomycin increased cytoplasmic $[Ca^{2+}]$ quantitatively more than nuclear $[Ca^{2+}]$. They found that the difference of $[Ca^{2+}]$ in the two compartments depended on the stimulus, and that the restoration of extracellular $[Ca^{2+}]$ after depletion of the internal store resulted only in increased cytosolic $[Ca^{2+}]$ without increase in nuclear $[Ca^{2+}]$. Taken together, these results imply that in a number of cells $[Ca^{2+}]_N$ can remain significantly below $[Ca^{2+}]_C$ for a period of many seconds. One possible explanation for these data is that the nuclear pores shut more or less completely when $[Ca^{2+}]_C$ is high. Alternatively, the nucleoplasm may contain a calcium buffer of low affinity but high capacity that takes up calcium when $[Ca^{2+}]_N$ rises above 300 nM, delaying equilibration. Whichever explanation is correct, this filtering of cytosolic calcium changes is likely to be important in determining the relationship between the pattern of electrical activity and gene expression. Some hint as to the complexity of control possible when calcium filtering is combined with multiple potential targets for raised $[Ca^{2+}]$ is suggested by the pattern of expression of the immediate early genes *zif-268* and c-*fos*. Although exceptions have been reported, the consensus view is that action potentials spaced at intervals too wide apart for any summation of $[Ca^{2+}]_i$ to occur do not up-regulate either gene. Strong stimulation of the type that will occur during induction of long-term potentiation up-regulates *zif-268*, while c-*fos* up-regulation seems to occur in intact brains only during pathologically sustained depolarization, such as occurs during epilepsy or ischaemia (Mackler et al., 1992; Labiner et al., 1993; Worley et al., 1993; but see Sheng et al., 1993). Figure 20.3 offers a pictorial summary of the traffic of $[Ca^{2+}]$ across the nuclear envelope. The diagram shown in Fig. 20.2 has been completed by adding to it the $[Ca^{2+}]$ transporters of the envelope discussed in the previous section. The diagram also visualizes the pores in both the closed and the fully permeable configuration.

The Biochemical Components of a Nuclear Calcium-Signalling Pathway are In Place

The classical pathway by which extracellular agonists activate intracellular calcium-dependent events is through the breakdown of phosphatidyl inositol bisphosphate (PIP_2) at the plasmalemma. PIP_2 breakdown produces diacyglycerol (DAG), which can activate protein kinase C, and $InsP_3$ which activates the $InsP_3$ receptor to release calcium from the endoplasmic reticulum. Some of this calcium may then diffuse into the nucleus, but, as discussed above, $InsP_3$ receptors on the inner nuclear membrane appear to play no role in the response of the cell to these agonists. Experiments on fibroblasts are beginning to hint at the presence of a second signalling pathway that will produce $InsP_3$ in the nucleoplasm, where it can activate $InsP_3$ receptors on the inner nuclear membrane. In fibroblasts, plasmalemmal PIP_2 breakdown can be activated by a number of agonists, including bombesin, which acts on G-protein-coupled receptors (Plevin et al., 1990). However, IGF-1, acting on a tyrosine kinase receptor, causes breakdown of PIP_2 to DAG in a nuclear fraction, with no PIP_2 breakdown at the plasmalemma (Divecha et al., 1991; Martelli et al. 1992). One of the downstream effects of nuclear PIP_2 breakdown is that phospholipase C is recruited to the nucleus (Divecha et al., 1991). It is therefore possible that the $InsP_3$ produced at the same time acts upon $InsP_3$ receptors on the inner nuclear membrane and causes a release of calcium from the intermembrane space into the nucleoplasm. Raised $[Ca^{2+}]_N$ could then act upon calmodulin-dependent processes, or in concert with DAG, to activate phospholipase C. At present, there is no direct evidence for a specifically nuclear $[Ca^{2+}]$ signal in response to any agonist, with one exception: Shankar et al. (1993), using AM-loaded dyes to study osteoclasts, found that integrin binding synthetic peptides and bone sialoprotein itself caused a $[Ca^{2+}]$ increase that was all but restricted to the nucleus, and which elicited cellular responses distinct from those produced by agonists such as calcitonin that raise $[Ca^{2+}]_C$ (Shankar et al., 1993, 1995). As discussed above, the interaction of the hormone 1-methyladenine with the plasma membrane of prophase-arrested starfish oocytes also induces a nuclear Ca^{2+} transient; however, this transient is not isolated, but coupled to other transients in the cytosol. We are at present investigating whether these phenomena do truly represent specific targeting of the calcium signal, or whether they are simply an example of the nuclear calcium amplification artifact described above.

Interest in the signalling pathway mediated by cADP-ribose is rapidly gaining momentum. The findings of Gerasimenko et al. (1995) on liver, and of Santella and Kyozuka (1997) on oocytes, have shown that the final effector of the pathway, the receptor sensitive to cADP-ribose, is present and functional on both membranes of the envelope (Fig. 20.4). Whether the nucleus also contains the meta-bolic pathway that transforms NAD into cADP-ribose is, however, still unknown.

Conclusions

The field of nuclear calcium is a young one. Biochemistry and molecular genetics are discovering that more and more processes within the nucleus are influenced by calcium. The control of $[Ca^{2+}]_N$ is not yet completely understood. Under most circumstances, $[Ca^{2+}]_N$ appears to follow $[Ca^{2+}]_C$ passively, but filtering at the nuclear envelope means that the time course of calcium changes, and therefore the particular downstream processes activated may be very different in cytosol and the nucleus. Recent work from fibroblasts suggests that a hitherto unsuspected signalling pathway may target calcium signals to the nucleus.

References

Abdelmajid, H., C. Leclerc-David, M. Moreau, P. Guerrier, and A. Ryazanov (1993) Release from the metaphase I block in invertebrate oocytes: possible involvement of Ca^{2+}/calmodulin-dependent kinase III. *Int. J. Dev. Biol.* 37: 279–290.

Al-Mohanna, F. A., K. W. T., Caddy, and S. R. Bolsover (1994) The nucleus is insulated from large cytosolic calcium ion changes. *Nature* 367: 745–750.

Allbritton, N. L., E. Oancea, M. A. Kuhn, and T. Meyer (1994) Source of nuclear calcium signals. *Proc. Natl. Acad. Sci. USA* 91: 12458–12462.

Bachs, O. and N. Agell (1995) *Calcium and Calmodulin Function in the Cell Nucleus.* R. G. Landes, Austin, TX.

Bachs, O., L. Lanini, J. Serratosa, M. J. Coll, R. Bastos, R. Aligué, E. Rius, and E. Carafoli (1990) Calmodulin-binding proteins in the nuclei of quiescent and proliferatively activated rat liver cells. *J. Biol. Chem.* 265: 18595–18600.

Bading, H., D. D. Ginty, and M. E. Greenberg (1993) Regulation of gene expression in hippocampal neurons by distinct calcium signaling pathways. *Science* 260: 181–186.

Badminton, M. N., A. K. Campbell, and C. M. Rembold, (1996) Differential regulation of nuclear and cytosolic Ca^{2+} in HeLa cells. *J. Biol. Chem.* 271: 31210–31214.

Badminton, M. N., J. M. Kendall, G. Sala-Newby, and A. K. Campbell (1995) Nucleoplasmin targeted aequorin provides evidence for a nuclear calcium barrier. *Exp. Cell Res.* 216: 236–243,.

Baitinger, C., J. Alderton, M. Poenie, H. Schulman, and R. A. Steinhardt. (1990) Multifunctional Ca^{2+}/calmodulin-dependent protein kinase is necessary for nuclear envelope breakdown. *J. Cell Biol.* 111: 1763–1773.

Baudier, J., C. Delphin, D. Grunwald, S. Khochbin, and J. J. Lawrence (1992) Characterization of the tumour suppressor protein p53 as a protein kinase C substrate and a S100b-binding protein. *Proc. Natl. Acad. Sci. USA* 89: 11627–11631.

Bellomo, G., M. Perotti, F. Taddei, F. Mirabelli, G. Finardi, P. Nicotera, and S. Orrenius (1992) Tumor necrosis factor induces apoptosis in mammary adenocarcinoma cells by an increase in intranuclear Ca^{2+} concentration and DNA fragmentation. *Cancer Res.* 52: 1342–1346.

Berezney, R. (1991) The nuclear matrix: a heuristic model for investigating genomic organization and function in the cell nucleus. *J. Cell Biochem.* 47: 109–124.

Berridge, M. J. (1993) Inositol trisphosphate and calcium signalling. *Nature* 361: 315–325.

Bosser, R., R. Aligué, D. Guerini, N. Agell, E. Carafoli, and O. Bachs (1993) Calmodulin can modulate the phosphorylation of nuclear proteins. *J. Biol. Chem.* 268: 15477–15483.

Bosser, R., M. Faura, J. Serratosa, J. Renau-Piqueras, M. Pruschy, and O. Bachs (1995) Phosphorylation of rat liver heterogeneous nuclear ribonucleoproteins A2 and C can be modulated by calmodulin. *Mol. Cell Biol.* 15: 661–670.

Boyko, V., O. Mudrak, M. Svetlova, Y. Negishi, H. Ariga, and N. Tomilin. (1994) A major cellular substrate for proteine kinases annexin II is a DNA-binding protein. *FEBS Lett.* 345: 139–142.

Bremer, J. W., H. Busch, and L. C. Yeoman (1981) Evidence for a species of nuclear actin distinct from cytoplasmic and muscle actins. *Biochemistry* 20: 2013–2017.

Brini, M., M. Murgia, L. Pasti, D. Picard, T. Pozzan, and R. Rizzuto (1993) Nuclear calcium concentration measured with specifically targeted recombinant aequorin. *EMBO J.* 12: 4813–4819.

Brown, E. M., G. Gamba, D. Riccardi, M. Lombardi, R. Butters, O. Kifor, A. Sun, M. A. Hediger, J. Lytton, and S. C. Hebert (1993) Cloning and characterization of an extracellular Ca^{2+}-sensing receptor from bovine parathyroid. *Nature* 366: 575–580.

Brüne, B., P. Hartzell, P. Nicotera, and S. Orrenius (1991) Spermine prevents endonuclease activation and apoptosis in thymocytes. *Exp. Cell Res.* 195: 323–329.

Bustamante, J. O. (1992) Nuclear ion channels in cardiac myocytes. *Pflugers Archiv.* 421: 473–485.

Cao, Q. P., C. A. McGrath, E. F. Baril, P. J. Quesenberry, and G. P. V. Reddy (1995) The 68 kDa calmodulin-binding protein is tightly associated with the multiprotein DNA polymerase α-primase complex in HeLa cells. *Biochemistry* 34: 3878–3883.

Chafouleas, J. G., L. Lagace, W. E. Bolton, A. E. Boyd, and A. R. Means (1984) Changes in calmodulin and its mRNA accompany reentry of quiscent (G_0) cells into the cell cycle. *Cell* 36: 73–81.

Chelsky, D., R. Ralph, and G. Jonak (1989) Sequence requirements for synthetic peptide-mediated translocation to the nucleus. *Mol. Cell. Biol.* 9: 2487–2492.

Chrivia, J. C., R. P. S. Kwok, N. Lamb, M. Hagiwara, M. R. Montminy, and R. H. Goodman (1993) Phosporylated CREB binds specifically to the nuclear protein CBP. *Nature* 365: 855–859,.

Ciapa, B., D. Pesando, M. Wilding, and M. J. Whitaker (1994) Cell-cycle calcium transients driven by cyclic changes in inositol trisphosphate levels. *Nature* 368: 875–878.

Clark, T. G. and J. L. Rosenbaum (1979) An actin filament matrix in hand-isolated nuclei of *Xenopus laevis* oocytes. *Cell* 18: 1101–1108.

Clipstone, N. A. and G. R. Crabtree (1992) Identification of calcineurin as a key signalling enzyme in T-lymphocyte activation. *Nature* 357: 695–697.

Clipstone, N. A., D. F. Fiorentino, and G. R. Crabtree (1994) Molecular analysis of the interaction of calcineurin with drug-immunophilin complexes. *J. Biol. Chem.* 269: 26431–26437.

Colas, P., C. Launay, A. E. Van-Loon, and P. Guerrier (1992) Regulation of cyclin proteolysis in *Patella vulgata* oocytes. *Biol. Cell* 76: 217(8A).

Colomer, J., A. López-Girona, N. Agell, and O. Bachs (1994) Calmodulin regulates the expression of cdks, cyclins and replicative enzymes during proliferative activation of human T lymphocytes. *Biochem. Biophys. Res. Commun.* 200: 306–312.

Connor, J. A. (1993) Intracellular calcium mobilization by inositol 1,4,5-trisphosphate: intracellular movements and compartmentalization. *Cell Calcium* 14: 185–200.

Corneliussen, B., M. Holm, Y. Waltersson, J. Onions, B. Hallberg, A. Thornell, and T. Grundström (1994) Calcium/calmodulin inhibition of basic–helix–loop–helix transcription factor domains. *Nature* 368: 760–764.

Crabtree, G. R. (1989) Contingent genetic regulatory events in T lymphocytes activation. *Science* 243: 355–361.

Dale, B., L. J. DeFelice, K. Kyozuka, L. Santella, and E. Tosti (1994) Voltage clamp of the nuclear envelope. *Proc. R. Soc. London Ser. B* 255: 119–124.

Dash, P. K., K. A. Karl, M. A. Colicos, R. Prywes, and E. R. Kandel (1991) cAMP response element-binding protein is activated by Ca^{2+}/calmodulin as well as cAMP-dependent protein kinase. *Proc. Natl. Acad. Sci. USA* 88: 5061–5065.

Davis, L. I. (1995) The nuclear pore complex. *Annu. Rev. Biochem.* 64: 865–896.

Davis, T. N. (1992) A temperature-sensitive calmodulin mutant loses viability during mitosis. *J. Cell Biol.* 118: 607–617..

De Boni, U. (1994) The interphase nucleus as a dynamic structure. *Int. Rev. Cytol.* 150: 149–171.

DeFelici, M., S. Dolci, and G. Siracusa (1991) An increase of intracellular free Ca^{2+} is essential for spontaneous meiotic resumption by mouse oocytes. *J. Exp. Zool.* 260: 401–405.

Deguchi, R. and K. Osanai (1994) Repetitive intracellular Ca^{2+} increases at fertilization and the role of Ca^{2+} in meiosis reinitiation from the first metaphase in oocytes of marine bivalves. *Dev. Biol.* 163: 162–174.

Divecha, N., H. Banfic, and R. R. Irvine (1991) The poly-phosphoinositide cycle exists in the nucleus of Swiss 3T3 cells under the control of a receptor (for IGF-I) in the plasma membrane, and stimulation of the cycle increases nuclear diacylglycerol and apparently induces translocation of protein kinase C to the nucleus. *EMBO J.* 10: 3207–3214.

DiVirgilio, F., T. H. Steinberg, and S. C. Silverstein (1990) Inhibition of fura-2 sequestration and secretion with organic anion transport blockers. *Cell Calcium* 11: 57–62.

Dou, Q. P., P. J. Markell, and A. B. Pardee (1992) Thymidine kinase transcription is regulated at the G1/S phase by a complex that contains retinoblastoma-like protein and a cdc2 kinase. *Proc. Natl. Acad. Sci. USA* 89: 3256–3260.

Dowd, D. R., P. N. MacDonald, B. S. Komm, M. R. Haussler, and R. Miesfeld (1991) Evidence for early induction of calmodulin gene expression in lymphocytes undergoing glucocorticoid-mediated apoptosis. *J. Biol. Chem.* 266: 18423–18426.

Dreyfuss, G., M. J. Matunis, S. Pinôl-Roma, and C. Burd (1993) hnRNP proteins and the biogenesis of RNA. *Annu. Rev. Biochem.* 62: 289–321.

Dubé, F. and L. Dufresne (1990) Release of metaphase arrest by partial inhibition of protein synthesis in blue mussel oocytes. *J. Exp. Zool.* 256: 323–332.

Egly, J. M., N. G. Miyamoto, V. Moncollin, and P. Chambon (1984) Is actin a transcription initiation factor for RNA polymerase B? *EMBO J.* 3: 2363–2371.

Faulk, E. A., J. D. McCully, T. Tsukube, N. C. Hadlow, I. B. Krukenkamp, and S. Levitsky (1995) Myocardial mitochondrial calcium accumulation modulates nuclear calcium accumulation and DNA fragmentation. *Ann. Thorac. Surg.* 60: 338–344.

Fluck, R. A., A. L. Miller, and L. F. Jaffe (1991) Slow calcium waves accompany cytokinesis in medaka fish eggs. *J. Cell Biol.* 115: 1259–1265.

Fruman, D. A., P. E. Mather, S. J. Burakoff, and B. E. Bierer (1992) Correlation of calcineurin phosphatase activity and programmed cell death in murine T cell hybridomas. *Eur. J. Immunol.* 22: 2513–2517.

Furuichi, T., K. Kohda, A. Miyawaki, and K. Mikoshiba (1994) Intracellular channels. *Curr. Opin. Neurobiol.* 4: 294–303.

Gaido, M. L. and J. A. Cidlowski (1991) Identification, purification, and characterization of a calcium-dependent endonuclease (NUC18) from apoptotic rat thymocytes. *J. Biol. Chem.* 266: 18580–18585.

Galione, A., H. C. Lee, and W. B. Busa (1991) Ca^{2+}-induced Ca^{2+} release in sea urchin egg homogenates: modulation by cyclic ADP-ribose. *Science* 253: 1143–1146.

Gardner, P. (1989) Calcium and T lymphocyte activation. *Cell* 59: 15–20.

Geiser, J. R., H. A. Sundberg, B. H. Chang, E. G. D. Muller, and T. N. Davis (1993) The essential mitotic target of calmodulin is the 110-kilodalton component of the spindle pole body in *Saccharomices cerevisiae*. *Mol. Cell. Biol.* 13: 7913–7924.

Geiser, J. R., D. van-Tuinem, S. E. Brockerhoff, M. M. Neff, and T. N. Davis (1991) Can calmodulin function without binding calcium? *Cell* 65: 949–959.

Genestier, L., M. T. Dearden-Badet, N. Bonnefoy-Berard, G. Lizard, and J. P. Revillard (1994) Cyclosporin A and FK506 inhibit activation-induced cell death in the murine WEHI-231 B cell line. *Cell. Immunol.* 155: 283–291.

Gerasimenko, O. V., J. V. Gerasimenko, A. V., Tepikin, and O. H. Petersen (1995) ATP dependent accumulation and inositol trisphosphate- or cyclic ADP-ribose-mediated release of calcium from the nuclear envelope. *Cell* 80: 439–444.

Gillot, I. and M. Whitaker (1993) Imaging calcium waves in eggs and embryos. *J. Exp. Biol.* 184: 213–219.

Gillot, I. and M. Whitaker (1994) Calcium signals in and around the nucleus in sea urchin eggs. *Cell Calcium* 16: 269–278.

Gonzales, G. A. and M. R. Montminy (1989) Cyclic AMP stimulates somatostatin gene transcription by phosphorylation of CREB at serine 133. *Cell* 59: 675–680.

Grandin, N. and M. Charbonneau (1991) Intracellular free calcium oscillates during cell division in *Xenopus* embryos. *J. Cell Biol.* 112: 711–718.

Greber, U. F. and L. Gerace (1995) Depletion of calcium from the lumen of the endoplasmic reticulum reversibly inhibits passive diffusion and signal-mediated transport into the nucleus. *J. Cell Biol.* 128: 5–14.

Greenberg, M. E., E. B. Ziff, and L. A. Greene (1986) Stimulation of neuronal acetylcholine receptors induces rapid gene transcription. *Science* 234: 80–83.

Grynciewicz, G., M. Poenie, and R. Y. Tsien (1985) A new generation of Ca^{2+} indicators with greatly improved fluorescence properties. *J. Biol. Chem.* 260: 3440–3450.

Hennager, D. J., M. J. Welsh, and S. DeLisle (1995) Changes in either cytosolic or nucleoplasmic inositol 1,4,5-trisphosphate levels can control nuclear calcium concentration. *J. Biol. Chem.* 270: 4959–4962.

Hepler, P. K. (1992) Calcium and mitosis. *Int. Rev. Cytol.* 138: 239–268.

Hepler, P. K. (1994) The role of calcium in cell division. *Cell Calcium* 16: 322–330.

Hernandez-Cruz, A., F. Sala, and P. R. Adams (1990) Subcellular calcium transients visualized by confocal microscopy in a voltage-clamped neuron. *Science* 247, 858–862.

Himpens, B., H. DeSmedt, and R. Casteels (1994) Subcellular calcium gradients in A7r5 vascular smooth muscle. *Cell Calcium* 15: 55–65.

Himpens, B., H. DeSmedt, G. Droogmans, and R. Casteels (1992) Difference in regulation between nuclear and cytoplasmic Ca^{2+} in cultured smooth muscle cells. *Am. J. Physiol.* 263: C95–C105.

Hiraga, K., K. Suzuki, E. Tsuchiya, and T. Miyakawa (1993). Identification and characterization of nuclear calmodulin-binding proteins of *Saccaromices cerevisiae*. *Biochem. Biophys. Acta* 1177: 25–30.

Hirai, S., H. Kawasaki, M. Yaniv, and K. Suzuki (1991) Degradation of transcription factors, c-*jun* and c-*fos*, by calpain. *FEBS Lett.* 287: 57–61.

Hoeffler, J. P., T. E. Meyer, Y. Yun, J. L. Jameson, and J. F. Habener (1988) Cyclic AMP-responsive DNA-binding protein: structure based on a cloned placental cDNA. *Science* 242: 1430–1433.

Humbert, J. P., N. Matter, J. C. Artault, P. Koppler, and A. N. Malviya (1996) Inositol 1,4,5-trisphosphate receptor is located to the inner nuclear membrane vindicating regulation of nuclear calcium signalling by inositol 1,4,5-trisphosphate. *J. Biol. Chem.* 271: 478–485.

Ishida, A. T., V. P. Bindokas, and R. Nuccitelli (1991) Calcium ion levels in resting and depolarized goldfish retinal ganglion cell somata and growth cones. *J. Neurophysiol.* 65: 968–979.

Jaffe, L., C. J. Gallo, R. H. Lee, Y. K. Ho, and T. L. Z. Jones (1993) Oocyte maturation in starfish is mediated by β-subunit complex of G-protein. *J. Cell Biol.* 121: 775–783.

James, J. L., G. A. Clawson, C. H. Chan, and E. A. Smuckler (1981) Analysis of the phospholipid of the nuclear envelope and endoplasmic reticulum of liver cells by high pressure liquid chromatography. *Lipids* 16: 541–545.

Jensen, K. F., C.-A. Ohmstede, R. S. Fisher, and N. Sayoun (1991) Nuclear and axonal localization of Ca^{2+}/calmodulin-dependent protein kinase type Gr in rat cerebellar cortex. *Proc. Natl. Acad. Sci. USA* 88: 2850–2853.

Jindal, H. K., W. G. Chaney, C. W. Anderson, R. G. Davis, and J. K. Vishwanatha. (1991) The protein-tyrosine kinase substrate, calpactin Y heavy chain (p36), is part of the primer recognition protein complex that interacts with DNA polymerase alpha. *J. Biol. Chem.* 266: 5169–5176.

Jones, D. P., D. J. McConkey, P. Nicotera, and S. Orrenius (1989) Calcium-activated DNA fragmentation in rat liver nuclei. *J. Biol. Chem.* 264: 6398–6403.

Jurgens, M., L. H. Hepler, B. A. Rivers, and P. K. Hepler (1994) BAPTA-calcium buffers modulate cell plate formation in stamen hair cells of Tradescantia: evidence for calcium gradients. *Protoplasma* 183: 86–89.

Kanatani, H., H. Shirai, K. Nakanishi, and T. Kurosawa (1969) Isolation and identification of meiosis-inducing substance in starfish *Asterina amurensis. Nature* 221: 273–274.

Kao, J. P. Y., J. M. Alderton, R. Y. Tsien, and R. A. Steinhardt (1990) Active involvement of calcium ion in mitotic progression of Swiss 3T3 fibroblast. *J. Cell Biol.* 111: 183–196.

Kasai, H. and O. H. Petersen (1994) Spatial dynamics of second messengers: IP_3 and cAMP as long-range and associative messengers. *Trends in Neurosci.* 17: 95–101.

Kaufman, M. L. and S. T. Homa (1993) Defining a role for calcium in the resumption and progression of meiosis in pig oocytes. *J. Exp. Zool.* 265: 69–76.

Kirschner, R. H., M. Rusli, and T. E. Martin (1977) Characterization of the nuclear envelope, pore complexes, and dense lamina of mouse nuclei by high reso-

lution scanning electron microscopy. *J. Cell Biol.* 72: 118–132.

Konishi, M., A. Olson, S. Hollingworth, and S. M. Baylor (1988) Myoplasmic binding of fura-2 investigated by steady-state fluorescence and absorbance measurements. *Biophys. J.* 54: 1089–1104.

Kubo, M., R. L. Kincaid, D. R. Webb, and J. T. Ransom (1994) The Ca^{2+}/calmodulin activated, phosphoprotein phosphatase calcineurin is sufficient for positive transcriptional regulation of the mouse IL-4 gene. *Int. Immunol.* 6: 179–188.

Kubota, H. Y., Y. Yoshimoto, and Y. Hiramoto (1993) Oscillation of intracellular free calcium in cleaving and cleavage-arrested embryos of *Xenopus laevis. Dev. Biol.* 160: 512–518.

Kumar, A., Raziuddin, T. H. Finlay, J. O. Thomas, and W. Szer (1984) Isolation of a minor species of actin from the nuclei of *Acanthamoeba castellanii. Biochemistry* 23: 6753–6757.

Labiner, D. M., L. S. Butler, Z. Cao, D. A. Hosford, C. Shin, and J. O. McNamara (1993) Induction of c-*fos* mRNA by kindled siezures: complex relationship with neuronal burst firing. *J. Neurosci.* 13: 744–751.

Lam, M., G. Dubyak, L. Chen, G. Nunez, R. L. Miesfeld, and C. W. Distelhorst (1994) Evidence that Bcl-2 represses apoptosis by regulating endoplasmic reticulum-associated Ca^{2+} fluxes. *Proc. Natl. Acad. Sci. USA* 91: 6569–6573.

Lane, R. D., D. M. Allan, and R. L. Mellgren (1992) A comparison of the intracellular distribution of μ-calpain, m-calpain, and calpastatin in proliferating human A431 cells. *Exp. Cell Res.* 203: 5–16.

Lanini, L., O. Bachs, and E. Carafoli (1992) The calcium pump of the liver nuclear membrane is identical to that of endoplasmic reticulum. *J. Biol. Chem.* 267: 11548–11552.

Larson, T. G. and D. L. Nuss (1994) Altered transcriptional response to nutrient availability in hypovirus-infected chestnut blight fungus. *EMBO J.* 13: 5616–5623.

Lee, A. S. (1987) Coordinated regulation of a set of genes by glucose and calcium ionophores in mammalian cells. *Trends Biochem. Sci.* 12: 20–23.

Lefèvre, B., A. Pesty, and J. Testat (1995) Cytoplasmic and nucleic calcium oscillations in immature mouse oocytes: evidence of wave polarization by confocal imaging. *Exp. Cell Res.* 218: 166–173.

Lin, C., G. Hajnoczky, and A. P. Thomas (1994) Propagation of cytosolic calcium waves into the nuclei of hepatocytes. *Cell Calcium* 16: 247–258.

Lorca, T., F. H. Cruzalegui, D. Fesquet, J. C. Cavadore, J. Méry, A. R. Means, and M. Dorée (1993) Calmodulin-dependent protein kinase II mediates inactivation of MPF and CSF upon fertilization of *Xenopus* eggs. *Nature* 366: 270–273.

Lu, K. P. and A. R. Means (1993) Regulation of the cell cycle by calcium and calmodulin. *Endocrine Rev.* 14: 40–48.

Mackler, S. A., B. P. Brooks, and J. H. Eberwine (1992) Stimulus-induced coordinate changes in mRNA abundance in single postsynaptic CA1 neurons. *Neuron* 9: 539–548.

Martelli, A. M., R. S. Gilmour, V. Bertagnolo, L. M. Neri, L. Manzoli, and L. Cocco (1992) Nuclear localization and signalling activity of phosphoinositidase C beta in Swiss 3T3 cells. *Nature* 358: 242–245.

Mazzanti, M., L. J. DeFelice, J. Cohen, and H. Malter (1990) Ion channels in the nuclear envelope. *Nature* 343: 764–767.

McConkey, D. J., P. Nicotera, P. Hartzell, G. Bellomo, and S. Orrenius (1989) Calcium-activated DNA fragmentation kills immature thymocytes. *FASEB J.* 3: 1843–1849.

Means, A. R. (1994) Calcium, calmodulin and cell cycle regulation. *FEBS Lett.* 347: 1–4.

Mellgren, R. L. (1991) Proteolysis of nuclear proteins by μ-calpain and m-calpain. *J. Biol. Chem.* 266: 13920–13924.

Mellgren, R. L. and Q. Lin. (1994) Selective nuclear transport of μ-calpain. *Biochem. Biophys. Res. Commun.* 204: 544–550.

Mellgren, R. L., K. Song, and M. T. Mericle (1993) m-Calpain requires DNA for activity on nuclear proteins at low concentrations. *J. Biol. Chem.* 268: 653–657.

Miller, A. L., R. A. Fluck, J. A. McLaughlin, and L. F. Jaffe (1993) Calcium buffer injections inhibit cytokinesis in *Xenopus* eggs. *J. Cell Sci.* 106: 523–534.

Minami, H., S. Inoue, and H. Hidaka (1994) The effect of KN-62, Ca^{2+}/calmodulin dependent protein kinase inhibitor on cell cycle. *Biochem. Biophys. Res. Commun.* 199: 241–248.

Miyamoto, S., M. Maki, M. J. Schmitt, M. Hatanaka, and I. M. Verma (1994) Tumor necrosis factor alpha-induced phosphorylation of I kappa B alpha is a signal for its degradation but not dissociation from NF-kappa B. *Proc. Natl. Acad. Sci. USA.* 91: 12740–12744.

Molinari, M., J. Anagli, and E. Carafoli (1995) PEST sequences do not influence substrate susceptibility to calpain proteolysis. *J. Biol. Chem.* 270: 2032–2035.

Montague, J. W., M. L. Gaido, C. Frye, and J. A. Cidlowski (1994) A calcium-dependent nuclease from apoptotic rat thymocytes is homologous with cyclophilin. Recombinant cyclophilins A, B, and C have nuclease activity. *J. Biol. Chem.* 269: 18877–18880.

Montminy, M. R. and L. M. Bilezikjian (1987) Binding of a nuclear protein to the cyclic-AMP response element of the somatostatin gene. *Nature* 328: 175–178.

Morgan, J. I. and T. Curran (1986) Role of ion flux in the control of c-*fos* expression. *Nature* 322: 552–555.

Nairn, A. C., B. Bhagat, and H. C. Palfray (1985) Identification of a calmodulin dependent protein kinase II and its major M_r 100,000 substrate in mammalian tissues. *Proc. Natl. Acad. Sci. USA* 82: 7939–7943.

Neylon, C. B., J. Hoyland, W. T. Mason, and R. F. Irvine (1990) Spatial dynamics of intracellular calcium in agonist-stimulated vascular smooth muscle cells. *Am. J. Physiol.* 259: C675-C686.

Nicotera, P. and A. D. Rossi (1994) Nuclear calcium: physiological regulation and role in apoptosis. *Mol. Cell Biochem.* 135: 89–98.

Nicotera, P., D. J. McConkey, D. P. Jones, and S. Orrenius (1989) ATP stimulates Ca^{2+} uptake and increases the free Ca^{2+} concentration in isolated liver nuclei. *Proc. Natl. Acad. Sci. USA* 86: 453–457.

Nicotera, P., B. Zhivotovsky, and S. Orrenius (1994) Nuclear transport and the role of calcium in apoptosis. *Cell Calcium* 16: 279–288.

Nowycky, M. C. and M. J. Pinter (1993) Time courses of calcium and calcium-bound buffers following calcium influx in a model cell. *Biophys. J.* 64, 77–91.

O'Keefe, S. J., J. Tamura, R. L. Kincaid, M. J. Tocci, and E. A. O'Neill (1992) FK-506- and CsA-sensitive activation of the interleukin promoter by calcineurin. *Nature* 357: 692–694.

O'Malley, D. M. (1994) Calcium permeability of the neuronal nuclear envelope: evaluation using confocal volumes and intracellular perfusion. *J. Neurosci.* 14: 5741–5758.

Ohta, Y., Y. Ohba, and E. Miyamoto (1990) Ca^{2+}/calmodulin-dependent protein kinase II: localization in the interphase nucleus and the mitotic apparatus of mammalian cells. *Proc. Natl. Acad. Sci. USA* 87: 5341–5345.

Ohya, Y. and D. Botstein. (1994) Diverse essential functions revealed by complementing yeast calmodulin mutants. *Science* 263: 963–966.

Osheroff, N. and E. L. Zechiedrich (1987) Calcium-promoted DNA cleavage by eukaryotic topoisomerase II: trapping the covalent enzyme-DNA complex in an active form. *Biochemistry* 26: 4303–4309.

Paine, P. L., L. C. Moore, and S. B. Horowitz (1975) Nuclear envelope permeability. *Nature* 254: 109–114.

Paliogianni, F., R. L. Kincaid, and D. T. Boumpas (1993) Prostaglandin E2 and other cyclic AMP elevating agents inhibit interleukine 2 gene transcription by counteracting calcineurin-dependent pathways. *J. Exp. Med.* 178: 1813–1817.

Parkinson, N. A., W. T. Mason, and S. R. Bolsover (1996) Calcium movement through the nuclear envelope in rat DRG cells. *J. Physiol.* 495p: 65–66.

Patel, T., G. J. Gores, and S. H. Kaufmann (1996) The role of proteases during apoptosis. *FASEB J.* 10: 587–597.

Pearson, B. E., H. P. Nasheuer, and T. S. F. Wang (1991) Human DNA polymerase α gene: sequences controlling expression in cycling and serum-stimulated cells. *Mol. Cell. Biol.* 11: 2081–2095.

Perez-Terzic, C., J. Pyle, M. Jaconi, L. Stehno-Bittel, and D. E. Clapham (1997) Conformational states of the nuclear pore complex induced by depletion of nuclear Ca^{2+} stores. *Science* 273: 1875–1877.

Perry, M. E. and A. J. Levine (1993) Tumor-suppressor p-53 and the cell cycle. *Curr. Opin. Genet. Dev.* 3: 50–54.

Pinõl-Roma, S. and G. Dreyfuss (1991) Transcription-dependent and transcription-independent nuclear transport of hnRNP proteins. *Science* 253: 312–314.

Planas-Silva, M. D. and A. R. Means (1992) Expression of a constitutive form of calcium/calmodulin dependent kinase II leads to arrest of the cell cycle in G_2. *EMBO J.* 11: 507–517.

Plevin, R., S. Palmer, S. D. Gardner, and M. J. Wakelam (1990) Regulation of bombesin-stimulated inositol 1,4,5-trisphosphate generation in Swiss 3T3 fibroblasts by a guanine-nucleotide-binding protein. *Biochem. J.* 268: 605–610.

Prendergast, G. C. and E. B. Ziff (1991) Mbh1: a novel gelsolin/severin related protein which binds actin *in vitro* and exhibits nuclear localization *in vivo*. *EMBO J.* 10: 757–766.

Pruschy, M., Y. Ju, L. Spitz, E. Carafoli, and D. S. Goldfarb (1994) Facilitated nuclear transport of calmodulin in tissue culture cells. *J. Cell Biol.* 127: 1527–1536.

Pujol, M. J., R. Bosser, M. Vendrell, J. Serratosa, and O. Bachs (1993) Nuclear calmodulin-binding proteins in rat neurons. *J. Neurochem.* 60: 1422–1428.

Rand, M. N., T. Leinders-Zufall, S. Agulian, and J. D. Kocsis (1994) Calcium signals in neurons. *Nature* 371: 291–292.

Rao, A. (1991) Signaling mechanisms in T cells. *Crit. Rev. Immunol.* 10: 495–519.

Rasmussen, C. D. and A. R. Means (1989) Calmodulin is required for cell cycle progression during G_1 and mitosis. *EMBO J.* 8: 73–82,.

Rasmussen, C. D., C. Garen, S. Brining, R. L. Kincaid, R. L. Means, and A. R. Means (1994) The calmodulin-dependent protein phosphatase catalytic subunit (calcineurin A) is an essential gene in *Aspergillus nidulans*. *EMBO J.* 13: 3917–3924.

Rasmussen, C. D., K. P. Lu, R. L. Means, and A. R. Means (1992) Calmodulin and cell cycle control. *J. Physiol.* 86: 83–88.

Ratan, R. R., F. R. Maxfield, and M. L. Shelanski (1988) Long lasting and rapid calcium changes during mitosis. *J. Cell Biol.* 107: 993–999.

Reddy, G. P., W. C. Reed, E. Sheehan, and D. B. Sacks (1992) Calmodulin-specific monoclonal antibodies inhibit DNA replication in mammalian cells. *Biochemistry* 31: 10426–10430.

Richard, I., O. Broux, V. Allamand, F. Fougerousse, N. Chiannilkulchai, N. Bourg, L. Brenguer, C. Devaud, P. Pasturaud, C. Roudaut, D. Hillaire, M. R. Passos-Bueno, M. Zatz, J. A. Tischfield, M. Fardeau, C. E. Jackson, D. Cohen, and J. S. Beckmann (1995) Mutations in the proteolytic enzyme calpain 3 cause limb-girdle muscular dystrophy type 2A. *Cell* 81: 27–40.

Riley, D. J., E. V.-H. P. Lee, and W.-H. Lee (1994) The retinoblastoma protein: more than a tumour suppressor. *Annu. Rev. Cell Biol.* 10: 1–29.

Ryazanov, A. and A. S. Spirin (1990) Phosphorylation of elongation factor 2: a key mechanism regulating gene expression in vertebrates. *New Biologist* 2: 843–850.

Sahlas, D. J., K. Milankov, P. C. Park, and U. De Boni (1993) Distribution of snRNPs, splicing factor SC-35 and actin in interphase nuclei: immunocytochemical evidence for differential distribution during changes in functional states. *J. Cell Sci.* 105: 347–357.

Sahyoun, N., H. LeVine III, and P. Cuatrecasas (1984) Ca^{2+}/calmodulin-dependent protein kinases from the neuronal nuclear matrix and postsynaptic density are structurally related. *Proc. Natl. Acad. Sci. USA* 81: 4311–4315.

Saido, T. C., H. Sorimachi, and K. Suzuki (1994) Calpain: new perspectives in molecular diversity and physiological–pathological involvement. *FASEB J.* 8: 814–822.

Santella, L. (1996) The cell nucleus: an Eldorado to future calcium research? *J. Membr. Biol.* 153: 83–92.

Santella, L. and K. Kyozuka (1994) Reinitiation of meiosis in starfish oocytes requires an increase in nuclear Ca^{2+}. *Biochem. Biophys. Res. Commun.* 203: 674–680.

Santella, L. and K. Kyozuka (1997) Effects of 1-methyladenine on nuclear Ca^{2+} transients and meiosis resumption in starfish oocytes are mimicked by the nuclear injection of inositol 1,4,5-trisphosphate and cADP-ribose. *Cell Calcium* 22: 11–20.

Scheer, U., H. Hinssen, W. W. Franke, and B. M. Jockusch (1984) Microinjection of actin-binding proteins and actin antibodies demonstrates involvement of nuclear actin in transcription of lampbrush chromosomes. *Cell* 39: 111–122.

Schindler, M. and L. W. Jiang (1990) *A Dynamic Analysis of the Nuclear Pore Complex. The Intelligent Tunnel* (Wang, E., Wang, J. L., Chien, S., Cheung, W. Y., and Wu, C. W. eds.). Academic Press, San Diego, pp. 249–263.

Schollmeyer, J. E. (1988) Calpain II involvement in mitosis. *Science* 240: 911–913.

Schontal, A., J. Sugarman, J. H. Brown, M. R. Hanley, and J. R. Feramisco (1991) Regulation of c-*fos* and c-*jun* protooncogene expression by the Ca^{2+}-ATPase inhibitor thapsigargin. *Proc. Natl. Acad. Sci. USA* 88: 7096–7100.

Shankar, G., I. Davison, M. H. Helfrich, W. T. Mason, and M. A. Horton (1993) Integrin receptor-mediated mobilization of intranuclear calcium in rat osteoclasts. *J. Cell Sci.* 105: 61–68.

Shankar, G., T. R. Gadek, D. J. Burdick, I. Davison, W. T. Mason, and M. A. Horton (1995) Structural determinants of calcium signalling by RGD peptides in rat osteoclasts: integrin dependent and independent actions. *Exp. Cell Res.* 219: 364–371.

Sheng, H. Z., R. D. Fields, and P. G. Nelson (1993) Specific regulation of immediate early genes by patterned neural activity. *J. Neurosci. Res.* 35: 459–467.

Sheng, M., S. T. Dougan, G. McFadden, and M. E. Greenberg (1988) Calcium and growth factor pathway of c-*fos* transcriptional activation require distinct upstream regulatory sequences. *Mol. Cell. Biol.* 8: 2787–2796

Sheng, M., G. McFadden, and M. E. Greenberg (1990) Membrane depolarization and calcium induce c-*fos* transcription via phosphorylation of transcription factor CREB. *Neuron* 4: 571–582.

Sheng, M., M. A. Thompson, and M. E. Greenberg (1991) CREB: a Ca^{2+}-regulated transcription factor phosphorylated by calmodulin-dependent kinases. *Science* 252: 1427–1430.

Shioda, M., K. Okuhara, H. Murofushi, A. Mori, H. Sakai, K. Murakami-Murofushi, M. Suzuki, and S. Yoshida (1991) Stimulation of DNA polymerase α activity by microtubule-associated proteins. *Biochemistry* 30: 11403–11412.

Silver, R. A., M. Whitaker, and S. R. Bolsover (1992) Intracellular ion imaging using fluorescent dyes: artefacts and limits to resolution. *Pflugers Archiv.* 420: 595–602.

Silver, R. B. (1989) Nuclear envelope breakdown and mitosis in sand dollar embryos is inhibited by microinjection of calcium buffers in a calcium-reversible fashion, and by antagonists of intracellular Ca^{2+} channels. *Dev. Biol.* 131: 11–26.

Simmen, R. C. M., B. S. Dunbar, V. Guerriero, J. G. Chafouleas, J. H. Clark, and A. R. Means (1984) Estrogen stimulates the transient association of calmodulin and myosin light chain kinase with the chicken liver nuclear matrix. *J. Cell Biol.* 99: 588–593.

Smeyne, R. J., M. Vendrell, M. Hayward, S. J. Baker, G. G. Miao, K. Shilling, L. M. Robertson, T. Curran, and J. I. Morgan (1993) Continuous c-*fos* expression precedes programmed cell death in vivo. *Nature* 363: 166–169.

Snow, P. and R. Nuccitelli (1993) Calcium buffer injections delay cleavage in *Xenopus laevis* blastomeres. *J. Cell Biol.* 122: 387–394.

Sorimachi, H., S. Imajoh-Ohmi, Y. Emori, H. Kawasaki, S. Olmo, Y. Minami, and K. Suzuki (1989) Molecular cloning of a novel mammalian calcium-dependent protease distinct from both m- and μ-types. Specific expression of the mRNA in skeletal muscle. *J. Biol. Chem.* 264: 20106–20111.

Sorimachi, H., N. Toyoma-Sorimachi, H. T. C. Saido, H. Kawasaki, H. Sugita, M. Miyasaka, K. Arahata, S. Ishiura, and K. Suzuki (1993) Muscle-specific calpain, p94, is degraded by autolysis immediately after translation, resulting in disappearance from muscle. *J. Biol. Chem.* 268: 10593–10605.

Squier, M. K. T., A. C. K. Miller, A. M. Malkinson, and J. J. Cohen (1994) Calpain activation in apoptosis. *J. Cell Physiol.* 159: 229–237.

Srinivasan, M., C. F. Edman, and H. Schulman (1994) Alternative splicing introduces a nuclear localization signal that targets multifunctional CaM kinase to the nucleus. *J. Cell Biol.* 126: 839–852.

Stehno-Bittel, L., A. Luckhoff, A. and D. E. Clapham (1995) Calcium release from the nucleus by $InsP_3$ receptor channels. *Neuron* 14, 163–167.

Stehno-Bittel, L., C. Perez-Terzic, and D. E. Clapham (1995) Diffusion across the nuclear envelope inhibited by depletion of the nuclear Ca^{2+} stores. *Science* 270: 1835–1838.

Steinhardt, R. A. and J. Alderton (1988) Intracellular free calcium rise triggers nuclear envelope breakdown in the sea urchin embryo. *Nature* 332: 364–366.

Stricker, S. A. (1995) Time-lapse confocal imaging of calcium dynamics in starfish embryos. *Dev. Biol.* 170: 496–518.

Stricker, S. A., V. E. Centonze, and R. F. Melendez (1994) Calcium dynamics during starfish oocyte maturation and fertilization. *Dev. Biol.* 166: 34–58.

Subramanyam, C., S. C. Honn, W. C. Reed, and G. P. V. Reddy (1990) Nuclear localization of 68 kDa calmodulin-binding protein is associated with the onset of DNA replication. *J. Cell Physiol.* 144: 423–428.

Sullivan, K. M. C., W. B. Busa, and K. L. Wilson (1993) Calcium mobilization is required for nuclear vesicle fusion in vitro: implications for membrane traffic and IP_3 receptor function. *Cell* 73: 1411–1422.

Sun, G. H., Y. Hirata, Y. Ohya, and Y. Anraku (1992) Mutations in yeast calmodulin cause defects in spindle body functions and nuclear integrity. *J. Cell Biol.* 119: 1625–1639.

Takahashi, K. (1990) Calpain substrate specificity. In *Intracellular-Dependent Proteolysis* (Mellgren R. L. and T. Murachi, eds.). CRC Press, Boca Raton, pp. 55–74.

Takuwa, N., W. Zhou, M. Kumada, and Y. Takuwa (1992) Ca^{2+}/calmodulin is involved in growth-induced retinoblastoma gene product phosphorylation in human vascular endothelial cells. *FEBS Lett.* 306: 173–175.

Takuwa, N., W. Zhou, M. Kumada, and Y. Takuwa (1993) Ca^{2+}-dependent stimulation of retinoblastoma gene product phosphorylation and $p34^{cdc2}$ kinase activation in serum-stimulated human fibroblast. *J. Biol. Chem.* 268: 138–145.

Tökés, Z. A. and G. A. Clawson (1990) Proteolytic activity associated with the nuclear scaffold. The effect of self-digestion on lamins. *J. Biol. Chem.* 264: 15059–15065.

Tombes, R. M. and G. G. Borisy (1989) Intracellular free calcium and mitosis in mammalian cells: anaphase onset is calcium modulated, but is not triggered by a brief transient. *J. Cell Biol.* 109: 627–636.

Tombes, R. M., C. Simerly, G. G. Borisy, and G. Schatten (1992) Meiosis, egg activation, and nuclear envelope breakdown are differentially reliant on Ca^{2+}, whereas germinal vesicle breakdown is Ca^{2+} independent in the mouse oocyte. *J. Cell Biol.* 117: 799–811.

Traub, P., A. Scherbarth, J. Willingale-Theune, M. Paulin-Levasseur, and R. Shoeman (1988) Differential sensitivity of vimentin and nuclear lamins from Ehrlich ascites tumour toward Ca^{2+}-activated neutral thiol proteinase. *Eur. J. Cell Biol.* 46: 478–482.

Tsien, R. Y. (1981) A non-disruptive technique for loading calcium buffers and indicators into cells. *Nature* 290: 527–528.

Tsuboi, A., E. S. Masuda, Y. Naito, H. Tokimitsu, K. Arai, and N. Arai (1994a) Calcineurin potentiates activation of the granulocyte-macrofage colony-stimulating factor gene in T cells: involvement of the conserved lymphokine element 0. *Mol. Biol. Cell.* 5: 119–128.

Tsuboi, A., M. Muramatsu, A. Tsutsumi, K. Arai, and N. Arai (1994b) Calcineurin activates transcription from the GM-CSF promoter in synergy with either protein kinase C or NF-κB/AP-1 in T cells. *Biochem. Biophys. Res. Commun.* 199: 1064–1072.

Twigg, J., R. Patel, and M. Whitaker (1988) Translational control of InsP₃-induced chromatin condensation during the early cell cycle of sea urchin embryos. *Nature* 332: 366–369.

Wahl, M., R. G. Sleight, and E. Gruenstein (1992) Association of cytoplasmic free Ca²⁺ gradients with subcellular organelles. *J. Cell Physiol.* 150: 593–609.

Wang. T. (1991) DNA polymerases. *Annu. Rev. Biochem.* 60: 513–552.

Welsh, M. J., J. R. Dedman, B. R. Brinkley, and A. R. Means (1978) Calcium-dependent regulator protein: localization in mitotic apparatus of eukaryotic cells. *Proc. Natl. Acad. Sci. USA* 75: 1867–1871.

Werlen, G., D. Belin, B. Conne, E. Roche, D. P. Lew, and M. Prentki (1993) Intracellular Ca²⁺ and the regulation of early response gene expression in HL-60 myeloid leukemia cells. *J. Biol. Chem.* 268: 16596–16601.

Whitaker, M. (1995) Regulation of the cell division cycle by inositol trisphosphate and the calcium signaling pathway. In *Advances in Second Messenger and Phosphoprotein Research* (Means, A. R., ed.). Raven Press, New York, pp. 299–310.

Whitaker, M. and R. Patel (1990) Calcium and cell cycle control. *Development* 108: 525–542.

White, B. A. (1985) Evidence for a role of calmodulin in the regulation of prolactin gene expression. *J. Biol. Chem.* 260: 1213–1217.

Williams, D. A., K. E. Fogarty, R. Y. Tsien, and F. S. Fay (1985) Calcium gradients in single smooth muscle cells revealed by the digital imaging microscope using fura-2. *Nature* 318: 558–561.

Williamson, J. R. and J. R. Monck (1989) Hormone effects on cellular Ca²⁺ fluxes. *Annu. Rev. Physiol.* 51: 107–124.

Witchel, H. J. and R. A. Steinhardt (1990) 1-methyladenine can consistently induce a fura-detectable transient calcium increase which is neither necessary nor sufficient for maturation in oocytes of the starfish *Asterina miniata. Dev. Biol.* 141: 393–398.

Woodrow, M., N. A. Clipstone, and D. Cantrell (1993) p21ʳᵃˢ and calcineurin synergize to regulate the nuclear factor of activated T cells. *J. Exp. Med.* 178: 1517–1522.

Worley, P. F., R. V. Bhat, J. M. Baraban, C. A. Erickson, B. L. McNaughton, and C. A. Barnes (1993) Thresholds for synaptic activation of transcription factors in hippocampus: correlation with long-term enhancement. *J. Neurosci.* 13: 4776–4786.

Yamamoto, M., T. Igarashi, M. Muramatsu, M. Fukagawa, T. Motokura, and E. Ogata (1989) Hypocalcemia increases and hypercalcemia decreases the steady-state level of parathyroid hormone messenger RNA in the rat. *J. Clin. Invest.* 83: 1053–1056.

Yano, S., Fukunaga, K., Y. Ushio, and E. Miyamoto (1994) Activation of Ca²⁺/calmodulin-dependent protein kinase II and phosphorylation of intermediate filament proteins by stimulation of glutamate receptors in cultured rat cortical astrocytes. *J. Biol. Chem.* 269: 5428–5439.

Yoshikuni, M., K. Ishikawa, M. Isobe, T. Goto, and Y. Nagahama (1988) Characterization of 1-methyladenine binding in starfish oocytes cortices. *Proc. Natl. Acad. Sci. USA.* 85: 1874–1877.

Yoshimoto, Y., T. Iwamatsu, and Y. Hiramoto (1985) Cyclic changes in intracellular free calcium levels associated with cleavage cycles in echinoderms and medaka eggs. *Biomed. Res.* 6: 387–394.

Zhang, D. H., P. Wadsworth, and P. K. Hepler (1992) Modulation of anaphase spindle microtubule structure in stamen hair cells of *Tradescantia* by calcium and related agents. *J. Cell Sci.* 102: 79–89.

21

Traversing the Cell Cycle: The Calcium/Calmodulin Connection

Anthony R. Means
Christina R. Kahl
Donna G. Crenshaw
Jennifer S. Dayton

A great deal has been written about the requirements for Ca^{2+} and Ca^{2+}-binding proteins, such as calmodulin, in cell growth and proliferation. The universal consensus is that both Ca^{2+} and calmodulin are required for cells to grow and divide. Much less is currently understood about specific signaling pathways by which these substances influence growth. We view Ca^{2+} as a primary intracellular messenger that acts via receptor proteins, such as calmodulin, to regulate signal transduction pathways. Given this premise, the important events that lead to cell growth and/or proliferation are generation of the Ca^{2+} signal, activation of the appropriate Ca^{2+} receptor, and recognition of the target protein or enzyme that is a prerequisite for the physiological event being investigated. In some cases, these primary events could initiate a cascade of subsequent reactions, some of which could be influenced by other message systems, such as that for cAMP, steroid hormones, or peptide hormones. In other scenarios, activation of the Ca^{2+}/calmodulin target could be the penultimate reaction required to realize the response. Recent reviews, including some from this laboratory, have focused on Ca^{2+} generation, Ca^{2+} signaling, Ca^{2+}-binding proteins, or Ca^{2+}/calmodulin action (Whitaker and Patel, 1990; Lu and Means, 1993; Means, 1994; Reddy, 1994; Berridge, 1995; Clapham, 1995; Takuwa et al., 1995; Whitfield et al., 1995). In this chapter, we emphasize what is known about specific pathways by which Ca^{2+}, calmodulin, or Ca^{2+}/calmodulin regulate cell progression through the cell cycle check points at G_1/S, G_2/M, and metaphase/anaphase.

Calcium Is Required for Cell Proliferation

Calcium is required for proliferation of all eukaryotic cells. For cells from multicellular organisms and many unicellular organisms, Ca^{2+} must be present in the culture media and levels of free intracellular Ca^{2+} must rapidly respond to external cues. When asynchronous cultures of proliferating human fibroblasts are placed in media that contains low Ca^{2+}, they cease dividing and accumulate in G_1 (Hazelton et al., 1979). Indeed, such cells experience two periods in which they are very sensitive to the depletion of extracellular Ca^{2+}. The first is early in G_1, before the restriction point where Ca^{2+} deprivation markedly delays entry into S phase. The second is late in G_1, close to the G_1/S boundary, where removal of Ca^{2+} results in greatly reduced numbers of cells that initiate DNA synthesis (Boynton et al., 1977). In the latter case, readdition of Ca^{2+} induces a synchronous entry of cells into S phase within 1 hour. Both G_1-sensitive block points are fully reversible. Both Ca^{2+}-dependent block points are also observed upon mitogenic stimulation of quiescent (G_0) cells. Data of this kind reveal that even for progression through G_1, Ca^{2+} must play multiple roles (Boynton et al., 1977; Hazelton et al., 1979).

Cell cycle progression can also be halted by depletion of the intracellular inositol trisphosphate (IP_3)-sensitive Ca^{2+} store in response to such pharmacological agents as thapsigargin, cyclopiazonic acid, or 2,5-di-*tert*-butylhydroquinone (BHQ) (Short et al., 1993). These compounds all work by selectively inhibiting the Ca^{2+}-pumping ATPase present in the endoplasmic reticulum. Such arrest is due to pool depletion rather than to changes in the levels of free cytosolic Ca^{2+} ($[Ca^{2+}]_i$). The consequences of emptying the intracellular Ca^{2+} pool are remarkable and include inhibition of DNA and protein synthesis, as well as passive diffusion and signal-mediated transport into the nucleus (Short et al., 1993; Greber and Gerace, 1995; Hussain et al., 1995). Proliferation arrest of NIH 3T3 cells is accompanied by a resetting of the cell cycle to a G_0-like state (F. Ribeiro-Neto and A. R. Means, unpublished data). This occurs when depletion is executed at any time from G_0 through G_1 through S phase. Even cells with partially replicated DNA behave, upon reversal of the arrest, as if they have retreated to G_0. If the thapsigargin-induced arrest is not reversed, the cells can eventually die due to apoptosis. Finally, Ca^{2+}-pool depletion triggers an influx of Ca^{2+} through voltage-independent Ca^{2+} channels in a process that has been termed capacitive Ca^{2+} entry (Putney, 1990). This process of depleting and repleting the stores by the action of Ca^{2+}-ATPases serves to increase the frequency of the Ca^{2+}-release cycle (Camacho and Lechleiter, 1993). However, in thapsigargin-treated cells, stimulation of Ca^{2+} entry appears to be a futile attempt to replete the intracellular Ca^{2+} stores. Thus, not only is extracellular Ca^{2+} required for cell cycle progression but also the presence of Ca^{2+} in the endoplasmic reticulum pool is absolutely required for proliferation.

Whether Ca^{2+} enters the cytoplasm via Ca^{2+} channels in the plasma membrane or by release from the IP_3-sensitive intracellular Ca^{2+} pools, the diffusion of this ion in the cell is very limited (Allbritton et al., 1992). Thus, repeated Ca^{2+} transients or spikes are the predominant mechanism used to generate a Ca^{2+} signal (Allbritton and Meyer, 1993; Berridge, 1995; Clapham, 1995). Calcium release can be restricted to a local area, proceed throughout the cell in waves, or may even be restricted to a particular organelle such as the nucleus, although the latter is controversial (Allbritton et al., 1994; Stehno-Bittel et al., 1995). And, regardless of how the Ca^{2+} signal is generated, it must be transduced. This is the role of the intracellular Ca^{2+} receptors, such as calmodulin.

Calmodulin is the only high-affinity Ca^{2+} receptor shown to be present in all eukaryotic cells. A variety of organisms such as the yeasts *Saccharomyces cerevisiae* and *Schizosaccharomyces pombe*, the ascomycete fungus *Aspergillus nidulans*, and the fruit fly *Drosophila melanogaster*, contain a unique calmodulin gene and, in all these organisms, the gene is essential (Davis et al., 1986; Takeda and Yamamoto, 1987; Rasmussen et al., 1990; Heiman et al., 1996). Mammals such as mouse, rat, and human contain three calmodulin genes (Bender et al., 1988; Fischer et al., 1988; Danchin et al., 1989; Nojima, 1989). In each case, the genes are located on different chromosomes but encode the same amino acid sequence (Berchtold, 1993). Remarkably, a number of cell types express all three gene products (see, e.g., Slaughter and Means, 1989; Zhang et al., 1993). To carefully evaluate the individual contribution of a single gene would require selective disruption by homologous recombination. In a chicken cell line, where two genes seem to be expressed, disruption of one produces no obvious cell proliferation defects, although the calmodulin content is decreased severalfold. However, these cells do show increased sensitivity to ionomycin-induced apoptosis (Qunrui Ye and Martin Berchtold, personal communication). Thus, most of what we know about the function of calmodulin in mammalian cell proliferation comes from a variety of pharmacological studies.

Calmodulin is Important for Progression into S Phase

Chafouleas et al. (1982) first showed that calmodulin levels in CHO cells doubled at the G_1/S boundary due to an increase in the rate of protein synthesis. Addition of the anticalmodulin drug W13 caused an arrest of the cell cycle in late G_1 and the block was readily reversed by removal of the drug. A 2-fold increase in the concentration of W13 caused cell death, so it was suggested that the G_1 concentration of calmodulin was required for survival whereas the extra calmodulin synthesized in late G_1 was required for an orderly progression into and through DNA synthesis. These observations were confirmed by Sasaki and Hidaka (1982) and have now been extended to a wide variety of cells in culture. Calmodulin is also required for re-entry of G_0 cells into the proliferative cycle in response to a mitogen (Chafouleas et al., 1984). In this case, calmodulin seems to be required at two points during re-entry. The first is soon after addition of the mitogen and the second is at the G_1/S boundary. These observations have also been proven to be universal in mammalian cells and extend to circumstances in which cells, such as human peripheral lymphocytes, normally rest in G_0 awaiting the appropriate environmental cue to proliferate (Colomer et al., 1993).

The importance of calmodulin for the G_1/S transition was confirmed by Rasmussen and Means

(1989a) who generated stable lines of BPV-transformed mouse C127 cells that harbored an inducible vector that expressed calmodulin antisense RNA. Accumulation of antisense RNA correlated with a reversible cell cycle arrest in late G_1. In addition, Reddy et al. (1992a) showed that treatment of permeabilized fibroblasts with calmodulin-specific monoclonal antibodies inhibited DNA synthesis. Perhaps more intriguing was the earlier observation of Rasmussen and Means (1987), who showed that increasing the concentration of calmodulin in C127 cells shortened the length of G_1 without affecting the duration of the other cell cycle compartments. This was the first example of overexpression of a protein that led to a decrease in the length of G_1. Subsequently, overexpression of four additional proteins has been reported to shorten G_1. These proteins are cyclins E, D, and A, as well as the proto-oncogene c-*myc* (Ando et al., 1993; Kato and Sherr, 1993; Ohtsubo and Roberts, 1993; Quelle et al., 1993; Resnitzky et al., 1994; 1995; Roussel et al., 1995; Wimmel et al., 1994). These latter four proteins have each been implicated in growth control and assigned roles in the progression from G_1 into S. Whereas the cyclin and c-*myc* studies were carried out in immortalized but nontransformed mammalian cell lines, Rasmussen and Means (1987) used BPV-transformed mouse C127 cells. Thus, it was possible that the effect of calmodulin on G_1 required the transformed phenotype. More recently, Lu et al. (1992) revealed that overexpression of calmodulin in *A. nidulans* also shortened the duration of the nuclear division cycle. It seems unlikely that the results obtained by overexpressing calmodulin are any less significant than the effects of overexpressing c-*myc* or the G_1 cyclins.

What is clear is that calmodulin-dependent pathways may influence progression past the restriction point. The order of appearance of the cyclins during G_1 is D before E before A (Heichman and Roberts, 1994; Reddy, 1994; Sherr, 1994; Resnitzky et al., 1995). However, both cyclin expression and activity of the cyclin/cyclin-dependent kinase (cdk) complexes overlap significantly as cells progress through G_1 into S phase. The activity of cdk complexes is regulated by cyclin-dependent kinase inhibitors (Sherr and Roberts, 1995). The p21 family regulates all G_1 cdk complexes, while the p16 family regulates cyclin D/cdk complexes. One critical event in G_1 progression is phosphorylation of the tumor suppressor retinoblastoma protein (Rb). Lundberg and Weinberg (1998) demonstrate that phosphorylation of Rb occurs in a sequential manner, first by cyclin D/cdk4/6 complexes and then by cyclin E/cdk2 complexes. Phosphorylation of Rb leads to release of the E_2F family of transcription factors, which regulate the expression of a number of genes involved in S phase

progression and/or growth control, including the genes encoding cyclin A, p34^{cdc2}, PCNA, and DNA polymerase α. DeGregori et al. (1995) found that overexpression of E_2F in cells arrested in G_1 by γ-irradiation is sufficient to overcome this block. These authors suggest that E_2F activation may constitute the mammalian cell G_1 restriction point. Calmodulin-dependent pathways may be required in G_1 prior to Rb phosphorylation, after Rb phosphorylation, or both. One possibility is that calmodulin and calmodulin-binding proteins influence the activity of cyclin/cdk complexes, either through direct effects on the cyclins, cdks, or cdk inhibitors or indirectly through upstream pathways. Another possibility is regulation of transcription by calmodulin-binding proteins because both calcineurin and calmodulin-dependent protein kinases (CaMKs) have established functions in the regulation of transcription factors.

Is it possible that calmodulin might influence G_1 progression in a similar way as used by the G_1 cyclins? At this time, it is not possible to state this conclusively but correlative data are provocative. The first implication that calmodulin might be involved in regulation of DNA synthesis was by Chafouleas et al. (1984), who showed that this protein was required for recovery of CHO cells from potentially lethal DNA damage induced by bleomycin. They speculated that calmodulin might be required for DNA repair and/or replication by regulating the activity of a DNA polymerase. Reddy et al. (1992b) identified a 68 kDa calmodulin-binding protein (CaM-BP68) whose expression and nuclear localization were associated with mitogen-dependent progression of mammalian cells from G_1 to S. This protein also fractionated with the "replitase complex" of Reddy and Pardee (1980) that contained DNA polymerase α and was implicated in the ability of the cell to enter S phase (Reddy and Fager, 1993). It was suggested by Reddy et al. (1994) that purified CaM-BP68 was capable of stimulating DNA synthesis when introduced into permeabilized hematopoietic progenitor cells. Cao et al. (1995) found that CaM-BP68 was tightly associated with the DNA polymerase α/primase complex isolated from HeLa cells, supporting an earlier observation that calmodulin-binding proteins were associated with immunopurified DNA polymerase α from a number of mammalian cell lines (Hammond et al., 1988). Unfortunately, no studies have been reported that show a role for calmodulin or CaM-BP68 in regulation of DNA polymerase α activity. However, it is intriguing that the cyclin D-Rb-E_2F pathway results in the synthesis of DNA polymerase α. It is certainly possible that activity of the enzyme may also require regulatory factors, one of which could be calmodulin. At any rate, the reagents now exist to elucidate how calmodulin is involved at the G_1/S boundary in mammalian cells.

The Quest for Identification of Calmodulin Targets

Calcineurin

Perhaps easier to interpret is the role for calmodulin in G_1 in unicellular organisms. When the single calmodulin gene is deleted or disrupted in the three fungal systems, *S. cerevisiae, S. pombe*, and *A. nidulans*, at least some of the cells arrest in G_1. Treatment of *S. cerevisiae* with low concentrations of the calmodulin inhibitor trifluoperazine (TFP) — concentrations that did not produce membrane effects — prevented cells from initiating DNA synthesis upon release from α-factor arrest (Eilam and Chernichovsky, 1988). A similar phenotype is observed when all three genes that encode subunits of the Ca^{2+}/calmodulin-dependent protein phosphatase, calcineurin are deleted (Cyert et al., 1991; Kuno et al., 1991; Liu et al., 1991; Cyert and Thorner, 1992). Since α-factor arrests the cell cycle before START, such studies are compatible with a role for calcineurin in the resumption of cell proliferation from a G_0/G_1 arrest. However, under normal growth conditions, yeast cells null for all three calcineurin subunits (two catalytic A genes and one Ca^{2+}-binding B gene) show no abnormalities in G_1 progression when cells of the opposite mating type are absent. Similarly, when the unique calcineurin A gene is deleted from *S. pombe*, no effects on DNA synthesis were observed even though null cells displayed a number of phenotypes, including impairment of polarity control, and delayed cytokinesis and mating (Yoshida et al., 1994). It has been possible to construct yeast strains in which calcineurin is essential, but such mutants are defective in cation homeostasis due to deletion of plasma membrane or vacuolar cation ATPases (Breuder et al., 1994; Cunningham and Fink, 1994; Mendoza et al., 1994; Garrett-Engele et al., 1995). The general conclusions have been that calcineurin is normally not essential and is not required for yeast cells to progress from M through G_1 and into S.

Results have been more satisfying in *A. nidulans*. This organism has a single calcineurin A gene which is essential (Rasmussen et al., 1994). The gene is transcribed in a cell cycle-dependent manner with maximal levels in late G_1, and analysis of the arrest point in the null strain suggest that it is early in the cell cycle, probably in G_1. We have now succeeded in preparing, by homologous recombination, a strain that is conditional for the expression of calcineurin A by replacing the endogenous gene with a copy regulated by the alcohol dehydrogenase (*alc* A) promoter (N. Nanthakumar and A. Means, unpublished observations). This promoter is regulated by the carbon source in the media. It is induced by threonine, repressed by glucose and, in the presence of glycerol,

allows low-level transcription. Germination of spores in repressing media or shifting exponentially growing cells into glucose results in arrest of the nuclear division cycle in G_1. However, because spores or exponentially growing cells must contain some calcineurin to survive, and it takes a few hours to deplete that calcineurin by turnover upon transfer to media that repress the calcineurin A gene, it is not possible to rule out additional roles for calcineurin in the nuclear division cycle. The immunosuppressive drug FK506 complexes with its target protein FKBP-12 and inhibits calcineurin in yeast and mammalian cells (Schreiber, 1992; Griffith et al., 1995). This drug is also lethal to *A. nidulans* displaying an LD_{50} of 50 ng/ml. An analog of FK506 called L-685,818 does not affect the growth of *A. nidulans*. This analog also binds to FKBP-12 and inhibits its prolyl isomerase activity but does not inhibit calcineurin (Dumont et al., 1992; Becker et al., 1993). These experiments suggest that the lethal effect of FK506 on the fungus is due to inhibition of calcineurin. The rapidity of the FK506 effect allowed us to more thoroughly examine whether calcineurin might be important for the requirement for calmodulin in G_2 and M. This was evaluated by questioning the effect of FK506 on the progression of cells released from arrest in S phase with hydoxyurea or in G_2 by means of a temperature-sensitive mutation in the $nimT^{cdc25}$ gene. In both cases, released cells proceeded through M before arresting, suggesting that calcineurin is not required to complete either G_2 or M (N. Nanthakumar and A. Means, unpublished observations). Together, these results suggest that a primary target for calmodulin regulation of G_1/S in *A. nidulans* is calcineurin. Possible substrates for calcineurin involved in G_1 progression are currently being sought.

Is calcineurin involved in the regulation of G_1 progression in mammalian cells? Certainly, this enzyme is required in the activation of T-cells in response to occupancy of the T-cell receptor (Crabtree and Clipstone, 1994; Cardenas and Heitman, 1995). Since these cells are quiescent prior to stimulation, they might be considered to be in G_0. It has long been known that Ca^{2+} is required for activation and that this process is sensitive to the immunosuppressive agents FK506 and cyclosporin A (CsA). Of course, the target of these drugs and a target of the Ca^{2+} signal is now known to be calcineurin (Schreiber, 1992; Crabtree and Clipstone, 1994; Cardenas and Heitman, 1995). Calcineurin dephosphorylates a cytoplasmic component of the transcription factor NF-AT that participates in inducing transcription from the interleukin-2 (IL-2) gene. The IL-2 promoter contains binding sites for NF-AT, NF-κB, AP-1, and Oct-1. Some evidence exists to suggest that all of the cognate transcription factors require Ca^{2+} and/or calcineurin (Ullman et al., 1990).

The current, conventional wisdom is that calcineurin-mediated dephosphorylation of the cytoplasmic NF-AT subunit results in its movement into the nucleus where it interacts with other binding components to form the transcriptionally active complex. IL-2 and other cytokines act as mitogens in an autocrine manner. Thus, this effect of calcineurin is more like the early G_0 requirement for Ca^{2+}/calmodulin in response to mitogenic signals (Chafouleas et al., 1984) and occurs long before the activated T-cells reach the G_1/S boundary. Once mitogenesis has been initiated by IL-2, cell proliferation is no longer sensitive to FK-506 but is blocked at the G_1/S boundary by another immunosuppressive drug, rapamycin (Crabtree and Clipstone, 1994; Cardenas and Heitman, 1995). Whereas rapamycin also binds FKBP-12, this complex works through a mechanism distinct from that of FK506/FKBP-12. Thus, at least in proliferating T-cells, the Ca^{2+}/calmodulin requirement for G_1/S is not likely to be mediated by calcineurin. Calcineurin also plays a role in cell cycle, progression in fibroblasts. When growth-arrested Swiss 3T3 cells are stimulated to enter the cell cycle, using several different growth factors, in the presence of cyclosporin A or FK506, there is a marked inhibition of DNA synthesis (Tomono et al., 1996). Furthermore, Tomono et al. (1998) demonstrate that these compounds block the expression of cyclins A and E, but not D_1, in fibroblast growth factor stimulated Swiss 3T3 cells. These results suggest calcineurin is required in late G_1 after cyclin D expression but prior to cyclin E or A expression. Determination of cdk activity and Rb phosphorylation status in this system is necessary to understand where in G_1 these cells are arrested in the presence of cyclosporin and FK506. The inescapable conclusion is that the role of calcineurin as the primary Ca^{2+}/calmodulin target at G_1/S cannot be universal. Since the intricacies of cell cycle regulation involving cyclins and cyclin-dependent kinases also vary considerably between organisms, maybe this conclusion should not come as a surprise.

Calmodulin-Dependent Protein Kinase

In G_1

Provocative evidence also exists to suggest that a calmodulin-dependent protein kinase could be a primary target for Ca^{2+}/calmodulin in G_1. Rasmussen and Rasmussen (1995) have shown that the CaM kinase-selective antagonist KN-93 (Sumi et al., 1991), arrests HeLa cells in G_1. These authors quantified histone H1 kinase activity precipitable with $p13^{suc1}$ and found a 4-fold increase at the arrest point. As the $p13^{suc1}$ protein will bind cyclin A/cdk2 and cyclin E/cdk2 (Dorée and Galas, 1994),

and both of these complexes are active in G_1 close to the S-phase boundary (Sherr, 1994), it was suggested that activation of cdk activity is not dependent on CaM kinase. Rasmussen and Rasmussen (1995) further concluded that the action of CaM kinase must be required after cdk activation in late G_1 but prior to DNA synthesis. Such a conclusion is compatible with the temporal requirement for calmodulin at G_1/S. However, it should be pointed out that neither individual cyclins nor cdk isoforms were examined and that the mitotic cdk, $p34^{cdc2}$, is always present during the proliferative cycle and also interacts with $p13^{suc1}$.

In a comparable study, Tombes et al. (1995) examined the effects of KN-93 on proliferation of NIH 3T3 fibroblasts. Two days after addition of KN-93 to exponentially growing cells, 95% of the cells were arrested in G_1. In addition, KN-93 prevented the mitogenic effects of basic FGF, PDGF, EGF, or IGF-1 added to serum-starved quiescent cells considered to be in G_0. If the drug was removed from the arrested cells within 1 or 2 days, the cell cycle block was readily reversible. However, after 3 days in the presence of KN-93 a large proportion of the arrested cells died with several characteristics that resembled apoptosis. Tombes et al. (1995) concluded that CaM kinase was required for progression of cells through G_1 and operated at a site common to the cascade of events triggered by both competence and progression factors. In subsequent experiments, Morris et al. (1998) found that KN-93 blocked both activation of cdk and phosphorylation of the retinoblastoma gene product, Rb, at the G_1 arrest point. At this point, cyclin D_1 levels are significantly reduced with an associated reduction in cdk4 complex activity. Additionally, cdk2 complex activity is reduced, even though cyclin A and cyclin E levels are constant. The authors conclude that the reduction in cdk2 is due to p27 binding to cyclin E/cdk2 complexes. However, the authors did not examine cyclin E/cdk2 or cyclin A/cdk2 activities individually and they did not determine the association of other cdk inhibitors with cdk2 complexes. These results differ significantly from Rasmussen and Rasmussen (1995), who found elevated H1 kinase activity in HeLa cells arrested at G_1 with KN-93. Taken at face value, these collective results suggest that KN-93 might arrest HeLa and 3T3 cells at different points in G_1 or by different mechanisms. Whereas HeLa cells are transformed and cannot be growth-arrested by serum starvation, 3T3 cells are immortalized but not transformed and can be arrested in G_0. So, the apparently different results could reflect the properties of normal vs. transformed cells. Alternatively, as previous studies have shown that calmodulin is required early after release of cells from G_0 and also close to the G_1/S boundary, it is possible that the 3T3 cells are blocked at the first arrest point whereas the HeLa cells are blocked at the second. To test this possibility, one could add KN-93 at various times after addition of a mitogen to 3T3 cells in G_0 and determine whether a second block point exists that is close to the G_1/S boundary. If a second block point does exist, it would be interesting to assess the state of Rb phosphorylation and individual cyclin/cdk activities. To further complicate the discussion, Agell et al. (1998) demonstrate that addition of the anti-calmodulin compound W-13 5 hours after proliferative stimulation of NRK cells, does not affect cyclin D levels nor its associated kinase activity, but blocks the nuclear localization of cyclin D/cdk 4 complexes. One possibility is that CaMK is required during early after release from the resting state and that by adding W-13 5 hours after the proliferative stimuli, Agell et al. (1998) bypassed this early requirement for CaMK in G_1 progression. Indeed, calmodulin is required at two distinct points in G_1 progression and may be required at multiple steps during G_1 progression (Chafouleas et al., 1984).

Although the studies described above implicate a CaM kinase in $G_0 \rightarrow G_1/S$ progression, pharmacological approaches can only be suggestive. We know neither the identity of the in vivo KN-93 target nor, if it is a CaM kinase, which enzyme isoform is important. Most mammalian cells contain multiple CaM kinases and HeLa and 3T3 cells possess at least two enzymes, CaM kinase II and CaM kinase I, that are subject to KN-93 inhibition in vitro. The studies in *S. cerevisiae* that deal with CaM kinases do not shed any light on the problem since three genes have been cloned and no effects on cell cycle were noted even upon deletion of all three genes (Ohya et al., 1991; Pausch et al., 1991). *Aspergillus nidulans* does contain a single CaM kinase gene (Kornstein et al., 1992). We have attempted to create a strain conditional for expression of the gene, and to disrupt the gene. Collectively, the experiments indicate that CaM kinase is essential. However, because expression of the gene is leaky in the conditional strain, and deletion must be done in either diploids or heterokaryons, it is most difficult to be certain where the cells become arrested (J. Dayton and A. Means, unpublished observations). The next important step is to identify the responsible calmodulin kinase targets at the arrest point. This is of obvious importance in the mammalian cells arrested by KN-93 as well. Recently, a second CaM kinase gene in *Aspergillus nidulans* was cloned in our laboratory (J. Joseph and A. R. Means, unpublished observations). This gene is also essential and experiments are being conducted to determine the cell cycle arrest point. If this CaM kinase is required prior to DNA synthesis, then studies into its function may help elucidate the pathways

involved in the G_1 arrest caused by KN-93 in mammalian cells.

In G_2

At first glance, the data that suggest a role for calmodulin kinase in G_1 are difficult to reconcile with previous results that suggest a role in G_2. In sea urchin embryos, microinjection of antibodies to CaM kinase IIα or synthetic peptides modeled on the autoinhibitory region of this enzyme prevent nuclear envelope breakdown (Baitinger et al., 1990). Since the sea urchin embryonic cell cycle is governed primarily by M-phase checkpoints, it is possible that in this system the targets for CaM kinase differ from those present in G_1 mammalian cells. On the other hand, overexpression of a constitutively active form of CaM kinase IIα in mouse C127 cells or in *S. pombe* causes a G_2 arrest (Planas-Silva and Means, 1992; Rasmussen and Rasmussen, 1994). In mouse C127 cells, the arrested cells exhibit elevated levels of H1 kinase activity, suggesting that the block is downstream of MPF activation. However, in *S. pombe* the arrest occurs with low H1 kinase activity. Since, in these experiments, constitutive CaM kinase was generated by two different mutations, the former by truncation and the latter by point mutation, the precise cause of the arrest may be different. A similar G_2 arrest is seen in an *A. nidulans* strain that harbors an inducible constitutively active CaM kinase. When germlings that contain exponentially dividing nuclei are induced to express this enzyme, the nuclear division cycle is blocked in G_2. In addition, if spores (in G_0) are germinated in inducing media, the cell cycle is blocked prior to entry into the first S phase (M. Sumi, J. Dayton, and A. Means, unpublished data). At this arrest point, H1 kinase activity is elevated. The components of the complex responsible for the H1 kinase activity measured at these arrest points have not been identified, but at present only one cyclin-dependent kinase (cdk) homolog, NIMX (Osmani et al., 1994), and one cyclin B homolog, NIME (O'Connell et al., 1992), have been described in *A. nidulans*.

Can we suggest a mechanism to explain the existence of two arrest points? One could envision that the constitutively active CaM kinase overrides a critical dephosphorylation event or phosphorylates a critical cdk substrate, possibly on the cdk complex or supercomplex, in an inappropriate manner. The result of this event would be nonproductive activation of H1 kinase leading to induction of a checkpoint. Precedent exists for induction of checkpoints in the presence of H1 kinase activity at both G_1 (Stueland et al., 1993; Tang and Reed, 1993) and G_2 (Amon et al., 1992; Sorger and Murray, 1992), in *S. cerevisiae*. Premature activation of cdk in G_1 could lead to inhibition of DNA synthesis. In a recent review, Su et al. (1995) propose a possible mechanism to prevent re-replication. Cumulative evidence suggests that cdk activity may prevent formation of a functional origin of replication. If preinitiation complexes are not present in *A. nidulans* spores, it is possible that the premature H1 kinase activity prevents their formation and thus entry into S phase. Constitutive CaM kinase activity in G_2 may also invoke a checkpoint due to unregulated phosphorylation. Here, the checkpoint could be mediated through failure to activate and/or stabilize the second serine/threonine kinase, NIMA, that is required for mitotic progression in *A. nidulans* (Oakley and Morris, 1983; Osmani et al., 1988; see also the discussion by Sorger and Murray, 1992).

Indeed, at the G_2 arrest point in *A. nidulans* caused by low extracellular Ca^{2+} or depletion of intracellular calmodulin, $p34^{cdc2}$ remains tyrosine phosphorylated and NIMA is inactive (Lu et al., 1993). So, precedent exists for a role for Ca^{2+}/calmodulin in the activation of both $p34^{cdc2}$ and NIMA. One candidate target for Ca^{2+}/calmodulin and calmodulin kinase is the tyrosine phosphatase cdc25, which is induced at G_2 and required to activate $p34^{cdc2}$/cyclin B by dephosphorylation of threonine 14 and tyrosine 15 (Millar and Russell, 1992; Dunphy, 1994; Maller, 1994). The argument is that inactive cdc25 exists in G_2 and must be activated by Ser/Thr phosphorylation. An unknown protein kinase begins the cascade that results in a small activation of the phosphatase to a low level which results in an incremental activation of maturation-promoting factor (MPF). This active MPF then hyperphosphorylates cdc25 which increases its activity to further activate MPF (see Izumi and Maller, 1995). Interestingly, Whitaker (1995) suggests that CaM kinase II can phosphorylate largely inactive cdc25 in vitro and increase its activity. The role of this kinase in vivo was investigated by synchronizing HeLa cells in S phase and releasing into KN-93. The cells arrested at G_2/M without detectable cdc25 phosphorylation, suggesting that a Ca^{2+}/calmodulin/CaM kinase step might be required in the activation of cdc25 at G_2. Regardless of whether this conclusion is correct, Whitaker (1995) and Rasmussen and Rasmussen (1995) found different KN-93-induced arrest points in the same cell type. The primary difference in the two studies is the proliferative state of the cells at the time of addition of KN-93. Rasmussen and Rasmussen (1995) found a G_1 block in exponentially growing cells and an elevated H1 kinase activity. Whitaker (1995) found a G_2 block in cells released from an arrest at S phase. One possibility is that cells in G_1 contain half the levels of calmodulin compared with cells in other compartments of the cell cycle (Chafouleas et al., 1982), so it is possible that these cells are more sensitive to KN-93.

Collectively, these observations indicate that HeLa cells, and perhaps other cells, require calmodulin at both G_1/S and G_2/M.

What Is Downstream of Calmodulin Kinase?

Clearly, more work needs to be done to determine how and by what means CaM kinase affects each transition in the cell cycle. At the moment, the only candidate as a direct target of the kinase is cdc25. Even this candidate enzyme remains to be proven conclusively to be the relevant substrate by definitive experiments in any cell type. Provocatively, at least in mammalian cells, multiple isoforms of cdc25 exist. Jinno et al. (1994) presented evidence that one of these isoforms, cdc25A, functions early in the cell cycle in normal rat kidney cells. The protein was found to be predominantly expressed in late G_1 and immunodepletion of the phosphatase effectively blocked entry into S phase. These authors speculated that cdc25A could be a regulator of the cdks that function in late G_1. Further evidence for this possibility was presented by Galaktionov et al. (1995), who showed that the proto-oncogene Raf1 associates with cdc25A. As Ras and Raf1 are involved in transducing signals from membrane-associated receptors that lead to mitogenesis, it was suggested that activation of the cell cycle by such pathways could involve cdc25A. Indeed, cdc25A, Ras, and Raf1 were shown to colocalize to the plasma membrane and, at least in vitro, the Raf1 kinase could phosphorylate cdc25A in a manner that led to an increase in its phosphatase activity by use of a synthetic substrate. Although formal evidence that cdc25A will activate a specific cdk/cyclin complex in G_1 has yet to be reported, it is tempting to speculate on a similar role for CaM kinase in both G_1 and G_2 progression.

Overexpression of an enzyme can frequently lead to a phenotype but it is infrequent that the phenotype is easily interpreted. In the case of enzymes that phosphorylate many substrates and can phosphorylate sites usually modified by a different kinase, results are even more difficult to interpret. Sadly, this is the case for CaM kinase. It is possible that the events that lead to a G_2 arrest are entirely fortuitous and of no physiological significance. Not only does CaM kinase phosphorylate many substrates, presumably with physiological significance, but also it can phosphorylate sites recognized by PKA or PKC. So, it might be misleading to interpret the G_2 block produced by overexpression of a constitutively active CaM kinase as indicating a role for calmodulin kinase in G_2 progression. It is interesting to note that, at least in *A. nidulans*, overexpression of the wild-type CaM kinase produces no obvious effects on the nuclear division cycle. Of course, as calmodulin is rate-limiting for growth of the fungus, this lack of effect could be due to the lack of availability of calmodulin required for activation. On the other hand, overexpression of constitutively active CaM kinase or addition of KN-93, which presumably inhibits endogenous CaM kinase, can arrest cell proliferation at the same points. Maybe this is another example of the requirement for orderly phosphorylation/dephosphorylation of a single substrate being required for proper cell cycle progression past a checkpoint.

Calcineurin and Calmodulin Kinase May Be Involved at Multiple Checkpoints

The question of the identity of the calmodulin target(s) that is/are required for mammalian cells to progress from G_0 to G_1 to S is complicated. Candidates discussed above include the 68 kDa calmodulin-binding protein that might influence the activity of DNA polymerase α, a multifunctional calmodulin-dependent kinase (or kinases), and the protein phosphatase 2B, calcineurin. If all three of these proteins are targets and all are required for entry into S phase, attempting to dissect the individual pathways and identify downstream targets will be a daunting task. Certainly, KN-93 and FK506 (which inhibits calcineurin) will arrest the progression of mitogenically activated mammalian cells into S phase. The possibility exists that FK506 could be working through a target other than calcineurin, but preliminary data in vascular smooth muscle cells suggest that a synthetic peptide that represents the autoinhibitory region of calcineurin A will also inhibit proliferation prior to S phase (Samuel E. George, personal communication). If both CaM kinase and calcineurin are required, could they be acting on a common substrate? At this time, the answer is unknown but there is precedent for a protein substrate that can be activated by phosphorylation by CaM kinase and inactivated by dephosphorylation by calcineurin. This protein is a neuronal Ca^{2+}-channel (Armstrong et al., 1991). One could question why a common second messenger would be employed to stimulate and inactivate a single physiological response. However, it must be kept in mind that Ca^{2+} transients in cells can be of very different magnitudes and the amount of Ca^{2+} (and calmodulin) required to activate various calmodulin-dependent enzymes can vary by at least 2 orders of magnitude. Indeed, CaM kinase II isoforms require more than 100-fold more Ca^{2+} than calcineurin for half-maximal activation (see Hanson and Schulman, 1992, and references therein). The diffusion of Ca^{2+} in cytoplasm is also quite limited (Allbritton et al., 1992). So, it is possible that temporally rapid Ca^{2+} transients of different amplitudes coupled with dis-

crete areas of release (or entry) into the cytoplasm could, in theory, result in an activation of CaM kinase followed by calcineurin, or vice versa. Such a complicated scenario points to the importance of identifying the relevant substrates for both kinase and phosphatase. Based on the current information, this may be most readily achieved in A. nidulans.

The question of why it is possible to achieve different arrest points in the same cell type in response to KN-93 could be due to the fact that calmodulin is required at multiple points in the cell cycle. The prominent block could well be influenced by the first such point encountered by the majority of the cells. Eilam and Chernichovsky (1988) described such a situation in S. cerevisiae in response to low concentrations of the anticalmodulin drug trifluoperazine (TFP). When the drug was added before START, prior to release from α-factor arrest (more or less equivalent to G_0 in mammalian cells), after release from a temperature-sensitive block caused by mutation of the cdc28 gene or at the transition from stationary phase to vegetative growth, cell growth, bud formation and DNA synthesis were inhibited. These results were interpreted to indicate that TFP exerted a cell cycle-specific inhibitory effect in G_1 before or at START. However, when TFP was added after execution of spindle pole body duplication, cell division was inhibited and the cells arrested with buds at G_2/M. The precise point of this arrest was determined to be just before the medial nuclear division. These results clearly showed a second TFP-sensitive arrest point in late G_2 or M. Thus, the position of the cells in the cell cycle at the time of exposure to TFP greatly influenced the nature of the arrest point.

Essential Roles for Calmodulin in *Saccharomyces cerevisiae*

Davis et al. (1986) showed that S. cerevisiae contained a single calmodulin gene (CMD1) and that this gene was essential. It is now known that calmodulin has at least four essential functions in yeast. The first two to be identified genetically were chromosome segregation and the polarized growth required for bud formation (Davis, 1992; Sun et al., 1992). These essential points in the cell cycle could explain the two TFP-sensitive points in cell proliferation observed by Eilam and Chernichovsky (1988). Subsequently, Ohya and Botstein (1994) individually mutated the codons for the phylogenetically conserved phenylalanine residues in the CMD1 gene, replaced the endogenous CMD1 gene with each mutant gene, and screened for temperature-sensitive growth. The resultant mutant strains were classified into four complementation groups which indicated that there were potentially as many essential targets of calmodulin. The four complementation groups had mutations that affected actin organization, calmodulin localization, bud emergence, or nuclear division. The latter two phenotypes are also compatible with the two TFP-sensitive points observed by Eilam and Chernichovsky (1988), as well as the previous genetic analysis by Davis (1992) and Sun et al. (1992).

The Davis laboratory has now identified two of the essential targets for calmodulin in *S. cerevisiae*. The essential mitotic target is the 100 kDa component of the spindle pole body called Nuf1p or SPC110 (Geiser et al., 1993). The second essential gene is that encoding the nonconventional myosin Myo2P. This calmodulin-binding protein is required at sites of cell growth (Brockerhoff et al., 1994). A homolog of Myo2P has also been shown to be essential in *A. nidulans* (McGoldrick et al., 1995). The startling fact about all of the essential targets of calmodulin found in *S. cerevisiae* is that they do not require calcium binding to calmodulin. In fact, Geiser et al. (1991) showed that calmodulin with all the Ca^{2+}-binding sites rendered inactive by mutation could support cell proliferation equally as well as the wild-type calmodulin. These observations, coupled with the fact that neither CaM kinase nor calcineurin is essential in *S. cerevisiae*, suggest that budding yeast has evolved in such a manner to maintain multiple essential requirements for calmodulin but to exclude the involvement of Ca^{2+} as an intracellular second messenger to support these requirements.

A number of observations suggest that the situation in *S. cerevisiae* is unique even among other fungal strains. First, the Ca^{2+}-dependent calmodulin-binding proteins CaM kinase and calcineurin are essential in the filamentous fungus *A. nidulans* (Rasmussen et al., 1994; Lu et al., 1995). Second, the *A. nidulans* calmodulin gene in which all four Ca^{2+}-binding sites have been silenced by mutation only partially complements, at best, whereas the wild-type *S. cerevisiae CMD1* gene fails to fulfill the essential roles of calmodulin in *A. nidulans* (J. Joseph and A. Means, unpublished observations). Moser et al. (1995) have shown that at least one Ca^{2+}-binding site must be present in *S. pombe* calmodulin to perform its essential function and that wild-type *S. cerevisiae* calmodulin fails to support the growth of fission yeast even when overproduced. On the other hand, the calcineurin A gene, although unique, is not essential in *S. pombe* (Yoshida et al., 1994). Indeed, at the time of this writing, no Ca^{2+}-dependent target of calmodulin has been shown to be essential in fission yeast. It will be very interesting to learn if the homolog of Nuf1p is essential in *S. pombe* or *A. nidulans* and, if so, whether these calmodulin-dependent events require Ca^{2+}.

Calmodulin Is Required for the G₂/M Transition

Whereas considerable circumstantial evidence exists to support a role for Ca^{2+} and calmodulin in the G_2 phase of proliferating cells, the best direct experimental confirmation is probably in *A. nidulans*. Lu et al. (1992) created a strain of this fungus that was conditional for the expression of calmodulin. When calmodulin levels were held low, this strain could not grow and characterization of the cells showed that about 80% were arrested in G_2. As the expression of the calmodulin gene was under control of the inducible *alc*A promoter, it was possible to induce calmodulin in the arrested cells and show that the block was fully reversible. Subsequently, Lu et al. (1993) substituted the *alc* A-driven calmodulin gene for the endogenous one in a strain of *A. nidulans* that also had a temperature-sensitive mutation in the *NIMT*^{cdc25} gene. At the restrictive temperature, the cells were arrested at G_2 prior to the activation of NIMX^{cdc2}, regardless of the level of Ca^{2+} or calmodulin. However, when the temperature-sensitive block was released and either extracellular Ca^{2+} or intracellular calmodulin had been depleted, the cells remained arrested in G_2. At this block point, NIMT^{cdc25} was probably inactive, since NIMX^{cdc2} remained tyrosine phosphorylated and the NIMA protein kinase did not become active. It was suggested that the status of intracellular Ca^{2+} pools was important. When EGTA was added only 10-15 minutes prior to the temperature shift, the cells synchronously proceeded into and through mitosis. However, if EGTA was present for an hour or more before shifting to the permissive temperature, the cells remained blocked in G_2. If only the entry of Ca^{2+} from the extracellular medium was important, simply removing extracellular Ca^{2+} should have been sufficient to maintain the cell cycle block and a time-dependence would not be anticipated. The conclusions of this study were that a threshold level of Ca^{2+}/calmodulin was required for the NIMT^{cdc25}-mediated tyrosine dephosphorylation of NIMX^{cdc2}, as well as for the activation of NIMA as a protein kinase.

At this time, exactly how Ca^{2+}/calmodulin regulates the activity of NIMT^{cdc25} and NIMA and thereby is involved in G_2 progression in *A. nidulans* is unknown. Neither enzyme directly binds calmodulin nor is the activity of either enzyme influenced by Ca^{2+} and/or calmodulin in vitro. Both NIMT^{cdc25} and NIMA are phosphoproteins and phosphorylation of each is required for activity (Lu et al., 1993; Izumi and Maller, 1995). Calmodulin kinase can phosphorylate both enzymes and increase their enzyme activities in vitro (Lu et al., 1993; Whitaker, 1995). At least in HeLa cells, an inhibitor of CaM kinase can produce a G_2 arrest, and in *A. nidulans* the unique CaM kinase gene is essential. But whether phosphorylation of either NIMT^{cdc25} or NIMA by CaM kinase represents a rate-limiting event in G_2 progression remains a mystery. To answer this question, the sites phosphorylated by CaM kinase must be identified. These residues must then be altered by mutation of the genes and the mutant forms must be used to replace the endogenous *A. nidulans* gene to determine if they are of physiological relevance. The problem is that both NIMT^{cdc25} and NIMA become hyperphosphorylated in vivo and it is clear that multiple protein kinases are involved (Izumi and Maller, 1995; Ye et al., 1995). Finally, although progression through mitosis from G_2 in *A. nidulans* requires both NIMX^{cdc2} and NIMA activities, the relationship between these two enzymes is uncertain. It is possible that they function via separate but parallel signaling pathways. Alternatively, the enzymes could represent two activities in a single pathway. Ca^{2+}/calmodulin (and possibly the CaM kinase) could be rate-limiting for the convergence of the two enzymes in the sequence of events that govern the G_2/M progression.

Progression Through Mitosis Requires Calmodulin

Calmodulin is not only required for the entry of cells into mitosis but also for the progression of cells through mitosis (Lu and Means, 1993). Calmodulin is found in the mammalian cell nucleus where it is associated with the centrosomes during mitosis (Welsh et al., 1978). It has been suggested that calmodulin moves into the nucleus in response to a sustained increase in intracellular Ca^{2+} (Luby-Phelps et al., 1995), and the translocation appears to be by an active mechanism (Pruschy et al., 1994). Several lines of evidence suggest that Ca^{2+} and calmodulin may be involved in the breakdown of the nuclear membrane that signals the initiation of mitosis. First transient increases in $[Ca^{2+}]_i$ have been associated with nuclear envelope breakdown in sea urchin embryos (Poenie et al., 1985). This event can be induced prematurely by the IP_3-stimulated release of Ca^{2+} from intracellular stores (Twigg et al., 1988). Alternatively, progression through mitosis can be arrested prior to dissolution of the nuclear membrane by microinjection of the Ca^{2+}-chelator EGTA (Steinhardt and Alderton, 1988; Twigg et al., 1988) or by expression of the Ca^{2+}-binding protein parvalbumin (Rasmussen and Means, 1989b). Baitinger et al. (1990) have shown that one calmodulin-dependent enzyme required for this process is CaM kinase II. Since nuclear lamins are phosphorylated in early mitosis and this modification leads to depolymerization, it is possible that CaM kinase is one of the enzymes responsible.

However, the mechanism by which this enzyme is involved has yet to be elucidated.

Second, Ca^{2+} and calmodulin have been suggested to play a role in the process that controls the transition from metaphase to anaphase. In the specialized case of the release, after fertilization, of *Xenopus* oocytes from metaphase arrest during the second meiotic division, the case is clear. Calcium is both necessary and sufficient to trigger release from this block (Whitaker and Patel, 1990). The block is due to the presence of a "cytostatic factor" activity (CSF) that includes the proto-oncogene c-*mos* and maintains the activity of maturation-promoting factor (MPF) which is composed of p34^{cdc2} and cyclin B. A rise in intracellular Ca^{2+} triggers the degradation of cyclin in metaphase-arrested *Xenopus* oocytes. Lorca et al. (1991) showed that introduction of a synthetic calmodulin-binding peptide, modeled on the calmodulin-binding region of smooth muscle myosin light chain kinase (MLCK), prevents this Ca^{2+}-induced cyclin degradation and they concluded that both Ca^{2+} and calmodulin are required for this event. Furthermore, it was shown that degradation of c-*mos* was not required and, in fact, did not occur. Subsequently, the target of Ca^{2+}/calmodulin was shown to be CaM kinase II. Lorca et al. (1993) demonstrated that introduction of a Ca^{2+}/calmodulin-independent form of CaM kinase II was sufficient for cyclin degradation and release from metaphase. This effect of the constitutively active enzyme fragment was independent of Ca^{2+} and occurred in the presence of an excess of the MLCK calmodulin-binding peptide. On the other hand, the response to the CaM kinase II fragment was effectively inhibited by a synthetic peptide analog of the autoinhibitory domain of mammalian CaM kinase IIα. The fertilization-induced rise in $[Ca^{2+}]_i$ was shown to activate endogenous CaM kinase and both the activity of the enzyme and release from metaphase were blocked by the autoinhibitory peptide. A third paper from the Dorée group (Morin et al., 1994) demonstrated that CaM kinase II is the only Ca^{2+} target required for the induction of anaphase in CSF extracts that contain metaphase spindles. This effect did not require direct phosphorylation of metaphase spindle proteins. However, anaphase did occur in the absence of CaM kinase II when spindles were added to extracts in which the ubiquitin-dependent proteolysis pathway had been previously activated. Morin et al. (1994) concluded that CaM kinase II is required both for the release from metaphase arrest and for sister chromatid exchange.

As mentioned previously, calmodulin is a component of the mammalian mitotic spindle apparatus throughout mitosis and is concentrated in the centrosomal regions at metaphase (Welsh et al., 1978). This observation led to experiments by Marcum et al.

(1978), who suggested that calmodulin might be involved in the depolymerization of microtubules that preceded the movement of chromosomes from the metaphase plate to the spindle poles during the metaphase–anaphase transition. The functional equivalent of the centrosome in cells that do not undergo nuclear envelope breakdown at mitosis is the spindle pole body (SPB). In *S. cerevisiae*, the 110 kDa protein component of the SPB is essential and, as mentioned previously, is a calmodulin-binding protein (Geiser et al., 1993). However, this interaction does not require Ca^{2+} and the 110 kDa protein apparently plays a structural rather than catalytic role. It has been suggested by Stirling et al. (1994) that calmodulin plays a role in controlling the stability of the 110 kDa protein. In mammalian cells, micromolar Ca^{2+} is sufficient to initiate microtubule depolymerization in the metaphase spindle (Cande, 1980). So, even if mammalian cells do contain a homolog to the yeast 110 kDa SPB protein and it binds calmodulin in the absence of Ca^{2+}, it is likely that a Ca^{2+}-dependent event is also required for the metaphase–anaphase transition. One Ca^{2+}/calmodulin-dependent enzyme that has also been localized to the centrosomal region of mitotic cells is CaM kinase II (Ohta et al., 1990). This enzyme has been shown to be a component of isolated centrosomes and to be enzymatically active (Pietromonaco et al., 1995). At least in sea urchins, a CaM kinase II substrate associated with the mitotic apparatus has been shown to exist (Dinsmore and Sloboda, 1988), although neither its identity nor function is known. Thus, the precise role of calmodulin and/or CaM kinase II in the metaphase–anaphase transition in metazoan cells remains to be determined.

Is Calmodulin Required for Cytokinesis?

The third event in mitosis that likely requires calmodulin is cytokinesis. The first suggestion of a role in cytokinesis was made by Guerriero et al. (1981), who localized MLCK to the nucleus and found it in the cleavage furrow along with calmodulin. Fishkind et al. (1991) microinjected a catalytically active but Ca^{2+}/calmodulin-independent form of MLCK into dividing cells, which resulted in a delay in the time between nuclear envelope breakdown and anaphase. This delay seemed to be coupled to disruption of spindle structure. However, the injected cells still formed functional cleavage furrows at the same rate as control cells. Fishkind et al. (1991) concluded that MLCK activity might affect mitosis by a control mechanism different than that involved in regulation of cortical activity by MLCK. The only residue on the regulatory myosin light chain (MLC) that is phosphorylated by MLCK is Ser^{19}. Satterwhite et

al. (1992) found that $p34^{cdc2}$ phosphorylated MLC on Ser^1 and Ser^2 in vitro and suggested that this modification might influence myosin II activity in mitosis. It was hypothesized that the $p34^{cdc2}$-mediated phosphorylation might keep myosin II from becoming activated until anaphase. Subsequently, Yamakita et al. (1994) found that in vivo, phosphate incorporation into MLC was 6–12 times greater in mitosis than in interphase. Analysis of the phosphorylated residues showed that in mitotic cells, MLC was phosphorylated predominantly on Ser^1 and Ser^2, with a lesser amount on Ser^{19}. In interphase cells, phosphorylation was exclusively on Ser^{19}. After cells were released from a mitotic arrest, phosphorylation of Ser^{19} increased by a factor of 20 whereas that of Ser^1 and Ser^2 was decreased. It was concluded that the change in phosphorylation site could be important in signaling the onset of cytokinesis. If this was true, the introduction of a constitutively active MLCK, as done by Fishkind et al. (1991), would have heavily favored Ser^{19} phosphorylation of MLC and therefore would have been predicted to accelerate the kinetics of cytokinesis and formation of the cleavage furrow. Collectively, these observations indicate that a role for MLCK in the regulation of cytokinesis remains to be proven. Certainly, it is equally possible that other protein kinases might productively phosphorylate MLC on Ser^{19}. Since cytokinesis is Ca^{2+}-dependent and both calmodulin and CaM kinase II are associated with the mitotic apparatus, it is possible that the latter enzyme might participate in the regulation of cytokinesis. Indeed, several groups have shown that CaM kinase II can phosphorylate MLC on Ser^{19} (Tuazon and Traugh, 1984; Chou and Rebhun, 1986; Satterwhite et al., 1992). This is another area worthy of further investigation.

Perspectives

The earliest paper that we found to support an absolute requirement for Ca^{2+} in the proliferation of animal cells was published in 1971 by Balk. The first indication that calmodulin might also be required was made 10 years later when Hidaka et al. (1981) reported that a calmodulin antagonist inhibited cell proliferation in a population of CHO cells synchronized in mitosis by mitotic selection. This was followed in 1986 by the first demonstration that calmodulin is an essential gene in yeast (Davis et al., 1986). Probably the most specific demonstration for the role of a Ca^{2+}/calmodulin-dependent enzyme in a cell cycle event was the Lorca et al. (1993) observation that CaM kinase II was required for the release of metaphase-arrested *Xenopus* eggs. Finally, the catalytic subunit of protein phosphatase 2B or calcineurin in *A. nidulans* was the first Ca^{2+}/calmodulin-dependent enzyme shown to be essential for cell cycle progression in any organism (Rasmussen et al., 1994). It has taken nearly 25 years for us to advance to the point where many scientists agree that Ca^{2+}-, calmodulin-, and Ca^{2+}/calmodulin-dependent enzymes are essential components of signaling cascades that are involved in the regulation of the cell cycle. Yet we still have not identified a single essential target for any calmodulin-dependent enzyme, much less elucidated all the steps in a Ca^{2+}-initiated cascade of events that regulates any given cell cycle transition. Due to the remarkably important roles for the cyclin-dependent protein kinases in checkpoint controls, it seems likely that one or more of these enzyme complexes will be influenced by Ca^{2+}/calmodulin. Evidence exists that activation of members of the cdk family at G_1/S requires extracellular Ca^{2+} and can be blocked by calmodulin antagonists (Takuwa et al., 1995). Similarly, a role of Ca^{2+}/calmodulin in the activation of cdc25 has been implicated at G_2/M (Lu et al., 1993; Whitaker, 1995). Surely, such observations will stimulate a number of groups to address these important and intriguing questions. The model systems, technology, and biological reagents now exist so that future progress should proceed at a much more rapid pace.

Acknowledgments We are very grateful to the members of our laboratory who allowed us to cite unpublished results. These individuals are Nanda Nanthakumar, James Joseph, and a former student, Mariko Sumi. Thanks are also extended to Samuel George and Martin Berchtold for providing us with preprints of unpublished manuscripts. The authors' work cited in this review was supported, in part, by research grants from the NIH and American Cancer Society.

References

Agell, N., R. Aligué, V. Alemany, A. Castro, M. Jaime, M. J. Pujol, E. Rius, J. Serratosa, M. Taulés, and O. Bachs (1998) New nuclear functions for calmodulin. *Cell Calcium* 23: 115–121.

Allbritton, N. L. and T. Meyer (1993) Localized calcium spikes and propagating calcium waves. *Cell Calcium* 14: 691–697.

Allbritton, N. L., T. Meyer, and L. Stryer (1992) Range of messenger action of calcium ion and inositol 1, 4, 5-trisphosphate. *Science* 258: 1812–1815.

Allbritton, N. L., E. Oancea, M. A. Kuhn, and T. Meyer (1994) Source of nuclear calcium signals. *Proc. Natl. Acad. Sci. USA* 91: 12458–12462.

Amon, A., U. Surana, I. Muroff, and K. Nasmyth (1992) Regulation of p34[cdc28] tyrosine phosphorylation is not required for entry into mitosis in *S. cerevisiae*. *Nature* 355: 368–371.

Ando, K., F. Ajchenbaum-Cymbalista, and J. D. Griffin (1993) Regulation of G_1/S transition by cyclins D2 and D3 in hematopoietic cells. *Proc. Natl. Acad. Sci. USA* 90: 9571–9575.

Armstrong, D. L., M. F. Rossier, A. D. Shcherbatko, and R. E. White (1991) Enzymatic gating of voltage-activated calcium channels. *Ann. N. Y. Acad. Sci.* 635: 26–34.

Baitinger, C., J. Alderton, M. Poenie, H. Schulman, and R. A. Steinhardt (1990) Multifunctional Ca^{2+}/calmodulin-dependent protein kinase is necessary for nuclear envelope breakdown. *J. Cell Biol.* 111: 1763–1773.

Balk, S. D. (1971) Calcium as a regulator of the proliferation of normal, but not of transformed, chicken fibroblasts in a plasma-containing medium. *Proc. Natl. Acad. Sci. USA* 68: 271–275.

Becker, J. W., J. Rotonda, B. M. McKeever, H. K. Chan, A. I. Marcy, G. Wiederrecht, J. D. Hermes, and J. P. Springer (1993) FK-506-binding protein: Three-dimensional structure of the complex with the antagonist L-685, 818. *J. Biol. Chem.* 268: 11335–11339.

Bender, P. K., J. R. Dedman, and C. P. Emerson Jr. (1988) The abundance of calmodulin mRNAs is regulated in phosphorylase kinase-deficient skeletal muscle. *J. Biol. Chem.* 263: 9733–9737.

Berchtold, M. W. (1993) Evolution of EF-hand calcium-modulated proteins. V. The genes encoding ER-hand proteins are not clustered in mammalian genomes. *J. Mol. Evol.* 36: 489–496.

Berridge, M. J. (1995) Calcium signaling and cell proliferation. *BioEssays* 17: 491–500.

Boynton, A. L., J. F. Whitfield, R. J. Isaacs, and R. Tremblay (1977) The control of human WI-38 cell proliferation by extracellular calcium and its elimination by SV-40 virus-induced proliferative transformation. *J. Cell. Physiol.* 92: 241–247.

Breuder, T., C. S. Hemenway, N. R. Movva, M.E. Cardenas, and J. Heitman (1994) Calcineurin is essential in cyclosporin A- and FK506-sensitive yeast strains. *Proc. Natl. Acad. Sci. USA* 91: 5372–5376.

Brockerhoff, S. E., R. C. Stevens, and T. N. Davis (1994) The unconventional myosin, Myo2p, is a calmodulin target at sites of cell growth in *Saccharomyces cerevisiae*. *J. Cell Biol.* 124: 315–323.

Camacho, P. and J. Lechleiter (1993) Increased frequency of calcium waves in *Xenopus laevis* oocytes that express a calcium- ATPase. *Science* 260: 226–229.

Cande, W. Z. (1980) A permeabilized cell model for studying cytokinesis using mammalian tissue culture cells. *J. Cell Biol.* 87: 326- 335.

Cao, Q. P., C. A. McGrath, E. F. Baril, P. J. Quesenberry, and G. P. V. Reddy (1995) The 68 kDa calmodulin-binding protein is tightly associated with the multiprotein DNA polymerase α–primase complex in HeLa cells. *Biochemistry* 34: 3878–3883.

Cardenas, M E. and J. Heitman (1995) Role of calcium in T-lymphocyte activation. In *Advances in Second Messenger and Phosphoprotein Research*, Vol. 30 (Means, A. R., ed.) Raven Press, New York, pp. 281–298.

Chafouleas, J. G., W. E. Bolton, H. Hidaka, A. E. Boyd III, and A. R. Means (1982) Calmodulin and the cell cycle: involvement in regulation of cell cycle progression. *Cell* 28: 41–50.

Chafouleas, J. G., L. Lagace, W. E. Bolton, A. E. Boyd III, and A. R. Means (1984) Changes in calmodulin and its mRNA accompany reentry of quiescent (G_0) cells into the cell cycle. *Cell* 36: 73–81.

Chou, Y.-H. and L. Rebhun (1986) Purification and characterization of a sea urchin egg Ca^{2+}-calmodulin-dependent kinase with myosin light chain phosphorylating activity. *J. Biol. Chem.* 261: 5389–5395.

Clapham, D. E. (1995) Calcium signaling. *Cell* 80: 259–268.

Colomer, J., N. Agell, P. Engel, J. Alberola-Ila, and O. Bachs (1993) Calmodulin expression during proliferative activation of human T lymphocytes. *Cell Calcium* 14: 609–618.

Crabtree, G. R. and N. A. Clipstone (1994) Signal transmission between the plasma membrane and nucleus of T lymphocytes. *Annu. Rev. Biochem.* 63: 1045–1083.

Cunningham, K. W. and G. R. Fink (1994) Calcineurin-dependent growth control in *Saccharomyces cerevisiae* mutants lacking PMC1, a homolog of plasma membrane Ca^{2+} ATPases. *J. Cell Biol.* 124: 351–363.

Cyert, M. S. and J. Thorner (1992) Regulatory subunit (CNB1 gene product) of yeast Ca^{2+}/calmodulin-dependent phosphoprotein phosphatases is required for adaptation to pheromone. *Mol. Cell. Biol.* 12: 3460–3469.

Cyert, M. S., R. Kunisawa, D. Kaim, and J. Thorner (1991) Yeast has homologs (CNA1 and CNA2 gene products) of mammalian calcineurin, a calmodulin-regulated phosphoprotein phosphatase. *Proc. Natl. Acad. Sci. USA* 88: 7376–7380.

Danchin, A., O. Sezer, P. Glaser, P. Chalon, and D. Caput (1989) Cloning and expression of mouse-brain calmodulin as an activator of *Bordetella pertussis* adenylate cyclase in *Escherichia coli*. *Gene* 80: 145–149.

Davis, T. N. (1992) A temperature-sensitive calmodulin mutant loses viability during mitosis. *J. Cell Biol.* 118: 607–617.

Davis, T. N., M. S. Urdea, F. R. Masiarz, and J. Thorner (1986) Isolation of the yeast calmodulin gene: calmodulin is an essential protein. *Cell* 47: 423–431.

DeGregori, J., T. Kowalik, and J. R. Nevins (1995) Cellular targets for activation by the E2F1 transcription factor include DNA synthesis- and G_1/S regulatory genes. *Mol. Cell. Biol.* 15: 4215–4224.

Dinsmore, J. H., and R. D. Sloboda (1988) Calcium and calmodulin dependent phosphorylation of a 62 kDa protein induces microtubule depolymerization in sea urchin mitotic apparatuses. *Cell* 53: 769–780.

Dorée, M. and S. Galas (1994) The cyclin-dependent protein kinases and the control of cell division. *FASEB J.* 8: 1114–1121.

Dumont, F. J., M. J. Staruch, S. L. Koprak, J. J. Siekierka, C. S. Lin, R. Harrison, T. Sewell, V. M. Kindt, T. R. Beattie, M. Wyvratt, and N. H. Sigal (1992) The immunosuppressive and toxic effects of FK-506 are mechanistically related: pharmacology of a novel antagonist of FK-506 and rapamycin. *J. Exp. Med.* 176: 751- 760.

Dunphy, W. G. (1994) The decision to enter mitosis. *Trends Cell Biol.* 4: 202–207.

Eilam, Y. and D. Chernichovsky (1988) Low concentrations of trifluoperazine arrest the cell division cycle of *Saccharomyces cerevisiae* at two specific stages. *J. Gen. Microbiol.* 134: 1063–1069.

Fischer, R., M. Koller, M. Flura, M. Mathews, M. A. Strehler-Page, J. Krebs, J. T. Penniston, E. Carafoli, and E. E. Strehler (1988) Multiple divergent mRNAs code for a single human calmodulin. *J. Biol. Chem.* 263: 17055–17062.

Fishkind, D. J., L. Cao, and Y. Wang (1991) Microinjection of the catalytic fragment of myosin light chain kinase into dividing cells: effects on mitosis and cytokinesis. *J. Cell Biol.* 114: 967–975.

Galaktionov, K., C. Jessus, and D. Beach (1995) Raf1 interaction with cdc25 phosphatase ties mitogenic signal transduction to cell cycle activation. *Genes Dev.* 9: 1046–1058.

Garrett-Engele, P., B. Moilanen, and M. S. Cyert (1995) Calcineurin, the Ca^{2+}/calmodulin-dependent protein phosphatase, is essential in yeast mutants with cell integrity defects and in mutants that lack a functional vacuolar H^+-ATPase. *Mol. Cell. Biol.* 15: 4103–4114.

Geiser, J. R., H. A. Sundberg, B. H. Chang, E. G. D. Muller, and T. N. Davis (1993) The essential mitotic target of calmodulin is the 110- kilodalton component of the spindle pole body in *Saccharomyces cerevisiae*. *Mol. Cell. Biol.* 13: 7913–7924.

Geiser, J. R., D. van Tuinen, S. E. Brockerhoff, M. M. Neff, and T. N. Davis (1991) Can calmodulin function without binding calcium? *Cell* 65: 949–959.

Greber, U. F. and L. Gerace (1995) Depletion of calcium from the lumen of endoplasmic reticulum reversibly inhibits passive diffusion and signal-mediated transport into the nucleus. *J. Cell Biol.* 28: 5–14.

Griffith, J. P., J. L. Kim, E. E. Kim, M. D. Sintchak, J. A. Thomson, M. J. Fitzgibbon, M. A. Fleming, P. R. Caron, K. Hsiao, and M. A. Navia (1995) X-ray structure of calcineurin inhibited by the immunophilin-immunosuppressant FKBP12–FK506 complex. *Cell* 82: 507–522.

Guerriero, V., D. R. Rowley, and A. R. Means (1981) Production and characterization of an antibody to myosin light chain kinase and intracellular localization of the enzyme. *Cell* 27: 449–458.

Hammond, R. A., K. A. Foster, M. W. Berchthold, M. Gassmann, A. M. Holmes, U. Hubscher, and N. C. Brown (1988) Calcium- dependent calmodulin-binding proteins associated with mammalian DNA polymerase-α. *Biochem. Biophys. Acta.* 951: 315–321.

Hanson, P. I. and H. Schulman (1992) Neuronal Ca^{2+}/calmodulin-dependent protein kinases. *Annu. Rev. Biochem.* 61: 559–601.

Hazelton, B., B. Mitchell, and J. Tupper (1979) Calcium, magnesium, and growth control in the WI-38 human fibroblast cell. *J. Cell Biol.* 83: 487–498.

Heichman, K. A. and J. M. Roberts (1994). Rules to replicate by. *Cell* 79: 557–562.

Heiman, R. G., R. C. Atkinson, B. F. Andruss, C. Bolduc, G. E. Kovalick, and K. Beckingham (1996) Spontaneous avoidance behavior in *Drosophila* null for calmodulin expression. *Proc. Natl. Acad. Sci. USA* 93(6): 2420–2425.

Hidaka, H., Y. Sasaki, T. Tanaka, T. Endo, S. Ohno, Y. Fujii, and T. Nagata (1981) *N*-(6-aminohexyl)-5-chloro-1-naphthalenesulfonamide, a calmodulin antagonist, inhibits cell proliferation. *Proc. Natl. Acad. Sci. USA* 78: 4354–4357.

Hunter, T. (1993) Braking the cycle. *Cell* 75: 839–841.

Hussain, A., C. Garnett, M. G. Klein, J.-J. Tsai-Wu, M. F. Schneider, and G. Inesi (1995) Direct involvement of intracellular Ca^{2+} transport ATPase in the development of thapsigargin resistance by chinese hamster lung fibroblasts. *J. Biol. Chem.* 270: 12140–12146.

Izumi, T. and J. L. Maller (1995) Phosphorylation and activation of the *Xenopus* cdc25 phosphatase in the absence of cdc2 and cdk2 kinase activity. *Mol. Biol. Cell.* 6: 215–226.

Jinno, S., K. Suto, A. Nagata, M. Igarashi, Y. Kanaoka, H. Nojima, and H. Okayama (1994) Cdc25A is a novel phosphatase functioning early in the cell cycle. *EMBO J.* 13: 1549–1556.

Kato, J.-Y. and C. J. Sherr (1993) Inhibition of granulocyte differentiation by G_1 cyclins D2 and D3 but not D1. *Proc. Natl. Acad. Sci. USA* 90: 11513–11517.

Kornstein, L. B., M. L. Gaiso, R. L. Hammell, and D. C. Bartelt (1992) Cloning and sequence determination of a cDNA encoding *Aspergillus nidulans* calmodulin-dependent multifunctional protein kinase. *Gene* 113: 75–82.

Kuno, T., H. Tanaka, J. Mukai, C. Chang, K. Hiraga, T. Miyakawa, and C. Tanaka (1991) cDNA cloning of a calcineurin B homolog in *Saccharomyces cerevisiae*. *Biochem. Biophys. Res. Commun.* 180: 1159–1163.

Liu, Y., S. Ishii, M. Tokai, H. Tsutsumi, O. Ohke, R. Akada, K. Tanaka, E. Tsuchiya, S. Fukui, and T. Miyakawa (1991) The *Saccharomyces cerevisiae* genes (CMP1 and CMP2) encoding calmodulin-binding proteins homologous to the catalytic subunit of mammalian protein phosphatase 2B. *Mol. Gen. Genet.* 227: 52–59.

Lorca, T., F. H. Cruzalegui, D. Fesquet, J.-C. Cavadore, J. Méry, A. Means, and M. Dorée (1993) Calmodulin-dependent protein kinase II mediates inactivation of MPF and CSF upon fertilization of *Xenopus* eggs. *Nature* 366: 270–273.

Lorca, T., S. Galas, D. Fesquet, A. Devault, J.-C. Cavadore, and M. Dorée (1991) Degradation of the proto-onco-

gene product p39^{mos} is not necessary for cyclin proteolysis and exit from meiotic metaphase: requirement for a Ca^{2+}-calmodulin dependent event. *EMBO J.* 10: 2087–2093.

Lu, K. P. and A. R. Means (1993) Regulation of the cell cycle by calcium and calmodulin. *Endocrine. Rev.* 14: 40–58.

Lu, K. P., N. Nanthakumar, J. S. Dayton, and A. R. Means (1995) Calcium and calmodulin regulation of the nuclear division cycle of *Aspergillus nidulans*. In *Advances in Molecular and Cell Biology*. Vol. 13 (Whitaker, M., ed.). JAI Press, Greenwich, CN, pp. 89–136.

Lu, K. P., S. A. Osmani, A. H. Osmani, and A. R. Means (1993) Essential roles for calcium and calmodulin in G_2/M progression in *Aspergillus nidulans*. *J. Cell Biol.* 121: 621–630.

Lu, K. P., C. D. Rasmussen, G. S. May, and A. R. Means (1992) Cooperative regulation of cell proliferation by calcium and calmodulin in *Aspergillus nidulans*. *Mol. Endocrinol.* 6: 365- 374.

Luby-Phelps, K., M. Hori, J. M. Phelps, and D. Won (1995) Ca^{2+}-regulated dynamic compartmentalization of calmodulin in living smooth muscle cells. *J. Biol. Chem.* 270: 21532- 21538.

Lundberg, A. S. and R. A. Weinberg (1998) Functional inactivation of the retinoblastoma protein requires sequential modification by at least two distinct cyclin-cdk complexes. *Mol. Cell. Biol.* 18(2): 753–761.

Maller, J. L. (1994) Biochemistry of cell cycle checkpoints at the G_2/M and metaphase/anaphase transitions. *Semin. Dev. Biol.* 5: 3061–3068.

Marcum, J. M., J. R. Dedman, B. R. Brinkley, and A. R. Means (1978) Control of microtubule assembly-disassembly by calcium-dependent regulator protein. *Proc. Natl. Acad. Sci. USA* 75: 3771–3775.

McGoldrick, C. A., C. Gruver, and G. S. May (1995) *myo*A of *Aspergillus nidulans* encodes an essential myosin I required for secretion and polarized growth. *J. Cell Biol.* 128: 577–587.

Means, A. R. (1994) Calcium, calmodulin and cell cycle regulation. *FEBS Lett.* 347: 1–4.

Mendoza, I., F. Rubio, A. Rodriguez-Navarro, and J. M. Pardo (1994) The protein phosphatase calcineurin is essential for NaCl tolerance of *Saccharomyces cerevisiae*. *J. Biol. Chem.* 269: 8792–8796.

Millar, J. B. and P. Russell (1992) The cdc25 M-phase inducer: an unconventional protein phosphatase. *Cell* 68: 7–10.

Morin, N., A. Abrieu, T. Lorca, F. Martin, and M. Dorée (1994) The proteolysis-dependent metaphase to anaphase transition: calcium/calmodulin-dependent protein kinase II mediates onset of anaphase in extracts prepared from unfertilized *Xenopus* eggs. *EMBO J.* 13: 4343–4352.

Morris, T. A., R. J. De Lorenzo, and R. M. Tombes (1998) CaMK-II reduces cyclin D1 levels and enhances the association of p27^{kip1} with cdk2 to cause G1 arrest in NIH 3T3 cells. *Exp. Cell Res.* 240: 281–227.

Moser, M. J., S. Y. Lee, R. E. Klevit, and T. N. Davis (1995) Ca^{2+} binding to calmodulin and its role in *Schizosaccharomyces pombe* as revealed by mutagenesis and NMR spectroscopy. *J. Biol. Chem.* 270: 20643–20652.

Nojima, H. (1989) Structural organization of multiple rat calmodulin genes. *J. Mol. Biol.* 208: 269–282.

Oakley, B. R. and N. Morris (1983) A mutation in *Aspergillus nidulans* that blocks the transition from interphase to prophase. *J. Cell Biol.* 96: 1155–1158.

O'Connell, M. J., A. H. Osmani, N. R. Morris, and S. A. Osmani (1992) An extra copy of $nimE^{cyclinB}$ elevates pre-MPF levels and partially suppresses mutation of $nim\ T^{cdc25}$ in *Aspergillus nidulans.*, 1989 *EMBO J.* 11: 2139–2149.

Ohta, Y., T. Ohba, and E. Miyamoto (1990) Ca^{2+}/calmodulin-dependent protein kinase II: localization in the interphase nucleus and the mitotic apparatus of mammalian cells. *Proc. Natl. Acad. Sci. USA* 87: 5341–5345.

Ohtsubo, M. and J. M. Roberts (1993) Cyclin-dependent regulation of G_1 in mammalian cells. *Science* 259: 1908–1912.

Ohtsubo, M., A. M. Theodoras, J. Schumacher, J. M. Roberts, and M. Pagano (1995) Human cyclin E, a nuclear protein essential for the G_1-S phase transition. *Mol. Cell. Biol.* 15: 2612–2624.

Ohya, Y. and D. Botstein (1994) Diverse essential functions revealed by complementing yeast calmodulin mutants. *Science* 263: 963–966.

Ohya, Y., H. Kawasaki, K. Suzuki, J. Londesborough, and Y. Anraku (1991) Two yeast genes encoding calmodulin-dependent protein kinases: isolation, sequencing and bacterial expressions of CMK1 and CMK2. *J. Biol. Chem.* 266: 12784–12794.

Osmani, A. H., N. van Peij, M. Mischke, M. J. O'Connell, and S. A. Osmani (1994) A single $p34^{cdc2}$ protein kinase (encoded by $nimX^{cdc2}$) is required at G_1 and G_2 in *Aspergillus nidulans. J. Cell Sci.* 107: 1519–1528.

Osmani, S. A., R. T. Pu, and N. R. Morris (1988) Mitotic induction and maintenance by overexpression of a G2-specific gene that encodes a potential protein kinase. *Cell* 53: 237–244,.

Pausch, M. H., D. Kaim, R. Kunisawa, A. Admon, and J. Thorner (1991) Multiple Ca^{2+}/calmodulin-dependent protein kinase genes in a unicellular eukaryote. *EMBO J.* 10: 1511–1522.

Pietromonaco, S. F., G. A. Seluja, and L. Elias (1995) Identification of enzymatically active Ca^{2+}/calmodulin-dependent protein kinase in centrosomes of hemopoietic cells. *Blood Cells, Molecules, and Diseases* 21: 34–41.

Planas-Silva, M. D. and A. R. Means (1992) Expression of a constitutive form of calcium/calmodulin dependent protein kinase II leads to arrest of the cell cycle in G_2. *EMBO J.* 11: 507–517.

Poenie, M., J. Alderton, R. Y. Tsien, and R. A. Steinhardt (1985) Changes in free calcium levels with stages of the cell division cycle. *Nature* 315: 147–149.

Pruschy, M., Y. Ju, L. Spitz, E. Carafoli, and D. S. Goldfarb (1994) Facilitated nuclear transport of calmodulin in tissue culture cells. *J. Cell Biol.* 127: 1527–1536.

Putney, J. W. J. (1990) Capacitative calcium entry revisited. *Cell Calcium* 11: 611–624.

Quelle, D. E., R. A. Ashmun, S. A. Shurtleff, J. Kato, D. Bar-Sagi, M. F. Roussel, and C. J. Sherr (1993) Overexpression of mouse D-type cyclins accelerates G_1 phase in rodent fibroblasts. *Genes Dev.* 7: 1559–1571.

Rasmussen, C. D. and A. R. Means (1987) Calmodulin is involved in regulation of cell proliferation. *EMBO J.* 6: 3961–3968.

Rasmussen, C. D. and A. R. Means (1989a) Calmodulin is required for cell cycle progression during G_1 and mitosis. *EMBO J.* 8: 73–82.

Rasmussen, C. D. and A. R. Means (1989b) The presence of parvalbumin in a nonmuscle cell line attenuates progression through mitosis *Mol. Endocrinol.* 3: 588–596.

Rasmussen, C. and G. Rasmussen (1994) Inhibition of G_2/M progression in *Schizosaccharomyces pombe* by a mutant calmodulin kinase II with constitutive activity. *Mol. Biol. Cell.* 5: 785–795.

Rasmussen, G. and C. Rasmussen (1995) Calmodulin-dependent protein kinase II is required for G_1/S progression in HeLa cells. *Biochem. Cell Biol.* 73: 201–207.

Rasmussen, C., C. Garen, S. Brining, R. L. Kincaid, R. L. Means, and A. R. Means (1994) The calmodulin-dependent protein phosphatase catalytic subunit (calcineurin A) is an essential gene in *Aspergillus nidulans. EMBO J.* 13: 2545–2552.

Rasmussen, C. D., R. L. Means, K. P. Lu, G. S. May, and A. R. Means (1990) Characterization and expression of the unique calmodulin gene of *Aspergillus nidulans. J. Biol. Chem.* 265: 13767- 13775.

Reddy, G. P. V. (1994) Cell Cycle: regulatory events in $G_1 \rightarrow S$ transition of mammalian cells. *J. Cell. Biochem.* 54: 379–386.

Reddy, G. P. V. and S. Fager (1993) Replitase: a complex integrating dNTP synthesis and DNA replication. *Crit. Rev. Euk. Gene Exp.* 3: 255–277.

Reddy, G. P. V. and A. B. Pardee (1980) A multienzyme complex for metabolic channeling in mammalian DNA replication. *Proc. Natl. Acad. Sci. USA* 77: 3312–3316.

Reddy, G. P. V., D. Deacon, W. C. Reed, and P. J. Quesenberry (1992a) Growth factor-dependent proliferative stimulation of hemopoietic cells is associated with the nuclear localization of 68 kDa calmodulin-binding protein. *Blood* 79: 1946–1955.

Reddy, G. P. V., W. C. Reed, D. H. Deacon, and P. J. Quesenberry (1994) Growth factor modulated calmodulin-binding protein stimulates nuclear DNA synthesis in hemopoietic progenitor cells. *Biochemistry* 33: 6605–6610.

Reddy, G. P. V., W. C. Reed, E. Sheehan, and D. B. Sacks (1992b) Calmodulin-specific monoclonal antibodies inhibit DNA replication in mammalian cells. *Biochemistry* 31: 10426–10430.

Resnitzky, D., M. Gossen, H. Bujard, and S.I . Reed (1994) Acceleration of the G_1/S phase transition by expression of cyclin D1 and E with an inducible system. *Mol. Cell. Biol.* 14: 1669–1679.

Resnitzky, D., L. Hengst, and S. I. Reed (1995) Cyclin A-associated kinase activity is rate limiting for entrance into S phase and is negatively regulated in G_1 by p27^{Kip1}. *Mol. Cell. Biol.* 15: 4347–4352.

Roussel, M. F., A. M. Theodoras, M. Pagano, and C. J. Sherr (1995) Rescue of defective mitogenic signaling by D-type cyclins. *Proc. Natl. Acad. Sci. USA* 92: 6837–6841.

Sasaki, Y. and H. Hidaka (1982) Calmodulin and cell proliferation. *Biochem. Biophys. Res. Commun.* 104: 451–456.

Satterwhite, L. L., M. J. Lohka, K. L. Wilson, T. Y. Scherson, L. J. Cisek, J. L. Corden, and T. D. Pollard (1992) Phosphorylation of myosin-II regulatory light chain by cyclin-p35cdc2: a mechanism for the timing of cytokinesis. *J. Cell Biol.* 118: 595–605.

Schreiber, S. L. (1992) Immunophilin-sensitive protein phosphatase action in cell signaling pathways. *Cell* 70: 365–368.

Sherr, C. J. (1994) G_1 phase progression: cycling on cue. *Cell* 79: 551–555.

Sherr, C. J. and J. M. Roberts (1995) Inhibitors of mammalian G_1 cyclin-dependent kinases. *Genes and Development* 9: 1149–1163.

Short, A. D., J. Bian, T. K. Ghosh, R. T. Waldron, S. L. Rybak, and D. L. Gill (1993) Intracellular Ca^{2+} pool content is linked to control of cell growth. *Proc. Natl. Acad. Sci. USA* 90: 4986–4990.

Slaughter, G. R. and A. R. Means (1989) Analysis of expression of multiple genes encoding calmodulin during spermatogenesis. *Mol. Endocrinol.* 3: 1569–1578.

Sorger, P. K. and A. W. Murray (1992) S-phase feedback control in budding yeast independent of tyrosine phosphorylation of p34^{cdc28}. *Nature* 355: 365–368.

Stehno-Bittel, L., A. Lückhoff, and D. E. Clapham (1995) Calcium release from the nucleus by InsP$_3$ receptor channels. *Neuron* 14: 163–167.

Steinhardt, R. A. and J. Alderton (1988) Intracellular free calcium rise triggers nuclear envelope breakdown in sea urchin embryo. *Nature* 332: 364–366.

Stirling, D. A., K. A. Welch, and M. J. R. Stark (1994) Interaction with calmodulin is required for the function of Spc110p, an essential component of the yeast spindle pole body. *EMBO J.* 13: 4329–4342.

Stueland, C. S., D. J. Lew, M. J. Cismowski, and S. I. Reed (1993) Full activation of p34^{cdc28} histone H1 kinase activity is unable to promote entry into mitosis in checkpoint-arrested cells of the yeast *Saccharomyces cerevisiae*. *Mol. Cell. Biol.* 13: 3744–3755.

Su, T. T., P. J. Follette, and P. H. O'Farrell. (1995) Qualifying for the license to replicate. *Cell* 81: 825–828.

Sumi, M., K. Kazutoshi, T. Ishikawa, A. Ishii, M. Hagiwara, T. Nagatsu, and H. Hidaka (1991) The newly synthesized selective Ca^{2+}/calmodulin dependent protein kinase II inhibitor KN-93 reduces dopamine content

in PC12h cells. *Biochem. Biophys. Res. Commun.* 181: 968–975.

Sun, G.-H., A. Hirata, Y. Ohya, and Y. Anraku (1992) Mutations in yeast calmodulin cause defects in spindle pole body functions and nuclear integrity. *J. Cell Biol.* 119: 1625–1639.

Takeda, T. and M. Yamamoto (1987) Analysis and in vivo disruption of the gene encoding for calmodulin in *Schizosaccharomyces pombe*. *Proc. Natl. Acad. Sci. USA* 84: 3580–3584.

Takuwa, N., W. Zhou, and Y. Takuwa (1995) Calcium, calmodulin and cell cycle progression. *Cell. Signalling* 7: 93–104.

Tang, Y. and S. I. Reed (1993) The Cdk-associated protein Cks1 functions both in G_1 and G_2 in *Saccharomyces cerevisiae*. *Genes and Dev.* 7: 822–832.

Tombes, R. M., S. Grant, E. H. Westin, and G. Krystal (1995) G_1 cell cycle arrest and apoptosis are induced in NIH 3T3 cells by KN-93, an inhibitor of CaMK-11 (the multifunctional Ca^{2+}/CaM kinase). *Cell Growth Diff.* 6: 1063–1070.

Tomono, M., K. Toyoshima, M. Ito, and H. Amano (1996) Calcineurin is essential for DNA synthesis in Swiss 3T3 fibroblasts. *Biochem. J.* 317: 675–680.

Tomono, M., K. Toyoshima, M. Ito, H. Amano, and Z. Kiss (1998) Inhibitors of calcineurin block expression of cyclins A and E induced by fibroblast growth factor in Swiss 3T3 fibroblasts. *Arch. Biochem. Biophys.* 352(2): 374–378.

Tuazon, P. T. and J. A. Traugh (1984) Activation of actin-activated ATPase in smooth muscle by phosphorylation of myosin light chain with protease-activated kinase I. *J. Biol. Chem.* 259: 541–546.

Twigg, J., P. Rajnikant, and M. Whitaker (1988) Translational control of InsP$_3$-induced chromatin condensation during the early cell cycles of sea urchin embryos. *Nature* 332: 366–369..

Ullman, K. S., J. P. Northrop, L. C. Verweij, and G. R. Crabtree (1990) Transmission of signals from the T-lymphocyte antigen receptor to the genes responsible for cell proliferation and immune function: the missing link. *Annu. Rev. Immunol.* 8: 421–452.

Welsh, M. J., J. R. Dedman, B. R. Brinkley, and A. R. Means (1978) Calcium-dependent regulator protein: localization in the mitotic spindle of eukaryotic cells. *Proc. Natl. Acad. Sci. USA* 75: 1867–1871

Whitaker, M. (1995) Regulation of the cell division cycle by inositol trisphosphate and the calcium signalling pathway. *Advances in Second Messenger and Phosphoprotein Research*, Vol. 30 (Means, A. R. ed.). Raven Press, New York, pp. 299–310.

Whitaker, M. and R. Patel (1990) Calcium and cell cycle. *Development* 108: 525–542.

Whitfield, J. F., R. P. Bird, B. R. Chakravarthy, R. J. Isaacs, and P. Morley (1995) Calcium — cell cycle regulator, differentiator, killer, chemopreventor, and maybe, tumor promoter. *J. Cell. Biochem.* 22: 74–91.

Wimmel, A., F. C. Lucibello, A. Sewing, S. Adolf, and R. Müller (1994) Inducible acceleration of G_1 progression through tetracycline-regulated expression of human cyclin E. *Oncogene* 9: 995- 997.

Yamakita, Y., S. Yamashiro, and F. Matsumura (1994) *In vivo* phosphorylation of regulatory light chain of myosin II during mitosis of cultured cells. *J. Cell Biol.* 124: 129–137.

Ye, X. S., G. Xu, R. T. Pu, R. R. Fincher, S .L. McGuire, A. H. Osmani, and S. A. Osmani (1995) The NIMA protein kinase is hyperphosphorylated and activated downstream of p34^{cdc2}/cyclin B: coordination of two mitosis promoting kinases. *EMBO J.* 14: 986–994.

Yoshida, T., T. Toda, and M. Yanagida (1994) A calcineurin-like gene *ppb1$^+$* in fission yeast: mutant defects in cytokinesis, cell polarity, mating and spindle pole body positioning. *J. Cell Science* 107: 1725–1735.

Zhang, S.-P., N. Natsukari, G. Bai, R.A. Nichols, and B. Weiss (1993) Localization of the multiple calmodulin messenger RNAs in differentiated PC12 cells. *Neuroscience* 55: 571–582.

22

Calcium in the Regulation of Intramitochondrial Enzymes

James G. McCormack
Richard M. Denton

Most of the chapters in this book are concerned with the signalling roles of Ca^{2+} within the cytosol of cells, or with the various transport proteins involved in controlling the concentration of Ca^{2+} in this major subcellular compartment. However, it is now clear that Ca^{2+} ions also fulfil an important second-messenger role within the mitochondrial matrix of mammalian, and perhaps all vertebrate, cells (Denton and McCormack, 1980; Hansford, 1991; McCormack and Denton, 1994a). Consequently, it is also now appreciated that a main function of the Ca^{2+}-transport processes which exist within the mitochondrial inner membrane of these cells is to regulate the concentration of Ca^{2+} within this important cellular subcompartment (McCormack et al., 1990, 1992). There is one major advantage which was presumably the main reason underlying the evolutionary acquisition of this role of mitochondrial Ca^{2+}. This would appear to be to allow stimulated cells to maintain energetic homeostasis when energy-requiring events are activated by raised cytosolic Ca^{2+}, and when there is also increased energy-requiring movement of Ca^{2+}. This is because it is key energy-producing events which are stimulated by raised intramitochondrial Ca^{2+} (McCormack et al., 1990; Hansford, 1991).

The purpose of this chapter is to review what is known about how Ca^{2+} extends its signal-transducing role to within the mitochondrial matrix to stimulate energy-producing events. It is fair to say at the outset that, as yet, we do not know as much molecular detail about intramitochondrial Ca^{2+}-sensitive processes (Nichols and Denton, 1995), or about mitochondrial Ca^{2+}-transport itself (Gunter and Gunter, 1994), as is known about the many analagous systems covered elsewhere in this book. This is probably because of some of the earlier misconceptions about mitochondrial calcium and its role, and hence the slower realisation, in comparison with that in the cytosol, of the second-messenger role of intramitochondrial Ca^{2+}. However, there have recently been some major advances in this area, and the increasing pace and impetus of the ongoing research suggests that this missing molecular information will soon be forthcoming.

Historical Perspective

The studies carried out by such distinguished scientists as Lehninger, Chance, and many others in the early 1960s on calcium uptake by isolated mitochondria (see Gunter and Gunter, 1994) formed an important part of the formulation of Mitchell's chemiosmotic hypothesis of mitochondrial energy transduction (Mitchell, 1966). However, they also showed that in vitro isolated mitochondria had the potential to accumulate massive amounts of calcium (in excess of 1000 nmol/mg of mitochondrial protein) (see Carafoli and Sottocasa, 1984). This led to the idea that mitochondria may act as reservoirs or sinks of calcium within cells, and that within the mitochondrial matrix, calcium had a largely inert or passive role. Then, as the importance of the signalling role of cytosolic Ca^{2+} was increasingly realised, and it became clear that some or all of this could be derived from within the cell, mitochondria were, for a time, strongly advocated as the intracellular source of messenger Ca^{2+} which would be released on receiving some signal that resulted from the binding of a Ca-mobilising hormone to its extracellular receptor (see Williamson et al., 1981; Fiskum and Lehninger, 1982; Nicholls and Åkerman, 1982). Tied in with this was also the concept that mitochondria played a key role in buffering or setting the cytosolic concentration of Ca^{2+} (Williamson et al., 1981; Fiskum and Lehninger, 1982; Nicholls and Åkerman, 1982).

These ideas have now diminished with (1) the discovery of Ca^{2+}-releasing molecules and mechanisms such as that induced by trigger Ca^{2+} (Fabiato and Fabiato, 1979), inositol-1,4,5-trisphosphate ($InsP_3$) (Berridge, 1988), and cyclic ADP-ribose (Sitsapesan et al., 1995), and the full realisation of the dominant role of reticular membrane systems as the intracellular structures involved in the control of cytosolic $[Ca^{2+}]$ (Becker et al., 1980); and (2) with the accumulation of the evidence that Ca^{2+} within the mitochondrial matrix actually has an important messenger role in itself (Denton and McCormack, 1980, 1985; McCormack et al. 1990; McCormack and Denton, 1994a). Thus, it is now thought that a primary purpose of the mitochondrial Ca^{2+}-transport system is to relay changes in cytosolic Ca^{2+} to within the matrix, where it plays an important role in the control of energy metabolism, and where its concentration normally exists within the same sub- to low micromolar range as in the cytosol (McCormack et al., 1990).

This concept originated in the 1970s with the discovery that there are three intramitochondrial dehydrogenases that play a key role in oxidative metabolism and that can be activated by Ca^{2+} within this concentration range (see Denton and McCormack, 1980). In the 1980s, supporting evidence accumulated from studies largely with isolated mitochondria (see McCormack et al., 1990, and now, in the 1990s, techniques have recently become available with which to study mitochondrial $[Ca^{2+}]$ and its role in intact cells (see below).

The Intramitochondrial Ca^{2+}-Sensitive Dehydrogenases

These are the pyruvate (PDH), NAD^+-isocitrate (NAD-ICDH), and 2-oxoglutarate (OGDH) dehydrogenases (Denton and McCormack, 1980, 1985, 1990) (Fig. 22.1). These enzymes are found exclusively within the mitochondrial matrix in mammalian cells. All three are key sites of NADH production for the respiratory chain, as well as being the major sites of production of respiratory CO_2 through their catalytic mechanisms which all involve essentially irreversible oxidative decarboxylations.

Pyruvate Dehydrogenase

This exists as a large multienzyme complex (M_R 7–10 MDa) containing multiple copies of three different enzymes (see Reed, 1981; Yeaman, 1989). The core

of the complex is made up of dihydrolipoate acetyl transferase (E2) units to which are attached the pyruvate decarboxylase (E1) and dihydrolipoate dehydrogenase (E3) units. The reaction catalysed by E1 is the irreversible step and this enzyme is composed of two different subunits and exists as a tetramer, $\alpha_2\beta_2$. The E1α-subunits can be phosphorylated by a specific kinase and dephosphorylated by a specific phosphatase (Reed, 1981; Yeaman, 1989). There are three sites of phosphorylation, and one of these controls the activity of the whole complex such that when it is phosphorylated, PDH is essentially inactive. The role of the phosphorylation of the other two sites is not yet fully established.

The PDH represents the point of commitment of carbohydrate in mammalian tissues and controls the supply of acetyl coenzyme A (CoA) from carbohydrate to the citrate acid cycle and fatty acid synthesis, and it is thus subject to much regulation (see Yeaman, 1989; Randle, 1995; Denton et al. 1996) (Fig. 22.2). The catalytic activity of PDH can be end-product inhibited by increases in the ratios of acetyl CoA/CoA and $NADH/NAD^+$, but, more important, the amount of active enzyme (PDHa) itself is regulated by the relative activities of PDHa kinase and PDH phosphate phosphatase (PDHP-Pase). The PDHa kinase is activated by increases in these same end-products (increased ratios of acetyl CoA/CoA and $NADH/NAD^+$), as well as by increased ATP/ADP, whereas it is inhibited by pyruvate. This type of "intrinsic" regulation achieved through sensing concentrations of local metabolites ensures that the flux through PDH is matched to the needs of the cell. Longer term regulation also appears to be achieved by regulation of the amount of PDHa kinase: for instance, in starvation its activity is increased to ensure conservation of limited carbohydrate stocks by switching more PDH into the inactive form (Yeaman, 1989; Randle, 1995).

In contrast, regulation of PDHP-phosphatase appears to be the means by which hormones may signal their acute requirements for changes in flux through PDH (Rutter et al., 1989; Denton et al., 1989, 1996). In this way, hormones or other agents can change flux whilst maintaining or increasing the key metabolite ratios mentioned above. In particular, PDHP-Pase can be activated several-fold by increases in $[Ca^{2+}]$ within the sub- to low micromolar range (Denton et al., 1972), and it is in this way that Ca^{2+}-mobilising hormones increase the amount of active enzyme (McCormack et al., 1990; Denton et al., 1996). This enzyme can also be activated as the result of insulin treatment of lipogenic tissues, whereby more acetyl CoA is formed to be used in stimulated fatty acid synthesis rather than in enhanced oxidative metabolism. The means by

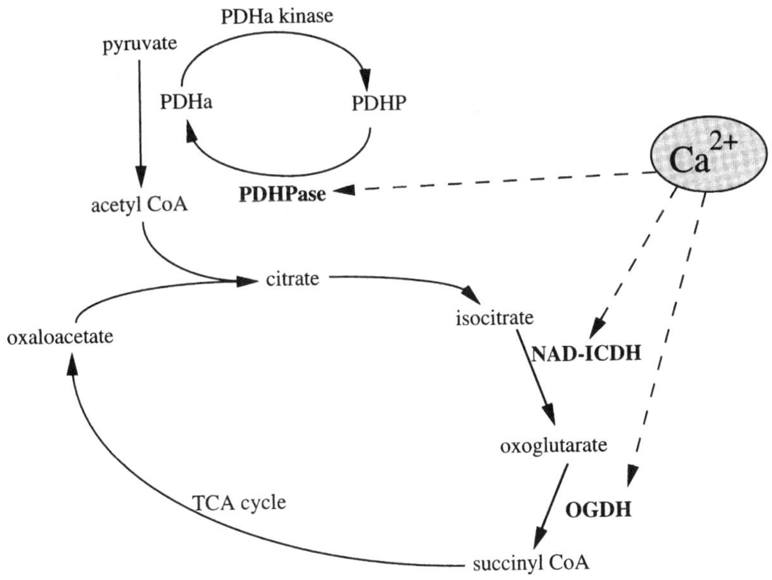

PDHa kinase
pyruvate
PDHa
PDHP
Ca2+
acetyl CoA
PDHPase
citrate
isocitrate
oxaloacetate
NAD-ICDH
oxoglutarate
TCA cycle
OGDH
succinyl CoA

Figure 22.1 The intramitochondrial dehydrogenases activated by Ca^{2+} (see text for definition of abbreviations).

which insulin activates this enzyme is not yet known (Denton et al., 1989, 1996).

Ca^{2+} appears to activate PDHP-Pase by causing a decrease in its K_m for Mg^{2+}, which itself is essential for its activity; the polyamine spermine also activates the enzyme (Thomas and Denton, 1986; Thomas et al., 1986; Midgley et al., 1987). Purified PDHP-Pase consists of two subunits of M_R 97 kDa and 50 kDa, respectively, with catalytic activity residing in the latter (Teague et al., 1982). The function of the other subunit is not known, but it contains one bound flavin adenine dinucleotide (FAD) molecule. Studies by Reed and coworkers with purified enzymes showed that Ca^{2+} binds to PDHP-Pase with a stoichiometry of one Ca^{2+} per 147 kDa unit (Lawson et al., 1993). They also showed that the Pase bound to the E2 of PDH, but only when Ca^{2+} was present, and that when this occurred a second

Pyruvate + NAD⁺ + CoA
Acetyl CoA + NADH + CO2
PDHa
End-product inhibitors
NADH
Acetyl CoA
Activators
Mg2+
Ca2+
spermine
PDHa
(active)
Activators
increases in:-
acetylCoA/CoA
NADH/NAD+
phosphatase
kinase
Inhibitors
pyruvate
ADP
Dichloroacetate
PDHP
(inactive)

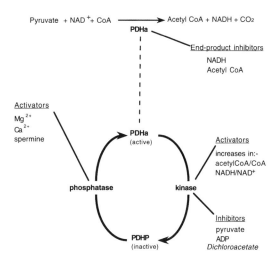

Figure 22.2 The regulation of mammalian pyruvate dehydrogenase (see text for description).

equivalent and non-interacting Ca^{2+} binding site became apparent. This has led to the idea that Ca^{2+} may act as a bridging ligand between PDHP-Pase and E2 (Lawson et al., 1993). The presence of E2 was also required for the activity of PDHP-Pase on E1 to be stimulated by Ca^{2+}; however, the activity of the PDHP-Pase against phosphopeptide substrates could not be stimulated by Ca^{2+} in either the absence or presence of E2, suggesting that other components on E1 may also be involved (Davis et al., 1977). The dissociation constants for the two binding sites of PDHP-Pase/E2 vary under different assay conditions but are in the range 5–40 μM which is somewhat higher than the $K_{0.5}$ for Ca^{2+} activation of the enzyme (Lawson et al., 1993). The $K_{0.5}$ for Ca^{2+} activation of the purified phosphatase varies from about 1 to 10 μM depending on the Mg^{2+} concentration (Thomas et al., 1986). Sr^{2+} ions can mimic the action of Ca^{2+} on PDHP-Pase, and also its effects on NAD-ICDH and OGDH, but at around 10-fold higher concentration ranges (McCormack and Osbaldeston, 1990). Ba^{2+} also acts as a mimic on NAD-ICDH and OGDH (at similar concentrations as Sr^{2+}), but interestingly does not appear to have any effect on PDHP-Pase (McCormack and Osbaldeston, 1990).

The cDNA for the 50 kDa subunit of PDHP-Pase has been cloned. The recombinant protein expressed in *Escherichia coli* has been shown to dephosphorylate PDH in a Ca^{2+}-sensitive manner (Lawson et al., 1993; P. Burnett, G. A. Rutter, and R. M. Denton, unpublished observations). The deduced amino acid sequence for the 50 kDa catalytic subunit contains a putative EF-hand Ca^{2+}-binding motif. However, there are a number of differences between this and the well-characterised EF-hand sequences of cytosolic Ca^{2+}-sensitive proteins (see Persechini et al., 1989; Strynadka and James, 1989). In particular, although the actual Ca^{2+} coordinating residues in the loop area of the EF-hand are present in PDHP-Pase, other residues which appear to be critical for correct folding are not there. For instance, residue 4 in the loop of EF-hands is almost exclusively hydrophilic whereas in PDHP-Pase it is an isoleucine; and residue 6 in the loop, which is a glycine in just about all known EF-hands is a leucine in PDHP-Pase. Substitution of this glycine residue may not allow the correct formation of the calcium-bound loop. Moreover, functional high-affinity EF-hands are usually found in pairs (Persechini et al., 1989; Strynadka and James, 1989). Therefore, at present there is some doubt as to whether this motif will be involved in Ca^{2+}-binding, although if it is, it is perhaps more likely the one involved in Ca^{2+} binding to PDHP-Pase when it is present alone. However, it has not yet been shown whether or not the 50 kDa subunit on its own will bind Ca^{2+}. There remains also

the identification of a site which may form part of a Ca^{2+}-binding pocket with the other part on the E2 subunits. The derived amino acid sequence from E2 itself (Koike et al., 1988) shows no homologous regions to Ca^{2+}-binding regions of other known Ca^{2+}-binding proteins, such as those containing EF-hands (Persechini et al., 1989; Strynadka and James, 1989) or the annexins (Klee, 1988; Burgoyne and Geisow, 1989).

There is also a report of a small inhibitory effect of Ca^{2+} on PDHa kinase which would serve to reinforce the effects of Ca^{2+} in increasing PDHa via PDHP-Pase activation (Cooper et al., 1974). However, the deduced amino acid sequence of the cDNA for PDHa kinase does not reveal any putative Ca^{2+}-binding domains (Popov et al., 1993), although it is of interest that this sequence and that for the only other intramitochondrial kinase, the branched-chain α-ketoacid dehydrogenase kinase (Popov et al., 1992) appear to establish a novel family of eukaryotic protein kinases which show most homology to the prokaryotic histidine kinase family.

NAD$^+$-Isocitrate Dehydrogenase

Mammalian NAD-ICDH probably exists as a hetero-octamer of composition $2(\alpha_2\beta\gamma)$ (Ramachandran and Colman, 1980; Ehrlich and Colman, 1983). Each of the different subunit types has a M_R around 40 kDa; however, the functions of the different subunits have not yet been fully established (but see below).

Ca^{2+} activates NAD-ICDH by causing a marked decrease in its K_m for its main substrate, *threo*-D$_S$-isocitrate (Denton et al., 1978). The presence of ADP or ATP is an absolute requirement for Ca^{2+}-sensitivity to be shown for this enzyme, in which it differs from PDH-Pase and OGDH. The Ca^{2+}-sensitivity of this enzyme (and also OGDH) can also be influenced by the ATP/ADP ratio, becoming more sensitive at lower ratios, with $K_{0.5}$ values for Ca^{2+} activation from 5 to 50 μM being obtained for NAD-ICDH (Rutter and Denton, 1988, 1989a). However, this means that the Ca^{2+}-sensitivity of NAD-ICDH appears to be expressed over a Ca^{2+} range which is about an order of magnitude greater than that for PDH and OGDH. The enzyme can also be inhibited by increases in ATP/ADP in themselves through increases in its K_m for isocitrate, and it can be end-product inhibited by increases in NADH/NAD$^+$ (McCormack et al., 1990). Both NAD-ICDH and OGDH, but not PDH, can also be inhibited by increases in pH over the physiological range (McCormack et al., 1990).

Studies with purified NAD-ICDH have shown that it requires the presence of isocitrate, Mg^{2+}, and AD(T)P for it to bind Ca^{2+} (Rutter and

Denton, 1989; Rutter, 1990). This dependence on other molecules for Ca^{2+} binding is not shown by either OGDH or PDHP-Pase/PDH. The dissociation constant of NAD-ICDH for Ca^{2+} is about 15 μM, which is consistent with its Ca^{2+}-sensitivity. The stoichiometry of Ca^{2+} binding appears to be 1 Ca per $\alpha_2\beta\gamma$ tetramer, suggesting that it may bind to either the β- or γ-subunit; however, binding at the interface between subunits is also a possibility. The former possibility would fit with the evidence that the purified α-subunit has a higher catalytic activity than purified forms of the other two subunits, which might therefore suggest that the β- and γ-subunits have more regulatory roles.

Each of the subunits of NAD-ICDH has now been cloned from mammalian sources (Nichols et al., 1993, 1995; Kim et al., 1995; Zeng et al., 1995). Comparison of these sequences with those for the enzyme from yeast (Cupp and McAlister-Henn, 1991, 1992), which has two subunits and does not show Ca^{2+}-sensitivity (Nichols et al., 1994), would suggest that the mammalian α-subunit, but not the β- or -subunits, has all of the appropriate residues and in the proper alignment for isocitrate binding and catalysis, again suggesting that it is the catalytic subunit. Putative adenine nucleotide binding sites have also been identified on the mammalian α- and γ-subunits. However, there was no significant similarity between sequences of any of the subunits of mammalian NAD-ICDH and those of other known Ca^{2+}-binding proteins (see above). There was also no obvious domain where the sequence of one of the mammalian subunits was dramatically different to either of the yeast subunits. Therefore, either the binding site for Ca^{2+} on one of the subunits of NAD-ICDH is novel, or is shared between subunits, or there is an as yet undetected Ca^{2+}-binding subunit in the purified preparations. Clearly, functional expression of the cDNA clones may yield some light on this.

2-Oxoglutarate Dehydrogenase

OGDH, like PDH, is a member of the 2-oxoacid dehydrogenase family and is also a multienzyme complex made up of multiple copies of three different enzymes, in this case 2-oxoglutarate decarboxylase (E1), dihydrolipoamide succinyl transferase (E2), and dihydrolipoamide dehydrogenase (E3), the last of these being the same enzyme that exists in PDH (Yeaman, 1989; Perham, 1991). Like PDH, the catalytic activity of the enzyme is end-product inhibited by increases in the ratios of succinyl CoA/CoA and NADH/NAD$^+$. However, unlike PDH, OGDH is not phosphorylated. It is activated directly by Ca^{2+} which causes a marked decrease in the K_m of the enzyme for 2-oxoglutarate (McCormack and

Denton, 1979). This K_m of the enzyme for this substrate is also decreased by a decrease in the ATP/ADP ratio, and decreases in this ratio also, as for NAD-ICDH, increase the sensitivity of the enzyme to Ca^{2+}, but this time the $K_{0.5}$ values for Ca^{2+} are altered within the 0.2–2 μM range and the presence of AD(T)P is not required for Ca^{2+}-sensitivity to be shown (McCormack and Denton, 1979; Rutter and Denton, 1988).

Studies with purified OGDH indicate that between 2.5 and 5 Ca^{2+} bind to each multienzyme complex (Rutter and Denton, 1989b). The dissociation constant for Ca^{2+} binding is in the range 1–7 μM, which is similar to its range of $K_{0.5}$ values for Ca^{2+} activation. The E1 step is again likely to be the irreversible step, and studies where this has been purified away from the rest of the complex and assayed with artificial electron acceptors would suggest that its reaction may also be the Ca^{2+}-sensitive step (Lawlis and Roche, 1981). However, functional OGDH consists of 12 copies of E1 subunits, 12 of E3, and 24 of E2 (Yeaman, 1989) and, given the above stoichiometry of Ca^{2+} binding, it is unlikely that Ca^{2+} binds to a site on any of the individual subunits. Nevertheless, isolated E1 and E3 subunits have been shown to form homodimers (Koike and Koike, 1976), and given that E3 is also a constituent of PDH and the other member of the 2-oxo acid dehydrogenase family, the branched-chain ketoacid dehydrogenase (Yeaman, 1989), this makes the most likely proposal to be that each Ca^{2+} binds to two E1 subunits.

The cDNA sequences are available for both the yeast (Repetto and Tzagoloff, 1989) and mammalian (Koike et al., 1992) E1 subunits of OGDH. However, the deduced amino acid sequence of the mammalian E1 again does not reveal any homology to other known Ca^{2+}-binding proteins (see above). The yeast enzyme is not Ca^{2+}-sensitive (Nichols et al., 1994), and comparison of this with the mammalian sequence shows that they are generally similar, but also reveals one insertion domain of about 20 amino acids in the mammalian sequence where there is no identity between them. This region contains 10 serine or threonine residues but no acidic negatively charged residues so it is difficult to see how this could form a Ca^{2+}-binding site. However, it is clearly a region which could be targeted in site-directed mutagenesis studies.

In summary, although there is now a lot of new sequence information on all of these Ca^{2+}-sensitive intramitochondrial dehydrogenases, the molecular details as to how Ca^{2+} may regulate these enzymes is not yet known. Of great interest, however, is that it appears from the present information described above that the intramitochondrial Ca^{2+}-sensitive dehydrogenases may have evolved different means of sensing the Ca^{2+} signal at the molecular level than other cellular Ca^{2+}-sensitive proteins. Indeed, there might well be different means of molecular interaction with Ca^{2+} for each of the different dehydrogenases.

Other Mitochondrial Ca^{2+}-Sensitive Enzymes

a-Glycerophosphate Dehydrogenase

The mitochondrial α-glycerophosphate dehydrogenase (GPDH) is not an intramitochondrial dehydrogenase, but it does pass electrons on to the respiratory chain as part of the α-glycerol phosphate shuttle for the oxidation of cytosolic NADH (see Hansford, 1991; McCormack and Denton, 1994b). This enzyme exists on the outside face of the inner mitochondrial membrane. However, it shares the striking similarity to the intramitochondrial dehydrogenases described above in being able to be activated by increases in Ca^{2+} within the low micromolar range — except that in the case of GPDH it is exposed to cytosolic and not mitochondrial matrix Ca^{2+}. Nevertheless, the concept of Ca^{2+} being able to increase the supply of respiratory chain substrate still pertains. This is of interest from an evolutionary point of view, as insect mitochondria also have a similar Ca^{2+}-sensitive GPDH on the external face of their inner mitochondrial membranes (Hansford and Chappell, 1967). However, the matrix dehydrogenases which are Ca^{2+}-sensitive in mitochondria from tissues of vertebrates are not Ca^{2+}-sensitive in insect mitochondria (McCormack and Denton, 1981). Interestingly, and consistent with this, insect mitochondria also do not appear to share the ability of their mammalian counterparts to transport sub- to low micromolar concentrations of Ca^{2+} (see McCormack and Denton, 1986).

In similar manner to both NAD-ICDH and OGDH, Ca^{2+} activates GPDH by causing a marked decrease in the K_m of the enzyme for its principal substrate, α-glycerophosphate (Hansford and Chappell, 1967; Fisher et al., 1973; Wernette et al., 1981; Rutter et al., 1992). However, unlike the matrix enzymes PDH, NAD-ICDH, and OGDH, the available sequence information on GPDH does indicate how it may be regulated by Ca^{2+}.

The purified mammalian GPDH exists as a single subunit of M_R 75 kDa which contains one molecule of bound FAD (Cole et al., 1978; Garrib and McMurray, 1986). The binding of Ca^{2+} to the purified enzyme does not appear to have been investigated. However, the deduced amino acid sequence from the cDNA for the mammalian enzyme reveals two clear EF-hand motifs at the carboxy-terminal end of the protein (Brown et al., 1994). Moreover, the more C-terminal of the two domains contains all of the necessary structural features to form a functional EF-hand; however, the other potential domain has a lysine residue in a position that is critical for Ca^{2+} binding, suggesting that it is unlikely to bind Ca^{2+} with high affinity. Pairings of high and low-affinity Ca^{2+}-binding domains are seen in other EF-hand-containing proteins, such as cardiac troponin C and parvalbumin (Persechini et al., 1989; Marsden et al., 1990).

Within the EF-hand consensus sequence, residue 9 has been implicated in controlling Ca^{2+}-dissociation rates (Renner et al., 1993). In general, proteins which have residues with short side-chains (e.g. Asn, Ser, Asp) at this position, such as calmodulin, exhibit rapid Ca^{2+} dissociation. In contrast, parvalbumin, which has the slowest Ca^{2+}-dissociation rate measured of any EF-hand protein (Breen et al., 1985), has a glutamate at this position. It is noteworthy, therefore, that GPDH also has a glutamate at this position (Brown et al., 1994), suggesting that this may allow the enzyme to sense time-averaged changes in cytosolic Ca^{2+} rather than respond to rapid oscillations, which is also likely to be the situation for the matrix Ca^{2+}-sensitive dehydrogenases (see later).

Interestingly, the yeast GPDH is around 13 kDa smaller that its mammalian counterpart and the amino acid sequence deduced from the yeast cDNA does not reveal any EF-hands (Roennow and Keilland-Brandt, 1993). It is not yet known whether or not the yeast enzyme is Ca^{2+}-sensitive, though clearly, from the above, the likelihood is that it will not be.

NAD(P)H Oxidoreductases

Plant mitochondria do not appear to contain intramitochondrial dehydrogenases which are activated by Ca^{2+} (see McCormack and Denton, 1981, 1986). However, plant mitochondria do contain NAD(P)H dehydrogenases within their mitochondrial inner membranes that are linked to the respiratory chain via ubiquinone, and which oxidise extramitochondrial NADH and NADPH (Moller et al., 1993). These dehydrogenases are activated by extramitochondrial Ca^{2+} with $K_{0.5}$ values of 1–3 μM (Moore and Åkerman, 1982; Moller et al., 1993). Again, as with α-glycerophosphate dehydrogenase in insect mitochondria, this appears to be a means whereby increases in cytosolic Ca^{2+} can lead to increases in respiration in mitochondria which do not appear to have the ability to take up Ca^{2+} into the matrix when the cytosolic concentration of Ca^{2+} is in the physiological 0.1-1 μM range.

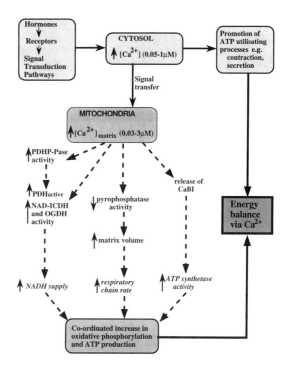

Figure 22.3 The postulated coordinated effect of Ca^{2+} on the overall process of oxidative phosphorylation (see text for description).

Matrix Pyrophosphatase

The activation by Ca^{2+} of the dehydrogenases described above will increase the supply of reducing equivalents to the respiratory chain. Although this in itself, by increasing the redox potential drive, may stimulate respiration and ATP production, it would obviously be more efficient and useful if the whole process of oxidative phosphorylation could be stimulated in concert (McCormack et al., 1990). Therefore, the topics of this and the next section are of great interest as they describe means by which the second-messenger signalling role of intramitochondrial Ca^{2+} may be extended to both the respiratory chain and the ATP synthetase (Fig. 22.3).

The mitochondrial matrix pyrophosphatase is implicated in the control of the mitochondrial volume which, in turn, is thought to play a key role in the control of the rate of the respiratory chain, particularly at the level of electron flow into the ubiquinone (UQ) pool (Halestrap, 1989). Moreover, it is known how Ca^{2+} affects this enzyme at the molecular level. The substrate for this enzyme is Mg^{2+} pyrophosphate, and Ca^{2+} causes an inhibition of this enzyme by Ca^{2+} pyrophosphate being a direct competitive inhibitor of the catalytic event (Halestrap, 1989; Davidson and Halestrap, 1989; Dubnova and Baykov, 1992). What is more, under likely physiological concentrations of matrix Mg^{2+} (Rutter et al.,

1990), the K_i and effective concentration range for Ca^{2+} is within the same sub- to low micromolar range as is effective in stimulating the matrix dehydrogenases as described above. However, it may be that this mechanism is more important in the mitochondria of some tissues, particularly liver and kidney, than in others such as heart, and this may be related to differences in the mitochondrial matrix Mg^{2+} concentration (Griffiths and Halestrap, 1993). In the liver, it is known that inorganic pyrophosphate is located primarily in the mitochondria (Davidson and Halestrap, 1988).

The molecular and sequence data that is available on pyrophosphatases (Cooperman et al., 1992) would suggest that, although their overall sequences differ substantially through evolution, they do show a remarkable level of conservation of both an extended active site structure, which has the character of a mini-mineral, and a catalytic mechanism. They require three or four Mg^{2+} ions at the active site and many of the 15–17 fully conserved active site residues are directly involved in their binding. Ca^{2+} is likely to directly interfere with the binding of Mg^{2+} to these residues. In mammalian tissues, it is not yet known if the mitochondrial form of the enzyme is different from the cytosolic form; however, it is likely that this will be the case as the yeast mitochondrial enzyme is a different protein to the cytosolic enzyme

(Lundin et al., 1991). The cytosolic pyrophosphatases are known to be soluble enzymes; however, there is evidence that the mitochondrial enzyme can also bind membranes, and may also be energy-linked (Lundin et al., 1992).

The consequent increases in matrix pyrophosphate concentration are thought to lead to an increase in the inward K^+ permeability of the mitochondrial inner membrane, perhaps via an interaction with the adenine nucleotide translocator (see Halestrap, 1989, 1994). The increase in matrix $[K^+]$ results in compensatory anion and water movement and an increase in matrix volume. Increases in matrix volume within the physiological range have a profound stimulatory effect on the rate of electron flow along the respiratory chain, particularly when fatty acids are used as respiratory substrate (Halestrap, 1989). In addition, such increases in matrix volume have also been shown to lead to the stimulation of pyruvate carboxylation, citrulline synthesis, and glutaminase activity (see Halestrap, 1989; McCormack et al., 1990). Hormones which stimulate these processes have been shown to lead to increases in matrix volume in liver (Davidson and Halestrap, 1988).

The ATP Synthetase

There have been a number of intriguing reports, particularly from the laboratories of Yamada and Harris (Yamada and Huzel, 1988; Harris and Das, 1991), which suggest that increases in matrix $[Ca^{2+}]$ may also lead to activation of the ATP synthetase itself. Thus, overall, the increases in matrix $[Ca^{2+}]$ may allow the whole process of oxidative phosphorylation — reducing power supply, the respiratory chain, and the ATP production machinery — to work at a higher rate but without the need to drop the key ATP/ADP or $NADH/NAD^+$ concentration ratios to achieve this (Fig. 22.3).

The mechanism by which it is proposed that Ca^{2+} may stimulate the ATP synthetase is by causing the release of a small inhibitory subunit (Yamada and Huzel, 1988, 1989). This inhibitory subunit has been shown to be distinct from, and its effects additive to those of, the well-known ATPase inhibitory subunit first identified by Pullman and Monroy (1963) (Yamada and Huzel, 1988). The Ca^{2+}-binding inhibitory subunit is heat-stable and has a minimal M_R of 6390 with 62 amino acids and an isoelectric point of 4.6, and there appears to be 5–10 mol/mol of synthetase in mammalian mitochondria (Yamada and Huzel, 1988). Little is known of the structure of this protein (see Harris and Das, 1991), or whether or not it can bind Ca^{2+} directly, or its site of interaction with the membrane-bound ATP synthetase. There is some information that it appears to be functionally inhibitory as a 12.5 kDa dimer, with Ca^{2+}

preventing inhibition by dissociating the protein into monomers (Yamada et al., 1981).

Although the concept of a coordinated regulation of oxidative phosphorylation by Ca^{2+} is obviously attractive, it is frustrating that, at present, no further information is available on this protein. For instance, it is possible that the alterations observed in synthetase activity are not due to direct effects of Ca^{2+}, but rather are indirect consequences of the activation of the dehydrogenases via, for example, increases in $NADH/NAD^+$.

The Mitochondrial Ca^{2+}-Transport System

Clearly, the concentrations of Ca^{2+} within mitochondria depend on the concentrations of Ca^{2+} in the cytoplasm and on the rates of uptake and egress of Ca^{2+} across the mitochondrial inner membrane (see Fiskum and Lehninger, 1982; Carafoli and Sottacasa, 1984; Crompton, 1985; Gunter and Gunter, 1994) (Fig. 22.4).

The Mitochondrial Ca^{2+}-Uniporter

Uptake of Ca^{2+} into the mitochondrial matrix is catalysed by an electrophoretic uniporter which is driven by the membrane potential component of the respiratory protonmotive gradient (which is about 180 mV, negative inside) (Fig. 22.4). It should be noted that, at present, Ca^{2+} is the only known second-messenger molecule which has access to the mitochondrial matrix (recent work, however, has shown that the NO radical may also act inside mitochondria).

This uptake pathway can be inhibited by physiological concentrations of Mg^{2+} (Fiskum and Lehninger, 1982; Carafoli and Sottacasa, 1984; Crompton, 1985; Gunter and Gunter, 1994), which acts in a competitive manner against Ca^{2+} (but is not itself transported) and confers a large degree of sigmoidicity upon the kinetics of Ca^{2+} uptake. At physiological extramitochondrial Mg^{2+} and Ca^{2+} concentrations, the Ca^{2+}-uptake pathway operates well below ($< 10\%$) its maximal capacity, which is about 150 nmol/min/mg of mitochondrial protein at 30°C [which is of similar maximal capacity to the mitochondrial inner membrane Na^+/H^+ exchanger (Garlid, 1988; see below)]. Under these conditions, its K_m for extramitochondrial Ca^{2+} is also in excess of 10 μM. The uptake pathway can be activated by physiological concentrations of the polyamine spermine (Nicchitta and Williamson, 1984; Lenzen et al., 1986; McCormack, 1989), whose effects appear to be independent of those of Mg^{2+}.

The uniporter can also transport Sr^{2+}, Mn^{2+}, Ba^{2+}, and Pb^{2+} in a mimetic manner and in that order of selectivity (see Åkerman and Nicholls,

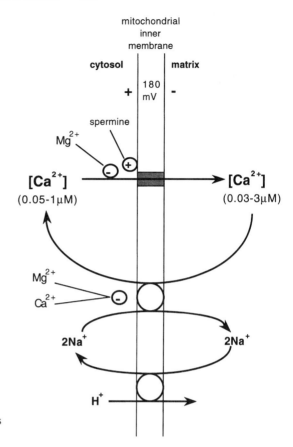

Figure 22.4 The Ca^{2+}-transport system of the mammalian inner mitochondrial membrane and its effectors (see text for description).

1983; Gunter and Gunter, 1994). The uptake pathway can also be inhibited nonphysiologically by the glycoprotein stain ruthenium red (Moore, 1971), or at least a component of this (Emerson et al., 1993), and by lanthanide ions (Mela, 1968).

Over the years, there have been many attempts by many different laboratories to identify the Ca^{2+}-uniporter (see Gunter and Pfeiffer, 1990; McCormack et al., 1992). However, as yet no consensus has emerged as to its identity or nature (carrier or channel), and certainly no cDNA or amino acid sequence data is available. The central problem is the lack of known blockers or other organic ligands of this putative protein of high enough specificity for tight binding and purification. Reconstitution assay are obviously also difficult for transport activities. Although the involvement of a ruthenium red-binding glycoprotein (which would also bind Ca^{2+} and perhaps Mg^{2+}) has been advocated, a variety of different entities with M_R values over the 15,000–42,000 range have been proposed, together with various additional proposed elements of 2,000–6,000 M_R for an associated channel (Gunter and Pfeiffer, 1990; Gunter and Gunter, 1994).

Mitochondrial Ca^{2+}-Egress Systems

The major mechanism for the egress of mitochondrial matrix Ca^{2+} operating under normal physiological conditions is a Ca^{2+}/Na^+ antiporter (Carafoli et al., 1974, and see Crompton, 1985; McCormack et al., 1990, 1992) (Fig. 22.4). There has been a consensus that this putative transporter catalysed a passive electroneutral $2Na^+/1Ca^{2+}$ exchange reaction, with this process then being driven by the subsequent (and 10-fold more active) Na^+/H^+ exhanger activity and thus the protonmotive force (see Brand, 1988). This conclusion was supported by reconstitution studies carried out as part of the purification procedure for this protein (see below) (Li et al., 1992). However, more recently others have provided evidence that it may, at least under some conditions, operate as a $Ca^{2+}/3Na^+$ exchanger or an active $Ca^{2+}/2Na^+$ mechanism (see Baysal et al., 1994; Gunter and Gunter, 1994).

The Na^+-dependent egress pathway can be inhibited by increases in the extramitochondrial concentrations of Ca^{2+} and Mg^{2+} within the physiological range (Hayat and Crompton, 1982, 1987). This effect of Ca^{2+} is thought to allow for increases in extramitochondrial Ca^{2+} to lead to larger changes in intramitochondrial Ca^{2+} (see McCormack et al., 1989). This transporter can also be inhibited by several non-physiological agents, and in particular by benzo-diazepines and other related drugs which are more commonly known as blockers of the plasma membrane Ca^{2+} channel (Vaghy et al., 1982), although some novel compounds developed from these chemical roots have much greater selectivity for the mitochondrial carrier (see Cox and Matlib, 1993). Both Ba^{2+} and Sr^{2+} mimicked the inhibitory effects of extramitochondrial Ca^{2+}, but only Sr^{2+} could act as a transport surrogate for Ca^{2+} (McCormack and Osbaldeston, 1990), whereas Li^+ could act as a transport substitute for Na^+ (Crompton et al., 1978).

The availability of organic ligands, such as diltia-zem, against this antiporter from which photoaffinity ligands could be made, together with the development of reconstitution assay systems using fluorescent ion-binding ligands, led to the identification of the Na^+/Ca^{2+} exchanger as a 110 kDa protein (Li et al., 1992). However, as yet no cDNA or other sequence has been published for this protein (Garlid, 1994).

There is also thought to be a Na^+-independent pathway for the egress of matrix Ca^{2+} that is as yet poorly characterised but which is thought to be perhaps a direct $Ca^{2+}/2H^+$ antiporter and which may also be an active mechanism (see Gunter and Gunter, 1994). It has been argued that the Na^+-dependent pathway may predominate in the mitochondria of excitable tissues such as heart, brain and muscle, whereas the Na^+-independent pathway would be dominant in mitochondria from tissues such as liver and kidney (see Gunter and Pfeiffer, 1990; Gunter and Gunter, 1994). However, others have argued that under physiological levels of Ca^{2+}-loading of mitochondria, the Na^+-dependent pathway dominates in these latter tissues also (see McCormack, 1985; McCormack et al., 1988), and that the Na^+-independent pathway may be artifactual in origin and is unlikely to operate physiologically (see Crompton, 1990a; McCormack and Denton, 1993).

Clearly, much work still needs to be done to identify, characterise, and sequence the proteins of the mitochondrial transport system. This remains one of the key challenges in this field and advances here would dramatically advance this field as a whole.

The Ca^{2+}-Operated Mitochondrial Pore

The transport systems described above are those which are thought to operate under normal cellular conditions for calcium. It can be calculated that under such conditions the energy cost of cycling Ca^{2+} across the mitochondrial inner membrane would always be less than 1% of mitochondrial respiratory capacity (Crompton, 1990b). Clearly, this cost is outweighed by the gains that the cell obtains in terms of energetic homeostasis. However, the uptake pathway still has a potential capacity that is about 10-fold that of the egress pathways, and therefore there remains the potential for mitochondria to accumulate calcium if the egress pathways become saturated (Nicholls, 1978). This may occur under conditions of cellular calcium overload, for example in disease conditions such as reperfusion following ischaemic episodes (Crompton, 1990b; Halestrap et al., 1993). Mitochondria may then take over the role of buffering cytosolic $[Ca^{2+}]$, albeit at a higher level than would normally be found (McCormack et al., 1990). This may then allow the cell some time to rectify the situation and recover, by pumping calcium out of the cell and into other storage systems, by limiting the rise in cytosolic Ca^{2+} and thus preventing activation of proteolysis and other Ca^{2+}-dependent cell destructive systems (McCormack et al., 1990).

It is also becoming evident that the concentration of Ca^{2+} in the mitochondrial matrix of animal cells may normally reach $10\,\mu M$, or even higher, for short transient periods following stimulation of cells by hormones or other agents which act through the mobilisation of intracellular calcium stores (Rizzuto et al., 1992, 1994; Rutter et al., 1996). These transient high concentrations greatly exceed the apparent increases in cytosolic Ca^{2+}. They are thought to be the result of the close juxtaposition of a subpopulation of mitochondria with the InsP3-activated release channels on the endoplasmic reticulum (Rizzuto et al., 1994). These mitochondria may thus be exposed to localised high concentrations of Ca^{2+} for a short period of time after cell stimulation, and during this period the rate of Ca^{2+} uptake exceeds that of Ca^{2+} egress, leading to the observed transient high increases in intramitochondrial Ca^{2+}.

However, the capacity of mitochondria to retain large loads of calcium is not limitless and can lead to their own damage. In particular, it is now realised that if mitochondrial matrix $[Ca^{2+}]$ reaches supra-physiological levels, it leads to the opening of a specific pore in the inner mitochondrial membrane (Al-Nasser and Crompton, 1986; Crompton et al., 1987, 1988; Gunter and Pfeiffer, 1990; Halestrap et al., 1993). The concentrations of matrix Ca^{2+} which lead to pore opening are in excess of the ranges in which it affects the enzyme systems described earlier. Pore opening is reversible on removal of Ca^{2+}, but open pores allow the passage of molecules up to about 2000 M_R and therefore the inner membrane becomes de-energised, and critical small molecules such as ATP, acetyl CoA and NADH may be lost. Therefore, opening of this pore has, not surprisingly, been postulated to be a potentially critical event in the processes leading to apoptosis and cell death. What is more, opening of the pore by matrix Ca^{2+} is promoted by other factors which are likely to arise under conditions of cell stress, such as increased inorganic phosphate and oxidative stress (Crompton et al., 1978, 1988). It has also been found that the immunosuppressant cyclosporin A is an inhibitor of pore opening (Crompton et al., 1988), and there is some evidence that this agent may protect cardiac cells and tissue from ischaemic or hypoxic insult, but only over a narrow concentration range (Nazareth et al., 1991; Griffiths and Halestrap, 1995). This latter report also provided some direct evidence, for the first time, for pore opening in vivo..

The mitochondrial binding site for cyclosporin A is a cyclophilin (peptidyl-prolyl *cis-trans* isomerase) which is distinct, and probably a different gene product, from the cytosolic equivalent and has a M_R of around 18 kDa (Connern and Halestrap, 1992; Andreeva et al., 1995). What is more, cyclophilin binding to mitochondrial membranes was found to be enhanced under conditions which would lead to pore opening in a process that was inhibitable by cyclosporin A (Connern and Halestrap, 1994). It has been proposed, on the basis of inhibitor studies, that the mitochondrial cyclophilin may bind to the adenine nucleotide translocase, and, in the presence of excess calcium, trigger a conformational change in this protein that induces pore opening (Halestrap et al., 1993). However, others have presented evidence that a membrane protein of approximately 10 kDa may also be part of the cyclosporin-A receptor of the Ca^{2+}-activated pore (Andreeva and Crompton, 1994).

Concluding Remarks

Clearly, much work remains to be done to establish the molecular basis for the interaction of Ca^{2+} ions with their intramitochondrial target proteins, and to identify and describe in molecular terms the mitochondrial Ca^{2+}-transport proteins. However, the growing realisation of the importance of the signalling role of intramitochondrial Ca^{2+} gives a strong impetus for this work to be carried out. Moreover, sequence information is now available for many of the mitochondrial matrix targets and, excitingly, it is beginning to look as if novel means of Ca^{2+}-protein interaction may be uncovered.

A number of approaches have been developed which demonstrate the importance of Ca^{2+} in the regulation of the intramitochondrial dehydrogenases within intact cell and tissue preparations (see Denton and McCormack, 1990; McCormack and Denton, 1994a, for reviews of this topic). Three approaches have been especially useful:

(1) The use of ruthenium red, which inhibits Ca^{2+} uptake into mitochondria when applied to some tissues. For example, rat hearts perfused with medium containing ruthenium red still show increases in contraction and O_2 uptake when exposed to β-adrenergic agonists, but the activation of PDH, and full maintainance of the ATP/ADP ratio, is no longer evident (McCormack and England, 1983; Unitt et al., 1989). Under these conditions, increases in citrate cycle activity thus appear to be brought about by the increases in the ADP/ATP ratio rather than by increases in intramitochondrial Ca^{2+}.

(2) The rapid preparation of mitochondria under conditions where both uptake and egress of Ca^{2+} are blocked. Enhanced activities of both PDH and OGDH can be demonstrated if mitochondria are prepared in this way from stimulated tissues, and these enhanced activities can be shown to be due to increased intramitochondrial concentrations of Ca^{2+} (McCormack and Denton, 1984; McCormack et al., 1990).

(3) Direct measurement of Ca^{2+} within mitochondria of intact cell preparations. This work began with the observations of Hansford and colleagues that when single heart cells were loaded with indo-1 or fura-2, a proportion of the dye ends up in the mitochondrial matrix, and that the cytosolic dye could be selectively quenched with low concentrations of Mn^{2+} (Miyata et al., 1991). This enabled them to demonstrate, as predicted on the basis of the known kinetics of the mitochondrial Ca^{2+}-transport system (Crompton 1990b), that mitochondrial matrix Ca^{2+} did not respond to "beat-to-beat" changes in cytosolic Ca^{2+}, but did increase as the result of "time-averaged" increases in cytosolic Ca^{2+}. Some Ca^{2+}-respondent dyes, such as rhod-2, are preferentially located in mitochondria in cells because of their positive charge, resulting in a rather straightforward means of measuring intramitochondrial Ca^{2+}, at least in some cell types (Hajnoczky et al., 1995; Rutter et al., 1996).

A further elegant approach has been the development of chimaeric proteins containing Ca^{2+}-reporting units, such as aequorin, which can be selectively expressed within the mitochondrial matrix (Pozzan et al., 1994; Rizzuto et al., 1992, 1994, 1995; Rutter et al., 1992, 1993, 1996). This approach has revealed much new information on the kinetics of mitochondrial Ca^{2+} uptake and egress, and on the concentrations of mitochondrial matrix Ca^{2+}, in intact cells. Attempts are also being made to link measurements of intramitochondrial Ca^{2+} with measurements of the physiological enzyme responses, and also to explore differential changes of intramitochondrial Ca^{2+} within single cells (Rutter et al., 1996; Robb-Gaspers et al., 1998). It is hoped and anticipated that this will lead to much new information on this important topic within the next few years.

Overall, there is now excellent evidence that increases in intramitochondrial Ca^{2+} are of central importance in the stimulation of mitochondrial respiration and ATP synthesis in a number of tissues, including heart and liver (McCormack et al., 1990; Robb-Gaspers et al. 1998). However, it is important to emphasise that these processes may also be stimulated under some circumstances by increases in the ADP/ATP ratio within mitochondria (Robb-Gaspers et al., 1998). One of the major challenges of the future is to develop techniques which will allow the relative roles of these two potential mechanisms to be established in a large range of conditions and cell types. In particular, techniques for the measurement of changes in the ATP/ADP ratio within mitochondria of intact cells are needed.

Acknowledgments Studies carried out in the authors' laboratories were supported by grants from the Medical Research Council, the British Diabetic Association, the British Heart Foundation, and the Lister Institute of Preventive Medicine.

References

Åkerman, K. E. O. and D. G. Nicholls (1983) Physiological and bioenergetic aspects of mitochondrial calcium transport. *Rev. Physiol. Biochem. Pharmacol.* 95: 149–201.

Al-Nasser, I. and M. Crompton (1986) The entrapment of the Ca^{2+}-indicator Arsenazo III in the matrix space of rat liver mitochondria by permeabilisation and resealing. *Biochem. J.* 239: 31–40

Andreeva, L. and M. Crompton (1994) An ADP-sensitive cyclosporin-A-binding protein in rat liver mitochondria. *Eur. J. Biochem.* 221: 261–8.

Andreeva, L., A. Tanveer, and M. Crompton (1995) Evidence for the involvement of a membrane-associated cyclosporin-A-binding protein in the Ca^{2+}-activated inner membrane pore of heart mitochondria. *Eur. J. Biochem.* 230: 1125–32.

Baysal, K., D. W. Jung, K. K. Gunter, T. E. Gunter, and G. P. Brierley (1994) Na^+-dependent Ca^{2+} efflux mechanism of heart mitochondria is not a passive $Ca^{2+}/2Na^+$ exchanger. *Am. J. Physiol.* 266: C800–8.

Becker, G. L, G. Fiskum, and A. L. Lehninger (1980) Regulation of free Ca^{2+} by liver mitochondria and endoplasmic reticulum. *J. Biol. Chem.* 255: 9009–12.

Berridge, M (1988) Inositol lipids and calcium signalling. *Proc. Roy. Soc. London, Ser. B Biol. Sci.* 234: 359–78.

Brand, M. D. (1988) The signalling role of mitochondrial calcium transport. *ISI Atlas Sci. Biochem.* 1: 350–4.

Breen, P. J., K. A. Johnson, and W. D. Horrocks (1985) Stopped-flow kinetic studies of metal ion dissociation or exchange in a tryptophan containing parvalbumin. *Biochemistry* 24: 4997–5004.

Brown, L. J., M. J. MacDonald, D. A. Lehn, and S. M. Moran (1994) Sequence of rat mitochondrial glycerol-3-phosphate dehydrogenase cDNA. Evidence for EF-hand calcium-binding domains. *J. Biol. Chem.* 269: 14363–6.

Burgoyne, R. D. and M. J. Geisow (1989) The annexin family of calcium binding proteins." *Cell Calcium* 10: 1–10.

Carafoli, E. and G. Sottocasa (1984) The uptake and release of calcium by mitochondria. *New Comprehen. Biochem.* 9: 269–89.

Carafoli, E., R. Tiozzo, G. Lugli, F. Crovetti, and C. Kratzing (1974) The release of calcium from heart mitochondria by sodium. *J. Mol. Cell Cardiol.* 6: 361–371.

Cole, E. S., A. L. Cyrus, P. D. Holohan, and T. L. Fondy (1978) Isolation and characterisation of flavin linked glycerol 3-phosphate dehydrogenase from rabbit skeletal muscle mitochondria and comparison with the enzyme from rabbit brain. *J. Biol. Chem.* 253: 7952–9.

Connern, C. P. and A. P. Halestrap (1992) Purification and N-terminal sequencing of peptidyl- prolyl *cis-trans*-isomerase from rat liver mitochondrial matrix reveals the existence of a distinct mitochondrial cyclophilin. *Biochem. J.* 284: 381–5.

Connern, C. P. and A. P. Halestrap (1994) Recruitment of mitochondrial cyclophilin to the mitochondrial inner membrane under conditions of oxidative stress that enhance the opening of a calcium-specific non-specific channel. *Biochem. J.* 302: 321–4,.

Cooper, R. H., P. J. Randle, and R. M. Denton (1974) Regulation of heart muscle pyruvate dehydrogenase kinase. *Biochem. J.* 143: 625–41.

Cooperman, B. S., A. A. Baykov, and R. Lahti (1992) Evolutionary conservation of the active site of soluble inorganic pyrophosphatase. *Trends Biochem. Sci.* 17: 262–6.

Cox, D. A. and M. A. Matlib (1993) Modulation of intramitochondrial free Ca^{2+} by antagonists of Na^+-Ca^{2+} exchange." *Trends Pharmacol. Sci.* 14: 408–13.

Crompton, M. (1985) The regulation of mitochondrial calcium transport in heart. *Curr. Top. Membr. Transp.* 25: 231–76.

Crompton, M. (1990a) Mitochondrial calcium transport. In *Intracellular Calcium Regulation* (Bronner, F., ed.), Alan R. Liss, New York, pp. 181–209.

Crompton, M. (1990b) The role of Ca^{2+} in the function and dysfunction of heart mitochondria. In *Calcium and the Heart* (Langer, G. A., ed.) Raven Press, New York, pp. 167–98.

Crompton, M., R. Moser, H. Ludi, and E. Carafoli (1978) The interrelations between the transport of sodium and calcium in mitochondria of various mammalian tissues. *Eur. J. Biochem.* 82: 25–31.

Crompton, M., A. Costi, and L. Hayat (1987) Evidence for the presence of a reversible Ca^{2+}- dependent pore activated by oxidative stress in heart mitochondria. *Biochem. J.* 245: 915–18.

Crompton, M., H. Ellinger, and A. Costi (1988) Inhibition by cyclosporin A of a Ca^{2+}-dependent pore in heart mitochondria activated by inorganic phosphate and oxidative stress. *Biochem. J.* 255: 357–60.

Cupp, J. R. and L. McAlister-Henn (1991) NAD-isocitrate dehydrogenase: cloning, disruption and nucleotide sequence of the IDH2 gene from *Saccharomyces cerevisiae*. *J. Biol. Chem.* 266: 22199–205.

Cupp, J. R. and L. McAlister-Henn (1992) Cloning and characterisation of the gene encoding IDH1 subunit of NAD-isocitrate dehydrogenase from *Saccharomyces cerevisiae*. *J. Biol. Chem.* 267: 16417–23.

Davidson, A. M. and A. P. Halestrap (1988) Inorganic pyrophosphate is located primarily in the mitochondria of the hepatocyte and increases in parallel with the decrease in light- scattering induced by gluconeogenic hormones, butyrate and ionophore A23187. *Biochem. J.* 254: 379–84,.

Davidson, A. M. and A. P. Halestrap (1989) Inhibition of mitochondrial-matrix pyrophosphatase by physiological [Ca^{2+}], and its role in the hormonal regulation of mitochondrial matrix volume. *Biochem. J.* 258: 817–21.

Davis, P. F., F. H. Pettit, and L. J. Reed (1977) Peptides derived from pyruvate dehydrogenase as substrates for PDH kinase and phosphatase. *Biochem. Biophys. Res. Commun.* 49: 563–71.

Denton, R. M. and J. G. McCormack (1980) On the role of the calcium transport cycle in heart and other mammalian mitochondria. *FEBS Lett.* 119: 1–8.

Denton, R. M and J. G. McCormack (1985) Ca^{2+} transport by mammalian mitochondria and its role in hormone action. *Am. J. Physiol.* 249: E543–54.

Denton, R. M. and J. G. McCormack (1990) Ca^{2+} as a second messenger within mitochondria of the heart and other tissues. *Annu. Rev. Physiol.* 52: 451–66.

Denton, R. M, P .J. Randle, and B. R. Martin (1972) Stimulation by calcium ions of pyruvate dehydrogenase phosphate phosphatase. *Biochem. J.* 128: 161–3.

Denton, R. M., D. A. Richards, and J. G. Chin (1978) Calcium ions in the regulation of NAD-isocitrate dehydrogenase from the mitochondria of rat heart and other tissues. *Biochem. J.* 176: 899–906.

Denton, R. M., P. J. W. Midgley, G. A. Rutter, A. P. Thomas, and J. G. McCormack (1989) Studies into the mechanism whereby insulin activates pyruvate dehydrogenase complex in adipose tissue. *Ann. N.Y. Acad. Sci.* 573: 285–96.

Denton, R. M., J. G. McCormack, G. A. Rutter, P. Burnett, N. J. Edgell, S. K. Moule, and T. A. Diggle (1996) Hormonal regulation of pyruvate dehydrogeanse complex. *Adv. Enzym. Regul.* 36: 183–198.

Dubnova, E. B. and A. A. Baykov (1992) Catalytic properties of the inorganic pyrophosphatase in rat liver mitochondria. *Arch. Biochem. Biophys.* 292: 16–19.

Ehrlich, R. S. and R. F. Colman (1983) Dissimilar subunits of DPN-isocitrate dehydrogenase. *J. Biol. Chem.* 249: 1848–56.

Emerson, J., M. J. Clarke, W.-L. Ying, and D. R. Sanadi (1993) The component of "ruthenium red" responsible for inhibition of mitochondrial calcium ion transport. Spectra, electrochemistry, and aquation kinetics. Crystal structure of μ-O-[(HCO$_2$)(NH$_3$)$_4$Ru]$_2$Cl$_3$. *J. Am. Chem. Soc.* 115: 11799–805.

Fabiato, A. and Fabiato, F. (1979) Calcium and cardiac excitation-contraction coupling. *Annu. Rev. Physiol.* 41: 473–84.

Fisher, A. B., A. Scarpa, K. F. LaNoue, D. Basset, and J. R. Williamson (1973) Respiration of rat lung mitochondria and the influence of Ca^{2+} on substrate utilisation. *Biochemistry* 23: 1438–46.

Fiskum, G. and A. L. Lehninger (1982) Mitochondrial regulation of intracellular calcium. In: *Calcium and Cell Function*, Vol. 2 (Cheung, W. Y., ed.). Academic Press, New York, pp. 39–80.

Garlid, K. D. (1988) Mitochondrial volume control. In *Integration of Mitochondrial Function* (Lemasters, J. J., Hackenbrock, C. R., Thurman, R. G., and Westerhoff, H. V., eds.). Plenum Press, New York, pp. 257–76.

Garlid, K. D. (1994) Mitochondrial cation transport: a progress report. *J. Bioenerg. Biomembr.* 26: 537–42.

Garrib, A. and W. C. McMurray (1986) Purification and characterisation of glycerol 3-phosphate dehydrogenase flavin-linked from rat liver mitochondria. *J. Biol. Chem.* 261: 8042-8.

Griffiths, E. J. and A. P. Halestrap (1993) Pyrophosphate metabolism in the perfused heart and isolated heart mitochondria and its role in regulation of mitochondrial function by calcium. *Biochem. J.* 290: 489–95.

Griffiths, E. J. and A. P. Halestrap (1995) Mitochondrial non-specific pores remain closed during cardiac ischaemia but open upon reperfusion. *Biochem. J.* 307: 93–8.

Gunter, K. K. and T. E. Gunter (1994) Transport of calcium by mitochondria. *J. Bioenerg. Biomembr.* 26: 471–85.

Gunter, T. E. and D. R. Pfeiffer (1994) Mechanisms by which mitochondria transport calcium. *Am. J. Physiol.* 258: C755–86.

Hajnoczky, G., L. D. Robb-Gaspers, M. B. Seitz, and A. P. Thomas (1995) Decoding of cytoplasmic calcium oscillations in the mitochondria. *Cell* 82: 415–24.

Halestrap, A. P. (1989) The regulation of the matrix volume of mammalian mitochondria *in vivo* and *in vitro* and its role in the control of mitochondrial metabolism. *Biochim. Biophys. Acta* 973: 355–82.

Halestrap, A. P. (1994) Interactions between oxidative stress and calcium overload on mitochondrial function. In *Mitochondria: DNA, Proteins and Disease.* (Darley-Usmar, V. and Shapira, A. H. V., eds.). Portland Press, London, ρp. 113–42.

Halestrap, A. P., E. J. Griffiths, and C. P. Connern. (1993) Mitochondrial calcium-handling and oxidative stress. *Biochem. Soc. Trans.* 21: 353–8.

Hansford, R. G. (1991) Dehydrogenase activation by Ca^{2+} in cells and tissues. *J. Bioenerg. Biomembr.* 23: 823–54.

Hansford, R. G. and J. B. Chappell (1967) The effect of Ca^{2+} on the oxidation of glycerol 3- phosphate by blowfly flight muscle mitochondria. *Biochem. Biophys. Res. Commun.* 27: 686–92.

Harris, D. A. and A. M. Das (1991) Control of mitochondrial ATP synthesis in the heart. *Biochem. J.* 280: 561–73.

Hayat, L. H. and M. Crompton (1982) Evidence for the existence of regulatory sites for Ca^{2+} on the Na^+/Ca^{2+} carrier of cardiac mitochondria. *Biochem. J.* 202: 509–18.

Hayat, L. H. and M. Crompton (1987) The effects of Mg^{2+} and adenine nucleotides on the sensitivity of the heart mitochondrial Na^+/Ca^{2+} carrier to extramitochondrial Ca^{2+}. A study using arsenazo III-loaded mitochondria. *Biochem. J.* 244: 533–8.

Kim, Y. O., I.-U. Oh, J. Jeng, B. J. Song, and T.-L. Huh (1995) Characterisation of a cDNA clone for human NAD^+-specific isocitrate dehydrogenase α-subunit and structural comparison with its isoenzymes from different species. *Biochem. J.* 308: 63–8.

Klee, C. B. (1988) Calcium dependent phospholipid and membrane binding proteins. *Biochemistry* 27: 6645–50.

Koike, K., Y. Urata, S. Ohta, Y. Kawa, and M. Koike (1988) Cloning and sequencing of cDNAs for the beta and alpha subunits of human pyruvate dehydrogenase. *Proc. Natl. Acad. Sci. USA* 85: 41–5.

Koike, K., Y. Urata, and S. Goto (1992) Cloning and nucleotide sequence of the cDNA encoding human 2-oxoglutarate dehydrogenase (lipoamide). *Proc. Natl. Acad. Sci. USA* 89: 1963–7.

Koike, M. and K. Koike (1976) Structure, assembly and function of mammalian α-keto acid dehydrogenase complexes. *Adv. Biophys.* 9: 187–227.

Lawlis, V. B. and T. E. Roche (1981) Inhibition of bovine kidney α-ketoglutarate dehydrogenase by NADH in the presence or absence of calcium ion and the effect of ADP on NADH inhibition. *Biochemistry* 20: 2523–7.

Lawson, J. E., X.-D. Niu, K. S. Browning, H. Le Trong, J. Yan, and L. J. Reed (1993) Molecular cloning and expression of the catalytic subunit of pyruvate dehydrogenase phosphatase and sequence similarity with protein phosphatase 2C. *Biochemistry* 32: 8987–93.

Lenzen S., R. Hickethier, and V. Panten (1986) Interactions between spermine and Mg^{2+} on mitochondrial Ca^{2+} transport. *J. Biol. Chem.* 261: 16478–83.

Li, W., Z. Shariat-Madar, M. Powers, X. Sun, R. D. Lane, and K. D. Garlid (1992) Reconstitution, identification, purification, and immunological characterization of the 110 kDa Na^+/Ca^{2+} antiporter from beef heart mitochondria. *J. Biol. Chem.* 267: 17983–9.

Lundin, M., H. Baltscheffsky, and H. Ronne (1991) Yeast PPA2 gene encodes a mitochondrial pyrophosphatase that is essential for mitochondrial function. *J. Biol. Chem.* 266: 12168–72.

Lundin, M., S. W. Deopujari, L. Lichko, L. Pereira-da-Silva, and H. Baltscheffsky (1992) Characterization of a mitochondrial inorganic pyrophosphatase in *Saccharomyces cerevisiae*. *Biochim. Biophys. Acta* 1098: 217–23.

Marsden, B. J., G. S. Shaw, and B. D. Sykes (1990) Calcium binding proteins. Elucidating the contributions to calcium affinity from an analysis of species variants and peptide fragments. *Biochem. Cell Biol.* 68: 257–62.

McCormack, J. G. (1985) Characterisation of the effects of Ca^{2+} on the intramitochondrial Ca^{2+}-sensitive enzymes from rat liver and within rat liver mitochondria. *Biochem. J.* 231: 581–95.

McCormack, J. G. (1989) Effects of spermine on mitochondrial Ca^{2+} transport and the ranges of extramitochondrial Ca^{2+} to which the matrix Ca^{2+}-sensitive dehydrogenases respond. *Biochem. J.* 264: 167–74.

McCormack, J. G. and R. M. Denton (1979) The effects of calcium ions and adenine nucleotides on the activity of pig heart 2-oxoglutarate dehydrogenase complex. *Biochem. J.* 180: 533–44.

McCormack, J. G. and R. M. Denton (1981) A comparative study of the regulation by Ca^{2+} of the activities of the 2-oxoglutarate dehydrogenase complex and NAD-isocitrate dehydrogenase from a variety of sources. *Biochem. J.* 196: 619–24.

McCormack, J. G. and R. M. Denton (1984) Role of Ca^{2+} ions in the regulation of intramitochondrial metabolism in the rat heart. Evidence from studies with isolated mitochondria that adrenaline activates the pyruvate and 2-oxoglutarate dehydrogenase complexes by increasing the intramitochondrial concentrations of Ca^{2+}. *Biochem. J.* 218: 235–47.

McCormack, J. G. and R. M. Denton. (1986) Ca^{2+} as a second messenger within mitochondria. *Trends Biochem. Sci.* 11: 258–62.

McCormack, J. G. and R. M. Denton (1993) The role of intramitochondrial Ca^{2+} in the regulation of oxidative phosphorylation in mammalian tissues. *Biochem. Soc. Trans.* 21: 793-8.

McCormack, J. G. and R. M. Denton (1994a) Signal transduction by intramitochondrial Ca^{2+} in mammalian energy metabolism. *News Physiol. Sci.* 9: 71–6.

McCormack, J. G. and R. M. Denton (1994b) Mammalian mitochondrial metabolism and its regulation. In *Mitochondria: DNA, Proteins and Disease*. (Darley-Usmar, V. and Shapira, A. H. V., eds.). Portland Press, London, pp. 81–112.

McCormack, J. G. and P. J. England (1983) Ruthenium red inhibits the activation of pyruvate dehydrogenase caused by positive inotropic agents in the perfused rat heart. *Biochem. J.* 214: 581–5.

McCormack, J. G. and N. J. Osbaldeston. (1990) The use of the Ca^{2+}-sensitive intramitochondrial dehydrogenases and entrapped fura-2 to study Sr^{2+} and Ba^{2+} transport across the inner membrane of mammalian mitochondria. *Biochem. J.* 192: 239–44.

McCormack, J. G., E. S. Bromidge, and N. J. Dawes (1988) Characterization of the effects of Ca^{2+} on the intramitochondrial Ca^{2+}-sensitive dehydrogenases within intact rat-kidney mitochondria. *Biochim. Biophys. Acta* 934: 282–92.

McCormack, J. G., H. M. Browne, and N. J. Dawes (1989) Studies on mitochondrial Ca^{2+}- transport and matrix Ca^{2+} using fura-2-loaded rat heart mitochondria. *Biochim. Biophys. Acta* 973: 420–7.

McCormack, J. G., A. P. Halestrap, and R. M. Denton (1990) The role of calcium ions in the regulation of mammalian intramitochondrial metabolism. *Physiol. Rev.* 70: 391–425.

McCormack, J. G., R. L. Daniel, N. J. Osbaldeston, G. A. Rutter, and R. M. Denton (1992) Mitochondrial Ca^{2+} transport and the role of matrix Ca^{2+} in mammalian tissues. *Biochem. Soc. Trans.* 20: 153–9.

Mela, L. (1968) Interactions of La and local anaesthetic drugs with mitochondrial Ca and Mn uptake. *Arch. Biochem. Biophys.* 123: 286–93.

Midgley, P. J. W., G. A. Rutter, A. P. Thomas, and R. M. Denton (1987) Effects of Ca^{2+} and Mg^{2+} on the activity of pyruvate dehydrogenase phosphate phosphatase within toluene- permeabilized mitochondria. *Biochem. J.* 241: 371–7.

Mitchell, P. (1966) *Chemiosmotic Coupling in Oxidative and Photosynthetic Phosphorylation.* Glynn Research, Bodmin, England.

Miyata, H., H. S. Silverman, S. J. Sollot, E. G. Lakatta, M. D. Stern, and R. G. Hansford (1991) Measurement of mitochondrial free Ca^{2+} in living single rat cardiac myocytes. *Am. J. Physiol.* 261: H1123–34.

Moller, I. A., A. G. Rasmussen, and K. M. Fredlund (1993) NAD(P)H-ubiquinone oxidoreductases in plant mitochondria. *J. Bioenerg. Biomembr.* 25: 377–84.

Moore, A. L. and K. E. O. Åkerman (1982) Ca^{2+} stimulation of the external dehydrogenase in Jerusalem artichoke mitochondria. *Biochem. Biophys. Res. Commun.* 109: 513–7.

Moore, C. L. (1971) Specific inhibition of mitochondrial calcium uptake by ruthenium red. *Biochem. Biophys. Res. Commun.* 42: 298–305.

Nazareth, W., N. Yafei, and M. Crompton (1991) Inhibition of anoxia-induced injury in heart myocytes by cyclosporin A. *J. Mol. Cell. Cardiol.* 23: 1351–4.

Nicchitta, V. C. and J. R. Williamson (1984) Spermine: a regulator of mitochondrial calcium cycling. *J. Biol. Chem.* 254: 12978–83.

Nicholls, D. G. (1978) The regulation of extramitochondrial free calcium ion concentration by rat liver mitochondria. *Biochem. J.* 176: 463–74.

Nicholls, D. G. and K. E. O. Åkerman (1982) Mitochondrial calcium transport. *Biochim. Biophys. Acta* 683: 57–88.

Nichols, B. J. and R. M. Denton (1995) Towards the molecular basis for the regulation of mitochondrial dehydrogenases by calcium ions. *Mol. Cell. Biochem.* 149: 203–12.

Nichols, B. J., L. Hall, A. C. F. Perry, and R. M. Denton (1993) Molecular cloning and deduced amino acid sequences of the γ-subunits of rat and monkey NAD^+-isocitrate dehydrogenase. *Biochem. J.* 295: 347–50.

Nichols, B. J., M. Rigoulet, and R. M. Denton (1994) Comparison of the effects of Ca^{2+}, adenine nucleotides and pH on the kinetic properties of mitochondrial NAD^+-isocitrate dehydrogenase and oxoglutarate dehydrogenase from the yeast *Saccharomyces cerevisiae* and rat heart. *Biochem. J.* 303: 461–5.

Nichols, B. J., A. C. F. Perry, L. Hall, and R. M. Denton (1995) Molecular cloning and deduced amino acid sequences of the α- and β- subunits of mammalian NAD^+-isocitrate dehydrogenase. *Biochem. J.* 310: 917–22.

Perham, R. N. (1991) Domains, motifs and linkers in 2-oxo acid dehydrogenase multienzyme complexes: a paradigm in the design of a multifunctional enzyme. *Biochemistry* 30: 8501–12.

Persechini, A., N. D. Moncrief, and R. H. Kretsinger (1989) The EF-hand family of calcium- modulated proteins. *Trends Neurosci.* 12: 462–8.

Popov, K. M., Y. Zhao, Y. Shimomura, M. J. Kuntz, and R. A. Harris (1992) Branched chain α-ketoacid dehydrogenase kinase. Molecular cloning, expression, and sequence similarity with histidine protein kinases. *J. Biol. Chem.* 267: 13127–30.

Popov, K. M., N. Y. Kedishvili, Y. Zhao, Y. Shimomura, D. W. Crabb, and R. A. Harris (1993) Primary structure of pyruvate dehydrogenase kinase establishes a new family of eukaryotic protein kinases. *J. Biol. Chem.* 268: 26602–6.

Pozzan, T., R. Rizzuto, P. Volpe, and J. Meldolesi (1994) Molecular and cellular physiology of intracellular calcium stores. *Physiol. Rev.* 74: 595–636.

Pullman, M. E. and G. C. Monroy (1963) A naturally occurring inhibitor of mitochondrial adenosine triphosphatase. *J. Biol. Chem.* 238: 3762–9.

Ramachandran, N. and Colman, R. F. (1980) Characterisation of distinct subunits of pig heart DPN-specific isocitrate dehydrogenase. *J. Biol. Chem.* 255: 8859–64.

Randle, P. J. (1995) Metabolic fuel selection: general integration at the whole body level. *Proc. Nutr. Soc.* 54: 317–27.

Reed, L. J. (1981) Regulation of mammalian pyruvate dehydrogenase complex by a phosphorylation-dephosphorylation cycle. *Curr. Top. Cell Regul.* 18: 95–106.

Renner, M., M. A. Danielson, and J. F. Falke (1993) Kinetic control of Ca^{2+} signalling: tuning the ion dissociation rates of EF-hand Ca^{2+} binding sites. *Proc. Natl. Acad. Sci. USA* 90: 6493–7.

Repetto, B. and A. Tzagoloff (1989) Structure and regulation of KGD1, the structural gene for yeast α-ketoglutarate dehydrogenase. *Mol. Cell. Biol.* 9: 2695–705.

Rizzuto, R., A. W. M. Simpson, M. Brini, and T. Pozzan (1992) Rapid changes of mitochondrial Ca^{2+} revealed by specifically targetted recombinant aequorin. *Nature* 358: 325–8.

Rizzuto, R., C. Bastianutto, M. Brini, M. Murgai, and T. Pozzan (1994) Mitochondrial Ca^{2+} homeostasis in intact cells. *J. Cell Biol.* 1183–94.

Rizzuto, R., M. Brini, P. Pizzo, M. Murgai, and T. Pozzan (1995) Chimeric green fluorescent protein as a tool for visualising subcellular organelles in living cells. *Curr. Biol.* 5: 635–42.

Robb-Gaspers, L. D., P. Burnett, G. A. Rutter, R. M. Denton, R. Rizzuto, and A. P. Thomas (1998) Integrating cytosolic calcium signals into mitochondrial metabolic responses. *EMBO J.* 17: 4987–5000.

Roennow, B. and M. C. Keilland-Brandt (1993) GUT2, a gene for mitochondrial glycerol 3- phosphate dehydrogenase from *Saccharomyces cerevisiae*. *Yeast* 9: 1121–30.

Rutter, G. A. (1990) Calcium binding to citrate cycle dehydrogenases. *Int. J. Biochem.* 22: 1081-8.

Rutter, G. A. and R. M. Denton (1988) Regulation of NAD^+-linked isocitrate dehydrogenase and 2-oxoglutarate dehydrogenase by calcium ions within toluene permeabilised rat mitochondria. Interactions with regulation by adenine nucleotides and $NADH/NAD^+$ ratios. *Biochem. J.* 252: 181–9.

Rutter, G. A. and R. M. Denton (1989a) Rapid purification of pig heart NAD-isocitrate dehydrogenase. Studies on the regulation of activity by Ca^{2+}, adenine nucleotides, Mg^{2+} and other metal ions. *Biochem. J.* 263: 445–52.

Rutter, G. A. and R. M. Denton (1989b) The binding of Ca^{2+} ions to pig heart NAD-isocitrate dehydrogenase and the 2-oxoglutarate dehydrogenase complex. *Biochem. J.* 263: 453–62.

Rutter, G. A., J. G. McCormack, P. J. W. Midgley, and R. M. Denton (1989) The role of Ca^{2+} in the hormonal regulation of the activities of pyruvate dehydrogenase and oxoglutarate dehydrogenase complexes. *Ann. N.Y. Acad. Sci.* 573: 206–17.

Rutter, G. A., N. J. Osbaldeston, J. G. McCormack, and R. M. Denton (1990) Measurement of matrix free Mg^{2+} concentration in rat heart mitochondria by using entrapped fluorescent probes. *Biochem. J.* 271: 627–34.

Rutter, G. A., W.-F. Pralong, and C. B. Wollheim (1992) Regulation of mitochondrial glycerol- phosphate dehydrogenase by Ca^{2+} within electropermealised insulin secreting cells INS-1. *Biochim. Biphys. Acta* 1175: 107–13.

Rutter, G. A., J.-T. Theler, M. Murgai, C. B. Wollheim, T. Pozzan, and R. Rizzuto (1993) Stimulated Ca^{2+} influx raises mitochondrial free Ca^{2+} to supramicromolar levels in a pancreatic cell line. Possible role in glucose and agonist-induced insulin secretion. *J. Biol. Chem.* 268: 22385–90.

Rutter, G. A., P. Burnett, R. Rizzuto, M. Brini, M. Murgai, T. Pozzan, J. M. Tavare, and R. M. Denton (1996) Subcellular imaging of intramitochondrial Ca^{2+} with

recombinant targetted aequorin. Significance for the regulation of pyruvate dehydrogenase activity. *Proc. Natl. Acad. Sci. USA*, 93: 5489–5494.

Sitsapesan, R., S. J. McGarry, and A. J. Williams (1995) Cyclic ADP-ribose, the ryanodine receptor and Ca^{2+} release. *Trends Pharmacol. Sci.* 16: 386–91.

Strynadka, N. C. J., and M. N. G. James (1989) Crystal structure of the helix-loop-helix calcium binding proteins. *Annu. Rev. Biochem.* 58: 951–98.

Teague, W. M., F. H. Pettit, T.-L. Wu, S. L. Silberman, and L. J. Reed (1982) Purification and properties of pyruvate dehydrogenase phosphatase form bovine heart and kidney. *Biochemistry* 21: 5585–92.

Thomas, A. P. and Denton, R. M. (1986) Use of toluene permeabilised mitochondria to study the regulation of adipose tissue pyruvate dehydrogenase *in situ*. *Biochem. J.* 238: 83–91.

Thomas, A. P., T. A. Diggle, and R. M. Denton (1986) Sensitivity of pyruvate dehydrogenase phosphate phosphatase to magnesium ions. Similar effects of spermine and insulin. *Biochem. J.* 238: 93–101.

Unitt, J. F., J. G. McCormack, D. Reid, L. K. MacLachlan, and P. J. England (1989) Direct evidence for the role of intramitochondrial Ca^{2+} in the regulation of oxidative phosphorylation in the stimulated rat heart. Studies using ^{31}P n.m.r. and ruthenium red. *Biochem. J.* 262: 293–301.

Vaghy, P. L., J. D. Johnson, M. A. Matlib, T. Wang, and A. Schwartz (1982) Selective inhibition of Na^+-induced Ca^{2+} release from heart mitochondria by diltiazem and certain other Ca^{2+} antagonist drugs. *J. Biol. Chem.* 257: 6000–2.

Wernette, M. E., R. S. Ochs, and H. A. Lardy (1981) Ca^{2+} stimulation of rat liver mitochondrial glycerophosphate dehydrogenase. *J. Biol. Chem.* 256: 12767–71.

Williamson, J. R., R. H. Cooper, and J. B. Hoek (1981) Role of calcium in the hormonal control of liver metabolism. *Biochim. Biophys. Acta* 639: 243–95.

Yamada, E. W. and N. J. Huzel (1988) The calcium-binding ATPase inhibitor protein from bovine heart mitochondria. Purification and properties. *J. Biol. Chem.* 263: 11498–503.

Yamada, E. W. and N. J.Huzel (1989) Calcium-binding ATPase inhibitor protein of bovine heart mitochondria. Role in ATP synthesis and effect of Ca^{2+}. *Biochemistry* 28: 9714–8.

Yamada, E. W., N. J. Huzel, and J. C. Dickison (1981) Reversal by uncouplers of oxidative phosphorylation and by Ca^{2+} of the inhibition of mitochondrial ATPase activity by the ATPase inhibitor protein of rat skeletal muscle. *J. Biol. Chem.* 256: 10203–7.

Yeaman, S. J. (1989) 'The 2-oxo acid dehydrogenase complexes: recent advances. *Biochem. J.* 257: 625–32.

Zeng, Y., C. Weiss, T.-T. Yao, J. Huang, L. Siconolfi-Baez, P. Hsu, and J. I. Rushbrook (1995) Isocitrate dehydrogenase from bovine heart: primary structure of subunit 3/4. *Biochem. J.* 310: 507–16.

23

Calcium Signaling in Neurons: A Case Study in Cellular Compartmentalization

Samuel S.-H. Wang
George J. Augustine

As in other cell types, virtually every facet of the function of neurons is regulated by calcium ions. The specific actions of calcium within neurons include many events common to all cell types. These include regulation of gene expression (Ghosh and Greenberg, 1995), membrane traffic, cell division, growth and motility (Zheng et al. 1994), development (Spitzer, 1995; Shatz, 1996), and cell death (Choi, 1994), as well as a large complement of actions that are characteristic to neurons, such as electrical signaling, rapid triggering of exocytosis, and synaptic plasticity. Furthermore, the thousands of types of neurons within the brain make it likely that certain calcium-signaling events are specialized for individual types of neurons. In sum, a remarkably rich abundance of neuronal calcium-signaling phenomena confers upon calcium the control of an extremely broad functional repertoire.

Neurons succeed in carrying out these diverse functions partly by being experts in compartmentalizing calcium. Accordingly, we have focused on mechanisms that underlie this spatial and temporal compartmentalization. Perhaps paradoxically, we believe that understanding such extreme examples of cellular calcium compartmentalization will provide general insights into calcium signaling in all cells. Because describing all known examples of calcium signaling within all types of neurons would yield an overwhelming yet underinformative list, we have instead tried to define some principles that are relevant to all types of neuronal calcium signaling. Our conclusion is that the information-processing functions of neurons rely both on the polarized anatomical organization of these cells and the strategic submicroscopic organization of calcium sources and targets within specific cell domains. Space limitations necessarily prohibit an exhaustive list of all relevant papers and we refer readers to other reviews for additional perspectives on the topic of neuronal calcium signaling (Regehr and Tank, 1994; Ghosh and Greenberg, 1995; Berridge, 1997) These space limitations even preclude any discussion of calcium signaling in glial cells, the predominant cell in the brain (e.g., Cornell-Bell et al., 1990; Cornell-Bell and Finkbeiner, 1991; Jahromi et al., 1992; Nedergaard, 1994; reviewed by Verkhratsky and Kettenmann, 1996).

Calcium Sources and Sinks in Neurons

In order to appreciate how neurons compartmentalize calcium signals, we begin with a brief overview of sources of calcium within neurons, as well as the physical properties of neuronal cytoplasm that serve to limit calcium movement (Fig. 23.1).

Ion Channels Deliver Calcium into Neuronal Cytoplasm

The best-characterized sources of neuronal calcium are voltage-gated ion channels. These plasma membrane proteins are pores that selectively allow calcium to diffuse from the extracellular medium into the cytoplasm. Whether these pores are open or closed is regulated by the potential of the plasma membrane, with the pores having a higher probability of opening when the membrane potential is depolarized. The properties of voltage-gated calcium channels are summarized in detail in the chapter by

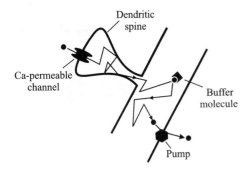

Figure 23.1 Cytoplasmic calcium sources and sinks. Upon entering neuronal cytoplasm via channels, a calcium ion can diffuse freely until it encounters a buffer, which reversibly binds calcium ions, or pumps that translocate calcium ions across membranes.

Tsien and Wheeler (Chapter 8) in this volume. Features of particular importance for considering neuronal calcium signaling are (1) voltage-dependent gating allows electrical signals, such as action potentials, to regulate calcium influx into neurons; (2) calcium channels exist in great variety; (3) calcium channels are targeted to distinct domains within the neuronal plasma membranes and (4) mechanisms exist for linking these channels to intracellular effector molecules.

Neurotransmitter receptors also serve as sites of Ca influx across the neuronal plasma membrane. These receptors are integral membrane proteins that bind chemical signals — neurotransmitters — secreted by neurons at synapses. The binding of transmitters to their receptors is the basic mechanism of communication between neurons and their target cells. Neurotransmitters cause Ca to flow across the plasma membrane in two ways. First, they can indirectly regulate Ca influx through the voltage-gated Ca channels as described above. They can do so either by changing the neuronal membrane potential, which determines the probability that these channels are open, or by activating intracellular signaling pathways, such as G-proteins and protein kinases and phosphatases, that regulate Ca-channel gating. Second, certain neurotransmitter receptors mediate Ca fluxes across the plasma membrane directly, by acting as Ca-permeable ion channels that are gated by a neurotransmitter. The most prominent examples of Ca-permeant neurotransmitter receptors/channels are the receptors for acetylcholine (Bregestrovski et al., 1979; Role, 1992) and glutamate (McDermott et al., 1986; Hollmann et al., 1991; Burnashev et al., 1992; Schneggenburger et al., 1993). Receptors for other neurotransmitters, such as ATP (Brake et al., 1994) and serotonin (reviewed by Gasic and Heinemann, 1992), also may be permeable to calcium. Ca-permeable glutamate receptors have received special attention because these receptors appear to mediate certain forms of synaptic plasticity (Bliss and Collingridge, 1993) and because their Ca permeability can also be dynamically regulated by

cellular membrane potential, impermeant ions, phosphorylation state, and RNA-editing mechanisms (Keller et al., 1992; Seeburg et al., 1995; Seeburg, 1996).

Calcium also can enter neurons via gap junctions connected to the cytoplasm of neighboring cells. Gap junctional calcium entry contributes to intercellular calcium signaling in neurons and glia and is necessary for the phenomenon of spreading depression (Yuste et al., 1992; Nedergaard et al., 1995; Verkhratsky and Kettenmann, 1996).

A separate source of Ca in neurons is release of intracellular "stored" Ca from the endoplasmic reticulum (ER). The Ca flows from the ER into neuronal cytoplasm through both inositol 1,4,5-trisphosphate (IP_3) receptors and ryanodine receptors (Berridge, 1993a). Both of these receptors are Ca-permeable ion channels in the ER membrane; their properties are summarized in detail in Chapter 9 in this volume. Ca release from IP_3 receptors is initiated by neurotransmitter receptor coupling to G-proteins which causes the production of IP_3 (Berridge, 1993a). While these receptors are known to play a prominent role in Ca signaling in most cells, their functions in neurons have received much less attention. Ca release from the ER has been implicated as a necessary step in inducing long-term synaptic plasticity in hippocampal (Bottolotto et al., 1994; Reyes and Stanton, 1996) and cerebellar (Berridge, 1993b; Khodakhah and Armstrong, 1997; Finch and Augustine, 1998) neurons. In nonneuronal cells, positive feedback of cytosolic Ca on ryanodine receptors and IP_3 receptors can cause regenerative Ca release from the ER (see Chapter 3, this volume). Feedback by Ca has been experimentally demonstrated to induce Ca release at ryanodine receptors in sympathetic neurons (Friel and Tsien, 1992), cerebellar Purkinje cells (Llano et al., 1994; Kano et al., 1995), and at IP_3 receptors in neuroblastoma cells (Wang and Thompson, 1995). However, the physiological context of regenerative Ca release in neurons remains to be characterized.

Calcium Removal Mechanisms Limit the Spatial Range of Neuronal Calcium Signals

Two mechanisms serve to localize Ca actions to the vicinity of its sources. First, dilution passively limits the action of calcium ions as they diffuse into progressively larger volumes within the neuron. Second, damping mechanisms actively restrict the movement of calcium ions. Several such damping mechanisms exist and can be distinguished by their underlying mechanisms (Baker et al., 1971; Baker and DiPolo, 1984; Allbritton et al., 1992). *Pumping* removes calcium from the cytoplasm before it can diffuse very far. *Binding* to buffer molecules retards the rate of movement of Ca, thereby confining Ca action to a smaller volume. Finally, *compartmentation* by membrane barriers prevents calcium from moving freely into and out of subcellular structures. This compartmentation enhances the ability of pumping and binding to sculpt the spatial and temporal range of Ca signals. We next summarize briefly what is known about each of these processes in neurons.

Pumping

After cytoplasmic Ca is elevated, neurons more slowly pump Ca out of their cytoplasm (Brinley et al., 1978; Ahmed and Connor, 1979; Ross et al., 1986; for review, see Ross, 1989). The transporters that contribute to neuronal Ca homeostasis are similar to those found in other cells, and are reviewed by Guerini and Carafoli (Chapter 10) and by Philipson (Chapter 11) elsewhere in this volume. Active transport removes Ca across the plasma membrane by the plasma membrane family of Ca/ATPases and by the Na–Ca exchanger; into the endoplasmic reticulum, by the ER family of Ca pumps; and into mitochondria. At rest and for low calcium loads, the major calcium sinks are thought to be the endoplasmic reticulum (Blaustein et al., 1978) and the plasma membrane, including Na–Ca exchange (Thompson, 1994; Gleason et al., 1995). For larger, micromolar rises in Ca concentration, mitochondria play a major role in ending the Ca signal (Barish and Thompson, 1983; Werth and Thayer, 1994; Herrington et al., 1996). More recently, it has been appreciated that, under certain conditions, mitochondria can also serve as calcium sources (Alneas and Rahamimoff, 1975; Herrington et al., 1996; Tang and Zucker, 1997).

Smaller diameter cellular structures, such as the dendrites of neurons, clear calcium much faster than large structures, such as cell bodies. For example, under optimal conditions Ca signals evoked by electrical activity can return to baseline with time constants of ~ 50 msec in the dendrites of both cerebellar Purkinje cells (Ross and Werman, 1987; Lev-Ram et al., 1992) and neocortical pyramidal neurons (Schiller et al., 1995; Markram et al., 1995). In contrast, somatic Ca signals usually decay over time scales of seconds (Ahmed and Connor, 1979; Ross et al., 1986; for review, see Ross, 1989). This difference is likely to be due, at least in part, to the much higher surface-to-volume ratio of dendrites, which can be less than $2\,\mu m$ wide, compared with neuron somata, which are tens or, occasionally, hundreds of micrometers in diameter.

Binding

Only a small fraction of the total Ca in cytoplasm exists as freely diffusing ions because most calcium is bound to relatively immobile buffer molecules within the cell. Calcium binds to such buffers rapidly, with equilibrium apparently reached within milliseconds (Smith and Zucker, 1980; Neher, 1986; Neher and Augustine, 1992; Klinghauf and Neher, 1997). Neuronal Ca buffering was first identified in the giant axon (Hodgkin and Keynes, 1957; Baker and Crawford, 1972) and then in nerve cell bodies (Tillotson and Gorman, 1980; Smith and Zucker, 1980). Heavy Ca buffering has subsequently been observed in a wide variety of neurons and neuroendocrine cells (for a thorough review, see Neher, 1995). Endogenous buffers appear to be highly selective for Ca over Mg, since Mg diffusion is not slowed (Baker and Crawford, 1972) and since free Mg in the cytoplasm reaches millimolar levels (Baker and DiPolo, 1984).

For the most part, cytoplasmic calcium buffers appear to be immobile (Neher and Augustine, 1992; Zhou and Neher, 1993), suggesting that they are of very high molecular weight or are anchored to intracellular structures. Binding to such immobile buffers greatly slows the effective rate of diffusion of Ca, because this ion can move only when it is not bound to the buffers. As a result, the average rate of diffusion of cytoplasmic Ca ions is much lower than the rate of diffusion when calcium is free in solution. For example, cytoplasmic buffering reduces the effective diffusion constant of Ca from $300\,\mu m^2/$sec to $<20\,\mu m^2/$sec (Hodgkin and Keynes, 1957; Gabso et al., 1997). When combined with active extrusion mechanisms, buffering can sharply limit the spatial range of a Ca signal. One simple measure of the range of Ca action is the mean distance moved by Ca in one dimension before it is extruded. This distance can be calculated to be $(2D_{eff}\,t)^{1/2}$, where D_{eff} is the effective diffusion coefficient for Ca and t is the uptake time constant (Baker and DiPolo, 1984; Allbritton et al., 1992). In a mammalian dendrite, the range of buffered Ca is on the order of $[(2)(10\,\mu m^2/sec)(0.05\,sec)]^{1/2} = 0.6\,\mu m$. Thus, homeostatic mechanisms play a critical role in regulating the spatial and temporal dimensions of calcium signaling in neurons and probably all other cells.

Despite extensive physiological characterization of cytoplasmic Ca buffers in neurons, in no case has Ca buffer capacity been decisively attributed to specific proteins. However, a number of Ca-binding proteins have been identified in neurons, including parvalbumin and calbindin (reviewed by Bainbridge et al., 1992). Some progress in identification has been made in hair cells, where the 28 kDa calbindin-D is abundant and can account for the exceptionally strong Ca buffering properties of these cells (Oberholtzer et al., 1988; Roberts, 1993). In cerebellar Purkinje cells, genetic knockout of this protein appears to reduce cytoplasmic calcium buffering properties (Airaksinen et al., 1997). Kindling reduces both levels of this protein and cytoplasmic calcium buffering properties in hippocampal granule cells (Kohr et al., 1991).

Membrane Compartmentation

The movement of Ca can also be restricted by the membrane topology of a neuron, such as the shape and outline of its plasma membrane and internal organelles. These shapes can trap Ca ions by forcing them to take indirect, tortuous paths, thereby both concentrating Ca and slowing its apparent rate of diffusion. The most prominent example of this is the dendritic spine, which forms "irregular contours and specialized appendages" extruding from the dendritic shaft (Peters et al., 1991). Spines assume a wide variety ∽f shapes, ranging from directly apposed to the dendritic shaft to being separated by a long neck (for a review, see Harris and Kater, 1994). Theoretical modeling of movement of messenger molecules, such as calcium, suggests that spines can act to compartmentalize chemical signals (Harris and Stevens, 1988; Wickens, 1988; Koch and Zador, 1993). The effectiveness of this compartmentalization has been directly measured by focal uncaging and photobleach techniques (Svoboda et al., 1996). These experiments revealed that equilibration times into and out of spines were on the order of tens of milliseconds, consistent with theoretical predictions of restricted diffusion.

Compartmentation by spines may also affect Ca as it diffuses along dendritic shafts (Wang and Augustine, 1995b). We estimate, based on electron micrographic reconstructions, that spine volume can be 30% of total dendritic volume in hippocampal pyramidal neuron dendrites and as much as 70% in Purkinje cell spiny dendrites (Harris and Stevens, 1988; 1989). As a result, diffusion may be impeded along the length of a spiny dendrite if molecules can pass into spines and be trapped there. Finally, diffusion in dendrites may also be slowed by the presence of internal blocking structures, such as cytoskeleton and internal organelles. Dendrites share with the cell body the property of having internal organelles (Peters et al., 1991), and these organelles, in addition to their other functions, may block diffusion.

Microanatomy of Neurons: Structural Polarization as a Source of Signaling Compartmentalization

In addition to the limitations imposed by the homeostatic mechanisms just described, still more of the calcium compartmentalization properties of neurons arise from the remarkable microscopic structure of these cells. Neurons are highly polarized and can compartmentalize calcium signals in two ways. First, calcium sources — such as voltage-gated calcium channels, neurotransmitter receptors, and intracellular calcium-release channels — are distributed nonuniformly among these compartments. Second, the autonomy of electrical signaling within these compartments will lead to segregated calcium influx through these channels. This section briefly summarizes the anatomical organization of various neurons and their channels and describes how this anatomy can lead to calcium compartmentalization.

Polarization of Neurons

Polarization of Neuronal Structure

Neurons are the most anatomically sophisticated and highly polarized cells in the body. Nearly all neurons possess the specialized structures of dendrites, cell body or soma, and axon (Fig. 23.2). Dendrites are elaborately branched structures, typically no more than 1 or $2 \mu m$ in diameter, possibly extending over tens or even hundreds of micrometers. Classically, the dendrites are viewed as areas of neurons specialized for receipt of electrical and chemical signals via synapses coming from other neurons. As will be seen below, neuronal dendrites also contain a rich diversity of calcium signaling. The soma or cell body of a neuron possesses the nucleus, endoplasmic reticulum, and Golgi apparatus, and thus is responsible for the majority of protein synthesis that goes on within the cell. Because calcium can regulate the expression of many neuronal genes, somatic and even nuclear calcium signaling have received considerable attention, as will be seen below. But perhaps the most distinctive part of a neuron is its axon. Axons are very thin neuronal processes, typically $1 \mu m$, or less in diameter, that can extend over distances as great as several meters. The function of the axon is to propagate a regenerative action potential from the somatic region along the axon to specialized structures at its end. These structures, termed presynaptic terminals, are the sites of secretion of neurotransmitter chemi-

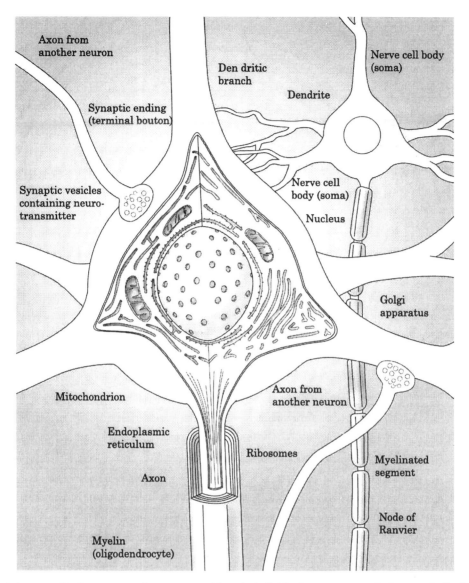

Figure 23.2 Generalized structure of a neuron. Although individual neurons can deviate markedly from this sterotype, all neurons possess a large number of distinctive subcellular compartments. (From Purves et al., 1997.)

cals and may also receive chemical signals responsible for retrograde signaling between neurons (Jessell and Kandel, 1993). Because secretion of neurotransmitters is regulated by calcium ions, presynaptic calcium signaling pathways have received extensive attention and have yielded much insight into calcium compartmentalization mechanisms in cells. In developing axons, the termination of an axon is a motile growth cone whose movement is targeted toward its future site of synaptic contact.

While most neurons possess dendrites and an axon, it is important to emphasize that neurons are extremely diverse in structure. Indeed, a vast number of types of neurons can be identified simply on the basis of their microanatomical structure (Ramon y Cajal, 1909–1911). For example, the structure of dendrites varies remarkably between different types of neurons, depending partly on the types of synaptic input that these dendrites are to receive (Bullock et al., 1977).

Polarization of Ca Sources in Neurons

All of the channels that act as sources of Ca are distributed in a nonrandom way within neurons. Among these sources, the polarized distribution of voltage-gated Ca channels within the neuron membrane has received the most attention. These channels are found throughout most neurons, though the density of Ca channels generally seems lower in axons than elsewhere (e.g., Smith et al., 1993). This is presumably a consequence of the fact that axons typically use voltage-gated channels that are not Ca-permeable to propagate electrical signals. Studies of the distribution of Ca channels within mammalian neurons have shown that these neurons target different types of channels to different regions. Certain types of Ca channels seem to be found exclusively in neuronal somata and dendrites, while others are most abundant in presynaptic terminals (Westenbroek et al., 1992; 1995). Because these channel types differ in their gating properties, such segregation presumably allows differential activation of Ca influx by various electrical signals.

Although neurotransmitter receptors are widely distributed in the membrane of neurons (e.g., Pettit et al., 1997), it is clear that these receptors are present at highest density in the immediate vicinity of sites of synaptic transmitter release (Hall and Sanes, 1993; Burns and Augustine, 1995). Such localization of receptors that are Ca-permeant or of those that generate IP_3 can lead to local Ca signals within subsynaptic regions (Müller and Connor, 1991; Murphy et al., 1994; Denk et al., 1995; Finch and Augustine, 1998).

Intracellular channels that mediate Ca release from the ER also are heterogeneously distributed within and among neurons. While both types of channels are found throughout most neurons, their densities vary markedly among neuron types (Sharp et al., 1993). Cerebellar Purkinje cells possess exceptionally high densities of both types of channels and distribute them in different patterns. While IP_3 receptors are found on ER throughout the Purkinje cell, ryanodine receptors are not found in dendritic spines (Walton et al., 1991; Takei et al., 1992). Such an arrangement presumably allows differential regulation of Ca release in response to synaptic and nonsynaptic signals.

Polarization of Neuronal Electrical Signaling

Because neurons possess Ca sources (voltage-gated channels, neurotransmitter receptors, gap junctions) that are sensitive to electrical signals, the distribution of electrical signals within the polarized structure of the neuron can also contribute importantly to compartmentalization of Ca signals. The classical view of electrical signaling in neurons is that a tug-of-war between excitatory and inhibitory synaptic transmission leads to the all-or-none production of an action potential at the axon hillock; the action potential then propagates down the axon to presynaptic terminals to release transmitter from the synaptic terminals of the neuron. While the action potential is therefore capable of spreading throughout neurons, many other electrical signals, such as the potential changes resulting from synaptic transmission, may be more restricted in their spatial range. As a result, regions of a neuron can operate semiautonomously with respect to both their electrical- and Ca-signaling capabilities.

Local Electrical Signaling Associated with Synaptic Transmission The initial sites of synaptic electrical signaling are dendritic spines. Current flow through a spine neck can cause a larger voltage change in the spine than in the rest of the dendrite, allowing the possibility that spines may act, to a degree, as separate electrogenic compartments (Koch and Zador, 1993; Harris and Kater, 1994). While experimental tests of this idea are largely negative (Harris and Steven, 1988; 1989), there is some evidence that synaptic electrical signals can be restricted to single spines (Denk et al., 1995).

Beyond the spines, synaptic potential changes clearly can spread locally throughout dendrites without triggering an action potential in the rest of the neuron. In the fly, low levels of visual input activate single dendritic branches of motion-sensitive cells (Borst and Egelhaaf, 1992). In Purkinje cells, calcium-electrogenic spikes can be generated in the dendritic arbor independently of the cell body (Llinás and Sugimori, 1980), and synaptic potentials can also be confined to single branches (Miyakawa et al., 1992) and small dendritic branchlets (Eilers et al., 1995).

Spatial Range of Action Potentials When the balance of synaptic excitation and inhibition exceeds threshold, an action potential is initiated in the axon (Eccles, 1964; for recent reviews, see Johnston et al., 1996 and Stuart et al., 1997). The subsequent spread of this action potential both forward along axonal branches to presynaptic terminals and back to the dendrites is governed largely by anatomical and electrical parameters, so that in some cases the action potential can fail to invade parts of a neuron.

In most neurons, action potentials are initiated in the soma or axon and can propagate retrogradely into the dendritic arbor (reviewed by Stuart et al., 1997). Two exceptions are the cerebellar Purkinje cell, where potential spread is largely passive (Llinás and Sugimori, 1980; Stuart and Häusser,

1994), and olfactory mitral cells, where action potentials can originate in dendrites and be propagated both forward and backward (Chen et al., 1997). Although neurons are capable of propagating action potentials backwards into their dendrites, this back-propagation is highly variable. For example, in pyramidal neurons, whether action potentials invade the entire dendritic arbor or attenuate significantly over distance depends upon the amount of recent electrical activity (Callaway et al. 1995, Spruston et al., 1995). When action potentials do not invade dendrites completely, propagation failure often occurs at points where the dendritites branch (Spruston et al., 1995). Recently, strong attenuation of action potentials has been demonstrated in vivo in cortical layer II/III neurons (Svoboda et al., 1997). Factors known to block back-propagation include inhibitory synaptic input, which hyperpolarizes and decreases the input resistance of cells (Callaway et al., 1995); transient potassium currents, which also prevent action potential initiation in dendrites (Hoffman et al., 1997); and acetylcholine receptor activation, which enables the back-propagation of trains of action potentials in hippocampal CA1 pyramidal cells (Tsubokawa and Ross, 1997).

Because action potentials do not always invade every branch of an axon, presynaptic terminals can fire separately. Action potential propagation has been observed to fail at branch points and swellings, with failures occurring in from one to all of the daughter axonal branches (Parnas, 1972; Grossman et al., 1979). Propagation failure can be caused by accumulation of potassium in extracellular space (Smith and Hatt, 1976), myelination (Swadlow et al., 1980), axon geometry and distribution of ion channels involved in action potential production (Joyner et al., 1980), and by recent changes in membrane potential (Debanne et al., 1997).

In sum, though neuronal electrical signaling can be quite widespread in its spatial extent, many types of neuronal electrical signals are spatially restricted and help to limit the range of Ca entry that is regulated by membrane potential.

Microanatomical Compartmentation of Neuronal Ca Signals

As a result of this polarization of neuronal structures and their calcium sources, subcellular Ca signals within neurons can be quite restricted. With the development of sophisticated methods for measuring these calcium signals in neurons and other cells (see Chapter 2, this volume), numerous interesting examples of calcium compartmentation have been observed in neurons.

Growth Cones

Calcium influx has been observed to be confined to single exploring growth cones (Freeman et al., 1985; Connor, 1986; Lipscombe et al., 1986). Also, Ca increases have been localized to subregions within growth cones, and are correlated with increases in filopodial number and length (Davenport and Kater, 1992). Application of neurotransmitter has been observed to cause a turning toward the stimulus that is accompanied by a localized increase in Ca (Zheng et al., 1994). For these reasons, Ca is believed to regulate their extension (for reviews, see Kater and Mills, 1991 and Doherty and Walsh, 1994; but see also Garyantes and Regehr, 1992).

Presynaptic Terminals

Action potentials have been observed to cause Ca increases that are spatially limited to presynaptic terminals. First, many experiments document that such Ca signals appear in presynaptic terminals but are absent in the axons that lead to the terminals (Miledi and Parker, 1981; Llinás, 1984; Stockbridge and Ross, 1984; Smith et al., 1993; Llano et al., 1997). This presumably reflects a higher density of Ca channels in the terminals than in the axons. Even within individual presynaptic terminals, pronounced Ca gradients can be observed. In the unusually large giant presynaptic terminals of squid, pronounced gradients can be observed between terminal regions containing active zones and transmitter release sites, and nearby regions with few of these release sites (Smith et al., 1993). The use of a low-affinity form of recombinant aequorin has allowed visualization of even more restricted Ca signals of less than $1 \mu m^2$ in area (Llinás et al., 1992). It is thought that these signals arise from "microdomains" of clustered Ca channels opening in synchrony, presumably within the active zone. As discussed in the subsequent section on "Nanoanatomical Ca-Signaling Arrangements", this organization of Ca channels and release sites permits the rapid triggering of transmitter secretion.

Dendrites

As mentioned before, spatial localization of calcium signaling occurs on a variety of scales in dendrites. Because Ca-dependent processes are known to be necessary for both short- and long-term synaptic plasticity, the spatial extent of a Ca signal in a dendritic arbor can strongly influence where that plasticity is expressed.

When electrical signals are actively propagated, Ca signals can invade part or all of the dendritic arbor (Jaffe et al., 1992; Regehr and Tank, 1992;

Markram et al., 1995; but see Svoboda et al., 1997). Dendritic localization has been found to have an in vivo functional correlate in fly interneurons, where the pattern of localization of Ca signals depends on the type and orientation of motion in a presented visual stimulus (Borst and Egelhaaf, 1992). In Purkinje neurons, synaptic activity causes Ca transients that can be localized to single branches of the dendritic arbor, and these transients can be inhibited selectively by inhibitory potentials (Miyakawa et al., 1992). Signals on this level have been reviewed elsewhere (Regehr and Tank, 1994; Johnston et al., 1996; Koch, 1997). Activation of parallel fiber synapses causes even more sharply localized Ca transients that stay within a single spiny branchlet of Purkinje cells (Eilers et al., 1995; Hartell, 1996). Synaptic transmission has been observed to elevate Ca in dendritic shafts adjoining the postsynaptic spine in hippocampal pyramidal cells (Murphy et al., 1994); sometimes, the rise in Ca is excluded from nearby spines (Guthrie et al., 1991).

On a very local scale, synaptic transmission and/or postsynaptic action potentials can elevate postsynaptic Ca concentration in single spines of hippocampal pyramidal neurons (Müller and Connor, 1991; Jaffe et al., 1994) and cerebellar Purkinje cells (Denk et al., 1995). These single-spine signals have been used to monitor success and failure of synaptic transmission in cultured cortical neurons (Murphy et al., 1994) and hippocampal slices (Yuste and Denk, 1995). Such work is consistent with the idea that spines can act to compartmentalize calcium and other chemical signals (Harris and Stevens, 1988; Wickens, 1988; Koch and Zador, 1993; Svoboda et al., 1996).

Although the endoplasmic reticulum of neurons appears to be a network that extends continuously throughout the neuronal cytoplasm (Terasaki et al., 1994), dendritic Ca-release signals from these stores can still be quite local. For example, local photolysis of caged IP_3 in Purkinje cells causes a local Ca-release transient, with no evidence for active spread of release beyond the uncaging spot (Wang and Augustine, 1995a; Finch and Augustine, 1998). Since the expression of long-term depression depends on postsynaptic calcium signals (Sakurai, 1990; Konnerth et al., 1992), local Ca transients could account for the input specificity of long-term synaptic depression (Finch and Augustine, 1998). Similar responses have also been found when IP_3 is produced by metabotropic glutamate receptors activated by caged glutamate (S. S.-H. Wang and G. J. Augustine, unpublished observations).

In contrast, in hippocampal pyramidal cell dendrites, activation of metabotropic glutamate receptors causes propagated Ca release (Jaffe and Brown, 1994). Likewise, synaptic stimulation has been shown to evoke spatially complex Ca release in hippocampal CA3 pyramidal neurons (Miller et al., 1996). In culture, histamine and muscarinic acetylcholine receptor agonists evoke propagating Ca waves in neuroblastoma cells (Wang and Thompson, 1995). The fact that both local and global Ca release have been observed suggests that whether or not Ca-release waves arise depends upon factors such as IP_3 production and the degree of regenerative feedback by Ca.

Nucleus

One recently proposed compartment for Ca signals is the nucleus itself, which has been suggested to be functionally separated from the cytoplasm. However, this idea is still controversial (see Chapter 20, this volume). It has been contradicted by recent experiments using fluorescent indicators targeted to the nucleus (e.g., Allbritton et al., 1994) which have supported the idea that Ca diffuses freely into the nucleus from outside. Others, however, showed fluorescence signals that were faster and larger in the nucleus than in the cytoplasm, leading to the suggestion that the nuclear envelope might act as a diffusional barrier between nucleus and cytoplasm (Przywara et al., 1991; Birch et al., 1992). Furthermore, carefully calibrated measurements in neurons and astrocytes show that the indicator dye is sequestered into small organelles in the cytoplasm, so that a Ca signal of uniform magnitude causes a smaller fluorescence change in the cytoplasm than in the nucleus (Connor, 1993; al-Mohanna et al., 1994). In principle, this phenomenon would be sufficient to account for differences in signal amplitude between nucleus and cytoplasm.

Sub-Microscopic Spatial Organization of Neuronal Calcium Signals

Although the examples described above make clear the compartmentalization of neuronal calcium signaling associated with somata, dendrites, axons, and synapses, the local arrangement of calcium channels and calcium-regulated effector proteins within any one of these compartments can provide intracellular signaling with even more spatial complexity (Augustine and Neher, 1992; Kasai, 1993). Such local signaling occurs over distances much less than a micrometer, spatial dimensions too small to be resolved by a light microscope. To distinguish calcium signals associated with microscopic neuronal structures from those restricted to smaller spatial dimensions, we will refer to the sub-microscopic organization of calcium signaling components by the term *nanoanatomy*. The key feature

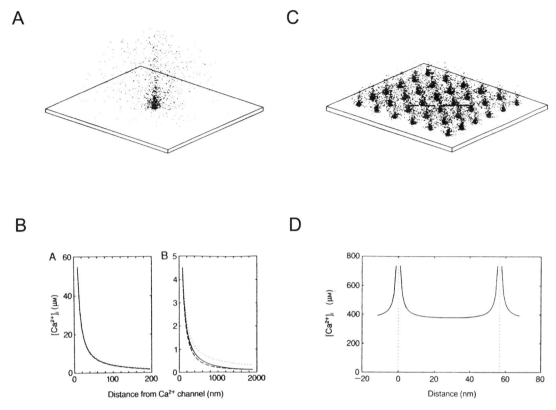

Figure 23.3 Predictions of calcium diffusion in the immediate vicinity of open channels. (A) Calcium entering through a single open channel procedures steep gradients over distances of nanometers. (B) Mathematical calculation of calcium profiles near a single open channel (positioned at 0 nm). Graphs A and B are the same predictions illustrated over two spatial ranges. Dotted lines indicate steady-state predictions, solid lines indicate predicted profiles 1 msec after calcium has started diffusing, and dashed lines indicate the same but after including 100 mM of a calmodulin-like calcium buffer. (C) Calcium entering through many open channels (40 nm spacing shown here) can summate. (D) Mathematical calculation of calcium profile near a pair of open channels (positioned at points shown by dotted vertical lines) from the array shown in (C). Summation of calcium entering from many open channels can elevate calcium concentrations to very high levels within the array of open channels. (From Smith and Augustine, 1988).

of this nanoanatomical calcium signaling is the precise spatial arrangement of calcium channels relative to each other and to the calcium-binding proteins that are activated by calcium fluxes through the channels.

Nanoanatomical Ca Signaling Arrangements

An initial understanding of Ca signaling at the nanoanatomical level arises by considering the diffusion of calcium ions away from the mouth of an open Ca channel. This diffusion can be calculated to yield a highly localized accumulation of calcium within tens of nanometers of the channel (Fig. 23.3A; see Chad and Eckert, 1984a; Simon and Llinás, 1985;

Smith and Augustine, 1988; Stern, 1992; Yamada and Zucker, 1992). If multiple calcium channels in close spatial proximity are opened simultaneously, then calcium entering from adjacent channels can sum together to make even larger local rises in calcium concentration. Depending upon the spatial relationship of calcium channels to each other and to intracellular calcium-binding proteins, three nanoanatomical arrangements could occur (Fig. 23.4; Schweizer et al., 1995).

Nanodomains

The first case occurs if the calcium receptor proteins are within a few nanometers of open calcium chan-

Nanodomain

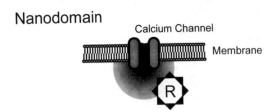

Microdomain

Radial Shell

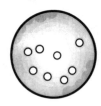

Figure 23.4 Three different geometric reltionships between local calcium signals and calcium receptor protein (R). In a nanodomain (top), calcium entering through a single open channel influences the activity of a calcium-binding protein positioned within nanometers of the channel. In a microdomain (center), summation of calcium entering through many channels (as in Fig. 23.3C, D) can bind to a calcium receptor positioned within a micrometer of any open channel. Calcium gradients shaped like radial shells (bottom) are relevant when calcium enters through many channels and diffuses distances of many microns before binding to its molecular target. (After Schweizer et al., 1995.)

nels (Fig. 23.4, top). This calcium signal has been referred to as a calcium domain (Chad and Eckert, 1984) or, in reference to the spatial dimensions involved, a nanodomain (Kasai, 1993). Present technology prevents accurate detection of such calcium nanodomains. However, theoretical estimates indicate that local calcium concentrations may be as high as hundreds of micromolar within the very small volume of a single nanodomain (Fig. 23.3B; Simon and Llinás, 1985; Smith and Augustine, 1988; Roberts et al., 1990; Yamada and Zucker, 1992).

Microdomains

A second type of local calcium signal arises when multiple calcium channels are clustered together within a membrane area of approximately $1 \mu m^2$ (Fig. 23.4, center). When these channels open, a Ca-binding protein positioned within the cluster will sense a microdomain of calcium (Llinás et al., 1992; Kasai, 1993). Theoretical calculations (Smith and Augustine, 1988; Roberts, 1994) predict that free Ca concentration within such signals can be as high as hundreds of micromolar, due to summation of calcium coming from multiple calcium channels (Fig. 23.3C,D). Initial empirical measurements are

in line with such estimates (Roberts et al., 1990; Llinás et al., 1992).

Radial Calcium Gradients

A third type of calcium signal will result when calcium channels are far ($1 \mu m$ or more) from target Ca-binding proteins (Fig. 23.4, bottom). In this case, *radial* concentration gradients will arise as calcium diffuses from the plasma membrane into the interior of the cell (e.g., Stockbridge and Moore, 1984). Such gradients are readily detected in optical imaging experiments (e.g., Lipscombe et al., 1988a; Hernandez-Cruz et al., 1990; Augustine and Neher, 1992). Because of the resulting dilution, calcium levels at the target proteins will be much lower than in the previous two cases, and may be within the 1–$10 \mu M$ range typically reported for calcium signaling events in nonneuronal cells.

Examples of Ca Signaling at the Nanoanatomical Level

The theoretical arguments that suggested the existence of nanoanatomical Ca signaling have now been bolstered by experimental evidence. We next consider a couple of cases where local nanoanatomy seems to be a critical determinant of the characteristics of Ca-dependent phenomena. However, we emphasize that these signals have been documented to varying degrees of rigor and in no case is a complete set of experimental data in hand.

Triggering of Transmitter Release from Presynaptic Terminals

Probably the best-documented action of calcium in neurons is the rapid triggering of neurotransmitter secretion, and investigation of such Ca-signaling events in presynaptic terminals has led the way in our appreciation of the local range of cellular Ca signaling. Decades of research indicate that action potentials that invade presynaptic terminals cause voltage-gated calcium channels to open, yielding a rise in calcium concentration that leads to the exocytotic secretion of neurotransmitters (Katz, 1969; Augustine et al., 1987; Zucker, 1996). Accumulating evidence indicates that nanodomains, microdomains, and radial shells are all involved in secretion at one or another synapse.

Nanodomains There are several indications that nanodomain-type calcium signals are responsible for triggering neurotransmitter release from some presynaptic terminals. First, presynaptic injection of the slowly binding calcium buffer EGTA does not block transmitter release, while a rapidly binding homolog, BAPTA, efficiently blocks release (Adler et al., 1991; von Gersdorff and Matthews, 1994). Knowledge of the kinetics of calcium binding by EGTA suggests that the Ca-binding protein that triggers release must bind calcium within tens of nanometers of the calcium channel (Neher, 1986; Stern 1992; Naraghi and Neher, 1997). Second, use of "caged" calcium compounds to raise presynaptic calcium concentration indicates that high calcium concentrations, on the order of 100–200 μM, are necessary to produce physiological rates of secretion (Heidelberger et al., 1994; Hsu et al., 1996). Titration of presynaptic calcium concentrations by calcium buffers with varying affinity yields similar estimates of calcium levels during transmitter release (Adler et al., 1991), as does measurement of local calcium concentration with a low-affinity version of aequorin (Llinás et al., 1992). Such high concentrations can be produced only within the immediate vicinity of calcium channels, implicating micro- or nanodomains in transmitter release. Third, pharmacologically modifying the number of open calcium channels causes changes in transmitter release that are consistent with nanodomain models of calcium signaling (Yoshikami et al., 1989; Augustine, 1990; reviewed by Augustine et al., 1991). Fourth, measurement of the opening of individual calcium channels suggests that only one channel needs to be open in order to

cause transmitter release (Stanley, 1993). Taken together, the evidence points toward a role for nanodomains in presynaptic terminals that secrete transmitters very rapidly (Stanley, 1997).

Microdomains There is also some evidence that microdomain-type calcium signals mediate secretion of transmitters at other synapses. In these cases, the microdomain organization seems important for distributing secretory control over a number of Ca channels. Particularly compelling are experiments done in auditory hair cells — mechanosensory cells that also secrete neurotransmitters and are tuned to respond to vibrational stimulation at kilohertz frequencies. By using calcium-sensitive ion channels as assays of local calcium levels in the vicinity of secretory sites, it was possible to document that calcium concentrations as high as the millimolar range result following the opening of voltage-gated calcium channels (Roberts et al., 1990; Roberts 1994). Such high calcium levels can only arise from the summation of calcium entry from many open calcium channels, one of the defining characteristics of microdomains. High-speed imaging of local calcium signals provides support for microdomains following depolarization of these cells (Tucker and Fettiplace, 1995). Additional support for a microdomain arrangement of calcium signaling comes from electron microscopic examination of the plasma membrane of these cells. Freeze-fracture microscopy reveals arrays of intramembranous particles that are present in the correct location and number to be ion channels. These presumptive channels are arranged in a tight cluster, surrounded by the sites of exocytosis (Roberts et al., 1990). Such an arrangement allows both the rapid rise and dissipation of Ca signals and seems optimally designed to produce fast, microdomain calcium signaling to cause exocytosis in these cells.

Less complete evidence implicates microdomain calcium signaling in the triggering of transmitter secretion at mammalian central synapses. For example, EGTA is almost as potent as BAPTA in blocking transmitter release at an auditory synapse, suggesting a substantial distance between the open calcium channels and Ca-binding proteins (Borst et al., 1996). Quantitative analysis of the dynamics of Ca diffusion led to the conclusion that influx from multiple calcium channels is needed for transmitter release at this synapse. The synergistic actions of calcium channel toxins at hippocampal and cerebellar synapses also indicates that calcium entering through several different types of calcium channels sums together to trigger transmitter release (Luebke et al., 1993; Takahashi and Momiyama, 1993; Wheeler et al., 1994; Mintz et al., 1995; Doroshenko et al., 1997). Such results are most readily explained by transmitter release being triggered by clusters that contain multiple types of calcium channels, consistent with the microdomain type of calcium signaling.

Radial Shells   Still other experiments point toward the importance of radial shell-type calcium signals in secretion at certain presynaptic terminals. The most obvious example is in peptide-secreting terminals. In these terminals, relatively modest levels of intracellular Ca are needed to produce exocytosis (Verhage et al., 1991; Peng and Zucker, 1993). In addition, influx of calcium through voltage-gated ion channels appears to be no more effective than influx through randomly distributed membrane pores induced by ionophores, indicating no preferential localization of the calcium channels near peptide-release sites (Verhage et al., 1991). The lack of tight spatial coupling between calcium channels and exocytotic sites is consistent with the fact that these terminals secrete peptides relatively slowly in contrast to the examples described above. In endocrine cells, such as adrenal chromaffin cells, diffuse calcium signals can also trigger exocytosis (Knight and Baker, 1982; Heinemann et al., 1994), although there is some evidence for a small component of rapid release that may be triggered by micro- or nanodomains (Horrigan and Bookman, 1994).

Even in presynaptic terminals where Ca nanodomains or microdomains rapidly trigger exocytosis, it is likely that more diffuse Ca signaling has other regulatory functions. In these cases, it is likely that the slow, more widespread calcium signaling events arise from gradual diffusion of calcium away from nanodomains and microdomains into much larger presynaptic volumes (Fig. 23.5). For example, a form of synaptic plasticity known as facilitation is thought to arise from the persistent, diffuse presence of calcium (Katz and Miledi, 1968; Rahamimoff, 1968). Diffuse calcium signals may produce other forms of plasticity, such as augmentation and post-tetanic potentiation (Fig. 23.3; Delaney et al., 1989; Swandulla et al., 1991; Kamiya and Zucker, 1994; Regehr et al., 1994; Tang and Zucker, 1997). Diffuse Ca signaling may also activate Ca-dependent ion channels in certain presynaptic terminals (Augustine, 1990).

In summary, spatial coupling between voltage-gated Ca channels and the Ca-binding proteins that trigger transmitter release allow diverse forms of local calcium signaling to regulate secretion of transmitters at different presynaptic terminals.

Regulation of Ion Channel Gating

Calcium signals also regulate the opening of ion channels in the plasma membrane. The two most-studied examples of Ca regulation are a type of potassium channel that opens when cytoplasmic Ca concentration is high and the feedback inhibition (or inactivation) of certain types of voltage-gated Ca channels. For both types of channels, substantial arguments can be made for the involvement of local Ca signaling in these regulatory events.

The Ca regulation of Ca-dependent potassium channels has received the most critical experimental attention. It is well established that these channels are found in presynaptic terminals and other places where voltage-gated Ca channels occur in high density (Augustine, 1990; Roberts et al., 1990; Knaus et al., 1996; Yazejian et al., 1997). Several experiments point toward a very close spatial coupling between the sources of Ca — the Ca channels — and the Ca effectors, the Ca-dependent K channels. First, K channels distant from the Ca channels are not activated when Ca channels open (Gola and Crest, 1993). Second, both freeze-fracture electron microscopy of hair cells (Roberts et al., 1990) and fluorescent staining studies of neuromuscular synapses (Robitaille et al., 1993) suggest that the Ca-activated K channels colocalize within 100 nm of Ca channels. Third, BAPTA-type Ca buffers are much more effective than EGTA in preventing activation of K channels following Ca influx through Ca channels (Roberts, 1993; Robitaille et al., 1993). Fourth, comparison of the Ca-sensitivity of K channel opening to the degree of opening produced by influx of Ca through Ca channels suggests that cytoplasmic Ca levels are on the order of $100\,\mu M$ (Roberts et al., 1990; Markwardt and Isenberg, 1992). These data are most consistent with the idea that Ca-dependent K channels are activated by microdomain or nanodomain Ca signals.

Opening of voltage-gated Ca channels causes a buildup of intracellular Ca, a recurring theme in this review. For certain types of Ca channels, this intracellular Ca signal can lead to inactivation and subsequent closure of the Ca channel (Eckert and Chad, 1984). There is some evidence that this form of Ca signaling occurs on a very local spatial range (Chad and Eckert, 1984; Imredy and Yue, 1992), such that the opening of single Ca channels is capable of causing self-inactivation. This inactivation is reduced by BAPTA and depends upon the number of Ca ions flowing through the channel, as predicted by a nanodomain form of Ca signaling. However, other observations are more readily explained by a microdomain Ca signal regulating these channels. For example, opening of nearby Ca channels enhances the rate of inactivation of a given Ca channel and enhances the sensitivity of inactivation to BAPTA. Further, EGTA is capable of reducing the rate of inactivation. Thus, while it is not yet certain whether nanodomains or microdomains are the

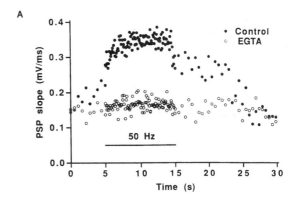

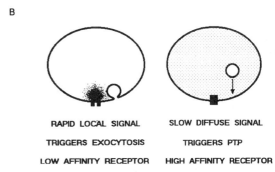

Figure 23.5 Calcium regulation of post-tetanic potentiation (PTP), a form of synaptic plasticity that enhances
neurotransmitter release. (A) PTP evoked by a train of
presynaptic stimuli (at bar) is blocked following presynaptic microinjection of EGTA. However, this buffer
has no effect upon transmitter released by low-frequency action potentials (at left, before stimulus train).
(B) Diagram of calcium signaling during neurotransmitter release and PTP. When calcium channels are
open, local rises in calcium concentration trigger release.
After the channels close, a more diffuse calcium signal
leads to PTP. (From Augustine et al., 1994).

dominant regulatory motif, it is clear that local Ca
signals regulate Ca-channel inactivation.

Regulation of Gene Expression by Local Signaling between Plasma Membrane and Nucleus

It is well known that Ca and many other intracellular
second messengers regulate neuronal gene expression. Primary targets of Ca in such actions are
immediate-early genes (IEGs; Morgan and Curran,
1995). One mechanism for Ca regulation of IEGs is
Ca-sensitive phosphorylation of a transcription factor, CREB (cAMP-responsive element-binding protein; Ginty, 1997). CREB is phosphorylated by
Ca/calmodulin-dependent protein kinases; for example, CaM kinase phosphorylation of Ser[133] of CREB
is critical in turning on CREB activity (Gonzalez and
Montminy, 1989). Because phosphorylation does not
require the synthesis of new proteins, CREBs and
IEGs have been suggested to play key roles in the
first stages of establishing long-term changes in
synaptic strength and even memory formation
(Frank and Greenberg, 1994; Bailey et al., 1996;
Yin and Tully, 1996). The CREB acts by binding to
CRE, the cAMP-response element of genes.
Activation of CRE by CREB then leads to transcrip-
tion of many genes, including the IEG c-*fos*.

It has been established that the source of Ca entry
determines the influence of Ca upon gene expression.
A good example of this differential Ca regulation of
gene expression comes from studies of hippocampal
neurons. In these cells, the Ser[133] activation site of
CREB can be phosphorylated by both voltage-gated
Ca entry and NMDA receptor activation, leading in
both cases to enhanced c-fos expression (Cole et al.,
1989; Lerea and McNamara, 1993; Thompson et al.,
1995; Deisseroth et al., 1996). However, these two
signals act through two different molecular pathways. Ca influx through L-type Ca channels clearly
activates transcription via CRE (Bading et al., 1993;
Ginty et al., 1993), as expected from the ability of this
signal to phosphorylate CREB. Remarkably, even
though NMDA receptor activation also phosphorylates CREB, expression of c-fos and other IEGs is
mediated by the Ca-responsive serum-response element (SRE). The lack of activation of CRE-depen-
dent transcription implies that NMDA receptor
activity may inactivate CREB through a second
pathway, perhaps at another phosphorylation site

(Ghosh and Greenberg, 1995). This dependence upon the nature of the Ca-influx pathway suggests mechanisms that detect differences in signal localization (Gallin and Greenberg, 1995; Ginty, 1997).

Spatial aspects of the coupling of Ca influx through L-type Ca channels to expression of IEGs has been examined in some detail. Local Ca signaling is indicated by the observation that BAPTA, but not EGTA, blocks Ca-mediated phosphorylation of CREB (Bito et al., 1996). Comparison of the actions of these and other Ca buffers suggests that Ca works within 1–$2\,\mu m$ of its site of entry, consistent with the microdomain or nanodomain models of local Ca action (Fig. 23.2). Preliminary experiments suggest that Ca-bound calmodulin diffuses into the nucleus to activate CREB phosphorylation following Ca entry (Deisseroth and Tsien, 1997). Although calmodulin is found throughout the cytoplasm of all cells, close spatial coupling may be provided by the preferential binding of calmodulin to L-type Ca channels (R. W. Tsien, personal communication). Differential effects of calmodulin upon nuclear CaM kinase (type IV) and protein phosphatases (such as calcineurin) allow the duration of the neuronal Ca signal to set the level of CREB phosphorylation and IEG expression (Bito et al., 1996).

Molecular Topology Yields Nanoanatomical Arrangements

The most crucial requirement for nanoanatomical Ca signaling is a tight spatial coupling between Ca sources and Ca targets. The mechanisms responsible for such molecular coupling in neurons are largely unknown but are likely to be fruitful grounds for future exploration. Here, we consider in detail the molecular underpinnings of nanoanatomical Ca signaling in presynaptic terminals, the only case where substantial progress has been made toward elucidating the mechanisms that ensure such spatial coupling.

Nanoanatomical Ca signaling in presynaptic terminals results from spatial linkage between voltage-gated Ca channels, the sources of Ca, and synaptotagmin, the Ca-binding protein that triggers exocytotic transmitter release. Substantial anatomical evidence indicates that calcium channels are clustered in presynaptic membranes in close spatial proximity to sites of neuronal exocytosis (Pumplin et al., 1981; Walrond and Reese, 1985; Robitaille et al., 1990; Engel, 1991; Smith et al., 1993; Haydon et al., 1994). While the molecular underpinnings of this clustering remain to be established, it is likely to involve the same cytoskeletal mechanisms that are involved in targeting of other voltage-gated ion channels (e.g., Srinivasan et al., 1988). Several lines of evidence indicate that Ca entering through these channels binds to synaptotagmin, an integral protein

of the synaptic vesicle membrane (Sudhof, 1995; Augustine et al., 1996). For example, microinjection of fragments of this protein inhibit transmitter release (Bommert et al., 1993; Elferink et al., 1993). Genetic knockout of this protein disrupts synaptic transmission in *Caenorhabditis elegans* (Nonet et al., 1993), *Drosophila* (Broadie et al., 1994), and mouse (Geppert et al., 1994). Particularly interesting is the observation that one *Drosophila* synaptotagmin mutant displays an altered Ca-dependence of transmitter release (Littleton et al., 1994). Genetic knockout of a single synaptotagmin gene does not completely eliminate all neurotransmitter release evoked by presynaptic action potentials. For example, in mice, loss of synaptotagmin I causes a selective loss of one rapid kinetic component of release while a slower component is spared (Geppert et al., 1994). The remaining release has been attributed to the residual presence of additional synaptotagmin isoforms (Li et al., 1995).

At least three possible mechanisms can serve to keep synaptotagmin nearby the clustered Ca channels. First, synaptic vesicles are physically attached to the plasma membrane via docking mechanisms that are still poorly defined (Bennett and Scheller, 1994; Sudhof, 1995; Augustine et al., 1996). In this arrangement, synaptotagmin is no more than one vesicle diameter (50 nm) away from the plasma membrane. Thus, as long as the docking process attaches the vesicle to the plasma membrane near the clusters of Ca channels, synaptotagmin should at least be within microdomain dimensions of the channels. Second, synaptotagmin appears to bind directly to presynaptic Ca channels in vitro (Leveque et al., 1992; Charvin et al., 1997; Sheng et al., 1997; Wiser et al., 1997). If this interaction occurs within the presynaptic terminal, it will keep the two proteins within a couple of nanometers of each other, well within the range of nanodomain Ca signaling. Finally, synaptotagmin also binds to syntaxin, an integral plasma membrane protein that binds both to calcium channels and to vesicle proteins (Bennett et al., 1992; Yoshida et al. 1992; Sheng et al., 1994, 1996). The binding of syntaxin to Ca channels seems to be important for transmitter release (Mochida et al., 1996). This interaction may also regulate the gating of Ca channels (Bezprozvanny et al., 1995; Wiser et al., 1996), though an in vivo function remains to be demonstrated (Mochida et al., 1995; Stanley and Mirotznik, 1997). While the relative importance of these three mechanisms is presently unknown, it is clear that direct and indirect interactions among the protein components of the Ca-signaling cascade are involved in constructing the nanoanatomy that underlies exocytotic secretion of transmitters.

Conclusions

Our survey indicates that neuronal Ca signaling works over spatial ranges that span several orders of magnitude. The most global signals act over entire cells or even networks of cells coupled by gap junctions. These signals rely upon widespread injection of Ca into cytoplasm through channels distributed throughout the neuron. Calcium signals of intermediate, microscopic range encompass anatomically defined regions of the cell, such as the dendrites or presynaptic terminals. They are mediated by the differential distribution of ion channels and electrical signals within these cellular regions, are limited by stringent homeostasis of Ca by neuronal buffers and pumps, and are contained by morphological specializations. Calcium signals of nanoanatomical dimensions result from the colocalization of ion channels and Ca-binding effector proteins, such as synaptotagmin and calmodulin, within localized domains within cellular regions. These mechanisms confer upon neurons the ability to use Ca for many purposes in a single cell, and in a wide dynamic temporal range of signaling properties, ranging from the sub-millisecond range required for efficient synaptic coupling of neurons to the essentially permanent changes in gene expression required for neuronal connectivity and learning and memory processes. While much has been learned about the early stages of neuronal Ca signaling, many further insights will emerge from future efforts to define better the identity and localization of neuronal Ca-effector pathways.

References

Adler, E. M., G. J. Augustine, S. N. Duffy, and M. P. Charlton (1991) Alien intracellular calcium chelators attenuate neurotransmitter release at the squid giant synapse. *J Neurosci.* 11:1496–507.

Ahmed, Z. and J. A. Connor, (1979) Measurement of calcium influx under voltage clamp in molluscan neurones using the metallochromic dye arsenazo III. *J. Physiol. (Lond.)* 286:61–82.

Airaksinen, M. S., J. Eilers, O. Garaschuk, H. Thoenen, A. Konnerth, and M. Meyer (1997) Ataxia and altered dendritic calcium signaling in mice carrying a targeted null mutation of the calbindin D28k gene. *Proc. Natl. Acad. Sci. USA* 94: 1488–93.

Allbritton N. L., T. Meyer, and L. Stryer (1992) Range of messenger action of calcium ion and inositol 1,4,5-trisphosphate. *Science* 258:1 812–1815.

Allbritton, N. L., E. Oancea, M. A. Kuhn, and T. Meyer (1994) Source of nuclear calcium signals. *Proc. Natl. Acad. Sci. USA* 91: 12458–12462.

al-Mohanna, F. A., K. W. Caddy, and S. R. Bolsover (1994) The nucleus is insulated from large cytosolic calcium ion changes. *Nature* 367:745–750.

Alneas, E. and R. Rahamimoff (1975) On the role of mitochondria in transmitter release from motor nerve terminals. *J. Physiol.* 248: 285–306.

Augustine, G. J. (1990) Regulation of transmitter release at the squid giant synapse by presynaptic delayed rectifier potassium current. *J. Physiol.* 431: 343–364.

Augustine, G. J. and E. Neher (1992) Neuronal Ca^{2+} signalling takes the local route. *Curr. Opin. Neurobiol.* 2: 302–307.

Augustine, G .J., M. P. Charlton, and S.J. Smith (1987) Calcium action in synaptic transmitter release. *Annu. Rev. Neurosci.* 10: 633–693.

Augustine, G. J., E. M. Adler, and M. P. Charlton (1991) The calcium signal for transmitter secretion from presynaptic nerve terminals. *Ann. N.Y. Acad. Sci.* 635: 365–381.

Augustine, G. J., H. Betz, K. Bommert, M. P. Charlton, W. M. DeBello, M. Hans, and D. Swandulla (1994) Molecular pathways for presynaptic calcium signaling. *Adv. Second Mess. Protein Phos. Res.* 29: 139–154.

Augustine, G. J., M. E. Burns, W. M. DeBello, D. L. Pettit, and F. E. Schweizer (1996) Exocytosis: proteins and perturbations. *Annu. Rev. Pharmacol. Toxicol.* 36: 659–701.

Bading, H., D. D. Ginty, and M. E. Greenberg (1993) Regulation of gene expression in hippocampal neurons by distinct calcium signaling pathways. *Science* 260: 181–186.

Bailey, C. H., D. Bartsch, and E. R. Kandel (1996) Toward a molecular definition of long-term memory storage. *Proc. Natl. Acad. Sci. USA* 93: 13445–13452.

Bainbridge, K. G., M. G. Celia, and J. H. Rogers (1992) Calcium-binding proteins in the nervous system. *Trends Neurosci.* 15: 303–308.

Baker, P. F. and A. C. Crawford (1972) Mobility and transport of magnesium in squid giant axons. *J. Physiol. (London)* 227: 855–874.

Baker, P. F. and R. DiPolo (1984) Axonal calcium and magnesium homeostasis. In *Current Topics in Membranes and Transport*, Vol. 22. The Squid Axon (Baker, P. F., ed.). Academic Press, pp. 195–263.

Baker, P. F., A. L. Hodgkin, and E. B. Ridgway (1971) Depolarization and calcium entry in squid giant axons. *J. Physiol. (London)* 218: 709–755.

Barish, M. E. and S .H. Thompson (1983) Calcium buffering and slow recovery kinetics of calcium-dependent outward current in molluscan neurones. *J. Physiol. (London)* 337: 201–219.

Bennett, M. K. and R. H. Scheller (1994) A molecular description of synaptic vesicle membrane trafficking. *Annu. Rev. Biochem.* 63: 63–100.

Bennett, M. K. N. Calakos, and R. H. Scheller (1992) Syntaxin: a synaptic protein implicated in docking of synaptic vesicles at presynaptic active zones. *Science* 257: 255–259.

Berridge, M. J. (1993a) Inositol trisphosphate and calcium signalling. *Nature* 361: 315–325.

Berridge, M.J . (1993b) A tale of two messengers. *Nature* 365: 388–389.

Berridge, M. J. (1997) Elementary and global aspects of calcium signaling. *J. Exp. Biol.* 200: 315–319.

Bezprozvanny, I., R. H. Scheller, and R. W. Tsien (1995) Functional impact of syntaxin on gating of N-type and Q-type calcium channels. *Nature* 378: 623–626.

Birch, B. D., D. L. Eng, and J. D. Kocsis (1992) Intranuclear Ca^{2+} transients during neurite regeneration of an adult mammalian neuron. *Proc. Natl. Acad. Sci. USA* 89: 7978–7982.

Bito, H., K. Deisseroth, and R. W. Tsien (1996) CREB phosphorylation and dephosphorylation: a Ca^{2+}- and stimulus duration-dependent switch for hippocampal gene expression. *Cell* 87: 1203–1214.

Blaustein, M. P., R. W. Ratzlaff, and E. S. Schweitzer (1978) Calcium buffering in presynaptic nerve terminals. II. Kinetic properties of the nonmitochondrial Ca sequestration mechanism. *J. Gen. Physiol.* 72: 43–66.

Bliss, T. V. and G. L. Collingridge (1993) A synaptic model of memory: long-term potentiation in the hippocampus. *Nature* 361: 31–9.

Bommert, K., M. P. Charlton, W. M. DeBello, G. J. Chin, H. Betz, and G. J. Augustine (1993). Inhibition of neurotransmitter release by C2-domain peptides implicates synaptotagmin in exocytosis. *Nature* 363: 163–165.

Borst, A. and M. Egelhaaf (1992) In vivo imaging of calcium accumulation in fly interneurons as elicited by visual motion stimulation. *Proc. Natl. Acad. Sci. USA* 89: 4139–4143.

Borst, J. G., F. Helmchen, and B. Sakmann (1996) Calcium influx and transmitter release in a fast CNS synapse. *Nature* 383: 43143–434.

Bottolotto, Z .A., Z. I. Bashir, C. H. Davies, G. L. Collingridge (1994) A molecular switch activated by metabotropic glutamate receptors regulates induction of long-term potentiation. *Nature* 368: 740–743.

Brake, A. J., M. J. Wagenbach, and D. Julius (1994) New structural motif for ligand-gated ion channels defined by an ionotropic ATP receptor. *Nature* 371: 519–523.

Bregestrovski, P. D., R. Miledi, and I. Parker (1979) Calcium conductance of acetylocholine-induced endplate channels. *Nature* 279: 638–639.

Brinley, F. J. Jr., T. Tiffert, and A. Scarpa (1978) Mitochondria and other calcium buffers of squid axon studied in situ. *J. Gen. Physiol.* 72: 101–127.

Broadie K., H. J. Bellen, A. DiAntonio, J. T. Littleton, and T. L. Schwarz (1994). Absence of synaptotagmin disrupts excitation-secretion coupling during synaptic transmission. *Proc. Natl. Acad. Sci. USA* 91: 10727–10731.

Bullock, T. H., R. Orkand, and A. Grinnell (1977) *Introduction to Nervous Systems.* W. H. Freeman, New York.

Burnashev, N., R. Schoepfer, H. Monyer, J. P. Ruppersberg, W. Gunther, P. H. Seeburg, and B. Sakmann (1992) Control by asparagine residues of calcium permeability and magnesium blockade in the NMDA receptor. *Science* 257: 1415–1419.

Burns, M. E. and G. J. Augustine (1995) Synaptic structure and function: dynamic organization yields architectural precision. *Cell* 83: 187–194.

Callaway, J. C., N. Lasser-Ross, and W. N. Ross (1995) IPSPs strongly inhibit climbing fiber-activated $[Ca^{2+}]_i$ increases in the dendrites of cerebellar Purkinje neurons. *J. Neurosci.* 15: 2777–2787.

Chad, J E. and R. Eckert (1984). Calcium domains associated with individual channels can account for anomalous voltage relations of CA-dependent responses. *Biophys. J.* 45: 993–999.

Charvin, N., C. Leveque, D. Walker, F. Berton, C. Raymond, M. Kataoka, Y. Shoji-Kasai, M. Takahashi, M. DeWaard, and M. J. Seagar (1997) Direct interaction of the calcium sensor protein synaptotagmin I with a cytoplasmic domain of the 1A subunit of the P/Q-type calcium channel. *EMBO J.* 16: 4591–4596.

Chen, W. R., J. Midtgaard, and G. M. Shepherd (1997) Forward and backward propagation of dendritic impulses and their synaptic control in mitral cells. *Science* 278: 463–466 .

Choi, D. W. (1994) Calcium and excitotoxic neuronal injury. *Ann. N.Y. Acad. Sci.* 747: 162–171.

Cole, A. J., D. W. Saffen, J. M. Baraban, and P.F. Worley (1989) Rapid increase of an immediate early gene messenger RNA in hippocampal neurons by synaptic NMDA receptor activation. *Nature* 340: 474–476.

Connor, J. A. (1986) Digital imaging of free calcium changes and of spatial gradients in growing processes in single, mammalian central nervous system cells. *Proc. Natl. Acad. Sci. USA* 83: 6179–6183.

Connor J. A. (1993) Intracellular calcium mobilization by inositol 1,4,5-trisphosphate: intracellular movements and compartmentalization. *Cell Calcium* 14: 185–200.

Cornell-Bell, A. H. and S. M. Finkbeiner (1991) Ca^{2+} waves in astrocytes. *Cell Calcium* 12: 185–204.

Cornell-Bell, A.H., S.M. Finkbeiner, M.S. Cooper, and S.J. Smith (1990) Glutamate induces calcium waves in cultured astrocytes: long-range glial signaling. *Science* 247:470–473.

Davenport, R. W. and S. B. Kater (1992) Local increases in intracellular calcium elicit local filopodial responses in Helisoma neuronal growth cones. *Neuron* 9: 405–416.

Debanne, D., N. C. Guerineau, B. H. Gahwiler, and S. M. Thompson (1997) Action-potential propagation gated by an axonal I_A-like K^+ conductance in hippocampus. *Nature* 389: 286–289.

Deisseroth, K. and R. W. Tsien, (1997) Calmodulin translocation to the nucleus mediates rapid synaptic control of CREB phosphorylation in hippocampal neurons. *Soc. Neurosci. Abstr.* 23: 310.

Deisseroth, K., H. Bito, and R.W. Tsien (1996. Signaling from synapse to nucleus: Postsynaptic CREB phosphorylation during multiple forms of hippocampal synaptic plasticity. *Neuron* 16: 89–101.

Delaney, K. R., R. S. Zucker, and D. W. Tank (1989) Calcium in motor nerve terminals associated with posttetanic potentiation. *J. Neurosci.* 9: 3558–3567.

Denk, W., M. Sugimori, and R. Llinás (1995) Two types of calcium response limited to single spines in cerebellar Purkinje cells. *Proc. Natl. Acad. Sci. USA* 92: 8279–8282.

Doherty, P. and F. S. Walsh (1994) Signal transduction events underlying neurite outgrowth stimulated by cell adhesion molecules. *Curr. Opin. Neurobiol.* 4: 49–55.

Doroshenko, P. A., A. Woppmann, G. Miljanich, and G. J. Augustine (1997) Pharmacologically distinct presynaptic calcium channels in cerebellar excitatory and inhibitory synapses. *Neuropharmacology* 36: 865–872.

Eccles, J. C. (1964) *The Physiology of Synapses.* Springer-Verlag, New York.

Eckert, R. and J. E. Chad (1984) Inactivation of Ca channels. *Prog. Biophys. Mol. Biol.* 44: 215–267.

Eilers, J., G. J. Augustine, and A. Konnerth (1995) Subthreshold synaptic Ca^{2+} signalling in fine dendrites and spines of cerebellar Purkinje neurons. *Nature* 373: 155–158.

Elferink, L. A., M. R. Peterson, and R. H. Scheller (1993) A role for synaptotagmin (p65) in regulated exocytosis. *Cell* 72: 153–159.

Engel, A. G. (1991) Review of evidence for loss of motor nerve terminal calcium channels in Lambert-Eaton myasthenic syndrome. *Ann. N.Y. Acad. Sci.* 635: 246–58.

Finch, E. A. and G. J. Augustine (1998) Local postsynaptic calcium signaling by inositol trisphosphate in Purkinje cell dendrites. *Nature,* submitted.

Frank, D. A. and M. E. Greenberg (1994) CREB: a mediator of long-term memory from mollusks to mammals. *Cell* 79: 5–8.

Freeman, J. A., P. B. Manis, G. J. Snipes, B. N. Mayes, P. C. Samson, J. P. Wilswo Jr., and D. B. Freeman (1985) Steady growth cone currents revealed by a novel circularly vibrating probe: a possible mechanism underlying neurite growth. *J. Neurosci. Res.* 13: 257–283.

Friel, D. D. and R. W. Tsien (1992) A caffeine- and ryanodine-sensitive Ca^{2+} store in bullfrog sympathetic neurones modulates effects of Ca^{2+} entry on $[Ca^{2+}]_i$. *J. Physiol. (London)* 450: 217–246.

Gabso, M., E. Neher, and M. E. Spira (1997) Low mobility of the Ca^{2+} buffers in axons of cultured *Aplysia* neurons. *Neuron* 18: 473–481.

Gallin, W. J. and M. E. Greenberg (1995) Calcium regulation of gene expression in neurons: the mode of entry matters. *Curr. Opin. Neurobiol.* 5: 367–374.

Garyantes, T. K. and W. G. Regehr (1992) Electrical activity increases growth cone calcium but fails to inhibit neurite outgrowth from rat sympathetic neurons. *J. Neurosci.* 12: 96–103.

Gasic, G. P. and S. Heinemann (1992) Determinants of the calcium permeation of ligand-gated cation channels. *Curr. Opin. Cell Biol.* 4: 670–677.

Geppert, M., Y. Goda, R. E. Hammer, C. Li, T. W. Rosahl, C. F. Stevens, and T.C. Südhof (1994) Synaptotagmin I: a major Ca^{2+} sensor for transmitter release at a central synapse. *Cell* 79: 717–727.

Ghosh, A. and M. E. Greenberg (1995) Calcium signaling in neurons: molecular mechanisms and cellular consequences. *Science* 268: 239–247.

Ginty, D. D. (1997) Calcium regulation of gene expression: isn't that spatial? *Neuron* 18: 183–186.

Ginty, D. D., J. M. Kornhauser, M. A. Thompson, H. Bading, K. E. Mayo, J. S. Takahashi, and M. E. Greenberg (1993) Regulation of CREB phosphorylation in the suprachiasmatic nucleus by light and a circadian clock. *Science* 260: 238–241.

Gleason, E., S. Borges, and M. Wilson (1995) Electrogenic Na-Ca exchange clears Ca^{2+} loads from retinal amacrine cells in culture. *J. Neurosci.* 15: 3612–3621.

Gola, M. and M. Crest (1993) Colocalization of active K_{Ca} channels and Ca^{2+} channels within Ca^{2+} domains in *Helix* neurons. *Neuron* 10: 689–699.

Gonzalez, G. A. and M. R. Montminy (1989) Cyclic AMP stimulates somatostatin gene transcription by phosphorylation of CREB at Ser-133. *Cell* 59: 675–680.

Grossman, Y., I. Parnas, and M. E. Spira (1979) Differential conduction block in branches of a bifurcating axon. *J. Physiol. (London)* 295: 283–305.

Guthrie, P. B., M. Segal, and S. B. Kater (1991) Independent regulation of calcium revealed by imaging dendritic spines. *Nature* 354: 76–80.

Hall, Z. W. and J. R. Sanes (1993) Synaptic structure and development: the neuromuscular junction. *Cell* 72: 99–121.

Harris, K. M. and S. B. Kater (1994) Dendritic spines: cellular specializations imparting both stability and flexibility to synaptic function. *Annu. Rev. Neurosci.* 17: 341–371.

Harris, K. M. and J. K. Stevens (1988) Dendritic spines of rat cerebellar Purkinje cells: serial electron microscopy with reference to their biophysical characteristics. *J. Neurosci.* 8: 4455–4469.

Harris, K. M. and J. K. Stevens (1989) Dendritic spines of CA1 pyramidal cells in the rat hippocampus: serial electron microscopy with reference to their biophysical characteristics. *J. Neurosci.* 9: 2982–2997.

Hartell, N. A. (1996) Strong activation of parallel fibers produces localized calcium transients and a form of LTD that spreads to distant synapses. *Neuron* 16: 601–610.

Haydon, P. G., E. Henderson, and E. F. Stanley (1994) Localization of individual calcium channels at the release face of a presynaptic nerve terminal. *Neuron* 13: 1275–1280.

Heidelberger R., and C. Heinemann, E.. Neher, and G. Matthews (1994). Calcium dependence of the rate of exocytosis in a synaptic terminal. *Nature.* 371: 513–515.

Heinemann, C., R. H. Chow, E. Neher, and R. S. Zucker (1994) Kinetics of the secretory response in bovine chromaffin cells following flash photolysis of caged calcium. *Biophys. J.* 67: 2546–2557.

Hernandez-Cruz, A., F. Sala, and P. R. Adams (1990). Subcellular calcium transients visualized by confocal microscopy in a voltage-clamped vertebrate neuron. *Science* 247: 858–862.

Herrington, J., Y. B. Park, D. F. Babcock, and B. Hille (1996) Dominant role of mitochondria in clearance of large Ca^{2+} loads from adrenal chromaffin cells. *Neuron* 16: 219–228.

Hodgkin, A. L. and R. D. Keynes (1957) Movements of labelled calcium in squid axons. *J. Physiol. (London)* 138: 253–281.

Hoffman, D. A., J. C. Magee, C. M. Colbert, and D. Johnston (1997) K^+ channel regulation of signal propagation in dendrites of hippocampal pyramidal neurons. *Nature* 387: 869–875.

Hollmann, M., M. Hartley, and S. Heinemann (1991) Ca2+ permeability of KA-AMPA-gated glutamate receptor channels depend on subunit composition. *Science* 252: 851–853.

Horrigan, F. T. and R. J. Bookman (1994) Releasable pools and the kinetics of exocytosis in adrenal chromaffin cells. *Neuron* 13: 1119–1129.

Hsu, S.-F., G. J. Augustine, and M. B. Jackson (1996) Adaptation of calcium-triggered exocytosis in presynaptic terminals. *Neuron* 17: 501–512.

Imredy, J. P. and D. T. Yue (1992) Submicroscopic Ca^{2+} diffusion mediates inhibitory coupling between individual Ca^{2+} channels. *Neuron* 9: 197–207.

Jaffe, D. B. and T. H. Brown (1994) Metabotropic glutamate receptor activation induces calcium waves within hippocampal dendrites. *J. Neurophysiol.* 72: 471–474.

Jaffe, D. B., D. Johnston, N. Lasser-Ross, J. E. Lisman, H. Miyakawa, and W. N. Ross (1992) The spread of Na^+ spikes determines the pattern of dendritic Ca^{2+} entry into hippocampal neurons. *Nature* 357: 244–246.

Jaffe, D. B., S. A. Fisher, and T. H. Brown (1994) Confocal laser scanning microscopy reveals voltage-gated calcium signals within hippocampal dendritic spines. *J. Neurobiol.* 25: 220–233.

Jahromi, B. S., R. Robitaille, and M. P. Charlton (1992) Transmitter release increases intracellular calcium in perisynaptic Schwann cells in situ. *Neuron* 8: 1069–1077.

Jessell, T. M. and E. R. Kandel (1993) Synaptic transmission: a bidirectional and self-modifiable form of cell–cell communication. *Cell* 72: 1–30.

Johnston, D., J. C. Magee, C. M. Colbert, and B. R. Christie (1996) Active properties of neuronal dendrites. *Annu. Rev. Neurosci.* 19: 165–186.

Joyner, R. W., M. Westerfield, and J. W. Moore (1980) Effects of cellular geometry on current flow during a propagated action potential. *Biophys. J.* 31: 183–194.

Kamiya, H. and R. S. Zucker (1994) Residual Ca^{2+} and short-term synaptic plasticity. *Nature* 371: 603–606.

Kano, M., O. Garaschuk, A. Verkhrtatsky, and A. Konnerth (1995) Ryanodine receptor-mediated intracellular calcium release in rat cerebellar Purkinje neurones. *J. Physiol. (London)* 487: 1–16.

Kasai H. (1993) Cytosolic Ca^{2+} gradients, Ca^{2+} binding proteins and synaptic plasticity. *Neurosci. Res.* 16: 1–7.

Kater, S. B. and L.R. Mills (1991) Regulation of growth cone behavior by calcium. *J. Neurosci.* 11: 891–899.

Katz, B. (1969) *The Release of Neural Transmitter Substances.* Liverpool University Press, Liverpool, UK.

Katz B. and R. Miledi (1968) The role of calcium in neuromuscular facilitation. *J. Physiol. (London)* 195: 481–492.

Keller, B. U., M. Hollmann, S. Heinemann, and A. Konnerth (1992) Calcium influx through subunits GluR1/GluR3 of kainate/AMPA receptor channels is regulated by cAMP dependent protein kinase. *EMBO J.* 11: 891–896.

Khodakah, K. and C. M. Armstrong (1997) Induction of long term depression and rebound potentiation by inositol trisphosphate in cerebellar Purkinje neurons. *Proc. Natl. Acad. Sci.* 94: 14009–14014.

Klinghauf, J. and E. Neher (1997) Modeling buffered Ca^{2+} diffusion near the membrane: implications for secretion in neuroendocrine cells. *Biophys. J.* 72: 674–690.

Knaus, H. G., C. Schwarzer, R. O. A. Koch, A. Eberhart, G. J. Kachorowski, H. Glossman, F. Wunder, O. Pongs, M. L. Garcia, and S. Sperk (1996) Distribution of high-conductance Ca^{2+} -activated K^+ channels in rat brain: targeting to axons and nerve terminals. *J. Neurosci.* 16: 955–963.

Knight, D. E. and P. F. Baker (1982) Calcium-dependence of catecholamine release from bovine adrenal medullary cells after exposure to intense electric fields. *J. Membr. Biol.* 68: 107–140.

Koch, C. (1997) Computation and the single neuron. *Nature* 385: 207–210.

Koch, C. and A. Zador (1993) The function of dendritic spines: devices subserving biochemical rather than electrical compartmentalizaiton. *J. Neurosci.* 13: 413–422.

Kohr, G., C. E. Lambert, and I. Mody (1991) Calbindin-D28K (CaBP) levels and calcium currents in acutely dissociated epileptic neurons. *Exp. Brain Res.* 85: 543–551.

Konnerth, A., J. Dreessen, and G. J. Augustine (1992) Brief dendritic calcium signals initiate long-lasting synaptic depression in cerebellar Purkinje cells. *Proc. Natl. Acad. Sci. USA* 89: 7051–7055.

Lerea, L. S. and J. O. McNamara, (1993) Ionotropic glutamate receptor subtypes activate c-fos transcription by distinct calcium-requiring intracellular signaling pathways. *Neuron* 10: 31–41.

Leubke, J. I., K. Dunlap, and T. J. Turner (1993) Multiple calcium channel types control glutamatergic synaptic transmission in the hippocampus. *Neuron* 11: 895–902.

Leveque, C., T. Hoshino, P. David, Y. Shoji-Kasai, K. Leye, A. Omori, B. Lang, O. El Far, K. Sato, N.

Martin-Moutot, M. Takahashi, and M. Seagar (1992) The synaptic vesicle protein synaptotagmin associates with Ca^{2+} channels and is a putative Lambert-Eaton myasthenic syndrome antigen. *Proc. Natl. Acad. Sci. USA* 89: 3625–3629.

Lev-Ram, V., H. Miyakawa, N. Lasser-Ross, and W. N. Ross (1992) Calcium transients in cerebellar Purkinje neurons evoked by intracellular stimulation. *J. Neurophysiol.* 68: 1167–1177.

Li, C., B. Ullrich, J. Z. Zhang, R. G. Anderson, N. Brose, and T. C. Sudhof (1995) Ca^{2+}-dependent and -independent activities of neural and non-neural synaptotagmins. *Nature* 375: 594–599.

Lipscombe, D., D. V. Madison, M. Poenie, H. Reuter, R. W. Tsien, and R. Y. Tsien (1988). Imaging of cytosolic Ca2+ transients arising from Ca2+ stores and Ca2+ channels in sympathetic neurons. *Neuron* 1: 355–365.

Lipscombe, D., D. V. Madison, M. Poenie, H. Reuter, R. Y. Tsien, and R. W. Tsien (1988) Spatial distribution of calcium channels and cytosolic calcium transients in growth cones and cell bodies of sympathetic neurons. *Proc. Natl. Acad. Sci. USA* 85: 2398–2402.

Littleton J. T., M. Stern, M. Perin, and H. J. Bellen (1994) Calcium dependence of neurotransmitter release and rate of spontaneous vesicle fusions are altered in Drosophila synaptotagmin mutants. *Proc. Natl. Acad. Sci. USA.* 91: 10888–10892.

Llano, I., R. DiPolo, and A. Marty (1994) Calcium-induced calcium release in cerebellar Purkinje cells. *Neuron* 12: 663–673.

Llano, I., Y. P. Tan, and C. Caputo (1997) Spatial heterogeneity of intracellular Ca^{2+} signals in axons of basket cells from rate cerebellar slices. *J. Physiol. (London)* 502: 509–519.

Llinás, R. R. (1984) The squid giant synapse. In: *Current Topics in Membranes and Transport, Vol. 22, The Squid Axon* (Baker, P. F., ed.). Academic Press, pp. 519–546.

Llinás, R. and M. Sugimori (1980) Electrophysiological properties of in vitro purkinje cell dendrites in mammalian cerebellar slies. *J.Physiol. (Lond.)* 305: 197–213.

Llinás, R., M. Sugimori, and R. B. Silver (1992). Microdomains of high calcium concentration in a presynaptic terminal. *Science* 256: 677–9.

MacDermott, A. B., M. L. Mayer, G. L. Westbrook, S. J. Smith, and J. L. Barker (1986) NMDA-receptor activation increases cyutoplasmic calcium concentration in cultured spinal cord neurones. *Nature* 321: 519–522.

Markram, H., P. J. Helm, and B. Sakmann (1995) Dendritic calcium transients evoked by single back-propagating action potentials in rat neocortical pyramidal neurons. *J. Physiol. (London)* 485: 1–20.

Markwardt, F. and G. Isenberg (1992). Gating of maxi K^+ channels studied by Ca^{2+} concentration jumps in excised inside-out multi-channel patches. *J. Gen Physiol.* 99: 841–862.

Miledi, R. and I. Parker (1981) Calcium transients recorded with arsenazo III in the presynaptic terminal of the squid giant synapse. *Proc. R. Soc. Lond. Ser.* B. 212: 197–211.

Miller, L. D., J. J. Petrozzino, G. Golarai, and J. A. Connor (1996) Ca^{2+} release from intracellular stores induced by afferent stimulation of CA3 pyramidal neurons in hippocampal slices. *J. Neurophysiol.* 76: 554–562.

Mintz, I. M., B. L. Sabatini, and W. G. Regehr (1995) Calcium control of transmitter release at a cerebellar synapse. *Neuron* 15: 675–688.

Miyakawa, H., V. Lev-Ram, N. Lasser-Ross, and W. N. Ross (1992) Calcium transients evoked by climbing fiber and parallel fiber synaptic inputs in guinea pig cerebellar Purkinje neurons. *J. Neurophysiol.* 68: 1178–1189.

Mochida, S., H. Saisu, H. Kobayashi, and T. Abe (1995) Impairment of syntaxin by botulinum neurotoxin C1 or antibodies inhibits acetylcholine release but not Ca2+ channel activity. *Neuroscience* 65: 905–915.

Mochida, S., Z. H. Sheng, C. Baker, H. Kobayashi, and W. A. Catterall (1996) Inhibition of neurotransmission by peptides containing the synaptic protein interaction site of N-type Ca^{2+} channels. *Neuron* 17: 781–788.

Morgan J. I. and T. Curran (1995) Immediate-early genes: ten years on. *Trends Neurosci.* 18: 66–67.

Müller, W. and J. A. Connor (1991) Dendritic spines as individual neuronal compartments for synaptic Ca^{2+} responses. *Nature* 354: 73–76.

Murphy, T. H., J. M. Baraban, W. G. Wier, and L. A. Blatter (1994) Visualization of quantal synaptic transmission by dendritic calcium imaging. *Science* 263: 529–532.

Naraghi, M. and E. Neher (1997) Linearized buffered Ca^{2+} diffusion in microdomains and its implications for calculation of $[Ca^{2+}]$ at the mouth of a calcium channel. *J. Neurosci.* 17: 6961–6973.

Nedergaard, M. (1994) Direct signaling from astrocytes to neurons in cultures of mammalian brain cells. *Science* 263: 1768–1771.

Nedergaard, M., A. J. Cooper, and S.A. Goldman (1995) Gap junctions are required for the propagation of spreading depression. *J. Neurobiol.* 28: 433–444.

Neher, E. (1986) Concentration profiles of intracellular calcium in the presence of a diffusible chelator. *Exp. Brain Res.* 14: 80–96.

Neher, E. (1995) The use of fura-2 for estimating Ca buffers and Ca fluxes. *Neuropharmacology* 34: 1423–1442.

Neher, E. and G. J. Augustine (1992) Calcium gradients and buffers in bovine chromaffin cells. *J. Physiol. (London)* 450: 273–301.

Nonet, M. L., K. Grundahl, B. J. Meyer, and J. B. Rand (1993) Synaptic function is impaired but not eliminated in *C. elegans* mutants lacking synaptotagmin. *Cell.* 73: 1291–1305.

Oberholtzer, J. C., C. Buettiger, M. C. Summers, and F. M. Matchinsky (1988) The 28-kDa calbindin-D is a major calcium-binding protein in the basilar papilla of the chick. *Biochemistry* 85: 3387–3390.

Parnas, I. (1972) Differential block at high frequency of branches of a single axon innervating two muscles. *J. Neurophysiol.* 35: 903–914.

Peng, Y. Y. and R. S. Zucker (1993) Release of LHRH is linearly related to the time integral of presynaptic Ca^{2+} elevation above a threshold level in bullfrog sympathetic ganglia. *Neuron* 10: 465–473.

Peters, A., S .L. Palay, and H. deF. Webster (1991) *The Fine Structure of the Nervous System: Neurons and Their Supporting Cells*, 3rd ed. Oxford University Press, New York.

Pettit, D. L., S. S.-H. Wang, K. R. Gee, and G. J. Augustine. (1997) Chemical two-photon uncaging: a novel approach to optical mapping of glutamate receptors. *Neuron 19: 465–471.*

Przywara, D. A., S. V. Bhave, A. Bhave, T. D. Wakade, and A. R. Wakade (1991) Stimulated rise in neuronal calcium is faster and greater in the nucleus than the cytosol. *FASEB J.* 5(2): 217–222.

Pumplin, D. W., T. S. Reese, and R. Llinas (1981) Are the presynaptic membrane particles the calcium channels? *Proc. Natl. Acad. Sci. USA* 78: 7210–7213.

Purves, D., G. J. Augustine, D. Fitzpatrick, L.C . Katz, A.-S. LaMantia, and J. O. McNamara (1997) *Neuroscience*. Sinauer and Associates, Sunderland, MA.

Rahamimoff R. (1968) A dual effect of calcium ions on neuromuscular facilitation. *J. Physiol (London)* 195: 471–480.

Ramon y Cajal, S. (1909–1911*) Histologie du Système Nerveux de l'Homme et des Vertébrés.* Translated by L. Azoulay. Consejo Superior de Investigaciones Cientificas, Madrid.

Regehr, W. G. and D. W. Tank (1992) Calcium concentration dynamics produced by synaptic activation of CA1 hippocampal pyramidal cells. *J. Neurosci.* 12: 4202–4223.

Regehr, W. G. and D. W. Tank (1994) Dendritic calcium dynamics. *Curr. Opin. Neurobiol.* 4: 373–382.

Regehr, W. G., K. R. Delaney, and D. W. Tank (1994) The role of presynaptic calcium in short-term enhancement at the hippocampal mossy fiber synapse. *J. Neurosci.* 14: 523–537.

Reyes, M. and P. K. Stanton (1996) Induction of hippocampal long-term depression requires release of Ca^{2+} from separate presynaptic and postsynaptic intracellular stores. *J. Neurosci.* 16: 5951–5960.

Roberts, W. M. (1993) Spatial calcium buffering in saccular hair cells. *Nature* 363: 74–76.

Roberts, W. M. (1994) Localization of calcium signals by a mobile calcium buffer in frog saccular hair cells. *J. Neurosci.* 14: 3246–3262.

Roberts, W. M., R. A. Jacobs, and A. J. Hudspeth (1990) Colocalization of ion channels involved in frequency selectivity and synaptic transmission at presynaptic active zones of hair cells. *J. Neurosci.* 10: 3664–3684.

Robitaille, R., E. M. Adler, and M. P. Charlton (1990) Strategic location of calcium channels at transmitter release sites of frog neuromuscular synapses. *Neuron* 5: 773–779.

Robitaille, R., M. L. Garcia, G. J. Kaczorowski, and M. P. Charlton (1993) Functional colocalization of calcium and calcium-gated potassium channels in control of transmitter release. *Neuron* 11: 645–655.

Role, L. W. (1992) Diversity in primary structure and function of neuronal nicotinic acetylcholine receptor channels. *Curr. Opin. Neurobiol.* 2: 254–262.

Ross, W. N. (1989) Changes in intracellular calcium during neuron activity. *Annu. Rev. Physiol.* 51: 491–506.

Ross, W. N. and R. Werman (1987) Mapping calcium transients in the dendrites of Purkinje cells from the guinea-pig cerebellum in vitro. *J. Physiol. (London)* 389: 319–336.

Ross, W. N., L. L. Stockbridge, and N. L. Stockbridge (1986) Regional properties of calcium entry in barnacle neurons determined with Arsenazo III and a photodiode array. *J. Neurosci.* 6: 1148–1159.

Sakurai, M. (1990) Calcium is an intracellular mediator of the climbing fiber in induction of cerebellar long-term depression. *Proc. Natl. Acad. Sci. USA* 87(9): 3383–3385.

Schiller, J., F. Helmchen, and B. Sakmann (1995) Spatial profile of dendritic calcium transients evoked by action potentials in rat neocortical pyramidal neurones. *J. Physiol. (London)* 487: 583–600.

Schneggenburger, R., Z. Zhou, A. Konnerth, and E. Neher (1993) Fractional contribution of calcium to the cation current through glutamate receptor channels. *Neuron* 11: 133–143.

Schweizer, F. E., H. Betz, and G. J. Augustine (1995) From vesicle docking to endocytosis: intermediate reactions of exocytosis. *Neuron* 14: 689–696.

Seeburg, P. H. (1996) The role of RNA editing in controlling glutamate receptor channel properties. *J. Neurochem.* 66: 1–5.

Seeburg, P. H., N. Burnashev, G. Kohr, T. Kuner, R. Sprengel, and H. Monyer (1995) The NMDA receptor channel: molecular design of a coincidence detector. *Recent Prog. Horm. Res.* 50: 19–34.

Sharp, A. H., P. S. McPherson, T. M. Dawson, C. Aoki, K. P. Campbell, and S. H. Snyder (1993) Differential immunohistochemical localization of inositol 1,4,5-trisphosphate- and ryanodine-sensitive Ca^{2+} release channels in rat brain. *J. Neurosci.* 13: 3051–3063.

Shatz, C. J. (1996) Emergence of order in visual system development. *Proc. Natl. Acad. Sci. USA* 93: 602–608.

Sheng, Z. H., J. Rettig, M. Takahashi, and W. A. Catterall (1994) Identification of a syntaxin-binding site on N-type calcium channels. *Neuron* 13: 1303–1313.

Sheng, Z. H., J. Rettig, T. Cook, and W. A. Catterall (1996) Calcium-dependent interaction of N-type calcium channels with the synaptic core complex. *Nature* 379: 451–454.

Sheng, Z. H., C. T. Yokoyama, and W. A. Catterall (1997) Interaction of the synprint site of N-type Ca^{2+} channels with the C2B domain of synaptotagmin I. *Proc. Natl. Acad. Sci. USA* 94: 5405–5410.

Simon S. M. and R. R. Llinás (1985) Compartmentalization of the submembrane calcium activity during calcium influx and its significance in transmitter release. *Biophys. J.* 48: 485–498.

Smith, D. O. and H. Hatt (1976) Axon conduction block in a region of dense connective tissue in crayfish. *J. Neurophysiol.* 39(4): 794–801.

Smith, S. J. and G. J. Augustine (1988) Calcium ions, active zones and synaptic transmitter release. *Trends Neurosci.* 11: 458–464.

Smith, S. J. and R. S. Zucker (1980) Aequorin response facilitation and intracellular calcium accumulation in molluscan neurones. *J. Physiol. (London)* 300: 167–196.

Smith, S. J., J. Buchanan, L. R. Osses, M. P. Charlton, and G. J. Augustine (1993) The spatial distribution of calcium signals in squid presynaptic terminals. *J. Physiol. (London)* 472: 573–593.

Spitzer, N. C. (1995) Spontaneous activity: functions of calcium transients in neuronal differentiation. *Perspect. Dev. Neurobiol.* 2: 379–386.

Spruston, N., Y. Schiller, G. Stuart, and B. Sakmann (1995) Activity-dependent action potential invasion and calcium influx into hippocampal CA1 dendrites. *Science* 268: 297–300.

Srinivasan, Y., L. Elmer, J. Davis, V. Bennett, and K. Angelides (1988) Ankryn and spectrin associate with voltage-dependent sodium channels in brain. *Nature* 333: 177–180.

Stanley, E. F. (1993) Single calcium channels and acetylcholine release at a presynaptic nerve terminal. *Neuron* 11: 1007–1011.

Stanley, E. F. (1997) The calcium channel and organization of the presynaptic transmitter release face. *Trends Neurosci.* 20: 404–409.

Stanley, E. F. and R. R. Mirotznik (1997) Cleavage of syntaxin prevents G-protein regulation of presynaptic calcium channels. *Nature* 385: 340–343.

Stern, M. D. (1992) Buffering of calcium in the vicinity of a channel pore. *Cell Calcium.* 13: 183–192.

Stockbridge, N. and J. W. Moore (1984) Dynamics of intracellular calcium and its possible relationship to phasic transmitter release and facilitation at the frog neuromuscular junction. *J. Neurosci.* 4: 803–811.

Stockbridge, N. and W. N. Ross (1984) Localized Ca^{2+} and calcium-activated potassium conductances in terminals of a barnacle photoreceptor. *Nature* 309: 266–268.

Stuart, G. and M. Hausser (1994) Initiation and spread of sodium action potentials in cerebellar Purkinkje cells. *Neuron* 13: 703–712.

Stuart, G., N. Spruston, B. Sakmann, and M. Häusser (1997) Action potential initiation and backpropagation in neurons of the mammalian CNS. *Trends Neurosci.* 20(3): 125–131.

Sudhof, T. C. (1995) The synaptic vesicle cycle: a cascade of protein–protein interactions. *Nature* 375:645–653.

Svoboda, K., D. W. Tank, and W. Denk (1996) Direct measurement of coupling between dendritic spines and shafts. *Science* 272: 716–719.

Svoboda, K., W. Denk, D. Kleinfeld, and D. W. Tank (1997) In vivo dendritic calcium dynamics in neocortical pyramidal neurons. *Nature* 385: 161–165.

Swadlow, H. A., J. D. Kocsis, and S. G. Waxman (1980) Modulation of impulse conduction along the axonal tree. *Annu. Rev. Biophys. Bioeng.* 9: 143–179.

Swandulla, D., M. Hans, K. Zipser, and G. J. Augustine (1991) Role of residual calcium in synaptic depression and posttetanic potentiation: fast and slow calcium signaling in nerve terminals. *Neuron* 7: 915–926.

Takahashi, T. and A. Momiyama (1993) Different types of calcium channels mediate central synaptic transmission. *Nature* 366: 156–158.

Takei, K., H. Stukenbrok, A. Metcalf, G. A. Mignery, T. C. Sudhof, P. Volpe, and P. DeCamilli (1992) Ca^{2+} stores in Purkinje neurons: endoplasmic reticulum subcompartments demonstrated by the heterogeneous distribution of the InsP$_3$ receptor, Ca^{2+}-ATPase, and calsequestrin. *J. Neurosci.* 12: 489–505.

Tang, Y. T. and R. S. Zucker (1997) Mitochondrial involvement in post-tetanic potentiation of synaptic transmission. *Neuron* 18: 483–491.

Terasaki, M., N. T. Slater, A. Fein, A. Schmidek, and T. S. Reese (1994) Continuous network of endoplasmic reticulum in cerebellar Purkinje neurons. *Proc. Natl. Acad. Sci. USA* 91: 7510–7514.

Thompson, M. A., D. D. Ginty, A. Bonni, and M. E. Greenberg (1995) L-type voltage-sensitive Ca^{2+} channel activation regulates c-fos transcription at multiple levels. *J. Biol. Chem.* 270: 4224–4235.

Thompson, S. H. (1994) Facilitation of calcium-dependent potassium current. *J Neurosci.* 14: 7713–7725.

Tillotson, D. and A. L. F. Gorman (1980) Non-uniform Ca^{2+} buffer distribution in a nerve cell body. *Nature* 286: 816–817.

Tsubokawa, H. and W. N. Ross (1997) Muscarinic modulation of spike backpropagation in the apical dendrites of hippocampal CA1 pyramidal neurons. *J. Neurosci.* 17(15): 5782–5791.

Tucker, T. and R. Fettiplace (1995) Confocal imaging of calcium microdomains and calcium extrusion in turtle hair cells. *Neuron* 15: 1323–1335.

Verhage, M., H. T. McMahon, W. E. J. M. Ghijsen, F. Boomsma, V. M. Wiegant, and D.G. Nicholls (1991) Differential release of amino acids, neuropeptides and catecholamines from isolated nerve endings. *Neuron* 4: 577–584.

Verkhratsky, A. and H. Kettenmann (1996) Calcium signalling in glial cells. *Trends Neurosci.* 19(8): 346–352.

von Gersdorff, H. and G. Matthews (1994) Inhibition of endocytosis by elevated internal calcium in a synaptic terminal. *Nature* 370: 652–655.

Walrond, J. P. and T. S. Reese (1985) Structure of axon terminals and active zones at synapses on lizard twitch and tonic muscle fibers. *J. Neurosci.* 5: 1118–1131.

Walton, P. D., J. A. Airey, J. L. Sutko, C. F. Beck, G. A. Mignery, T. C. Sudhof, T. J. Deerinck and M. H. Ellisman (1991) Ryanodine and inositol trisphosphate receptors coexist in avian cerebellar Purkinje neurons. *J. Cell Biol.* 113: 1145–1157.

Wang, S. S.-H. and G. J. Augustine (1995a) Confocal imaging and local photolysis of caged compounds: dual probes of synaptic function. *Neuron* 15: 755–760.

Wang, S. S.-H. and G. J. Augustine (1995b) Localized photolysis of caged compounds in the dendritic arbors of cerebellar Purkinje cells. *Biophys. J.* 68: A291.

Wang, S. S.-H. and S. H. Thompson (1995) Local positive feedback by calcium in the propagation of intracellular calcium waves. *Biophys. J.* 69: 1683–1697.

Werth, J. L. and S. A. Thayer (1994) Mitochondria buffer physiological calcium loads in cultured rat dorsal root ganglion neurons. *J. Neurosci.* 14: 348–356.

Westenbroek, R. E., J. W. Hell, C. Warner, S. J. Dubel, T. P. Snutch, and W. A. Catterall (1992) Biochemical properties and subcellular distribution of an N-type calcium channel $\alpha 1$ subunit. *Neuron* 9: 1099–1115.

Westenbroek, R. E., T. Sakurai, E. M. Elliott, J. W. Hell, T. V. B. Starr, T. P. Snutch, and W. A. Catterall (1995) Immunochemical identification and subcellular distribution of the α_{1A} subunits of brain calcium channels. *J. Neurosci.* 15: 6403–6418.

Wheeler, D. B., A. Randall, and R. W. Tsien (1994) Roles of N-type and Q-type Ca^{2+} channels in supporting hippocampal synaptic transmission. *Science* 264: 107–111.

Wickens, J. (1988) Electrically coupled but chemically isolated synapses: dendritic spines and calcium in a rule for synaptic modification. *Prog. Neurobiol.* 31: 507–528.

Wiser, O., M. K. Bennett, and D. Atlas (1996) Functional interaction of syntaxin and SNAP-25 with voltage-sensitive L- and N-type Ca^{2+} channels. *EMBO J.* 15: 4100–4110.

Wiser, O., D. Tobi, M. Trus, and D. Atlas (1997) Synaptotagmin restores kinetic properties of a syntaxin-associated N-type voltage sensitive calcium channel. *FEBS Lett.* 404: 203–207.

Yamada, W. M. and R. S. Zucker (1992) Time course of transmitter release calculated from simulations of a calcium diffusion model. *Biophys. J.* 61: 671–682.

Yazejian, B., D. A. DiGregorio, J. L. Vergara, R. E. Poage, S. D. Meriney, and A. D. Grinnell (1997) Direct measurements of presynaptic calcium and calcium-activated potassium currents regulating neurotransmitter release at cultured *Xenopus* nerve-muscle synapses. *J. Neurosci.* 17: 2990–3001.

Yin, J. C. and T. Tully (1996) CREB and the formation of long-term memory. *Curr. Opin. Neurobiol.* 6: 264–268.

Yoshida, A., C. Oho, A. Omori, R. Kuwahara, T. Ito, and M. Takahashi (1992). HPC-1 is associated with synaptotagmin and omega-conotoxin receptor. *J. Biol. Chem.* 267: 24925–24928.

Yoshikami, D., Z. Bagabaldo, and B. M. Olivera (1989) The inhibitory effects of omega-conotoxins on Ca channels and synapses. *Ann. N.Y. Acad. Sci.* 560: 230–248.

Yuste, R. and W. Denk (1995) Dendritic spines as basic functional units of neuronal integration. *Nature* 375: 682–684.

Yuste, R., A. Peinado, and L. C. Katz (1992) Neuronal domains in developing neocortex. *Science* 257: 665–669.

Zheng, J. Q., M. Felder, J. A. Connor, and M. M. Poo (1994) Turning of nerve growth cones induced by neurotransmitters. *Nature* 368(6467): 140–144.

Zhou Z. and E. Neher (1993) Mobile and immobile calcium buffers in bovine adrenal chromaffin cells. *J. Physiol. (London)* 469: 245–273.

Zucker, R. S. (1996) Exocytosis: a molecular and physiological perspective. *Neuron* 17: 1049–1055.

24

The Roles of Calcium and Calmodulin in Photoreceptor Cell Functions

Zvi Selinger
Baruch Minke

For many organisms, light is the major sensory information that they receive from the environment. The first stage of processing this information is carried out by the photoreceptor cells, which are light-sensitive neurons that have specialized in one particular function. Photoreceptors translate and amplify incoming light stimuli through a cascade of chemical reactions and produce finally an electrical signal that can be communicated to other cells of the nervous system. The light-induced chemical reactions, from isomerization of the photosensor molecule rhodopsin, to the electrical signal at the plasma membrane, constitute the phototransduction cascade. Due to the large amplification which is generated by this process, photoreceptor cells can detect a single photon, the smallest quantity of light. While possessing this utmost sensitivity, the photoreceptors can also process light stimuli, which have several orders of magnitude greater intensity. To operate reliably over this enormously wide range of light intensities, the photoreceptors must strictly avoid saturation of the phototransduction cascade. This remarkable performance is achieved by a rapid and efficient adjustment of sensitivity in accordance with the ambient light intensity. The adjustment of sensitivity of the photoreceptor cells, which occurs both during, as well as following, background illuminations is known as light and dark adaptation. The process of light adaptation expands the operational range of the photoreceptors from 10^2, the common range that saturates receptors, to at least 10^7, a unique achievement among signaling systems.

It is now generally accepted that vertebrate and invertebrate photoreceptors have evolved two different phototransduction cascades and generate fundamentally different output signals. The vertebrate photoreceptors use the cyclic GMP phototransduction cascade and hyperpolarize in response to light, while the invertebrate ones use the inositol-lipid signaling as the major phototransduction pathway and depolarize in response to light. For this reason, vertebrate and invertebrate will be treated separately with special emphasis on *Drosophila* phototransduction, as the study of this latter system has been greatly advanced in recent years.

Invertebrate Phototransduction

The phosphoinositide cascade of vision

Invertebrate phototransduction is a prototype system of Ca^{2+}-mediated signaling, operating via the phosphoinositide cascade. For many years, it served as a model system for cell signaling, having the advantages of single photon analysis and advanced techniques of Ca^{2+} measurements provided by the giant ventral photoreceptors of the *Limulus*. In recent years, however, the pioneering genetic dissection of phototransduction through isolation of *Drosophila* visual mutants (Pak, 1979) has brought the *Drosophila* visual system to the forefront. This has become evident with the introduction of molecular genetics, germ line transformation, and enhancer trap methodologies combined with whole-cell voltage clamp and biochemical analysis that all converged on the *Drosophila* visual system (Selinger et al., 1993; Ranganathan et al., 1995; Hardie and Minke, 1995). These new methodologies harnessed the unparalleled genetic potential of *Drosophila* and provided invaluable and comprehensive information that could not have been predicted by comparison with other phosphoinositide signaling systems.

Invertebrate phototransduction begins with the absorption of a photon by the photosensor molecule rhodopsin, which is converted by photochemical reactions to the active state of metarhodopsin. Invertebrate metarhodopsin does not bleach and does not regenerate spontaneously to rhodopsin. As will be discussed below, to provide the essential temporal resolution of the visual system, the activity of metarhodopsin is terminated by active processes and the photopigment is regenerated to unphosphorylated rhodopsin by a Ca^{2+}-dependent process that ensures the fidelity and maintenance of the photoreceptor cells. The next step of invertebrate phototransduction is activation of a Gq-protein by metarhodopsin, which catalyzes the exchange of GDP bound to the Gq protein for free GTP. The Gq protein charged with GTP then activates a phosphoinositide-specific phospholipase C enzyme which catalyses the breakdown of phosphotidylinositol-bisphosphate into two second messengers: diacylglycerol and inositol trisphosphate ($InsP_3$). Diacylglycerol activates protein kinase C, while $InsP_3$ releases Ca^{2+} from internal Ca^{2+} stores. The final step in the light response of invertebrate photoreceptors is the opening of cation channels at the plasma membrane to bring about depolarization of the photoreceptor cells. Genetic evidence in *Drosophila* demonstrated that phospholipase C activity is absolutely required for phototransduction, as null alleles of the *norp A* gene, which encodes for the eye-specific phospholipase C enzyme, do not respond to light (Fig. 24.1).

Second Messengers of Excitation

The genetic evidence that phospholipase C (PLC) is required for excitation of *Drosophila* photoreceptors implicates the involvement of the second messenger $InsP_3$ and the ensuing Ca^{2+} release from internal stores in light excitation. Diacylglycerol, which is also generated by PLC activity and activates protein kinase C (PKC), is apparently not involved in light excitation. This notion is based on studies of a *Drosophila* mutant, characterized by inactivation but no afterpotential C (*inaC*), which lacks eye-specific PKC. In this mutant, excitation is unaffected although, as will be discussed separately, there are several defects in this mutant that lead to abnormal response termination and light adaptation (Hardie et al., 1993). The question of whether Ca^{2+} is the messenger of excitation was directly addressed by photorelease of caged Ca^{2+} loaded through the recording pipette into blind mutant photoreceptors that do not respond to light but have normal light-activated channels. Although the photolysis raised the cellular Ca^{2+} from $<1\,\mu M$ to about $10-50\,\mu M$ within 1 ms, it failed to activate any light-sensitive channels.

Positive controls demonstrated that the photoreleased Ca^{2+} activated the electrogenic Na^+/Ca^{2+} exchanger and profoundly modulated the light-induced current (Hardie, 1995). Furthermore, measurements of cellular Ca^{2+} with fluorescent indicators in Ca^{2+}-free external medium failed to detect light-induced Ca^{2+} release from internal stores (Peretz et al., 1994b; Ranganathan et al., 1994). In the *Limulus* and honeybee drone photoreceptors (Baumann and Walz, 1989a), however, the situation is different as in both species $InsP_3$ was found to release Ca^{2+} from internal stores and, in *Limulus*, photoreceptor injections of Ca^{2+} do indeed activate a current with similar properties to the light-activated conductance (Payne et al., 1986).

Despite the large body of evidence implicating a phosphoinositide cascade in the invertebrate phototransduction process, it came as a surprise that in the *Limulus* photoreceptors there is also evidence for cGMP as a second messenger of light excitation. Injection of hydrolysis-resistant analogs of cGMP into limulus photoreceptors mimicked light excitation, and in excised patches it is cGMP but not Ca^{2+} that activates the light-sensitive channels (Bacigalupo et al., 1991). In *Drosophila*, a cyclic nucleotide-gated channel (CNGC) has been cloned from a cDNA library and was reported to be expressed in eye tissue, making it possible that cGMP participates in light excitation (Baumann et al., 1994). The genetic evidence in *Drosophila*, however, indicates that if cGMP is a cellular transmitter of visual excitation, then this signal must be generated subsequent to, and be absolutely dependent on, phospholipase C activity. An obligatory role for a rise of cytosolic Ca^{2+} followed by elevation of cGMP has been suggested for excitation of the *Limulus* photoreceptors (Shin et al., 1993).

Ca^{2+} Release from Internal Stores

A rise in $InsP_3$ levels which induces Ca^{2+} release from internal stores is the common sequence of events in phosphoinositide signaling. In the *Limulus* and in the honeybee drone, there is clear evidence that both light and $InsP_3$ release Ca^{2+} from internal stores (Baumann and Walz, 1989a). In microvillar photoreceptors of invertebrates, the Ca^{2+} stores have been localized to a specialized compartment of smooth endoplasmic reticulum vesicles, the submicrovillar cisternas (SMC), that have been found to accumulate Ca-precipitate in the presence of oxalate (Baumann and Walz, 1989b). The SMC that can be detected by electron microscopy are much smaller in *Drosophila* than in the honeybee drone or in *Limulus*. In *Drosophila*, light induces a massive rise in intracellular Ca^{2+}; however, this signal is completely abolished when the photoreceptors are incubated in Ca^{2+}-free

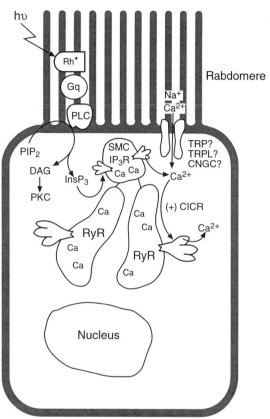

Figure 24.1 Phototransduction in the invertebrate photoreceptors. Photoexcitation of rhodopsin activates the inositol lipid transduction pathway leading to production of two second messengers: inositol trisphosphate (InsP$_3$) and diacylglycerol (DAG). The second messenger DAG activates an eye-specific protein kinase C that functions in response termination and light adaptation. On the other hand, InsP$_3$ causes release of Ca^{2+} from internal stores (SMC). The information that the InsP$_3$-sensitive stores are depleted of Ca^{2+} is relayed to channels at the cell membrane that are gated by an unknown mechanism putatively related to capacitative Ca^{2+} entry and the TRP protein. The identity of the light-dependent channels and their subunit composition has not been unequivocally determined; however, genetic evidence implicates TRP and TRPL as major components. Amplification and spatial spread of the Ca^{2+} signal apparently operate by Ca^{2+}-induced Ca^{2+} release from ryanodine-sensitive stores that have been identified in the honeybee drone photoreceptors.

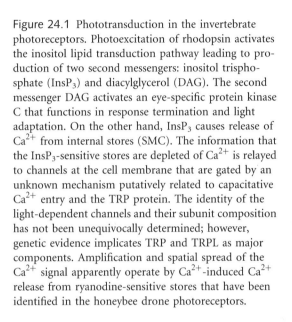

medium, indicating that the light-induced rise in cytosolic Ca^{2+} is due to influx of external Ca^{2+} (Peretz et al., 1994b; Ranganathan et al., 1994). Since it is unlikely that *Drosophila* has evolved a profoundly different mechanism of phosphoinositide signaling than other insects, it has been argued that the inability to detect Ca^{2+} release from internal stores in *Drosophila* is due to the particularly small size of the Ca^{2+} stores and the presence of a highly localized Ca^{2+} release that could not be detected by spatial averaging of the Ca^{2+} indicator fluorescence over the whole cell.

Despite the inability to demonstrate Ca release from internal stores in *Drosophila* photoreceptors, extensive studies of the *trp* and *inaC* mutants are consistent with an essential role of Ca^{2+} stores in *Drosophila* phototransduction. In wild-type *Drosophila*, the receptor potential evoked by intense light is characterized by a peak response which rapidly declines to a lower steady state (representing adaptation) that is maintained as long as the light is turned on. The major phenotype of the *trp* mutant is the inability to maintain sustained excitation in response to continuous intense illumination. Detailed studies

have indicated that the decline of the *trp* light response to baseline was due to exhaustion of excitation rather than excessive adaptation (Minke, 1982). Furthermore, application of low concentration of the Ca^{2+}-blocker La^{3+} conferred a *trp* phenotype on wild-type *Drosophila* while it had no effect on the transient response of the *trp* mutant (Suss-Toby et al., 1991). Finally, whole-cell recording from *trp* and wild-type photoreceptors revealed that the *trp* mutant is missing a major component of current with high permeability to Ca^{2+} (Hardie and Minke, 1992). Taken together, these results are most economically explained by an essential role of the TRP protein in a channel that introduces Ca^{2+} into the cell to refill the internal Ca^{2+} stores. The phenotype of the *inaC* mutant which lacks the eye-PKC enzyme has been ascribed to a primary defect revealed at the level of the single photon responses (Hardie et al., 1993). In wild type, these single photon responses are characterized by a sharp rise and a rapid decay while in the *inaC* mutant the single photon responses fail to terminate, demonstrating an abnormally slow decline of each response. On the assumption that the individual responses represent quantal Ca^{2+} release from internal stores, the multifacets phenotype of inaC was explained as follows:

1. The receptor potential of *inaC* is inactivated during intense illumination since the internal stores are depleted of Ca^{2+} due to ineffective termination of Ca release from these stores.
2. In the absence of external Ca^{2+}, when replenishment of Ca^{2+} in the stores is compromised, the receptor potential of *inaC* is inactivated by much dimmer lights than in the presence of external Ca^{2+}.
3. On the other hand, *inaC* is more sensitive to dim light than wild type because each photon releases more Ca^{2+} due to the inability to terminate the release.
4. The slow termination of the photoresponse of *inaC* after a short pulse of light is explained by the slow termination of the individual responses as these sum up to produce the receptor potential.

From these considerations, the site of action of PKC on the *Drosophila* phototransduction pathway is likely to be on Ca^{2+} sequestration by the internal stores or on the negative feedback of Ca^{2+} on Ca^{2+} release from the internal stores (Hardie et al., 1993; Selinger et al., 1993).

The Light-Sensitive Conductance

One of the major unresolved questions in invertebrate phototransduction is the nature of the light-activated channels and their mechanism of gating. Detailed analysis of the light-activated channels of *Drosophila* is still lacking. The available data comes from measurements of the effects of ionic substitutions on the reversal potential, using whole-cell voltage-clamp recordings from photoreceptors of isolated ommatidia (Hardie et al., 1993). These studies revealed that the light-activated conductance has a large fraction of Ca^{2+} current with the permeability of ratio of $P_{Ca} : P_{Mg} : P_{Na} = 40:8:1$. A large light-induced Ca^{2+} influx was detected with indicator dyes and Ca^{2+} selective microelectrodes (Peretz et al., 1994a,b). Identification of the components that are involved in the light-sensitive conductance has greatly benefited from the availability of *Drosophila* visual mutants. Perhaps the most useful mutant in this line of investigation is the transient receptor potential mutant *trp*. The photoreceptors of the *trp* mutant are unable to maintain sustained excitation in response to continuous intense illumination. Unlike the receptor potential of wild type in which the peak amplitude declines to a lower steady-state level that represents adaptation, in the *trp* mutant the response

decays back to baseline and, consequently, the photoreceptors are inactivated as long as the light is turned on. Two main findings suggested that the *trp* gene product is a Ca^{2+}-channel subunit that is gated by the inositol lipid transduction cascade. First, application of La^{3+}, a blocker of Ca^{2+}-binding sites, rapidly and efficiently converted the response of a wild-type fly into a *trp* phenotype (Suss-Toby et al., 1991). Second, whole-cell voltage-clamp recording from wild-type and *trp* photoreceptors showed that the reversal potential of *trp* is considerably more negative than in wild type and is much less dependent on extracellular Ca^{2+}, indicating that the relative Ca^{2+} permeability of *trp* is reduced about 10-fold (Hardie and Minke 1992). Independent measurements of Ca^{2+} influx using Ca^{2+} indicator dyes or Ca^{2+}-selective microelectrode also showed that, compared with wild type, Ca^{2+} influx in *trp* is greatly reduced (Peretz et al., 1994a, 1994b). Furthermore, sequence comparison showed that the transmembrane domains of *trp* reveal weak but significant homology to the vertebrate voltage-gated Ca^{2+} channel α-subunits (Phillips et al., 1992). Taken together, these results led to a model which suggests that the *trp* gene product is a subunit of a channel with high permeability to Ca^{2+} that serves to replenish Ca^{2+} in the internal Ca^{2+} stores. This model assumes that release of Ca^{2+} from internal stores is an essential component of *Drosophila* phototransduction and that the normal response of *trp* to dim light or to a short pulse of light is due to less efficient mechanism of Ca^{2+} replenishment, which suffices when the release of Ca^{2+} from the internal stores is moderate or minimal. In screening for calmodulin-binding proteins, another eye-specific gene was discovered whose predicted amino acid sequence showed 40% homology to *trp* and was therefore called *trp*-like (*trp*l) (Phillips et al., 1992). The *trp*l gene product could be a subunit of another channel that operates in null *trp* mutant. The gating mechanism of *trp* and *trp*l is still obscure and the hypothetical gating mechanism will be discussed in the context of the widespread but poorly understood phosphoinositide-mediated capacitative Ca^{2+} entry (CCE). A powerful approach to dissect the components of the light-induced current in *Drosophila* photoreceptors was demonstrated by the recent isolation of a *trp*l mutant (Niemeyer et al., 1996). The somewhat surprising finding was that the *trp*l mutant had no phenotype. However, the *trp* and *trp*l double-mutant was almost totally unresponsive to light, indicating that the TRP and TRPL proteins accounted for most if not all of the light-induced current. These results were interpreted as indicating that TRP and TRPL each constitute a separate channel that can be activated on its own but that have some overlapping functions. While other possibilities have not been entirely ruled out, the isolation of the *trp*l mutant is an important step toward dissection of the light-induced current into its components.

Capacitative Ca^{2+} Entry

Studies of the fruit fly *Drosophila* often provide essential guidelines needed to advance the understanding of similar phenomena in higher organisms. The *Drosophila* visual mutant *trp* that featured significantly in studies of "capacitative Ca^{2+} entry" is, indeed, no exception to this rule. The term "capacitative Ca^{2+} entry" (Putney, 1990) is a metaphor that helps explain the somewhat peculiar relationship between Ca^{2+} release from internal stores and extracellular Ca^{2+} entry into the cell, in that both are observed in almost every inositol lipid signaling system (Berridge, 1995). Drawing an analogy to a capacitor in an electrical circuit, fully charged internal Ca^{2+} stores prevent Ca^{2+} entry but immediately begin to promote entry when, Ca^{2+} is discharged from the stores. While not ascribing a precise mechanism, the "capacitative Ca^{2+} entry" model implies a cause-and-effect relationship between Ca^{2+} store depletion and extracellular Ca^{2+} entry. It should be pointed out, however, that while capacitative Ca^{2+} entry has been observed in many mammalian cells, its signaling pathway and gating mechanism are largely unknown and candidate mammalian participants in this pathway have been only recently identified. As already discussed in previous sections, the transient receptor potential mutant (*trp*) was suggested to be the first putative capacitative Ca^{2+}-entry mutant (Minke and Selinger, 1991). The evidence for that is based on the following findings:

1. The *trp* photoreceptors are unable to maintain sustained excitation on prolonged intense illumination. In fact, during the inactivation of the *trp* response the number of single photon events progressively decreases and completely disappears when the response is decayed to baseline. On the assumption that single photon events represent the release of Ca^{2+} from internal stores these findings are consistent with depletion of the internal Ca^{2+} stores as part of the *trp* phenotype.
2. Detailed studies of wild-type (WT) and the *trp* mutant indicate that the light-induced conductance of WT consists of at least two components that differ in their ion selectivity and La^{3+}-sensitivity. The *trp* mutation was found to remove a major component of the light-induced current which has a high Ca^{2+} permeability.
3. Direct measurements of the light-induced increase in cellular Ca^{2+} showed that in the *trp* mutant there is a substantial decrease in the light-induced Ca^{2+} influx.
4. Prolonged Ca^{2+} deprivation confers a *trp* phenotype on the wild type.

At an early stage of these studies, it was already suggested that the *trp* gene product is a putative channel subunit that is gated by Ca^{2+}-store depletion (Minke and Selinger, 1991). Heterologous expression of the TRP and TRPL proteins showed that expression of TRP gives rise to a current that is activated by Ca^{2+}-store depletion while expression of TRPL results in a constitutively active current. Furthermore, construction of TRP/TRPL chimeras established that the C-terminal domains of TRP are responsible for the store operated current (Sinkins et al., 1996). More recently, the sequence data provided by the human genome project led to cloning of three human *trp* genes (Wes et al., 1996; Zhu et al., 1996) and identification of as many as six mouse *trp* homolog genes. Convincing evidence that the *trp* homolog genes participate in capacitative Ca^{2+} entry was provided by a recent study, demonstrating that expression of two of the human *trp* homolog genes in COS-M6 cells enhances the store-operated Ca^{2+} entry beyond the endogenous Ca^{2+} entry of untransfected cells (Zhu et al., 1996). Furthermore, mouse fibroblasts transfected with fragments of the mouse *trp* homolog genes in antisense orientation suppressed the endogenous store operated Ca^{2+} entry observed in nontransfected cells (Zhu et al., 1996). On the basis of this information, it appears that *trp* is an archetypal member of a multigene family, whose products are likely to play a major role in Ca^{2+} mobilization via capacitative Ca^{2+} entry. It can therefore be expected that identification of the target for capacitative Ca^{2+} entry in mammalian cells will lead to the identification of other participants in this pathway and to the elucidation of the underlying mechanism of this process in the not too distant future.

Calmodulin

The Ca^{2+}-receptor protein calmodulin is widely used as a mediator of Ca^{2+} signaling in a variety of eukaryotic cells (Klee, 1988). It is among the most highly conserved proteins, differing between vertebrate and *Drosophila* in only 3 out of 148 amino acids. Many target proteins, including protein kinases, protein phosphatase, ion channels, adenylyl cyclase, cyclic nucleotide phosphodiesterase, and nitric oxide synthetase are regulated by intracellar Ca^{2+} transients operating through calmodulin. In general, the effects of calmodulin have been studied in vitro as it is not easy to manipulate the levels of calmodulin in whole cells. In *Drosophila*, however, genetic studies identified the *ninaC* mutant that made it possible to manipulate the levels of calmodulin and its subcellular distribution in the photoreceptor cells in vivo. The *ninaC* locus encodes two unconventional myosins, p132 and p174, consisting of fused protein kinase and myosin head domains that are expressed in *Drosophila* photoreceptor cells (Montell and Rubin, 1988). The p174 protein is localized to the rhabdomeral membranes whereas the p132 protein is present in the cytosol (Porter et al., 1992). The ninaC proteins are the major calmodulin-binding proteins in the *Drosophila* photoreceptor cell. Differential expression of the ninaC proteins in a null *ninaC* background showed that the p174 protein is responsible for the presence of calmodulin in the rhabdomer, the dense microvillar compartment that houses the transducing proteins, while the p132 is responsible for localization of calmodulin to the cytosol. The rhabdomer contains a much higher concentration of calmodulin than the cytosol. The calmodulin concentration in the rhabdomer has been estimated to be 0.5 mM (Porter et al., 1993). Identification of the two calmodulin-binding sites on the ninaC proteins led to the creation of a transgenic fly that expresses the p174 protein, containing the myosin and kinase domains, but that lacks the calmodulin-binding sites. This transgenic fly had a reduced level of calmodulin in the retina and most of it was present in the central extracellular matrix. Unlike the $ninaC^{P235}$ null allele, the transgenic fly p[$ninaC^{\Delta B}$] does not undergo light- and age-dependent retinal degeneration, thus allowing a more meaningful interpretation of its phenotype. The electrophysiological light response of p[$ninaC^{\Delta B}$] was similar to that of a newly emerged null $ninaC^{235}$ mutant fly, consisting of a larger response to the first light pulse that returns to baseline more slowly than the response of wild type. Surprisingly, the response of p[$ninaC^{\Delta B}$] to a second consecutive light pulse was much smaller than the response to the first stimulus and the responses of wild type, the latter of which do not change between the first and second light pulses (Porter et al., 1995). These results suggest that calmodulin participates in light adaptation and response termination but its target is as yet unknown.

Termination of the Photoresponse

Both vertebrate and invertebrate photoreceptors have the general requirement for speed and sensitivity. The invertebrates' photosensor rhodopsins, however, are thermostable, and do not bleach or regenerate spontaneously to rhodopsin. For this reason, the activity of metarhodopsin must be stopped by an effective termination reaction. On the other hand, there is also a need to restore the inactivated photopigment to an excitable state, in order to keep a sufficient number of photopigment molecules available for excitation. The first step of photopigment inactivation takes place by phosphorylation of metarhodopsin (the active form of the photopigment) by rhodopsin kinase. The rhodopsin kinase is absolutely specific to metarhodopsin and does not phosphorylate the inactive rhodopsin (Doza et al., 1992). Phosphorylation of metarhodopsin considerably decreases its ability to activate the guanine nucleotide-binding protein but this inactivation is not complete. Further inactivation of phosphorylated metarhodopsin is achieved upon binding of 49 kDa arrestin. *Drosophila* photoreceptors have two arrestins: a 39 kDa protein which is membrane bound and is present at low concentrations and a major form of 49 kDa arrestin that associates and dissociates from membranes containing metarhodopsin and rhodopsin, respectively. The *Drosophila* arrestins undergo light-dependent phosphorylation. This phosphorylation is carried out by Ca^{2+}/calmodulin-dependent protein kinase (Byk et al., 1993; Matsumoto et al., 1994), but the physiological consequences of this reaction are not clear. Detailed studies revealed that the 49 kDa arrestin plays a dual role in the photopigment cycle. On the one hand, it quenches the ability of phosphorylated metarhodopsin to activate the G-protein and on the other hand, it protects it from dephosphorylation. Photoregeneration of metarhodopsin to the inactive state of rhodopsin results in dissociation of arrestin to the cytosol and exposes the phosphorylated rhodopsin to phosphatase activity. The rhodopsin phosphatase is activated directly by Ca^{2+} and it is encoded by the *rdgC* gene (Steele et al., 1992). Binding and release of arrestin to metarhodopsin and rhodopsin respectively, confines the rdgC phosphatase to the inactive rhodopsin, thereby ensuring the return of phosphorylated metarhodopsin to the rhodopsin pool without initiating transduction in the dark, thus helping to maintain the fidelity of the photoreceptors' response to light.

Both biochemical and genetic studies of the arrestin cycle helped explain the following observations that remained obscure for many years. (1) Large net conversion of rhodopsin to metarhodopsin results in prolonged depolarizing after potential (PDA), representing sustained excitation long after the light is turned off. Throughout this period, the photoreceptors are partially desensitized (inactivated) and are less sensitive to subsequent test light. The ability to manipulate the concentration of arrestin in the membrane enabled direct biochemical experiments which clearly showed that, whenever the concentration of metarhodopsin exceeds the concentration of available arrestin, metarhodopsin remains persistently capable of activating the G-protein long after the light is turned off. In contrast, in the presence of high arrestin levels, the activity of metarhodopsin rapidly declines to baseline. Elegant genetic experiments and isolation of arrestin-deficient mutants convincingly showed that this holds true also in the intact photoreceptor in vivo. In the mutant with low arrestin level, suffice it to convert a small amount of rhodopsin to metarhodopsin to elicit prolonged depolarizing afterpotential (PDA), again indicating that the PDA response arises by the presence of metarhodopsin in excess over arrestin (Byk et al., 1993; Dolph et al., 1993). (2) For many years, it has been observed that a variety of phototransduction mutants eventually develop a light- and age-dependent retinal degeneration. What came as a surprise was that strong alleles of the phototransduction mutants which did not show any electrical light response revealed much more pronounced light-dependent retinal degeneration than weak alleles with a feeble light response that were protected from retinal degeneration. The discovery that in the retinal degeneration C mutant the *rdgC* locus encodes a protein phosphatase which is Ca^{2+}-dependent and that phosphorylated rhodopsin is a major substrate for the rdgC phosphatase (Byk et al., 1993) suggested that hyperphosphorylation of rhodopsin, as indeed occurs in the *rdgC* flies, leads to retinal degeneration. On the basis of these findings it was proposed that in the phototransduction mutants, light-dependent conversion of rhodopsin to metarhodopsin results in phosphorylation of the photopigment which is not followed by dephosphorylation, since blockade of the transduction prevents the rise in cellular Ca^{2+} necessary for the rdgC rhodopsin phosphatase activity (Byk et al., 1993). These predictions were subsequently confirmed by genetic experiments in which suppressor mutants of *rdgC* were also found to suppress retinal degeneration of the no receptor potential A (*norpA*) transduction mutant, indicating that in both mutants (*rdgC* and *norpA*) retinal degeneration is caused by a similar mechanism (Kurada and O'Tousa, 1995).

Light Adaptation

Summation of relatively large and fast single photon responses (quantum bumps) that form the receptor potential give rise to two undesirable features. (1) It produces large fluctuations, thereby resulting in a poor signal-to-noise ratio. (2) Saturation of the response is expected even with dim light due to the limited voltage range which is available in every excitable cell. These problems are solved in an ingenious way by the mechanism of light and dark adaptation. Results obtained from fluctuation analysis of the light-induced macroscopic current led to the suggestion of the "adapting bump model" (Wong et al., 1982). According to this model, the major effect of light adaptation is to decrease the size of the quan-

tum bumps by several orders of magnitude and to shorten their duration by a smaller factor (approx 2–4). Adaptation is therefore a gain control mechanism operating by a negative feedback loop that reduces the amplification of one or more steps in the phototransduction cascade. Of physiological significance is the ability of adaptation to prevent the receptor potential from reaching saturation and to produce a large reduction in the noise created by individual bumps. Furthermore, the decrease in the lag time and duration of individual bumps results in an improved temporal resolution.

Experiments in the *Limulus* and flies have shown that introduction of GTPγS or InsP$_3$ both excite and adapt the photoreceptor cells. These experiments were the first indication that inositol lipid signaling produces the internal transmitters that both excite and adapt the invertebrate photoreceptor cells. In all of these experiments, GTPγS and InsP$_3$ produced quantal bumps similar in shape but of different amplitude than those produced by light. Surprisingly, the question of how these nonquantal stimuli can give rise to quantal responses was not dealt with. A straightforward explanation for these findings, however, is that the mechanism which generates the quantal responses operates beyond the site of action of GTPγS and InsP$_3$. Consistent with this interpretation are experiments on *Limulus* photoreceptors that tested the ability of Ca^{2+} to excite and adapt the photoreceptor cells (Payne et al., 1986). It was found that pressure injection of Ca^{2+}, creating a large Ca^{2+} transient, elicited effective excitation of the *Limulus* photoreceptors. On examining the effect of pressure injection of Ca^{2+} on adaptation, it was found that Ca^{2+} injection effectively adapted the photoreceptors to subsequent stimuli of light or InsP$_3$ but had little effect on a second injection of Ca^{2+}. The most obvious conclusion from these experiments is that the site of adaptation must lie between the sites of action of InsP$_3$ and Ca^{2+}.

In the invertebrate photoreceptors, a subpopulation of the endoplasmic reticulum, the submicrovillar cisternae (SMC), constitute the internal Ca^{2+} stores that release Ca^{2+} upon binding of InsP$_3$ to its receptor. It therefore follows that the SMC is the site of light adaptation and that release of packets of Ca^{2+} from these stores underlies the generation of quantum bumps. Findings consistent with this notion are that while light and InsP$_3$ injection caused noisy transient events comprising quantum bumps, the response to Ca^{2+} injection was a smooth inward current, indicating that Ca^{2+} excites the photoreceptor at a site distal to the site that generates the quantum bumps. Studies of the *Limulus* ventral photoreceptors demonstrated that injection of InsP$_3$ elevates the concentration of intracellular Ca^{2+} and consequently depolarizes the photoreceptor cells. This $InsP_3$-induced elevation of cellular Ca^{2+}, was inhibited by a prior injection of Ca^{2+} or $InsP_3$ delivered 1 sec earlier (Payne et al., 1988, 1990). These results thus indicate that $InsP_3$-induced Ca^{2+} release from internal stores can inhibit further release of Ca^{2+} from the $InsP_3$-sensitive Ca^{2+} stores. If this mechanism operates in other invertebrate photoreceptors, it may account, at least in part, for light adaptation of invertebrate phototransduction at the level of the Ca^{2+} stores.

Vertebrate Photoreceptors

The Cyclic GMP Cascade of Vision

The vertebrate photoreceptor cells, rods and cones, have evolved a unique transduction system. Unlike other neurons or the invertebrate photoreceptors, the rods and cones are depolarized in the dark because the light-sensitive channels are open in the dark, producing the so called "dark current," while light closes these channels thereby eliminating the "dark current." Photoexcitation begins with absorption of a photon by the visual pigment rhodopsin. The photoexcited rhodopsin catalyzes the exchange of guanosine diphosphate (GDP) for guanosine triphosphate (GTP) on the G-protein transducin (T). The GTP-bound form of transducin then activates a cyclic guanosine monophosphate (cGMP)-specific phosphodiesterase that catalyzes the rapid breakdown of cGMP. The light-activated signal is thereby translated into a decrease in the level of cytosolic cGMP. This leads to dissociation of cGMP from binding sites on the cGMP-gated channels at the plasma membrane, the channels close, and the photoreceptor cells hyperpolarize (Kaupp and Koch, 1992; Lagnado and Baylor, 1992; Pugh and Lamb, 1993; Yau, 1994). Many of the components involved in activation and regulation of this signaling pathway have been identified, the individual reaction has been reconstituted with purified components, and this cascade has served as a model system for G-protein-coupled signaling. This system is geared for rapid on and off mechanisms which are required for the high temporal resolution of the light signals and it has the ability to modulate its sensitivity, operating reliably both in dim and intense light without reaching saturation. This means that each step in the transduction cascade should have an efficient "turn-on" mechanism, as well as an equally effective "turn-off" mechanism. Furthermore, as the photoreceptor cells need to change their sensitivity in line with the intensity of the ambient light, they rely on an efficient mechanism to adjust the sensitivity and kinetics of the photoreceptors. As will be discussed below,

while cGMP is the second messenger of light excitation, Ca^{2+} ions change the sensitivity and response kinetics of the photoreceptors by processes that are collectively referred to as light and dark adaptation (Fig. 24.2).

A Delicate Balance between Ca^{2+} Fluxes Generates a Ca^{2+} Signal

The Na^+ Ca^{2+} Exchanger

Early observations have indicated the presence of a significant Ca^{2+} component in the dark current of vertebrate photoreceptors. This implies that in the dark, under steady-state conditions, there must be a mechanism to remove Ca^{2+} from the cell in order to avoid toxic concentrations of cellular Ca^{2+}. An obvious candidate for this task is the Na^+/Ca^{2+} exchanger that exists in photoreceptors and many other cells. In the dark, there is a balance between Ca^{2+} influx through the light-dependent channels and Ca^{2+} extrusion through the Na^+/Ca^{2+} exchanger. Illumination causes closure of the light-dependent channels, leading to a net Ca^{2+} efflux and a decrease in cellular Ca^{2+}. In terms of Ca^{2+} levels, the light-evoked signal is a fall in photoreceptor free Ca^{2+} concentrations. This is a unique response since excitable cells usually respond to stimulation by a rise in cytosolic Ca^{2+}. Indeed, during the reign of the "Ca^{2+} hypothesis" of phototransduction, the claim that a rise in cytosolic Ca^{2+} is the final signal for closing of the light-dependent channels was based on the observation that light causes an efflux of Ca^{2+} from the rod photoreceptors. This efflux was misinterpreted as an indication for a rise in cytosolic Ca^{2+} followed by extrusion of the excess of Ca^{2+} through the Na^+/Ca^{2+} exchanger. As this model was based on measurements of extracellular Ca^{2+}, we know now that the light-induced increase in extracellular Ca^{2+} concentrations was the result of eliminating the Ca^{2+} influx into the cells due to closure of the light-dependent channels, while the Na^+/Ca^{2+} exchanger was unaffected and continued to extrude Ca^{2+} from the cells. In view of the great importance of curbing the rise of cytosolic Ca^{2+} during dark periods, the photoreceptor's Na^+/Ca^{2+} exchanger also uses K^+ movement with the stochiometry of four Na^+ against one Ca^{2+} and one K^+ to extrude Ca^{2+} from the cell (Cervetto et al., 1989). This property enables the exchanger to use the concentration gradients of both Na^+ and K^+ to extrude Ca^{2+}, thereby reaching much lower intracellular Ca^{2+} concentration during light conditions than those dictated by the Na^+ concentrations across the plasma membrane and the membrane potential.

Figure 24.2 Phototransduction in vertebrate rod photoreceptors. The photosensor rhodopsin (Rh) is turned on (excited) by light and is turned off by a phosphorylation reaction mediated by rhodopsin kinase (RhK), which is sequentially followed by binding of arrestin. Photoexcited rhodopsin activates the G-protein transducin (T) which, in turn, activates the enzyme phosphodiesterase (PDE) that hydrolyses cGMP to GMP. This excitation pathway brings about a decrease in the level of cGMP, leading to dissociation of cGMP from the cyclic nucleotide-gated channel (CNGC) which shuts off and, as a result, the photoreceptor cell hyperpolarizes. The excitation pathway also generates a second signal manifested by a decrease in cellular Ca^{2+}. The signal of a fall in cellular Ca^{2+} concentration operates by the proteins S-modulin/recoverin, calmodulin, and guanylyl cyclase-activating protein (GCAP), exerting their effects on rhodopsin kinase, the β-subunit of the CNG channel and the reticular guanylyl cyclase, respectively. All these Ca^{2+}-mediated effects participate in photorecovery and light adaptation. (This figure is reproduced with permission from K. W. Koch, *Cell Calcium* (1995) 18: 314-321.

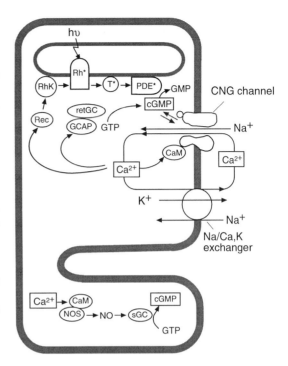

The Cyclic GMP-Gated Channels

It is now well established that the light-dependent channels of vertebrate photoreceptors are gated by cGMP and not by Ca^{2+}. On the other hand, Ca^{2+} has a modulatory effect on the light-dependent cGMP-gated channel (CNGC) (Kaupp, 1995). As already mentioned, the photoreceptor CNG channels are nonselective cation channels permeable to Ca^{2+}, in addition to Na^+. Under physiological conditions, approximately 70% of the cGMP-sensitive inward current is carried by Na^+, 15% by Ca^{2+}, 5% by Mg^{2+}, and 8% by Na^+ flux through the electrogenic Na^+/Ca^{2+}, K^+ exchanger. The Ca^{2+} ions interact with the CNG channel at three different sites. First, it blocks the channel with high affinity by binding at the extracellular side to glutamate residue of the CNGC at position 363 of the putative pore region (Root and Mackinon, 1993; Eismann et al., 1994). The single channel conductance of rod cells in the absence of Ca^{2+} or Mg^{2+} is 20–25 pS. At physiological levels of extracellular Ca^{2+} or Mg^{2+} (1 mM), the single channel conductance is dramatically decreased to 0.1 pS. This decrease in conductance has important physiological consequences as it greatly improves the signal-to-noise ratio that is needed for single photon detection. Second, Ca^{2+} and Mg^{2+} also block the rod's CNG channels from the cytoplasmic side but at much higher concentrations (Ki ∼ 1 mM),

which are hardly attainable under physiological conditions. Third, calcium-calmodulin binding to the β-subunit of the rod CNG channel decreases the affinity of the channel to cGMP. When the concentrations of Ca^{2+} fall in response to light, calmodulin loses its Ca^{2+} and comes off the channel, rendering the channel more sensitive to cGMP and thus enhancing the photorecovery and contributing to light adaptation. Furthermore, the β subunit confers on the hetero-oligomeric CNG channel flickering behavior and inhibition by L-*cis*-diltiasem (Chen et al., 1993; Körschen et al., 1995).

The Guanylyl Cyclase

In response to light, the vertebrate photoreceptor's enzyme cGMP-phosphodiesterase is activated. As a result, the level of cGMP falls and the cGMP-gated channels close. The receptor potential returns to the dark baseline when the cGMP-gated channels (CNGC) are reopened. This requires that the concentration of cGMP would be restored to the high dark levels needed to reopen the CNG channels. The task of restoring the cGMP level is mediated by the enzyme guanylyl cyclase (GC) that converts GTP to cGMP. From this consideration it follows that the level of cGMP is determined by a balance between the activities of the phosphodiesterase that breaks down cGMP (cGMP-PDE) and the GC which synthesizes cGMP.

The notion that photoexcitation of rhodopsin is coupled to activation of cGMP-PDE by the G-protein transducin has been established more than a decade ago. In contrast, the regulation of the guanylyl cyclase activity by light has been unraveled only in the last few years. It turned out that the GC activity is regulated by the level of Ca^{2+} in the photoreceptor cells. However, the regulation of GC activity by Ca^{2+} is indirect and is mediated by acidic Ca^{2+}-binding proteins (Palczewski et al., 1994, Dizhoor et al., 1995) that are referred to as guanylyl cyclase-activating proteins (GCAP). The GCAP, in the presence of low Ca^{2+} concentration, causes a substantial activation of the guanylyl cyclase and restores the Ca^{2+}-sensitive regulation of GC in reconstituted system. Furthermore, dialysis of GCAP into functionally intact rod outer segments decreased the sensitivity, time to peak, and recovery time of the electrical response to light (Gorczyca et al., 1994). These properties are hallmarks of light adaptation.

Several lines of evidence indicate that the GCAP molecules are the Ca^{2+} sensors which mediate termination of the receptor potential and light adaptation of the photoresponse. Purified GCAP changes its fluorescence in response to a decrease in Ca^{2+} concentration and, like calmodulin, migrates differently on SDS-PAGE in the presence and absence of Ca^{2+}. The amino acid sequence of GCAP shows the presence of three EF-hand Ca^{2+}-binding domains, consistent with the Ca^{2+}-sensitive regulation of GC by GCAP. Furthermore, a myristoylated 27 amino acid peptide of the N-terminus of GCAP and monoclonal antibodies raised against recombinant GCAP both efficiently inhibited the activation of GC by GCAP. These results attest to the specificity of guanylyl cyclase activation by GCAP and suggest that its N-terminal is involved in the interaction with the enzyme (Eismann et al., 1994).

The Rhodopsin Kinase

Phosphorylation of rhodopsin and binding of arrestin inactivate the photoexcited rhodopsin. Thereby, the gain of the entire process is determined by the life time of the excited rhodopsin and, consequently, the number of transducin molecules that it activates. While binding of arrestin is very fast, an increase in rhodopsin kinase activity is translated into a decrease in the gain of the phototransduction cascade. The proteins S-modulin and recoverin are Ca^{2+}-sensor proteins with a structure similar to calmodulin, having typical EF-hand motifs and myristoylated binding to the surface membrane. At high Ca^{2+} concentration, these proteins prolong the life time of the activated cGMP phosphodiesterase. As this effect was dependent on ATP, it was found to be mediated by inhibition of rhodopsin kinase (Kawamura, 1993; Kawamura et al., 1993). Since the light response results in a decrease in cellular Ca^{2+} in the photoreceptor cells, this causes disinhibition (i.e., activation) of rhodopsin kinase, thereby enhancing photopigment-dependent light adaptation. An important question is the relevance of rhodopsin phosphorylation in the in vitro biochemical studies to the reactions that take place in the rod cell under physiological conditions. This question has been recently addressed, using a transgenic mouse expressing C-terminal-truncated rhodopsin in the rod photoreceptor. The C-terminal truncation eliminated three phosphorylation sites that the biochemical experiments implicated in the shut-off process. The transgene rod photoreceptors contained a normal amount of rhodopsin, of which 10% was truncated. A similar fraction of the single photon responses recorded from the transgenic rods failed to terminate normally, lasting on an average 20 times longer than the normal responses (Chen et al., 1995). These experiments strongly suggest that phosphorylation of one or more of the three phosphorylation sites that had been deleted is involved in rhodopsin shut-off under physiological conditions.

Conclusions

While much remains to be discovered, the picture which emerges is as follows. Vertebrate photoreceptors use two different second messengers: the nucleotide cGMP that mediates light excitation and Ca^{2+} ions that mediate photoresponse termination and light adaptation (Koch, 1995). Since Ca^{2+} ions enter the photoreceptor cell mainly or exclusively through the light-dependent channels, the decrease in the intracellular level of Ca^{2+} is a reliable register for the number of these channels that have been closed by light excitation. When light excitation brings about closure of CNG channels, this is immediately and reliably translated into a fall in cytosolic Ca^{2+}. The decrease in cellular Ca^{2+} is the signal for GCAP to increase the rate of cGMP synthesis which, in turn, sets the sensitivity and the kinetics of light excitation. Two additional tiers of Ca-mediated light adaptation operate at the level of the photopigment and the cGMP-gated channel:

(1) The inhibition of rhodopsin kinase by S-modulin and recoverin is relieved when the concentration of Ca^{2+} falls, thereby the life time of the photoexcited rhodopsin is shortened, and the gain of the entire phototransduction process is decreased.

(2) A decrease in cellular Ca^{2+} leads to dissociation of calmodulin from the CNGC β-subunit. As a result, the affinity of the channel to cGMP is increased and the channel opens at lower cGMP levels, making it necessary to further decrease the level of cGMP to close the cGMP-gated channel.

References

Bacigalupo, J., Johnson, E. C., Vergara, C., and Lisman, J. E. (1991) Light-dependent channels for excised patches of *Limulus* ventral photoreceptors are opened by cGMP. *Proc. Natl. Acad. Sci. USA* 88: 7938–7942.

Baumann, A., Frings, S., Godde, M., Seifert, R., and Kaupp, U. B. (1994) Primary structure and functional expression of a *Drosophila* cyclic nucleotide-gated channel present in eyes and antenna. *EMBO J.* 13: 5040–5050.

Baumann, O. and Walz, B. (1989a) Calcium and inositol polyphosphate-sensitivity of the calcium-sequestering endoplasmic reticulum in the photoreceptor cells of the honeybee drone. *J. Comp. Physiol.* 165: 627–636.

Baumann, O. and Walz, B. (1989b) Topography of the Ca^{2+} sequestering endoplasmic reticulum in photoreceptors and pigmented glial cells in the compound eye of the honeybee drone. *Cell Tissue Res.* 255: 511–522.

Berridge, M. J. (1995) Capacitative calcium entry. *Biochem. J.* 312: 1–11.

Byk, T., Bar-Yaacov, M., Doza, Y. N., Minke, B., and Selinger, Z. (1993) Regulatory arrestin cycle secures the fidelity and maintenance of the fly photoreceptor cell. *Proc. Natl. Acad. Sci. USA* 90: 1907–1911.

Cervetto, L., Lagnado, L., Perry, R. J., Robinson, D. W., and McNaughton, P. A. (1989) Extrusion of calcium from rod outer segments is driven by both sodium and potassium gradients. *Nature* 337: 740–743.

Chen, J., Makino, C. L., Peachey, N. S., Baylor, D. A,. and Simon, M. I. (1995) Mechanism of rhodopsin inactivation *in vivo* as revealed by a COOH-terminal truncation mutant. *Science*, 267: 374–377.

Chen, T. Y., Peng, Y. W., Dhallan, R. S., Ahamed, B., Reed, R. R., and Yau, K. W. (1993) A new subunit of the cyclic nucleotide-gated cation channel in retinal rods. *Nature*, 362: 764–767.

Dizhoor, A. M., Olshevskia, E. V., Henzel, W. J., Wong, S. C., Stults, J. T., Ankoudinova, I. N., and Hurley, J. B. (1995) Cloning, sequencing and expression of a 24-kDa Ca^{2+}-binding protein activating photoreceptor guanylyl cyclase. *J. Biol. Chem.* 270: 25200–25206.

Dolph, P. J., Ranganathan, R., Colley, N. J., Hardy, R. W., Socolich, M., and Zuker, C. S. (1993) Arrestin function in inactivation of G-protein coupled receptor rhodopsin in vivo. *Science* 260: 1910–1916.

Doza, Y. N., Minke, B., Chorev, M., and Selinger Z. (1992) Characterization of fly rhodopsin kinase. *Eur. J. Biochem.* 209: 1035–1040.

Eismann, E., Muller, F., Heinemann, S. H., and Kaupp, U. B. (1994) A single charge within the pore region of a cGMP gated channel controls rectification, Ca^{2+} blockage and ionic selectivity. *Proc. Natl. Acad. Sci. USA* 91: 1109–1113.

Gorczyca, W. A., Gray-Keller, M. P., Detwiler, P. B., and Palczewski, K. (1994) Purification and physiological evaluation of a guanylate cyclase activating protein from retinal rods. *Proc. Natl. Acad. Sci. USA* 91: 4014–4018.

Hardie, R. C. (1995) Photolysis of caged Ca^{2+} facilitates and inactivates but does not directly excite light-sensitive channels in *Drosophila* photoreceptors. *J. Neurosci.* 15: 889–902.

Hardie, R. C. and Minke, B. (1992) The *trp* gene is essential for a light-activated Ca^{2+} channel in *Drosophila* photoreceptors. *Neuron*, 8: 643–651.

Hardie, R. C. and Minke, B. (1995) Phosphoinositide mediated phototransduction in *Drosophila* photoreceptors: the role of Ca^{2+} and *trp*. *Cell Calcium* 18: 256–274.

Hardie, R. C., Peretz, A., and Suss-Toby, E., et al. (1993) Protein kinase C is required for light adaptation in *Drosophila* photoreceptors. *Nature* 363: 634–637.

Kaupp, U. B. (1995) Family of cyclic nucleotide gated ion channels. *Curr. Opin. Neurobiol.* 5: 434–442.

Kaupp, U. B. and Koch, K. W. (1992) Role of cGMP and Ca^{2+} in vertebrate photoreceptor excitation and adaptation. *Annu. Rev. Physiol.* 54: 153–175.

Kawamura, S. (1993) Rhodopsin phosphorylation as a mechanism of cyclic GMP phosphodiesterase regulation by S-modulin. *Nature* 362: 855–857.

Kawamura, S., Hisatomi, O., Kayada, S., Tokunaga, F., and Kuo, C. H. (1993) Recoverin has S-modulin activity in frog rods. *J. Biol. Chem.* 268: 14579–14582.

Klee, C. B. (1988) Interactions of calmodulin with Ca^{2+} and target proteins. In *Calmodulin*, (Klee, C. B. and Cohen, P., eds.). Elsevier, New York, pp. 35–89.

Koch, K. W. (1995) Control of photoreceptor proteins by Ca^{2+}. *Cell Calcium* 18: 314–321.

Körschen, H. G., Illing, M., Seifert, R., Sesti, F., et al. (1995) A 240 kDa protein represents the complete β subunit of the cyclic nucleotide-gated channel from rod photoreceptors. *Neuron* 15: 627–636.

Kurada, P. and O'Tousa, E. (1995) Retinal degeneration caused by dominant rhodopsin mutations in *Drosophila*. *Neuron* 14: 571–579.

Lagnado, L. and Baylor, D. A. (1992) Signal flow in visual transduction. *Neuron* 8: 995–1002.

Matsumoto, H., Kurien, B. T., Takagi, Y., and Kahn, E. S., et al. (1994) Phosrestin I undergoes the earliest light-induced phosphorylation by a calcium/calmodulin-dependent protein kinase in *Drosophila* photoreceptors. *Neuron* 12: 997–1010.

Minke, B. (1982) Light-induced reduction in excitation efficiency in the *trp* mutant of *Drosophila*. *J. Gen. Physiol.* 79: 361–385.

Minke, B. and Selinger, Z. (1991) Inositol lipid pathway in fly photoreceptors: excitation, calcium mobilization and retinal degeneration. In *Progress in Retinal Research*, (Osborne, N. N. and Chader, G. J., eds.). Pergamon Press, Oxford, pp. 99–124.

Montell, C. and Rubin, G. M. (1988) The *Drosophila* ninaC locus encodes two photoreceptor specific proteins with domains homologous to protein kinases and the myosin heavy chain head. *Cell* 52: 757–772.

Niemeyer, B. A., Suzuki, E., Scott, K., Jalnik, K., and Zuker, C. S. (1996) The *Drosophila* light activated conductance is composed of the two channels *TRP* and *TRP*L. *Cell* 85: 651–659.

Pak, W. L. (1979) Study of photoreceptor function using *Drosophila* mutants. In *Neurogenetics, Genetic Approaches to the Nervous System* (Breakfeld, X. O., ed.) New York, Elsevier, pp. 67–99.

Palczewski, K., Subbaraya, I., Gorczyca, W. A., Helekar, B. S., et al. (1994) Molecular cloning and characterization of retinal photoreceptor guanylyl cyclase-activating protein, *Neuron* 13: 395–404.

Payne, R., Corson, D. W., and Fein, A. (1986) Pressure injection of calcium both excites and adapts *Limulus* ventral photoreceptors. *J. Gen. Physiol.* 88: 107–126.

Payne, R., Flores, T. M., and Fein, A. (1990) Feedback inhibition by calcium limits the release of calcium by Inositol trisphosphate in *Limulus* ventral photoreceptors. *Neuron* 4: 547–555.

Payne, R., Walz, B., Levy, S., and Fein, A. (1988) The localization of calcium release by inositol trisphosphate in *Limulus* photoreceptors and its control by negative feedback. *Philos. Trans. R. Soc. Lond. (Biol.)* 320: 359–379.

Peretz, A., Sandler, C., Kirschfeld, K., Hardie, R. C., and Minke, B. (1994a) Genetic dissection of light-induced Ca^{2+} influx into *Drosophila* photoreceptors. *J. Gen. Physiol.* 104: 1057–1077.

Peretz, A., Suss-Toby, E., Rom-Glas, A., Arnon, A., Payne, R., and Minke, B. (1994b) The light response of *Drosophila* photoreceptors is accompanied by an increase in cellular calcium: effects of specific mutations. *Neuron* 12: 1257–1267.

Phillips, A M., Bull, A., and Kelly, L. (1992) Identification of a *Drosophila* gene encoding a calmodulin binding protein with homology to the *trp* phototransduction gene. *Neuron* 8: 631–642.

Porter, J. A., Hicks J. L., Williams, D. S., and Montell, C. (1992) Differential localizations of and the requirements for the two *Drosophila* ninaC kinase/myosins in photoreceptor cells. *J. Cell Biol.* 116: 683–693.

Porter, J. A., Minke, B., and Montell, C. (1995) Calmodulin binding to *Drosophila* ninaC required for termination of phototransduction. *EMBO J.* 14: 4450–4459.

Porter, J. A., Yu, M., Doberstein, S. K., Pollard, T. S., and Montell, C. (1993) Dependence of calmodulin localization in the retina on the ninaC unconventional. *Myosin Science*, 262: 1038–1042.

Pugh, E. N. Jr. and Lamb, T. D. (1993) Amplification and kinetics of the activation steps in phototransduction. *Biochim. Biophys. Acta* 1141: 111–149.

Putney, J .W. Jr. (1990) Capacitative calcium entry revisited. *Cell Calcium* 116: 611–624.

Ranganathan, R., Bacskai, B. J., Tsien, R. Y., and Zuker, C. S. (1994) Cytosolic calcium transients: spatial localization and role in *Drosophila* photoreceptor cell function. *Neuron* 13: 837–848.

Ranganathan, R., Malicki, D. M., and Zuker, C. S. (1995) Signal transduction in *Drosophila* photoreceptors. *Annu. Rev. Neurosci.* 18: 283–317.

Root, M. J. and Mackinnon, R. (1993) Identification of an external divalent cation-binding site in the pore of a cGMP-activated channel. *Neuron* 11: 459–466.

Selinger, Z., Doza, Y. N., and Minke, B. (1993) Mechanisms and genetics of photoreceptors desensitization in *Drosophila* flies. *Biochim. Biophys. Acta* 1179: 283–299.

Shin, J., Richard, E. A., and Lisman, J. E. (1993) Ca^{2+} is an obligatory intermediate in the excitation cascade of *Limulus* photoreceptors. *Neuron* 11: 845–855.

Sinkins, W. G., Vaca, L., Hu, Y., Kunze, D. L., and Schilling, W. P. (1996) The COOH-terminal domain of *Drosophila* TRP channels confers thapsigargin sensitivity. *J. Biol. Chem.* 271: 2955–2960.

Steele, F. R., Washburn, T., Rieger, R., and O'Tousa, J. E. (1992) *Drosophila* retinal degeneration C (rdgC) encodes a novel serine/threonine protein phosphatase. *Cell* 69: 669–676.

Suss-Toby, E., Selinger, Z., and Minke, B. (1991) Lanthanum reduces the excitation efficiency in fly photoreceptors. *J. Gen. Physiol.* 98: 849–868.

Wes, P. D., Chevesich, J., Jeromin, A., Rosenberg, C., Stetten, G., and Montel, C. (1996) *TRP*C1, a human homolog of a *Drosophila* store-operated channel. *Proc. Natl. Acad. Sci. USA* 92: 9652–9659.

Wong, F., Knight, B. W., and Doge, F. A. (1982) Adapting bump model for entral photoreceptors of *Limulus*. *J. Gen. Physiol.* 79: 1089–1113.

Yau, K. W. (1994) Phototransduction mechanism in retinal rods and cones. *Invest. Ophthalmol. Vis. Sci.* 35: 9–32.

Zhu, X., Jiang, M., Peyton, M., Boulay, G., Hurst, R., Stefani, E., and Birnbaumer, L. (1996) *trp*, a novel mammalian gene family essential for agonist activated capacitative Ca^{2+} entry. *Cell* 85: 661–671.

25

Calcium in Muscle Contraction

Setsuro Ebashi
Makoto Endo
Iwao Ohtsuki

Overview: A Brief Historical Sketch

In this chapter, the mechanism of how muscle contraction is regulated by Ca^{2+} will be described with special references to vertebrate skeletal muscle, which is one of the most differentiated tissues and, therefore, equipped with a very subtle device for this purpose.

Muscle has a special position in the Ca^{2+} research. As will be described below, the thoughts which eventually led to today's Ca^{2+} concept (cf. Ebashi and Endo, 1968) have been fostered exclusively in the research on skeletal muscle. Scientists, including most muscle researchers, believed until around 1970 that Ca^{2+} would be the factor specific to muscle contraction. Now Ca^{2+} is the most popular factor thought to regulate or control intracellular processes of almost all tissues and, as a result, muscle has become nothing more than one of numerous subjects of Ca^{2+} research. At this stage, it may be useful to look back on the progress of Ca^{2+} research in the muscle field. Emphasis will be laid on the reason that the role of this ubiquitous factor had remained unrecognized for a long time.

Ringer's discovery (1883) of the indispensability of Ca^{2+} for cardiac contractility is a famous story in the history of biological sciences, embellished by plausible anecdotes. However, this finding did not straightforwardly develop to today's Ca^{2+} concept. It was then explained by Ringer's followers that the excitability of cardiac and smooth muscles was based on the balance between excitatory Ca^{2+} and inhibitory K^+ in the bathing solution (cf. Stiles, 1901). The fact that the removal of Ca^{2+} from the outer medium induced repetitive twitches of skeletal muscle, which could be suppressed by Ca^{2+} (Ringer, 1886), gave an impression that each tissue would have its own way of excitation and the modification of its excitability

by some ingredients of the outer medium, such as inorganic ions, should be made in a way specific to that tissue. In other words, there was no idea that Ca^{2+} would have a role common to every tissue. One of the reasons that the interpretation of Ringer's finding was thus distorted might be the lack of the concept of the membrane. For the contemporary scientists, excitation was the matter of the whole cell, not the specialized function of the cytoplasmic membrane.

The pioneering work, which eventually led to the present Ca^{2+} concept, was carried out by Heilbrunn (1940). Although he appreciated Ringer's finding from a physiological point of view, he did not accept Ringer as his forerunner and esteemed Ringer's Ca^{2+} as one of the environmental conditions (Heilbrunn, 1937). Whether or not this explanation be reasonable, he was undoubtedly the first person who considered Ca^{2+} as the factor exerting its effect on the intracellular system, namely, the contractile elements.

Heilbrunn's conviction about the essential role of Ca^{2+} in muscle contraction might be derived from the general atmosphere of scientists engaged in research on cell motility, in which Heilbrunn himself had been involved. They believed that cell motility was under the control of Ca^{2+}, but the evidence shown by them was dubious and sometimes erroneous; in other words, they reached the conclusion based on a priori belief in Ca^{2+}. Since they were dealing with unicellular organisms, they were compelled to recognize the presence of the plasma membrane that demarcated the cytoplasm from its environment. This must have incited Heilbrunn to shape a proper idea.

In spite of his epoch-making idea and experiments, however, Heilbrunn could not immediately open a new frontier of research. The reasons for

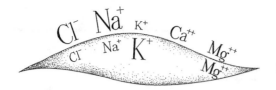

Figure 25.1 Distributions of inorganic ions outside and inside the cell. (Modified from Ebashi, 1974.)

this are as follows. First, his experiment that showed the contracture of injured muscle was carried out in isotonic CaCl₂ solution, which seemed far from being physiological. Even his later work, i.e., the demonstration of the shortening of an intact single muscle fiber upon injection of Ca^{2+} through a micropipette (Heilbrunn and Wiercinski, 1947), was not enough to persuade other people to believe the role of Ca^{2+} (prior to this, Ca^{2+} injection experiments were carried out by two groups, Keil and Sichel, 1936, and Kamada and Kinosita 1943; the latter was the only group who immediately agreed with Heilbrunn's proposal). Second, the actomyosin–ATP system, the dramatic accomplishment of Szent-Györgyi and his school (cf. Szent-Györgyi, 1951), could not notice the effect of Ca^{2+} (the Ca^{2+} concentration necessary for activation was so low that they performed all experiments in a solution containing sufficient Ca^{2+}). Third, and perhaps the most important, the elegant comment of Hill (1948) that denied the possibility that Ca^{2+} would be the triggering factor for contraction was accepted by many physiologists as a reasonable counterevidence against Heilbrunn's idea. Skeletal muscle, however, is equipped with a trick, which has eventually proved the apparently unreasonable idea of Heilbrunn to be true (see section "Excitation–Contraction Coupling").

Today's Ca^{2+} concept was not directly originated from the Heilbrunn proposal but was derived from the work on the intrinsic relaxing factor, work which was pioneered by Marsh (1951) and followed up by several others. Subsequent identification of the relaxing factor as the granular or microsome ATPase preparation (Kumagai et al., 1955) tempted many researchers to search for the "true" relaxing factor, an imaginary soluble substance which should have been produced by the microsomes with the aid of ATP. This might reflect the contemporary biochemists' hope to find out a new active substance something like cyclic AMP. The interesting finding that EDTA mimicked the action of the natural relaxing factor (Bozler, 1954; Watanabe, 1955) extended substantial help in reaching the relevant conclusion. Further progress of the research that eventually established the Ca^{2+} concept of today can be found in already published review articles (cf. Weber, 1964;

Ebashi and Endo, 1968; Ebashi et al., 1969; Ebashi, 1993).

Figure 25.1, in which emphasis is laid on the virtual absence of Ca^{2+} in the cytoplasm (Ebashi, 1968), is now accepted as the general feature of all kinds of cells, including prokaryotes (Gangola and Rosen, 1987). It is worthy of note, however, that no evidence for the presence of Ca^{2+} regulation has been shown in the prokaryote. In other words, the prokaryote has remained in its birth state for nearly 4×10^9 years so far as Ca^{2+} regulation is concerned. In the meantime, the eukaryote, provided with a Ca^{2+}-regulatory system, has achieved a rapid evolution since its appearance. In this way, the skeletal muscle of mammalians has obtained the most elegant device for Ca^{2+} regulation, the details of which will be described in the following chapters.

The guiding principle of this chapter is to present the most fundamental but still attractive aspects of Ca^{2+} regulation of muscle contraction in a lucid way. Consequently, the description will concentrate its focus on the mechanism in vertebrate skeletal muscle. This tendency is particularly pronounced in the part of regulatory mechanisms (sections "Troponin" and "Nontroponin Systems (Myosin-Related Regulation)"), disregarding some important matters in other kinds of muscle. On the other hand, the processes to link excitation with contraction (section "Excitation–Contraction Coupling" inseparably entangled in the intricate mechanisms of the cell. As a result, considerable attention will be paid to the event in cardiac and smooth muscle.

Excitation–Contraction Coupling

Ca^{2+} as the Intracellular Messenger

As described in the previous section, Ca^{2+} is now accepted as one of the most crucial factors for controlling or regulating intracellular processes of almost all kinds of cells. This was first established around 1960 in skeletal muscle contraction, where Ca^{2+} acts as the messenger in 'excitation–contraction coupling" (ECC), the link between excitation, namely electrical phenomenon at the surface membrane, and consequent activation of contractile elements (cf. Ebashi and Endo, 1968). Since then, similar roles of Ca^{2+} were demonstrated not only in cardiac and smooth muscle, but also in a vast number of other cells that are capable of responding to stimuli.

To utilize Ca^{2+} in this manner, cells normally keep cytoplasmic Ca^{2+} concentration at a very low level, such as 0.1 μM, less than 1/10,000 of that in the extracellular fluid (Figure 25.1). Inside the cells, there are tubular or vesicular organelles for storing Ca^{2+}, in which Ca^{2+} is usually heavily accumulated; although most Ca^{2+} is loosely bound inside the lumen of the store, free Ca^{2+} concentration in the store is of the order of that in the extracellular fluid.

Upon stimulation of the cells, Ca^{2+} can be mobilized either from the extracellular fluid or from the intracellular store. Both the cytoplasmic membrane and the membrane of the Ca^{2+} store have Ca^{2+} channels, opening of which causes Ca^{2+} mobilization, a passive movement of Ca^{2+} down the electrochemical potential gradient. When the cellular responses cease, mobilized Ca^{2+} must be either extruded from the cell or reaccumulated by the store. In these processes, Ca^{2+} has to move against the gradient of electrochemical potential and, therefore, active transport is required.

While Ca^{2+} is the universal messenger, each cell utilizes it in its own way with a specific time course, intensity, and so on. For this purpose, each cell selects suitable Ca^{2+} source(s), extracellular fluid or intracellular Ca^{2+} store or both, and uses the most suitable molecular species of Ca^{2+} channel(s) and active transport systems.

In sharp contrast to most of other messengers that are enzymatically formed after the arrival of the stimulus to cells, Ca^{2+} pre-exists on the spot and is simply separated from the cytoplasm by membranes. Therefore, when rapid and massive mobilization of a messenger is required, Ca^{2+} is more suitable than other messengers, because a large amount of Ca^{2+} can quickly be supplied simply by opening Ca^{2+} channels. The fact that the role of Ca^{2+} as the messenger was first discovered in skeletal muscle is not without relations with this characteristic.

Another feature of Ca^{2+} as an intracellular messenger is the fact that there are many high-affinity Ca^{2+}-binding sites in the cytoplasm, which apparently retard diffusion of Ca^{2+} and tend to localize it. On the other hand, the Ca^{2+} store usually forms a network throughout the cell (Fig. 25.2), and Ca^{2+}-release channels of the store are activated by Ca^{2+} itself as described later (p. 583). This enables Ca^{2+} release from the store to propagate throughout the cell; this property is particularly important in smooth muscle (p. 585).

Excitation–Contraction Coupling in Skeletal Muscle

Role of T-tubules

The physiological activation of skeletal muscle must produce tension to bear a load and/or shortening against the load, with rapid enough transition from the relaxed to the contracted state.

The unitary structure of tension-generating apparatus of skeletal muscle is a single half-sarcomere, many thousands of which are connected in series in a muscle fiber (Fig. 25.2). Since a relaxed sarcomere cannot hold tension, all the sarcomeres in series should be activated simultaneously to attain effective tension. Action potential must propagate quickly enough over the entire length of fibers to serve as a signal for rapid simultaneous activation in the longitudinal direction.

Thousands of myofibrils arranged in parallel in a single fiber should also be simultaneously activated to obtain rapid and full activation of the whole fiber. As referred to in the previous section (p. 580), diffusion of any activating substance from the cell surface was too slow to cause the rapid transition (Hill, 1948). T-tubules are the very device for the rapid and uniform activation in the transverse direction. they form a network running in the transverse direction of a fiber and surrounding each myofibril throughout the whole-fiber cross section at a particular level of the sarcomere (Fig. 25.2). Since their lumen directly continues to the extracellular space (Endo, 1964; Huxley, 1964) and since there are Na channels in the T-tubule membrane (Costantin, 1970), action potential can propagate through the T-tubule from the surface deep into the fiber, securing a rapid simultaneous activation in the transverse direction. Indeed, electrical interruption of the T-tubule at its every orifice by glycerol treatment abolishes ECC (Gage and Eisenberg, 1969).

The sarcoplasmic reticulum (SR) is the Ca^{2+} store and its end is expanded, called terminal cisternae. The T-tubule is sandwiched between a pair of terminal cisternae in the longitudinal direction, forming the "triad"[(Fig. 25.2). The action potential or depolarization of the T-tubule membrane sends a signal to the terminal cisternae to open its Ca^{2+}-release channels, resulting in Ca^{2+} mobilization.

Ca^{2+}-Release Channels of the SR, the Ryanodine Receptor

The Ca^{2+} giving rise to contraction of skeletal muscle derives exclusively from intracellular Ca^{2+} store, the SR. Even if Ca^{2+} is thoroughly removed from the extracellular fluid, or Ca^{2+} influx through voltage-dependent Ca^{2+} channels is completely

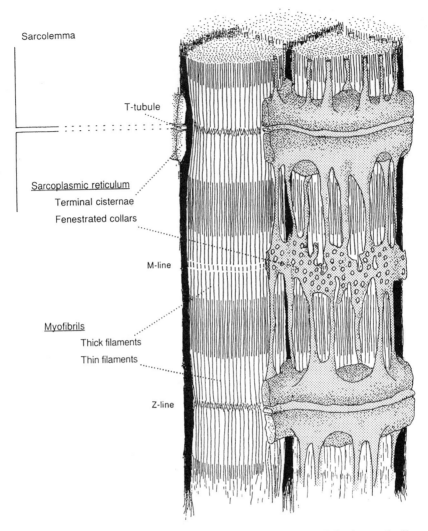

Figure 25.2 T-tubule, sarcoplasmic reticulum, and myofibrils of frog fast skeletal muscle fiber. Sarcoplasmic reticulum consists of terminal cisternae, longitudinal tubules, and fenestrated collars between two adjacent T-tubules which transversely surround myofibrils at the level of Z-lines. The lumen of T-tubules is open to the extracellular space. (From Peachey, 1965, but modified for this article, by permission of L. Peachey.)

inhibited, excitation can still evoke contraction of essentially the same magnitude. On the other hand, in dyspedic mice, in which the Ca^{2+}-release channel of the SR of skeletal muscle was genetically knocked out, ECC was abolished (Takeshima et al., 1994).

The Ca^{2+}-release channel protein of the SR is called the ryanodine receptor. For many years, ryanodine was known to induce contracture of skeletal muscle but to inhibit cardiac contraction (Jenden and Fairhust, 1969). The mechanism of this puzzling action was elucidated rather recently: ryanodine acts on the Ca^{2+}-release channel only when the channel is open, and fixes it in an open state (Fleischer et al., 1985). Utilizing the specific binding of ryanodine, the channel protein of rabbit muscle was isolated, purified (Imagawa et al., 1987; Inui et al, 1987; Lai et al., 1988), and sequenced (Takeshima et al., 1989). It contains about 5000 amino acid residues (molecular weight, about 565 kDa) and a membrane-spanning (channel) region towards its C-terminus. It forms a tetramer and shows quatrefoiled structure under the electron microscope. Skeletal muscle of avian and lower animals has two kinds of ryanodine receptors in equal amounts; one is similar to the mammalian skeletal muscle type and the other is brain type (Oyamada et al., 1994).

In 1968, a phenomenon called Ca^{2+}-induced Ca^{2+} release (CICR) was reported in the skeletal muscle SR independently by two groups (Endo et al., 1970; Ford and Podolsky, 1970), i.e., Ca^{2+} in the micromolar range causes Ca^{2+} to be released from the SR. The CICR is enhanced by ATP and related adenine compounds, as well as caffeine and other xanthine derivatives. It is inhibited by Mg^{2+}, ruthenium red, procaine, and some other local anesthetics (cf. Endo, 1985).

When incorporated into the lipid bilayer membrane, the purified ryanodine receptor protein shows Ca^{2+}-channel activity with all the properties of CICR (Lai et al., 1988). Further, the enhancing or inhibitory conditions for ryanodine binding to the ryanodine receptor are exactly the same as the enhancing or inhibitory conditions for CICR, respectively (Oyamada et al., 1993). These results clearly indicate that the ryanodine receptor is the CICR channel itself. However, CICR is not the mechanism of physiological Ca^{2+} release in skeletal muscle, that is to say, Ca^{2+} is not the physiological mediator to open Ca^{2+}-release channels of skeletal muscle in ECC, although CICR operates in contractures in a pathological state known as malignant hyperthermia or contractures induced by caffeine (cf. Endo, 1985).

Nevertheless, the ryanodine receptor appears to be identical with the physiological Ca^{2+}-release channel, because, as already mentioned, dyspedic mice lack ECC and also because morphologically the ryanodine receptor is very similar to the foot structure in the triad region that is considered to be the physiological Ca^{2+}-release channel (Franzini-Armstrong, 1970). Therefore, it must be assumed that the ryanodine receptor/Ca^{2+}-release channel has two different modes of opening: the physiological mode and the CICR mode.

The T–SR Coupling

How can an action potential in skeletal muscle fiber open the Ca^{2+}-release channel of the SR? Although action potential is accompanied by Na^+ influx, entry of Na^+ is not required, but depolarization of the T-tubule membrane is the necessary and sufficient condition for physiological Ca^{2+} release. Depolarization must be detected by a voltage-sensor molecule in the T-tubule membrane, which appears to be the same molecule as the L-type Ca^{2+}-channel sensitive to dihydropyridine (DHP) Ca^{2+} antagonists as well as D600 and diltiazem; these drugs tend to fix the voltage sensor in an inactivated state and abolish the effect of depolarization to cause Ca^{2+} release in the

same way as the effect of depolarization to cause Ca^{2+} release in the same way as they affect the Ca^{2+} channel. Dysgenic mice, which genetically lack the α_1-subunit of the DHP receptor, are devoid of both ECC and Ca^{2+} current, and upon introduction of cDNA of the α_1-subunit to the defective cell, both ECC and Ca^{2+} current are recovered (Tanabe et al., 1988). Thus, the DHP receptor is essential for ECC, and, as already mentioned, Ca^{2+} current is not necessary for the coupling. Consequently, it is almost certain that DHP receptor molecules serve as the voltage sensor molecules, which have been known to be situated just opposite to Ca^{2+}-release channels (Fig. 25.3).

The way in which information of the voltage sensor is conveyed to the Ca^{2+}-release channel of the SR is considered to be a protein–protein interaction between the DHP receptor and the ryanodine receptor, with the signal being transmitted either directly or through (an) unknown third protein(s). Unlike skeletal muscle, Ca^{2+} release from the SR in cardiac cells requires Ca^{2+} influx through voltage-dependent Ca^{2+} channels of the plasma membrane. Using dysgenic mice, Tanabe et al. (1990) showed that when the α_1-subunit of cardiac DHP receptor is expressed in dysgenic cells, ECC is recovered, but it is now the cardiac type and requires Ca^{2+} current. They then performed a chimeric study between skeletal and cardiac DHP receptors and found that cytoplasmic loop II/III (Fig. 25.3) is essential for determining skeletal or cardiac type of ECC. However, further detailed mechanism of the T–SR coupling is entirely unknown.

Excitation–Contraction Coupling in Cardiac and Smooth Muscle

Cardiac Muscle

Control of the contractile strength of skeletal muscle is exercised basically by motoneurons which adjust the number of fibers excited; each muscle fiber responds to the order of the nerve with essentially a fixed magnitude of contraction for each action potential, more or less in all-or-none manner. In cardiac muscle, however, all cardiac cells contract at every beat and, therefore, control of the strength of cardiac contraction to meet the physiological requirements must be made by the cardiac cells themselves, adjusting the magnitude of contraction of each cell. To do this, cardiac cells must widely and precisely determine the amount of Ca^{2+} mobilized by an action potential, unlike skeletal muscle cells.

Action potential of ventricular cardiac myocytes has a prominent plateau, during which a substantial amount of Ca^{2+} current flows into the cell. The magnitude of the Ca^{2+} influx is roughly in parallel with

A

B

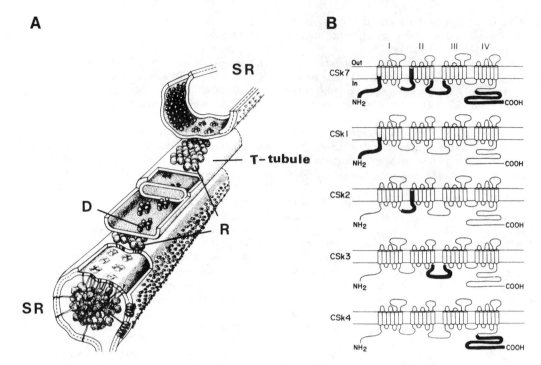

Figure 25.3 (A) Three-dimensional reconstruction of a triad, showing relative positions of tetramers of ryanodine receptors (R) and T-tubule tetrads (DHP receptors: D). (From Block et al., 1988, by the permission of Prof. C. Franzini-Armstrong.) (B) Schematic representation of the structures of chimeric DHP receptors composed of skeletal muscle (darkly shaded areas) and cardiac DHP-receptor (the remaining part). CSk7 and CSk3 showed skeletal muscle-type and CSk1, 2, and 4 cardiac-type T-SR coupling. (From Tanabe et al., 1990, by permission of T. Tanabe.)

the magnitude of the resulting Ca^{2+} transient and contractile strength. However, the amount of Ca^{2+} influx is not sufficient to cause contraction, and most of the Ca^{2+} acting on the contractile system of the mammalian heart comes from the SR (Mitchell et al., 1984).

A problem then arises of why there is a parallelism between Ca^{2+} influx and Ca^{2+} transient. CICR (p. 583) is the most plausible way to explain the parallelism: Ca^{2+} entering from the extracellular space during an action potential may cause Ca^{2+} release from the SR, i.e., the more Ca^{2+} that enters, the larger the amount of Ca^{2+} released, and hence the stronger contraction. CICR in cardiac muscle was demonstrated many years ago (Fabiato and Fabiato, 1972), and it has also been shown that cardiac SR has a similar ryanodine receptor/Ca^{2+}-release channel with CICR activity to that in skeletal muscle SR (Nakai et al., 1990; Ohtsu et al., 1990).

To accept the CICR theory of cardiac ECC, however, we have to clear up some complicated problems.

First, the CICR in cardiac skinned fibers in physiological condition is not sensitive enough to Ca^{2+}

(cf. Endo, 1985). Second, the rate of Ca^{2+} release should uniquely be determined by the intracellular Ca^{2+} concentration at the moment. The experimental results are rather at variance with this expectation (e.g., Ca^{2+} release, begun upon depolarization, is immediately stopped by repolarization, when Ca^{2+} concentration has already reached the level that should evoke sufficient Ca^{2+} release; Cannell et al., 1987). Third, why could Ca^{2+} release be graded in spite of the positive feedback nature of CICR?

These difficulties could be overcome if each Ca^{2+}-release channel (or each group of channels) behaves independently of the other channels, the channel activity being controlled by the local Ca^{2+} concentration that is different from the average of the whole cell. The local Ca^{2+} concentration is assumed to reach effective level only by Ca^{2+} influx through (a) voltage-dependent Ca^{2+} channel(s) of the sarcolemma situated just opposite to the Ca^{2+}-release channel. It is possible that Ca^{2+} released through (a) ryanodine receptor(s) cannot effectively support the open state of its own because channels spontaneously close (although they reopen) even in the presence of opening stimulus and Ca^{2+} rapidly dissipates from the small local space around the channel(s), which prevents regeneration of Ca^{2+} release (Stern, 1992; Wier et al., 1994). Indeed, unit activity of Ca^{2+}-release channels was detected with confocal microscopy as local stochastic changes in Ca^{2+} concentration, termed Ca^{2+} sparks; they either occur spontaneously (Cheng et al., 1993) or are evoked by a small Ca^{2+} influx under voltage-clamp condition (López-López et al., 1994).

Since the CICR theory is very attractive and there is no alternative, researchers in this field have thus made an elaborate effort to overcome the counter-evidence against this theory. However, simple questions raised from a pharmacological point of view, i.e., inhibitors of CICR, procaine, and adenine (which could effectively suppress CICR from the SR of cardiac skinned fibers) do not inhibit cardiac contraction, (cf. Endo, 1985) should be solved soon.

Smooth Muscle

In most smooth muscles, action potentials are Ca^{2+} spikes and bring about a sufficient amount of Ca^{2+} influx to cause contraction (it must be noted that in most visceral smooth muscles, this type of mechanism is playing a substantial role in physiological movement). In this sense, ECC in smooth muscle is entirely different from that in striated muscles, in which the main source of Ca^{2+} for contraction is the intracellular store, although in some smooth muscles, Ca^{2+} secondarily released from the store by the Ca^{2+} influx may contribute to contraction (Ganitkevich and Isenberg, 1992).

In agonist-induced contraction of smooth muscle, however, Ca^{2+} comes essentially from the intracellular store. The initial phase of contraction induced by the agonist is exactly the same irrespective of the presence or absence of extracellular Ca^{2+}. There are two kinds of Ca^{2+}-release channels of smooth muscle Ca^{2+} store: the ryanodine receptor, similar to those in the SR of striated muscles, and the channel activated by inositol trisphosphate (IP_3 receptor). In agonist-induced contraction, the lattern channels play the major role (cf. Berridge, 1993).

Iino (1990) found that opening of IP_3-receptor channel is enhanced by a sub-micromolar concentration of Ca^{2+} so that it apparently shows a property of CICR, although in the absence of IP_3 Ca^{2+} cannot open this Ca^{2+} channel. The IP_3-receptor channels have other similarities with the ryanodine receptor in several respects, although quantitatively they are different to a large extent. In accordance with these results, homologous sequences are demonstrated in the primary structures of both types of Ca^{2+}-release channels (Furuichi et al., 1989).

The apparent CICR nature of the IP_3 receptor plays a very important role in agonist-induced contraction. Iino et al. (1993) found that agonist-induced Ca^{2+} release in an individual smooth muscle cell occurs in an all-or-none-like manner. With increase in agonist concentration, the response suddenly jumps from no release to the maximum release. When the amount of Ca^{2+} in the store is decreased in the Ca^{2+}-free medium, agonist response suddenly disappears in spite of the fact that a substantial amount of Ca^{2+} still remains in the store as evidenced by caffeine-induced Ca^{2+} release. During the all-or-none-like Ca^{2+} release, the wave of Ca^{2+} release propagates throughout the cell by a positive feedback mechanism so that Ca^{2+} released at a site enhances Ca^{2+} release in the neighboring site (Fig. 25.4; Iino and Endo, 1992; Iino et al., 1993). As stated before, muscle should have a mechanism to activate a certain length of contractile elements more or less simultaneously for the effective tension production (see p. 581 for skeletal muscle). In smooth muscle, the propagation of Ca^{2+} release in this way is the very device for this purpose. Similar propagation of Ca^{2+} waves through IP_3 receptors has been demonstrated upon fertilization of an egg (Miyazaki et al., 1992).

The role of ryanodine receptor in smooth muscle, which is less abundantly distributed than IP_3 receptor in the store membrane (Yamazawa et al., 1992) is not known at present except the case of secondary Ca^{2+} release described above (Ganitkevich and Isenberg, 1992).

Excitation–Metabolism (Glycogenolysis) Coupling

Since skeletal muscle, particularly fast skeletal muscle, is a machine to make a rapid transition from relaxed state to contracted state and, therefore, consumes an enormous amount of ATP in a short time (roughly speaking, the ATP breakdown parallels the shortening speed), there must be a device to link the excitation or contraction with the process to retrieve lost ATP.

It has been well known that the onset of skeletal muscle contraction is accompanied with rapid glycogenolysis. This suggests that (a) step(s) in glycogenolytic reactions would be responsible for this link.

A large part of ADP derived from ATP breakdown is restored to ATP by creatine kinase, but a significant amount of ADP is converted into ATP and AMP via adenylate kinase. Since AMP has a strong activating effect on phosphorylase, a key enzyme in glycogenolysis, it was once considered as a plausible candidate for this link. However, it was soon realized that AMP was immediately converted into inactive IMP by adenylic acid deaminase.

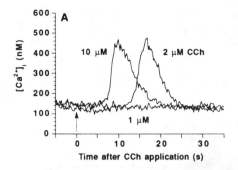

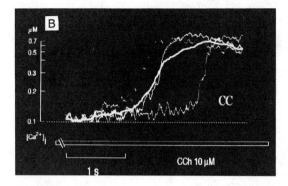

Figure 25.4 (a) All-or-none $[Ca^{2+}]_i$ rise due to car-bachol (CCh)-induced Ca^{2+} release. (B) Averaged time course of $[Ca^{2+}]_i$ of a whole cell (thick line) and within three regions in the cell (thin lines) showing spread of $[Ca^{2+}]_i$ wave within the cell. CC, critical $[Ca^{2+}]_i$ for the transition to rapid upstroke. (From Iino et al., 1993.)

In view of the success of the Ca story, it was rather natural to suppose that Ca^{2+} might play some role. Indeed, it was noticed that phosphorylase b kinase, which converts phosphorylase b, the inactive form, into phosphorylase a, the active form, requires micro-molar Ca^{2+} (Ozawa et al., 1967). This finding has revealed three important aspects. First, the motivator of the link is not the matter associated with contrac-tion, but the excitation itself that mobilizes Ca^{2+}. Second, since this enzyme is known to be activated by cyclic AMP, it becomes a question of which reac-tion, Ca^{2+}-dependent or cyclic AMP-dependent, would be of primary importance. The answer is that Ca^{2+} is more fundamental; in the absence of Ca^{2+}, cyclic AMP does not activate this enzyme (Ozawa and Ebashi, 1967). Third, there was an impression up to that time that Ca^{2+} was an agent specific only to muscle contraction. However, the finding that Ca^{2+} can exert its effect on a common metabolic process dissipated such a prejudice and has opened the way to the Ca^{2+} era of today (p. 579).

Troponin

General View

The investigation into the molecular mechanism of the Ca^{2+} regulation of muscle contraction began from the finding that superprecipitation of actomyo-

sin is sensitized to Ca^{2+} by a protein factor called native tropomyosin (Ebashi, 1963; cf. Ebashi and Endo, 1968). This factor was then found to be a complex of two proteins. One was tropomyosin, which had already been known, and the other was a new globular protein named troponin (Ebashi and Kodama, 1965; cf. Ebashi and Endo, 1968; Ebashi, 1974). The interaction of myosin and actin of verte-brate striated muscle in the presence of ATP exhibits its contractile response without any aid from other proteins (Fig. 25.5a). The fundamental role of the troponin–tropomyosin complex is to depress this contractile interaction. This inhibition is removed by Ca^{2+} through its binding to troponin and, as a result, the contraction takes place (Fig. 25.5a).

Troponin is distributed along the entire length of the thin filament at regular intervals of about 40 nm (Ohtsuki et al., 1967; Ohtuski, 1974). This periodicity is based on the end-to-end arrangement of fibrous tropomyosin molecules, which lie along the grooves of the double-stranded actin filament (Fig. 25.6) (cf. Ebashi et al., 1969). The effect of Ca^{2+} on troponin is transmitted to the actin filament in a process in which tropomyosin plays an essential role.

Troponin Components

Troponin is composed of three different components: troponins C, I, and T (Greaser and Gergely, 1971; cf.

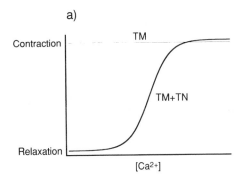

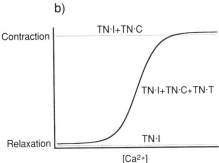

Figure 25.5 Ca^{2+}-regulatory mechanism of troponin and tropomyosin in muscle contraction. (a) Contractile interaction of myosin and actin in the presence of troponin (TN) and/or tropomyosin (TM). (b) Contractile interaction of myosin–actin–tropomyosin in the presence of troponin components (troponin, C, TN · C; troponin I, TN · I; troponin T, TN · T). (From Ohtsuki et al., 1986.)

Ebashi, 1974; cf. Perry, 1979; cf. Leavis and Gergely, 1984; cf. Ohtsuki et al., 1986). In the absence of troponins C, I, and T, the contractile interaction of myosin and actin–tropomyosin is activated irrespective of Ca^{2+} concentrations. Troponin I inhibits the contractile interaction and this inhibition is removed by troponin C even in the absence of Ca^{2+}. The inhibitory action of troponin I and the deinhibitory action of troponin C are integrated by troponin T into the Ca^{2+}-dependent processes in the thin filament (Fig. 25.5b).

Troponin I

The contractile interaction of myosin–actin is abolished by troponin I in the presence of tropomyosin, while in the absence of tropomyosin, it is hardly, or only slightly depressed by troponin I. The maximum inhibition is obtained when tropomyosin is present

in an amount equimolar to troponin I. This inhibition is neutralized by troponin C regardless of Ca^{2+} concentrations (Fig. 25.5b), unless troponin T is present.

Most of the inhibitory action of troponin I (composed of 178 amino acid residues) resides in a small cyanogen bromide fragment (residues 96–116) of troponin I from rabbit skeletal muscle, called CN4 fragment or inhibitory fragment (Syska et al., 1976). The amino acid sequence of the inhibitory region of troponin I is highly conserved among various muscles (Kobayashi et al., 1989).

Troponin C

Troponin C is the Ca^{2+}-binding component of troponin. The interaction of troponin C with troponin I is potentiated by Ca^{2+} (Potter and Gergely,

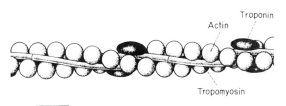

Figure 25.6 Structure of thin filament. (From Ebashi, 1974.)

10 nm

1974; Ohnishi et al., 1975). The binary complex of troponins C and I binds to actin–tropomyosin in the absence of Ca^{2+}, but it does not in the presence of Ca^{2+}. The complex by itself cannot confer the Ca^{2+}-sensitivity on the contractile interaction of myosin and actin-tropomyosin (Fig. 25.5b).

Troponin has four Ca^{2+}-binding sites; two with high affinity and two with low affinity (Ebashi et al., 1968). This is also the case for separated troponin C. These sites also show more or less affinities for Mg^{2+} (Potter and Gergely, 1975; Ogawa, 1985; Morimoto, 1991b). The Ca^{2+}-binding to troponin C in myofibrils and its relationship to the contraction are discussed in the section "Ca^{2+} Binding to Troponin C in Myofibrils and its Relation to Contraction."

Troponin T

Troponin T is the tropomyosin-binding component and modifies the interaction of tropomyosin and the complex of troponins C and I. The function of troponin T is localized in the C-terminal chymotryptic subfragment region (troponin T_2), which interacts with tropomyosin, troponin I, and troponin C. The N-terminal subfragment (troponin T_1) has a stronger affinity for tropomyosin but shows no function (Fig. 25.7) (Ohtsuki, 1979; cf. Ohtsuki, 1980).

Both the regulating action and the tropomyosin-binding activity of troponin T_2 are greatly reduced by the removal of the C-terminal 17 residues, while the interaction with troponin I or C is not affected. Thus, the tropomyosin-binding activity of troponin T_2 has primary importance from a physiological point of view (cf. Ohtsuki et al., 1986).

Ca^{2+} Regulation in the Thin Filament

In the absence of Ca^{2+}, the thin filament—the complex of actin, tropomyosin, and troponin—is prevented from interacting with myosin. This is due to the firm binding of troponin I to actin–tropomyosin; under this condition, troponin I binding to troponin C is weak. The Ca^{2+} potentiates the binding of troponin C to troponin I with the concurrent depression of the inhibitory actin of troponin I on actin–tropomyosin. This releases the contractile interaction of actin–tropomyosin with myosin. This Ca^{2+}-dependent competition in binding to the inhibitory region of troponin I between troponin C and actin–tropomyosin is the key process of Ca^{2+} regulation (fig. 25.8).

Even in the absence of Ca^{2+}, the complex of troponins C and I (without troponin T) does not inhibit the interaction of myosin and actin–tropomyosin, though it shows some affinity for actin–tropomyosin (see section on "Troponin C"). The inhibitory action

of the troponin I coupled with troponin C becomes apparent only when troponin T coexists (Figure 25.5b). The principal action of troponin T is to strengthen the affinity of the inhibitory region of troponin I for actin–tropomyosin (Fig. 25.8).

In this connection, it is interesting that calmodulin antagonizes the inhibitory action of troponin I Ca^{2+}-dependently in the absence of troponin T, but does not affect it in the presence of troponin T (Amphlett et al., 1976; Yamamoto, 1983; Morimoto and Ohtsuki, 1987). This somewhat puzzling effect is due to the weaker affinity of calmodulin than that of troponin C for troponin I; the interaction of calmodulin and troponin I is more effectively attenuated by troponin T than that of troponins C and I.

The filamentous structure of tropomyosin to cover a certain length of the actin alignment is essential for Ca^{2+} regulation, but it is not necessary that two adjoining tropomyosin molecules are conjugated tightly at their ends (Tawada et al., 1975).

As the mechanism involved in the Ca^{2+} regulation of the contractile interaction of myosin and actin, the "steric block hypothesis" has been proposed (Huxley, 1972; Wakabayashi et al., 1975). It is claimed that the position of tropomyosin in the absence of Ca^{2+} does not allow myosin access to actin. A time-resolved x-ray diffraction study demonstrated that tropomyosin movement precedes the myosin cross-bridge attachment to the thin filament or the force generation (Kress et al., 1986). Though very attractive, the final approval of this hypothesis still remains to be seen (Ishikawa and Wakabayashi, 1994).

Removal and Replacement of Troponin Components in Myofibrils

The Ca^{2+}-regulatory mechanisms concerning the roles of three troponin components have mostly been studied in vitro by using isolated proteins (cf. Ohtsuki et al., 1986). These studies could not exclude the possibility that they might not represent the actual roles of troponin components under physiological conditions. The crucial experiment to dispel such an apprehension is to remove and replace each component in myofibrils or skinned fibers and to compare the mode of Ca^{2+} regulation of preparations thus treated with that of original ones. A considerable success has been achieved along this line, which has opened a new frontier of troponin research.

Troponin C

Troponin C is removed completely from myofibrils (Morimoto and Ohtsuki, 1987, 1988) and skinned fibers (Morimoto et al., 1988; Morimoto and

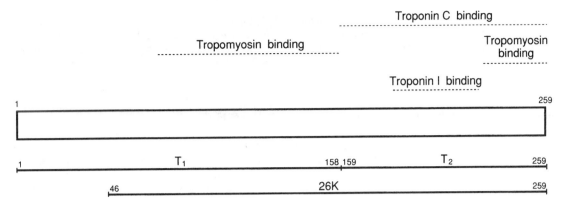

Figure 25.7 Schematic representation of amino acid sequence of troponin T from rabbit skeletal muscle. (From Ohtsuki et al., 1986, but modified for this article.)

Ohtsuki, 1994a) by treatment with CDTA (CyDTA), a strong chelator both for Ca^{2+} and Mg^{2+}. This treatment completely suppresses the Ca^{2+}-activation of myofibrillar ATPase or tension development of skinned fibers, and the subsequent troponin substitution recovers the Ca^{2+}-activated contraction.

Troponin C·I·T Complex

The troponin $C \cdot I \cdot T$ complex in myofibrils or skinned fibers is replaced by troponin T from outside in an excess amount at slightly acidic conditions (Hatakenaka and Ohtsuki, 1992; Shiraishi et al., 1992). As a consequence, troponins C and I are removed from the myofibrils or skinned fibers together with the original troponin T, which is replaced by added troponin T; the contraction is then activated regardless of Ca^{2+} concentrations (troponin T alone scarcely affects the contractile response of actomyosin to ATP) (Fig. 25.5b). Addition of troponin I to the myofibrils treated in

this way completely depresses their contraction, and addition of troponin C restores their Ca^{2+}-sensitivity.

Replacement Studies

Plenty of information has been accumulated from the replacement experiment of troponin C (cf. Moss, 1992; Ohtsuki, 1995). The studies with troponin C-depleted myofibrils or skinned fibers have shown that the well-known difference in the Sr^2-sensitivity (relative to Ca^{2+}-sensitivity), or the difference in the cooperativity of Ca^{2+}-activation between fast skeletal and cardiac muscles, is dependent only on the troponin C species (Morimoto and Ohtsuki, 1987, 1988, 1994a), in accord with the results of previous studies (Yamamoto, 1983).

Concerning the ability of Ca^{2+}-activation, troponin C is exchangeable among three kinds of striated muscles: i.e., fast skeletal, slow skeletal, and cardiac (Nakamura et al., 1994). However, the troponin C in the vertebrate myofibrils cannot be replaced by inver-

Figure 25.8 Ca^{2+}-dependent competition between troponin C and actin–tropomyosin for troponin I in the thin filament. The inhibitory region (dotted portion) of troponin I binds to actin–tropomyosin in the absence of Ca^{2+}. In the presence of Ca^{2+}, the inhibitory region of troponin I is dissociated from actin–tropomyosin and binds to Ca^{2+}-bound troponin C (see section on "Troponin, Ca^{2+} Regulations in the Thin Filament"). These interactions control the entire regulatory unit composed of 2 × 7 actin molecules, 2 tropomyosin molecules, and 2 troponin molecules. C, I and T, troponins C, I and T; TM, tropomyosin.

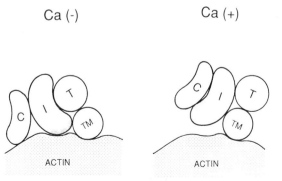

tebrate troponin C and vice versa. Crustacean myofibrils are activated only by crustacean troponin C, not by molluscan troponin C. The species specificity of troponin C in invertebrates thus seems to be more strict than invertebrates.

A number of troponin T isoforms, the variety of which is located in the N-terminal amino acid residues, are produced by the alternative splicing (Wilkinson et al., 1984; Briggs et al., 1987). Myofibrillar preparations, each containing only one type among various troponin T isoforms prepared by the troponin T exchange procedure, however, showed no difference from one another in the mode of Ca^{2+}-activation profiles (Hatakenaka, 1996). This is in accord with the incapability of troponin T_1 in regulation (see p. 588) and is consistent with a previous finding that the proteolytic subfragment of troponin T (26 k fragment), devoid of the N-terminal variable region (Fig. 25.7), shows the same Ca^{2+}-regulating action as native troponin T (Ohtsuki et al., 1984; Pan et al., 1991; Shiraishi et al., 1992).

The amino acid sequences of troponin I from vertebrate striated muscles are very similar (see section "Troponin I"), but cardiac troponin I has 20–30 additional residues at the N-terminal end. Phosphorylation of Ser residues in this N-terminal region by cAMP-dependent protein kinase depresses the interaction with troponin C and consequently decreases the Ca^{2+}-affinity of troponin C in myofibrils (Yamamoto, 1983). The roles of this N-terminal additional region have also been examined by using recombinant truncated troponin I that lacks N-terminal residues and troponin T-treated myofibrillar preparations (Guo et al., 1994; Wattanapermpool et al., 1995). The troponin exchange procedure has also revealed that troponin I and/or troponin C are/is responsible for different pH-sensitivities of fast and slow types of chicken skeletal muscles (Kawashima et al., 1995) and of bovine cardiac and rabbit skeletal muscles (Ball et al., 1994). A definite increase in the Ca^{2+}-sensitivity of contraction has been found in the skinned filters hybridized with two troponin T mutants of hypertrophic cardiomyopathy (Morimoto et al., 1988).

It should be noted that troponins I and C are removed from cardiac skinned muscle preparations by vanadate treatment (Strauss et al., 1992), though this procedure is applicable only to a particular species.

Ca^{2+} Binding to Troponin C in Myofibrils and its Relation to Contraction

Ca^{2+} binding to isolated troponin C does not simply reflect the situation in the myofibrils, because the Ca^{2+}-affinity of troponin C is often altered if it forms complexes with other thin filament proteins

(cf. Leavis and Gergely, 1984; cf. Ebashi and Ogawa, 1988). Thus, it is necessary to measure the Ca^{2+} binding specific to troponin C within the myofibrillar lattice in the presence of ATP. The measurement of Ca^{2+} binding of the whole myofibrils, however, involves the Ca^{2+} binding to other Ca^{2+}-binding proteins, such as myosin. This difficulty is overcome by using troponin C-depleted myofibrillar preparations (see section "Troponin, Replacement Studies").

The Ca^{2+}-binding to troponin C in fast skeletal muscle myofibrils thus determined is composed of four Ca^{2+}-binding sites, i.e., two with high affinities and two with low affinities (Morimoto and Ohtsuki, 1989), in accordance with the results of isolated troponin C (section "Troponin C"). The Ca^{2+}-activation of the myofibrillar contraction is caused by the Ca^{2+} binding to the low-affinity sites of troponin C. Among the two low-affinity sites, the second site (II) from the N-terminal is crucial according to the analysis with the probe fluorescence (Morimoto, 1991a).

The Ca^{2+}-binding to the low-affinity site of troponin C is potentiated cooperatively by the interaction of myosin with actin in the presence of ATP, but it is potentiated noncooperatively by the rigor interaction of myosin and actin in the absence of ATP (Morimoto, 1991a). the contractile interaction of myosin with actin, once activated by Ca^{2+} binding to the low-affinity site of troponin C, potentiates, in turn, the affinity of troponin C for Ca^{2+} through a certain feedback mechanism, which gives rise to the cooperative Ca^{2+} regulation of contraction in skeletal muscle. The same feedback mechanism is also present, though weakly, in the Ca^{2+}-activated myofibrillar contraction of cardiac muscle (Morimoto and Ohtsuki, 1994b).

Nontroponin Systems (Myosin-Related Regulation)

In the section on "Troponins", troponin was described as if it were the model of the Ca^{2+}-regulating system. It is true that troponin is the most advanced device, but this does not mean that it should represent the regulatory systems. The remarkable characteristic of the troponin system is its association with the actin filament. Consequently, the troponin system is often called the "actin-linked system." This is in sharp contrast with other systems described below, in which myosin plays more or less substantial roles in regulatory processes (Table 25.1). Tropomyosin is an inseparable partner of troponin, but it is not essential for regulatory systems, though it plays a role in some cases.

Table 24.1 Ca^{2+} Regulation in Different Actomyosin Systems

Type of Ca^{2+} Regulation	Proteins Involved in Ca^{2+} Regulation			Mode of Ca^{2+} Regulation
	Ca^{2+}-Receptor	Essential Proteins	Contractile Protein Responsible for Regulation	
Troponin-regulated	Troponin (troponin C)	Tropomyosin	Actin	Deinhibition
Myosin-linked	Myosin (ELC)[a]	RLC	Myosin	(Deinhibition?)
Smooth muscle[a]	Calmodulin	MLCK-RLC	Myosin	Activation
Slime mold	Myosin (ELC)		Myosin	Inhibition

ELC, essential light chain of myosin; RLC, regulatory light chain of myosin; MLCK, myosin light chain kinase.

[a]See the text.

There are three kinds of myosin-dependent regulatory systems. The first one is generally called the "myosin-linked system," and is widely distributed in the animal kingdom (Kendrick-Jones et al., 1970). The Ca^{2+}-receptive site of this system has not yet been identified; perhaps it resides in some small domain formed by the essential light chain (ELC) and a part of the heavy chain (Fromherz and Szent-Györgyi, 1995). Its regulatory light chain (RLC) is easily removed by EDTA treatment (often called EDTA-light chain). This removal deprives myosin of its ability to bind Ca^{2+}, and its contractile interaction with actin becomes independent of Ca^{2+} concentrations. The myosin molecule thus treated restores its function of recombining the light chain (Szent-Györgyi et al., 1973). Because of this interesting property, this light chain has been studied extensively from various points of view.

First, there was an impression that troponin would be the device of higher animals, including higher protostomias such as arthropods. However, troponin has now been isolated from the muscles of nematodes (Endo and Obinata, 1981) and even molluscs (Ojima and Nishita, 1986), from which the myosin-linked system was originally discovered (in these lower animals, troponin is not simply a de-depressor but acts partly as an activator). Perhaps troponin is distributed more widely than is supposed at present. On the other hand, the myosin-linked system has not been recognized in higher animals. The distribution of both systems in the animal kingdom is thus an interesting evolutionary problem.

Second, there is the regulatory system of smooth muscle. In sharp contrast to skeletal muscle, myosin and actin of smooth muscle cannot make a contractile response to ATP by themselves and need the support of some activating system. According to widely accepted views, the most crucial support is

rendered by the phosphorylation of the regulatory light chain. This process is catalyzed by a calmodulin-dependent kinase (myosin light chain kinase, MLCK) (cf. Harshorne, 1987). The Ca^{2+} regulation is thus carried out by a nonmember of the contractile system (there is a minor opinion that MLCK activates the contractile process through its structural effect on actin, something like an activating type of troponin; cf. Ebashi and Kuwayama, 1994). Dephosphoryla- tion of once phosphorylated light chain is a necessary condition for relaxation, but not a sufficient one. In addition to this mechanism, there have been reports of a few systems that modify the contractile processes through the actin filament (cf. Kohama and Saida, 1995). The activity of smooth muscle cells is profoundly influenced by various extrinsic humoral agents and the mode of signal transduction is very complicated. These make the situation in this muscle far more complex than that in skeletal and cardiac muscle.

Third, there is the regulatory system in the slime mold (*Physarum polycephalum*). The vigorous protoplasmic streaming of its plasmodial type and the movement of its ameboid type are both carried out by the actin–myosin–ATP interaction. Phosporylation of myosin heavy chain is a prerequisite for its active state, but the phosporylation–dephosphylation cycle itself is not involved in ordinary movements, which are under the direct control of Ca^{2+} derived from intrinsic Ca^{2+} stores. The most characteristic point of this strange creature is that Ca^{2+} does not activate its streaming and ameboid movement, but depresses them. Responses of extracted actomyosin to ATP are also inhibited by Ca^{2+} (Kohama et al., 1980; cf. Kohama, 1987). This "Ca^{2+} inhibition" has so far not clearly been recognized in higher creatures, but it is worthy of further investigation because it will provide important information for the evolutionary development of Ca^{2+} regulation.

All three of these regulation systems are very attractive subjects, but each may require an independent article even for brief description of its outline; the reader is referred to relevant review articles on each system.

References

Amphlett, G. W., T. C. Vanaman, and S. V. Perry (1976) Effect of the troponin C-like protein from bovine brain on the Mg^{2+} stimulated ATPase of skeletal muscle actomyosin. *FEBS Lett.* 72: 163–168.

Ball, K. L., M. D. Johnson, and R. J. Solaro (1994) Isoform specific interactions of troponin I and troponin C determine pH sensitivity of myofibrillar Ca^{2+} activation. *Biochemistry* 33: 8464–8471.

Berridge, M. J. (1993) Inositol trisphosphate and calcium signalling. *Nature* 361; 315–325.

Block, B. A., T. Imagawa, K. P. Campbell, and C. Franzini-Armstrong (1988) Structural evidence for direct interaction between the molecular components of the transverse tubule sarcoplasmic reticulum junction in skeletal muscle. *J. Cell Biol.* 107: 2587–2600.

Bozler, E. (1954) Binding calcium and magnesium by the contractile elements. *J. Gen Physiol.* 38: 735–742.

Briggs, M. M., J. J.-C. Lin, and F. H. Schachat (1987) The extent of amino-terminal heterogeneity in rabbit fast skeletal muscle troponin T. *J. Muscle Res. Cell Motil.* 8: 1–12.

Cannell, M. B., J. R. Berlin, and W. J. Lederer (1987) Effect of membrane potential changes on the calcium transient in single rat cardiac muscle cells. *Science* 238: 1419–1423.

Cheng, H., W. J. Lederer, and M. B. Cannell (1993) Calcium sparks: elementary events underlying excitation–contraction coupling in heart muscle. *Science* 262: 740–743.

Costantin, L. L. (1970) The role of sodium current in the radial spread of contraction in frog muscle fibers. *J. Gen. Physiol.* 55: 703–715.

Ebashi, S. (1963) Third component participating in the superprecipitation of "Natural actomyosin". *Nature* 200: 1010–1012.

Ebashi, S. (1968) Molecular mechanism of muscle contraction. *J. Jpn. Med. Doctors Soc.* 59; 239–257.

Ebashi, S. (1974) Regulatory mechanism of muscle contraction with special reference to the Ca–troponin–tropomyosin system. *Essays Biochem.* 10: 1–36.

Ebashi, S. (1993) From the relaxing factor to troponin. *Biomed. Res.* 14: (Suppl. 2): 1–7.

Ebashi, S. and M. Endo (1968) Calcium ions and muscle contraction. *Prog. Biophys. Mol. Biol.* 18: 123–183.

Ebashi, S. and K. Kodama (1965) A new protein factor promoting aggregation of tropomyosin. *J. Biochem.* 58: 107–108.

Ebashi, S. and H. Kuwayama (1994) Is phosphorylation the main physiological action of myosin light chain kinase? *Can. J. Physiol. Pharmacol.* 72: 1377–1379.

Ebashi, S. and Y. Ogawa (1988) Troponin C and calmodulin as calcium receptors: mode of action and sensitivity to drugs. In *Handbook of Experimental Pharmacology*, Vol. 83 (Baker, P. F., ed.). Springer-Verlag, Berlin pp. 31–56.

Ebashi, S., A. Kodama, and F. Ebashi (1968) Troponin, I. Preparation and physiological function. *J. Biochem.* 64: 465–477.

Ebashi, S., M. Endo, and I. Ohtsuki (1969) Control of muscle contraction. *Q. Rev. Biophys.* 2: 351–384.

Endo, M. (1964) Entry of a dye into the sarcotubular system of muscle. *Nature* 202: 1115–1116.

Endo, M. (1985) Calcium release from sarcoplasmic reticulum. *Curr. Top. Membr. Transp.* 25: 181–230.

Endo, M., M. Tanaka, and Y. Ogawa (1970) Calcium induced release of calcium from the sarcoplasmic reticulum of skinned skeletal muscle fibres. *Nature* 228: 34–36.

Endo, T. and T. Obinata (1981) Troponin and its components from ascidian smooth muscle. *J. Biochem.* 89: 1599–1608.

Fabiato, A. and F. Fabiato (1972) Excitation–contraction coupling of isolated cardiac fibers with disrupted or closed sarcolemmas. Calcium-dependent cyclic and tonic contractions. *Circ. Res.* 31: 293–307.

Fleischer, S. E., E. M. Ogunbunmi, M. C. Dixon, and E. A. Fleer (1985) Localization of Ca^{2+} release channels with ryanodine in junctional terminal cisternae of sarcoplasmic reticulum of fast skeletal muscle. *Proc. Natl. Acad. Sci. USA* 82: 7256–7259.

Ford, L. E. and R. J. Podolsky (1970) Regenerative calcium release within muscle cells. *Science* 167: 58–59.

Franzini-Armstrong, C. (1970) Studies on the triad. I. Structure of the junction of frog twitch fibers. *J. Cell Biol.* 47: 488–499.

Fromherz, S. and A. G. Szent-Györgyi (1995) Role of essential light chain EF hand domains in calcium binding and regulation of scallop myosin. *Proc. Natl. Acad. Sci. USA* 93: 7652–7656.

Furuichi, T. S., Yoshikawa, A. Miyawaki, K. Wada, N. Maeda, and K. Mikoshiba (1989) Primary structure and functional expression of the inositol 1,4,5-trisphosphate-binding protein P_{400}. *Nature* 342: 32–38.

Gage, P. W. and R. S. Eisenberg (1969) Action potentials, after potentials, and excitation-contraction coupling in frog sartorious fibers without transverse tubules. *J. Gen. Physiol.* 53: 298–310.

Gangola, P. and B. P. Rosen (1987) Maintenance of intracellular calcium in *Esherichia coli*. *J. Biol. Chem.* 262: 12750–12754.

Ganitkevich, V. Ya. and G. Isenberg (1992) Contribution of Ca^{2+} induced Ca^{2+} release to the $[Ca^{2+}]_i$ transients in myocytes from guinea-pig urinary bladder. *J. Physiol.* 458: 119–137.

Greaser, M. L. and J. Gergely (1971) Reconstitution of troponin activity from three protein components. *J. Biol. Chem.* 246: 4226–4233.

Guo, X., J. Wattanapermpool, K. A. Palmiter, A. M. Murphy, and R. J. Solaro (1994) Mutagenesis of cardiac troponin I: role of the unique NH_2-terminal peptide in myofilament activation. *J. Biol. Chem.* 269: 15210–15216.

Hartshorne, D. J. (1997) Biochemistry of the contractile proteins in smooth muscle. In *Physiology of the Gastrointestinal Tract*, Vol. 1, (Johnson, L. R. ed.). Raven Press, New York, pp. 423–482.

Hatakenaka, M. (1996) The Ca^{2+}-activation profile of rabbit fast skeletal myofibrils is not affected by troponin T isoforms. *Biomed. Res.* 17: 95–100.

Hatakenaka, M. and I. Ohtsuki (1992) Effect of removal and reconstitution of troponins C and I on the Ca^{2+}-activated tension development of single glycerinated rabbit skeletal muscle fibers. *Eur. J. Biochem.* 205: 985–993.

Heilbrunn, L. V. (1937) *An Outline of General Physiology.* W. B. Saunders, Philadelphia.

Heilbrunn, L. V. (1940) The action of calcium on muscle protoplasm. *Physiol. Zool.* 13: 88–94.

Heilbrunn, L. V. and F. J. Wiercinski (1947) The action of various cations on muscle protoplasm. *J. Cell. Comp. Physiol.* 29: 15–32.

Hill, A. V. (1948) On the time required for diffusion and its relation to processes in muscle. *Proc. R. Soc. (London)* Ser. B 135: 446–453.

Huxley, H. E. (1964) Evidence for continuity between the central elements of the triads and extracellular space in frog sartorious muscle. *Nature* 202: 1067–1071.

Huxley, H. E. (1972) Structural changes in the actin- and myosin-containing filaments during contraction. *Cold Spring Harbor Symp. Quant. Biol.* 37: 361–376.

Iino, M. (1990) Biphasic Ca^{2+} dependence of inositol 1,4,5-trisphosphate-induced Ca release in smooth muscle cells of the guinea pig taenia caeci. *J. Gen. Physiol.* 95: 1103–1122.

Iino, M. and M. Endo (1992) Calcium-dependent immediate feedback control of inositol 1,4,5-trisphosphate-induced Ca^{2+} release. *Nature* 360: 76–78.

Iino, M., T. Yamazawa, Y. Miyashita, M. Endo, and H. Kasai (1993) Critical intracellular Ca^{2+} concentration for all-or-none Ca^{2+} spiking in single smooth muscle cells. *EMBO J.* 12: 5287–5291.

Imagawa, T., J. S. Smith, R. Coronado, and K. P. Campbell (1987) Purified ryanodine receptor from skeletal muscle sarcoplasmic reticulum is the Ca^{2+} permeable pore of the calcium release channel. *J. Biolo. Chem.* 262: 16636–16643.

Inui, M., A. Saito, and S. Fleischer (1987) Purification of the ryanodine receptor and identity with feet structures of junctional terminal cisternae of sarcoplasmic reticulum from fast skeletal muscle. *J. Biol. Chem.* 262: 1740–1747.

Ishikawa, T. and T. Wakabayashi (1994) Calcium induced change in three-dimensional structure of thin filaments of rabbit skeletal muscle as revealed by cryo-electron microscopy. *Biochem. Biophys. Res. Commun.* 203: 951–958.

Jenden, D. J. and A. S. Fairhurst (1969) The pharmacology of ryanodine. *Pharmacol. Rev.* 21: 1–25.

Kamada, T. and H. Konosita (1943) Disturbances initiated from naked surface of muscle protoplasm. *Jpn. J. Zool.* 10: 469–493.

Kawashima, A., S. Morimoto, A. Suzuki, F. Shiraishi, and I. Ohtsuki (1995) Troponin isoform dependent pH dependence of Ca^{2+}-activated myofibrillar ATPase activity of avian slow/fast skeletal muscles. *Biochem. Biophys. Res. Commun.* 207: 585–592, 1995.

Keil, E. M. and E. J. M. Sichel (1936) The injection of aqueous solutions, including acetylcholine, into the isolated muscle fiber. *Biol. Bull.* 71: 402.

Kendrick-Jones, J., W. Lehman, and A. G. Szent-Györgyi (1970) Regulation in molluscan muscles. *J. Mol. Biol.* 54: 313–326.

Kobayashi, T., T. Takagi, K. Konishi, and J. A. Cox (1989) Amino acid sequence of crayfish troponin I. *J. Biol. Chem.* 264: 1551–1557.

Kohama, K. (1987) Ca-inhibitory myosins: their structures and function. *Adv. Biophys.*, 23: 149–182.

Kohama, K. and K. Saida (1995) *Smooth Muscle Contraction, New Regulatory Modes* (Kohama, K. and Saida K. eds). Japanese Scientific Society Press, Tokyo and Karger, Basel.

Kohama, K., K. Kobayashi, and S. Mitani (1980) Effects of Ca ion and ADP on superprecipitation of myosin B from slime mold, *Physarum policephalum. Proc. Jpn. Acad.* 56B: 591–596.

Kress, M., H. E. Huxley, A. R. Faruqi, and J. Hendrix (1986) Structural changes during activation of frog muscle studied by time-resolved X-ray diffraction. *J. Mol. Biol.* 188: 325–342.

Kumagai, H., S. Ebashi, and F. Takeda (1955) Essential relaxing factor in muscle other than myokinase and creatine phosphokinase. *Nature* 176: 166–168.

Lai, F. A., H. P. Erickson, E. Rousseau, Q.-Y. Liu, and G. Meissner (1988) Purification and reconstitution of the calcium release channel from skeletal muscle. *Nature* 331: 315–319.

Leavis P. C. and J. Gergely (1984) Thin filament proteins and thin filament-linked regulation of vertebrate muscle contraction. *CRC Crit. Rev. Biochem.* 16: 235–305.

López-López, J. R., P. S. Shacklock, C. W. Balke, and W. G. Wier (1994) Local stochastic release of Ca^{2+} in voltage-clamped rat heart cells: visualization with confocal microscopy. *J. Physiol.* 480: 21–29.

Marsh, B. B. (1951) A factor modifying muscle fibre synaeresis. *Nature* 167: 1065–1066.

Mitchell, M. R., T. Powell, D. A. Terrar, and V. W. Twist (1984) Ryanodine prolongs Ca-currents while suppressing contraction in rat ventricular muscle cells. *Br. J. Pharmacol.* 81: 13–15.

Miyazaki, S., M. Yuzaki, K. Nakada, H. Shirakawa, S. Nakanishi, S. Nakade, and K. Mikoshiba (1992) Block of Ca^{2+} wave and Ca^{2+} oscillation by antibody to the inositol 1,4,5-trisphosphate receptor in fertilized hamster eggs. *Science* 257: 251–255.

Morimoto, S. (1991a) Effect of myosin cross-bridge interaction with actin on the Ca^{2+}-binding properties of troponin C in fast skeletal myofibrils. *J. Biochem.* 109: 120–126.

Morimoto, S. (1991b) Effect of Mg^{2+} on the Ca^{2+}-binding to troponin C in rabbit fast skeletal myofibrils. *Biochim. Biophys. Acta* 1073: 336–340.

Morimoto, S. and I. Ohtsuki (1987) Ca^{2+}- and Sr^{2+}-sensitivity of the ATPase activity of rabbit skeletal myofibrils: effect of the complete substitution of troponin C with cardiac troponin C, calmodulin, and parvalbumins. *J. Biochem.* 101: 291–301.

Morimoto, S. and I. Ohtsuki (1988) Effect of substitution of troponin C in cardiac myofibrils with skeletal troponin C or calmodulin on the Ca^{2+}- and Sr^{2+}-sensitive ATPase activity. *J. Biochem.* 104: 149–154.

Morimoto, S. and I. Ohtsuki (1989) Ca^{2+} binding to skeletal muscle troponin C in skeletal and cardiac myofibrils. *J. Biochem.* 105: 435–439.

Morimoto, S. and I. Ohtsuki (1994a) Role of troponin C in determining the Ca^{2+}-sensitivity and cooperativity of the tension development in rabbit skeletal and cardiac muscles. *J. Biochem.* 115; 144–146.

Morimoto, S. and I. Ohtsuki (1994b) Ca^{2+} binding to cardiac troponin C in the myofilament lattice and its relation to the myofibrillar ATPase activity. *Eur. J. Biochem.* 226: 597–602.

Morimoto, S., T. Fujiwara, and I. Ohtsuki (1988) Restoration of Ca^{2+}-activated tension of CDTA-treated single skeletal muscle fibres by troponin C. *J. Biochem.* 104: 873–874.

Morimoto, S., F. Yanaga, R. Minakami, and I. Ohtsuki (1998) Ca^{2+}-sensitizing effects of the mutations at Ile-79 and Arg-92 in hypertrophic cardiomyopathy. *Am. J. Physiol.* 275: C200–C207.

Moss, R. L. (1992) Ca^{2+} regulation of mechanical properties of striated muscle: mechanistic studies using extraction and replacement of regulatory proteins. *Circ. Res.* 70: 865–884.

Nakai, J., T. Imagawa, Y. Hakamata, M. Shigekawa, H. Takeshima, and S. Numa (1990) Primary structure and functional expression from cDNA of the cardiac ryanodine receptor/calcium release channel. *FEBS Lett.* 271: 169–177.

Nakamura, Y., F. Shiraishi, and I. Ohtsuki (1994) The effect of troponin C substitution on the Ca^{2+}-sensitive ATPase activity of vertebrate and invertebrate myofibrils by troponin Cs with various numbers of Ca^{2+}-binding sites. *Comp. Biochem. Physiol.* 108B: 121–133.

Ogawa, Y. (1985) Calcium binding to troponin C and troponin: effects of Mg^{2+}, ionic strength and pH. *J. Biochem.* 97: 1011–1023.

Ohnishi, S., K. Maruyama, and S. Ebashi (1975) Calcium-induced conformational changes and mutual interactions of troponin components as studied by spin labeling. *J. Biochem.* 78: 73–81.

Ohtsu, K., H. F. Willard, V. K. Kanna, F. Zorzato, N. M. Green, and D. H. MacLennan (1990) Molecular cloning of cDNA encoding the Ca^{2+} release channel (ryanodine receptor) of rabbit cardiac muscle

sarcoplasmic reticulum. *J. Biol. Chem.* 265: 13472–13483.

Ohtsuki, I. (1974) Localization of troponin in thin filament and tropomyosin paracrystal. *J. Biochem.* 75: 753–765.

Ohtsuki, I. (1979) Molecular arrangement of troponin-T in the thin filament. *J. Biochem.* 86: 491–497.

Ohtsuki, I. (1980) Functional organization of the troponin–tropomyosin system. In *Muscle Contraction; Its Regulatory Mechanism* (Ebashi, S. et al., eds). Japanese Scientific Society Press, Tokyo and Springer-Verlag, Berlin, pp. 237–240.

Ohtsuki, I. (1995) Troponin components and calcium-ion regulation of myofibrillar contraction in skeletal muscle. In *Ca as Cell Signal.* Igaku-syoin, Tokyo, pp. 36–42.

Ohtsuki, I., T. Masaki, Y. Nonomura, and S. Ebashi (1967) periodic distribution of troponin along the thin filament. *J. Biochem.* 61: 817–819.

Ohtsuki, I., F. Shiraishi, N. Suenaga, T. Miyata, and M. Tanokura (1984) A 26 K fragment of troponin T from rabbit skeletal muscle. *J. Biochem.* 95: 1337–1342.

Ohtsuki, I., K. Maruyama, and S. Ebashi (1986) Regulatory and cytoskeletal proteins of vertebrate skeletal muscle. *Adv. protein Chem.* 38: 1–67.

Ojima, T. and K. Nishita (1986) Troponin from akazara scallop striated adductor muscles. *J. Biol. Chem.* 261: 16749–16754.

Oyamada, H., M. Iino, and M. Endo (1993) Effects of ryanodine on the properties of Ca^{2+} release from the sarcoplasmic reticulum in skinned skeletal muscle fibres of the frog. *J. Physiol.* 470: 335–348.

Oyamada, H., T. Murayama, T. Takagi, M. Iino, N. Iwabe, T. Miyata, Y. Ogawa, and M. Endo (1994) Primary structure and distribution of ryanodine-binding protein isoforms of the bullfrog skeletal muscle. *J. Biol. Chem.* 269: 17206-17214.

Ozawa, E. and S. Ebashi (1967) Requirement of Ca ion for the stimulating effect of cyclic $3',5'$-AMP on muscle phosphorylase *b* kinase. *J. Biochem.* 62: 285–286.

Ozawa, E., K. Hosoi, and S. Ebashi (1967) Reversible stimulation of muscle phophorylase *b* kinase by low concentrations of calcium ions. *J. Biochem.* 61: 531–533.

Pan, B.-S., A. M. Gordon, and J. D. Potter (1991) Deletion of the first 45 NH_2-terminal residues of rabbit skeletal troponin T strength binding of troponin to immobilized tropomyosin. *J. Biol. Chem.* 266: 12432–12438.

Peachey, L. D. (1965) The sarcoplasmic reticulum and transverse tubules of the frog's sartorius. *J. Cell Biol.* 25: 209–231.

Perry, S. V. (1979) The regulation of contractile activity in muscle. *Biochem. Soc. Trans.* 7: 593–617.

Potter, J. D. and J. Gergely (1974) Troponin, tropomyosin, and actin interaction in the Ca^{2+} regulation of muscle contraction. *Biochemistry* 13: 2697–2703.

Potter, J. D. and J. Gergely (1975) The calcium and magnesium binding sites on troponin and their role in the regulation of myofibrillar adenosine triphosphatase. *J. Biol. Chem.* 250: 4628–4633.

Ringer, S. (1883) A further contribution regarding the influence of the different constituents of the blood on the contraction of the heart. *J. Physiol.* 4: 29–42.

Riner, S. (1886) A further contribution regarding the effect of minute quantities of inorganic salts on organised structures. *J. Physiol.* 7: 118–127.

Shiraishi, F., M. Kambara, and I. Ohtsuki (1992) Replacement of troponin components in myofibrils. *J. Biochem.* 111: 61–65.

Stern, M. D. (1992) Theory of excitation–contraction coupling in cardiac muscle. *Biophys. J.* 63: 497–517.

Stiles, P. G. (1901) On the rhythmic activity of the oesophagus and the influence upon it of various media. *Am. J. Physiol.* 5: 338–357.

Strauss, J. D., C. Zeugner, J. E. Van Eyk, C. Bletz, M. Troschka, and J. C. Ruegg (1992) Troponin replacement in permeabilized cardiac muscle. Reversible extraction of troponin I by incubation with vanadate. *FEBS Lett.* 310: 229–234.

Syska, H., J. M. Wilkinson, R. J. A. Grand, and S. V. Perry (1976). The relationship between biological activity and primary structure of troponin I from white skeletal muscle of rabbit. *Biochem. J.* 153: 375–387.

Szent-Györgyi, A. (1951) *Chemistry of Muscular Contraction*, 2nd edn. Academic Press, New York.

Szent-Györgyi, A. G., E. M. Szentkiralyi, and J. Kendrick-Jones (1973) The light chains of scallop myosin as regulatory subunits. *J. Mol. Biol.* 74: 179–203.

Takeshima, H., S. Nishimura, T. Matsumoto, H. Ishida, K. Kangawa, N. Minamino, H. Matsuo, M. Ueda, M. Hanaoka, T. Hirose, and S. Numa (1989) Primary structure and expression from complementary DNA of skeletal muscle of ryanodine receptor. *Nature* 339: 439–445.

Takeshima, H., M. Iino, H. Takekura, M. Nishi, J. Kuno, O. Minowa, H. Takano, and T. Noda (1994) Excitation–contraction uncoupling and muscular degeneration in mice lacking functional skeletal muscle ryanodine-receptor gene. *Nature* 369: 556–559.

Tanabe, T., K. G. Beam, J. A. Powell, and S. Numa (1988) Restoration of excitation–contraction coupling and slow calcium current in dysgenic muscle by dihydropyridine receptor complementary DNA. *Nature* 336: 134–139.

Tanabe, T., K. G. Beam, B. A. Adams, T. Niidome, and S. Numa (1990) Regions of the skeletal muscle dihydropyridine receptor critical for excitation–contraction coupling. *Nature* 346: 567–569.

Tawada, Y., H. Ohara, T. Ooi, and K. Tawada (1975) Non-polymerizable tropomyosin and control of the superprecipitation of actomyosin. *J. Biochem.* 78: 65–72.

Wakabayashi, T., H. E. Huxley, L. A. Amos, and A. Klug (1975) Three-dimensional image reconstruction of actin–tropomyosin complex and actin-tropomyosin-troponin T-troponin I complex. *J. Mol. Biol.* 93: 477–497.

Watanabe, S. (1955) Relaxing effects of EDTA on glycerol-treated muscle fibers. *Arch. Biochem. Biophys.* 54: 559–562.

Wattanapermpool, J., X. Guo, and R. J. Solaro (1995) The unique amino-terminal peptide of cardiac troponin I regulates myofibrillar activity only when it is phosphorylated. *J. Mol. Cell. Cardiol.* 27: 1383–1391.

Weber, A. (1964) Energized calcium transport and relaxing factors. *Current Topics in Bioenergetics* (Sanadi, D. R. ed.). Academic Press, New York, pp. 203–254.

Wier, C. W., T. M. Egan, J. López-López, and C. W. Balke (1994) Local control of excitation–contraction coupling in rat heart cells. *J. Physiol.* 474; 463–471.

Wilkinson, J. M., A. J. Moir, and M. D. Waterfield (1984) The expression of multiple forms of troponin T in chicken-fast-skeletal muscle may result from differential splicing of a single gene. *Eur. J. Biochem.* 143: 47–56.

Yamamoto, K. (1983) Sensitivity of actomyosin ATPase to calcium and strontium ions. Effect of hybrid troponins. *J. Biochem.* 93: 1061–1069.

Yamazawa, T., M. Iino, and M. Endo (1992) Presence of functionally different compartments of the Ca^{2+} store in single intestinal smooth muscle cells. *FEBS Lett.* 301: 181–184.

26

Calcium in Programmed Cell Death

David J. McConkey
Sten Orrenius

Apoptosis (programmed cell death) is a highly regulated process of selective cell deletion involved in development, normal cell turnover, hormone-induced tissue atrophy, cell-mediated immunity, tumor regression, and a growing number of pathological disorders (typified by AIDS and Alzheimer's disease) (Wyllie et al., 1980; Thompson, 1995). The response is characterized by stereotyped morphological alterations, including plasma and nuclear membrane blebbing, organelle relocalization and compaction, chromatin condensation, and the formation of membrane-enclosed structures termed "apoptotic bodies" that are extruded into the extracellular milieu (Wyllie et al., 1980; Savill et al., 1993). Uptake of apoptotic debris is carefully controlled, as apoptotic cells and bodies are specifically recognized and cleared by neighboring epithelial cells and professional phagocytic cells (macrophages) before their contents can be released into the extracellular milieu, thereby allowing for cell death to occur in the absence of inflammation (Savill et al., 1993).

Apoptosis has historically been characterized biochemically by endogenous endonuclease activation, resulting first in the production of domain-sized large (50–300 kilobase) DNA fragments (Filipski et al., 1990; Brown et al., 1993; Oberhammer et al., 1993) and subsequently in the generation of oligonucleosomal cleavage products commonly referred to as "DNA ladders" (Wyllie, 1980). It appears that DNA ladders are derived from the larger DNA fragments but that the two events may be mediated by different enzymatic activities than can be distinguished by their divalent cation requirements (Sun and Cohen, 1994). In addition to endonuclease activation, more recent work has demonstrated that a family of cysteine proteases homologous to the *Caenorhabditis elegans* cell death gene *ced-3* and human interleukin-1β-converting enzyme (ICE) are also critically involved in the response, as inhibitors of these enzymes block both endonuclease activation and cell death (Yuan et al., 1993; Fernandes-Alnemri et al., 1994, 1995; Kumar et al., 1994; Wang et al., 1994; Martin and Green, 1995; Nicholson et al., 1995; Tewari et al., 1995). The important substrates for these proteases remain largely unidentified but include poly(ADP-ribose) polymerase (PARP), the lamins, and a viral inhibitor of their activity (baculovirus p35) (Kaufmann et al., 1993; Lazebnik et al., 1993; Bump et al., 1995; Tewari et al., 1995; Xue and Horvitz, 1995).

At the molecular level, apoptosis is regulated by many familiar oncogenes (*bcl-2*, *myc*, *ras*, *abl*, *fos*) and tumor suppressor genes (*p53*, *Rb*) (Thompson, 1995). However, the family of polypeptides homologous to *bcl-2* appear collectively to be the most important. Regulation of apoptosis susceptibility is their only established function to date. One class of *bcl-2* homologs (including *bcl-2*, *bcl-x$_L$*, *mcl-1*, and several viral proteins) suppress apoptotic cell death, while another group (*bax*, *bcl-x$_s$* and *bak*) promote apoptosis sensitivity (Oltvai and Korsmeyer, 1994). Thus, overexpression of BCL-2 or BCL-X$_L$ blocks apoptosis induced by very diverse stimuli, including growth factor withdrawal, tumor necrosis factor, engagement of the Fas antigen, ionizing radiation, oncogenes such as *myc*, and chemotherapeutic agents (Thompson, 1995). Moreover, the apoptosis-regulatory functions of BCL-2 and its homologs are evolutionarily conserved, as the *C. elegans* cell death suppressor *ced-9* is a structural and functional homolog of human *bcl-2* (Hengartner et al., 1992; Hengartner and Horvitz, 1994). Together, these observations have suggested to many investigators that BCL-2 regulates a central biochemical signal in the pathway to cell death, the identity of which is still unknown. However, several strong candidates are emerging. Work from our laboratory and others over the past 5 years has demonstrated that alterations in intracellular Ca^{2+} homeostasis are commonly involved in promoting apoptosis, and more recent work suggests that one aspect of BCL-2 function involves preventing these alterations. Here, we will discuss the role of Ca^{2+} in regulating apoptosis and compare it with other general mediators of apoptotic cell death.

Calcium Alterations in Apoptosis

Early studies by Kaiser and Edelman (1977) demonstrated that glucocorticoid-stimulated apoptosis is associated with enhanced Ca^{2+} influx. We have since confirmed that glucocorticoids induce cytosolic Ca^{2+} elevations in thymocytes via Ca^{2+} influx across the plasma membrane (McConkey et al., 1989b), work that provided the first evidence that increases in intracellular Ca^{2+} might be involved in triggering apoptosis. This has received strong support by the recent observation by Snyder's group of an increased level of type 3 inositol 1,4,5-trisphosphate (InsP$_3$) receptor in lymphocytes undergoing apoptosis (Khan et al., 1996). However, intracellular Ca^{2+} storage sites also appear to be affected, as the Ca^{2+} pool located in the endoplasmic reticulum is depleted in a lymphoid cell line in response to glucocorticoid treatment (Lam et al., 1993), and a similar phenomenon has been documented in an interleukin 3 (IL-3)-dependent myeloid cell line undergoing apoptosis following IL-3 withdrawal (Baffy et al., 1993). Circumstantial evidence suggests that the mitochondrial Ca^{2+} pool may also be affected (Richter, 1993), as mitochondrial membrane potential drops very early during apoptosis (Petit et al., 1995; Zamzani et al., 1995a, 1995b), and it is well known that the maintenance of mitochondrial Ca^{2+} homeostasis is dependent upon mitochondrial membrane potential (Richter, 1993). Apoptosis in other systems also appears to involve elevations in the cytosolic Ca^{2+} concentration. For example, rapid, sustained Ca^{2+} increases precede the cytolysis of the targets of cytotoxic T-lymphocytes (Allbritton et al., 1988) and natural killer (NK) cells (McConkey et al., 1990). In developing T-lymphocytes, high-affinity engagement of the T-cell receptor induces apoptosis (McConkey et al., 1989a; Shi et al., 1989; Smith et al., 1989; Murphy et al., 1990) that involves a sustained Ca^{2+} elevation (McConkey et al., 1989a; Nakagama et al., 1992).

Both second messenger- and damage-mediated mechanisms can be involved in promoting Ca^{2+} increases in apoptotic cells. In an example of the former, T-cell receptor engagement on thymocytes leads to a sustained increase in the cytosolic Ca^{2+} concentration that involves protein tyrosine kinase activation, phosphorylation of the γ isoform of phospholipase C, phosphoinositide hydrolysis that leads to the production of (InsP$_3$), and mobilization of Ca^{2+} from the endoplasmic reticulum and extracellular milieu that promote cell death (McConkey et al., 1989a, 1994). Similarly, surface antigen receptor engagement on B-cells leads to Ca^{2+} increases that promote cell death (Tsubata et al., 1993; Yao and Scott, 1993; Parry et al., 1994; Norvell et al., 1995). Thus, in these examples of apoptosis, Ca^{2+} increases occur via a controlled, physiological mechanism that is also utilized in alternative responses, such as cellular activation leading to proliferation.

Work from our laboratories and others has revealed another mechanism that is involved in promoting sustained cytosolic Ca^{2+} increases in apoptotic cells. It is well known that cytosolic Ca^{2+} concentration is maintained at roughly 100 nM in resting cells, whereas the concentrations in the extracellular milieu, and the ER, are much higher (in the millimolar range). Early work on the biochemical mechanisms underlying the cytotoxocity of agents that generate reactive oxygen species in cells (oxidative stress) indicated that the Ca^{2+}-transport systems localized to the ER, mitochondria, and plasma membrane can be damaged by oxygen radicals (Orrenius et al., 1989). This leads to diffusion of Ca^{2+} down its concentration gradient, a disruption of intracellular Ca^{2+} homeostasis, and sustained Ca^{2+} increases. Oxidative stress is now know to be commonly involved in apoptosis (Lennon et al., 1990; Hockenbery et al., 1993; Buttke and Sandstrom, 1994; Dypbukt et al., 1994; Fang et al., 1995), and it is therefore possible that oxidative disruption of intracellular Ca^{2+} homeostasis is involved in these systems. In support of this idea, we have recently shown that the glucocorticoid-induced Ca^{2+} increase observed in thymocytes is blocked by antioxidants (Fernandez et al., 1995), and Kroemer's laboratory has presented evidence that oxidative stress leads to disruption of mitochondrial Ca^{2+} stores (Zamzani et al., 1995a).

Direct evidence that Ca^{2+} increases can mediate apoptotic endonuclease activation and cell death has been obtained from experiments with intracellular Ca^{2+} buffering agents and extracellular Ca^{2+} chelators. We (McConkey et al., 1989a, 1989b, 1990, 1994; Aw et al., 1990; Bellomo et al., 1992; Zhivotovsky et al., 1993) and others (Story et al., 1992; Robertson et al., 1993) have shown that these agents can inhibit both DNA fragmentation and death in apoptotic cells. The Ca^{2+}-dependent regulatory cofactor calmodulin may link these Ca^{2+} alterations to the effector machinery, as we and others have shown that calmodulin antagonists can interfere with apoptosis in some of these systems (McConkey et al., 1989b, Dowd et al., 1991) and increases in calmodulin expression are linked to apoptosis in glucocorticoid-treated thymoma cells (Dowd et al., 1991) and in prostatic epithelial cells following withdrawal of androgen (Furuya and Isaacs, 1993). Independent evidence for the involvement of Ca^{2+} influx in the triggering of apoptosis has come from studies with specific Ca^{2+}-channel blockers, which abrogate apoptosis in the regressing prostate following testosterone withdrawal (Martikainen and Isaacs, 1990) and in pancreatic β-cells treated with serum from patients with type I diabetes (Juntti-Berggren et al., 1993).

Other support for the involvement of Ca^{2+} in apoptosis comes from the observation that agents which directly mobilize Ca^{2+} can trigger apoptosis in diverse cell types. Early work by Kaiser and Edelman (1978) demonstrated that the cytolytic effects of glucocorticoids on lymphoid cells can be mimicked by treating the cells with Ca^{2+} ionophores. Subsequently, Wyllie et al. (1984) demonstrated that Ca^{2+} ionophores cause endonuclease activation, as well as many of the morphological changes that are typical of apoptosis in thymocytes. Calcium ionophores also trigger apoptosis in prostate tumor cells (Martikainen and Isaacs, 1990) and in nonmetastatic melanoma lines (D. J. McConkey, unpublished observations). Independent evidence for the general relevance of this mechanism has come from studies with the endoplasmic reticular Ca^{2+}-ATPase inhibitor thapsigargin, the product of the plant *Thapsa garganica*, that can also trigger all of the morphological and biochemical events of apoptosis in thymocytes (Jiang et al., 1994) and some other cell types (Kaneko and Tsukamoto, 1994; Choi et al., 1995; Levick et al., 1995).

A final argument for a central role for Ca^{2+} in regulating apoptosis comes from recent and ongoing work on the biochemical mechanisms of apoptosis suppression by the BCL-2 oncoprotein. The possible relationship between Ca^{2+} and BCL-2 was first suggested by work by Baffy et al. (1993), who showed that BCL-2 can block the depletion of the endoplasmic reticular Ca^{2+} pool in transfectants of an interleukin 3-dependent cell line (32D). Interestingly, these authors also demonstrated that constitutive levels of Ca^{2+} in mitochondria (measured following treatment with an uncoupler that promotes rapid and selective depletion of this intracellular Ca^{2+} store) were significantly lower in BCL-2-expressing cells compared with vector control transfectants, consistent with the notion that BCL-2 may also regulate Ca^{2+} compartmentalization in mitochondria. More recently, Lam et al. (1994) have shown that overexpression of BCL-2 interferes with thapsigargin-induced Ca^{2+} mobilization from the ER in the WEHI7.2 T lymphoma cell line, an effect that is associated with preservation cell viability. Precisely how BCL-2 regulates intracellular Ca^{2+} is still unclear, although given its colocalization with Ca^{2+}-transport sites in mitochondria, the ER, and the nuclear envelope (deJong et al., 1994), a direct effect of BCL-2 on Ca^{2+} channel(s) is possible. Alternatively, given the tight interrelationship between Ca^{2+} and oxidative stress, BCL-2 could be influencing Ca^{2+} homeostasis via effects on cellular redox status (Hockenbery et al., 1993; Kane et al., 1993).

Calcium Coupling to the Effector Pathway

An important aspect of ongoing research involves defining the biochemical consequences of Ca^{2+} mobilization in apoptotic cells, and at present there are two models to explain how these alterations might trigger apoptosis. In one, depletion of intracellular stores and possibly influx of Ca^{2+} across the plasma membrane promote a sustained Ca^{2+} increase that acts as a signal for apoptosis, perhaps, in part, by activating key catabolic enzymes that make up the effector machinery. In the second, it is not the Ca^{2+} increase but the emptying of intracellular Ca^{2+} stores that triggers apoptosis, perhaps by disrupting intracellular architecture and allowing key elements of the effector machinery to gain access to their substrates. These models are certainly not mutually exclusive. Evidence for both models will be presented below, but it should be emphasized at the outset that definitive proof for either one is lacking at present.

Possible Targets for Ca^{2+} Elevations

As illustrated schematically in Fig 26.1, there are many potential targets for Ca^{2+} signalling in apoptosis. Some of them are discussed below.

Signal Transduction Intermediates

Activation of Ca^{2+}-dependent protein kinases and/or phosphatases that lead to alterations in gene transcription represents one possible way that Ca^{2+} might regulate apoptosis. The most convincing support for this hypothesis has come from experiments with the immunosuppressant cyclosporin A, a compound that binds a family of cytosolic receptors termed cyclophilins and, in so doing, forms a composite molecular surface that binds to and inhibits a Ca^{2+}/calmodulin-dependent protein serine/threonine phosphatase, calcineurin (Liu et al., 1991). Studies by several independent laboratories have shown that cyclosporin A can block Ca^{2+}-dependent apoptosis in lymphoid model systems (Shi et al., 1989;

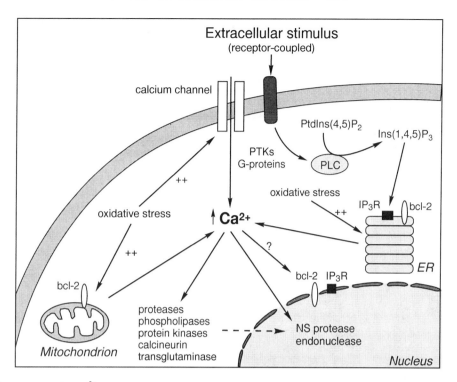

Figure 26.1 Targets for Ca^{2+} in the regulation of apoptosis. Extracellular agonists, calcium ionophores, or the endoplasmic reticular Ca^{2+}-ATPase antagonist thapsigargin are capable of triggering sustained Ca^{2+} increases that mediate apoptosis in diverse model systems. Alternatively, recent work suggests that oxidative stress can disrupt intracellular Ca^{2+} homeostasis, possibly via oxidation of critical sulfhydryls present in the Ca^{2+} translocases located in the plasma membrane, mitochondria, and ER. Of interest is the fact that members of the BCL-2 family of apoptosis suppressors are localized to several intracellular Ca^{2+} regulatory sites, including the mitochondria, ER, and nucleus. Calcium elevations most likely exert their effects via activation of both cytosolic and nuclear targets, the latter of which include a Ca^{2+}-dependent protease associated with the nuclear scaffold (NS protease) and the endonuclease. Abbreviations: $PtdIns(4,5)P_2$, phosphatidyl inositol-(4,5)-bisphosphate; $Ins(1,4,5)P_3$, phosphatidyl inositol-(1,4,5)-trisphosphate; IP_3R, receptor for $Ins(1,4,5)P_3$.

Amendola et al., 1994; Bonnefoy-Berard et al., 1994; Makrigiannis et al., 1994), indicating that calcineurin activation may be required for these responses. Our recent work confirms that cyclosporin A and FK506 block Ca^{2+}-stimulated apoptosis in T-cell hybridomas but they are without effect on Ca^{2+}-dependent apoptosis in immature rodent thymocytes in vitro (S. Jiang, S. C. Chow, and S. Orrenius, unpublished observations). Induction of the orphan steroid receptor Nur77 and the Fas ligand represent at least two of the confirmed molecular targets of calcineurin in mature T-cells and T-cell hybridomas (Anel et al., 1994; Yazdanbakhsh et al., 1995). The involvement of calcineurin in Ca^{2+}-stimulated apoptosis could also potentially explain, in part, the sensitivity of various apoptotic pathways to inhibition by calmodulin antagonists.

Ca^{2+}-Activated Proteases

There is some evidence that Ca^{2+}-sensitive protease(s) might represent direct targets for Ca^{2+} elevations in apoptosis. Recent work has shown that the Ca^{2+}-dependent neutral protease calpain is rapidly activated in T-lymphocytes following treatment with glucocorticoids or exposure to γ-irradiation, and that calpain antagonists can block DNA fragmentation associated with the response (Squier et al., 1994). Similarly, Henkart's group has demonstrated that some (but not all) pathways of apoptosis in mature T-lymphocytes can be inhibited by calpain antagonists (Sarin et al., 1993, 1995). The cytoskeletal protein fodrin is at least one substrate for calpain that is cleaved in T-cells following treatment with glucocorticoids or engagement of the Fas antigen (Martin et al., 1995). Similarly, we have recently obtained evidence that the cytoskeletal protein vimentin is also cleaved in apoptotic cells by a calpain-sensitive mechanism (J. Kiefer et al., unpublished observation). Precisely how fodrin and vimentin cleavage participate in the apoptosis effector mechanism is unclear, although they may be involved in cellular shrinkage, membrane blebbing or other structural alterations associated with apoptosis.

Other Ca^{2+}-activated proteases may also participate in the process. Previous work has demonstrated that incubation of isolated nuclei in the presence of Ca^{2+} promotes the rapid degradation of a family of nuclear matrix proteins, the lamins (Clawson et al., 1992; Tokes and Clawson, 1989). The protease responsible for lamin cleavage in this system is directly associated with the nuclear matrix and is activated by Ca^{2+}. In parallel, independent work from several laboratories has shown that lamins are also degraded in cells undergoing apoptosis (Kaufmann, 1989; Lazebnik et al., 1993; Oberhammer et al., 1994; Neamati et al., 1995), and in thymocytes, lamin cleavage occurs via a Ca^{2+}-dependent mechanism (Neamati et al., 1995), inspiring investigators to investigate the possible involvement of the lamin protease in the response. Thus, we have found that a specific peptide inhibitor of the putative lamin protease (also known as "nuclear scaffold (NS) protease") blocks cellular shrinkage and DNA fragmentation in thymocytes exposed to antibodies to the T-cell receptor, thapsigargin, and glucocorticoids, but not in cells treated with etoposide, a cancer chemotherapeutic agent that acts via induction of DNA damage (Zhivotovsky et al., 1995; McConkey, 1996). We have also found that inhibitors of this protease selectively block both lamin cleavage and DNA fragmentation in isolated nuclei incubated with Ca^{2+}, while inhibitors of the ICE family of proteases do not. Importantly, however, the NS protease is not the only protease that can cleave the lamins, as indicated by recent work by Lazebnik et al. (1995), who have shown that an inhibitor of the ICE/ced-3 family of cysteine proteases blocks lamin cleavage in another isolated nuclei system, whereas treatment with inhibitors of the NS protease does not (Lazabnik et al., 1995). Moreover, in this system, lamin cleavage, chromatin condensation, and DNA fragmentation do not require exogenous Ca^{2+}. Whether Ca^{2+} is involved in generating the apoptosis-promoting activity found within the extracts that these authors use to promote nuclear lamin cleavage is not clear at present. In addition, whether BCL-2 is capable of blocking the nuclear alterations induced by these extracts is also not known, and it is therefore possible that the activity present in them is an irreversibly activated component of the effector machinery (i.e., an ICE protease) that is no longer subject to Ca^{2+} regulation.

Ca^{2+}-Activated Endonuclease(s)

As introduced above, endonuclease activation that results in the formation of oligonucleosome-length DNA fragments (DNA ladders) remains the most characteristic biochemical feature of apoptotic cell death. Early work by Hewish and Burgoyne (1973) and later by Vanderbilt et al. (1982) demonstrated that a Ca^{2+}/Mg^{2+}-dependent enzyme activity capable of generating characteristic apoptotic chromatin cleavage patterns is constitutively present within nuclei of a variety of different cell types. Subsequent work by Cohen and Duke (1984), and Wyllie et al. (1984), demonstrated the involvement of this activity in the DNA fragmentation observed in thymocytes undergoing apoptosis, and it is now thought that it mediates DNA fragmentation in a variety of other model systems as well. The search for and purification of potential Ca^{2+}-dependent apoptotic nucleases have subsequently been undertaken by several laboratories. Thus, Gaido and Cidlowski (1991) have described a low-molecular-weight nuclease (NUC18) with Ca^{2+}- and Mg^{2+}-dependence activity in apoptotic lymphoid cells in response to several kinds of apoptotic stimuli. (Interestingly, the purified NUC18 shares amino acid sequence homology with cyclophilin, and human recombinant cyclophilin A has biochemical and pharmacological properties identical to native NUC18 [Montague et al., 1994].) NUC18 is also present in untreated thymocytes in precursor form or as part of a higher molecular-weight complex (> 100 kDa), suggesting that the enzyme is maintained in an inactive complex from which the nuclease dissociates in response to apoptotic signals. Although the precise mechanism of liberation of active enzyme from its precursors is unknown, an attractive possibility is that it may involve proteolysis.

The Ca^{2+}-dependent endonuclease DNase I is another excellent candidate apoptotic nuclease (Preitsch et al., 1993). Addition of the enzyme to isolated nuclei and other reconstitution systems promotes the formation of DNA strand breaks that possess the same $5'-PO_4$ and $3'-OH$ end groups found in DNA fragments isolated from apoptotic cells. Although the enzyme is localized within the rough endoplasmic reticulum, the Golgi complex, and small (secretory) vesicles in viable cells, it is also found within the perinuclear space of apoptotic cells, and it is possible tht structural alterations in the ER and/or nuclear envelope associated with apoptosis may promote the entry of DNase I into the nucleus (see below). A similar mechanism may promote entry of an ER-localized fraction of the NS lamin protease into the nucleus. Several other proteins with Ca^{2+}/Mg^{2+} endonuclease activity have been isolated (Ishida et al., 1974; Wyllie et al., 1992; Nikonova et al., 1993; Ribeiro and Carson, 1993), but to date proof that any one of these activities is directly involved in oligonucleosomal DNA fragmentation in apoptosis is lacking.

Transglutaminase Activation

Transglutaminases are a group of Ca^{2+}-dependent enzymes that catalyze the post-translational coupling of amines (including polyamines) into proteins and the cross-linking of proteins via γ-glutamyl lysine bridges when the amine is a peptide-bound lysine residue. Tissue transglutaminase has been implicated in a number of physiological processes, including cross-linking of integral plasma membrane proteins with the cytoskeleton. Recent work indicates that tissue transglutaminase is also involved in induction of apoptosis (Fesus et al., 1987, 1989). Expression of transglutaminase mRNA and protein levels increase markedly in dying cells. The enzyme appears to be activated by elevations of the cytosolic Ca^{2+} concentration, which are involved in apoptosis in many different systems. Isolation of apoptotic bodies from a number of different tissues has shown that they are resistant to dissolution by detergents and chaotrophic agents; this may, in part, be explained by the fact that surface polypeptides in these structures are cross-linked via γ-glutamyl lysine isopeptide bonds (Taresa et al., 1992) The resistance of these structures to proteolysis may allow them to accumulate, and they can be detected in the media of cell cultures that have high rates of apoptotic cell death (Fesus et al., 1991). Isodipeptide can also be detected in normal plasma, and its concentration increases following induction of apoptosis in various organs, including the thymus and liver.

The role of transglutaminase in promoting cell death and/or phagocytosis is still poorly understood. One possibility is that protein cross-linking stabilizes apoptotic cells and bodies, preventing leakage of intracellular contents into the extracellular milieu (which can trigger inflammation). Alternatively, transglutaminase modification may target proteins for subsequent degradation. Intriguingly, overexpression of the enzyme has been reported to trigger apoptotic cell death (Melino et al., 1994), suggesting that transglutaminase may be a component of the death effector pathway. Further efforts are required to identify the substrates for transglutaminase in apoptotic cells and to determine the consequences of their modification.

Exposure of Phosphatidyl Serine and Macrophage Recognition

Recent work indicates that the movement of phosphatidyl serine (PS) from the inner to the outer surface of the plasma membrane, a process that functions in the removal of apoptotic cells and bodies by both professional phagocytic cells (macrophages) and neighboring cells in tissues (Fadok et al., 1992a, 1992b; Savill et al., 1993), is another component of apoptosis that is regulated by alterations in cytosolic Ca^{2+}. Phospholipids are known to be distributed asymmetrically across the plasma membrane, with phosphatidyl choline and sphingomyelin localized primarily to the outer leaflet and PS and phosphatidyl ethanolamine restricted almost exclusively to the inner leaflet under normal conditions (Verkleij et al., 1973; Zwaal et al., 1975). Most of the work on plasma membrane lipid asymmetry has been conducted with red blood cells, where it is known that PS localization is regulated by energy-dependent processes that involve specific lipid transporters ("flipases" and "flopases") that move PS to the outside or inside surface, respectively. Transport in both directions is ATP-dependent and sensitive to sulfhydryl modifying agents. Interestingly, inhibition of lipid movement with reagents that abrogate the activities of both the flipase and flopase does not result in loss of membrane asymmetry (Connor and Schroit, 1990), suggesting that these enzymes may primarily function to restore lipid asymmetry following its disruption. However, transport can also be inhibited by increasing the cytosolic Ca^{2+} concentration, which does result in rapid nonspecific redistribution of all phospholipids (Bevers et al., 1990). The mechanism of Ca^{2+}-mediated PS exposure is not yet clear, but in red blood cells it closely parallels formation of cytoskeleton-free lipid microvesicles (Sims et al., 1989) and other phenomena, such as calpain activation (Fox et al., 1991) and the formation of phosphatidyl inositol(4,5)-bisphosphate-Ca^{2+} complexes (Sulpice et al., 1994). However, neither direct Ca^{2+} effects, nor calpain-mediated protein degradation (Basse et al., 1993; Vanags et al., 1996), nor $Ins(4,5)P_2$ accumulation (Bevers et al., 1995), nor inhibition of the flipase can singularly accommodate the membrane rearrangements that occur. More recent work suggests that Ca^{2+}-mediated redistribution is mediated by Ca^{2+}- and sulfhydryl-sensitive, energy-dependent lipid scramblase (Williamson et al., 1995). Therefore, elevations in the cytosolic Ca^{2+} concentration probably promote PS exposure and allow for macrophage recognition primarily by inactivating the PS translocase and by activating the scramblase (Verhoven et al., 1995). Interestingly, PS exposure on aged red blood cells is associated with increased cytosolic Ca^{2+} levels and increased cell density, suggesting that these events may be mechanistically related in both red cells and apoptotic cells. In addition, given its regulation by Ca^{2+}- and calpain-dependent mechanisms, it is tempting to speculate that red blood cell microvesiculation may be functionally related to the formation of apoptotic bodies by nucleated cells undergoing apoptosis.

Plasma membrane lipid asymmetry is critically involved in several other important physiological functions. For example, surface-exposed PS serves as the point of assembly for the coagulation factors Va and Xa to enter into the prothrombinase complex (Rosing et al., 1985). In addition, PS exposure enhances membrane fusion events and appears to be involved in the initiation of microvesiculation in red blood cells (Schewe et al., 1992). Finally, surface PS is also detectable on certain tumor cells (Utsugi et al., 1991). Although the mechanisms underlying the latter have not been defined, it is possible that PS exposure is involved in the prominent macrophage infiltration observed in most solid tumors and that surface PS may represent a potential target for anti-tumor therapies.

Possible Consequences of Intracellular Ca^{2+}-Pool Depletion

In some cellular systems, extracellular or intracellular Ca^{2+} chelators can actually promote DNA fragmentation, even though other triggers of apoptosis in these systems (i.e., glucocorticoids, growth factor withdrawal) have been shown to deplete the ER Ca^{2+} store. These observations have led Baffy et al. (1993) and Lam et al. (1993) to propose that depletion of the ER Ca^{2+} store may itself serve as a signal for apoptosis. How could this occur? At least two of the catabolic enzymes proposed to be involved in the effector mechanism of apoptosis (DNase I and an extranuclear pool of the NS protease) are localized to the ER, and it is therefore possible that loss of Ca^{2+} leads to release of these factors into the perinuclear region or into the nuclear matrix itself. In addition, it is known that ER Ca^{2+}-pool depletion results in the release of a small biomolecule that participates in a retrograde signal for plasma membrane Ca^{2+} influx, and it is possible that it or another molecule released in a similar fashion can also promote cell death.

Depletion of mitochondrial Ca^{2+} stores may also participate in the signal for apoptosis. Mitochondrial Ca^{2+} uptake is driven by mitochondrial membrane potential ($\Delta\Psi$) (Richter, 1993). In de-energized mitochondria, Ca^{2+} can be released by a reversal of the uptake pathway. Under conditions of oxidative stress, mitochondrial Ca^{2+} cycling can reach critical levels, leading to increased energy expenditure and a dramatic fall in $\Delta\Psi$. Recent work has shown that a fall in mitochondrial $\Delta\Psi$ is an early event in apoptosis (Petit et al., 1995; Zamzani et al., 1995a, 1995b), and ruthenium red, an inhibitor of the mitochondrial Ca^{2+}-uptake pathway, blocks apoptosis in L929 fibroblasts (Hennet et al., 1993) and inhibits the progression of apoptosis in glucocorticoid-treated splenocytes (Zamzani et al., 1995a), suggesting that mitochondrial Ca^{2+} release is involved. Again, further efforts are required to determine the relationship between this event and the activation of the proteases and nucleases of the effector pathway.

Alternative Signals for Apoptosis

Oxidative Stress

Several lines of evidence indicate that reactive oxygen species are involved in promoting apoptosis in diverse model systems. Treatment of cells with low to moderate doses of exogenous oxidants (i.e., hydrogen peroxide, tert-butyl peroxide, menadione) can trigger apoptosis (McConkey, et al., 1988; Lennon et al., 1990, 1991; Hennet et al., 1993; Hockenbery et al., 1993; Kane et al., 1993; Buttke and Sandstrom, 1994). Moreover, apoptosis induced by agents that are not direct oxidants (TNF, glucocorticoids, thapsigargin, chemotherapeutic agents) is associated with oxygen radical production and depletion of intracellular antioxidants (i.e, reduced glutathione) (Hennet et al., 1993; Hockenbery et al., 1993; Kane et al., 1993; Richter, 1993; Mayer and Noble, 1994; Wolfe et al., 1994; Fernandez et al., 1995; Slater et al., 1995; Zamzani et al., 1995a). Although the identities of the oxygen radicals involved in each system are still under active investigation, roles for superoxide (Hannet et al., 1993; Zamzani et al., 1995b), lipid peroxides (Hockenbery et al., 1993), nitric oxide (Xie et al., 1995), and hydroxyl radicals (Wolfe et al., 1994) have been proposed. In these systems, exogenous antioxidants such as N-acetyl cysteine and free radical scavengers block DNA fragmentation and cell death (Mayer and Noble, 1994; Wolfe et al., 1994; Fernandez et al., 1995; Slater et al., 1995). Finally, several laboratories have now presented strong evidence that BCL-2 and BCL-X_L possess antioxidant properties that may be involved in their abilities to inhibit cell death. Thus, oxidative stress is another good candidate for a central cell death signal, one which may affect intracellular Ca^{2+} homeostasis. However, it should be noted that BCL-2 and BCL-X_L can still inhibit apoptosis under conditions of low oxygen (Jacobson and Raff, 1995; Shimizu et al., 1995), which has been raised as an argument against a universal role for oxidative stress in the response.

Intracellular Acidification

Eastman and coworkers have proposed another alterntive to Ca^{2+} that may serve as a general signal for cell death. In their efforts to characterize the biochemical mechanisms underlying chemotherapy-induced apoptosis, they determined that isolated nuclei from their cellular models possessed an endonuclease activity that was stimulated by acidic pH (Dnase II) (Barry and Eastman, 1993). Interestingly, Mary Collins and her coworkers (personal communication) have recently identified an acidic nuclease in IL-3-dependent hematopoietic cells that can also be activated by Ca^{2+} under the appropriate conditions, suggesting that intracellular acidification and alterations in intracellular Ca^{2+} homeostasis may represent independent ways of arriving at the same endpoint (endonuclease activation) in apoptotic cells. Other efforts have shown that a drop in cytoplasmic pH precedes the morphological and biochemical features of apoptosis in certain models (Barry and Eastman, 1992; Li and Eastman, 1995), observations that have since been confirmed by other investigators in other systems (Gottlieb et al., 1995). Moreover, the protective effects of certain survival factors and agents that activate protein kinase C have been linked to activation of the Na^+/H^+ antiporter and intracellular alkylinization (Rajotte et al., 1992; Gottlieb et al., 1995). Finally, the drop in pH is prevented by overexpression of BCL-2 (A. Eastman, personal communication), suggesting that acidification may represent another cellular target for this family of apoptosis suppressors. It will be interesting to determine whether intracellular acidification is related to oxidative stress and intracellular Ca^{2+} alterations in these systems.

Ceramide Production

The generation of bioactive signal transduction regulators via the hydrolysis of plasma membrane phospholipids is emerging as an important general means of regulating apoptotic cell death. Of all the second messengers shown to be involved, most recent attention has been focused on ceramide as a possible ubiquitous trigger of apoptosis (Obeid et al., 1993; Pushkareva et al., 1995). Two major mechanisms have been identified that appear to contribute to the formation of ceramide under different circumstances. The most common pathway involves activation of the enzyme sphingomyelinase, which catalyzes the hydrolysis of sphingomyelin to form ceramide and diacylglycerol, a response that is involved in apoptosis induced by tumor necrosis factor, engagement of the Fas antigen, and ionizing radiation (Obeid et al., 1993; Cifone et al., 1994; Haimovitz-Friedman et al., 1994; Jarvis et al., 1994b; Gulbins et

al., 1995; Pushkareva et al., 1995; Tepper et al., 1995). Ceramide can also be formed de novo via activation of ceramide synthase, and it has been shown that the cytotoxic effects of the chemotherapeutic agent daunorubicin on leukemic cell lines are mediated by this pathway (Bose et al., 1995). Notably, exogenous hydrolysis-resistant ceramide analogs or sphingomyelinase can mimic the effects of TNF and the other apoptosis-inducing agents to trigger endonuclease activation and cell death, indicating that ceramide production is sufficient to induce apoptosis (Obeid et al., 1993; Cifone et al., 1994; Haimovitz-Friedman et al., 1994; Jarvis et al., 1994b; Bose et al., 1995; Gulbins et al., 1995; Tepper et al., 1995). In addition, the possibility that ceramide is an evolutionarily conserved trigger of apoptosis is suggested by the fact that overexpression of the *Drosophila* cell death protein, Reaper, leads to upregulation of cellular ceramide levels (J. M. Abrams, personal communication). Interestingly, phorbol esters and diacylglycerol (DAG) antagonize the death-promoting effects of ceramide in diverse models (Obeid et al., 1993; Jarvis et al., 1994a), suggesting that a balance between ceramide and DAG may determine the outcome of death signals. The downstream targets for ceramide remain unclear, but candidates include a ceramide-dependent protein phosphatase (CAPP) (Wolff et al., 1994), a ceramide-activated protein kinase (CAPK) (Kolesnick and Golde, 1994), the ζ isoform of protein kinase C (Lozano et al., 1994), and the H-*ras* proto-oncogene (Gulbins et al., 1995). Whether ceramide acts upstream, downstream, or independently of the other candidate biochemical mediators of apoptosis remains to be determined.

Conclusions and Future Directions

The independent efforts of many laboratories over the past several years have established that alterations in intracellular Ca^{2+} homeostasis are commonly involved in initiating apoptosis. Ongoing work suggests that activation of Ca^{2+}-stimulated signaling networks and catabolic enzymes represents one way these signals are translated into responses In addition, preliminary evidence indicates that the depletion of intracellular Ca^{2+} pools can itself serve as a signal for cell death, perhaps by promoting relocalization of some of the key catabolic enzymes involved and by enhancing oxidative stress in mitochondria. In some systems, other signals have been defended as key regulators of the response, and further efforts are therefore required to determine if and how all of these signals are interrelated.

An important component of the defense of Ca^{2+} as a central cell death regulator is the idea that BCL-

2 and its homologs specifically regulate intracellular Ca^{2+} compartmentalization, an idea that has strong preliminary support but requires a good deal of additional investigation. In particular, further efforts are required to determine how BCL-2 exerts its effects on Ca^{2+} and whether these effects are required for its cell death-suppressing function. Similarly, it is maintained at present that any central cell death signal would directly or indirectly activate one or more members of the ICE family of cysteine proteases, and at present there is no evidence available that any of the candidate central cell death signals are capable of this. Elucidation of these relationships over the next few years should provide new targets for therapeutic intervention that may aid in the treatment of the expanding number of diseases, including cancer, AIDS, and neurodegenerative diseases, in which apoptosis is thought to play a central role in their pathologies.

References

Allbritton, N. L., Verret, C. R., Wolley, R. C., and Eisen, H. N. (1988) Calcium ion concentrations and DNA fragmentation in target cell destruction by murine cloned cytotoxic T lymphoscytes. *J. Exp. Med.* 167: 514–527.

Amendola, A., Lombardi, G., Oliverio, S., Colizzi, V., and Piacentini, M. (1994) HIV-1 gp 120-dependent induction of apoptosis in antigen-specific human T cell clones is characterized by "tissue" transglutaminase expression and prevented by cyclosporin A *FEBS Lett.* 339: 258–264.

Anel, A., Buferne, M., Boyer, C., Schmitt-Verhulst, A. M., and Golstein, P. (1994) T cell receptor-induced Fas ligand expression in cytotoxic T lymphocyte clones is blocked by protein tyrosine kinase inhibitors and cyclosporin A. *Eur. J Immunol.* 24: 2469–2476.

Aw, T. Y., Nicotera, P., Manzo, L., and Orrenius, S. (1990) Tributyltin stimulates apoptosis in rat thymocytes. *Arch. Biochem. Biophys.* 283: 46–50.

Baffy, G., Miyashita, T., Williamson, J. R., and Reed, J. C. (1993) Apoptosis induced by withdrawal of interleukin-3 (IL-3) from an IL-3-dependent hematopoietic cell line is associated with repartitioning of intracellular calcium and is blocked by enforced BCL-2 oncoprotein production. *J. Biol. Chem.* 268: 6511–6519.

Barry, M. A. and Eastman, A. (1992) Endonuclease activation during apoptosis: the role of cytosolic Ca^{2+} and pH. *Biochem. Biophys. Res. Commun.* 186: 782–789.

Barry, M. A. and Eastman, A. (1993) Identification of deoxyribonuclease II as an endonuclease involved in apoptosis. *Arch Biochem. Biophys.* 300: 440–450.

Basse, F., Gaffet, P., Rendu, F., and Bienvenue, A. (1993) Translocation of spin-labeled phospholipids through the plasma membrane during thrombin- and ionophore A23187-induced platelet aggregation. *Biochemistry* 32: 2337–2344.

Bellomo, G., Perotti, M., Taddei, F., Mirabelli, F., Finardi, G., Nicotera, P., and Orrenius, S. (1992) Tumor necrosis factor α induces apoptosis in mammary adenocarcinoma cells by an increase in intranuclear free Ca^{2+} concentration and DNA fragmentation. *Cancer Res.* 52: 1342–1346.

Bevers, E. M., Verahllen, P. F., Visser, A. J., Comfurius, P., and Zwaal, R. F. (1990) Bidirectional transbilayer lipid movement in human platelets as visualized by the fluorescent membrane probe 1[4-(trimethylammonio)-phenyl]-6-phenyl-1,3,5-hexatriene. *Biochemistry* 29: 5132–5137.

Bevers, E. M., Wiedmer, T., Comfurius, P., Zhao, J., Smeets, E. F, Schlegel, R. A., Schroit, A. J., Weiss, H. J., Williamson, P., Zwaal, R. F., and Sims, P. J. (1995) The complex of phosphatidylinositol 4,5-bisphosphate and calcium ions is not responsible for Ca^{2+}-induced loss of phospholipid aymmetry in the human erythrocyte: a study of Scott syndrome, a disorder of calcium-induced phospholipid scrambling. *Blood* 86: 1983–1991.

Bonnefoy-Berard, N., Genestier, L., Flacher, M., and Revillard, J. P. (1994) The phosphoprotein phosphatase calcineurin controls calcium-dependent apoptosis in B cell lines. *Eur. J. Immunol.* 24: 325–329.

Bose, R., Verheij, M., Haimovitz-Friedman, A., Scotto, K., Fuks, Z., and Kolesnick, R. (1995) Ceramide synthase mediates daunorubicin-induced apoptosis: an alternative mechanism for generating death signals. *Cell* 82: 405–414.

Brown, D. G., Sun, X. M., and Cohen, G. M. (1993) Dexamethasone-induced apoptosis involves cleavage of DNA to large fragments prior to internucleosomal fragmentation. *J. Biol. Chem.* 268: 3037–3039.

Bump, N. J., Hackett, M., Hugunin, M., Seshagiri, S., Brady, K., Chen, P., Ferenz, C., Franklin, S., Ghayur, T., Li, P., Licari, P., Mankovich, M., Shi, L., Greenberg, A. H., Miller, L. K., and Wong, W. W. (1995) Inhibition of ICE family proteases by baculovirus antiapoptotic protein p35. *Science* 269: 1885–1888.

Buttke, T. M. and Sandstrom, P. A. (1994) Oxidative stress as a mediator of apoptosis *Immunol. Today* 15: 7–10.

Choi, M. S., Boise, L. H., Gottschalk, A. R., Quintans, J., and Thompson, C. B. (1995) The role of BCL-XL in CD40-mediated rescue from anti-mu-induced apoptosis in WEHI-231 B lymphoma cells. *Eur. J. Immunol.* 25: 1352–1357.

Cifone, M. G., Maria, R. D., Roncaioli, P., Rippo, R. M., Azuma, M., Lanier, L. L., Santoni, A., and Testi, R. (1994) Apoptotic signaling through CD95 (Fas/Apo-1) activates acidic sphingomyelinase. *J. Exp. Med.* 180: 1547–1552.

Clawson, G. A., Norbeck, L. L., Hatem, C. L., Rhodes, C., Amiri, P., McKerrow, J. H., Patierno, S. R. and Fiskum, G. (1992) Ca^{2+}-regulated serine protease associated with the nuclear scaffold. *Cell Growth Differ.* 3: 827–838.

Cohen, J. J. and Duke, R. C. (1984) Glucocorticoid activation of a calcium-dependent endonuclease in thymocyte nuclei leads to cell death *J. Immunol.* 132: 38–42.

Connor, J. and Schroit, A. J. (1990) Aminophospholipid translocation in erythrocytes: evidence for the involvement of a specific transporter and an endofacial protein. *Biochemistry* 29: 37–43.

deJong, D., Prins, F. A., Mason, D. Y., Reed, J. C., Ommen, G. B.v., and Kluin, P. M. (1994) Subcellular localization of the bcl-2 protein in malignant and normal lymphoid cells. *Cancer Res.* 54: 256–260.

Dowd, D. R., MacDonald, P. N., Komm, B. S., Maussler, M. R., and Miesfeld, R. (1991) Evidence for early induction of calmodulin gene expression in lymphocytes undergoing glucocorticoid-mediated apoptosis. *J. Biol. Chem.* 266: 18423–18426.

Dypbukt, J. M., Ankarcrona, M., Burkitt, M., Sjoholm, A., Strom, K., Orrenius, S., and Nicotera, P. (1994) Different prooxidant levels stimulate growth, trigger apoptosis, or produce necrosis of insulin-secreting RINm5F cells. The role of intracellular polyamines. *J. Biol. Chem.* 269: 30553–30560.

Fadok, V. A., Savill, J. S., Haslett, C., Bratton, D. L., Doherty, D. E., Campbell, P. A., and Henson, P. M. (1992a) Different populations of macrophages use either the vitronectin receptor or the phosphatidylserine receptor to recognize and remove apoptotic cells. *J. Immunol.* 149: 4029–4035.

Fadok, V. A., Voelker, D. R., Campbell, P. A., Cohen, J. J., Bratton, D. L., and Henson, P. M. (1992b) Exposure of phosphatidylserine on the surface of apoptotic lymphoces triggers specific recognition and removal by macrophages. *J. Immunol.* 148: 2207–2216.

Fang, W., Rivard, J J., Ganser, J. A., LeBien, T. W., Nath, K. A., Mueller, D. L., and Behrens, T. W. (1995) BCL-xL rescues WEHI 231 B lymphocytes from oxidant mediated death following diverse apoptotic stimuli. *J. Immunol.* 155: 66–75.

Fernandes-Alnemri, T., Litwack, G., and Alnemri, E. S. (1994) CPP32, a novel human apoptotic protein with homology to *Caenorhabditis elegans* cell death protein ced-3 and mammalian interleukin-1b-converting enzyme. *J. Biol. Chem.* 269: 30761–30764.

Fernandes-Alnemri, T., Litwack, G., and Alnemri, E. S. (1995) Mch2, a new member of the apoptotic Ced-3/Ice cysteine protease gene family. *Cancer Res.* 55: 2737–2742.

Fernandez, A., Kiefer, J., Fosdick, L., and McConkey, D. J. (1995) Oxygen radical production and thiol depletion are required for Ca^{2+}-mediated endogenous endonuclease activation in apoptotic thymocytes. *J. Immunol.* 155: 5133–5139.

Fesus, L., Thomazy, V., and Falus, A. (1987) Induction and activation of tissue transglutaminase during programmed cell death. *FEBS Lett.* 224: 104–108.

Fesus, L., Thomazy, V., Autuori, F., Ceru, M. P., Tarcsa, E., and Piacentini, M. (1989) Apoptotic hepatocytes become insoluble in detergents and chaotropic agents as a result of transglutaminase action. *FEBS Lett.* 245: 150–154.

Fesus, L., Tarcsa, E., Kedei, N., Autuori, F., and Piacentini, M. (1991) Degradation of cells dying by apoptosis leads to accumulation of epsilon(gamma-glutamyl)lysine isodipeptide in culture fluid and blood. *FEBS Lett.* 284: 109–112.

Filipski, J., Leblanc, J., Youdale, T., Sikorska, M., and Walker, P. R. (1990) Periodicity of DNA folding in higher order chromatin structures. *EMBO J.* 9: 1319–1327.

Fox, J. E., Austin, C. D., Reynolds, C. C., and Steffen, P. K. (1991) Evidence that agonist-induced activation of calpain causes the shedding of procoagulant-containing microvesicles from the membrane of aggregating platelets. *J. Biol. Chem.* 266: 13289–13295.

Furuya, Y. and Isaacs, J. T. (1993) Differential gene regulation during programmed death (apoptosis) versus proliferation of prostatic glandular cells induced by androgen manipulation. *Endocrinology* 133: 2660–2666.

Gaido, M. L. and Cidlowski, J. A. (1991) Identification, purification, and characterization of a calcium-dependent endonuclease (NUC 18) from apoptotic rat thymocytes. *J. Biol. Chem.* 266: 18580–18585.

Gottlieb, R. A., Giesing, H. A., Zhu, J. Y., Engler, R. L., and Babior, B. M. (1995) Cell acidification in apoptosis: granulocyte colony-stimulating factor delays programmed cell death in neutrophils by up-regulating the vacuolar H^+-ATPase. *Proc. Natl. Acad. Sci. USA* 92: 5965–5968.

Gulbins, E., Bissonnette, R., Mahboubi, A., Martin, S., Nishioka, W., Brunner, T., Baier, G., Baier-Bitterilch, G., Byrd, C., Lang, F., Kolesnick, R., Altman, A., and Green, D. (1995) FAS-induced apoptosis is mediated via a ceramide-initiated RAS signaling pathway. *Immunity* 2: 341–351.

Haimovitz-Friedman, A., Kan, C. C., Ehleiter, D., Persaud, R. S., McLoughlin, M., Fuks, Z., and Kolesnick, R. N. (1994) Ionizing radiation acts on cellular membranes to generate ceramide and initiate apoptosis. *J. Exp. Med.* 180: 525–535.

Hengartner, M. O. and Horvitz, H. R. (1994) *C. elegans* cell death gene ced-9 encodes a functional homolog of mammalian proto-oncogene bcl-2. *Cell* 76: 665–676.

Hengartner, M. O., Ellis, R. E., and Horvitz, H. R. (1992) *Caenorhabditis elegans* gene ced-9 protects cells from programmed cell death. *Nature* 356: 494–499.

Hennet, T., Richter, C., and Peterhans, E. (1993) Tumour necrosis factor-alpha induces superoxide anion production in mitochondria of L929 cells. *Biochem. J.* 289: 587–592.

Hewish, D. R. and Burgoyne, L. A. (1973) Chromatin substructure. The digestion of chromatin DNA at regularly spaced sites by a nuclear deoxyribonuclease. *Biochem. Biophys. Res. Commun.* 52: 504–510.

Hockenbery, D. M., Oltvai, Z. N., Yin, X. M., Milliman, C. L., and Korsmeyer, S. J. (1993) BCL-2 functions in an antioxidant pathway to prevent apoptosis. *Cell* 75: 241–251.

Ishida, R., Akiyoshi, H., and Takahashi, T. (1974) Isolation and purification of calcium and magnesium dependent endonuclease from rat liver nuclei. *Biochem. Biophys. Res. Commun.* 56: 703–708.

Jacobson, M. D. and Raff, M. C. (1995) Programmed cell death and Bcl-2 protection in very low oxygen. *Nature* 374: 814–816.

Jarvis, W. D., Fornari, F. A., Browning, J. L., Gewirtz, D. A., Kolesnick, R. N., and Grant, S. (1994a) Attenuation of ceramide-induced apoptosis by digly-ceride in human myeloid leukemia cells. *J. Biol. Chem.* 269: 31685–31692.

Jarvis, W. D., Kolesnick, R. N., Fornari, F. A., Traylor, R. S., Gewirtz, D. A., and Grant, S. (1994b) Induction of apoptotic DNA damage and cell death by activation of the sphingomyelin pathway. *Proc. Natl. Acad. Sci. USA* 91: 73–77.

Jiang, S., Chow, S. C., Nicotera, P., and Orrenius, S. (1994) Intracellular Ca^{2+} signals activate apoptosis in thymo-cytes: studies using the Ca^{2+} ATPase inhibitor thapsi-gargin. *Exp. Cell Res.* 212: 84–92.

Juntti-Berggren, L., Larsson, O., Rorsman, P., Ammala, C., Bokvist, K., Wahlander, K., Nicotera, P., Dybukt, J. M., Orrenius, S., Hallberg, A., and Berggren, P. (1993) Increased activity of L-type Ca^{2+} channels exposed to serum from patients with type I diabetes. *Science* 261: 86–90.

Kaiser, N. and Edelman, I. S. (1977) Calcium dependence of glucocorticoid-induced lymphocytolysis. *Proc. Natl. Acad. Sci. USA* 74: 638–642.

Kaiser, N. and Edelman, I. S. (1978) Further studies on the role of calcium in glucocorticoid-induced lymphocyto-lysis. *Endocrinology* 103: 936–942.

Kane, D. J., Sarafian, T. A., Anton, R., Hahn, H., Gralla, E. B., Valentine, J. S., Ord, T., and Bredesen, D. E. (1993) BCL-2 inhibition of neural death: decreased generation of reactive oxygen species. *Science* 262: 1274–1277.

Kaneko, Y. and Tsukamoto, A. (1994) Thapsigargin-induced persistent intracellular calcium pool depletion and apoptosis in human hepatoma cells. *Cancer Lett.* 79: 147–155.

Kaufmann, S. H. (1989) Induction of endonucleolytic DNA cleavage in human acute myelogenous leukemia cells by etoposide, camptothecin, and other cytotoxic anti-cancer drugs: a cautionary note. *Cancer Res.* 49: 5870–5878.

Kaufmann, S. H., Desnoyers, S., Ottaviano, Y., Davidson, N. E., and Poirier, G. G. (1993) Specific proteolytic cleavage of poly(ADP-ribose) polymerase: an early marker of chemotherapy-induced apoptosis. *Cancer Res.* 53: 3976–3985.

Khan, A. A., Soloski, M. J., Sharp, A. H., Schilling, G., Sabatini, D. M., Li, S.-H., Ross, C. A., and Snyder, S. H. (1996) Lymphocyte apoptosis: mediation by

increased type 3 inositol 1,4,5-trisphosphate receptor. *Science* 273: 503–507.

Kolesnick, R. and Golde, D. W. (1994) The sphingomyelin pathway in tumor necrosis factor and interleukin-1 signaling. *Cell* 77: 325–328.

Kuman, S., Kinoshita, M., Noda, M., Copeland, N. G., and Jenkins, N. A. (1994) Induction of apoptosis by the mouse Nedd2 gene, which encodes a protein similar to the product of *Caenorhabditis elegans* cell death gene ced-3 and the mammalian IL-1b-converting enzyme. *Genes Dev.* 8: 1613–1626.

Lam, M., Dubyak, G., and Distelhorst, C. W. (1993) Effect of glucocorticoid treatment on intracellular calcium homeostasis in mouse lymphoma cells. *Mol. Endocrinol.* 7: 686–693.

Lam, M., Dubyak, G., Chen, L., Nunez, G., Miesfeld, R. L., and Distelhorst, C. W. (1994) Evidence that bcl-2 represses apoptosis by regulating endoplasmic reticu-lum-associated Ca^{2+} fluxes. *Proc. Natl. Acad. Sci. USA* 91.

Lazebnik, Y. A., Cole, S., Cooke, C. A., Nelson, W. G., and Earnshaw, W. C. (1993) Nuclear events of apoptosis in vitro in cell-free mitotic extracts: a model system for analysis of the active phase of apoptosis. *J. Cell. Biol.* 123: 7–22.

Lazebnik, Y. A., Takahashi, A., Moir, R. D., Goldman, R. D., Poirier, G. G., Kaufmann, S. H., and Earnshaw, W. C. (1995) Studies of the lamin proteinase reveal multi-ple parallel biochemical pathways during apoptotic execution. *Proc. Natl. Acad. Sci. USA* 92: 9042–9046.

Lennon, S. V., Martin, S. J., and Cotter, T. G. (1990) Induction of apoptosis (programmed cell death) in tumour cells by widely diverging stimuli. *Biochem. Soc. Trans.* 18: 343–345.

Lennon, S. V., Martin, S. J., and Cotter, T. G. (1991) Dose-dependent induction of apoptosis in human tumour cell lines by widely divergent stimuli. *Cell Prolif.* 24: 203–204.

Levick, V., Coffey, H., and D'Mello, S. R. (1995) Opposing effects of thapsigargin on the survival of developing cerebellar granule neurons in culture. *Brain Res.* 676: 325–335.

Li, J. and Eastman, A. (1995) Apoptosis in an interleukin-2-dependent cytotoxic T lymphocyte cell line is asso-ciated with intracellular acidification. *J. Biol. Chem.* 270: 3203–3211.

Liu, J., Farmer, J. D., Lane, W. S., Friedman, J., Weissman, I., and Schreiber, S. L. (1991) Calcineurin is a common target of cyclophilin–cyclosporin A and FKBP–FK506 complexes. *Cell* 66: 807–815.

Lozano, J., Berra, E., Municio, M. M., Diaz-Meco, M. T., Dominguez, I., Sanz, L., and Moscat, J. (1994) Protein kinase C zeta isoform is critical for kB-dependent pro-moter activation by sphingomyelinase. *J. Biol. Chem.* 269: 19200–19202.

Makrigiannis, A. P., Blay, J., and Hoskin, D. W. (1994) Cyclosporin A inhibits 2-chloroadenosine-induced DNA cleavage in mouse thymocytes. *Int. J. Immunopharmacol.* 16: 995–1001.

Martikainen, P. and Isaacs, J. (1990) Role of calcium in the programmed cell death of rat ventral prostatic glandular cells. *Prostate* 17: 175–187.

Martin, S. J. and Green, D. R. (1995) Protease activation during apoptosis: death by a thousand cuts? *Cell* 82: 349–352.

Martin, S. J., O'Brien, G. A., Nishioka, W. K., McGahon, A. J., Mahboubi, A., Saido, T. C., and Green, D. R. (1995) Proteolysis of fodrin (non-erythroid spectrin) during apoptosis. *J. Biol. Chem.* 270: 6425–6428.

Mayer, M. and Noble, M. (1994) N-acetyl-L-cysteine is a pluripotent protector against cell death and enhancer of trophic factor-mediated cell survival in vitro. *Proc. Natl. Acad. Sci. USA* 91: 7496–7500.

McConkey, D. J. (1996) Calcium-dependent, interleukin 1β-converting enzyme inhibitor-insensitive degradation of lamin B₁ and DNA fragmentation in isolated thymocyte nuclei. *J. Biol. Chem.* 271: 22398–22406.

McConkey, D. J., Hartzell, P., Nicotera, P., Wyllie, A. H., and Orrenius, S. (1988) Stimulation of endogenous endonuclease activity in hepatocytes exposed to oxidative stress. *Toxicol. Lett.* 42: 123–130.

McConkey, D. J., Hartzell, P., Amador-Perez, J. F., Orrenius, S., and Jondal, M. (1989a) Calcium-dependent killing of immature thymocytes by stimulation via the CD3/T cell receptor complex. *J. Immunol.* 143: 1801–1806.

McConkey, D. J., Nicotera, P., Hartzell, P., Bellomo, G., Wyllie, A. H., and Orrenius, S. (1989b) Glucocorticoids activate a suicide process in thymocytes through an elevation of cytosolic Ca²⁺ concentration. *Arch. Biochem. Biophys.* 269: 365–370.

McConkey, D. J., Chow, S. C., Orrenius, S., and Jondal, M. (1990) NK cell-induced cytotoxicity is dependent on a Ca²⁺ increase in the target. *FASEB J.* 4: 2661–2664.

McConkey, D. J., Fosdick, L., D'Adamio, L., Jondal, M., and Orrenius, S. (1994) Co-receptor (CD4/CD8) engagement enhances CD3-induced apoptosis in thymocytes. Implications for negative selection. *J. Immunol.* 153: 2436–2443.

Melino, G., Annicchiarico-Petruzzeli, M., Piredda, L., Candi, E., Gentile, V., Davies, P. J., and Piacentini, M. (1994) Tissue transglutaminase and apoptosis: sense and antisense tranfection studies with human neuroblastoma cells. *Mol. Cell. Biol.* 14: 6584–6596.

Montague, J. W., Gaido, M. L., Frye, C., and Cidlowski, J. A. (1994) A calcium-dependent nuclease from apopototic rat thymocytes is homologous with cyclophilin. Recombinant cyclophilins A, B, and C have nuclease activity. *J. Biol. Chem.* 269: 18877–18880.

Murphy, K., Heimberger, A., and Loh, D. Y. (1990) Induction by antigen of intrathymic apoptosis of CD4+CD8+ thymocytes in vivo. *Science* 250: 1720–1723.

Nakagama, T., Ueda, Y., Yamada, H., Shores, E. W., Singer, A., and June, C. H. (1992) In vivo calcium elevations in thymocytes with T cell receptors that are specific for self ligands. *Science* 257: 96–99.

Neamati, N., Fernandez, A., Wright, S., Kiefer, J., and McConkey, D. J. (1995) Degradation of lamin B1 precedes oligonucleosomal DNA fragmentation in apoptotic thymocytes and isolated thymocyte nuclei. *J. Immunol.* 154: 3788–3795.

Nicholson, D. W., Ali, A., Thornberry, N. A., Vaillancourt, J. P., Ding, C. K., Gallant, M., Gareau, Y., Griffin, P. R., Labelle, M., Lazebnik, Y. A., Munday, N. A., Raju, S. M., Smulson, M. E., Yamin, T. T., Yu, V. L., and Miller, D. K. (1995) Identification and inhibition of the ICE/ced-3 protease necessary for mammalian apoptosis. *Nature* 376: 37–43.

Nikonova, L. V., Beletsky, I. P., and Umansky, S. R. (1993) Properties of some nuclear nucleases of rat thymocytes and their changes in radiation-induced apoptosis. *Eur. J. Biochem.* 215: 893–901.

Norvell, A., Mandik, L., and Monroe, J. G. (1995) Engagement of the antigen receptor on immature murine B lymphocytes results in death by apoptosis. *J. Immunol.* 154: 4404–4413.

Obeid, L. M., Linardic, C. M., Karolak, L. A., and Hannun, Y. A. (1993) Programmed cell death induced by ceramide. *Science* 259: 1769–1771.

Oberhammer, F., Wilson, J. W., Dive, C., Morris, I. D., Hickman, J. A., Wakeling, A. E., Walker, P. R., and Sikorska, M. (1993) Apoptotic death in epithelial cells: cleavage of DNA to 300 and/or 50 kb fragments prior to or in the absence of internucleosomal fragmentation. *EMBO J.* 12: 3679–3684.

Oberhammer, F. A., Hochegger, K., Froschl, G., Tiefenbacher, R., and Pavelka, M. (1994) Chromatin condensation during apoptosis is accompanied by degradation of lamin A+B without enhanced activation of cdc2 kinase. *J. Cell Biol.* 1267: 827–837.

Oltvai, Z. N. and Korsmeyer, S. J. (1994) Checkpoints of dueling dimers foil death wishes. *Cell* 79: 189–192.

Orrenius, S., McConkey, D. J., Bellomo, G., and Nicotera, P. (1989) Role of Ca²⁺ in toxic cell killing. *Trends Pharmacol. Sci.* 10: 281–285.

Parry, S. L., Holman, M. J., Hasbold, J., and Klaus, G. G. B. (1994) Plastic-immobilized anti-m or anti-d antibodies induce apoptosis in mature murine B lymphocytes. *Eur. J. Immunol.* 24: 974–979.

Peitsch, M. C., Polzar, B., Stephan, H., Crompton, T., MacDonald, H. R., Mannherz, H. G., and Tschopp, J. (1993) Characterization of the endogenous deoxyribonuclease involved in nuclear DNA degradation during apoptosis (programmed cell death). *EMBO J.* 12: 371–377.

Petit, P. X., Lecoeur, H., Zorn, E., Dauguet, C., Mignotte, B., and Gougeon, M. L. (1995) Alternations in mitochondrial structure and function are early events of dexamethasone-induced thymocyte apoptosis. *J. Cell Biol.* 130: 157–167.

Pushkareva, M., Obeid, L. M., and Hannun, Y. A. (1995) Ceramide: an endogenous regulator of apoptosis and growth suppression. *Immunol. Today* 16: 294–297.

Rajotte, D., Haddad, P., Haman, A., Cragoe, Jr. E. J., Hoang, T. (1992) Role of protein kinase C and the

Na^+H^+ antiporter in suppression of apoptosis by granulocyte macrophage colony-stimulating factor and interleukin-3. *J. Biol. Chem.* 26: 9980–9987.

Ribeiro, J. M. and Carson, D. A. (1993) Ca^{2+}/Mg^{2+}-dependent endonuclease from human spleen: purification, properties, and role in apoptosis. *Biochemistry* 32: 9129–9136.

Richter, C. (1993) Pro-oxidants and mitochondrial Ca^{2+}: their relationship to apoptosis and oncogenesis. *FEBS Lett.* 325: 104–107.

Robertson, L. E., Chubb, S., Meyn, R. E., Story, M., Ford, R., Hittelman, W. N., and Plunkett, W. (1993) Induction of apoptotic cell death in chronic lymphocytic leukemia by 2-chloro-2′-deoxyadenosine and 9-beta-D-arabinosyl-2′-fluoradenine. *Blood* 81: 143–150.

Rosing, J., van Rijn, J. L., Bevers, E. M., van Dieijen, G., Comfurius, P., and Zwaal, R. F. (1985) The role of activated human platelets in prothrombin and factor X activation. *Blood* 65: 319–332.

Sarin, A., Adams, D. H., and Henkart, P. A. (1993) Protease inhibitors selectively block T cell receptor-triggered programmed cell death in a murine T cell hybridoma and activated peripheral T cells. *J. Exp. Med.* 178: 1693–1700.

Sarin, A., Nakajima, H., and Henkart, P. A. (1995) A protease-dependent TCR-induced death pathway in mature lymphocytes. *J. Immunol.* 154: 5806–5812.

Savill, J., Fadok, V., Henson, P., and Haslett, C. (1993) Phagocytic recognition of cells undergoing apoptosis. *Immunol. Today* 14: 131–136.

Schewe, M., Muller, P., Korte, T., and Herrmann, A. (1992) The role of phospholipid asymmetry in calcium phosphate-induced fusion of human erythrocytes. *J. Biol. Chem.* 267: 5910–5915.

Shi, Y., Sahai, B. M., and Green, D. R. (1989) Cyclosporin A inhibits activation-induced cell death in T-cell hybridomas and thymocytes. *Nature* 339: 625–626.

Shimizu, S., Eguchi, Y., Kosaka, H., Kamiike, W., Matsuda, H., and Tsujimoto, Y. (1995) Prevention of hypoxia-induced cell death by Bcl-2 and Bcl-xL. *Nature* 374: 811–813.

Sims, P. J., Wiedmer, T., Esmon, C. T., Weiss, H. J., and Shattil, S. J. (1989) Assembly of the platelet prothrombinase complex is linked to vesiculation of the platelet plasma membrane. Studies in Scott syndrome: an isolated defect in platelet procoagulant activity. *J. Biol. Chem.* 264: 17049–17057.

Slater, A. F., Nobel, C. S., Maellaro, E., Bustamante, J., Kimland, M., and Orrenius, S. (1995) Nitrone spin traps and a nitroxide antioxidant inhibit a common pathway of thymocyte apoptosis. *Biochem. J.* 306: 771–778.

Smith, C. A., Williams, G. T., Kingston, R., Jenkinson, E. J., and Owen, J. J. T. (1989) Antibodies to CD3/T cell receptor complex induce death by apoptosis in immature T cells in thymic cultures. *Nature* 337: 181–184.

Squier, M. K. T., Miller, A. C. K., Malkinson, A. M., and Cohen, J. J. (1994) Calpain activation in apoptosis. *J. Cell Physiol.* 159: 229–237.

Story, M. D., Stephens, L. C., Tomosovic, S. P., and Meyn, R. E. (1992) A role for calcium in regulating apoptosis in rat thymocytes irradiated in vitro. *Int. J. Radiat. Biol.* 61: 243–251.

Sulpice, J. C., Zachowski, A., Devaux, P. F., and Giraud, F. (1994) Requirement for phosphatidylinositol 4,5-bisphosphate in the Ca^{2+}-induced phospholipid redistribution in the human erythrocyte membrane. *J. Biol. Chem.* 269: 6347–6354.

Sun, X. M. and Cohen, G. M. (1994) Mg^{2+}-dependent cleavage of DNA into kilobase pair fragments is responsible for the initial degradation of DNA in apoptosis. *J. Biol. Chem.* 269: 14857–14860.

Taresa, E., Kedei, N., Thomazy, V., and Fesus, L. (1992) An involucrin-like protein in hepatocytes serves as a substrate for tissue transglutaminase during apoptosis. *J. Biol. Chem.* 267: 25648–25651.

Tepper, C. G., Jayadev, S., Liu, B., Bielawska, A., Wolff, R., Yonehara, S., Hannun, Y. A., and Seldin, M. F. (1995) Role for ceramide as an endogenous mediator of Fas-induced cytotoxicity. *Proc. Natl. Acad. Sci. USA* 92: 8443–8447.

Tewari, M., Quan, L. T., O'Rourke, K., Desnoyers, S., Salvesen, G. S., and Dixit, V. M. (1995) Yama/CPP32b, a mammalian homolog of ced-3, is a crmA-inhibitable protease that cleaves the death substrate poly(ADP-ribose) polymerase. *Cell* 81: 801–809.

Thompson, C. B. (1995) Apoptosis in the pathogenesis and treatment of disease. *Science* 267: 1456–1462.

Tokes, Z. A. and Clawson, G. A. (1989) Proteolytic activity associated with the nuclear scaffold. The effect of self-digestion on lamins. *J. Biol. Chem.* 264: 15059–15065.

Tsubata, T., Wu, J., and Honjo, T. (1993) B-cell apoptosis induced by antigen receptor crosslinking is blocked by a T-cell signal through CD40. *Nature* 364: 645–648.

Utsugi, T., Schroit, A. J., Connor, J., Bucana, C. D., and Fidler, I. J. (1991) Elevated expression of phosphatidylserine in the outer membrane leaflet of human tumor cells and recognition by activated human blood monocytes. *Cancer Res.* 51: 3062–3066.

Vanags, D. M., Pörn-Ares, M. I., Coppola, S., Burgess, D. H., and Orrenius, S. (1996) Protease involvement in fodrin cleavage and phosphatidylserine exposure in apoptosis. *J. Biol. Chem.* 271: 31075–31085.

Vanderbilt, J. N., Bloom, K. S., and Anderson, J. N. (1982) Endogenous nuclease: properties and effects on transcribed genes in chromatin. *J. Biol. Chem.* 257: 13009–13017.

Verhoven, B., Schlegel, R. A., and Williamson, P. (1995) Mechanisms of phosphatidylserine exposure, a phagocyte recognition signal, on apoptotic T lymphocytes. *J. Exp. Med.* 182: 1597–1601.

Verkleij, A. J., Zwaal, R. F., Roelofsen, B., Comfurius, P., Kastelijn, D., and van Deenen, L. L. (1973) The asymmetric distribution of phospholipids in the human red cell membrane. A combined study using phospholipases and freeze-etch electron microscopy. *Biochem. Biophys. Acta* 323: 178–193.

Wang, L., Miura, M., Bergeron, L., Zhu, H., and Yuan, J. (1994) Ich-1, an ICE/ced-3-related gene, encodes both positive and negative regulators of programmed cell death. *Cell* 78: 739–750.

Williamson, P., Bevers, E. M., Smeets, E. F., Comfurius, P., Schlegel, R. A., and Zwaal, R. F. (1995) Continuous analysis of the mechanism of activated transbilayer lipid movement in platelets. *Biochemistry* 34: 10448–10455.

Wolfe, J. T., Ross, D., and Cohen, G. M. (1994) A role for metals and free radicals in the induction of apoptosis in thymocytes. *FEBS Lett.* 352: 58–62.

Wolff, R. A., Dobrowsky, R. T., Bielawska, A., Obeid, L. M., and Hannun, Y. A. (1994) Role of ceramide-activated protein phosphatase in ceramide-mediated signal transduction. *J. Biol. Chem.* 269: 19605–19606.

Wyllie, A. H. (1980) Glucocorticoid-induced thymocyte apoptosis is associated with endogenous endonuclease activation. *Nature* 284: 555–556.

Wyllie, A. H., Kerr, J. F. R., and Currie, A. R. (1980) Cell death: the significance of apoptosis. *Int. Rev. Cytol.* 68: 251–305.

Wyllie, A. H., Morris, R. G., Smith, A. L., and Dunlop, D. (1984) Chromatin cleavage in apoptosis: association with condensed chromatin morphology and dependence on macromolecular synthesis. *J. Pathol.* 142: 67–77.

Wyllie, A. H., Arends, M. J., Morris, R. G., Walker, S. W., and Evan, G. (1992) The apoptosis endonuclease and its regulation. *Sem. Immunol.* 4: 389–397.

Xie, K., Huang, S., Dong, Z., Juang, S. H., Gutman, M., Xie, Q. W., Nathan, C., and Fidler, I. J. (1995) Transfection with the inducible nitric oxide synthase gene suppresses tumorigenicity and abrogates metastasis by K-1735 murine melanoma cells. *J. Exp. Med.* 181: 1333–1343.

Xue, D. and Horvitz, H. R. (1995) Inhibition of the *Caenorhabditis elegans* cell-death protease CED-3 by a CED-3 cleavage site in baculovirus p35 protein. *Nature* 377: 248–251.

Yao, X. R. and Scott, D. W. (1993) Antisense oligodeoxynucleotides to the blk tyrosine kinase prevent anti-m-mediated growth inhibition and apoptosis in a B-cell lymphoma. *Proc. Natl. Acad. Sci. USA* 90: 7946–7950.

Yazdanbakhsh, K., Choi, J. W., Li, Y., Lau, L. F., and Choi, Y. (1995) Cyclosporin A blocks apoptosis by inhibiting the DNA binding activity of the transcription factor Nur77. *Proc. Natl. Acad. Sci. USA* 92: 437–441.

Yuan, J., Shaham, S., Ledoux, S., Ellis, H. M., and Horvitz, H. R. (1993) The *C. elegans* cell death gene ced-3 encodes a protein similar to mammalian interleukin-1b-converting enzyme. *Cell* 75: 641–652.

Zamzani, N., Marchetti, P., Castedo, M., Decaudin, D., Macho, A., Hirsch, T., Susin, S. A., Petit, P. X., Mignotte, B., and Kroemer, G. (1995a) Sequential reduction of mitochondrial transmembrane potential and generation of reactive oxygen species in early programmed cell death. *J. Exp. Med.* 182: 367–377.

Zamzani, N., Marchetti, P., Castedo, M., Zanin, C., Vayssiere, J. L., Petit, P. X., and Kroemer, G. (1995b) Reduction in mitochondrial potential constitutes an early irreversible step of programmed lymphocyte death in vivo. *J. Exp. Med.* 181: 1661–1672.

Zhivotovsky, B., Nicotera, P., Bellomo, G., Hanson, K., and Orrenius, S. (1993) Ca^{2+} and endonuclease activation in radiation-induced lymphoid cell death. *Exp. Cell Res.* 207: 163–170.

Zhivotovsky, B., Gahm, A., Ankarcrona, M., Nicotera, P., and Orrenius, S. (1995) Multiple proteases are involved in thymocyte apoptosis. *Exp. Cell Res.* 221: 404–412.

Zwaal, R. F., Roelofsen, B., Comfurius, P., and van Deenen, L. L. (1975) Organization of phospholipids in human red cell membranes as detected by the action of various purified phospholipases. *Biochem. Biophys. Acta* 406: 83–96.

27

Pathology of Calcium-Transporting Membrane Systems

David H. MacLennan
Julian Loke
Alex Odermatt

Muscle contraction is regulated by the concentration of free Ca^{2+} in the sarcoplasm (Ebashi et al., 1969). In skeletal muscle, sarcoplasmic Ca^{2+} concentrations are controlled almost exclusively by the sarcoplasmic reticulum, so that skeletal muscle can be stimulated to contract repeatedly in the absence of extracellular Ca^{2+}. Although the sarcoplasmic reticulum contains many proteins (Lytton and MacLennan, 1992), Ca^{2+} uptake in fast-twitch skeletal muscle is carried out by only one, the sarco- or endoplasmic reticulum Ca^{2+}-ATPase isoform, SERCA1, encoded by the *ATP2A1* gene. Uptake of Ca^{2+} by slow-twitch skeletal muscle, cardiac muscle, smooth muscle, and nonmuscle tissues is through one or the other of two alternatively spliced SERCA2 isoforms encoded by the *ATP2A2* gene. Ca^{2+} release from fast- and slow-twitch skeletal muscle is through the Ca^{2+}-release channel (sometimes referred to as the ryanodine receptor) isoform 1, encoded by the *RYR1* gene, while Ca^{2+} release from cardiac muscle is through the ryanodine receptor isoform 2 (*RYR2*). Defects in any of these genes would be expected to result in altered Ca^{2+} regulation and tissue-specific abnormalities. In the case of *RYR1*, gene knockouts are lethal (Takeshima et al., 1994), reflecting its expression in more than one muscle type. It would be predicted that knockout of *ATP2A2* would also be lethal, since it is expressed in many tissues. In this review chapter, we describe mutations in *RYR1* that are not lethal, but which lead to malignant hyperthermia and central-core disease, and mutations in *ATP2A1* that lead to Brody disease (BD).

Malignant Hyperthermia

Although malignant hyperthermia (MH) was first recorded in 1900 (Gibson et al., 1900), it was not until six decades later that Denborough et al. (1962) reported 10 cases in a single Australian family, demonstrating that MH is inherited as an autosomal dominant abnormality. By 1970, enough evidence had accumulated to assign the site of the primary defect to the skeletal muscle (Britt and Kalow, 1968; Berman et al., 1970). In the early 1970s, Kalow et al. (1970) showed that contracture of muscle fascicles from MH-susceptible (MHS) patients were more sensitive than normal to caffeine and Ellis et al. (1972) demonstrated similar high sensitivity of MH muscle to halothane. These abnormal responses provided a diagnostic test for MH — the caffeine–halothane contracture test (CHCT) — that is still widely used. In 1966, it was discovered that pigs afflicted with porcine stress syndrome (PSS) also developed MH reactions that were virtually identical to human MH reactions, thus providing an animal model for human MH (Hall et al., 1966). In 1975, dantrolene was demonstrated to be an antidote for MH reactions, permitting full recovery in many cases (Harrison, 1975). In recent years, genetic analysis has shown that a large fraction of human MH cases, but not all, are associated with dominant mutations in the *RYR1* gene that encodes the Ca^{2+}-release channel of skeletal muscle sarcoplasmic reticulum (the ryanodine receptor) (MacLennan and Phillips, 1992). Current research is aimed at the identification of all genes and all mutations that cause human MH. Excellent comprehensive reviews of clinical aspects of MH and its physiological, biochemical, and genetic basis have been published elsewhere (Britt, 1991; Mickelson and Louis, 1996).

Human Malignant Hyperthermia

Human MH is a genetic abnormality, inherited as an autosomal dominant mutation (Denborough et al., 1962). Britt and Kalow (1970) estimated the incidence to be about 1 case in 15,000 anesthetics in children and about 1 case in 50,000 to 1 in 100,000 anesthetics in adults. Ørding (1985) found the incidence of MH episodes to be about 1 case in 65,000 anesthetics in Denmark. These reported incidences are probably marked underestimates of the true incidence of MH susceptibility because of the incomplete penetrance of the gene and the difficulty in defining mild reactions. Many MH-susceptible patients are never identified because they are never anesthetized or are not anesthetized enough times to develop an MH reaction. An MH-like episode may occur in individuals who have inherited muscle diseases with deleterious phenotypes, such as central-core disease (CCD, Denborough et al., 1973), King-Denborough syndrome (King and Denborough, 1973), Duchenne muscular dystrophy (Brownell et al., 1983), myotonia fluctuans (Vita et al., 1995), and possibly other myopathies (Brownell, 1988; Heiman-Patterson et al., 1988). It is, therefore, important to differentiate true MH from those muscle diseases that give rise to MH-like reactions (Iaizzo and Lehmann-Horn, 1995).

Individuals with MH susceptibility may respond to the administration of potent inhalational anesthetics and depolarizing skeletal muscle relaxants with a rising end tidal carbon dioxide, skeletal muscle rigidity, tachycardia, unstable and rising blood pressure, hyperventilation, cyanosis, a falling arterial oxygen tension, an increasing arterial carbon dioxide tension, lactic acidosis, and, eventually, fever (Britt, 1991). Muscle cell damage brings about electrolyte imbalance, with early elevation of serum K^+ and Ca^{2+} and a later elevation of muscle enzymes, such as creatine kinase, and muscle proteins, such as myoglobin in the serum and urine. If therapy is not initiated immediately, the patient may die within minutes from ventricular fibrillation, within hours from pulmonary edema or coagulopathy, or within days from postanoxic neurological damage and cerebral edema or obstructive renal failure, resulting from the release of muscle proteins into the circulation (Britt, 1991). During convalescence, muscle soreness and muscle edema may develop. In some instances, the reactions induced in susceptible individuals may be much milder (Britt, 1991), being char-

acterized by masseter muscle rigidity alone (Kosko et al., 1992), by fever alone, or by reactions of intermediate severity.

Malignant hyperthermia crises are rare below the age of 3 years and the incidence of crises declines progressively above the age of 30 years. Males are more affected, both in terms of incidence of reactions and of positive biopsies (Britt, 1991), possibly because of their greater musculature or because of hormonal differences. Many individuals who have had MH reactions have had multiple previous uneventful general anesthetics, often of long duration and with known triggering agents. The reason for the failure of a crisis to be precipitated on exposure to potent inhalational anesthetics and/or to succinylcholine is not clear, but it may be that an additional environmental trigger, such as strenuous exercise or extreme emotional stress and agitation (both of which might affect circulating hormonal levels), a major muscle injury, drugs, or a viral or bacterial infection associated with fever, is needed to initiate a reaction.

Anesthesiologists identify many patients at risk through case histories, which include information on their kinship to individuals who have had an MH reaction and/or a combined history of a persistently elevated creatine kinase with chronic and incapacitating muscle pain and cramps. For patients known or suspected to have malignant hyperthermia, anesthetic routines are changed to any desired combination of nontriggering anesthetics, such as barbiturates, tranquilizers, narcotics, propofol, ketamine, nitrous oxide, and local anesthetics (Britt, 1991; McKenzie et al., 1992).

Under present-day standard anesthetic practice, heart rate, blood pressure, body temperature, end tidal carbon dioxide production, and arterial oxygen saturation are monitored during the course of anesthesia (Britt, 1991). An increase in several of these factors may lead to the clinical diagnosis of MH (Larach et al., 1994). When this occurs, the administration of the MH-triggering anesthetic is stopped immediately, the gas machine is changed and the patient is hyperventilated with 100% oxygen, and the antidote — dantrolene — is administered until muscles relax, fever, heart rate and respiratory rate decline significantly toward normal, and blood gases normalize. Sodium bicarbonate is also given to correct metabolic acidosis, insulin is given if serum potassium and blood glucose are elevated, and furosemide and mannitol are given to prevent the onset of acute renal failure or muscle and brain edema. These practices have lowered the death rate from MH episodes from over 80% to less than 7% in recent years, but neurological, muscle, and kidney damage still contribute to the morbidity that results from MH reactions (Britt, 1991).

Diagnostic Tests for Malignant Hyperthermia

In humans, MH susceptibility is seldom associated with ill health. Accordingly, the most important goals of MH research are to provide therapy during an acute MH episode and to prevent MH episodes through identification of MHS individuals in advance of anesthesia.

The caffeine–halothane contracture test (CHCT) and the in vitro contracture test (IVCT) were developed on the premise that the muscle from MHS individuals might be abnormally sensitive to agents that induce contractures. Therefore, MHS muscle fascicles might contract in the presence of lower amounts of either caffeine or halothane, or, conversely, have larger contractures in the presence of these agents than the muscle from normal individuals. The North American test protocol (Larach, 1989) and the European test protocol (European MH Group, 1984) have been standardized on the basis of this premise. In both protocols, a positive response to both caffeine and halothane results in the diagnosis of MHS. A positive response to either caffeine or halothane, but not both, is considered to be MHS in the North American protocol, but as MH-equivocal (MHE) in the European test. The MHE patients are treated as MHS patients.

The CHCT is a useful clinical test (Larach, 1993), distinguishing those at risk for MH from unaffected patients. As a clinical test, the CHCT ensures that appropriate anesthetics are administered to those patients who are MHS, while those diagnosed as normal can be treated with routine anesthetics. Since failure to detect MH susceptibility can result in a serious or fatal outcome, sensitivity (the ability of the test to detect MH when it is present) approaching 100% is more important for a clinical diagnosis than is specificity (the ability of the test to exclude false positive results in normal patients). Strict, objective criteria have been developed to define when an MH reaction has almost certainly occurred (D6), when it has very likely occurred (D5), when it is unlikely to have occurred (D2), or likely not to have occurred (D1) (Larach et al., 1994). The clear definition of an acute MH reaction has permitted case findings for determination of the sensitivity of the CHCT and IVCT. Testing of control patients has allowed the determination of specificity. The North American CHCT currently achieves 97% sensitivity and 78% specificity (95% CI, 70–85%) based on 32 MH patients with a D6 MH reaction and 120 controls (95% CI, 84–100%) (Larach et al., 1992a, 1992b; Allen et al., 1998). The European IVCT offers 99% sensitivity (95% CI, 84-100%) based on investigation of 17 patients who survived a D6 MH reaction and 29 controls and 93.6% specificity (Ørding et al., 1997). Combined results from the European MH group suggest that the IVCT has 100% sensitivity and 87% specificity (95% CI, 82–92%) on the basis of studies of 95 patients with a D6 reaction and 200 controls.

It is most likely that limitations in sensitivity and specificity of the CHCT arise because the outcome of the test depends on the interplay among a very large number of biochemical reactions within muscle cells. In a multifactorial system, the potentially deleterious effect of an abnormal Ca^{2+}-release channel, the most common cause of MH, may be compensated for in some individuals by overactivity of systems such as the sarcoplasmic reticulum Ca^{2+} pump, the plasma membrane Ca^{2+} pump, the Na^+/Ca^{2+} exchangers, or the Ca^{2+}-uptake system of the mitochondria, all of which might act to bring about Ca^{2+} homeostasis before contracture is triggered. These factors might be altered further by the fitness of the muscle analyzed.

Porcine Malignant Hyperthermia

Individuals among herds of lean, heavily muscled swine are susceptible to fatal crises characterized by muscle rigidity, mottling of the skin, and tachycardia (O'Brien, 1987). This syndrome is referred to as the porcine stress syndrome (PSS). Crises are brought on by physical and emotional stresses, including overheating, exercise, mating, transportation to market, and fear. When given halothane and succinylcholine, or halothane alone, susceptible pigs frequently develop MH reactions similar to those suffered by human MHS patients (Hall et al., 1966).

Stress-induced deaths in swine occur predominantly with homozygous MH animals and rarely in heterozygotes. In a study of the response of homozygotes and heterozygotes to a 3-minute halothane challenge, none of 197 heterozygotes responded, while 168 of 179 homozygotes responded to the halothane challenge (Otsu et al., 1991). In an early study (Britt et al., 1978), continued halothane administration was found to bring about increased heart rate, cyanosis, rigidity, and fever in both MH homozygotes and heterozygotes. The response was graded from normal through MH heterozygotes to MH homozygotes. In a more recent study (Fletcher et al., 1993), however, heterozygotes did not respond to continued halothane administration, even in the presence of succinylcholine, raising the possibility that the genetic background on which the MH gene is expressed may affect the MH response. The MH heterozygotes gain about 5% more lean meat than normal animals (Simpson and Webb, 1989; O'Brien et al., 1994), they have better feed conversion and fat is redistributed from muscle to backfat.

The Physiological Basis for Malignant Hyperthermia

Examination of the early and late events in the development of a halothane-induced MH reaction in swine have proved very informative (Berman et al., 1970). Studies of blood chemistry revealed that lactic acidosis, presumably originating in glycogenolysis and glycolysis, and the release of K^+, Mg^{2+}, and Ca^{2+} from muscle occurs within seconds after halothane administration. Observed rises of serum K^+ to as high as 7–8 mM have been sufficient to cause cardiac arrest in some cases. A rise in body temperature usually occurs only after 15 or 20 minutes, and perhaps even substantially longer. On the basis of measurements of oxygen uptake, the major source of heat production has been shown to be anaerobic, arising from the breakdown of creatine phosphate and ATP (Berman et al., 1970). The neutralization of bicarbonate in the blood by lactic acid and the direct production of carbon dioxide by the tricarboxylic acid cycle produces large amounts of CO_2 in the blood and exhaled air. The release of Ca^{2+} into the blood from the muscle was the first indication that a disturbance in Ca^{2+} homeostasis in muscle was the primary defect in MH.

Studies of the mechanisms controlling changes in muscle tension have revealed that the interaction of actin and myosin is regulated by Ca^{2+} and that Ca^{2+} regulation is mediated through the Ca^{2+}-binding protein, troponin (Zot and Potter, 1987). Muscle Ca^{2+} concentrations are regulated by the activities of Ca^{2+} pumps and channels located in the sarcoplasmic reticulum and transverse tubular systems. The Ca^{2+} also plays a role in the control of glycolysis in muscle through its activation of phosphorylase kinase (Brostrom et al., 1971). A defect in Ca^{2+} regulation, leading to the chronic elevation of Ca^{2+} within the sarcoplasm, could induce muscle contracture, extensive glycolysis, and enhanced mitochondrial oxidation of glycolytic end products (Fig. 1). These various reactions, leading to high turnover of ATP, could be responsible for the elevated temperatures associated with MH episodes (Britt 1991; MacLennan and Phillips, 1992).

Abnormalities in the Ca^{2+} pump were ruled out in biochemical studies of MH swine (Nelson, 1988). It is of interest, however, that muscle from Brody patients with Ca^{2+}-pump deficiency gave a positive in vitro CHCT result (Karpati et al., 1986), even though fulminant MH reactions have not been reported in Brody disease. On the other hand, higher rates of Ca^{2+}-induced Ca^{2+} release, particularly at low levels of inducing Ca^{2+}, were observed in membrane vesicle preparations from both human (Endo et al., 1983) and porcine (Ohnishi et al., 1983) muscle, and closure of single porcine MH channels at high Ca^{2+} concen-

trations was shown to be inhibited (Fill et al., 1990; Shomer et al., 1993). In comparable studies of human muscle, Ca^{2+}-release channels with abnormally high caffeine sensitivity were detected in MH individuals (Fill et al., 1991). In sarcoplasmic reticulum from MHS swine, ryanodine binding, which is dependent on the open state of the Ca^{2+}-release channel, was found to be enhanced (Mickelson et al., 1988). Digestion with trypsin revealed an alteration in the amino acid sequence of the Ca^{2+}-release channel in MHS animals (Knudson et al., 1990). Thus, comparative biochemical and physiological studies implicated the Ca^{2+}-release channel as a potential causal factor for MH.

The Ca^{2+}-release channels have been cloned from skeletal (*RYR1*) (Takeshima et al., 1989; Zorzato et al., 1990; Fujii et al., 1991), cardiac (*RYR2*) (Nakai et al., 1990; Otsu et al., 1990), and nonmuscle sources (*RYR3*) (Giannini et al., 1992; Hakamata et al., 1992). These proteins contain from 4872 to 5037 amino acids with masses between 550,000 and 564,000 Da. The Ca^{2+}-release channel has been isolated from skeletal muscle and shown to form a tetrameric complex (Inui et al., 1987). Transmembrane sequences are located at the COOH-terminal end of the molecule (Takeshima et al., 1989; Zorzato et al., 1990). Predicted ATP-, Ca^{2+}-, and calmodulin-binding sites are clustered in two sites in the molecule. One lies in the proposed transmembrane sequences in the COOH-terminal end of the protein, encompassing the region between amino acid residues 4253 and 4499 (Takeshima et al., 1989). A second regulatory region has been proposed to lie between amino acid residues 2600 and 3000 (Otsu et al., 1990). This sequence would contain residues 2809 and 2843 in cardiac and skeletal isoforms, respectively, the major phosphorylation sites in these proteins (Witcher et al., 1991; Suko et al., 1993). A predicted ATP-binding sequence begins at residue 2652 and the region between residues 2800 and 3050 contains predicted calmodulin-binding sites (Otsu et al., 1990).

A defect in the Ca^{2+}-release channel, giving rise to abnormal Ca^{2+} regulation within skeletal muscle, could account for all of the signs of MH (MacLennan and Phillips, 1992). If the Ca^{2+}-release channels had longer open times in the presence of anesthetic agents, intracellular Ca^{2+} might be chronically elevated, resulting in muscle contracture and activation of the first steps in glycogenolysis through activation of phosphorylase kinase. Muscle contracture and the pumping of excessive amounts of cytoplasmic Ca^{2+} back to the lumen of the sarcoplasmic reticulum would consume large amounts of ATP, thus generating large quantities of heat. The ADP formed would stimulate glycolysis and the mitochondrial oxidation of pyruvate derived from glucose. These hypermetabolic responses would lead to deple-

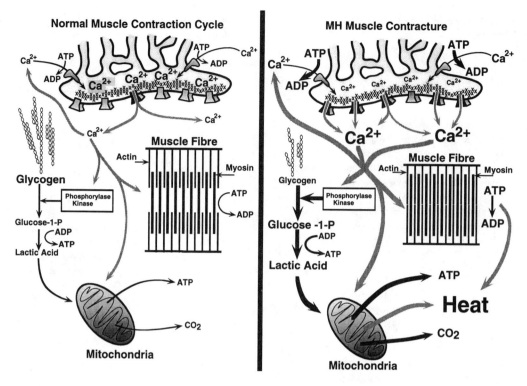

Figure 27.1 A proposed mechanism for the induction of malignant hyperthermia caused by abnormalities in the Ca^{2+}-release channel of skeletal muscle sarcoplasmic reticulum. In a normal relaxation/contraction cycle (left), Ca^{2+} is pumped into the sarcoplasmic reticulum by a Ca^{2+} pump to initiate relaxation, stored within the lumen in association with calsequestrin, and released through a Ca^{2+}-release channel to initiate contraction. Glycolytic and aerobic metabolism proceed only rapidly enough to maintain the energy balance in the cell. The Ca^{2+} release is highly regulated and, even when stimulated, has a relatively short open time. Abnormal MH or CCD Ca^{2+}-release channels (right) are sensitive to lower concentrations of activators of opening, releasing Ca^{2+} at enhanced rates, and not closing readily. The activity of the mutated channel floods the cell with Ca^{2+} and overpowers the Ca^{2+} pump that would normally lower cytoplasmic Ca^{2+}. Sustained muscle contraction (contracture) accounts for rigidity, while sustained glycolytic and aerobic metabolism account for the generation of excess lactic acid, CO_2 and heat, and for enhanced oxygen uptake. Damage to cell membranes and imbalances of ionic concentrations can account for the life-threatening systemic problems that arise during a malignant hyperthermia episode. (Adapted from MacLennan, D. H. and Phillips, M. S. (1992) *Science* 256: 789, with permission, © AAAS.)

tion of ATP, glycogen, and oxygen; to the production of excess lactic acid, carbon dioxide, and heat; and, ultimately, to the disruption of intracellular and extracellular ion balances, with consequent muscle cell damage.

Linkage of the *RYR1* Gene to Malignant Hyperthermia

In early studies of porcine MH, Andersen and Jensen (1977) demonstrated linkage between inheritance of porcine MH and polymorphisms in the gene that encodes glucose phosphate isomerase (*GPI*). Later studies established a linkage group for the porcine *HAL* gene (the designation of the MH gene that gives rise to halothane sensitivity), *GPI*, and the gene for 6-phosphogluconate dehydrogenase (*PGD*) (Archibald and Imlah, 1985; Gahne and Juneja, 1985), located near the centromere of pig chromosome 6 (Davies et al., 1988; Chowdhary et al., 1989; Harbitz et al., 1990). The homologous region around the human *GPI* locus was found to be on the long arm of chromosome 19 (Lusis et al., 1986), making this a candidate region for human MH. Cloning of the human skeletal muscle ryanodine receptor (*RYR1*) cDNA (Zorzato et al., 1990) led to the localization of *RYR1* to human chromosome 19q13.1 (MacKenzie et al., 1990) and permitted the identification of several restriction fragment length polymorphisms (RFLPs) in the human *RYR1* gene (MacLennan et al., 1990). In a study of linkage between inheritance of MH and these *RYR1* polymorphisms and flanking markers, cosegregation with *RYR1* markers was found in 23 meioses in nine families, leading to a lod score (the logarithm of the odds favoring linkage vs. nonlinkage) of 4.2 for a recombinant fraction of 0.0. The probability of linkage of more than 10,000:1 identified *RYR1* as a candidate gene for MH in humans (MacLennan et al., 1990). In an independent investigation of linkage of human MH to a series of chromosome 19q markers, the MH locus was also assigned to the region of human chromosome 19 where *RYR1* was localized (McCarthy et al., 1990).

Malignant Hyperthermia Mutation in Swine

The demonstration of linkage of *RYR1* to MH led to parallel searches in both swine and humans for sequence differences in the *RYR1* gene between MH and normal individuals (Fujii et al., 1991; Otsu et al., 1991; Gillard et al., 1992a, 1992b). In a comparison of the cDNA sequences of MH (Pietrain) and normal (Yorkshire) pigs, the substitution of T for C at nucleotide 1843, leading to the substitution of Cys for Arg at amino acid residue 615, was the only amino acid sequence alteration found (Fujii et al., 1991). Initial studies showed an association between inheritance of the mutation and inheritance of MH in some 80 animals from five different breeds. Tight linkage was established in a study of backcrosses between British Landrace heterozygous animals of the N/n genotype and homozygous MH animals of the n/n genotype (Otsu et al., 1991) in which 376 animals were tested, including 338 that represented informative meioses. Cosegregation of the phenotype with the Cys for Arg[615] substitution was observed, leading to a lod score favoring linkage of 102 for a recombinant fraction of 0.0.

The porcine MH mutation seems to have been selected because it contributes to dressed carcass weight, to leanness and heavy muscling, and to the redistribution of fat from muscle to backfat (Simpson and Webb, 1989; O'Brien et al., 1994). A possible explanation for these phenomena is that a leaky or hypersensitive Ca^{2+}-release channel could give rise to spontaneous muscle contractures and this continual toning of the muscle might lead to muscle hypertrophy (MacLennan and Phillips, 1992). The utilization of ATP for spontaneous contractures would limit the deposition of fat.

Malignant Hyperthermia Mutations in Humans

The demonstration of linkage between MH and a substitution of Cys for Arg[615] in the *RYR1* gene in swine led to a search for the corresponding mutation in human MH families. The equivalent human mutation, Cys for Arg[614], was first found in a single family of five members in which the mutation segregated with MH (Gillard et al., 1992a). This mutation has been found in about 4% of MH families worldwide. In most families where it has been found, it segregates with individuals who have been diagnosed by the CHCT as being MHS (Gillard et al., 1992a; Hogan et al., 1992), but in a few families where it is found, linkage was not observed (Deufel et al., 1995; Serfas et al., 1996). The questions that arise, therefore, are whether the Arg[614] to Cys mutation is really causal of MH, whether the phenotypic diagnosis based on CHCT is incorrect, or whether other MH-causing mutations are also cosegregating in these families (MacLennan, 1995).

The assessment of the Arg[614] to Cys mutation as causative of human MH has a strong genetic and biochemical basis. The corresponding mutation is linked to porcine MH with a lod score of 102 favoring linkage for a recombinant fraction of 0.0 (Otsu et al., 1991). The mutation has been found across a species barrier between swine and humans, where it segregates with MH in all but a few individuals analyzed (Gillard et al., 1992a). In biochemical studies of the porcine channel, a measurable defect has been observed in the closing of the Ca^{2+}-release channel (Fill et al., 1990; Shomer et al., 1993). When expressed in heterologous cell culture, the mutant form of the ryanodine receptor has been shown to release Ca^{2+} in response to lower levels of added caffeine or halothane than the normal channel (Otsu et al., 1994; Treves et al., 1994). All of these test results support the view that the Arg[614] to Cys mutation is causal of MH.

The phenotypic assay (CHCT) has an inherent error, defined by its imperfect sensitivity and specificity. The limited specificity can probably be attributed to the multifactorial nature of the contracture response. Accordingly, the occasional discordance between the CHCT- and DNA-based diagnoses, at least for the Arg[614] to Cys mutation, is most likely to arise from the CHCT test. Another possibility is that discordant results arise from the segregation of a second MH mutation in the family under study. These problems are complicating factors in proving the causal nature of other MH mutations through linkage of inheritance of MH and mutant *RYR1* alleles (Phillips et al., 1994; MacLennan, 1995; Serfas et al., 1996).

Table 27.1 Mutations Associated with MH, CCD, or BD

Mutation	Region	Association	Reference(s)
RYR1 Mutations Associated with MH and CCD			
Cys35Arg	Exon 2	MH	Lynch et al. (1997)
Arg163Cys	Exon 6	MH, CCD	Quane et al. (1993)
Gly248Arg	Exon 9	MH	Gillard et al. (1992b)
Gly341Arg	Exon 11	MH	Quane et al. (1994b)
Ile403Met	Exon 12	MH, CCD	Quane et al. (1993)
Tyr522Ser	Exon 14	MH, CCD	Quane et al. (1994a)
Arg552Trp	Exon 15	MH	Keating et al. (1997)
Arg614Leu	Exon 17	MH	Quane et al. (1997)
Arg614Cys	Exon 17	MH	Gillard et al. (1992a)
Arg2162Cys	Exon 39	MH	Manning et al. (1998b)
Arg2162His	Exon 39	MH CCD	Manning et al. (1998b)
Val2167Met	Exon 39	MH	Manning et al. (1998b)
Thr2205Met	Exon 39	MH	Manning et al. (1998b)
Gly2434Arg	Exon 45	MH	Keating et al. (1994) Phillips et al. (1994)
Arg2435His	Exon 45	MH, CCD	Zhang et al. (1993)
Arg2458Cys	Exon 46	MH	Manning et al. (1998a)
Arg2458His	Exon 46	MH	Manning et al. (1998a)
ATP2A1 Mutations Associated with BD			
Arg198stop	Exon 7	BD	Odermatt et al. (1996)
Cys675stop	Exon 15	BD	Odermatt et al. (1996)
Defective splice donor	Intron 3	BD	Odermatt et al. (1996)
Frameshift at Pro[147]	Exon 5	BD	Odermatt et al. (1997)

Continued analysis of the DNA sequence of ryanodine receptors from probands from MH families has now associated MH with at least 15 *RYR1* mutations, accounting for the abnormality in about 25% of MH families (Table 27.1). Thus, the causal MH mutations in about 75% of MH families have yet to be found.

The search for additional MH mutations in *RYR1* is hampered by the large size of the gene. The cDNA is over 15,000 nucleotide base pairs (bp) long and the gene from which it is derived is 159,000 bp long (Phillips et al., 1996). The gene contains 106 exons, of which two are alternatively spliced into mRNA to create different forms of the protein. The MH mutations found to date have clustered in exons 2–18 and 39–46 (Table 27.1). If additional mutations continue to cluster in a few exons, mutant searches will be much less onerous than might be anticipated, considering the length of the coding sequences in the gene.

Searching for Additional MHS Loci

The *RYR1* gene is a very strong candidate gene for MH, but it is not the only one. There are cases where diagnosis appears to be accurate and where no linkage between *RYR*1 and MH can be defined (Ball and Johnson, 1993; MacLennan, 1995). In at least some of these cases, both false positive and false negative diagnoses would have to be invoked to prove linkage. Linkage of MH to chromosome 17q in several families has been reported (Levitt et al., 1992), making candidates of the sodium channel α-subunit gene (*SCN4A*) and two subunits of the dihydropyridine receptor, *CACNLB1* and *CACNLG*, located on chromosome 17q11.2–q24 (Olckers et al., 1992). In subsequent studies, however, linkage of MH to chromosome 17q and to the candidate genes was ruled out in other non-chromosome-19-linked European families (Sudbrak et al., 1993). It is possible that myotonia fluctuans and other demonstrated defects of the sodium channel α-subunit protein might have been associated with abnormal responses to succinylcholine, including muscle rigidity (Iaizzo and Lehmann-Horn, 1995; Vita et al., 1995).

Malignant hyperthermia has also been linked to chromosome 7q, with a lod score less than 3 in a single family (Iles et al., 1994). The presence of the gene that encodes the α_2/δ subunit of the dihydropyridine receptor (CACNL2A) on chromosome 7q21–q22 (Powers et al., 1994) establishes it as a candidate gene for MH. Sequencing of this candidate has not revealed a causal mutation.

Several large, nonchromosome-19-linked European MH families have been included in a systematic linkage study using a set of polymorphic microsatellite markers that cover the entire human genome (Sudbrak et al., 1995). A single family was linked to chromosome 3q13.1 with a lod score of 3.22. The high lod score found in this linkage analysis suggests that true heterogeneity exists for MH. Thus, this chromosome-3-linked family offers an excellent opportunity for identifying an additional causal gene in MH.

In continued analysis, two more candidate loci were identified (Robinson et al., 1997). The first is on human chromosome 1q, the site of a candidate gene, CACNL1A3, that encodes the α1-subunit of the dihydropyridine receptor. The second is on human chromosome 5p, where no candidate gene has been mapped. A third family provides evidence for at least one other unspecified locus.

Analysis of the chromosome-1q-linked family led to the discovery of the mutation of G3333 to A in the CACNL1A3 gene, leading to the mutation of Arg[1086] to His in the α1-subunit of the L-type voltage-dependent Ca^{2+} channel (Monnier et al., 1997). This study clearly demonstrates the involvement of the dihydropyridine receptor in the regulation of the ryanodine receptor and illustrates how defects in dihydropyridine receptor function can manifest as defects in Ca^{2+} regulation, leading to susceptibility to malignant hyperthermia.

Central-Core Disease

Central-core disease (CCD) is a rare, nonprogressive myopathy, presenting in infancy, which is characterized by hypotonia and proximal muscle weakness (Shy and Magee, 1956). Additional variable clinical features include pes cavus, kyphoscoliosis, foot deformities, congenital hip dislocation, and joint contractures (Shuaib et al., 1987). Although signs may be severe, up to 40% of patients who demonstrate central cores may be clinically normal (Shuaib et al., 1987). Diagnosis is made on the basis of the lack of oxidative enzyme activity in central regions of skeletal muscle cells (Dubowitz and Pearse, 1960), observed upon histological examination of muscle biopsies. Electron microscopic analysis shows disintegration of the contractile apparatus, ranging from blurring and streaming of the Z-lines to total loss of myofibrillar structure (Dubowitz and Roy, 1970; Isaacs et al., 1975; Hayashi et al., 1989). The sarcoplasmic reticulum and transverse tubular systems are greatly increased in content and are, in general, less well structured. The NADH-tetrazolium reductase reactions reveal pale circular areas referred to as central cores. Mitochondria are depleted in the cores, but may be enriched around the surfaces of the cores. The inheritance of CCD has been linked closely to RYR1 (Kausch et al., 1991; Mulley et al., 1993) and to four RYR1 mutations in several CCD families (Quane et al., 1993; Zhang et al., 1993). The physiological changes in muscle cells that give rise to the presence of central cores is of current research interest.

The Genetic Basis of Central-Core Disease

Genetic analysis indicates that the disorder is inherited as an autosomal dominant trait with variable penetrance (Isaacs et al., 1975; Eng et al., 1978; Byrne et al., 1982; Hayashi et al., 1989). An important feature of CCD is its close association with susceptibility to malignant hyperthermia (Denborough et al., 1973; Eng et al., 1978; Frank et al., 1980; Brownell, 1988). This association led investigators to establish a linkage between CCD and markers in chromosome 19q in large Australian (Haan et al., 1990; Mulley et al., 1993) and European (Kausch et al., 1991) CCD pedigrees. Zhang et al. (1993) linked the substitution of Arg[2434] with His to CCD in a Canadian family, obtaining a lod score of 4.8 favoring linkage with a recombinant fraction of 0.0, while Quane et al. (1993) associated CCD with three other RYR1 mutations (Table 27.1).

Mutations corresponding to fifteen human MH and CCD mutations were made in a full length rabbit RYR1 cDNA, and wild type and mutant cDNAs were transfected into HEK-293 cells (Tong et al., 1997). After about 48 hours, intact cells were loaded with the fluorescent Ca^{2+} indicator, fura-2, and intracellular Ca^{2+} release, induced by caffeine or halothane, was measured by photometry. Ca^{2+} release in cells expressing MH or CCD mutant ryanodine receptors was invariably significantly more sensitive to low concentrations of caffeine and halothane than Ca^{2+} release in cells expressing wild-type receptors or receptors mutated in other regions of the molecule. Abnormal sensitivity in the Ca^{2+} photometry assay provides supporting evidence for a causal role in MH and CCD for each of fifteen single amino acide mutations in the ryanodine receptor.

It is of interest that 8 of the 12 codons that have been shown to give rise to 15 MH or CCD mutations (Table 27.1) encode amino acids lying between Cys^{35} and Arg^{614}, while the other four codons encode amino acids lying between Arg^{2162} and Arg^{2458}. Thus, there are two clusters of MH mutations in the 5000-amino acid protein that may constitute MH regulatory domains. In each of these two regions, MH and CCD mutations are interspersed, demonstrating that there is no preferential site in the ryanodine receptor where mutation might give rise to the formation of central cores (Table 27.1). An interesting feature of the known MH and CCD mutations is that 13 of the 15 involve either loss or gain of an Arg residue. This suggests that positive charges within the two proposed MH regulatory domains are critical to regulatory function.

Arginine-614, which gives rise to two MH mutants (Table 27.1), lies in a region of the ryanodine receptor that is homologous to the inositol trisphosphate (IP_3)-binding region in the IP_3 receptor (Mignery and Südhof, 1990; Miyawaki et al., 1991). This may indicate that the cluster of amino acids between residues 35 and 614 forms a regulatory domain in the Ca^{2+}-release channel that is concerned with ligand activation of the channel. The MH and CCD mutations between Arg^{2162} and Arg^{2458} may form a second MH regulatory domain. The boundaries of these two proposed regulatory domains might be deduced from analysis of the alignments of the amino acid sequences of the ryanodine and IP_3 receptors. They are contiguous between amino acids 1 and 668 (of the ryanodine receptor), but the next 967 amino acids in the ryanodine receptor are deleted from the IP_3 receptor. The amino acid sequences are also contiguous between amino acids 1637 and 2650 of the ryanodine receptor, but, at this point, another 1045 amino acids are deleted from the IP_3 receptor, so that contiguity is re-established only after amino acid 3695 of the ryanodine receptor. The MH and CCD mutations lying between Cys^{35} and Arg^{614} fit nicely into the first conserved regulatory domain (amino acids 1–668) and mutations lying between Arg^{2162} and Arg^{2458} fit nicely into the second conserved domain (amino acids 1637–2650). Thus, the boundaries of these two conserved domains may prove to be the boundaries within which MH and CCD mutations, which affect regulation of the Ca^{2+}-release channel, will be found. The sequence that encompasses amino acid residues 2600–3000, previously proposed to include ligand-binding sites (Otsu et al., 1990), may represent a third regulatory domain that is unique to ryanodine receptors.

The Physiological Basis for Central-Core Disease

The Arg^{615} to Cys mutation is associated with muscle hypertrophy in swine, while the Arg^{2434} to His mutation, for example, is associated with variable degrees of muscle atrophy, metabolically inert cores, proximal muscle weakness, and MH (Table 27.1). If both of these mutations were to lead to poorly regulated Ca^{2+} release into the muscle cell, they could trigger spontaneous muscle contractions. Such spontaneous contractions could lead to the muscle hypertrophy observed in swine. In this case, the system of pumps and exchangers in the plasma membrane and organellar systems of mitochondria and sarcoplasmic reticulum within the cell could remove excess Ca^{2+} from the sarcoplasm without deleterious effects on the muscle cell. The CCD mutations might be more severe, leading to disorganization of the contractile proteins in the central core, a proliferation of sarcoplasmic reticulum and transverse tubules, and a loss of functional mitochondria, leading to damage to the interior of the cell and to loss of mitochondrial function and structural abnormalities in the central core (Fig. 2). These losses could, in turn, lead to muscle weakness and atrophy.

The physical alterations in the central core may be the result of physiological adaptation to functional alterations in the Ca^{2+}-release channel that lead to elevated Ca^{2+} levels within the muscle cell. Muscle cells regulate Ca^{2+} through at least four systems: plasma membrane Ca^{2+} pumps (PMCAs), sarco(endo)plasmic reticulum Ca^{2+} pumps (SERCAs), Na^+/Ca^{2+} exchangers, and mitochondria. The Ca^{2+} pumps and Na^+/Ca^{2+} exchangers in the plasma membrane can remove Ca^{2+} from the muscle cell, depositing it in extracellular spaces (Carafoli, 1987). Of these two systems, the Ca^{2+} pump has the higher affinity for Ca^{2+}, while the Na^+/Ca^{2+} exchanger is more active with elevated levels of intracellular Ca^{2+}. The sarcoplasmic reticulum is the major regulator of Ca^{2+} within the muscle cell, removing it from the cytoplasm, storing it, and releasing it again to initiate muscle contraction. If Ca^{2+} concentrations are elevated, the mitochondria can transport Ca^{2+} to matrix spaces, thereby protecting the cell from Ca^{2+}-induced damage (Wrogemann and Pena, 1976). If the Ca^{2+}-release channel were to release excessive amounts of Ca^{2+} within the muscle cell, then the Na^+/Ca^{2+} exchanger and mitochondria might play a more important role in Ca^{2+} regulation in a CCD cell than in a normal cell. Extrusion of excess Ca^{2+} from the cell might, itself, have deleterious effects on the skeletal muscle cell, which is believed to carry out intracellular cycling of a constant level of Ca^{2+} without uptake and expulsion of external Ca^{2+}.

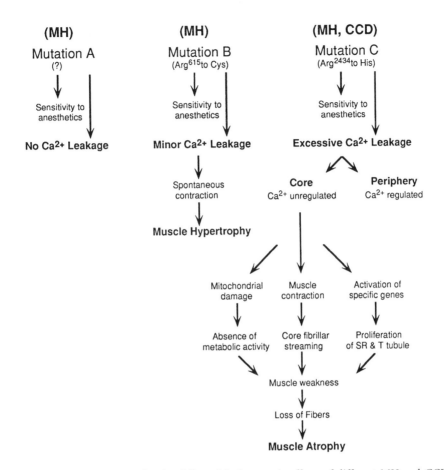

Figure 27.2 A proposed mechanism for the differential phenotypic effects of different MH and CCD mutations. (A) The MH mutations in the ryanodine receptor lead to the common phenotype of sensitivity to anesthetics. (B) Some mutations may also lead to spontaneous Ca^{2+} release sufficient to trigger spontaneous contractions. If this trigger Ca^{2+} were readily regulated, the major phenotypic effect would be spontaneous exercise-induced muscle hypertrophy. The Arg^{615} to Cys mutation was selected in swine because it leads to increased lean muscle mass. (C) Mutations leading to excessive spontaneous Ca^{2+} release may have no phenotypic effect on the periphery of the cell, but be deleterious to the central core. The Ca^{2+} released from the sarcoplasmic reticulum can be regulated by four systems, including the organellar sarcoplasmic reticulum and mitochondria and the plasma membrane Ca^{2+} pumps and Na^+/Ca^{2+} exchangers. Under normal circumstances, the bulk of the Ca^{2+} is cycled only through the sarcoplasmic reticulum. If enhanced Ca^{2+} release occurred spontaneously, in CCD muscle, the additional Ca^{2+} regulatory systems might be co-opted to regulate Ca^{2+}. The plasma membrane exchangers and pumps would be effective in regulating Ca^{2+} near the periphery, but, in the core, mitochondria may be forced to accumulate excessive Ca^{2+}, destroying themselves in the process. The degeneration of mitochondria could lead to diminished ATP production and to the degeneration of a central, possibly compartmented, core. The disorganization of the central core might be brought about by higher core levels of Ca^{2+}, which could cause contraction at the core, but not at the periphery, and lead to myofibrillar streaming and membrane disorganization. Elevated Ca^{2+} may stimulate the transcription of genes that encode proteins of the sarcoplasmic reticulum and transverse tubules that would be required to re-establish Ca^{2+} homeostasis. This may stimulate proliferation of internal membrane proteins at the transcriptional level. The phenotypic effects would be the formation of a disorganized, metabolically deficient core, which could lead to cell death and muscle atrophy. (Adapted from MacLennan, D. H. and Phillips, M. S. (1995) In *Ion Channels and Genetic Diseases* (Dawson, D. C. and Frizzell, R. A., eds.). Rockefeller University Press, New York, pp. 89–100, with permission from Rockefeller University Press.)

Pumps and exchangers in the plasma membrane might be more effective in protecting the periphery of the cell than the interior of the cell where the full burden of regulation of excess Ca^{2+} would fall on the sarcoplasmic reticulum and mitochondria. Mitochondria, which have a high capacity for Ca^{2+} uptake, would, undoubtedly, participate in removal of excess Ca^{2+} from the central core of the muscle cell and might destroy themselves in an effort to protect the cell from Ca^{2+}-induced necrosis (Wrogemann and Pena, 1976). Loss of mitochondria from the center of the cell would, in turn, lead to lower ATP synthesis and might be an underlying cause of the disorganization of the central core, leading to muscle weakness and muscle atrophy. Elevated Ca^{2+} in the interior of the muscle cell might have the same effects on the core of the muscle cell as MH would have on the whole cell. Of most interest would be its effects on muscle contraction. The differential contraction of the core of the muscle, in relation to the periphery, could lead to the disorganization and "streaming" of both fibers and membrane systems that is observed in the central cores. The profusion of sarcoplasmic reticulum and transverse tubules might be induced at the gene level by high local Ca^{2+} concentrations. Thus, mutations in $RYR1$ can lead to a spectrum of pathophysiological responses that range from muscle hypertrophy to muscle atrophy.

When CCD mutations were introduced into full-length rabbit Ca^{2+} release channel cDNA and expressed transiently in HEK-293 cells, resting Ca^{2+} concentrations were higher than in cells expressing wild-type or MH mutant RyR1 proteins, suggesting that the CCD mutants are exceptionally permeable (Tong et al., 1999). Under these conditions, HEK-293 cells expressing both MH and CCD mutant RyR1 exhibited lower maximal peak amplitudes of caffeine-induced Ca^{2+} release than cells expressing wild-type RyR1, suggesting that both MH and CCD mutants were more leaky so that Ca^{2+} stores were reduced in size. The content of endogenous sarco(endo)plasmic reticulum Ca^{2+} ATPase isoform 2b (SERCA2B) was increased in HEK-293 cells expressing wild-type or mutant RyR1, supporting the view that sarco- or endoplasmic reticulum Ca^{2+} storage capacity is increased as a compensatory response to an enhanced Ca^{2+} leak.

Brody Disease

Brody disease (Brody, 1969) was first defined as a disorder of muscle function characterized by painless muscle cramping and exercise-induced impairment of muscle relaxation. In a normal muscle contraction/relaxation cycle, Ca^{2+} is released from the sarcoplas-

mic reticulum into the cytoplasm, where it binds to troponin in the thin filament, releasing constraints on the interaction between actin and myosin and inducing muscle contraction (Zot and Potter, 1987). Then, Ca^{2+} is pumped back into the lumen of the sarcoplasmic reticulum by a Ca^{2+} pump to initiate relaxation. In his studies of a skeletal muscle biopsy, Brody (1969) showed that the sarcoplasmic reticulum from his patient was deficient in both Ca^{2+} uptake and Ca^{2+}-ATPase activity. Consequently, his report focused attention on the possibility that defects in the Ca^{2+} pump might underlie the disease. This possibility was supported by studies in which fast-twitch skeletal muscle fibers from four Brody patients were shown to be deficient in Ca^{2+}-ATPase (Karpati et al., 1986; Danon et al., 1988), but it was undermined by studies of Benders et al. (1994) in which a 50% decrease in the activity of the fast-twitch Ca^{2+}-ATPase isoform, SERCA1, was recorded, even though no reduction in protein content was observed. Recent analysis of the $ATP2A1$ gene that encodes SERCA1 has also provided conflicting results. In two Brody families, both $ATP2A1$ alleles were shown to code for truncated, inactive forms of SERCA1 (Odermatt et al., 1996), but, in four other Brody families, $ATP2A1$ was unaltered (Zhang et al., 1995; Odermatt et al., 1996). Thus, current research is aimed at the understanding of the heterogeneous genetic origins of Brody disease (or syndrome).

The diagnosis of Brody disease (Odermatt et al., 1996) is based on a lifelong history of difficulty in performing sustained, strenuous muscular activities, such as running upstairs, because the muscles stiffen during exercise and, temporarily, cannot be used. The exercise-induced delay in muscle relaxation involves the legs, arms, and eyelids, and may be worse in cold weather. Clinical examination demonstrates normal strength with a single effort, normal sensation, and normal deep tendon reflexes, but progressive difficulty in relaxing muscles during repeated forceful contraction. Percussion myotonia is absent. Standard needle electromyography shows normal spontaneous, insertion, and voluntary activity, but no electrical activity nor myotonic discharges in the muscles after exercise and during the delayed relaxation. The disease is not life threatening and no specific therapy is indicated.

There are no recorded estimates of the incidence of Brody disease, partly because the syndrome is difficult to diagnose and partly because of its rarity. Inheritance of Brody disease is autosomal recessive in the two cases where a genetic diagnosis has been possible (Odermatt et al., 1996), but it is also heterogeneous in origin (Zhang et al., 1995) and, in some families, inheritance might be dominant. A rough estimate of the incidence of the disease might be 1 in 10,000,000 births. While this figure seems rare indeed, the incidence of carriers required to achieve this incidence would be 1 in 1600, if the disease were invariably autosomal recessive.

Physiological Basis for Brody Disease

All of those who have investigated the role of the Ca^{2+}-ATPase in Brody disease have found diminished Ca^{2+}-ATPase activity, but reports of the fraction of Ca^{2+}-ATPase that remains have varied dramatically (Karpati et al., 1986; Danon et al., 1988; Taylor et al., 1988; Benders et al., 1994). Brody (1969) reported that ATP-dependent Ca^{2+} uptake in the microsomal fraction of skeletal muscle from his patient was reduced to 25% of control values. Studies by Taylor et al. (1988) showed that the Ca^{2+}-ATPase activity in skeletal muscle microsomes of a single Brody patient was reduced to about 10% of the activity of control samples. Karpati et al. (1986) measured Ca^{2+} uptake in microsomal fractions from Brody muscle and found it to be reduced to 2% or less of control values.

In mammalian tissues, three different *ATP2A* genes encode five different sarco(endo)plasmic reticulum Ca^{2+}-ATPase (SERCA) proteins (MacLennan et al., 1985; Brandl et al., 1986, 1987; Lytton and MacLennan, 1988; Burk et al., 1989; Lytton et al., 1989). The SERCA1a protein is encoded by the *ATP2A1* gene located on human chromosome 16p12.1–p12.2 (Callen et al., 1995), the *ATP2A2* gene that encodes SERCA2 is located on chromosome 12q23–q24.1 (Otsu et al., 1993), and the *ATP2A3* gene that encodes SERCA3 is located on chromosome 17q13.3 (Dode et al., 1996). The human *ATP2A1* gene contains 23 exons, spanning about 26 kb of genomic DNA (Zhang et al., 1995). Exon 22 is retained to form the adult SERCA1a protein, ending in Gly, while alternative splicing, leading to the removal of exon 22, leads to the formation of the neonatal SERCA1b protein, terminating in the sequence Glu-Asp-Pro-Glu-Asp-Glu-Arg-Arg-Lys (Brandl et al., 1986, 1987). The SERCA1a protein accounts for more than 99% of SERCA isoforms expressed in adult fast-twitch skeletal muscle (type II) fibers, while SERCA1b predominates in neonatal fibers (Brandl et al., 1987; Wu et al., 1995). A similar, but not identical, splicing occurs in the *ATP2A2* gene, creating the cardiac/slow-twitch isoform SERCA2a and the smooth muscle/nonmuscle isoform, SERCA2b (Lytton and MacLennan, 1988). The SERCA1 isoforms are not expressed to any significant extent in any tissue other than fast-twitch skeletal muscle, while SERCA2 isoforms are expressed to some extent in virtually all other tissues (Wu et al., 1995). SERCA3 also has complex splicing at the C-terminal end (Poch et al., 1998).

Supporting evidence for the view that defects in the Ca^{2+}-ATPase are involved in Brody disease was presented by Karpati et al. (1986) and by Danon et al. (1988), who showed that polyclonal and monoclonal antibodies against the Ca^{2+}-ATPase from chicken fast-twitch skeletal muscle sarcoplasmic reticulum reacted only poorly, if at all, with histochemical type 2 (fast-twitch) fibers in sections of the skeletal muscle of four Brody patients. The antibodies did, however, react with the Ca^{2+}-ATPase in type 1 fibers. These studies suggested that the Ca^{2+}-ATPase protein was either absent from type 2 fibers or was present in a form with altered antigenicity. In cross sections of muscle biopsies, type 2 fibers appeared angular and atrophied, whereas type 1 fibers appeared normal. Since the SERCA1 isoform predominates in type 2 (fast-twitch) fibers, while the SERCA2 isoform predominates in type 1 (slow-twitch) fibers (Brandl et al., 1987), the studies of Karpati et al. (1986) and of Danon et al. (1988) imply that defects in the SERCA1 isoform of the Ca^{2+}-ATPase might be directly involved in the manifestation of Brody disease.

Benders et al. (1994) analyzed both Ca^{2+}-ATPase activity and Ca^{2+}-ATPase protein content in 10 Brody patients. Using an antibody specific against SERCA1, they estimated that 83% of the total Ca^{2+}-ATPase in both Brody patient and control muscle homogenates was SERCA1. Moreover, the content of total Ca^{2+}-ATPase protein and SERCA1 protein, measured both by immunoreactivity and by phosphorylation, was identical in the two sample populations. The Ca^{2+}-stimulated ATPase activity in Brody muscle homogenates, however, was reduced to only 50% of the activity found in comparable samples from normal patients. A similar pattern was observed when ATPase activities and protein contents were measured in cells cultured from normal and Brody disease muscle.

The findings of Karpati et al. (1986) and those of Benders et al. (1994) implicate SERCA1 in Brody disease, but support very different views of its etiology. The results of Karpati et al. (1986) are consistent with mutations in *ATP2A1* that cause virtually 100% loss of SERCA1 protein and Ca^{2+}-ATPase activity. Such mutations might affect translation or they might result in a protein that is degraded before it is functionally incorporated into the bilayer. The studies of Benders et al. (1994) are consistent with mutations in *ATP2A1* that do not affect SERCA1 expression, but diminish its V_{max} to about 50% of control values. Studies of mutagenesis of the Ca^{2+}-ATPase (MacLennan et al., 1992) demonstrate many examples of mutations or small deletions that affect expression and stability of the Ca^{2+}-ATPase and many examples of mutations that reduce Ca^{2+}-ATPase activity by 10–90%.

Genetic Analysis of Brody Disease

In order to determine whether genetic defects exist in the *ATP2A1* gene, sequencing of *ATP2A1* cDNA or genomic DNA has now been carried out on six unrelated Brody disease probands (Zhang et al., 1995; Odermatt et al., 1996, 1997). Amplification of the full-length *ATP2A1* cDNA from two of these patients did not reveal any deletion or apparent alternative splicing, nor were any defects detected in the analysis of the *ATP2A1* cDNAs from these two patients (Zhang et al., 1995). Although quantitative analysis of mRNA levels was not carried out, it was evident from the ease with which PCR amplification was accomplished that the mRNA was relatively abundant. In a third Brody proband, the 23 individual *ATP2A1* exons, plus the flanking intron sequence and several hundred base pairs of upstream sequence in the *ATP2A1* gene, were amplified and sequenced. Again, no evidence for a mutation in *ATP2A1* was found (Zhang et al., 1995).

In a study of two more Brody disease families, clinical evaluation identified three individuals in one family and two individuals in a second family with typical signs of Brody disease. A 50% reduction in Ca^{2+}-ATPase and Ca^{2+}-transport activities had been reported in the first family (Benders et al., 1994), while a complete loss in SERCA1 had been reported in the second family (Karpati et al., 1986). Since the *ATP2A1* gene that encodes SERCA1 has been localized to chromosome 16 in the interval between *D16S297* and *D16S288* (Callen et al., 1995), haplotype analysis was carried out on these families using genetic markers between *D16S295* and *D16S304* (Odermatt et al., 1996). In the first family, both parents inherited the same haplotype for the 6.6 cM interval between markers *D16S288* and *D16S304*. Two affected patients (one affected family member was unavailable) inherited this common haplotype from each parent, but a crossover between markers *D16S298* and *D16S300* in the maternal chromosome inherited by one of them limited the region of haplotype identity so that the two were homozygous only for the *D16S288/D16S298* haplotype. The unaffected parents and an unaffected sibling were heterozygous for this haplotype and two other unaffected siblings did not carry this haplotype in either chromosome. These results were consistent with autosomal recessive inheritance of Brody disease.

In the second family, the parents also shared an identical haplotype on most of one chromosome. Both affected patients inherited the maternal copy of this common haplotype and both inherited the different paternal haplotype. Inheritance by the two patients of the same two *ATP2A1* intervals was, again, consistent with the autosomal recessive inheritance of Brody disease, suggesting that, for these two

families, a defect was located within the interval flanked by *D16S297* on the telomeric side and by *D16S300* on the centromeric side (containing the *ATP2A1* gene).

Sequencing of exon 7 in the proband from the first family and comparison with normal *ATP2A1* sequence (Zhang et al., 1995) revealed the homozygous mutation of C592 to T, resulting in the mutation of the Arg^{198} codon, CGA, to the stop codon, TGA (Table 27.1). The truncated product would be devoid of phosphorylation, nucleotide-binding, and Ca^{2+}-binding domains and, almost certainly, of activity (MacLennan et al., 1992). Genetic data were fully consistent with the recessive inheritance of the C592 to T mutation as the causal factor for Brody disease in the first family.

Sequencing of amplified genomic DNA from the proband in the second family revealed two different mutations in *ATP2A1*, consistent with the inheritance of the two different haplotypes. The mutation of C2025 to A in the TGC codon for Cys^{675} in exon 15 of the paternally inherited chromosome created the stop codon, TGA, predicted to lead to a truncated protein of 674 amino acids (Table 27.1). The truncated gene product would contain phosphorylation and nucleotide-binding domains, but the Ca^{2+}-binding domain would be disrupted (MacLennan et al., 1992). The mutation of the invariant GT dinucleotide to CT at the splice donor site of intron 3 in the maternally inherited chromosome (Table 27.1) was predicted to lead preferentially to skipping of exon 3 and less frequently to partial retention of intron 3 (Krawczak et al., 1992). If exon 2 of *ATP2A1* were spliced to exon 4 and transcribed, the product would be truncated, consisting of 45 normal amino acids, followed by 5 novel amino acids. If intron 3 were retained partially and transcribed, the product would also be truncated, consisting of 73 normal amino acids, followed by 49 novel amino acids. Both potential gene products would be missing phosphorylation, nucleotide-binding and Ca^{2+}-binding domains, and activity (MacLennan et al., 1992). Thus, in the second family, the inheritance of Brody disease was autosomal recessive and was associated with compound heterozygosity for the two *ATP2A1* mutations, leading to the prediction of total loss of SERCA1 protein and activity.

Since one of the patients in the original study of Zhang et al. (1995) had been included in the immunohistochemical analysis of Karpati et al. (1986) and had been scored as SERCA1-deficient, the sequence of the *ATP2A1* gene in this family was re-evaluated (Odermatt et al., 1997). Two affected brothers in this family were found to have the identical haplotype in the segment of chromosome 16p12 where *ATP2A1* is located. Resequencing of the *ATP2A1* gene then led to the discovery of the homozygous deletion of a C in

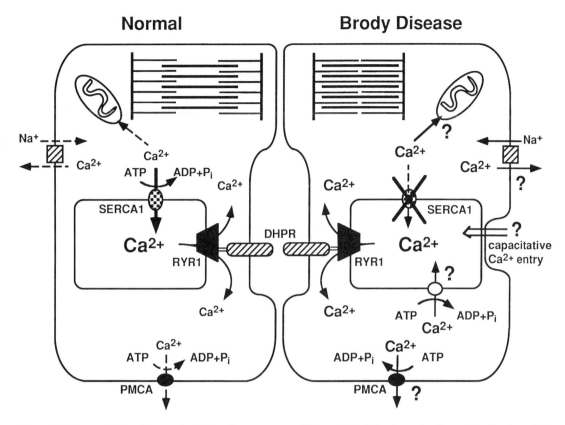

Figure 27.3 A model for the exercise-induced contracture of fast-twitch skeletal muscle observed in Brody patients and for the activation of compensatory relaxation mechanisms. In a normal contraction/relaxation cycle (left), Ca^{2+} is released from the sarcoplasmic reticulum through a Ca^{2+}-release channel (*RYR1*) to initiate contraction. The Ca^{2+} is pumped back into the lumen of the sarcoplasmic reticulum by a Ca^{2+}-ATPase (SERCA1), rapidly lowering cytoplasmic Ca^{2+} concentrations and leading to relaxation. In those Brody patients who lack functional SERCA1 (right), Ca^{2+} removal from the cytoplasm is delayed and prolonged exercise induces higher cytoplasmic Ca^{2+} concentrations, leading to contracture. The Ca^{2+} is removed from the cytoplasm, eventually, permitting relaxation, and the store of Ca^{2+} in the sarcoplasmic reticulum is replenished, since subsequent contractions are unaffected (Benders et al., 1994). These observations suggest that other Ca^{2+} regulatory systems are activated. The simplest compensation would involve expression of SERCA2 or SERCA3 as replacements for inactive SERCA1. In this way, the Ca^{2+} store would be replenished directly. A more circuitous compensatory route would involve activation of plasma membrane Ca^{2+}-ATPases (PMCA) or Na^+/Ca^{2+} exchangers, which would eject Ca^{2+} from the cell. In this case, the sarcoplasmic reticulum would have to be reloaded with Ca^{2+} from the extracellular space, perhaps by capacitative Ca^{2+} entry (Berridge, 1994).

a series of three Cs in exon 5, leading to a frameshift at Pro^{147} in both affected brothers that would truncate and inactivate both copies of SERCA1. Analysis of the *ATP2A1* sequence in a sixth Brody family did not uncover any defect (Odermatt et al., 1997).

The three families in which *ATP2A1* mutations truncate and inactivate *SERCA1* demonstrate recessive inheritance of Brody disease. For those families in which Brody disease shows dominant inheritance, *ATP2A1* is probably not the causal gene (Odermatt et al., 1997). It is, however, conceivable that defects might occur in a gene that encodes a protein which either directly or indirectly modifies the specific activity of SERCA1. In a study of the sarcolipin (SLN) gene, which encodes a potential SERCA1 inhibitor, Odermatt et al. (1997) were unable to detect mutations that might be linked to Brody disease. Wevers et al. (1992) and Bender et al. (1994) showed that dantrolene, which reduces myofibrillar Ca^{2+} concentration by blocking Ca^{2+} release from the sarcoplasmic reticulum (Ohta et al., 1990; Nelson and Lin, 1993), is effective in the treatment of some, but not all, Brody patients. Therefore, the possibility must be considered that mutations in other components of muscle Ca^{2+} regulation might be responsible for the clinical manifestation of some forms of Brody disease.

In spite of the predicted absence of SERCA1 in the several Brody patients who inherited *ATP2A1* gene defects, all are able to relax their fast-twitch skeletal muscles, although at a significantly reduced rate. In cultured muscle cells from both patients in one of these families and controls, the sarcoplasmic Ca^{2+} concentration at rest and the increase in intracellular Ca^{2+} concentration after addition of acetylcholine were found to be the same (Benders et al., 1994). However, the time required to reach resting intracellular Ca^{2+} levels after Ca^{2+} release was increased several-fold in cells derived from these patients. Thus, the phenotype was consistent with reduced SERCA1 activity, but not with complete loss of SERCA1 function.

Concentrations of Ca^{2+} in Brody muscle might be lowered through a combination of other mechanisms which regulate Ca^{2+} in muscle (Carafoli, 1987). There might be compensatory Ca^{2+} removal by plasma membrane Ca^{2+}-ATPases (PMCAs), by Na^+/Ca^{2+} exchangers in the plasma membrane, by mitochondrial Ca^{2+} uptake, or by the proliferation of sarcoplasmic or endoplasmic reticulum that contains compensating levels of SERCA2 (MacLennan et al., 1985) or SERCA3 (Burk et al., 1989) isoforms (Fig. 3). Of these possible compensatory processes, only the latter would be predicted to result in Ca^{2+} loading of the sarcoplasmic reticulum, a process necessary for subsequent muscle contraction. As an alternative, refilling of Ca^{2+}-depleted sarcoplasmic reticulum might be possible through some form of capacitative Ca^{2+} entry (Berridge, 1995).

The literature on the content of Ca^{2+}-ATPase protein in Brody patients is conflicting. Benders et al. (1994) reported a normal Ca^{2+}-ATPase protein content and a 50% reduction in Ca^{2+}-ATPase activity in two patients from one *ATP2A1*-deficient family, while genetic study would predict a complete loss of SERCA1 protein and function (Odermatt et al., 1996). Further evaluation of these discrepancies will be necessary. Present findings, however, provide clear genetic evidence that *ATP2A1* is a candidate gene for at least one autosomal recessive form of Brody disease and support the earlier view that the loss of SERCA1 function underlies the manifestation of this form of the disease. Further investigation will also be necessary to determine the cause of the identical syndrome in patients who do not have an *ATP2A1* defect and who seem to inherit an autoso-mal dominant form of the disease (Odermatt et al., 1997).

Acknowledgments   We thank our many colleagues for advice and discussion in the preparation of this review chapter. Research grants to D. H. M., supporting original work from our laboratory, were from the Medical Research Council of Canada (MRCC), the Muscular Dystrophy Association of Canada (MDAC), the Heart and Stroke Foundation of Ontario (HSFO), and the Canadian Genetic Diseases Network of Centers of Excellence. J. L. and A. O. are postdoctoral fellows of the MRCC.

References

Allen, G. C., M. G. Larach, and A. R. Kunselman (1998) The sensitivity and specificity of the caffeine halothane contracture test: a report from the North American Malignant Hyperthermia Registry. *Anesthesiology* 88: 579–88.

Andersen, E. and P. Jensen (1977) Close linkage established between the HAL locus for halothane sensitivity and the PHI (phosphohexose isomerase) locus in pigs of the Danish Landrace breed. *Nord. Vet. Med.* 29: 502–4.

Archibald, A. L. and P. Imlah (1985) The halothane sensitivity locus and its linkage relationships. *Anim. Blood Groups Biochem. Genet.* 16: 253–63.

Ball, S. P. and K. J. Johnson (1993). The genetics of malignant hyperthermia. *J. Med. Genet.* 30: 89–93.

Benders, A. A. G. M., J. H. Veerkamp, A. Oosterhof, P. J. H. Jongen, R. J. M. Bindels, L. M. E. Smit, H. F. M. Busch, and R. A. Wevers (1994) Ca^{2+} homeostasis in Brody's disease. A study in skeletal muscle and cultured muscle cells and the effects of dantrolene and verapamil. *J. Clin. Invest.* 94:741–48.

Berman, M. C., G. G. Harrison, A. B. Bull, and J. E. Kench (1970) Changes underlying halothane-induced malignant hyperthermia in Landrace pigs. *Nature* 225: 653–55.

Berridge, M. J. (1995) Capacitative calcium entry. *Biochem. J.* 312: 1–11.

Brandl, C. J., N. M. Green, B. Korczak, and D. H. MacLennan (1986) Two Ca^{2+}-ATPase genes: homologies and mechanistic implication of deduced amino acid sequences. *Cell* 44: 597–607.

Brandl, C. J., S. DeLeon, D. R. Martin, and D. H. MacLennan (1987) Adult forms of the Ca^{2+}-ATPase of sarcoplasmic reticulum: expression in developing skeletal muscle. *J. Biol. Chem.* 262: 3768–74.

Britt, B. A. (1991) Malignant hyperthermia: a review. In *Thermoregulation: Pathology, Pharmacology and Therapy* (Schonbaum, E. and P. Lomax, eds.). Pergamon Press, New York. pp. 179–292.

Britt, B. A. and W. Kalow (1968) Hyperrigidity and hyperthermia associated with anaesthesia. *Ann. N.Y. Acad. Sci.* 151: 947–58.

Britt, B. A. and W. Kalow (1970) Malignant hyperthermia: a statistical review. *Can. Anaesth. Soc. J.* 17: 293–315.

Britt, B. A., W. Kalow, and L. Endrenyi (1978) *Malignant Hyperthermia—Patterns of Inheritance in Swine* (Second International Symposium on Malignant Hyperthermia) New York, Grune and Stratton, pp. 195–211.

Brody, I. A. (1969) Muscle contracture induced by exercise. A syndrome attributable to decreased relaxing factor. *N. Engl. J. Med.* 281: 187–92.

Brostrom, C. O., F. L. Hunkeler, and E. G. Krebs (1971) The regulation of skeletal muscle phosphorylase kinase by Ca^{2+}. *J. Biol. Chem.* 246: 1961–67.

Brownell, A. K. W. (1988) Malignant hyperthermia: relationship to other diseases. *Br. J. Anaesth.* 60: 303–8.

Brownell, A. K. W., R. T. Paasuke, A. Elash, S. B. Fowlow, C. G. F. Seagram, R. J. Diewold, and C. Friesen (1983) Malignant hyperthermia in Duchenne muscular dystrophy. *Anaesthesiology* 58: 180–82.

Burk, S. E., J. Lytton, D. H. MacLennan, and G. E. Shull (1989) cDNA cloning, functional expression and mRNA tissue distribution of a third organellar Ca^{2+} pump. *J. Biol. Chem.* 264: 18561–68.

Byrne, E., P. C. Blumbergs, and J. F. Hallpike (1982) Central core disease. Study of a family of five affected generations. *J. Neurol. Sci.* 53: 77–83.

Callen, D. F., S. A. Lane, H. Kozman, G. Kremmidiotis, S. A. Whitmore, M. Lowenstein, N. A. Doggett, N. Kenmochi, D. C. Page, D. R. Maglott, W. C. Nierman, K. Murakawa, R. Berry, J. M. Sikela, R. Houlgatte, C. Auffray, and G. R. Sutherland (1995) Integration of transcript and genetic maps of chromosome 16 at near-1-Mb resolution: demonstrations of a "hot spot" for recombination at 16p12. *Genomics* 29: 503–11.

Carafoli, E. (1987) Intracellular calcium homeostasis. *Annu. Rev. Biochem.* 56: 395–433.

Chowdhary, B. P., I. Harbitz, A. Makinen, W. Davies, and I. Gustavvson (1989) Localization of the glucose phosphate isomerase gene to p12–q21 segment of chromosome 6 in pig by *in situ* hybridization. *Hereditas* 111: 73–78.

Danon, M. J., G. Karpati, J. Charuk, and P. Holland (1988) Sarcoplasmic reticulum adenosine triphosphatase deficiency with probable autosomal dominant inheritance. *Neurology* 38: 812–15.

Davies, W., I. Harbitz, R. Fries, G. Stranzinger, and J. G. Hauge (1988) Porcine malignant hyperthermia carrier detection and chromosomal assignment using a linked probe. *Anim. Genet.* 19: 203–12.

Denborough, M. A., J. F. A. Forster, and R. R. H. Lovell (1962) Anaesthetic deaths in a family. *Br. J. Anaesth.* 34: 395.

Denborough, M. A., X. Dennett, and R. McD. Anderson (1973) Central-core disease and malignant hyperpyrexia. *Br. Med. J.* 1: 272–73.

Deufel, T., R. Sudbrak, Y. Feist, B. Rubsam, I. Du Chesne, K.-L. Schafer, N. Roewer, T. Grimm, F. Lehmann-Horn, E. J. Hartung, and C. R. Muller (1995) Discordance in a malignant hyperthermia pedigree between *in vitro* contracture-test phenotypes and haplotypes for the MHS1 region on chromosome 19q12–13.2 comprising the C1840T transition in the *RYR1* gene. *Am. J. Hum. Genet.* 56: 1334–42.

Dode, L., F. Wuytack, P. F. Kools, F. Babia-Aissa, L. Raeymaekers, F. Brike, W. J. van de Ven and R. Casteels (1996) cDNA cloning, expression and chromosomal localization of the human sarco/endoplasmic reticulum Ca(2+)-ATPase 3 gene. *Biochem. J.* 318: 689–699.

Dubowitz, V. and A. G. E. Pearse (1960) Oxidative enzymes and phosphorylase in central-core disease of muscle. *Lancet*: 23–24.

Dubowitz, V. and S. Roy (1970) Central core disease of muscle: clinical histochemical and electron microscopic studies of an affected mother and child. *Brain* 93:133–46.

Ebashi, S., M. Endo, and I. Ohtsuki (1969) Control of muscle contraction. *Quart. Rev. Biophys.* 2: 351–84.

Ellis, F. R., N. P. Keaney, D. G. F. Harriman, D. W. Sumner, K. Kyei-Mensah, J. H. Tyrrell, J. B. Hargreaves, R. K. Parkh, and P. L. Mulrooney (1972) Screening for malignant hyperthermia. *Br. Med. J.* 3: 559–61.

Endo, M., S. Yagi, T. Ishizuka, K. Horiuti, Y. Koga, and K. Amaha (1983) Changes in the Ca-induced Ca release mechanism in sarcoplasmic reticulum from a patient with malignant hyperthermia. *Biomed. Res.* 4: 83–92.

Eng, G. D., B. S. Epstein, W. K. Engel, D. W. McKay, and R. McKay (1978) Malignant hyperthermia and central core disease in a child with congenital dislocating hips. *Arch. Neurol.* 35: 189–97.

European MH Group (1984) Malignant hyperpyrexia, a protocol for the investigation of malignant hyperthermia (MH) susceptibility. *Br. J. Anaesth.* 56: 1267–69.

Fill, M., R. Coronado, J. R. Mickelson, J. Vilven, J. Ma, B. A. Jacobson, and C. F. Louis (1990) Abnormal ryanodine receptor channels in malignant hyperthermia. *Biophys. J.* 50: 471–75.

Fill, M., E. Stefani, and T. E. Nelson (1991) Abnormal human sarcoplasmic reticulum Ca^{2+} release channels in malignant hyperthermia skeletal muscle. *Biophys. J.* 59: 1085–90.

Fletcher, J. E., P. A. Calvo, and H. Rosenberg (1993) Phenotypes associated with malignant hyperthermia susceptibility in swine genotyped homozygous or heterozygous for the ryanodine receptor mutation. *Br. J. Anaesth.* 71: 410–17.

Frank, J. P., Y. Harati, I. J. Butler, T. E. Nelson, and C. I. Scott (1980) Central core disease and malignant hyperthermia syndrome. *Ann. Neurol.* 7: 11–17.

Fujii, J., K. Otsu, F. Zorzato, S. de Leon, V. K. Khanna, J. Weiler, P. J. O'Brien, and D. H. MacLennan (1991) Identification of a mutation in porcine ryanodine receptor associated with malignant hyperthermia. *Science* 253: 448–51.

Gahne, B. and R. K. Juneja (1985) Prediction of the halothane (Hal) genotypes of pigs by deducing Hal, Phi, Po2, Pgd haplotypes of parents and offspring: results from a large-scale practice in Swedish breed. *Anim. Blood Groups Biochem. Genet.* 16: 265–83.

Giannini, G., E. Clementi, R. Ceci, G. Marziali, and V. Sorrentino (1992) Expression of a ryanodine receptor-Ca^{2+} channel that is regulated by TGF-β. *Science* 257: 91–94.

Gibson, C. L., A. B. Johnson, G. E. Brewer, J. P. Tuttle, and A. V. Moschcowitz (1900) Heat-stroke as a post operative complication. *J. Am. Med. Assoc.* 35: 1685.

Gillard, E. F., K. Otsu, J. Fujii, V. K. Khanna, S. de Leon, J. Derdemezi, B. A. Britt, C. L. Duff, R. G. Worton, and D. H. MacLennan (1992a) A substitution of cysteine for arginine-614 in the ryanodine receptor is potentially causative of human malignant hyperthermia. *Genomics* 11: 751–55.

Gillard, E. F., K. Otsu, J. Fujii, C. L. Duff, S. de Leon, V. K. Khanna, B. A. Britt, R. G. Worton, and D. H. MacLennan (1992b) Polymorphisms and deduced amino acid substitutions in the coding sequence of the ryanodine receptor (RYR1) gene in individuals with malignant hyperthermia. *Genomics* 13:1247–54.

Haan, E. A., C. J. Freemantle, J. A. McCure, K. L. Friend, and J. C. Mulley (1990) Assignment of the gene for central core disease to chromosome 19. *Hum. Genet.* 86: 187–90.

Hakamata, Y., J. Nakai, H. Takeshima, and K. Imoto (1992) Primary structure and distribution of a novel ryanodine receptor/calcium release channel from rabbit brain. *FEBS Lett.* 312: 229–35.

Hall, L. W., N. Woolf, J. W. Bradley, and D. W. Jolly (1966) Unusual reaction to suxamethonium chloride. *Br. Med. J.* 2: 1305.

Harbitz, L., B. Chowdhary, P. Thomsen, W. Davies, U. Kaufman, S. Kran, I. Gustavsson, K. Christensen, and J. Hauge (1990) Assignment of the porcine calcium release channel gene, a candidate for the malignant hyperthermia locus, to the 6p11-q21 segment of chromosome 6. *Genomics* 9: 243–48.

Harrison, G. G. (1975) Control of the malignant hyperpyrexic syndrome in MHS swine by dantrolene sodium. *Br. J. Anaesth.* 47: 62.

Hayashi, K., R. G. Miller, and A. K. W. Brownell (1989) Central core disease: ultrastructure of the sarcoplasmic reticulum and T-tubules. *Muscle Nerve* 12: 95–102.

Heiman-Patterson, T., H. Rosenberg, J. E. Fletcher, and A. J. Tahmoush (1988) Malignant hyperthermia in myotonia congenita, halothane–caffeine contracture testing in neuromuscular disease. *Muscle Nerve* 11: 453–57.

Hogan, K., F. Couch, and P. A. Powers (1992) A cysteine-for-arginine substitution (R614C) in the human skeletal muscle calcium release channel cosegregates with malignant hyperthermia. *Anesth. Analg.* 75: 441–48.

Iaizzo, P. A. and F. Lehmann-Horn (1995) Anesthetic complications in muscle disorders. *Anesthesiology* 82: 1093–96.

Iles, D. E., F. Lehmann-Horn, S. W. Scherer, L. C. Tsui, D. O. Weghuis, R. F. Suijkerbuijk, L. Heytens, G. Mikala, A. Schwartz, F. R. Ellis, A. D. Stewart, and B. Wieringa (1994) Localization of the gene encoding the a2/d-subunits of the L-type voltage-dependent calcium channel to chromosome 7q and analysis of the segregation of flanking markers in malignant hyperthermia susceptible families. *Hum. Mol. Genet.* 3: 969.

Inui, M., A. Saito, and S. Fleischer (1987) Purification of the ryanodine receptor and identity with feet structures of junctional terminal cisternae of sarcoplasmic reticulum from fast skeletal muscle. *J. Biol. Chem.* 262: 1740–47.

Isaacs, H., J. J. A. Heffron, and M. Badenhorst (1975) Central core disease. A correlated genetic, histochemical, ultramicroscopic and biochemical study. *J. Neurol. Neurosurg. Psychiat.* 38: 1177–86.

Kalow, W., B. A. Britt, and M. E. Terreau (1970) Metabolic error of muscle metabolism after recovery from malignant hyperthermia. *Lancet* ii: 895–98.

Karpati, G., J. Charuk, S. Carpenter, C. Jablecki, and P. Holland (1986) Myopathy caused by a deficiency of Ca^{2+}-adenosine triphosphatase in sarcoplasmic reticulum (Brody's disease). *Ann. Neurol.* 20: 38–49.

Kausch, K., F. Lehmann-Horn, M. Janka, B. Wieringa, T. Grimm, and C. R. Müller (1991) Evidence for linkage of the central core disease locus to the proximal long arm of human chromosome 19. *Genomics* 10: 765–69.

Keating, K. E., K. A. Quane, B. M. Manning, M. Lehane, E. Hartung, K. Censier, A. Urveyler, M. Klausnetger, C. R. Müller, J. J. A. Heffron, and T. V. McCarthy (1994) Detection of a novel RYR1 mutation in four malignant hyperthermia pedigrees. *Hum. Mol. Genet.* 3: 1855–58.

Keating, K. E., L. Giblin, P. J. Lynch, K. A. Quane, M. Lehane, J. J. A. Heffron, and T. V. McCarthy (1997) Detection of a novel mutation in the ryanodine receptor gene in an Irish malignant hyperthermia pedigree: correlation of the IVCT response with the affected and unaffected haplotypes. *J. Med. Genet.* 34(4): 291–96.

King, J. O. and M. A. Denborough (1973) Anaesthetic-induced malignant hyperpyrexia in children. *J. Pediatr.* 83: 37–40.

Knudson, C. M., J. R. Mickelson, C. F. Louis, and K. P. Campbell (1990) Distinct immunopeptide maps of the sarcoplasmic reticulum Ca^{2+} release channel in malignant hyperthermia. *J. Biol. Chem.* 265: 2421–30.

Kosko, J. R., B. W. Brandom, and K. H. Chan (1992) Masseter spasm and malignant hyperthermia: a retrospective review of a hospital-based pediatric otolaryngology practice. *Int. J. Pediatr. Otorhinolaryngol.* 23: 45–50.

Krawczak, M., J. Reiss, and D. N. Cooper (1992) The mutational spectrum of single base-pair substitutions in messenger RNA splice junctions of human genes — causes and consequences. *Hum. Genet.* 90: 41–54.

Larach, M. G., for the North American Malignant Hyperthermia Group (1989) Standardization of the caffeine halothane muscle contracture test. *Anesth. Analg.* 69: 511–15.

Larach, M. G. (1993) Should we use muscle biopsy to diagnose malignant hyperthermia susceptibility? *Anesthesiology* 79: 1–4.

Larach, M. G., J. R. Landis, J. S. Bunn, and M. Diaz (1992a) Prediction of malignant hyperthermia susceptibility in low-risk subjects; an epidemiologic investigation of caffeine halothane contracture responses. *Anesthesiology* 76: 16–27.

Larach, M. G., J. R. Landis, B. S. Shirk, and M. Diaz (1992b) Prediction of malignant hyperthermia susceptibility in man, improving sensitivity of the caffeine halothane contracture test. *Anesthesiology* 77: A1052.

Larach, M. G., A. R. Localio, G. C. Allen, M. A. Denborough, F. R. Ellis, G. A. Gronert, R. F. Kaplan, S. M. Muldoon, T. E. Nelson, H. Ording, H. Rosenberg, B. E. Waud, and D. J. Wedel (1994) A clinical grading scale to predict malignant hyperthermia susceptibility. *Anesthesiology* 80: 771.

Levitt, R. C., A. Olckers, S. Meyers, J. E. Fletcher, H. Rosenberg, H. Isaacs, and D. A. Meyers (1992) Evidence for the localization of a malignant hyperthermia susceptibility locus (MHS2) to human chromosome 17q. *Genomics* 14: 562–66.

Lusis, A. J., C. Heinzmann, R. S. Sparkes, J. Scott, T. J. Knott, R. Geller, M. C. Sparkes, and T. Mohandas (1986) Regional mapping of human chromosome 19: organization of genes for plasma lipid transport (APOC1, -C2 and -E and LDLR) and the genes C3, PEPD and GPI. *Proc. Natl. Acad. Sci. USA* 83: 3929–33.

Lynch, P. J., R. Krivosic-Horber, H. Reyford, N. Monnier, K. Quane, P. Adnet, G. Haudecoeur, I. Krivosic, T. McCarthy, and J. Lunardi (1997) Identification of heterozygous and homozygous individuals with the novel RYR1 mutation Cys35Arg in a large kindred. *Anesthesiology* 86: 620–26.

Lytton, J. and D. H. MacLennan (1988) Molecular cloning of cDNAs from human kidney coding for two alternatively spliced products of the cardiac Ca^{2+} ATPase gene. *J. Biol. Chem.* 263: 15024–31.

Lytton, J. and D. H. MacLennan (1992) Sarcoplasmic reticulum. In Heart and Cardiovascular System: Scientific Foundations, Vol. 2, 2nd Edn. (Fozzard, H. A., Haber, E., Jennings, R. B., Katz A. M., and Morgan, H. E., eds.). Raven Press, New York, pp.1203–22.

Lytton, J., A. Zarain-Herzberg, M. Periasamy, and D. H. MacLennan (1989) Molecular cloning of the mammalian smooth muscle sarco(endo)plasmic reticulum Ca^{2+}-ATPase. *J. Biol. Chem.* 264: 7059–65.

MacKenzie, A. E., R. G. Korneluk, F. Zorzato, J. Fujii, M. Phillips, D. Iles, B. Wieringa, S. Le Blond, J. Bailly, H. F. Willard, C. Duff, R. G. Worton, and D. H. MacLennan (1990) The human ryanodine receptor gene: its mapping to 19q13.1, placement in a chromosome 19 linkage group and exclusion as the gene caus-

ing myotonic dystrophy. *Am. J. Hum. Genet.* 46:1082–89.

MacLennan, D. H. (1995) Discordance between phenotype and genotype in malignant hyperthermia. *Curr. Opin. Neurol.* 8: 397–401.

MacLennan, D. H. and M. S. Phillips (1992) Malignant hyperthermia. *Science* 256: 789–94.

MacLennan, D. H., C. J. Brandl, B. Korczak, and N. M. Green (1985) Amino-acid sequence of a Ca^{2+}–Mg^{2+}-dependent ATPase from rabbit muscle carcoplasmic reticulum, deduced from its complementary DNA sequence. *Nature (London)* 316: 696–700.

MacLennan, D. H., C. Duff, F. Zorzato, J. Fujii, M. Phillips, R. G. Korneluk, W. Frodis, B. A. Britt, and R. G. Worton (1990) Ryanodine receptor gene is a candidate for predisposition to malignant hyperthermia. *Nature* 343: 559–61.

MacLennan, D. H., D. M. Clarke, T. W. Loo, and I. Skerjanc (1992) Site-directed mutagenesis of the Ca^{2+} ATPase of sarcoplasmic reticulum. *Acta Physiol. Scand.* 146: 141–50.

Manning, B. M., K. A. Quane, P. J. Lynch, A. Urwyler, V. Tegazzin, R. Krivosic-Horber, K. Censier, G. Comi, P. Adnet, W. Wolz, J. Lunardi, C. R. Muller, and T. V. McCarthy (1998a) Novel mutations at a CpG dinucleotide in the ryanodine receptor in malignant hyperthermia. *Hum. Mutat.* 11: 45–50.

Manning, B. M., K. A. Quane, H. Ording, A. Urwyler, V. Tegazzin, M. Lehane, J. O. Halloran, E. Hartung, L. M. Giblin, P. J. Lynch, P. Vaughan, K. Censier, D. Bendizen, G. Comi, L. Heytens, T. Fagerlund, W. Wolz, J. J. A. Heffron, C. R. Muller, and T. V. McCarthy (1998b) Identification of novel mutations in the ryanodine recptor gene (RYR1) in malignant hyperthermia: genotype phenotype correlation. *Am. J. Hum. Genet.* 62: 599–609.

McCarthy, T. V., J. M. S. Healy, J. J. A. Heffron, M. Lehane, M. Deufel, F. Lehmann-Horn, M. Faralli, and K. Johnson (1990) Localization of the malignant hyperthermia susceptibility locus to human chromosome 19q12–13.2. *Nature* 343: 562–64.

McKenzie, A. J., K. G. Couchman, and N. Pollock (1992) Propofol is a "safe" anaesthetic agent in malignant hyperthermia susceptible patients. *Anesth. Intens. Care* 20: 165–68.

Mickelson, J. R. and C. F. Louis (1996) Malignant hyperthermia: excitation–contraction coupling, Ca^{2+} release channel, and cell Ca^{2+} regulation defects. *Physiol. Rev.* 76(2): 537–92.

Mickelson, J. R., E. M. Gallant, L. A. Litterer, K. M. Johnson, W. E. Rempel, and C. F. Louis (1988) Abnormal sarcoplasmic reticulum ryanodine receptor in malignant hyperthermia. *J. Biol. Chem.* 263: 9310.

Mignery, G. A. and T. C. Südhof (1990) The ligand binding site and transduction mechanism in the inositol-1,4,5-triphosphate receptor. *EMBO J.* 9: 3893–98.

Miyawaki, A., T. Furuichi, Y. Ryou, S. Yoshikawa, T. Nakagawa, T. Saitoh, and K. Mikoshiba (1991) Structure–function relationships of the mouse inositol

1,4,5-trisphosphate receptor. *Proc. Natl. Acad. Sci. USA* 88: 4911–15.

Monnier, N., V. Procaccio, P. Stieglitz and J. Lunardi (1997) Malignant-hyperthermia susceptibility is associated with a mutation of the α 1-subunit of the human dihydropyridine-sensistive L-type voltage-dependent calcium-channel receptor in skeletal muscle. *Am. J. Hum. Genet.* 60: 1316–25.

Mulley, J. C., H. M. Kozman, H. A. Phillips, A. K. Gedeon, J. A. McCure, D. E. Iles, R. G. Gregg, K. Hogan, F. J. Couch, J. L. Weber, D. H. MacLennan, and E. A. Haan (1993) Refined genetic localization for central core disease. *Am. J. Hum. Genet.* 52: 398–405.

Nakai, J., T. Imagawa, Y. Hakamata, M. Shigekawa, H. Takeshima, and S. Numa (1990) Primary structure and functional expression from cDNA of the cardiac ryanodine receptor/calcium release channel. *FEBS Lett.* 271: 169–77.

Nelson, T. E. (1988) SR function in malignant hyperthermia. *Cell* 9: 257–65.

Nelson T. E. and M. Lin (1993) Dantrolene activates and then blocks the ryanodine receptor Ca²⁺ release channel in a planar lipid bilayer. *Biophys. J.* 64: A380.

O'Brien, P. J. (1987) Etiopathogenetic defect of malignant hyperthermia: hypersensitive calcium-release channel of skeletal muscle sarcoplasmic reticulum. *Vet. Res. Comm.* 11: 527–59.

O'Brien, P. J., R. O. Ball, and D. H. MacLennan (1994) Effects of heterozygosity for the mutation causing porcine stress syndrome on carcass quality and live performance characteristics. Proceedings of the 13th International Pig Veterinarian Society Congress, Bangkok. p. 481.

Odermatt, A., P. E. M. Taschner, V. K. Khanna, H. F. M. Busch, G. Karpati, C. K. Jablecki, M. H. Bruning, and D. H. MacLennan (1996) Mutations in the gene encoding SERCA1, the human fast-twitch skeletal muscle sarcoplasmic reticulum Ca²⁺ ATPase, are associated with Brody disease. *Nat. Genet.* 14: 191–94.

Odermatt, A., P. E. M. Taschner, S. W. Scherer, B. Beatty, V. K. Khanna, D. R. Cornblath, V. Chaudhry, W.-C. Yee, B. Schrank, G. Karpati, M. H. Breuning, N. Knoers, and D. H. MacLennan (1997) Characterization of the gene encoding human sarcolipin, a proteolipid associated with SERCA1; absence of structural mutations in five patients with Brody disease. *Genomics* 45: 541–53.

Ohnishi, S. T., S. Taylor, and G. A. Gronert (1983) Calcium-induced Ca²⁺ release from sarcoplasmic reticulum of pigs susceptible to malignant hyperthermia. The effects of halothane and dantrolene. *FEBS Lett.* 161: 103–7.

Ohta, T., S. Ito, and A. Ohga (1990) Inhibitory action of dantrolene on Ca²⁺-induced Ca²⁺ release from sarcoplasmic reticulum in guinea pig skeletal muscle. *Eur. J. Pharmacol.* 178: 11–19.

Olckers, A., D. A. Meyers, S. Meyers, E. W. Taylor, J. E. Fletcher, H. Rosenberg, H. Isaacs, and R. D. Levitt (1992) Adult muscle sodium channel α-subunit is a gene candidate for malignant hyperthermia susceptibility. *Genomics* 14: 829–31.

Ørding, H. (1985) Incidence of malignant hyperthermia in Denmark. *Anesth. Analg.* 64:700–4.

Ørding, H., V. Brancadoro, S. Cozzolino, F. R. Ellis, V. Glauber, E. F. Gonano, P. J. Halsall, E. Hartung, J. J. A. Heffron, L. Heytens, G. Kozak-Ribbens, H. Kress, R. Krivosic-Horber, F. Lehmann-Horn, W. Mortier, Y. Nivochey, E. Ranklev-Twetman, S. Sigurdsson, M. Snoeck, P. Stieglitz, V. Tegazzin, A. Urwyler, and F. Wappler (1997) *In vitro* contracture test for diagnosis of malignant hyperthermia following the protocol of the European MH Group: results of testing patients surviving fulminant MH and unrelated low-risk subjects. *Acta Anaesthesiol. Scand.* 41: 955–66.

Otsu, K., H. F. Willard, V. K. Khanna, F. Zorzato, N. M. Green, and D. H. MacLennan (1990) Molecular cloning of cDNA encoding the Ca²⁺ release channel (ryanodine receptor) of rabbit cardiac muscle sarcoplasmic reticulum. *J. Biol Chem.* 265: 13472–83.

Otsu, K., V. K. Khanna, A. L. Archibald, and D. H. MacLennan (1991) Co-segregation of porcine malignant hyperthermia and a probable causal mutation in the skeletal muscle ryanodine receptor gene in backcross families. *Genomics* 11: 744–50.

Otsu, K., J. Fujii, M. Periasamy, M. Difilippantonio, M. Uppender, D. C. Ward, and D. H. MacLennan (1993) Chromosome mapping of five human cardiac and skeletal muscle sarcoplasmic reticulum genes. *Genomics* 17: 507–9.

Otsu, K., N. Nishida, Y. Kimura, T. Kuzuya, M. Hori, T. Kamada, and M. Tada (1994) The point mutation Arg⁶¹⁵ to Cys in the Ca²⁺ release channel of skeletal muscle sarcoplasmic reticulum is responsible for hypersensitivity to caffeine and halothane in malignant hyperthermia. *J. Biol. Chem.* 269: 9413.

Phillips, M. S., V. K. Khanna, S. de Leon, W. Frodis, B. A. Britt, and D. H. MacLennan (1994) The substitution of Arg for Gly²⁴³³ in the human skeletal muscle ryanodine receptor is associated with malignant hyperthermia. *Hum. Mol. Genet.* 3: 2181–86.

Phillips, M. S., J. Fujii, V. K. Khanna, S. deLeon, K. Yakabata, P. J. deJong, and D. H. MacLennan (1996) The structural organization of the human skeletal muscle ryanodine receptor (*RYR*1) gene. *Genomics* 34: 24–41.

Poch, E., S. Leach, S. Snape, T. Cacic, D. H. MacLennan, and J. Lytton (1998) Functional characterization of alternatively spliced human SERCA3 transcripts *Am. J. Physiol.* in press.

Powers, P. A., S. W. Scherer, L.-C. Tsui, R. G. Gregg, and K. Hogan (1994) Localization of the gene encoding the α2/s subunit (CACNL2A) of the human skeletal muscle voltage-dependent Ca²⁺ channel to chromosome 7q21–q22 by somatic cell hybrid analysis. *Genomics* 19: 192–93.

Quane, K. A., J. M. S. Healy, K. E. Keating, B. M. Manning, F. J. Couch, F. J. Palmucci, C. Doriguzzi, T. H. Fagerlund, K. Berg, H. Ording, D. Bendixen, W.

Mortier, U. Linz, C. R. Müller, and T. V. McCarthy (1993) Mutations in the ryanodine receptor gene in central core disease and malignant hyperthermia. *Nat. Genet.* 5: 51–55.

Quane, K. A., K. E. Keating, J. M. S. Healy, B. M. Manning, R. Krivosic-Horber, I. Krivosic, N. Monnier, J. Lunardi, and T. V. McCarthy (1994a) Mutation screening of the RYR1 gene in malignant hyperthermia: Detection of a novel Tyr to Ser mutation in a pedigree with associated central cores. *Genomics* 23: 236–39.

Quane, K. A., K. E. Keating, B. M. Manning, J. M. S. Healy, K. Monsieurs, J. J. A. Heffron, M. Lehane, L. Heytens, R. Krivosic-Horber, P. Adnet, F. R. Ellis, N. Monnier, J. Lumardi, and T. V. McCarthy (1994b) detection of a novel common mutation in the ryanodine receptor gene in malignant hyperthermia: implications for diagnosis and heterogeneity studies. *Hum. Mol. Genet.* 3: 471.

Quane, K. A., H. Ording, K. E. Keating, B. M. Manning, R. Heine, D. Bendixen, K. Berg, R. Krivosic-Horber, F. Lehmann-Horn, T. Fagerlund, and T. V. McCarthy (1997) detection of a novel mutation at amino acid position 614 in the ryanodine receptor in malignant hyperthermia. *Br. J. Anaesth.* 79: in press.

Robinson, R. L., N. Monnier, W. Wolz, M. Jung, A. Reis, G. Nuernberg, J. L. Curran, K. Monsieurs, P. Stieglitz, L. Heytens, R. Fricker, C. van Broeckhoven, T. Deufel, P. M. Hopkins, J. Lunardi, and C. R. Mueller (1997) A genome wide search for susceptibility loci in three European malignant hyperthermia pedigrees. *Hum. Mol. Genet.* 6(6): 953–61.

Serfas, K., P. Bose, L. Patel, K. Wrogemann, M. S. Phillips, D. H. MacLennan, and C. Greenberg (1996) Comparison of the segregation of the *RYR*1 C1840T mutation with segregation of the caffeine/halothane contracture test results for malignant hyperthermia susceptibility in a large Manitoba Mennonite family. *Anesthesiology* 84: 322–29.

Shomer, N. H., C. F. Louis, M. Fill, L. A. Litterer, and J. R. Mickelson (1993) Reconstitution of abnormalities in the malignant hyperthermia-susceptible pig ryanodine receptor. *Am. Physiol. Soc.* C125–35.

Shuaib, A., R. T. Paasuke, and K. W. Brownell (1987) Central core disease: clinical features in 13 patients. *Medicine* 66: 389–96.

Shy, G. M. and K. R. Magee (1956) A new congenital non-progressive myopathy. *Brain* 79: 610–21.

Simpson, S. P. and A. J. Webb (1989) Growth and carcass performance of British Landrace pigs heterozygous at the halothane locus. *Anim. Prod.* 49: 503–9.

Sudbrak, R., A. Golla, P. Powers, R. Gregg, G. Du, I. Chesne, F. Lehmann-Horn, and T. Deufel (1993) Exclusion of malignant hyperthermia susceptibility (MHS) from a putative MHS2 locus on chromosome 17q and of the alpha 1, beta 1, gamma subunits of the dihydropyridine receptor calcium channel as candidates for the molecular defect. *Hum. Mol. Genet.* 2: 857–62.

Sudbrak, R., V. Procaccio, M. Klausnitzer, J. L. Curran, K. Monsieurs, C. Van Broeckhoven, R. Ellis, L. Heyetens, E. J. Hartung, G. Kozak-Ribbens, D. Heilinger, J. Weissenbach, F. Lehman-Horn, C. R. Mueller, T. Deufel, A. D. Stewart, and J. Lunardi (1995) Mapping of a further malignant hyperthermia susceptibility locus to chromosome 3q13.1. *Am. J. Hum. Genet.* 56: 684–91.

Suko, J., I. Maurer-Fogy, B. Plank, O. Bertel, W. Wyskovsky, M. Hohenegger, and G. Hellmann (1993) Phosphorylation of serine 2843 in ryanodine receptor-calcium release channel of skeletal muscle by cAMP-, cGMP- and CaM-dependent protein kinase. *Biochim. Biophys. Acta* 1175: 193–206.

Takeshima, H., S. Nishimura, T. Matsumoto, H. Ishida, K. Kangawa, N. Minamino, H. Matsuo, M. Ueda, M. Hanaoka, T. Hirose, and S. Numa (1989) Primary structure and expression from complementary DNA of skeletal muscle ryanodine receptor. *Nature* 339: 439–45.

Takeshima, H., M. Iino, H. Takekura, M. Nishi, J. Kuno, O. Minowa, H. Takano, and T. Noda (1994) Excitation–contraction uncoupling and muscular degeneration in mice lacking functional skeletal muscle ryanodine receptor gene. *Nature* 369: 556–59.

Taylor, D. J., M. J. Brosnan, D. L. Arnold, P. J. Bore, P. Styles, J. Walton, and G. K. Radda (1988) Ca^{2+}-ATPase deficiency in a patient with an exertional muscle pain syndrome. *J. Neurol. Neurosurg. Psychiat.* 51: 1425–33.

Tong, J., H. Oyamada, N. Demaurex, S. Grinstein, T. V. McCarthy, and D. H. MacLennan (1997) Caffeine and halothane sensitivity of intracellular Ca^{2+} release is altered by fifteen calcium release channel (ryanodine receptor) mutations associated with malignant hyperthermia and/or central core disease *J. Biol. Chem.* 272: 26332–39.

Tong, J., T. V. McCarthy, and D. H. MacLennan (1999) Measurement of resting cytosolic Ca^{2+} concentrations and Ca^{2+} store size in HEK-293 cells transfected with malignant hyperthermia or central core disease mutant Ca^{2+} release channels. *J. Biol. Chem.* 274: in press.

Treves, S., F. Larini, P. Menegazzi, T. H. Steinberg, M. Koval, B. Vilsen, J. P. Andersen, and F. Zorzato (1994) Alteration of intracellular Ca^{2+} transients in COS-7 cells transfected with the cDNA encoding skeletal-muscle ryanodine receptor carrying a mutation associated with malignant hyperthermia. *Biochem. J.* 301: 661–65.

Vita, G. M., A. Olckers, A. E. Jedlicka, A. L. George, T. Heiman-Patterson, H. Rosenberg, and J. E. Fletcher (1995) Masseter muscle rigidity associated with glycine 1306-to-alanine mutation in the adult muscle sodium channel α-subunit gene. *Clin. Invest.* 82: 1097–103.

Wevers, R. A., P. J. E. Poels, E. M. G. Joosten, G. G. H. Steenbergen, A. A. G. M. Benders, and J. H. Veerkamp (1992) Ischaemic forearm testing in a patient with Ca^{2+}-ATPase deficiency. *J. Inher. Metab. Dis.* 15: 423–25.

Witcher, D. R., R. J. Kovacs, H. Schulman, D. C. Cefali, and L. R. Jones (1991) Unique phosphorylation site on the cardiac ryanodine receptor regulates calcium channel activity. *J. Biol. Chem.* 266: 11144–52.

Wrogemann, K. and S. D. J. Pena (1976) Mitochondrial calcium overload: a general mechanism of cell necrosis in muscle diseases. *Lancet* 1: 672–73.

Wu, K.-D., W.-F. Lee, J. Wey, D. Bungard, and J. Lytton (1995) Localization and quantification of endoplasmic reticulum Ca^{2+} ATPase isoform transcripts. *Am. J. Physiol.* 269: C775–84.

Zhang, Y., H. S. Chen, V. K. Khanna, S. de Leon, M. S. Phillips, K. Schappert, B. A. Britt, A. K. W. Brownell, and D. H. MacLennan (1993) Identification of a mutation in human ryanodine receptor associated with central core disease. *Nat. Genetics* 5: 61–65.

Zhang, Y., J. Fujii, M. S. Phillips, H.-S. Chen, G. Karpati, W.-C. Yee, B. Schrank, D. R. Cornblath, K. B. Boylan, and D. H. MacLennan (1995) Characterization of cDNA and genomic DNA encoding SERCA1, the Ca^{2+}-ATPase of human fast-twitch skeletal muscle sarcoplasmic reticulum, and its elimination as a candidate gene for Brody disease. *Genomics* 30: 415–25.

Zorzato, F., J. Fujii, K. Otsu, M. Phillips, N. M. Green, F. A. Lai, G. Meissner, and D. H. MacLennan (1990) Molecular cloning of cDNA encoding human and rabbit forms of the Ca^{2+} release channel (ryanodine receptor) of skeletal muscle sarcoplasmic reticulum. *J. Biol. Chem.* 265: 2244–56.

Zot, A. S. and J. D. Potter (1987) Structural aspects of troponin–tropomyosin regulation of skeletal muscle contraction. *Ann. Rev. Biophys. Chem.* 16: 535–59.

Index